PRECALCULUS

A Study of Functions and their Applications

PRECALCULUS

A Study of Functions and their Applications

Todd Swanson
Hope College

Janet Andersen
Hope College

Robert Keeley
Calvin College

Harcourt College Publishers

Fort Worth Philadelphia San Diego
New York Orlando Austin San Antonio
Toronto Montreal London Sydney Tokyo

Publisher: Emily Barrosse
Executive Editor: Angus McDonald
Acquisitions Editor: Liz Covello
Marketing Strategist: Julia Downs Conover
Developmental Editor: Jay Campbell
Project Management: Progressive Publishing Alternatives
Production Manager: Alicia Jackson
Art Director: Carol Bleistine
Cover Credit: Mackinac Bridge photo courtesy of Tim Burke, Michigan Department of Transportation

Precalculus: A Study of Functions and their Applications 1/e

ISBN: 0-03-024964-3
Library of Congress Catalog Card Number: 00-105279

Address for domestic orders:
Harcourt College Publishers, 6277 Sea Harbor Drive, Orlando, FL 32887-6777
1-800-782-4479
e-mail: collegesales@harcourt.com

Address for international orders:
International Customer Service, Harcourt Inc.
6277 Sea Harbor Drive, Orlando, FL 32887-6777
(407) 345-3800
Fax: (407) 345-4060
e-mail: hbintl@harcourt.com

Address for editoral correspondence:
Harcourt College Publishers, Public Ledger Building, Suite 1250,
150 S. Independence Mall West,
Philadelphia, PA 19106-3412

Web Site Address
http://www.harcourtcollege.com

Printed in the United States of America

0123456789 048 10 987654321

To Debra, Emma, and Tyler

To Jim, John, and Joy

To Laura, Bethany, Meredith, Bryan, and Lynnae

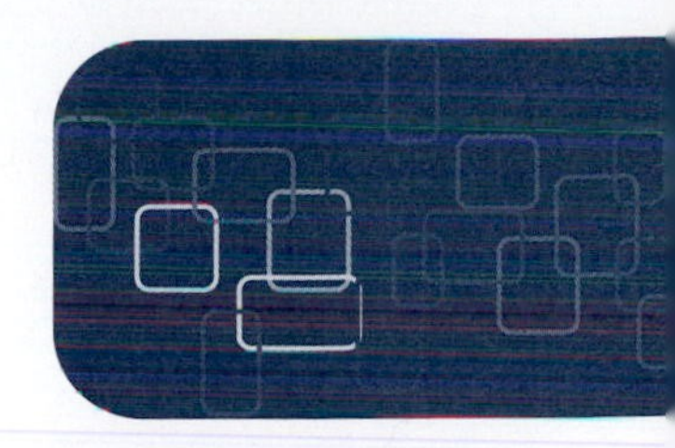

Preface

PHILOSOPHY

This book treats functions as the object of study while focusing on important mathematical concepts. In contrast to the organization of most precalculus texts, we introduce students to linear, exponential, logarithmic, periodic, and power functions in the second chapter and develop the properties of these functions throughout the remainder of the text as mathematical concepts are explored. This allows us to emphasize the connections between mathematical ideas as we further the student's knowledge about a particular type of function.

Building the study of functions throughout the book, rather than isolating the study of a single type of function to one or two chapters, leads to a natural review of the properties of various functions throughout the text. This organization also addresses a common student misconception that mathematics consists of unrelated bits of information which can be forgotten once a particular topic is finished.

Another common student misconception is that mathematics is useless and dry. To address this, we have made the study of applications an integral part of our text. Applications occur in the readings in three different ways: the focus of a section, an example used to illustrate a concept, and the motivation for a mathematical concept. Applications also occur in homework exercises, investigations, and projects. Every application has been researched and care has been taken to insure that the data given is accurate.

Our underlying premise is that students learn best when they are engaged with the material. We have designed this text so that students will read, write, discuss, and explore mathematical concepts. It is written in a conversational format with reading questions interspersed throughout each section. Each homework set includes investigations where students are asked to extend and apply the material presented in that section. Each chapter includes a project, which is a larger scale exploration or application. All of these features invite students to be actively engaged in their study of precalculus.

This textbook is designed for a one-semester college course in precalculus (although we believe that it is equally well suited for a one-year high school precalculus course). It incorporates writing, projects, multiple representations of mathematical concepts, and the use of technology in the form of graphing calculators. This makes it an excellent preparation for a modern calculus course.

This text is also well suited for students going on to a traditional calculus course as well as various science courses. Throughout the text we have tried to illustrate how mathematicians think by asking the kinds of questions that are interesting to mathematicians. We feel that this text meets the recommendations of the Mathematical Association of America, and reflects the spirit of the standards set by the National Council of Teachers of Mathematics and the American Mathematical Association of Two-Year Colleges.

KEY FEATURES

Precalculus: A Study of Functions and their Applications has several important features:

- **Coverage of Functions.** Each of the basic types of functions (linear, exponential, logarithmic, periodic, and power) is introduced early in the book. Properties of these functions are developed throughout the remainder of the text. This leads to a natural review as students build on their understanding of these functions. It also emphasizes the connections of mathematical ideas by unifying the study of functions around mathematical concepts rather than around types of functions.
- **Writing Style.** Students are expected to read the text. It is written in a conversational format addressed to the student. Reading questions are incorporated throughout each section. In these questions, students are asked to write about key concepts and do simple problems.
- **Exercises & Investigations.** Each exercise set includes investigations. These are extended problems in which students explore either a mathematical concept or an application. Several topics covered in the reading first appear as an investigation in an earlier section. This foreshadows mathematical ideas and allows the students to gain some firsthand experience with a concept before encountering it formally. Investigations can be used as part of a regular homework assignment, group homework, a small group in-class activity, or as a class discussion.
- **Projects.** A project is included with each chapter. These incorporate a variety of topics and can be used to reinforce and extend the topics presented in that chapter. A typical project requires one or two class periods and approximately five additional hours of student work outside class. The projects are best completed by students working in groups. These projects were written with the partial support of a grant from the National Science Foundation. This material was previously published under the title *Projects for Precalculus* by Saunders College Publishing.
- **Applications.** The inclusion of applications is an integral part of the text. All of the applications included in the text involve real situations with actual data. These applications occur in the reading, the exercises, the investigations, and the projects. Every application has been

researched and care has been taken to insure that the data given is accurate.

- **Historical Anecdotes.** Included in the reading are historical anecdotes. We also include some recent developments in mathematics so students realize that mathematical knowledge continues to evolve and that interesting, applicable mathematical questions are still being explored. We want students to see mathematics as an ongoing human endeavor with roots in the past.
- **Approach.** Topics are presented using multiple representations (symbolic, numerical, graphical, and verbal).
- **Graphing Calculator.** It is assumed that students have daily access to graphing calculators. We emphasize ways in which the calculator can be used to gain mathematical understanding, but rarely give specific keystrokes.

ANNOTATED TABLE OF CONTENTS

CHAPTER 1: An Introduction to Functions. Various representations of functions as well as the language and notation associated with functions are introduced. This chapter also illustrates how graphing calculators can be used and misused in the study of functions.

CHAPTER 2: Families of Functions. Linear, exponential, logarithmic, periodic, and power functions are introduced. Students are shown how to recognize these functions in their various representations. Students are also shown how to obtain a formula when given a linear, exponential, or power function either numerically or graphically. This lays the groundwork for these functions and for their use throughout the remainder of the book. Examples of real world situations are included for each type of function.

CHAPTER 3: New Functions from Old. The basic functions from Chapter 2 are transformed to form new functions in a variety of ways. In particular, the relationship between a transformed function in its symbolic form is compared with its graphical form. Transformations include addition and multiplication as well as composition. The relationship between a function and its inverse is also explored.

CHAPTER 4: Polynomial and Rational Functions. Polynomials, introduced as transformations of particular power functions, are important enough to study as independent objects. We look at their properties as well as how they can be combined through division to form rational functions. Applications are given throughout the chapter.

CHAPTER 5: Trigonometric Functions. The periodic functions of sine and cosine, introduced in Chapter 2, are reviewed and other trigonometric functions are introduced in this chapter. These functions are introduced by using the unit circle definitions. The geometry of a circle, including arc length and area, is also explored. The transformations, introduced in Chapter 3, are applied to the trigonometric functions. Trigonometric identities are introduced throughout the chapter and are the focus of Section 5.4.

CHAPTER 6: Applications of Trigonometric Functions. Using trigonometric functions to model situations in the world is the focus of this chapter. It begins by looking at problems involving triangles. Then combinations of the periodic functions with other periodic functions as well as with non-period functions are explored. Doing this allows us to expand the areas in which we can use trigonometric functions to model applications.

CHAPTER 7: Solving Equations and Fitting Functions to Data. Different methods for solving equations and inequalities are introduced. This structure gives students a review of the functions first introduced in Chapter 2. The techniques of linear, exponential, and power regression are introduced as methods of fitting functions to data.

CHAPTER 8: Getting Ready for Calculus. This chapter serves as an introduction to calculus by exploring the concepts of limit, the derivative, and the integral. These topics are meant to help students prepare for the study of calculus and serve only as introductions.

CHAPTER 9: Additional Topics. The text concludes with a look at parametric equations, vectors, and multivariable functions. A property of a conic section is the focus of the project in this chapter. While these topics are not vital for a preparation for the study of calculus, they are of interest because many applications use these mathematical concepts.

SUPPLEMENTS TO THE TEXT

The following supplements are available for instructors and students from the publisher:

Instructor's Resource Manual. This manual contains solutions to all the reading questions, exercises, investigations, and projects. It also outlines methods for ways to effectively use the book.

Printed Test Bank and Prepared Tests. The test bank consists of multiple-choice and short-answer test items organized by chapter and section. The prepared tests include nine sets of ready-to-copy tests, one set for each chapter. Each set comprises of one multiple-choice and two show-your-work tests. Answers for every test item are provided.

ESATEST 2000 TM Computerized Test Bank. A flexible, powerful computerized testing system, the ESATEST 2000 TM Computerized Test Bank contains all the test bank questions and allows instructors to prepare quizzes and examinations quickly and easily. It offers teachers the ability to select, edit, and create not only test items but algorithms for test items as well. Teachers can tailor tests according to a variety of criteria, scramble the order of test items, and administer tests on-line, either over a network or via the Web. ESATEST 2000 TM also includes full-function grade book and graphing features. This software is available in Windows and Macintosh formats.

Student Solutions Manual. This manual contains complete solutions to every odd-numbered problem in the reading questions and exercises and to all chapter review exercises.

Graphing Calculator Manual. The Graphing Calculator Manual contains programming ideas for various Texas Instruments graphing calculators and investigates the application of those programming operations to solve problems.

Projects for Precalculus. The authors of the text previously published *Projects for Precalculus*, a popular and successful NSF-sponsored program in reform precalculus. This manual contains 26 carefully prepared and tested activities that promote conceptual understanding and active learning. The projects included at the end of each chapter in the text were originally published in this supplement.

ACKNOWLEDGMENTS

We are grateful for the support that we have received throughout this project from Hope College and Calvin College. In particular, we are thankful to our colleagues for their encouragement and advice throughout this long process. A special thanks goes to Donatella Delfino and John Van-Iwaarden, both from Hope College, for field-testing our manuscript.

The projects and some of the investigations in this book were written with support from the National Science Foundation (Grant DUE-9354741). We are thankful for this support.

We are very thankful to our student assistants, Rhonda Pardue, John Krueger, Dana Horner, Shannon See, Andy DeYoung, and Muhammed Hameed for their outstanding work and assistance. Their help has been a valuable part of this project.

We appreciate the comments and positive support we received from those that reviewed portions of the manuscript for this book,

Laurie Burton, *Central Washington University*
Marcia Drost, *Texas A&M University*
Barbara Edwards, *Oregon State University*
Susan Forman, *Bronx Community College*
Carol Lynn Hancock, *Appalachian State University*
Richard J. Maher, *Loyola University Chicago*
Nancy L. Matthews, *University of Oklahoma*
Katherine McGivney, *University of Arizona*
Ed Nichols, *Chattanooga State Technical Community College*
Greg Rhoads, *Appalachian State University*
Barbara Sausen, *Fresno City College*
David Slavit, *Washington State University, Vancouver*
Rebecca G. Wahl, *University of South Dakota*
Pamela J. Wells, *Grand Valley State University*
Sandra Wray-McAfee, *University of Michigan–Dearborn*

We are grateful to the accuracy reviewers,

Nancy Matthews, *Austin Peay State University*
Margaret Donlan, *University of Delaware*

who helped this book through the production process by finding and correcting mistakes that were made along the way.

Thanks also go to the staff at Harcourt College Publishing. In particular, we wish to thank Marc Sherman, Alexa Epstein, Liz Covello, and Jay Campbell for their guidance in bringing our manuscript to publication.

Finally, we are very thankful to our families for their patience, understanding, and encouragement throughout this project.

Todd Swanson
Janet Andersen
Robert Keeley

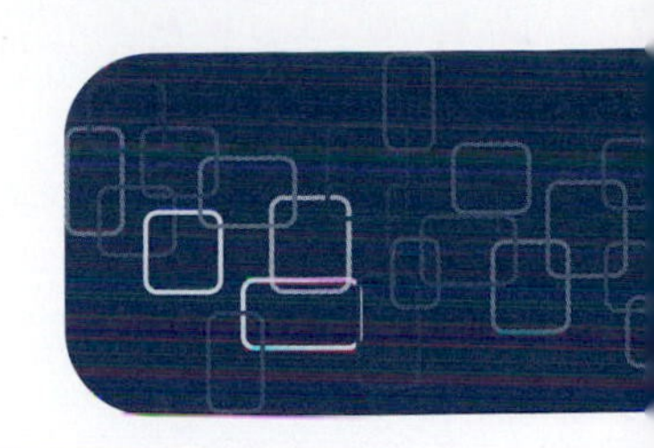

Contents

PRECALCULUS

A Study of Functions and their Applications

CHAPTER OUTLINE

CHAPTER OVERVIEW

- The idea of a function
- How functions are represented and evaluated
- How functions can be represented effectively on a calculator
- How functions can be used to mathematically model a physical situation

AN INTRODUCTION TO FUNCTIONS

Various representations of functions as well as the language and notation associated with functions are introduced in this chapter. Also included are illustrations of how graphing calculators can be used and misused in the study of functions. The chapter concludes with a section on mathematical modeling, which shows how functions can be developed in order to solve various problems.

1.1 FUNCTIONS

Functions are all around us. They are useful in modeling physical, financial, and even sociological situations in addition to being interesting to study from a mathematical perspective. By understanding the behavior of functions, we can better understand how to choose functions to describe real-world situations. In this section, we consider four ways to represent functions, how to determine if something is a function, and how to evaluate functions.

FOUR REPRESENTATIONS OF A FUNCTION

Functions can be represented in a variety of ways. The first thing that probably comes to your mind when you see the word *function* is a mathematical formula such as the formula for the area of a circle:

$$A = \pi r^2.$$

This is a *symbolic* representation of a function, but there are other ways of representing functions. Often you see functions represented graphically, especially in magazines and newspapers. Figure 1 is an example of a *graphical* representation of a function that describes the average salary of a National Basketball Association (NBA) player with respect to time.[1]

Another way that functions can be represented is as a table of numbers such as Table 1, which describes the U.S. population over time.[2] This table is known as a *numerical* representation of a function. Symbols, graphs, and numbers are all mathematical objects, so it should not surprise you that all three can be used in different ways to represent functions. There is, however, a fourth way of commonly representing functions, one that does

[1] USA Today, *15 November 1996, p. 16C.*

[2] *Mark S. Hoffman, ed.,* The World Almanac *(New York: Pharos Books, 1990), pp. 552–53.*

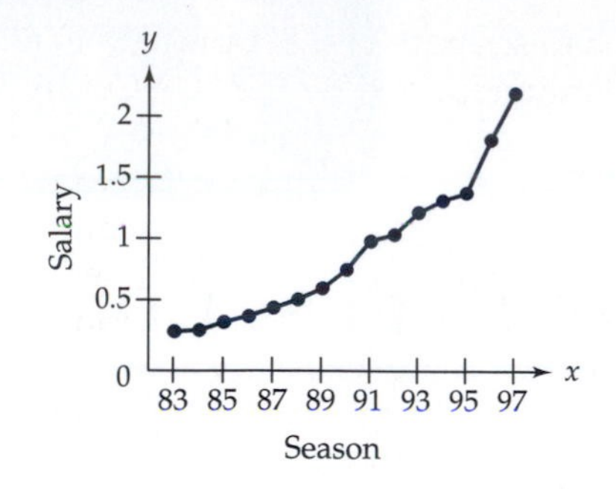

Figure 1 Graphical representation of the average salary of an NBA player (in millions of dollars).

TABLE 1 U.S. Population by Official Census

Year	Population
1800	5,308,483
1820	9,638,453
1840	17,063,353
1860	31,443,321
1880	50,189,209
1900	76,212,168
1920	106,021,537
1940	132,164,569
1960	179,323,175
1980	226,542,203
1990	248,709,873

not seem quite so mathematical: using words to describe a function. We call this a *verbal* representation of a function. For example, you might read the cost of tuition in a college catalog as

> Tuition for undergraduate students who are Michigan residents taking anywhere from 12 to 16 credit hours is the same, a total of $1,390 per semester. Tuition is $124 per credit hour for fewer than 12 credits and for each credit over 16.[3]

This description can be thought of as a function in which the input is the number of credit hours a student takes and the output is the total cost of tuition.

Understanding how functions can be represented in these four ways (symbolically, graphically, numerically, and verbally) and learning how these representations are related to one another is something you will see over and over again throughout this book. All four representations are equally valid, but each representation has its own advantages and disadvantages. A symbolic representation is best if you want to evaluate your function for a specific input. It is also the best representation if you want to manipulate or transform your function in some way. A graphical representation is best for quickly seeing the behavior of your function. It is easy to see where the function is changing when looking at a graph. A numerical representation is best if you want to convey information about your function at a few selected points, such as size of the U.S. population every 20 years from 1800 to 1980. This method is also the most common way to represent a function when you are collecting experimental data. Verbally is often the best way to describe a function to someone else and, in some sense, is the most natural way to represent many common functions. So, all four representations are equally important, and which one you want to use for a particular function depends on the function and what you are trying to do with it. For this reason, it is important to know how to change from one representation to another.

Definition of Function

Although we have talked about functions and looked at examples of functions, we have not yet defined exactly what we mean by the word *function*.[4] A **function** is a rule that assigns an input to at most one output. To see how this definition works, we look at each of the four examples we mentioned earlier.

The formula for the area of a circle is $A = \pi r^2$. Think about why this is a function. If you are given a radius, how many choices do you have for the area? Just one. This is the definition of a function: for each input, there is at most *one* output. For example, a circle of radius 9 centimeters has an area of 81π square centimeters. The area cannot possibly be anything else. Look at the function given in Figure 1, which describes the average salary for an NBA player. If you are given a season, there is at most *one* possible average salary. Again, the definition of a function is that there is at

[3]Grand Valley State University Undergraduate and Graduate Catalog *(Allendale, MI: Grand Valley State University, 1996–97), 1996, p. 42.*

[4]*In mathematics, words have very specific meanings, and one of the keys to doing well in mathematics is to learn and understand the definitions.*

most one possible output. Table 1, which describes the U.S. population, is a function since there is only one population reported by the U.S. Census Bureau for any given year. The tuition description is a function since, given the number of credit hours you are taking, there is only one tuition. Notice that although a function must have at most one output for any given input, the opposite is not true. It is possible to have a function where different inputs result in the same output. For example, it costs $1390 in tuition if you enroll for 12, 13, 14, 15, or 16 credits.

The definition of a function is that every input has at most *one* output. This is true whether a function deals with numbers or other objects. For example, consider the rule

What is the color of clothing worn by this person?

This rule is *not* a function because a person may be wearing more than *one* color of clothing.[5] The rule

What is the blood type of this person?

is a function, however, because each person has one and only one blood type.

READING QUESTIONS

1. What is the definition of a function?
2. Give an example not mentioned in the text of a function represented
   (a) Symbolically.
   (b) Graphically.
   (c) Numerically.
   (d) Verbally.
3. Can you have a function where an input is repeated with different outputs? Why or why not?
4. Can you have a function where an output is repeated with different inputs? Why or why not?

HOW TO DETERMINE WHETHER A RULE IS A FUNCTION

Remember that the definition of a function is that, no matter how it is represented, there will be at most one output for each input. We use this definition to determine if a rule is a function. This determination is not necessarily easy. When determining whether a rule represented symbolically is a function, you need to decide if any single input can give more than one answer. For example, suppose the relationship between the input, x, and the output, y, is given by

$$y = x^2.$$

This is a function because for any input, you have a single output, namely the square of your input. For example, there is one and only one number

[5] *Mathematicians call these types of rules relations.*

that represents 3^2. Suppose, however, the relationship between the input, x, and the output, y, is given by

$$y = \pm\sqrt{x}.$$

This is not a function because there is a single input that has more than one output. For example, the input 9 has both the output $+3$ and -3.

It is relatively easy to determine whether or not a rule represented graphically is a function. Since the horizontal axis of a graph represents the input and the vertical axis represents the output, you can determine if you have a function by using the vertical line test. The **vertical line test** states that if there exists a vertical line that crosses the graph at more than one point, then the graph does not represent a function. This statement is true because if any vertical line crosses the graph at more than one point, then there is more than one output for that single input. The graph given in Figure 2(a) is a function since every input (x-value) has only one corresponding output (y-value). The graph given in Figure 2(b), however, is not a function since there is an input that has three outputs as shown by the vertical line.

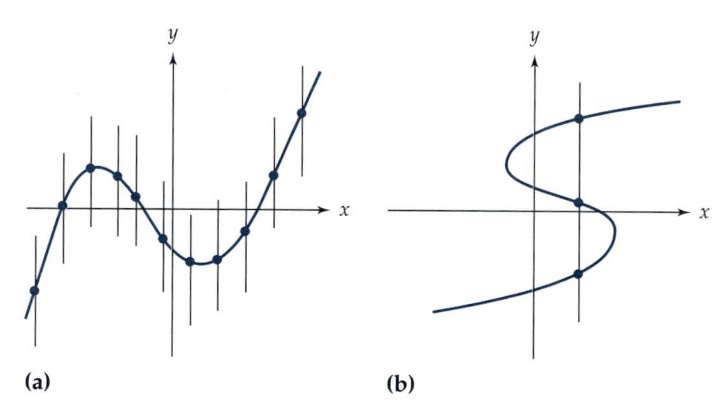

Figure 2 An example of the vertical line test.

TABLE 2 Table 2(a) Is a Function and Table 2(b) Is Not

Input	1	3	4	6	9
Output	3	6	8	6	5

(a)

Input	1	3	3	6	9
Output	3	6	8	6	5

(b)

It is also relatively easy to determine if a table of data represents a function. In this case, you need to determine if any of the inputs are repeated with different outputs. If so, then the table of data does not represent a function since a single input has more than one output. The data in Table 2(a) represent a function since every input has only one output. In Table 2(b), however, the input 3 has two different outputs, namely 6 and 8, so this does not represent a function. Notice that Table 2(b) is not a function even though the inputs 1, 6, and 9 each have only a single output. These three inputs follow the definition of a function, whereas a single input, 3, does not. To be a function, though, the definition has to hold true for *every* input. So, if a single input has more than one output, you do not have a function no matter what happens to the other inputs.

In verbal representations, similar to symbolic representations, it is sometimes difficult to determine if your rule represents a function. You need to determine if there can be more than one output for any single input. For example, the rule

What is the license plate number of this car?

is a function since a given car has only one license plate number. The rule

What kind of car does this person own?

however, is not a function since some people own more than one car. In each of the four representations, the key to determining if a rule is a function is to know the definition of a function.

READING QUESTIONS

5. Explain how you would determine whether or not a rule is a function if it is represented
   (a) Symbolically.
   (b) Graphically.
   (c) Numerically.
   (d) Verbally.
6. Explain why each of the following do not represent functions.
   (a) $y^2 = 4 - x^2$

   (b)

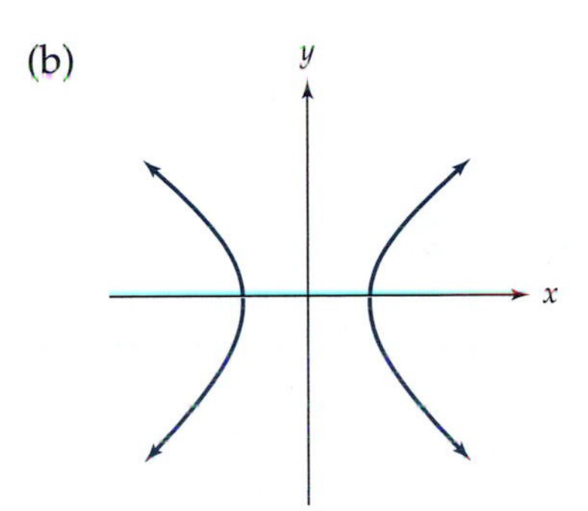

   (c)

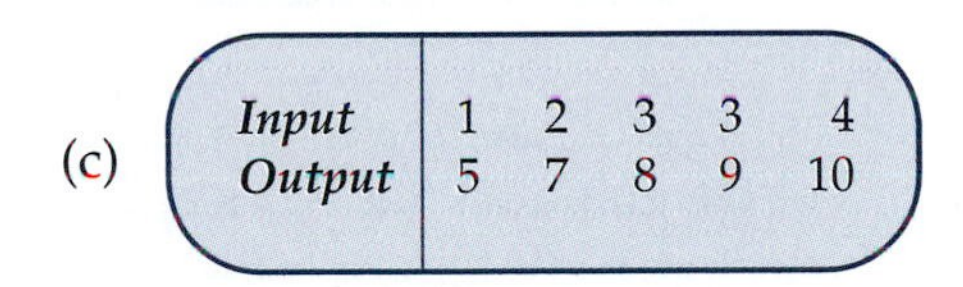

Input	1	2	3	3	4
Output	5	7	8	9	10

   (d) What is the name of your sister?
7. Why does the vertical line test work?

DOMAIN AND RANGE

Two important words associated with functions are domain and range. The **domain** of a function is the set of *valid* inputs. The **range** is the set of *actual* outputs. For example, suppose you had the following function:

What is the blood type of someone in our mathematics class?

The domain is the set of people in your mathematics class and the range is all the blood types of the people in your class, namely some subset of {A+, A−, B+, B−, AB+, AB−, O+, O−}.

You need to be careful when determining the domain and range of a function. Notice that the domain is limited to the *valid* inputs. In the blood type rule, it would be wrong to say that the input is "people." According to the description of the function, the input needs to be the name of a person in your mathematics class, so "Richard Nixon" or "Napoleon Bonaparte" are probably not valid inputs. Even though every person has a blood type, the function indicates that you are only choosing people in your mathematics class.

When a function is represented symbolically, it is natural to assume that the domain is the set of real numbers. Sometimes the domain is explicitly restricted, but sometimes the character of the function itself restricts the domain. The two most common things that make an input invalid are if

- The input causes division by zero.
- The input causes a negative number under a square root sign.

Example 1 Find the domain and range of $y = \sqrt{x-2}$.

Solution Since we must have a nonnegative number under the square root sign, we need to have $x - 2 \geq 0$, which implies $x \geq 2$. The domain of this function is therefore $x \geq 2$. The range is $y \geq 0$ since the square root of a number will be either zero or positive.

Example 2 Find the domain and range of $y = 1/(x-3)$.

Solution The function $y = 1/(x-3)$ is undefined when $x = 3$ since this would cause division by zero. So, the domain is all real numbers not equal to 3. The range is all real numbers except for zero since it is impossible to divide 1 by a number and get zero for an answer, but you can get any other answer you wish.

When a function is represented graphically, the domain is the projection of the graph on the x-axis and the range is the projection of the graph on the y-axis. In Figure 3, the domain of the graph is $-5 \leq x \leq 5$, and the range of the graph is $-6 \leq y \leq 4$. When a function is given symbolically, it is often helpful to graph the function as an aid in determining the domain and range.

The graphs of the two functions we did earlier, $y = \sqrt{x-2}$ and $y = 1/(x-3)$, are shown in Figure 4. You can see by looking at the graph, for example, that the domain of $y = \sqrt{x-2}$ is $x \geq 2$. Although graphs like these are helpful in determining a function's domain and range, it is usually best to make sure your domain and range are correct by also analyzing the function in its symbolic form.

Suppose you want to find the domain and range of a function that is represented numerically. In this case, the domain is all the numbers in the input row or column (usually the first), and the range is all the numbers in the output row or column (usually the second). Look at the function given in Table 3. The domain is $\{1, 3, 6, 9, 12, 15\}$, and the range is $\{0, 3, 5, 6, 8\}$.

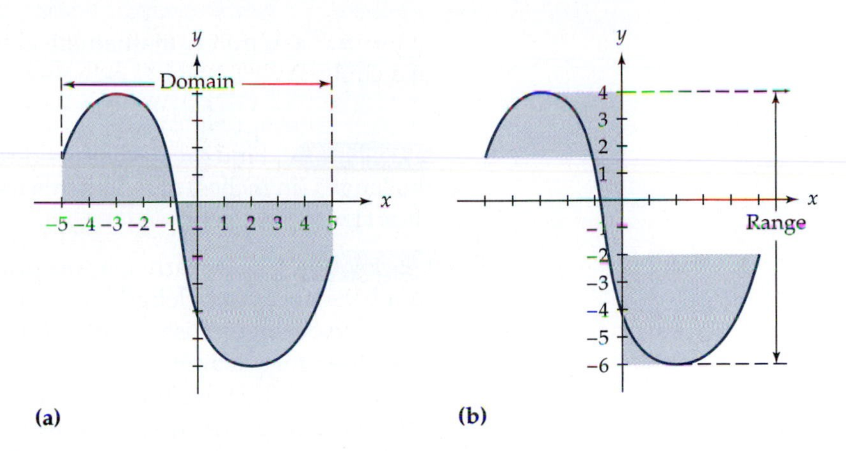

Figure 3 Domain and range.

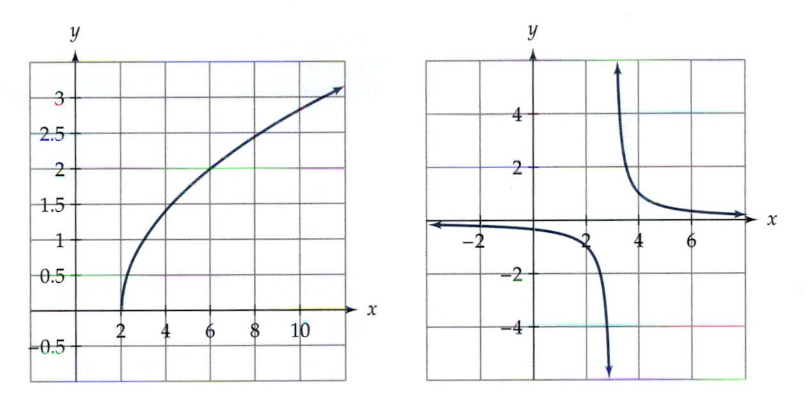

Figure 4 Graphs of $y = \sqrt{x-2}$ and $y = 1/(x-3)$. You can use the graphical representation of a function to help determine its domain and range.

TABLE 3 This Table Represents a Function

Input	1	3	6	9	12	15
Output	3	8	6	5	0	5

When a function is described verbally or if the function is modeling a physical situation, it is important to take into account the context of the function when giving the domain and range. For example, consider the function $y = \pi x^2$. The domain of this function is all real numbers, and the range is $y \geq 0$ (since squaring a number means the output cannot be negative). Suppose, however, you were told this function represents the area of a circle, y, given the radius of the circle, x. In this case, the domain is $x > 0$ since the radius of a circle cannot be negative or zero. The range is $y > 0$ since the square of a positive number will always be positive. So, the domain and the range differ depending on whether you are thinking of $y = \pi x^2$ as a purely mathematical function or as the formula for the area of a circle.

Example 3 Find the domain and range for the function whose input is the length (in inches) of a person's foot and whose output is his or her shoe size.

Solution It is clearly inappropriate to say that the domain is all real numbers since a foot length of -2 in. or even of 100 in. is *not* a valid size for a person's foot. Instead, the domain is (approximately) 2 in. to 16 in.[6] Also, the range is limited to actual outputs. It is inappropriate to say the range of our shoe size function consists of all the real numbers between 0 and 26 because shoe sizes are always multiples of $\frac{1}{2}$. So, the range is $\{1, 1\frac{1}{2}, 2, 2\frac{1}{2}, \ldots, 24\frac{1}{2}, 25, 25\frac{1}{2}, 26\}$.

READING QUESTIONS

8. What is meant by the *domain* of a function?

9. What is meant by the *range* of a function?

10. What are two things that limit the domain of a function when it is represented symbolically?

11. Why is the context of a function important when determining domain and range?

12. Give the domain and range for each of the following.

(a) $y = \sqrt{x - 4}$

(b)

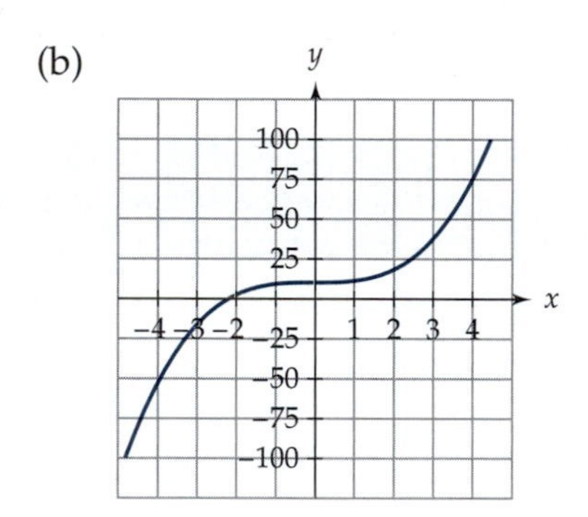

(c)

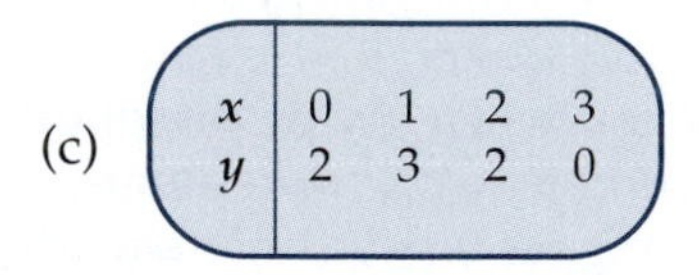

x	0	1	2	3
y	2	3	2	0

(d) What is the highest temperature in Detroit, Michigan, for a given day?

[6] *According to the* 1997 Guiness Book of World Records, *Matthew McGory has the biggest feet currently known. He wears a size 26 shoe, which corresponds to a foot length of 16 in.*

EVALUATING FUNCTIONS

Function Notation

The notation $f(x)$ is commonly used to denote the output of a function. This notation is pronounced "f of x." The letter f is typically used to remind us that this object is a $f$unction. The letter x is typically used to denote the input. The f is the *label* representing the function (like your name is the label representing you), and the x is a *placeholder* representing the input. To evaluate the function, you substitute your input in place of the letter x and perform the designated calculation. For example, $f(x) = x + 2$ means "my function, called f, is the rule that adds two to a given input." The letter x should be thought of as a space for your input. So, $f(x) = x + 2$ should be thought of as $f(\_\_) = \_\_ + 2$. If you want to evaluate your function, substitute your input for x and add 2 to get your answer. For example, $f(5) = 5 + 2 = 7$, $f(-1) = -1 + 2 = 1$, and $f(\pi) = \pi + 2$. In fact, $f(y) = y + 2$ and $f(x + 1) = (x + 1) + 2$ because the function f is the rule "add two to your input" regardless of the input.

When a function is modeling a physical situation, it is common to use letters for your function and input that remind you of the physical quantities. For example, the function for the area of a circle is typically written as $A = \pi r^2$ rather than $f(x) = \pi x^2$ to remind us that the input is the *r*adius of the circle and the output is the *a*rea of the circle.

USING LETTERS IN MATHEMATICS

It is customary to reserve particular letters to denote certain things in mathematics. For example, you probably spent most of your first-year algebra course trying to find x. It is customary to use letters at the end of the alphabet, such as x, y, and z, for variables. Letters at the beginning of the alphabet are often used to represent constants. For example, the standard form of a quadratic equation is $y = ax^2 + bx + c$, where a, b, and c are the coefficients. The letters f, g, and h are usually used for function labels. If letters are used to represent specific physical quantities, then these conventions do not apply.

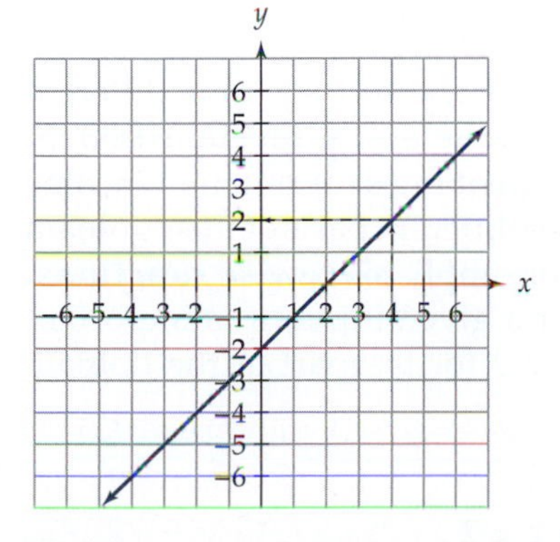

Figure 5 Evaluating this function at $x = 4$.

You do not need to have a function in symbolic form to evaluate it. On a graph, the inputs are on the horizontal axis and the outputs are on the vertical axis. So, to evaluate your function for a given input, find the appropriate value on the horizontal axis, go up (or down) to your graph, and read across to find the output on your vertical axis. For example, in the graph in Figure 5, the function evaluated at $x = 4$ gives an output of $y = 2$.

To evaluate a function represented numerically, find your input in the first row or column and read the corresponding output in the second row or column. For example, in Table 4, the function evaluated at $g = 10$ gives an output of $C(10) = \$12.10$.

TABLE 4 A Function Represented Numerically

g	0	5	10	15
$C(g)$	0	\$6.05	\$12.10	\$18.15

Changing Input Versus Changing Output

When evaluating functions, you need to be careful of the order in which you perform your calculations. For example, there is a difference between $f(x)+2$ and $f(x+2)$. Let $f(x)=x^2$. Then, for $f(x)+2$, we would evaluate the function at x (square x) and add 2 to our answer to get $f(x)+2=x^2+2$. For $f(x+2)$, we would evaluate the function at $x+2$ (square $(x+2)$), which would result in $f(x+2)=(x+2)^2=x^2+4x+4$. As you can see, the results are quite different. The difference between $f(x)+2$ and $f(x+2)$ is that, in the first case, you are adding 2 to the *output*, whereas in the second case, you are adding 2 to the *input*. This is especially obvious when your function represents a physical situation.

Example 4 Let C be the function whose input is the number of gallons of gasoline, g, and whose output is cost (in dollars). What is the significance of $C(g)+2$ versus $C(g+2)$?

Solution $C(g)+2$ says that you pay for g gallons of gas and then decide to give the cashier an extra \$2. So, you are paying \$2 more than the price of the gasoline you purchased; that is, you are adding 2 to your output. On the other hand, $C(g+2)$ means that you purchase $g+2$ gallons of gasoline. You are only paying for the gasoline—no additional tip this time!—but you decided to buy 2 gallons more of gas than you had originally intended. This time, you added 2 to your input. Notice that both of these functions, $C(g)+2$ and $C(g+2)$, give money as an answer, but $C(g)+2$ is adding \$2 to the output, whereas $C(g+2)$ is adding 2 gallons to the input.

Piecewise Defined Functions

Some functions may be difficult to evaluate because they have different rules for different parts of their domain. An example is the function used for determining income tax since there are different rates for different income levels. Functions that have different rules for different parts of their domain are called **piecewise defined functions,** or simply **piecewise functions.** When evaluating a piecewise function for a given input, it is important to make sure you are using the rule defined for the part of the domain containing your input.

For example, suppose

$$f(x)=\begin{cases}-x+6, & \text{if } x<2\\ x^2, & \text{otherwise}\end{cases}$$

and you are asked to evaluate $f(x)$ for $x = -4,\ 0,\ 2,\ 4$. Evaluating at these points requires that we use different rules, depending on the input.

$$\begin{aligned} f(-4) &= -(-4) + 6 = 10 \\ f(0) &= -0 + 6 = 6 \\ f(2) &= 2^2 = 4 \\ f(4) &= 4^2 = 16. \end{aligned}$$

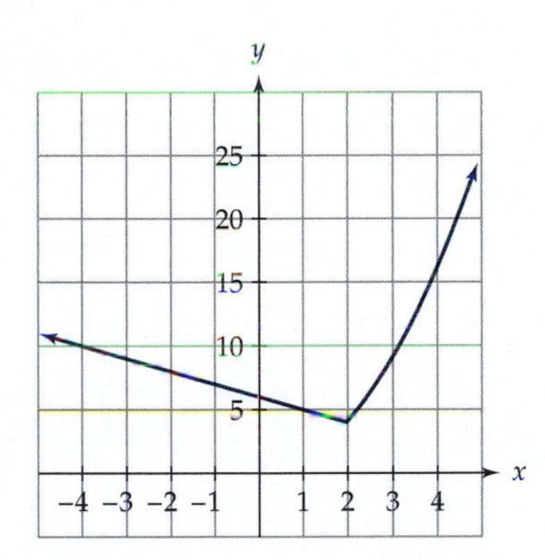

Figure 6 The graph of the piecewise function $f(x) = \begin{cases} -x+6, & \text{if } x < 2 \\ x^2, & \text{otherwise.} \end{cases}$

The decision whether to use the rule, "Find the opposite of the input and add six" or the rule, "Square the input" depends on whether your input is less than 2. The graph of this function is shown in Figure 6.

You may be familiar with the absolute value function, $|x|$, which is defined as a piecewise function.

$$|x| = \begin{cases} -x, & \text{if } x < 0 \\ x, & \text{otherwise.} \end{cases}$$

The graph of this function is shown in Figure 7.

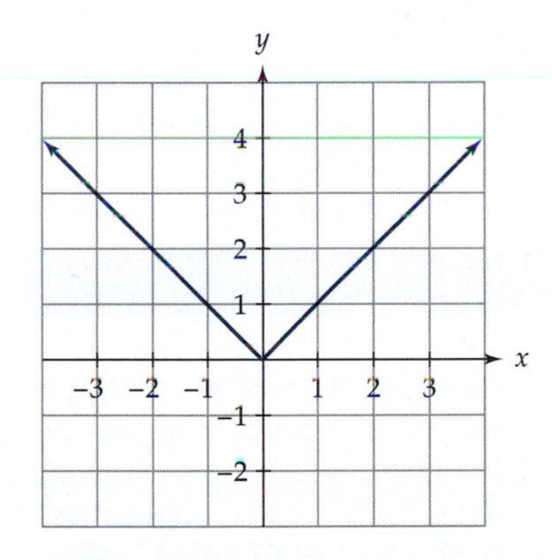

Figure 7 A graph of $f(x) = |x|$.

Notice the abrupt change in the graph at $x = 0$. There was also an abrupt change of the graph in Figure 6 at $x = 2$. Graphs of piecewise functions often (but not always) have an abrupt change of behavior at the point where the rule changes.

Many functions that model physical situations are piecewise functions. For example, F & H SoftHouse, an Internet server, has a cost of \$10 for the first 45 hr and \$0.70 for each hour used above 45.[7] This is a piecewise function since the rule we use depends on whether we use less than or more than 45 hr. If we let c represent the cost of using the Internet server for t hours, then a symbolic representation of this function is:

$$c(t) = \begin{cases} 10, & \text{if } 0 \le t \le 45 \\ 0.7(t-45) + 10, & \text{if } t > 45. \end{cases}$$

The cost of 10 hr of Internet service, or $c(10)$, would be \$10, $c(30) = \$10$, and $c(50) = 0.7(50 - 45) + 10 = \$13.50$.

READING QUESTIONS

13. In the notation $f(x)$, what does x represent?

14. Evaluate each of the following at $x = 2$.

(a) $f(x) = \dfrac{x-3}{x+2}$

[7] *F & H SoftHouse*, SoftHouse Home Pages *(visited 8 August 1997) (http://www.softhouse.com)*.

(b)

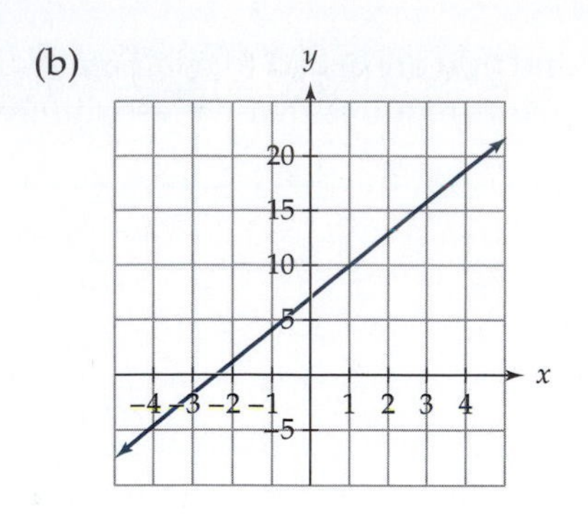

(c)

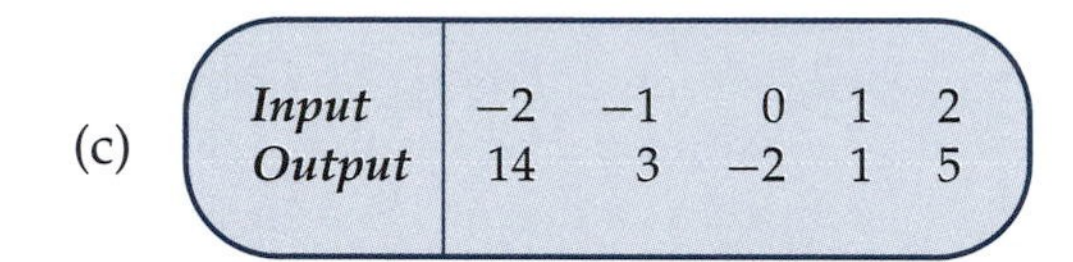

Input	−2	−1	0	1	2
Output	14	3	−2	1	5

15. What is the difference between $f(x+2)$ and $f(x)+2$?

16. What is a piecewise function?

17. Let

$$g(x) = \begin{cases} -x+3, & \text{if } x < 0 \\ 2x+3, & \text{if } 0 \le x \le 4 \\ x^2, & \text{otherwise.} \end{cases}$$

Find $g(-4)$, $g(0)$, $g(4)$, and $g(10)$.

SUMMARY

Functions are useful in modeling physical, financial, and even sociological situations in addition to being interesting to study from a mathematical perspective. They can be represented symbolically (as a formula), graphically, numerically (as a table of numbers), and verbally. Specifically, a **function** is a rule that assigns an input to at most one output. We can use the **vertical line test** to determine if a graph is a function.

Two important words associated with functions are domain and range. The **domain** of a function is the set of *valid* inputs. The **range** is the set of *actual* outputs.

The notation $f(x)$, pronounced "f of x," is commonly used to denote the output of a function. The letter f is typically used to remind us that this object is a $f$unction. The letter x is typically used to denote the input. The f is the *label* representing the function (like your name is the label representing you), and the x is a *placeholder* representing the input.

Functions that have different rules for different parts of their domain are called **piecewise defined functions,** or simply **piecewise functions.** When evaluating a piecewise function for a given input, it is important to make sure you are using the rule defined for the part of the domain containing your input.

EXERCISES

1. The function for the area of a circle is $A = \pi r^2$. Another (although less common) way of writing this is $A(r) = \pi r^2$. What is the significance of the (r) in $A(r)$?
2. Evaluate each of the following functions at the points indicated.
   (a) Evaluate $A = \pi r^2$ for $r = 4$.
   (b) Evaluate $f(x) = 3x^2 - 4$ for $x = 1$.
   (c) Evaluate the average annual salary, given in millions of dollars, for an NBA player for the 1995 season.

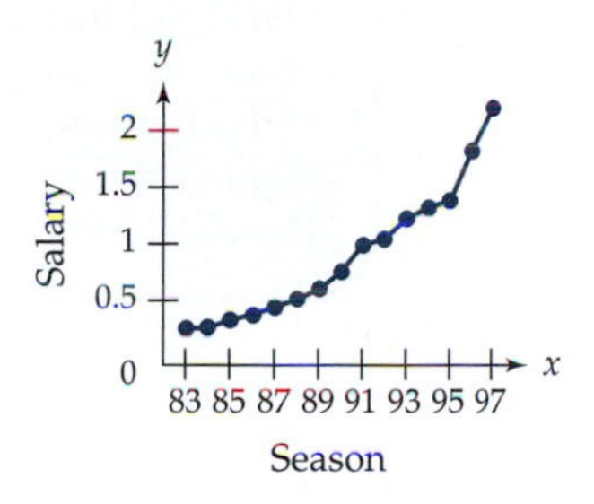

   (d) Evaluate the following graph for $x = 3$.

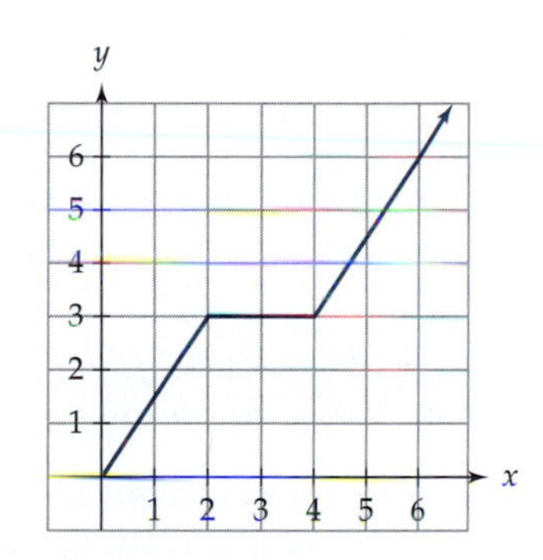

   (e) Evaluate the U.S. population by the Bureau of the Census for 1940.

Year	Population
1800	5,308,483
1820	9,638,453
1840	17,063,353
1860	31,443,321
1880	50,189,209
1900	76,212,168
1920	106,021,537
1940	132,164,569
1960	179,323,175
1980	226,542,203
1990	248,709,873

(f) Evaluate the "Top Salaries of Professional Basketball Players in 1996"[8] for Shaquille O'Neal.

Name	Yearly Salary
Michael Jordan	$30,140,000
Horace Grant	$14,857,000
Reggie Miller	$11,250,000
Shaquille O'Neal	$10,714,000
Gary Payton	$10,212,000
David Robinson	$9,952,000
Juwan Howard	$9,750,000
Hakeem Olajuwon	$9,655,000
Alonzo Mourning	$9,380,000
Dennis Rodman	$9,000,000

(g) Evaluate the tuition cost for 11 credits.

Tuition for undergraduate students who are Michigan residents taking from 12 to 16 credit hours is the same, a total of $1,390 per semester. Tuition is $124 per credit hour for fewer than 12 credits and for each credit over 16.

(h) Evaluate the function given below where the input is yourself.

What are the last two numbers of your Social Security number?

3. Why is the following rule a function?

Given the number of gallons of a particular grade of gasoline at a certain gas station, determine the cost of the gasoline.

4. Determine if the following rules are functions. If the rule is a function, give the domain and range. If it is not, give an example of an input that has more than one output.
   (a) A listing of the outdoor temperature every hour for a 24-hr period during spring break in Tampa, Florida.
   (b) A listing of the number of people swimming at a certain beach given the temperature of the water.
   (c) $A = 6s^2$, where s is the length of the side of a cube and A is the surface area.

[8] USA Today, *15 November 1996, p. 16C.*

(d)

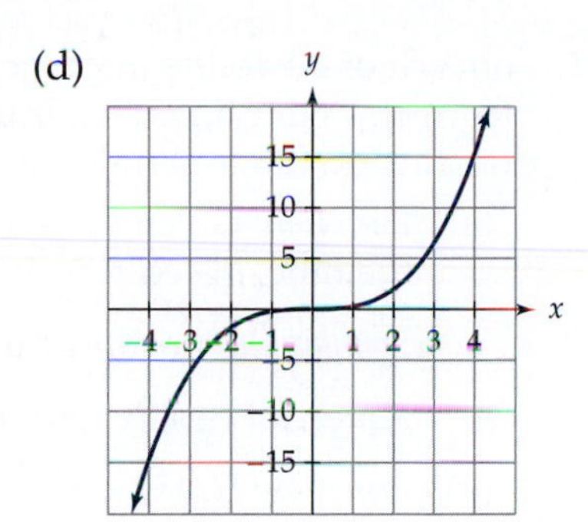

(e)

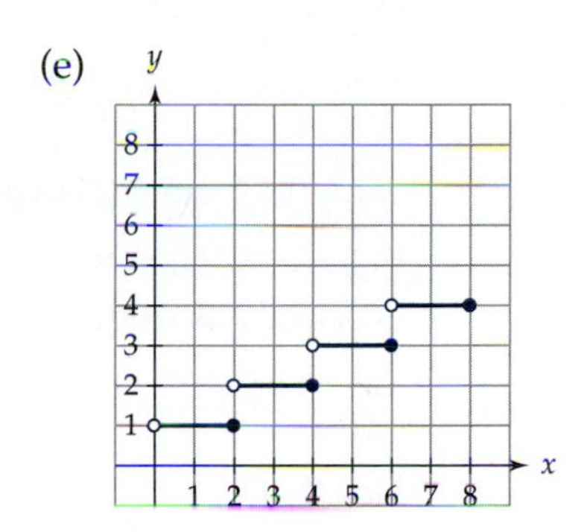

(f) $y = 3$

(g) $y = \sqrt{x - 4}$

(h) $y = \dfrac{1}{x + 7}$

(i) $y = \dfrac{-2 \pm \sqrt{x + 14}}{6}$

(j)

Input	20	21	22	23	24
Output	16	18	19	20	20

(k)

Input	16	18	19	20	20
Output	20	21	22	23	24

5. Compare $y = \frac{4}{3}\pi x^3$ and $V = \frac{4}{3}\pi r^3$.
   (a) What are the domain and the range for the function $y = \frac{4}{3}\pi x^3$?
   (b) The formula to calculate the volume of a sphere is given by $V = \frac{4}{3}\pi r^3$, where r represents the radius and V represents the volume. What are the domain and the range?
   (c) Your answer for part (b) should differ from your answer for part (a) even though the formulas look the same except for different letters. What is causing this difference?

6. For each of the following functions, the domain, the range, or both are incorrect. Determine what is incorrect and describe what a reasonable domain and range should be.
   (a) *Function*: Given the speed of a car in miles per hour, determine the time, in seconds, it takes to stop.

   Domain: all positive numbers

   Range: all positive numbers
   (b) *Function*: $f(x) = \sqrt{x^2 - 4}$

   Domain: $-2 \leq x \leq 2$

   Range: all numbers greater than or equal to zero
   (c) *Function*: You are buying a new car at a dealership. Given the amount of money borrowed, determine the monthly payment.

   Domain: \$10,000 $\leq m \leq$ \$40,000, given that m is the amount of money borrowed

   Range: \$10,000 $\leq p \leq$ \$40,000, given that p is the monthly car payment
   (d) *Function*:

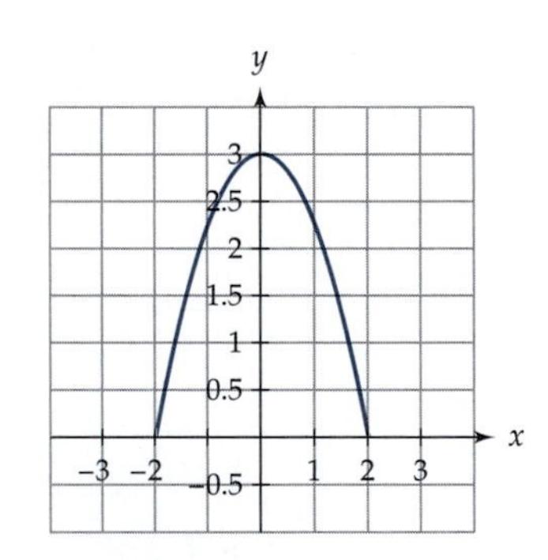

   Domain: $x \geq 0$

   Range: $0 \leq y \leq 3$

7. Evaluate each of the following functions at the given inputs.
   (a) If $f(x) = 1/x$, determine $f(10)$ and $f(\frac{1}{10})$.
   (b) If $h(z) = 3z - 4$, determine $h(\pi)$ and $h(\pi + 1)$.
   (c) If $g(n) = n + 1$, determine $g(2)$ and $g(x^2)$.
   (d) If $f(c) = \frac{9}{5}c + 32$, determine $f(20)$ and $f(40)$.
   (e) If

$$f(x) = \begin{cases} 2x + 3, & \text{if } x < 3 \\ x^2, & \text{otherwise,} \end{cases}$$

   determine $f(-5)$ and $f(5)$.

8. In general, what is the difference between $f(x+a)$ and $f(x)+a$, where a is a constant?
9. Let n be the function used to determine the number of hamburgers produced from a certain amount of ground beef, where x is the weight of ground beef in pounds and $n(x)$ is the number of hamburgers you can make. In terms of pounds of ground beef and numbers of hamburgers, what is the meaning of
   (a) $n(x+2)$
   (b) $n(x)+2$
10. Let s be the function used to determine the shoe size of someone with a certain length foot, where x is the length of the person's foot in inches and $s(x)$ is the shoe size. In terms of shoe size and the length of someone's foot, what is the meaning of
   (a) $s(x-1)$
   (b) $s(x)-1$

INVESTIGATIONS

INVESTIGATION 1: COMPARING INTERNET PACKAGES

Internet companies offer a variety of different packages. Some are designed to be attractive to the casual user and some to the frequent user. Two popular national companies, America Online and Prodigy, provide similar services to their customers. We are interested in investigating the cost of the following packages provided by these companies.[9]

Internet Provider	America Online	Prodigy
Cost for unlimited use	$21.95	$19.95
Cost of basic service	$4.95	$9.95
Hours included in basic service	3	10
Cost per hour of additional hours	$2.50	$1.50

1. Fill in the following table with the cost of using the number of hours specified for each of the four different plans.

[9] *America Online,* Pricing Plans *(visited 24 February 2000) ⟨http://www.aol.com/info/pricing.html⟩. Prodigy,* Prodigy Internet *(visited 29 February 2000) ⟨http://www.prodigy.com/pcom/prodigy-internet/splash-index.html⟩.*

Hours	America Online Basic	Prodigy Basic	America Online Unlimited	Prodigy Unlimited
3				
6				
9				
12				
15				

2. When is it cheaper to buy the unlimited service for America Online instead of the basic? Answer the same question for Prodigy.
3. Suppose you were writing a review of these services for a newspaper. Write a brief analysis of the costs of these services with recommendations for which one an individual should purchase if cost is the overriding factor. Since different people will spend different amounts of time using the Internet, your report should contain recommendations depending on different use patterns (e.g., if a person intends to use the service for fewer than 5 hr a month, between 5 and 10 hr a month, etc.).

INVESTIGATION 2: INPUT AND OUTPUT CHANGES

In this section, we briefly looked at the difference between $f(x + a)$ and $f(x) + a$, where a is a constant. This investigation explores this same concept graphically.

1. As mentioned in the reading, the key difference between $f(x + a)$ and $f(x) + a$ is that one changes the input, whereas the other changes the output. Which is which? How do you know?
2. Let $f(x) = x^2$.
   (a) Determine a symbolic representation for each of the following functions.

$$y = f(x + 2)$$
$$y = f(x) + 2$$
$$y = f(x - 2)$$
$$y = f(x) - 2$$

   (b) Sketch $f(x) = x^2$ and each of the functions from part (a).
3. Repeat question 2, this time using the function $f(x) = 3x$.
4. Describe in words the difference between the graphs of $y = f(x)$ and $y = f(x + a)$, where a is a constant. Be sure to mention what happens when a is positive versus what happens when a is negative.

5. Describe in words the difference between the graphs of $y = f(x)$ and $y = f(x) + a$, where a is a constant. Be sure to mention what happens when a is positive versus what happens when a is negative.
6. Let f be the graph of the following function.

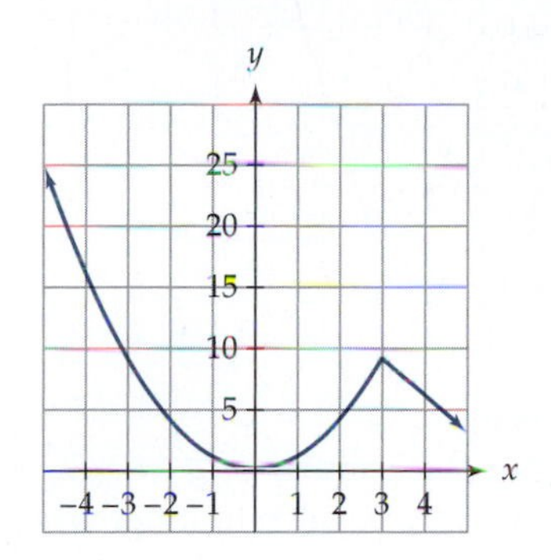

Match the graphs to the following functions.

$y = f(x - 1)$

$y = f(x) - 1$

$y = f(x + 1)$

$y = f(x) + 1$

(a)

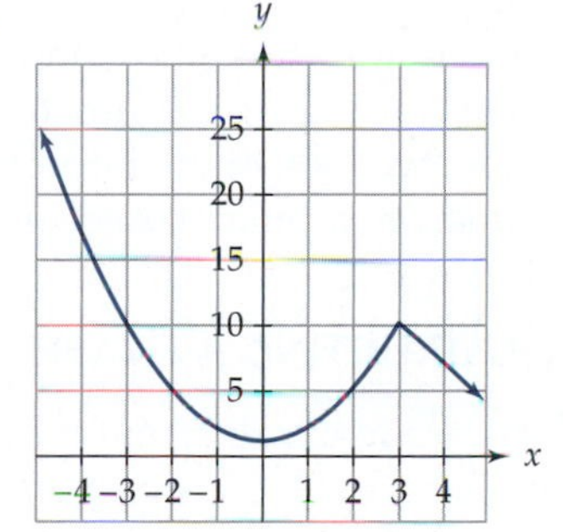

(b)

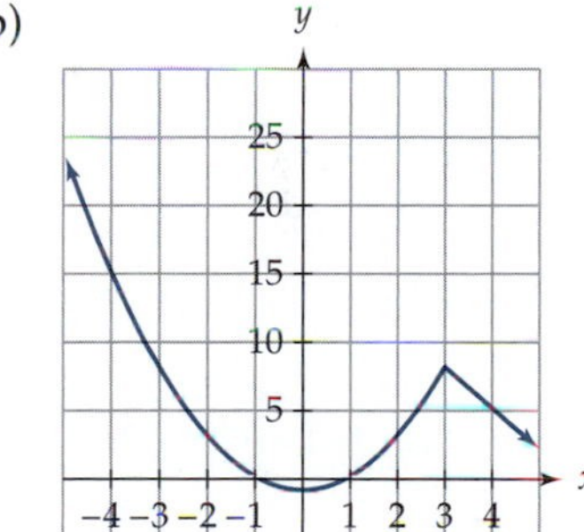

(c)

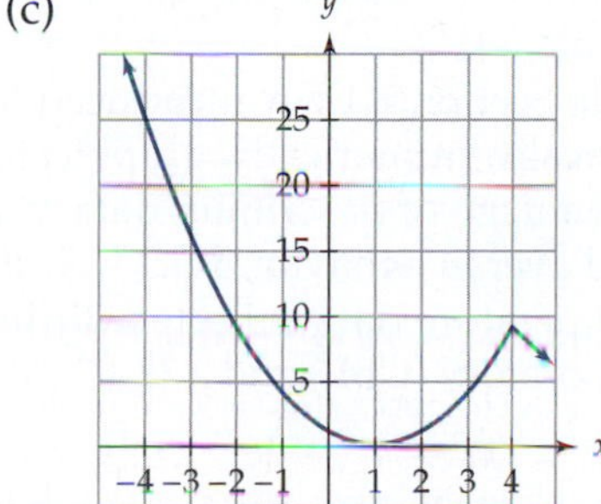

(d)

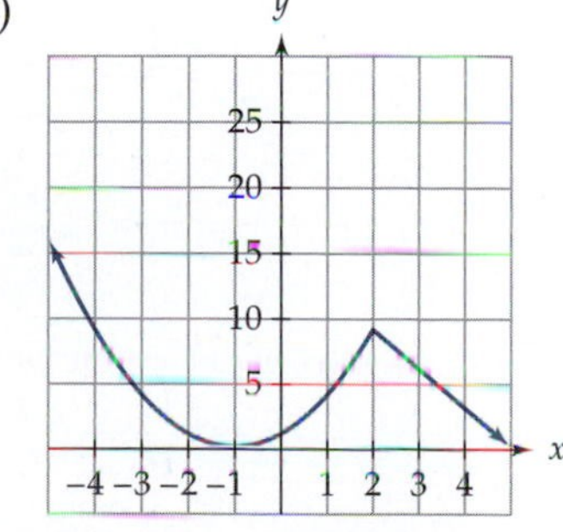

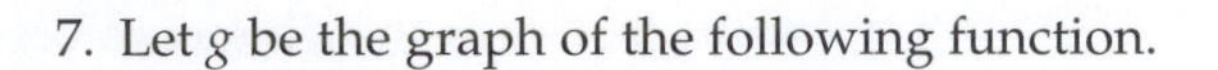

7. Let g be the graph of the following function.

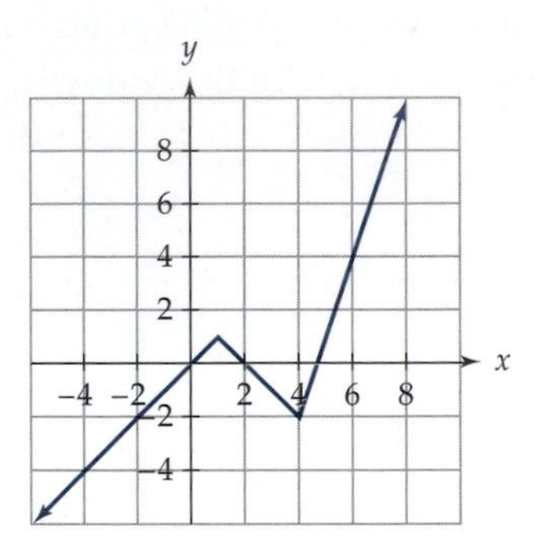

Draw the graphs of $y = g(x - 1)$, $y = g(x + 1)$, $y = g(x) - 1$, and $y = g(x) + 1$. Be sure to label which is which.

1.2 GRAPHICAL REPRESENTATIONS OF FUNCTIONS

Someone once said that a picture is worth a thousand words. In many ways, this is true. You can read about seeing the Grand Canyon, the New York City skyline, or Mount Rainier, but it takes a lot of words to gain the same understanding you get from just looking at a picture. The same principle is true with the relationship between the graphical and symbolic representations of a function. Looking at a graph often gives you a better understanding of the function than just looking at the symbols. In this section, we look at the importance of graphs, how to interpret them, some common vocabulary used in analyzing them, and scatterplots.

CONSTRUCTING A GRAPH

There are many different types of graphs, such as pie graphs, bar graphs, and line graphs. Newspapers and magazines publish graphs daily to track the stock market, express information about the weather, and present other types of information. In this book, we focus our attention on graphs of functions plotted in the Cartesian[10] coordinate system (which is the familiar xy-plane).

In Section 1.1 we introduced four ways of representing functions (symbolically, numerically, graphically, and verbally). We mentioned that one advantage of describing data in graphical form is that you can see trends and overall behavior quickly and easily. We illustrate this by using outside temperature data collected during a 24-hr period with a Calculator-Based Laboratory (CBL) and a TI-83 calculator. The program we used first stored

[10] *Named after René Descartes, who first used the two-axis system to plot points. He is the same person who philosophized about his existence by saying, "I think, therefore I am."*

the data in the calculator numerically and then graphed the data. Part of the hourly temperatures collected are given in Table 1. This numerical listing gives a lot of information, but it takes some time to see any relationships or trends. With the calculator graph of these data, however, shown in Figure 1, we can quickly and easily see the relationship between time and temperature. We can see that the temperature initially decreased, leveled off for a short time, then increased fairly rapidly before leveling off again toward the end, except for a brief increase and decrease.

TABLE 1 Temperature Data in Degrees Fahrenheit over a 24-hr Period

Time	4 P.M.	5 P.M.	6 P.M.	7 P.M.	8 P.M.	9 P.M.	10 P.M.	11 P.M.	12 A.M.	1 A.M.	2 A.M.	3 A.M.
Temperature	72°	69°	65°	63°	62°	62°	61°	59°	58°	57°	56°	56°
Time	4 A.M.	5 A.M.	6 A.M.	7 A.M.	8 A.M.	9 A.M.	10 A.M.	11 A.M.	12 P.M.	1 P.M.	2 P.M.	3 P.M.
Temperature	57°	62°	66°	72°	76°	80°	81°	82°	84°	83°	83°	85°

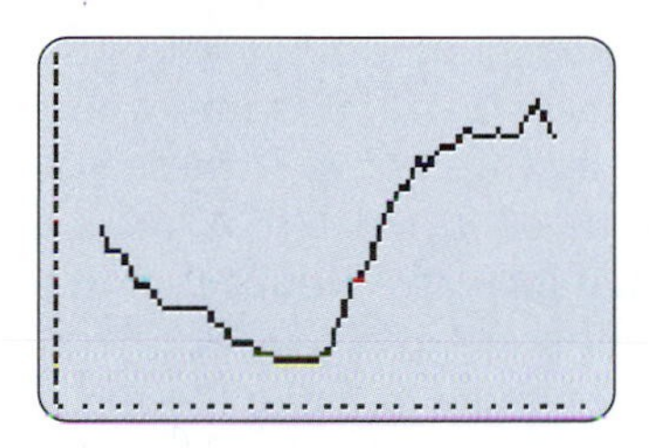

Figure 1 A relationship between time and temperature (without labels).

The graph shown in Figure 1 is not as useful at it could be. Because this graph has no labels for the horizontal and vertical axes, the information it provides is limited. Without labels, the only information we can obtain from the graph is its shape. Looking at Figure 1, we see that the temperature cooled down and then warmed up. Sometimes, that may be all we are interested in knowing about a graph. In this case, however, we would like to be able to read the graph to determine the temperature at a given time of day. This will combine the specific type of information you can get from Table 1 along with the ease of seeing trends that graphs provide. Figure 2 contains another graph of the same situation. This time we carefully labeled our axes to show that the input represented time and the output represented temperature. We also gave the scales we used to measure each of these, which allows us to gain additional information from the graph. We can now see when the temperature decreased, what the minimum temperature was during this period, and so on directly from the graph.

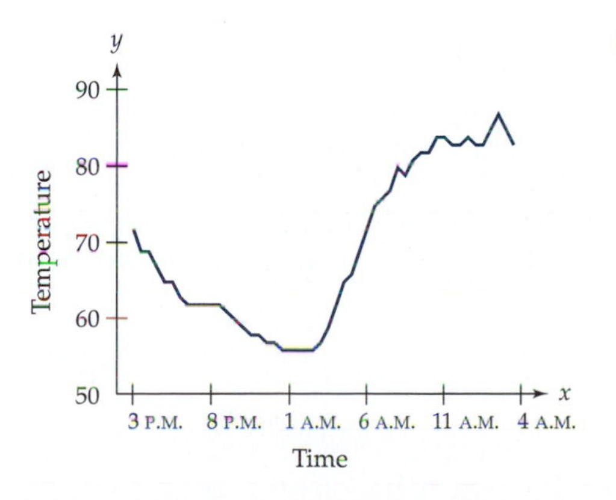

Figure 2 A relationship between time and temperature (with labels).

Notice in Figure 2 that time was placed on the horizontal axis and temperature was placed on the vertical axis. In a function that relates time and temperature, we want time to be the input and temperature to be the output because the temperature *depends* on the time of day. (*Note:* The temperature will also depend on your location and many other variables, but in all our examples, we focus on just one input variable at a time.) When constructing graphs, we place the input (or the independent variable) on the horizontal axis and the output (or the dependent variable) on the vertical axis. So, if b depends on a, we place a on the horizontal axis and b on the vertical axis. This allows us to observe the change in b as a varies. For example, the cost of a used car depends on the number of miles it has been driven, the cost of tuition depends on how many hours of credit you are taking, and how many pages of a book you read depends on how much time you spend reading. If you were to graph these relationships, the number of miles driven, the number of credit hours, and the time spent reading would be the inputs and would go on the horizontal axis. Recognizing and quantifying these types of relationships correctly is an important part of doing science and mathematics.

Notice that the tick marks on the x-axis in Figure 2 are evenly spaced. For example, the distance between 3 P.M. and 4 P.M. is the same as the distance of any other 1 h portion along the horizontal axis. Likewise, there is the same distance between 70° and 72° as there is between any other 2° segment along the vertical axis. When drawing graphs by hand, it is important to have your tick marks on an axis the same distance apart. If these scales were not done uniformly, the graph would be distorted. What looks like a quick increase might, in reality, be a gradual increase. Notice that in our temperature example, we do not have the same scale on both the horizontal and vertical axes. In this case, it hardly makes sense to make these scales the same because our input and output are in different units. Even if the input and output were given in the same units, the scales do not have to be the same. Not using the same scales on both axes might cause the graph to appear stretched or compressed. There are times, however, when not having the scales the same is advantageous.

Some care should be used when determining a proper domain and range in which to graph your function. If the domain or range you choose is too small, you will not see your entire graph and you may be omitting the part that you are most interested in seeing. If the domain or range you choose is too large, you may not be able to see the information you want from your graph. For example, you might not be able to see small increases or decreases in your function. In choosing a proper domain for the temperature graph, we needed to make sure we included our entire 24-hr period.[11] We also needed to make sure our range included the lowest temperature recorded, 56°, and the highest temperature recorded, 87°. As you can see from our graph, our window was a little bit wider than that to allow a sense of "margins" for the graph. Notice that the origin[12] was not

[11] *The calculator originally stored these as the numbers 19 through 43, and we converted them to the actual time of day.*

[12] *The origin is the point* $(0, 0)$ *and is located at the intersection of the two axes.*

included in our temperature graph. If the origin is not part of the domain and range, it should not be included on the graph unless there is a good reason to do so.

READING QUESTIONS

1. What are the advantages of a graphical representation of a function?
2. Suppose "your pay depends on the number of years you have been working at a company." Which of the two variables, pay or number of years worked, would you put on the horizontal axis and which on the vertical axis? Why?
3. Why is it important to label the axes of your graphs?
4. What would be a consequence of not having equally spaced tick marks on an axis?

WORDS USED TO DESCRIBE GRAPHS

A few words are commonly used by mathematicians to describe graphs. Most of you have heard before and seem like natural choices for the concepts they describe. Some of these words may be new to you. It is important to understand their definitions since discussing graphs without a common vocabulary is difficult and leads to confusion.

Increasing and Decreasing

A graph of a function is said to be **increasing** over a particular interval if, for each point in that interval, every point to the right has a greater output. Symbolically, we say a function is increasing on an interval if for any two numbers a and b on the interval with $b > a$, then $f(b) > f(a)$. Thus, the larger the input, the larger the output. Increasing graphs are going "uphill" as the input increases. This does not have to happen over the entire graph since increasing is a property of intervals. In our temperature example (Figure 2), the graph is increasing over the interval from 4 A.M. to 9 A.M.

A graph is **decreasing** over an interval if, for each point in that interval, every point to the right has a smaller output. Symbolically, we say a function is decreasing on an interval if for any two numbers a and b on the interval with $b > a$, then $f(b) < f(a)$. In other words, the larger the input, the smaller the output. Decreasing graphs are going "downhill" as the input increases. In our temperature example (Figure 2), the graph is decreasing over the interval from 9:30 P.M. to 11:30 P.M.

Concavity

Another word used in describing graphs is *concavity*. Informally, a graph is **concave up** if it bends upward, like a bowl that can hold water. A graph is **concave down** if it bends downward, like an umbrella when it is being used properly. The concavity of a graph and whether it is increasing or decreasing are two independent ideas. An increasing function can either

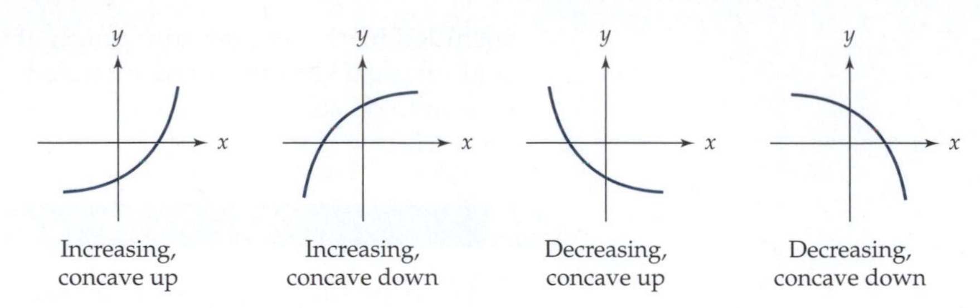

Figure 3 Increasing and decreasing functions with different types of concavity.

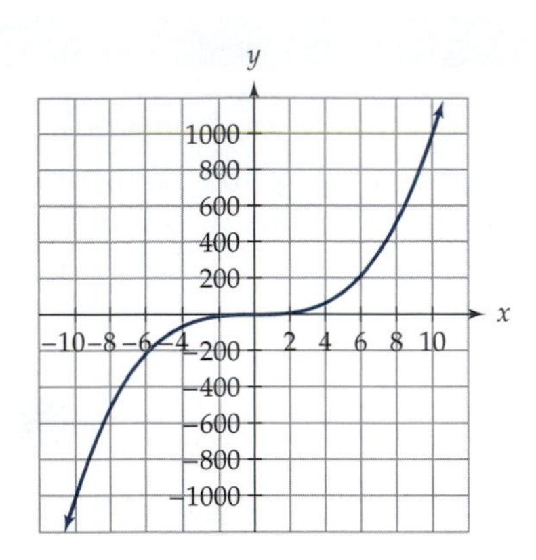

Figure 4 The graph of $y = x^3$ is concave down when $x < 0$ and concave up when $x > 0$.

be concave up or concave down. Similarly, a decreasing function can either be concave up or concave down. (See Figure 3.)

The difference between increasing/decreasing and concavity is that increasing and decreasing describe how the output of the function is changing, whereas the concavity of a function tells how the *rate* of increase or decrease is changing. This is how concavity is defined. A graph is **concave up** on intervals where the rate of change increases. A graph is **concave down** on intervals where the rate of change decreases. For example, the graph in Figure 4 is always increasing. Where it is concave down (when $x < 0$), it is increasing at a slower and slower rate. So, its rate of change is decreasing, which is shown by the graph being less steep near the origin. Where it is concave up (when $x > 0$), it is increasing at a faster and faster rate. So, its rate of change is increasing, which is shown by the graph getting steeper as you move to the right.

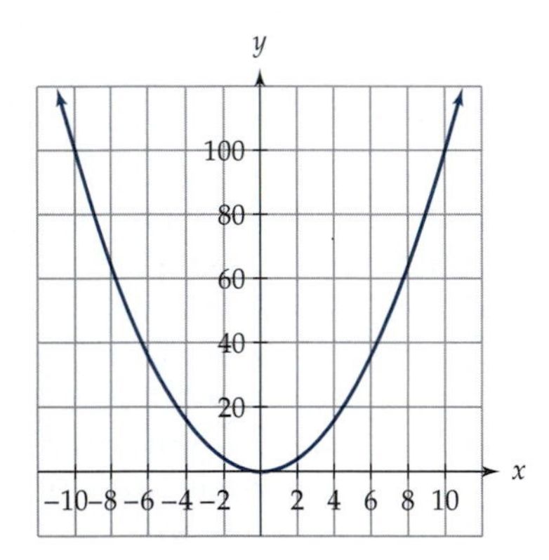

Figure 5 The graph of $y = x^2$.

Example 1 The graph of $y = x^2$ is shown in Figure 5. Describe where this function is decreasing, increasing, concave up, and concave down.

Solution The function is decreasing when $x < 0$ and increasing when $x > 0$. It is always concave up and never concave down.

X- and Y-Intercepts

Some important points on a graph are the x- and y-intercepts. The ***y*-intercept** is the value of the function where the graph crosses the y-axis. This is the point where the input (or x-coordinate) of the function is zero. An ***x*-intercept** is the x-coordinate of a point where the function crosses the x-axis. It is also called a *root* or *zero* of the function. These are the places where the output (or y-coordinate) of the function is zero. Whereas a function can have at most one y-intercept, it can have any number of x-intercepts. The graph shown in Figure 6 has a y-intercept of -6 since it crosses the y-axis at the point $(0, -6)$. It has x-intercepts of -3 and 2 since it crosses the x-axis at the points $(-3, 0)$ and $(2, 0)$.

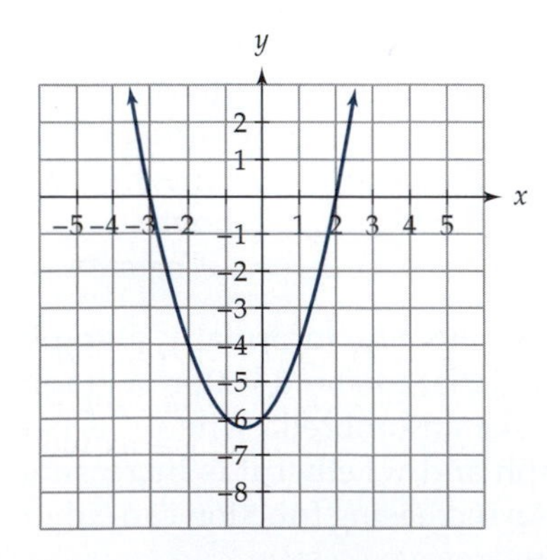

Figure 6 The graph of $y = x^2 + x - 6$.

Symmetry

Another characteristic of graphs is symmetry. Graphs in the Cartesian plane have two common types of symmetry: symmetry about the y-axis and symmetry about the origin. A graph is **symmetric about the y-axis** if the part of the graph to the left of the y-axis is a reflection of the part to the right. If f is a function that is symmetric about the y-axis, then f has the property $f(x) = f(-x)$. Thus, an input and the negative of that input will give the same output.[13] The graph of $f(x) = x^2$ shown in Figure 5 is symmetric about the y-axis because $f(-x) = (-x)^2 = x^2 = f(x)$.

A graph is **symmetric about the origin** if, when the graph is rotated 180°, with the origin being the center of rotation, it looks the same as it did before it was rotated. If f is a function that is symmetric about the origin, then f has the property $f(x) = -f(-x)$. Thus, any input and the negative of any input will give opposite outputs.[14] The graph of $f(x) = x^3$, that is shown in Figure 4, is symmetric about the origin because $-f(-x) = -(-x)^3 = x^3 = f(x)$.

READING QUESTIONS

5. What does it mean when we say that a function is increasing?
6. Are the golden arches at a McDonald's restaurant concave up or concave down?
7. If the point $(0, 8)$ was on the graph of a function, would it be located at an x-intercept or the y-intercept? How do you know?
8. If f is symmetric about the origin and $f(1) = 2$, what is $f(-1)$?

USING GRAPHS TO REPRESENT PHYSICAL SITUATIONS

Graphical representations of functions are only useful if you are able to correctly interpret the information they present. Therefore, it is important to have the ability to convert a verbal description of a physical situation to a graph and vice versa. Often, we are primarily interested in determining only the shape of the graph. In this case, the only labels we need put on the axes are to describe the inputs and outputs. The examples here do not include any sort of units or specific numbers; we are primarily interested in the shape of the graph and want to concentrate on the general properties such as where it is increasing, where it is decreasing, and its concavity. For example, suppose a person was walking away from you down a hallway at a constant speed. In this situation, distance depends on time. Therefore, a graph of this situation would have time on the x-axis and the distance from you on the y-axis. The graph representing the function would look like a straight line with positive slope. The graph is increasing because the person is walking away from you. The graph is a line because the person is walking at a constant speed. If the graph were a horizontal line, then the output would not be changing so the person would not be moving. If the graph were a line with a negative slope, then the person would be moving closer to you.

What if the person was not walking at a constant rate? In modeling physical situations, you need to consider not only whether the function is increasing or decreasing, but the rate of increase or decrease as well.

[13] *Functions that have this property, $f(x) = f(-x)$, are also called* even *functions.*

[14] *Functions that have this property, $f(x) = -f(-x)$, are also called* odd *functions.*

Concavity is going to be important since it describes the rate of increase or decrease. Figure 7 shows two graphs that represent the distance a person is away from you as he or she walks down a hallway. In Figure 7(a), the person starts out quickly walking away from you and then slows down. This gives us a graph that is always increasing but concave down since the person's rate is decreasing. In Figure 7(b), the person starts out slowly walking away from you but then speeds up. This gives us a graph that is also increasing but concave up since the person's rate is increasing. Remember that concavity shows how the person's *rate* of walking is changing. A graph that is concave up will have a rate that is increasing, whereas a graph that is concave down will have a rate that is decreasing. Notice that both graphs are increasing over the entire interval, showing that the person is always walking away from you.

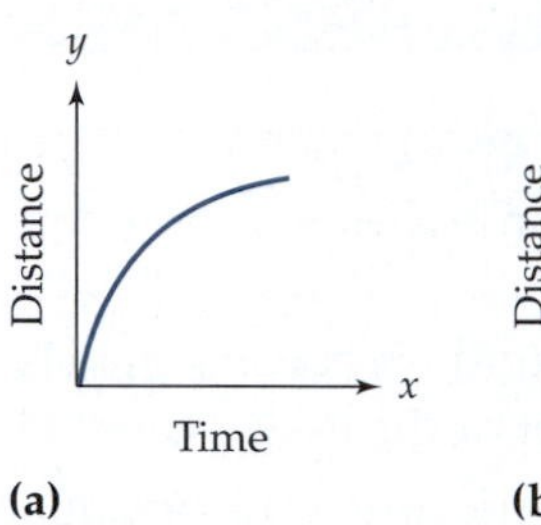

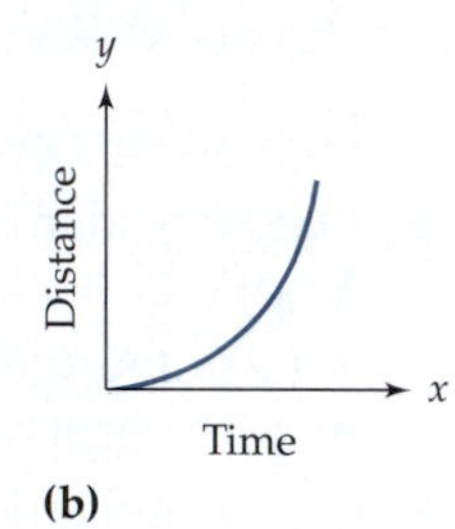

Figure 7 Two graphs showing a person's position as he or she walks down a hallway.

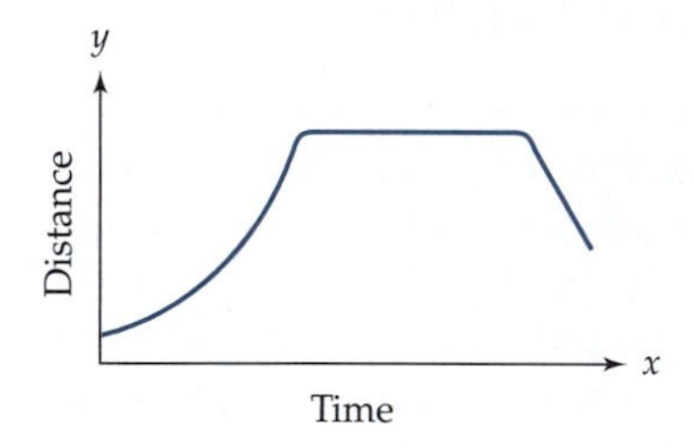

Figure 8

Example 2 The graph shown in Figure 8 represents a function where time is the input and the distance a person is away from you as he or she walks down a hall is the output. Describe how that person is walking.

Solution Because at the beginning the graph is increasing and concave up, the person is walking away from you and the rate of walking is increasing. So, the person starts out walking slowly and then speeds up. Where the graph is horizontal, the person has stopped. Where the graph is decreasing, the person is walking back toward you at a constant rate. Since the graph stops at a higher point than where it began, the person stopped before returning to the starting point.

Example 3 Suppose you threw a baseball straight up in the air so that its maximum height was 30 ft and you caught it as it returned to the ground. Describe how the height of the ball changed over time. Sketch a graph of this situation where time is the input and the height of the ball is the output.

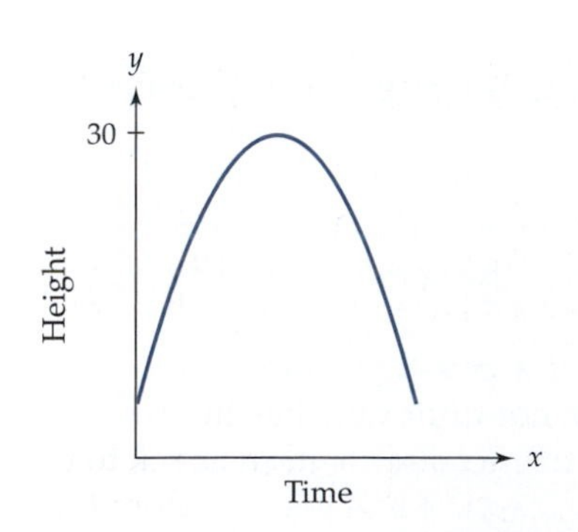

Figure 9

Solution To get the ball to a height of 30 ft, you would have to throw it fairly hard. Initially it would travel quite quickly. Gravity, however, would cause it to slow down until, at 30 ft, it would actually stop. This is the vertex of the graph. The ball will then start to travel back toward the ground. As it does, gravity will cause it to speed up until you catch it. A graph of this situation, with time as the input and height above the ground as the output, is shown in Figure 9. Notice that this is not a picture of the flight of the ball; that would look like a vertical line. It is a graph of the height of the ball versus time. The shape of the graph corresponds to our verbal description. When the graph is steep, it indicates that the ball is traveling rapidly, which occurs soon after the ball is thrown and just before it is caught. In the middle, the graph is not as steep, which indicates that the ball was traveling slowly.

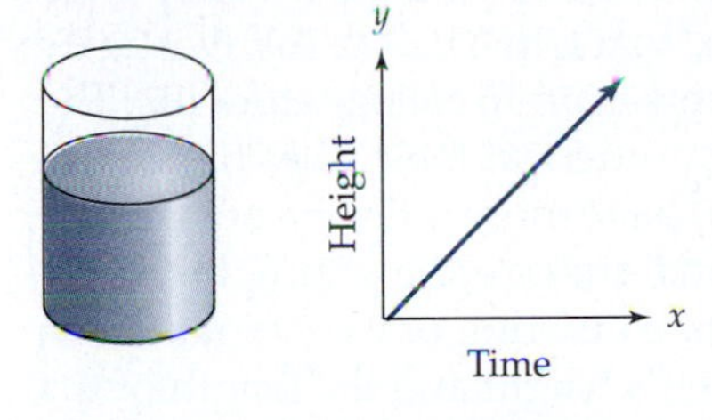

Figure 10 If water were added to a cylinder at a constant rate, then the graph of this function would be linear.

The last two examples involved graphing moving objects. Now we look at something a little different. Suppose you were filling a typical cylindrical drinking glass with water from a faucet. The water level in the glass rises at a fairly constant rate. The graph of this event, with time as an input and height of water as an output, looks like an increasing line since lines have a constant rate of increase. (See Figure 10.) The more interesting questions about water levels occur when you are filling containers that are not cylinders.

Figure 11

Example 4 Suppose you are filling a cone-shaped glass from a water faucet, where the water is flowing at a constant rate. (See Figure 11.) Sketch a graph of this function, where time is the input and the level of water is the output.

Solution The graph looks like Figure 12. Since the radius of the cone near the bottom is very small, it does not take much water to make the water level in the glass go up by a fixed distance. Therefore, the water level would rise very quickly at first. As you add more water, however, the rate at which the level is rising slows down because it takes more water for the level to go up by that same distance near the top of the cone than it did at the bottom. The graph is increasing since the height of the water increases as the glass is filled. The graph starts out quite steep (since the water level is rising quickly at first) and gets less steep as the top of the glass is approached. This gives a graph that is *concave down* for the entire interval. Our graph will start at the origin, since the glass is empty when we begin and the domain of our function ends when the glass is full.

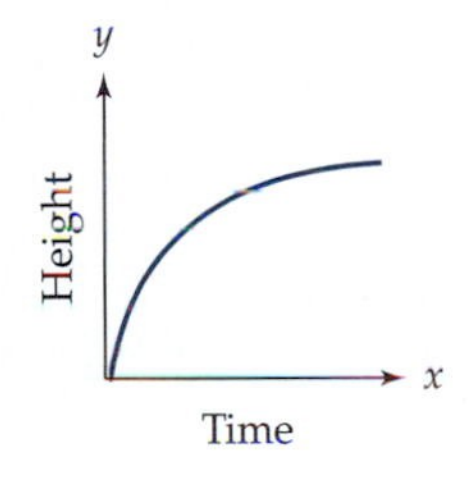

Figure 12

READING QUESTIONS

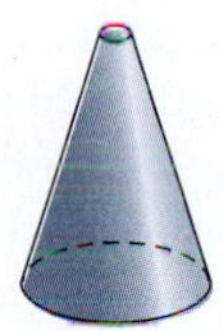

9. Suppose you have a container that is an inverted cone (similar to the accompanying picture) that is being filled with water flowing at a constant rate. Sketch a graph of this event, where the input is time and the output is the height of the water in the container. How does this graph differ from the one in Example 4?

10. Suppose a person left an office, walked slowly down a hall at a constant rate, stopped to talk to someone for a few seconds, and ran back to the office when the phone rang. Sketch a graph of this event, where time is the input and the distance the person is from the office is the output.

SCATTERPLOTS

All the graphs we have looked at so far were functions. Sometimes, however, we are interested in graphs that are not functions, which can occur when you plot data points. For example, a rule of thumb says that a person's height is about the same as his or her arm span (the distance from fingertip to fingertip with arms outstretched). If this rule is true, then when these data are plotted, the points should be close to the line $y = x$. To investigate this relationship, we collected measurements from 16 college students. The data are shown in Table 2. Note that this numerical table does not represent a function since some inputs (like 168) have more than one output. We can still graph the data as a scatterplot and analyze the graph, however. A **scatterplot** is a graph used to plot data consisting of two variables. In our example, the two variables are a person's height and the length of the person's arm span. A scatterplot of the data from Table 2 is shown in Figure 13. In this example, one of these variables does not depend on the other; rather, both variables depend on other things. Therefore, we do not have an independent variable and a dependent variable. This does not mean that height and arm span are not related. It just means that one is not the *cause* of the other.[15] In cases like these, it does not matter which variable is on the horizontal axis and which is on the vertical axis.

TABLE 2 Heights and Arm Spans (in Centimeters) of College Students

Height	Arm Span	Height	Arm Span
152	159	173	170
156	155	173	169
160	160	173	176
163	166	179	183
165	163	180	175
168	176	182	181
168	164	183	188
173	171	193	188

Looking at the scatterplot, the data appear to be fairly linear. Although the points do not all lie in a line, they produce a linear pattern. To test the idea that a person's height is about the same as the length of his or her arm span, we graphed the line $y = x$ along with the scatterplot. Since the input and output are the same on this line, it serves as a reference for those whose height is exactly the same as the arm span. As you can see from Figure 14,

[15] *This error is common in interpreting correlational studies, studies that demonstrate that two things are related. It is often tempting to assume that one of the variables causes the change in the other one when often they are both caused by something else.*

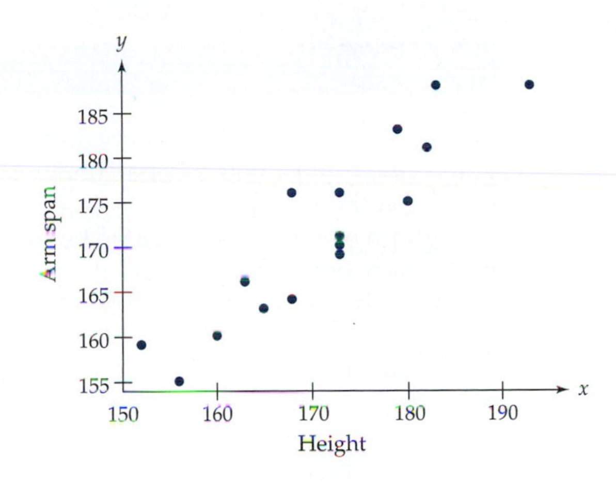

Figure 13 A scatterplot of students' heights and arm spans.

this line[16] does a fairly good job of approximating our data. The points above the line in Figure 14 have larger outputs than inputs. They represent people whose arm spans are longer than their height. The opposite is true for people whose points lie below the line.

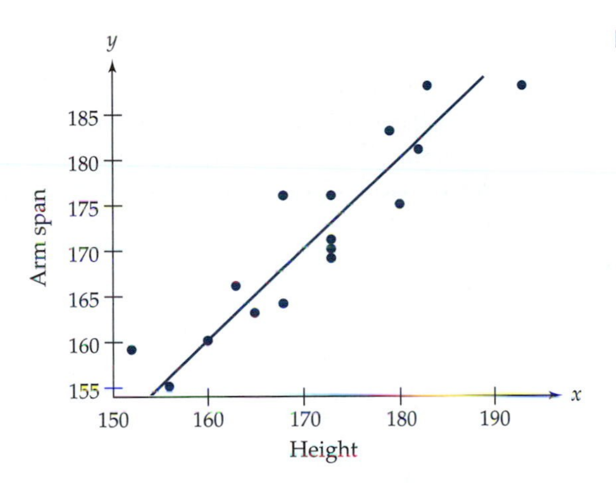

Figure 14 A scatterplot of students' heights and arm spans with the line $y = x$.

READING QUESTIONS

11. What is a scatterplot?

12. Why does the scatterplot shown in Figure 13 not represent a function?

13. What can we say about the height and arm span of people whose points lie above the line $y = x$ in Figure 14? What can we say about the height and arm span of people whose points lie below the line $y = x$ in Figure 14?

[16] *Although the line $y = x$ fits our data quite nicely, it is not the best-fitting line. The line that best fits a set of data is called a* ***regression line.*** *We discuss this concept in Section 7.4.*

SUMMARY

Graphs are very useful in understanding functions because looking at the graph often gives you a better understanding of the function than just looking at the symbols.

A graph of a function is said to be **increasing** over a particular interval if, for each point in that interval, every point to the right has a greater output. Symbolically, we say a function is increasing on an interval if for any two numbers a and b on the interval with $b > a$, then $f(b) > f(a)$. A graph is **decreasing** over an interval if, for each point in that interval, every point to the right has a smaller output. Symbolically, we say a function is decreasing on an interval if for any two numbers a and b on the interval with $b > a$, then $f(b) < f(a)$.

Although increasing and decreasing describe how the output of the function is changing, the concavity of a function tells how the *rate* of increase or decrease is changing. A graph is **concave up** on intervals where the rate of change increases. A graph is **concave down** on intervals where the rate of change decreases.

Some important points on a graph are the x- and y-intercepts. The **y-intercept** is the value of the function where the graph crosses the y-axis. This is the point where the input (or x-coordinate) of the function is zero. An **x-intercept** is the x-coordinate of a point where the function crosses the x-axis. It is also called a *root* or *zero* of the function. These are the places where the output (or y-coordinate) of the function is zero.

Another characteristic of graphs is symmetry. A graph is **symmetric about the y-axis** if the part of the graph to the left of the y-axis is a reflection of the part to the right. If f is a function that is symmetric about the y-axis, then f has the property $f(x) = f(-x)$. A graph is **symmetric about the origin** if, when the graph is rotated 180° with the origin being the center of rotation, it looks the same as it did before it was rotated. If f is a function that is symmetric about the origin, then f has the property $f(x) = -f(-x)$.

All graphs are not functions. One such graph that may not be a function is that of a scatterplot. A **scatterplot** is a graph used to plot data consisting of two variables.

EXERCISES

1. For each pair of variables listed below, indicate which is the independent and which is the dependent variable.
   (a) The temperature of an oven after you turn it on and set it for 350°F and time.
   (b) The number of gallons of gasoline that you buy and total cost of gasoline.
   (c) The amount of daylight in a day and the date.

(d) The time it takes to finish a running race and the length of a race.

(e) The month of the year and the cost of heating your home.

(f) The age and the price of a used car.

2. The graph of this function appears to be that of a line, but it was graphed inappropriately. Determine what is wrong with the graph and sketch the function properly by estimating points on the graph.

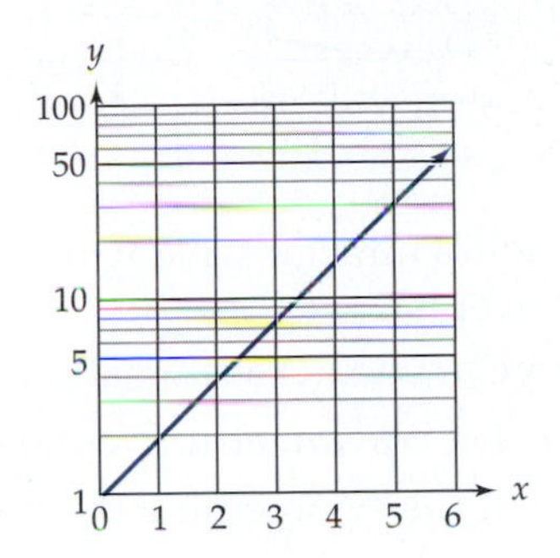

3. Indicate an appropriate domain and range for each of the following relationships. Be sure to indicate which variable is your input and which is your output.
   (a) The total meal cost for a wedding reception depends on the number of guests in attendance.
   (b) The amount of daylight on January 1 depends on your latitude in the Northern Hemisphere.
   (c) The average high daily temperature in Chicago depends on the day of the year.
   (d) The time it takes to drive to work depends on the distance you live from your work.
   (e) The speed of a car accelerating as it enters a freeway depends on time.

4. Use the terms *y-intercept, x-intercept, increasing, decreasing, concave up, concave down, symmetric about the y-axis,* and *symmetric about the origin* to describe the following graphs.

(a)

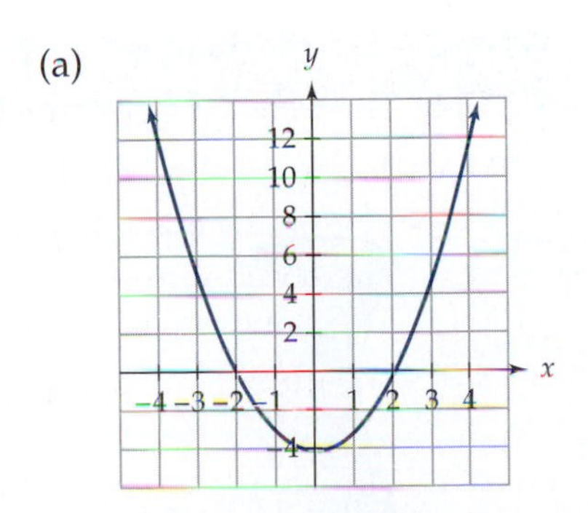

(b)

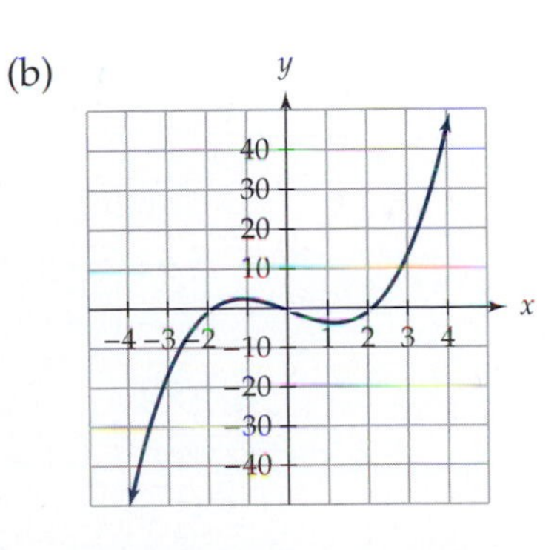

(c)

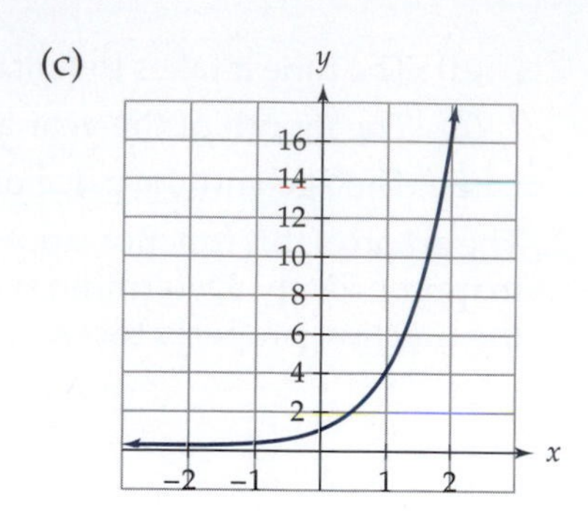

(d)

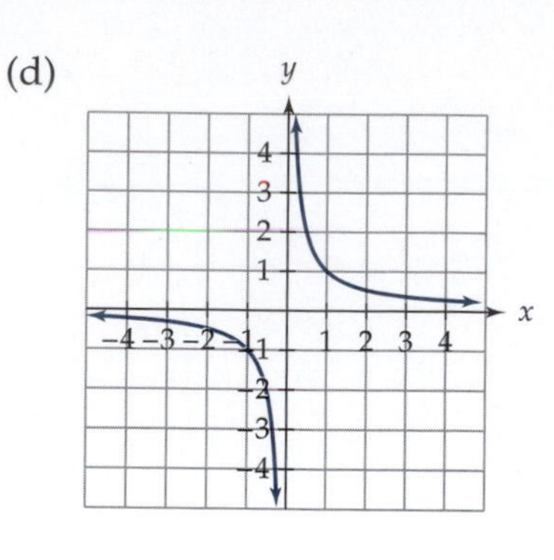

5. Why will a function have at most one y-intercept but could have several x-intercepts?
6. Let f be a function such that $f(1) = 2$.
   (a) If f is symmetric about the y-axis, what is $f(-1)$?
   (b) If f is symmetric about the origin, what is $f(-1)$?
7. Suppose f is a function that is defined at $x = 0$. What, if anything, do you know about $f(0)$ if
   (a) f is symmetric about the origin?
   (b) f is symmetric about the y-axis?
8. Many types of containers can be filled with water. The shape of the container impacts how much liquid the container holds and how quickly the level of water rises as it is being filled with water flowing at a constant rate. For each of the following containers, graph the height of the water as a function of time and give a reasonable domain and range. Be sure to put time on the horizontal axis and the height of the water level on the vertical axis.
   (a) The following container is 5 in. tall and will take 15 sec to fill.

   (b) The following container is 8 in. tall and will take 20 sec to fill.

9. The following graphs were created using a calculator attached to a motion detector. The graphs give time (in seconds) as the input and the distance (in meters) the person was away from the motion detector as the output. For each of the following graphs, describe in words the motion of the person walking. Indicate when the person was walking away or toward the motion detector and when the person was speeding up or slowing down.

(a)

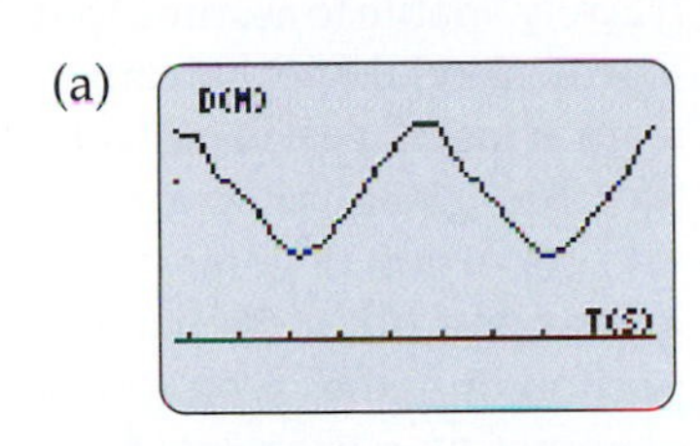

(b)

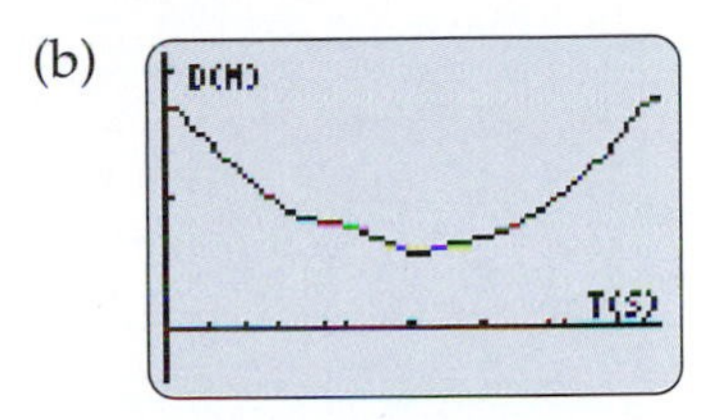

(c)

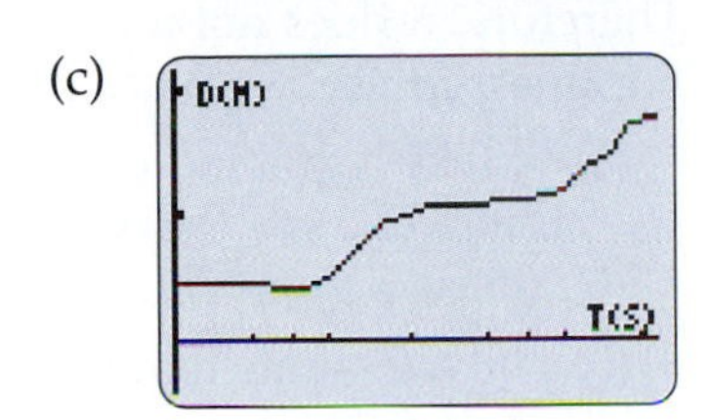

(d)

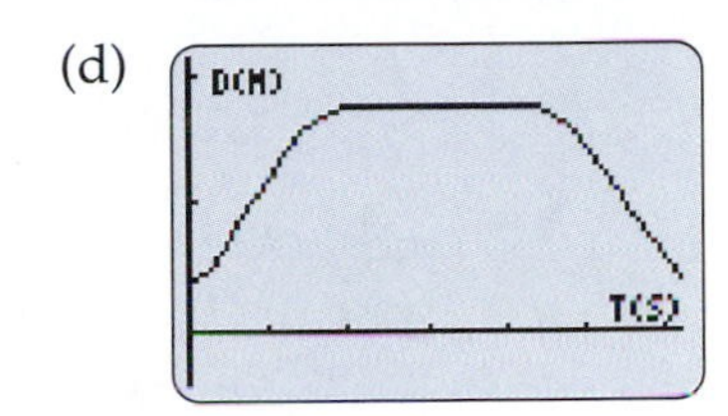

10. The accompanying graph shows the distance traveled by a car from 0 to 7 sec. Answer the following questions using this graph.

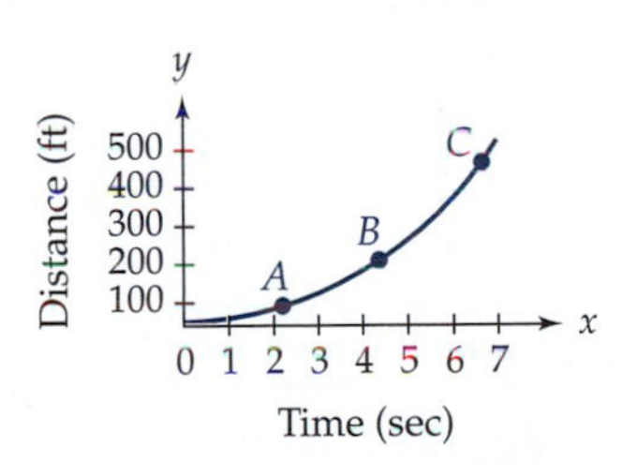

(a) Is the graph concave up or concave down?

(b) How does the distance of the car from its starting point compare at points A, B, and C?

(c) How does the velocity of the car compare at points A, B, and C?

(d) If a car is accelerating, its velocity is increasing. If a car is decelerating, its velocity is decreasing. Is the car accelerating or decelerating? Briefly justify your answer.

(e) What is the relationship between the concavity of the graph and the acceleration or deceleration of the car?

11. If someone does not want to emphasize a change in a particular item graphically, sometimes he or she will set the vertical scale so that it includes more than the entire range of the function. Suppose a school wanted to show a graph of students' average math SAT scores over

the last 10 years, which have declined from 500 to 460. The graph has a vertical scale that goes from 0 to 600. Explain why the decrease has less of an impact on the viewer if the graph shows a vertical scale of 0 to 600 instead of 460 to 500.

12. The scatterplot of students' heights and arm spans given in Figure 14 from this section contained the line $y = x$. What does that line represent?

13. It is reasonable to assume that there is a relationship between how much a professional sports team pays its players and how many games the team wins in a given season. In the 1996–1997 NBA basketball season, the Chicago Bulls had the highest payroll ($58.27 million) and the highest proportion of games won (0.841). The Vancouver Grizzlies had the lowest payroll in the league ($18.64 million) and also the lowest proportion of games won (0.171).[17] Although the payroll and proportion of games won are related, one is not necessarily the cause of the other. Therefore, it does not matter which variable is placed on which axis. For this exercise, we place the payroll on the horizontal axis.
    (a) Make a graph and place two points, one for the Chicago Bulls and one for the Vancouver Grizzlies, on the graph. Connect these points with a line.
    (b) If we assume that the proportion of games won and the payroll have a linear relationship, we expect points representing other teams in the league to be on or near this line. What can we say about a team that was below the line? What can we say about a team that was above the line?
    (c) The Boston Celtics had a payroll of $25.99 million and a percentage of games won of 0.183. Plot this point on your graph. What can we say about their winnings compared with their payroll?
    (d) The Utah Jazz had a payroll of $25.11 million and a proportion of games won of 0.780. Plot this point on your graph. What can we say about their winnings compared with their payroll?
    (e) Do you think the other teams in the league would be close to this line? Why or why not?

14. To determine a function's concavity, you need to determine if its rate of change is increasing or decreasing. A function's rate of change can be described as the change in output divided by the change in input, or change in output/change in input. Use the accompanying table of data to answer the following questions.

x	−1	0	1	2	3	4
$f(x)$	1	2	5	10	17	26

    (a) Determine the rate of change (or change in output/change in input) for each consecutive pair of inputs.

[17] USA Today, *15 November 1996, p. 16C.*

(b) Is the rate of change increasing or decreasing?

(c) Make a scatterplot of the data and connect the points with a smooth curve. Is your graph concave up or concave down? How is your answer to this question related to part (b)?

INVESTIGATIONS

INVESTIGATION 1: HUMAN ANATOMY

Vitruvius Pollio (who lived in the first century B.C.) wrote that people were designed by nature in such a way that 4 fingers make one palm, 4 palms make one foot, 6 palms make one cubit (the length from the elbow to the fingertips), 4 cubits make one person's height, and 24 palms make a person's height.[18] We asked college students to measure some of the attributes that Vitruvius discusses. They generated the following data by measuring their palm widths, foot lengths, distance from their elbows to fingertips (cubit), and heights.

Palm (cm)	Foot (cm)	Cubit (cm)	Height (cm)
7.5	21.5	43.0	152.4
7.5	22.0	39.0	156.2
7.0	21.5	42.3	160.0
7.0	24.5	43.0	162.6
7.8	23.7	40.5	165.1
7.4	22.5	41.5	167.6
7.5	25.0	43.5	172.7
8.3	24.0	45.4	172.4
6.5	22.5	43.9	172.7
8.0	24.5	45.5	172.9

1. Make a scatterplot of the palm and foot data. Put palm width on the horizontal axis and foot length on the vertical axis. Superimpose the line $y = 4x$ on your scatterplot.
2. What do the points on the line $y = 4x$ represent in terms of palm width and foot length? What would points that lie above the line represent?
3. In the same manner, make a scatterplot of the foot length and the height of the 10 people represented here.
4. According to Vitruvius, four palms make one foot and 24 palms make one person's height. If this were true, how many lengths of a person's foot would make that person's height? Superimpose the line that would represent this on your scatterplot from question 3.
5. Measure your palm, foot, and height and those of four other people. Place this information on the graphs.

[18] *H. Arthur Klein,* The World of Measurements *(New York: Simon and Schuster, 1974), p. 68.*

6. Based on the points on your graph, do you think Vitruvius was correct in his assessment of human anatomy? Why or why not?

INVESTIGATION 2: DISTANCE VERSUS VELOCITY

A common function that models the vertical position of a ball as it is thrown up in the air is $h(t) = -16t^2 + v_0t + h_0$, where $h(t)$ is the height (in feet) after t seconds, v_0 is the initial velocity (in feet per second), and h_0 is the initial height (in feet). The function that models the velocity of the same situation is $v(t) = -32t + v_0$, where $v(t)$ is the velocity in feet per second after t seconds and v_0 is again the initial velocity. Suppose we toss a ball up in the air at 48 ft/sec from a height of 6 ft.

1. Rewrite h and v to include an initial velocity of 48 ft/sec and an initial height of 6 ft.
2. Graph h and v on two separate set of axes using appropriate domains and ranges.
3. Your graph of v should be decreasing. What does this tell you about the concavity of h? Why?
4. You should be able to read the initial height on the graph of h and the initial velocity on the graph of v. Where do each of these occur?
5. Identify the point on your velocity graph and on your height graph where $v(t) = 0$. What is happening to the height of the ball at that point?
6. Identify the point on your velocity graph and on your height graph where $h(t) = 0$. What is happening to the velocity of the ball at that point?
7. Describe in words what is happening to both the height of the ball and the velocity of the ball. Does this make sense? Explain.

1.3 CALCULATOR GRAPHICS

As we saw in Section 1.2, graphs are a good way of quickly conveying information about a function. Graphing complicated functions by hand can be a laborious process. Graphing these same functions using technology can often be accomplished with a few keystrokes. The technology that makes our lives easier comes with a price (aside from what you paid for your calculator). Sometimes, with the ease of making graphs with a calculator, we do not think about the results we get and simply accept what we see on the screen as a good representation of our function. This is not always the case, however. In this section, we give you some pointers on how to best use a graphing calculator and some cautions on how to interpret the information you get from your calculator correctly.

CONNECTING THE DOTS

When you first learned to graph functions by hand, you probably evaluated the function at different inputs to obtain a number of points. You then plotted these points and connected them with a smooth curve. This method is essentially how a graphing calculator or a computer graphs a function. Graphing calculators plot points by darkening pixels on the screen. The pixels are then connected by short lines.[19] If your calculator evaluates the "right" points, then you obtain a fairly accurate representation of the graph of your function. If these points do not demonstrate the key properties of the function, however, then your graph may be inaccurate.

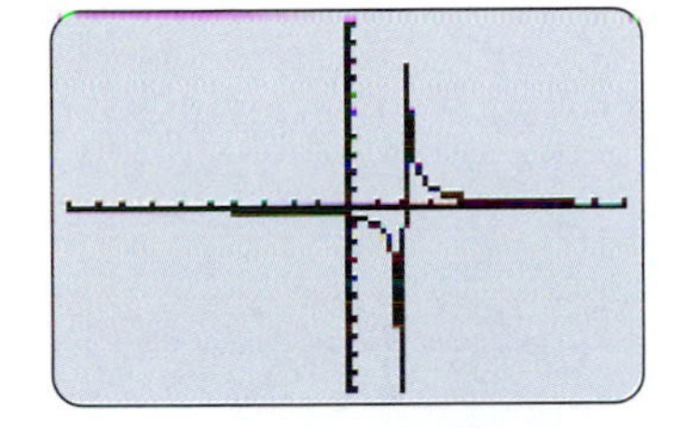

Figure 1 A TI-83 calculator graph of $f(x) = 1/(x-2)$ in the default window.

Because a graph contains an infinite number of points and because there are only a finite number of pixels in any calculator, the graph will never be completely accurate. Most of the time, however, the graph conveys enough information for you to analyze the function correctly. Unfortunately, this is not always the case. At times, the inaccuracies of a calculator-generated graph will lead you to misinterpret the function. For example, Figure 1 shows the screen from a graphing calculator for the function $f(x) = 1/(x-2)$. Although this graph shows many of the important features of this function, it gives an incorrect view of what happens when $x = 2$. The graph appears to be connected, when actually the function is undefined for $x = 2$. It is important to know when you can and cannot depend on a calculator graph to analyze a function.

Let's see why the calculator graphed $f(x) = 1/(x-2)$ incorrectly. Calculators generally have a default window that uses a domain of $-10 \le x \le 10$. With this domain, the points the calculator uses to evaluate the function are not points you would typically choose. For example, in the graph shown in Figure 1, the calculator evaluated the function at zero and then moved one pixel to the right to the value $x = 0.21276596$. It then drew a line from $(0, -0.5)$ to $(0.21276596, -0.5595238)$. Usually, the actual points chosen for evaluation do not matter; we do not care exactly what points are evaluated as long as we get an accurate picture of the shape of the graph. In this case of $f(x) = 1/(x-2)$, however, not evaluating the function at $x = 2$ causes inaccuracies. The calculator evaluated the function and plotted the points $(1.9148936, -11.75)$ and $(2.1276596, 7.8333336)$. It then connected these two points even though this function is undefined between them. Since the calculator never evaluated the function at $x = 2$, it did not recognize that the function was undefined and instead mindlessly connected successive points.

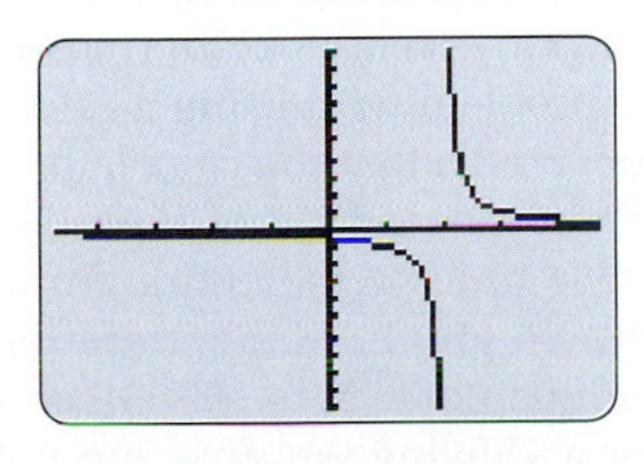

Figure 2 A TI-83 calculator graph of $f(x) = 1/(x-2)$ on the domain of $-4.7 \le x \le 4.7$. This graph was obtained by using the zoom-decimal feature.

There are two solutions to this problem. One is to realize that the calculator graph is inaccurate and, by understanding your function, compensate for the inaccuracies. For example, once you realize that calculators will often draw nearly vertical lines where the function is undefined, you learn to analyze your function to determine what the graph should be. The other solution is to change the viewing window to force the calculator to evaluate the function at $x = 2$. (We discuss how to do this later in this section.) This strategy causes the calculator to realize that $f(x) = 1/(x-2)$ is undefined at $x = 2$, and the resulting graph will not be connected. Figure 2 shows a calculator graph that more accurately depicts the function $f(x) = 1/(x-2)$.

Because a calculator graphs by plotting and connecting points, additional problems can result. Features of a graph may not be depicted properly. For example, suppose you wanted to use your calculator to determine

[19] *Many calculators have a feature that allows you to leave these points unconnected.*

the x-intercepts of $f(x) = x^3 + 0.2x^2$. A calculator graph of this function in the default screen is shown in Figure 3(a). From this graph, it appears that there is one x-intercept, and it occurs at approximately (0, 0). In Figure 3(b), we zoomed in at this point. It appears that there is still one x-intercept for this function, but this is not the case. There are two x-intercepts, one at $(-0.2, 0)$ and one at (0, 0). The reason we do not see both of them is that they occur very close together, and between -0.2 and 0, the graph goes above the x-axis by a very small amount. Do not rely on your calculator for all the answers. The function $f(x) = x^3 + 0.2x^2$ can easily be evaluated symbolically to determine both of the x-intercepts.

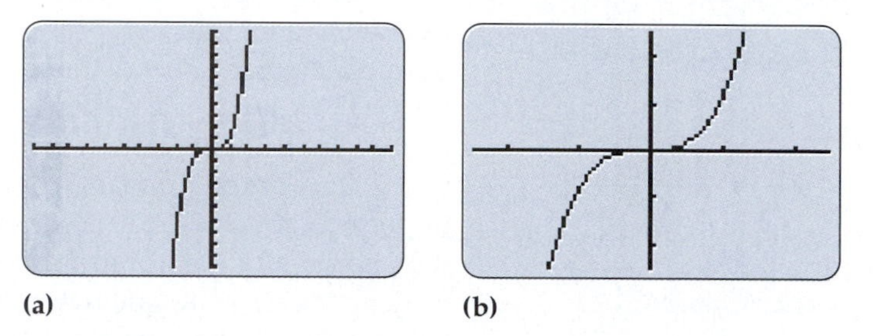
(a) (b)

Figure 3 Two views of a calculator graph of $f(x) = x^3 + 0.2x^2$.

READING QUESTIONS

1. How does a graphing calculator produce a graph of a function?
2. Why might a graphing calculator produce a poor graphical representation of the function $f(x) = 1/(x + 3)$?
3. Graph $f(x) = 4x^3 - 0.4x^2$. How many x-intercepts do you see? How many x-intercepts does $f(x)$ have?

THE VIEWING WINDOW

We are often interested in the shape of the graph ("the big picture"). For example, we may want to see where the function increases or decreases, if it is connected or not, or its concavity. The problem with using a calculator graph to determine the shape of the function is that the calculator's viewing window limits what we see. In fact, if the domain of the function includes all real numbers, it is impossible to view the entire graph either with a calculator graph or a hand-drawn graph. Many graphing calculators have a default window with a domain of $-10 \leq x \leq 10$ and a range of $-10 \leq y \leq 10$. These choices for a viewing window, are not always good. It is important to select a good viewing window for your function *before* you graph it. Do not get in the habit of always relying on the default window. To determine a proper viewing window, you should first analyze the function, which includes determining an appropriate domain and range and knowing the general behavior of various types of functions. Also, it is sometimes a good idea to use more than one viewing window.

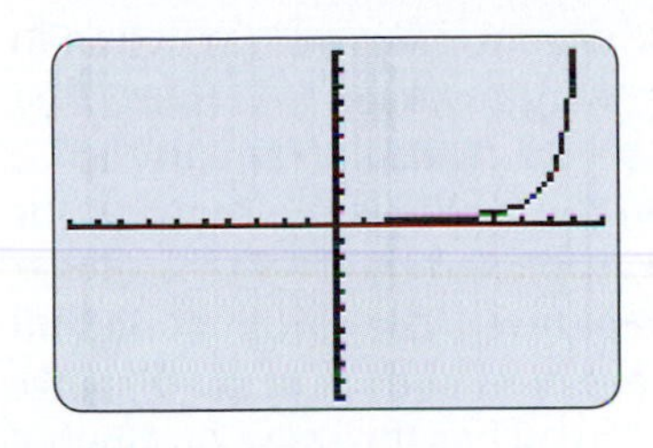

Figure 4 A poor viewing window for the function $f(x) = 10/(x - 10)^2$.

Figure 4 shows a graph of $f(x) = 10/(x - 10)^2$ in the calculator's default window. This viewing window is a poor choice for seeing the graph's overall shape: The default window was used instead of analyzing the function to determine an appropriate viewing window. By viewing the function in this window, it appears that $f(x) = 10/(x - 10)^2$ is constantly increasing and always concave up. This is not the case.

Let's determine a good viewing window by analyzing our function. The function $f(x) = 10/(x - 10)^2$ is undefined when $x = 10$ since this causes division by zero. Because the function is undefined at $x = 10$, it is important to have a graph that includes values for x that are greater than 10 as well as those that are less than 10 before making conclusions about the shape of $f(x) = 10/(x - 2)^2$. Thus, the default window of $-10 \leq x \leq 10$ is a poor choice for this function. We need to also analyze the range. Since the numerator is positive and the denominator will always be positive (the square of any nonzero number is positive), the range of this function is $f(x) > 0$. One good viewing window for this function is a domain of $0 \leq x \leq 20$ and a range of $-2 \leq y \leq 10$. Our domain is centered around the point where the function is undefined. Even though the function has no negative outputs, we extended our viewing window down to $y = -2$ so that if we traced the function, the coordinates of the points would be displayed below the graph. A specific upper bound for the viewing window is not very important for this function. We chose 10 simply because it was used in the original default window. Figure 5 shows a graph of $f(x) = 10/(x - 10)^2$ using the revised viewing window. With this viewing window, we see the function is always concave up, increasing when $x < 10$, decreasing when $x > 10$, and not connected at $x = 10$.

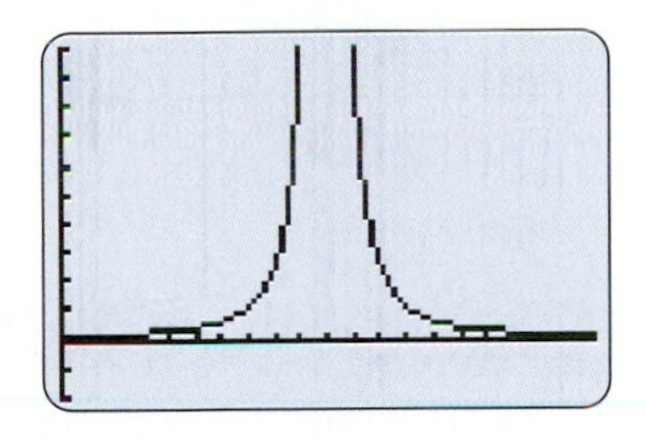

Figure 5 A graph of $f(x) = 10/(x - 10)^2$ in a viewing window of $0 \leq x \leq 20$ and $-2 \leq y \leq 10$.

Many functions, such as $f(x) = 10/(x - 10)^2$, have a domain and range that either include all real numbers or have an upper or lower limit of positive or negative infinity. Obviously, you cannot use these limits as the domain and range for your viewing window. Instead, you need to determine a viewing window that will allow you to see the behavior you are interested in analyzing. At times, however, determining the domain and range of your function does give the proper viewing window, particularly when functions are being used to model real-world phenomenon. In this case, clues from the context of the problem can help us determine a proper domain and range for our viewing window. For example, the equation $t = 1 + (s/20) + (70/s)$ can be used to determine the length of time needed for a traffic light to stay yellow. In this equation, t is time (in seconds) and s is the speed limit (in feet per second). For this situation, you would use your driving experience to determine the proper viewing window. A reasonable domain for this function is 36.7 ft/sec $\approx$ 25 mph to 80.7 ft/sec $\approx$ 55 mph (the range of speed limits on streets where there are traffic signals). A reasonable range might be 3 to 6 sec (a guess of how long most lights stay yellow). Figure 6 shows what this function looks like in the window we have chosen. Although the function could be defined at speeds less than 25 mph and greater than 55 mph, it really does not make sense for this application.

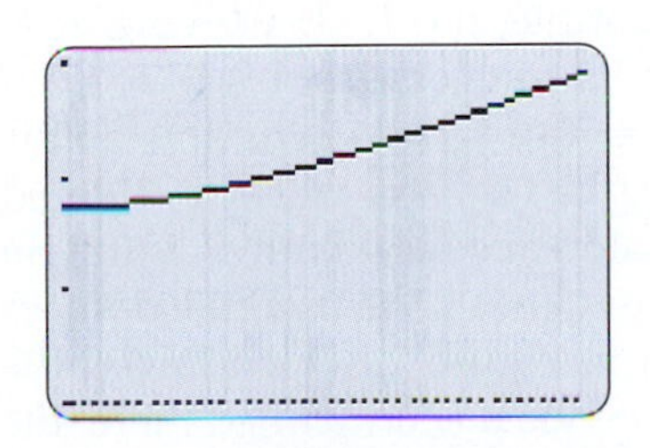

Figure 6 A graph of $t = 1 + s/20 + 70/s$ on the domain of $37.6 \leq s \leq 80.7$.

Be aware that your choice of a viewing window may distort the appearance of your graph. For example, the semicircle $f(x) = \sqrt{5 - x^2}$ looks like a portion of an ellipse in the viewing window used for Figure 7(a), whereas it looks more like a semicircle in the viewing window used for Figure 7(b). Some calculators have a feature that allows you to easily "square up" a graph so that the distance between units on the x-axis is the same as the distance between units on the y-axis. (For example, on the TI-83 calculator, squaring up is done using the zoom-square feature.) The common default screen of $-10 \le x \le 10$ and $-10 \le y \le 10$ does not have the same distance between units on the two axes. This default screen has the same number of *units* on the x-axis as are on the y-axis. Since most calculators have screens that are wider than they are tall, however, the same number of units on each axis means that the distance between units will be different for the two axes, which causes the graphs in the default screen to appear distorted.

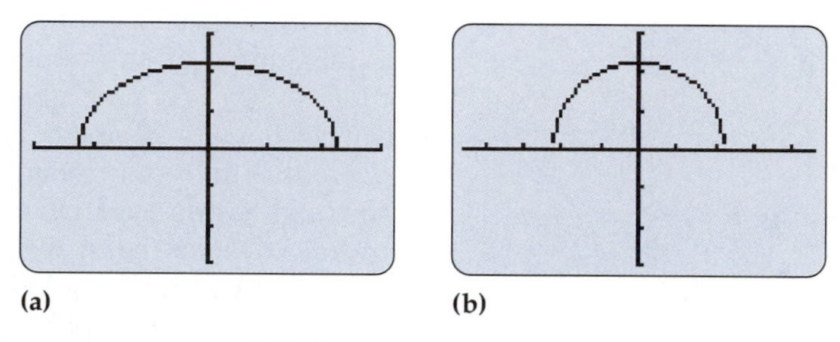

Figure 7 Graphs of $f(x) = \sqrt{5 - x^2}$ using two different viewing windows.

Finally, at times you may want to adjust your viewing window so that your calculator will evaluate your function at points a specified distance apart, such as at each unit or each tenth of a unit. This distance can be thought of as the "distance" between pixels. There are different ways of creating such viewing windows. Some calculators have a zoom-integer or a zoom-decimal feature that adjusts the screen so that each pixel represents an integer or a tenth of an integer. You can also change the domain and range for your viewing window to force the calculator to make this adjustment. In fact, you can adjust your viewing window so that your function is evaluated at points any certain distance apart you desire. This option is important if, for example, you want the function $f(x) = 1/(x - 2)$ to be evaluated at $x = 2$ so that the calculator will show that the graph is disconnected. To choose a viewing window that will force your calculator to evaluate your function at points a specified distance apart, you need to know how wide your calculator screen is in terms of pixels and adjust the viewing window appropriately. For example, the TI-83 viewing window has a width of 95 pixels.[20] There are 95 integers from -47 to 47 if you include -47 and 47. Therefore, if you set your viewing window on the domain $-47 \le x \le 47$ and trace your graph, the function will be evaluated at every integer value. If you want it evaluated at every tenth of a unit on the TI-83, set your domain from -4.7 to 4.7 or 0 to 9.4. We used the domain $-4.7 \le x \le 4.7$ to graph $f(x) = 1/(x - 2)$ in Figure 2, which forced the calculator to evaluate the function at $x = 2$. Since it is undefined there, the calculator showed the graph as it should be: disconnected.

[20] *To find the number of pixels in the screen for the calculator you are using, look in the book that came with your calculator.*

READING QUESTIONS

4. Why is it important to graph a function in a proper viewing window?
5. What should you do to determine an appropriate viewing window?
6. What does it mean to say your calculator screen is "squared up"?
7. What would be a good viewing window for the function $f(x) = 3/(x-20)^2$?
8. Graph the function $f(x) = x^2$ and adjust your viewing window so that, when tracing, the x-coordinates increase by 0.1. (*Hint:* Check to see if your calculator has a zoom-decimal feature.)

WARNING: THE RESULTS GIVEN BY YOUR CALCULATOR ARE USUALLY APPROXIMATIONS

A graphing calculator is a powerful tool that can be used to help you understand mathematical concepts and solve problems. As mentioned earlier, however, do not treat it as a magic box that will give you all the answers. Most "answers" you get by analyzing a calculator graph are approximations. Sometimes approximations are fine, but sometimes the exact answer is desired. When using a calculator, it is important to know when an answer is an approximation and when it is not. For example, suppose you want to solve the equation $x^2 + 2x - 6 = 0$. One method is to graph the function $f(x) = x^2 + 2x - 6$ and use your calculator to find the values of x where $f(x) = 0$. You can trace or use a built-in feature that some calculators have to determine zeros. This feature is usually more accurate than just tracing, but the zeros are still likely to be approximations. Using this feature on a TI-83,[21] we determined that one of the zeros of $f(x) = x^2 + 2x - 6$ is 1.6457513. (See Figure 8.) We know this is not the exact zero because $f(1.6457513) \approx -5.85483 \times 10^{-8} \neq 0$. Using the quadratic formula, you can determine that the exact positive zero is $\sqrt{7} - 1$. Since this number is irrational,[22] it is impossible for this particular calculator to give the exact answer. Sometimes an approximate answer is fine, but remember that answers obtained from your calculator are usually approximations.

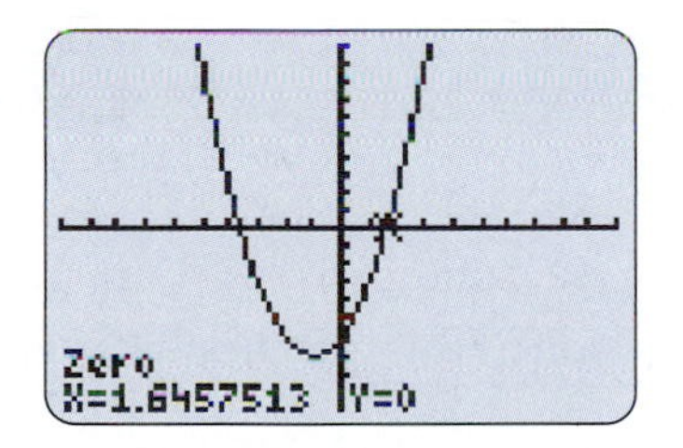

Figure 8 A TI-83 calculator screen showing an approximation of a zero of $f(x) = x^2 + 2x - 6$.

Example 1 Use your calculator to determine the x-intercepts or zeros of $f(x) = x^3 + 4x^2 - 2x - 8$. Are these zeros exact or approximations?

Solution Using the TI-83, we determined that the zeros are -4, -1.414214, and 1.4142136. Since $f(-4) = 0$, the -4 is exact. The other two are approximations because $f(-1.414214) \approx 3.2006756 \times 10^{-6}$ and $f(-1.414214) \approx 5.76208 \times 10^{-7}$.

[21] *Choose the* ***2:zero*** *option on the* ***CALC*** *menu on the TI-83.*

[22] *The decimal form of an irrational number goes on indefinitely.*

READING QUESTIONS

9. If you are using a calculator graph to determine the zeros of a function, what is one way to tell if these zeros are approximations?

10. Use your calculator to determine the x-intercepts or zeros of $f(x) = x^3 + 5x^2 + 4x - 4$. Are these zeros exact or approximations?

SUMMARY

Graphing functions using technology can often be accomplished with a few keystrokes. Sometimes, however, we have to make sure the graph given by a calculator accurately reflects the real graph of the function. Graphing calculators plot points by darkening pixels on the screen. The pixels are then connected by short lines. By this method, a calculator may connect a graph that should not be connected.

When looking at a function on a calculator, choosing a proper viewing window is important. The standard viewing window will not always be the best. In fact, you may not even see your graph in that window.

When calculators are used to find things such as zeros and points of intersection, the numbers given are often not exact. It is important to understand when these numbers are exact and when they are not.

EXERCISES

1. Let $f(x) = 0.004x^3 - 0.13x^2 + 10$.
   (a) Graph f with your viewing window set at $-10 \leq x \leq 10$ and $-10 \leq y \leq 10$. Describe the behavior of this function as it appears in this viewing window.
   (b) Graph f with your viewing window set at $-10 \leq x \leq 40$ and $-10 \leq y \leq 10$. Describe the behavior of this function as it appears in this viewing window.
   (c) Which of your two descriptions from parts (a) and (b) best describes the shape of the function? Why?
2. Graph the function $f(x) = 0.1x^2 + x - 30$ in a window that allows you to see its overall shape.
   (a) Sketch the graph of the function. What are the dimensions of your viewing window?
   (b) Use your graph to determine the roots of the function. Are your answers approximations, or are they exact?
   (c) Use your graph to estimate the minimum value of the function.
3. A proper graph of the function $f(x) = 2/(x - 3)$ will be disconnected at the line $x = 3$.

(a) Why is this function disconnected at $x = 3$?

(b) Graph this function on your calculator so that it is disconnected. (*Hint:* Set your viewing window so that one of the pixels has an x-value of 3.)

(c) What are the dimensions of your viewing window from part (b)? How are these numbers related to the number of pixels on your screen horizontally?

4. A standard $-10 \leq x \leq 10$ calculator screen is not an appropriate window to view the general shape of the graphs of the functions given below. Analyze these functions to determine an appropriate viewing window that shows the graph's general shape and make a sketch of your calculator graph. Be sure to indicate the dimensions of the viewing window you used.

   (a) $f(x) = 2x + 20$

   (b) $g(x) = 4x^2 - 3x + 12$

   (c) $h(x) = 8 \cdot 10^x$

   (d) $j(x) = \dfrac{10}{(x - 15)^3}$

5. The lines $f(x) = 3x/2$ and $g(x) = -2x/3 + 6$ are perpendicular.

   (a) Graph both lines in the same graph using the calculator's default window. Do they look perpendicular? Why or why not?

   (b) Find a viewing window so that the lines appear to intersect at right angles. What are the dimensions of your viewing window? How are these numbers related to the number of pixels on your screen horizontally and vertically?

6. Use your calculator to graph the function $f(x) = x$.

   (a) In a "squared-up" graph of this function, the line $f(x) = x$ makes a 45° angle with the x-axis. Adjust your viewing window so that the graph shows this angle. Remember that when the screen is "square," the distance between units on the x-axis and y-axis are the same. You will probably not have the same number of units on each axis since your viewing screen is probably not square.

   (b) Adjust your viewing window so that the line appears to have a very steep slope. What are the dimensions of your viewing window?

   (c) Adjust your viewing window so that the line appears to have a very gradual slope. What are the dimensions of your viewing window?

7. Use your calculator to determine the zeros of $f(x) = x^2 + 2x - 7$. Are these zeros exact or approximations? Explain.

8. Use your calculator to determine the zeros of $f(x) = 2x^3 - x^2 - 5x + 3$. Are these zeros exact or approximations? Explain.

9. Use your calculator to determine the point of intersection of $f(x) = 0.4x - 2$ and $g(x) = 5x + 3$. Is your point exact or an approximation? Explain.

10. Let $f(x) = x^3 - x^2 - 2x + 2$.
    (a) Graph f using the default window $-10 \leq x \leq 10, -10 \leq y \leq 10$. Sketch your calculator graph.
    (b) Find a viewing window so that the graph of f starts in the bottom left-hand corner and ends in the top right-hand corner of the screen.
    (c) Find a viewing window so that the graph of f looks like an increasing line.
    (d) Find a viewing window so that the graph of f looks like a decreasing line.
    (e) Find a viewing window so that the graph of f looks like a horizontal line.

INVESTIGATIONS

INVESTIGATION 1: PIXELS

As stated, a viewing window on a calculator can be adjusted in such a way that the calculator will evaluate your function every tenth of a unit, every unit, or any other fixed distance that you desire. If your calculator has a zoom-decimal or zoom-integer feature, with the push of a couple of buttons you can get points one unit or one tenth of a unit apart. Knowing the width of your screen in terms of the number of pixels and adjusting the viewing window appropriately, however, allow you to obtain any distance between points you wish. For example, the viewing window on a TI-83 is 95 pixels wide. If we graph on a domain of $0 \leq x \leq 94$, the calculator will evaluate our function at every unit; there are 95 integers from 0 to 94 if you include both 0 and 94. Therefore, each pixel will correspond to one of these integers.

1. Let n be the number of pixels in the width of your screen.
    (a) What is the value of n for your calculator?
    (b) If you want your calculator to evaluate your function at every integer starting with zero, what should your domain be in terms of n?
    (c) If you want your calculator to evaluate your function at every half a unit starting with zero, what should your domain be in terms of n?
2. Graph $f(x) = 0.01x^2 - 3$ on your calculator and adjust your viewing window so that it evaluates the function at $0, 0.5, 1, 1.5, \ldots$ when you trace on it.
    (a) What is the domain for your viewing window?
    (b) Trace your graph to find $f(21.5)$.
    (c) Graph $g(x) = 1/(x - 30)$ using the same window. By tracing on your graph, how can you tell that the function is undefined at $x = 30$?

3. Graph $f(x) = x^2 - 4x + 3$ on your calculator and adjust your viewing window so that it evaluates the function at $0, 0.2, 0.4, 0.6, \ldots$ when you trace it.
   (a) What is the domain of your viewing window?
   (b) Trace your graph to find the x-intercepts and the y-intercept.
   (c) Are your answers to part (b) exact or approximations? How do you know?
   (d) Adjust your viewing screen so each pixel is still 0.2 unit apart but starts at -5 instead of 0. How does this change your domain?
4. Write a formula to determine the maximum value for a viewing window, X_{max}, on your calculator given that it starts at X_{min} and that the distance between pixels is d.
5. In this section, we mentioned that a TI-83 calculator set on the default screen, $-10 < x < 10$, evaluated at an input of 0 and then at an input of 0.21276596. Why did the calculator use the input 0.21276596?

1.4 MATHEMATICAL MODELING

Have you ever wondered how formulas that you have seen in mathematics and science texts were developed? For example, how would you go about deriving the formula for the area of a circle, $A = \pi r^2$, or the height of an object in free fall, $h(t) = -16t^2 + v_0 t + h_0$? Taking a physical situation and developing a mathematical formula to describe it is called **mathematical modeling.** In this section, we demonstrate the process of creating a mathematical model. In doing so, we define what our model should contain, look at our model using a specific situation, generalize our model, and analyze our model by connecting it back to the physical situation. We first go through this process for a few examples and then look at an application where we model the time needed for a traffic light to stay yellow.

DEFINING OUR MODEL

When presented with a situation for which you want to develop a mathematical model, it is important to take a few moments to try to understand the factors involved in setting up an appropriate model. If the situation is fairly well defined, then this process may be simple. Often, however, the situation is not well defined and several assumptions must be made to set up an appropriate model. Once appropriate assumptions are made, drawing a picture of the situation is usually a good idea.

Consider an example that may be familiar to you if you have been around small children. Think of puzzles, often made out of wood or heavy cardboard, for little children. To match the shapes properly, the pieces are fit nicely into the shallow holes on the board. For example, a child has to fit the triangular shape in the triangular hole, the square shape in the square hole, and so on. For some children, this process is, as the old saying goes, "As difficult as trying to put a square peg in a round hole." Or does the saying go, "As difficult as trying to put a round peg in a square hole?" Which is more difficult, or, alternatively, which one fits better? This is an example of a question that is not well defined. We need to make some assumptions as to what it means to "fit better." One way of defining "fit better" is "a smaller area between the puzzle piece and the hole." We will assume that the two holes (square and round) have the same area. Finally, we will assume that we have the largest square puzzle piece that will fit in the round (or circular) hole on the board and the largest round (or circular) puzzle piece that will fit in the square hole on the board. With these assumptions, the question is transformed into the following: "Which has greater area, the difference between a square inscribed in a circle of area A or the difference between a circle inscribed in a square of area A?" To answer this mathematical question generated by looking at a children's toy, we need to draw the two figures, generate the appropriate formulas, and compare the answers. (We come back to this later.)

Mathematical models are derived to answer a variety of questions, some generated by pure curiosity and some generated by necessity. The next example, although perhaps motivated by curiosity, has more of a practical use than our previous one.

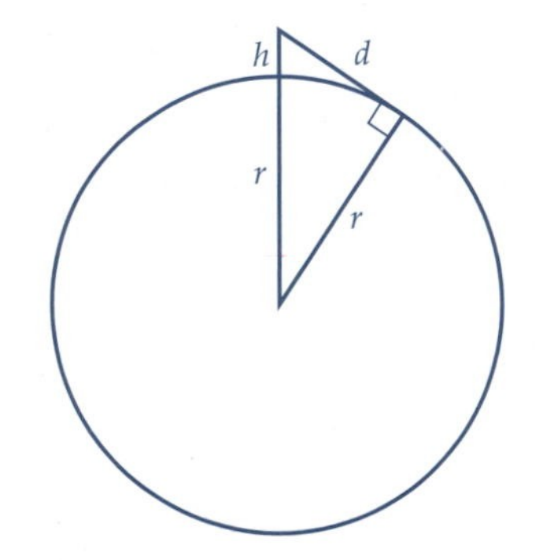

Figure 1 A cross-sectional view of Earth showing distance to the horizon, d, as a function of a person's height above Earth, h.

If you have ever been to the top of a skyscraper, you know that you can see much farther from there than when you are standing on the ground. The higher you are above Earth, the farther you can see out to the horizon. But how much farther? We can set up a mathematical model to answer this question by developing a function where the height someone is above Earth is the input and the distance they can see out to the horizon is the output. The first step in developing our model is to make a drawing of the situation. Figure 1 is a picture of the distance to the horizon. We have made some assumptions in drawing this figure. We assumed that Earth is a perfect sphere, that there is nothing in our way when we look out to the horizon, and that our line of sight to the horizon is a straight line. The circle in Figure 1 represents the circumference of Earth. The line segment labeled h is the distance someone is above Earth. The line segments labeled r are the radii of Earth. The line segment labeled d is the distance to the horizon. Since d is tangent to the circle and r is a radius, the angle where they intersect is a right angle. To develop our model, we need to write a function where h is the input and d is the output.

READING QUESTIONS

1. If you want to develop a mathematical model, explain why it is important to transform the question, "Which is more difficult, putting a square puzzle piece into a round hole or putting a round puzzle piece into a square hole?" to "Which has greater area, the difference between a square inscribed in a circle of area A or the difference between a circle inscribed in a square of area A?"

2. Why is it important to have the area of the circular hole and the area of the square hole the same when we were developing the puzzle-piece model?
3. What assumptions would you have to make to develop a mathematical model that gives the price of a pizza for a given diameter?

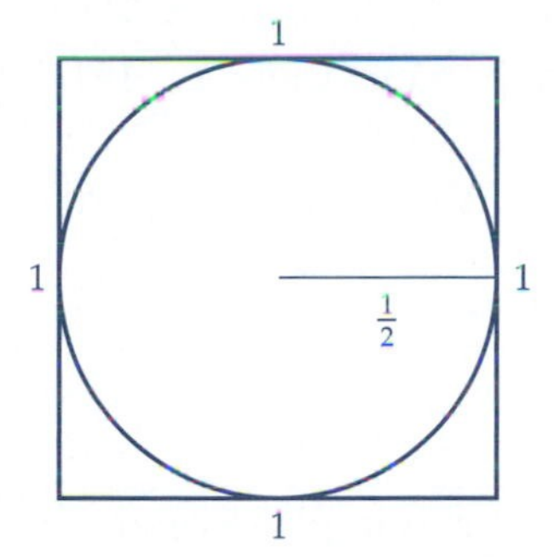

Figure 2 A circle inscribed in a square of area 1.

GETTING SPECIFIC

Once you have made assumptions and your model is well defined, it is helpful to test the model with specific examples. Looking at specific examples will help you better understand your model as well as test the validity of the model by comparing it with what you know about the situation. This step will often include algebraic manipulation. Working with functions in their symbolic form is a valuable tool in setting up mathematical models.

Let's return to the question about a square puzzle piece in a round hole versus a round puzzle piece in a square hole. Remember that we transformed this to the mathematical question, "Which has greater area, the difference between a square inscribed in a circle of area A or the difference between a circle inscribed in a square of area A?" Instead of starting with general circles and squares of area A, we start by looking at a specific example. Suppose we have a circle inscribed in a square of area 1. We need to find the area of the inscribed circle. If the area of the square is 1, then the length of each side is $\sqrt{1} = 1$. So, the radius of the circle inscribed in this square is $\frac{1}{2}$. (See Figure 2.) The area of the circle is then $\pi\left(\frac{1}{2}\right)^2 = \pi/4$. Therefore, the difference between the area of the outer square and the area of the inner circle is $1 - \pi/4 \approx 0.2146$.

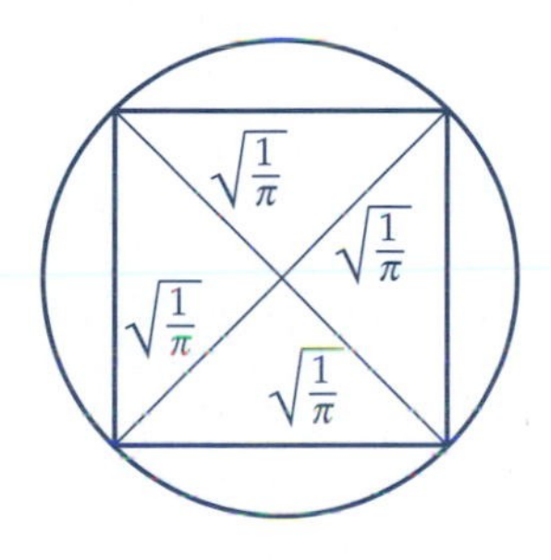

Figure 3 A square inscribed in a circle of area 1.

Suppose we have a square inscribed in a circle of area 1. We need to find the area of the inscribed square. If the area of the circle is 1, then $\pi r^2 = 1$. Solving for r, we get $r = \sqrt{1/\pi}$. We can use this radius to find the length of the side of a square and use that information to find the area of the square, or we can use the radius to directly determine the area. We choose the latter method. If we think of the square as four right triangles as shown in Figure 3, then the area of each triangle is $A = \frac{1}{2}bh = \frac{1}{2}\sqrt{1/\pi}\sqrt{1/\pi} = 1/(2\pi)$. So the area of the square is $4 \times 1/(2\pi) = 2/\pi$. The difference in the area of the circle and the area of the square is $1 - 2/\pi \approx 0.3634$. For this specific example, we see that a round puzzle piece fits in a square hole "better" than a square peg fits into a round hole since the difference in area is smaller.

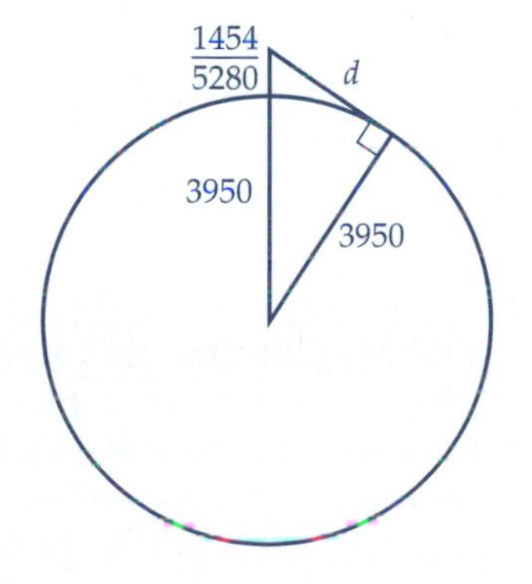

Figure 4 Distance to horizon from the top of the Sears Tower in Chicago.

Let's return to our example of determining the distance to the horizon given the distance you are above Earth. The Sears Tower in Chicago, at 1454 ft, is the tallest building in the United States. If you were standing on top of this building, how far could you see out to the horizon? Figure 4 shows another view of the circumference of Earth with some specific numbers included. (*Note:* To show the height above Earth, Figure 4 is not drawn to scale.) The radius of Earth is approximately 3950 mi. Converting the height of the Sears Tower to miles, gives $h = 1454/5280$. We have a right triangle with legs of length d and 3950, and a hypotenuse of length $3950 + 1454/5280$. Using the Pythagorean theorem, we have $d^2 + 3950^2 = (3950 + 1454/5280)^2$. Solving for d, gives $d = \sqrt{(3950 + 1454/5280)^2 - 3950^2} \approx 47$ mi. Therefore, when standing on top of the Sears Tower, one can see approximately 47 mi out to the horizon.

READING QUESTIONS

4. Why is "getting specific" a good strategy?
5. Which has greater area, the difference between a square inscribed in a circle of area 2 or the difference between a circle inscribed in a square of area 2? What are the two differences in area?
6. The Sears Tower is 1454 ft tall. Why is this the same thing as 1454/5280 mi?
7. Solving the equation $x^2 = 9$ gives $x = \pm 3$. We had a similar situation when we were solving for d in our "distance we could see out to the horizon" example. Why didn't we include the $\pm$ sign?

GENERALIZING

Once you have worked through enough specific examples to gain an adequate understanding of your model, it is time to generalize. How many examples you should do before generalizing depends on your understanding of the model. Examples only allow you to reach definitive conclusions for a few cases, so only doing examples is not sufficient. To have something that is always true or versatile enough to be used in many situations, we must generalize. Using a mathematical model to answer your question in general is what makes mathematical modeling so powerful. Generalizing typically means using variables instead of numbers. This is algebra. In fact, one should think of algebra as generalized arithmetic. Algebra is such a powerful tool because it allows us to answer questions of the form, "What will always happen?" instead of limiting our questions to specific cases.

Let's return one more time to the puzzle piece question. We are now ready to answer our mathematical question, "Which has greater area, the difference between a square inscribed in a circle of area A or the difference between a circle inscribed in a square of area A?" The process is almost identical to that in the specific example done earlier, which is usually the case. Most of the time, the steps done using variables are identical to the ones done using numbers. This is another reason why getting specific first is a good strategy.

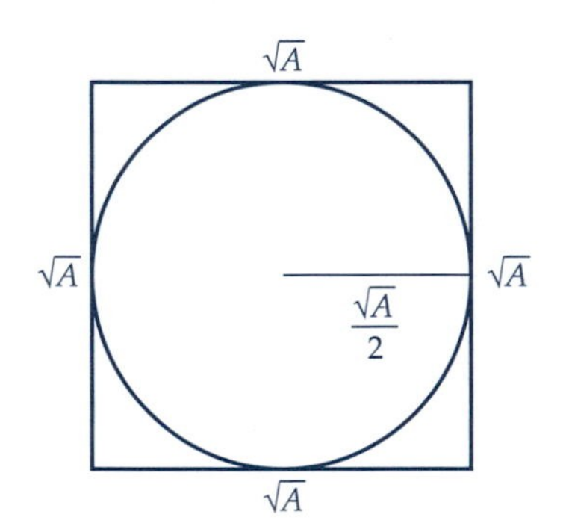

Figure 5 A circle inscribed in a square of area A.

Suppose we have a circle inscribed in a square of area A. We need to find the area of the circle. If the area of a square is A, then the length of each side is $\sqrt{A}$. The radius of the circle inscribed in this square will then be $\sqrt{A}/2$. (See Figure 5.) So, the area of the circle is $\pi(\sqrt{A}/2)^2 = \pi A/4$. The difference between the area of the square and the area of the circle is $A - \pi A/4 = A(1 - \pi/4) \approx 0.2146A$.

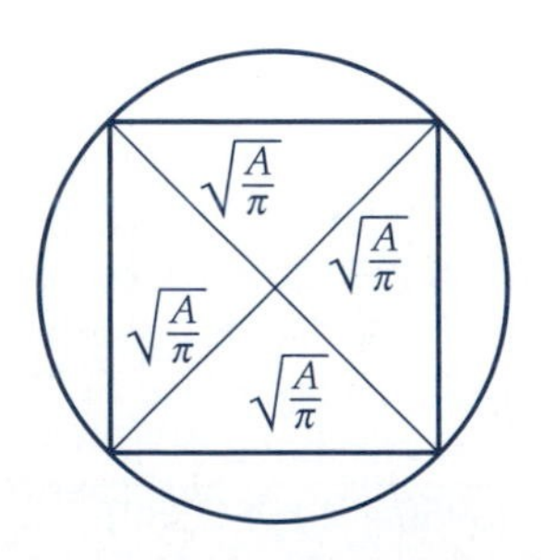

Figure 6 A square inscribed in a circle of area A.

Now suppose we have a square inscribed in a circle of area A. We need to find the area of the square. If the area of a circle is A, then $\pi r^2 = A$. Solving for r, gives $r = \sqrt{A/\pi}$. We use this radius to find the area of the square, as in our specific example. We again think of the square as four right triangles (as shown in Figure 6). The area of each triangle is $A = \frac{1}{2}bh = \frac{1}{2}\sqrt{A/\pi}\sqrt{A/\pi} = A/(2\pi)$, so the area of the square is $4 \cdot A/(2\pi) = 2A/\pi$. The difference between the area of the circle and the area of the square is $A - 2A/\pi = A(1 - 2/\pi) \approx 0.3634A$. Because $0.2146A < 0.3634A$ for every positive value of A, we see that a round puzzle piece will *always* fit

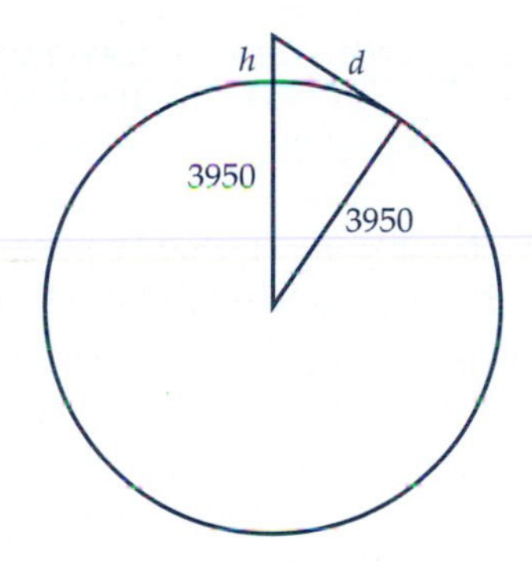

Figure 7 A cross-sectional view of Earth showing distance to the horizon, d, as a function of a person's height above Earth, h.

into a square hole "better" than a square puzzle piece will fit into a round hole.

Let's generalize the second example we looked at earlier: "If you are standing h miles above Earth, how far can you see out to the horizon?" The process of writing the distance one can see out to the horizon as a function of height above Earth is similar to that of our specific example with the Sears Tower. The only difference in our general procedure versus our specific example is using h in place of 1454/5280. Using Figure 7 as a guide and the Pythagorean theorem, we get $d^2 + 3950^2 = (3950 + h)^2$. Solving for d gives $d = \sqrt{(3950 + h)^2 - 3950^2}$. Simplifying gives $d = \sqrt{3950^2 + 2(3950)h + h^2 - 3950^2} = \sqrt{7900h + h^2}$. Therefore, the distance (in miles) one can see out to the horizon, d, while standing h miles above Earth is given by the function $d = \sqrt{7900h + h^2}$.

READING QUESTIONS

8. Why is it important to find general formulas as opposed to only doing specific examples?

9. Why is algebra important in generalizing?

10. In the conclusion to our puzzle piece question, we stated that $0.2146A < 0.3634A$ for every *positive* value of A. Why is is safe to assume that A is always positive?

11. Using the general model for the distance you could see to the horizon, how far could you see out to the horizon if you were on a 555-ft-tall building?

ANALYZING

Once you have set up a mathematical model, it is important to analyze your model and the answers it gives. In doing so, you should do three things: connect your model back to the physical situation, determine any limitations of your model, and ask yourself other questions that could lead to developing a different (and perhaps better) model. You want to connect the model back to the original situation to make sure your model makes sense. If it does not correspond with your perception of reality, then there is a problem with your model. You want to determine if there are any limitations of your model so you know when it is appropriate to use it. For example, you need to determine an appropriate domain for your function. Finally, analyzing your model can often lead to other questions. You may want to change one or more of your original assumptions and develop an alternative model. You can analyze your model by examining the function not only symbolically, but also numerically and graphically.

Let's return to our distance to the horizon example. The function that gives the distance one can see out to the horizon, d, for a given height above Earth, h, was $d = \sqrt{7900h + h^2}$. The graph of this function is shown in Figure 8. Notice that the graph is increasing and concave down. Does this make sense? To relate this graph back to the physical situation, suppose you are going up in an elevator located on the outside of a tall building.

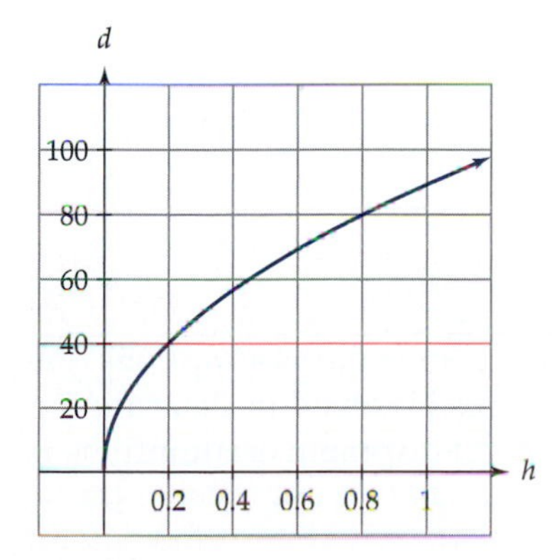

Figure 8 A graph of $d = \sqrt{7900h + h^2}$.

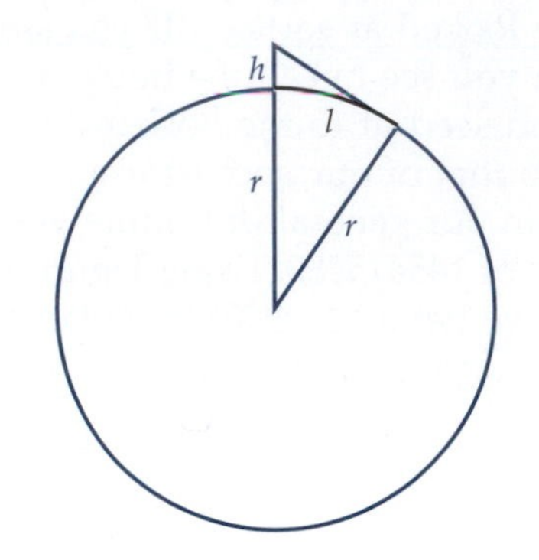

Figure 9 The arc length, l, could also be used as the distance to the horizon.

Since the graph is increasing, our model says that the higher you travel up in the elevator, the farther you can see out to the horizon. That should make sense. Since the graph is concave down, our model says that as you travel higher in the elevator, the rate at which you can see farther is decreasing.

Figure 8 shows what our function looks like for $0 \leq h \leq 1$, that is, for inputs within a mile of Earth. Is our model limited to distances relatively close to Earth's surface, or will it work for any distance above Earth? It would seem logical that if you were high enough, you could see almost a quarter of the way around Earth, but no farther. The function $d = \sqrt{7900h + h^2}$ increases without bound, however. In other words, you can make the output as large as you want. Our model finds the straight line distance from the observer to the horizon and not the distance along Earth. Perhaps we should find a new model to determine how far you can see out to the horizon. Instead of finding the straight line distance from the observer to the horizon, we can find the arc length from the surface of Earth directly below the observer out to the horizon. (See Figure 9.)

READING QUESTIONS

12. What should you do when you analyze a mathematical model?

13. If we wanted to use our model $d = \sqrt{7900h + h^2}$ to determine the distance to the horizon from the tops of buildings, what would be an appropriate domain and an appropriate range?

AN APPLICATION: CROSSROADS

Have you ever wondered how traffic engineers determine how long a traffic signal should stay yellow before it turns red? If it is yellow for too short of a time, a vehicle can get caught in a zone where there is neither enough time to brake safely nor enough time to drive through the intersection before the light turns red. If it is yellow for too long, drivers will slow down and stop, thinking the light will soon turn red, and traffic will be held up. We want to develop a mathematical model that will find the optimal time that a traffic light should stay yellow.[23] In doing so, we follow the same steps as earlier:

- Define our model.
- Look at a specific example.
- Develop a generalized model.
- Analyze our model.

Defining Our Model

As before, the first step in defining our model is to make assumptions. We begin by making assumptions about the factors involved in determining an appropriate time for the yellow light. One of our main assumptions is the existence of a *critical point*. We assume that as you approach a traffic

[23] *Arthur Eisenkraft and Larry Kirkpatrick, "Stop on Red, Go on Green …"* Quantum, *January/February 1994, pp. 34–36.*

light, there is a critical point, shown in Figure 10, such that if you are ahead of the point, then you should drive through the intersection, and if you are behind the point, then you should stop. There is a little room for error around this critical point, but not much. Determining the distance from this point to the intersection is the first step in developing our model.

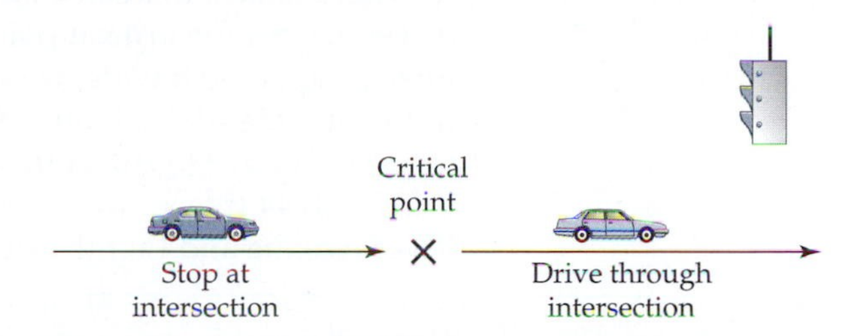

Figure 10 When the light turns yellow, cars to the left of the critical point should have enough distance to stop safely before the intersection and cars to the right of the critical point should have enough time to drive through the intersection before the light turns red.

If drivers to the left of the critical point should stop and drivers to the right of the point should drive through the light, then drivers at the critical point when the light turns yellow should be able to do both. To determine the time needed for the yellow light, we need to find the distance from the critical point to the intersection, which can be done by finding the minimum distance that a driver needs to react to a yellow light and stop safely. Since a driver at the critical point also should be able to drive through the intersection, the time needed for the yellow light should be the same as the time needed for the driver to drive from the critical point through the intersection.

Getting Specific

Now that our model is defined, let's consider an example. We now determine the time needed for a yellow light at the intersection of River Avenue and Eighth Street in Holland, Michigan, for vehicles traveling on River Avenue.

To determine the location of our critical point, we need to make some assumptions to determine the distance it takes a car to stop safely. These assumptions are the speed of cars traveling on River Avenue, how much time it takes a driver to react to a yellow light, and how fast a car can safely slow to a stop. The speed limit on River Avenue is 30 mph (which equals 44 ft/sec), so we assume that our vehicle is going that speed. We assume the reaction time (the time it takes to apply the brake from when a driver sees the light turn yellow) is 1 sec. We also assume that a safe braking deceleration is 10 ft/sec^2 (which is an acceleration of -10 ft/sec^2).[24] We also measured the width of the intersection and determined that it was 70 ft wide.

[24] *We determined these numbers through experimentation with one author at the controls of a car while braking at 30 mph (without spilling his coffee!) and another author timing with a stopwatch.*

We are now ready to calculate the time needed for the yellow light. Remember that we want a driver at the critical point to have the option of either stopping safely before the intersection or traveling through the intersection before the light turns red. Using our assumptions of reaction time, speed of the vehicle, and braking acceleration, we determined that the distance needed for a driver to react and stop was approximately 140.8 ft. (The details of this calculation are left for Investigation 1.) Therefore, the distance from the critical point to the intersection is 140.8 ft. Because the intersection is 70 ft wide, we know that the distance from the critical point to the far side of the intersection is 140.8 ft + 70 ft = 210.8 ft. The time it takes for the car to travel this distance (maintaining a speed of 44 ft/sec) is 210.8 ft/44 ft/sec = 4.79 sec. So, the time needed for the yellow light at River Avenue and Eighth Street is 4.79 sec.

Generalizing

All intersections with traffic lights are not identical to the one at River and Eighth in Holland, Michigan, however. To answer the question for a variety of intersections, we need to generalize. In our example, what assumptions were dependent on the intersection being at River and Eighth and what assumptions are true for most intersections? It is probably realistic to assume that the reaction time of drivers is 1 sec and the acceleration of braking is -10 ft/sec^2 no matter what intersection we consider. The width of intersections and the speed limit of the roads where these intersections are located, however, will vary. Because we want to develop a function with only one input, we cannot let both speed and intersection width be variables. We assume that the intersection is 70 ft wide and let the speed limit be our variable. We want to develop a function where the posted speed limit on a road (or the velocity of the car), v, is the input and the time needed for a yellow light, t, is the output. This function will have three terms: one involving the reaction time, one involving the acceleration of braking, and one involving the width of the intersection.

The process of deriving the formula is similar to the calculations done in our specific example. We now use stopping distance to find the distance from the critical point to the far side of the intersection and divide this value by the velocity to determine the time needed for a car to travel that far. This result is the time needed for the yellow light. The equation we obtain is $t = 1 + v/20 + 70/v$. (Details on how the equation is obtained are left for Investigation 1.)

Analyzing

Now that we have a mathematical model that can be used to determine the time needed for a yellow traffic light, let's connect it back to our original situation and assumptions. The function we obtained is $t = 1 + v/20 + 70/v$, where t is the time in seconds and v is the velocity in feet per second. Notice that this function consists of three terms. The term involving reaction time is the 1 because we assumed that reaction time was 1 sec. This value is not dependent on the speed of the car. The term involving acceleration of braking is $v/20$. This term makes our function increase because as v increases, $v/20$ also increases because the faster you travel, the more time it takes to slow to a stop safely. The last term, $70/v$, involves the width of the intersection. This term makes our function decrease because as v increases, $70/v$ decreases. The faster you travel, the less time it takes to travel through the intersection.

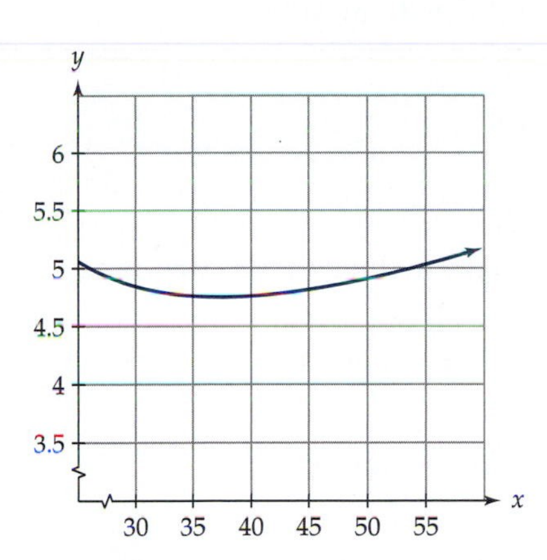

Figure 11 A graph of the yellow light function.

Let's determine a reasonable domain and range for our yellow light function. Since the domain of this function should include all the speed limits on roads with traffic lights, a domain of 25 to 55 mph is appropriate. The range should include all the times of yellow lights. Using the answer (4.79 sec) obtained earlier from our specific example as a guide, our guess for a range is 3 to 6 sec. The graph of the yellow light function with the specified domain and range is shown in Figure 11. Notice that this function both decreases and increases, which makes sense because, as pointed out earlier, one term is causing the function to decrease and one is causing the function to increase.

Once you obtain a model and analyze it, your work may not be done. You may want to develop a different model based on different assumptions. For example, we could let the width of the intersection instead of the speed of the car be the variable. One of our original assumptions was that the cars had to drive through the intersection before the light turned red. We could change this assumption to allow the car to just make it halfway through or to start into the intersection before the light turns red. Making these changes in our assumptions leads to the development of different models. It is then interesting to compare models to determine which best models actual traffic lights.

READING QUESTIONS

14. What is the purpose of the *critical point* in defining our model for the length of time for a yellow traffic light?

15. Our general model, $t = 1 + v/20 + 70/v$, contains three terms. What does each term represent?

16. According to our model, how does the width of the intersection affect the time needed for a yellow light? How does the velocity of the cars affect the time needed for a yellow light?

SUMMARY

Taking a physical situation and developing a mathematical formula to describe it is called **mathematical modeling.** To develop a mathematical model, first define the model, which may involve restating the question in a way that a mathematical formula can more easily be obtained. After the terms are better defined, you may develop a formula for a specific situation, which gives you a feel for the problem before getting too abstract. Once you understand the problem and how to solve it for some specific cases, you need to generalize. Generalizing often leads to a formula that will work in many situations. Finally, once your abstract model is obtained, you can analyze the model by connecting it back to the physical situation by asking additional questions or adjusting some of the original variables.

EXERCISES

1. Solve the following formulas for the indicated variable.
   (a) Solve $F = \frac{9}{5}C + 32$ for C.
   (b) Solve $A = 2\pi r^2 + 2\pi rh$ for h.
   (c) Solve $V = \pi r^2 h$ for r.
   (d) Solve $A = (\sqrt{3}/4)a^2$ for a.
2. What assumptions would you have to make to develop a mathematical model that gives
   (a) the time needed for someone to drive to work.
   (b) the price of a used car.
   (c) the time needed for someone to mow a lawn.
   (d) the time needed to study for a test.
3. In our puzzle piece example from this section, we made the assumption that the outside figures had the same area. In this exercise, we make the assumption that the inside figures have the same area.
   (a) Suppose a square of area A is inscribed in a circle. What is the area of the circle?

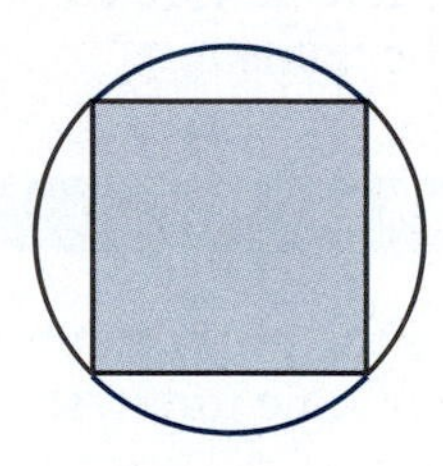

   (b) Suppose a circle of area A is inscribed in a square. What is the area of the square?

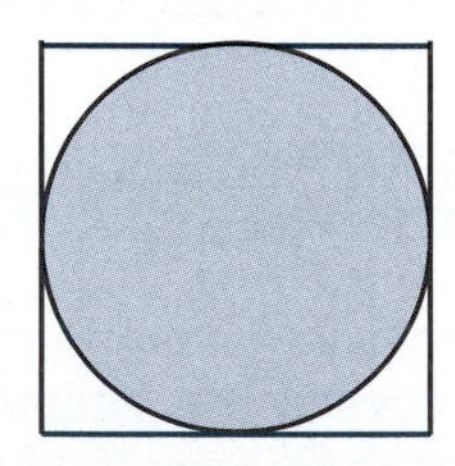

   (c) Which has greater area, the difference between a square of area A inscribed in a circle or the difference between a circle of area A inscribed in a square?
   (d) In our puzzle piece example from this section, we concluded that the round puzzle piece always fits into a square hole better than a square puzzle piece fits into the round hole. How does this conclusion compare with your conclusion in part (c)?

4. Here are two pictures of circles inscribed in squares of area 1.

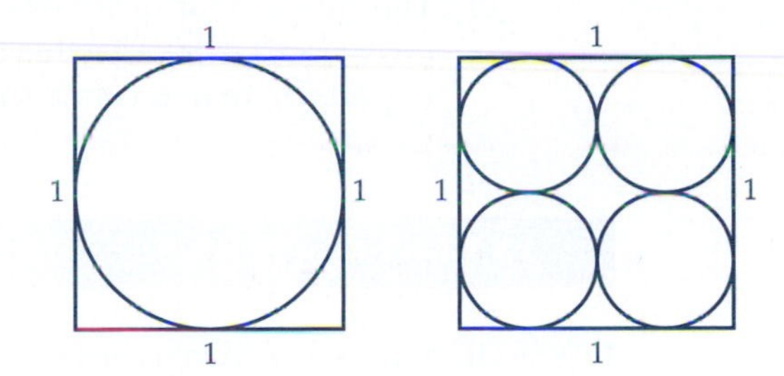

 (a) Find the area of the one large circle.
 (b) Find the sum of the areas of the four small circles.
 (c) Which is larger, the area of the one large circle or the sum of the areas of the four small circles?
 (d) Instead of four circles, suppose there were nine circles in the square (arranged in three rows and three columns). What is the sum of the areas of the nine circles?
 (e) Suppose there were n^2 circles arranged in the square (n rows and n columns of circles). What is the area of each circle? What is the sum of the areas of all n^2 circles?

5. In this section, we developed the formula $d = \sqrt{7900h + h^2}$, where h is the height above Earth and d is the distance you can see to the horizon. In this formula, why must the height above Earth be given in miles?
6. If you were standing on top of the Eiffel Tower in Paris (984 ft tall), how far could you see out to the horizon?
7. If you were standing on top of a building and could see 25 mi out to the horizon, how tall is the building in miles? How tall is the building in feet?
8. Suppose there were tall buildings on Mars (which has a radius of approximately 2100 mi) and you were standing on one with a height of h miles. Determine a function where h is the input and d, the distance you can see out to the horizon, is the output.
9. In the application at the end of this section, we found that the function $t = 1 + v/20 + 70/v$, where v was the velocity of a car in feet per second, gave the time, t (in seconds), needed for a yellow light. The unit on the 1 is seconds, the unit on the 20 is feet per second per second (ft/sec^2), and the unit on the 70 is feet. For the output to be time in seconds, each term of the function has to be in terms of seconds. Show that this is true.
10. A common mathematical model used to determine the height of an object after it is thrown into the air is $h(t) = -16t^2 + v_0 t + h_0$, where $h(t)$ is the height after t seconds, v_0 is the initial velocity (in feet per second), and h_0 is the initial height (in feet). Suppose a water balloon is thrown straight up into the air at 30 ft/sec from a height of 6 ft.
 (a) How long will it take for the balloon to hit the ground?

(b) What is the velocity of the balloon when it reaches its maximum height?

(c) The model that describes the velocity of the same balloon is $v(t) = -32t + 30$. Use this information, along with your answer to part (b), to determine the maximum height of the balloon.

INVESTIGATIONS

INVESTIGATION 1: CROSSROADS

We left out a number of the algebraic details in our application from this section. In this investigation, you are asked to provide these details. We begin by finding the time needed for a yellow light at the intersection at River Avenue and Eighth Street in Holland, Michigan. We then generalize this situation and develop an alternate model.

1. We start with our specific example. The first step in that process was to find the distance our vehicle needs to stop safely. So, we need an equation relating distance to both velocity and acceleration. We get this equation by manipulating and combining two other equations. In these equations, a = acceleration, v_f = final velocity, v_0 = initial velocity, t = time, and d = distance traveled. We use the fact that acceleration is change in velocity divided by time, or

$$a = \frac{(v_f - v_0)}{t}. \tag{1}$$

Next, we need to know that average velocity is distance traveled divided by time (a form of the equation $d = rt$), or

$$\frac{(v_0 + v_f)}{2} = \frac{d}{t}. \tag{2}$$

The following summarizes the specific data for the intersection at River Avenue and Eighth Street mentioned in our application:

Initial velocity, v_0	$= 44$ ft/sec
Width of intersection	$= 70$ ft
Reaction time to brake	$= 1$ sec
Acceleration of braking, a	$= -10$ ft/sec^2

(a) Since we will be braking to a complete stop, the final velocity in our equations is zero. Using this, combine equations (1) and (2) to get $d = -v_0^2/(2a)$.

(b) Once the brake is applied, how much distance will it take for our car on River Avenue to stop?

(c) How much total distance, including reaction time, will it take for our car to stop once the light turns yellow? This distance is from the beginning of the intersection to the critical point.

(d) From this critical point, how much time is needed for the front of our car (maintaining a speed of 44 ft/sec) to make it to the far side of the intersection safely? This time will be the minimum time needed for a yellow light.

(e) The yellow light at River and Eighth stays on for 4.2 sec. Is this what you obtained? If not, what might explain the difference?

2. Now let's consider the more general situation. As in our application, we assume the intersection is still 70 ft wide, the reaction time is still 1 sec, and the acceleration of braking is still -10 ft/sec^2. Let the initial velocity vary. The final velocity, v_f, continues to be zero since we are interested in stopping the car.
   (a) Find the function whose input is the initial speed of the vehicle, v_0, in feet per second and whose output is the distance from the critical point to the far side of the intersection.
   (b) Using your function from part (a), find the function whose input is the initial speed of the vehicle, v_0, in feet per second and whose output is the time needed for a yellow light, t, in seconds.

3. We did some analysis of this model in our application. We also mentioned other assumptions we could make to develop different models. One assumption was instead of having the car make it to the far side of the intersection before the light turned red, it only had to make it halfway through.
   (a) Determine the function needed for a yellow light if all our original assumptions are the same except the car only has to make it halfway through the intersection before the light turns red.
   (b) Graph your function from part (a) along with a graph of the original model. How is the new graph different from the original graph?
   (c) How does the time needed for a yellow light for a car traveling at 44 ft/sec compare with the actual time of the yellow light at River and Eighth (4.2 sec)? Which model seems to work best?

PROJECTS 1.5

CRICKETS: NATURE'S THERMOMETER

Crickets are one of nature's more interesting insects, partly because of their musical ability. In England, the chirping or singing of a cricket was once considered a sign of good luck. In China and Japan, crickets were kept in fancy cages in the house so the residents could enjoy their singing. Many of us are so used to hearing this sound on a summer evening that we would probably think that something was wrong if it were missing. The male cricket "sings" to attract the female cricket—not just to keep you up at night—by rubbing his two front wings together.[25]

An interesting fact about crickets is that their activity depends on temperature. As a result, they can be thought of as "natural" thermometers.

[25] Ross E. Hutchins, *Insects* (Englewood Cliffs, NJ: Prentice-Hall, 1966), pp. 54–57.

The rate of a cricket's chirp increases as the temperature increases; it also depends on the type of cricket. So, if you know the right formula and the type of cricket you hear chirping, you can estimate the temperature by counting the chirps. Changes in humidity and different crickets of the same type also produce variations in a cricket's chirping rate. The dominant factor, however, is temperature, so formulas relating temperature to the number of chirps are fairly accurate. The following are rules for finding the temperature, in degrees Fahrenheit, for three different types of crickets.[26]

- The field cricket is the black cricket commonly found in the United States. For a field cricket, count the number of chirps in 15 sec and add 38 to obtain the temperature.
- The tree cricket is small and pale green and is usually found on trees. For this cricket, the temperature can be obtained by counting the number of chirps in 7 sec and adding 46.
- The snowy tree cricket is the species whose music is most in tune with that of the temperature since it is believed to be the most accurate. For this cricket, count the number of chirps in 14 sec and add 42.

1. Find a function relating temperature to the number of chirps by following these steps.
   (a) Notice that each "rule" involves counting the number of chirps in a different predetermined period: 15 sec for the field cricket, 7 sec for the tree cricket, and 14 sec for the snowy tree cricket. The input for our function will be *number of chirps per minute*.
      i. For the field cricket, if n is the number of chirps per minute, what is the number of chirps in 15 sec?
      ii. For the tree cricket, if n is the number of chirps per minute, what is the number of chirps in 7 sec?
      iii. For the snowy tree cricket, if n is the number of chirps per minute, what is the number of chirps in 14 sec?
   (b) Translate the "rule" for each cricket into a function where the input is n, number of chirps per minute, and the output is T, temperature in degrees Fahrenheit.
   (c) If each type of cricket chirps 120 times in 1 min, what are the three different temperatures?
2. In the following questions, various properties and characteristics of the functions relating temperature to the number of chirps are examined.
   (a) Explain how you know, just from looking at your functions from question 1, part (b), that these equations are linear.
   (b) For each function, specify a domain (inputs that are reasonable for this situation) and the corresponding range. (*Note:* Assume these functions are not valid when the temperature is above 70°F.)
   (c) Graph each of the three functions on the same set of axes.

[26] *Lucy Clausen,* Insect Fact and Folklore *(New York: MacMillan, 1958), pp. 62–63.*

(d) Explain the physical meaning of the y-intercept in terms of temperature and number of cricket chirps.

(e) Remembering that slope is (change in output) $\div$ (change in input), explain the physical meaning of slope for these functions in terms of temperature and number of cricket chirps.

3. We now want to find functions where the input is temperature and the output is the number of chirps per minute; that is, we wish to find the inverse functions for the three functions found in question 1 part (b).

    (a) For each of the three functions from question 1, part (b), solve for n in terms of T; that is, find the inverse function.

    (b) What is the domain and the range of each of your inverse functions? How do they compare with the domain and the range for the three original functions?

    (c) What is the slope of each of the inverse functions? How does it compare with the slope of the original functions?

    (d) If it is 70°F, how many chirps will each cricket produce in 1 min?

    (e) Graph the three inverse functions on the same set of axes.

    (f) We wish to determine which cricket chirps at the highest rate. So, we are looking for the line with the largest y-values, or the line that is "on top."

        i. Examine your graph for the temperature range from 40°F to 70°F. Notice that the line that has the highest y-values depends on where you are in this interval. Find the point of intersection for these two lines.

        ii. Which cricket chirps at the highest rate and at which temperatures when $40 < T < 70$?

    (g) The slope of a line tells how "steep" the line is. If you have two lines, is it always true that the line with the largest slope gives the largest output? Explain your answer. How does this relate to your answer to question 3, part (f)?

4. Suppose we wish to find temperature in degrees Celsius[27] rather than degrees Fahrenheit. For this question, consider only the field cricket rather than all three crickets.

    (a) Use the function found in question 1, part (b) for the *field cricket* and convert it to a new function relating number of chirps per minute to temperature in degrees Celsius. (*Hint:* Use $T_C = \frac{5}{9}(T_F - 32)$, where T_F is temperature in degrees Fahrenheit and T_C is temperature in degrees Celsius.)

    (b) Convert this function to a verbal rule, such as one of those originally given, that can be used easily to convert chirps to temperature in degrees Celsius. Your rule should be written in the following form: Count the number of chirps in ___ seconds and add ___. Refer back to question 1, part (a), to remind yourself how to convert from words to symbols. (*Note:* For this rule to

[27] *This would give us a metric cricket.*

be easily used, it is necessary to round quantities to the nearest whole number.)

REVIEW EXERCISES

1. Evaluate each of the following functions.
   (a) Evaluate $A = (\sqrt{3}/4)a^2$ for $a = 5$.
   (b) Find $f(3)$ if $f(x) = 2x^2 - 7x$.
   (c) In the accompanying graph of g, find $g(2)$ and x such that $g(x) = 3$.

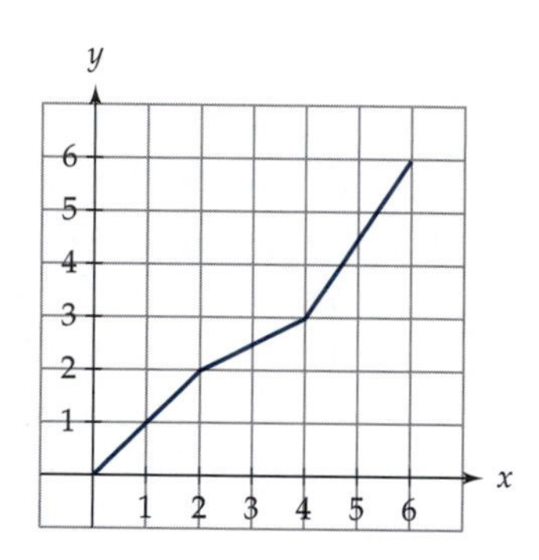

   (d) Evaluate the mean SAT math score for 1980.

Year	1975	1980	1985	1990	1995
SAT math mean score (recentered scale)	498	492	500	501	506

   (e) Evaluate the cost of a 5 min phone call. Evaluate the cost of a 30 min phone call.

   The cost of a long distance phone call is just 99 cents for the first 20 min and then just 10 cents a minute for each minute over 20 min.

2. Determine if the following rules are functions. If the rule is a function, then give the domain and the range. If it is not a function, then give an example of an input that has more than one output.
   (a) A listing of the number of hours of daylight per day of the year in Las Vegas, Nevada

(b)

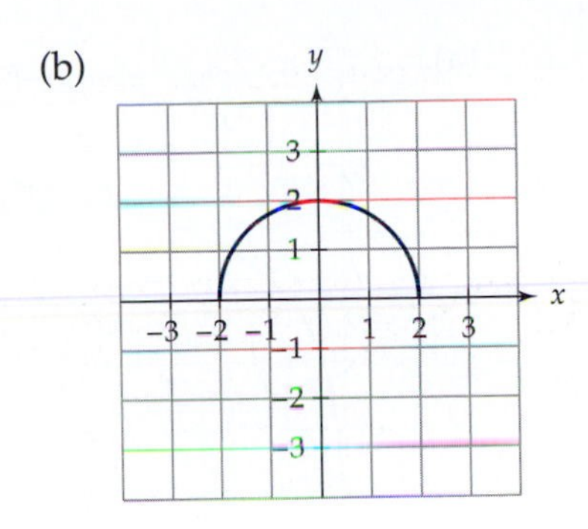

(c) $y = \sqrt{x^2 - 5}$

(d) $y = \dfrac{1}{2x - 1}$

(e) $y = x^{2/3}$

(f)

Input	1	2	3	3	5
Output	5	5	6	7	8

3. Evaluate each of the following functions at the given inputs.
   (a) If $f(x) = 1/x^2$, determine $f(5)$ and $f(\frac{1}{5})$.
   (b) If $g(w) = w^2 + 1$, determine $g(2)$ and $g(x+1)$.
   (c) If $h(n) = 2n - 5$, determine $h(2)$ and $h(x^2)$.
   (d) If $c(f) = \frac{5}{9}(f - 32)$, determine $c(68)$ and $c(-40)$.
   (e) If

$$g(x) = \begin{cases} x - 7, & \text{if } x < 0 \\ x^2 + 1, & \text{otherwise,} \end{cases}$$

   determine $g(-5)$ and $g(5)$.
4. Let f be the function used to determine the temperature of your house in the winter such that t is the time of day and $f(t)$ is the temperature of the house in degrees Fahrenheit. In terms of time of day and temperature of the house, what is the meaning of the following?
   (a) $f(t+2)$
   (b) $f(t) + 2$
5. Use the terms *y-intercept, x-intercept, increasing, decreasing, concave up, concave down, symmetric about the y-axis,* and *symmetric about the origin "as appropriate"* to describe the following graphs.

(a)

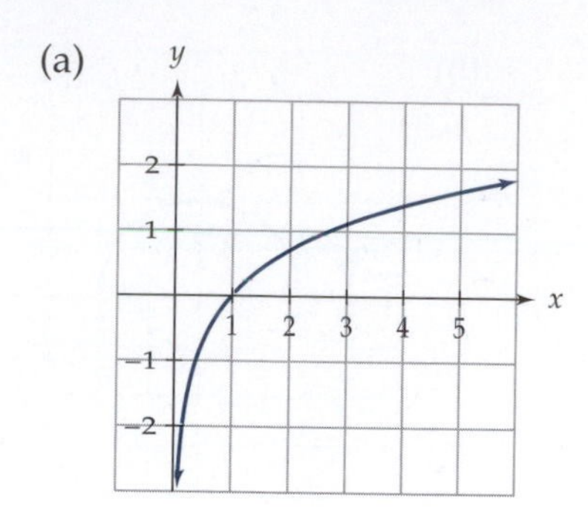

(b)

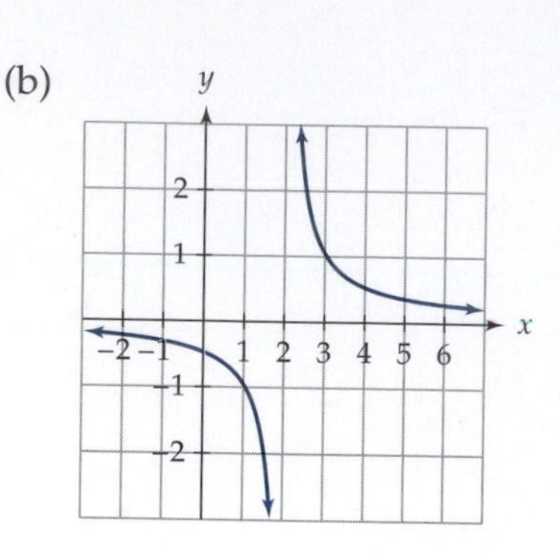

(c)

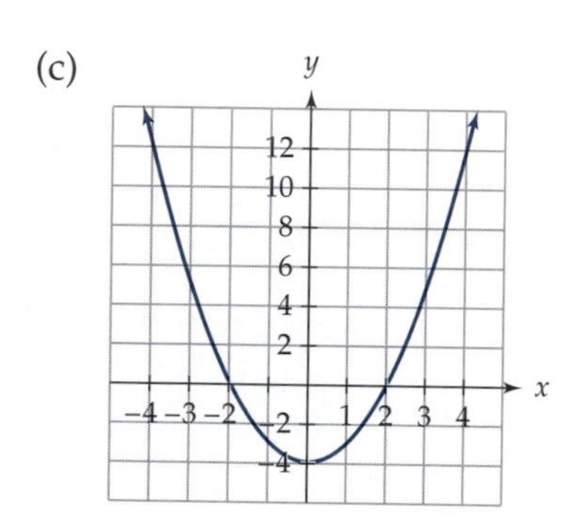

6. Let f be a function such that $f(4) = -5$.
   (a) If f is symmetric about the y-axis, what is $f(-4)$?
   (b) If f is symmetric about the origin, what is $f(-4)$?
7. Many types of containers can be filled with water. The shape of the container has an impact on how much liquid the container holds and how quickly the level of water rises as it is being filled with water flowing at a constant rate. For each of the following containers, graph the height of the water as a function of time and give a domain and range. Be sure to put time on the horizontal axis and the height of the water level on the vertical axis.
   (a) The following container is 5 in. tall and will take 15 sec to fill.

(b) The following container is 8 in. tall and will take 20 sec to fill.

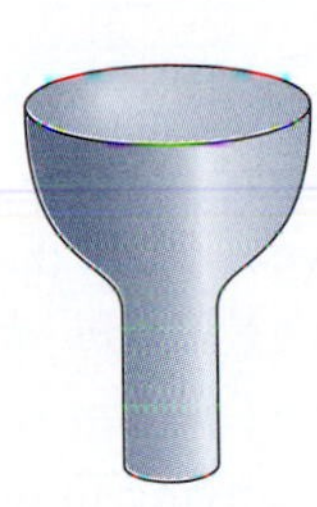

8. Describe each of the following functions as concave up or concave down.

(a) $f(x) = -2x^2 + 3x - 4$

(b)

x	-1	0	1	2	3	4
$g(x)$	1	6	10	13	15	16

(c)

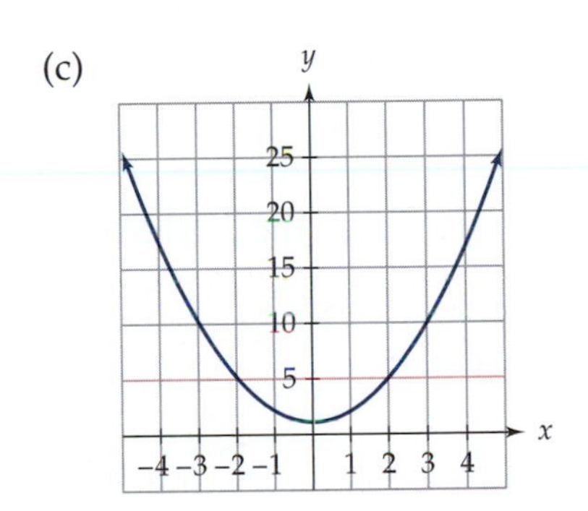

9. A standard $-10 \le x \le 10$ calculator screen is not an appropriate window to view the general shape of the graphs of the following functions. Analyze these functions to determine an appropriate viewing window that shows the graph's general shape and make a sketch of your calculator graph. Be sure to indicate the dimensions of the viewing window you used.

(a) $g(x) = x^2 - 3x + 20$

(b) $f(x) = \dfrac{15}{(x-20)^2}$

(c) $h(x) = 5\left(\frac{1}{4}\right)^{x+20}$

10. Use your calculator to determine the zeros of $f(x) = 8x^4 + 34x^3 - 3x^2 - 85x + 25$. Are these zeros exact or approximations? Explain.

11. Use your calculator to determine the point of intersection of $f(x) = x + 2$ and $g(x) = 6x + 7$. Is your point exact or an approximation? Explain.
12. Solve the following formulas for the indicated variable.
    (a) Solve $C = \frac{5}{9}(F - 32)$ for F.
    (b) Solve $A = 2\pi r(r + h)$ for h.
    (c) Solve $F = (Gm_1m_2/d^2)$ for d.
    (d) Solve $F = (mv^2/r)$ for r.
13. The accompanying figure is of a window. It consists of a semicircle of radius r on top of a square.

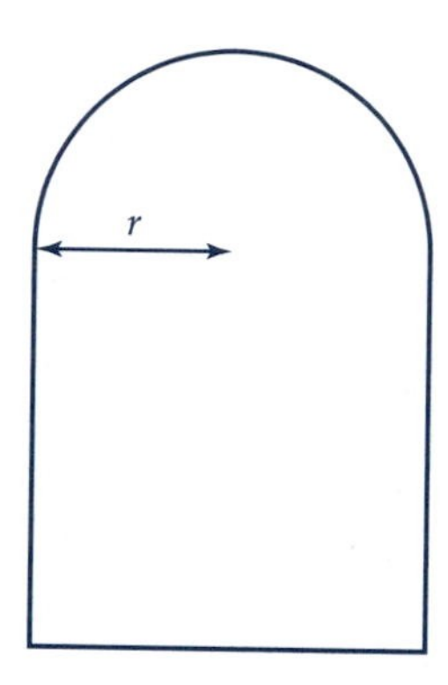

    (a) Find the area of the window in terms of r.
    (b) Find the perimeter of the window in terms of r.
14. If you were looking out over the ocean on top of Mauna Loa in Hawaii (13,680 ft above sea level), how far could you see out to the horizon? (Use the formula developed in Section 1.4.)
15. Suppose a tank holds 1000 gal of water that will drain from the bottom of the tank in 10 min. The volume of the water, V, remaining in the tank after t minutes can be modeled by

$$V(t) = 1000\left(1 - \frac{t}{10}\right)^2.$$

    (a) What is the domain for V?
    (b) What is the range for V?
    (c) What is the volume of the tank after 5 min?
    (d) How long will it take for the tank to be half full?

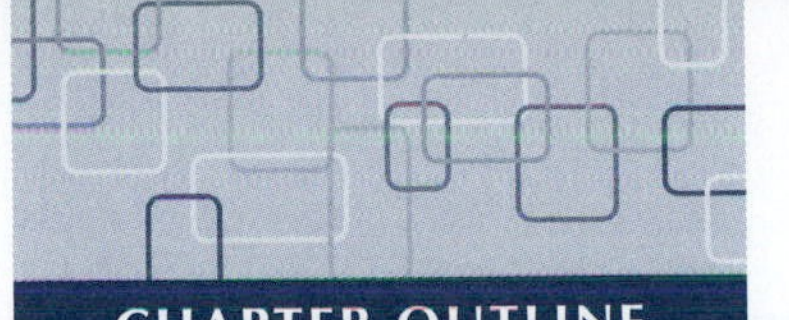

CHAPTER OUTLINE

CHAPTER OVERVIEW

- Recognizing linear, exponential, logarithmic, periodic, and power functions in their various representations
- The properties of these functions as well as the similarities and differences between these functions
- Finding equations for linear, exponential, and power functions
- The sine and cosine functions are defined as periodic functions

FAMILIES OF FUNCTIONS

Linear, exponential, logarithmic, periodic, and power functions are introduced in this chapter. Students learn to recognize these functions in their various representations. Students also learn how to obtain a formula when a given a linear, exponential, or power function either numerically or graphically. This knowledge lays the groundwork for these functions and for their use throughout the remainder of the book.

2.1 LINEAR FUNCTIONS

Linear functions are the simplest type of function. They are also the most commonly used functions. Examples of linear functions include finding distances traveled, wages for hourly employees, prices of long-distance telephone calls, and sales tax. Most functions that convert one unit of measurement to another—such as the conversion of feet to meters, dollars to francs, and temperature on the Fahrenheit scale to the Celsius scale—are also linear. In this section, we explore properties of linear functions and how to determine if a function is linear.

DEFINITIONS AND SYMBOLIC REPRESENTATIONS OF LINEAR FUNCTIONS

All linear functions have the same type of relationship between the input and the output. To explore this relationship, Table 1 looks at men's shoe sizes and the length of men's feet in inches.

Notice that as the shoe size column increases by 1, the length column *always* increases by $\frac{1}{3}$. Another way of saying this is (change in output)/(change in input) $= \frac{1}{3}$. This constant rate of change is what defines a linear function. A **linear function** is one in which the rate of change is constant. The **rate of change** is the change in output divided by the change in input. We can see why these functions are called linear when we plot the points from Table 1; see Figure 1(a). The constant rate of change is shown by all the points lying on the same line; see Figure 1(b).

y-Intercept and Slope

The linear function relating shoe size to foot length in inches (and every other linear function, for that matter) can be described with only two pieces of information: the y-intercept and the slope. The ***y*-intercept** is where the graph crosses the y-axis. It can also be thought of as the output when

TABLE 1 Comparing U.S. Men's Shoe Size to the Length of a Man's Foot in Inches

Shoe Size	5	6	7	8	9	10	11
Average Foot Length (in.)	9	$9\frac{1}{3}$	$9\frac{2}{3}$	10	$10\frac{1}{3}$	$10\frac{2}{3}$	11

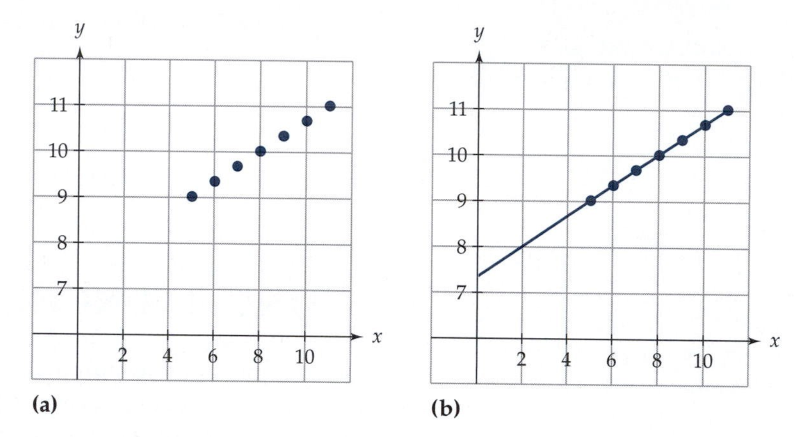

Figure 1 The data from Table 1 lie on the same line.

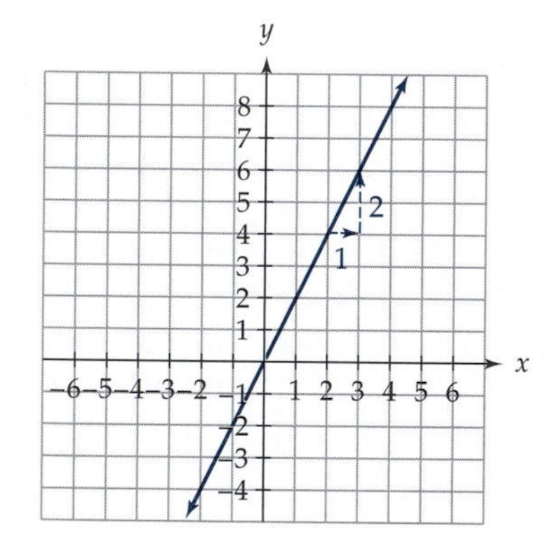

Figure 2 A graph of a linear function with a slope of 2.

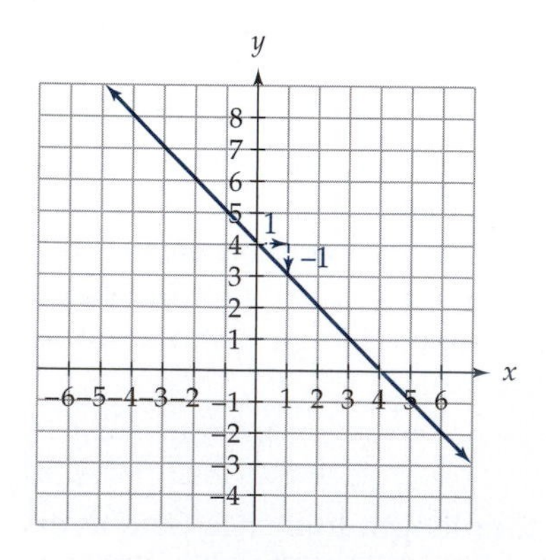

Figure 3 A graph of a linear function with a slope of −1.

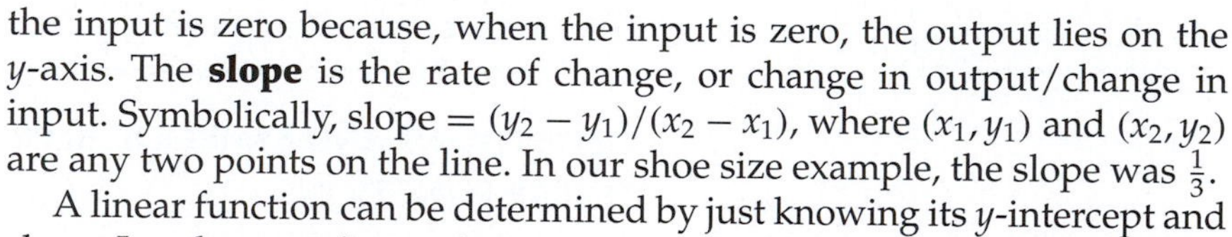

the input is zero because, when the input is zero, the output lies on the y-axis. The **slope** is the rate of change, or change in output/change in input. Symbolically, slope $= (y_2 - y_1)/(x_2 - x_1)$, where (x_1, y_1) and (x_2, y_2) are any two points on the line. In our shoe size example, the slope was $\frac{1}{3}$.

A linear function can be determined by just knowing its y-intercept and slope. In other words, we do not need a symbolic formula. For example, suppose we had a linear function whose y-intercept was 0 and whose slope was 2. Then, for every one unit of increase horizontally, there will be two units of increase vertically, or for every one unit of increase in the input, the output will increase by two. So, $f(0) = 0$ (the y-intercept), $f(1) = 2$, $f(2) = 4$, $f(3) = 6$, and so on. Every time you move to the right one unit, you must go up two units. (See Figure 2.)

What happens, however, if the slope is negative? A linear function with a negative slope is a decreasing function. Suppose we have a linear function whose y-intercept is 4 and whose slope is -1. Then for every one unit of increase in the input, the output will *decrease* by 1. Therefore, $f(0) = 4$ (the y-intercept), $f(1) = 3$, $f(2) = 2$, $f(3) = 1$, and so on. Every time you move to the right one unit, you go *down* one unit. (See Figure 3.)

In summary,

- A linear function with a positive slope is an increasing function.
- A linear function with a negative slope is a decreasing function.

Slope-Intercept Form

Knowing the slope and the y-intercept for a linear function is sufficient for finding the symbolic representation of the function, that is, the equation for that function. The most common way to write a linear equation is using the **slope-intercept form:**

$$\text{output} = \text{slope} \times \text{input} + y\text{-intercept}$$

or

$$y = mx + b,$$

where y is the output, m is the slope, x is the input, and b is the y-intercept. Let's give the equations for the linear functions we have already seen. The equation for the linear function shown in Figure 2 with a y-intercept of 0 and slope of 2 is $y = 2x + 0$ or $y = 2x$. The equation for the linear function shown in Figure 3 with a y-intercept of 4 and slope of -1 is $y = -x + 4$.

Example 1 Determine the equation for the linear function representing the conversion from shoe size to inches given in Table 1.

Solution We saw earlier that the slope of this function was $\frac{1}{3}$. We can estimate the y-intercept from the graph in Figure 1(b) as being something between 7 and 8. To determine its exact value, we can start with the first entry in Table 1 and work backwards to determine $f(0)$. We know that $f(5) = 9$, and with a slope of $\frac{1}{3}$, $f(4) = 8\frac{2}{3}$, $f(3) = 8\frac{1}{3}$, $f(2) = 8$, $f(1) = 7\frac{2}{3}$, and $f(0) = 7\frac{1}{3}$. Therefore, the equation is $f(x) = \frac{1}{3}x + 7\frac{1}{3}$, or to use variable labels that are more descriptive, $l = \frac{1}{3}s + 7\frac{1}{3}$, where l is the length of a foot in inches and s is the shoe size.

Horizontal and Vertical Lines

We have seen that linear functions with positive slopes are increasing and those with negative slopes are decreasing. What about linear functions with a slope of zero? A linear function with a slope of zero is neither increasing nor decreasing. A slope of zero means (change in output)/(change in input) $= 0$, so there is no change in the output. Therefore, represented graphically, these lines are horizontal. Symbolically, a linear function with a slope of zero is written as $y = 0x + b$, or simply $y = b$. These functions give the same output regardless of the input. For example, $y = 3$ means that no matter what the input is, the output is always 3. The graph of $y = 3$ is shown in Figure 4(a).

On a vertical line, the change in input is zero, which means that the formula for slope results in division by zero. So, vertical lines have slopes that are undefined. Although vertical lines are not functions, they can still be represented symbolically. Vertical lines have constant x-values. Therefore, these lines can be represented by the equation $x = k$, where k is a constant. For example, the graph of the line $x = 5$ is shown in Figure 4(b).

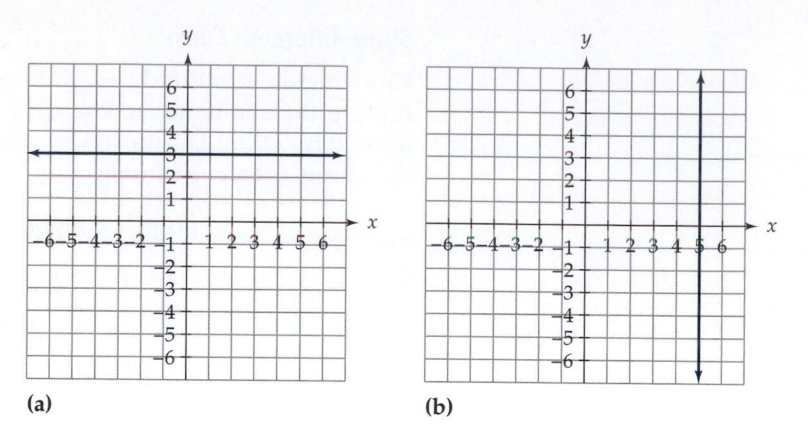

Figure 4 Graphs of (a) the horizontal line $y = 3$ and (b) the vertical line $x = 5$.

READING QUESTIONS

1. What is the definition of a linear function?
2. Why are the two ways of thinking about the y-intercept actually the same?
3. What is the slope of the line containing the points in the following table?

x	1	2	3	4
y	3	6	9	12

4. Suppose you have a linear function which decreases. What do you know about the slope of this line?
5. What is the y-intercept of the linear function $f(x) = 3x + 4$?
6. Graph each of the following.
   (a) $y = -2$
   (b) $x = 4$

GRAPHICAL REPRESENTATIONS OF LINEAR FUNCTIONS

Linear functions have a constant rate of change, so they look like lines when represented graphically. Hence, they are called *linear* functions. If you are given the graph of a line (and therefore know you have a linear function), you can determine its equation by estimating the slope and the y-intercept. One way to estimate the slope is to find the change in y when x increases one unit. To estimate the y-intercept, find where the graph intersects the y-axis.

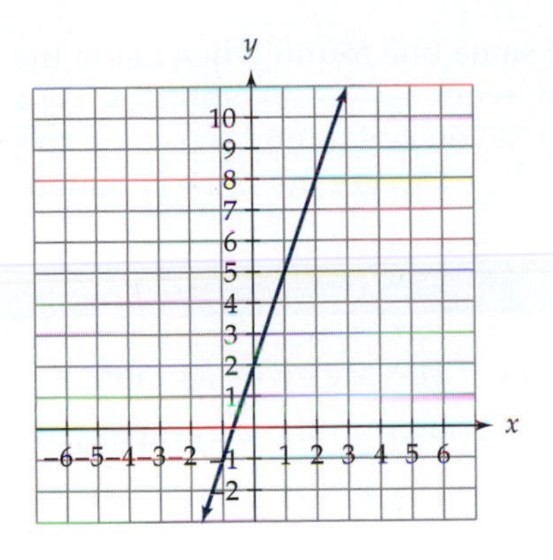

Figure 5

Example 2 Give the equation for the linear function shown in Figure 5.

Solution The graph increases three units in the vertical direction for each increase of one unit in the horizontal direction, which gives a slope of $\frac{3}{1} = 3$. (See Figure 6.) The y-intercept for this function is 2. Therefore, the equation of this line is $y = 3x + 2$.

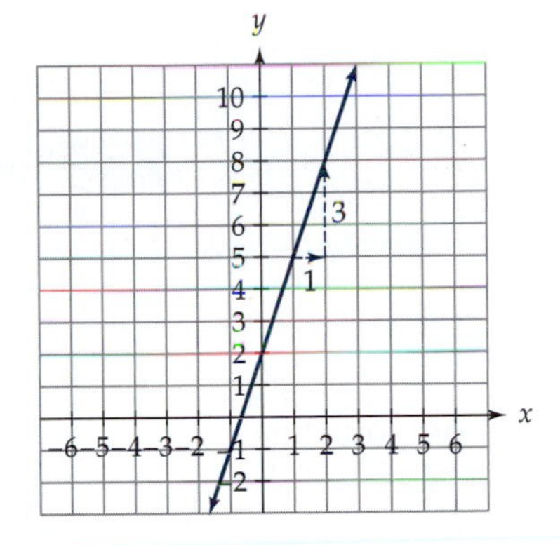

Figure 6

If the y-axis is not included in the graphical representation of the function, making it impossible to estimate the y-intercept directly from the graph, you can still find the equation of the line by finding the coordinates of any two points on the line, (x_1, y_1) and (x_2, y_2). Once you have the two points, do the following to find the equation.

- Use the two points to find the slope,

$$m = \frac{y_2 - y_1}{x_2 - x_1}.$$

- Substitute the value of the slope for m and the coordinates of one of your points for x and y into the equation $y = mx + b$.
- Solve for b.
- Write your equation in slope-intercept form.

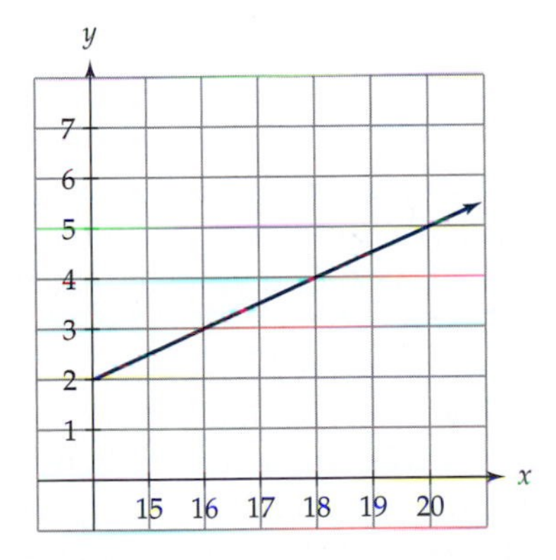

Figure 7

Example 3 Find the equation of the linear function shown in Figure 7.

Solution The graph of this function does not include the y-axis. (The vertical line on the left side of the graph is $x = 14$.) We can, however, use the coordinates of *any* two points on this line to determine its equation. For convenience, we use the points $(14, 2)$ and $(20, 5)$, because they have integer coordinates. Using these points, the slope is

$$m = \frac{(5-2)}{(20-14)} = \frac{3}{6} = \frac{1}{2}.$$

We use the point $(14, 2)$ and the slope we just found to solve for b, the y-intercept.

$$\begin{aligned} 2 &= \tfrac{1}{2} \cdot 14 + b \\ 2 &= 7 + b \\ -5 &= b. \end{aligned}$$

Writing our equation in slope-intercept form, we get $y = \frac{1}{2}x - 5$.

Suppose in Example 3 we had used the point $(20, 5)$ instead of the point $(14, 2)$ to solve for b. Then we would have

$$\begin{aligned} 5 &= \tfrac{1}{2} \cdot 20 + b \\ 5 &= 10 + b \\ -5 &= b. \end{aligned}$$

The resulting equation is $y = \frac{1}{2}x - 5$, the same one found when using the point (14, 2). The equation of a line has to work for *all* the points on your line. Hence, regardless of which two points you use to find the slope and which point you use to solve for b, you will *always* get the same equation.

READING QUESTIONS

7. Explain how to find the equation of a line if you are given a graph.
8. Find the equation of the linear function shown in the accompanying graph.

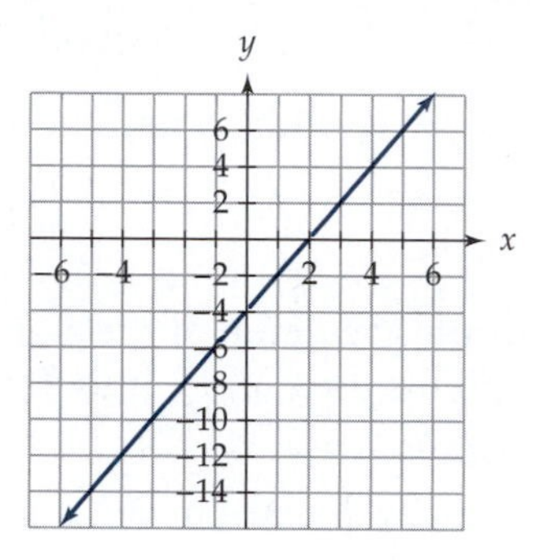

9. Suppose you know that two of the points on your line are (2, 3) and (4, 0). You find that the slope is $(0 - 3)/(4 - 2) = -\frac{3}{2}$. Using the point (2, 3), you find that the equation of this line is $y = -\frac{3}{2}x + 6$. If you used the point (4, 0) instead of the point (2, 3), what equation would you get? How do you know?

NUMERICAL REPRESENTATIONS OF LINEAR FUNCTIONS

Let f be a function represented numerically. How do we determine if f is linear by looking at the data? Once again, we rely on the definition of linear functions; that is, only linear functions have *constant* slope. Table 2 represents two functions. One is linear and one is not. Can you tell which function is linear?[1]

The linearity is easy to spot because the function given by Table 2(a) has a constant rate of change, whereas the function given by Table 2(b) does not. Remember, the definition of a linear function is that the rate of change must be constant. If the slope m is constant, then one unit of change in the input will produce m units of change in the output no matter what the input is. Notice in Table 2(a) that as x increases by 1, y always increases by 2. Thus, Table 2(a) is a linear function.

Suppose we decide that a function represented numerically is linear and we wish to find its equation. Again, we can do this with two pieces of

[1] *The symbol Δ (the uppercase Greek letter delta) is commonly used to denote a change in a variable. So, the change in x is commonly denoted as Δx (pronounced "delta x"), and the change in y is denoted as Δy (pronounced "delta y").*

TABLE 2 One of These Tables Is a Numerical Representation of a Linear Function

(a)

x	y	$\frac{\Delta y}{\Delta x}$
−1	−1	
		$\frac{2}{1}=2$
0	1	
		$\frac{2}{1}=2$
1	3	
		$\frac{4}{2}=2$
3	7	
		$\frac{4}{2}=2$
5	11	

(b)

x	y	$\frac{\Delta y}{\Delta x}$
−1	2	
		$\frac{-2}{1}=-2$
0	0	
		$\frac{2}{1}=2$
1	2	
		$\frac{16}{2}=8$
3	18	
		$\frac{32}{2}=16$
5	50	

information: the constant slope and the y-intercept. It is easy to determine the y-intercept if the input value of zero is included in the table. Remember that the y-intercept is where the graph crosses the y-axis or the output when the input is zero. If you are given a table of data and zero is one of your inputs, then you already have the y-intercept. For example, the formula for the linear function given by Table 2(a) must be $y = 2x + 1$ because the slope is 2 and y is equal to 1 when x is equal to 0.

If the data you are given do not include $x = 0$, you can use any two points to determine the equation of the line the same way you did in the graphical examples.

TABLE 3

x	y	$\frac{\Delta y}{\Delta x}$
7	21	
		$\frac{-4}{2}=-2$
9	17	
		$\frac{-6}{3}=-2$
12	11	
		$\frac{-6}{3}=-2$
15	5	
		$\frac{-10}{5}=-2$
20	−5	

Example 4 Find the equation of the linear function given in Table 3.

Solution The slope of the function is −2. Although the y-intercept is not given, we can still find the equation of the line by using any point in the table. Using the point (7, 21), we get

$$21 = -2 \cdot 7 + b$$
$$21 = -14 + b$$
$$35 = b.$$

So, our equation is $y = -2x + 35$.

READING QUESTIONS

10. How do you determine if your function is linear if you are given a table of data?

11. Determine if the following tables represent linear functions. For those that do, give the equation.

(a)

x	-2	-1	0	1	2
y	10	14	18	22	26

(b)

x	-2	-1	1	4	6
y	-3	-1	3	9	13

VERBAL REPRESENTATIONS OF LINEAR FUNCTIONS

Many situations are best modeled with a linear function. Not surprisingly, any situation with a *constant* rate of change can be described with a linear function. For example, if you travel at a *constant* velocity of 60 mph, then the distance you travel is given by the linear function

$$d(t) = 60t,$$

where $d(t)$ is distance (in miles) and t is time (in hours). In this situation, the y-intercept is zero, or $d(0) = 0$. At 0 hr, the distance traveled is 0 mi. The slope is 60, which means that (change in distance)/(change in time), or velocity, is 60 mph.

The key to determining if a verbal representation of a physical situation is best modeled by a linear function is no different than for graphical or numerical representations: You need to determine if there is a constant rate of change. Once again, you can find the symbolic representation by determining two pieces of information, the slope and the y-intercept. In a physical (or verbal) situation, the y-intercept will be your starting point (i.e., your output when your input is zero) and the slope will be your constant rate of change. (*Note:* Because slope is a *rate* of change, the units of slope will always be *something per something*, such as miles per hour, dollars per hour, or inches per shoe size.)

Example 5 One of the authors recently had his house connected to cable television. The cost of installation was \$38.00 and the fee is \$10.05 per month for limited basic service. Find the equation for the function whose input is time (in months) and whose output is cost (in dollars).

Solution This example is a linear function because the rate of change—in this case, the monthly fee (in dollars per month)—is constant. The y-intercept is the installation cost of \$38.00. So, the equation for this function is

$$C(m) = 10.05m + 38,$$

where $C(m)$ is the total cost of cable television for m months.

Another familiar example of linear functions is converting units of measurement. We measure length, weight, time, velocity, sound levels, intelligence, and so forth in a variety of units. Length, for example, can be measured in inches, feet, meters, miles, or cubits, to name just a few units. Most conversions from one unit to another can be represented by a linear function because the rate of change for measurement is *constant;* that is, there are always 12 in. per foot, 2.54 cm per inch, and 4 qt per gallon. These types of conversions are often given verbally by such statements as "There are 12 inches in a foot," or "There are 5280 feet in a mile." We are interested in converting these verbal descriptions to an equation.

Let's start with some simple examples of converting from inches to feet. To find how many feet are in 24 in., you write

$$24 \text{ in.} \times \frac{1 \text{ ft}}{12 \text{ in.}} = 2 \text{ ft}.$$

To find how many feet are in 36 in., you write

$$36 \text{ in.} \times \frac{1 \text{ ft}}{12 \text{ in.}} = 3 \text{ ft}.$$

To find how many feet are in n inches, you write

$$n \text{ in.} \times \frac{1 \text{ ft}}{12 \text{ in.}} = \frac{1}{12}n \text{ ft}.$$

In general, this process can be written as

$$\text{number of feet} = \tfrac{1}{12} \times \text{number of inches} \quad \text{or} \quad f(n) = \tfrac{1}{12}n.$$

Notice that no matter how many feet you have in your measurement, changing the number of inches by 1 means changing the number of feet by $\frac{1}{12}$. The rate of change of this function is constant, which is why this function is linear. Notice that the slope, $\frac{1}{12}$, is the conversion factor; that is, there is $\frac{1}{12}$ of a foot per inch. The y-intercept is zero because if the input (inches) is zero, then the output (feet) is also zero.

Example 6 Give the formula that can be used to convert gallons to quarts (4 qt = 1 gal). What is the physical meaning of the slope and the y-intercept in your formula?

Solution Because

$$\text{number of gallons} \times \frac{4 \text{ qt}}{1 \text{ gal}} = \text{number of quarts},$$

the formula is $q(g) = 4g$, where g is the number of gallons and $q(g)$ is the number of quarts. The slope, 4, is the conversion factor; that is, there are

4 qt per gallon. The y-intercept is zero because if you have zero gallons (the input), then you have zero quarts (the output).

Not all functions that model measurement have a y-intercept of zero. For example, to convert from degrees Celsius to degrees Fahrenheit, you can use the linear equation

$$F = \frac{9}{5}C + 32.$$

In this case, the y-intercept is 32 because when the input (degrees Celsius) is zero, then the output (degrees Fahrenheit) is 32; that is, 0°C = 32°F. The slope of $\frac{9}{5}$ means that there are $\frac{9}{5}$ degrees Fahrenheit for every one degree Celsius (or 9 degrees Fahrenheit for every 5 degrees Celsius).

Measurements can be in units you may not typically consider. For example, when describing the length of your foot to someone, you probably would not give the length as inches or centimeters; you would probably give your shoe size. Earlier in this section, we derived the following formula for determining the length of a person's foot in inches when given that person's (U.S. men's) shoe size:

$$\text{foot length in inches} = \frac{1}{3} \times \text{shoe size} + 7\frac{1}{3}$$

The slope of $\frac{1}{3}$ means that as the shoe size increases by 1, the feet that best fit that shoe size are $\frac{1}{3}$ of an inch longer. Another perspective is that slope = (change in output)/(change in input) = $\frac{1}{3}$, so when the input (shoe size) changes by 3, the output (length of foot) changes by 1 in. The y-intercept also has meaning. Because the y-intercept is the output when the input is 0, a y-intercept of $7\frac{1}{3}$ means that a 0 shoe size corresponds to a $7\frac{1}{3}$ in. foot (if size 0 men's shoes existed).

HISTORICAL NOTE

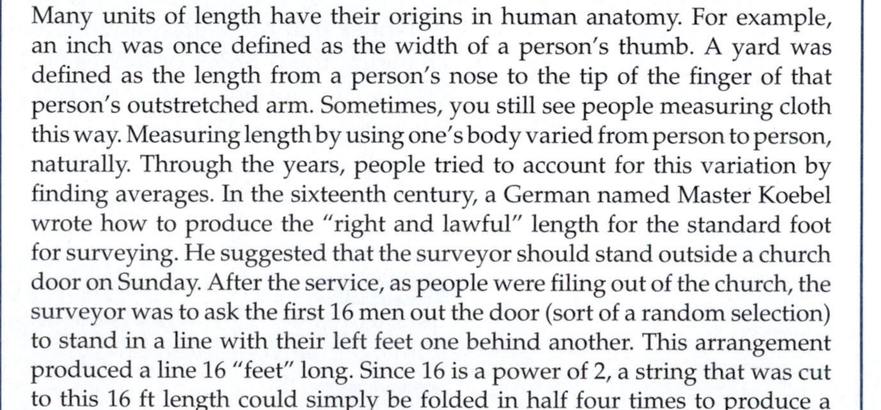

Many units of length have their origins in human anatomy. For example, an inch was once defined as the width of a person's thumb. A yard was defined as the length from a person's nose to the tip of the finger of that person's outstretched arm. Sometimes, you still see people measuring cloth this way. Measuring length by using one's body varied from person to person, naturally. Through the years, people tried to account for this variation by finding averages. In the sixteenth century, a German named Master Koebel wrote how to produce the "right and lawful" length for the standard foot for surveying. He suggested that the surveyor should stand outside a church door on Sunday. After the service, as people were filing out of the church, the surveyor was to ask the first 16 men out the door (sort of a random selection) to stand in a line with their left feet one behind another. This arrangement produced a line 16 "feet" long. Since 16 is a power of 2, a string that was cut to this 16 ft length could simply be folded in half four times to produce a length of 1 ft.

Source: H. Arthur Klein, *The World of Measurements* (New York: Simon and Schuster, 1974), p. 66.

READING QUESTIONS

12. How do you determine if a verbal description of a physical situation represents a linear function?
13. Write a symbolic formula for each of the following.
    (a) Total wages if you earn $6.25 per hour
    (b) Conversion of inches to yards
14. The equation that converts from degrees Fahrenheit to degrees Celsius is $C = \frac{5}{9}F - 17\frac{7}{9}$.
    (a) What is the physical significance of the $\frac{5}{9}$?
    (b) What is the physical significance of the $-17\frac{7}{9}$?

LINEARITY IS NOT ALWAYS OBVIOUS

Functions Represented Graphically

Determining if a function is linear is not necessarily as easy as it might seem. If the function is given graphically, it seems as though you just have to see if it looks like a straight line. The problem is that a function might appear to be a straight line when it actually is not because the slope might change very little in a certain interval. Figure 8 shows two graphs. They both look like straight lines, so it might seem that they are both linear functions. The function in Figure 8(a), however, is the graph of the rational function $y = 20/x$ on the interval $19 \leq x \leq 20$. The function in Figure 8(b) is a graph of the linear function $y = -0.05x + 2$, also on the interval $19 \leq x \leq 20$.

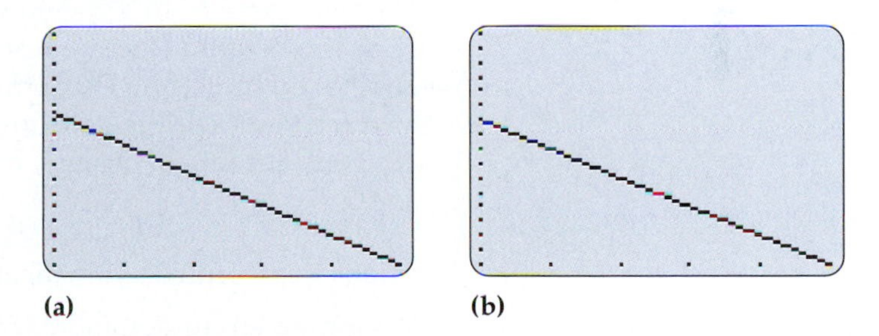

Figure 8 One of these functions is linear and one is not.

The problem with looking at a graph and determining that it represents a linear function is that the viewing window has a tremendous impact on the shape of the graph. In fact, using a carefully chosen viewing window can cause almost any function to look linear.[2] Figure 9 shows another view of the graphs in Figure 8, but this time with the viewing window $5 \leq x \leq 20$. With these graphs, it is clear which function is not linear. Be careful when concluding, from a single viewing window, that a function is linear.

[2] *Try it yourself. Use your graphing calculator and input some nonlinear function. Then use the zoom-in function of your calculator over and over until a section of the graph looks like a line.*

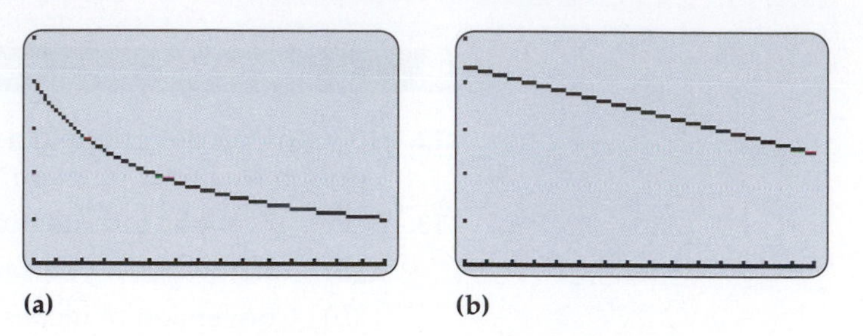

(a) (b)

Figure 9 Same graphs as in Figure 8, but with viewing windows from 5 to 20.

Functions Represented Numerically

TABLE 4

x	-1	0	1
$f(x)$	-2	0	2

If you are given numerical data and determine that $\Delta y/\Delta x$ is constant for each pair of points given, a linear function will model that data, but it does not necessarily mean that the data are best represented by a linear function. For example, you could have used several different equations to generate the three points in Table 4. The functions $f(x) = 2x$, $f(x) = 2x^3$, or even something as complicated as $f(x) = (4/\sqrt{3})\sin(\pi x/3)$ all have $f(-1) = -2, f(0) = 0$, and $f(1) = 2$. In fact, there are an infinite number of functions that contain these three points. So, how do you know which equation best represents your data? It depends. If your function is only these three points, then you do not need an equation; you already have the entire function. If you want a symbolic representation that contains these three points and there is no other information about this function, then any legitimate representation is valid. In particular, the equation $f(x) = 2x$ works for these three points. If these three points originated from a physical situation, then knowing the behavior of your system can help you determine which symbolic formula to use. For example, suppose these three points came from a system where there was a *constant* rate of change (such as a constant velocity). Knowing that the rate of change is constant means that this must be a linear function, so the equation must be $f(x) = 2x$. If the three points originated from a physical system that did not always have a constant rate of change, however, then the function clearly cannot be linear.

Functions Represented Symbolically

If you are given an equation written in slope-intercept form, then it is clear the equation represents a linear function. Not every linear function, however, will be given to you in this form. You may have to do some algebra and solve for y to decide whether or not the function is linear. Consider the function

$$\frac{y-4}{7} = x.$$

This function does not look like the equation for a line, but if we solve for y, we get

$$y = 7x + 4.$$

Now this equation is in the familiar slope-intercept form and is easy to recognize as a linear function.

Example 7 Determine if the equation

$$\frac{y-4}{7} = (x+1)x$$

represents a linear function.

Solution When we solve

$$\frac{y-4}{7} = (x+1)x$$

for y by multiplying both sides by 7, distributing $7x$, and adding 4, we get

$$y = 7x^2 + 7x + 4,$$

which is *not* a linear function since it has an x^2 term.

If a function in symbolic form is not given in slope-intercept form, then you must be careful when deciding if it is a linear function. It is often helpful to solve for y before deciding if your function is linear.

READING QUESTIONS

15. Why do you have to be careful about concluding that a function is linear when given a graphical representation? A numerical representation?
16. What should you do to determine if a symbolic representation is a linear function?
17. Determine if the following represent linear functions. For those that do, give the equation. For those that do not, briefly explain why.

(a)

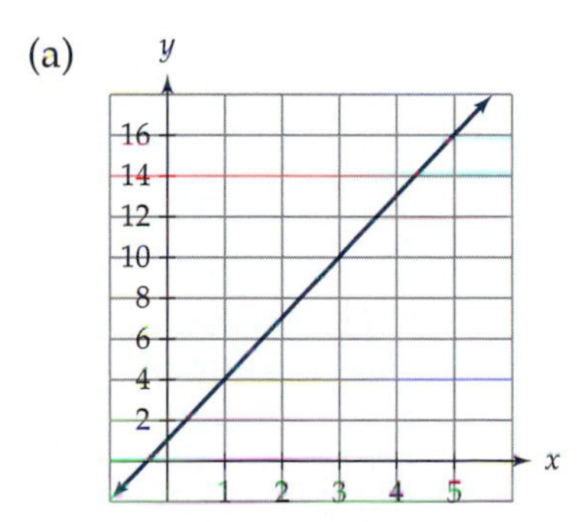

(b)

x	-1	2	5
y	3	4	5

(c) $\dfrac{8(y+3)}{4} = x$

SUMMARY

A **linear function** is one in which the rate of change is constant. The **rate of change** is the change in output divided by the change in input. The **y-intercept** is where the graph crosses the y-axis. It can also be thought of as the output when the input is zero because, when the input is zero, the output lies on the y-axis. The **slope** is the rate of change, or (change in output)/(change in input). Symbolically, slope $= (y_2 - y_1)/(x_2 - x_1)$, where (x_1, y_1) and (x_2, y_2) are any two points on the line. A linear function with a positive slope is an increasing function, and a linear function with a negative slope is a decreasing function.

The most common way to write a linear equation is using the **slope-intercept form,** $y = mx + b$, where y is the output, m is the slope, x is the input, and b is the y-intercept.

Symbolically, a linear function with a slope of 0 (a horizontal line) is written as $y = b$. These functions give the same output regardless of the input. Vertical lines have slopes that are undefined. They can still be represented by the equation $x = k$, where k is a constant.

You can find the equation of the line by finding the coordinates of any two points on the line, (x_1, y_1) and (x_2, y_2). Once you have the two points, you can use the two points to find the slope, substitute the value of the slope for m and the coordinates of one of your points for x and y into the equation $y = mx + b$, solve for b, and then write your equation in slope-intercept form.

EXERCISES

1. Let f be a function such that, whenever the input changes by 2, the output changes by 3. Is f a linear function? If so, give the slope. If not, explain why.
2. Let f be a linear function whose slope is 3.
   (a) Suppose $f(1) = 2$. What is $f(2)$? What is $f(5)$?
   (b) Suppose $f(1) = a$. What is $f(2)$? What is $f(5)$?
3. Let f be the linear function represented by the accompanying graph. Is the slope positive or negative? Is the y-intercept positive or negative? Briefly justify your answers.

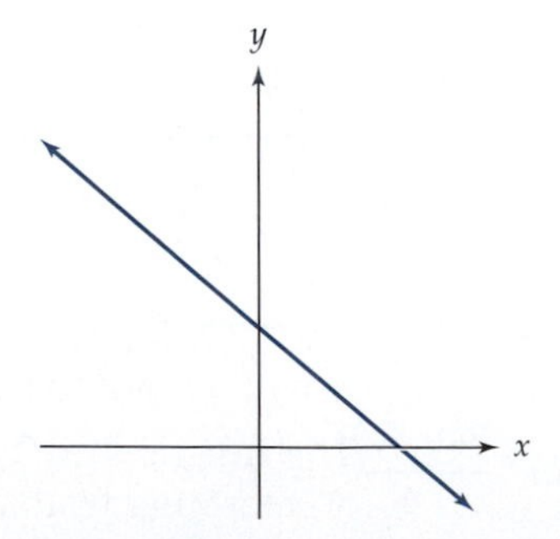

4. Let f be the linear function represented by the accompanying table. Is the slope positive or negative? Is the y-intercept positive or negative? Briefly justify your answers.

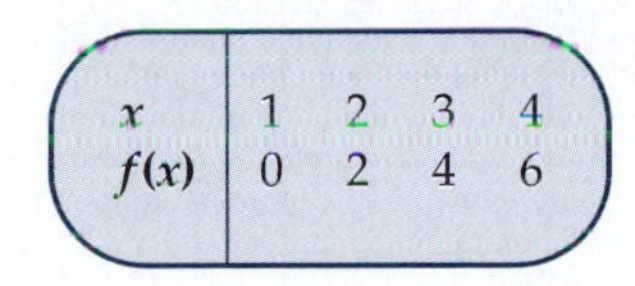

x	1	2	3	4
$f(x)$	0	2	4	6

5. Answer the following about the domain and range of linear functions.
   (a) Give the domain and range of the following linear functions.
      i. $f(x) = 2x + 7$
      ii. $f(x) = -\frac{2}{3}x - 5$
      iii. $f(x) = 7$
   (b) In general, what can be said about the domain and range of linear functions?

6. For each of the following pairs of functions, one is linear and one is not. Which is which? Explain why.
   (a) i. $f(x) = 3x + 4$
      ii. $f(x) = (x + 3)x + 4$

   (b) i.

x	0	1	2	3
$f(x)$	2	5	7	8

   ii.

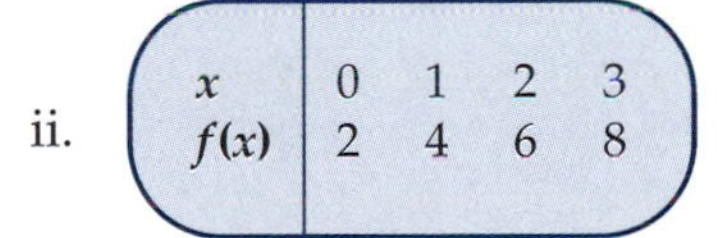

x	0	1	2	3
$f(x)$	2	4	6	8

   (c) i.

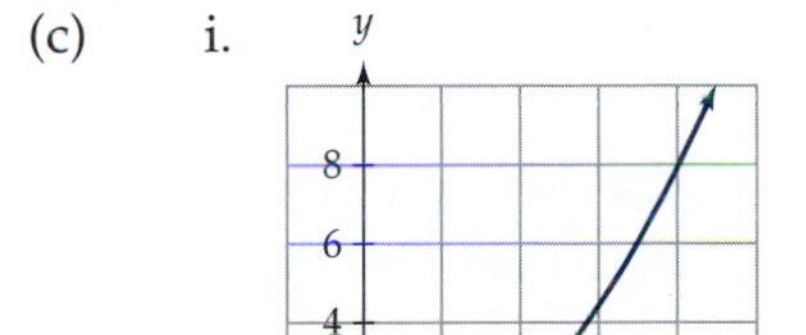

   ii.

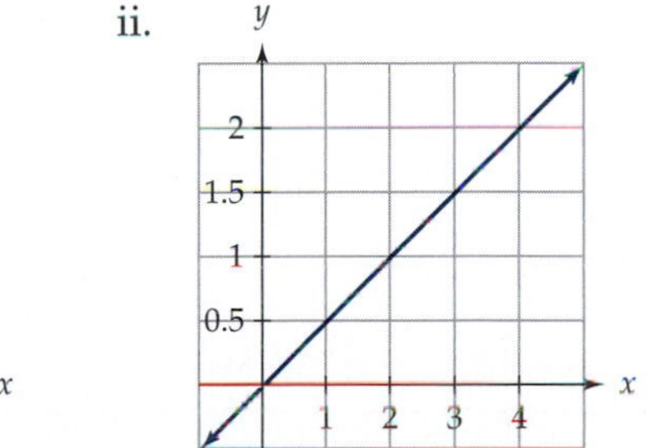

   (d) i. The function giving the distance traveled by a car that is traveling at 70 mph
      ii. The function giving the distance traveled by a car that is accelerating from 0 mph to 45 mph

7. Give the equations for each of the following linear functions.

(a)

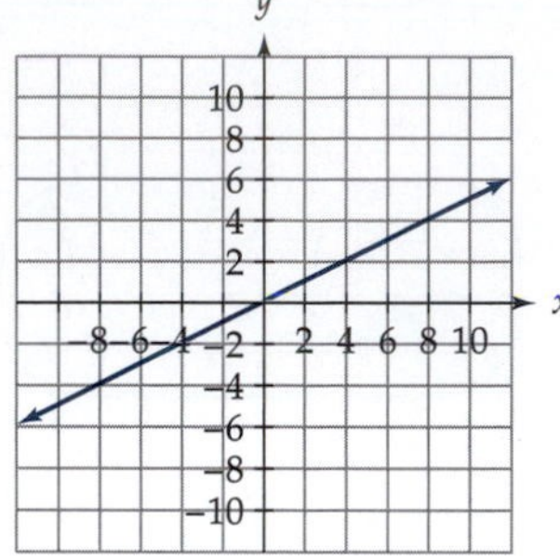

(b)

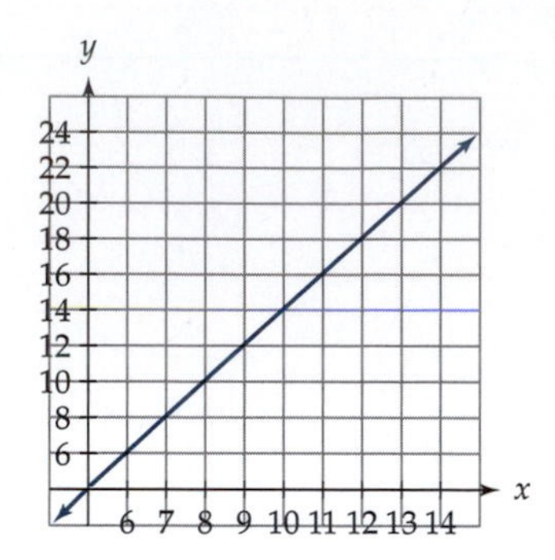

(c)

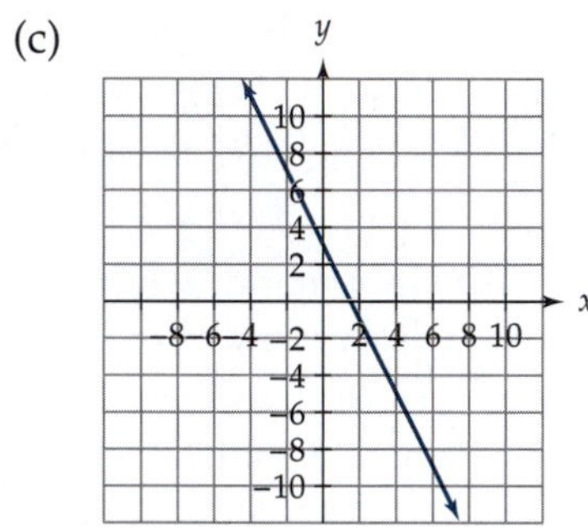

(d)

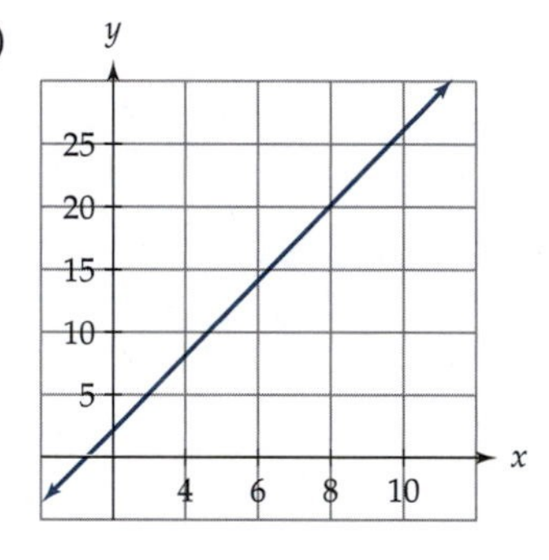

(e)

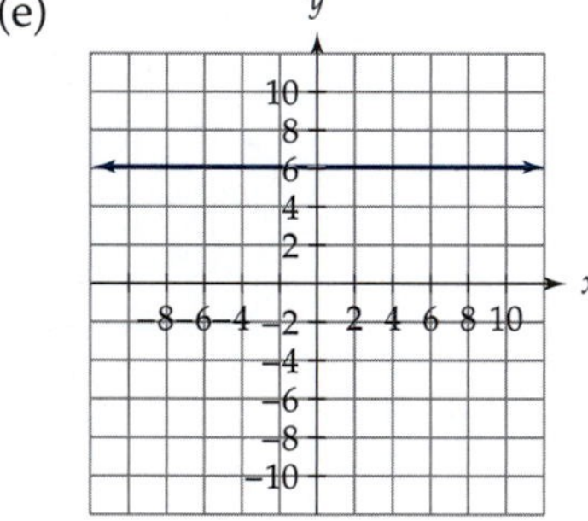

8. For each of the following, determine if the data represent a linear function. If so, find the equation of the line. If not, briefly explain why.

(a)

x	0	1	2	3	4
$f(x)$	5	9	13	17	21

(b)

x	2	5	8	11	14
$f(x)$	4	−2	−8	−14	−20

(c)

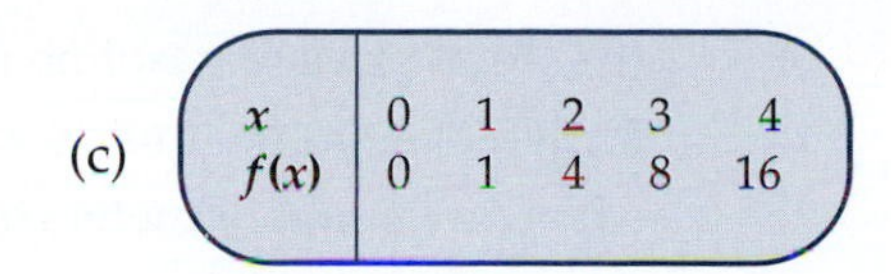

x	0	1	2	3	4
$f(x)$	0	1	4	8	16

(d)

x	−2	−1	1	4	8
$f(x)$	2	4	6	8	10

(e)

x	0	1	2	4	8
$f(x)$	−3	0	3	9	21

9. Determine which of the following functions are linear. If a function is linear, then give its slope and y-intercept.
   (a) $2y = 3x + 7$
   (b) $2(y + 2) = 5(x - 7)$
   (c) $\dfrac{y-2}{x+4} = 3$
   (d) $x = \dfrac{y+2}{3}$
   (e) $\dfrac{y}{x} = x + 2$
   (f) $\dfrac{x+2}{y-7} = x + 3$
   (g) $y = 8$
   (h) $x = 3$
10. Look at the following table of gasoline prices.[3] It has been claimed that gas prices are a linear function of the octane level. Is this claim true or false with respect to these data? Justify your answer.

Octane Level	**Mobil**	**Amoco**	**Shell**
87	\$1.11	\$1.20	\$1.06
89	\$1.20	\$1.29	\$1.17
93	\$1.29	\$1.38	\$1.28

11. Each of the following represents a linear function. Give the equation.
   (a) Total cost of a long-distance phone call if you are paying \$0.15 a minute

[3] *Gasoline prices in Holland, Michigan, on 29 July 1997.*

(b) Total price of gasoline if you are paying \$1.18 a gallon

(c) Conversion of quarts to gallons (4 qt = 1 gal)

(d) Conversion of meters to centimeters (1 m = 100 cm)

(e) Conversion of centimeters to inches (2.54 cm = 1 in.)

(f) Conversion of French francs to U.S. dollars using the exchange rate of 4.83 francs = 1 U.S. dollar[4]

12. Suppose a cable television service advertised the cost of installation as \$25 dollars with a monthly fee of \$12. Let C be the function whose input, m, is the number of months and whose output, $C(m)$, is the total cost of cable television.

    (a) Explain why this function is linear.

    (b) Write the equation for this function.

    (c) What is the total cost after 3 months?

    (d) Instead of getting cable television, a person could install an antenna system to receive the broadcast channels clearly. Suppose the cost of an installed antenna system is \$395. At what month does the total cost of the cable television service exceed the cost of the antenna system?

13. Look at the copy of an electric bill for June 1997 from Consumers' Energy. Notice that there are three different rates. Let c be the function whose input, e, is the amount of electricity used in kilowatt-hours (kWh) and whose output, $c(e)$, is the cost of this electricity.

First 300 kWh × \$0.090906		\$27.27
Next 300 kWh × \$0.076206		\$22.86
Next 49 kWh × \$0.090906		\$4.45
649 kWh total	Total electric	\$54.58

    (a) Explain why c is a piecewise function.

    (b) Let c_1 be the function where the domain is restricted to the first 300 kWh.

        i. Explain why c_1 is a linear function.

        ii. Find an equation for c_1 and graph it for the appropriate domain.

        iii. Explain the meaning of the slope and y-intercept for c_1 in terms of the amount of electricity used and its cost.

    (c) Let c_2 be the function where the domain is restricted to 300 to 600 kWh total.

        i. Explain why c_2 is a linear function.

        ii. Find an equation for c_2 and graph it for the appropriate domain on the same graph as c_1. (*Note:* The two lines should connect at $e = 300$.)

        iii. Explain the meaning of the slope and y-intercept for c_2 in terms of the amount of electricity used and its cost.

[4] *American Express Financial Services,* Exchange Rate *(visited 4 November 1997)* ⟨*http://www.register.com/amex/p0000108.htm*⟩.

(d) Let c_3 be the function where the domain is restricted to greater than 600 kWh.

i. Explain why c_3 is also a linear function.

ii. Find an equation for c_3 and graph it for the appropriate domain on the same graph as c_1 and c_2. (*Note:* Again, your lines should connect.)

iii. Explain the meaning of the slope and y-intercept of c_3 in terms of the amount of electricity used and its cost.

14. The speed of a ship is often given in knots. One knot is equal to 1 nautical mile per hour. A nautical mile differs from the common statute mile, which is equal to 5280 ft. A nautical mile, rather than being defined in terms of feet, is defined to be one minute of one degree of a great circle on Earth. Becaue Earth is not a perfect sphere, different lengths have been given for a nautical mile. There are approximately 1.15 statute miles in 1 nautical mile.

(a) Write an equation that can be used to convert from statute miles to nautical miles.

(b) Use your equation from part (a) to convert 20 statute miles into nautical miles.

(c) Use your equation from part (a) to convert 55 mph into knots (nautical miles per hour).

15. The history of the English measurement system is quite interesting. For example, the well-known nursery rhyme,

Jack and Jill went up the hill
To fetch a pail of water.
Jack fell down and broke his crown,
And Jill came tumbling after.

was originally written as a protest against King Charles I of England. In the rhyme, Jack and Jill represent measures of volume rather than children. A *jackpot* or simply a *jack* is the volume of a small glass. A *jill* is equivalent to two jacks. In seventeenth-century England, liquor was often sold by the jack. Charles, being one of those nasty English kings, imposed a tax on a jack of liquor. He also reduced its size. The citizens of England were therefore getting less for more money, which caused widespread dissatisfaction with the king. Eventually, Charles lost more than his crown. In 1649, he lost his head.[5]

The accompanying graph can be used to convert from cups, c, to jacks, j.

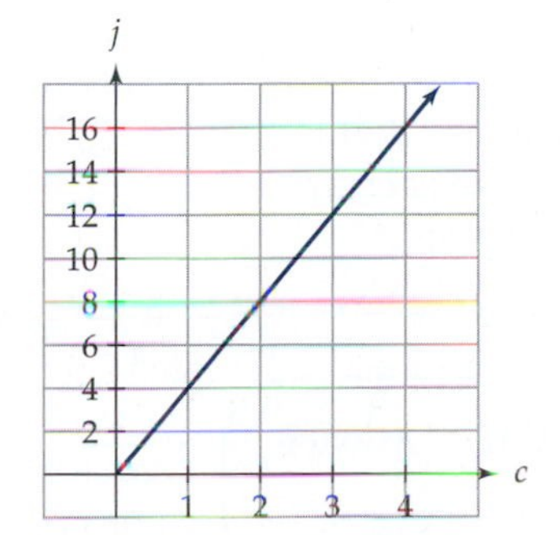

(a) Why does it "make sense" that the conversion from cups to jacks is a linear function?

(b) According to the graph, 2 cups is approximately how many jacks?

(c) According to the graph, 10 jacks is approximately how many cups?

[5] *Klein,* The World of Measurements, *p. 39.*

(d) Find the equation for the function that converts from cups (the input) to jacks (the output).

16. There are approximately 0.6 mi in 1 km.
    (a) Write a formula that converts from miles to kilometers.
    (b) What is the physical meaning of the slope?
    (c) What is the physical meaning of the y-intercept?

17. The following table compares the length of a foot in inches to U.S. men's shoe sizes. This table is similar to one from earlier in this section, except the roles of the input and output are reversed. In this table, the length of a person's foot is the input and the shoe size is the output.

Length (in.)	8	9	10	11	12
Shoe Size	2	5	8	11	14

Because the ratio $\Delta s/\Delta l$ (where s is shoe size and l is the length of a person's foot) is constant, we can conjecture that the function that best describes these data is linear. We want to find a function where length of a man's foot in inches is the input and his shoe size is the output.

(a) What is the slope of this function? What does this number mean in terms of shoe size and length of a person's foot?
(b) What is the y-intercept of this function? What does this number mean in terms of shoe size and length of a person's foot?
(c) What is the equation for this function?
(d) How is this similar to and different from the function whose input is shoe size and whose output is length ($l = \frac{1}{3}s + 7\frac{1}{3}$) given in Example 1?

18. Shoe-sizing scales differ around the world. An international standard, the Mondo Point sizing scale, is often used with sporting equipment such as ski boots or in-line skates. This system uses a straight centimeter scale, so a Mondo Point size shoe of 26 would fit a foot 26 cm long. Other scales are not as simple, but most are similar. For example, the scales for U.S. women, U.S. men, the United Kingdom (U.K.), and Mexico all show that an increase of 1 in. in foot length results in an increase of three shoe sizes. If you have a foot that is 10 in. long, however, you would wear a different size shoe in each system, as shown in the table.

Inches	**U.S. Women**	**U.S. Men**	**U.K.**	**Mexico**
10	9.5	8	7.5	7

We want to determine functions that convert from foot length (the input) to shoe size (the output) for each of these different systems.

(a) These functions all have the same slope. What is it?

(b) What are the y-intercepts for each sizing system? What are the equations for these four systems?

(c) Graph your four functions from part (b). What do these four graphs have in common?

(d) Measure your foot in inches. What size would you wear in the four different systems?

19. The temperature of Earth increases with an increase in depth from the surface. This increase is known as the geothermal gradient. In the outer few thousand feet of Earth, the temperature increases at a rate of approximately 1°F for every 80-ft increase in depth. The initial temperature is based on the yearly average surface temperature of the location rather than the surface temperature at a specific time.[6]

(a) How do you know that the function that determines the temperature beneath the surface of Earth is linear?

(b) Assuming the average surface temperature of a given location is 50°F, give a function where the input is the depth from the surface of Earth and the output is the temperature in degrees Fahrenheit.

(c) The Quincy Mine is an abandoned copper mine in Michigan's upper peninsula. At one time it was the deepest mine in North America with a depth of 6200 ft.[7] The average surface temperature in Quincy is 50°F. What is the temperature at the bottom of the mine?

20. One of the most popular toys of the past 30 years is the Barbie doll. Barbie was designed as a fashion doll, and her dimensions do not necessarily match those of a real person. For example, her head is disproportionately large. If she were full size, she would wear a size $9\frac{1}{4}$ hat; that's xxxxxxx-large![8] We measured a Barbie and found the following information.

Barbie's Features	Measurement (mm)
Height	278
Waist	72
Head circumference	115
Foot length	29

[6] *Chauncey D. Holmes,* Introduction to College Geology, 2d ed. *(New York: MacMillan, 1962), p. 20.*

[7] *William H. Pyne, "The Quincy Mine: The Old Reliable,"* Michigan History, *June 1957, p. 238.*

[8] *Based on size $7\frac{1}{2}$ hat being x-large and $7\frac{7}{8}$ being xx-large.*

We are interested in finding a function, $l(d)$, whose input is a measurement on a Barbie doll and whose output is the measurement of a "life-size" Barbie. Because Barbie is meant to be a model and because the average model is taller than the average woman, let's assume that if Barbie were life size, her height would be 5 ft 10 in., or 70 in. Thus, 70 in./278 mm is the conversion factor from doll-size to adult-size Barbie measurements.

(a) Explain why this function, $l(d)$, will be linear.

(b) Find an equation where your input is the Barbie doll measurement (in millimeters) and the output is the life-size Barbie's measurement (in inches).

(c) Use your equation from part (b) to find the foot, head, and waist size of the full-size Barbie. How does this compare with most women of similar height?

21. When you gather data, you seldom find a perfect linear relationship, even if the situation involves a constant rate of change. In working with measured data, you often are interested in finding a function that is a good fit for the data, although not a perfect fit. Students from a college class were asked to measure their foot length and the length from their wrist to their elbow. We selected eight of these students and entered their data (rounded to the nearest centimeter) in the following table.

Foot Length (cm)	21	22	22	23	24	25	25	27
Length from Wrist to Elbow (cm)	23	23	24	22	23	26	28	29

(a) Why are these data not a function?

(b) Make a scatterplot of these data with foot length on the horizontal axis and length from wrist to elbow on the vertical axis.

(c) The points appear to be fairly linear, but not perfectly. Why is this?

(d) Draw a line on your graph that best fits the data.

(e) By picking two points that lie on your line (not necessarily those in the table), find a linear equation that models these data.

(f) What is the physical meaning of the slope of your line? What is the physical meaning of the y-intercept of your line?

(g) It has been said the length of a person's foot is the same as the length from a person's wrist to his or her elbow. Does the equation of your line support this statement? Why or why not?

(h) Measure your foot length (in centimeters) and use your linear model to predict the length from your wrist to your elbow. Then measure this length to check your prediction.

INVESTIGATIONS

INVESTIGATION 1: FLAT TAX

U.S. income tax laws are quite complicated. There have been various proposals to simplify the system. Many of these proposals are referred to as a flat tax. To most people, a flat tax implies a single tax rate, but this is usually not the case in most of the proposals. For example, a tax proposal endorsed by 1996 presidential candidate Steve Forbes proposed that a person be taxed at a rate of 17% only on the income earned above \$13,300. The first \$13,300 would not be taxed at all.

1. Why is Steve Forbes's flat tax function a piecewise function?
2. Give an equation for Steve Forbes's flat tax where income is the input and the amount of the federal income tax is the output. What are the domain and the range for your function?
3. How much would a person earning \$20,000 per year pay in federal income tax? What percentage of that person's income is used to pay federal income tax?
4. How much would a person earning \$40,000 per year pay in federal income tax? What percentage of that person's income is used to pay federal income tax?
5. You should see from your answers to questions 3 and 4 that the percentage of income paid in federal income tax is not the same for everyone since people are only taxed on what they make above \$13,300. Give a formula for the function where the input is the income of an individual and the output is the percentage of that income that goes toward the federal income tax. What are the domain and the range of this function?

INVESTIGATION 2: PARALLEL AND PERPENDICULAR LINES

When René Descartes developed the coordinate system that we use for graphing equations, he was connecting algebra with geometry. Many of the proofs that once required creative thought in geometry could now be done algorithmically using algebra. In other words, you can just manipulate the symbols and get your proof. Some elementary relationships between lines in our coordinate system illustrate this connection.

1. On graph paper, sketch a graph of $y = 2x + 3$. What are the slope and the y-intercept of your line?
2. On the same graph, sketch a line parallel to the line you just made. What are the slope and the y-intercept of the new line?
3. Think about what would be true about two parallel lines in general. Would the y-intercept (starting point) have to be the same? Would the slope (rate of change) have to be the same? Why?
4. On a new graph, again sketch the line $y = 2x + 3$.
5. We want to find the equation of a line perpendicular to $y = 2x + 3$. Will the slope of the perpendicular line be positive or negative? Briefly justify your answer.

6. Carefully sketch the graph of a line that is perpendicular to $y = 2x + 3$ and that intersects the line at the point $(0, 3)$.
7. Use your graph to find the slope and y-intercept of this perpendicular line.
8. Draw the line perpendicular to $y = 2x + 3$, but this time, have the lines intersect at the point $(1, 5)$. Find the equation of this line.
9. What is the same about the equations of your two perpendicular lines? What is different? How could you have predicted this based on your answer to question 3?
10. The slope of a perpendicular line is always the negative reciprocal of the original slope.
    (a) Compare the slope of the line $y = 2x + 3$ with the slope of the perpendicular line you found in question 7. Are they negative reciprocals?
    (b) Why must the slopes of perpendicular lines have opposite signs?
11. Find the equation of the line perpendicular to each of the following at the given point.
    (a) $y = \frac{1}{2}x - 4$ at $(0, -4)$
    (b) $y = 2x$ at $(2, 4)$
    (c) $y = -\frac{1}{2}x - 4$ at $(0, -4)$
    (d) $y = -2x$ at $(2, -4)$
    (e) $y = mx + 1$ at $(0, 1)$
    (f) $y = mx + b$ at $(0, b)$

2.2 EXPONENTIAL FUNCTIONS

Money in a savings account, the cost of college tuition, population, and decay of a radioactive substance are all examples of situations that increase (or decrease) by a fixed percentage. For example, the interest earned is a percentage of the money in the account. Next year's tuition increase is a percentage of this year's tuition. Functions that increase (or decrease) by a fixed percentage are exponential functions. This type of function has the property of eventually increasing (or decreasing) very rapidly, something that is fortunate when you have money in the bank! In this section, we look at properties of exponential functions, compare exponential and linear functions, and show how to determine if something is an exponential function.

DEFINITION OF EXPONENTIAL FUNCTIONS

A New Look at an Old Story

Our variation on an old story involves the owner of a baseball team who, with 10 games left in the regular season, tries to sign a young baseball player to a contract. The player convinces the team owner to pay him in an unorthodox manner. Rather than a simple annual salary, the player asks the owner to give him one dollar for signing his contract and to double his salary each time the team plays a game. The owner, seeing a chance to secure a reasonably good player for what seems like an insignificant wage, readily agrees. Given that the player will only receive \$2 for the first game and there are only 10 games left in the regular season, how expensive could this agreement be? Table 1 shows how much the player received for each of the 10 games. Remember that each game's payment is double the previous payment. The payment for game 0 represents the \$1 bonus he received for signing the contract.

TABLE 1 How Much the Player Received for Each of the 10 Games Left in the Regular Season

Game	0	1	2	3	4	5	6	7	8	9	10
Payment	\$1	\$2	\$4	\$8	\$16	\$32	\$64	\$128	\$256	\$512	\$1024

Altogether, the player received $\$1 + \$2 + \$4 + \$8 + \cdots + \$1024 = \$2047$ for playing the 10 games. Notice that for playing the 10th game, the player received \$1024, quite an improvement over the \$2 he received after the first game! The real impact of this method of payment, however, would come if there were more games. To understand, look at that pattern generated. Let $P(n)$ be the function representing the payment the player received on the nth game. Then $P(0) = 1$ because he received one dollar for signing the contract. To find $P(1)$, we double the amount he received for signing. So, $P(1) = 2P(0) = 2 \cdot 1 = 2$. Following this pattern, we have

$$P(2) = 2P(1) = 2 \cdot 2 = 2^2 = 4$$
$$P(3) = 2P(2) = 2 \cdot 2^2 = 2^3 = 8$$
$$P(4) = 2P(3) = 2 \cdot 2^3 = 2^4 = 16.$$

In general, the nth game's pay is double the previous game's pay, so

$$P(n) = 2P(n-1) = 2 \cdot 2^{n-1} = 2^n.$$

This method of payment is not too bad for the owner when the player only plays 10 games, but the story does not end here. With the young player's help, the team made the playoffs and went on to win the World Series, playing a total of 19 additional games. The owner quit bragging to his friends about how he got a star player for such cheap wages when a friend pointed out that he would owe the young player $\$2^{29} = \$536{,}870{,}912$ simply for playing the final game. In fact, the player earned a total of \$1,073,741,823 for the 10 regular games and 19 playoff games.

The baseball player's payment function is an example of an exponential function. Exponential functions always have a constant growth factor. The **growth factor** is the ratio, $f(x+1)/f(x)$. A constant growth factor simply means that the $(x+1)$th output is a multiple of the xth output. In our baseball example, the growth factor was 2 because the next game's salary was two times the previous game's salary.

Comparing Linear and Exponential Functions

Exponential functions are similar to linear functions in that they both have a constant rate. What these rates measure, however, is different and leads to different types of behavior. Recall that a linear function has a constant *rate of change*, $[f(x_2) - f(x_1)]/(x_2 - x_1)$. Hence, every time the input increases by one, a fixed amount will be *added* to the previous output. For example, let L be the linear function where $L(0) = 1$ and the rate of change is 2 (i.e., the change in the output is twice the change in the input). Table 2 gives the values of L for the first 10 integers.

TABLE 2 Linear Function Where $L(0) = 1$ and the Rate of Change Is 2

x	0	1	2	3	4	5	6	7	8	9	10
$L(x)$	1	3	5	7	9	11	13	15	17	19	21

An exponential function, on the other hand, has a constant growth factor, $f(x+1)/f(x)$. Hence, every time 1 is added to the input, a fixed amount will be *multiplied* by the previous output. For example, let E be the exponential function where $E(0) = 1$ and the growth factor is 2 (i.e., $E(x+1)$ is double $E(x)$). Table 3 gives the values of E for the first 10 integers. It does not take long to see that multiplication produces a much faster increase than addition!

TABLE 3 Exponential Function Where $E(0) = 1$ and the Growth Factor Is 2

x	0	1	2	3	4	5	6	7	8	9	10
$E(x)$	1	2	4	8	16	32	64	128	256	512	1024

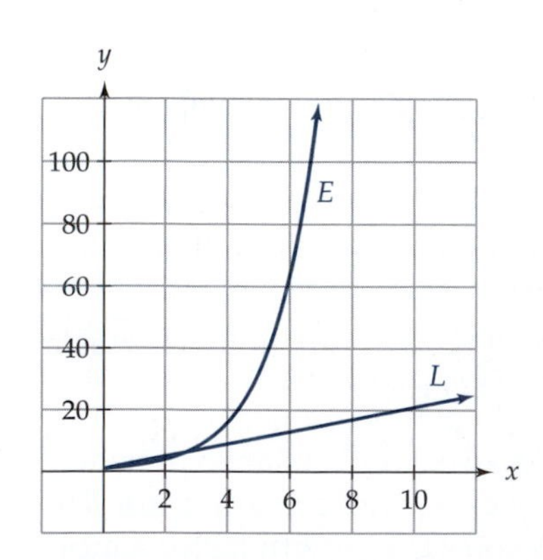

Figure 1 Graphs of E and L.

A graph of the two functions, E and L, are shown in Figure 1. Even though the linear function starts out larger for small values of x, the exponential function soon becomes much larger. A constant growth factor is significantly more powerful than a constant rate of change if you are interested in functions that increase quickly. In fact, any exponential function (with a growth rate larger than 1) will eventually become larger than any linear function. Thankfully, exponential functions are used to model money accumulated through interest. Unfortunately, exponential functions are also used to model accumulated debt.

We saw in the baseball example that $P(n) = 2^n$ was the exponential function that started with \$1 and had a growth factor of 2. Let's generalize this result by finding the formula for the exponential function, E, where $E(0) = b$ and E has a growth factor of a. We assume that $a > 0$ and that $a \neq 1$. $E(0)$ was given to us as b. Because the growth rate is a, the next output has to be a times the previous output, or

$$E(1) = a \cdot E(0) = a \cdot b.$$

Similarly,

$$E(2) = a \cdot E(1) = a \cdot ab = a^2 b.$$

In general,

$$E(n) = a^n b = ba^n.$$

(*Note:* Even though we have derived the formula for $E(n)$ using only integers, the same formula works for any real number.) Our values for $E(n)$ are summarized in Table 4.

TABLE 4 Exponential Function Where $E(0) = b$ and the Growth Factor Is a

x	0	1	2	3	4	5	6	...	n
$E(x)$	b	ab	a^2b	a^3b	a^4b	a^5b	a^6b	...	ba^n

We are now ready to give the formal definition of an exponential function. An **exponential function** is a function where the growth factor, $f(x+1)/f(x) = a$, is constant.[9] Symbolically, an exponential function is

$$f(x) = ba^x,$$

where $a > 0$ and $a \neq 1$. The y-intercept (or starting value) is b, and the growth factor is a. So, $b = f(0)$ and $a = f(x+1)/f(x)$. The growth factor, a, is also called the **base.**

READING QUESTIONS

1. If our baseball player participated in 10 regular season games and 12 playoff games before reaching the World Series, how much did he earn for playing the first game of the World Series?
2. How is a linear function similar to an exponential function? How is a linear function different from an exponential function?
3. Let f be the exponential function $f(x) = 3 \cdot 4^x$. What is the y-intercept? What is the growth factor?

[9] *The value of x in $f(x+1)/f(x)$ does not have to be an integer.*

PROPERTIES OF EXPONENTIAL FUNCTIONS

Now that we have the definition and a general formula for exponential functions, we are interested in exploring their properties. To make things simpler, we assume that the starting amount, b, is equal to 1. In other words, we only look at functions of the form $E(x) = a^x$.

Growth Factors Greater Than One

Let's compare the three exponential functions $f(x) = 2^x$, $g(x) = 3^x$, and $h(x) = 4^x$. Some values for these functions are given in Table 5. Graphs of f, g, and h are shown in Figure 2.

TABLE 5 Values for $f(x) = 2^x$, $g(x) = 3^x$, and $h(x) = 4^x$

x	0	1	2	3	4
$f(x) = 2^x$	1	2	4	8	16
$g(x) = 3^x$	1	3	9	27	81
$h(x) = 4^x$	1	4	16	64	256

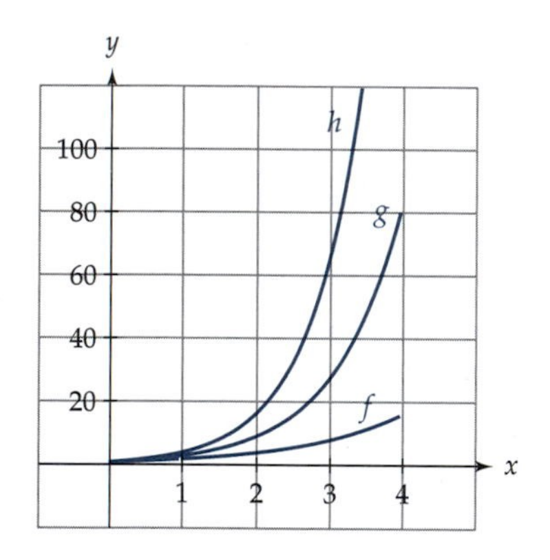

Figure 2 Graphs of $f(x) = 2^x$, $g(x) = 3^x$, and $h(x) = 4^x$ on a domain of $0 \leq x \leq 4$.

It is fairly obvious that an exponential function with a larger growth factor will give a larger output (as long as $x > 0$). Mathematically, if $a > c$ and $x > 0$, then $a^x > c^x$. The larger the growth factor, the larger the outputs.

Also notice that all three functions contain the points $(0, 1)$ and $(1, a)$. This property (as well as the other properties of exponential functions) is a consequence of the properties of exponents. Looking at this symbolically, if $E(x) = a^x$, then $E(0) = a^0 = 1$ and $E(1) = a^1 = a$. Both of these properties should also make sense when you remember that an exponential function of the form $E(x) = a^x$ has a starting value of 1 and a growth rate of a.

Another observation is that all the outputs are positive. Because $a > 0$, $E(x) = a^x$ will always be positive. In other words, starting with one unit and multiplying by a positive growth factor guarantees that your output will be positive.

All three of these functions are increasing and concave up. The graphs are increasing because, for a growth factor greater than 1, the larger the input, the larger the output. Symbolically, if $x_2 > x_1$, then $a^{x_2} > a^{x_1}$. In words, the more days something has been growing, the more you have. The graphs are concave up because the rate of increase is increasing. Observe in Table 5 that the difference between successive outputs is increasing. If you multiply a large amount by a growth factor of a, your increase is greater than if you multiply a small amount by a growth factor of a.

So far, we have only looked at positive inputs. Exponential functions, however, are also defined for negative values of x. Table 6 gives the same three functions, but this time with negative inputs. Approximations are rounded to three decimal places. A graph of f, g, and h using negative inputs is shown in Figure 3.

TABLE 6 Values for $f(x) = 2^x$, $g(x) = 3^x$, and $h(x) = 4^x$ When $-4 \leq x \leq 0$

x	0	−1	−2	−3	−4
$f(x) = 2^x$	1	$\frac{1}{2} = 0.5$	$\frac{1}{4} = 0.25$	$\frac{1}{8} = 0.125$	$\frac{1}{16} \approx 0.063$
$g(x) = 3^x$	1	$\frac{1}{3} \approx 0.333$	$\frac{1}{9} \approx 0.111$	$\frac{1}{27} \approx 0.037$	$\frac{1}{81} \approx 0.012$
$h(x) = 4^x$	1	$\frac{1}{4} = 0.25$	$\frac{1}{16} \approx 0.063$	$\frac{1}{64} \approx 0.016$	$\frac{1}{256} \approx 0.004$

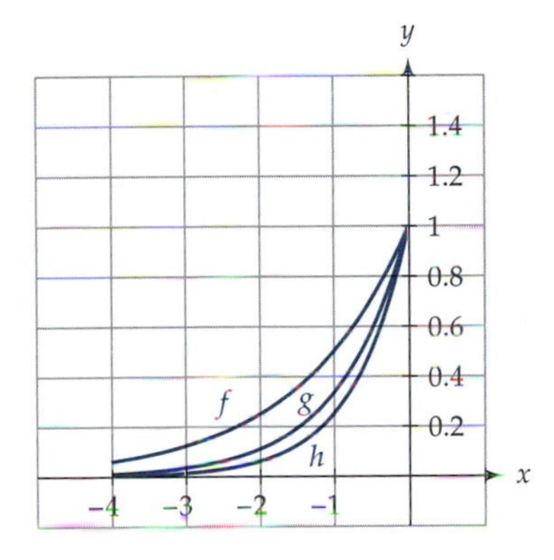

Figure 3 Graphs of $f(x) = 2^x$, $g(x) = 3^x$, $h(x) = 4^x$ on a domain of $-4 \leq x \leq 0$.

In some sense, we get the opposite behavior for negative inputs. This time, the larger the growth factor, the *smaller* the output. Symbolically, if $a > c$ and $x < 0$, then $a^x < c^x$. Let's think about why this behavior occurs. Recall that a number raised to a negative exponent means that you raise the *reciprocal* of the number to a positive exponent. In other words, $2^{-1} = \frac{1}{2}$, $2^{-2} = \left(\frac{1}{2}\right)^2$, and $2^{-3} = \left(\frac{1}{2}\right)^3$. For negative inputs, the functions 2^x, 3^x, and 4^x are giving you powers of $\frac{1}{2}$, $\frac{1}{3}$, and $\frac{1}{4}$. Larger growth factors give larger denominators, causing the reciprocal to be smaller. Therefore, the larger the growth factor, the smaller the output for negative values of x. Notice that our outputs continue to be positive because we are still multiplying positive numbers.

Although our functions are still increasing and still concave up, all three functions get closer to the x-axis as x approaches negative infinity. These functions all have the x-axis as a horizontal asymptote. In general, the graph of a function has a **horizontal asymptote,** $y = c$, if the values of the function get closer and closer to c as x approaches either infinity or negative infinity. For example, $f(x) = 2^x$ has the x-axis as a horizontal asymptote because, as x approaches negative infinity, $f(x)$ gets closer and closer to zero.

Growth Factors Less Than One (and Greater Than Zero)

So far, we have only looked at functions where a is greater than 1. Let's look at a table of data (Table 7) and graphs (Figure 4) of the functions $f(x) = \left(\frac{1}{2}\right)^x$, $g(x) = \left(\frac{1}{3}\right)^x$, and $h(x) = \left(\frac{1}{4}\right)^x$. Each of these functions has a positive growth factor that is less than 1. Because our growth factor is smaller than 1, the next output will be *smaller* than the previous output. In other words, the amount is decaying rather than growing.

TABLE 7 Values for $f(x) = \left(\frac{1}{2}\right)^x$, $g(x) = \left(\frac{1}{3}\right)^x$, and $h(x) = \left(\frac{1}{4}\right)^x$ When $-4 \leq x \leq 4$

x	−4	−3	−2	−1	0	1	2	3	4
$f(x) = \left(\frac{1}{2}\right)^x$	16	8	4	2	1	$\frac{1}{2} = 0.5$	$\frac{1}{4} = 0.25$	$\frac{1}{8} = 0.125$	$\frac{1}{16} \approx 0.063$
$g(x) = \left(\frac{1}{3}\right)^x$	81	27	9	3	1	$\frac{1}{3} \approx 0.333$	$\frac{1}{9} \approx 0.111$	$\frac{1}{27} \approx 0.037$	$\frac{1}{81} \approx 0.012$
$h(x) = \left(\frac{1}{4}\right)^x$	256	64	16	4	1	$\frac{1}{4} = 0.25$	$\frac{1}{16} \approx 0.063$	$\frac{1}{64} \approx 0.016$	$\frac{1}{256} \approx 0.004$

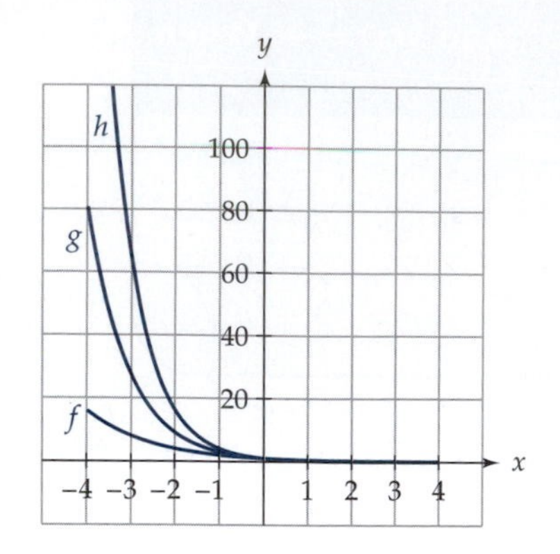

Figure 4 Graphs of $f(x) = \left(\frac{1}{2}\right)^x$, $g(x) = \left(\frac{1}{3}\right)^x$, and $h(x) = \left(\frac{1}{4}\right)^x$ on a domain of $-4 \le x \le 4$.

In comparing these functions, we again see that a larger growth factor gives a larger output when $x > 0$ and a smaller output when $x < 0$. Just like when the growth factors were greater than 1, these functions still contain the points $(0, 1)$ and $(1, a)$, where a is the growth factor. The outputs are still positive. These properties are the same because the properties of exponents are not affected by the size of the growth factor.

These functions are different, however, in that this time they are decreasing. If you raise a number between 0 and 1 to a larger exponent, you get a smaller value. Your amount is decaying rather than growing. For example, suppose the baseball player had agreed to a salary of \$10,000 for the first game and a growth factor of $\frac{1}{2}$. For the second game, he would have received \$5000; for the third game, he would receive \$2500; and so on. By the 10th game, his payment would only be \$19.53, and by the time he finished the World Series, he would be playing for less than a penny!

The functions are still concave up, which means that the rate of change is increasing. Because the rate of change is increasing and the function is decreasing, the larger the input, the smaller the decrease. The functions still have a horizontal asymptote at $y = 0$, but this time it is as x approaches positive infinity. These graphs are just like the graphs of $f(x) = 2^x$, $g(x) = 3^x$, and $h(x) = 4^x$ except that they have been "flipped" over the y-axis.

Let's summarize the properties of $f(x) = a^x$, where $a > 0$ and $a \neq 1$. Even though we only looked at a few examples to observe these properties, they are true for all exponential functions. In particular, these properties are true even when a is not an integer.[10]

PROPERTIES

Comparing exponential functions, $f(x) = a^x$ and $g(x) = c^x$, where $a > c$:

1. The affect of the size of the growth factor:
   - $a^x > c^x$ if $x > 0$. The larger the growth factor, the larger the output if your input is positive.
   - $a^x < c^x$ if $x < 0$. The larger the growth factor, the smaller the output if your input is negative.
2. Points in common:
   - $(0, 1)$
   - $(1, a)$
3. Domain and range:
   - The domain is all real numbers.
   - The range is $f(x) > 0$.

continued on page 95

[10] *All these properties are consequences of the properties of exponents.*

4. Increasing/decreasing:
   - $f(x) = a^x$ is increasing for all values of x if the growth factor, a, is greater than 1.
   - $f(x) = a^x$ is decreasing for all values of x if the growth factor, a, is between 0 and 1.
5. Concavity:
   - $f(x) = a^x$ is always concave up.

Why the Growth Factor Cannot Be Negative or Equal to One

We keep repeating that the growth factor, a, must be greater than 0 and not equal to 1. Let's see why. If the growth factor were 1, each day you have one times the amount you had the previous day. Thus, you have exactly the same amount each day. Symbolically, your function would be $f(x) = 1^x = 1$ since 1 raised to any power is still 1. This function is the horizontal line $y = 1$. This horizontal line does not have many of the properties listed earlier, so we exclude this as an exponential function.

Why can't the growth factor be negative? The domain of exponential functions is all real numbers. Thus, if a were negative, we would have to worry about things such as $a^{1/2}$. Recall that $a^{1/2} = \sqrt{a}$ is undefined for negative values of a. In fact, there are an infinite number of inputs where a^x would be undefined if we allowed a to be negative. Instead of worrying about this issue (and others), we simply exclude the possibility of negative values for our growth factor, a. So, exponential functions are only defined for $a > 0, a \neq 1$.

HISTORICAL NOTE

The asymptotic behavior of exponential functions was written about by Zeno the Eleatic (fl. ca. 450 B.C.)[11] in a series of paradoxes. In one of them, the *Dichotomy*, Zeno argues that to go from point A to point B, you must first cover half that distance, then half the remaining distance, and so forth. Notice that the function that gives the distance from you to point B is $d(t) = \left(\frac{1}{2}\right)^t$. The process Zeno describes continues until one concludes that an infinite number of steps must be taken to reach point B, which is the same as saying that $d(t) = \left(\frac{1}{2}\right)^t$ has the horizontal asymptote $y = 0$. If there are finite pieces of time (as some in Zeno's time claimed), then it would be impossible to get to point B. Although, while this is clearly not the case—we do eventually get places—Zeno's philosophical argument is a demonstration of one of the properties of exponential functions.

Source: Carl B. Boyer, *A History of Mathematics* (Princeton, NJ: Princeton University Press, 1985), p. 82.

[11] *The abbreviations fl. and ca. stand for "flourished circa," which means flourished about. This date gives the approximate year that Zeno did his most significant work.*

READING QUESTIONS

4. Why will every exponential function pass through the point (0, 1)?
5. Explain why, for an exponential function, it is necessary that the growth factor be positive.
6. Let $f(x) = a^x$ such that $f(1) = 6$. What is the value of a? How do you know?
7. Why is 1 excluded as a growth factor for exponential functions?
8. Let $f(x) = a^x$. In the accompanying graph of f, is a greater than 1 or less than 1? How do you know?

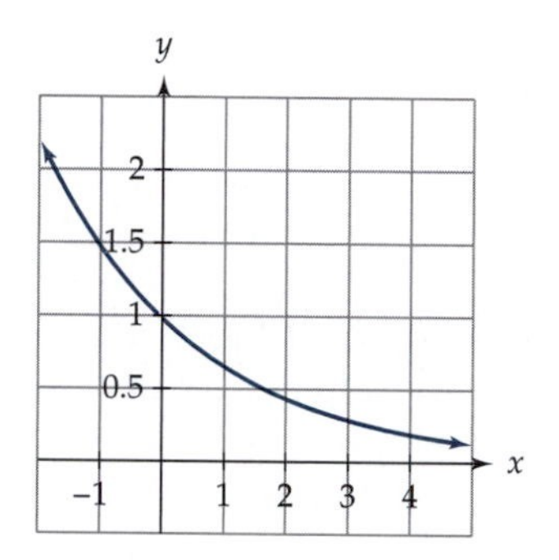

THE SIGNIFICANCE OF *b*

We looked at exponential functions of the form $f(x) = a^x$ for various values of a. Now let's return to exponential functions of the form $f(x) = ba^x$ and explore the significance of b. For simplicity, we restrict b to be positive.[12] As before, we start with a table, this time using the functions $f(x) = \frac{1}{3} \cdot 2^x$, $g(x) = 2^x$, and $h(x) = 3 \cdot 2^x$. (See Table 8.) Approximations are rounded to three decimal places. Figure 5 is a graph of these three functions for $-4 \le x \le 4$.

Notice that the shape of the graph has not changed. All three graphs are similar to the shape of an exponential function whose growth factor is greater than 1. The most noticeable change is that the functions do not all cross the y-axis at the point (0, 1). Rather, the functions go through the point $(0, b)$ because b is the starting point, or y-intercept. Symbolically, if $f(x) = ba^x$, then $f(0) = ba^0 = b \cdot 1 = b$. Changing the y-intercept will not change the growth factor, but it will change the actual output you have at any particular point.

It is also no longer true that the functions contain the point $(1, a)$. Instead, the functions now contain the point $(1, ba)$. You can easily find the values of both b and a from a graph or table. Recall that b is the y-intercept, or $f(0)$, and a is the growth factor, $a = f(x+1)/f(x)$. Because the growth factor is constant, you can use any points that are one unit apart to find a. For

[12] *Unlike our restrictions on the growth factor, a, it is possible to have negative values for b. We look at what happens when b is negative in Chapter 3.*

TABLE 8 Values for $f(x) = \frac{1}{3} \cdot 2^x$, $g(x) = 2^x$, and $h(x) = 3 \cdot 2^x$

x	$f(x) = \frac{1}{3} \cdot 2^x$	$g(x) = 2^x$	$h(x) = 3 \cdot 2^x$
−4	$\frac{1}{48} \approx 0.021$	$\frac{1}{16} \approx 0.063$	$\frac{3}{16} \approx 0.188$
−3	$\frac{1}{24} \approx 0.042$	$\frac{1}{8} = 0.125$	$\frac{3}{8} = 0.375$
−2	$\frac{1}{12} \approx 0.083$	$\frac{1}{4} = 0.25$	$\frac{3}{4} = 0.75$
−1	$\frac{1}{6} \approx 0.167$	$\frac{1}{2} = 0.5$	$\frac{3}{2} = 1.5$
0	$\frac{1}{3} \approx 0.333$	1	3
1	$\frac{2}{3} \approx 0.667$	2	6
2	$\frac{4}{3} \approx 1.333$	4	12
3	$\frac{8}{3} \approx 2.667$	8	24
4	$\frac{16}{3} \approx 5.333$	16	48

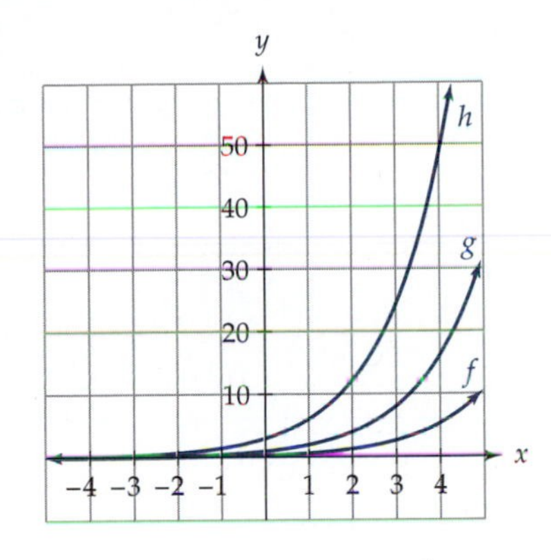

Figure 5 Graphs of $f(x) = \frac{1}{3} \cdot 2^x$, $g(x) = 2^x$, and $h(x) = 3 \cdot 2^x$.

example, using data from Table 8, the ratio of $h(4)/h(3) = \frac{48}{24} = 2$, which tells us that the growth factor for h is 2.

READING QUESTIONS

9. Why is b the value of the y-intercept?

10. Give the equation for the exponential function whose y-intercept is 2 and whose growth factor is 5.

11. Let f be the exponential function whose graph is shown. What is the value of b? How do you know?

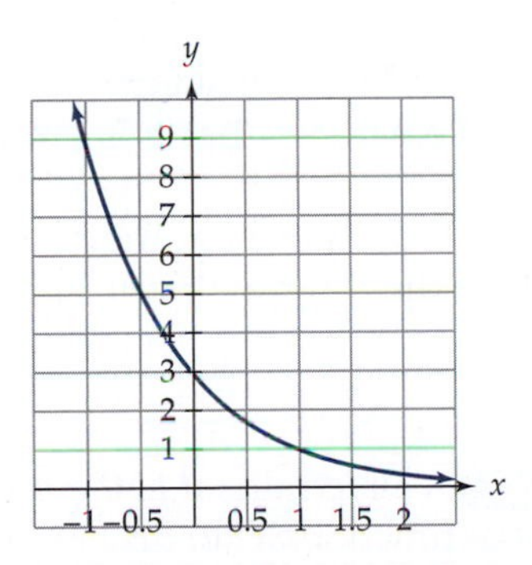

12. Suppose the baseball player negotiated a contract where he received \$500 for signing the contract and his pay was doubled for each game he played. What is the equation for the exponential function where the number of games played is the input and the output is his payment for that game?

HOW DO YOU KNOW IF YOUR FUNCTION IS EXPONENTIAL?

Suppose you are given a function symbolically, graphically, numerically, or verbally. How do you know if it is an exponential function?

Symbolic Representations

Let's start with a symbolic representation. In this case, make sure you have it in the form "$y =$" and see if it can be written as $y = ba^x$ for some values of a and b.

Example 1 Determine if $3^x y = 2$ is an exponential function. If so, give the growth factor and the starting value (i.e., the values of a and b). If not, briefly explain why.

Solution Solving $3^x y = 2$ for y, we get $y = 2/3^x = 2\left(\frac{1}{3}\right)^x$. This is an exponential function since it is in the form $y = ba^x$. The growth factor is $\frac{1}{3}$, and the starting value is 2.

Example 2 Determine if $x^3 y = 2$ is an exponential function. If so, give the growth factor and the starting value (i.e., the values of a and b). If not, briefly explain why.

Solution Solving $x^3 y = 2$ for y, we get $y = 2/x^3 = 2(1/x)^3$. This is not an exponential function since it is not in the form $y = ba^x$. All exponential functions have x in the exponent, and this one does not.

Numerical Representations

Now let's look at functions represented numerically. The key to determining if you have an exponential function is to remember that exponential functions have a constant growth factor. So, you need to determine the ratios $f(x+1)/f(x)$ and see if they are all the same. If so, your function is exponential and this constant is your value of a. To find b, find the output for $x = 0$. Remember that b is your starting value. If that output is not given, you can determine b by solving $f(x) = ba^x$ for b. Doing so, we get $b = f(x)/a^x$. You can then find b using the value of a, any input, x, and its corresponding output, $f(x)$. This process is illustrated in the next few examples.

Example 3 Determine if f, the function shown in Table 9, is an exponential function. If so, find its equation. If not, briefly justify why.

TABLE 9

x	0	1	2	3	4
$f(x)$	8	1.6	0.32	0.064	0.0128

Solution To decide if f is an exponential function, we need to see if the growth factor is constant.

$$\frac{f(1)}{f(0)} = \frac{1.6}{8} = 0.2, \qquad \frac{f(2)}{f(1)} = \frac{0.32}{1.6} = 0.2,$$
$$\frac{f(3)}{f(2)} = \frac{0.064}{0.32} = 0.2, \qquad \frac{f(4)}{f(3)} = \frac{0.0128}{0.064} = 0.2.$$

Because the ratio $f(x+1)/f(x)$ is constant, this is an exponential function with $a = 0.2$. To find the value of b, notice that $f(0) = 8$. So, this is the exponential function $f(x) = 8 \cdot 0.2^x$.

Example 4 Determine if f, the function shown in Table 10, is an exponential function. If so, find its equation. If not, briefly justify why.

TABLE 10

x	3	4	5	6	7
$f(x)$	6.75	20.25	60.75	182.25	546.75

Solution To see if f is an exponential function, we need to see if the growth factor is constant.

$$\frac{f(4)}{f(3)} = \frac{20.25}{6.75} = 3, \qquad \frac{f(5)}{f(4)} = \frac{60.75}{20.25} = 3,$$
$$\frac{f(6)}{f(5)} = \frac{182.25}{60.75} = 3, \qquad \frac{f(7)}{f(6)} = \frac{546.75}{182.25} = 3.$$

Because the ratio $f(x+1)/f(x)$ is constant, this is an exponential function with $a = 3$. We do not have the value of $f(0)$, so we cannot read the value of b directly from the table. For any value of x, however, $b = f(x)/a^x$. Using the point $(3, 6.75)$, we get $b = f(3)/a^3 = 6.75/a^3$. We already discovered that $a = 3$, so $b = 6.75/3^3 = 6.75/27 = 0.25$. This is the exponential function $f(x) = 0.25 \cdot 3^x$.

Example 5 Determine if f, the function shown in Table 11, is an exponential function. If so, find its equation. If not, briefly justify why.

TABLE 11

x	1	2	3	4	5
$f(x)$	1	8	27	64	125

Solution To see if f is an exponential function, we need to see if the growth factor is constant.

$$\frac{f(2)}{f(1)} = \frac{8}{1} = 8, \qquad \frac{f(3)}{f(2)} = \frac{27}{8} = 3.375,$$
$$\frac{f(4)}{f(3)} = \frac{64}{27} \approx 2.370, \qquad \frac{f(5)}{f(4)} = \frac{125}{64} \approx 1.953.$$

Because the ratio $f(x+1)/f(x)$ is not constant, this is not an exponential function.

Graphical Representations

Determining if a graphical representation is an exponential function is a matter of considering whether it has the right shape, using a couple of points to find the values of a and b, and then graphing the exponential function to see if it is identical to the one you were given. This process is illustrated in the following examples.

Example 6 Determine if the graph in Figure 6 is an exponential function. If so, find its equation. If not, briefly justify why.

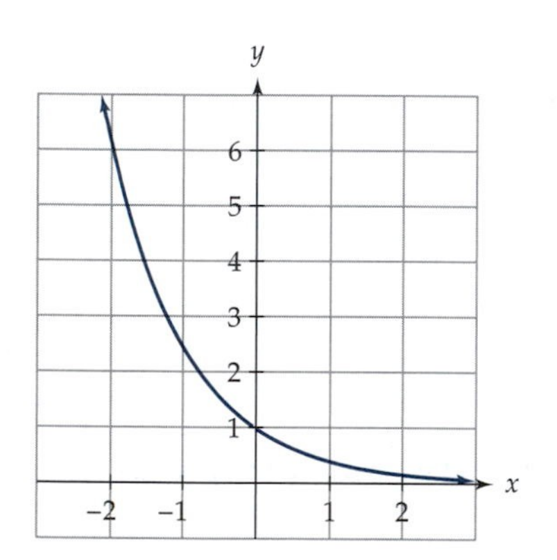

Figure 6

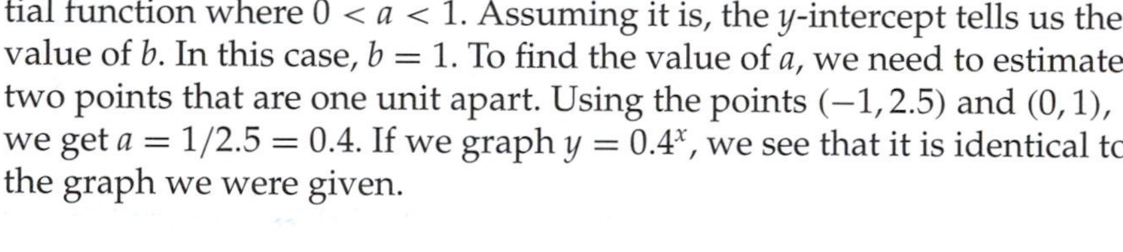

Solution The function is concave up, decreasing, and seems to have a horizontal asymptote as x gets large. It is possible that it is an exponential function where $0 < a < 1$. Assuming it is, the y-intercept tells us the value of b. In this case, $b = 1$. To find the value of a, we need to estimate two points that are one unit apart. Using the points $(-1, 2.5)$ and $(0, 1)$, we get $a = 1/2.5 = 0.4$. If we graph $y = 0.4^x$, we see that it is identical to the graph we were given.

Example 7 Determine if the graph in Figure 7 is an exponential function. If so, find its equation. If not, briefly justify why.

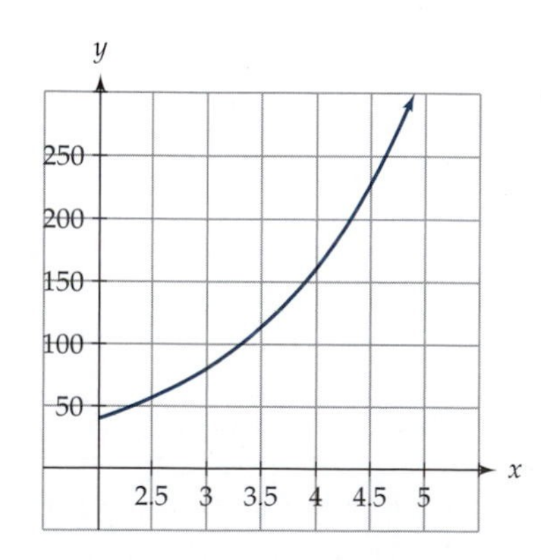

Figure 7

Solution The function is concave up, increasing, and seems to be getting large rapidly. It is possible that it is an exponential function where $a > 1$. Assuming it is, we can find the value of a by choosing two points that are one unit apart. Using the points $(2, 40)$ and $(3, 80)$, we get $a = \frac{80}{40} = 2$. We cannot find the value of b by using the y-intercept since the y-axis is not included in this graph. Using $a = 2$ and the point $(2, 40)$, however, we get $b = f(x)/a^x = 40/2^2 = \frac{40}{4} = 10$. If we graph $y = 10 \cdot 2^x$, we see that it is identical to the graph we were given.

Example 8 Determine if the graph in Figure 8 is an exponential function. If so, find its equation. If not, briefly justify why.

Solution The graph is increasing and concave up, but the y-intercept is $(0, 0)$. If this were an exponential function, the starting value would be

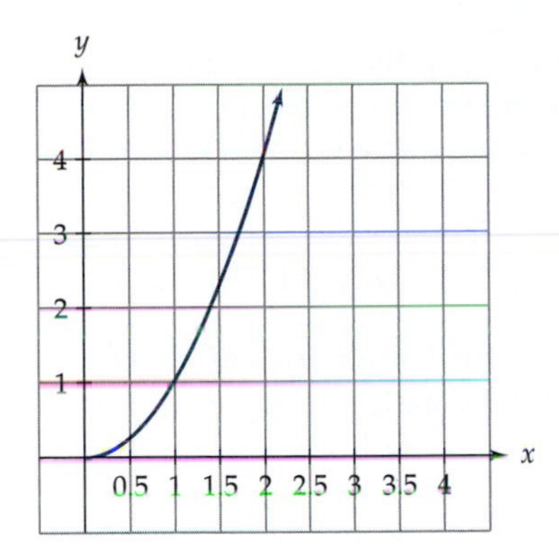

Figure 8

zero, which means that all the outputs would have to be zero because $0 \cdot a^x = 0$ for all values of x. So, this cannot be an exponential function. (*Note:* Recall that the exponential function $f(x) = ba^x$ has a y-intercept of $(0, b)$.)

Example 9 Determine if the graph in Figure 9 is an exponential function. If so, find its equation. If not, briefly justify why.

Solution The graph is increasing and concave up. It is possible that it is an exponential function where $a > 1$. Assuming it is, the value of b is 1 since the y-intercept is $(0, 1)$. Using the points $(0, 1)$ and $(1, 2)$, we get $a = \frac{2}{1} = 2$. If we graph $y = 2^x$, however, we see that the only places where the graph matches our function are the two points we used to find a. So, it is not an exponential function. (See Figure 10.)

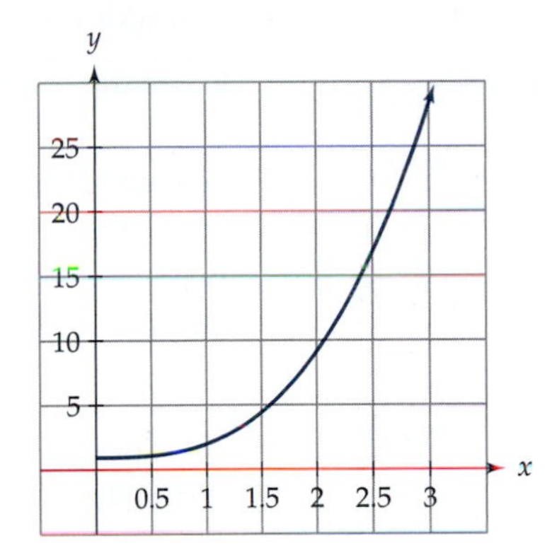

Figure 9

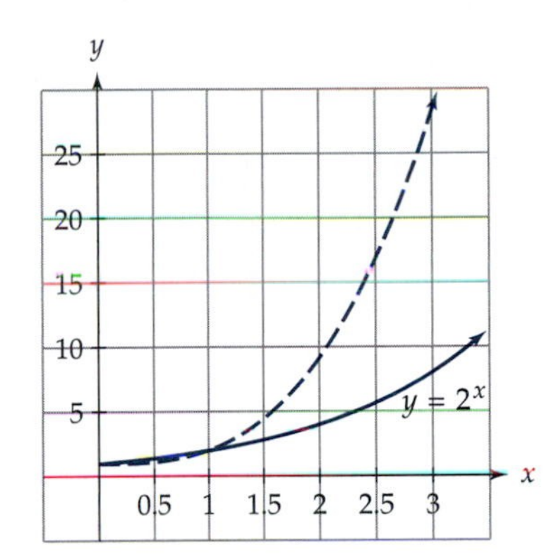

Figure 10

Verbal Representations

To determine if a verbal description of a function is exponential, ask yourself if this situation has a constant growth factor. Many things grow exponentially, such as money, human populations, and bacteria. For example, your savings account may be earning 3% interest, the population of your town might be increasing by 5% each year, or the population of an endangered species might be decreasing by 10% each year. Each of these situations has a constant growth factor, so we can use exponential functions to describe them.

Example 10 In Michigan, state universities increased their tuition by an average of 7.9% per year from 1986 to 1996. In 1986, the average yearly tuition in the state was approximately \$1800. Let C be the function whose input, t, is the number of years since 1986 and whose output, $C(t)$, is the estimated cost of tuition that year. Is this an exponential function? If so, find its equation. If not, explain why.

Solution This is an exponential function since the growth factor is constant. It might seem as though the growth factor is 0.079, but the tuition *increase* from 1986 to 1987 in dollars was 1800×0.079. Therefore, the new tuition in 1987 was

$$1800 + 1800 \times 0.079 = 1800(1 + 0.079) = 1800(1.079).$$

Thus, the 1987 tuition was 107.9% as much as the tuition in 1986. With the same reasoning, we say that the tuition in 1988 was 107.9% as much as the tuition in 1987 and so on. So, the growth factor is 1.079. The starting value is $1800 if we use 1986 as our starting point. The exponential function that describes the yearly tuition would then be

$$C(t) = 1800(1.079)^t,$$

where t is the number of years since 1986 and $C(t)$ is the cost of the average yearly tuition in dollars at a Michigan public university.

In Example 10, we found that the growth factor was 1.079. The 7.9% = 0.079 is referred to as a growth rate. A **growth rate** is always 1 less than the corresponding growth factor. So, if a population is increasing in such a way that the growth factor is 1.1, then the growth rate is 0.1 or 10%.

Example 11 In 1980, there were approximately 1,200,000 African elephants. In 1990, there were approximately 700,000 African elephants.[13] Assuming the population growth factor is constant, find the exponential function that models these data and use it to predict the number of African elephants in 2000.

Solution Assuming a constant population growth factor means that this will be an exponential function. Let P be the function whose input is time (in years since 1980) and whose output is the estimated number of elephants. We are given that $P(0) = 1{,}200{,}000$, so $P(t) = 1{,}200{,}000a^t$ because $b = P(0)$. We are not given the growth factor, but instead we know that $P(10) = 700{,}000$. So, $700{,}000 = 1{,}200{,}000a^{10}$. Dividing by 1,200,000 and taking the 10th root of both sides gives $a \approx 0.95$. Thus, each year, the elephant population is approximately 95% of what it was the previous year. Our function is $P(t) = 1{,}200{,}000 \cdot 0.95^t$. The population in 2000 (assuming the growth factor remains constant) would be $P(20) = 1{,}200{,}000 \cdot 0.95^{20} \approx 430{,}000$.

Notice in Example 11 that we were never given the growth factor. As long as you know that your function is exponential, knowing any two

[13] *John A. Burton, ed.,* The Atlas of Endangered Species *(New York: Macmillan Publishing, 1991), p. 138.*

points on your function allows you to algebraically solve for both b and a, as illustrated in Example 11.

Example 12 An employee was offered the choice of receiving either a $1000 bonus added to her base pay every year or a 2.3% cost of living increase added to her base pay every year. She chose to take the $1000 bonus. Her annual salary in 1997 was $33,500. Let S be the function whose input is the number of years since 1997 and whose output is the employee's base salary for that year. Is S an exponential function? If so, find its equation. If not, explain why.

Solution The employee's salary is being increased at a constant rate of $1000 per year. This amount, however, is *added* to her salary, which means that the 1000 is a constant rate of change rather than a growth factor. So, S is a linear function, not an exponential function.

Application: Chicken Bacteria

The information in Table 12 was found on the label of a package of chicken. It was also noted on the label that at 40°F, the bacteria double in number every 6 hr. Because the bacteria are growing at a constant rate, it is an exponential function. Let $B(t) = ba^t$ be the number of bacteria on the chicken after t days. Looking at the table, we see that $b = 360$. To find the growth factor, notice that $B(1)/B(0) = 5800/360 \approx 16$. (If we look at the other ratios, we see that they are also approximately 16.) So, the equation is $B(t) = 360 \cdot 16^t$.

TABLE 12 Bacteria Count on Chicken (per Square Centimeter in a 40°F Refrigerator)

Time	Bacteria	Condition
Day 0	360	OK
Day 1	5,800	OK
Day 2	92,000	Fair
Day 3	1,475,000	Poor
Day 4	23,600,000	Odor
Day 5	377,500,000	Slime

Notice, however, that the label identified the growth factor as "bacteria double in 6 hr," whereas we have the growth factor as "16 times as many bacteria in 1 day." The reason for the difference in the two numbers is the units of time. When dealing with a physical situation, it is important to pay attention to units. Our growth factor, a, equals the ratio of outputs whose inputs are one unit apart. The question becomes, What do we mean by "one unit apart"? In terms of bacteria, we could use one unit to represent 1 hr, 1 day, or even 1 min. We chose to have "one unit" mean 1 day. In this case, we cannot simply use 2 as our value of a because the bacteria double in 6 hr, not 1 day. If the bacteria double in 6 hr, however, then there are four times as much bacteria in 12 hr, eight times as much in 18 hr, and 16 times as much in 24 hr. So, our growth factor for 1 day is 16, which agrees with the value of a we determined by using the data in Table 12.

Although the function describing growth of bacteria is similar to the example of growth of tuition in Example 10, there is a fundamental difference in these two examples. It makes sense to use our chicken bacteria function to determine the number of bacteria after 1.5 days, but it does not make sense to use our tuition function to find the average tuition cost after 1.5 years. Bacteria grow almost continuously, so using a function that allows for any input is fine. Tuition does not grow continuously; it usually changes just once a year. Therefore, the domain of our tuition function is restricted to nonnegative integers. Another limitation of this model is that we cannot assume that tuition will continue to grow at this rate just because it has over the past 10 years. We need to be especially careful when using models of this sort to predict future growth because many factors (economic and social) impact tuition.

READING QUESTIONS

13. Determine if each of the following is an exponential function. If so, give its equation. If not, briefly explain why.

(a) $3y = \left(\frac{1}{4}\right)^x$

(b)

x	-2	-1	0	1	2
$f(x)$	-7	0	1	2	9

(c)

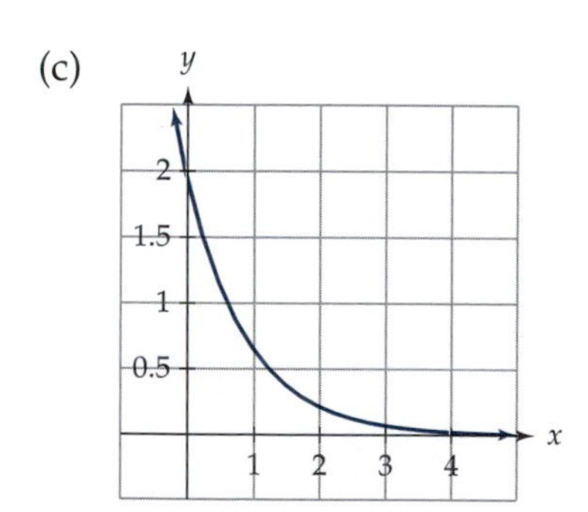

(d) The function describing the amount of money received by returning soda bottles if the deposit is $0.10 per bottle.

14. Explain why bacteria doubling every 6 h is equivalent to 16 times as many bacteria after each day.

15. Why is it not reasonable to use the formula for our tuition function in Example 10 to find the tuition after $3\frac{1}{2}$ years?

SUMMARY

Functions that increase (or decrease) by a fixed percentage are exponential functions. An **exponential function** is a function where the growth factor is constant.

Symbolically, an exponential function is

$$f(x) = ba^x.$$

where $a > 0$ and $a \neq 1$. The y-intercept (or starting value) is b, and the **growth factor** is a. So, $b = f(0)$ and $a = f(x+1)/f(x)$. The growth factor, a, is also called the **base.**

Exponential functions of the form $f(x) = ba^x$ contain the points $(0, b)$ and $(1, ba)$. If b is positive, then the domain is all the real numbers and the range is $f(x) > 0$. Again assuming b is positive, an exponential function will be increasing if the growth factor is greater than 1 and decreasing if the growth factor is between 0 and 1. In either case, the function is always concave up with the x-axis as a **horizontal asymptote.**

When exponential functions are given verbally, they often include a growth rate. A **growth rate** is always 1 less than the corresponding growth factor.

EXERCISES

1. Let $h(x) = 2^x$. Find each of the following.
   (a) $h(-2)$
   (b) $h(-1)$
   (c) $h(0)$
   (d) $h(\frac{1}{2})$
   (e) $h(10)$
2. Let f be an exponential function, $f(x) = ba^x$.
   (a) What is the domain of f? Why is this true?
   (b) What is the range of f? Why is this true?
3. Determine if each of the following is an exponential function. If so, give the growth factor and the starting value.
   (a) $y = 3^x$
   (b) $x^3y = 1$
   (c) $3y = (\sqrt{2})^x$
   (d) $y = \pi^x$
   (e) $y = x^\pi$

4. Determine if each of the graphs given is an exponential function. If so, give the equation for the function. If not, briefly explain why.

(a)

(b)

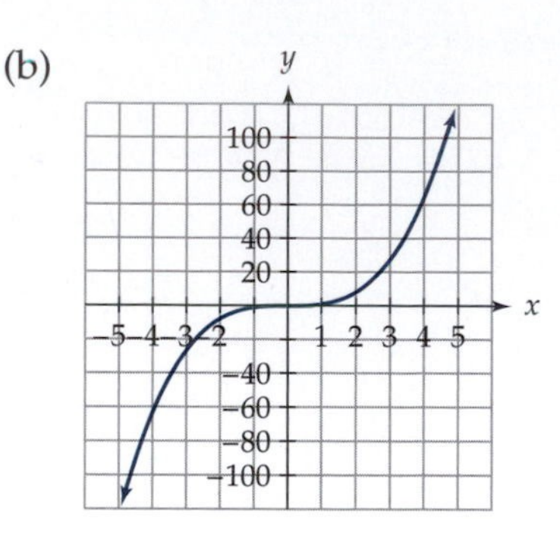

(c)

(d)

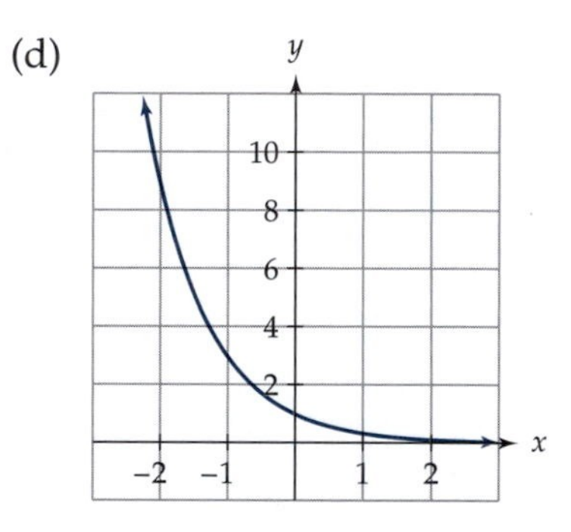

5. Determine if each of the following tables is an exponential function. If so, give the equation for the function. If not, briefly explain why.

(a)

x	-2	-1	0	1	2
$f(x)$	$\frac{1}{9}$	$\frac{1}{3}$	1	3	9

(b)

x	-2	-1	0	1	2
$f(x)$	$\frac{1}{2}$	$1/\sqrt{2}$	1	$\sqrt{2}$	2

(c)

x	-2	-1	0	1	2
$f(x)$	5	2	1	2	5

(d)

x	-2	-1	0	1	2
$f(x)$	0.694	0.833	1	1.2	1.44

(e)

x	-2	-1	0	1	2
$f(x)$	-8	-1	0	1	8

6. Determine if each of the following describes an exponential function. If so, give the growth factor and the starting value. If not, briefly explain why.
   (a) The cost of gasoline is $1.23 per gallon.
   (b) The number of basketball teams playing in the nth round is $\frac{1}{2}$ the number that played in the previous round.
   (c) A radioactive substance is decaying at a rate where the amount left after 1 year is 98.7% of the original amount.
   (d) The total distance you travel in your car is increasing by 50 miles every hour.
   (e) The population of Expo, U.S.A., is increasing at a rate of 3.5% per year.
7. Let f be an exponential function where the starting value is 1. We are interested in comparing $f(x)$ and $f(-x)$.
   (a) Suppose $f(x) = 2^x$. Compute $f(-3)$, $f(-2)$, $f(-1)$, $f(1)$, $f(2)$, $f(3)$. What is the relationship between $f(x)$ and $f(-x)$ for this function and these inputs?
   (b) Let $f(x) = a^x$. Compute $f(-3)$, $f(-2)$, $f(-1)$, $f(1)$, $f(2)$, $f(3)$. What is the relationship between $f(x)$ and $f(-x)$?
   (c) What property of exponents is causing this behavior?
   (d) Is $f(x) = a^x$ symmetric about the y-axis, the origin, or neither? Briefly justify your answer.
8. Explain what is wrong with the following statement:

   If college tuition is increasing at a constant rate of 7.9% per year and knowing that linear functions have a constant rate of change, I can conclude that college tuition is increasing linearly.

9. Let $P(n)$ represent the amount of money paid to our hypothetical baseball player who received $1 for signing his contract and whose pay was doubled for every game the team played.
   (a) For what game did he first earn more than $1,000,000?
   (b) We stated that he would earn $536,870,912 for playing the last game of the World Series. What did he earn for playing the second-to-last game?
   (c) Suppose he received $1 for signing his contract and agreed to receive 1.5 times as much for the next game. What function

represents his pay for the nth game? What would he have earned for the 10th game? What would he have earned for the last game of the World Series (the 29th game)?

(d) Suppose he wanted to receive $1,000,000 for playing the last game of the World Series (a total of 29 games). If he received $1 for signing his contract, what growth factor should he have negotiated?

10. Let $f(x) = a^x$. Label each of the following as true, sometimes true, or false. Briefly justify your answer.

(a) The domain of f includes negative inputs.

(b) The range of f includes negative outputs.

(c) The growth factor, a, can be positive or negative.

(d) If $a > 0$, then f will be increasing.

(e) f always has the x-axis as a horizontal asymptote.

(f) If $a > c$ (where a and c are positive numbers not equal to 1), then $a^x > c^x$ for all x.

(g) The x-intercept for f is $x = 1$.

(h) The y-intercept for f is $y = 1$.

(i) f is concave up.

11. An employee was offered the choice of receiving either a $1000 bonus added to his base pay every year or a 2.7% cost of living increase added to his base pay every year. His current annual salary is $33,500.

(a) Let B be the function representing the annual pay if he chooses to take the $1000 bonus option. Is B a linear or exponential function? Why?

(b) Find the formula for B.

(c) Let C be the function representing the annual pay if he chooses to take the 2.7% cost of living increase. Is C a linear or an exponential function? Why?

(d) Find the formula for C.

(e) If he works for the company for 5 more years, which is the better choice?

(f) If he works for the company for 20 more years, which is the better choice?

12. For each of the following, convert the doubling rate to a growth factor.

(a) If a substance doubles in 6 months, what is its growth factor for 1 year?

(b) If a substance doubles in 6 min, what is its growth factor for 1 h?

(c) If a substance doubles in 2 days, what is its growth factor for 1 day?

(d) If a substance doubles in 1 h, what is its growth factor for 1 day?

13. Let L be the linear function through the points $(0, 1)$ and $(10, 50)$. Let E be the exponential function through the points $(0, 1)$ and $(10, 50)$.

(a) Which will be greater, $L(5)$ or $E(5)$? How do you know?
(b) Which will be greater, $L(20)$ or $E(20)$? How do you know?
(c) Let $0 < c < 10$. Which is greater, $L(c)$ or $E(c)$? How do you know?
(d) Let $d > 10$. Which is greater, $L(d)$ or $E(d)$? How do you know?

14. Let $P(t)$ be the population of whosits at year t on the Island Chaos. For each of the following, describe what is happening to the population.
 (a) $P(t) = 10 \cdot 1.5^t$
 (b) $P(t) = 150t + 10$
 (c) $P(t) = 10 \cdot 0.5^t$
 (d) $P(t) = 50t + 10$

15. Suppose there are 16 croquet teams competing for the championship. Consider two different play-off strategies. In the first strategy, a round consists of every team playing every other team and the five worst teams overall are eliminated. This process continues until there is one team left. In the second strategy, teams are paired and only the winner continues on to the next round. Again, this process continues until there is one team left.
 (a) How many total rounds will be played in the first strategy? How many total rounds will be played in the second strategy?
 (b) What type of function models the first strategy? What type of function models the second strategy?
 (c) Give an equation for the function that models the first strategy, being sure to specify the domain.
 (d) Give an equation for the function that models the second strategy, being sure to specify the domain.

16. For each of the following, find the equation of the exponential function that contains that pair of points.
 (a) $(0, 2)$ and $(1, 3)$
 (b) $(0, 1)$ and $(2, 4)$
 (c) $(1, 5)$ and $(2, 7)$
 (d) $(1, 3)$ and $(5, 2)$

17. Let B be the function representing the number of bacteria on the chicken. Assume that the bacteria double every 6 h and that initially there are 360 bacteria on the chicken.
 (a) When will there be 720 bacteria on the chicken?
 (b) Table 12 stated that there were 1,475,000 bacteria on the chicken on the third day. When were there $1{,}475{,}000/2 = 737{,}500$ bacteria on the chicken?

18. Let f be an exponential function whose growth factor is 2. Suppose $f(10) = 3072$.
 (a) For what value of x does $f(x) = 1536$?
 (b) For what value of x does $f(x) = 6144$?
 (c) For what value of x does $f(x) = 384$?

INVESTIGATIONS

INVESTIGATION 1: POPULATION

Population growth is a concern of many people around the world because of its impact on the environment, the food supply, and other limited resources. Although this growth, particularly long-term growth, is very difficult to predict, it is nonetheless necessary to make predictions to plan for the future. Mathematical models are often used to make reasonable predictions. These models can be quite complicated since population growth patterns are dependent on many variable factors. Relatively accurate information, however, can be generated from simple mathematical models, particularly when their use is restricted to a short period.

To help understand how growth rates affect the size of a population, we compare population growth rates from three different countries. A document produced by the United Nations Population Fund[14] gives an annual growth rate of approximately 1% for Ireland, a growth rate of approximately 2% for Mexico, and a growth rate of approximately 3% for Ethiopia from 1990 to 1995. For comparison purposes, assume each of these countries had an initial population of 1,000,000.

1. All the rates given are growth *rates*, not growth factors. A growth rate is used to determine the additional population, not the total population. Similarly, interest rates are used to determine additional money added to your account, not the total money in your account.
   (a) Convert each of the growth rates to a growth factor.
   (b) Give an exponential function that models the population growth for each of the three countries, assuming the growth factor is constant.[15]

2. Use the exponential functions you found in question 1(b) to complete the following table.

Country	Growth Rate	Population after 10 Years	Population after 100 Years
Ireland	1%		
Mexico	2%		
Ethiopia	3%		

3. Graph the population for these three countries, on the same set of axes, with time on the horizontal axis and population on the vertical axis. First use a domain of 0 to 10 years, then, on another graph, use a domain of 0 to 100 years.

[14] *United Nations Population Fund,* Population and the Environment: The Challenges Ahead *(London: Banson Productions, 1991), pp. 39–43.*

[15] *In reality, population growth rates do not stay constant. We assume they do to simplify this problem.*

4. Describe the differences and similarities between the graphs with the 10-year domain and those with the 100-year domain. How are the shapes of the graphs in these two windows related to the resulting differences in populations of the three countries after 10 years and after 100 years?
5. The populations of each of these countries in 1990 was not 1,000,000 but was actually much larger. In 1990, Mexico had a population of about 90,000,000 and Ethiopia had a population of about 50,000,000. If the growth rates of these two countries stay constant, use a graph of the appropriate functions to estimate the year that Ethiopia's population would exceed that of Mexico.
6. If the growth rate of one country is double that of another country, describe what "double" means. Does it mean that the growth factor is double? Does it mean that the population of the second country will always be double that of the first? Does it mean that the increase in population of the second will always be double that of the first? Exactly what is doubled? Use examples from previous questions or make up other examples to support your conclusions.

INVESTIGATION 2: THE IMPORTANCE OF e

When mathematicians think of exponential functions, they often think about the particular exponential function, e^x, whose growth factor is $e \approx 2.718281828$. You may have noticed that your calculator has a button for e^x but does not have a similar button for 2^x or for any other exponential function. In this investigation, we explore one of the reasons e is such an important growth factor.

1. We are interested in exploring the rate of change of $f(x) = a^x$ close to its starting value, $(0, 1)$. Let h represent a small positive number.
   (a) Let $f(x) = 2^x$. Write an expression that represents the slope between the points $(0, 1)$ and $(h, 2^h)$.
   (b) Using your expression from part (a), fill in the following table for the values of h given.

h	0.1	0.01	0.001	0.0001
Slope				

The slopes in the table represent the slope between the point $(0, 1)$ and a point on the graph of your function that is quite close to the point $(0, 1)$. As you continue this process with smaller and smaller values of h, you can start to think of each slope as representing the slope of the function at the point $(0, 1)$ (rather than thinking of it as the slope between two points). Using the values from your table, estimate the "slope" of the function $f(x) = 2^x$ at the point $(0, 1)$.

2. Repeat question 1, this time using the function $g(x) = 3^x$.
3. Repeat question 1, this time using the function $h(x) = e^x$.
4. Which of the three functions has the smallest slope at the point $(0, 1)$? Which of these has the largest slope at the point $(0, 1)$? Why is this happening?
5. What do you know about the slope of $k(x) = a^x$ at the point $(0, 1)$ if $a < 1$? What do you know about the slope of $k(x) = a^x$ at the point $(0, 1)$ if $a > 1$?
6. What appears to be special about the slope of the function $h(x) = e^x$ at the point $(0, 1)$?
7. Let $f(x) = e^x$.
   (a) Write the expression that finds the slope between the points $(c, f(c)) = (c, e^c)$ and $(c + h, f(c + h)) = (c + h, e^{c+h})$.
   (b) Assuming h is a small positive number, your expression in part (a) can be thought of as the "slope" of the function $f(x) = e^x$ at the point (c, e^c). Using properties of exponents, rewrite your expression from part (a) so that it is equal to $(e^h - 1)/h$ times something.
   (c) Knowing that the slope of $f(x) = e^x$ at the point $(0, 1)$ equals 1 and that the slope at $(0, 1)$ is approximated by $(e^h - 1)/h$ for small values of h, what can you say about the slope of $f(x) = e^x$ at the point (c, e^c)? The only growth factor for which this will be true is e, just one of the things that makes e so special!

2.3 LOGARITHMIC FUNCTIONS

In sections 2.1 and 2.2, we looked at linear and exponential functions. All these functions are similar in that an input is multiplied by some number, added to some number, or raised to some exponent; these are all basic algebraic operations. Logarithmic functions are different. These functions do not just use basic arithmetic operations to convert an input to an output. In this section, we give the definition of the logarithmic function, explore some of its properties, and consider the type of applications where logarithms are used.

MAGNITUDES

Let's start by reviewing the metric system, something that turns out to be a "natural" use of the logarithm function. Recall that there are 10 meters in a dekameter, 100 meters in a hectometer, and 1000 meters in a kilometer. Each of these units is 10 times larger than its preceeding unit. The term *magnitude* is used to describe this type of relationship. If we use meters as our base, we say that a kilometer has magnitude 3 because 10^3 meters equals 1 kilometer. Similarly, a hectometer has magnitude 2 and a dekameter has magnitude 1. We could also say that a meter has magnitude 0 because a meter $= 10^0$ meter.

Magnitude is defined in the dictionary as the "greatness of size" of something. For our purposes, we define the **magnitude** of a number a as the exponent b such that $10^b = a$. Magnitudes can be defined using numbers besides powers of 10, but here we choose 10 because our number system is in powers of 10. Notice the relationship between magnitudes and exponents. A magnitude is just another word asking for an exponent with a given base.[16] With this definition of magnitude, the magnitude of 10,000 is 4 because $10{,}000 = 10^4$. This definition is similar to the way we defined magnitude when looking at the metric system because both our definition of magnitude and the metric system are based on powers of 10. It is clear that the magnitude of 10 is 1 and the magnitude of 100 is 2. What, then, is the magnitude of 50? Because 50 is about halfway between 10 and 100, it is reasonable to start with a guess of 1.5. We can check our guess by going back to our definition of magnitudes and using our calculator to find that $10^{1.5} \approx 31.6$. That answer is too low, so we increase our guess to 1.6. Because $10^{1.6} \approx 39.8$, this guess is still too low. Trying 1.7, we find that $10^{1.7} \approx 50.1$. Therefore, the magnitude of 50 is approximately 1.7. Notice that magnitude is not a linear function.

Example 1 Estimate the magnitude of 45,876 to two decimal places.

Solution Since 45,867 is between 10,000 (magnitude 4) and 100,000 (magnitude 5), it will have a magnitude between 4 and 5. To estimate the magnitude to two decimal places, we use the same "guess and check" method used earlier.

Guess	4.60	4.65	4.66	4.67
Check	$10^{4.60} \approx 39{,}811$	$10^{4.65} \approx 44{,}668$	$10^{4.66} \approx 45{,}709$	$10^{4.67} \approx 46{,}774$

Because 45,867 is closer to $10^{4.66}$ than to $10^{4.67}$, the magnitude of 45,876 is approximately 4.66.

There is a similarity between writing the magnitude of a number and writing that number in scientific notation. In Example 1, we saw that the magnitude of 45,867 is approximately 4.66. If we write 45,867 in scientific notation, we find that $45{,}867 \approx 4.59 \times 10^4$. The exponent on the 10 is simply the whole-number part of the magnitude of 45,867. Because both magnitudes and scientific notation are based on exponents of 10, this result is not surprising.

READING QUESTIONS

1. What is meant by the *magnitude* of a number?
2. What is the magnitude of one million?
3. Estimate the magnitude of 6,132,678 to two decimal places.

[16] *Recall that with c^d, the value of c is called the* base *and the value of d is called the exponent.*

THE LOGARITHM FUNCTION

The function whose input is a number and whose output is its magnitude is called the **logarithm** function or, simply, the **log** function. In this section, our logarithm function is always of base 10. Recall that *base* refers to the number you are raising to a power. So,

$$\log x = y \qquad \text{is equivalent to} \qquad 10^y = x.$$

Thus, y is the exponent that, when 10 is raised to that power, gives us x.[17] The output of the logarithm function is called either the magnitude or the logarithm. Table 1 shows the logarithm (or magnitude) of some numbers between 0 and 10.

TABLE 1 The Logarithm Function for Values of x, Where $0 < x \le 10$

x	0.1	0.5	1	2	3	4	5	6	7	8	9	10
$\log x$	-1	≈ -0.30	0	≈ 0.30	≈ 0.48	≈ 0.60	≈ 0.70	≈ 0.78	≈ 0.85	≈ 0.90	≈ 0.95	1

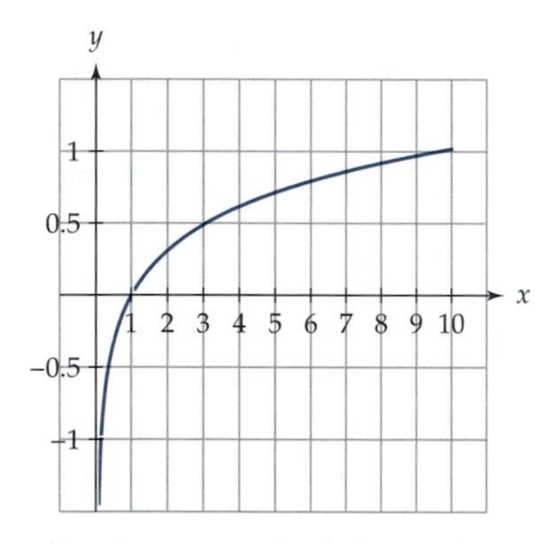

Figure 1 The logarithm function on a domain of $0 < x \le 10$.

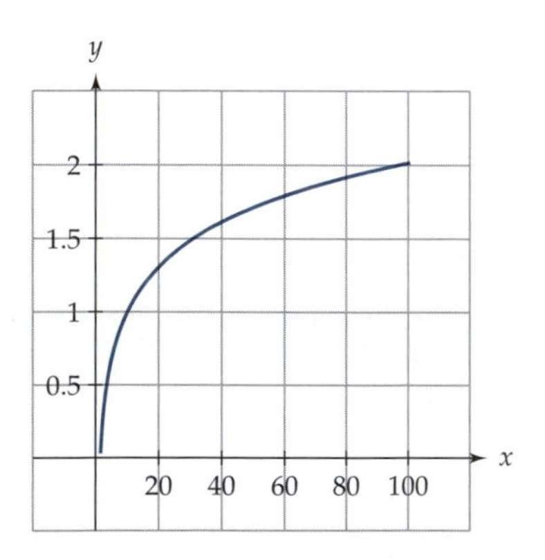

Figure 2 The logarithm function on a domain of $1 \le x \le 100$.

Note that the points $(1, 0)$ and $(10, 1)$ are included in Table 1. Thus, $\log(1) = 0$ and $\log(10) = 1$. Remember that the logarithm function is telling us the magnitude (or exponent) of the number, and $1 = 10^0$, whereas $10 = 10^1$.

The graph in Figure 1 is the logarithm (or magnitude) function, $\log x$, for $0 < x \le 10$. Even though the graph is of $f(x) = \log x$ with 10 as the base, we see later that this is the basic shape for the log function with *any* positive base greater than 1.

Notice that the graph is increasing and concave down. It is increasing because the larger the number, the larger its magnitude. It is concave down because the rate of change is decreasing, as we see in Table 1. Notice that when the input increased from 1 to 2, the output increased by 0.30. When the input increased from 9 to 10, the output increased by only 0.05.

The graph of $f(x) = \log x$ will continue to be concave down for $10 \le x \le 100$. (See Figure 2.) Because $\log 10 = 1$ and $\log 100 = 2$, the increase in the output will be one unit, the same as the increase in the output between $\log 1 = 0$ and $\log 10 = 1$. Yet for $1 \le x \le 10$, the change in the input is $10 - 1 = 9$ units and the change in the output is one unit. For $10 \le x \le 100$, however, the change in the input is $100 - 10 = 90$ units and the change in the output is again only one unit. The same increase in height (or change in outputs) is spread over a much larger interval, which causes the graph to be concave down.

In general, the logarithm function's graph will be steep at the beginning but will quickly become flatter and flatter as your input increases. This logarithm function does not have an asymptote as in the function $f(x) = \left(\frac{1}{2}\right)^x$ because the logarithm function never approaches a horizontal line. The logarithm function continues to grow, although *very* slowly!

[17] *Later in the book, we look at powers of other numbers besides 10.*

READING QUESTIONS

4. What is the range of the logarithm function for $1 \leq x \leq 1000$? What is happening to the graph of $\log x$ for $100 \leq x \leq 1000$?

5. Why is the logarithm function concave down?

PROPERTIES OF THE LOGARITHM FUNCTION

Basic Properties

Looking at Figures 1 and 2, we notice several properties of the logarithm (or magnitude) function. Some of these properties of $f(x) = \log x$ are the following.

PROPERTIES

Basic properties of the logarithm functions are:

- $\log(1) = 0$.
- $\log x$ is negative for $0 < x < 1$ and positive for $x > 1$.
- $f(x) = \log x$ has a domain of $x > 0$ and a range of all real numbers.
- The graph of $f(x) = \log x$ is increasing and concave down.

These properties are true because logarithms are merely magnitudes or exponents. As with exponential functions, the properties of these functions are just consequences of the properties of exponents. Knowing the properties of exponents allows you to find the characteristics of either logarithmic functions or exponential functions. For example, the magnitude of 1 is 0 because $10^0 = 1$, which is analogous to saying that 1 m has magnitude 0. The magnitude of something less than 1 (but greater than 0) is negative because 10 has to be raised to a negative exponent to get a number smaller than 1. For example, because $1/10^2 = 10^{-2}$, $\log(\frac{1}{100}) = -2$, which is similar to saying that 1 cm has magnitude -2. The magnitude of anything greater than 1 is positive. If you want to raise 10 to a power to get a number greater than 1, your exponent will be positive. For example, the magnitude of 1000 or log 1000 is 3.

So far, we have looked at finding magnitudes of positive numbers. What about finding magnitudes of negative numbers? Notice in Figures 1 and 2 that $f(x) = \log(x)$ does not exist for negative values of x because the logarithm of a negative number is undefined. For example, to what exponent does 10 need to be raised to get -100? There is no such exponent. Ten raised to *any* power will always be positive. Similarly, log 0 is undefined. There is no exponent p such that $10^p = 0$. In terms of the metric system, there is no unit to describe a length of zero or a negative length. Hence, the domain of the log function is restricted to the set of positive real numbers.

Adding Logarithms

It was mentioned earlier that the output of the the logarithm function increases by one unit as the input increases from 1 to 10. The output again increases by 1 when the input increases from 10 to 100. This idea was helpful when observing the shape of the graph of the logarithm function. Because the logarithm function is defined as magnitudes, the output of the logarithm function will always increase by 1 whenever the input changes by a factor of 10. For example,

$$\log(2) \approx 0.3010 \qquad \text{because} \qquad 10^{0.3010} \approx 2$$

and

$$\log(20) \approx 1.3010 \qquad \text{because} \qquad 10^{1.3010} = 10 \cdot 10^{0.3010} \approx 10 \cdot 2 = 20.$$

Multiplying a number by 10 will always increase the magnitude by one unit. Symbolically, if $\log c = q$, then $\log(10c) = q + 1$.

Let's do another example using different numbers. Let a and b be numbers such that $\log(a) = 3$ and $\log(b) = 4$. Thus, $a = 10^3$ and $b = 10^4$. Notice that $\log(a) + \log(b) = 3 + 4 = 7$. What input would give an output of 7? It would be 10^7, the *product* of the original two inputs, 10^3 and 10^4. Try it with any pair of numbers. This property always works. In general, we can say that adding the outputs of the logarithm function corresponds to multiplying the inputs. Symbolically, we summarize this property as follows.

PROPERTIES

The multiplication property:

$\log(ab) = \log(a) + \log(b)$ for any positive numbers a and b.

Even though this property may seem like a new concept, it is merely a consequence of the properties of exponents. Whenever you multiply two numbers with exponents that have the same base, you add the exponents. Symbolically, if $\log a = p$ and $\log b = q$, then $10^p = a$ and $10^q = b$. So, $ab = 10^p 10^q = 10^{p+q}$. Thus, $\log(ab) = p + q = \log(a) + \log(b)$.

Logarithms were originally developed for computational purposes (see the Historical Note later in this section). This property of converting a multiplication problem to an addition problem is one reason logarithms were so useful before electronic calculators were available. Addition of large numbers is much easier than multiplication! To use logarithms to multiply two very large numbers, we use a log table (readily available in the precalculator days) to find the logs of our two numbers and then add the two logarithms. By taking this sum and reading our log table "backwards," we can find the result of our multiplication without actually doing any multiplication.

Multiplying Logarithms by Constants

The property of being able to add the logarithms of numbers to find the logarithm of a product leads to another property of logarithms. Let's start by using $\log(a) + \log(b) = \log(ab)$ to find $\log a^2$. Because

$$\log(a^2) = \log(aa),$$

we know that

$$\log(aa) = \log(a) + \log(a)$$
$$= 2\log(a).$$

Similarly,

$$\log(a^3) = \log(a) + \log(a^2)$$
$$= \log(a) + 2\log(a)$$
$$= 3\log(a).$$

In general, for all positive integer values of n,

$$\log(a^n) = \log(a) + \log(a^{n-1})$$
$$= \log(a) + (n-1)\log(a)$$
$$= n\log(a).$$

So, for integer exponents, the logarithm of a number raised to an exponent is the same as the value of the exponent multiplied by the logarithm of the number. This property, however, is also true for noninteger exponents, again as a consequence of the properties of exponents. Suppose $\log(b) = p$. Then $b = 10^p$. Therefore, $b^r = (10^p)^r = 10^{pr}$. So, $\log(b^r) = pr = r\log(b)$. This gives us the following property.

PROPERTIES

The power property:

$\log(a^r) = r\log(a)$ for any real number r and positive number a.

This second property of logarithms again made logarithms an easy way of doing calculations before the era of electronic calculators.[18] Finding a number raised to an exponent r was much easier using logarithms since, instead of repeated multiplication, you could merely multiply the log of the number by r. It also gave a way of finding values such as $2^{1/2} = \sqrt{2}$ by using $r = \frac{1}{2}$. For example, finding the value of 12^{10} by hand would be difficult, but using logarithm tables, we look up log 12 and find that it is approximately 1.079. We know that $\log 12^{10} = 10 \log 12$, so multiplying 1.079 by 10 gives us 10.79. The antilog of 10.79 (i.e., the number whose log is 10.79) is found in a logarithm table as $\approx 6.166 \times 10^{10}$. Thus, $12^{10} \approx 6.166 \times 10^{10}$.

READING QUESTIONS

6. If $\log r$ is negative, what do you know about r?
7. Why is it impossible to find a number p such that $10^p = -4$?
8. If $\log 14 \approx 1.146$, what is $\log 140$?
9. Given that $\log 4 = 0.602$, describe how to find $\log 16$.
10. Given that $\log 4 = 0.602$, describe how to find $\log 2$.

[18] *The slide rule—a simple computational device popular before calculators—is one such device that used logarithms for doing calculations.*

HISTORICAL NOTE

John Napier, who is usually credited with the invention of logarithms, was not a professional mathematician; he was a Scottish laird who wrote on many different topics while managing his large estates. Napier was seeking a procedure for making computation of large numbers easier. He published his system of logarithms in 1614. This book, whose title translates as *A Description of the Marvelous Rule of Logarithms,* was immediately admired by Henry Briggs, a professor of geometry at Oxford. In 1615, Napier and Briggs met and agreed that 10 should be used as the base for his logarithms. In 1617, the year of Napier's death, Briggs published a table of logarithms of the numbers 1 through 1000, each to 14 places! This table was extended in 1624 to include the logarithms (again to 14 places) of the numbers 1 through 20,000 and 90,000 through 100,000.

The English word *logarithm* comes from two Greek words: *logos,* meaning "reckoning or computing," and *arithmos,* meaning "number." From their inception, logarithms were meant to be a way to reckon or calculate numbers. Only later were many other uses for Napier's interesting discovery, the logarithm function, found.

Sources: Carl B. Boyer, *A History of Mathematics* (Princeton, NJ: Princeton University Press, 1985), pp. 342–45; Steven Schwartzman, *The Words of Mathematics: An Etymological Dictionary of Mathematical Terms Used in English* (Washington, DC: Mathematical Association of America, 1994), p. 128.

WHY USE MAGNITUDES?

Many scientific measurements are in terms of magnitudes. For example, the pH scale, the way earthquakes are measured, distances to the stars, and sound intensity are all measured in terms of magnitudes and therefore use a logarithmic scale. One reason is that it is easy to distinguish between magnitudes of change, whereas it is often difficult to distinguish change measured in differences. An illustration of magnitudes and differences is shown in Figure 3.

Figure 3 contains 12 boxes of dots. The number of dots in each box is given at the bottom of each box. Notice that each box in the second row contains one more dot than each corresponding box in the first row and that each box in the third row contains 10 more dots than each corresponding box in the first row. As you look down the first column of boxes, it is very easy to see the increase in dots between each row. In the second column, this same increase is more difficult to see. In the third and fourth columns, these same increases are almost impossible to detect. By looking across any of the three rows, however, it is very easy to see the increase in the number of dots per box. The increase across the rows goes from a magnitude of 0 ($10^0 = 1$) to a magnitude of 1 ($10^1 = 10$) to a magnitude of 2 ($10^2 = 100$) and finally to a magnitude of 3 ($10^3 = 1000$). The human eye perceives this change of magnitudes better than it perceives the change in terms of individual dots, especially when the number of dots is large.

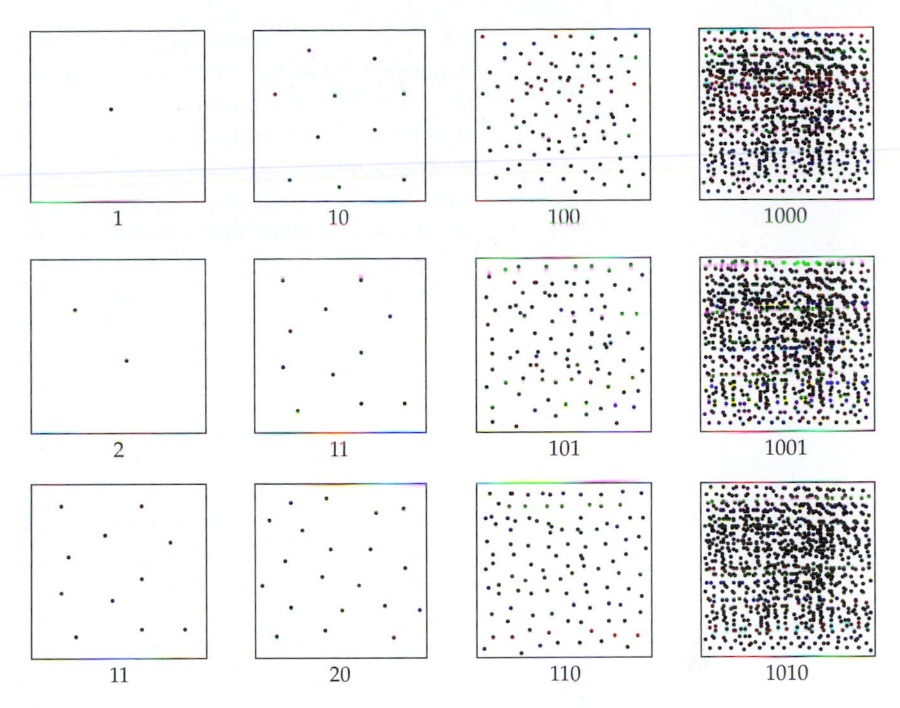

Figure 3 These boxes help visualize the difference between changes in magnitudes and differences.

Another situation modeled with magnitudes is the way sound is perceived. It has been suggested that we do not perceive changes in sound intensity linearly, but, rather, logarithmically.[19] The smallest change in sound intensity that the human ear can detect is about 26%, which corresponds to a change of 1 decibel (dB). The volume control on a stereo or television compensates for the human ear's nonlinearity by using a "logarithmic-taper potentiometer." This device is designed to give the impression of a linear increase as the volume is raised.[20] Look again at the dots in Figure 3. Let's imagine that these represent sound. If your stereo had a top volume of 1000 dots (like the box on the top right) and your volume control was linear, then turning the volume dial one quarter of the way would be the equivalent of going from 1 dot to 250 dots in that quarter turn. You would go from no sound to a great deal of sound very quickly. This change would represent a change of about 24 dB.

Let's see why increasing our sound level from 1 to 250 is equal to an increase of about 24 dB. Remember that 1 dB is defined as a change in sound intensity of about 26%. Because 1 dB is a 26% increase, the function modeling decibels has a growth rate and so is an exponential function. An increase of 26% means that the level of sound after an increase of 1 dB is

[19] *Some scientists have proposed a power function to model the way humans perceive sound. The logarithmic model, however, has been accepted for quite some time, and it serves to illustrate the properties of logarithms for our purposes.*

[20] *Stan Gibilisco and Neil Sclater, eds.,* Encyclopedia of Electronics, *2d ed. (Blue Ridge Summit, PA: TAB Professional and Reference Books, 1990), p. 69.*

$1 + 0.26 = 1.26\%$ of the original sound level. So, decibels are an exponential function with a growth factor of 1.26. An increase of 24 dB is $(1.26)^{24} \approx 250$, the increase in the sound level when we turned our linear volume control. It is no accident that logarithms that measure magnitudes are connected with exponential functions that have a constant growth factor. Both of these functions deal with exponents. For exponential functions, the exponent is the input. For logarithmic functions, the exponent is the output. We explore this connection further when we look at inverse functions in Section 3.5.

Let's return to the sound system with the linear volume control. The next quarter turn, from 250 dots to 500 dots, still represents a change in sound. This turn, however, only represents a change of about 3 dB since $(1.26)^{24+3} = (1.26)^{27} \approx 500$. The next quarter turn, from 500 to 750, represents a change of only about 2 dB. In the last quarter turn of our linear volume control, you can barely hear any difference in sound level. Without the logarithmic-taper potentiometer, it would be difficult to control the volume because almost all the change would be in the first quarter of the volume control. Your volume level would go from too soft to too loud too quickly, like the graph of the logarithm function that increases rapidly at first and then increases slowly.

The shape of the logarithm graph makes it a good modeling function for any behavior that increases rapidly at first but then slows down dramatically. For example, think about the way you learn a new computer program. You quickly go from knowing nothing about that particular program to learning enough to start using the program. Your knowledge of the program has increased quickly in a short period. Soon, however, your rate of learning about the program slows down. You continue to learn as you encounter new situations, but you only learn a few new things now and then. So, your learning went from a initial rapid growth to a slow, gradual growth. Some type of logarithmic function would best model your learning of the computer program.

READING QUESTIONS

11. Why is measuring in magnitudes useful when trying to perceive the difference in large quantities?

12. What is the growth factor for the exponential function whose input is the number of decibels and whose output is the sound level?

13. What do log functions and exponential functions have in common?

14. Describe a physical situation (not mentioned in this section) for which a logarithm function might be a good model.

SUMMARY

The **magnitude** of a number a is the exponent b such that $10^b = a$. Magnitudes can be defined using other numbers besides powers of 10, but for this section, we chose 10 because our number system is in powers of 10. A magnitude is just another word asking for an exponent with a given base.

The function whose input is a number and whose output is its magnitude is called the **logarithm** function or, simply, the **log** function. In this section, we only looked at logarithm functions with a base of 10. Therefore,

$$\log x = y \qquad \text{is equivalent to} \qquad 10^y = x.$$

Some basic properties of $f(x) = \log x$ are the following.

- $\log(1) = 0$.
- $\log x$ is negative for $0 < x < 1$ and positive for $x > 1$.
- $f(x) = \log x$ has a domain of $x > 0$ and a range of all real numbers.
- The graph of $f(x) = \log x$ is increasing and concave down.

Two other important properties of logarithms are the following.

- $\log(ab) = \log(a) + \log(b)$ for any positive numbers a and b.
- $\log(a^r) = r\log(a)$ for any real number r and positive number a.

EXERCISES

1. The following table gives the values of $\log x$ for some values of x. Answers are rounded to two decimal places.

x	0.1	0.5	1	2	3	4	5	6	7	8	9	10
$\log x$	−1	−0.30	0	0.30	0.48	0.60	0.70	0.78	0.85	0.90	0.95	1

   Explain why this table does not represent each of the following.
   (a) A linear function
   (b) An exponential function
2. Why is the graph of the logarithm function concave down?
3. The function $f(x) = \log x$ increases without bound. Thus, no matter what number M you choose, you can find a value of x such that $\log x > M$. Why is this property true?
4. The magnitude function m cannot be a linear function because $m(10) = 1$ and $m(100) = 2$, yet $m(55) \approx 1.74$. Explain why this information shows that magnitude is not linear.
5. Using $\log 5 \approx 0.69897$, find each of the following.
   (a) $\log 50$
   (b) $\log 5000$
   (c) $\log 25$
   (d) $\log 125$
6. Estimate p to two decimal places.
   (a) Find p such that $10^p \approx 27$.

(b) Find p such that $10^p \approx 600$.

(c) Find p such that $10^p \approx 3500$.

7. Suppose $\log(a) = 0.4$. Find each of the following.
   (a) $\log(10a)$
   (b) $\log(100a)$
   (c) $\log(1000a)$
   (d) $\log(a^2)$
   (e) $\log(a^{1/2})$
   (f) $\log(a^3)$
8. Suppose $\log(20) = b$. Find each of the following.
   (a) The value of c such that $\log(c) = 2b$
   (b) The value of c such that $\log(c) = b + 3$
   (c) The value of c such that $\log(c) = b/4$
9. Using the properties of logarithms given in this section, show that $\log(a/b) = \log a - \log b$ for $a > 0, b > 0$. (*Hint:* Use that $a/b = a \cdot 1/b$.)
10. Using the properties of logarithms given in this section, show that $\log \sqrt[n]{a} = (\log a)/n$ for any positive integer n and $a > 0$.
11. Graph $f(x) = \log x$ and $g(x) = \log 10x$ on the same set of axes.
    (a) What is the relationship between these two graphs? Why is this relationship true?
    (b) Without graphing, explain how a graph of $h(x) = \log 100x$ would compare with a graph of $f(x) = \log x$.
    (c) Without graphing, explain how a graph of $j(x) = \log 0.1x$ would compare with a graph of $f(x) = \log x$.
12. Determine whether each of the following is true or false. Justify your answers.
    (a) $\log ab = \log a + \log b$
    (b) $\log ab = \log a \cdot \log b$
    (c) $\log(a + b) = \log a + \log b$
    (d) $\log a^b = (\log a)^b$
    (e) $\log a^b = b \cdot \log a$
13. In the example involving decibel level, we stated that if the sound level went from 500 to 700, the decibels would increase by less than 2. (The decibel level at 500 is approximately 27.) Show that this is true.
14. The pH of a solution is defined as $\text{pH} = -\log [H^+]$, where H^+ is the hydrogen ion content in moles per liter of a given solution. The solution is said to be acidic if the pH is less than 7 and basic if the pH is greater than 7.
    (a) Determine the pH of a solution that has a hydrogen ion content of 0.0004 mole per liter.
    (b) Determine the hydrogen ion content of a solution that has a pH of 5.
    (c) Which has a greater hydrogen ion content, an acidic solution or a basic solution?

INVESTIGATIONS

INVESTIGATION 1: CREATING A LOG TABLE

How do you suppose the first logarithm tables were made? Certain values were easy to find, such as log 10. How would you go about finding the logarithm of values that are not powers of 10? Trial and error could give you some of them, especially if you had *lots* of patience. For instance, finding log 2 is just like solving the equation $2 = 10^x$.

Guessing different solutions to this problem and checking (possibly using tables made by people with lots of spare time) allows us to get fairly close. You do not have to guess and check for every number, however. Using the properties of logarithms from this section and knowing a few logarithms, you can easily estimate the logarithms of other numbers.

The following table gives the logarithms for the first five integers. Answers are rounded to three decimal places.

x	1	2	3	4	5
log x	0	0.301	0.477	0.602	0.699

1. Use the properties of logarithms to find the logarithms of as many numbers as possible from 6 to 30. For example, $\log 6 = \log(2 \cdot 3) = \log 2 + \log 3 \approx 0.301 + 0.477 = 0.778$. You cannot use the values given in the table to find log 7, however, because 7 is not the product of any number besides itself and 1. Also, 7 is not the quotient, root, or power of any of the numbers given in the table.
2. In general, if you were computing a table of logarithms for the numbers 1 through 100 (without the aid of a calculator log key), which ones could you find easily by knowing the logarithms of 1 through 5 and using properties of logarithms? Keep in mind that once you generate another log, you can use it to find more. For example, once you know that $\log 6 \approx 0.778$, you can use it to find $\log 6^2 = \log 36$. Are there any numbers you cannot find? Why or why not?

INVESTIGATION 2: LOGS OF OTHER BASES

Suppose we were to use a base other than 10 for our magnitude or logarithm function. If we choose base 2, our magnitudes represent powers of 2 rather than powers of 10. We denote the logarithm function with base 2 as $\log_2 x$. It is easy to see that $\log_2 2 = 1$, $\log_2 4 = 2$ and $\log_2 8 = 3$. Answer the following questions about $\log_2(x)$.

1. What is $\log_2 32$? What is $\log_2 1024$?
2. Complete the following table for $f(x) = \log_2(x)$ by using the "guess and check" method.

x	1	2	3	4	5	6	7	8	9	10
$f(x)$										

3. It is no longer true in base 2 that increasing the input by a factor of 10 results in an increase of 1 in the output. Why not? How much do you have to increase the input before getting an increase of 1 in the output?
4. Plot the points from your table and connect them with a smooth curve. Compare this graph with Figure 2 in this section. How are the two graphs similar? How are they different?
5. Explain why each of the following properties of logarithms is true for base 2.
   (a) $\log_2 a$ is negative for $0 < a < 1$.
   (b) $\log_2 1 = 0$.
   (c) $\log_2 a$ is positive for $a > 1$.
   (d) $\log_2 a$ is undefined if a is negative or equal to zero.
   (e) $\log_2 a + \log_2 b = \log_2(ab)$.
   (f) $b \log_2 a = \log_2(a^b)$.

2.4 PERIODIC FUNCTIONS

Many things in the world repeat in a cyclic pattern. For example, the amount of daylight changes from day to day, but the amount of daylight on March 1 is much the same from year to year. The position of the second hand on a clock repeats itself every minute. The position of a child swinging on a swing is also repetitious. Repetitious behavior can be described mathematically as being periodic.[21] None of the functions discussed so far in Chapter 2 describes these periodic phenomena, because linear functions, exponential functions, and logarithmic functions continue to increase or decrease in the long run. These types of functions do not go through the same set of outputs repeatedly, so none is appropriate for describing periodic behavior. In this section, we show how periodic functions are defined, give situations that can be defined using periodic functions, and introduce the sine and cosine functions.

WHAT ARE PERIODIC FUNCTIONS?

Imagine watching the second hand move around the face of a clock. The second hand is in the same position every 60 sec. Because this repetition occurs at regular intervals, the behavior is periodic. To be periodic, something must not only repeat itself, but repeat itself at regular intervals. A **periodic function** is a function that gives the same output for inputs a

[21] *Magazines published at regular intervals are called* periodicals *for the same reason.*

fixed distance apart. Symbolically, we say that f is periodic if for some real number p, $f(x) = f(x+p)$ for all x. The **period** of this function is the smallest value of p for which this relationship is always true; that is, the period is the *minimum* fixed distance where the outputs are always the same. If the period of f is 5, then $f(x)$ and $f(x+5)$ are equal for *every* value of x.

If a function is represented graphically, it is fairly easy to determine whether or not it is periodic. Four graphs are shown in Figure 1. Graph (a) is not periodic. Although it is going up and down on equal intervals, it does not give the same output on each of these intervals. Graph (b) is not periodic because the intervals on which it repeats are not regular. Graph (c) is a periodic function because it gives the same output for inputs a fixed distance apart. Although graph (d) repeats itself on equal intervals, it is not the graph of a function.

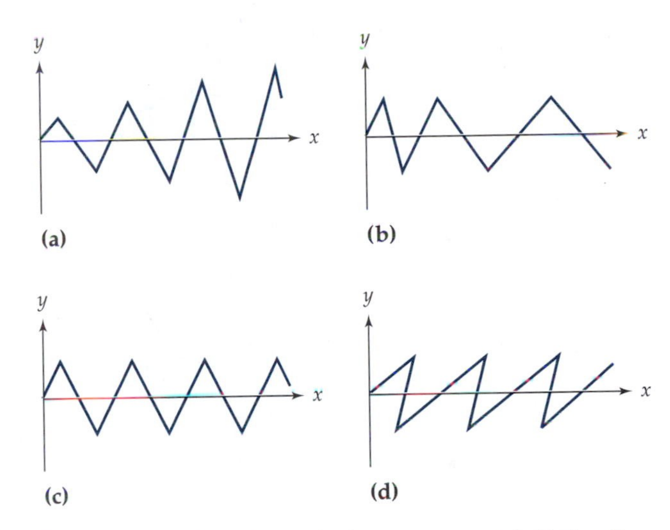

Figure 1 Of these four graphs, only graph (c) represents a periodic function.

READING QUESTIONS

1. What is a periodic function?
2. Why is graph (d) in Figure 1 not a function?
3. Give an example of something that can be described using periodic functions.
4. Suppose f is a periodic function whose period is 4. If $f(1) = -2$, what is $f(5)$?

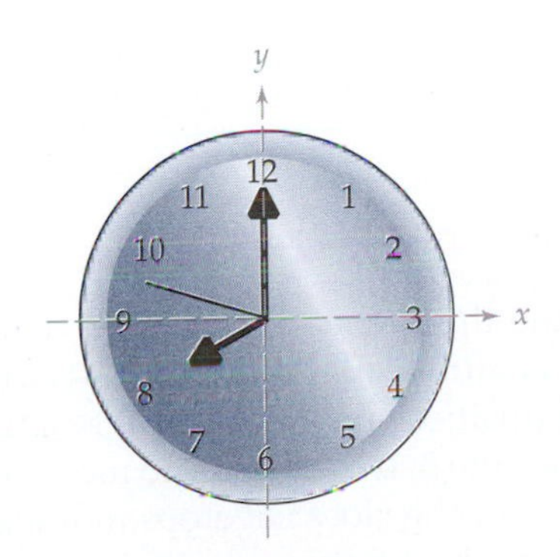

Figure 2 A clock centered on an xy-coordinate system.

GRAPHING A PERIODIC FUNCTION

Imagine the face of a clock centered on an xy-coordinate system. (See Figure 2.) The position of the tip of the second hand can be described in terms of its position on this coordinate system in two ways. We use one function to describe the *vertical* position of the point on the end of the second hand and another to describe the point's *horizontal* position. We use time as an input for both of these functions. The horizontal axis of our graph is labeled 0 to 120 sec because it gives two cycles of the second hand.

First, consider the vertical position function. We assume the second hand is 6 in. long and starts straight up, in the 12 o'clock position. Therefore, at time 0, the point on the end of the hand is as high above the horizontal axis as possible (which is 6 in.). As the second hand moves clockwise, the point goes down for 30 sec (reaching a low point of 6 in. below the horizontal axis), then up for another 30 sec, repeating this process over and over. The points on the graph in Figure 3 represent the vertical distance of the tip of the second hand. They are given for 120 sec (or two revolutions of the second hand) in 5-sec intervals. Notice that the graph reflects the second hand's behavior of being straight up at 0 sec and straight down at 30 sec.

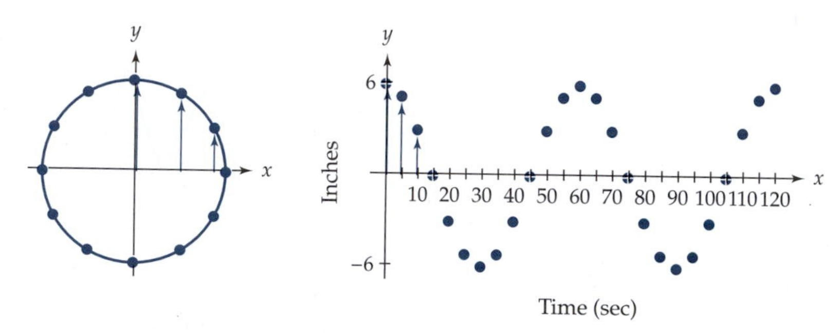

Figure 3 A graphical representation of the vertical position function.

To complete our graph we just need to connect these points with a smooth curve. (See Figure 4.)

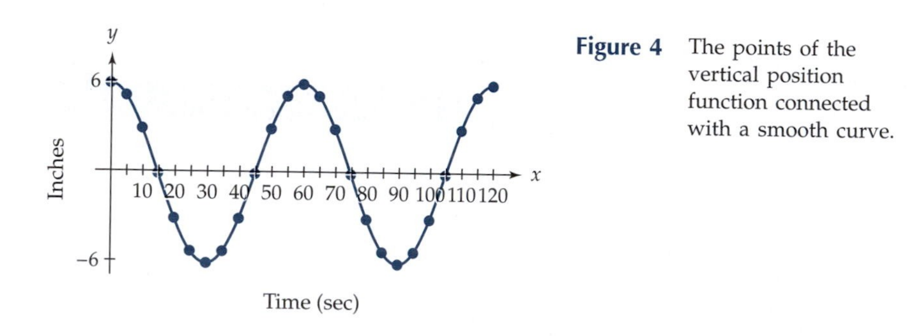

Figure 4 The points of the vertical position function connected with a smooth curve.

Describing the horizontal position of the tip of the second hand on a clock is similar to describing the vertical position. As the hand starts on the 12 and moves clockwise, its horizontal position moves to the right and continues to move to the right until it reaches the 3. It then starts to move to the left and continues past the 6 (the middle of the clock); it stops moving left when it reaches the 9. From the 9 back to the 12, it moves to the right again.

Graphing the function representing the horizontal position is similar to graphing the vertical position. The input is the same, but the output is the horizontal distance the points on the circle are to the right (or left) of the vertical axis. Because we graph this *horizontal* distance on the clock as *vertical* distance on our graph, this process can be tricky. As you can see in Figure 5, the graph of the horizontal distance is similar to the graph of the vertical distance.

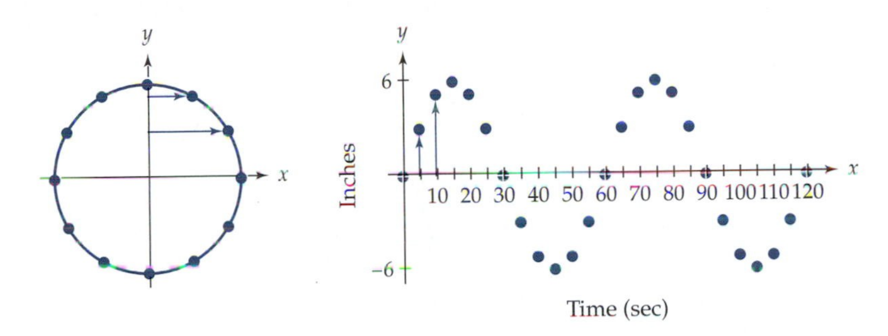

Figure 5 A graphical representation of the horizontal position function.

Graphs of both the vertical and horizontal positions of the tip of a second hand represent periodic functions. These graphs repeat themselves every 60 sec, so the period of both the horizontal position function and the vertical position function is 60. In other words, for any two inputs that are 60 sec apart, the outputs are the same. Later in this section we see how to develop formulas to describe this type of behavior.

READING QUESTIONS

5. What is the period of the function describing the vertical position of the second hand of a clock?
6. Sketch the graph of a vertical position function for a clock if the second hand starts on the 3 instead of the 12.
7. How would the graph of a vertical position function for a clock change if the second hand was 4 in. long instead of 6 in. long?
8. How would the graph of a vertical position function for a clock change if we used the minute hand instead of the second hand?

AMPLITUDE AND PERIOD

Two important words associated with periodic functions are *period* and *amplitude*. Remember that the **period** is the *minimum* fixed distance where the outputs are always the same. The period of a second hand on a clock is 60 sec, the time it takes to complete one revolution. The period is also the time (or distance) it takes the graph to complete one cycle. The start of this cycle can occur at any point on the graph. Regardless of where it starts, however, 60 sec later it returns to the same point. Whereas the period is associated with the input of a function, the amplitude is associated with the output. Suppose a horizontal line is drawn through a graph of a periodic function halfway between the function's maximum and minimum values; the **amplitude** is the distance from this horizontal line to the maximum (or minimum) value. (See Figure 6.) The amplitude can also be defined as $\frac{1}{2}(M - m)$, where M is the function's maximum value and m is the function's minimum value. The amplitude of the second hand function is 6 in., which is the length of the second hand and hence the maximum distance the graph was above and below the horizontal axis.

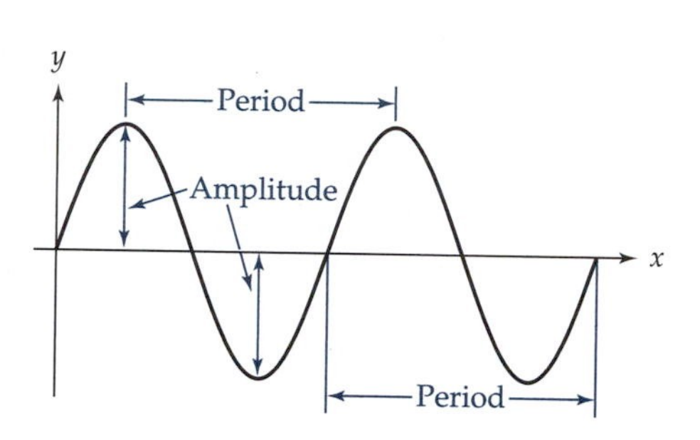

Figure 6 Amplitude and period.

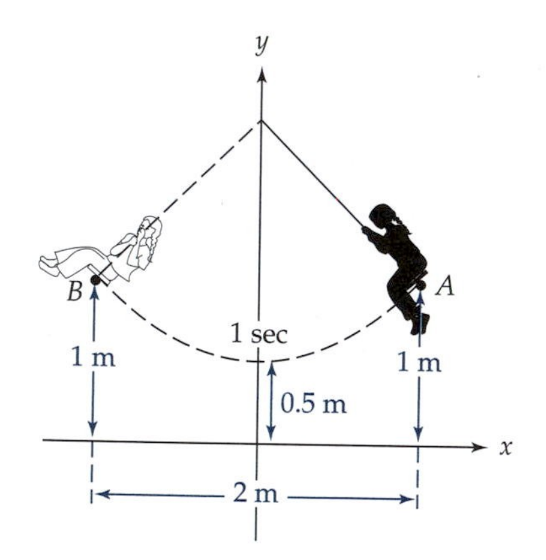

Figure 7 The swing function.

In the clock example, both the horizontal and the vertical functions had the same period and amplitude, which is not always the case. Emma, a daughter of one of the authors, has a small swing in her backyard. As she swings, the functions that describe her position are periodic. Figure 7 shows a drawing of the dimensions of a typical swing path. The lowest point of the swing is 0.5 m, the highest point is 1 m,[22] the horizontal distance traveled is 2 m, and the length of time it takes to swing from point A to point B is 1 sec. An xy-coordinate system is included in Figure 7. The x-axis is at ground level, and the y-axis is where the swing hangs straight down.

In describing our position functions, let's choose our starting point as point A, the point at which the swing is at its highest and farthest to the right. First, we determine the period and amplitude for the vertical position function. As Emma swings forward, she starts 1 m above the ground (the highest position), then is at 0.5 m above the ground (the lowest position), then is back up to 1 m at point B. Because she went from the highest position to the lowest and back to the highest, she completed one period for the vertical position function. This process repeats as she continues to swing. The period for the vertical position function is therefore 1 sec. Because the difference between the swing's highest height and lowest height is $1 - 0.5 = 0.5$ m, the amplitude is half of 0.5, or 0.25 m. The graph of the vertical position function is shown in Figure 8.

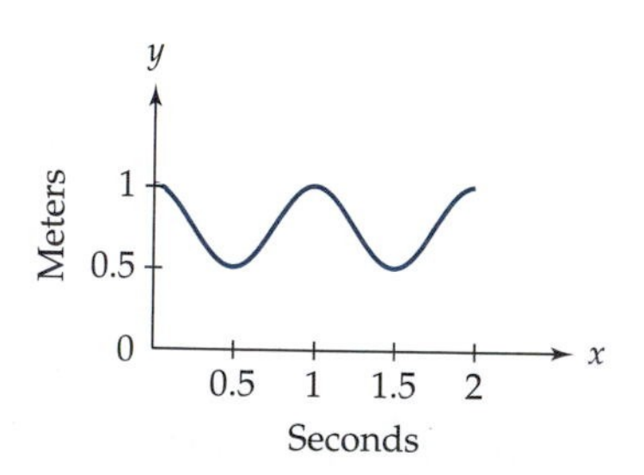

Figure 8 A graphical representation of the vertical position function of the swing.

Both the period and the amplitude are different for the horizontal position function. Instead of looking at the swing's minimum and maximum heights, we need to see when the swing is farthest to the right and farthest to the left. The starting point is again point A. As Emma swings forward, she eventually reaches point B, and as she swings backwards, she returns to point A, constituting one period. This period is 2 sec, twice as long as

[22] *Emma might enjoy going higher, but her parents are concerned about excessive swing amplitude.*

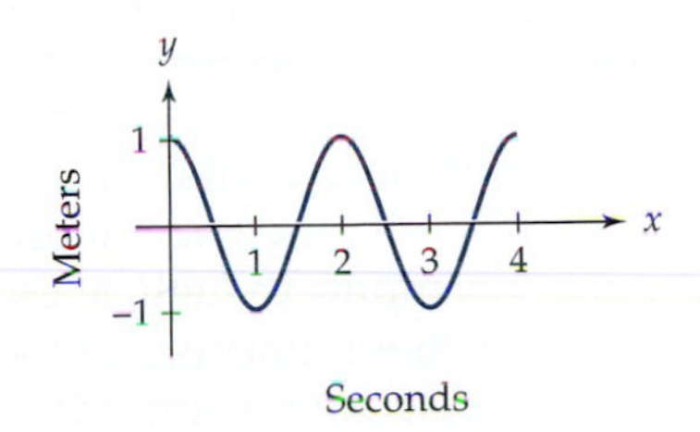

Figure 9 A graphical representation of the horizontal position function of the swing.

that of the vertical function. Because the horizontal distance between when the swing was farthest to the right (or left) and when it is at the center is 1 m, the amplitude for the horizontal position is 1 m. The amplitude always equals one-half of the total distance; in this case, $\frac{1}{2} \cdot 2 = 1$. The horizontal position function is shown in Figure 9.

Graphs of both the vertical and horizontal positions of Emma's swing represent periodic functions. In many respects, they are similar to those in the clock example. For instance, all four graphs have the same basic shape. Both clock graphs, however, are centered at the x-axis, whereas the vertical swing position graph is centered at $y = 0.75$. Also, in the clock example, both the horizontal and vertical position functions have the same amplitude and period, whereas in the swing example, the amplitude and period of the two functions are different.

READING QUESTIONS

9. Explain why the horizontal and vertical position functions of Emma's swing have different periods.
10. What could cause the amplitude of the horizontal position function for Emma's swing to change?
11. What could cause the period of the horizontal position function for Emma's swing to change?

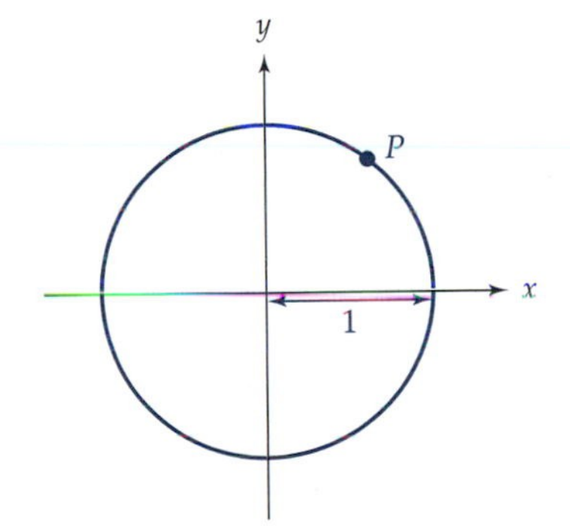

Figure 10 A unit circle with an arbitrary point P.

Figure 11 A unit circle showing the point $(\cos x, \sin x)$.

SINE AND COSINE

We use the sine and cosine functions to describe periodic behavior symbolically. You may be familiar with the right triangle definitions of these functions as $\sin\theta = \text{opposite}/\text{hypotenuse}$ and $\cos\theta = \text{adjacent}/\text{hypotenuse}$. Another way to define them is in terms of a unit circle (a circle with radius 1). These definitions are similar to the vertical and horizontal position functions of our earlier clock example. Hence, the sine and cosine functions are periodic.

Consider an arbitrary point P on a unit circle, centered at the origin as shown in Figure 10. The sine and cosine functions are defined in terms of the position of point P. The input is the distance on the circle from $(1, 0)$ *counterclockwise* to P. In other words, x is the arc length from $(1, 0)$ to the point P. **Sine** of x (abbreviated $\sin x$) is the vertical position of P, and **cosine** of x ($\cos x$) is the horizontal position of P. The coordinates of P are therefore $(\cos x, \sin x)$. (See Figure 11.)

One complete revolution around the circle is 2π. Remember that the formula for the circumference of a circle is $C = 2\pi r$ and because our radius is 1, the circumference of this circle is 2π. The input is the distance counterclockwise from $(1, 0)$, so we often see inputs for sine and cosine as fractions or multiples of π. One-half way around the circle is π, one-quarter way around the circle is $\pi/2$, and so on. Inputs larger than 2π mean that you have gone around the circle more than one revolution. Negative inputs represent arc lengths that proceed clockwise from $(1, 0)$ rather than counterclockwise. Because all the values of the coordinates of the points begin to repeat when we have gone around the circle a distance of 2π, the period of the sine function and of the cosine function is 2π.

Constructing graphs of the sine and cosine functions is similar to constructing graphs for the vertical and horizontal position functions for our clock we referred to earlier. Instead of time as an input, though, these functions have arc length as an input. We also start these functions at the point (1, 0) (3 o'clock on a clock) and proceed counterclockwise rather than clockwise. Because arc length is the input, it makes sense to have the circumference of the unit circle transformed into the x-axis for our graph, as shown in Figure 12.

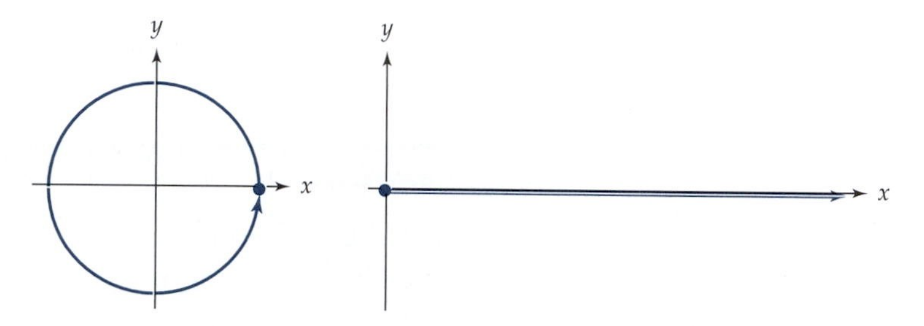

Figure 12 A unit circle transformed into the x-axis.

To graph $y = \sin x$, remember that on a unit circle, $\sin x$ was defined as the y-coordinate of point P. So, $\sin x$ tells us how far P is above (or below) the x-axis. To illustrate, let's plot eight points around the unit circle, which subdivides the distance around the circle, 2π, into eight equal lengths of $\pi/4$. We then transfer the height of these points to our graph and connect the points with a smooth curve, as shown in Figure 13.

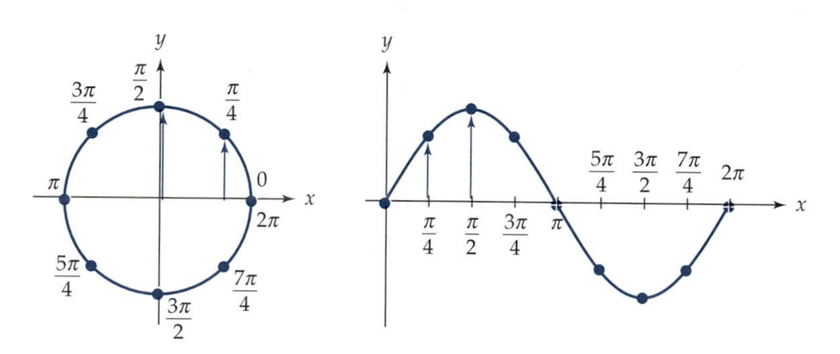

Figure 13 The graphical form of $y = \sin x$.

Similarly, we can determine what a graph of $y = \cos x$ looks like. We again plot eight points around the unit circle. Then we transfer the horizontal distance of each of these points from the y-axis to our graph and connect the points with a smooth curve. (See Figure 14.) The amplitude of $y = \cos x$ is 1 because it is both the maximum and the minimum horizontal coordinate of P. Also, the amplitude equals one-half the difference between the maximum and minimum values, or $\frac{1}{2}(1 - (-1)) = 1$. For the same reasons, the amplitude of $y = \sin x$ is also 1.

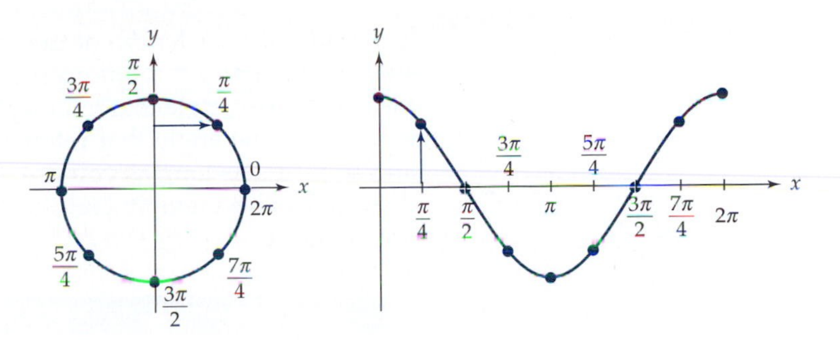

Figure 14 The graphical form of $y = \cos x$.

Notice that all the graphs of periodic behavior are very similar. In fact, the graph of $y = \sin x$ is itself very similar to the graph of $y = \cos x$. In some sense, they are the same graph except that one starts at the origin and the other one has a y-intercept of $(0, 1)$. The shape and behavior of these two functions, $y = \sin x$ and $y = \cos x$, are what makes them the appropriate choice for modeling periodic behavior.

READING QUESTIONS

12. Why is one revolution around a unit circle a distance of 2π?

13. What is the definition of $\sin x$ in terms of a unit circle?

14. What is the definition of $\cos x$ in terms of a unit circle?

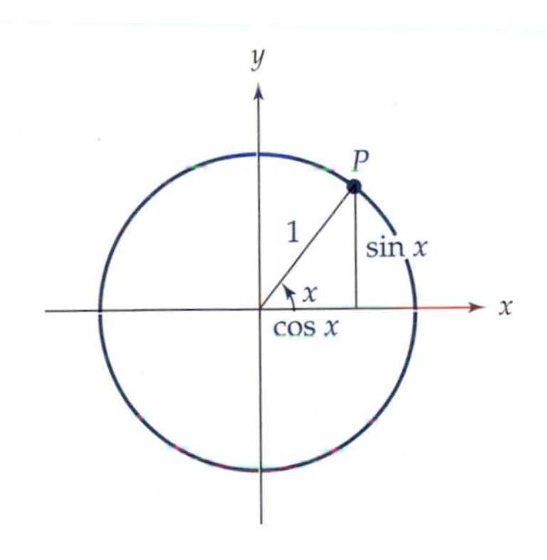

Figure 15 The sine and cosine functions on a right triangle in a unit circle.

THE RELATIONSHIP BETWEEN THE RIGHT TRIANGLE AND THE UNIT CIRCLE DEFINITIONS OF SINE AND COSINE

We just defined sine and cosine as periodic functions that give the coordinates of points on the unit circle. In previous courses, however, these same functions were probably defined in terms of right triangles, where $\sin x$ = opposite/hypotenuse and $\cos x$ = adjacent/hypotenuse. Obviously, the two ways of defining these functions must be related or they would not be called the same thing. To see this relationship, study Figure 15.

The length of the vertical side of the triangle is equal to $\sin x$ (because the length of the hypotenuse is 1), and the length of the horizontal side of the triangle is equal to $\cos x$ (again because the hypotenuse is equal to 1). The lengths of these two lines also give the coordinates of the point P. Obviously, these two definitions are the same. Yet is it really obvious?

Recall that, when thinking of $\sin x$ and $\cos x$ in terms of right triangles, the input is the angle. When thinking of $\sin x$ and $\cos x$ in terms of the unit circle, however, the input is the arc length. So, Figure 15 leads us to believe that the two ways of defining these functions are the same, until we realize that the inputs do not correspond.

The solution is to have the angle equal the arc length on the unit circle, forcing the inputs to be the same. This is exactly why we use radians when working with sine and cosine functions. The **radian** measure of an angle is the length of the arc on the unit circle that is intercepted by that angle. (See Figure 16.)

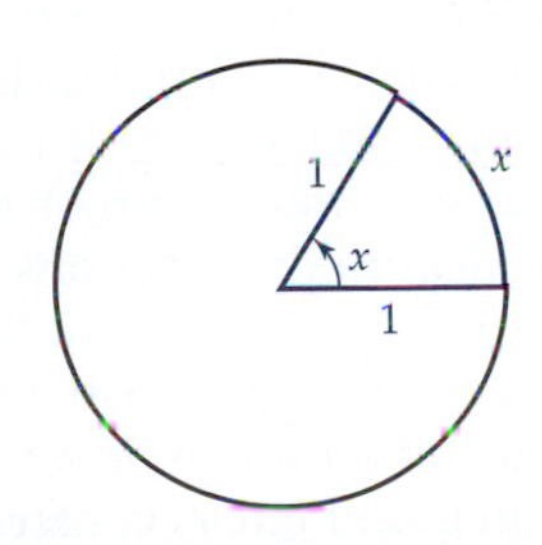

Figure 16 The measure of angle x is x radians.

Because the arc length of the unit circle is $2\pi \cdot 1 = 2\pi$, the radian measure of the angle that intercepts the entire circle is also 2π. The radian measure of the angle that intercepts half of the circle must be π, the radian measure of the angle that intercepts a quarter of the circle must be $\pi/2$, and so forth. As long as you use radians for your input, the right triangle definitions and the unit circle definitions of the sine and cosine functions are the same.

READING QUESTIONS

15. To show that the right triangle definition gives the same function as the unit circle definition of $\sin x$, why it is important to use a *unit* circle and not, for instance, a circle of radius 2?

16. What is meant by the *radian* measure of an angle?

17. Why are radians important?

18. What is the radian measure of the angle shown?

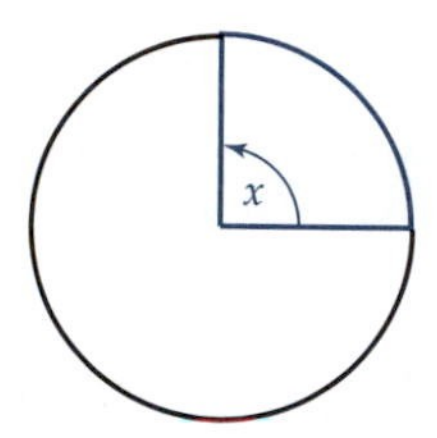

SUMMARY

Repetitious behavior can be described mathematically as being periodic. To be periodic, something must not only repeat itself, but repeat itself at regular intervals. A **periodic function** is a function that gives the same output for inputs a fixed distance apart. Symbolically, we say that f is periodic if for some real number p, $f(x) = f(x+p)$ for all x. The **period** of this function is the smallest value of p for which this relationship is always true; that is, the period is the *minimum* fixed distance where the outputs are always the same.

The period is associated with the input of a function, whereas the amplitude is associated with the output. Suppose a horizontal line is drawn through a graph of a periodic function halfway between the function's maximum and minimum values; the **amplitude** is the distance from this horizontal line to the maximum (or minimum) value. The amplitude can also be defined as $\frac{1}{2}(M - m)$, where M is the function's maximum value and m is the function's minimum value. Because the maximum or minimum value may not be some finite number, not all periodic functions have amplitude.

We often use the sine and cosine functions to describe periodic behavior symbolically. These functions can be defined using a unit circle (a circle with radius 1). Consider an arbitrary point P on a unit circle, centered at the origin. The sine and cosine functions are defined in terms of the position of point P. The input is the distance on the circle from $(1, 0)$ *counterclockwise* to P. In other words, x is the arc length from $(1, 0)$ to the point P. **Sine** of x (abbreviated $\sin x$) is the vertical position of P, and **cosine** of x ($\cos x$) is the horizontal position of P. The coordinates of P are therefore $(\cos x, \sin x)$.

The input for sine and cosine functions are often given as angle measures in radians. The **radian** measure of an angle is the length of the arc on the unit circle intercepted by that angle. One complete revolution around the circle is 2π radians.

EXERCISES

1. Determine if each of the following graphs represents a periodic function. If so, give the period and the amplitude of the function. If not, briefly explain why.

(a)

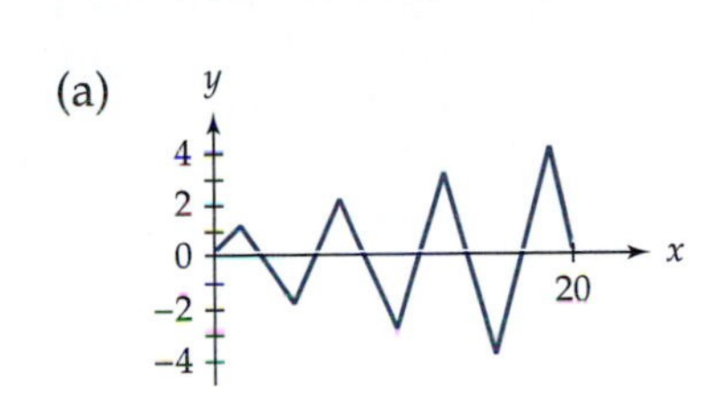

(b)

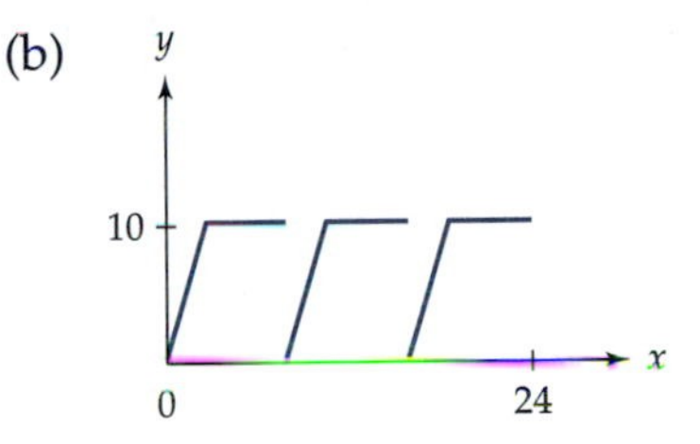

(c)

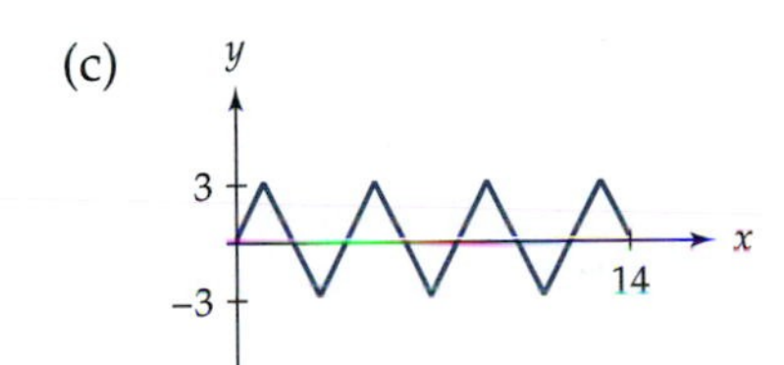

(d)

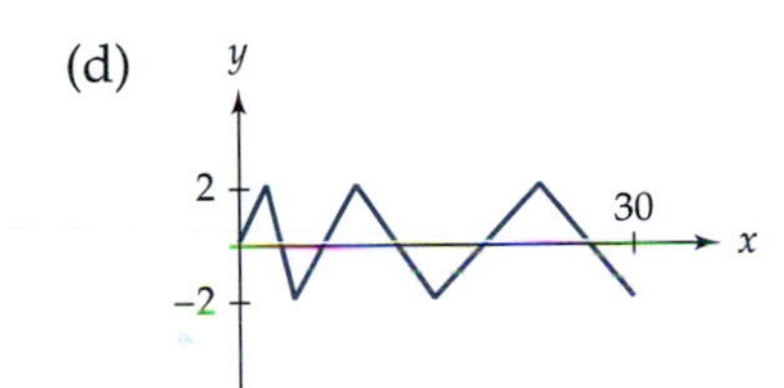

2. Let S be the function whose input is the day of the year and whose output is the total amount of possible daylight in Grand Rapids, Michigan. Explain why S is a periodic function. What is its period?
3. Let f be the periodic function, a portion of which is shown in the accompanying graph. Evaluate each of the following.
   (a) $f(1)$
   (b) $f(5)$
   (c) $f(10)$
   (d) $f(20)$
   (e) $f(-3)$

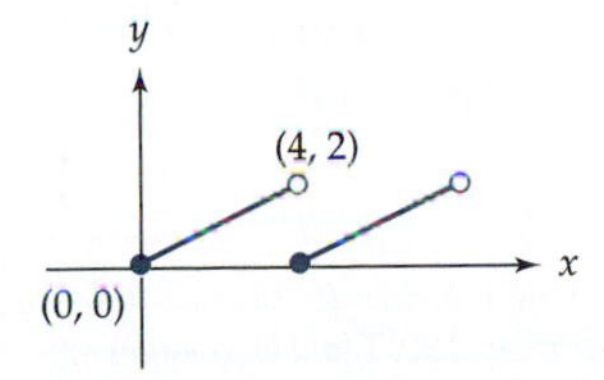

4. In July 1997, a huge Ferris wheel opened up at an amusement park in Osaka, Japan. The wheel is 112.5 m from the ground to the top and has a diameter of 100 m. It holds 60 cabins that seat up to 8 people each. One revolution of the wheel takes 15 min.[23] For the following situations, construct a graph with your height above the ground on the vertical axis and the minutes after the wheel begins to turn on the horizontal axis. Label the period and amplitude in each graph. Assume the wheel revolves continuously and goes slow enough so that you can hop on and off.
   (a) You get on the Osaka Ferris wheel at its lowest point (12.5 m above the ground) and it turns clockwise for two revolutions.
   (b) You get on the Osaka Ferris wheel at its lowest point and it turns clockwise for two revolutions. This time, however, it takes the wheel 20 min to complete one revolution.
   (c) Suppose they enlarge the Osaka Ferris wheel so that its highest point is 150 m from the ground and the wheel has a diameter of 140 m. It still completes one revolution in 15 min. You get on this new wheel at its lowest point and it turns clockwise for two revolutions.
5. How would the graphs of the functions from question 4 change if the Ferris wheel rotated counterclockwise instead of clockwise?
6. What are the period and the amplitude of the periodic function describing the vertical position of the point on the tip of a 4-in.-long minute hand on a clock?
7. What are the period and the amplitude of the periodic function describing the horizontal position of the point on the tip of a 2-in.-long hour hand on a clock?
8. In the text, the amplitude of a periodic function is defined as the distance from a horizontal line to the function's maximum (or minimum) value where the horizontal line lies halfway between the maximum and minimum values. Amplitude is also defined as $\frac{1}{2}(M - m)$, where M is the function's maximum value and m is the function's minimum value. Explain why these two definitions are the same.
9. The child in the swing in the accompanying figure takes 1.5 sec to swing forward and 1.5 sec to swing backward. The maximum height the child is off the ground is 5 ft, and the minimum height is 2 ft. The distance from when the child is farthest to the right to when he or she is farthest to the left is 10 ft.
   (a) Construct a graph of the vertical position of the swing as the child swings for 6 sec if the child starts at the farthest point to the right. What are the period and the amplitude for your graph?
   (b) Construct a graph of the horizontal position of the swing as the child swings for 6 sec if the child starts at the farthest point to the right. What are the period and the amplitude for your graph?

[23] *Japan Information Network, "World's Tallest Ferris Wheel in Osaka,"* Monthly News, August 1997 *(visited 4 November 1997)* ⟨*www.jinjapan.org/kidsweb/news/97-8/wheel.html*⟩.

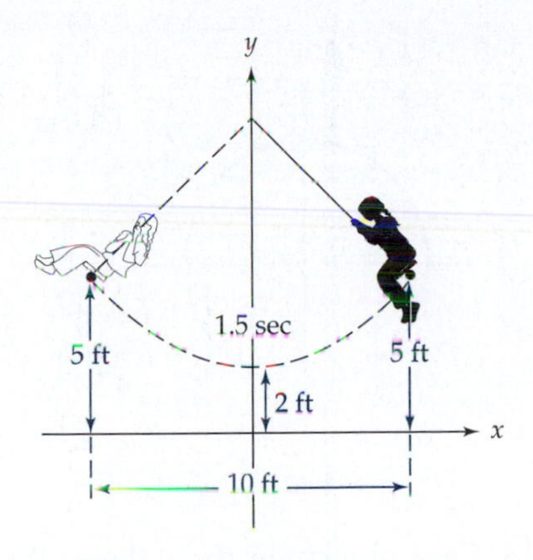

10. Let V be the function whose input is the position of a child in a swing and whose output is his or her height from the ground. If the amplitude of V is 3.5 ft and the period of V is 1, what do you know about the swing?
11. Use the accompanying figure to approximate the values of the sine and cosine functions. Each letter represents the length of the arc beginning at $(1, 0)$ and ending at the indicated point.
    (a) $\cos A$
    (b) $\sin B$
    (c) $\cos C$
    (d) $\sin D$
    (e) $\sin E$
    (f) $\cos F$

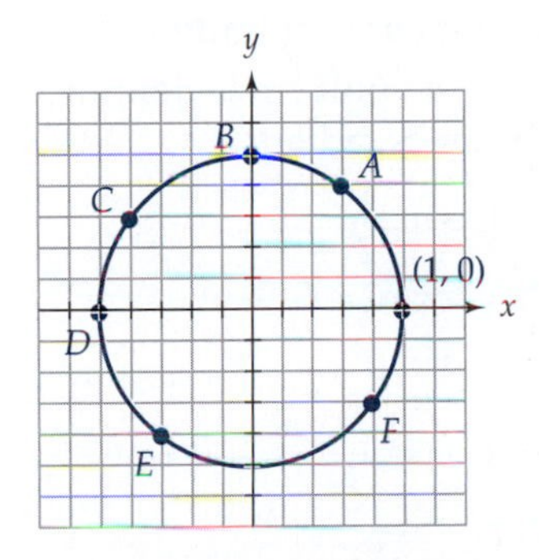

12. A set of axes divides a plane into four quadrants. The quadrants are labeled as shown in the accompanying figure on the next page. In which quadrants are the following statements true?
    (a) $\sin x > 0$ and $\cos x < 0$
    (b) $\sin x < 0$ and $\cos x < 0$
    (c) $\sin x > 0$ and $\cos x > 0$
    (d) $\sin x < 0$ and $\cos x > 0$

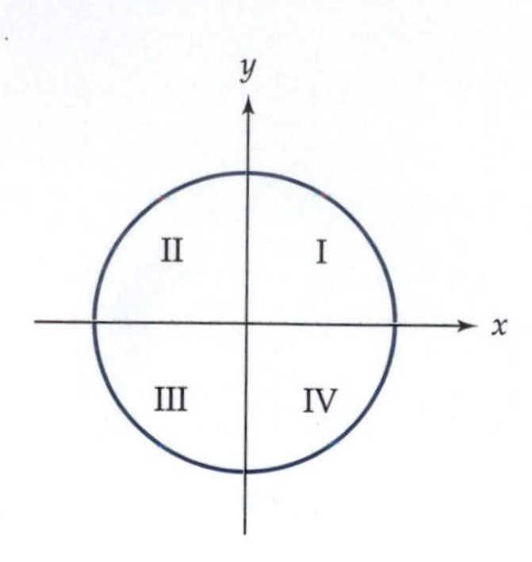

13. On a unit circle, draw the angle whose initial side is the positive x-axis and whose terminal side intersects the circle to form an angle with the radian measure given below.
    (a) π
    (b) $\frac{\pi}{4}$
    (c) $\frac{3\pi}{2}$
    (d) $\frac{5\pi}{4}$
    (e) $-\frac{\pi}{4}$
    (f) $\frac{7\pi}{3}$
14. Explain why it is important to measure an angle in radians when using the unit circle definitions of $\sin x$ and $\cos x$.
15. Using the unit circle definitions, evaluate the following functions. The input is in radians.
    (a) $\sin \frac{\pi}{2}$
    (b) $\cos 2\pi$
    (c) $\sin 0$
    (d) $\cos \frac{3\pi}{2}$
    (e) $\cos \pi$
16. Angles can be measured in either degrees or radians.
    (a) Explain why the function that converts from degrees to radians is a linear function.
    (b) There are 360° in the angle that goes completely around the unit circle and 2π radians in the angle that goes completely around the unit circle. Using this information, determine the conversion factor that converts from degrees to radians.
    (c) Give the equation of the function that converts from degrees to radians.

(d) Convert each of the following degrees to a radian measure.
   i. 90°
   ii. 45°
   iii. 60°
   iv. 180°

17. Using the unit circle definition of $\sin x$ and $\cos x$, explain why $(\sin x)^2 + (\cos x)^2 = 1$.

INVESTIGATIONS

INVESTIGATION: DAYLIGHT

One of the cyclic functions in this section has its input as the day of the year and its output as the maximum possible hours of daylight. We are interested in determining mathematical properties of this function and exploring how they are related to what you know about the seasons.

The following table gives the minutes of daylight for every 10th day (where Day 1 is January 1st) for Grand Rapids, Michigan.

Day	Hours	Minutes	Total Minutes	Day	Hours	Minutes	Total Minutes
1	9	5	545	191	15	11	911
11	9	16	556	201	14	56	896
21	9	33	573	211	14	37	877
31	9	55	595	221	14	14	854
41	10	20	620	231	13	49	829
51	10	47	647	241	13	23	803
61	11	13	673	251	12	54	774
71	11	42	702	261	12	26	746
81	12	11	731	271	11	57	717
91	12	40	760	281	11	28	688
101	13	9	789	291	10	59	659
111	13	36	816	301	10	32	632
121	14	3	843	311	10	6	606
131	14	27	867	321	9	43	583
141	14	49	889	331	9	24	564
151	15	5	905	341	9	9	549
161	15	16	916	351	9	2	542
171	15	21	921	361	9	2	542
181	15	19	919				

1. Graph the points from the table with the x-axis representing the day of the year (numbered 1 through 365) and the y-axis representing the number of minutes of daylight.
2. Why is this function periodic?

3. What day had the most daylight? What day had the least daylight? Do these answers correspond with your experience and your knowledge of the seasons? Explain.
4. Looking at the table, find the time of year when the amount of daylight is changing most rapidly. When is it changing the least? How does this answer correspond with your experience and your knowledge of the seasons? Explain.
5. What is the amplitude of your function? What does this number mean in terms of number of minutes of daylight?
6. What is the period of your function?

2.5 POWER FUNCTIONS

Functions that describe the position of an accelerating car or the formula for the area of a circle are examples of power functions. These types of functions have exponents, but unlike exponential functions, the exponent is not the input. In this section, we explore characteristics of power functions with positive exponents and how to determine if something is a power function.

DEFINITIONS AND EXAMPLES

If you are driving a car at a constant velocity, the function describing your distance traveled over time is a linear function because your rate is constant. Suppose, however, you are accelerating. Then, your velocity (rate of change) is not constant, so the function describing your distance traveled over time would *not* be linear. Now let's say you stop at a stop sign and then accelerate at a constant rate of 8 ft/sec^2 for 10 sec. Accelerating at this rate means that your velocity will increase 8 ft/sec for each second you continue traveling. (With this acceleration rate, you reach a velocity of approximately 55 mph after 10 sec.) The function that describes your distance from the stop sign over time can be represented by the equation

$$d(t) = 4t^2,$$

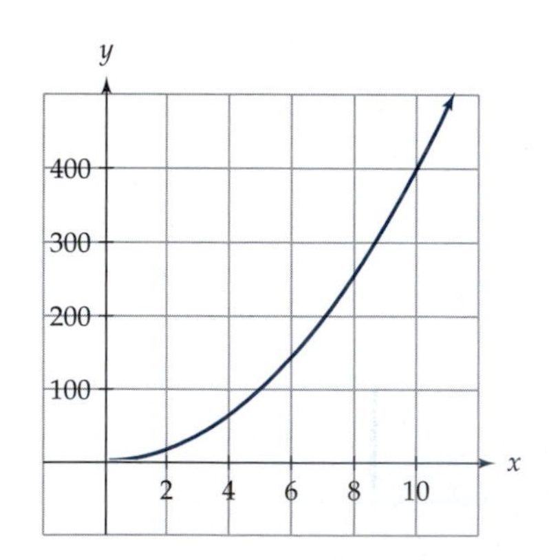

Figure 1 Distance traveled (in feet) for a given time (in seconds) with a constant acceleration of 8 ft/sec^2.

where $d(t)$ is the distance traveled (in ft) after t sec. The graph of this function is shown in Figure 1. It is obviously not a linear function. The rate of change is not constant, the graph does not look like a straight line, and the symbolic representation of the function is not of the form $f(t) = mt + b$. Instead, it is an example of a power function.

A **power function** is a function of the form $f(x) = bx^a$. In this section, we look only at power functions where the exponent is a positive rational number. We start by assuming that $b = 1$, so we first consider power functions of the form $f(x) = x^a$. You should already be familiar with some power functions, such as $f(x) = x^2$ and $g(x) = x^3$. The graphs of these two power functions are shown in Figure 2.

Compare the graphs of $f(x) = x^2$ and $g(x) = x^3$. Overall, these graphs look quite different, yet they have a similar shape in the first quadrant. We

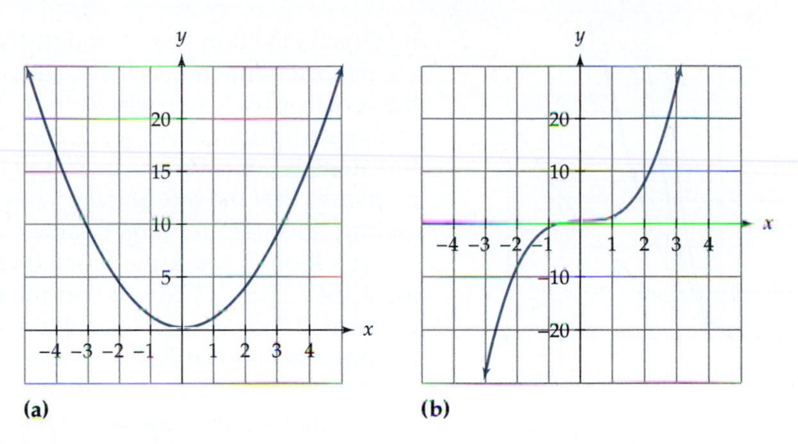

Figure 2 The basic power function shapes. (a) $f(x) = x^2$; (b) $g(x) = x^3$.

want to explore why the graphs look different (yet similar) and how the behavior of a power function depends on the value of its exponent. Power functions with negative exponents as well as polynomials are discussed in Section 4.3.

READING QUESTIONS

1. What is the main difference between something that can be modeled with a linear function and something that can be modeled with a power function?
2. How can you tell if a function given to you in symbolic form is a power function?

POSITIVE INTEGER EXPONENTS

Let's start by exploring the connections between the equation, the shape of the graph, and the behavior of the data of functions of the form $f(x) = x^a$, where a is a positive integer. First, we focus on positive (or more precisely, nonnegative) inputs. Table 1 contains data for four different power functions. What can we conclude from looking at Table 1? It's fairly noticeable

TABLE 1 Comparing Power Functions Using Nonnegative Inputs

x	$y = x^2$	$y = x^3$	$y = x^4$	$y = x^5$
0	0	0	0	0
1	1	1	1	1
2	4	8	16	32
3	9	27	81	243
4	16	64	256	1024

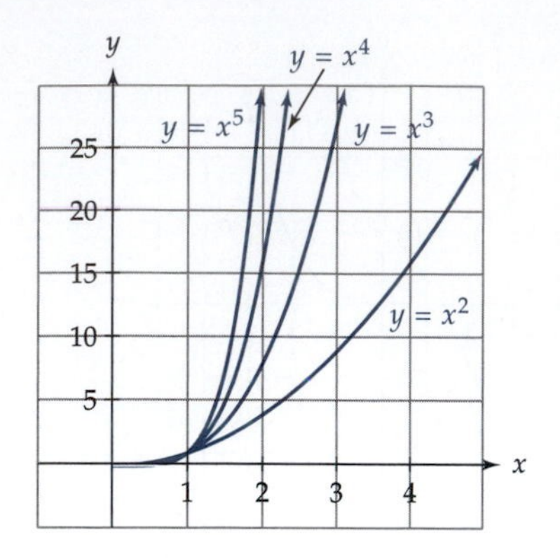

Figure 3 Comparing the graphs of power functions using nonnegative inputs.

that each function goes through the points $(0,0)$ and $(1,1)$. We can also see that each function is increasing, and the larger the exponent, the faster the function increases when $x > 1$. These properties are seen in the graph of these four functions; see Figure 3. It is also obvious that each of these functions is concave up, at least in the first quadrant. Recall that concave up means that the rate of increase is increasing. In this case, the graphs are getting "steeper" for larger values of x.

Let's look at the same four functions, this time using negative inputs for x. (See Table 2.) Notice that the behavior of the functions is no longer uniform. If the exponent is odd, a negative input gives a negative output. If the exponent is even, however, a negative input gives a positive output. The graphs of the four functions are shown in Figure 4. Notice that the functions with even exponents are above the x-axis and are concave up, whereas the functions with odd exponents are below the x-axis and are concave down, depending on whether the negative inputs were multiplied an even or an odd number of times. A graph of all four power functions with inputs of $-4 \leq x \leq 4$ is shown in Figure 5.

TABLE 2 Comparing Power Functions Using Negative Inputs

x	x^2	x^3	x^4	x^5
−4	16	−64	256	−1024
−3	9	−27	81	−243
−2	4	−8	16	−32
−1	1	−1	1	−1

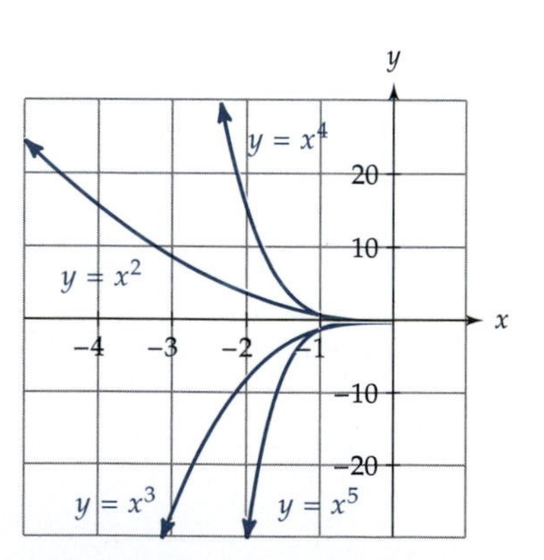

Figure 4 Comparing the graphs of power functions using negative inputs.

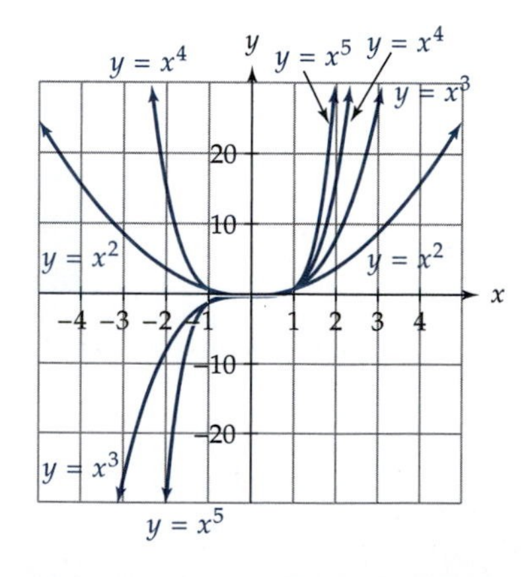

Figure 5 Comparing the graphs of power functions with integer exponents.

PROPERTIES

The following is a list of properties of $f(x) = x^a$, where a is a positive integer.

1. Points in common:
   - $f(0) = 0$
   - $f(1) = 1$
2. Domain and range:
   - If a is even, then the function has a domain of all real numbers and a range of $f(x) \geq 0$.
   - If a is odd, then the function has a domain of all real numbers and a range of all real numbers.
3. Concavity:
   - If a is even, then the function is always concave up.
   - If a is odd, then the function is concave down for $x < 0$ and concave up for $x > 0$.
4. Symmetry:
   - If a is even, then the function is symmetric with respect to the y-axis.[24]
   - If a is odd, then the function is symmetric with respect to the origin.[25]
5. For $|x| > 1$, the larger the exponent, the faster the function is increasing/decreasing and the steeper the graph.

Although we only looked at four functions to generate this list of properties, these properties are true for *all* functions of the form $f(x) = x^a$, where a is a positive integer. Each property here is a consequence of the properties of exponents. For example, $f(0) = 0^a = 0$ regardless of the value of a because zero multiplied by itself will always be zero.

READING QUESTIONS

3. The following graphs are that of $y = x^4$, $y = x^5$, $y = x^{10}$, and $y = x^{15}$. Identify which is which and briefly justify your answer.

(a)

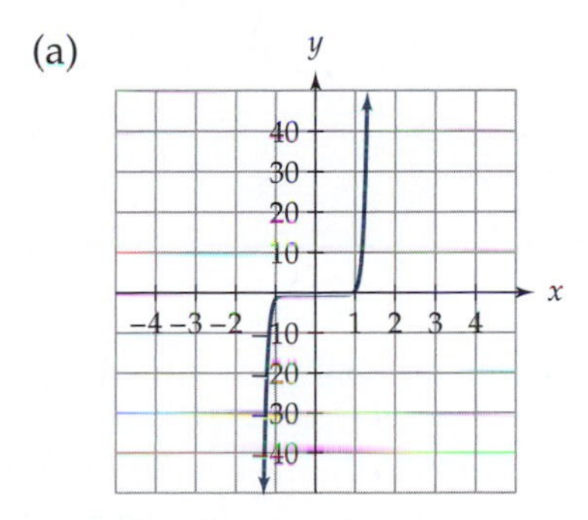

(b)

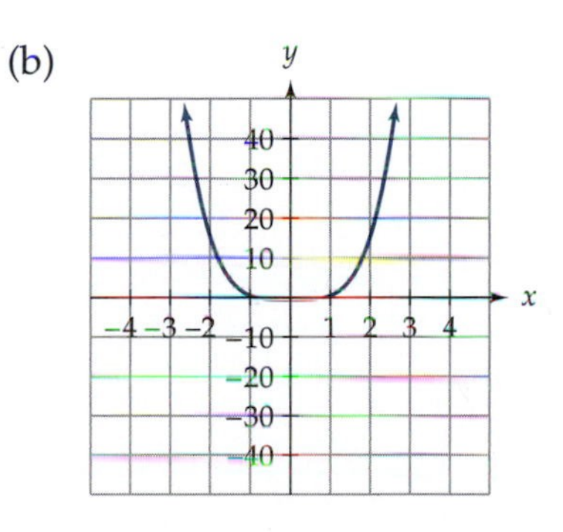

[24] *That is, if you fold the graph in half along the y-axis, the two pieces coincide.*

[25] *That is, if you rotate the graph 180° around the origin, it looks the same.*

(c)

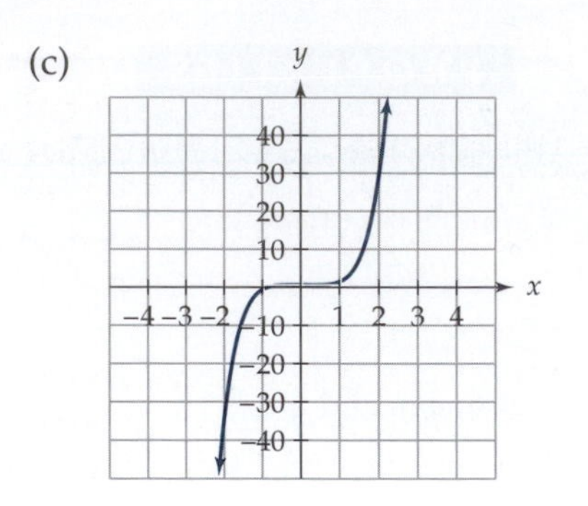

(d)

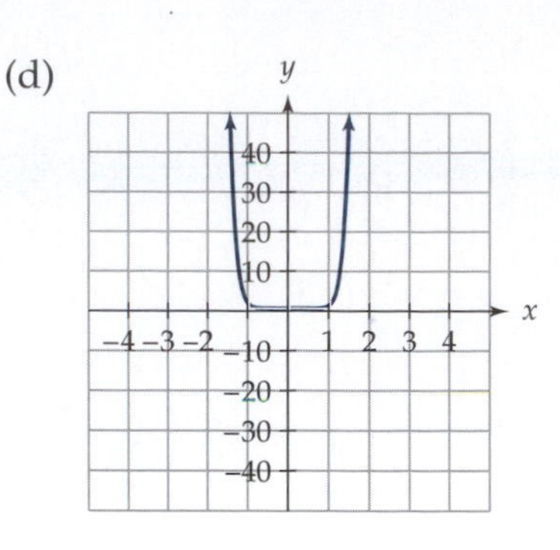

4. Let $f(x) = x^a$, where a is a positive integer. Explain why it is always true that $f(1) = 1$.
5. Let $f(x) = x^a$, where a is a positive integer. What is the relationship between the value of a and the value of $f(-1)$?
6. What happens to the concavity of power functions with an odd exponent at the point (0, 0)?
7. What happens to the concavity of power functions with an even exponent at the point (0, 0)?

POSITIVE RATIONAL EXPONENTS

We know how power functions behave when the exponent is a positive integer. Now let's allow the exponent to include positive rational numbers. Recall that a rational number is one that can be written as a fraction. We are again interested in making connections between the symbolic formula, the shape of the graph, and the behavior of the data. We start by using positive inputs. Table 3 contains data (rounded to two decimal places) for five power functions with rational exponents using positive inputs.

TABLE 3 Comparing Power Functions with Rational Exponents Using Nonnegative Inputs

x	$y = x^{1/3}$	$y = x^{1/2}$	$y = x^{2/3}$	$y = x^{3/2}$	$y = x^{5/3}$
0	0	0	0	0	0
1	1	1	1	1	1
2	1.26	1.41	1.59	2.83	3.17
3	1.44	1.73	2.08	5.20	6.24
4	1.58	2	2.52	8	10.08

Once again, all these functions have the points $(0, 0)$ and $(1, 1)$ in common. It is also true that these functions are increasing and that the larger the exponent, the faster the increase. The first three functions, $y = x^{1/3}$, $y = x^{1/2}$, and $y = x^{2/3}$, increase quite slowly. (Notice that their outputs are equal to or smaller than their inputs.) In fact, the increase for these three

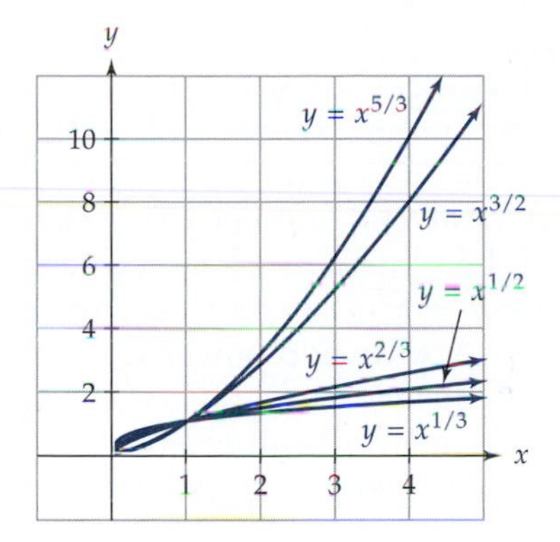

Figure 6 Comparing the graphs of power functions with rational exponents using nonnegative inputs.

seems to be "slowing down." For example, look at $y = x^{1/3}$; it increases by 0.26 unit between $x = 1$ and $x = 2$ and only increases 0.14 unit between $x = 3$ and $x = 4$. On the other hand, the graphs of $y = x^{3/2}$ and $y = x^{5/3}$ are getting steeper and steeper, the same behavior we noticed when we looked at exponents that were positive integers. The two different types of behavior are demonstrated in the graphs of the five functions shown in Figure 6.

Notice in Figure 6 that the three functions that were increasing slowly and whose increase was "slowing down" are concave down. Recall that the definition of concave down is a function whose rate of change is decreasing. The other two functions resemble the graphs of power functions with positive integer exponents. Both of these are concave up. Power functions with rational exponents are concave down in the first quadrant if the exponent is less than 1. This fact should be somewhat intuitive because these functions, in general, increase slower than the linear function $y = x$. (The function $y = x$ can be thought of as the power function $y = x^1$).

Now that we know what happens with power functions with rational exponents when the inputs are positive, let's look at these same functions using negative inputs. (See Table 4.) This time, a variety of things are happening. Some columns have negative outputs, one column has positive outputs, and some columns do not have any outputs at all. In the columns of Table 4 that do have outputs, notice that the numbers are the same (with the exception of some negative signs) as the outputs in Table 3. A graph of these functions is shown in Figure 7. Only three functions are shown in this figure because two of them, $y = x^{1/2}$ and $y = x^{3/2}$, do not have any outputs for negative inputs.

TABLE 4 Comparing Power Functions with Rational Exponents Using Negative Inputs

x	$y = x^{1/3}$	$y = x^{1/2}$	$y = x^{2/3}$	$y = x^{3/2}$	$y = x^{5/3}$
−1	−1	—	1	—	−1
−2	−1.26	—	1.59	—	−3.17
−3	−1.44	—	2.08	—	−6.24
−4	−1.58	—	2.52	—	−10.08

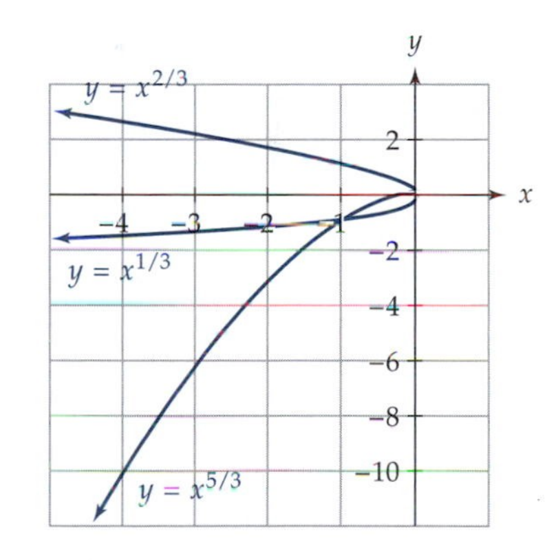

Figure 7 Comparing the graphs of power functions with rational exponents using negative inputs.

These functions are behaving this way because of, once again, properties involving exponents. For example, recall that $x^{1/2} = \sqrt{x}$. You cannot take a square root of a negative number because there is no real number that, when squared, gives a negative answer. In fact, any even root of a negative number does exist because there is no real number that, when raised to an even power, gives a negative answer. So, $y = x^{1/2}$ and $y = x^{3/2}$ are undefined for negative inputs, and their domains consist of $x \geq 0$.

Some roots, however, are defined for negative inputs. For example, cube roots are defined for negative values of x. Thus, functions like $y = x^{1/3}$, which can be written as $y = \sqrt[3]{x}$, are defined for all real numbers. This fact is true for any odd root. Recall that $x^{n/d}$ can be written as $(\sqrt[d]{x})^n$. Notice that $y = x^{5/3}$ gave negative outputs because $y = x^{5/3} = (\sqrt[3]{x})^5$; so, after taking the cube root, you are raising your answer to an odd power. On the other hand, $y = x^{2/3}$ gave positive outputs because $y = x^{2/3} = (\sqrt[3]{x})^2$; so, after taking the cube root, you are raising your answer to an even power. Remember that raising your answer to an even integer power makes your answer positive, whereas raising your answer to an odd integer power makes your answer negative.

Although we looked at only five functions to observe these properties, they are true for all power functions with positive rational exponents. Let's summarize what we know about these type of functions.

PROPERTIES

The following is a list of properties for $f(x) = x^a$ where $a = n/d$ is a positive rational number written in simplified form.

1. Points in common:
   - $f(0) = 0$
   - $f(1) = 1$
2. Domain:
   - If the denominator of the exponent is even, then the domain is $x \geq 0$.
   - If the denominator of the exponent is odd, then the domain is all real numbers.
3. Range:
   - If the numerator is even, then the range is $f(x) \geq 0$.
   - If the numerator is odd and the denominator is odd, then the range is all real numbers.
   - If the numerator is odd and the denominator is even, then the range is $f(x) \geq 0$.
4. Concavity for $x \geq 0$ (first quadrant):
   - If $a > 1$, then the function is concave up (and gets steeper).
   - If $a < 1$, then the function is concave down (and gets flatter).
5. Symmetry:
   - If the denominator of the exponent is even, then the function is only defined for $x \geq 0$ and there is no symmetry.
   - If the denominator of the exponent is odd and the numerator is even, then the function is symmetric with respect to the y-axis.
   - If the denominator and the numerator are both odd, then the function is symmetric with respect to the origin.

TECHNOLOGY TIP

Power functions with fractional exponents are sometimes defined on a domain that includes all the real numbers and sometimes on a domain that includes only positive values, depending on the exponent. Because of the way graphing calculators plot curves, they may give you an incorrect domain for a power function. For example, graph $y = x^{3/5}$ on your calculator with a domain of $-10 \le x \le 10$. Many calculators, like the TI-82, give you a graph similar to this:

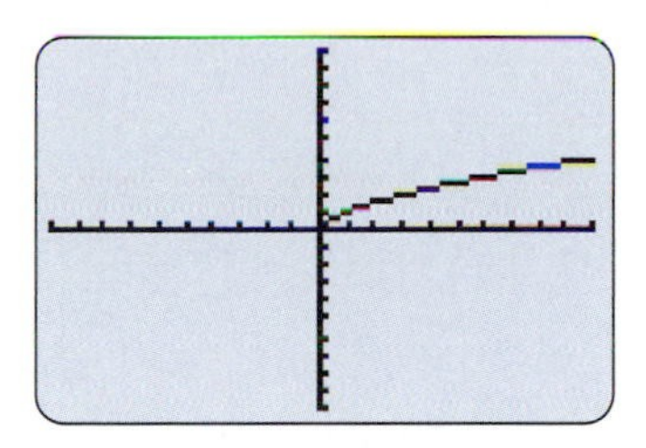

Your calculator has assumed the domain is $x \geq 0$. Yet, because the denominator of the exponent is odd, the domain is actually all real numbers. It is important to think about the appropriate domain before graphing with a machine. You may be able to correct the problem with the domain by entering the function into your calculator in the form $y = (x^{1/5})^3$. Give it a try!

READING QUESTIONS

8. The following graphs are of $y = x^{1/4}$, $y = x^{3/5}$, $y = x^{4/5}$, $y = x^{6/5}$, and $y = x^{9/5}$. Which is which?

(a)

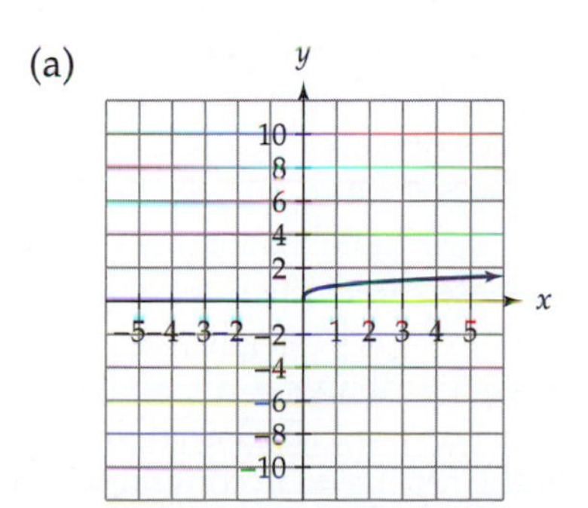

(b)

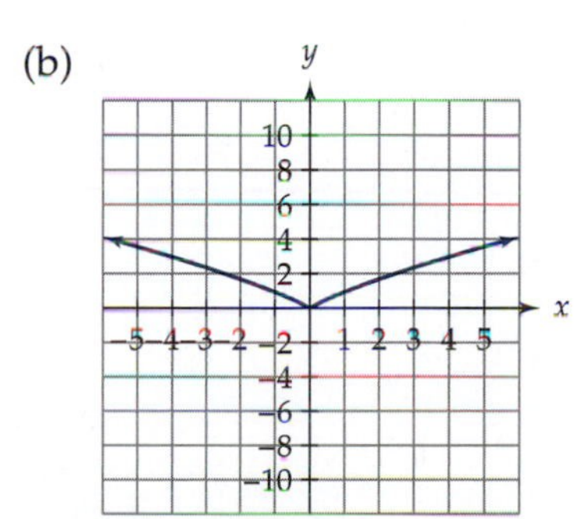

(c)

(d)

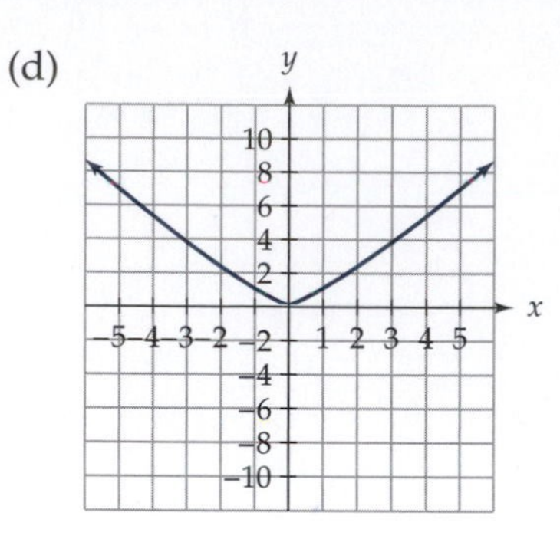

(e)

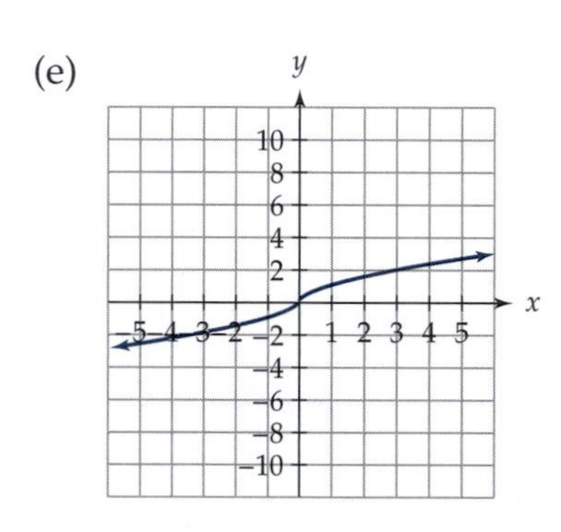

9. Let $f(x) = x^a$, where a is a positive rational number. Why is it true that $f(0) = 0$ and $f(1) = 1$?

10. Give the domain and range for each of the following.

(a) $y = x^{3/4}$

(b) $y = x^{9/8}$

(c) $y = x^{2/7}$

(d) $y = x^{13/7}$

11. You are given a power function whose graph is concave up in the first quadrant. What do you know about the exponent?

12. You are given a power function whose graph is concave down in the first quadrant. What do you know about the exponent?

OTHER VALUES FOR b

So far, we have only looked at power functions where $b = 1$, that is, power functions of the form $f(x) = x^a$. What happens, however, if $b \neq 1$, that is, if $f(x) = bx^a$? For now, we restrict ourselves to positive values of b. Negative values of b are explored in Section 3.1.

Let's start to answer this question by comparing $y = x^2$, $y = \frac{1}{2}x^2$, and $y = 4x^2$. Table 5 contains data for these three power functions. The graph of these three functions is shown in Figure 8. Notice that the basic shape of the three graphs is the same. They all look like parabolas, although $y = 4x^2$ is steeper than $y = x^2$, and $y = \frac{1}{2}x^2$ is not quite as steep. Notice what happens for $x = 1$. Earlier in this section, we observed that *all* power functions of the form $y = x^a$ contain the point $(1, 1)$. That is no longer true for power

TABLE 5 Comparing $y = x^2$, $y = \frac{1}{2}x^2$, and $y = 4x^2$

x	x^2	$\frac{1}{2}x^2$	$4x^2$
−2	4	2	16
−1	1	0.5	4
0	0	0	0
1	1	0.5	4
2	4	2	16

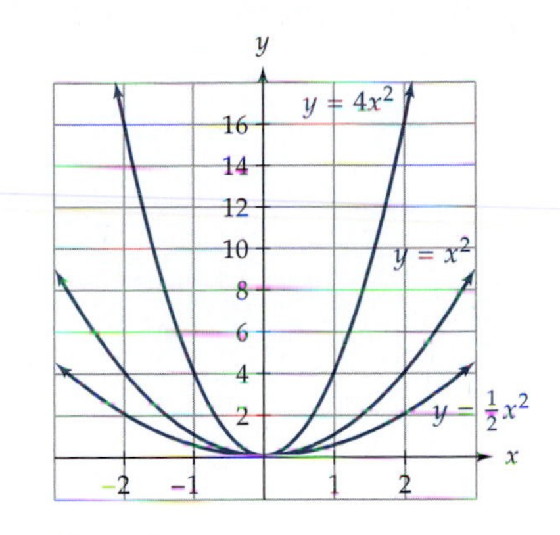

Figure 8 Comparing graphs of $y = x^2$, $y = \frac{1}{2}x^2$, and $y = 4x^2$.

functions of the form $y = bx^a$. The function $y = x^2$ still contains the point $(1, 1)$, but $y = \frac{1}{2}x^2$ contains the point $(1, \frac{1}{2})$ and $y = 4x^2$ contains the point $(1, 4)$. The behavior of the functions at $x = 0$, however, has not changed. All three functions still go through the point $(0, 0)$.

Let's look at why $f(x) = bx^a$ for $b \neq 1$ changes the behavior of the power function for $x = 1$ but does not affect the behavior for $x = 0$. The answer is quite simple: just evaluate $f(x) = bx^a$ for both $x = 1$ and $x = 0$. We have $f(1) = b \cdot 1^a = b \cdot 1 = b$ because 1 raised to any power is still 1. So, power functions of the form $f(x) = bx^a$ contain the point $(1, b)$, which makes finding the value of b from a graph or a table of numbers easy: just look at the output for $x = 1$. To see what happens when $x = 0$, notice that $f(0) = b \cdot 0^a = b \cdot 0 = 0$. So, *every* power function contains the point $(0, 0)$, regardless of the value of b.

Other properties of power functions are not affected by the value of b (as long as b is positive). Our earlier discussion of domain and range, concavity, and symmetry still holds true. The important difference in using a value of b other than 1 is the following.

PROPERTIES

Points in common for $f(x) = bx^a$:

- $f(1) = b$
- $f(0) = 0$

READING QUESTIONS

13. What impact does the b have on the graph of $y = bx^a$?

14. The tables and graphs on the next page represent power functions of the form $y = bx^a$. Give the value of b for each.

(a)

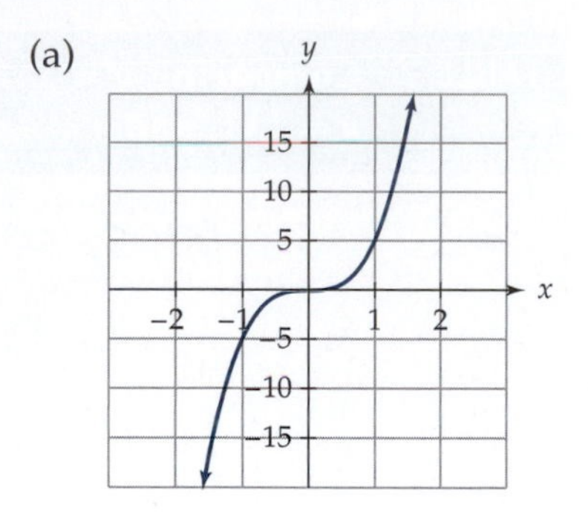

(b)

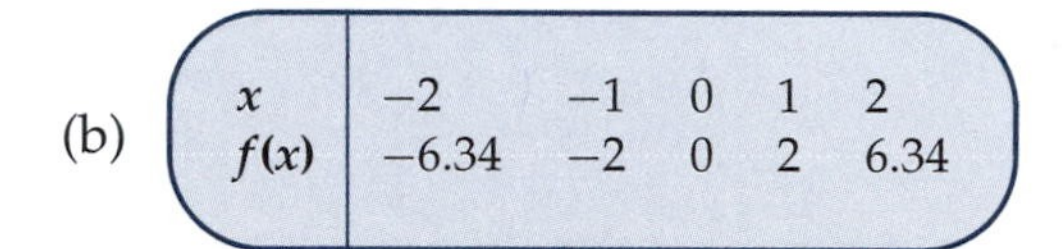

x	−2	−1	0	1	2
$f(x)$	−6.34	−2	0	2	6.34

(c)

x	−2	−1	0	1	2
y	—	—	0	3	8.49

DETERMINING IF A FUNCTION IS A POWER FUNCTION

Suppose we want to determine if a given function is a power function. How do we do this? If it is a symbolic formula, the answer is fairly simple: just see whether or not the formula is of the form $f(x) = bx^a$. For example, $f(x) = 32x^{3/2}$ is a power function, whereas the function $g(x) = 32^x$ is an exponential function because the input is an exponent.

When given a table of data or a graph, it may not be easy to determine if it represents power functions because the behavior of power functions depends on the exponent and can look similar to the behavior of other types of functions. The key to determining if a table of data or a graph does represent a power function is to remember the properties described earlier in this section. For example, Figure 9 consists of a function in numerical form (entries in the table are rounded to two decimal places) as well as a

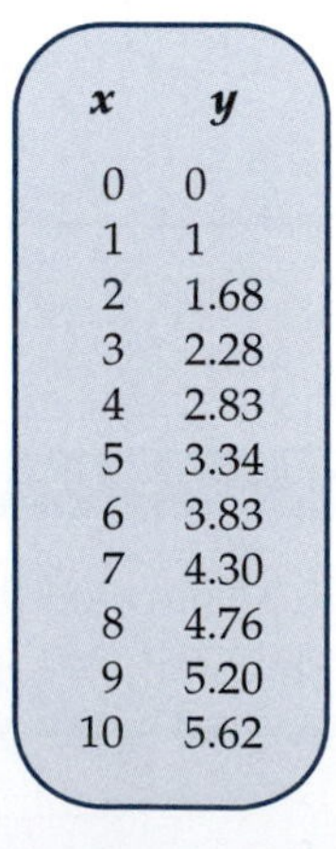

x	y
0	0
1	1
2	1.68
3	2.28
4	2.83
5	3.34
6	3.83
7	4.30
8	4.76
9	5.20
10	5.62

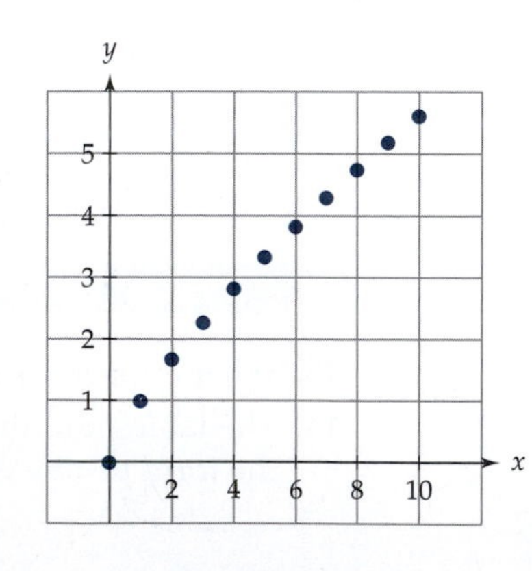

Figure 9 A table of data with a corresponding graph of the points.

graph of that function. We want to determine if this represents a power function.

The function shown in Figure 9 contains the two points $(0, 0)$ and $(1, 1)$, so it is plausible that it could be represented symbolically by $y = x^a$ for some appropriate value of a. Notice that the graph is concave down. If it is a power function, the exponent must be less than 1. Let's assume a power function of the form $f(x) = x^a$ and see how well our answer fits the data. To determine the value of the exponent, choose one point from your table of data. You do not want to choose either the point $(0, 0)$ or $(1, 1)$ because these points are common to every power function of the form $y = x^a$. Let's use the point $(2, 1.68)$. We need to determine what power of 2 gives us 1.68; that is, we need to find a such that $2^a = 1.68$. We can approximate this exponent by the "guess and check" method. Start by choosing a reasonable exponent, determine if your answer is too low or too high, and pick your next guess accordingly. Table 6 shows how we guessed the exponent for this problem.

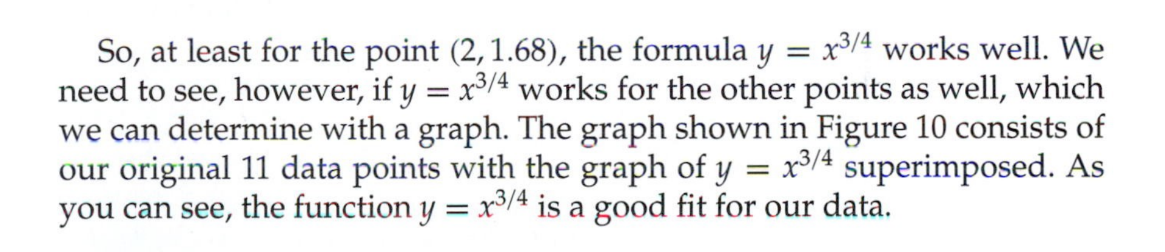

TABLE 6 Using the "Guess and Check" Method to Solve $2^a = 1.68$

Guess for a	2^a	Too Low/Too High
0.5	1.41	Too low
0.7	1.62	Too low
0.8	1.74	Too high
0.75	1.68	It works!

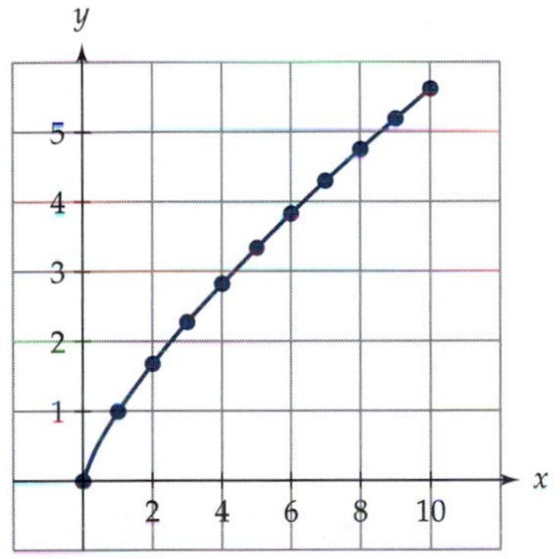

Figure 10 A graph of the original points with the function $y = x^{3/4}$ superimposed.

So, at least for the point $(2, 1.68)$, the formula $y = x^{3/4}$ works well. We need to see, however, if $y = x^{3/4}$ works for the other points as well, which we can determine with a graph. The graph shown in Figure 10 consists of our original 11 data points with the graph of $y = x^{3/4}$ superimposed. As you can see, the function $y = x^{3/4}$ is a good fit for our data.

Example 1 Determine if the data and graph in Figure 11 represent a power function. If so, give the formula. If not, explain why.

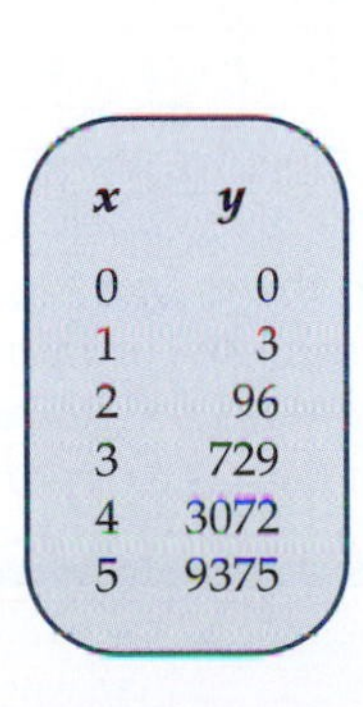

x	y
0	0
1	3
2	96
3	729
4	3072
5	9375

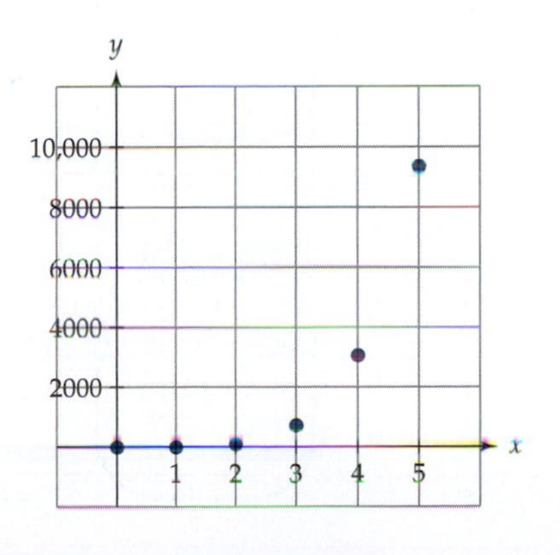

Figure 11

Solution At first glance, it appears that this function cannot be a power function because it does not contain the point $(1, 1)$. It certainly cannot be a function of the form $y = x^a$ because, regardless of a, this function *always* contains the point $(1, 1)$. It could, however, be of the form $y = bx^a$. This type of power function would still contain the point $(0, 0)$, but it would go through the point $(1, b)$ rather than $(1, 1)$. Because our function has the point $(1, 3)$, b would have to be 3 if we truly have a power function. Looking at the graph (or the table of data), it is obvious that the exponent will be greater than 1. Let's use the point $(2, 96)$ to find a. Once again, we use the "guess and check" method as shown in Table 7 to solve $3(2^a) = 96$.

TABLE 7

Guess for a	$3(2^a)$	Too Low/Too High
6	192	Too High
5	96	It works!

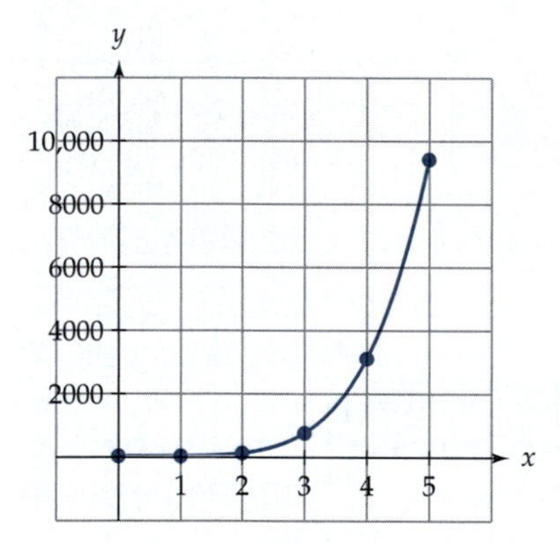

Figure 12

To see if this works for other points, we again plot our original points along with the graph of $y = 3x^5$. Our graph is shown in Figure 12. As you can see, the function $y = 3x^5$ is a good fit for our data.

Example 2 Determine if the data and graph in Figure 13 represent a power function. If so, give the formula. If not, explain why.

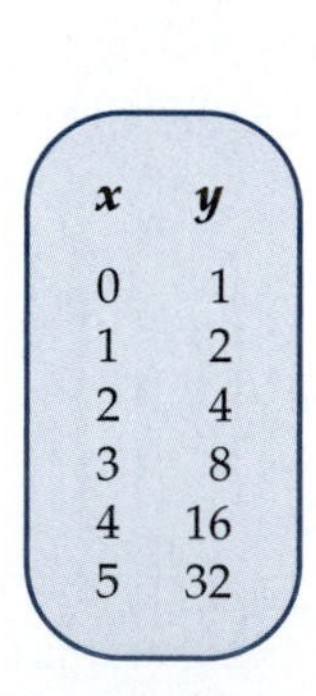

x	y
0	1
1	2
2	4
3	8
4	16
5	32

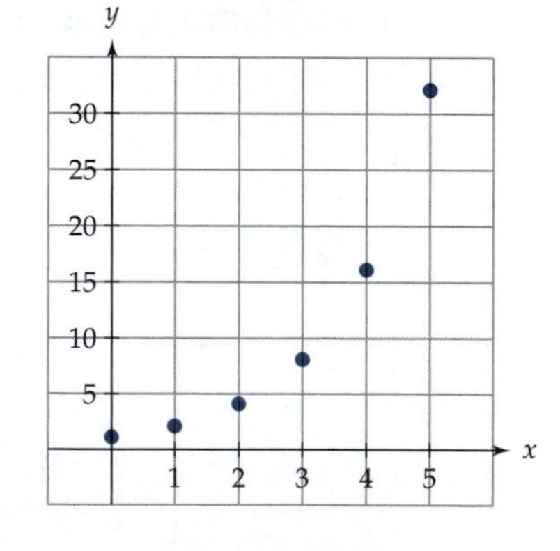

Figure 13

Solution Although the graph looks similar to a power function, it is suspicious because it does not contain the point $(0, 0)$. Any function of the form $f(x) = bx^a$ must contain the point $(0, 0)$, so this function cannot be a power function. It is, in fact, the exponential function $y = 2^x$.

In summary, to decide if a table of data or a graph represents a power function of the form, $f(x) = bx^a$, you should do the following.

- Check to be sure that it contains the point $(0, 0)$.
- Look at the output for $x = 1$, which gives the value of b.
- Use any point[26] to "guess and check" the value of a. (A look at the shape of the graph and the values of the table should enable you to find the value of a with only a few guesses.)

READING QUESTIONS

15. Let $f(x) = bx^a$.

(a) Suppose $f(1) = 4$. What is the value of b?

(b) Suppose $f(1) = 4$ and $f(2) = 32$. What is the value of a?

(c) Suppose $f(1) = 4$ and $f(2) = 4.76$ (rounded to two decimal places). What is the value of a?

VERBAL DESCRIPTIONS OF POWER FUNCTIONS: AN APPLICATION

As noted at the beginning of this section, power functions have a number of applications, from accelerating a car to finding a formula for area. Johannes Kepler used power functions to model the orbits of the planets. Using data similar to Table 8, he tried to find patterns in the movement of the planets. Table 8 shows the mean distance of each planet from the sun, given in astronomical units (AU), and the time, in Earth years, it takes each planet to orbit the sun. (*Note:* One astronomical unit is the mean distance from the sun to Earth, or approximately 1.496×10^8 km.)

TABLE 8 Planet Data*

Planet	Distance from the Sun (Astronomical units)	Orbit Time (Earth years)
Mercury	0.39	0.24
Venus	0.72	0.62
Earth	1.00	1.00
Mars	1.52	1.88
Jupiter	5.20	11.86
Saturn	9.59	29.46

**Eric Chaissen and Steve McMillan,* Astronomy Today, *2d ed. (Upper Saddle River, NJ: Prentice Hall, 1996), p. 42.*

As Kepler considered the distance of each planet from the sun and the time it takes each one to orbit the sun, it was clear that the relationship was

[26] *Make sure you use a point where the x-value is neither 0 nor 1.*

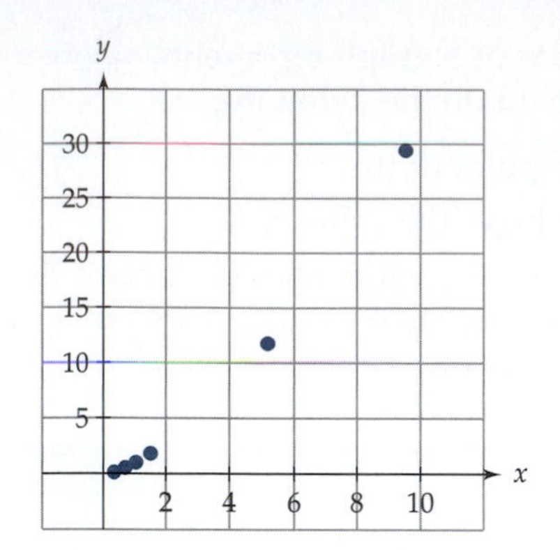

Figure 14 Orbit time (in Earth years) as a function of distance from the sun (in astronomical units).

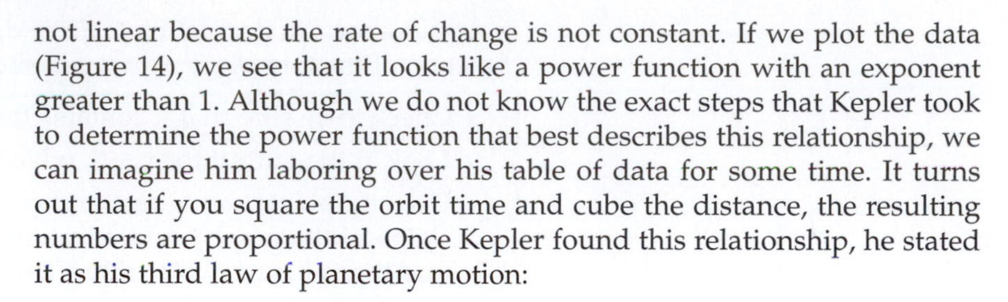

not linear because the rate of change is not constant. If we plot the data (Figure 14), we see that it looks like a power function with an exponent greater than 1. Although we do not know the exact steps that Kepler took to determine the power function that best describes this relationship, we can imagine him laboring over his table of data for some time. It turns out that if you square the orbit time and cube the distance, the resulting numbers are proportional. Once Kepler found this relationship, he stated it as his third law of planetary motion:

> The square of the length of time it takes any planet to orbit the sun is proportional to the cube of the planet's mean distance from the sun.

Recall that if two variables are proportional, then one is always equal to the other one times some constant. If w is proportional to z, for example, then we can say that w is equal to some constant (mathematicians often use k to represent the constant) times z. In this case, Kepler's third law indicates that

$$t^2 = k \cdot d^3,$$

where t is the time and d is the distance. This equation can be rewritten as

$$t = k_2 \cdot d^{3/2}$$

by taking the square root of both sides.[27]

To use this law to find the time of orbit for a given planet, we need to find the value of k_2. Notice that k_2 is playing the role of b when a power function is given as $f(x) = bx^a$. Recall that the value of b is the output for $f(1)$. So, our value of k_2 must be $f(1) = 1$ (using the data for Earth's measurements). If it seems too convenient that the value of k_2 results in an answer as simple as 1 realize that it is because of the units. Because the astronomical unit is based on the distance of Earth from the sun and the year is based on Earth's orbit, it is no wonder that the value of k_2 equals 1. (It is similar to using the length of your foot as a unit to measure your feet and getting 1 for an answer.) Figure 15 consists of a plot of the six data points along with the graph of $t = d^{3/2}$. Notice that this function is a very good fit for our data.

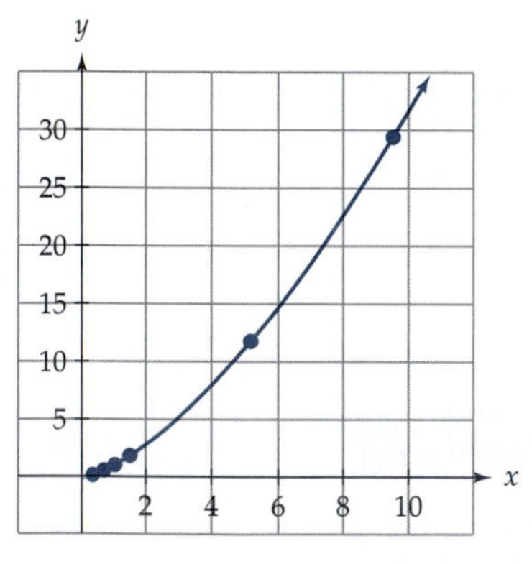

Figure 15 A plot of the six data points along with the graph of $t = d^{3/2}$.

When Kepler derived his law, he had only the data from the first six planets in our solar system. Table 9 contains data for the three planets discovered after the time of Kepler. If Kepler's law is correct, it should also work for these planets, even though he did not use this data to discover the relationship between time and distance. Using Kepler's law with the data from Uranus, the orbit should take about $19.19^{3/2} \approx 84.07$ years, the same answer given in Table 9.

[27] *k_2 is equal to the square root of the original constant. Mathematicians are often sloppy and call both constants k because there is not yet a value for our constant.*

TABLE 9 Planet Data for Uranus, Neptune, and Pluto

Planet	Date Discovered	Distance from the Sun d	Orbit Time t
Uranus	1781	19.19	84.07
Neptune	1846	30.06	164.82
Pluto	1930	39.53	248.6

HISTORICAL NOTE

Johannes Kepler was born in 1571 in a small town in Germany. His father was a mercenary soldier who disappeared without a trace, and his mother was once tried as a witch. In 1589, he attended the University of Tubinger, where he received straight A's in all subjects except one—Astronomy (in which he received an A−). Kepler became a mathematics teacher at Protestant High School in Braz, Austria, and quickly made a reputation for himself. His lectures were so disorganized that in his second year of teaching, no students signed up for his astronomy-mathematics class.

In 1601, Kepler was hired as an astronomer by Emperor Rudolph of Bohemia. Kepler replaced Tycho Brahe who, from 1576 to 1597, compiled a precise continuous record of planetary motion. With the aid of these data, Kepler was able to develop his three laws of planetary motion and was the first scientist to connect a theoretical model of astronomy to reality.

Source: Michael Zeilik, *Astronomy: The Evolving Universe* (New York: Harper and Row, 1976), pp. 54–64.

READING QUESTIONS

16. Explain why Kepler's third law describes a power function.

17. Use Kepler's law to find the orbit time for Neptune. Compare your answer with that given in Table 9.

SUMMARY

A **power function** is a function of the form $f(x) = bx^a$. In this section, we limited our look to power functions where a was a positive rational number.

For a power function of the form $f(x) = x^a$, where a is a positive integer, $f(0) = 0$ and $f(1) = 1$. If a is even, then the function has a domain of all real numbers and a range of $f(x) \geq 0$. If a is odd, then the function has a domain of all real numbers and a range of all real numbers. If a is even, then the function is symmetric with respect to the y-axis. If a is odd, then the function is symmetric with respect to the origin.

Suppose f is a power function of the form $f(x) = x^a$, where $a = n/d$ is a positive rational number written in simplified fractional form. If the denominator of the exponent is even, then the domain is $x \geq 0$. If the denominator of the exponent is odd, then the domain is all real numbers. If the numerator is even, then the range is $y \geq 0$. If the numerator is odd and the denominator is odd, then the range is all real numbers. If the numerator is odd and the denominator is even, then the range is $y \geq 0$.

For a power function of the form $f(x) = bx^a$, $f(0) = 0$ and $f(1) = b$. To decide if a table of data or a graph represents a power function of the form, $f(x) = bx^a$, you should check to be sure that it contains the point $(0, 0)$; look at the output for $x = 1$ (to give the value of b) and use any point to "guess and check" the value of a. A look at the shape of the graph and the values of the table should enable you to find the value of a in only a few guesses.

When power functions are described verbally, the two variables involved are often described as being proportional to one another. If two variables are proportional, then one is always equal to the other one times some constant. Sometimes, as in Kepler's third law, some power of one variable is proportional to some power of the other variable. These types of relationships can be described using power functions.

EXERCISES

1. Below is the function whose input is the length of a pendulum and whose output is the period of the pendulum (i.e., the length of time it takes to go back and forth through one complete swing).

Length (in.)	5	10	15	20	25
Period (sec)	0.75	1.05	1.28	1.46	1.62

   (a) Explain why this function is not a linear function.
   (b) This function is a power function, $P(x) = bx^a$. Will a be greater than or less than 1? Briefly justify your answer.

2. The formulas for each of the following are examples of power functions, $f(x) = bx^a$. Give the formula, identifying the value of a and the value of b.
   (a) The function to find the area of a circle
   (b) The function to find the area of a square
   (c) The function to find the volume of a cube
   (d) The function to find the volume of a sphere

3. Match the functions to the graphs.
   (a) $y = x^2$
   (b) $y = x^{10}$
   (c) $y = x^5$

(d) $y = x^9$

(e) $y = x^{2/3}$

(f) $y = x^{7/3}$

(g) $y = x^{3/4}$

(h) $y = x^{7/4}$

i.

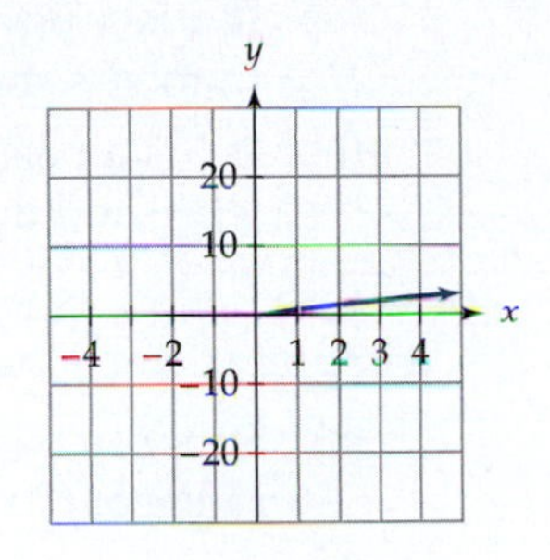

ii.

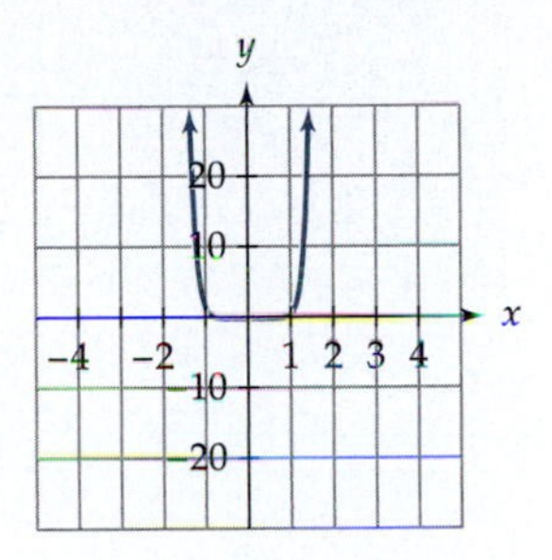

iii.

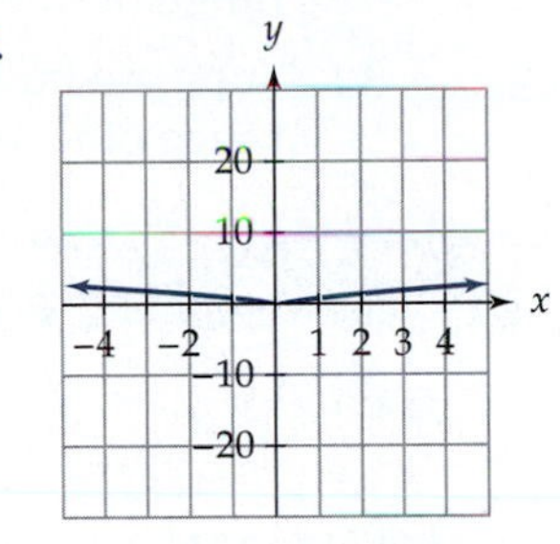

iv.

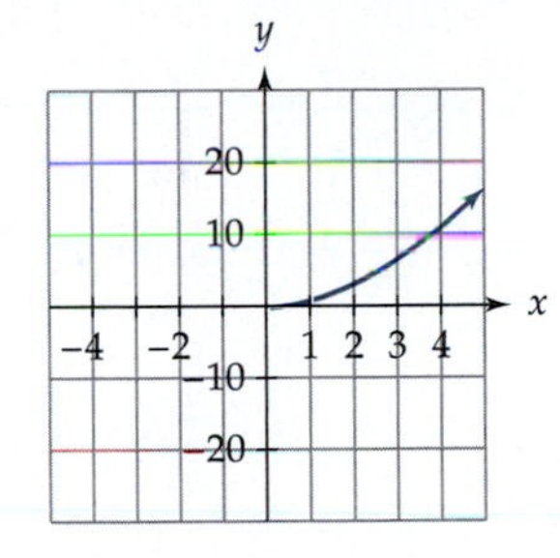

v.

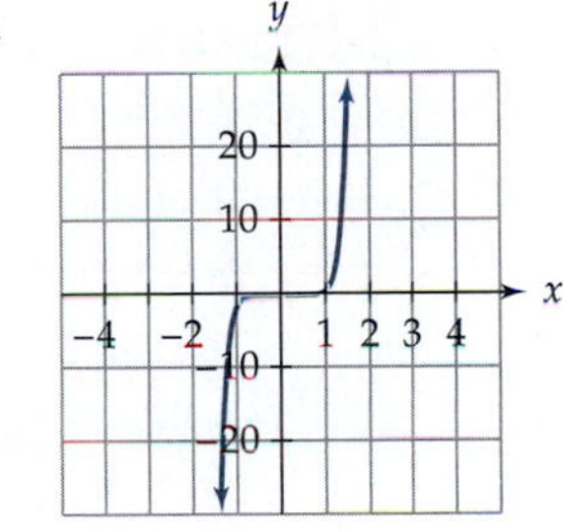

vi.

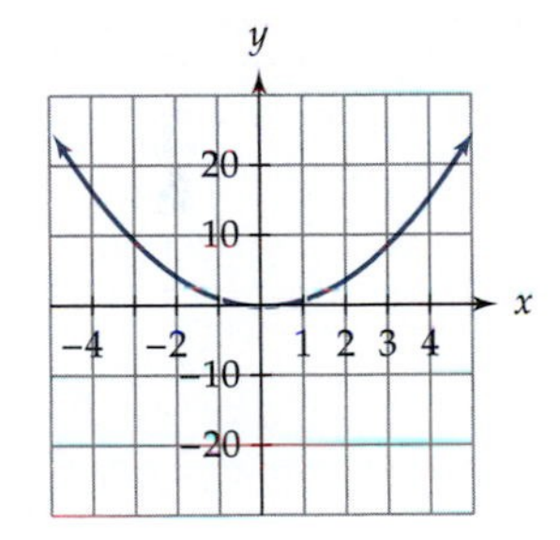

vii.

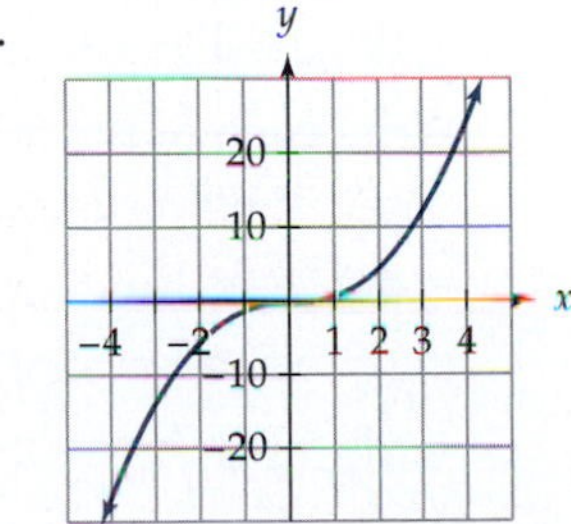

viii.

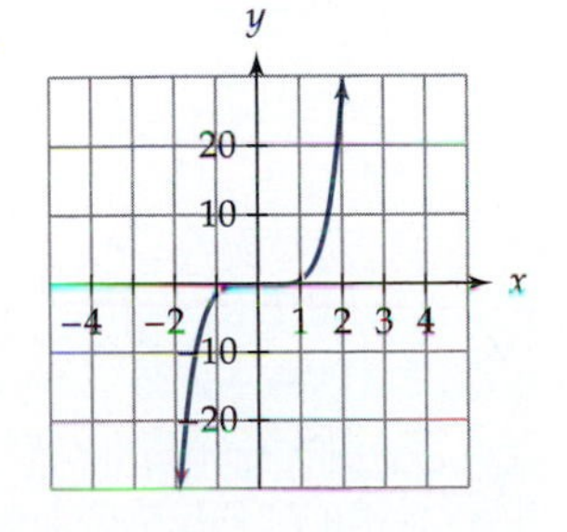

4. Let f be a function whose y-intercept is 4. How do you know that this function cannot be a power function?

5. Let $f(x) = x^a$. Label each of the following as always true, sometimes true, or never true. Briefly justify your answer.
   (a) If a is a positive rational number, then f will be concave up for $x \geq 0$.
   (b) If a is a positive integer, then f will be concave up for $x \geq 0$.
   (c) $f(0) = 0$ and $f(1) = 1$.
   (d) If a is a positive integer, then the domain of f will be all real numbers and the range will be $f(x) \geq 0$.
   (e) If a is a positive rational number, then the domain of f will be $x \geq 0$ and the range will be $f(x) \geq 0$.
6. Let $f(x) = x^a$, where a is a rational number, $a = n/d$. Assume a is in simplified fractional form.
   (a) Suppose n is even and d is odd. What are the domain and the range of f? Why?
   (b) Suppose n is odd and d is even. What are the domain and the range of f? Why?
   (c) Suppose both n and d are odd. What are the domain and the range of f? Why?
7. Let $f(x) = x^n$.
   (a) Let $n = 2$. Explain why $f(x) = f(-x)$.
   (b) Let n be an even integer. Explain why $f(x) = f(-x)$. What does this imply about the symmetry of f?
8. Let $f(x) = x^n$.
   (a) Let $n = 3$. Explain why $f(x) = -f(-x)$.
   (b) Let n be an odd integer. Explain why $f(x) = -f(-x)$. What does this imply about the symmetry of f?
9. Give the domain and range for each of the following.
   (a) $f(x) = x^{10}$
   (b) $f(x) = 3x^7$
   (c) $f(x) = x^{1/2}$
   (d) $f(x) = 4x^{7/11}$
   (e) $f(x) = x^{11/4}$
   (f) $f(x) = 15x^{14/9}$
10. Write each of the following using radicals (i.e., write $x^{1/2}$ as $\sqrt{x}$).
    (a) $3x^{7/4}$
    (b) $2x^{5/2}$
    (c) $x^{72/25}$
    (d) $16x^{1/5}$
11. Determine if each of the following is a power function. If so, give an equation for the function. If not, briefly explain why. All entries in the table are rounded to two decimal places.

(a)

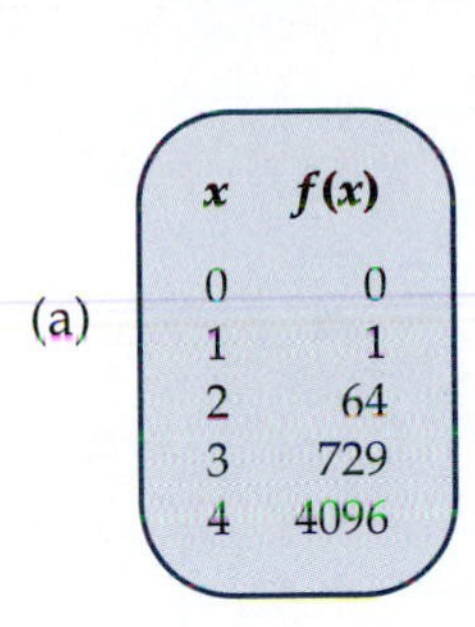

x	$f(x)$
0	0
1	1
2	64
3	729
4	4096

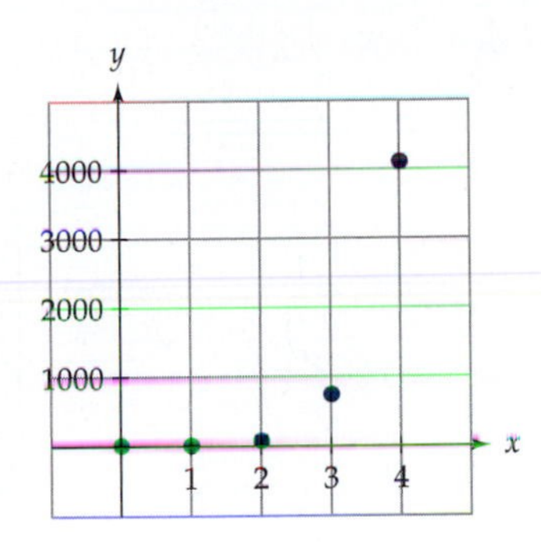

(b)

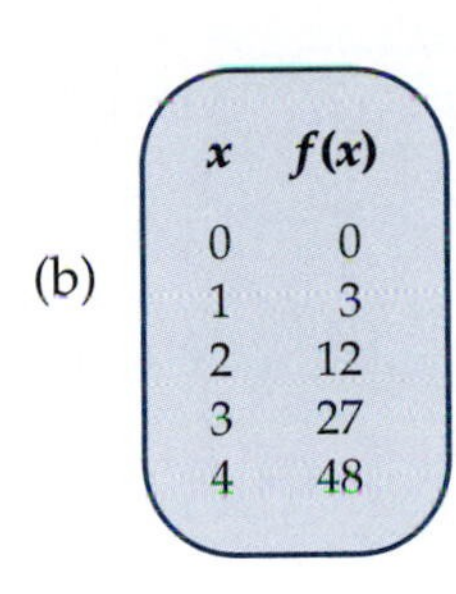

x	$f(x)$
0	0
1	3
2	12
3	27
4	48

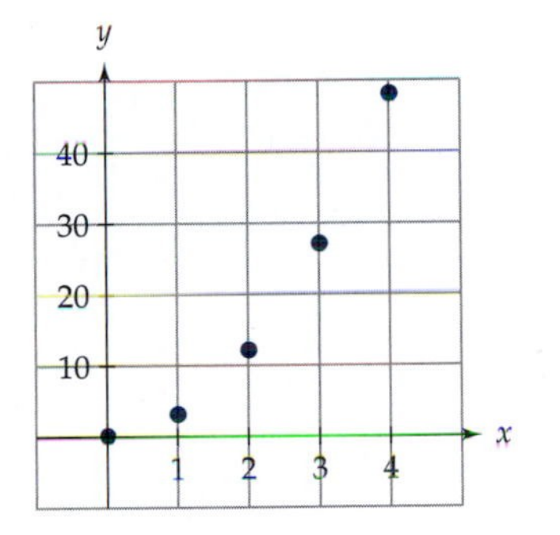

(c)

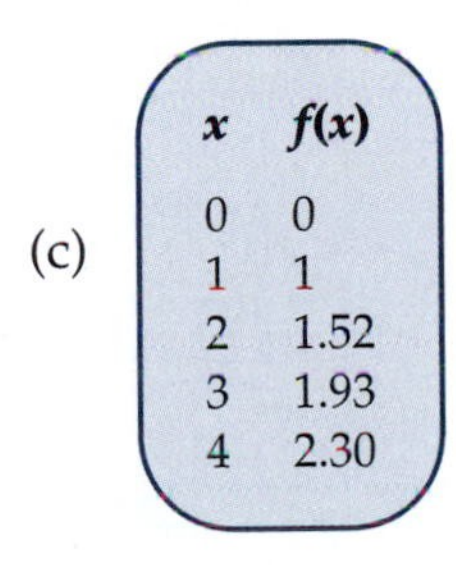

x	$f(x)$
0	0
1	1
2	1.52
3	1.93
4	2.30

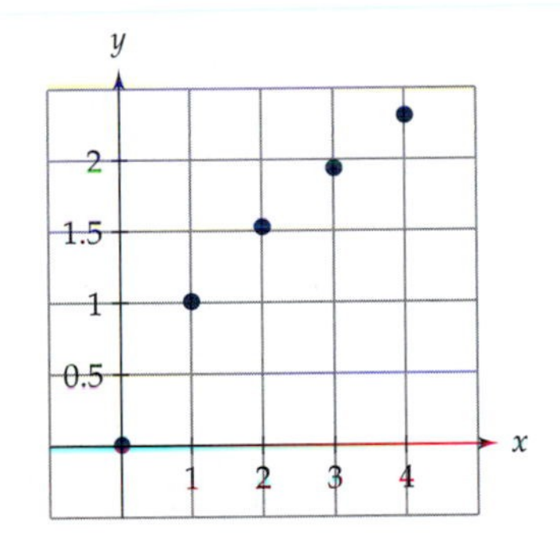

(d)

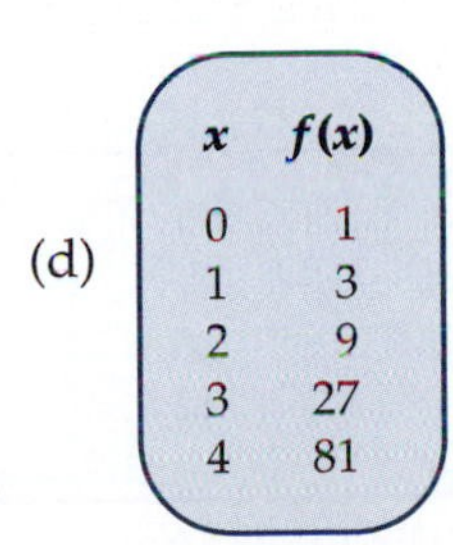

x	$f(x)$
0	1
1	3
2	9
3	27
4	81

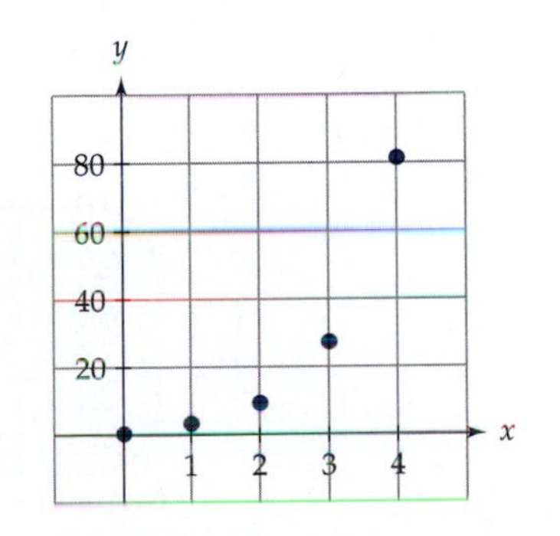

(e)

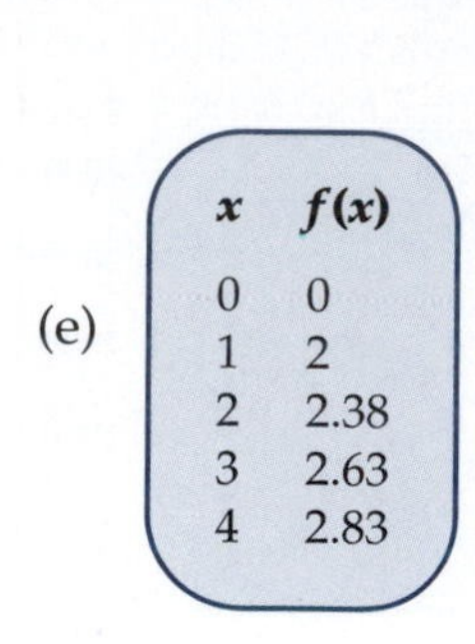

x	$f(x)$
0	0
1	2
2	2.38
3	2.63
4	2.83

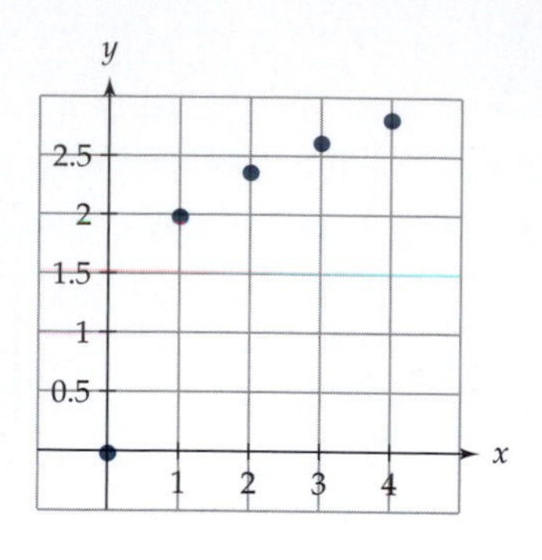

(f)

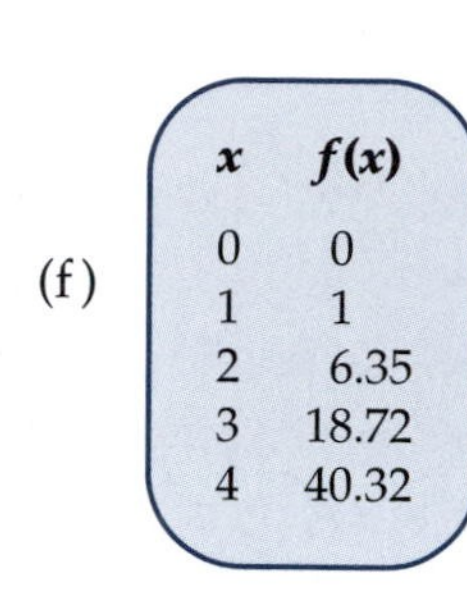

x	$f(x)$
0	0
1	1
2	6.35
3	18.72
4	40.32

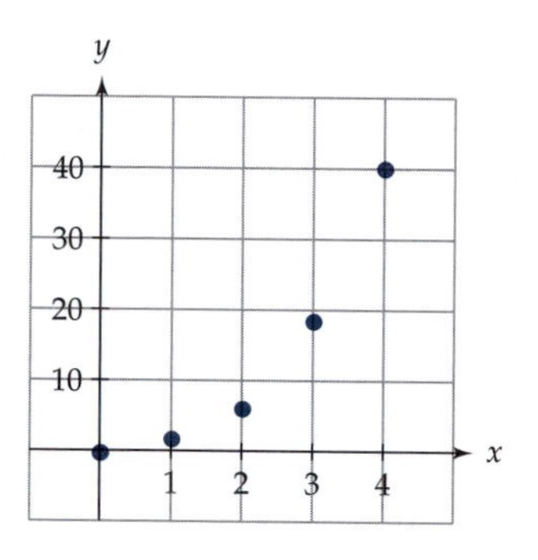

(g)

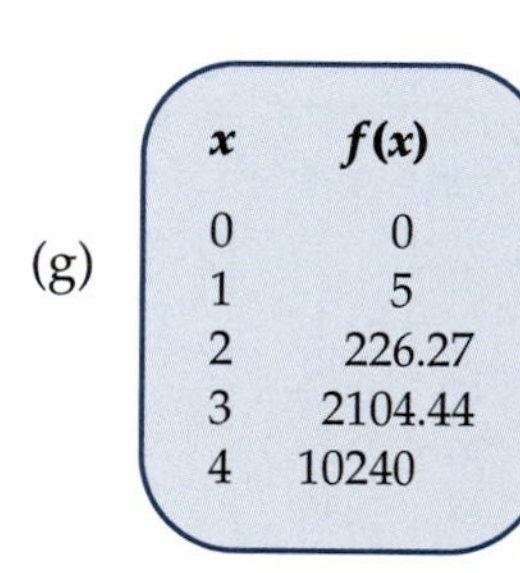

x	$f(x)$
0	0
1	5
2	226.27
3	2104.44
4	10240

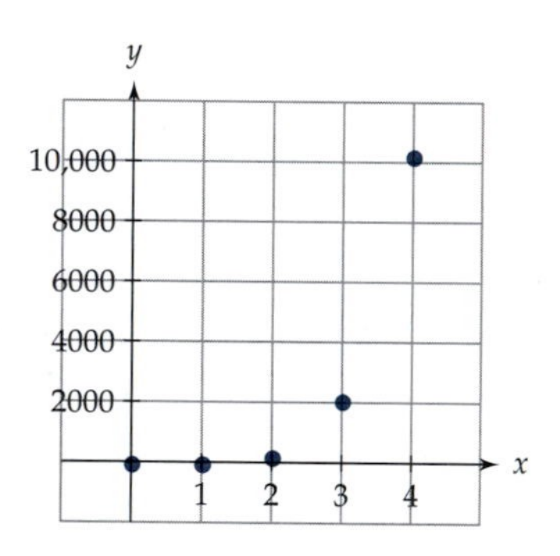

12. Determine formulas for each of the following functions. One is linear, one is exponential, and the other is a power function.

x	0	1	2	3	4
$f(x)$	−5	−3	−1	1	3
$g(x)$	4	8	16	32	64
$h(x)$	0	3	12	27	48

13. Determine formulas for each of following functions. One is linear, one is exponential, and the other is a power function.

x	0	1	2	3	4
$f(x)$	0	2.5	20	67.5	160
$g(x)$	4	6	9	13.5	20.25
$h(x)$	3	10.5	18	25.5	33

14. Can you have a power function, $f(x) = bx^a$, where $f(0) = 1$? Why or why not?
15. Let f be the power function $f(x) = bx^a$ such that each of the following is true.
    - The domain and the range are all real numbers.
    - The graph is concave down for $x < 0$ and concave up for $x > 0$.
    - The graph contains the point $(1, 1)$.

    What do you know about the value of a and the value of b?
16. How accurate is Kepler's formula in predicting the time of orbit for Neptune and Pluto?
17. The accompanying table contains the same planetary data as Table 9, except that the distance is in kilometers instead of astronomical units. (*Note:* 1 AU $\approx 1.496 \times 10^8$ km.)

Planet	**Distance from the Sun** d	**Orbit Time** t
Mercury	0.58344×10^8	0.24
Venus	1.07712×10^8	0.62
Earth	1.496×10^8	1.00
Mars	2.27392×10^8	1.88
Jupiter	7.7792×10^8	11.86
Saturn	14.34664×10^8	29.46

(a) Find a formula for this power function.

(b) What is the relationship between your formula and the formula we found using astronomical units, $t = d^{3/2}$? What is causing this relationship?

18. For each of the following, a proportion and a data point are given. Using this information, determine a constant of proportionality and the equation of the power function.
    (a) The input is directly proportional to the ninth power of the output. The point $(4, 18.66)$ satisfies this relationship.
    (b) The square of the input is directly proportional to the fifth power of the output. The point $(2, 9.24)$ satisfies this relationship.
    (c) The fourth power of the input is directly proportional to the output. The point $(0.6, 1.56)$ satisfies this relationship.
    (d) The square of the input is directly proportional to the cube of the output. The point $(1.3, 0.858)$ satisfies this relationship.

INVESTIGATIONS

INVESTIGATION: ACCELERATION

At the beginning of this section, it was stated that the function $d(t) = 4t^2$ models the distance traveled by a car accelerating at a rate of 8 ft/sec^2. In this function, $d(t)$ is the distance the car is from the stop sign after t seconds. Let's see why this equation might be correct.

1. Complete the table for the function $d(t) = 4t^2$. Recall that velocity = change in distance/change in time and acceleration = change in velocity/change in time.

Time (sec)	Distance (ft)	Velocity (ft/sec)	Acceleration (ft/sec^2)
0	0	—	—
1	4	4	—
2	16	12	8
3			
4			
5			
6			
7			
8			
9			
10			

2. What did you notice about the acceleration?

3. Fill out the following table, this time for the function $d(t) = 3t^2$.

Time (sec)	Distance (ft)	Velocity (ft/sec)	Acceleration (ft/sec^2)
0		—	—
1			—
2			
3			
4			
5			
6			
7			
8			
9			
10			

4. What did you notice about the acceleration this time?
5. Fill out the following table for the function $d(t) = bt^2$.

Time (sec)	Distance (ft)	Velocity (ft/sec)	Acceleration (ft/sec^2)
0		—	—
1			—
2			
3			
4			
5			
6			
7			
8			
9			
10			

6. Suppose the function $d(t) = 10t^2$ gives the distance traveled (in feet) from the stop sign after t seconds. What is the car's constant acceleration?
7. Give the equation that models the distance traveled by a car that starts from a complete stop and accelerates at a rate of c ft/sec^2.

8. The *second difference* for a function is defined as (change in slope)/(change in input). What does this investigation show about the second differences of a power function of the form $f(x) = bx^2$?
9. What is the relationship between the concavity of a function and second differences? What does your the table in question 5 tell you about the concavity of the function $f(x) = bx^a$?

PROJECTS 2.6 NEWTON: A REAL SWINGER

A pendulum consists of an object suspended from a fixed point so that it can freely swing back and forth. The period of a pendulum (i.e., the time it takes to go back and forth through one complete swing) is dependent only on the length of the string that is holding the mass, a fact first discovered by Galileo.[28] Although absolutely true only for an ideal pendulum (i.e., one in which the mass is concentrated at a point, the string has no mass, and there is no wind resistance on the mass as it swings), properties of ideal pendulums can still be used to mathematically determine the period of less than ideal pendulums.

More than 300 years ago Isaac Newton used a pendulum to estimate the speed of sound. This fact was first learned by one of the authors while on a tour of Cambridge University in England. The tour guide stopped at a colonnade in Neville's Court and clapped her hands, and a nice echo came back with a slight delay. She explained that this place was where Newton determined the speed of sound using an echo. He measured the length of the hallway and doubled it, which gave him the distance that the sound traveled. He then had to find the length of time between clapping his hands and hearing the echo. By dividing d, the total distance the sound traveled, by t, the time needed for the sound to travel there and back, he would be able to compute the speed of sound. There was one problem, however. The time difference between the clap and the echo returning was less than a second. Although there were clocks in Newton's day, there were no stopwatches that would measure to the accuracy needed. How did Newton measure the time? He used a pendulum. Newton knew the relationship that existed between the length of the string and the period of a pendulum. To measure the time, he varied the length of a pendulum until the period matched up with the time between the clap and its return. Through this experiment, he calculated the speed of sound to be between 920 and 1085 ft/sec,[29] which was a great feat with such simple instruments!

1. We made a slightly less than ideal pendulum with fishing line and a small metal weight and collected the data in Table 1. We use these data to determine a function for the period of a pendulum.

[28]*James R. Newman, ed.,* The Harper Encyclopedia of Science, *rev. ed. (New York: Harper and Row, 1967), p. 894.*

[29]*Richard S. Westfall,* Never at Rest: A Biography of Isaac Newton *(New York: Cambridge University Press, 1980), p. 456.*

TABLE 1 The Number of Periods per Minute for Various-Length Pendulums

Length of String (in.)	Number of Periods (per minute)	Length of Period (sec)
1	165	
3	101	
5	80	
10	57	
15	47	
20	41	
25	37	
30	33.5	
35	31	
40	29.5	
45	27	
50	26.5	

(a) Complete Table 1 by computing the time (in seconds) it takes for one period of the pendulum. Round your answer to two decimal places.

(b) Graph the data from Table 1 with length of string on the horizontal axis and length of the period on the vertical axis. Sketch a smooth curve to connect the points. Is your graph linear? If not, what is the shape of your graph? What does this tell you about the function for the period of a pendulum in terms of the length of the string?

(c) The formula for the length of time of the period of a pendulum P is of the form $P = c\sqrt{l}$, where c is some constant and l is the length of string.

i. Is this consistent with the shape of your graph in question 1(b)? Why or why not?

ii. Complete Table 2 by using the data from Table 1 and dividing the length of the period (in seconds) by the square root of the length of the string to show the value of c for each case. Find the average of these values to give an approximate value for the constant c. Explain how you obtained your approximation.

2. The formula for the period of an ideal pendulum is

$$P = \frac{2\pi}{\sqrt{g}}\sqrt{l}, \tag{1}$$

where P is the period of the pendulum, l is the length of the string, and g is the acceleration due to gravity.

TABLE 2 Find the Value of c Where c = length of period/$\sqrt{\text{length of string}}$

Length of String (in.)	Length of Period (sec)	c
1		
3		
5		
10		
15		
20		
25		
30		
35		
40		
45		
50		

(a) Give an estimate for the acceleration of gravity by using your approximation of c.

(b) The acceleration of gravity is approximately equal to 386 in./sec^2. How does this compare with your approximation of c you found in question 2(a)? How do you account for the discrepancy?

(c) The acceleration of gravity is not the same everywhere on Earth. For example, on a mountain, the value of g would be less than at sea level. How would this affect the period of a pendulum? Justify your answer.

For the rest of the project, use $P = (2\pi/\sqrt{g})\sqrt{l}$, *where* $g = 386$ in./sec^2.

3. Wall clocks are often constructed so that the pendulum has a period of 1 sec, whereas grandfather clocks are often constructed so that the pendulum has a period of 2 sec.

   (a) If a clock were to have a pendulum that had a period of 1 sec, how long should the pendulum be?

   (b) What is the length of a clock's pendulum that has a period of 2 sec? Is this length twice that of the 1-sec pendulum? What does this say about the relationship between the length of the string and the period of the pendulum?

4. The accuracy of a pendulum clock is dependant on the length of the pendulum. A clock could have a pendulum cut incorrectly during construction, or, through a change in temperature, a metal pendulum could shrink or expand.

   (a) Suppose you were building a clock with a pendulum that had a period of 1 sec.

i. If you accidently made the pendulum $\frac{1}{4}$ in. too long, how would that affect the period? Would your clock run fast or slow? By how much?

ii. How would making the pendulum $\frac{1}{4}$ in. too long affect the accuracy of the clock after 30 days? (*Hint:* Find the number of periods per second.)

(b) Would the clock be more or less accurate if you made the same $\frac{1}{4}$-in. error on a pendulum that had a period of 2 sec? Explain.

REVIEW EXERCISES

1. Let f be a linear function whose slope is 4. Suppose $f(1) = 3$. What is $f(3)$? What is $f(10)$?
2. Give the equations for each of the following linear functions.

(a)

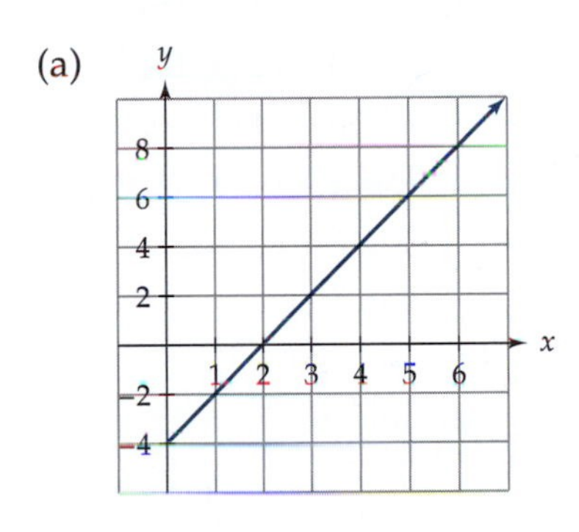

(b)

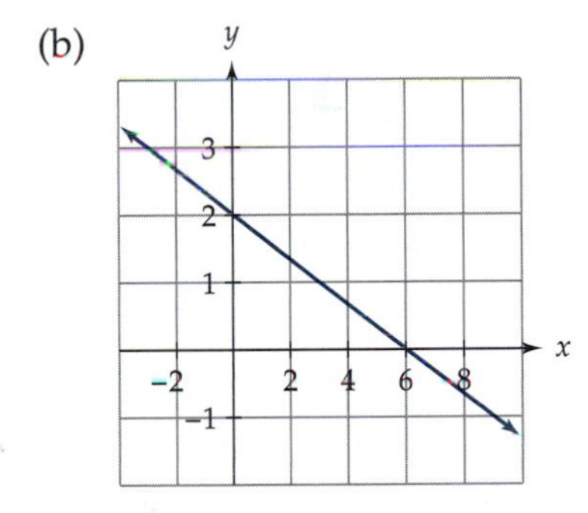

3. Each of the following represents a linear function. Give the equation.
   (a) Conversion of cubic inches to gallons (231 in.^{3} = 1 gal)
   (b) Conversion of miles to feet (5280 ft = 1 mi)
   (c) Total cost of a cell phone call if you are paying \$0.35 a minute
   (d) Total price of wire if you are paying \$0.15 per foot
4. Determine if each of the following describes a linear function, an exponential function, or neither. If it is a linear function, give the slope. If it is exponential, give the growth factor. If it is neither, briefly explain why.
   (a) The population of Here, U.S.A., is increasing at a rate of 2000 people per year.
   (b) The population of There, U.S.A., is increasing at a rate of 0.5% per year.
   (c) The cost of a telephone call is 99¢ for the first 20 min and 10¢ for each minute after 20 min.
   (d) An investment is doubling in value every year.
   (e) Your daughter's height is increasing 2 in. per year.

5. For each of the following, convert the doubling rate to a growth factor.
   (a) If a substance doubles in 3 months, what is its growth factor for 1 year?
   (b) If a substance doubles in 3 days, what is its growth factor for 1 day?
   (c) If a substance doubles in 30 min, what is its growth factor for 1 day?
6. Let $P(t)$ be the population of Whats at year t in Whatville. For each of the following, describe what is happening to the population of Whats.
   (a) $P(t) = 1000 \cdot 0.5^t$
   (b) $P(t) = 1500 - 100t$
   (c) $P(t) = 1500 \cdot 1.05^t$
   (d) $P(t) = 1000 + 105t$
7. For each of the following, find the equation of the exponential function that contains that pair of points.
   (a) $(0, 5)$ and $(1, 15)$
   (b) $(0, 5)$ and $(5, 15)$
   (c) $(2, 5)$ and $(4, 20)$
8. Using $\log 2 \approx 0.30103$ and $\log 3 \approx 0.47712$, find each of the following.
   (a) $\log 6$
   (b) $\log 8$
   (c) $\log 600$
   (d) $\log 36$
9. Suppose $\log(a) = \frac{3}{4}$. Find each of the following.
   (a) $\log(10a)$
   (b) $\log(100a)$
   (c) $\log(a^2)$
   (d) $\log(a^{4/3})$
10. Suppose $\log(13) = a$. Find each of the following.
   (a) The value of b such that $\log(b) = a/2$
   (b) The value of b such that $\log(b) = a + 2$
11. Sound is usually measured using decibels. To measure sound in decibels, the sound is compared to a standard, I_0. This standard, $I_0 = 10^{-16}$ W/cm^2 (watts per square centimeter), is approximately the lowest intensity sound audible to humans. The loudness of sound can then be calculated using the formula

$$\text{decibels} = 10 \log\left(\frac{I}{I_0}\right),$$

   where I is the sound's intensity.
   (a) Find the decibel measure of a typical conversation where I, the intensity of sound, is 10^{-10} W/cm^2.
   (b) Find the decibel measure of a large truck passing by where I, the intensity of sound, is 10^{-7} W/cm^2.

(c) How many times larger is the intensity of the sound of a jet taking off (120 dB) than the intensity of the sound of a busy street (80 dB)?

12. Let f be the periodic function, a portion of which is shown in the accompanying graph. Evaluate each of the following.
    (a) $f(1)$
    (b) $f(5)$
    (c) $f(28)$
    (d) $f(42)$
    (e) $f(-3)$

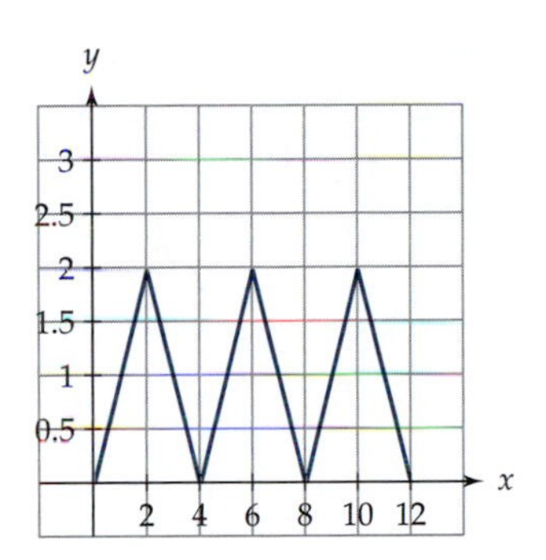

13. In 2000, the London Eye, the world's tallest Ferris wheel, opened along bank of the River Thames. The wheel has a diameter of 450 ft. The are 32 capsules, each capable of holding 25 people. One revolution of the wheel takes 30 min.[30] Suppose you get on the London Eye at ground level and it turns clockwise for two revolutions. Construct a graph with your height above the ground on the vertical axis and the minutes after the wheel begins to turn on the horizontal axis. Label the period and amplitude in the graph.
14. What are the period and the amplitude of the periodic function describing the vertical position of the point on the tip of a 5-in. second hand on a clock?
15. On a unit circle, draw the angle whose initial side is the positive x-axis and whose terminal side intersects the circle to form an angle with the radian measure given below.
    (a) $\frac{\pi}{3}$
    (b) $\frac{\pi}{2}$
    (c) $\frac{3\pi}{4}$
    (d) $-\frac{7\pi}{8}$

[30] *British Airways,* London Eye—Statistics, *(visited 20 December 1999) ⟨www.british-airways.com/millennium/docs/londoneye/stats.html⟩.*

(e) $\frac{9\pi}{2}$

(f) $-\frac{2\pi}{3}$

16. Convert each of the following degrees to a radian measure.
    (a) 30°
    (b) 135°
    (c) 240°
    (d) 270°
17. Match the functions to the graphs.
    (a) $y = x^{3/5}$
    (b) $y = x^{5/3}$
    (c) $y = x^{2/3}$
    (d) $y = x^{5/4}$
    (e) $y = x^{3/4}$
    (f) $y = x^{8/5}$

i.

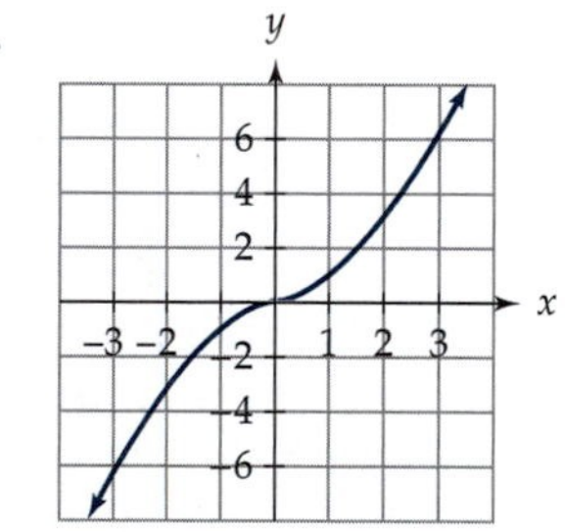

ii.

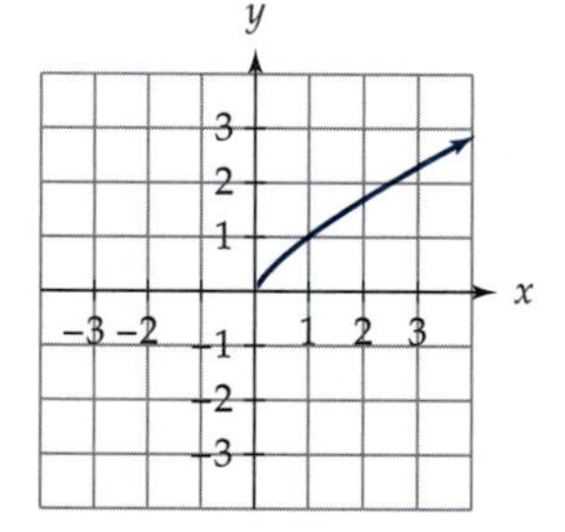

iii.

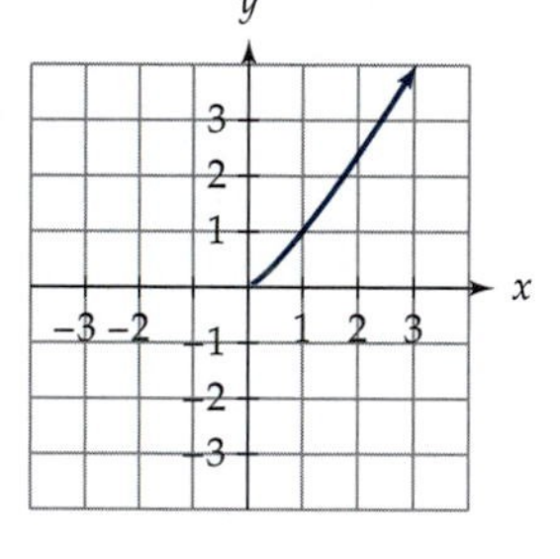

iv.

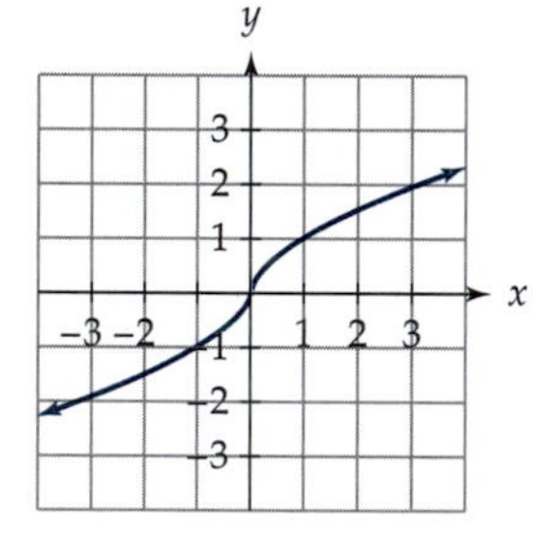

v.

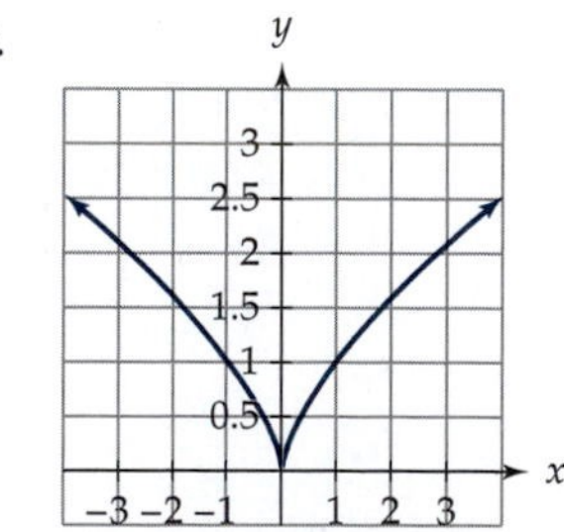

vi.

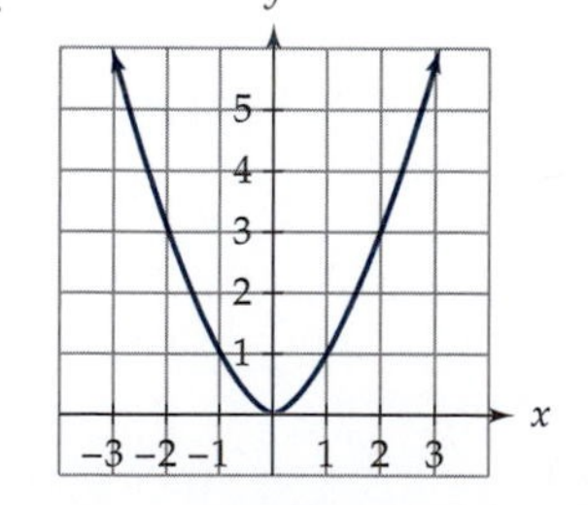

18. Give the domain and the range for each of the following.
    (a) $f(x) = 3x^5$
    (b) $f(x) = x^8$
    (c) $f(x) = x^{2/3}$
    (d) $f(x) = 2x^{5/6}$
    (e) $f(x) = 4x^{7/5}$
19. Write each of the following using radicals (i.e., write $x^{1/2}$ as $\sqrt{x}$).
    (a) $f(x) = x^{1/4}$
    (b) $f(x) = 4x^{7/11}$
    (c) $f(x) = x^{11/4}$
    (d) $f(x) = 15x^{14/9}$
20. Determine formulas for each of following functions. One is linear, one is exponential, and the other is a power function.

x	0	1	2	3	4
$f(x)$	1.5	2	2.5	3	3.5
$g(x)$	0	0.5	4	13.5	32
$h(x)$	5	10	20	40	80

21. For the following functions, one is linear, one is exponential, one is a power function, and one is none of these three. Determine formulas for the linear, exponential and power functions.

x	0	1	2	3	4
$f(x)$	0	0.15	0.6	1.35	2.4
$g(x)$	0.15	0.45	0.95	1.65	1.95
$h(x)$	20	8	3.2	1.28	0.512
$j(x)$	0.15	0.55	0.95	1.35	1.75

22. For each of the following, a proportion and a data point are given. Using this information, determine a constant of proportionality and the equation of the power function.
    (a) The input is directly proportional to the fifth power of the output. The point $(32, 20)$ satisfies this proportional relationship.
    (b) The square of the input is directly proportional to the third power of the output. The point $(27, 72)$ satisfies this proportional relationship.

CHAPTER OUTLINE

CHAPTER OVERVIEW

- Changing the input or the output of a function to shift, stretch, compress, or reflect the graph of the function
- Adding or multiplying two functions
- Composition of functions
- Finding the inverse of a function

NEW FUNCTIONS FROM OLD

In this chapter, we focus on changing and combining functions. We show how changing the input or the output of a function or how adding or multiplying functions together can create new functions with graphs that sometimes look very different than those shown in chapter 2. Such changes allow the modeling of many more complex situations.

3.1 FUNCTION TRANSFORMATIONS: CHANGES IN OUTPUT

In chapter 2, we introduced five basic types of functions: linear, power, exponential, logarithmic, and periodic. In this chapter, we look at transforming these functions. We first transform these functions by adding a constant to the output and multiplying the output by a constant. We see how changing the output by a constant affects the function numerically, graphically, symbolically, and verbally. Because the output appears on the vertical axis of a graph, the effect of changing the output will always be a vertical change.

ADDING OR SUBTRACTING A CONSTANT TO THE OUTPUT

The function that models the basic federal income tax system is composed of four piecewise linear functions, each based on the amount of income the taxpayer earns. Let's focus on one piece of one of these functions by looking at the tax for a married couple filing jointly for 1997 with a taxable income[1] of more than \$41,200 but not over \$99,600. The amount of their tax is 15% of the first \$41,200 in income plus 28% of the income earned above \$41,200.[2] If we let i be the amount of taxable income a couple earns and $t(i)$ be the amount of tax owed, then our function is

$$\begin{aligned} t(i) &= 0.15(41{,}200) + 0.28(i - 41{,}200) \\ &= 6180 + 0.28i - 11{,}536 \\ &= 0.28i - 5356. \end{aligned}$$

[1] *Taxable income is a person's total income less any deductions and exemptions that person is allowed to take.*

[2] *Department of the Treasury, Internal Revenue Service,* 1997 1040 Instructions, *(Washington, DC: GPO, 1997), p. 51.*

One way to change the amount of tax owed is to include tax credits. A tax credit is a direct reduction in the amount of tax one has to pay. For example, in 1997, the U.S. government allowed a $5000 tax credit for expenses related to the adoption of a child. Thus, the amount of tax a couple had to pay for 1997 would be reduced by $5000 if they had adopted a child that year. Notice that a tax credit is changing the output, that is, the amount of tax owed. Because the amount of tax owed is output, a tax credit is a change in output. First, let's look at this change numerically. Table 1 shows the amount of tax owed for couples with and without the tax credit.

TABLE 1 A Comparison of Tax Owed with and without an Adoption Credit

Taxable Income ($)	**$50,000**	**60,000**	**70,000**	**80,000**	**90,000**
Tax Owed without Adoption Credit ($)	$8644	11,444	14,244	17,044	19,844
Tax Owed with Adoption Credit ($)	$3644	6,444	9,244	12,044	14,844

For a given income, the amount of tax owed with the adoption credit is always $5000 less than the amount of tax owed without the adoption credit. If we let a be the function that determines the amount of tax owed for couples taking the adoption credit, then $a(i) = t(i) - 5000$. Therefore, $a(i) = 0.28i - 5356 - 5000 = 0.28i - 10{,}356$.

Let's compare the graphs of t and a. We start by plotting the points given in Table 1. In Figure 1(a), each point representing the amount of tax owed with the adoption credit, $a(i)$, is 5000 units lower than the corresponding point representing the amount of tax owed without the adoption credit,

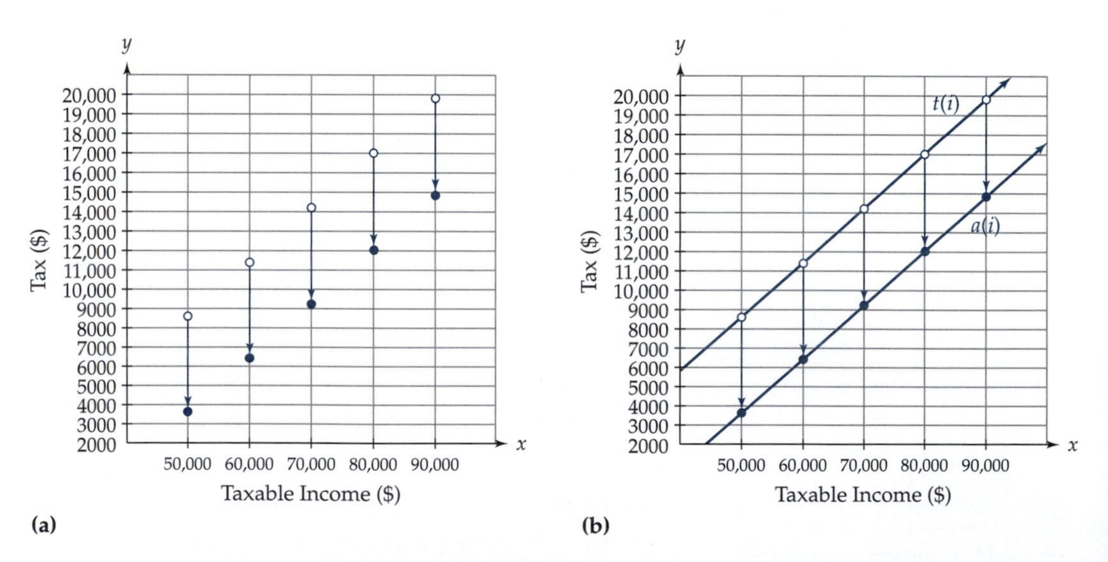

Figure 1 Graphical representations of the tax functions. In (a) each point is shifted down by $5000. In (b) the function that includes the adoption tax credit, $a(i)$, is $5000 lower than the function that does not, $t(i)$.

$t(i)$, because output is represented vertically on a graph. By connecting the points with a line, we see that subtracting 5000 from the output results in a vertical shift downward of 5000 units from the original linear function. See Figure 1(b).

Addition of a positive constant has the opposite effect of subtraction; it will result in a vertical shift upward. We can see this shift in Figure 1 if we remember that $a(i) = t(i) - 5000$ implies that $t(i) = a(i) + 5000$ and notice that $t(i)$ is shifted up from $a(i)$ by 5000 units. In general, the following is true.

PROPERTIES

If f is a function and c is a positive number, then

- The graph of $y = f(x) + c$ is the graph of $y = f(x)$ shifted up c units.
- The graph of $y = f(x) - c$ is the graph of $y = f(x)$ shifted down c units.

Example 1 Given the functions f and g in Table 2, answer the following questions.

(a) How are the values of $g(x)$ related to the values of $f(x)$?

(b) If $f(x) = x^2$, determine a formula for g.

(c) How is the graph of g related to the graph of f?

TABLE 2

x	0	1	2	3	4
$f(x)$	0	1	4	9	16
$g(x)$	−2	−1	2	7	14

Solution

(a) For a given input, the output $g(x)$ is 2 less than that of $f(x)$.

(b) Since $g(x) = f(x) - 2$, then $g(x) = x^2 - 2$.

(c) The graph of g would look like the graph of f shifted down two units.

Example 2 One of the graphs in Figure 2 is of the function $f(x) = x^3$. The other is a transformation of $f(x) = x^3$ resulting from a vertical shift. What is the formula for the transformed function?

Solution The graph of $f(x) = x^3$ is the one that passes through the origin. The other graph appears to be shifted up four units since it passes through the point $(0, 4)$. Therefore, its formula is $g(x) = x^3 + 4$.

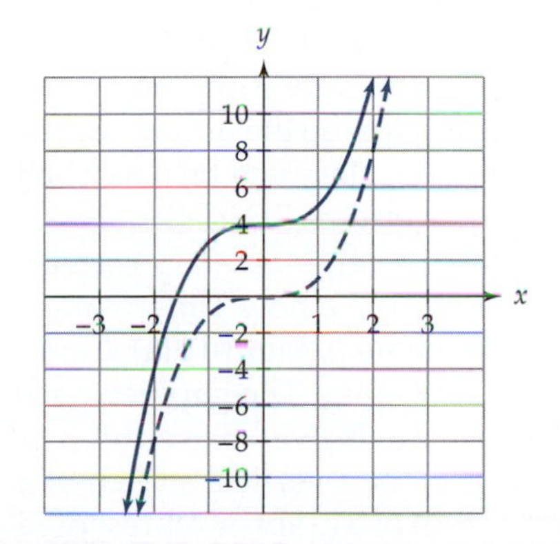

Figure 2

READING QUESTIONS

1. What is a tax credit? Why do we consider a tax credit a change in output?
2. Why does adding a constant to a function shift the function's graph *vertically*?
3. If $f(x) = 2x - 3$ and $g(x) = f(x) - 1$, what is the formula for g? How is the graph of g related to the graph of f?

MULTIPLYING THE OUTPUT BY A POSITIVE CONSTANT

In section 2.2, you read about a baseball player who convinced the team owner to pay him in an unorthodox manner. After receiving a signing bonus of \$1, the player was to get \$2 for the first game, \$4 for the second game, \$8 for the third game, and so on, doubling the payment for each additional game played. His salary is given by the function $P(n) = 2^n$, where $P(n)$ is the player's pay for the nth game. Let's extend this fictitious story to include a couple of other players who want more. When the owner of the team tried to sign Addison to a contract, he asked for \$5 more per game than our original player. When the owner tried to sign Timermann to a contract, he asked for five times as much per game as our original player. What do these new payment functions look like? Let's first look at them numerically. (See Table 3.)

TABLE 3 Three Unorthodox Payment Schemes for a Baseball Player

Game	0	1	2	3	4
Original Players Payment (\$)	1	2	4	8	16
Addison's \$5 More Payment (\$)	6	7	9	13	21
Timermann's Five Times as Much Payment (\$)	5	10	20	40	80

Think about these new payment schemes as functions and notice that the change is in the outputs. The function for Addison has outputs that are five more than the original function, and the function for Timermann has outputs that are five times the original. By letting $P(n)$ be the output for the function describing the original salary, Addison's salary can be described as $P(n) + 5$ and Timermann's salary can be described as $5P(n)$.

A graph of Addison's salary compared with the original salary is a vertical shift because each point is five units higher. To see how Timermann's salary compares, let's plot the points given in Table 3. Figure 3(a) shows that the distance between the two salaries is changing. The gap gets progressively larger. Each of the points for Timermann is five *times* as high as each corresponding point for the original player, or a vertical stretch of a *factor* of five. It is a vertical change since the output of a graph is represented

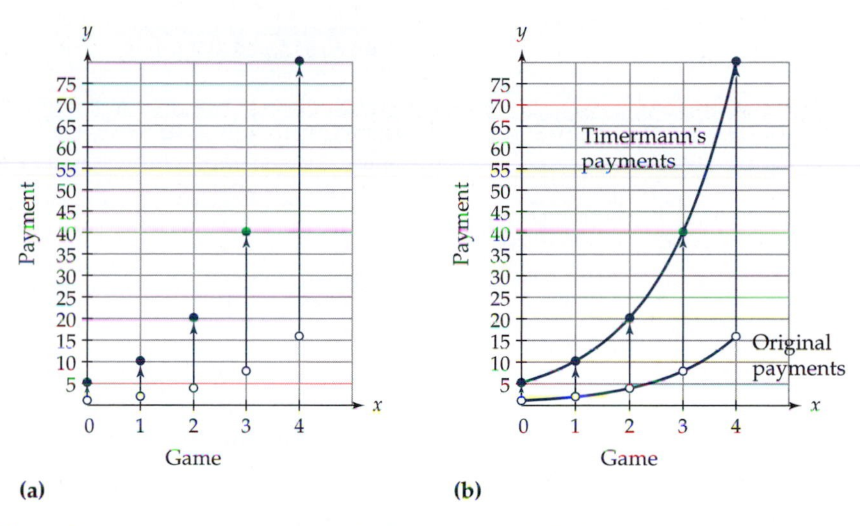

Figure 3 A graphical comparison of payments to baseball players. Each new point in (a) is five times farther away from the x-axis than its original point. This is shown in (b) by having the entire function stretched by a factor of 5.

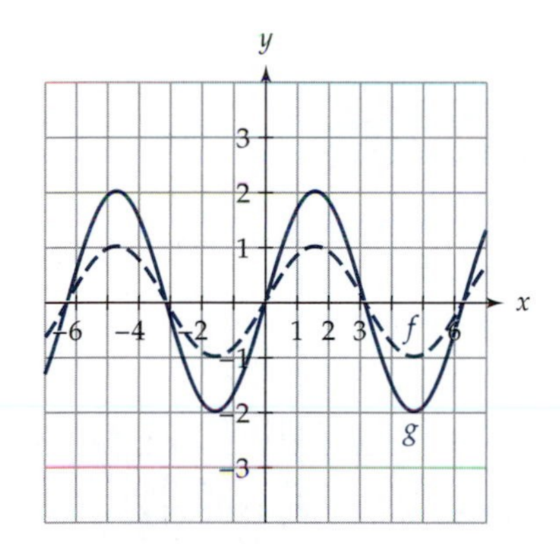

Figure 4 The graph of $f(x) = \sin x$ has amplitude 1, and the graph of $g(x) = 2\sin x$ has amplitude 2.

vertically. By connecting points in Figure 3(a) with a smooth curve, we see that multiplying the output of our function by 5 results in a vertical stretch of our original curve by a factor of 5. See Figure 3(b). Notice that a vertical stretch differs from a vertical shift. If a function is shifted vertically, then all the points are shifted the same amount. If a function is stretched vertically, however, then the points on the x-axis do not move and instead act as anchor points. All other points are moved away from the x-axis by the constant factor. For example, if a function is being stretched by a factor of 5 and the function contains the points $(2, 3)$ and $(1, -4)$, then the stretched function contains the points $(2, 15)$ and $(1, -20)$. Both points are now five times farther away from the x-axis. If a function is multiplied by a positive number greater than 1, then the stretching always occurs away from the x-axis, with outputs that are on the x-axis remaining the same and all others moving farther away from the x-axis. For example, Figure 4 shows the graphs of $f(x) = \sin x$ and $g(x) = 2\sin x$. Notice that g is stretched away from the x-axis. Points on the x-axis remain the same, positive outputs remain positive, and negative outputs remain negative. The function $g(x) = 2\sin x$ has an amplitude[3] of 2, whereas f has an amplitude of 1.

In our previous examples, we multiplied functions by a number, which caused their graphs to be stretched farther away from the x-axis. Will that always happen, or are there positive numbers that, after multiplying, give smaller answers? Let's compare $f(x) = \sin x$ and $h(x) = \frac{1}{3}\sin x$. Because the outputs, $h(x)$, will be one-third as large as the outputs, $f(x)$, the graph of h will be *closer* to the x-axis than the graph of f. Thus, when compared with f, the graph of h is compressed rather than stretched. The amplitude of h is $\frac{1}{3}$. (See Figure 5.) In general, the following is true.

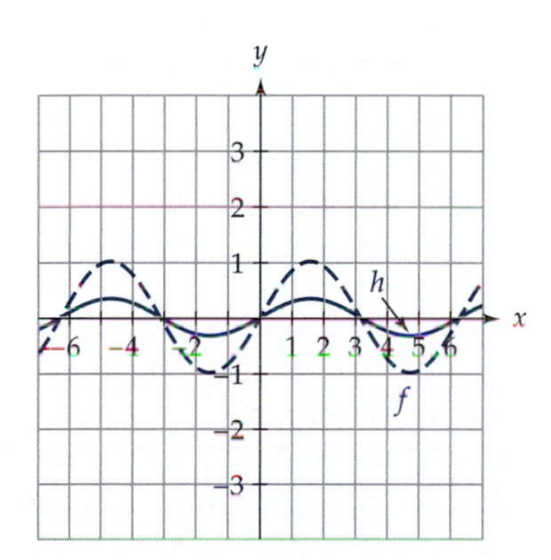

Figure 5 The graph of $f(x) = \sin x$ has amplitude 1, and the graph of $h(x) = \frac{1}{3}\sin x$ has amplitude $\frac{1}{3}$.

[3] *Remember that the amplitude of a periodic function is half the difference between the function's largest output and its smallest output.*

PROPERTIES

The graph of $y = c \cdot f(x)$ is the graph of $y = f(x)$ vertically stretched by a factor of c for $c > 0$. (If $0 < c < 1$, then the graph of $y = c \cdot f(x)$ can also be referred to as being compressed by a factor of $1/c$.)

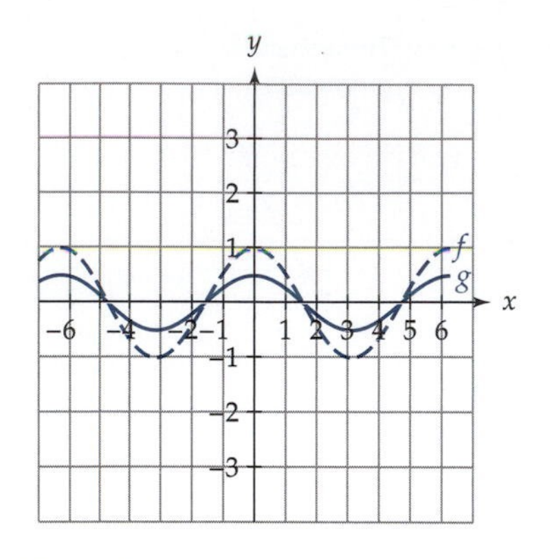

Figure 6

Example 3 Figure 6 shows the graphs of f and g. If $f(x) = \cos x$, determine a formula for g.

Solution From the graph, it appears that for a given input, the output for g is half that of f. For example, f contains the point $(0, 1)$ whereas g contains the point $(0, \frac{1}{2})$. Therefore, $g(x) = \frac{1}{2}f(x) = \frac{1}{2}\cos x$.

Example 4 How are f and g related in Table 4?

TABLE 4

x	−3	−2	−1	0	1	2	3
$f(x)$	−5	−4	−2	1	5	10	16
$g(x)$	−15	−12	−6	3	15	30	48

Solution Because each output, $g(x)$, is three times that of $f(x)$, $g(x) = 3f(x)$.

Example 5 The amount of money in a bank account can be calculated by using the formula $f(t) = P(1 + r)^t$, where P is the principal (the amount of money invested), r is the annual interest rate, and t is the length of time, in years, that the money is in the account. Suppose you have \$1000 invested in an account that is earning 4% per year. The function $f(t) = 1000(1.04)^t$ can be used to calculate the amount of money you have in your account. Which of the following situations could be represented by a function whose graph is a vertical stretch of f by a factor of 2? Justify your answer.

A. You invest \$1000 in an account that earns 8% per year.

B. You invest \$2000 in an account that earns 4% per year.

Solution Because a stretch of 2 means that we multiply the output by 2, the function we want has the form $2 \cdot f(t) = 2 \cdot 1000(1.04)^t = 2000(1.04)^t$. This function describes the amount of money in the account for situation B. The amount of money in the account for situation A can be calculated using the function $A(t) = 1000(1.08)^t$, which cannot be rewritten as two times the original function. We can also see that situation A is not correct by checking a specific point. For example, $f(1) = 1040$, whereas $A(1) = 1080$. Because $A(1) \neq 2f(1)$, situation A cannot be represented by a function whose graph is a vertical stretch of f by a factor of 2. So, the only situation that works is situation B.

READING QUESTIONS

4. How does a vertical stretch in a graph differ from a vertical shift?

5. How are f and g related in the following table?

x	−3	−2	−1	0	1	2	3
$f(x)$	−8	−4	−2	0	2	4	8
$g(x)$	−4	−2	−1	0	1	2	4

6. Given a function f, how does the graph of $y = a \cdot f(x)$, where $0 < a < 1$, differ from the graph of $y = b \cdot f(x)$, where $b > 1$?

7. What points on a graph do not change when the graph is stretched vertically?

MULTIPLYING THE OUTPUT BY A NEGATIVE CONSTANT

So far, we have only considered what happens when you multiply a function's output by a positive number. Let's see what happens when we multiply a function's output by a negative number. Let $f(x) = x^3$ and consider $y = -1 \cdot f(x) = -f(x) = -x^3$. Table 5 shows some outputs for both of these functions. Not surprisingly, the outputs $-f(x)$ just have the opposite sign of $f(x)$. Figure 7 compares the graphs of these two functions. You can see that $y = -f(x)$ looks like $y = f(x)$ except that it is reflected across the x-axis because the positive outputs, $f(x)$, became negative outputs, $-f(x)$, and vice versa. In general, the following is true.

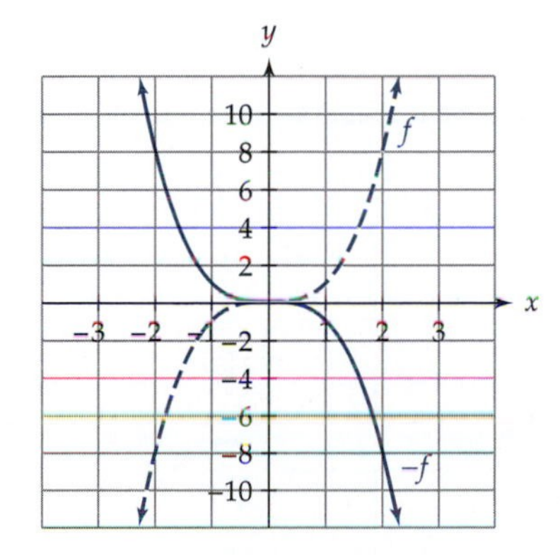

Figure 7 A graphical comparison of $f(x)$ and $-f(x)$ when $f(x) = x^3$.

TABLE 5 A Comparison of $f(x)$ and $-f(x)$ When $f(x) = x^3$

x	−3	−2	−1	0	1	2	3
$f(x)$	−27	−8	−1	0	1	8	27
$-f(x)$	27	8	1	0	−1	−8	−27

PROPERTIES

The graph of $y = -f(x)$ is a reflection of the graph of $y = f(x)$ across the x-axis.

Multiplying by a negative number other than -1 will reflect the graph over the x-axis as well as vertically stretch or compress it. For example, in Figure 8, the graph of $f(x) = \log x$ is transformed to $g(x) = -2\log x$.

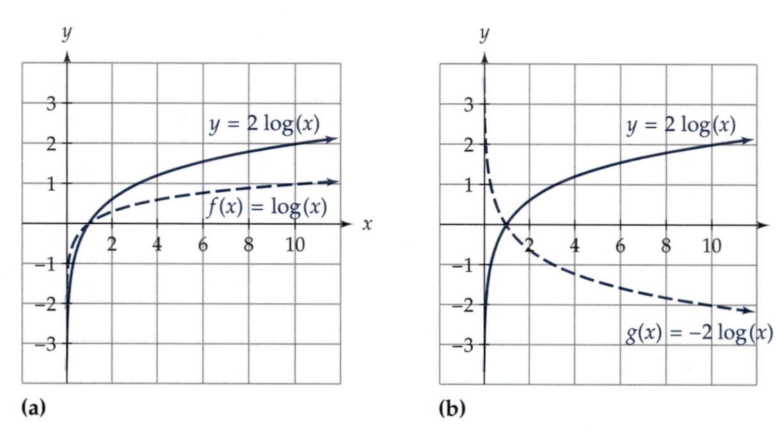

Figure 8 (a) $f(x) = \log x$ stretched by a factor of 2 to become $y = 2\log x$. (b) $y = 2\log x$ reflected across the x-axis to become $g(x) = -2\log x$.

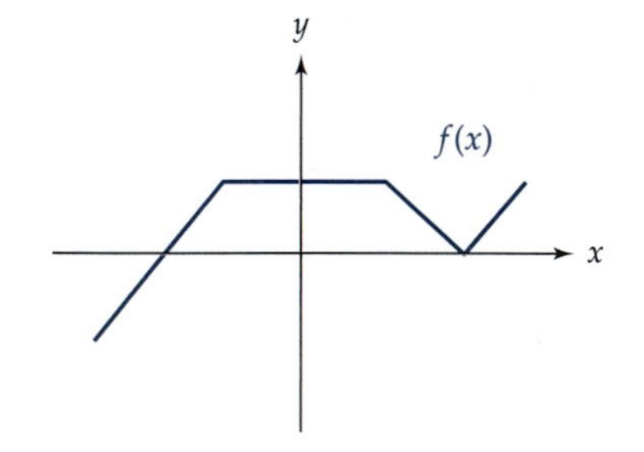

Figure 9

Example 6 Given the graph of $y = f(x)$ in Figure 9, sketch a graph of $y = -\frac{1}{2}f(x)$.

Solution The graph of $y = -\frac{1}{2}f(x)$ looks like the graph $y = f(x)$ stretched by a factor of $\frac{1}{2}$ and reflected over the x-axis. A sketch of $y = \frac{1}{2}f(x)$ is shown in Figure 10(a), and $y = -\frac{1}{2}f(x)$ is shown in Figure 10(b).

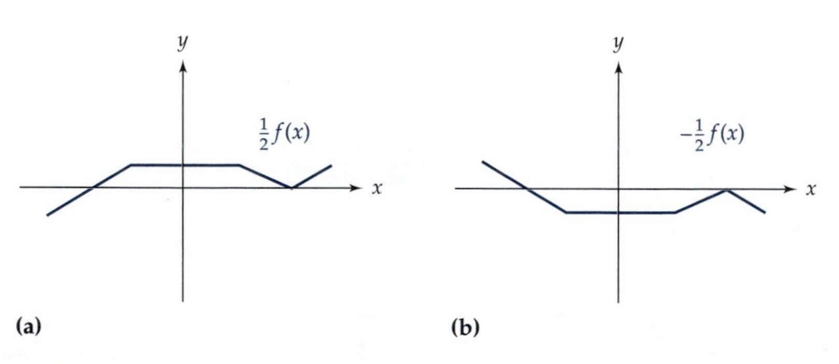

Figure 10

READING QUESTIONS

8. How is the graph of $y = -f(x)$ different from the graph of $y = f(x)$?

9. Shown are the graphs of f, $3f$, and $-3f$. Which is which?

(a)

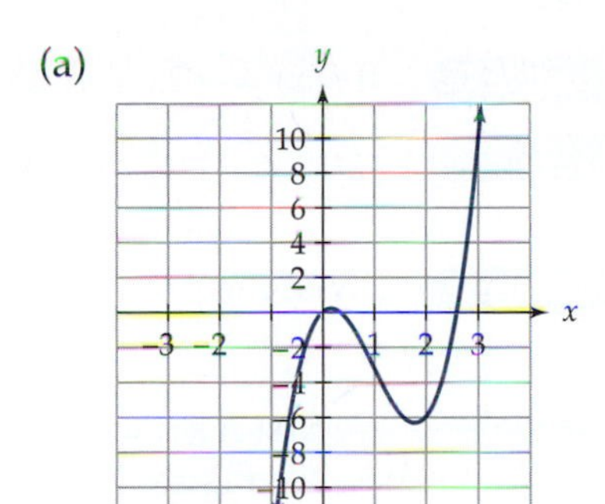

(b)

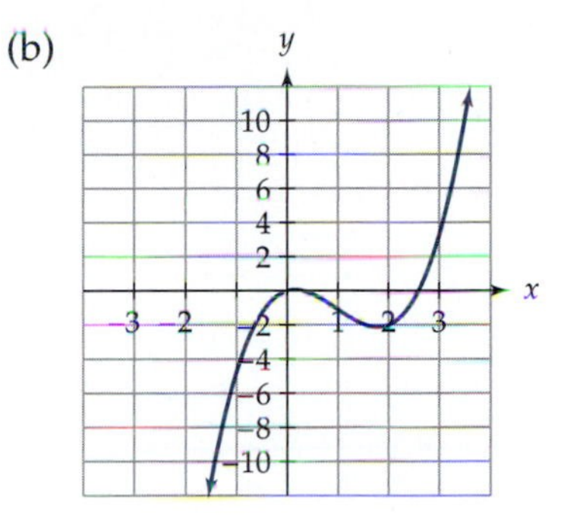

(c)

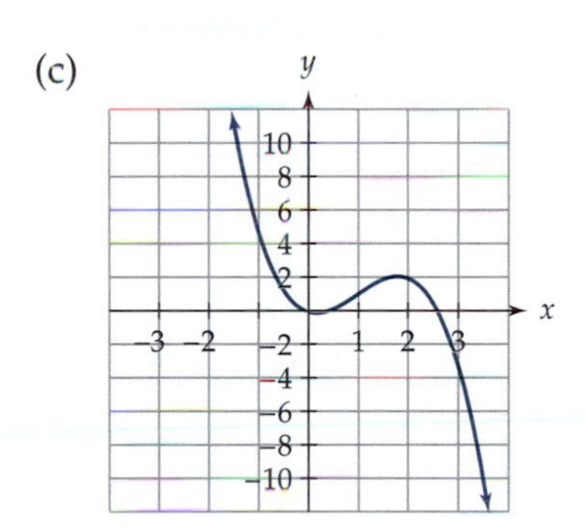

COMBINING OUTPUT CHANGES

Suppose we want to vertically shift *and* stretch a function. We need to add (or subtract) a constant to the function as well as multiply the function by a constant. Because all these operations deal with changes in output, they all result in vertical changes to a graph: vertical shifts, vertical stretches or compressions, or vertical reflections. When doing multiple changes to an output of a function, you need to follow the ordinary order of operations. That is, we need to perform operations in grouping symbols first, multiply and divide next, and add or subtract last.

Example 7 Let f be function such that $f(1) = 6$.

(a) Compute $-2f(1) + 3$ and $-2(f(1) + 3)$.

(b) How would the graph of $y = -2f(x) + 3$ differ from $y = -2(f(x) + 3)$.

Solution

(a) We have $-2f(1) + 3 = -2(6) + 3 = -12 + 3 = -9$ and $-2(f(1) + 3) = -2(6 + 3) = -2(9) = -18$.

(b) The graph of $y = -2f(x) + 3$ is the graph of $y = f(x)$ vertically stretched by a factor of 2, reflected across the x-axis, then shifted up three units. The graph of $y = -2(f(x) + 3)$ is the graph of $y = f(x)$ shifted up three units, then vertically stretched by a factor of 2 and reflected across the x-axis.

Example 8 If $f(x) = -2x + 4$, what is the formula for $\frac{1}{2}f(x) - 3$?

Solution $\frac{1}{2}f(x) - 3 = \frac{1}{2}(-2x + 4) - 3 = -x + 2 - 3 = -x - 1.$

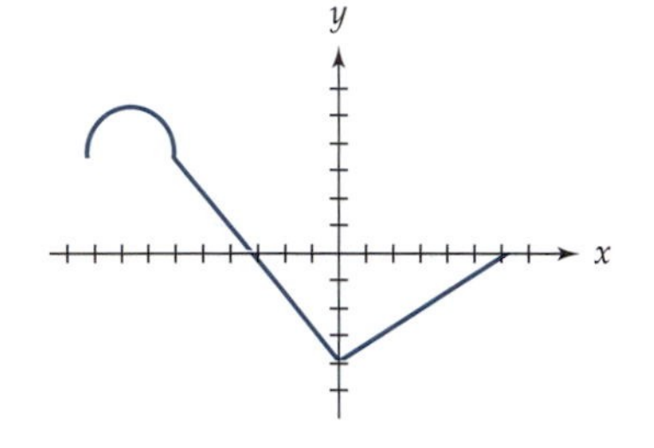

Figure 11

Example 9 Given the graph of $y = f(x)$ in Figure 11, sketch a graph of $y = -\frac{1}{2}f(x) + 4$.

Solution We first need to stretch $y = f(x)$ by a factor of $\frac{1}{2}$ and reflect it over the x-axis; see Figure 12(a). We then need to shift it up four units; see Figure 12(b).

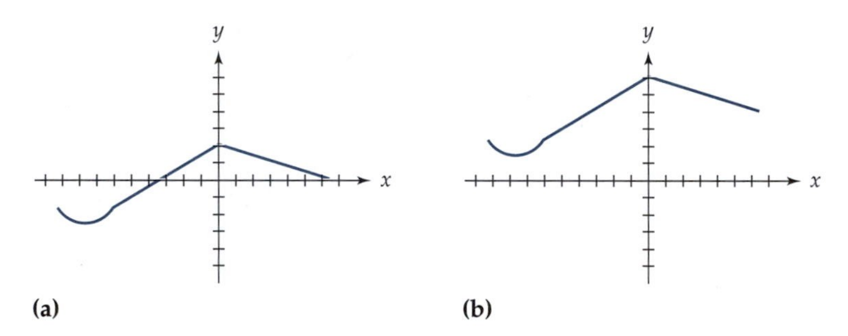

Figure 12

HISTORICAL NOTE

René Descartes is well-known as a philosopher who concluded "Cogito ergo sum" (I think, therefore I am). He is also known as the father of modern mathematics for making the connection between algebra and geometry and thus creating analytic geometry.

Descartes was born in 1596 into a French noble family. As a child in frail health, he spent much of his time in bed. During this time of quiet meditation, he developed some of his ideas for his philosophy and mathematics. After leaving school at the age of 18, Descartes went through a short period of gambling and carousing with friends in Paris. Quickly tiring of that, he decided to become a soldier. He enlisted under the Elector of Bavaria, who was then waging war against Bohemia. While celebrating St. Martin's Eve on November 10, 1619, Descartes had an experience that changed his life. After considerable drinking, Descartes fell asleep and had three vivid dreams. In his second dream, the magic key that would unlock the treasure house of nature—the exploration of natural phenomena through mathematics, including the application of algebra to geometry—was supposedly revealed to him.

continued on page 181

After spending a couple of years as a soldier, Descartes spent three years of "meditation" in Paris. He then went to Holland, where he continued to study mathematics and philosophy as well as optics, chemistry, physics, anatomy, embryology medicine, astronomy, and meteorology. In 1637, he finally published his work, *A Discourse on the Method of Rightly Conducting the Reason and Seeking Truth in the Sciences*. The third and last appendix of this book contains Descartes's new geometry, where he shows how algebraic equations can be graphed on an xy-plane, commonly called the Cartesian coordinate system. This connection between algebra and geometry was the beginning of modern mathematics.

Source: E. T. Bell, *Men of Mathematics* (New York: Simon and Schuster, 1937), pp. 35–55.

READING QUESTIONS

10. Complete the following table.

x	−2	−1	0	1	2	3
$f(x)$	−5	−3	−1	2	4	8
$-3f(x)+1$						
$2f(x)-3$						

11. Given the accompanying graph of $y = f(x)$, sketch a graph of $h(x) = -3(f(x) + 4)$.

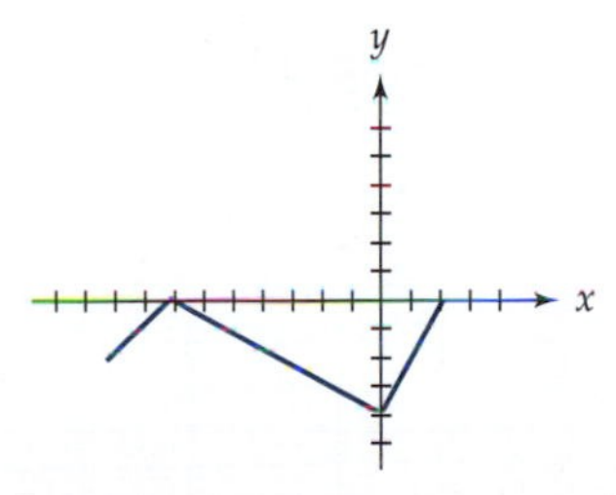

SUMMARY

Adding a positive constant to the output of a function causes a vertical upward shift in the graph of the function. Adding a negative constant causes a vertical downward shift. In particular, if f is a function and c is a positive number, then the graph of $y = f(x) + c$ is the graph of $y = f(x)$ shifted up

c units; the graph of $y = f(x) - c$ is the graph of $y = f(x)$ shifted down c units.

Multiplying the output of a function by a constant has a different affect. Multiplying by a positive constant greater than 1 causes the function to stretch vertically. Multiplying by a positive constant less than 1 causes the function to be compressed. In particular, the graph of $y = c \cdot f(x)$ is the graph of $y = f(x)$ vertically stretched by a factor of c for $c > 0$. (If $0 < c < 1$, then the graph of $y = c \cdot f(x)$ can also be referred to as being compressed by a factor of $1/c$.)

When multiplying by a negative number, the function is first reflected over the x-axis and then stretched or compressed.

You can also combine multiplying and adding constants to the output of a function. When doing these multiple changes to the output you need to follow the ordinary order of operations.

EXERCISES

1. Why do transformations in the output of a function always show themselves on a graph as vertical changes?
2. Complete the following table for the different transformations on f.

x	1	3	5	8	9
$f(x)$	−2	5	7	−3	10
$f(x) - 3$					
$-2f(x) - 3$					
$4(f(x) + 7)$					

3. Fill in the missing entries in the following table.

x	1	2	3	4	5
$f(x)$	3			−3	
$f(x) + 5$			10		6
$2f(x) - 7$		−15			

4. If $g(x) = 2x^4 + 3$, determine formulas for the following functions.
   (a) $y = 2g(x)$
   (b) $y = -g(x) - 4$
   (c) $y = 2g(x) + 7$

5. Use the accompanying graph f to sketch a graph of each of the following transformations.

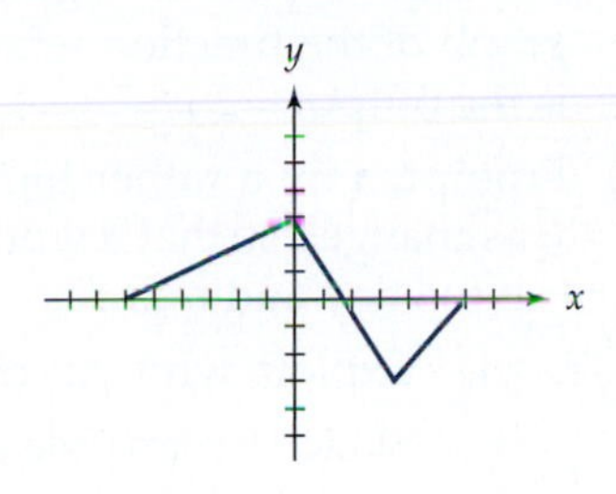

(a) $y = f(x) + 2$ (b) $y = 2f(x) + 1$ (c) $y = -2f(x) - 3$

6. The accompanying graph is of $h(x) = |x|$. Match the following functions to the following graphs. (Not all equations will be used.)

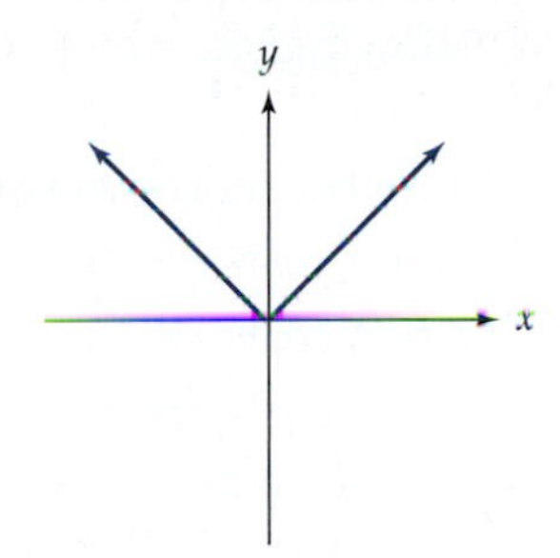

(a) $y = 1.5h(x)$ (b) $y = -h(x)$ (c) $y = h(x) - 3$
(d) $y = 1.5h(x) - 3$ (e) $y = -h(x) + 3$ (f) $y = h(x) + 3$
(g) $y = -1.5h(x) - 3$ (h) $y = 1.5h(x) + 3$

i.

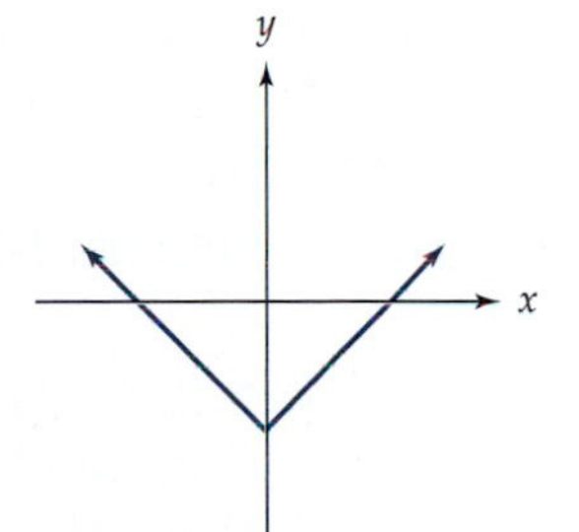

ii.

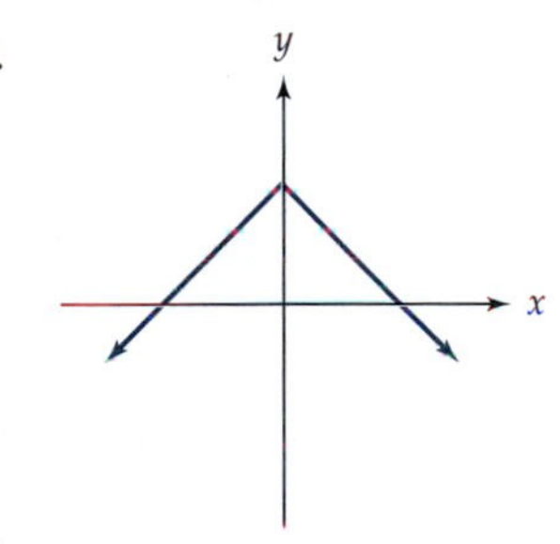

iii.

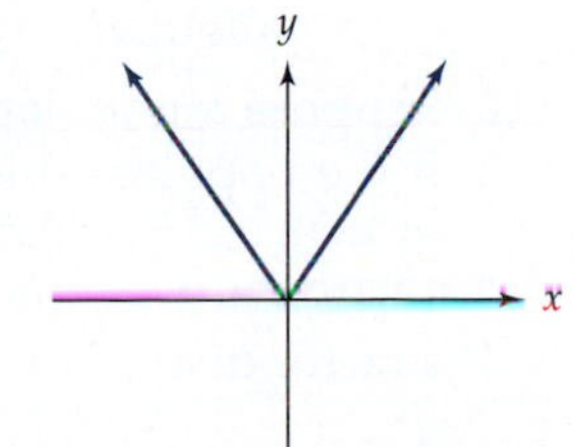

iv.

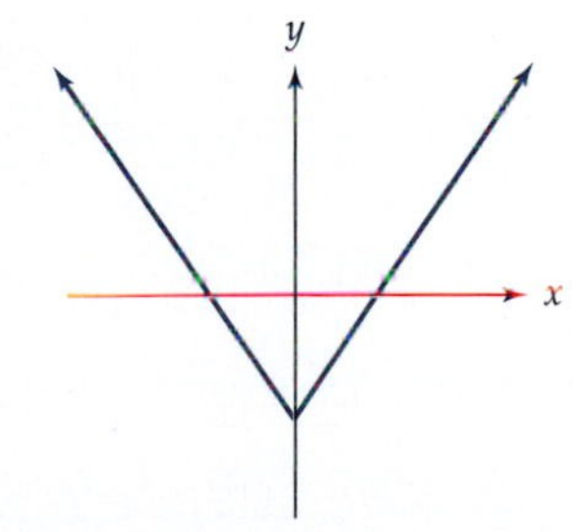

7. One of the authors had his thermostat on his house set at 70°F from 6 A.M. to 10 P.M. and 64°F from 10 P.M. to 6 A.M.
   (a) For a 24-hr period beginning at midnight, sketch a plausible graph of the function where time is the input and temperature is the output.
   (b) Prompted by a rather high gas bill one month, the thermostat was changed so that it would be 3°F cooler in the house throughout the day and night.
      i. Explain why this change is a change in output.
      ii. Sketch a plausible graph of the new time versus temperature function.
      iii. What could be done to get a change in input rather than a change in output?
8. Suppose c represents the function where x is a certain state college and $c(x)$ is the cost of tuition for the college in hundreds of dollars. Match the following situations to the formulas. (Not all functions will be used.)
   (a) All tuition costs are increased by $100.
   (b) All tuition costs are decreased by $100.
   (c) All tuition costs are increased by 10%.
   (d) All tuition costs are decreased by 10%.

i. $y = 0.1c(x)$	ii. $y = 0.9c(x)$	iii. $y = c(x) + 1$
iv. $y = 1.1c(x)$	v. $y = -0.1c(x)$	vi. $y = c(x) - 1$
vii. $y = c(x) - 10$	viii. $y = c(x) + 10$	ix. $y = c(x) + 0.1$

9. Suppose g is a vertical transformation of f and $f(3) = g(3)$.
   (a) Is g a vertical shift or a stretch of f? How do you know?
   (b) What is $f(3)$?
10. What is the amplitude of $y = \frac{3}{2}\sin x$?
11. Answer the following questions about transformations of f with the constant a.
   (a) Suppose $f(1) > 0$ and $0 < af(1) < f(1)$. What do you know about a?
   (b) Suppose $f(1) > 0$ and $af(1) > f(1)$. What do you know about a?
   (c) Suppose $f(1) > 0$ and $a + f(1) > f(1)$. What do you know about a?
   (d) Suppose $f(1) > 0$ and $a + f(1) < f(1)$. What do you know about a?
12. Suppose a population of a town is growing at a rate of 3% per year. If the population in 1970 was 2000, then the function describing the growth can be modeled by $p(t) = 2000(1.03)^t$, where t is the time since 1970. Which of the following situations describes a vertical stretch by a factor of 4?
   A. The population of a town in 1970 was 8000 and has increased by 3% per year.

B. The population of a town in 1970 was 2000 and has increased by 12% per year.

13. The functions g and h in the following table are transformations of the function f. Determine formulas for g and h in terms of f.

x	1	3	5	8	9
$f(x)$	−1	0	2	5	9
$g(x)$	−5	−4	−2	1	5
$h(x)$	1	3	7	13	21

14. In the United States, the function that converts the length of someone's foot to a man's shoe size is different than the one that converts foot length to a woman's shoe size. The table shows the functions that convert the length of a person's foot, in inches, to a man's shoe size and a woman's shoe size.

Foot Length (in.)	**8**	**9**	**10**	**11**	**12**
Men's Shoe Size	2	5	8	11	14
Women's Shoe Size	3.5	6.5	9.5	12.5	15.5

(a) Let m represent the function where the input, x, is the length of a person's foot and the output, $m(x)$, is the men's shoe size that person would wear. What kind of a function is it?

(b) Determine the formula where the input, x, is the length of a person's foot and the output, $m(x)$, is the men's shoe size that person would wear.

(c) Let w represent the function where the input, x, is the length of a person's foot and the output, $w(x)$, is the women's shoe size that person would wear. What is the relationship between $w(x)$ and $m(x)$? Use this relationship to determine a formula for w.

15. In this section, we stated that the functions used to determine someone's federal income tax are piecewise functions. We looked at one piece of one function. The following is a complete piecewise function for married couples filing jointly. The input for this function is taxable income, i, and the output is the amount of their federal tax, $t(i)$.

$$t(i) = \begin{cases} 0.15i & \text{if } \$0 < i \leq \$41{,}200 \\ 6180.00 + 0.28(i - 41{,}200) & \text{if } \$41{,}200 < i \leq \$99{,}600 \\ 22{,}532.00 + 0.31(i - 99{,}600) & \text{if } \$99{,}600 < i \leq \$151{,}750 \\ 38{,}698.50 + 0.36(i - 151{,}750) & \text{if } \$151{,}750 < i \leq \$271{,}050 \\ 81{,}646.50 + 0.396(i - 271{,}050) & \text{if } i > \$271{,}050 \end{cases}$$

For the following questions, use the piece of the function for couples with taxable income of $99,600 < i \le $151,750; that is $t(i) = 22{,}532.00 + 0.31(i - 99{,}600)$.

(a) What does the $(i - 99{,}600)$ represent in this function?

(b) How did the I.R.S. determine the 22,532.00 in this function?

(c) Suppose a couple takes the adoption tax credit of $5000. Why is this credit a change in *output*?

(d) Rewrite the piece of this function for those taking the adoption credit of $5000. Simplify your formula.

16. Suppose f is a linear function, $f(x) = mx + b$.

(a) If f is transformed by adding a constant to the output, will the slope or y-intercept change? Explain.

(b) If f is transformed by multiplying the output by a constant, will the slope or y-intercept change? Explain.

INVESTIGATIONS

INVESTIGATION: HOT WATER

The difference in temperature between a warm object and its surrounding temperature can be modeled by a decreasing exponential function. We took a cup of hot water, heated it in a microwave, and allowed it to cool on the kitchen counter. We recorded the temperature of the water (in degrees Celsius) every 5 min for 85 min. Table 6 represents the data. In this investigation, we determine the formula of the function that models these data.

TABLE 6

Time (min)	0	5	10	15	20	25	30	35	40
Temperature (°C)	64.9	58.8	54.0	50.1	46.8	44.0	41.6	39.4	37.5
Time (min)	45	50	55	60	65	70	75	80	85
Temperature (°C)	35.9	34.5	33.3	32.0	30.8	29.9	29.0	28.2	27.6

1. Because the *difference* in temperature between a warm object and its surrounding temperature can be modeled with an exponential function, we first need to transform our data. The surrounding temperature (the temperature of the kitchen) was 21.4°C.

(a) Transform Table 6 so that the output is the *difference* in temperature between the cup of water and the kitchen. (*Note:* The outputs should all be positive.)

(b) Plot your transformed data. Explain why your graph looks like a decreasing exponential function.

2. Exponential functions are of the form $f(x) = ba^x$, where b is the y-intercept and a is the growth factor.
   (a) By simply looking at the data, determine the y-intercept.
   (b) Use a data point other than the y-intercept to determine a value for the growth factor.
   (c) Give the equation for your exponential function.
   (d) Plot your transformed data along with your function. Does your function do a good job of modeling the data? (*Note:* If it does not fit very well, then you might want to redo parts (b) and (c) by choosing another point.)
   (e) If the function that models the cooling cup of water is exponential, then the temperature decreases by the same percentage amount each minute. What is that amount for the cooling water in question 2, part (c)?
3. Transform your exponential function so that it models the original data given in the table. Plot this function along with the original data.

3.2 FUNCTION TRANSFORMATIONS: CHANGES IN INPUT

In the last section, we transformed functions by adding and multiplying the output by a constant. In this section, we show the effects of adding and multiplying the *input* by a constant. Because the input appears on the horizontal axis of a graph, the effect on the function's graph of changing the input is a horizontal change. At the end of this section, we transform functions by combining both input and output changes.

ADDING AND SUBTRACTING A CONSTANT TO THE INPUT

In section 3.1, we looked at the function whose input was the amount of taxable income for a married couple and whose output was their federal income tax. We saw that a tax credit was a direct reduction in someone's tax. Because the amount of tax was the output of the function, we had a change in output. As a result, when we compared the graphs of the function before and after a tax credit, one was vertically shifted from the other. Another way that taxes are commonly reduced is with deductions and exemptions. These directly reduce the amount of a person's taxable income and are therefore a change in input. In 1997, a family of four was allowed to take a standard deduction of \$6900 and an exemption of \$10,600, which reduced their taxable income by \$17,500. If I is their total income and i is their taxable income, then $I - \$17{,}500 = i$. Let's assume the family of four has a taxable income between \$41,200 and \$99,600 (similar to the assumption made in section 3.1). To determine the amount of tax owed, use the function

$$t(i) = 0.28i - 5356 \quad \text{for} \quad \$41{,}200 < i \leq \$99{,}600,$$

where i represents the amount of taxable income and $t(i)$ represents the amount of tax owed. Let's also look at the function, T, for families in this same income category. In this function, the input, I, is the *total* income and the output is the amount of tax owed. Because $i = I - \$17{,}500$, we can rewrite the function, $t(i)$, so that the input is the total income by replacing i with $I - \$17{,}500$. Because we are changing the input, we are also changing the domain. Our new function is

$$\begin{aligned} T(I) &= 0.28(I - 17{,}500) - 5356 && \text{if} \quad \$41{,}200 < I - 17{,}500 \leq \$99{,}600 \\ &= 0.28I - 10{,}256 && \text{if} \quad \$58{,}700 < I \leq \$117{,}100. \end{aligned}$$

Notice that the endpoints of the domain for T are larger than the endpoints for t. Graphically, T is shifted to the right of t, which makes sense for two reasons. First, we made an input change, and because input is represented horizontally on a graph, we expect to see a horizontal shift. Second, considering deductions and exemptions, we need an income \$17,500 higher to pay the same tax, which forces the shift to be to the *right*. Both t and T are graphed in Figure 1. Adding a constant to the input of any function always results in a horizontal shift. In general, the following is true.

PROPERTIES

If $y = f(x)$ is a function and c is a positive number, then

- The graph of $y = f(x + c)$ is the graph of $y = f(x)$ shifted to the left c units.
- The graph of $y = f(x - c)$ is the graph of $y = f(x)$ shifted to the right c units.

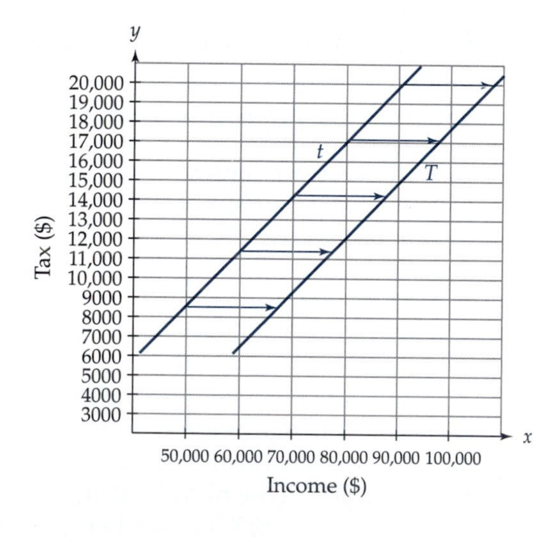

Figure 1 Graphical representations of federal tax functions. $T(I)$ represents the amount of tax paid given a total income of I dollars, and $t(i)$ represents the amount of tax paid given a taxable income of i dollars.

Example 1 Given the functions f and g in Table 1, answer the following questions.

(a) How are the values of $g(x)$ related to the values of $f(x)$?

(b) Knowing that $f(x) = x^3$, determine a formula for g.

(c) How is the graph of g related to the graph of f?

TABLE 1

x	−2	−1	0	1	2	3
$f(x)$	−8	−1	0	1	8	27
$g(x)$	−27	−8	−1	0	1	8

Solution

(a) The values for $g(x)$ are shifted one column to the right of those for $f(x)$. Thus, for outputs to be equal, the input of f must be one less than that of g.

(b) Symbolically, $g(x) = f(x - 1)$. Therefore, $g(x) = (x - 1)^3$.

(c) The graph of g would look like the graph of f shifted to the right one unit.

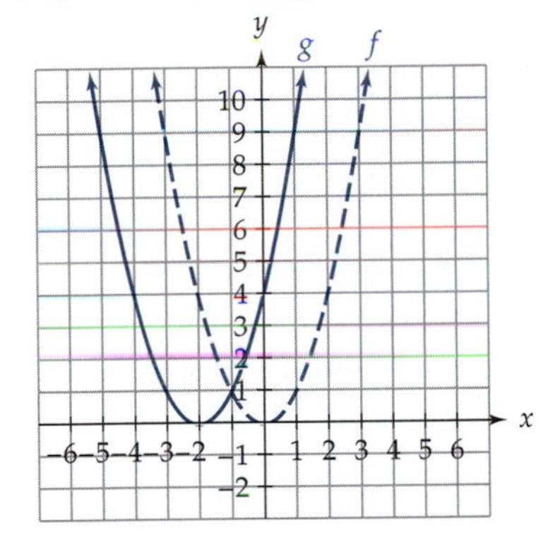

Figure 2

Example 2 One of the graphs in Figure 2 is of the function $f(x) = x^2$, and the other is a transformation of $f(x) = x^2$ resulting from a horizontal shift. What is the formula for the transformed function, g?

Solution The graph of $f(x) = x^2$ is the one that passes through the origin. Because the other graph passes through the point $(-2, 0)$, it is shifted to the left two units. So, for the outputs to be equal, the input of f must be two more than that of g. Hence, $g(x) = f(x + 2) = (x + 2)^2$.

READING QUESTIONS

1. Why is a tax deduction considered a change in input?
2. If $f(x) = 2x - 7$, determine the formula for $g(x) = f(x + 1)$. Describe the graph of g compared with that of f.
3. If g is represented by the accompanying graph, sketch a graph of $h(x) = g(x - 2)$.

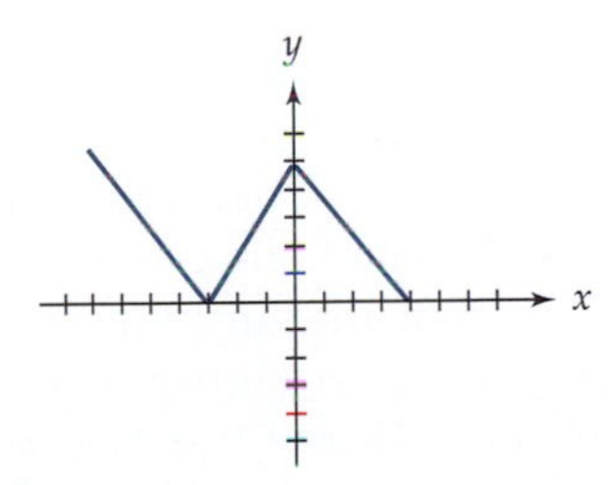

MULTIPLYING THE INPUT BY A POSITIVE CONSTANT

In section 3.1, we saw that adding a constant to the output of a function resulted in a vertical shift of the function's graph and that multiplying the output of a function by a constant resulted in a vertical stretch of the function's graph. In this section, we saw that adding a constant to the input results in a horizontal shift of the function's graph. Now let's see what happens when we multiply the input by a constant. Given the function $f(x)$ as defined by Table 2 and filling in the missing entries of the table, we can see how the function $g(x) = f(3x)$ looks.

TABLE 2 Given f, Determine Outputs for g Such That $g(x) = f(3x)$

x	−3	−1	0	1	3
$f(x)$	10	2	1	2	10
$g(x) = f(3x)$					

Because $g(-3) = f(3 \cdot -3) = f(-9)$, which is not defined in Table 2, we cannot fill in that entry of the table. Next, we have $g(-1) = f(3 \cdot -1) = f(-3) = 10$. To determine $g(0)$, we have $g(0) = f(3 \cdot 0) = f(0) = 1$. The next entry gets filled in a similar manner, and a number for the last entry, like the first, cannot be determined. Table 3 shows a completed version of Table 2.

TABLE 3 A Comparison of $f(x)$ and $g(x) = f(3x)$

x	−3	−1	0	1	3
$f(x)$	10	2	1	2	10
$g(x) = f(3x)$		10	1	10	

By comparing the outputs for g and f in Table 3, we see that each 10 has moved in toward the center of the table and the 1 that was in the middle stayed there. When we look at a graph of these points, we see the same thing. The symbolic function used for f in Table 2 is $f(x) = x^2 + 1$, which means that the formula for g is $g(x) = f(3x) = (3x)^2 + 1$. Both f and g are shown in Figure 3. You can see that $f(3) = g(1) = 10$ and $f(-3) = g(-1) = 10$ and that $f(0) = g(0) = 1$. The graph of g is the graph of f horizontally compressed by a factor of 3. If a function is horizontally compressed, then it is always a movement toward or away from the y-axis with the points on the y-axis not changing. We sometimes refer to a compression away from the y-axis as a horizontal stretch. In general, the following is true.

PROPERTIES

The graph of $y = f(cx)$ is the graph of $y = f(x)$ horizontally compressed by a factor of c for $c > 0$. (If $0 < c < 1$, then the graph of $y = f(cx)$ can also be referred to as being stretched by a factor of $1/c$.)

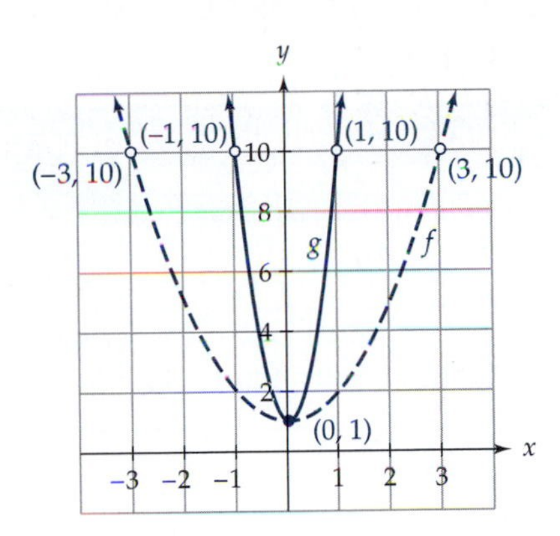

Figure 3 The graphs of $f(x) = x^2 + 1$ and $g(x) = (3x)^2 + 1$.

Example 3 In Figure 4, $f(x) = \cos x$. Determine a formula for $g(x)$.

Solution It appears from the graph that $g(x)$ is a horizontal stretch by a factor of 2. Thus, it is a horizontal compression by a factor of $\frac{1}{2}$. Therefore, $g(x) = f(x/2) = \cos(x/2)$.

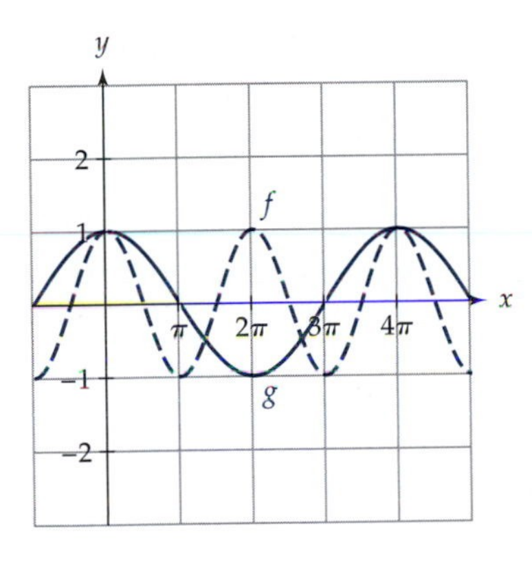

Figure 4

Example 4 Assume it costs \$42 to have cable television installed in your home and \$12 a month for limited basic service. The total cost of cable television can be calculated using the function $C(t) = 42 + 12t$, where t is the length of time, in months, that you have the service. Which of the following situations can be represented by a function whose graph is a horizontal compression of C by a factor of $\frac{3}{2}$? Explain your answer.

A. It costs \$63 to have cable television installed in your home and \$18 a month for basic service.

B. It costs \$42 to have cable television installed in your home and \$18 a month for basic service.

Solution A horizontal compression of C by a factor of $\frac{3}{2}$ is $C(\frac{3}{2}t) = 42 + 12(\frac{3}{2}t) = 42 + 18t$, which describes situation B. The total cost of cable television for situation A is $A(t) = 63 + 18t$, which is equal to $y = \frac{3}{2}C(t)$. Therefore, the correct answer is situation B.

Application: Chicken Bacteria

In section 2.2, we looked at an application involving bacteria on chicken as an example of an exponential function. We saw that bacteria on fresh chicken that was kept at 40°F increased by a factor of 16 every day, which the label noted meant that the bacteria was doubling every 6 hr. The label also stated that the number of bacteria would double every 12 hr if the chicken were kept at 36°F. Decreasing the temperature causes the bacteria to increase more slowly. Let's compare the functions that describe bacteria growth at these two different temperatures. If we start with an average of 100 bacteria per square centimeter on a chicken, then the functions representing bacteria growth over time can be represented by Table 4.

TABLE 4 The Bacteria Count on Chicken per Square Centimeter in a 36°F Refrigerator and a 40°F Refrigerator

Time (hr)	0	1	2	3	4	5	6	7	8	9	10	11	12
Bacteria at 40°F	100	112	126	141	159	178	200	224	252	283	317	356	400
Bacteria at 36°F	100	106	112	119	126	133	141	150	159	168	178	189	200

Notice that in Table 4 every other bacteria count in the 36°F row is the same as a bacteria count in the 40°F row, but at a different time. For example, the bacteria count at 36°F is 112 after 2 hr, whereas the bacteria count at 40°F is 112 after 1 hr. If the function used to determine the number of bacteria at 36°F after t hours is $y = N_{36}(t)$ and the function used to determine the number of bacteria at 40°F after t hours is $y = N_{40}(t)$, then $N_{36}(2) = N_{40}(1) = 112$. In general, it takes the bacteria growing at 40°F half as long to reach a certain size population when compared with the bacteria growing at 36°F. Thus, $N_{36}(x) = N_{40}(\frac{1}{2}x)$. Therefore, N_{36} is a horizontal compression of N_{40} by a factor of $\frac{1}{2}$, which is the same as a horizontal stretch by a factor of 2, as we can see in the graph of these two functions. We start by plotting the points given in Table 4. In Figure 5(a), points representing the number

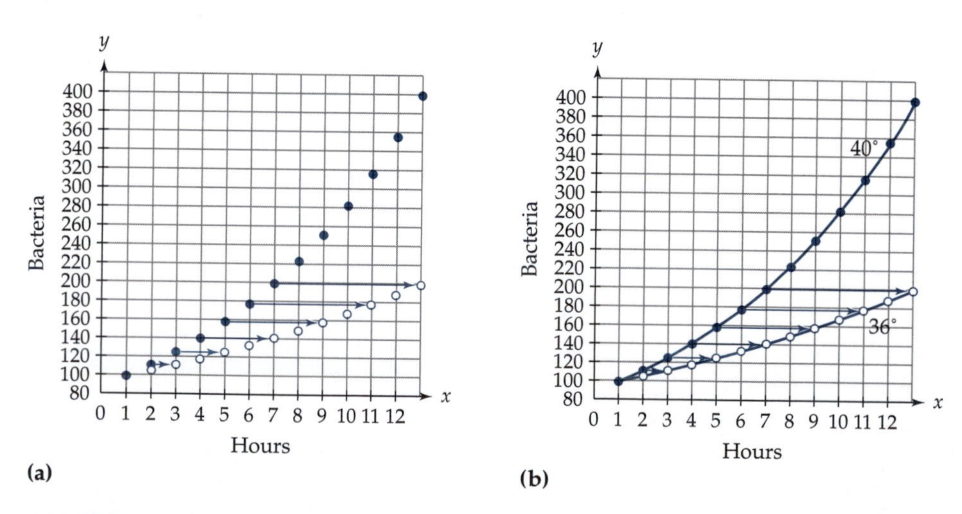

Figure 5 (a) A plot of the points from Table 4. (b) The same plot connected by a smooth curve. Notice that the time it takes the chicken in the 36°F refrigerator to reach a certain bacteria count is twice that of the chicken in the 40°F refrigerator.

of bacteria on the chicken in a 36°F refrigerator are moved to the right a total of twice their distance from the y-axis. By connecting the points in Figure 5(a) with a smooth curve, we again see that multiplying the input of a function by $\frac{1}{2}$ results in a horizontal stretch of the entire original curve. See Figure 5(b).

READING QUESTIONS

4. How is the graph of $y = f(3x)$ different from the graph of $y = f(x)$?

5. Complete as much of the following table as possible.

x	-2	-1	0	1	2
$f(x)$	−6	−4	−2	0	2
$f(2x)$					

6. At $t = 12$ hr, the number of bacteria on the chicken kept at 36°F is 200. When is the number of bacteria on the chicken kept at 40°F equal to 200?

MULTIPLYING THE INPUT BY A NEGATIVE CONSTANT

So far, we considered only what happens when you multiply the input of a function by a positive number. Let's see what happens when you multiply the input by a negative number. A function, f, is shown in Table 5. We are interested in determining the values for $f(-1 \cdot x) = f(-x)$. When $x = -3, f(-x) = f(-(-3)) = f(3)$. From the table, we can see that $f(3) = 9$. Therefore, the first entry in the table for $f(-x)$ is 9. To determine the second entry, we need to find $f(-(-2))$. Because $f(-(-2)) = f(2) = 7$, our second entry for $f(-x)$ is 7. Continuing this process, we obtain the answers given in Table 6. Notice that the values of $f(-x)$ are the same as those of $f(x)$ except in reverse order. Also note that when $x = 0, f(x) = f(-x)$ because $-0 = 0$. We see these same properties when we look at this function graphically.

The equation we used for f in Table 5 is $f(x) = 2x + 3$, which means that $f(-x) = 2(-x) + 3 = -2x + 3$. The graphs of these two functions are

TABLE 5 A Function f shown in Numerical form

x	-3	-2	-1	0	1	2	3
$f(x)$	−3	−1	1	3	5	7	9
$f(-x)$							

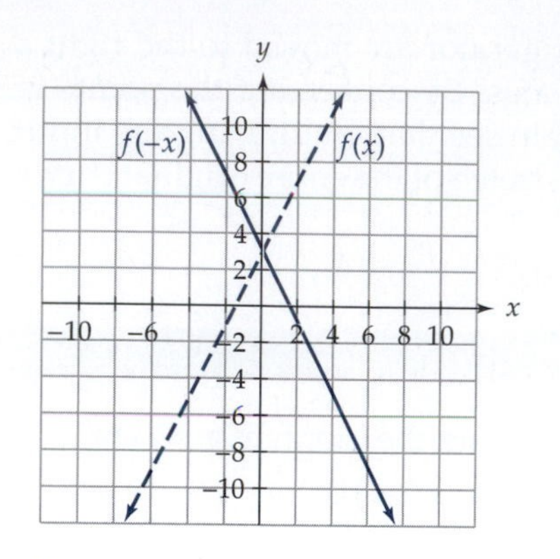

Figure 6 Graphs of $y = f(x)$ and $y = f(-x)$ when $f(x) = 2x + 3$.

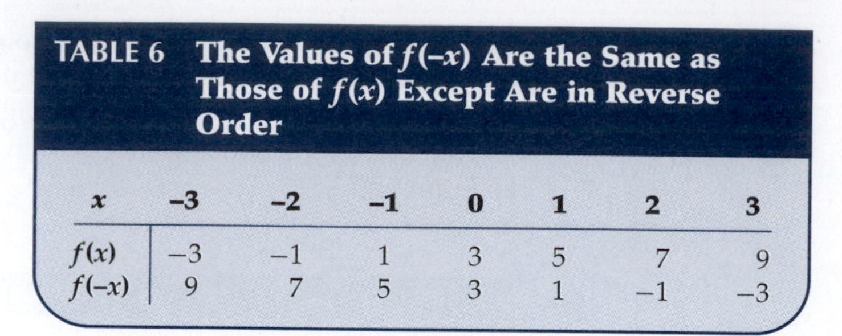

TABLE 6 The Values of $f(-x)$ Are the Same as Those of $f(x)$ Except Are in Reverse Order

x	−3	−2	−1	0	1	2	3
$f(x)$	−3	−1	1	3	5	7	9
$f(-x)$	9	7	5	3	1	−1	−3

shown in Figure 6. Notice that $y = f(-x)$ is a reflection of $y = f(x)$ across the y-axis. This "flip" could be predicted by noticing the reverse order of the outputs in Table 6. In general, the following is true.

PROPERTIES

The graph of $y = f(-x)$ is a reflection of the graph of $y = f(x)$ across the y-axis.

Multiplying the input by a negative number other than 1 results in not only a graph that is reflected across the y-axis, but also one that is horizontally compressed.

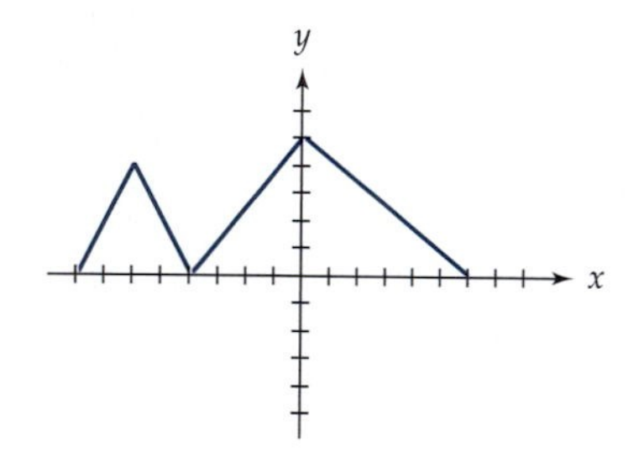

Figure 7

Example 5 Given the graph of $y = f(x)$ in Figure 7, sketch a graph of $y = f(-2x)$.

Solution The graph of $y = f(-2x)$ looks like the graph $y = f(x)$ reflected over the y-axis and compressed by a factor of 2. A sketch of $y = f(-x)$ is shown in Figure 8(a), and a sketch of $y = f(-2x)$ is shown in Figure 8(b).

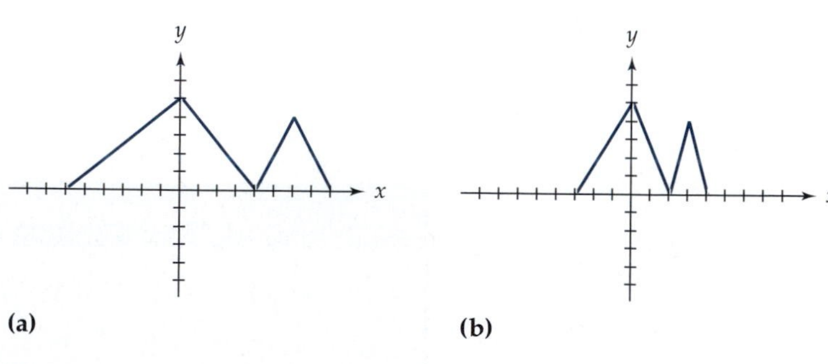

(a) (b)

Figure 8

READING QUESTIONS

7. How is the graph of $y = f(-x)$ different from the graph of $y = f(x)$?
8. The following graphs are of $y = f(x)$, $y = f(-x)$, and $y = f(-2x)$. Which is which?

(a)

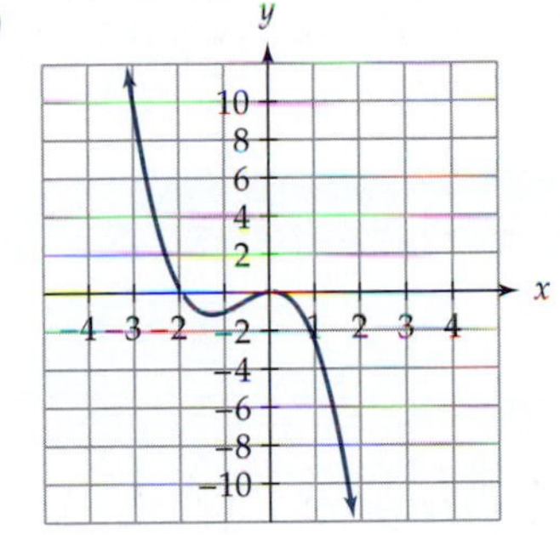

(b)

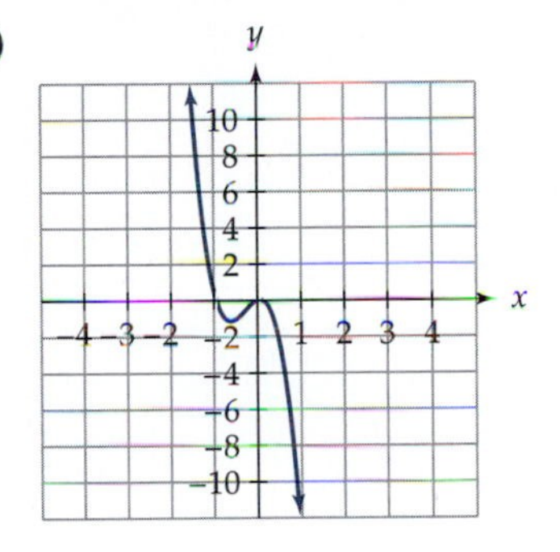

(c)

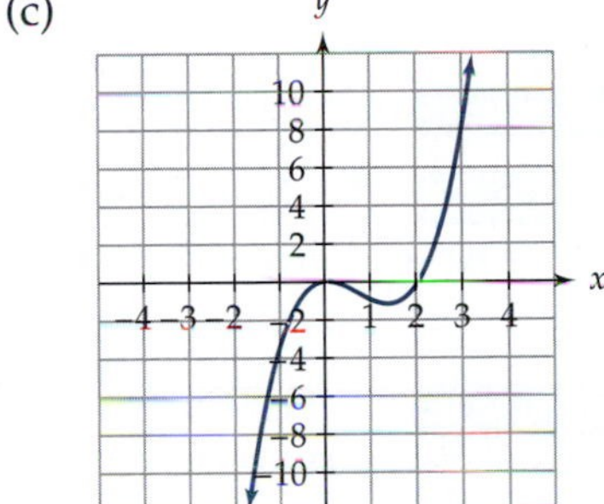

COMBINATIONS OF INPUT AND OUTPUT CHANGES

Let's look at what happens when we combine input changes together with output changes. In each case, it is important to first make the input changes before making the output changes.

Example 6 Complete as much of Table 7 as possible.

TABLE 7

x	−2	−1	0	1	2
$f(x)$	−7	−4	−1	6	5
$f(-x) + 3$					
$2f(x + 1) - 1$					

Solution For $x = -2$, we have $f(-(-2)) + 3 = f(2) + 3 = 5 + 3 = 8$. For $x = -1$, we have $f(-(-1)) + 3 = f(1) + 3 = 6 + 3 = 9$. The rest of the outputs are determined in a similar manner and are shown in Table 8.

For $x = -2$, we have $2f(-2+1) - 1 = 2f(-1) - 1 = 2(-4) - 1 = -9$. For $x = -1$, we have $2f(-1+1) - 1 = 2f(0) - 1 = 2(-1) - 1 = -3$. The rest of the outputs are determined in a similar manner except for $x = 2$. In that case, $2f(2+1) - 1 = 2f(3) - 1$. Because $f(3)$ is not defined in Table 7, we cannot determine a value for it.

TABLE 8

x	-2	-1	0	1	2
$f(x)$	−7	−4	−1	6	5
$f(-x) + 3$	8	9	2	−1	−4
$2f(x+1) - 1$	−9	−3	11	9	

Example 7 If $f(x) = x^2$, what is the formula for $y = \frac{1}{2}f(2x+1) - 5$?

Solution We have $y = \frac{1}{2}f(2x+1) - 5 = \frac{1}{2}(2x+1)^2 - 5 = \frac{1}{2}(4x^2 + 4x + 1) - 5 = (2x^2 + 2x + \frac{1}{2}) - 5 = 2x^2 + 2x - \frac{9}{2}$.

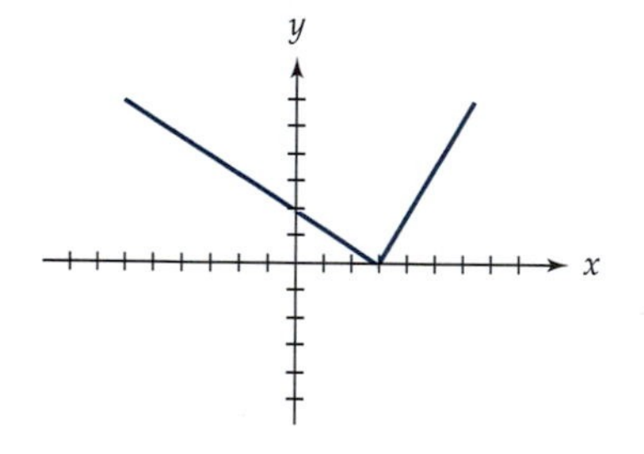

Figure 9

Example 8 Given the graph of $y = f(x)$ in Figure 9, sketch a graph of $y = 2f(-3x) - 4$.

Solution The graph of $y = f(-3x)$ looks like the graph $y = f(x)$ reflected over the y-axis and compressed by a factor of 3 and is shown in Figure 10(a). The graph of $y = 2f(-3x) - 4$ looks like the graph $y = f(-3x)$ vertically stretched by a factor of 2 and shifted four units down. This graph is shown in Figure 10(b).

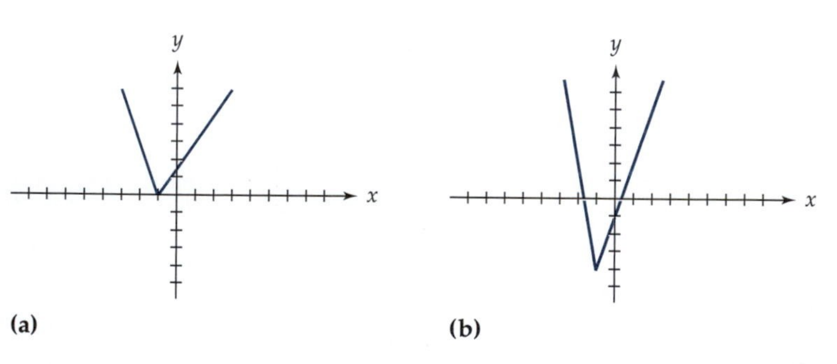

Figure 10

READING QUESTIONS

9. Complete as much of the following table as possible.

x	0	1	2	3	4
$f(x)$	2	4	8	11	13
$2f(x-1)+3$					

10. If $g(x) = 2^x$, determine a formula for $y = 3g(2x - 1) + 7$.

SUMMARY

Changing the input of a function is similar to changing the output except that the changes are horizontal rather than vertical. The following is a summary of all of the output and input changes from sections 3.1 and 3.2.

- If $y = f(x)$ is a function and c is a positive number, then
  - The graph of $y = f(x) + c$ is the graph of $f(x)$ shifted up c units.
  - The graph of $y = f(x) - c$ is the graph of $f(x)$ shifted down c units.
- The graph of $y = c \cdot f(x)$ is the graph of $y = f(x)$ vertically stretched by a factor of c for $c > 0$. (If $0 < c < 1$, then the graph of $y = c \cdot f(x)$ can also be referred to as being compressed by a factor of $1/c$.)
- The graph of $y = -f(x)$ is a reflection of the graph of $y = f(x)$ across the x-axis.
- If $y = f(x)$ is a function and c is a positive number then
  - The graph of $y = f(x + c)$ is the graph of $y = f(x)$ shifted to the left c units.
  - The graph of $y = f(x - c)$ is the graph of $y = f(x)$ shifted to the right c units.
- The graph of $y = f(cx)$ is the graph of $y = f(x)$ horizontally compressed by a factor of c for $c > 0$. (If $0 < c < 1$, then the graph of $y = f(cx)$ can also be referred to as being stretched by a factor of $1/c$.)
- The graph of $y = f(-x)$ is a reflection of the graph of $y = f(x)$ across the y-axis.
- Do input changes before doing output changes.

EXERCISES

1. Why do transformations of the input of a function always result in horizontal changes on a graph?
2. When a graph is compressed horizontally, what points do not change? Why?
3. Given the function f in the following table, complete as much of the table as possible for the different transformations of f.

x	−4	−3	−2	−1	0	1	2	3	4
$f(x)$	−2	5	7	−3	10	−1	6	0	8
$f(2x)$									
$f(x+3)$									
$4f(-x)+1$									

4. If $g(x) = x^2 + 3$, determine formulas for the following functions.
   (a) $y = 2g(x)$ and $y = g(2x)$
   (b) $y = -g(x)$ and $y = g(-x)$
   (c) $y = 2g(x+1) + 3$
5. Using the accompanying graph f, sketch a graph of each of the following transformations.

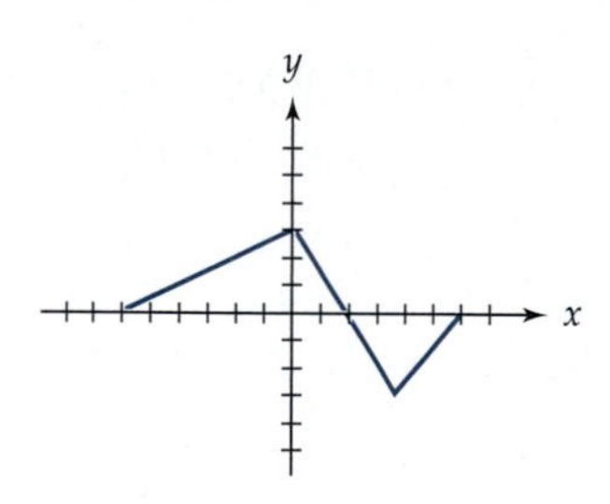

   (a) $y = f(x-2)$
   (b) $y = f(-2x)$
6. Use the accompanying graph of $h(x) = |x|$ to match the following functions to the graphs given. (Not all the functions will be used.)

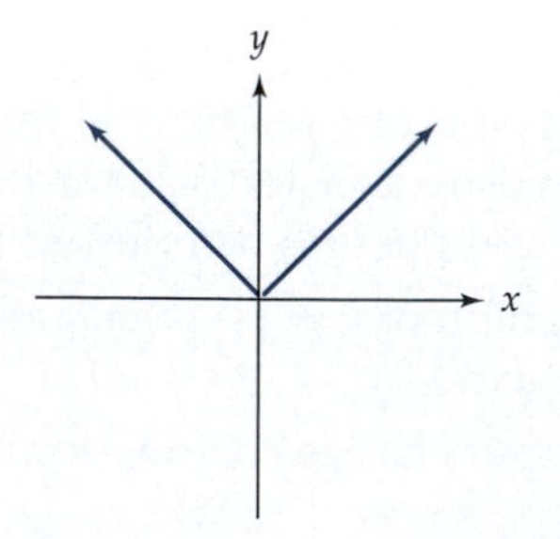

(a) $y = h(x+3)$ (b) $y = -h(x)$ (c) $y = h(x) - 3$
(d) $y = h(2x)$ (e) $y = h(x-3)$ (f) $y = h(-x)$
(g) $y = h(x) + 3$ (h) $y = h(\frac{1}{2}x)$

i. 

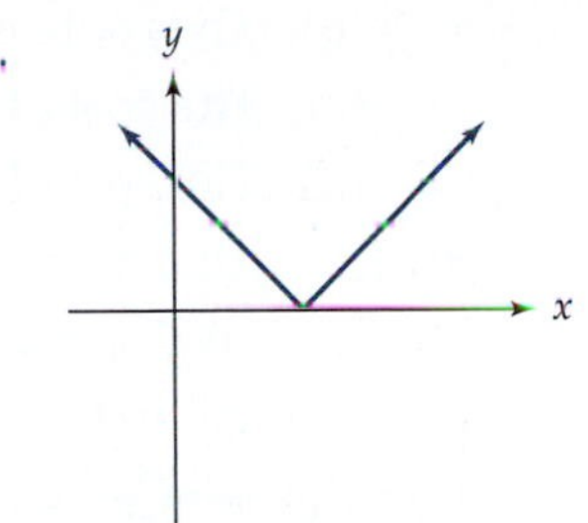

ii.

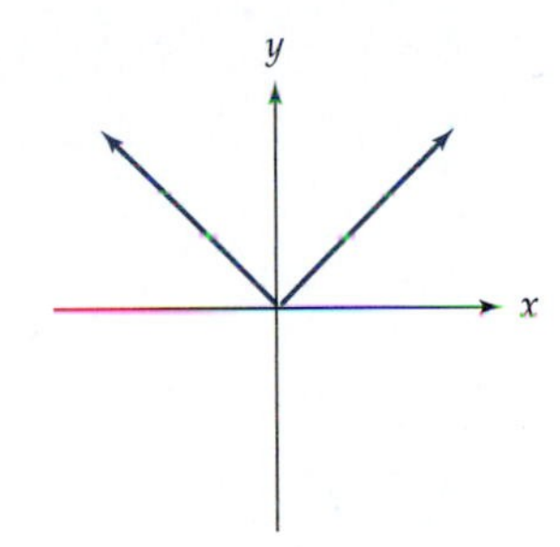

iii.

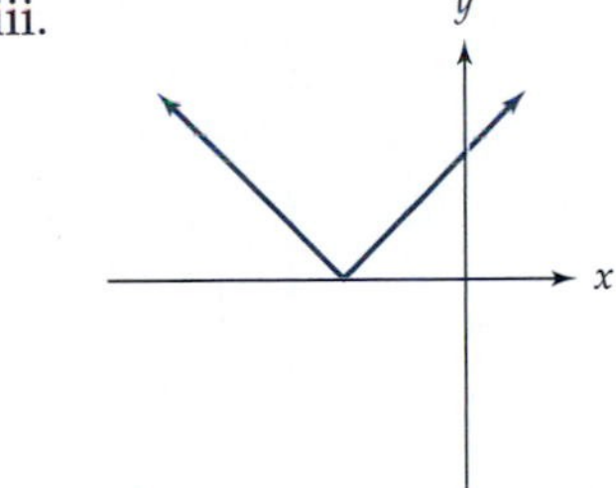

iv.

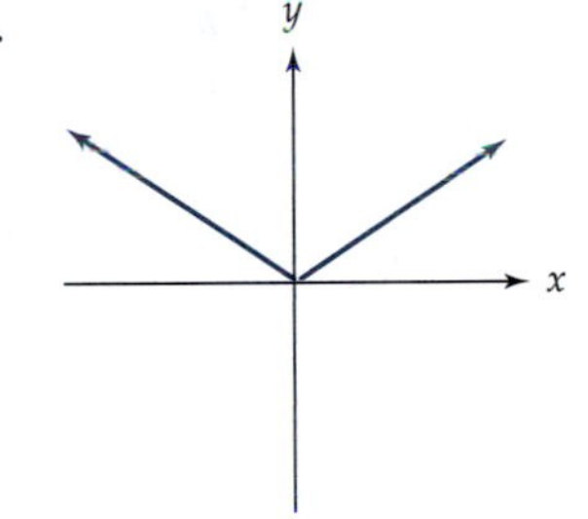

7. Assume it costs 10 cents a minute to make a long-distance telephone call. In addition, there is a \$0.90 surcharge per call. So, the total cost of a telephone call is $c(t) = 0.10t + 0.90$, where t is the length of the call (in minutes). Which of the following long-distance calling plans can be represented by a function whose graph is a horizontal compression of c by a factor of $\frac{5}{2}$?

   A. The call costs 4 cents a minute and there is a \$0.90 surcharge.

   B. The call costs 25 cents a minute and there is a \$0.90 surcharge.

   C. The call costs 25 cents a minute and there is a \$2.25 surcharge.

   D. The call costs 4 cents a minute and there is a \$0.36 surcharge.

8. Suppose you manufacture widgets. Assume $f(x) = x^2$ is the function that describes your profit per day in terms of x = number of widgets produced per day. To increase your profit as much as possible, which do you prefer? Justify your answer.

   A. Produce one more widget per day.

   B. Make one more dollar in profit per day.

9. Suppose c represents the function where n is the number of miles a rural mail carrier drives on a typical day and $c(n)$ is the cost of gasoline used on a typical day. Match the following situations to the following formulas. (Not all functions will be used.)
   (a) Due to road construction, the length of the mail carrier's route is increased by 10%.
   (b) The route is altered so that it is 10 mi longer than normal.
   (c) The cost of gasoline is increased by 10%.
   (d) The cost of gasoline is decreased by 10%.

   i. $y = 1.1c(n)$ ii. $y = 0.9c(n)$ iii. $y = c(n - 10)$
   iv. $y = c(n + 10)$ v. $y = c(n) + 10$ vi. $y = c(n) - 10$
   vii. $c(1.1n)$ viii. $c(0.1n)$ ix. $c(0.9n)$

10. In this section, we stated that the functions used to determine someone's federal income tax are piecewise functions. We looked at one piece of one function. Below is the complete piecewise function for married couples filing jointly. The input for this function is taxable income, i, and the output is the amount of their federal tax, $t(i)$.

$$t(i) = \begin{cases} 0.15i & \text{if } \$0 < i \leq \$41{,}200 \\ 6180.00 + 0.28(i - 41{,}200) & \text{if } \$41{,}200 < i \leq \$99{,}600 \\ 22{,}532.00 + 0.31(i - 99{,}600) & \text{if } \$99{,}600 < i \leq \$151{,}750 \\ 38{,}698.50 + 0.36(i - 151{,}750) & \text{if } \$151{,}750 < i \leq \$271{,}050 \\ 81{,}646.50 + 0.396(i - 271{,}050) & \text{if } i > \$271{,}050 \end{cases}$$

For the following questions, use the piece of the function for couples with taxable income of $\$99{,}600 < i \leq \$151{,}750$; that is $t(i) = 22{,}532.00 + 0.31(i - 99{,}600)$.
   (a) Suppose a couple takes \$15,000 in deductions from their income. Why would we consider this change a change in *input*?
   (b) Rewrite the piece of this function for those taking \$15,000 in deductions. The input should be total income and the output should be federal tax. Simplify your formula.
   (c) What is the new domain for the piece of the function you found in part (b)?

11. Suppose f is a linear function of the form $f(x) = mx + b$.
   (a) If f is transformed by adding a constant to the input, will the slope or y-intercept change? Explain.
   (b) If f is transformed by multiplying the input by a constant, will the slope or y-intercept change? Explain.

12. Given the functions $f(x) = \log(x)$, $g(x) = \log(10x)$, and $h(x) = \log(100x)$, do the following.
   (a) Graph the three functions on the same set of axes.
   (b) Describe how these three graphs appear to be related.
   (c) Although g and h are input transformations of f, the graphs are vertical shifts of each other. Why?

13. In this section, we looked at the function for bacteria growing on chicken kept at 40°F. In this case, the number of bacteria was doubling every 6 hr. This function can be represented by the formula $N_{40}(t) = 100 \cdot 2^{t/6}$, where t is the time in hours and $N_{40}(t)$ is the number of bacteria per square centimeter on chicken kept at 40°F. The number of bacteria after t hours for chicken kept at 70°F is represented by $N_{70}(t) = N_{40}(6t)$.
    (a) What is the number of bacteria after 4 h for chicken kept at 70°F?
    (b) Determine a formula for the number of bacteria after t hours for chicken kept at 70°F.
    (c) How often are bacteria doubling for chicken kept at 70°F?
14. In section 1.2, we stated that if f is symmetric about the y-axis, then $f(x) = f(-x)$. We also stated that if f is symmetric about the origin, then $f(x) = -f(-x)$.
    (a) With what you learned in this section about reflections, why does it make sense to say that if $f(x) = f(-x)$, then f is symmetric about the y-axis?
    (b) With what you learned in this section about reflections, why does it make sense to say that if $f(x) = -f(-x)$, then f is symmetric about the origin?
    (c) Is there a function that is both symmetric about the y-axis and the origin? If so, sketch it. If not, explain why.

INVESTIGATIONS

INVESTIGATION 1: TRANSFORMATIONS OF LINEAR AND EXPONENTIAL FUNCTIONS

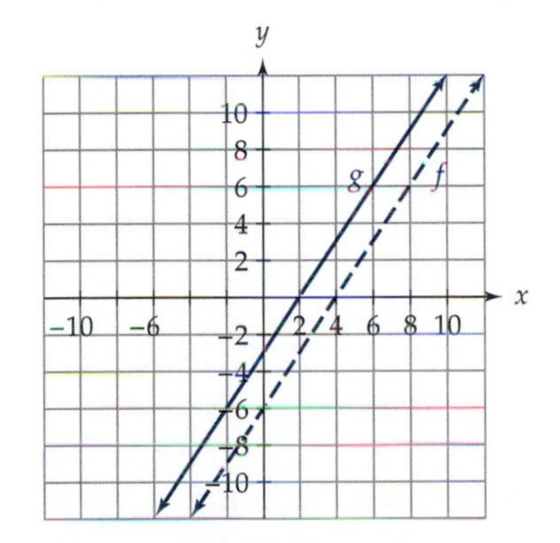

Figure 11

Some transformations can either be a horizontal or a vertical shift. For example, Figure 11 shows $f(x) = \frac{3}{2}x - 6$ and $g(x) = \frac{3}{2}x - 3$. The graph of g can be considered a horizontal shift of f by moving it two units to the left or a vertical shift of f by moving it three units up. When can a transformation be obtained by either a vertical or horizontal shift? We look at this question as it pertains to linear and exponential functions.

1. Let's start by looking at vertical and horizontal shifts of linear equations.
    (a) Let $f(x) = 2x + 3$.
        i. Determine the equation for $y = f(x + 4)$.
        ii. The horizontal shift of four units to the left for f is the same as a vertical shift of how many units and in what direction?
    (b) Let $f(x) = 2x + 3$.
        i. Determine the equation for $y = f(x - 4)$.
        ii. The horizontal shift of four units to the right for f is the same as a vertical shift of how many units and in what direction?

(c) Let $g(x) = -\frac{1}{3}x + 5$.

i. Determine the equation for $y = g(x - 2)$.

ii. The horizontal shift of two units to the right for g is the same as a vertical shift of how many units and in what direction?

(d) Let $h(x) = mx + b$.

i. Determine the equation for $y = h(x+c)$, where c represents a constant.

ii. The horizontal shift of c units for h is the same as a vertical shift of how many units and in what direction?

2. Let's explore what type of vertical transformation is equivalent to a horizontal shift for an exponential function.

(a) Let $f(x) = 2^x$.

i. Determine the equation for $y = f(x + 3)$. Simplify your equation so that the exponent is just x.

ii. The horizontal shift of three to the left for f is the same as a vertical stretch by what factor?

(b) Let $g(x) = 0.4^x$.

i. Determine the equation for $y = g(x - 2)$. Simplify your equation so that the exponent is just x.

ii. The horizontal shift of two to the right for g is the same as a vertical stretch by what factor?

(c) Let $h(x) = a^x$, where a is a positive real number not equal to 1.

i. Determine the equation for $y = h(x + c)$, where c is some real number. Simplify your equation so that the exponent is just x.

ii. The horizontal shift of c units for h is the same as a vertical stretch by what factor?

TABLE 9

Time (sec)	Position (cm)
0.00	0
0.05	18.2
0.10	33.7
0.15	46.9
0.20	57.6
0.25	65.0
0.30	70.8
0.35	74.1
0.40	74.6
0.45	72.7
0.50	68.3
0.55	61.7
0.60	52.1
0.65	40.1
0.70	25.5
0.75	8.6

INVESTIGATION 2: A SHOT IN THE DARK

Picture the arc made by a baseball after it is hit out into center field or the path of a stream of water in a drinking fountain. Both of these paths are approximately parabolic. A parabola is a transformation of the equation $y = x^2$. In a physics laboratory, we obtained data that describes the path made by a small steel ball launched at an angle.[4] With the aid of a camera and a strobe light, the path of the ball was photographed and the position of the ball at various times was determined. The data in Table 9 and in Figure 12 describe the vertical position of the ball (in centimeters) versus time (in seconds).

In this investigation, we want to determine a formula that fits Figure 12 data by transforming the power function $y = x^2$. Once we have the formula that describes the vertical position, we can use it to find a graph and formula for the velocity of the steel ball and finally the acceleration

[4] *Paul DeYoung, Physics Department, Hope College, provided us with this projectile motion data.*

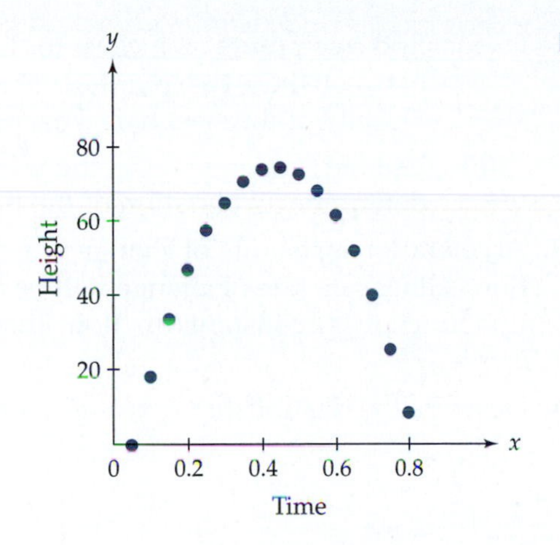

Figure 12

of the steel ball. Because the points given in the table come from an actual experiment, they do not all lie exactly on the parabola.

1. Let's examine the relationship between time and the vertical position of the projectile.
   (a) Determine the point at which the ball is at its highest position. We assume this point is the vertex of our parabola.
   (b) The vertex of $y = x^2$ is at point $(0, 0)$. Transform this function so that it has the vertex given by your answer from part (a). You will need to use both a horizontal and vertical shift to do so.
   (c) Your function should now be of the form $y = (x + b)^2 + c$. To make it fit the data, find an appropriate factor a so that your function is of the form $y = a(x + b)^2 + c$. To do so, use the last point from the table, $(0.75, 8.6)$, plug it into your function, and solve for a. Rewrite your function using the notation t (for time) as the input and $p(t)$ (for position) as the output. Graph your function along with the data.
2. Velocity is the rate of change of the position. If the vertical position of the steel ball can be described with a parabola, then its vertical velocity can be described with a line.[5] Because the velocity function is a line, we need only two points to determine a formula for velocity. We want to find the x- and y-intercepts.
   (a) When the ball is traveling upwards, its velocity is positive, and when the ball is traveling downwards, its velocity is negative. At what point during the flight of the ball will the velocity be 0? At what time does it occur? This point is the x-intercept of our line.
   (b) Finding the y-intercept is more difficult than finding the x-intercept. The y-intercept is when time is zero. In other words, the y-intercept is the initial velocity. To find the initial velocity, we need to find the rate of change in the position function close to the origin. Using your position function from question 1, part (c),

[5] *See Investigation 1 in section 2.5 to see why this is true.*

find two points *very* close to the origin. Find the rate of change between these two points and use it as your estimate of the initial velocity of the steel ball. This point is the y-intercept of our line.

(c) Use the x-intercept and the y-intercept from parts (a) and (b) to determine an equation of the line describing the velocity, $v(t)$.

3. Acceleration is the rate of change of velocity. Because the velocity function is linear, its rate of change will be constant. Therefore, the acceleration function is a constant function. Find an equation for the acceleration, $a(t)$.
4. Describe the relationship between your three functions, p, v, and a.

3.3 COMBINING FUNCTIONS

Earlier in this chapter, we saw what happened when we combined the input and the output of a function with constants. Constants, however, are not the only thing that can be combined with functions. We can create new functions from the original set of functions introduced in chapter 2 in a variety of ways. In this section, we explore what happens when we take two or more functions and combine their outputs using the algebraic operations of adding, subtracting, multiplying, and dividing.

ADDING AND SUBTRACTING FUNCTIONS

Tracking Population Growth

The U.S. Census Bureau is responsible for conducting the census taken every 10 years in the United States. In addition to that huge task, it also estimates the number of people in the country at any given time and posts this information on the Internet.[6] On the morning of 8 February 1998 (9:14:05 A.M. EST), the bureau estimated the population to be 269,126,342 and also gave some information about how this estimate was determined. On 8 February, the bureau was assuming the following statistics about the change in population:

- One birth every 8 sec
- One death every 12 sec
- One international migrant (net) every 39 sec
- One federal U.S. citizen (net) returning every 4473 sec

In this situation, four different functions (births, deaths, migrants, and returning citizens) combine to form one function (net gain in population). Because each function has a constant rate of change, they are linear functions with different slopes. The output of each function is population change, and the input is time (in seconds). Let the first function, b, represent the number of births in the United States, which is given verbally in

[6] *U.S. Bureau of Census,* POPClock Projection *(visited 8 February 1998) 〈http://www.census.gov/cgi-bin/popclock〉.*

the chart as "one birth every 8 sec." Therefore, the slope of this function is $\frac{1}{8}$ (change in population over change in time). Symbolically, this function is $b(t) = \frac{1}{8}t$, where the input, t, is the number of seconds since our starting time (8 February 1998 at 9:14:05 A.M. EST) and the output is the number of births since that time.[7] Similarly, the other functions are

$d(t) = \frac{1}{12}t$, where $d(t)$ is the number of deaths

$m(t) = \frac{1}{39}t$, where $m(t)$ is the number of international migrants

$c(t) = \frac{1}{4473}t$, where $c(t)$ is the number of returning citizens

Combining these functions gives one function for total population, p. Starting with our base population (269,126,342), we add the number of births to the number of international migrants and the number of U.S. citizens who return to the country. We also need to subtract the number of deaths. We now have

$$p(t) = 269{,}126{,}342 + b(t) + m(t) + c(t) - d(t)$$
$$p(t) = 269{,}126{,}342 + \frac{1}{8}t + \frac{1}{39}t + \frac{1}{4473}t - \frac{1}{12}t$$
$$p(t) \approx 269{,}126{,}342 + 0.0675t.$$

The formula for p is a linear equation with a slope of 0.0675 and a y-intercept of 269,126,342. The y-intercept represents the starting population. The slope represents the number of additional people added to the population each second. As a fraction with a numerator of 1, $0.0675 \approx \frac{1}{15}$. Because slope = change in population/change in time, a slope of $\frac{1}{15}$ means that there is a net gain of approximately one person every 15 sec.

Definitions and Examples

By adding and subtracting the functions associated with population, we were able to create one function that combined four pieces of information. The process of combining functions by adding or subtracting occurs frequently and in many contexts. For example, the cost of driving a car is the sum of gas costs plus the costs for repair and the costs for maintenance, which can all be expressed in average dollars per mile.

Let's formally define the concept of adding or subtracting two functions. We define the sum of two functions, f and g, as the function that adds the output $f(x)$ to the output $g(x)$. Symbolically, $(f+g)(x) = f(x) + g(x)$. Similarly, we define $(f-g)(x)$ as the difference of the outputs of f and g, or $(f-g)(x) = f(x) - g(x)$.

Example 1 Let $f(x) = x^2$ and $g(x) = \log x$. Find $(f+g)(2)$ and $(f-g)(2)$.

[7] *Population growth, if left unchecked by outside influences, is usually best modeled with an exponential function. Because this population estimate is revised on a monthly basis, however, linear equations do an acceptable job of estimating population over the short term. As mentioned earlier, most functions appear linear if the viewing window is small enough.*

Solution We have $(f+g)(2) = f(2) + g(2) = 2^2 + \log 2 = 4 + \log 2 \approx 4.301$. Similarly, we have $(f-g)(2) = 2^2 - \log 2 = 4 - \log 2 \approx 3.699$.

Example 2 Two functions, u and t, are given in Table 1. Let s be the sum of u and t. Find s.

TABLE 1

x	0	1	2	3	4
$u(x)$	6	9	13	18	24
$t(x)$	−3	−2	−1	0	1

Solution As you can see in Table 2, the values of $s(x)$ are obtained by merely adding the appropriate outputs.

TABLE 2

x	0	1	2	3	4
$u(x)$	6	9	13	18	24
$t(x)$	−3	−2	−1	0	1
$s(x) = (u+t)(x)$	3	7	12	18	25

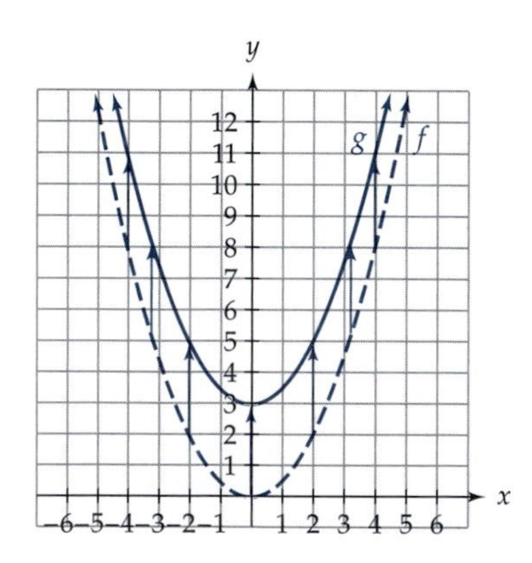

Figure 1 Graphs of $f(x) = 0.5x^2$ and $g(x) = 0.5x^2 + 3$. Notice that the vertical distance between the two functions is constant.

In section 3.1, we saw the impact on the graph of a function when we added a constant to the output. Because we were changing the output, the result was always a vertical shift in the graph. Not surprisingly, we also get a vertical change in the graph if we add the outputs of two or more functions to create a new function. This time, however, the vertical change will vary for different inputs because the shift is no longer constant. For example, let $f(x) = 0.5x^2$. Let's compare $f(x) = 0.5x^2$, $g(x) = 0.5x^2 + 3$, and $h(x) = 0.5x^2 + x$. Figure 1 is a graph of $f(x) = 0.5x^2$ and $g(x) = 0.5x^2 + 3$. You can see that g is shifted up three units from f. Graphs of $f(x) = 0.5x^2$ and $h(x) = f(x) + x = 0.5x^2 + x$ are shown in Figure 2. Notice that the vertical distance between the graphs of h and f varies. The two graphs are zero units apart at $x = 0$, two units apart at $x = 2$, four units apart at $x = 4$, and so on. As x gets larger, the distance between $0.5x^2$ and $0.5x^2 + x$ increases. Also notice that the graph of h is not always above the graph of f. For example, when $x = -2$, the graph of h is two units below the graph of f. In fact, whenever x is negative, the graph of $h(x) = 0.5x^2 + x$ is below the graph of $f(x) = 0.5x^2$ because you are adding a negative number to the output. The graph of h is a vertical transformation of the graph of f, but the amount (and even the direction) of the vertical shift varies.

Analyzing Functions Combined by Adding or Subtracting

Figure 2 showed that adding functions produces graphs that are not as simple as those produced by adding a constant to a graph. In Figure 2, it is fairly easy to see that the graph of h is similar to $f(x) = 0.5x^2$, but it is difficult to see the impact of $y = x$ by just looking at Figure 2. Let's look at another example. Combining the periodic function $y = \sin x$ with the power function $y = 0.1x^2$ gives the graph in Figure 3. Notice that this graph combines the general shape of the power function, $y = 0.1x^2$, with the wave effect of the periodic function, $y = \sin x$. As x gets farther from zero, the effect of $\sin x$ is less pronounced. Looking at the same function with a larger viewing window produces a graph that seems indistinguishable from the graph of $y = 0.1x^2$. (See Figure 4.) As x gets large, the values of $0.1x^2$ are much larger than the values of $\sin x$.[8] If the outputs of one of the functions is much larger than the other, it will dominate that portion of the graph.

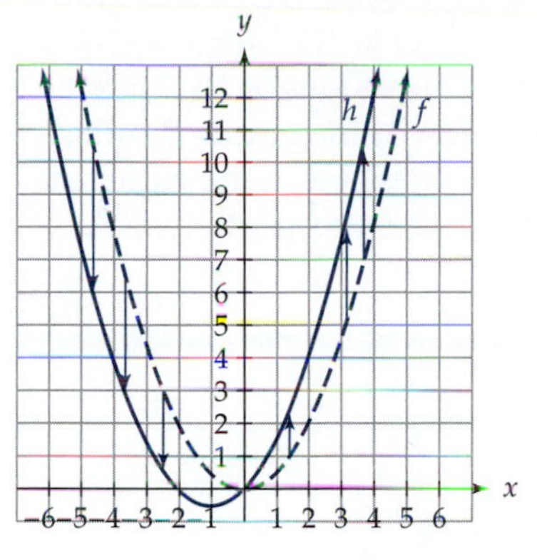

Figure 2 Graphs of $f(x) = 0.5x^2$ and $h(x) = 0.5x^2 + x$. Notice that the vertical distance between the two functions varies.

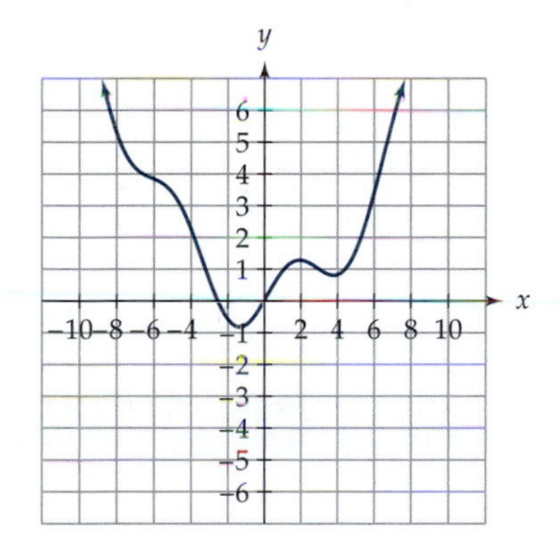

Figure 3 A graph of $f(x) = \sin x + 0.1x^2$.

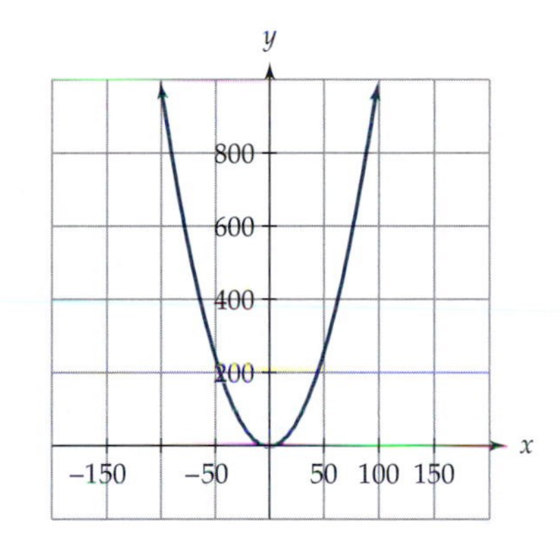

Figure 4 Another view of $f(x) = \sin x + 0.1x^2$. In this viewing window, the effects of adding $\sin x$ to $y = 0.1x^2$ are not noticeable.

Example 3 Let $f(x) = x$ and $g(x) = 2^x$. Let $h(x) = f(x) + g(x) = x + 2^x$. Graph h and analyze the graph in terms of f and g.

Solution Figure 5 shows $h(x) = x + 2^x$. When x is less than zero, the output of $g(x) = 2^x$ is quite small in absolute value compared with the output of $f(x) = x$, so the graph is similar to $y = x$. As x gets larger, the output of $g(x) = 2^x$ is much larger than the output of $y = x$, so the graph is similar to the graph of $y = 2^x$.

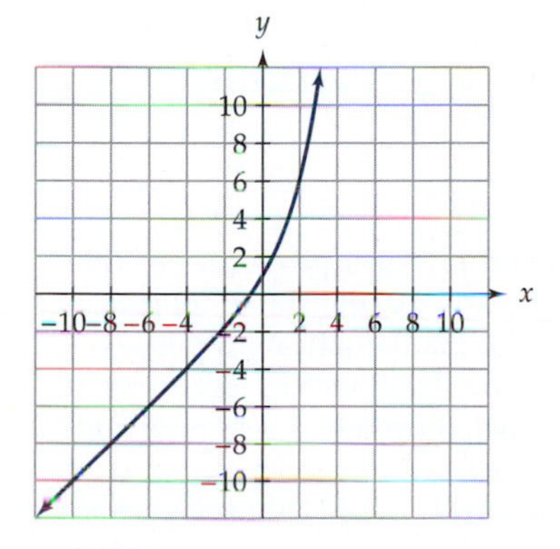

Figure 5

[8] *Recall that the output of* $\sin x$ *is always between* -1 *and* 1, *inclusive.*

Example 4 Let $f(x) = 10^x$ and $g(x) = 10^{-x}$. The graphs of f and g are shown in Figure 6. Let $h(x) = f(x) + g(x) = 10^x + 10^{-x}$. Graph h and analyze the graph in terms of f and g.

Solution The graph of h is shown in Figure 7. Notice that this graph is similar to $f(x) = 10^x$ for $x > 0$ and similar to $g(x) = 10^{-x}$ for $x < 0$. The values of 10^x are quite large for $x > 0$, whereas the values of 10^{-x} are quite small for $x > 0$. So, f dominates this portion of the graph. Similarly, the values of $g(x) = 10^{-x}$ are quite large for $x < 0$, whereas the values of $f(x) = 10^x$ are quite small for $x < 0$. So, g dominates this portion of the graph.

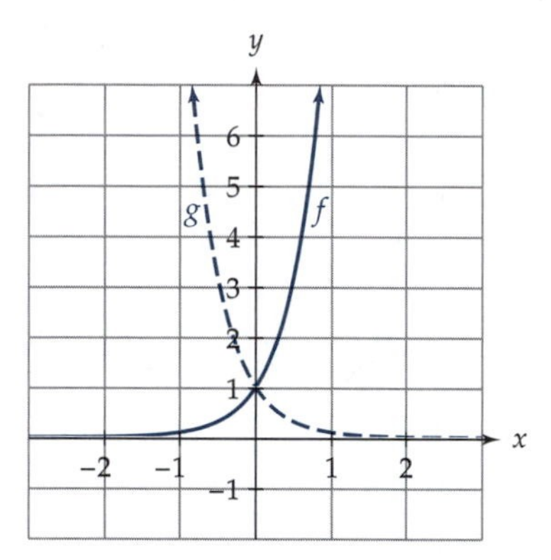

Figure 6

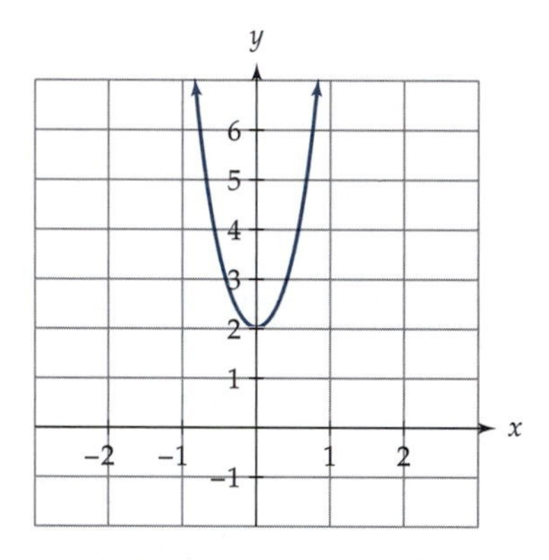

Figure 7

HISTORICAL NOTE

Conic sections had been studied for over 2000 years when, in the early 1600s, scientists found two new applications: Kepler's discovery that the planets move around the sun in elliptical orbits and Galileo's finding that the path of a projectile, disregarding air resistance, is a parabola. Galileo also thought that the curve that results when a chain, wire, or rope hangs under its own weight is in the shape of a parabola. This time Galileo was mistaken. The curve formed by a hanging chain is actually a catenary. This shape is defined symbolically as a combination of exponential functions. Gottfried Wilhelm Leibniz (1646–1716) discovered the equation for one type of catenary:

$$f(x) = \frac{a}{2}\left(e^{x/a} + e^{-x/a}\right),$$

where $a > 0$ and a depends on the tension and physical properties of the chain. The number e, used for the base of the exponential function, is approximately 2.718. This number has many useful mathematical properties.

The shape of a catenary can be seen in common items such as electric and telephone lines, necklaces, and jump ropes. In fact, the method of supporting trolley wire in a horizontal position using other wires is referred to as a catenary system. One of the more notable catenaries in the United States is the Gateway Arch in St. Louis. It is in the shape of a transformed catenary (as shown in the accompanying figure).

continued on page 209

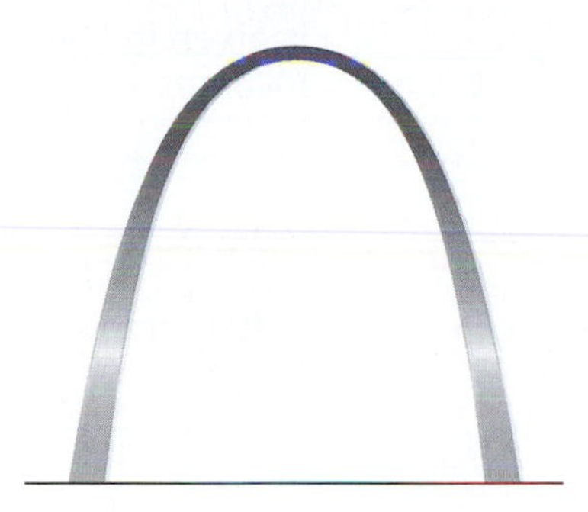

Sources: Carl Boyer, *A History of Mathematics* (Princeton, NJ: Princeton University Press, 1968), p. 358, 505–6; National Park Service, *Jefferson National Expansion Memorial—Arch History and Architecture* (visited 4 February 1998) (www.nps.gov/jeff/arch-home.htm); Steven Schwartzman, *The Words of Mathematics* (Washington, D.C.: Mathematics Association of America, 1994), p. 41.

READING QUESTIONS

1. Suppose the death rate was given by the Census Bureau as one death every 18 sec but the other information remained the same. Give the equation for p in this case.
2. Using the data from Table 1, make a row for $r(x) = (u - t)(x)$.
3. Let $f(x) = x^2$ and $g(x) = 2^x$. Find $(f + g)(3)$.
4. Why does adding two functions produce a *vertical* change in the graph?
5. Why was it difficult to observe the effect of adding $\sin x$ to $y = 0.1x^2$ in the viewing window given in Figure 4?

MULTIPLYING AND DIVIDING FUNCTIONS

Flat Tax

Section 2.1 on linear functions contained an investigation on the "flat tax," a tax system in which every American pays the same tax rate. The 1996 tax proposal of Steve Forbes, a presidential candidate, was that a single person with no dependents would pay 17% of his or her income over \$13,300, with the first \$13,300 not being taxed. Symbolically, the amount of tax one would pay is $0.17(x-13,300)$, where x is one's income. As you may have discovered in that investigation, this tax rate is not actually constant. Because you do not pay 17% of your entire income as tax, your actual tax rate is not 17%. Instead, the tax rate is going to be the amount of tax you pay, $0.17(x - 13,300)$, divided by your total income, x. The actual tax rate is given by the function $r(x) = 0.17(x - 13,300)/x$, where x is your income and $r(x)$ is the tax rate you are paying.

This function is only defined for incomes greater than \$13,300 because the first \$13,300 earned each year is not taxed. Therefore, the domain for $r(x)$ is $x \geq 13,300$. The graph of r for incomes between \$13,300 and \$100,000 is given in Figure 8. Notice that the actual tax rate is less than 17% for any income.

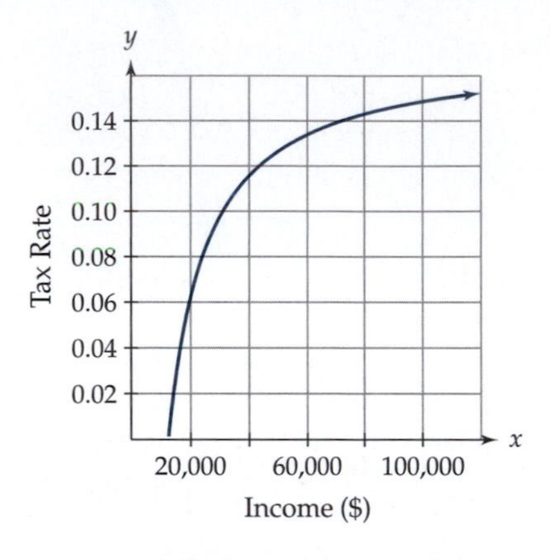

Figure 8 A graph of the tax rate, $r(x) = 0.17(x - 13,300)/x$.

As you can see from the graph, this rate is not flat after all, although as the income gets progressively greater, the rate gets closer to 17%. In fact, when looking at a different view of this function in Figure 9, the function is very close to a flat tax with a rate of 17% for the wealthiest Americans.

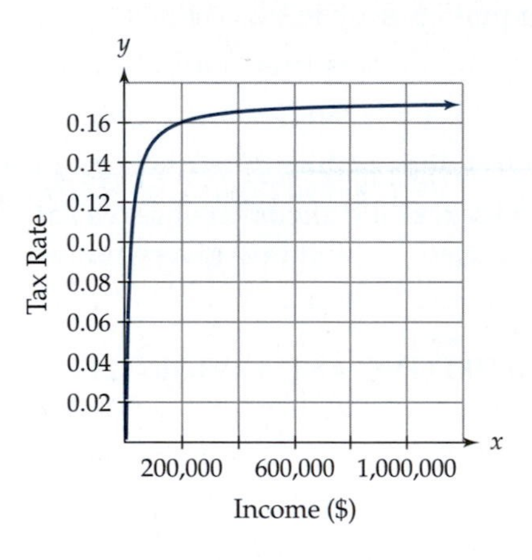

Figure 9 Another view of the tax rate function for very high incomes.

The function we used to investigate the flat tax plan combined two functions using division. The process of finding a rate of any sort, such as percent, gas mileage, or speed, always involves division. Rates are just one of many situations in which functions are combined by division or multiplication.

Definitions and Examples

Let's formally define the process of multiplying or dividing two functions. As with addition and subtraction, we define the product of f and g as the function that multiplies the output $f(x)$ and the output $g(x)$. Symbolically,

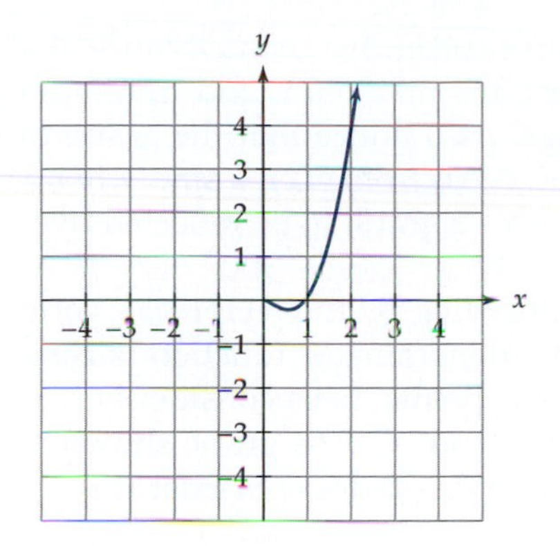

Figure 10 A graph of $h(x) = 3x^2 \cdot \log x$.

$(f \cdot g)(x) = f(x) \cdot g(x)$. Similarly, we define $(f/g)(x)$ as the quotient of the outputs of f and g, or $f(x)/g(x)$. For example, if $f(x) = 3x^2$ and $g(x) = \log(x)$, then $h(2) = (f \cdot g)(2) = (3 \cdot 2^2)(\log 2) = 12 \log 2$. The graph of h is shown in Figure 10.

It is not surprising that h has an x-intercept (or root) at $x = 1$, because $\log(x) = 0$ when $x = 1$. In fact, a function equal to the product of two or more functions will only have roots at values where one or more of its components has a root.[9]

Example 5 The functions s and t are given in Table 3. Let p be the product of s and t. Find p.

TABLE 3

x	0	1	2	3	4
$s(x)$	6	9	13	18	24
$t(x)$	−3	−2	−1	0	1

Solution Table 4 shows the functions s and t given along with their product, p. Notice that the values of $p(x)$ are obtained by multiplying the appropriate outputs.

TABLE 4

x	0	1	2	3	4
$s(x)$	6	9	13	18	24
$t(x)$	−3	−2	−1	0	1
$p(x) = (s \cdot t)(x)$	−18	−18	−13	0	24

Figure 11 Graphs of $f(x) = x$ and $g(x) = \sin x$. What happens when we multiply them?

Multiplying two functions given by graphs is difficult to analyze visually. When adding two functions, you can often envision the graph because it is possible to visualize addition of distances fairly easily. Unfortunately, it is much more difficult to visualize the graphical effect of multiplying two functions.

Figure 11 shows the linear function $f(x) = x$ and the periodic function $g(x) = \sin x$. What happens when we multiply them?

[9] *This fact holds because the only way to multiply two numbers and get zero for an answer is if one of the original numbers is zero.*

Figure 12 shows the functions $g(x) = \sin x$ and $h(x) = x \sin x$. Notice that the amplitude of h is not defined because the maximum and minimum values of the function continue to change. Also notice that the graph of $h(x) = x \sin x$ is on the opposite side of the x-axis from $g(x) = \sin x$ when x is negative; multiplying by a negative causes a graph to be reflected over the x-axis.

Dividing functions is the same as multiplying by the reciprocal, yet it is important to consider points where the denominator function is zero. The quotient function will not exist at these points. Let's consider $k(x) = 2^x/(x^2 - 4)$. The denominator has zeros at 2 and -2. The graph shown in Figure 13 shows that, as expected, $k(x) = 2^x/(x^2 - 4)$ does not exist at $x = 2$ and $x = -2$.

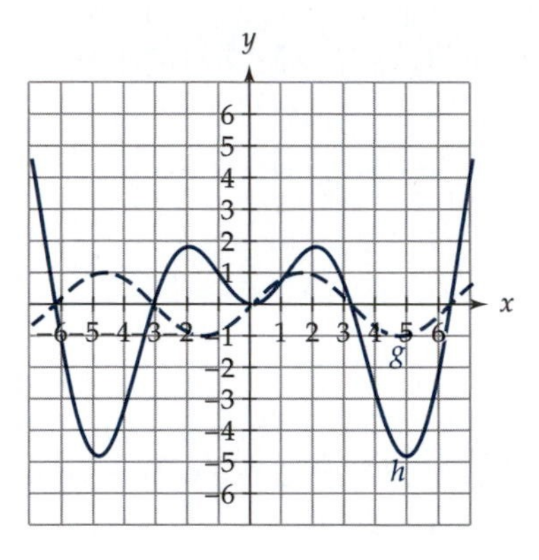

Figure 12 Graphs of $g(x) = \sin x$ and $h(x) = x \sin x$.

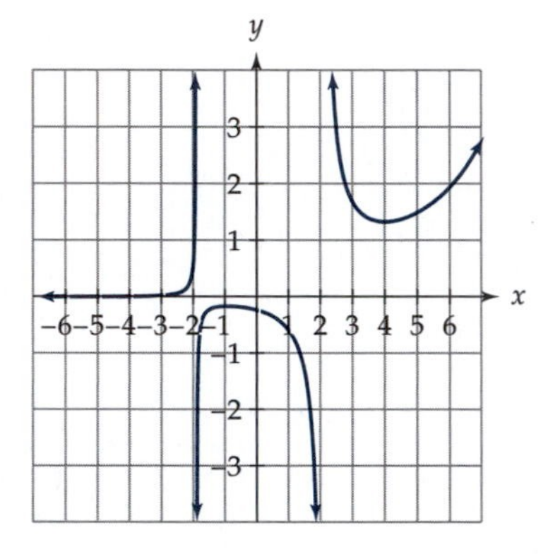

Figure 13 A graph of $k(x) = 2^x/(x^2 - 4)$.

READING QUESTIONS

6. What would your tax rate be under Steve Forbes's plan if you made $30,000 per year?
7. Using the data in the following table, fill in the row for $(s/t)(x)$ where it is defined.

x	0	1	2	3	4
$s(x)$	6	9	13	18	24
$t(x)$	−3	−2	−1	0	1
$(s/t)(x)$					

8. Does $f(x) = x^2/(x - 5)$ have any points where it is undefined? If so, where?
9. Let $f(x) = x^2$ and $g(x) = 2^x$. Find $(f \cdot g)(3)$.

10. Why are the graphs of $y = x \sin x$ and $y = \sin x$ sometimes on opposite sides of the x-axis?

THE DOMAIN OF COMBINATIONS OF FUNCTIONS

The domain of a function created by combining two or more functions is usually the intersection[10] of the domains of the original functions. For example, the functions f, g, and their sum are shown in Table 5. Notice that because f is not defined for $x = 5$ and g is not defined for $x = -1$, $h(x)$ is also not defined for those values. The domain of h is all values of x that are in the domain of both f and g.

TABLE 5 The Functions f, g, and $(f + g)$ in Table Form

x	−1	0	1	2	3	4	5
$f(x)$	4	6	9	13	18	24	
$g(x)$		−3	−2	−1	0	1	2
$h(x) = (f+g)(x)$		3	7	12	18	25	

Example 6 Find the domain of $f(x) = x^3 \log x$.

Solution Because f is an algebraic combination of $\log x$ and x^3, we need to consider the domains of both pieces. The domain of the power function $y = x^3$ is all real numbers. The domain of $\log x$ is only positive real numbers. Therefore, the domain of f is $x > 0$.

Functions involving division have an additional domain issue. The quotient function is not defined for points where the denominator equals zero.

Example 7 Find the domain of

$$f(x) = \frac{(x-5)(x+2)}{(x+3)(x-4)}.$$

Solution The numerator and the denominator are defined for all real numbers. The denominator function, however, is equal to zero when x is -3 or 4. Therefore, the domain of f is all real numbers except -3 and 4.

Whenever you combine two functions, be sure to consider the domain. It can have a huge impact on the behavior of the function, especially for quotient functions close to the values that cause the denominator to be

[10] *The intersection of the domains is all inputs for which* both *original functions are defined.*

zero. In section 3.1, when we added, subtracted, multiplied, and divided the output of our functions by constants, we did not indicate any changes in the domain because a constant has a domain of all real numbers. Whenever you take the intersection of any set of numbers with the set of all real numbers, you always end up with your original set. Only when you include functions with restricted domains or when you divide functions may the domain be affected.

In general, the following rules apply when finding the domain of a function formed by combining other functions.

- The domain of a function formed by adding, subtracting or multiplying functions is the intersection of the domains of the component functions.
- The domain of a function formed by dividing functions is the intersection of the domains of the component functions except for those values for which the denominator is equal to zero.

READING QUESTIONS

11. Find the domain of each of the following.

(a) $g(x) = \sqrt{x} + 6$

(b) $h(x) = \dfrac{1}{x-2} + \sqrt{x}$

12. What can we say about the domain of a function formed by combining a function with a constant?

13. Why are there additional domain considerations for functions with quotients?

SUMMARY

We can combine two or more functions by using algebraic operations. Adding or subtracting functions results in the outputs of the individual functions being combined. We defined $(f+g)(x)$ as $f(x)+g(x)$. Similarly, we defined $(f-g)(x)$ as $f(x)-g(x)$.

We can also combine functions by multiplying or dividing them in the same way we add or subtract functions, allowing us to define $(f \cdot g)(x) = f(x) \cdot g(x)$ and $(f/g)(x) = f(x)/g(x)$.

For function combinations involving addition, subtraction, or multiplication, the domain of the resultant function is the intersection of the domains of the original functions, that is, all values that are in *both* functions. For function combinations involving division, however, the domain is the intersection of the original functions except those values for which the denominator of the new function is zero.

EXERCISES

1. Describe a nonmathematical situation in which the outputs of two functions are algebraically combined to form a new function.

2. Under Steve Forbes's tax plan, will you ever reach a tax rate of exactly 17%? How could you tell by looking at the symbolic form of the function?
3. How could you change Steve Forbes's plan so that it would truly be a flat tax?
4. Let $f(x) = x^2$ and $g(x) = 2x + 4$. Find the following.
   (a) $(f+g)(2)$
   (b) $(f+g)$
   (c) $(f-g)(3)$
   (d) $(f-g)$
   (e) $(f \cdot g)(-1)$
   (f) $(f \cdot g)$
   (g) $\left(\frac{f}{g}\right)(4)$
   (h) $\left(\frac{f}{g}\right)$
   (i) $(2f-g)(3)$
   (j) $(2f-g)$
   (k) $\left(\frac{f+g}{f-g}\right)(-3)$
   (l) $\left(\frac{f+g}{f-g}\right)$
5. If the population function is given by $p(t) = 268{,}722{,}001 + 0.1044t$, what is the average number of seconds it takes for an additional person to be added to the population?
6. Assume the United States had a population of 300,000,000 and there is a net gain of one person every 11 sec. Give an equation representing p, where the input, t, is the amount of time that passed since the population was 300,000,000 and the output, $p(t)$, is the current population.
7. In the accompanying figure, the graph of $f(x) = 0.5x^2 + x$ is given. Sketch a graph of $g(x) = 0.5x^2 - x$. How is this similar to the graph of $f(x) = 0.5x^2 + x$? How is this different from the graph of $f(x) = 0.5x^2 - x$?

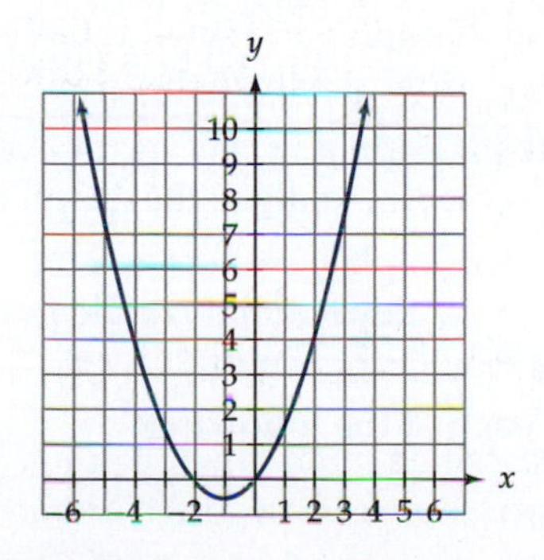

8. Complete the following table for those values where the function is defined.

x	-1	0	1	2	3	4	5
$f(x)$	2	5	7	6	8	3	7
$g(x)$	1	0	3	9	7	4	3
$(f+g)(x)$							
$(f \cdot g)(x)$							
$\left(\frac{f+g}{g}\right)(x)$							

9. Let f, g, and h be represented by the accompanying graphs.

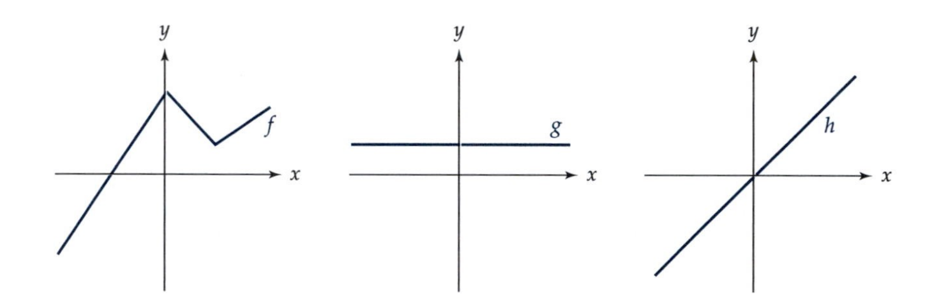

(a) Plot the graph of $y = (f + g)(x)$.
(b) Plot the graph of $y = (f + h)(x)$.

10. Let $f(x) = x/[(x-4)(x-6)]$, $g(x) = \log x$, $h(x) = (g + f)(x)$, and $k(x) = (f/g)(x)$.
    (a) Give the domain for f and g.
    (b) Give the domain for h and k. How are these related to your answer for part (a)?
11. Where will the graph of $f(x) = 3^{-x} + x$ look similar to $y = 3^{-x}$? Where will it look similar to $y = x$?
12. Let $f(x) = \sin x$. Be sure to choose your viewing window carefully to do each of the following.
    (a) Graph $y = f(x) + x$. Give the coordinates of your viewing window. Analyze the graph in terms of f and $y = x$.
    (b) Graph $y = f(x) + x^2$. Give the coordinates of your viewing window. Analyze the graph in terms of f and $y = x^2$.
    (c) Graph $y = f(x) + 2^{-x}$. Give the coordinates of your viewing window. Analyze the graph in terms of f and $y = 2^{-x}$.
13. Let $f(x) = \sin x$. Be sure to choose your viewing window carefully to do each of the following.

(a) Graph $y = xf(x)$. Give the coordinates of your viewing window. Analyze the graph in terms of f and $y = x$.

(b) Graph $y = x^2 f(x)$. Give the coordinates of your viewing window. Analyze the graph in terms of f and $y = x^2$.

(c) Graph $y = 2^{-x} f(x)$. Give the coordinates of your viewing window. Analyze the graph in terms of f and $y = 2^{-x}$.

14. Let $f(x) = x$, $g(x) = 2x$, and $h(x) = (g/f)(x)$.
    (a) Give the domain of f and g.
    (b) Give the domain of h.
    (c) How does the domain of h compare with the domain of $c(x) = 2$? Why is there a difference between the domain of h and the domain of c?

15. Operating a car can be expensive. Besides the obvious cost of gasoline are correlated expenses of repairs, oil, and maintenance. Some typical costs are:

    Average cost of gasoline: $1.05 per gallon

    Average gas mileage: 22 mi/gal

    Cost of oil change: $22.50

    Oil changes: one every 3000 mi

    Average frequency for repairs: once every 25,000 mi

    Average cost of each repair: $200

    (a) We are interested in finding a function that gives us the cost per mile of operating this car.
        i. Let g be the function whose input, m, is number of miles and whose output, $g(m)$, is cost of gasoline. Find g.
        ii. Let l be the function whose input, m, is number of miles and whose output, $l(m)$, is cost of oil changes. Find l. (*Note:* Although the cost of oil occurs in lump sums, your function should be written in terms of cost per mile.)
        iii. Let r be the function whose input, m, is number of miles and whose output, $r(m)$, is cost of repairs. Find r. (*Note:* Although the cost of repairs occurs in lump sums, your function should be written in terms of cost per mile.)
        iv. Combine g, l, and r to obtain a function whose input is number of miles and whose output is operating costs.
        v. Use your answer from part iv to find the cost per mile of operating the car.
    (b) In addition to operating costs are annual costs of about $500 per year for insurance and about $40 per year for license. Assume the car was driven 12,000 mi in 1 year. What is the expected total cost of the car (excluding car payments) for that year?

16. Suppose you throw an object into the air from a height of h_0 with an initial velocity of v_0. Then the function whose input is time (in seconds)

and whose output is distance of the object from the ground (in feet) is

$$h(t) = -16t^2 + v_0 t + h_0.$$

(a) Suppose you threw a ball into the air from a height of 5 ft and an initial velocity of 20 ft/sec.

i. How high was the ball at $t = 1$ sec?

ii. When did the ball hit the ground? (*Hint:* Set the distance equal to zero and solve for t.)

(b) Let's look at why the function for the height of a ball thrown in the air combines three components.

i. Why does it make sense that a function that measures the vertical distance of an object thrown into the air would include the term h_0?

ii. Why does this function include the term $v_0 t$? Why is the initial velocity multiplied by the time?

iii. The component $-16t^2$ can be rewritten as $-16t^2 = (-16t)(t)$. The -16 comes from the acceleration of gravity.[11] Knowing that acceleration gives you the change in velocity, explain why this function includes the term $-16t^2 = (-16t)(t)$. Why is the -16 multiplied by t^2?

INVESTIGATIONS

INVESTIGATION 1: VERTICES

In this section, you examined what happened when you added two functions together. Chapter 4 looks at combinations of some power functions in more detail. One such combination is of the form $f(x) = x^2 + bx$, where b is any real number. When $b = 0$, the function $y = x^2$ is a parabola whose vertex is at the origin.

Let's consider what happens to $y = x^2$ as we add the linear function $g(x) = bx$. We will concentrate on looking at what happens to the vertex.[12]

1. Sketch a graph of $y = x^2$, $y = x^2 + 2x$, and $y = x^2 + 4x$ on the same set of axis. Describe the relationship between the vertices of the three parabolas.

2. We are interested in finding the coordinates of the vertices for the three parabolas. Obviously, the vertex of $y = x^2$ is $(0, 0)$.

   (a) To find the vertex of $y = x^2 + 2x$, begin by completing the square. We get $y = x^2 + 2x = (x^2 + 2x + 1) - 1 = (x + 1)^2 - 1$.

   i. Use what was learned in sections 3.1 and 3.2 about vertical and horizontal shifts to describe the relationship between $y = x^2$ and $y = x^2 + 2x = (x + 1)^2 - 1$.

[11] *The acceleration of gravity is* -32 ft/sec^2.

[12] *The vertex is the lowest point of a parabola that is concave up.*

ii. Use your answer to question 2, part (a)(i), to give the coordinates of the vertex for $y = x^2 + 2x$.

(b) Find the vertex of the parabola $y = x^2 + 4x$.

3. Now we want to look at $y = x^2 - 2x$ and $y = x^2 - 4x$.

   (a) *Before* graphing these functions, predict what the graphs will look like. What do you think the coordinates of the vertices will be?

   (b) Graph the functions $y = x^2$, $y = x^2 - 2x$, and $y = x^2 - 4x$ on the same set of axes.

   (c) Find the coordinates of the vertices for the equations $y = x^2 - 2x$ and $y = x^2 - 4x$.

4. In general, describe the relationship between the graphs of $y = x^2$ and $y = x^2 + bx$. What are the coordinates of the vertex for $y = x^2 + bx$?

5. We want to describe the curve we obtain if we draw a curve through the vertices of $y = x^2 + bx$ for several different values of b.

   (a) Using your answer to question 4, describe the relationship between the x-coordinate of the vertex for $y = x^2 + bx$ and the y-coordinate of the vertex for $y = x^2 + bx$.

   (b) This relationship tells you that all the vertices will lie on the same curve. Give the equation for that curve.

   (c) Sketch that curve on the same set of axes as $y = x^2$, $y = x^2 + 2x$, $y = x^2 + 4x$, $y = x^2 - 2x$, and $y = x^2 - 4x$. Does your curve go through all five vertices?

INVESTIGATION 2: THE GATEWAY ARCH

The historical note in this section mentioned that the Gateway Arch (part of the Jefferson National Expansion Memorial in St. Louis, Missouri) is not a parabola but a shape known as a catenary. In this investigation, we look at the shape of the Gateway Arch and compare it with a parabola.

The word *catenary* comes from the Latin word for chain. A catenary is the shape formed by the graph of the hyperbolic cosine (abbreviated cosh). It is the shape of a uniform flexible cable or chain whose ends are supported from the same height. The hyperbolic cosine is defined as

$$\cosh x = \tfrac{1}{2}e^x + \tfrac{1}{2}e^{-x}.$$

The functions e^x and e^{-x} are exponential functions whose base is e, a number approximately equal to 2.718. The number e is an important constant that occurs in a variety of mathematical contexts and so is denoted by its own symbol (similar to the number π). The exponential function e^x is often the "nicest" exponential function to use, partly because it crosses the y-axis with a slope of 1.

1. We want to examine the relationship between the graph of $y = \frac{1}{2}\iota^x + \frac{1}{2}e^{-x}$ and the graphs of $f(x) = \frac{1}{2}e^x$ and $g(x) = \frac{1}{2}e^{-x}$.

   (a) On the same set of axes, sketch a graph of $f(x) = \frac{1}{2}e^x$ and $g(x) = \frac{1}{2}e^{-x}$.

(b) *Without* sketching $y = \frac{1}{2}e^x + \frac{1}{2}e^{-x}$, predict the shape of the graph by looking at the graphs from part (a) and using your knowledge of what happens when you add functions together.

(c) Graph $y = \frac{1}{2}e^x + \frac{1}{2}e^{-x}$. Was your prediction accurate?

2. Now let's look at the Gateway Arch. The equation that gives the shape of the arch[13] is

$$y = 693.8597 - 68.7672 \cosh 0.0100333x,$$

where y is the height above the ground and x is the distance from the line of symmetry through the middle of the arch.

(a) Rewrite $\cosh 0.0100333x$ as a sum of exponential functions noting that $\cosh x = \frac{1}{2}e^x + \frac{1}{2}e^{-x}$.

(b) Rewrite the Gateway Arch function in terms of exponential functions instead of using the $\cosh x$ function.

(c) The arch has a height of approximately 625 ft and a span of approximately 600 ft. Using an appropriate viewing window, graph the function that gives its shape.

3. We are interested in comparing the shape of the Gateway Arch with an appropriate parabola.

(a) The parabola used is of the form $y = -ax^2 + b$.

i. Why is the value of a negative?

ii. Why is our parabola of the form $-ax^2$ rather than $-x^2$?

iii. Why is it necessary to add a constant, b, to the equation for our parabola?

(b) We want a parabola that matches the Gateway Arch at the vertex and at ground level. Hence, our parabola must contain the points $(0, 625)$, $(-300, 0)$, and $(300, 0)$. Use these points to find the values of a and b.

(c) Graph the parabola on the same set of axes as the formula for the Gateway Arch.

(d) How do your two graphs compare? How would you describe the shape of the Gateway Arch compared with the shape of a parabola?

3.4 COMPOSITION OF FUNCTIONS

In section 3.3, we looked at algebraic operations on functions. All these operations involved combining outputs by adding, subtracting, multiplying, or dividing. The type of function operation in this section is distinctly different. This time, rather than combining outputs, we link functions by causing the output of one to be the input of the second. In this section, we

[13] *National Park Service,* Jefferson National Expansion Memorial-Arch History and Architecture *(visited 2 February 1998) (http://www.nps.gov/jeff/arch-ov.htm).*

look at the definition of composition, what it means in terms of various representations, and how to "reverse" composition.

DEFINITION OF COMPOSITION

At an arcade, you first buy tokens for the games you want to play. Next, you play any one of a set of games in which, if you are successful, the tokens are replaced with tickets. Finally, you redeem these tickets for a prize. This situation involves three functions. The first function (accomplished at the token machine) changes money to tokens. The second function (accomplished at the arcade game) changes tokens to tickets. The third function (accomplished at the prize counter) changes tickets to a prize. Notice the relationship between the inputs and outputs. The output from one function becomes the input for the next. The three functions are linked together. The composition (or linking) of these three functions allows you to walk into the arcade with money and walk out with prizes.

Let's look at another example. Let the first function, C, have an input, d, which is the day of the year in 1998 and an output which is the high temperature in degrees Celsius for that day in Holland, Michigan. For example, $C(12) = 0°\text{C}$ means that the high temperature (in Celsius) in Holland, Michigan, for the 12th day of the year (12 January 1998) was 0°C. Many of us prefer to see temperature in degrees Fahrenheit. For Fahrenheit temperatures, we need a second function that converts from degrees Celsius to degrees Fahrenheit. This second function is $F(C) = \frac{9}{5}C + 32$, where the input, C, is the temperature in degrees Celsius and the output, F, is the corresponding temperature in degrees Fahrenheit. Once again, the output from the first function is the input for the second function.

Let's formalize this process. The **composition** of two functions occurs when the functions are linked so that the output of one function is the input of the other. Symbolically, if two functions are denoted by f and g, then g composed with f is denoted as $g \circ f$, which is pronounced "g of f". Another way of writing this composition is $g(f(x))$. Writing it this way reminds us that the output of f is to be the input for g.

The following diagram shows how composition works:

$$x \xrightarrow{f} f(x) \xrightarrow{g} g(f(x)).$$

The f and g above the arrows remind us that a function is a *process* of taking an input, doing something to it, and producing an output. The first arrow shows that the input is x, the process is the function f, and the output is $f(x)$. The second arrow shows that the input is $f(x)$, the process is the function g, and the output is $g(f(x))$. We have linked f and g together to produce a new function, $g \circ f$, whose input is x and whose output is $g(f(x))$.

Think back to the arcade. The diagram representing this process is

$$\text{money} \xrightarrow{\text{token machine}} \text{tokens} \xrightarrow{\text{arcade game}} \text{tickets} \xrightarrow{\text{prize counter}} \text{prize.}$$

The function is the process of changing an input to an output. In each case, the output for one process becomes the input for the next.

The following diagram is for the temperature example:

$$\text{day} \xrightarrow{C(d)} \text{Celsius temperature} \xrightarrow{F(C)=\frac{9}{5}C+32} \text{Fahrenheit temperature.}$$

Notice that composition is very different from the type of operations seen in section 3.3. There, we were interested in algebraically combining outputs. For example, we found the function for total population by adding birth and immigration rates while subtracting death rates. Composition, on the other hand, links two functions by using outputs as inputs.

Example 1 Let $f(x) = x^2$ and $g(x) = x + 3$. Find $g(f(2))$ and $f(g(2))$.

Solution First, we need to find $f(2) = 2^2 = 4$, which then becomes our input for g. We have $g(4) = 4 + 3 = 7$. Therefore, $g(f(2)) = 7$. A shorter way of writing this is $g(f(2)) = g(4) = 7$. To compute $f(g(2))$, we first find $g(2) = 2 + 3 = 5$. Using 5 as the input for f, we get $f(5) = 5^2 = 25$. So, $f(g(2)) = f(5) = 25$.

Notice that when we reversed the order of composition, the answer changed.[14] That usually happens in composition. In fact, sometimes reversing the order of the functions does not make any sense. Think about the arcade. The output of the last function (the prize) is not an appropriate input for the other two functions (the token machine or the arcade game). When we try to reverse the order, the output no longer makes sense as the input for the next function. The same thing is true of our temperature example. If we start with degrees Celsius and convert to degrees Fahrenheit, we do not get an input we can use in $C(d)$ that converts day to temperature. The inputs and outputs no longer match when the order is reversed.

Let's return to the functions used in Example 1, $f(x) = x^2$ and $g(x) = x + 3$. Suppose we want to find a formula for $g(f(x))$. Because $f(x) = x^2$, we have $g(f(x)) = g(x^2)$ (replacing the notation $f(x)$ with its output, x^2). To evaluate $g(x^2)$, recall from section 1.1 that the x in the function notation is just a placeholder for the input. So, for $g(x^2)$ instead of $g(x)$, just replace the x in the formula for $g(x)$ with an x^2. We now have $g(f(x)) = g(x^2) = x^2 + 3$. If we compose the functions the other way, we get $f(g(x)) = f(x + 3) = (x + 3)^2$. Looking at these two formulas, it is obvious that $g(f(x)) \neq f(g(x))$.

READING QUESTIONS

1. A bread machine takes ingredients and turns them into bread dough, which it then bakes to form a loaf of bread. Explain why this process is a composition and draw an appropriate arrow diagram to illustrate.

[14] *Because reversing the order of composition sometimes gives us a different result, composition is said to be noncommutative.*

2. Let f and g be two functions such that the composition, $g(f(x))$, is defined. Explain why it is not necessarily true that the composition, $f(g(x))$, will also be defined.
3. Let $f(x) = (x - 3)/x$ and $g(x) = 2x^3$. Find $g(f(1))$.

COMPOSITION OF FUNCTIONS FOR DIFFERENT REPRESENTATIONS

Regardless of whether the functions are given as symbols, tables, or graphs, composing two functions always means using the output of the first function as the input of the second.

Example 2 Let $f(x) = 4x$ and $g(x) = x^3$. Find the formulas for $g \circ f$ and $f \circ g$.

Solution We have $g \circ f = g(4x) = (4x)^3 = 64x^3$. Similarly, we have $f \circ g = f(x^3) = 4x^3$. Notice that changing the order of composition gives a different answer.

Example 3 Let f and g be represented by Table 1. Find $g(f(3))$ and $f(g(3))$.

TABLE 1

x	−3	−2	−1	0	1	2	3
$f(x)$	4	−1	0	2	3	−2	1
$g(x)$	−2	0	2	4	2	1	−1

Solution Use the input $x = 3$ in the first row of Table 1. Under this input, we see that $f(3) = 1$. Then start over in the first line with the input $x = 1$ to find that $g(1) = 2$. So, $g(f(3)) = g(1) = 2$. Similarly, starting with the input $x = 3$, we see that $g(3) = -1$. Starting over with this input in the first row, we see that $f(-1) = 0$. So, $f(g(3)) = f(-1) = 0$.

Example 4 Let f and g be represented by the graphs in Figure 1. Find $g(f(2))$.

Solution To find $g(f(2))$, we first need to find $f(2)$. By looking at the graph of $f(x)$, we see that it contains the point $(2, 4)$, so $f(2) = 4$. By looking at the graph of $g(x)$ (using the input $x = 4$), we see that it contains the point $(4, 5)$, so $g(4) = 5$. Thus, $g(f(2)) = g(4) = 5$.

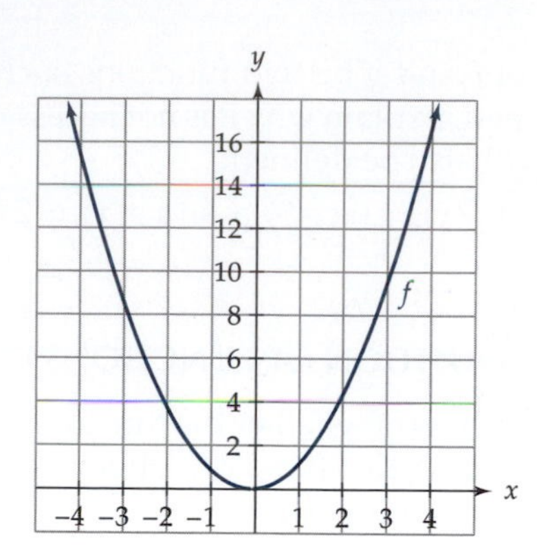

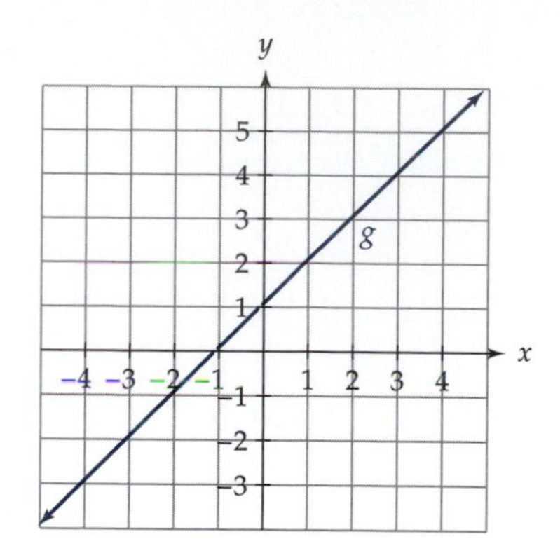

Figure 1

Composition does not have to be limited to two functions. To compose as many as you like, remember to use the output of the previous function as the input of the next. You are composing a string of functions "two at a time." We illustrate in the next example.

TABLE 2

x	−2	−1	0	1	2
$h(x)$	3	1	−1	0	2

Example 5 Let $f(x) = 3^x - 2$, $g(x) = \log x$, and h be given by Table 2. Find $h(g(f(1)))$.

Solution To find $h(g(f(1)))$, we first find that $f(1) = 3^1 - 2 = 3 - 2 = 1$. Using 1 as the input for g, we find that $g(1) = \log 1 = 0$. Using 0 as the input for h, we see from Table 2 that $h(0) = -1$. So, $h(g(f(1))) = h(g(1)) = h(0) = -1$.

READING QUESTIONS

4. Let $f(x) = \sqrt{x}$, g be given by the accompanying graph and h be given by the table. Find each of the following.

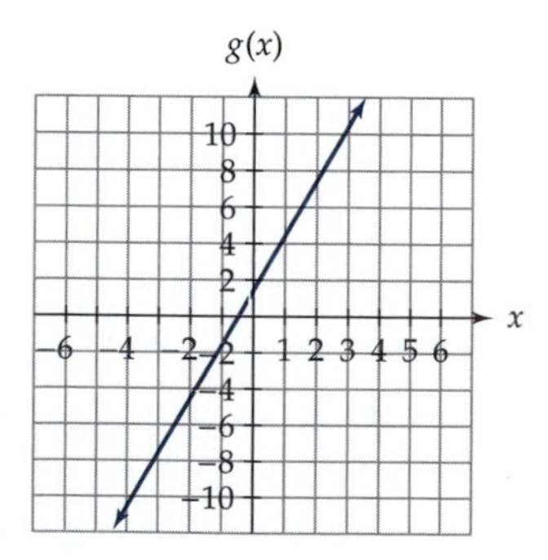

(a) $h(g(0))$

(b) $f(h(-2))$

(c) $g(f(9))$
(d) $g(f(h(-1)))$

x	-2	-1	0	1	2
$h(x)$	4	1	2	0	-1

HOW TO TELL IF A SITUATION INVOLVES COMPOSITION

One of the authors decided to wallpaper his daughter's room. He measured the room and discovered it was 492.75 in. around. He wanted to wallpaper the bottom 32 in. of the wall as a wainscoting. He knew that the wallpaper came in rolls that were 20.75 in. wide and 33 ft long. Using this information, he needed to calculate the number of rolls of wallpaper to buy.[15]

This example is one of composition. To calculate the number of rolls of wallpaper, he first needs to convert the measurement of the room to strips of wallpaper. He then needs to convert the number of strips to the number of rolls of wallpaper he needs to buy. Calculating the number of rolls of wallpaper needed for his daughter's room is a two-step process where the output from the first calculation, number of strips, becomes the input for the second calculation.

Let's look at another example. Table 3 gives some temperature data collected during a 24-hr period with a Calculator-Based Laboratory and a TI-83 calculator.

TABLE 3 Temperature Data in Degrees Fahrenheit over a 24-hr Period

Time	4 P.M.	5 P.M.	6 P.M.	7 P.M.	8 P.M.	9 P.M.	10 P.M.	11 P.M.	12 A.M.	1 A.M.	2 A.M.	3 A.M.
Temperature (°F)	72°	69°	65°	63°	62°	62°	61°	59°	58°	57°	56°	56°
Time	4 A.M.	5 A.M.	6 A.M.	7 A.M.	8 A.M.	9 A.M.	10 A.M.	11 A.M.	12 P.M.	1 P.M.	2 P.M.	3 P.M.
Temperature (°F)	57°	62°	66°	72°	76°	80°	81°	82°	84°	83°	83°	85°

Let F represent the function given by Table 3. The input, t, is time of day and the output, $F(t)$, is temperature in degrees Fahrenheit. Suppose we wanted the input to be time of day but wanted the output to be temperature in degrees Celsius. Again, we have an example of composition. What we need is a function that will convert from degrees Fahrenheit (the output of our current function) to degrees Celsius (the output we want). The output of the first function is the input of the second function. Let C represent the function whose input is time of day and whose output is the temperature in degrees Celsius. Then $C(t) = (C \circ F)(t) = C(F(t))$, where $F(t)$ is the function whose input is time of day and whose output is the temperature,

[15] *Assuming he did not need to match a pattern.*

TABLE 4 Temperature Data in Degrees Celsius over a 24-hr Period

Time	4 P.M.	5 P.M.	6 P.M.	7 P.M.	8 P.M.	9 P.M.	10 P.M.	11 P.M.	12 A.M.	1 A.M.	2 A.M.	3 A.M.
Temperature (°F)	22°	21°	18°	17°	17°	17°	16°	15°	14°	14°	13°	13°
Time	4 A.M.	5 A.M.	6 A.M.	7 A.M.	8 A.M.	9 A.M.	10 A.M.	11 A.M.	12 P.M.	1 P.M.	2 P.M.	3 P.M.
Temperature (°F)	14°	17°	19°	22°	24°	27°	27°	28°	29°	28°	28°	29°

whereas $C(F)$ converts degrees Fahrenheit to degrees Celsius. Note that $C(F) = \frac{5}{9}(F - 32)$. Table 4 represents our new function, C.

One of the ways to know if a physical situation involves the composition of functions is to ask if there is more than one function giving you information. Just knowing you are working with more than one function, however, does not necessarily mean you have a composition. The situation may involve algebraically combining two functions by adding, subtracting, multiplying, or dividing. The key to knowing that you have composition is whether or not you have a multistep process where the output from your first step becomes the input for your second step.

READING QUESTIONS

5. In the wallpaper example, explain why the first function the author will use is $S(m) = m/20.75$, where m is the measurement of the room (in inches) and $S(m)$ is the number of strips of wallpaper needed.
6. Let C be the function whose input is the time of day and whose output is temperature in degrees Celsius. Using Table 3, calculate $C(10 \text{ P.M.})$ and show it approximately equals 16°C.
7. Give an example of a physical situation involving composition.

DOMAIN AND RANGE

Let f and g be two functions such that $h(x) = (g \circ f)(x) = g(f(x))$. How are the domain and range of h related to the domain and range of f and g? Let's start with the domain. For an input to be valid for h, it must first be valid for f. At first glance, it might appear that the domain of h should equal the domain of f. We need to be careful, however, to make sure that the resulting inputs for g are also valid. That is, we need to be sure that all the outputs for f qualify as valid inputs for g. Otherwise, we will have a problem when we try to evaluate g. Thus, the domain for h is going to be a subset of the domain for f.[16] Let's look at an example.

[16] *The entire set is considered a subset of itself, so it is possible for the domain of h to equal the domain of f.*

Let $f(x) = x - 2$, $g(x) = \sqrt{x}$, and $h(x) = g(f(x))$. To find the domain for h, we start with the domain for f, which is all real numbers. The domain for g, however, is restricted to inputs that are greater than or equal to zero. Therefore, we need to restrict the inputs of f to those that give *outputs* that are positive or zero; otherwise, we will encounter a problem when we try to evaluate g. If $f(x) \geq 0$, then $x - 2 \geq 0$, giving $x \geq 2$. So, the domain of h is $x \geq 2$. If we look at the formula for h, we have $h(x) = g(x-2) = \sqrt{x-2}$. For this function to be defined, we need to restrict the values of x to those that give nonnegative numbers inside the square root sign. Thus, the domain is going to be $x \geq 2$, the same answer as when we found the domain of h by thinking of it as a composition.

Sometimes, knowing that a function, $h(x) = g(f(x))$, is a composition puts restrictions on the domain that do not seem obvious when you simply look at the formula for h. For example, suppose $f(x) = 1/(x+2)$ and $g(x) = 1/x$. The domain of f is all real numbers except for $x = -2$. The domain of g is all real numbers except 0, so we also have to eliminate inputs for f that give 0 as an output. In this case, there are none. Therefore, the domain of h is all real numbers except $x = -2$. If we look at the formula for h, we get $h(x) = g(f(x)) = g(1/(x+2)) = 1/(1/(x+2)) = x+2$, which appears to be defined for all real numbers. We can only see the reason for the restriction in the domain of h when we know that h is a composition.

Now that we have seen the domain, let's consider what happens to the range. In general, if $h(x) = g(f(x))$, then any output for h is also an output for g. At first glance, it might appear that the range of h should equal the range of g. Once again, however, we need to be careful. The range of h may not be equal to the entire range of g. The inputs of g must be *outputs of* f. This requirement may restrict what numbers can be inputs for g and, therefore, restrict what outputs you can get from g. So, the range of h is going to be a subset of the range of g.

For example, let $f(x) = x^2$, $g(x) = 3x + 1$, and $h(x) = g(f(x))$. The range of g is all real numbers. The range of h, however, is not all real numbers because the outputs for f will always be nonnegative. Because our inputs for g are restricted, our outputs for g will be restricted as well. The range for h is $y \geq 1$ because if you multiply a nonnegative number by 3 and add 1, you always get an answer greater than or equal to 1. If we look at the formula for h, then we have $h(x) = g(f(x)) = g(x^2) = 3x^2 + 1$. Because $3x^2 \geq 0$, the range of h will be $h(x) \geq 1$, the same answer we obtained when we found the range of h by thinking of it as a composition.

Example 6 Let f and g be represented by Table 5 and Table 6. Find the domain and range for $h = g(f(x))$.

TABLE 5

x	-3	-2	-1	0	1	2	3
$f(x)$	5	2	1	0	2	-1	-2

TABLE 6

x	0	1	2	3	4	5	6
$g(x)$	-1	4	3	0	-1	2	1

Solution The domain for f is $\{-3,-2,-1,0,1,2,3\}$, so our domain for h is a subset of these numbers. The inputs for g are restricted to being the integers from 0 to 6. Because $f(2) = -1$ and $f(3) = -2$, 2 and 3 are not valid inputs for h. Therefore, the domain of h is $\{-3,-2,-1,0,1\}$. To find the range of h, find the outputs for each of these inputs. We have

$$h(-3) = g(f(-3)) = g(5) = 2$$
$$h(-2) = g(f(-2)) = g(2) = 3$$
$$h(-1) = g(f(-1)) = g(1) = 4$$
$$h(0) = g(f(0)) = g(0) = -1$$
$$h(1) = g(f(1)) = g(2) = 3.$$

So, the range of h is $\{-1,2,3,4\}$. Notice that this range is a subset of the range for g.

In summary,

- The domain of $h(x) = g(f(x))$ is a subset of the domain of f. It includes only those values for which $f(x)$ is a valid input for g.
- The range of $h(x) = g(f(x))$ is a subset of the range of g. It includes only those values that are outputs of g for inputs given by $f(x)$.

READING QUESTIONS

8. Let $f(x) = x/(x+2)$ and $g(x) = 1/x$. Find the domain of $h(x) = g(f(x))$.
9. Let $f(x) = \sqrt{x}$ and $g(x) = x^2$. Find the domain and range of $h(x) = g(f(x))$.
10. Let f and g be represented by the tables given below. Find the domain and range for $k(x) = f(g(x))$.

x	−3	−2	−1	0	1	2	3
$f(x)$	5	2	1	0	2	−1	−2

x	0	1	2	3	4	5	6
$g(x)$	−1	4	3	0	−1	2	1

FUNCTION DECOMPOSITION

Sometimes, when looking at the final result of a composition, we are interested in "going backward" to determine the original functions used to form the composition. "Undoing" composition also helps to understand the process of composition. It is often difficult to reverse this process; it is close to impossible if the function is given numerically or graphically. If the function is given symbolically, however, we can usually find a way of "undoing" composition.

For example, suppose $h(x) = (3^x + 1)^2$. We want to rewrite h as the composition of two functions, $f(x)$ and $g(x)$. If $h(x) = g(f(x))$, then the output of f is the input of g. So, g is the "outside function" whose input is the result of f. In the formula for h, the last step was the "squaring function." Because we are squaring the quantity $3^x + 1$, this expression must be the output from f. So, using $f(x) = 3^x + 1$ and $g(x) = x^2$ (the squaring function) gives us $h(x) = g(f(x)) = g(3^x + 1) = (3^x + 1)^2$.

Let's look at some more examples. Keep in mind that composition is not unique. There is often more than one way to choose f and g so that $h(x) = g(f(x))$.

Example 7 Let $h(x) = \sin(3x^4)$. Find a function f and a function g so that $h(x) = g(f(x))$.

Solution The "outside" function is the $y = \sin x$ function. So, choose $g(x) = \sin x$. Because we are taking the sine of $3x^4$, it must be the output from f. So, $f(x) = 3x^4$. To check, we see that $g(f(x)) = g(3x^4) = \sin(3x^4)$, which is what we are given for h.

Example 8 Let $h(x) = 3x^2 + 1$. Find a function f and a function g so that $h(x) = g(f(x))$.

Solution The "outside" function (the operation performed last) could be "adding 1." So, we let $g(x) = x + 1$. Because the input is $3x^2$, we have $f(x) = 3x^2$. Finding the composition, we get $g(f(x)) = g(3x^2) = 3x^2 + 1 = h(x)$.

We could also have used "multiply by 3 and add 1" as the "outside" function. In this case, $g(x) = 3x + 1$. Because we multiplied x^2 by 3 and added 1, we must have $f(x) = x^2$. Finding the composition gives $g(f(x)) = g(x^2) = 3x^2 + 1 = h(x)$. In this case, there is more than one way to choose the functions f and g.

READING QUESTIONS

11. Find functions f and g so that $g(f(x)) = h(x) = 10^{x^2+1}$.

12. Find two different pairs of functions, f and g, so that $g(f(x)) = 3 - 1/x^2$.

SUMMARY

The **composition** of two functions occurs when the functions are linked so that the output of one function becomes the input of the other. For example, we use $g(f(x))$ to show that the output of f becomes the input of g. We can also show this composition by writing $(g \circ f)(x)$.

Regardless of the form of these functions—graphical, numerical, or symbolic—the procedure for evaluating these functions is the same. You find the output of the first function and use it as the input of the second function.

The domain of $g(f(x))$ is a subset of the domain of f. It includes only those values for which $f(x)$ is a valid input for g. The range of $g(f(x))$ is a subset of the range of g. It includes only those values which are outputs of g for inputs given by $f(x)$.

If given a function symbolically that is a composition of other functions, the functions originally used to form the composition can sometimes be determined. This process is not unique, so there is often more than one way to decompose these functions.

EXERCISES

1. Let S be the function representing the process of placing a letter in an envelope and sealing it. Let A be the function representing the process of writing the address on the envelope.
   (a) Describe what $S \circ A$ represents.
   (b) Is $A \circ S$ a valid process? If so, what does it represent?
2. What is the difference between algebraically combining two functions and composing two functions?
3. Let $f(x) = \cos x$ and $g(x) = 3x^2 + 4$.
   (a) Find $(g \circ f)(0)$.
   (b) Find $(g \cdot f)(0)$.
4. The process of doing laundry involves two steps, first washing and then drying the clothes. Explain why this process is a composition and draw an appropriate arrow diagram to illustrate it.
5. Let f and g be given by the following table. Where possible, fill in the third row, giving the outputs for $g(f(x))$.

x	−3	−2	−1	0	1	2	3
$f(x)$	4	−1	0	2	3	−2	1
$g(x)$	−2	0	2	4	2	1	−1
$g(f(x))$							

6. Let f and g be represented by the accompanying graphs. Find $f(g(2))$.

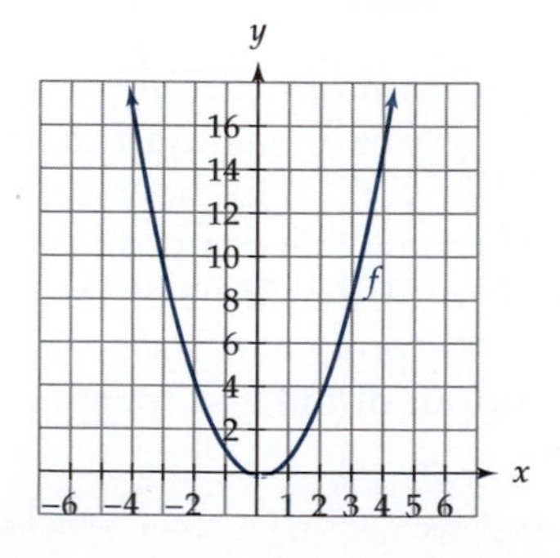

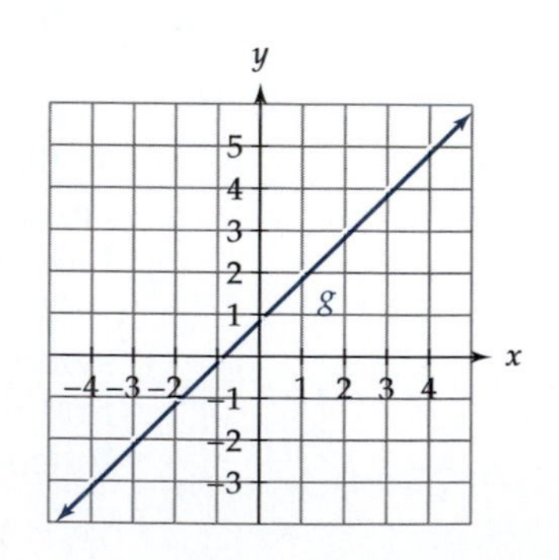

7. Let $f(x) = 2^x$, $g(x) = 3x - 4$, and $h(x) = \sqrt{x}$. Find the equation for each of the following.
   (a) $g \circ f$
   (b) $f \circ h$
   (c) $h \circ g$
   (d) $h \circ f \circ g$

8. In this section, we gave an example of one of the authors needing to buy wallpaper for his daughter's room. He needs to calculate the number of rolls of wallpaper he should buy.

   He intends to wallpaper the bottom 32 in. of the wall.

   The wallpaper comes in rolls that are 20.75 in. wide and 33 ft long.

   The wallpaper he is buying does not have a pattern that needs to be matched.

   (a) Explain why this calculation will be a composition.
   (b) Let the first function be S, where the input, m, is the perimeter of the room (not including doorways) in inches and the output, $S(m)$, is the number of strips of wallpaper that will need to be used. Find the formula for S.
   (c) Let the second function be $R(S)$, where the input, S, is the number of strips of wallpaper needed to paper the room and $R(S)$ is the number of rolls of wallpaper. Find the formula for $R(S)$.
   (d) Find the composition $R \circ S$.
   (e) If the perimeter of the room (not including doorways) is 492.75 in., find the number of rolls of wallpaper the author should buy.

9. You may have heard that it is possible to estimate the temperature by counting the number of cricket chirps. According to *Insect Fact and Folklore*,[17] the temperature in degrees Fahrenheit is approximately equal to the number of chirps in 15 sec plus 38.
   (a) Find the equation for the function whose input is number of chirps in 15 sec and whose output is temperature in degrees Fahrenheit.
   (b) Your answer to part (a) should be a linear function. Explain the physical meaning of the slope and y-intercept.
   (c) Suppose you wanted the temperature in degrees Celsius rather than degrees Fahrenheit. Explain why you would have a composition.
   (d) Find the equation for the function whose input is number of chirps and whose output is temperature in degrees Celsius. (*Note:* $C = \frac{5}{9}(F - 32)$.)
   (e) Your answer to part (d) should be a linear function. Explain the physical meaning of the slope and the y-intercept.

[17] *Lucy Clausen,* Insect Fact and Folklore *(New York: Macmillan, 1958), pp. 62–63.*

10. Given

$$f(x) = \begin{cases} x^2 + 1, & \text{if } x \leq 0 \\ 3x + 1, & \text{if } 0 < x \leq 5 \\ 2^{x-1}, & \text{if } x > 5, \end{cases}$$

find each of the following.
   (a) $f(f(-1))$
   (b) $f(f(0))$
   (c) $f(f(f(1)))$

11. Let f be given by the accompanying graph, $g(x) = x - 3$, and k be given by the table. Where possible, find each of the following.

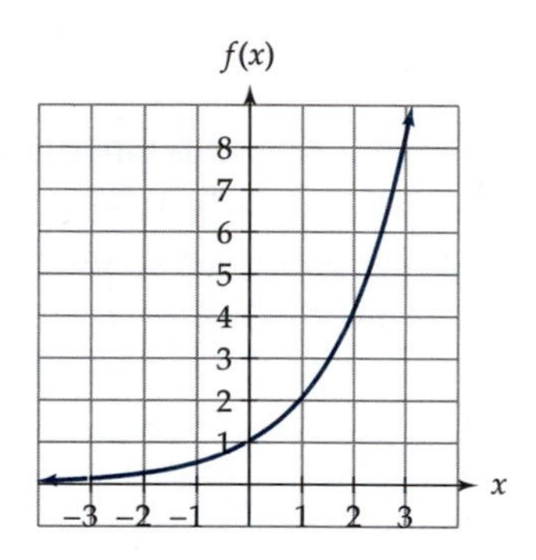

   (a) $k(f(0))$
   (b) $k(k(f(1)))$
   (c) $f(k(g(0)))$
   (d) $g(g(2))$
   (e) $f(k(g(1)))$

x	−4	−3	−2	−1	0	1	2	3	4
$k(x)$	−2	0	1	0	2	−1	3	4	2

12. Let $f(x) = x - 4$, $g(x) = \sqrt{x}$, $h(x) = g(f(x))$, and $k(x) = f(g(x))$.
   (a) Give the domain for f and g.
   (b) Give the domain for h and k. How is this related to your answer for part (a)?
   (c) Give the range for f and g.
   (d) Give the range for h and k. How is this related to your answer for part (c)?

13. Let f and g be represented by the following tables. Find the domain and range for $k(x) = f(g(x))$.

x	−3	−2	−1	0	1	2	3
$f(x)$	5	2	1	0	2	−1	−2

x	0	1	2	3	4	5	6
$g(x)$	−1	4	3	0	−1	2	1

14. For each of the following, find functions f and g so that $h(x) = g(f(x))$.
    (a) $h(x) = (x+3)^2$
    (b) $h(x) = x^2 + 3$
    (c) $h(x) = 10^{x^2}$
    (d) $h(x) = \log(x+1)$
    (e) $h(x) = 3x^4 - 2x^3$
15. For each of the following pairs of functions, find $g \circ f$. Simplify your answer.
    (a) $f(x) = 1/x$ and $g(x) = 1/x$
    (b) $f(x) = x + 1$ and $g(x) = x - 1$
    (c) $f(x) = 10^x$ and $g(x) = \log x$
    (d) $f(x) = \sqrt{x}$ and $g(x) = x^2$
    (e) $f(x) = 3x + 2$ and $g(x) = \frac{1}{3}(x-2)$
    (f) What do you notice about each of these examples?
16. The notation $f^n(x)$ means to compose a function with itself n times. For example, if $f(x) = x^2 + 1$, then $f^2(x) = f(f(x)) = f(x^2+1) = (x^2+1)^2 + 1 = x^4 + 2x^2 + 2$. For each of the following, find a pattern that allows you to predict $f^n(x)$ for a general value of n.
    (a) $f(x) = 1/x$
    (b) $f(x) = x + 1$
    (c) $f(x) = x^3$

INVESTIGATIONS

INVESTIGATION: WIND CHILL

If you have ever been outside on a cold, windy day, you have experienced the chilling effect of the wind. The wind chill equivalent temperature, commonly known as wind chill, is a number that represents the effect of the wind on your perception of cold. Equations for wind chill were first published in 1939 by Paul Siple and Charles Passel, based on experiments done in Antarctica.[18] Table 7 gives the wind chill equivalent temperature for air temperatures given in degrees Celsius and wind speeds given in meters per second.[19] Wind chill is a *multivariable* function, which means that there is more than one input (in this case, air temperature and wind speed) for this function.

[18] *John E. Oliver,* The Encyclopedia of Climatology *(New York: Van Nostrand Reinhold, 1987), pp. 928–29.*

[19] *Joseph M. Moran and Michael D. Morgan,* Meteorology: The Atmosphere and the Science of Weather, 2d ed. *(New York: Macmillan, 1989), p. 73.*

TABLE 7 Wind Chill Temperature (°C)

Wind Speed (m/sec)	Air Temperature (°C)															
	6	**3**	**0**	**−3**	**−6**	**−9**	**−12**	**−15**	**−18**	**−21**	**−24**	**−27**	**−30**	**−33**	**−36**	**−39**
3	3	−1	−4	−7	−11	−14	−18	−21	−24	−28	−31	−34	−38	−41	−45	−48
6	−2	−6	−10	−14	−18	−22	−26	−30	−34	−38	−42	−46	−50	−54	−58	−62
9	−6	−10	−14	−18	−23	−27	−31	−35	−40	−44	−48	−53	−57	−61	−65	−70
12	−8	−12	−17	−21	−26	−30	−35	−39	−44	−48	−53	−57	−62	−66	−71	−75
15	−9	−14	−18	−23	−27	−32	−37	−41	−46	−51	−55	−60	−65	−69	−74	−79
18	−10	−14	−19	−24	−29	−33	−38	−43	−48	−52	−57	−62	−67	−71	−76	−81
21	−10	−15	−20	−25	−29	−34	−39	−44	−49	−53	−58	−63	−68	−73	−77	−82
24	−10	−15	−20	−25	−30	−35	−39	−44	−49	−54	−59	−63	−68	−73	−78	−83

1. Find the wind chill if the temperature is −15°C and the wind speed is 12 m/sec.
2. An air temperature of −15°C is equivalent to 5°F. A wind speed of 12 m/sec is equivalent to approximately 27 mph. Find the wind chill if the air temperature is 5°F and the wind speed is 27 mph. Be sure to give your answer in degrees Fahrenheit rather than degrees Celsius.
3. The equation for wind chill is

$$W_C = 33 - \frac{(10.45 + 10\sqrt{V_{\text{m/sec}}} - V_{\text{m/sec}})(33 - C)}{22.04},$$

where W_C is the wind chill temperature in degrees Celsius, $V_{\text{m/sec}}$ is the speed of the wind in meters per second, and C is the air temperature in degrees Celsius.

   (a) Let $C = -15$°C and $V_{\text{m/sec}} = 12$ m/sec. Use the wind chill formula to compute the wind chill for these values. Does your answer agree with your answer to question 1?

   (b) We want to convert the wind chill formula from metric to English measurements.

      i. First, convert the speed from meters per second to miles per hour. There are approximately 1609 m in a mile and exactly 3600 sec in 1 h. Use this information to find the conversion factor, a, so that $V_{\text{m/sec}} = aV_{\text{mph}}$. Use this conversion factor to change from $V_{\text{m/sec}}$ (speed in meters per second) to V_{mph} (speed in miles per hour) in the wind chill formula.

      ii. Now using $C = \frac{5}{9}(F - 32)$, convert the temperature from degrees Celsius to degrees Fahrenheit in the formula. Also be sure to convert the resulting wind chill temperature to degrees Fahrenheit.

4. Use your answer from question 3, part (b)(ii) to find the wind chill (in degrees Fahrenheit) for an air temperature of 5°F and a wind speed of 27 mph. Does this agree with your answer to question 2?

3.5 INVERSE FUNCTIONS

In this chapter, we have looked at transforming functions by adding or multiplying a constant to the input or output of a function. We have also looked at algebraically combining two or more functions and determining the composition of functions. There is yet another type of function operation to explore. This operation involves having a function go "backward" by switching the roles of the input and the output. This type of function operation is known as finding a function's inverse. In this section, we look at what an inverse function is, see how to determine if a function has an inverse, and learn how to find it.

WHAT IS AN INVERSE?

You are familar with the word *inverse* in mathematics as it applies to the operations of arithmetic. For example, adding 3 to a number is the inverse of subtracting 3 from a number. Multiplying a number by 4 is the inverse of dividing a number by 4. These operations are called inverse operations because the net result is no change; one operation undoes the other. This same idea is true of inverse functions. Because functions can be an involved process, let's look at a physical process first.

Suppose you are sitting at your desk in your dorm room and a friend calls to borrow some of your class notes. You decide to take them to your friend's room. To do so, you get your notes from your folder, get up from your desk, walk to your friend's room, and give your friend the notes. To undo this process, you take the notes from your friend, walk back to your room, sit down at your desk, and put the notes back in your folder. Notice that to undo this process you had to reverse it. Not only did you have to reverse the order of things, but you also had to reverse each individual action. For example, instead of getting up from your desk, you sat down at your desk. Also notice that the input and the output switched roles. If you think of getting the notes to your friend as a function, then the input is the class notes in your folder and the output is the class notes in your friend's hands. The inverse function (or the process of taking your notes back) reversed the input and the output. The input of the inverse function is the class notes in your friend's hands, and the output of the inverse function is the class notes in your folder. These two things, reversing each process and reversing the role of the input and the output, always happen when we find an inverse.

The numbers in Table 1 represent the function where the input is the age of a baby in months and the output is the average weight of a baby in pounds. This table can be used to predict the weight based on the baby's

TABLE 1 The Average Weight of Babies Given Their Age

Age (months)	0	1	2	3	4	5	6
Weight (lb)	7.1	9.3	11.3	13.0	14.6	16.0	17.2

TABLE 2 The Age of Babies Given Their Average Weight

Weight (lb)	7.1	9.3	11.3	13.0	14.6	16.0	17.2
Age (months)	0	1	2	3	4	5	6

age. For example, if a child is 3 months old, you might expect his or her weight to be about 13.0 lb. You could, however, also use the weight of a child to predict his or her age. In this case, we want a function where the input is weight and the output is age. Such a function is shown in Table 2. Notice that Table 2 is identical to Table 1 except that the inputs and outputs are reversed. From Table 2, you can predict that a child who weighs 13.0 pounds is about 3 months old. Tables 1 and 2 are inverses of each other. Notice also that if you start with Table 2, then its inverse is Table 1. If one function is the inverse of a second, then the second is also the inverse of the first.

Let's look at another example. Two functions with which you are familiar are those that convert temperature in degrees Fahrenheit to degrees Celsius and vice versa:

$$C = \tfrac{5}{9}(F - 32)$$
$$F = \tfrac{9}{5}C + 32.$$

In the first equation, 32 is subtracted from the input, F. That difference is then multiplied by $\frac{5}{9}$. In the second equation, the process is reversed; both the order and the mathematical operations are reversed. The input, C, is multiplied by $\frac{9}{5}$ (the same as dividing by $\frac{5}{9}$). That product is then added to 32. The diagram illustrates these two processes. The first line is read left to right, and the second is read right to left.

$$\text{Subtract } 32 \Rightarrow \text{ multiply by } \tfrac{5}{9}.$$
$$\text{Add } 32 \Leftarrow \text{ divide by } \tfrac{5}{9}.$$

Another way to see that these two functions are inverses of each other is to look at a specific point. If 68°F is the input for the first function, then its output is 20°C. Using 20°C as the input for the second function, its output is 68°F. The 68 was converted to 20 in the first function, and the 20 was converted back to the 68 in the second function. This process of taking the output of one function and using it as the input of a second function is composition. When we take the composition of inverse functions, we get back what we started with. Hence, inverse functions "undo" each other. The formal definition of inverses is as follows.

DEFINITIONS

Two functions f and g are **inverses** of each other if

- $f(g(x)) = x$ for every x in the domain of g
- $g(f(x)) = x$ for every x in the domain of f

The functions $f(x) = \frac{5}{9}(x - 32)$ and $g(x) = \frac{9}{5}x + 32$ are the temperature conversion functions written in function notation. To see that they are inverses of each other, we find the composition of $g(f(x))$ and $f(g(x))$.

$$g(f(x)) = g\left(\tfrac{5}{9}(x-32)\right) = \tfrac{9}{5}\left(\tfrac{5}{9}(x-32)\right) + 32 = (x-32) + 32 = x$$

$$f(g(x)) = f\left(\tfrac{9}{5}x + 32\right) = \tfrac{5}{9}\left(\tfrac{9}{5}x + 32 - 32\right) = \tfrac{5}{9}\left(\tfrac{9}{5}x\right) = x.$$

In both cases, $g(f(x)) = x$ and $f(g(x)) = x$, which proves that f and g are inverses of each other.

READING QUESTIONS

1. Suppose the function f consisted of multiplying the input by 5 and then adding 12. What would be the inverse of this function?
2. What is the relationship between the inputs and outputs of a function and the inputs and outputs of its inverse?
3. Describe an inverse function in your own words.

HOW TO FIND AN INVERSE FUNCTION

Finding the inverse of a function can be simple or difficult, depending on both the type of function and form in which it is given. The notation for the inverse of a function, f, is f^{-1}. The -1 is used to denote the inverse and is not an exponent.[20] The temperature conversion functions are often written as

$$C = \tfrac{5}{9}(F - 32)$$
$$F = \tfrac{9}{5}C + 32.$$

They could also be written as

$$f(x) = \tfrac{5}{9}(x - 32)$$
$$f^{-1}(x) = \tfrac{9}{5}x + 32.$$

The nice thing about this notation is that the f and the f^{-1} quickly identify these functions as inverses of each other. The problem with this notation is that the labels F and C are lost. These labels are helpful in distinguishing the input and the output for each function.

Finding an Inverse Numerically

Finding an inverse numerically is simply a matter of reversing the inputs and outputs.

TABLE 3

x	0	1	2	3	4
$f(x)$	3	7	9	11	12

Example 1 Find the inverse of the function given in Table 3.

Solution To find the inverse, we simply need to switch the input and the output around. The inverse function, f^{-1}, is shown in Table 4.

TABLE 4

x	3	7	9	11	12
$f^{-1}(x)$	0	1	2	3	4

[20] *If you want the -1 to indicate an exponent, then write your function as $(f(x))^{-1} = 1/f(x)$.*

Finding an Inverse Symbolically

Because an inverse function reverses the role of the input and output, finding an inverse of a function that is represented symbolically is done by solving the equation for the input.

Example 2 The function used to determine the temperature in degrees Fahrenheit, T, given the number of chirps per minute for a field cricket, n, is given by the formula, $T = (n/4) + 38$. Find the inverse of this function and describe what the input and output represent.

Solution Solving $T = (n/4) + 38$ for n gives

$$\begin{aligned} T &= \frac{n}{4} + 38 \\ T - 38 &= \frac{n}{4} \\ 4(T - 38) &= n. \end{aligned}$$

Therefore, the inverse of $T = (n/4) + 38$ is $n = 4(T - 38)$. The input for the inverse function is temperature in degrees Fahrenheit, and the output is the number of chirps per minute for a field cricket.

If a function is given in function notation, f, the inverse, f^{-1}, is found in much the same way, with some notation changes. Manipulating a function with the notation $f(x)$ can be a bit cumbersome. Replacing $f(x)$ with y makes it easier. We also switch the letters x and y after finding the formula for the inverse to remind us that the roles of the input and the output have switched.

Example 3 Find the inverse of $f(x) = \frac{2}{3}x - 4$.

Solution Rewriting the function as $y = \frac{2}{3}x - 4$ and solving for the input, x, gives

$$\begin{aligned} y &= \tfrac{2}{3}x - 4 \\ y + 4 &= \tfrac{2}{3}x \\ \tfrac{3}{2}(y + 4) &= x. \end{aligned}$$

Our inverse function is $x = \frac{3}{2}(y + 4)$. Rewriting this function in inverse function notation (replacing the x with y and the y with x), gives $y = \frac{3}{2}(x + 4)$ or $f^{-1}(x) = \frac{3}{2}(x + 4)$.

In chapter 2, we saw the relationship between exponential functions with a growth factor of 10 and logarithmic functions. The logarithmic function was used to determine the magnitude of a number. For example, determining the magnitude or logarithm of 415 is the same as solving the equation $10^x = 415$. In this case, $x = \log 415 \approx 2.618$. (Remember that $10^a = b$ means that $\log b = a$.) So, the inverse of the logarithm is the exponential function with a growth factor of 10. Similarly, the inverse of $f(x) = 10^x$ is $f^{-1}(x) = \log x$.

Example 4 Find the inverse of $f(x) = 3 \cdot 10^x$.

Solution If we rewrite this function as $y = 3 \cdot 10^x$, then $y/3 = 10^x$. By taking the logarithm of both sides to "undo" the exponential function, we get $\log(y/3) = x$. Rewriting this in inverse function notation gives $f^{-1}(x) = \log(x/3)$.

The inverse of a power function, $f(x) = x^a$, is $f^{-1}(x) = x^{1/a}$ because $f^{-1}(f(x)) = f^{-1}(x^a) = (x^a)^{1/a} = x$. Not all power functions have inverses for their entire domain, as we soon see. Even on a restricted domain, however, it is still true that $f^{-1}(x) = x^{1/a}$.

Example 5 Find the inverse of $f(x) = 4x^{3/5}$.

Solution We first rewrite the function as $y = 4x^{3/5}$ and then solve the equation for x.

$$\frac{y}{4} = x^{3/5}$$

$$\left(\frac{y}{4}\right)^{5/3} = x.$$

Rewriting in inverse function notation gives $f^{-1}(x) = (x/4)^{5/3}$.

Finding an Inverse Graphically

The graph of a function and the graph of its inverse have a nice relationship. For example, Figure 1 shows the graphs of $f(x) = x^3$ and $f^{-1}(x) = x^{1/3}$. Also included is the line $y = x$. Notice that the graph of $f^{-1}(x)$ is a reflection of the graph of $f(x)$ across the line $y = x$.

To show that the graph of f^{-1} is always the graph of f reflected over the line $y = x$, consider a function that passes through an arbitrary point (a, b). The inverse of this function passes through the point (b, a) because an inverse reverses the role of inputs and outputs. We want to show that (b, a) is a reflection of (a, b) across the line $y = x$. In Figure 2, horizontal and vertical line segments have been drawn to form a square whose sides are of length $a - b$. One diagonal of this square is the line $y = x$, and the other diagonal is the line segment connecting (a, b) and (b, a). Because the diagonals of a square are perpendicular bisectors of one another, (b, a) must be a reflection of (a, b) across the line $y = x$.

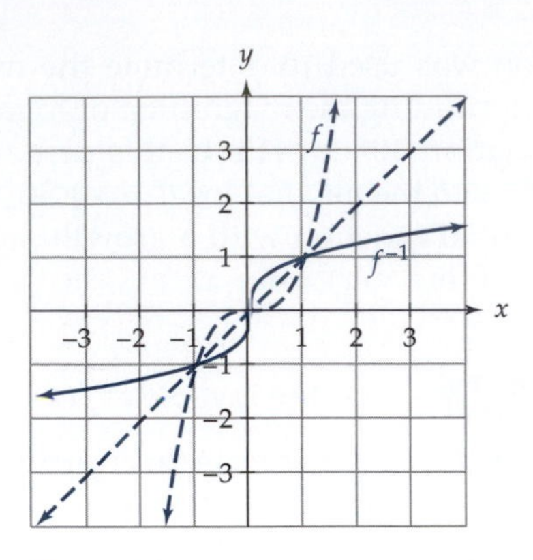

Figure 1 The graphs of $f(x) = x^3$ and $f^{-1}(x) = x^{1/3}$ along with the line $y = x$.

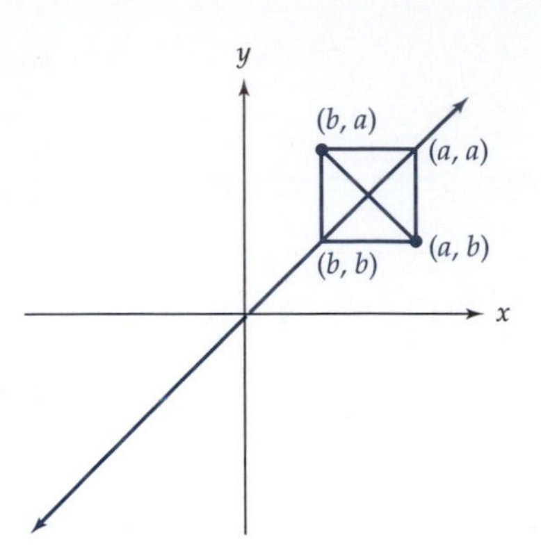

Figure 2 The point (b, a) is a reflection of the point (a, b) across the line $y = x$.

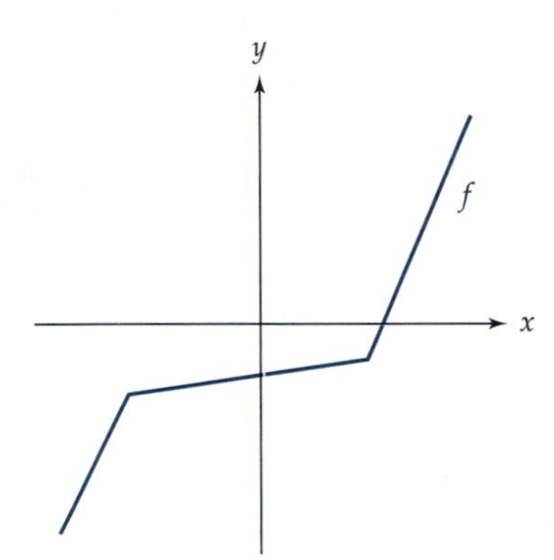

Figure 3

Example 6 Given the graph of the function f shown in Figure 3, sketch a graph of f^{-1}.

Solution The graph of f^{-1} is a reflection of f across the line $y = x$. To sketch the graph of f^{-1}, first draw in the line $y = x$ and then reflect points on the graph across this line. See Figure 4(a). To complete the graph, connect the points. See Figure 4(b).

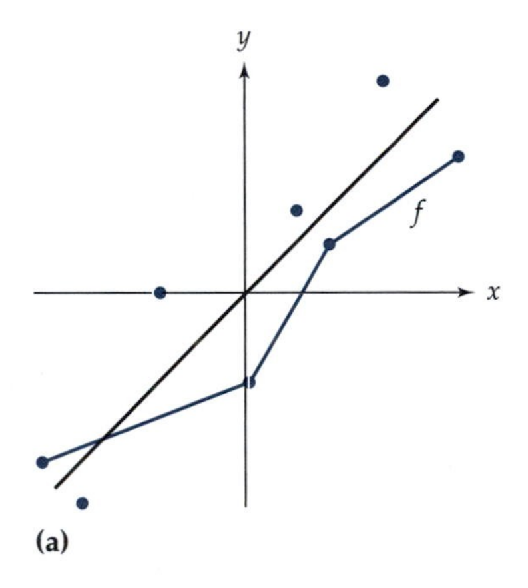

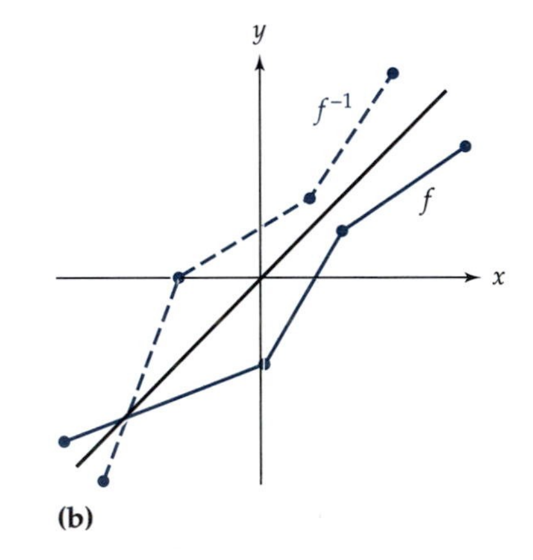

Figure 4

READING QUESTIONS

4. What does the -1 represent in the notation f^{-1}?

5. Find the inverse of $f(x) = x/3 - \frac{4}{3}$.

6. Use the table to find the inverse of $g(x)$.

x	-2	-1	0	1	2
$g(x)$	4	3	1	0	-2

7. What relationship does a function and its inverse have when they are represented graphically?

NOT ALL FUNCTIONS HAVE INVERSES

The reversing of a process cannot always be done. For example, the process of mixing ingredients for a loaf of bread and then baking it cannot be undone. You cannot take a loaf of bread and make flour, yeast, water, and butter out of it. The same thing is true for some mathematical processes. The process of finding an inverse of a function is not always possible. In particular, it is impossible if the inverse happens to be something that is not a function. For example, the function given in Table 5 represents the height of a ball thrown in the air (in feet) for given amounts of time (in seconds). Table 5 represents a function because for each input there is only one output. If the inputs and outputs are reversed, however, the result is not a function. (See Table 6.) In Table 6, you can see, for example, that the input of 6 ft has two outputs, 0 sec and 3 sec, which makes sense from our physical situation because what goes up must come down, but which violates the definition of a function. Therefore, the function described in Table 5 does not have an inverse.

TABLE 5 The Height of a Ball for Given Amounts of Time

Time (sec)	0	0.5	1	1.5	2	2.5	3
Height (ft)	6	26	32	42	32	26	6

TABLE 6 The Time a Ball Is in the Air (in seconds) for Given Heights of the Ball

Height (ft)	6	26	32	42	32	26	6
Time (sec)	0	0.5	1	1.5	2	2.5	3

In chapter 1, we defined a function as a rule that assigns an input to at most one output. A function, however, can have an output with more than one input, such as the function in Table 5. This situation causes problems when trying to find an inverse. The property that guarantees a function has an inverse is called one-to-one. A function is **one-to-one** if for each output there is only one corresponding input. If a function is given as a table of numbers, then it is easy to determine if it is one-to-one; merely check that none of the outputs is repeated. In Table 5, some of the outputs are repeated. Therefore, that function is not one-to-one, and hence it does not have an inverse.

Determining if a function represented graphically has an inverse is similar to determining if a graph is a function. Recall that we used the vertical line test to determine if a graph represented a function. If a vertical line crossed a graph at more than one point, then an input had more than one output and so the graph did not represent a function. A similar test is used to determine if a graph represents a one-to-one function. In this case, we want to check if each output has only one input. To do so, we use the horizontal line test. The **horizontal line test** states that if there exists a horizontal line that crosses a graph at more than one point, then the graph does not represent a one-to-one function. Hence, the graph will not have an inverse. The graph in Figure 5(a) represents a one-to-one function and therefore has an inverse. The graph in Figure 5(b), however, fails the horizontal line test. It is not a one-to-one function and therefore does not have an inverse.

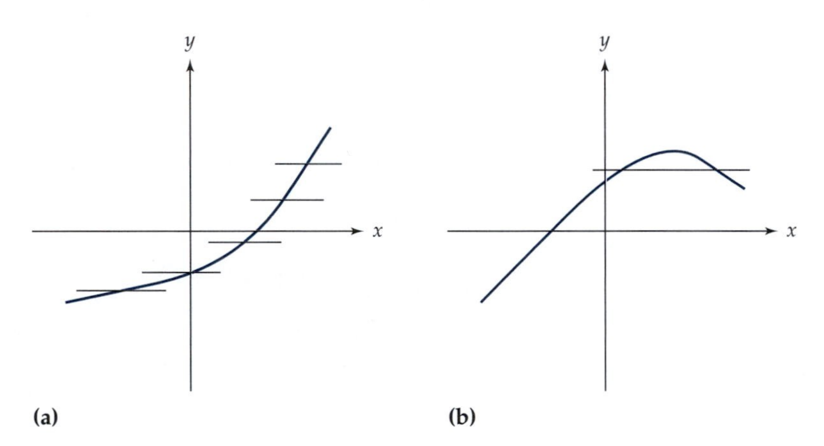

Figure 5 The graph in (a) passes the horizontal line test, and the graph in (b) fails the horizontal line test.

If a function is given symbolically, sometimes the easiest way to tell if it is one-to-one is to graph it. Some of the functions studied in chapter 2 are not one-to-one and so do not have inverses. For example, power functions with even exponents are not one-to-one. If $f(x) = x^2$, it is easy to see that $f(3) = 9$ and $f(-3) = 9$. Because two inputs give the same output, this function is not one-to-one and does not have an inverse.

READING QUESTIONS

8. Why don't some functions have inverses?

9. What is a one-to-one function?

10. How does the horizontal line test determine if a function has an inverse?

RESTRICTING THE DOMAIN OF A FUNCTION TO FIND ITS INVERSE

We stated earlier that functions that are not one-to-one do not have inverses. Often, however, we can take a function that is not one-to-one and, by restricting its domain, make it one-to-one on that interval. For example, we already mentioned that the function $f(x) = x^2$ is not one-to-one. See Figure 6(a). If we only consider nonnegative inputs, however, then we have a piece of the function that is one-to-one. We can write this function as $f(x) = x^2, x \geq 0$. To find the inverse, we again substitute y for $f(x)$. Solving $y = x^2$ for x gives $x = \pm\sqrt{y}$, which is not a function. Because we restricted x so that it could not be negative, however, $x \neq -\sqrt{y}$. Therefore, the inverse function is $x = \sqrt{y}$. Rewriting in function notation gives $f^{-1}(x) = \sqrt{x}$. The graphs of $f(x) = x^2, x \geq 0$, and its inverse, $f^{-1}(x) = \sqrt{x}$, are shown in Figure 6(b).

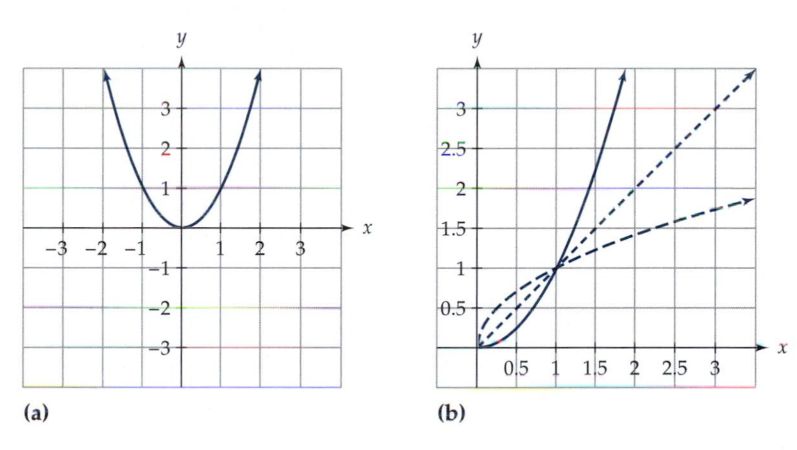

Figure 6 The graph in (a) shows that the inverse of $y = x^2$ does not exist because it is not one-to-one. The graph in (b) shows how restricting the domain of $f(x) = x^2$ to $x \geq 0$ allows us to find the inverse.

The trigonometric functions of sine and cosine studied in chapter 2 are also not one-to-one. Because they are periodic, the outputs are repeated over and over again. These functions, however, do have inverses if we restrict their domain. (See Figure 7.)

DEFINITIONS

These inverse functions are defined as follows:

$y = \sin^{-1} x$ if and only if $x = \sin y$ and $\dfrac{-\pi}{2} \leq y \leq \dfrac{\pi}{2}$

$y = \cos^{-1} x$ if and only if $x = \cos y$ and $0 \leq y \leq \pi$.

Because the range of the sine function is -1 to 1, the domain of $y = \sin^{-1} x$ is $-1 \leq x \leq 1$. Be aware that the -1 in $y = \sin^{-1} x$ is not an exponent. Just as with function notation, it refers to being an inverse. The function $y = \sin^{-1} x$ can also be written as $y = \arcsin x$. Similarly, the function

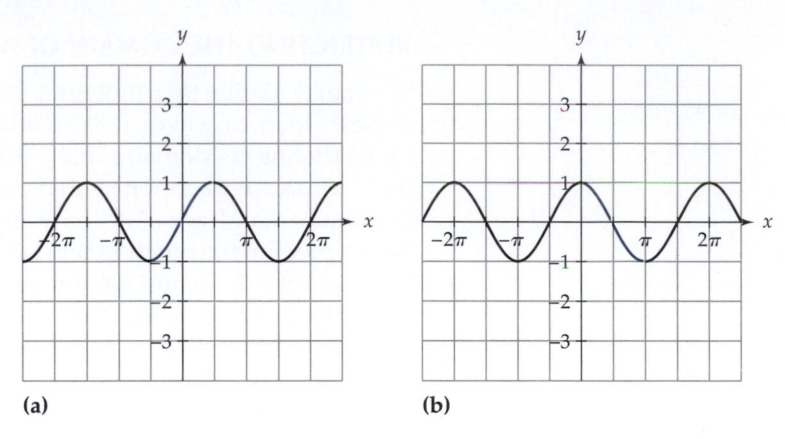

Figure 7 The graph in (a) shows how the domain of the sine function can be restricted so that it has an inverse. The graph in (b) shows how the domain of the cosine function can be restricted so that it has an inverse.

$y = \cos^{-1} x$ can also be written as $y = \arccos x$. Recall that the input for both $y = \sin x$ and $y = \cos x$ is the distance on a unit circle. So, for $y = \arcsin x$ and $y = \arccos x$, the output is the distance on a unit circle. The notation arcsine and arccosine reminds us that these functions are defined in terms of the arc length of a circle.

Example 7 Find the inverse, f^{-1}, of $f(x) = 3\sin x$. Give the domain and range of f^{-1}.

Solution To find f^{-1}, we first write f as $y = 3\sin x$. Solving for x gives

$$y = 3\sin x$$
$$\frac{y}{3} = \sin x$$
$$\sin^{-1}\left(\frac{y}{3}\right) = x.$$

Rewriting the inverse in function notation gives $f^{-1}(x) = \sin^{-1}(x/3)$. Our original function, $f(x) = 3\sin x$, has a range of $-3 \le f(x) \le 3$, and because we found its inverse, its domain is restricted to $-\pi/2 \le x \le \pi/2$. Because inverse functions switch the roles of the inputs and the outputs, f^{-1} has a domain of $-3 \le x \le 3$ and a range of $-\pi/2 \le f^{-1}(x) \le \pi/2$.

READING QUESTIONS

11. How can a function that is not one-to-one be changed so that it has an inverse?

12. In the definitions for inverse sine and cosine, the restrictions on the domains for sine and cosine are different. Why are they not the same?

13. Give an interval for the domain of the accompanying graph so that it will have an inverse on the restricted domain.

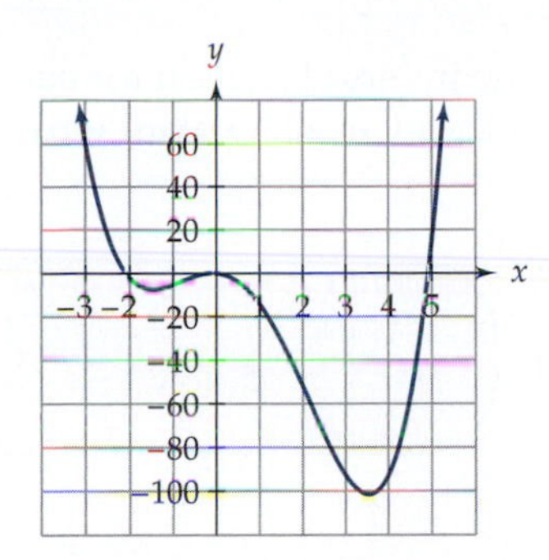

SUMMARY

Two functions, f and g, are **inverses** of each other if $f(g(x)) = x$ for every x in the domain of g and $g(f(x)) = x$ for every x in the domain of f.

Finding the inverse of a function given numerically is simply a matter of reversing the inputs and outputs. Finding the inverse of a function given symbolically is done by solving the equation for the input. Finding the inverse of a function given graphically is done by reflecting the function over the line $y = x$.

For a function to have an inverse, it must be one-to-one. A function is **one-to-one** if for each output there is only one corresponding input. A simple way to determine if a graph is one-to-one is to use the horizontal line test. The **horizontal line test** states that if there exists a horizontal line that crosses a graph at more than one point, then the graph does not represent a one-to-one function.

When a function is not one-to-one, we can restrict its domain (like the sine and cosine functions) so that it is one-to-one and thus has an inverse.

EXERCISES

1. Suppose you and your friends were playing a game of Simon Says in which Simon said, "Take your left shoe off, walk forward five paces, take your right shoe off, walk backward five paces, and put your left shoe back on." Now Simon tells you to do the inverse of the first set of instructions. What do you do?
2. Give an example of a physical situation that does not have an inverse.
3. If $f(x) = 2x + 3$, we can see that the input is first multiplied by 2 and then 3 is added. To undo, subtract 3 and then divide by 2. Therefore, $f^{-1}(x) = (x - 3)/2$. Use this same procedure to find the inverses of the following functions.

   (a) $g(x) = \dfrac{x}{2} - 5$

   (b) $h(x) = \dfrac{x+1}{3}$

   (c) $j(x) = \dfrac{x-7}{5} - 1$

4. Find the inverse function for each of the following. If the inverse function does not exist, explain why.

(a)

x	0	1	2	3	4	5
$f(x)$	3	5	10	5	3	1

(b)

x	0	1	2	3	4	5
$g(x)$	2	1	9	7	5	3

(c)

x	0	1	2	3	4	5
$h(x)$	2	6	9	6	2	1

5. Determine if each of the following functions is one-to-one. If not, give a restricted domain so that the function will be one-to-one on that interval.

(a) $f(x) = 3x^2$

(b) $g(x) = 3 \cdot 2^x$

(c) $y = \log(x + 3)$

(d) $h(x) = \sin(2x)$

(e)

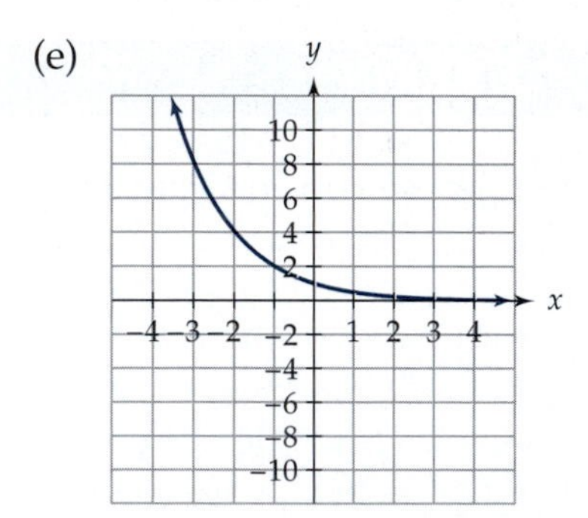

(f)

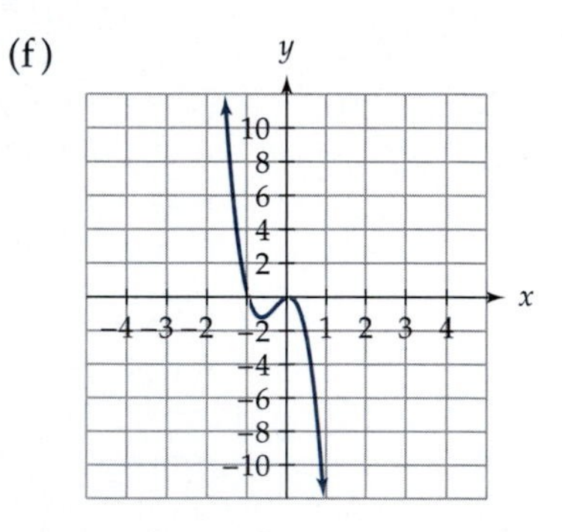

6. Find the inverse function for each of the following. Give the domain and range of the original function as well as the domain and range of the inverse function.

(a) $f(x) = 4(x - 7)$

(b) $g(x) = 2 \cdot 10^x$

(c) $h(x) = 3x^5$

(d) $j(x) = 2\log x$

(e) $k(x) = 4\cos x$

7. Given the accompanying graph of f, sketch the graph of f^{-1}.

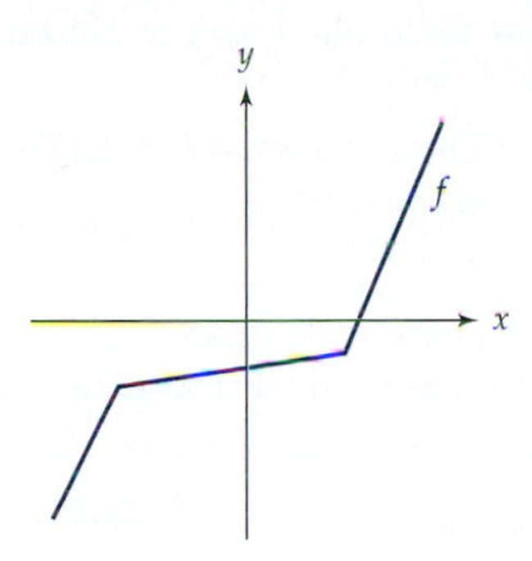

8. The period of a pendulum is the time it takes the pendulum to move in one direction and then back to the starting point. The period is given by the function

$$T = 2\pi\sqrt{\frac{L}{9.8}},$$

where L is the length of the pendulum in meters and T is the time in seconds.

   (a) Determine the domain and range of this function. (Include the appropriate units in your answer.)

   (b) Find the inverse of this function and give its domain and range.

9. Show that $g(x) = (x - 2)/3 + 6$ and $f(x) = 3(x - 6) + 2$ are inverses of one another by showing $f(g(x)) = x$ and $g(f(x)) = x$.

10. Explain why the function $f(x) = \pi x^2$ does not have an inverse, whereas the function $A = \pi r^2$, where r is the radius of a circle and A is the area of a circle, does have an inverse.

11. How is the slope of a linear function related to the slope of its inverse? (Assume the slope of the line is not zero.)

12. The volume, V, and surface area, A, of a sphere are given by the functions

$$V = \tfrac{4}{3}\pi r^3 \quad \text{and} \quad A = 4\pi r^2.$$

   (a) Write the formula for the volume of a sphere as a function of its surface area and simplify your formula. Your formula should not contain the letter r.

   (b) Determine the volume of a sphere whose surface area is 36π square inches.

13. The current federal income tax is modeled by a complicated piecewise function. Suppose the government came out with a new simplified

version,

$$f(x) = \begin{cases} 0.1x, & \text{if } 0 \leq x \leq 10{,}000 \\ 0.2x - 1000, & \text{if } x > 10{,}000, \end{cases}$$

where x is the amount of income earned and $f(x)$ is the amount of tax owed.

(a) Describe what $f^{-1}(2000)$ represents.

(b) Find f^{-1} and include appropriate domains for this piecewise function.

14. The Richter scale relates the energy of an earthquake in ergs, x, to the magnitude of an earthquake, $M(x)$, by the function

$$M(x) = \frac{\log x - 11.8}{1.5}.$$

(a) Find $M(2.5 \times 10^{20})$.

(b) Find $M^{-1}(x)$.

(c) Find $M^{-1}(5)$.

INVESTIGATIONS

INVESTIGATION: MIRROR IMAGE

Many interesting questions can be asked about inverse functions. One we consider in this investigation deals with functions that are their own inverses. In the function $f(x) = x$, it is obvious that our definition of an inverse is satisfied: $f(f(x)) = x$. What other functions will satisfy that definition?

1. Show that each of the following functions are their own inverses by showing that $f(f(x)) = x$.

   (a) $f(x) = \dfrac{1}{x}$

   (b) $f(x) = -\dfrac{1}{x}$

   (c) $f(x) = -x$

2. Graph the three functions from question 1. What do all these graphs have in common?

3. Let $g(x) = -x/(x+1)$.

   (a) Does this function appear to be its own inverse? Why or why not?

   (b) Pick any three values for x and plug them into the function. Take that output and plug it into the function again. What was the result? Does this result support the notion that g is its own inverse?

   (c) Find $g \circ g$. Is g its own inverse?

   (d) Graph g and explain how it is related to $f(x) = 1/x$.

4. For a function to be its own inverse, its graph must be symmetric about the line $y = x$. Suppose f is such a function. Let $g(x) = f(x - a) + b$.
   (a) What is the relationship between the graphs of f and g?
   (b) The function g will not necessarily be symmetric with respect to the line $y = x$. Why not?
   (c) What conditions on a and b guarantee that g is symmetric with respect to $y = x$?
   (d) How does the answer to part (c) relate to your answer to question 3, part (d)?
5. Sketch a function different from those you explored in this investigation that is its own inverse. Be sure it is also a function.

PROJECTS 3.6

SETTING THE TONE

In music, the frequency of a given sound determines the pitch. For example, a tone with a frequency of 220 Hz[21] makes the sound of the note called "A below middle C." The next A, one octave higher, has a frequency of 440 Hz. Doubling the frequency always results in a note that is one octave higher. So, doubling a pitch of 440 Hz to 880 Hz gives us yet another A, this time two octaves higher than the original. Similarly, halving 220 Hz gives us another A, this time with a frequency of 110 Hz, which is one octave below the original note.

The tricky part is finding the frequencies of the notes in between. By listening to music all our lives, we have been conditioned to hear certain notes as "right" or "in tune." Pythagoras first discovered that when two notes are played together and create a pleasant sound, the ratio of their frequencies can be written as the ratio of two small whole numbers.[22] For example, a C with a frequency of 264 Hz sounds very nice with a G of frequency of 396 Hz. The fraction $\frac{396}{264}$ simplifies to $\frac{3}{2}$, so the ratio of their frequencies is 3:2. Generally, the smaller the numbers in the simplified ratio, the more pleasant the sound.[23]

The most common type of scale, a major scale, consists of the eight notes we sometimes think of as do, re, mi, fa, sol, la, ti, and do. In the key of C, these notes are C, D, E, F, G, A, B, and C′, or just one octave of the white keys on the piano. (See Figure 1.) When counting from C to D, we move one whole step or two half steps, because C♯ is between the two. Because some of the notes in the major C scale do not have other notes in between them, it is easier to count in half steps rather than whole steps. From C to G, for example, is seven half steps.

[21] *Frequency is measured in the unit hertz, which is the number of cycles per second made by a sound wave.*

[22] *Boyer,* A History of Mathematics, *p. 60.*

[23] *Bill G. Aldridge, A. A. Strassenburg, and Gary S. Waldman,* The Guitar, A Module on Wave Motion and Sound *(New York: McGraw-Hill, 1975), pp. 63–66.*

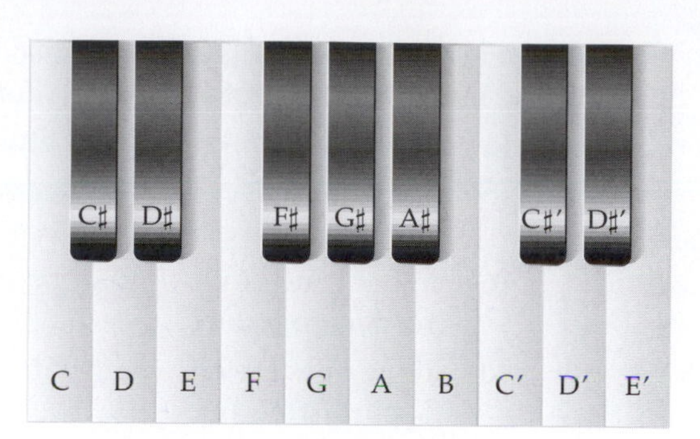

Figure 1 A portion of a piano keyboard.

1. If using the "nice ratios" that Pythagoras discovered, then comparing the frequencies of the different notes works well as long as you stay in a particular key. Some frequencies, however, do not match when comparing the same note in different keys. The ratios given in Tables 1 and 2 are those of the frequencies between a given note and C. For example, the ratio of the frequencies between D and C is $\frac{297}{264} = \frac{9}{8}$. Complete the tables, rounding the frequencies to the nearest 0.1 Hz.

You can see a problem immediately. What is called an E has a different frequency in the key of C than it has in the key of D. Being in tune in one key would then sound out of tune in another. This problem continues for every possible key, and there are 12 major scales alone! The impact of this problem on the music world was significant. Musicians were not always able to play with each other. If they started tuning with different notes, then they would sound out of tune. To have the instruments always in tune, they would need to play every piece in the same key. Besides boring the listeners, it would also be torture for singers, who would sometimes be forced to sing above or below their range.

One solution to the tuning dilemma is to redefine what is considered "in tune." Instead of making all the intervals nice ratios, around 1600 the idea

TABLE 1 Frequencies of Notes in the Key of C

Notes	Number of Half Steps above Starting Note	Ratio of Frequencies	Frequencies (Hz)
C	0	1:1	264
D	2	9:8	297
E	4	5:4	
F	5	4:3	
G	7	3:2	
A	9	5:3	
B	11	15:8	
C′	12	2:1	528

TABLE 2 Frequencies of Notes in the Key of D

Notes	Number of Half Steps above Starting Note	Ratio of Frequencies	Frequencies (Hz)
D	0	1:1	297
E	2	9:8	
F♯	4	5:4	
G	5	4:3	
A	7	3:2	
B	9	5:3	
C♯	11	15:8	
D′	12	2:1	594

to take the 12 notes of the scale and make the ratio of one note to the next the same came into practice. In other words, take the frequency of A (110 Hz)[24] and multiply it by some value r to get the frequency for A♯. Then take the frequency of A♯ and multiply it by that same r to get the frequency for B. Continue this process until you get to the next A, which is 220 Hz. You now have a geometric sequence with r as the constant ratio. This idea did not gain immediate acceptance. In fact, it took more than 200 years before it was universally accepted. J. S. Bach helped matters by writing a series of pieces, *The Well-Tempered Clavier*, in the early 1700s for the newly tuned keyboard. It wasn't until the nineteenth century that equal temperament was adopted by all countries and for all instruments.[25]

2. We will now determine exactly how all 12 notes in an octave can be equally tempered.
   (a) Doubling the frequency of any note will give the same note one octave higher. What is the ratio of the frequencies of a note and the note one octave lower? Assume that the ratio of frequencies r between each of the 12 half steps in each octave is the same. Determine the exact value of r.
   (b) Using the constant ratio r determined in part (a), complete Table 3 by finding the frequencies of the missing notes from A (110 Hz) to A′ (220 Hz). Round to the nearest 0.1 Hz.
   (c) Find a function in which the number of half steps, n, above the starting note, A, is the input and the frequency of the note, f, is the output. This function should be an exponential function.

[24] *This A is below the C that is one octave below middle C. We chose this A because it has the same frequency as the A string on a guitar, which is used later in the project.*

[25] *Karl Geiringer,* Instruments in the History of Western Music *(New York: Oxford University Press, 1978), p. 283.*

TABLE 3 Frequencies of an Octave of Notes Using an Exponential Function

Number of Half Steps above Starting Note	Notes	Frequencies
0	A	110
1	A♯	
2	B	
3	C	
4	C♯	
5	D	
6	D♯	
7	E	
8	F	
9	F♯	
10	G	
11	G♯	
12	A′	220

3. On different musical instruments, various adjustments are made to change the frequency of the notes. One common method is to change the length of a tube or a pipe, as in organs or trombones, or change the length of a string, as in violins or guitars. If you decrease the length of a guitar string by half, then the frequency of the note played is doubled. In this situation, there is a similar relationship between a note and its frequency as there is between a note and the length of the string on which it is played. This relationship, however, is backward. As a note gets higher, its frequency is given by a larger number, but its string length is given by a smaller number. We measured the length of the A string on a guitar and found it to be 61.2 cm.[26] When the string is played without reducing its length, the note is an A. When length is reduced by moving your finger on the fingerboard, the frequency of the tone increases. When the string is half as long as original, the frequency is doubled and the first octave, or A′, is heard.

   (a) Halving the length of any guitar string gives the same note one octave higher. What would be the ratio of string lengths of a note and the note one octave lower? Assume the ratio of string lengths r between each of the 12 half steps in each octave is the same. Determine the exact value of r.

   (b) Using the constant ratio r determined in part (a), complete Table 4 by finding the missing lengths of the A string. They should form

[26] *Not every guitar has the same string length.*

TABLE 4 Guitar String Lengths for an Octave of Notes Using an Exponential Function

Number of Half Steps above Starting Note	Notes	String Lengths (cm)
0	A	61.2
1	A♯	
2	B	
3	C	
4	C♯	
5	D	
6	D♯	
7	E	
8	F	
9	F♯	
10	G	
11	G♯	
12	A′	30.6

a geometric sequence. When building a guitar, these numbers determine the positions of the frets, which are the metal ridges along the neck of a guitar. When you push down a string on the neck of a guitar, a fret will "cut" the string off at just the right point to produce the note you want.

(c) Determine a function in which the number of half steps above the starting note, n, from Table 4 is the input and the length of the string, l, is the output. This function should be an exponential function.

4. The frequency of a specific guitar string is inversely proportional to its length. Let's show that this statement is true, at least for the A string on our guitar.

   (a) Add your data from Tables 3 and 4 to Table 5.

   (b) If the frequency of the guitar string is inversely proportional to its length, then the product of a string length and its corresponding frequency is a constant. Using a few data values from Table 5, approximate this constant.[27]

   (c) Let f represent your frequency function from question 2, part (c) and l represent your string length function from question 3, part

[27] *Because rounding occurred in determining the values in Table 5, you will not get exactly the same value for each product, but your values should be close.*

TABLE 5 Guitar String Lengths and Frequencies for an Octave of Notes Using an Exponential Function

Number of Half Steps above Starting Note	Notes	String Lengths (cm)	Frequencies
0	A	61.2	110
1	A♯		
2	B		
3	C		
4	C♯		
5	D		
6	D♯		
7	E		
8	F		
9	F♯		
10	G		
11	G♯		
12	A′	30.6	220

(c). Which composition of functions,

$$f \circ l, \quad f^{-1} \circ l, \quad f \circ l^{-1}, \quad f^{-1} \circ l^{-1},$$
$$l \circ f, \quad l^{-1} \circ f, \quad l \circ f^{-1}, \quad l^{-1} \circ f^{-1},$$

represent the function that would have the data for the string length column of Table 5 as the input and the data in the frequency column as the output? Explain your answer.

(d) Find the function composition that you determined was correct in part (c). Does your resulting function show that the frequency of the guitar string is inversely proportional to its length? Explain.

REVIEW EXERCISES

1. Given the function f in the following table, complete as much of the table as possible for the different transformations of f.

x	−4	−3	−2	−1	0	1	2	3	4
$f(x)$	−2	5	7	−3	10	−1	6	0	8
$f(-2x)$									
$f(x+4)$									
$f\left(\frac{x}{2}\right)+1$									

2. For $g(x) = 2^x + 5$, determine formulas for the following functions.
   (a) $y = 2g(x)$ and $y = g(2x)$
   (b) $y = 2g(4x) + 4$
3. Use the accompanying graph of f to sketch a graph of each of the following transformations.

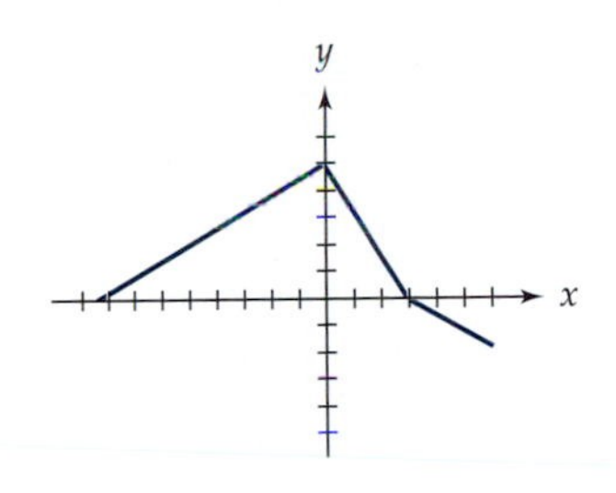

   (a) $y = 2f(x+1)$
   (b) $y = f(-x) + 3$
4. Suppose c represents the function where n is the number of course credits a student is taking and $c(n)$ is the cost of tuition in dollars. Match the following situations to the formulas given below. (Not all functions will be used.)
   (a) The cost of tuition after a 10% increase.
   (b) The cost of tuition after a $100 decrease.
   (c) The cost of tuition after there is a increase of 10% in the number of credits taken.
   (d) The cost of tuition after one more credit is taken.

i. $y = 1.1c(n)$	ii. $y = 0.1c(n)$	iii. $y = c(n) + 1$
iv. $y = c(n+1)$	v. $y = c(n) + 100$	vi. $y = c(n) - 100$
vii. $y = c(1.1n)$	viii. $y = c(0.1n)$	ix. $y = c(0.9n)$

5. Suppose f is an exponential function of the form $f(x) = ba^x$.
   (a) If f is transformed by adding a constant, c, to the input, will the growth factor or y-intercept change? Explain exactly how it will change.

(b) If f is transformed by multiplying the input by a constant, c, will the growth factor or y-intercept change? Explain exactly how it will change.

6. Let $f(x) = 3x - 4$ and $g(x) = x^2 + 2x - 1$. Find the following.
   (a) $(f - g)$
   (b) $(f - g)(2)$
   (c) $(f \cdot g)(-3)$
   (d) $(2f - g)$
   (e) $(2f - g)(-1)$
7. Complete the following table.

x	**0**	**1**	**2**	**3**	**4**	**5**	**6**
$f(x)$	2	4	6	1	3	5	0
$g(x)$	0	6	1	5	2	4	3
$(f+g)(x)$							
$(f \cdot g)(x)$							
$f(g(x))$							
$g(f(x))$							

8. Let f and g be represented by the accompanying graphs.

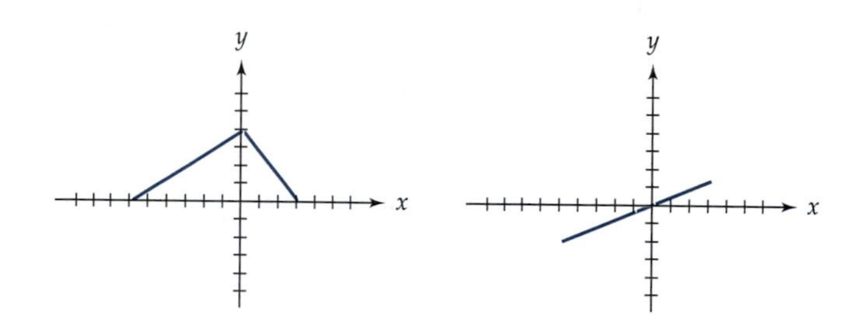

   (a) Plot the graph of $y = (f + g)(x)$.
   (b) Plot the graph of $y = (f - g)(x)$.
9. Let $f(x) = \sqrt{x+2}$, $g(x) = 1/x$, $h(x) = (g + f)(x)$, and $k(x) = (g/f)(x)$.
   (a) Give the domain for f and g.
   (b) Give the domain for h and k. How are these related to your answer for part (a)?
10. Let $f(x) = x^2 + 4$, $g(x) = \log x$, $h(x) = g(f(x))$, and $k(x) = f(g(x))$.
   (a) Give the domain for f and g.
   (b) Give the domain for h and k. How is this related to your answer for part (a)?

(c) Give the range for f and g.

(d) Give the range for h and k. How is this answer related to your answer for part (c)?

11. Let $f(x) = x^2 + 3x$, $g(x) = x^{3/2}$, and $h(x) = 5x - 8$. Find the equation for each of the following.

(a) $g \circ f$

(b) $f \circ g$

(c) $h \circ g$

(d) $h \circ f \circ g$

12. For each of the following, find functions f and g so that $h(x) = g(f(x))$.

(a) $h(x) = \sqrt{x - 2}$

(b) $h(x) = \sqrt{x} - 2$

(c) $h(x) = 2^{x+3}$

(d) $h(x) = 2^x + 3$

13. Determine if each of the following functions is one-to-one. If not, give a restricted domain so that the function will be one-to-one on that interval.

(a) $y = 3 + \log x$

(b) $h(x) = \cos(2x)$

(c) $f(x) = x^{4/5}$

(d) $g(x) = 5^x$

14. Find the inverse function for each of the following. Give the domain and range of the original function as well as the domain and range of the inverse function.

(a) $f(x) = \dfrac{x+3}{4}$

(b) $g(x) = 3\log(x + 2)$

(c) $h(x) = 16x^{4/5}$

(d) $j(x) = \dfrac{x-3}{x+5}$

15. Given the accompanying graph of f, sketch the graph of f^{-1}.

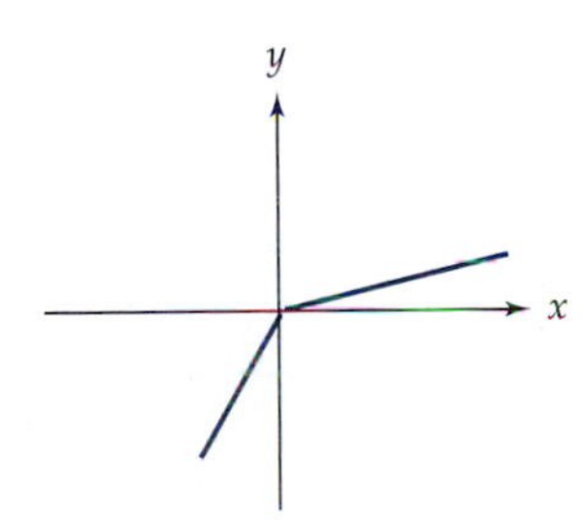

16. Suppose a tank holds 1000 gal of water that will drain from the bottom of the tank in 10 min. The volume of the water, $V(t)$, remaining in the

tank after t minutes can be modeled by

$$V(t) = 1000\left(1 - \frac{t}{10}\right)^2.$$

(a) Determine the domain and range of this function. (Include the appropriate units in your answer.)

(b) Find the inverse of this function and give its domain and range.

17. Show that $g(x) = (x+2)/(x-4)$ and $f(x) = (4x+2)/(x-1)$ are inverses of one another by showing $f(g(x)) = x$ and $g(f(x)) = x$.

18. Computer screens, like television sets, use the length of the diagonal as a measurement of the "area" of the screen. A computer screen is typically a rectangle where the height is three-fourths the width.

    (a) Show that the area of a computer screen can be written as

    $$A = \tfrac{3}{4}w^2,$$

    where w is the width of the screen.

    (b) Show that the diagonal of a computer screen can be written as

    $$d = \tfrac{5}{4}w,$$

    where w is the width of the screen.

    (c) Find the area of a computer screen in terms of its diagonal measure. (Your formula should have d as the only input and A as the output.

    (d) Determine the area of a computer screen that has a diagonal of 15 in.

19. Sound is usually measured using decibels (abbreviated dB). To measure sound in decibels, the sound is compared with a standard, I_0. This standard, 10^{-16} W/cm^2, is approximately the lowest intensity sound that is audible to humans. The loudness of sound, in decibels, can then be calculated using the formula

    $$D(x) = 10\log\left(\frac{x}{10^{-16}}\right),$$

    where x is the sound's intensity.

    (a) Find $D^{-1}(x)$.

    (b) Find $D^{-1}(50)$.

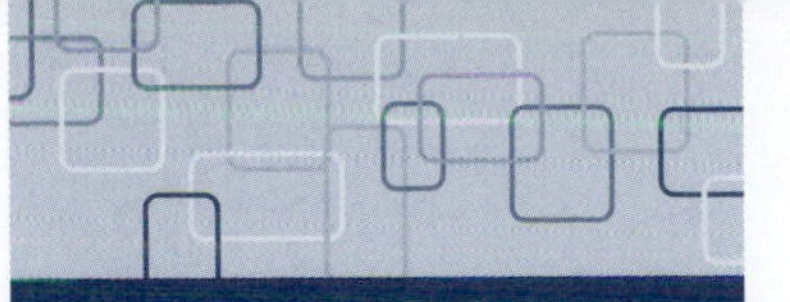

CHAPTER OUTLINE

CHAPTER OVERVIEW

- How to recognize and use quadratic functions in various settings
- The characteristics and behaviors of various polynomial functions, including their roots and local extrema
- Properties of power functions with negative exponents
- Asymptotes
- The behavior of rational functions

POLYNOMIAL AND RATIONAL FUNCTIONS

In this chapter, certain commonly used combinations of functions are introduced. Quadratic functions are introduced as transformations of $y = x^2$. Properties of quadratic functions as well as other polynomial functions, power functions with negative exponents, and rational functions are explored. We also show the use of these functions to model physical situations.

4.1 QUADRATIC FUNCTIONS

In chapter 3, we explored a variety of ways of combining basic functions to form new functions. In this chapter, we concentrate on certain combinations of functions known as polynomial and rational functions. We begin by looking at a special type of polynomial function called a quadratic function. In this section, we explore how quadratic functions can be expressed symbolically, we show how to find the vertex and roots, and we explore some applications.

DEFINING QUADRATIC FUNCTIONS

Imagine throwing a ball into the air. The velocity of the ball is continuously changing from the moment it leaves your hand until the moment you catch it. Because the velocity (which is the change in position divided by the change in time) is changing continuously, the slope of the function that has time as the input and height as the output is also changing continuously. Because the slope is changing, the function that models the height of the ball is not linear.

The graph of the function with time as the input and the height of a thrown ball as the output is shown in Figure 1. This shape is known as a parabola. The function that models the height of the ball is known as a quadratic function. A **quadratic function** is a function that can be written in the form $f(x) = ax^2 + bx + c$, where a, b, and c are real numbers such that $a \neq 0$.[1] Although we often use other forms for a quadratic function, the form $f(x) = ax^2 + bx + c$ is called the *standard form*. The graph of every quadratic function (where a is not equal to zero)is a parabola. As mentioned

[1] *The term* quadratic *comes from the Latin* quadratum, *which means square.*

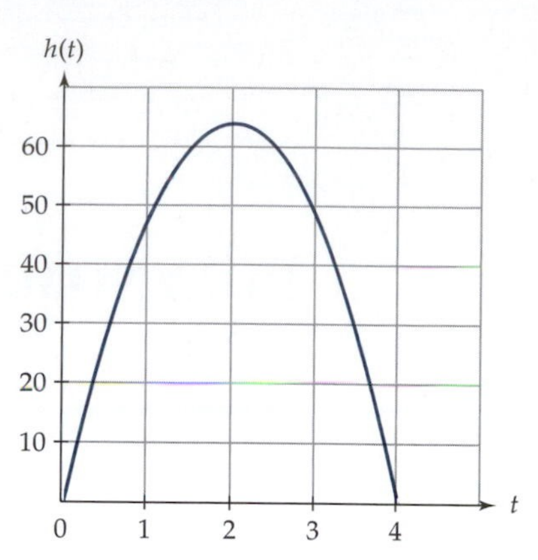

Figure 1 A graph of the height of a ball as a function of time.

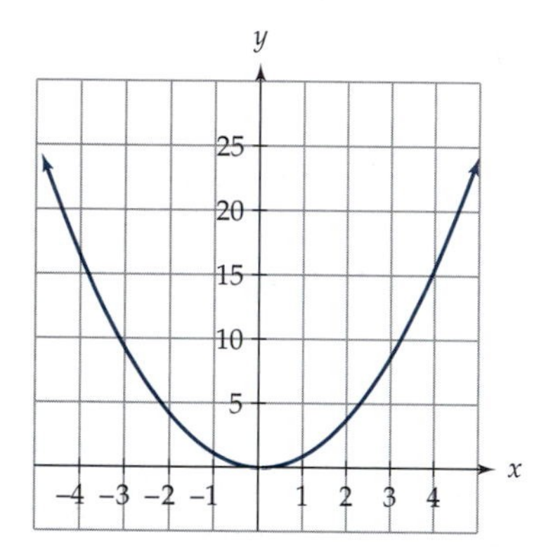

Figure 2 The function $y = x^2$ shows the shape of a parabola.

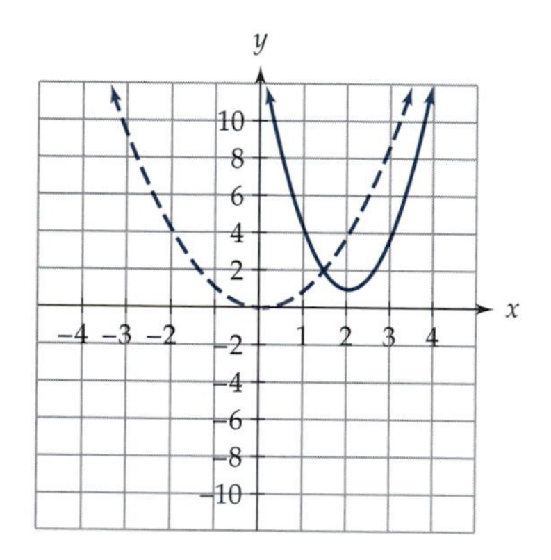

Figure 3 The function $f(x) = 3(x - 2)^2 + 1$ is a stretch and shift of the function $y = x^2$.

in chapter 2, a graph of the function $y = x^2$ is in the shape of a parabola. (See Figure 2.)

All the parabolas we look at in this section are the result of shifting, reflecting, or stretching the function $y = x^2$. As a result, they can be written as $f(x) = a(x - h)^2 + k$. This form is called the *vertex form* for a quadratic function. The vertex of this parabola has been shifted from the point $(0, 0)$ to the point (h, k), and the function has been vertically stretched by a factor of a. The function shown in Figure 1 is $h(t) = -16(t - 2)^2 + 64$. It has a vertex at $(2, 64)$, is reflected across the x-axis, and has been stretched by a factor of 16.

We can expand the symbolic form of a quadratic function in vertex form to put it in standard form. The function $f(x) = 3(x - 2)^2 + 1$, for example, is a parabola that has been vertically stretched by a factor of 3 and has its vertex at the point $(2, 1)$. (See Figure 3.) By multiplying and combining terms, we find that $f(x) = 3(x - 2)^2 + 1 = 3x^2 - 12x + 13$. Notice that it is no longer possible to "read" the vertex from the expanded form of f. Also notice that the leading coefficient, a, in both the standard form and vertex form of a quadratic function is the same. If a is positive, then the parabola opens up. If a is negative, then the parabola opens down.

Suppose we start with an equation in standard form and want to find the vertex. To do so, we need to complete the square. Completing the square is the process of taking a quadratic function in standard form and writing it to contain a perfect square binomial of the form $f(x) = a(x - h)^2 + k$, which is the vertex form for a quadratic function. For example, let $f(x) = x^2 + 6x + 2$. To complete the square, we take half of the coefficient of the x-term, square this result, and then both add and subtract it to the function. For $f(x) = x^2 + 6x + 2$, we add and subtract $\left(\frac{6}{2}\right)^2 = 3^2 = 9$. We start with

$$f(x) = x^2 + 6x + 2.$$

When we complete the square, we get

$$f(x) = (x^2 + 6x + 9) + 2 - 9,$$

which simplifies to

$$f(x) = (x + 3)^2 - 7.$$

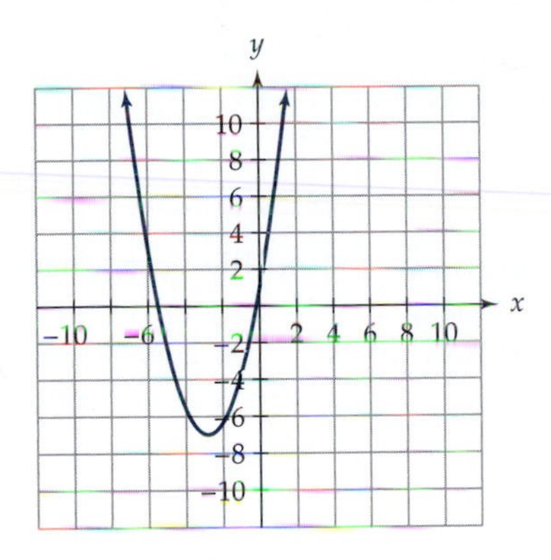

Figure 4 The graph of $f(x) = (x + 3)^2 - 7$.

When f is written in vertex form, we see that the graph of f is the same as the graph of $y = x^2$ shifted three units to the left and seven units down. (See Figure 4.)

Example 1 Find the vertex of $g(x) = x^2 + 12x + 9$.

Solution The coefficient of the x-term is 12, and $\left(\frac{12}{2}\right)^2$ is 36. To complete the square, we add (and subtract) 36 to our function, which gives

$$\begin{aligned} g(x) &= (x^2 + 12x + 36) + 9 - 36 \\ &= (x + 6)^2 - 27 \\ &= (x - (-6))^2 - 27. \end{aligned}$$

The vertex is the point $(-6, -27)$.

The symbolic form of the function in the previous example illustrates why the vertex is always going to be a maximum or minimum point for the function. Notice that because $(x + 6)$ is squared, the result of that part of the function is always nonnegative. Hence, the smallest value that part of the function can have is zero. This value will occur when $x = -6$, the x-value of the vertex. Because $g(-6) = -27$, the smallest output is -27. The vertex will always be a minimum point when the leading coefficient, a, is positive, and it will always be a maximum point when the leading coefficient, a, is negative.

So far, we have only completed the square in a situation in which the coefficient of x^2 was one. If the coefficient of x^2 is not 1, then you need to factor out this constant before completing the square. This step is illustrated in Example 2.

Example 2 Find the vertex of $h(x) = 3x^2 - 15x - 3$.

Solution We begin by factoring 3 out of the terms of this function, which gives $h(x) = 3(x^2 - 5x - 1)$. Completing the square and simplifying leads to

$$\begin{aligned} h(x) &= 3\left(\left(x^2 - 5x + \tfrac{25}{4}\right) - 1 - \tfrac{25}{4}\right) \\ &= 3\left(\left(x - \tfrac{5}{2}\right)^2 - \tfrac{29}{4}\right) \\ &= 3\left(x - \tfrac{5}{2}\right)^2 - 3\left(\tfrac{29}{4}\right) \\ &= 3\left(x^2 - \tfrac{5}{2}\right)^2 - \tfrac{87}{4}. \end{aligned}$$

The vertex is the point $\left(\frac{5}{2}, -\frac{87}{4}\right)$.

As you can see, there is an algorithm that will always allow you to put a quadratic function in vertex form if the function is given in standard form. These steps are as follows.

1. Factor out the leading coefficient.
2. Add and subtract the square of half the coefficient of x to complete the square.
3. Simplify.

Putting a function in vertex form then allows us to find the vertex easily. If we perform this algorithm on the general quadratic function $f(x) = ax^2 + bx + c$, then we see that the x-coordinate of the vertex is $-b/2a$. You can then find the y-coordinate by finding $f(-b/2a)$. So, the vertex of $f(x) = ax^2 + bx + c$ is $(-b/2a, f(-b/2a))$.[2]

PROPERTIES

Standard Form and Vertex Form

- $f(x) = ax^2 + bx + c$ is the standard form for a quadratic function.
- $f(x) = a(x - h)^2 + k$ is the vertex form for a quadratic function. The vertex is at the point (h, k).
- In both standard form and vertex form, a has the same value. If a is positive, then the parabola opens up; if a is negative, then the parabola opens down.

READING QUESTIONS

1. Rewrite $f(x) = 4(x - 6)^2 + 3$ in standard form.
2. Give the equation for the parabola that has been vertically stretched by a factor of 3 and whose vertex is $(2, 1)$.
3. Find the vertex of $f(x) = x^2 + 8x + 3$.

SYMMETRY AND ROOTS

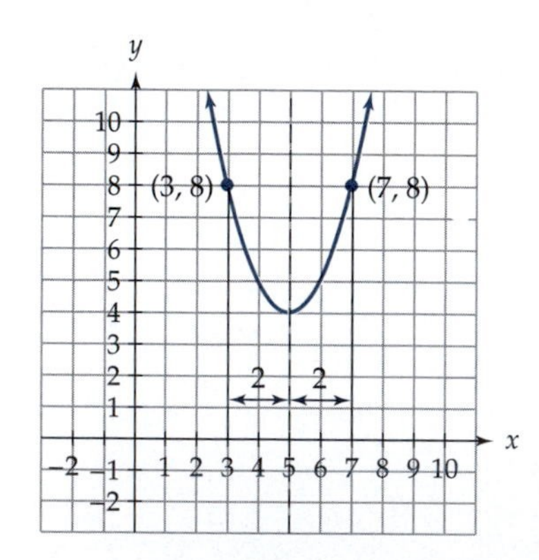

Figure 5 Inputs equidistant from the line of symmetry have the same output.

Knowing the vertex of a parabola and the stretch factor allows us to sketch the parabola and give its symbolic representation. Therefore, it is worthwhile to look at other ways to find the vertex. One way is to notice the symmetry of the parabola. The graph of $y = x^2$ is symmetric about the y-axis (which is the line $x = 0$), and its vertex is the point $(0, 0)$. Because graphs of quadratic functions are just shifted and stretched versions of $y = x^2$, they too are symmetric about the vertical line going through their vertex. For example, the function $f(x) = (x - 5)^2 + 4$ has a line of symmetry of $x = 5$ because the vertex is at $(5, 4)$. Evaluating this function at 3 and at 7 (two units away from the line of symmetry), we see that $f(3) = (3 - 5)^2 + 4 = (-2)^2 + 4 = 8$ and $f(7) = (7 - 5)^2 + 4 = (2)^2 + 4 = 8$. The output is the same because the points are equal units away from the line of symmetry. (See Figure 5.) Conversely, two points with the same output will always be equidistant from the line of symmetry (something you are asked to prove in the exercises). Thus, if we know any two points of a

[2] *You are asked to show this result in the exercises.*

quadratic function with the same y-values, then the x-value of the vertex must be exactly halfway between the x-values of these two points.

Example 3 Find the x-coordinate of the vertex of the quadratic function that includes the points $(1, 6)$ and $(4, 6)$.

Solution Because both these points have the same y-value, the x-value of the vertex must be halfway between the x-values of the two given points. That puts the x-value of the vertex at $x = (1 + 4)/2 = 2.5$.

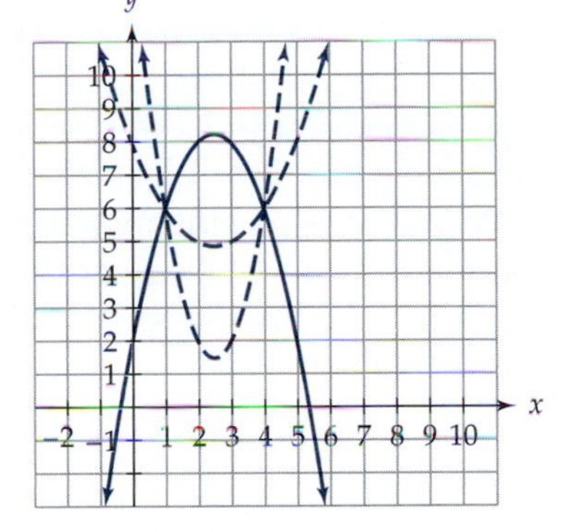

Figure 6 There are an infinite number of different parabolas that go through the points $(1, 6)$ and $(4, 6)$.

In Example 3, we unfortunately do not have enough information to give the y-value for the vertex. Figure 6 shows that there are many parabolas that contain the two points given. To find the vertex, we need more information, such as another point.

Points other than the vertex are also important. The x-intercepts, or the roots of a quadratic function, are important in many applications. The **roots** of a function, f, are all the values of x such that $f(x) = 0$. For example, $f(x) = (x + 2)(x - 4)$ has roots at $x = -2$ and $x = 4$. (See Figure 7.) In general, if $(x - r)$ is a factor of a function, then r is a root and vice versa. One way to find the roots of a function, then, is to factor the function.

Example 4 Find the roots of $f(x) = x^2 - x - 12$.

Solution To find the roots, we need to solve $0 = x^2 - x - 12$, which can be accomplished by factoring. Because factoring this function gives $0 = (x - 4)(x + 3)$, then $x - 4 = 0$, so $x = 4$ or $x + 3 = 0$ and thus $x = -3$. Therefore, the roots of this function are $x = 4$ and $x = -3$.

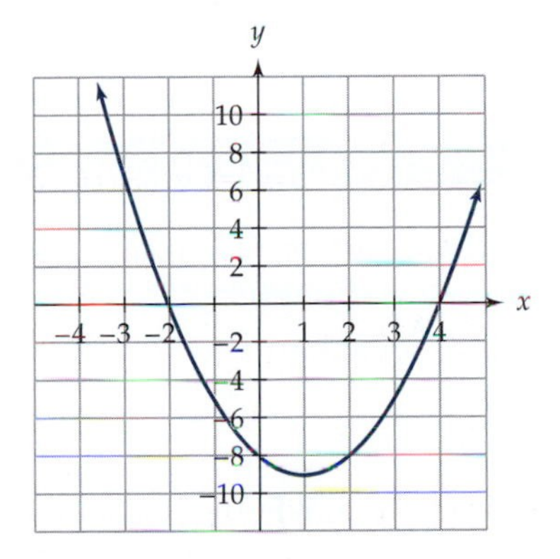

Figure 7 The graph of $f(x) = (x + 2)(x - 4)$ has roots at $x = -2$ and $x = 4$.

One reason to know the roots of a quadratic function is to determine when a projectile hits the ground. Suppose, for example, you fired a projectile into the air; also suppose the height of this projectile is modeled by the function $h(t) = -16t^2 + 500t$, where t is time (in seconds) and $h(t)$ is height (in feet). Factoring $h(t) = -16t^2 + 500t$ gives $h(t) = -16t(t - 31.25)$. The roots are $t = 0$ and $t = 31.25$. Thus, the projectile was at ground level at 0 sec (which makes sense because that is the time it was fired) and again at 31.25 sec. Because the vertex is halfway between any two points with equal output, the projectile must have reached its highest point at $t = (0 + 31.25)/2 = 15.625$ sec.

Sometimes, quadratics cannot be factored easily or cannot be factored at all. In such cases, one way of finding roots, if they exist, is to use the **quadratic formula**,

$$x = \frac{-b \pm \sqrt{b^2 - 4ac}}{2a}.$$

The quadratic formula is the solution[3] to $ax^2 + bx + c = 0$.

[3] *You are asked to show this solution in the exercises.*

Example 5 Find the roots and vertex of $f(x) = 3x^2 - 30x + 73$.

Solution To find the roots, we use the quadratic formula:

$$x = \frac{30 \pm \sqrt{(-30)^2 - 4(3)(73)}}{2(3)}$$
$$= \frac{30 \pm \sqrt{24}}{6}.$$

This result simplifies to $x = 5 + (\sqrt{6}/3)$ and $x = 5 - (\sqrt{6}/3)$. The x-coordinate of the vertex is halfway between the roots at $x = 5$. Because $f(5) = -2$, the vertex is at $(5, -2)$.

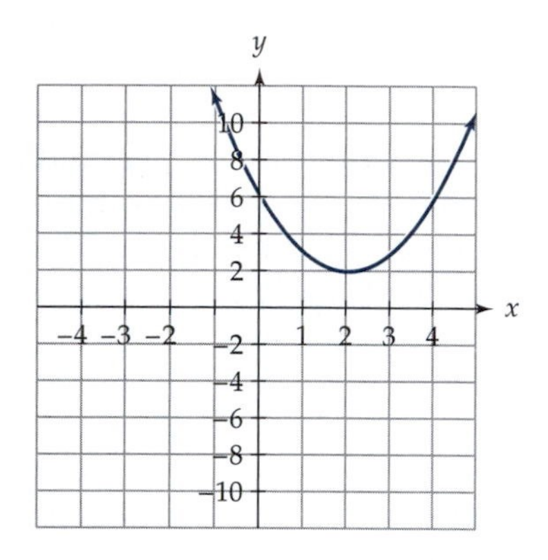

Figure 8 This quadratic function has no real roots.

Because it is possible for a parabola to never intersect the x-axis, not every quadratic function has real roots. For example, in $f(x) = x^2 - 4x + 6$ (shown in Figure 8), there are no real roots. What happens when we use the quadratic formula in this case? Using the quadratic formula gives the roots of f as

$$x = \frac{4 \pm \sqrt{(-4)^2 - 4(1)(6)}}{2}$$
$$= \frac{4 \pm \sqrt{16 - 24}}{2}$$
$$= \frac{4 \pm \sqrt{-8}}{2}.$$

The result is a negative number under the square root sign, indicating that there is no real root.[4]

READING QUESTIONS

4. What is the line of symmetry in a parabola that contains the points $(4, 7)$ and $(8, 7)$?
5. Given the two roots of a quadratic function, what can and cannot be determined about the vertex?
6. What are the roots of $f(x) = 3(x - 5)(x + 4)$?
7. How do you know from using the quadratic formula whether or not the parabola crosses the x-axis?

FINDING THE EQUATION

If you know (or have good reason to believe) a function can be best modeled by a quadratic function, you can find the symbolic representation of that function if you know the vertex and another point or any three arbi-

[4] *It is still possible to consider the two roots if we use the set of complex numbers. In this set, $\sqrt{-1}$ is defined as i. So, the roots of $f(x) = x^2 - 4x + 6$ are $2 \pm \sqrt{2}i$.*

trary points. Just as two points determine a line, three points determine a parabola.[5]

Suppose we know that the vertex of a parabola is $(2, 5)$ and that the parabola also includes the point $(0, 3)$. The vertex form can help us find the symbolic representation of this function. We know, using the vertex, that $f(x) = a(x-2)^2 + 5$. Now we need to find the value of a. To do so, substitute 0 for x and 3 for $f(x)$ into the equation:

$$\begin{aligned} f(x) &= a(x-2)^2 + 5 \\ 3 &= a(0-2)^2 + 5 \\ 3 &= 4a + 5 \\ a &= -\tfrac{1}{2}. \end{aligned}$$

Thus, our function is $f(x) = -\frac{1}{2}(x-2)^2 + 5 = -\frac{1}{2}x^2 + 2x + 3$.

One reason it is sufficient to have only two points, if one of them is the vertex, is that symmetry automatically gives us a third point. In our last example, we were given the vertex $(2, 5)$ and the point $(0, 3)$. We know that a point whose x-value is two away from the line of symmetry must have a y-value of 3. Thus, the point $(4, 3)$ is also on the parabola. Knowing one point and the vertex means we actually know three points.

Let's go back to our last example and find the symbolic representation for the function that passes through the points, $(0, 3)$, $(2, 5)$ and $(4, 3)$. This time, however, we use a different method, one that does not require that we know that one of these is the vertex. We substitute each of these points, one at a time, for x and $f(x)$ in the standard form $f(x) = ax^2 + bx + c$ to get three equations and three unknowns. We can then solve for each unknown, which allows us to find our original function.

Using the three points $(0, 3)$, $(2, 5)$ and $(4, 3)$ gives the three equations

$$0a + 0b + c = 3 \quad (1)$$

$$4a + 2b + c = 5 \quad (2)$$

$$16a + 4b + c = 3. \quad (3)$$

From equation (1), we see that $c = 3$. We substitute $c = 3$ into equations (2) and (3) to get

$$4a + 2b = 2 \quad (4)$$

and

$$16a + 4b = 0. \quad (5)$$

This process has reduced our problem to one of two equations and two unknowns. Multiplying equation (4) by -2 gives

$$-8a - 4b = -4. \quad (6)$$

Adding equations (5) and (6) together gives $a = -\frac{1}{2}$. By substituting $a = -\frac{1}{2}$ back into *either* equation (4) or (5), we find that $b = 2$. We now know all three variables. The symbolic representation f is therefore

$$f(x) = -\tfrac{1}{2}x^2 + 2x + 3,$$

[5] *The equation for a parabola can be determined when given three points that lie on the parabola. If given three arbitrary points on a plane, however, a unique parabola will contain them if they are not collinear.*

the same result obtained when we found f knowing that one of the points was the vertex. This second process, although not always simple, always works for finding the symbolic representation of a quadratic function when you are given three noncollinear points. It was easier to find the function using the vertex, so we use that method when possible.

Example 6 Find the symbolic representation for the quadratic function that contains the points $(1,3)$, $(2,6)$, and $(4,-2)$.

Solution The three equations that we obtain by using these three points are

$$3 = a + b + c \tag{7}$$

$$6 = 4a + 2b + c \tag{8}$$

$$-2 = 16a + 4b + c. \tag{9}$$

Subtracting equation (7) from equation (8) gives

$$3 = 3a + b, \tag{10}$$

and subtracting equation (8) from equation (9) gives

$$-8 = 12a + 2b. \tag{11}$$

Multiplying equation (10) by -2 gives

$$-6 = -6a - 2b. \tag{12}$$

By adding equations (11) and (12), we find that $a = -\frac{7}{3}$. We substitute $a = -\frac{7}{3}$ back into either equation (10) or (11) to get $b = 10$. By substituting both the values for a and b into any of the original equations, we find that $c = -\frac{14}{3}$. So,

$$f(x) = -\tfrac{7}{3}x^2 + 10x - \tfrac{14}{3}.$$

Let's return to the ball example from the beginning of this section. Suppose you throw a ball from ground level that stays in the air for 4 sec and reaches a maximum height of 64 ft above the point where you released it. Using this information, we can determine the function that describes the ball's position.[6] The time when the ball is at its highest position is always halfway between when you threw it and when you caught it, provided these two points are at the same height. Thus, at $t = 2$ sec, we know that $h = 64$ ft. Because this height is the highest position reached by the ball, the point $(2,64)$ is the vertex of our parabola. We know that the roots of this function are 0 and 4, so we can find the symbolic representation of the function. Because 0 and 4 are roots, $(t - 0)$ and $(t - 4)$ must be factors of the function. Thus, we can write $h(t) = a(t - 0)(t - 4) = at(t - 4)$. By

[6] *Throughout this section, we assume that gravity is the only force acting on an object after it is thrown. Of course, other forces such as air resistance act on the ball, but we ignore those.*

substituting the other point that we know, $(2, 64)$, for $(t, h(t))$ we can solve this equation for a:

$$\begin{aligned} h(t) &= at(t-4) \\ 64 &= a(2)(2-4) \\ a &= -16. \end{aligned}$$

We can now write our function as

$$h(t) = -16t(t-4)$$

or, putting it in standard form,

$$h(t) = -16t^2 + 64t,$$

where t is the time (in seconds) since the ball was thrown and h is the height above your hand. Through observation, scientists have determined that because of the constant force of gravity on Earth, the value of a for any object acted on by gravity is always -16 ft/sec^2. In standard form, the function is

$$h(t) = -16t^2 + v_0 t + h_0,$$

where v_0 is the initial velocity and h_0 is the initial height. From our example, we see that our ball had an initial height, h_0, of 0 ft and an initial velocity, v_0, of 64 ft/sec.

Example 7 In an investigation in section 3.2, a table of data and a graph showed the height of a steel ball as a function of time. Because gravity is the main force acting on the steel ball after it is shot into the air and because other forces acting on it were minimal, it is reasonable to expect that a quadratic function could model the vertical position of the projectile as a function of time. The data from the investigation are shown in Table 1 and Figure 9. Find a quadratic equation that models the data.

TABLE 1

Time (sec)	Position (cm)
0.00	0
0.05	18.2
0.10	33.7
0.15	46.9
0.20	57.6
0.25	65.0
0.30	70.8
0.35	74.1
0.40	74.6
0.45	72.7
0.50	68.3
0.55	61.7
0.60	52.1
0.65	40.1
0.70	25.5
0.75	8.6

Solution One way to find an equation to model these data is to pick three points and find the equation for the parabola containing those three points. We choose three points that are not close together and that seem

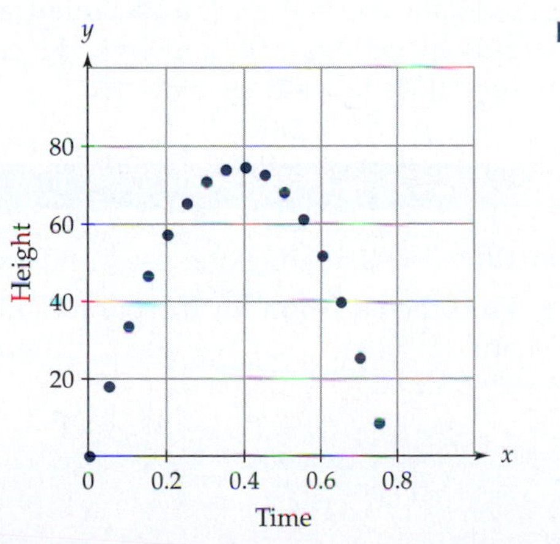

Figure 9

to "fit" the shape of the parabola in our graph. Given these criteria, we use the points $(0, 0)$, $(0.40, 74.6)$ and $(0.70, 25.5)$.[7] We obtain the following equations using the standard form $h(t) = at^2 + bt + c$:

$$0 = 0a + 0b + c$$
$$74.6 = 0.16a + 0.4b + c$$
$$25.5 = 0.49a + 0.7b + c.$$

Using these equations to solve for a, b, and c gives $a \approx -500.2$, $b \approx 386.6$, and $c = 0$. We now have an equation of $h(t) = -500.2t^2 + 386.6t$. Plotting this function on the graph with the data indicates that we have a fairly good model for this situation. (See Figure 10.)

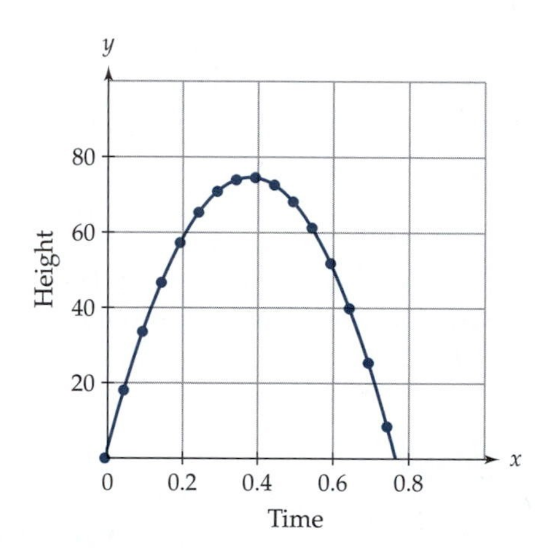

Figure 10

Notice that the leading coefficient in Example 7 was around -500. Earlier in this section, we mentioned that in all cases involving objects being thrown or dropped, the leading coefficient was -16. How can both of these be true? The answer is that it depends on the units. The examples at the beginning of this section were all in feet and seconds. In that case, the coefficient of x^2 is always -16 ft/sec^2. When using meters and seconds, the coefficient is -4.9 m/sec^2, and when using centimeters and seconds, it is -490 cm/sec^2. Considering that the data used in Example 7 are actual measured data and that we found a quadratic function to model this behavior simply by choosing three points, the coefficient of -500 that we obtained is very close to what we expected.

READING QUESTIONS

8. Why does knowing one point and a vertex actually give us three points?

9. Find the equation for the parabola containing the points $(-1, 2)$, $(0, 1)$ and $(2, 5)$.

[7] *Other choices are also reasonable and should give close to the same result.*

SUMMARY

A **quadratic function** is a function that can be written in the form $f(x) = ax^2 + bx + c$, where a, b, and c are real numbers and $a \neq 0$. The graph of every quadratic function is a parabola. All quadratic functions are shifts, reflections, and stretches of the function $y = x^2$, so all quadratic functions can also be written in vertex form, $f(x) = a(x - h)^2 + k$. In both forms, if a is positive, then the parabola opens up; if a is negative, then the parabola opens down. The vertex is always at the point (h, k).

Every quadratic function is symmetric with respect to a vertical line that passes through the vertex. The **roots** of a function are all values of x such that $f(x) = 0$. For a quadratic function in standard form, the roots can be found by using the **quadratic formula,**

$$x = \frac{-b \pm \sqrt{b^2 - 4ac}}{2a}.$$

You can find the equation for any quadratic function if you know the vertex and any other point or if you know any three points. If you know the vertex, substitute the values into the vertex form of the equation. If you do not know the vertex, substitute each point, one at a time, for x and $f(x)$ in the standard form for the quadratic equation and then solve the three equations for the three unknowns.

EXERCISES

1. Find the vertex of each of the following by rewriting each function in vertex form if it is not already given in vertex form.
   (a) $f(x) = (x - 5)^2 - 4$
   (b) $g(x) = x^2 - 4x + 7$
   (c) $h(x) = 2x^2 + 12x + 14$
2. Sketch, by hand, a reasonable graph of the following functions.
   (a) $f(x) = (x - 2)^2 + 3$
   (b) $g(x) = 2(x + 4)^2 - 3$
3. Rewrite the following quadratic functions in standard form.
   (a) $f(x) = 4(x - 2)^2 + 3$
   (b) $g(x) = -3(x + 6)^2 - 1$
   (c) $h(x) = 6(x + 1)^2 - 4$
4. In this section, you learned that the x-coordinate of the vertex of $f(x) = ax^2 + bx + c$ is $-b/2a$. To prove this property we need to write an arbitrary quadratic function in vertex form.
   (a) Factor a out of each term of $f(x) = ax^2 + bx + c$.
   (b) Find one-half of the middle term (after you have factored out a) and square it. What does this value equal?

(c) Complete the square on f and rewrite it in vertex form.

(d) Explain why this answer shows that the x-coordinate of the vertex is $-b/2a$.

5. We want to derive the quadratic formula for $p(x) = ax^2 + bx + c$.

 (a) Set $p(x) = 0$. Explain why solving this equation is equivalent to saying "find the roots."

 (b) Divide both sides of $p(x) = 0$ by a. Why can you assume that $a \neq 0$?

 (c) Complete the square and rewrite your equation in vertex form $(0 = (x-h)^2 + k)$.

 (d) Solve this equation for x.

 (e) Simplify your answer from part (d) to show that

$$x = \frac{-b \pm \sqrt{b^2 - 4ac}}{2a}.$$

6. It was shown in this section that when the leading coefficient of a quadratic function, a, is positive, the vertex of the function gives a minimum value. Given a quadratic function in vertex form, $f(x) = a(x-h)^2 + k$, where a is negative, explain why the vertex gives a maximum value for $f(x)$. Use only references to the function's symbolic form in your explanation.

7. The following table of data represents a quadratic function. What is the line of symmetry? Briefly justify your answer.

x	1	3	5	7
$f(x)$	5	5	13	29

8. We want to prove an assertion made in this section that any pair of inputs that are equidistant from the line of symmetry will have the same output. Let's begin this proof with a quadratic equation in vertex form, $f(x) = a(x-h)^2 + k$.

 (a) What is the line of symmetry?

 (b) Suppose two inputs are d units away from the line of symmetry. Write them in terms of h and d.

 (c) Substitute your answer from part (b) for x in $f(x) = a(x-h)^2 + k$ and compare their outputs.

 (d) Explain why the answer from part (c) proves that any two inputs equidistant from the line of symmetry have the same output.

9. In Exercise 8, you were asked to prove that any pair of inputs equidistant from the line of symmetry have the same output. In this exercise, you are asked to prove the converse, that any two points with the same output must be equidistant from the line of symmetry.

 (a) Suppose you have two points, (x_1, j) and (x_2, j), with the same output. Substitute both points into $f(x) = a(x-h)^2 + k$ to form two equations.

(b) Because the two equations have the same output set them equal to each other.

(c) Simplify the equation by subtracting k from both sides and dividing both sides by a.

(d) Take the square root of both sides of the function. Remember that $\sqrt{x^2} = |x|$.

(e) Explain how this result proves that the inputs must be equidistant from the line of symmetry.

10. Find the symbolic representation of a quadratic function with the following properties and sketch a graph.
    (a) The parabola opens up and has no real roots.
    (b) The parabola opens up and has one real root.
    (c) The parabola opens up and has two real roots.
    (d) The parabola opens down and has two real roots.
    (e) The parabola opens down and has one real root at $x = 5$.

11. Find the real roots of the following quadratic functions.
    (a) $f(x) = x^2 + 4x - 12$
    (b) $g(x) = 2x^2 + 4x - 6$
    (c) $h(x) = \frac{1}{2}x^2 - \frac{9}{2}x + 9$
    (d) $j(x) = x^2 - 8x + 16$
    (e) $k(x) = x^2 - 4$
    (f) $m(x) = 2x^2 - 20x + 100$
    (g) $n(x) = x^2 + 4$

12. Find the equation for the quadratic function that contains the following.
    (a) Vertex at $(4, 3)$ and the point $(1, 2)$
    (b) Vertex at $(5, -3)$ and the point $(0, 0)$
    (c) The three points $(2, 2)$, $(6, 8)$ and $(-2, 4)$
    (d) The three points $(-2, 0)$, $(6, 0)$ and $(0, 4)$

13. Give the range of the following functions.
    (a) $f(x) = x^2 - 4x + 6$
    (b) $g(x) = x^2 - 5x - 7$
    (c) $h(x) = 2x^2 + 3x - 4$

14. A projectile similar to the one in Example 7 is shot and the following data are recorded.

Time (sec)	0	0.1	0.2	0.3	0.4	0.5	0.6	0.7	0.8
Vertical Position (cm)	0	38.4	66.8	85.4	94.2	93.2	82.5	61.9	31.5

Find a quadratic function that models these data.

15. Find the equations for the following quadratic functions.
    (a) The function $y = x^2$ is shifted two units to the left and one unit down and is vertically stretched by a factor of -1.
    (b) A quadratic function has a vertex at $(0, -4)$, and the leading coefficient is -2.
    (c) A quadratic function has roots at $x = 0$ and $x = 1$ and contains the point $(\frac{1}{2}, 2)$.
    (d) A quadratic function is represented by the accompanying graph.

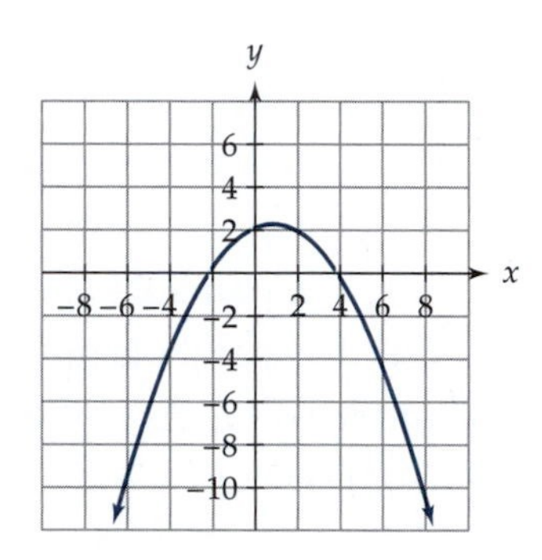

16. Let $p(x) = (x - h)^2 + k$. What restrictions, if any, on h and k guarantee that p will have
    (a) No real roots.
    (b) One real root.
    (c) Two real roots.
17. Let $p(x) = -(x - h)^2 + k$. What restrictions, if any, on h and k guarantee that p will have
    (a) No real roots.
    (b) One real root.
    (c) Two real roots.
18. Label each of the following as always true, sometimes true, or never true. Briefly justify your answers.
    (a) If a function's graph is in the shape of a parabola, then it is a shift and/or stretch and/or reflection of $y = x^2$.
    (b) $f(x) = (x + 1)(x - 2)$ has roots at $x = 1$ and $x = -2$.
    (c) If p contains the points $(-1, 2)$ and $(2, 2)$, then its line of symmetry is $x = 1$.
    (d) If $p(x) = ax^2 + bx + c$ contains the points $(-1, 2)$ and $(2, 2)$, then a is positive.
    (e) Given three noncollinear points, there is only one quadratic function that will contain them all.
19. A water balloon is thrown out of a window of a tall building such that its height can be described by the function $h(t) = -16t^2 + 80$, where t is time in seconds since the balloon was thrown and $h(t)$ is the height

of the balloon (in feet) above the sidewalk. Find the time it takes the balloon to hit the sidewalk.

20. A ball is thrown vertically into the air such that its height can be described by the function $h(t) = -16t^2 + 48t + 6$, where t is the time (in seconds) since the ball was thrown and $h(t)$ height of the ball (in feet) above the ground.
    (a) At what time does the ball reach its maximum height?
    (b) What is the maximum height?
    (c) How long does it take the ball to reach the ground?

INVESTIGATIONS

INVESTIGATION 1: HANG TIME

Two of the authors went to a math teachers' convention where they worked with some calculator-based laboratory equipment. One project involved calculating how high someone jumped based on how long the person was in the air. They set up a light source on the ground pointing horizontally and a light sensor opposite it. A jumper then stood in the path of the light. When the jumper jumped, the beam of line shined on the sensor until the jumper landed. The calculator then reported how high the jumper jumped. Needless to say, all the math teachers were fascinated and enjoyed seeing how the calculator performed this experiment and others. How was this calculator able to report the height of the jumper based only on the time the jumper was in the air?

1. The general equation for describing the height of an object that has been thrown is

$$h(t) = -16t^2 + v_0 t + h_0,$$

where t is time (in seconds), v_0 is the initial velocity (in feet per second), and h_0 is the initial height (in feet). With this equation, we can find the maximum height of the jumper or of any thrown object as long as we know how long it is in the air. Let's begin by thinking about what we know about the height of the object. We know the initial height, $h_0 = 0$. We also know that the height of the object at the time it finished its trip is also zero, so $h(s) = 0$, where s represents the time that the jumper landed on the ground again. Let's say that our jumper stayed in the air for 0.6 sec. That makes $s = 0.6$. With this information, we can find the initial velocity.

    (a) Substitute 0.6 for t, 0 for h_0, and 0 for $h(t)$ and determine the initial velocity, v_0.
    (b) Now that you know the initial velocity, you can find the height of the object at the halfway point.
        i. At what time did that maximum height occur?
        ii. Substitute your value for v_0 and the time of the maximum height for t to find the maximum height.

2. We can generalize this process so that we do not have to go through as many steps each time we take a reading on the time a jumper is in the air. Let's do this process again, but this time we just use s for the total time and $s/2$ for the time of the maximum height.
   (a) Substitute s for t, 0 for h_0, and 0 for $h(t)$ into the equation for h and determine the initial velocity, v_0.
   (b) Now that you have the initial velocity in terms of s, you can find the height of the object at the halfway point. Substitute your value for v_0 and $s/2$ for t into the equation for h and simplify. What is the maximum height?

INVESTIGATION 2: GOLDEN RATIO

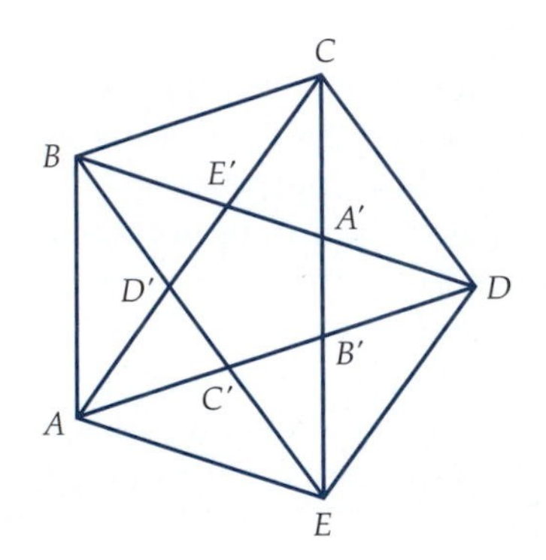

Figure 11

The Pythagoreans, a group of ancient Greeks, had a view of mathematics that put it at the very center of their understanding of life. The Pythagorean School was said to have a motto that read "all is number."[8] For the Pythagoreans, mathematics played an almost religious role. Consequently, certain relationships (like the famous Pythagorean theorem) were, to them, more than just nice theorems; they believed that these relationships held much deeper and more significant meaning.

Some historians believe that the pentagram (see Figure 11) held special significance for the Pythagorean School. The pentagram is simply a regular pentagon with the diagonals drawn in. One characteristic of this shape is that it is self-replicating; the diagonals form another pentagon that is similar to the original. You can then draw in the diagonals and make a pentagon again. That is not the only aspect of this shape that is self-replicating, however. If you look at one of the triangles formed by the sides of the original pentagon (such as $\triangle ABD'$), then you see an isosceles triangle that has interesting proportions. The ratio of one of the longer legs to the shorter leg is a well-known ratio known today as the *golden ratio*.

The golden ratio can be seen when dividing a line segment in a certain way. If a line segment is divided so that the ratio of the larger segment to the smaller is the same as the whole segment to the larger, then both these ratios are the golden ratio. (See Figure 12.)

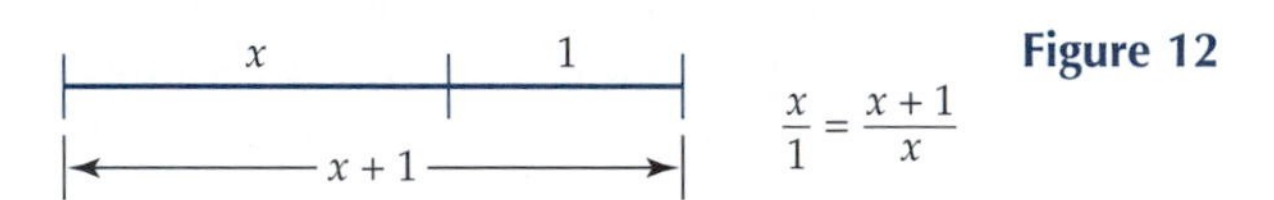

Figure 12

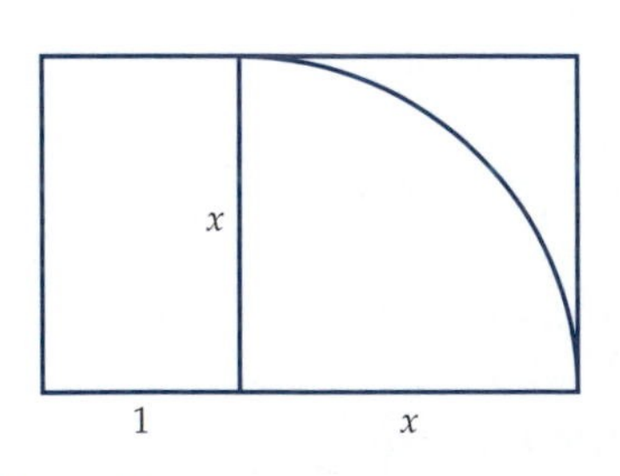

Figure 13

The golden ratio can be found by solving $(x+1)/x = x/1$. This ratio is shown in the proportions of the rectangle in Figure 13. If we start with a rectangle whose sides are length x and 1 and add a square (with sides of length x) to the side of this rectangle, then we have a new rectangle with sides x and $x+1$. If these rectangles are similar (that is, if the ratio of their

[8] *Carl B. Boyer,* A History of Mathematics *(Princeton, NJ: Princeton University Press, 1985), pp. 54–56.*

sides are the same), then $x/1$ is the golden ratio. This rectangle is then called the golden rectangle.

1. Solve the $(x+1)/x = x/1$ for x. Because $x/1$ is the ratio of one of these rectangles, x is the golden ratio.
2. Give a numerical approximation (to four decimal places) for the golden ratio.
3. Let's continue exploring this idea of the self-replicating nature of the golden ratio. Now that we have a number for this ratio, let's see if it really is self-replicating. Because the golden ratio is stated as $(x+1)/x = x/1$, it seems that we should be able to add 1 to our number, divide by the original number, and get the original number back again. Do this step by adding 1 to your numerical approximation of the golden ratio from question 2 and dividing that sum by your numerical approximation. That is, show that if x is your numerical approximation for the golden ratio, $(x+1)/x \approx x/1$.
4. When you solved for x in question 1, you should have obtained an exact value for the golden ratio, $(1+\sqrt{5})/2$. Show that if $x = (1+\sqrt{5})/2$, then $(x+1)/x = x/1$.
5. Geometrically, the golden rectangle is said to be particularly pleasing to the eye. It has been seen in many areas of art and architecture. Many Greek buildings, such as the Parthenon, supposedly have the proportions of the golden ratio. Some people have even found this relationship in various pieces of art (like the Mona Lisa) and in the human form as well. Although these uses for the golden ratio are interesting, they may or may not be true, but it is true that the Fibonacci sequence approximates the golden ratio. The Fibonacci sequence is formed by starting with a pair of 1s and then generating the next number in the sequence by adding the previous two numbers $(1, 1, 2, 3, 5, 8, \ldots)$.
   (a) Give the next five numbers in the Fibonacci sequence.
   (b) Use your last two numbers from the Fibonacci sequence to find their ratio. Is it close to the golden ratio?
   (c) The proportion we showed earlier, $x/1 = (x+1)/x$, can be thought of as a process of going from $x/1$ to $(x+1)/x$. We could think of doing this as, "Given a fraction, add the numerator and denominator and divide this sum by the numerator."
      i. Starting with the fraction $\frac{1}{1}$, add the numerator and denominator and divide this sum by the numerator. (Leave your answer in fraction form.)
      ii. Using your answer from part (i), add the numerator and denominator and divide this sum by the numerator.
      iii. Using your answer from part (ii), add the numerator and denominator, divide this sum by the numerator, and repeat this process a number of times. What do you notice happening?
   (d) Instead of starting with $\frac{1}{1}$, start with any number, a_1, add 1 to it, and divide the sum by your number, $(a_1+1)/a_1$. Now, use your

answer, a_2, add 1 to it, and divide this new sum by a_2: $(a_2+1)/a_2$. Repeat this process at least five more times. What do you notice happening?

6. Find five common rectangular items (like a television screen or a credit card or a piece of paper) and measure them. Are any of these the golden ratio? Do you think that this rectangle is particularly pleasing?

4.2 POLYNOMIAL FUNCTIONS

In section 4.1, we looked at quadratic functions. They are a special case of a more general class of functions known as polynomial functions. In this section, we define polynomial functions and examine their behavior in terms of shape and roots.

DEFINITION OF POLYNOMIAL FUNCTIONS

Pretend, for a moment, that you are president of STA,[9] a manufacturer of plastic bottle caps. You are interested in shipping a large quantity of bottle caps to the Wake-Up High-Energy Cola Company using as few boxes as possible. The largest box you can ship via parcel post through the U.S. Postal Service is "up to 70 pounds and a maximum of 108 inches in length and girth combined." After discovering that girth means the perimeter of a cross section, you need to determine what box size gives the largest volume, assuming your boxes will have a square base as in Figure 1.

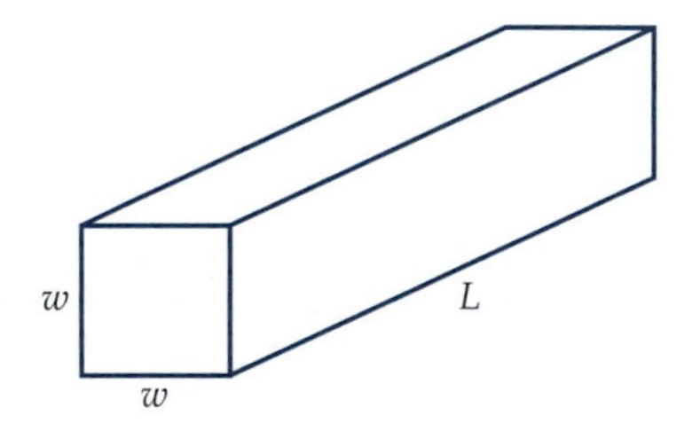

Figure 1 A box with a square base.

Because you are interested in maximizing volume, you decide to let the girth and length combined equal 108 in. Thus, $4w + L = 108$ or, equivalently, $L = 108 - 4w$. Therefore, the volume of your box is given by $V(w) = w \cdot w \cdot (108 - 4w)$. A graph of V for $0 \leq w \leq 27$ is shown in Figure 2.

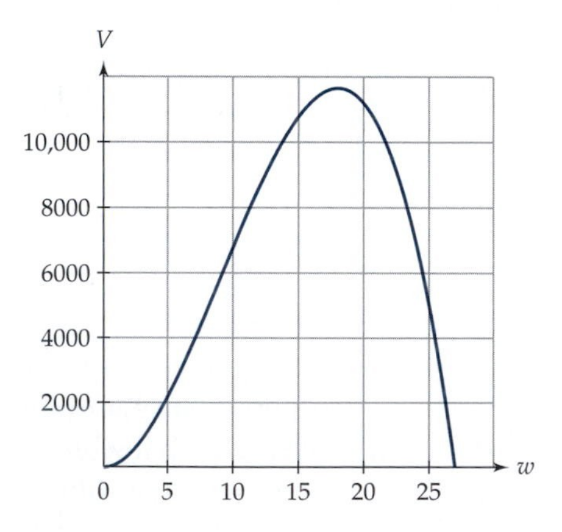

Figure 2 A graph of $V(w) = w^2(108 - 4w)$ giving the volume of the box for $0 \leq w \leq 27$.

Notice that although this graph is similar to a parabola, it is clear that something different is happening. There is obviously a "highest point" for

[9] *STA stands for "Sorry, Try Again," the famous slogan placed inside each bottle cap.*

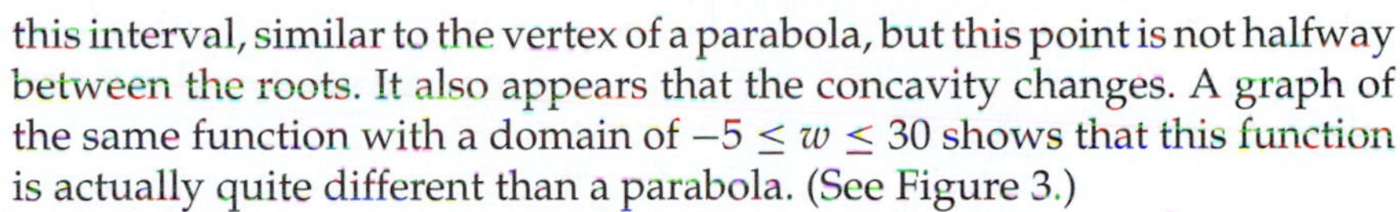

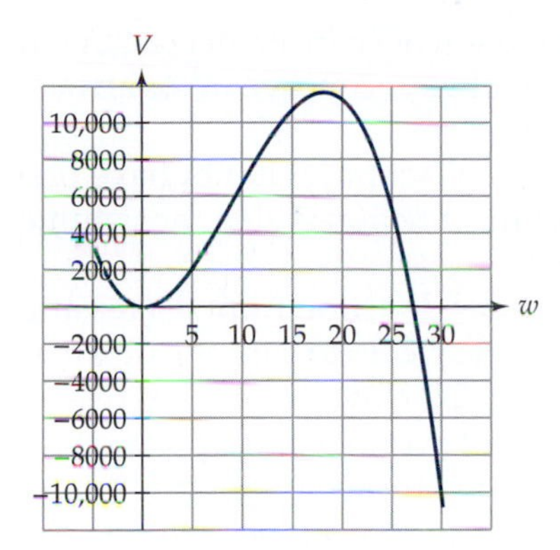

Figure 3 A graph of $V(w) = w^2(108 - 4w)$ giving the volume of the box for $-5 \le w \le 30$.

this interval, similar to the vertex of a parabola, but this point is not halfway between the roots. It also appears that the concavity changes. A graph of the same function with a domain of $-5 \le w \le 30$ shows that this function is actually quite different than a parabola. (See Figure 3.)

Let's look closer at this function. If we expand $V(w) = w^2(108 - 4w)$, we get $V(w) = -4w^3 + 108w^2$, which is an example of a polynomial function. A **polynomial function** is a function of the form

$$p(x) = a_n x^n + a_{n-1}x^{n-1} + \cdots + a_1 x + a_0,$$

where the exponents are nonnegative integers and the coefficients are real numbers. In other words, a polynomial function is a sum of power functions where each exponent is a nonnegative integer. Some examples of polynomial functions are

$$q(x) = 4x^3 - 80x^2 + 400x$$
$$s(x) = 10x^{10} - x^2 + 2$$
$$f(x) = x^9 - x^8 + x^7 - x^6.$$

The **degree** of a polynomial is the largest exponent when considering all terms with nonzero coefficients. Our volume function, V, is a polynomial function of degree 3.[10] The function $f(x) = x^9 - x^8 + x^7 - x^6$ has degree 9. Symbolically, the degree of the general polynomial function

$$p(x) = a_n x^n + a_{n-1}x^{n-1} + \cdots + a_1 x + a_0$$

is n (assuming $a_n \neq 0$).

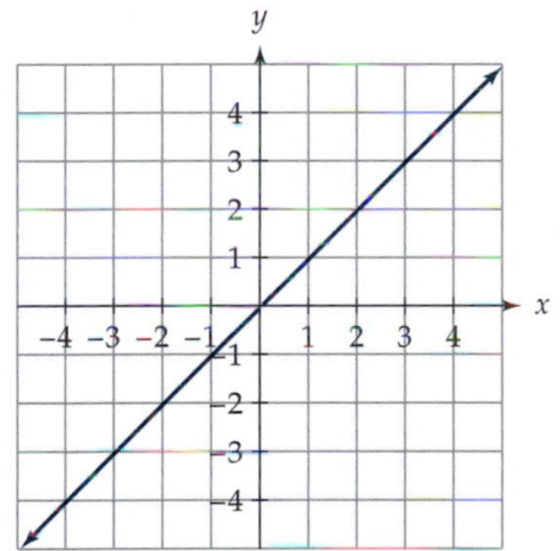
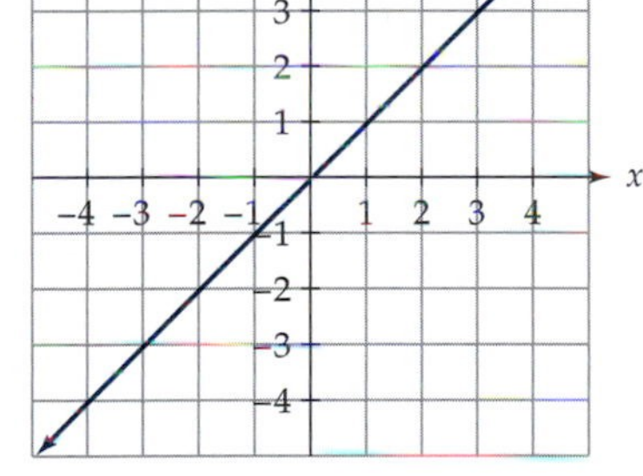

Figure 4 A graph of a first-degree polynomial function.

READING QUESTIONS

1. What is the degree of $p(x) = \frac{1}{2}x^4 - x^3 + 4$?
2. Why is $f(x) = x^3 + 2x - \sqrt{x}$ not a polynomial function?
3. For the function giving the volume of the box, V, why did we initially choose a domain of $0 \le w \le 27$?

GRAPHS OF POLYNOMIAL FUNCTIONS

Shape of the Graph

There is a connection between the degree of a polynomial function and the appearance of its graph. To explore this connection, let's start with lines. Linear functions are polynomial functions of degree 1. Every line looks similar to Figure 4. Lines may be shifted, reflected, or stretched, but they all have the same basic shape. Notice that a line always has at most one x-intercept (horizontal lines do not always have x-intercepts), never changes direction, and is neither concave up nor concave down.

Second-degree polynomial functions are quadratics, functions we closely studied in section 4.1. Quadratics, like lines, all have the same basic shape. They may be shifted, reflected, or stretched, but they all look similar to Figure 5. As you saw in section 4.1, a second-degree polynomial function may have zero, one, or two roots, depending on how the parabola

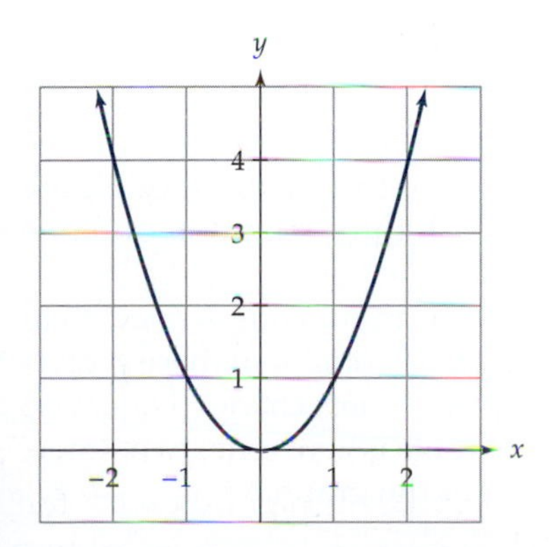

Figure 5 A graph of a second-degree polynomial function.

[10] *A polynomial of degree 3 is also known as a cubic polynomial.*

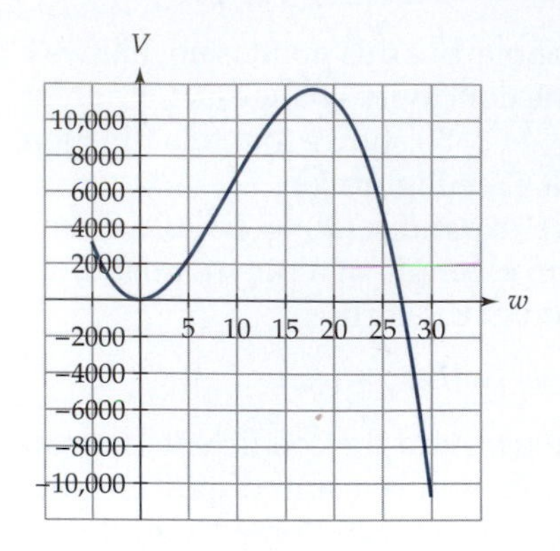

Figure 6 A graph of V, the function giving the volume of the parcel post box.

has been shifted. All quadratics change direction once (from decreasing to increasing or vice versa), and they are either always concave up or always concave down.

Graphs of third-degree polynomial functions (cubics) do not have one basic shape. Let's look again at the graph of the function giving the volume of the box, V, for $-5 \leq w \leq 30$ in Figure 6.

This graph has two "turns" where it changes direction from decreasing to increasing or vice versa. It changes from concave up to concave down. It also has two roots, one at $x = 0$ and one at $x = 27$. Not all cubics, however, look like this one. Figure 7 shows the graphs of two other cubic polynomial functions. These two cubic polynomial functions are always increasing. As a consequence, they always have exactly one real root. Notice that there is still a change in concavity from concave down (for $x < 0$) to concave up (for $x > 0$).

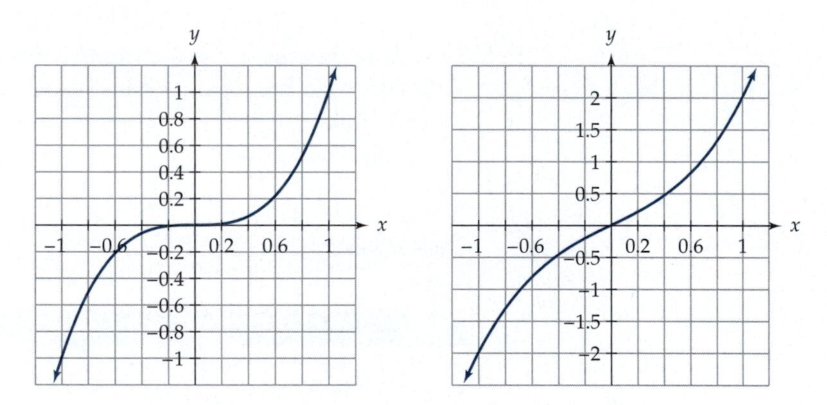

Figure 7 The graphs of two cubic polynomial functions that do not change direction.

These two shapes, a shape with two changes in direction and a shape with no changes in direction, are the only possibilities for a cubic polynomial function. The number of roots of a cubic depends both on the shape and on how the graph has been shifted. If the shape is always increasing (or always decreasing) such as the graphs in Figure 7, then there is always exactly one real root. If the shape has two changes in direction, such as the graph in Figure 6, then there could be one, two, or three real roots. Figure 8 shows the graphs of two more cubics, one with three real roots and one with one real root.

Fourth-degree polynomial functions are even more complicated than cubics. Four different shapes are shown in Figure 9. Notice that these graphs change direction either one time or three times. They change concavity either two times or zero times. The number of roots is zero, one, two, three, or four, depending on the shape and the shift of the graph.

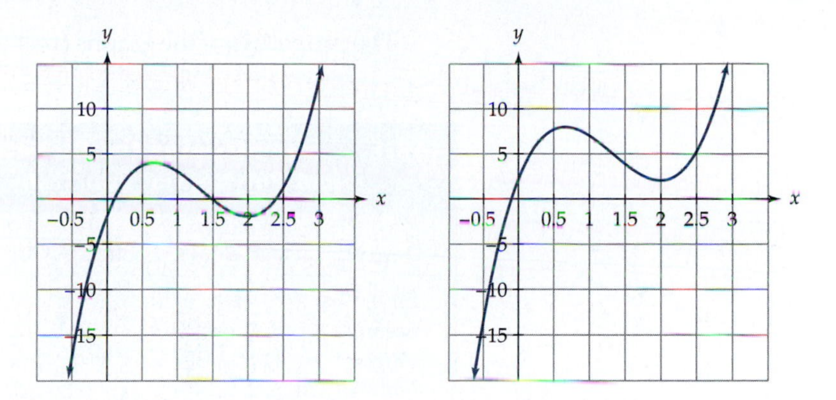

Figure 8 The graphs of cubic polynomial functions with three real roots and with one real root.

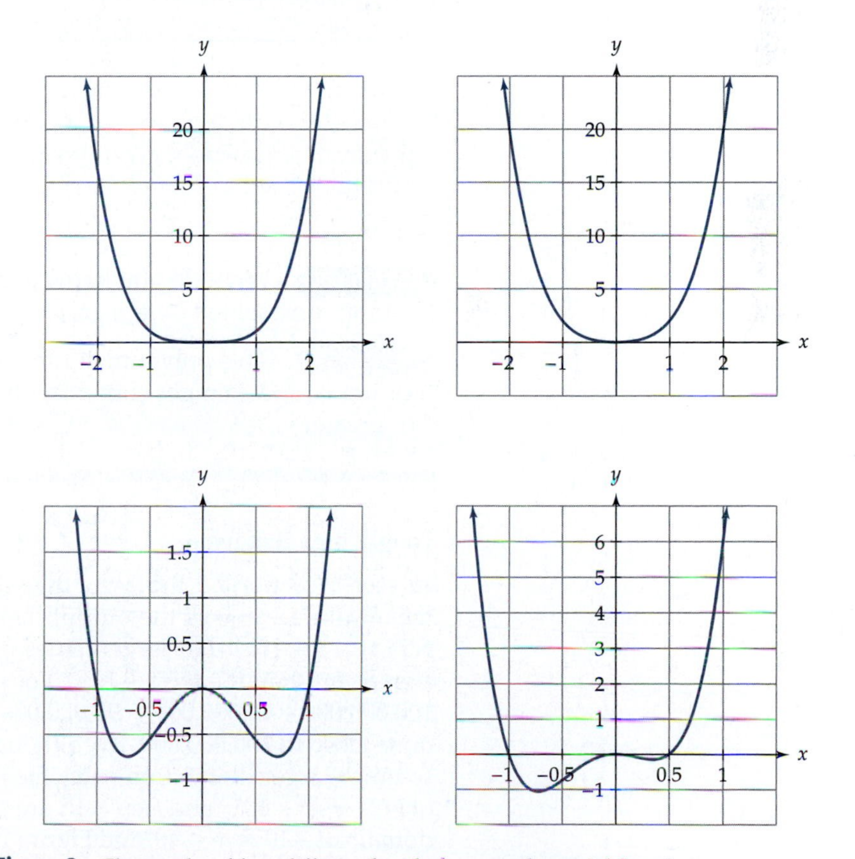

Figure 9 The graphs of four different fourth-degree polynomial functions.

The properties of the graphs considered so far are summarized in Table 1.

TABLE 1 Summarizing the Properties of the Graphs of Polynomial Functions

Degree	Real Roots	Changes in Direction	Changes in Concavity
First	1	0	—
Second	0, 1, or 2	1	0
Third	1, 2, or 3	0 or 2	1
Fourth	0, 1, 2, 3, or 4	1 or 3	0 or 2

PROPERTIES

In general, if p is a nth-degree polynomial function, then

- p has at most n real roots
- p changes direction at most $n-1$ times
- p changes concavity at most $n-2$ times

These facts can be used to gain information about the possible degree of a polynomial function, given the graph. The graph alone, however, is not sufficient for determining the degree.

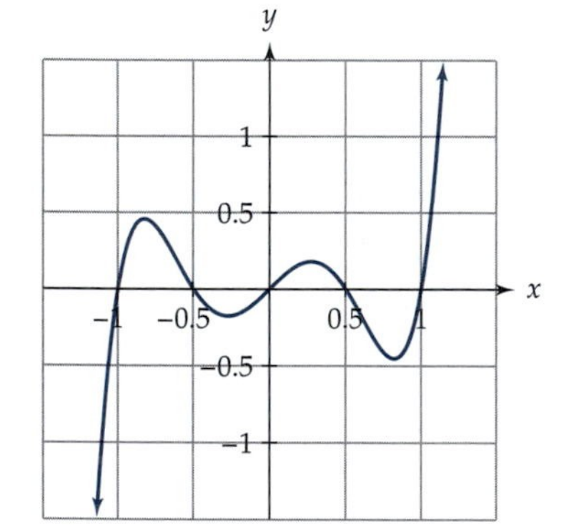

Figure 10

Example 1 What do you know about the degree of the polynomial function represented by the graph in Figure 10?

Solution This polynomial function has five roots, changes direction four times, and changes concavity three times. Thus, it must be of degree 5 or greater.

Long-Range Behavior

As seen in section 2.5, the larger the exponent of a power function, the larger the absolute value of the output. Let's consider the polynomial function $p(x) = x^4 - 10x^3$. For large values of x, the term x^4 is going to be more significant than the term $-10x^3$. For example, $p(100) = 100^4 - 10 \cdot 100^3 = 100{,}000{,}000 - 10{,}000{,}000 = 90{,}000{,}000$. This number is, relatively speaking, quite close to 100,000,000. Even though the x^3-term is large, the x^4-term is so much larger that it dominates the function for large values of x. Graphs of $p(x) = x^4 - 10x^3$ and $f(x) = x^4$ are shown in Figure 11. Figure 11(a) has a domain of $-10 \leq x \leq 10$, and Figure 11(b) has a domain of $-100 \leq x \leq 100$. In these figures, you can see that although the functions look different close to the origin, their long-range behavior is quite similar.

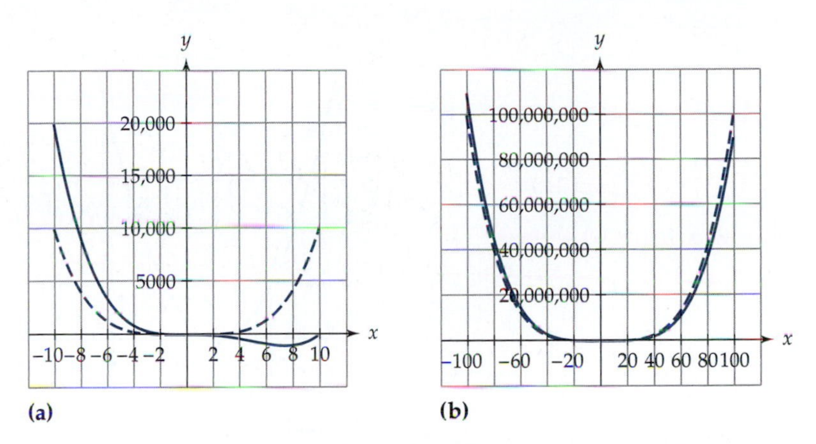

Figure 11 (a) The functions $p(x) = x^4 - 10x^3$ and $f(x) = x^4$ are compared when $-10 \le x \le 10$. (b) The same two functions are compared when $-100 \le x \le 100$.

As the absolute value of x gets large, the x^4-term in the function always dominates the behavior of the function. We denote x approaching larger and larger positive numbers as $x \to +\infty$, and we denote x approaching negative numbers with a larger and larger absolute value as $x \to -\infty$. The notation $x \to \pm\infty$ denotes the absolute value of x becoming large (for both positive and negative values of x).[11] In general, $p(x) = a_n x^n + a_{n-1}x^{n-1} + \cdots + a_1 x + a_0$ is approximately equal to $a_n x^n$ as $x \to \pm\infty$.

For example, as $x \to \pm\infty$,

$$q(x) = 4x^3 - 80x^2 + 400x \approx 4x^3$$
$$s(x) = 10x^{10} - x^2 + 2 \approx 10x^{10}$$
$$f(x) = x^9 - x^8 + x^7 - x^6 \approx x^9$$

We can determine the long-range behavior of any polynomial function simply by knowing the long-range behavior of power functions with positive integer exponents. In section 2.5, we saw that $f(x) = x^n$ looks similar to Figure 12(a) if n is even and looks similar to Figure 12(b) if n is odd.

Notice that if n is even, the output behaves similarly regardless of whether $x \to +\infty$ or $x \to -\infty$. In Figure 12(a), the output gets larger and larger as the absolute value of the input gets large. Symbolically, we write $f(x) \to +\infty$ as $x \to \pm\infty$. If n is even but the coefficient of x^n is negative, then the output still behaves the same regardless of whether $x \to +\infty$ or $x \to -\infty$. This time, however, $f(x) \to -\infty$. The output behaves the same because raising a negative number to an even power results in a positive answer. This behavior also corresponds to what we noticed in Table 1. If n is 2 or 4, then the function changes direction an odd number of times, which forces both ends of the graph to be in the same direction.

On the other hand, if n is odd, then the output does the opposite as $x \to +\infty$ when compared with $x \to -\infty$. In Figure 12(b), $f(x) \to +\infty$ as $x \to +\infty$ but does the opposite (i.e., $f(x) \to -\infty$) as $x \to -\infty$. If n is odd

[11] *The notation $x \to \pm\infty$ is only used if the behavior of the function is the same both as $x \to \infty$ and as $x \to -\infty$.*

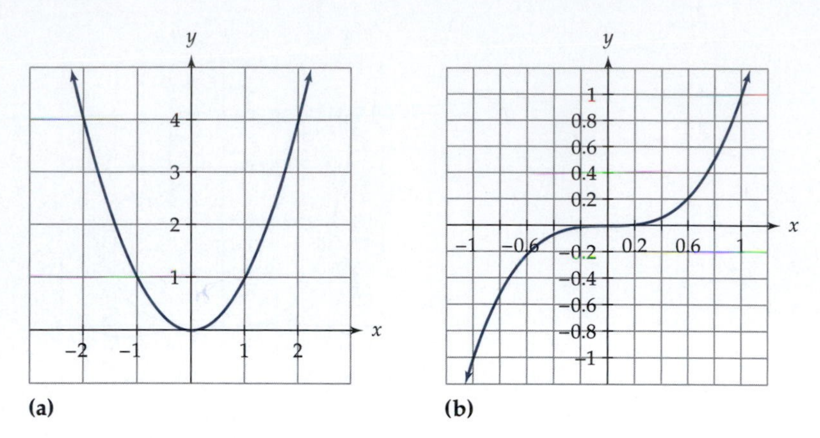

Figure 12 (a) The basic shape of $y = x^n$ when n is even; (b) the basic shape of $y = x^n$ when n is odd.

and the coefficient of x^n is negative, then the output still does the opposite but the order is reversed. This time, $f(x) \to -\infty$ as $x \to +\infty$, but $f(x) \to +\infty$ as $x \to -\infty$. Again, that the output does the opposite is a consequence of the properties of exponents because raising a negative number to an odd power results in a negative answer, which also corresponds to what we noticed in Table 1. If n is 1 or 3, the function changes direction an even number of times, which forces the two ends of the graph to be in opposite directions.

A summary of the possible long-range behaviors of a polynomial function, $p(x) = a_nx^n + a_{n-1}x^{n-1} + \cdots + a_1x + a_0$, is illustrated by the graphs in Figures 13 and 14.

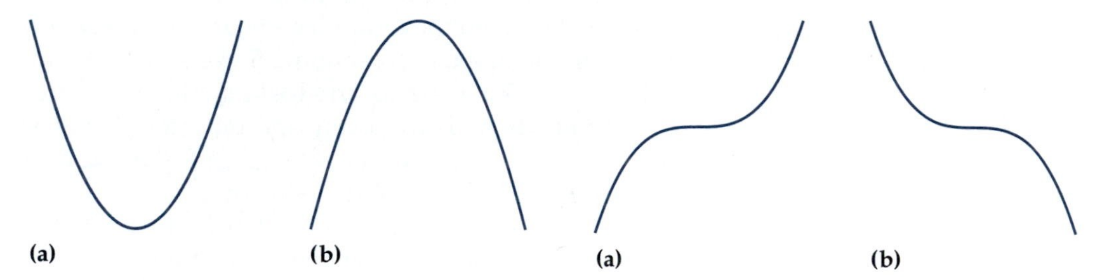

Figure 13 (a) The long-range behavior of p when n is even and a_n is positive; (b) the long-range behavior of p when n is even and a_n is negative.

Figure 14 (a) The long-range behavior of p when n is odd and a_n is positive; (b) the long-range behavior of p when n is odd and a_n is negative.

READING QUESTIONS

4. Give the smallest possible degree for the polynomial function represented by the accompanying graph.

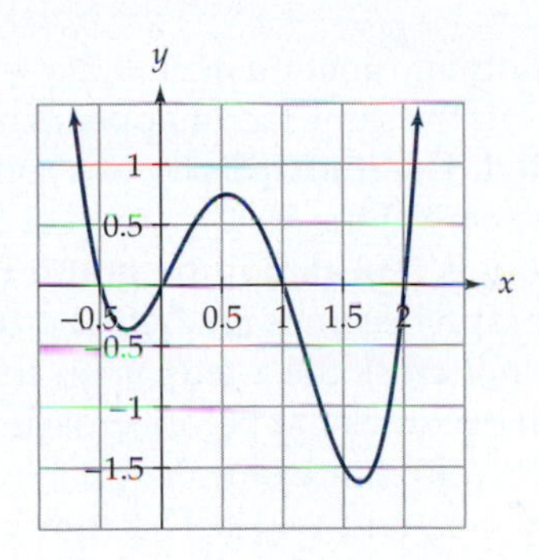

5. What do you know about the number of real roots of a fifth-degree polynomial function?
6. Suppose the graph of the polynomial function, p, changes concavity three times. What do you know about the degree of p?
7. Answer the following questions, assuming the graph of the accompanying polynomial function does not change direction outside of the viewing window.

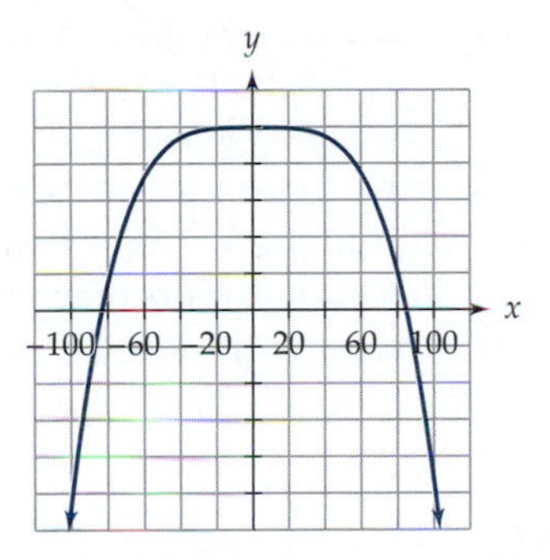

   (a) Is the exponent of the leading term of the function even or odd? Briefly justify your answer.
   (b) Is the coefficient of the leading term of the function positive or negative? Briefly justify your answer.
8. What do you know about a polynomial function, p, where $p(x) \to -\infty$ when $x \to \pm\infty$?

FINDING ROOTS

As we saw in section 4.1, we are often interested in finding the vertex and the roots of quadratic functions. For polynomial functions in general, we are similarly often interested in finding the points where the function changes direction and the points where it crosses the x-axis. Unfortunately, finding either of these points for a general polynomial function is significantly harder and often impossible.

As with quadratics, if r is a root of the polynomial function, p, then $(x - r)$ is a factor of the polynomial and vice versa. So, if we can factor a polynomial, we can find the roots and vice versa. For example, the roots of

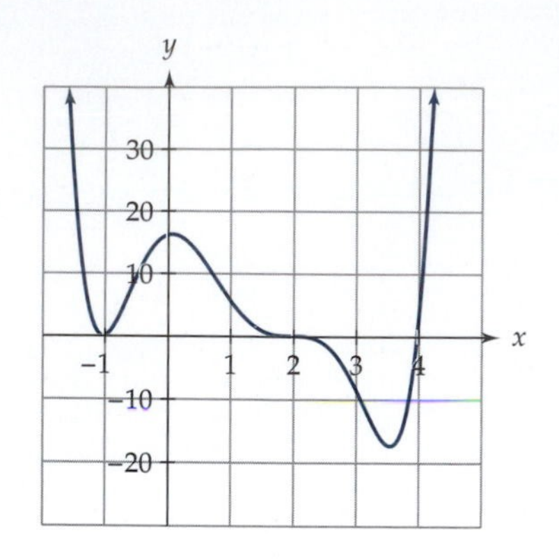

Figure 15 A graph of the polynomial function $g(x) = \frac{1}{2}(x+1)^2(x-2)^3(x-4)$.

the polynomial function $q(x) = 2(x+3)(x-1)(2x-5)$ are $x = -3$, $x = 1$, and $x = \frac{5}{2}$. If the same factor appears more than once, then the root is said to be repeated. The **multiplicity** of a root, r, is the number of times the factor $(x-r)$ appears in the factorization of the polynomial. Graphically, you can tell if a root is repeated if the graph touches but does not cross the x-axis or if the graph becomes flat before crossing the x-axis. If the graph touches but does not cross the x-axis, then the multiplicity of that root is even. If the graph becomes flat before crossing the x-axis, then the multiplicity of that root is odd. For example, $g(x) = \frac{1}{2}(x+1)^2(x-2)^3(x-4)$ has roots at $x = -1$, $x = 2$, and $x = 4$. The root $x = -1$ has multiplicity two, the root $x = 2$ has multiplicity three, and the root $x = 4$ has multiplicity one. A graph of g is shown in Figure 15.

From the graph, you can determine that $x = -1$ is a repeated root with an even multiplicity because the graph touches but does not cross the x-axis at $x = -1$. You can also determine that $x = 2$ is a repeated root with an odd multiplicity because the graph flattens before crossing the x-axis at $x = 2$. The root $x = 4$ is not a repeated root because the graph looks similar to a straight line when it crosses the x-axis at $x = 4$.

The fundamental theorem of algebra guarantees that an nth-degree polynomial function will always have exactly n roots (counting multiplicities). So, for example, the fifth-degree polynomial function $p(x) = 3(x-1)^2(x-2)^3$ has five roots if you count the root $x = 1$ twice (because it has multiplicity two) and the root $x = 2$ three times (since it has multiplicity three). Just as with quadratic functions, however, it is possible to have roots that are not real numbers.[12] Nonreal roots always come in pairs. When a polynomial function has roots that are not real, the graph changes direction twice without crossing the x-axis. Figure 16 is the graph of a third-degree polynomial function that has one real root and two nonreal (complex) roots.

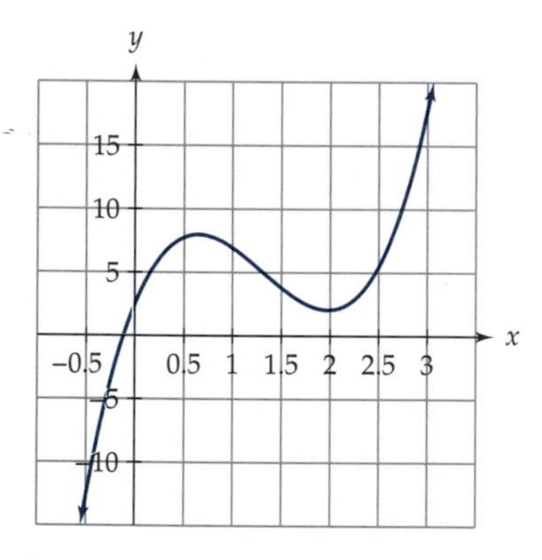

Figure 16 The graph of a cubic polynomial function with only one real root.

Example 2 Determine a plausible formula for the fourth-degree polynomial function, p, shown in Figure 17. Write your answer in factored form. Assuming the graph does not change direction outside the viewing window, determine if the leading coefficient for p is positive or negative.

Solution The roots appear to be $x = -1$ and $x = 3$. The root at $x = -1$ is not a repeated root because the function looks like a line when it crosses the x-axis here. The root at $x = 3$ has an odd multiplicity. So, the formula for p is $p(x) = a(x+1)(x-3)^s$, where s is an odd number. Because p is a fourth-degree polynomial function, s must equal 3. Therefore, $p(x) = a(x+1)(x-3)^3$. The leading coefficient, a, must be positive because $p(x)$ is approaching positive infinity as $x \to \pm\infty$.

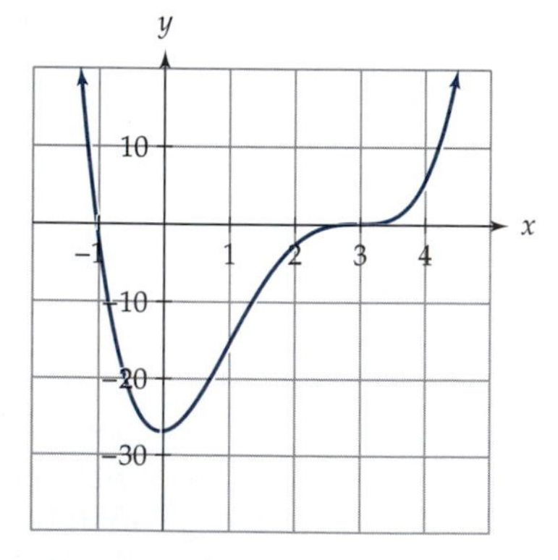

Figure 17

Example 3 Determine a plausible formula for the eighth-degree polynomial function, p, shown in Figure 18, and write your answer in factored form. Determine if the leading coefficient for p is positive or negative. Assume that all the roots are shown and that all the roots of p are real numbers. One of the roots is at $x = -2$.

[12] *Such roots will be complex numbers. As mentioned in section 4.1, complex numbers include the number i, which equals* $\sqrt{-1}$.

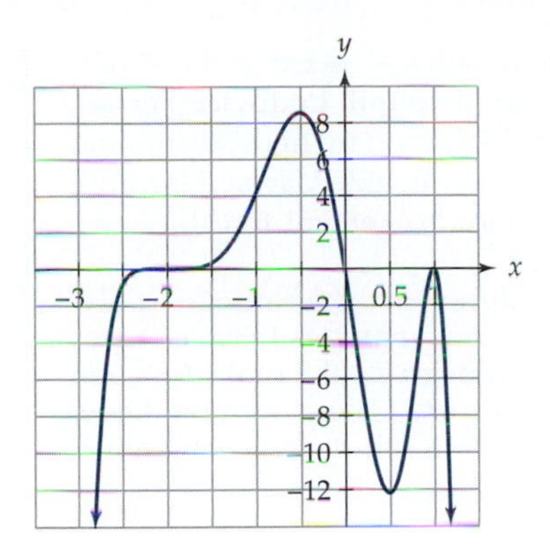

Figure 18

Solution We know that one of the roots is $x = -2$. The other roots appear to be $x = 0$ and $x = 1$. The root at $x = -2$ has an odd multiplicity, whereas the root at $x = 1$ has an even multiplicity. So, the formula for p should be $ax(x+2)^s(x-1)^t$, where s is odd and t is even. Because p is an eighth-degree polynomial function and all the roots are real numbers, we also know that $1 + s + t = 8$. We could either choose $s = 3$ and $t = 4$ or we could choose $s = 5$ and $t = 2$. Because the graph is quite flat near $x = -2$, we assume that $x = -2$ has multiplicity five. Using this assumption, $p(x) = ax(x+2)^5(x-1)^2$. The leading coefficient, a, must be negative because $p(x)$ is approaching negative infinity as $x \to \pm\infty$.

Finding the roots of a polynomial function given in factored form or finding the factored form of a polynomial function given the roots is relatively easy. Unlike the situation with quadratics where we have the quadratic formula, however, there is no formula for finding the roots of a general polynomial function. Thus, given the equation of a polynomial, it is often impossible to find the exact value of the roots.

HISTORICAL NOTE

There has been a long-standing interest in finding roots of polynomial functions. As early as 1550 B.C., the Egyptians knew how to find the zeros of a linear equation. By 300 B.C., the Greeks were geometrically finding roots of quadratic functions. The Hindus had derived a version of the quadratic formula by A.D. 1025. The search was now on to find a solution to the general cubic polynomial function.

Solutions of special cases of cubic equations began as early as A.D. 275 by the Greeks. It was the Italians, however, who in 1545 published a generic solution for solving a cubic. In those days, it was common for mathematicians to send each other "challenge problems." In 1530, Zuanne de Tonini da Coi, a teacher, sent a challenge to Tartaglia to find solutions to the cubics $x^3 + 3x^2 = 5$ and $x^3 + 6x^2 + 8x = 1000$. Tartaglia was a self-educated mathematician who was interested in applying mathematics to artillery science. His name, which means stammerer, was a nickname. He was given this name because he had difficulty speaking due to losing the use of some of his facial muscles as a child when he received a saber cut on his face during the battle at Brescia. For some time, Tartaglia was unable to solve these two cubics. Eventually, however, he not only found the solutions but also discovered how to solve a general cubic. He decided to keep the solution of the general cubic a secret, imparting the information to a fellow mathematician, Jerome Cardan, only under an oath of secrecy. Cardan was a man of curious contrasts who, in the course of his lifetime, was an astrologer, philosopher, gambler, algebraist, physicist, physician, father of a murderer, professor at the University of Bologna, and inmate of a poorhouse. In 1545, Cardan published *Ars Magna*, the first great Latin treatise devoted solely to algebra. In this book, breaking his oath to Tartaglia, he published the solution to the general cubic equation.

There was now a renewed interest in finding the solution for the general fourth-degree polynomial function. Again, it was the Italians in the sixteenth century who made the greatest headway. Another challenge problem created

continued on page 286

by Da Coi, this time given to Cardan, was to solve $x^4 + 6x^2 + 36 = 60x$. Unable to do so, Cardan gave the problem to his pupil, Lodovico Ferrari. Ferrari (who died at the age of 38 after being poisoned by his sister) found the solution. Cardan also published this solution in *Ars Magna*. Not until 1637 did Descartes succeed in finding a solution to general fourth-degree polynomial functions.

Next came an attempt to find the general solution to the fifth-degree polynomial function. After many attempts by famous mathematicians, it was shown in 1824 that a general solution to the fifth-degree polynomial function is impossible. Using a theory developed by Evariste Galois, we now know that there is no solution for finding the roots of a nth-degree polynomial function if n is 5 or greater.

Sources: D. T. Smith, *History of Mathematics, Vol. 1.* (New York: Dover, 1951), pp. 295–300; D. T. Smith, *History of Mathematics, Vol. 2.* (New York: Dover, 1953), pp. 435–36, 443–44, 455–70.

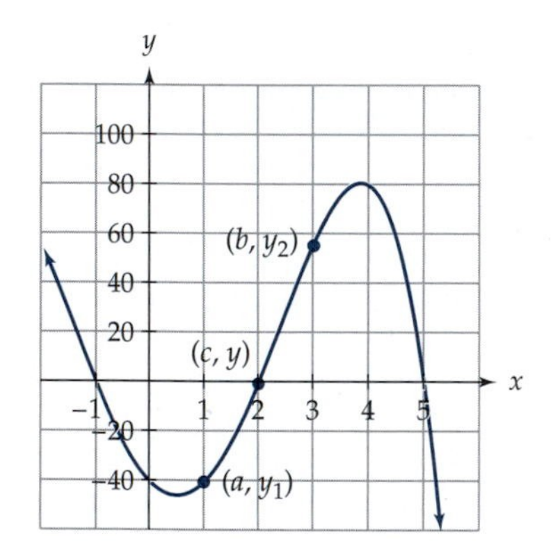

Figure 19 Graphical illustration of the intermediate value theorem.

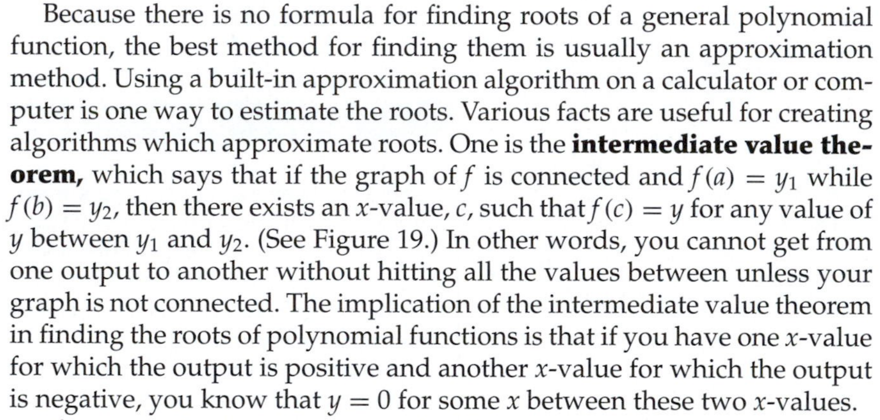
Because there is no formula for finding roots of a general polynomial function, the best method for finding them is usually an approximation method. Using a built-in approximation algorithm on a calculator or computer is one way to estimate the roots. Various facts are useful for creating algorithms which approximate roots. One is the **intermediate value theorem,** which says that if the graph of f is connected and $f(a) = y_1$ while $f(b) = y_2$, then there exists an x-value, c, such that $f(c) = y$ for any value of y between y_1 and y_2. (See Figure 19.) In other words, you cannot get from one output to another without hitting all the values between unless your graph is not connected. The implication of the intermediate value theorem in finding the roots of polynomial functions is that if you have one x-value for which the output is positive and another x-value for which the output is negative, you know that $y = 0$ for some x between these two x-values.

That a root lies between an x-value that gives a positive output and an x-value that gives a negative output (also known as Descartes rule of signs) is the basis for the approximation method known as the bisection method. Its algorithm is as follows.

1. Find two x-values, a and b, such that $f(a) < 0$ and $f(b) > 0$.
2. Find the midpoint of a and b, $m = (a + b)/2$.
3. Determine if $f(m)$ is positive or negative.
4. If $f(m) > 0$, then replace b with m. If $f(m) < 0$, then replace a with m.
5. Repeat the process, using your new values for a and b.

This process is continued until you have approximated the root to the desired accuracy. Notice that the bisection method will not work to find roots with even multiplicity because roots of this nature do not cross the x-axis. This method is illustrated in Example 4.

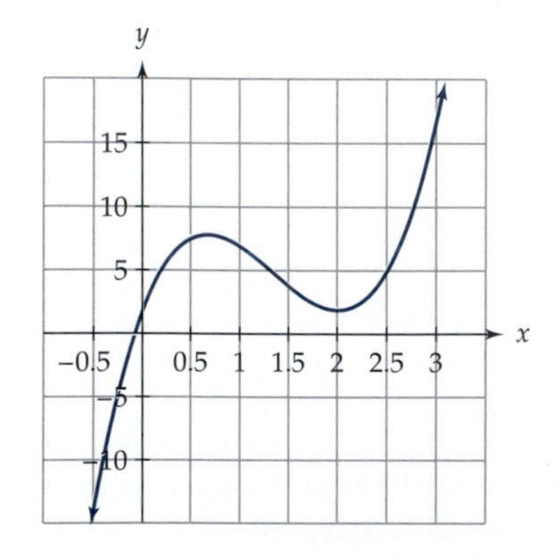

Figure 20

Example 4 Use the bisection method to approximate to two decimal places the root of the polynomial function $p(x) = 5x^3 - 20x^2 + 20x + 2$. The graph of p is shown in Figure 20.

Solution Looking at the graph, we start with $x = -0.5$ and $x = 0$. Thus, the first midpoint is $x = -0.25$. Subsequent calculations are shown in Table 2. Because the last two x-values, $-0.09375 \approx -0.09$ and $-0.08984375 \approx -0.09$, are the same when rounded to two decimal places, the root is $x \approx -0.09$.

TABLE 2

Two x-values		Midpoint	Sign of Output
−0.5	0	−0.25	Negative
−0.25	0	−0.125	Negative
−0.125	0	−0.0625	Positive
−0.125	−0.0625	−0.09375	Negative
−0.09375	−0.0625	−0.078125	Positive
−0.09375	−0.078125	−0.0859375	Positive
−0.09375	−0.0859375	−0.08984375	Positive
−0.09375	−0.08984375		

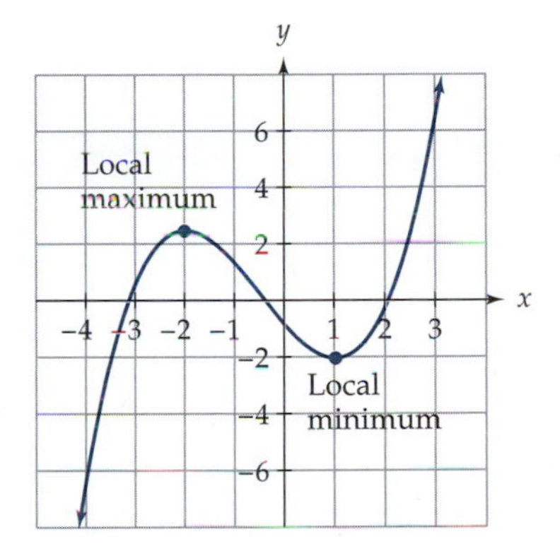

Figure 21 A graph illustrating local maximum and local minimum.

As with roots, there is no formula for finding local maxima or local minima of polynomial functions. A **local maximum** is a point $(c, f(c))$ such that $f(c) \geq f(x)$ for x close to c.[13] Similarly, a **local minimum** is a point $(b, f(b))$ such that $f(b) \leq f(x)$ for x close to b. (See Figure 21.)

With quadratics, we know that the maximum (or minimum) occurs at the vertex and that it is always possible to find the coordinates of the vertex. Unfortunately, there are no extensions of this technique to higher-degree polynomial functions (although calculus can sometimes be used to find local maxima and minima). As with finding roots, because there is no formula or algorithm, the best method for finding local maxima and minima is usually an approximation technique.

Example 5 Suppose (as in the beginning of this section) you are interested in finding the volume of the largest box with a square base. You want the length plus the perimeter of the cross section to equal 108 in.

(a) Find a formula for the volume of the box, V, in terms of the length of a side of the square base, w.

(b) Approximate the dimensions of the largest possible box and give its volume.

Solution

(a) Let w represent the length of a side of the square base and L represent the length of the box. Then $V = w \cdot w \cdot L$. Because the length plus the perimeter of the cross section equals 108 in., we

[13] *In calculus, you learn the precise meaning of "close."*

have $4w + L = 108$ or, equivalently, $L = 108 - 4w$. Substituting in place of L gives $V(w) = w^2(108 - 4w) = -4w^3 + 108w^2$.

(b) The domain of V is $0 < w < 27$ because both w and L must be positive. A graph of V for this domain is shown in Figure 22. By using a TI-83 to estimate the local maximum, we find that it occurs at $w = 18$. So, the box should have a square base whose width is 18 in. and the length of the box should be $L = 108 - 4w = 108 - 4 \cdot 18 = 36$ in. The volume of this box is 11,664 in.^{3}

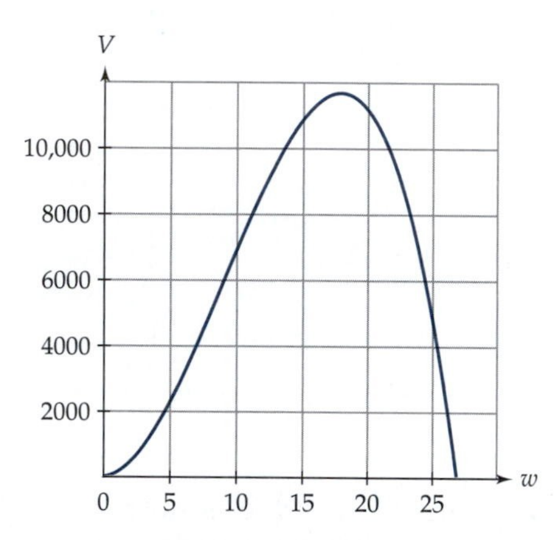

Figure 22

READING QUESTIONS

9. What is the relationship between the roots of a polynomial function and its factors?

10. Let $p(x) = -(x - 1)^3(x + 1)(x + 2)^2$.

(a) Give the degree of p.

(b) Give the multiplicity of each root of p.

(c) Without using a calculator or computer, sketch the shape of the graph of p.

11. Let $q(x) = x^4 + x^3 + x - 1$. How do you know that q has a root between $x = 0$ and $x = 1$?

12. What is meant by a local minimum?

SUMMARY

A **polynomial function** is a function of the form $p(x) = a_n x^n + a_{n-1}x^{n-1} + \cdots + a_1 x + a_0$, where the exponents are nonnegative integers and the coefficients are real numbers. The **degree** of a polynomial is the largest exponent when considering all terms with nonzero coefficients.

Graphs of polynomials do not have one basic shape, but if p is an nth-degree polynomial, then p has at most n real roots, p changes direction at most $n - 1$ times, and p changes concavity at most $n - 2$ times.

The long-range behavior of polynomial functions is similar to the behavior of the leading term when written in standard form.

The **multiplicity** of a root, r, is the number of times the factor $(x - r)$ appears in the factorization of the polynomial. If the graph of the function touches but does not cross the x-axis at roots, then the multiplicity is even. If the graph becomes flat before crossing the x-axis, then the multiplicity is greater than 2 and is odd.

The **intermediate value theorem** states that if the graph of f is connected and $f(a) = y_1$ while $f(b) = y_2$, then there exists some x-value, c, such that $f(c) = y$ for any value of y between y_1 and y_2. This theorem leads to the bisection method for finding roots. To use it, you need to find a value for which $f(x)$ is positive and another for which $f(x)$ is negative. Then you find the value of $f(x)$ for the point midway between those two points and repeat the process using one of the original points and the midpoint.

There is no formula for finding local maxima or minima of polynomial functions. A **local maximum** is a point $(c, f(c))$ such that $f(c) \geq f(x)$ for x close to c. In the same way, a **local minimum** is a point $(c, f(c))$ such that $f(c) \leq f(x)$ for x close to c. The best method for finding local maxima or minima in polynomials is usually an approximation technique.

EXERCISES

1. Pretend (one more time) that you are the president of STA, a manufacturer of plastic bottle caps, searching for the largest possible shipping box.
   (a) According to United Parcel Service, the largest box they accept is "up to 150 lbs with a maximum of 130 in. in length and girth combined and a maximum of 108 in. in length alone."
      i. Find the polynomial function whose output is the volume of a box with a square base and a combined girth and length of 130 in., V, and whose input is the width of the base, w.
      ii. Give a reasonable domain for this function.
      iii. Use a graph of this function to estimate the dimensions of the box with the largest volume.
   (b) According to Federal Express, the largest box they accept is "up to 150 lbs with a maximum of 165 in. in length and girth combined and a maximum of 119 in. in length alone."
      i. Find the polynomial function whose output is the volume of a box with a square base and a combined girth and length of 165 in., V, and whose input is the width of the base, w.
      ii. Give a reasonable domain for this function.
      iii. Use a graph of this function to estimate the dimensions of the box with the largest volume.

2. Explain why each of the following is not a polynomial function.
   (a) $y = x^3 - \dfrac{1}{x^2}$
   (b) $y = 3^x$
   (c) $y = x^4 + x^3 - 10x^2 + \pi x + x^{-1}$
   (d) $y = x^{11/7} - 2x^{7/11}$
3. If possible, give the degree of each of the following polynomial functions. If it is not possible to find the degree, then explain what you *can* conclude about the degree.
   (a) $y = 10x^3 + 100x^2$
   (b) $y = x^2 - x^3 + 14 + x^5 - 8x^4$
   (c) $p(x) = 23x(2x+1)^3(x-1)$
   (d) $q(x) = -8(x+2.3)(x-4)(x-10)^2$

(e)

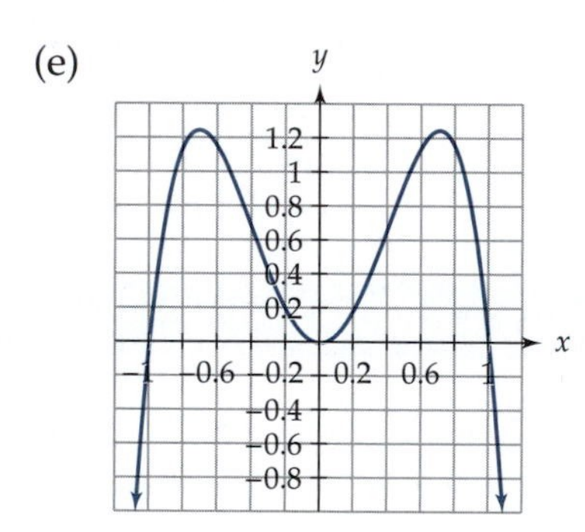

(f)

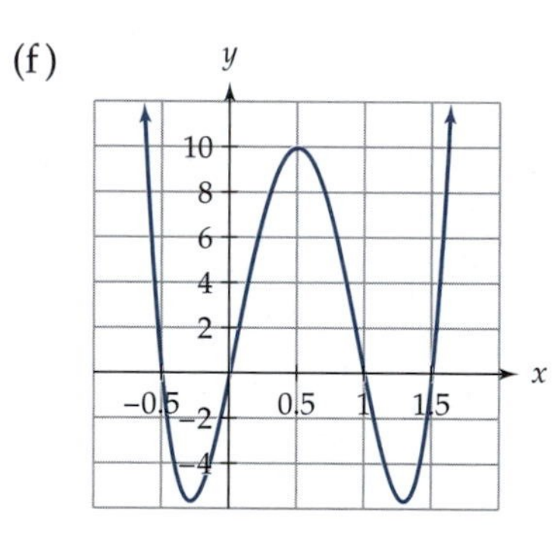

4. Give the smallest possible degree for each of the following polynomial functions.
   (a) The polynomial function p has two real roots, changes direction two times, and changes concavity three times.
   (b) The polynomial function q has three real roots, changes direction four times, and changes concavity three times.
   (c) The polynomial function r has seven real roots, changes direction six times, and changes concavity five times.
   (d) The polynomial function s has one real root, changes direction zero times, and changes concavity one time.
5. Assuming the graph does not change directions for x-values not shown, use each graph to give the smallest possible degree, determine if the leading coefficient is positive or negative, and classify the multiplicity of each root as 1, even, or an odd number greater than 1. Briefly justify your answers.

(a)

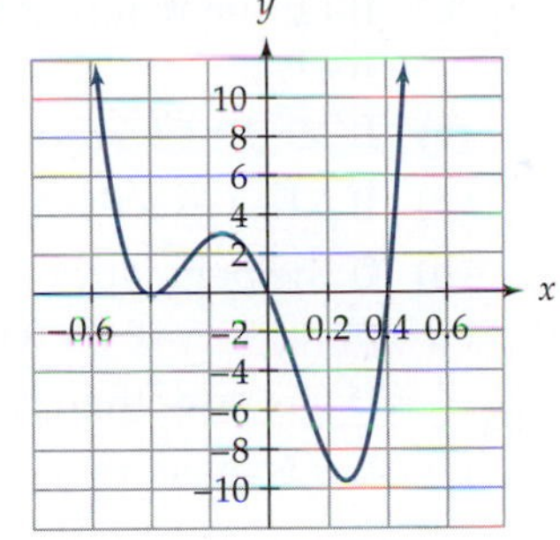

(b)

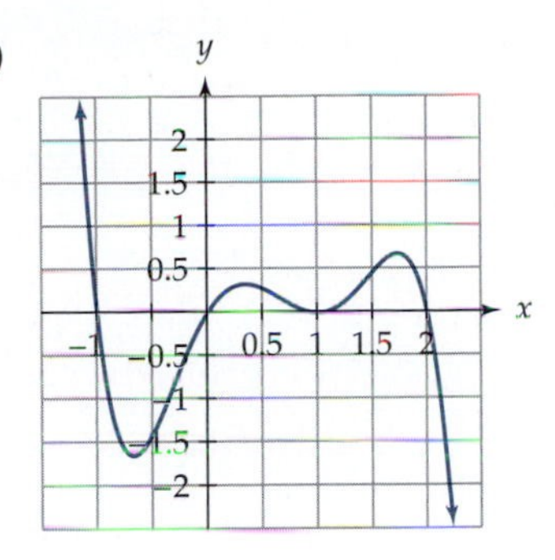

(c)

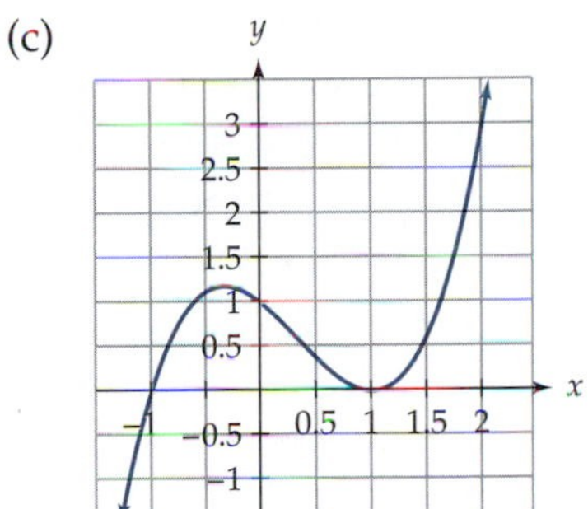

(d)

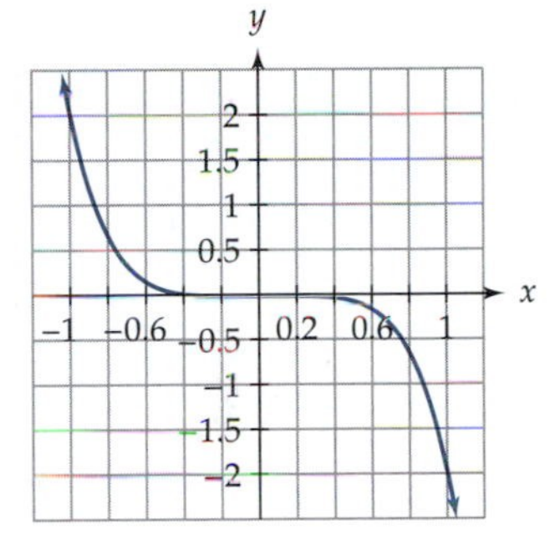

6. Let f be a polynomial function of odd degree. Explain (in words) why it is impossible for $f(x) \to +\infty$ both as $x \to +\infty$ and as $x \to -\infty$.
7. Sketch the graph of a polynomial function that has the following properties. There may be more than one possible answer.
   (a) The degree of p is 4, it has two real roots, it changes concavity two times, and $p(x) \to -\infty$ as $x \to +\infty$.
   (b) The degree of q is 5, it has a root at $x = 0$ with multiplicity three and a root at $x = 1$ with multiplicity two, and $q(x) \to +\infty$ as $x \to +\infty$.
   (c) The degree of r is 2, it has no real roots, and the vertex is at $(1, 2)$.
   (d) The degree of s is 7, it has a root at $x = -1$ with multiplicity four and a root at $x = 3$ with multiplicity one, it has no other real roots, and $s(x) \to -\infty$ as $x \to +\infty$.
8. Let $p(x) = a_n x^n + a_{n-1}x^{n-1} + \cdots + a_1 x + a_0$. Label each of the following as always true, sometimes true, or never true. Briefly justify your answer.
   (a) The function p has n real roots.
   (b) If n is even, then $p(x) \to +\infty$ as $x \to +\infty$.
   (c) If a_n is positive, then $p(x) \to +\infty$ as $x \to +\infty$.
   (d) If a_n is positive and n is odd, then $p(x) \to -\infty$ as $x \to +\infty$.

(e) If all the real roots of p have multiplicity one, then p has n real roots.

(f) If all the real roots of p have multiplicity two, then n is even.

(g) If p has exactly eight real roots, then $n = 8$.

(h) If the behavior of p is the same regardless of whether $x \to +\infty$ or $x \to -\infty$, then a_n is positive.

(i) If n is odd, then p has at least one real root.

(j) If n is even, then p has at least two real roots.

9. Explain (in words) why a polynomial function with even degree must change direction an odd number of times.

10. Match the functions to the graphs. Assume that the graphs do not change direction outside of the viewing window.

(a) $y = x^{10} - 1000x^5$

(b) $y = x^7 - 50x^6 - 40x^5 - 30x^4 - 20x^3 - 10x^2 - x$

(c) $y = -x^4 + 10x^3 - 10x^2$

(d) $y = -5x^3 + 50x^2 - x$

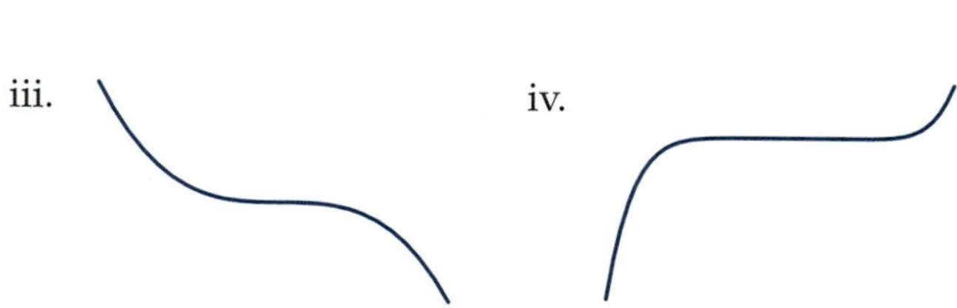

11. (a) Here is a graph of the function f.

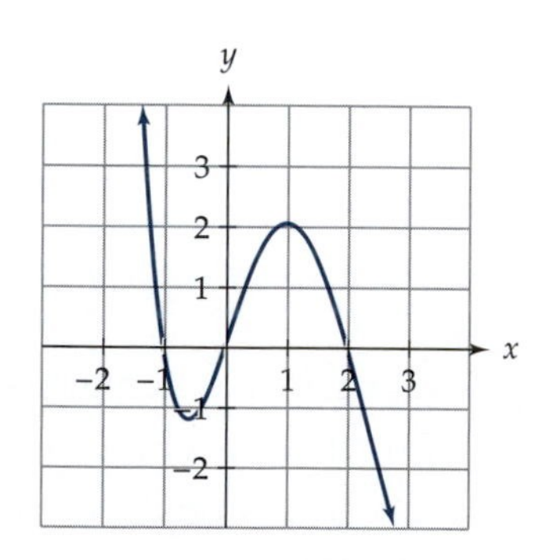

i. Give at least one reason why this graph cannot be of an exponential or logarithmic function.

ii. Assume this graph is of a polynomial function. By looking at the graph, what do you know about the degree of the function? What do you know about the roots of the function?

(b) By using the accompanying graph of the same function, f, in a different viewing window, what can you now conclude about the degree of the function?

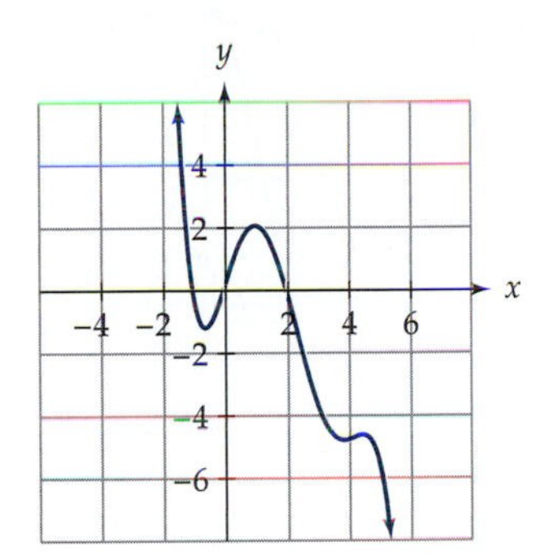

12. (a) Is knowing the number of real roots of a polynomial function sufficient for finding its degree? Explain.

    (b) Is knowing both the degree of a polynomial function and the locations of its real roots sufficient for finding the equation of the function? Explain.

13. Let $p(x) = a_n x^n + a_{n-1}x^{n-1} + \cdots + a_1 x + a_0$. What condition on n guarantees that p will have at least one real root? Justify your answer.

14. Find the exact values of the real roots for the following polynomial functions.

    (a) $y = \frac{1}{4}x(x+2)(3x+1)(5x-6)^2(x-8)$

    (b) $y = x^2 + x + 1$

    (c) $y = (x^2 - 9)(x+2)$

    (d) $y = x^3 - 8$

    (e) $y = x^3 - 2x^2 + x$

15. Give a possible equation for a polynomial function with the following properties. There is more than one correct answer.

    (a) The degree of p is 3, it has a root at $x = -\frac{1}{2}$ of multiplicity two and a root at $x = 1$, and $p(x) \to -\infty$ as $x \to \infty$.

    (b) The degree of q is 4, it has a root at $x = -2$ and a root at $x = 2$, both roots have even multiplicities, and $q(0) = 1$.

    (c) The degree of r is 2, it has no real roots, and $r(x) \to +\infty$ as $x \to +\infty$.

    (d) The degree of s is 5; it has roots at $x = -10$, $x = -\frac{7}{2}$, $x = 0$, $x = \sqrt{3}$, and $x = 5.46$; and $s(-1) > 0$.

16. Give a possible equation for the polynomial functions shown in the accompanying graphs. You may assume that the absolute value of the leading coefficient is 1.

(a)

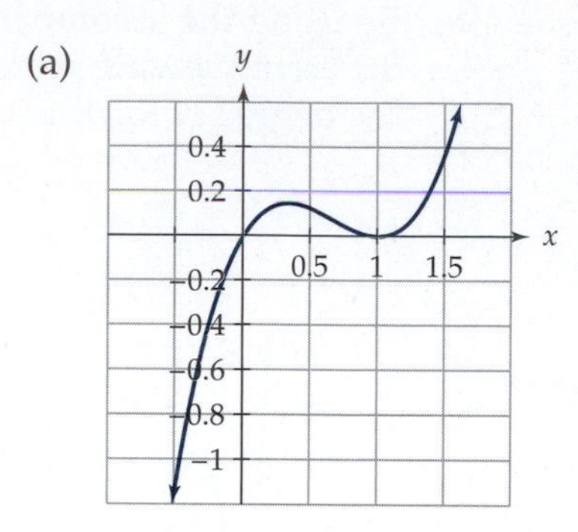

(b)

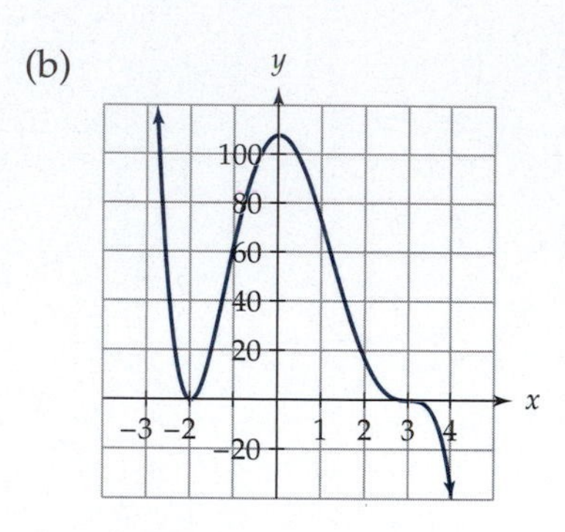

(c)

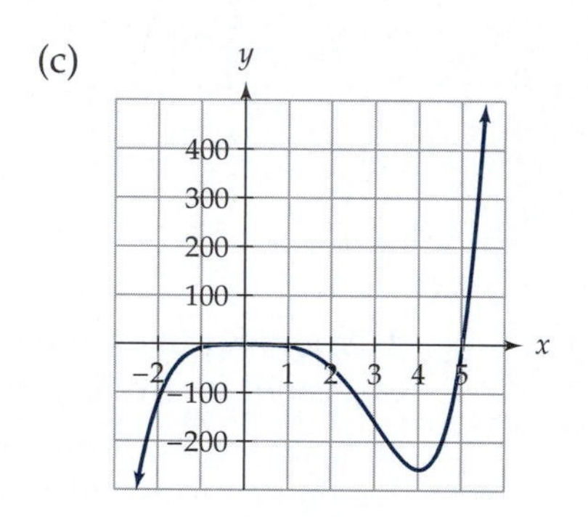

(d)

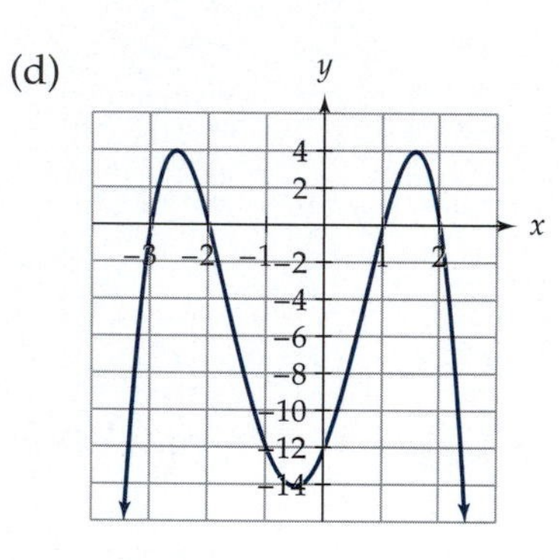

17. Explain why the function $p(x) = x^9 - x^8 + x^5 + x^3 - x^2 + 2$ must have a root between $x = -1$ and $x = 0$.
18. Use the bisection method to approximate to two decimal places the real root of each of the following polynomial functions.
    (a) $y = x^3 - x + 5$
    (b) $y = -x^3 + 6x^2 - 12x + 10$
19. Explain why the bisection method will not work for approximating the real root of $y = x^4 - x^3 + 1.25x^2 - x + 0.25$.
20. Approximate the roots, local maxima, and local minima of each of the following functions.
    (a) $p(x) = 2x^3 - 6x^2 + 2$
    (b) $q(x) = 3x^4 - 3x^2 - 1$
21. Explain why a periodic function can never be represented by a polynomial function.
22. Manufacturers of boxes for puzzles and many games construct the boxes by taking a flat piece of cardboard, cutting a square out of each corner, and folding the sides. See the accompanying figure. We measured an old Spirograph box and discovered that the original size of the cardboard was 15 in. by 12 in. Let a represent the length of the side of the square cut out of each corner.

(a) Give the formula for the function, V, representing the volume of the box in terms of a.

(b) What is the domain of V?

(c) Graph V for the appropriate domain.

(d) Using your graph, estimate the value of a that gives the maximum volume for the box. What are the dimensions and the volume of this box?

(e) The actual box was constructed using $a = 1$ in. What are the dimensions and the volume of the actual box?

(f) Why do you suppose the manufacturers did not use a box with the maximum volume?

23. A graph of the average speeds of the cars (in miles per hour) that won the Indianapolis 500 versus the year of the race (in years since 1900) is shown.[14] Also shown is the fourth-degree polynomial function that best fits the data.

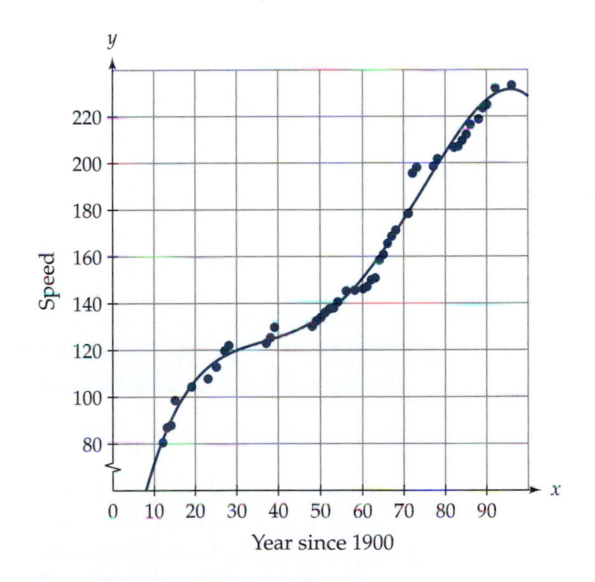

(a) What characteristics of a fourth-degree polynomial function make this function a good choice for a function that fits the data?

(b) Would you use this function to predict the speed of the car that wins the Indianapolis 500 in 2000? Why or why not?

[14] *USA Today*, 23 May 1997, p. 5E.

24. You decide to go into the business of manufacturing doodads, a new product that can be used as packaging. Assume you do not actually make the doodads until an order is placed so that you never have a surplus.
    (a) After careful analysis by your mathematical consultant, it is determined that the cost of manufacturing doodads is given by the function

    $$C(x) = 0.001x^3 - 0.030x^2 + 0.3x,$$

    where x is number of pounds of doodads and $C(x)$ is cost in dollars.
        i. How much does it cost to manufacture 1 lb of doodads?
        ii. How much does it cost to manufacture 10 lb of doodads?
    (b) Suppose $R(x) = 5x$ is the revenue function, where x is the number of pounds of doodads and $R(x)$ is revenue in dollars.
        i. What is the price of 1 lb of doodads?
        ii. What is the price of n pounds of doodads?
        iii. Fill in the blank: The function R tells you that you are selling doodads for ____ per pound.
        iv. Graph C and R on the same set of axes. Looking at this graph, are you making or losing money if you sell 50 lb of doodads?
    (c) Let $P(x) = R(x) - C(x)$.
        i. Graph P.
        ii. For what value of x do you start losing money? Justify your answer.
        iii. How could you use the graphs of C and R to determine when you start losing money?
        iv. Ideally, how many pounds of doodads should you sell? Briefly justify your answer.

INVESTIGATIONS

INVESTIGATION 1: SYMMETRY OF CUBICS

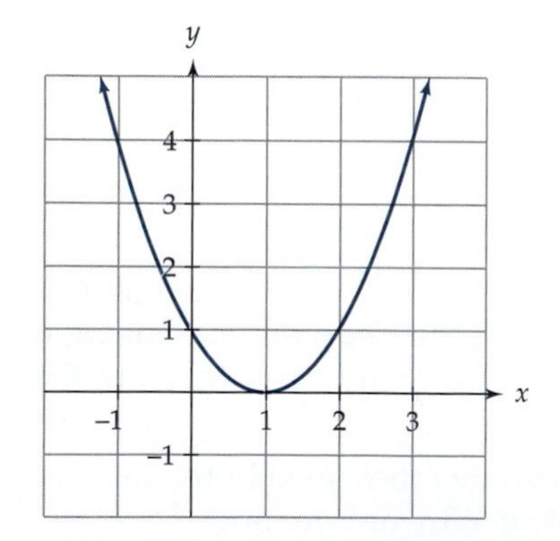

Figure 23

In section 4.1, we discovered that every second-degree (quadratic) polynomial function is symmetric about a vertical line through its vertex. (See Figure 23.) What, if anything, can we say about the symmetry of higher-degree polynomial functions? Let's start by considering third-degree (cubic) polynomial functions.

1. Graphs of four different third-degree polynomial functions are shown in Figure 24. In your own words, make a hypothesis about the symmetry of *any* cubic polynomial function.
2. Recall from section 1.2 that there are two kinds of symmetry: symmetry about a vertical line and symmetry about a point. Symmetry about a vertical line is when the part of the graph to the left of the vertical line

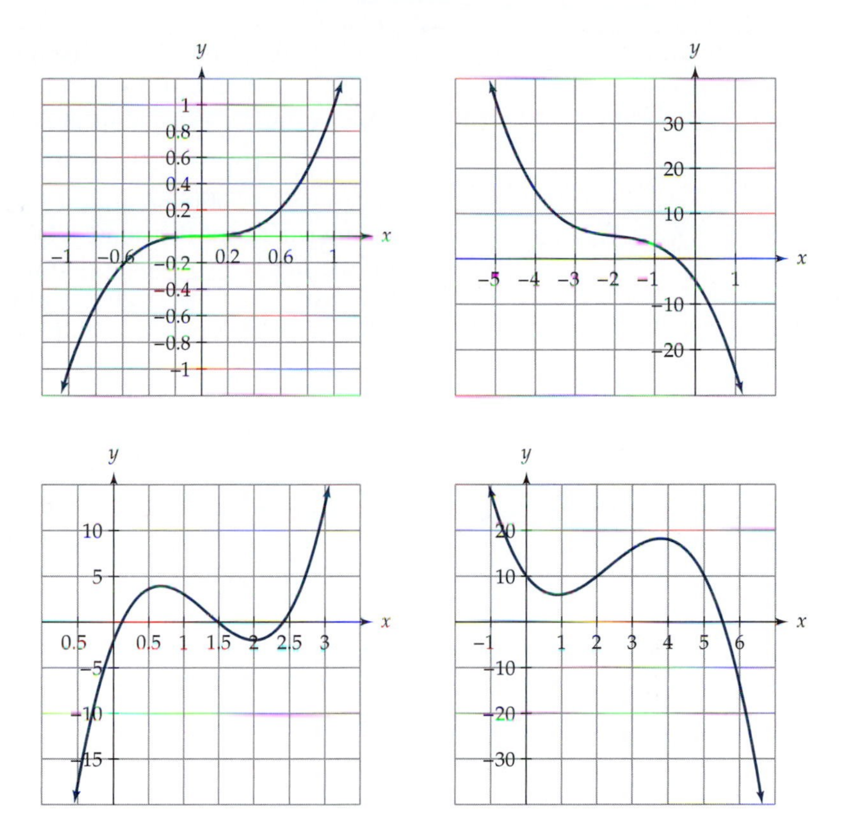

Figure 24

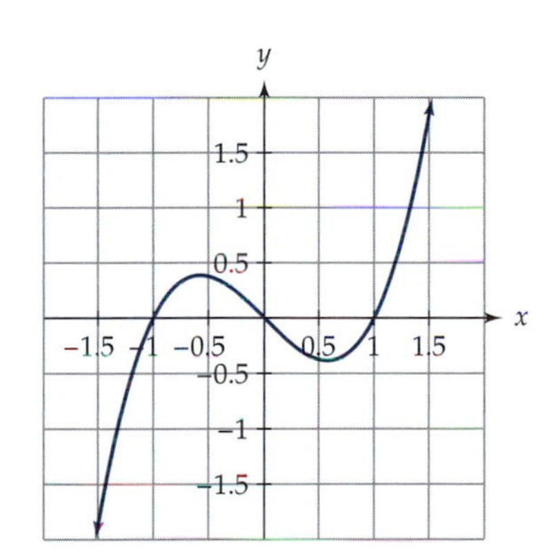

Figure 25

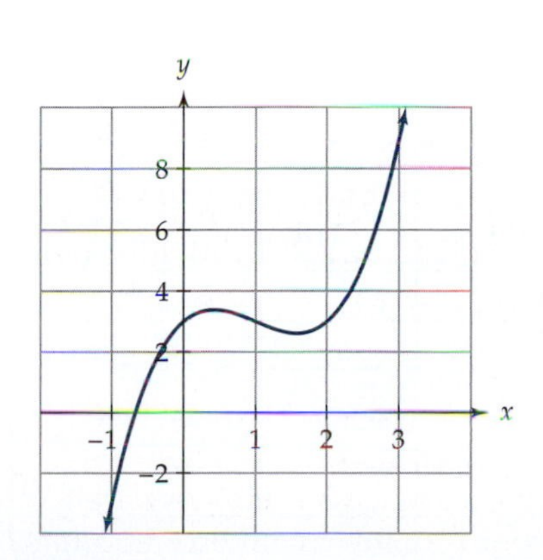

Figure 26

is a reflection of the part to the right of the vertical line (such as Figure 23). If a function, f, is symmetric about the y-axis, then $f(x) = f(-x)$. Symmetry about a point is when a graph rotated 180° (with the point being the center of rotation) looks the same after it was rotated as before. If a function, f, is symmetric about the origin, then $f(x) = -f(-x)$. It appears that a cubic polynomial function is symmetric about a point. The question is how to locate that point.

(a) Place a dot on each of the four graphs in Figure 24 at the point of symmetry. In general, describe the location of the point of symmetry of a third-degree polynomial function.

(b) Figure 25 is a graph of $s(x) = x^3 - x$. It appears that the point of symmetry is at the origin. As mentioned earlier (and in section 1.2), if a function, f, is symmetric about the origin, then $f(x) = -f(-x)$. Prove that s is symmetric about the origin by showing that $s(x) = -s(-x)$.

(c) Figure 26 is a graph of $q(x) = x^3 - 3x^2 + 2x + 3$. The point of symmetry for this function is $(1, 3)$. Verify this point of symmetry by doing the following.

  i. Shift q so that the point of symmetry is at the origin. Explain why $r(x) = q(x+1) - 3$ is the appropriate transformation of q.

  ii. Using $q(x) = x^3 - 3x^2 + 2x + 3$, rewrite $r(x) = q(x+1) - 3$ so that it is in the form $r(x) = ax^3 + bx^2 + cx + d$.

iii. Show that r is symmetric about the origin by showing that $r(x) = -r(-x)$.

iv. Explain why showing that r is symmetric about the origin proves that q is symmetric about the point $(1, 3)$.

3. In general, a cubic polynomial function, $p(x) = ax^3 + bx^2 + cx + d$, is symmetric about the point where it changes concavity (also known as an *inflection point*). Using calculus, one can show that the inflection point of p is at $x = -b/3a$. Algebraically, we can prove that $(-b/3a, p(-b/3a))$ is the point of symmetry by doing the following.
   (a) As in question 2, part (c)(i), we begin by shifting the function so that the point of symmetry is at the origin. Explain why $g(x) = p(x - b/3a) - p(-b/3a)$ is the appropriate shift of p.
   (b) Using $p(x) = ax^3 + bx^2 + cx + d$, rewrite $g(x) = p(x - b/3a) - p(-b/3a)$ so that it is in the form $g(x) = ax^3 + bx^2 + cx + d$.
   (c) Show that g is symmetric about the origin by showing that $g(x) = -g(-x)$.
4. Unfortunately, there is no guaranteed symmetry for polynomial functions of degree four or higher. Give an example of a fourth-degree polynomial function that is not symmetric about either a point or a line.

INVESTIGATION 2: SAVING FOR COLLEGE

Vinnie and Viola, upon the birth of their first child, Vera, received a card from Aunt Veronica (who lives in Vancouver) saying that she would be sending \$600 every year for Vera's college fund. Vinnie and Viola decide to invest Aunt Veronica's money in a certificate of deposit (CD) that earns 6% interest. They want to know how much will be in the fund by Vera's 18th birthday because they have their heart set on Vera's attending the University of Victoria Valley. They anticipate an annual tuition of \$10,000 per year (or total tuition of \$40,000 for 4 years) when Vera is 18 years old. If there will not be a sufficient amount, then they intend to add their own contribution to Aunt Veronica's money. For simplicity, we assume interest is only compounded at the end of the year.

1. Fill in the table, giving the amount of money in the CD for the first 5 years.

Year	Amount of Money Prior to the Addition of Current Year's Interest	Interest	Total Money in CD at End of the Year
0	\$600	\$36	\$636
1	\$1236	\$74.16	\$1310.16
2			
3			
4			
5			

2. Vinnie and Viola quickly become tired of completing a table to find out how much money will be available. They decide to use their algebra skills and attempt to find a formula, letting the interest rate equal r so that they can explore what happens for various investment strategies.
   (a) Explain why the amount of money available at the end of the first year can be written as $600(1+r)$.
   (b) Explain why the amount of money available at the end of the second year can be written as $[600+600(1+r)]\cdot(1+r) = 600[(1+r)+(1+r)^2]$.
   (c) Write an expression giving the amount of money available at the end of the third year.
   (d) Continuing this pattern, what is the amount of money available at the end of 18 years? Label this M.
   (e) Explain why M is a polynomial function.
3. The polynomial function M from question 2, part (d), is a geometric sum. Thus, it can be rewritten as a rational function.[15] A *rational function* is the quotient of two polynomial functions.
   (a) Recall that a sum of the form $S = v + v^2 + v^3 + \cdots + v^n$ is called a geometric sum. Show that $vS - S = v^{n+1} - v$. From this equation, solve for S to obtain
   $$S = \frac{v^{n+1} - v}{v - 1}.$$
   (b) Use your answer to part (a) to rewrite M as a rational function.
4. How much money will Vera have for her college fund when she reaches 18? Will it be enough?
5. Graph M. Estimate a value of r so that $M(r) \geq 40{,}000$.
6. Suppose Vinnie and Viola cannot find a CD with the needed interest rate and decide to contribute some of their own money to Vera's college fund. Assume the highest interest rate they can find is 6%.
   (a) They decide to contribute $100 per year (so that a total of $700 is invested each year). In this case, how much money is available when Vera turns 18?
   (b) Contributing $100 per year to Vera's college fund was not sufficient. Determine how much Vinnie and Viola need to add to Aunt Violet's contribution so that Vera has $40,000 in her college fund when she turns 18.

4.3 POWER FUNCTIONS WITH NEGATIVE EXPONENTS

In section 2.5, we considered the behavior of power functions with positive exponents. In the previous two sections, we considered combinations of power functions with nonnegative integer exponents. In this section, we

[15] *Rational functions are covered in detail in section 4.4.*

look at the behavior of power functions with negative exponents. We discover that these have vertical and horizontal asymptotes and learn how to describe asymptotic behavior. We also see how to find the inverse of a power function with a negative exponent.

VERTICAL AND HORIZONTAL ASYMPTOTES

A walk frequently taken by a father and his young daughter is about 3000 ft in length. Walking alone, the father can complete this route in about 10 min. Assuming his rate is constant, it is 3000 ft/10 min = 300 ft/min. If his daughter walks along with him and they walk at her pace, it takes about 30 min to complete the route. Assuming her rate is constant, it is 3000 ft/30 min = 100 ft/min. Notice that it takes three times as long to complete the route when walking at the daughter's pace and that the rate is one-third as much. Suppose, as a compromise, the father lets his daughter keep the pace for half the route and then sets the pace for the other half by carrying her on his shoulders. As a result, they complete the route in 20 min (halfway between the other two times of 10 min and 30 min). What is the average rate in this case? You might think that the average rate is 200 ft/min (halfway between the other two rates), but that is not the case. The average rate is actually 3000 ft/20 min = 150 ft/min. Why? First, notice that this rate is closer to the daughter's rate of 100 ft/min than the father's rate of 300 ft/min. This rate makes sense because although they were each traveling at their respective rates for half the *distance*, they were not each traveling at their respective rates for half the *time*. Because the daughter walks slower than the father, the two spend more time walking at her rate than at her father's.

This conclusion also makes sense because the function that describes this situation is not linear. Because speed (or rate) is distance divided by time, $r = d/t$ and we fixed the distance of our route at 3000 ft, the function where time is the input and rate is the output is $r = 3000/t$. Because this function is not linear, inputs of 10, 20, and 30 do not give three outputs that are the same distance apart. The function $r = 3000/t$ can be thought of as a power function with a negative exponent if we rewrite it as $r = 3000t^{-1}$. A graph of this function is shown in Figure 1.

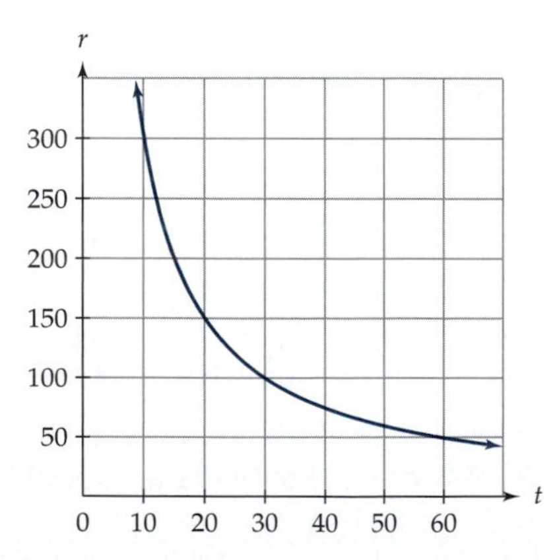

Figure 1 A graph of $r = 3000t^{-1}$.

In section 2.5, we looked at power functions, which are defined as having the form $f(x) = bx^a$, where a and b are constants. In that section, we only considered functions where the exponent, a, was a positive number. Now, however, we are interested in exploring what happens when a is a negative number. The most basic of these functions is $f(x) = x^{-1} = 1/x$. A graph of this function is shown in Figure 2(a). Notice that the graph of $f(x) = 1/x$ has the same basic shape in the first quadrant as the graph shown in Figure 1. This graph, however, looks quite different than those of power functions with positive exponents. See Figure 2(b). The most striking difference is that these graphs have asymptotes. The function shown in Figure 2(a) has two asymptotes. The horizontal asymptote is the x-axis, and the vertical asymptote is the y-axis.

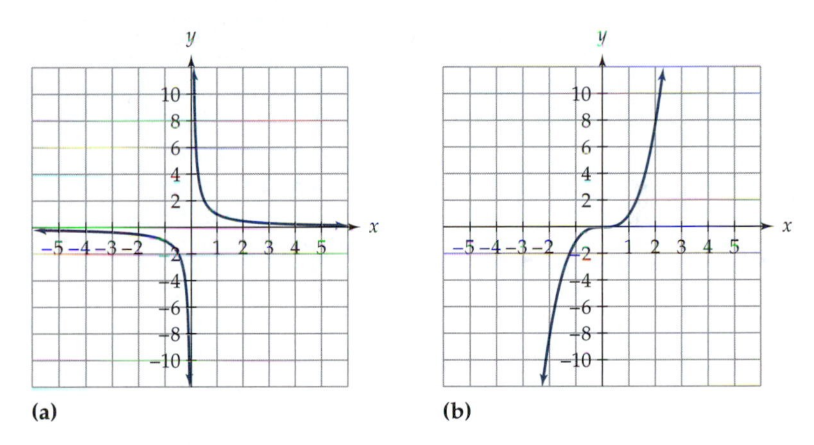

Figure 2 A comparison of power functions with negative and positive exponents. (a) The graph of $y = x^{-1}$; (b) the graph of $y = x^3$.

In general, a **horizontal asymptote** of a function, f, is the line $y = c$ where the outputs of f approach c as x approaches either negative or positive infinity. For $f(x) = 1/x$ shown in Figure 2, the outputs of f approach 0 both as x approaches negative infinity and as x approaches positive infinity. Therefore, f has a horizontal asymptote of $y = 0$ (which is the x-axis). A **vertical asymptote** of a function, f, is the line $x = c$ where the outputs of f approach either positive or negative infinity as x approaches c. Rational functions that have a vertical asymptote of $x = c$ are undefined when $x = c$. In the function $f(x) = 1/x$, shown in Figure 2(a), the outputs of f approach positive infinity as x approaches 0 from the right and negative infinity as x approaches 0 from the left. Therefore, f has a vertical asymptote of $x = 0$ (which is the y-axis).

READING QUESTIONS

1. Suppose you were going on a 150-mi trip in your car. Determine the function where rate (in miles per hour) is the input and time (in hours) is the output.

2. What is a horizontal asymptote?
3. Does $f(x) = 1/x$ have any x-intercepts? Why or why not?

NEGATIVE INTEGER EXPONENTS

The two basic shapes for graphs of power functions with negative integer exponents can be illustrated by looking at the graph of $f(x) = x^{-1}$ in Figure 3(a) and $g(x) = x^{-2}$ in Figure 3(b). Notice that both graphs have the y-axis as a vertical asymptote and the x-axis as the horizontal asymptote. Let's explore this behavior more closely by first focusing on the vertical asymptote. Table 1 shows outputs for $f(x) = x^{-1} = 1/x$ and $g(x) = x^{-2} = 1/x^2$ as x gets closer to zero from both the left and the right. Notice in both functions, f and g, that as the input gets closer to zero the absolute value of the output gets large because the reciprocal of numbers close to zero (such as 0.001) are large numbers. For $g(x) = 1/x^2$, the outputs are large positive numbers because raising either a positive or a negative number to an even exponent always gives a positive answer. For $f(x) = 1/x$, the outputs are large positive numbers when the inputs are positive and close to zero, but the outputs are negative numbers with large absolute value when the inputs are negative and close to zero. Also notice that both functions are undefined for $x = 0$ because division by zero is undefined.

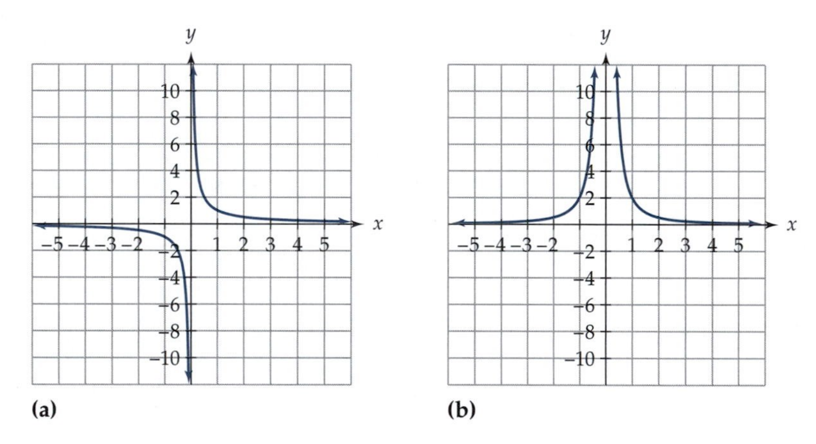

Figure 3 Graphs of (a) $f(x) = x^{-1}$ and (b) $g(x) = x^{-2}$, the two basic shapes of power functions with negative integer exponents.

TABLE 1 A Comparison $f(x) = 1/x$ and $g(x) = 1/x^2$ Close to the Point $x = 0$

x	1	0.1	0.01	0.001	0	−0.001	−0.01	−0.1	−1
$f(x) = \frac{1}{x}$	1	10	100	1,000	Undefined	−1,000	−100	−10	−1
$g(x) = \frac{1}{x^2}$	1	100	10,000	1,000,000	Undefined	1,000,000	10,000	100	1

The vertical asymptote in $g(x) = 1/x^2$ can be described by saying $g(x)$ *approaches positive infinity as x approaches zero.* This description can be written as $g(x) \to +\infty$ as $x \to 0$. The behavior of $f(x) = 1/x$ is more complicated because $f(x)$ is approaching positive infinity for small positive numbers and negative infinity for negative numbers with small absolute value. See Figure 3(a). We describe this behavior as *$f(x)$ approaches positive infinity as x approaches zero from the right* and *$f(x)$ approaches negative infinity as x approaches zero from the left.* This description can be written $f(x) \to +\infty$ as $x \to 0^+$ and $f(x) \to -\infty$ as $x \to 0^-$. (The notation $x \to 0^+$ indicates that x is approaching zero from the right, and the notation $x \to 0^-$ indicates x is approaching zero from the left.)

The vertical asymptotes occurred where our functions were undefined, but the horizontal asymptotes demonstrate the long-range behavior of the functions. In the case of our two functions, $f(x) = 1/x$ and $g(x) = 1/x^2$, the reciprocal of large numbers results in outputs close to 0. Therefore, the behavior of both $f(x) = 1/x$ and $g(x) = 1/x^2$ can be described the same way. That is, $f(x) \to 0$ as $x \to \pm\infty$ and $g(x) \to 0$ as $x \to \pm\infty$.

Power functions with negative integer exponents other than -1 and -2 still behave in a manner similar to $f(x) = x^{-1}$ and $g(x) = x^{-2}$. If the exponent is odd, then the function behaves in a manner similar to $f(x) = x^{-1}$. If the exponent is even, then the function behaves in a manner similar to $g(x) = x^{-2}$. Even with power functions of the form $f(x) = bx^a$, where b is some number other than 1, the behavior is similar because these functions are just vertical stretches of $f(x) = x^a$. We had the same result when we looked at power functions with positive integer exponents. Those with even exponents behaved like $y = x^2$, and those with odd exponents behaved like $y = x^3$.

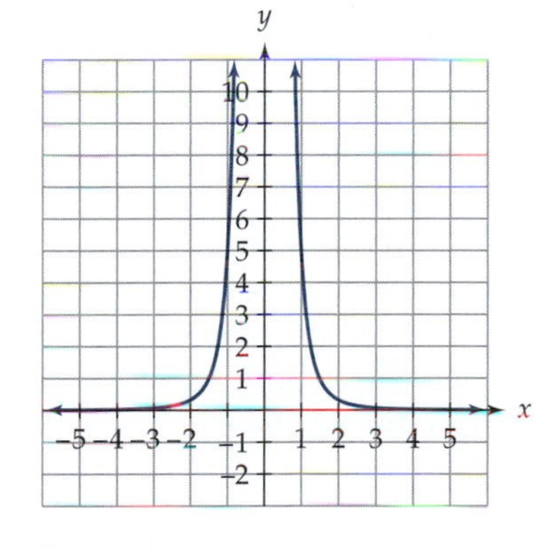

Figure 4

Example 1 Determine the vertical and horizontal asymptotes of $f(x) = 5x^{-4}$. Describe the function's behavior close to the vertical asymptote and describe its long-range behavior.

Solution Because the exponent is even, the outputs are all positive. For large positive or negative values of x, $f(x)$ gets close to zero, so $f(x) \to 0$ as $x \to \pm\infty$. As x gets close to zero, $f(x)$ gets large, so $f(x) \to +\infty$ as $x \to 0$. Thus, the function has a horizontal asymptote of $y = 0$ (the x-axis) and a vertical asymptote of $x = 0$ (the y-axis), which can be seen in Figure 4.

READING QUESTIONS

4. Describe in words what the following statement means.

$$f(x) \to +\infty \text{ as } x \to 0^+.$$

5. Is the following statement describing a vertical or horizontal asymptote?

$$f(x) \to 0 \text{ as } x \to \pm\infty.$$

6. Let $g(x) = 2x^{-3}$.

(a) What happens to $g(x)$ as $x \to +\infty$?

(b) What happens to $g(x)$ as $x \to -\infty$?

(c) What happens to $g(x)$ as $x \to 0^+$?

(d) What happens to $g(x)$ as $x \to 0^-$?

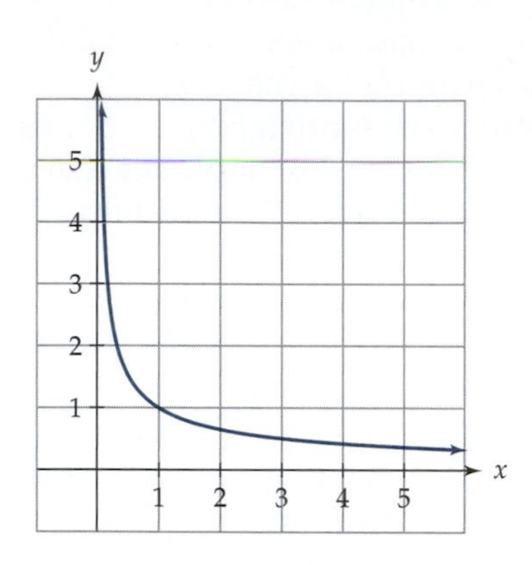

Figure 5 A graph of $f(x) = x^{-3/5}$ for positive inputs.

NEGATIVE RATIONAL EXPONENTS

The behavior of power functions with negative rational exponents (that are not integers) is very similar to the behavior of those with negative integer exponents. Some properties of power functions with negative exponents occur regardless of whether the exponent is an integer. For example, suppose $f(x) = x^{-a}$, where a is some positive rational number (not necessarily an integer). This function can be written $f(x) = 1/x^a$. As x gets larger, x^a also gets larger and therefore $1/x^a$ gets smaller. More precisely, as $x \to +\infty$, $x^a \to +\infty$ and therefore $1/x^a \to 0$, which is true regardless of whether or not a is an integer. Thus, the x-axis is a horizontal asymptote for all power functions with negative exponents. It is also true that the y-axis is a vertical asymptote for all power functions with negative exponents because, for $f(x) = 1/x^a$, as $x \to 0^+$, $x^a \to 0^+$ and therefore $1/x^a \to +\infty$.

Because all power functions with negative exponents have the same type of behavior for positive inputs, the graphs of these functions all have the same basic shape. Figure 5 is the graph of $f(x) = x^{-3/5}$ for positive inputs. Notice that this shape is similar to that of $f(x) = 1/x$ and $g(x) = 1/x^2$ in the first quadrant.

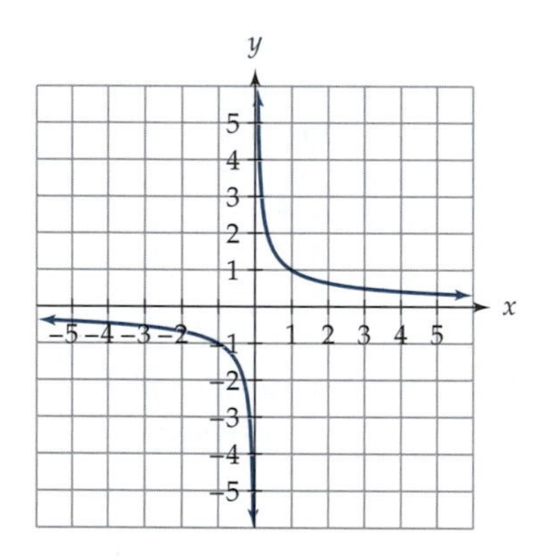

Figure 6 A graph of $f(x) = x^{-3/5}$.

For negative inputs, the behavior of power functions with integer exponents depends on whether or not the exponent is even or odd. We can similarly categorize power functions with negative rational exponents that are not integers. Because the exponent is not an integer, it will neither be even nor odd. We can, however, look at whether the numerator or denominator of the fraction in simplified form is even or odd, which will affect how the function behaves for negative inputs. Three different combinations are possible: the numerator can be even and the denominator can be odd, the numerator can be odd and the denominator can be even, or both the numerator and denominator can be odd.[16]

Suppose $f(x) = x^{-3/5}$. This function can be written as a composition of functions, $f(x) = \left(1/\sqrt[5]{x}\right)^3$. Because $\sqrt[5]{x}$ is an odd root, it is defined for negative inputs. Because the exponent, 3, is odd, the function resembles a power function with a negative odd integer exponent. A graph of f is shown in Figure 6.

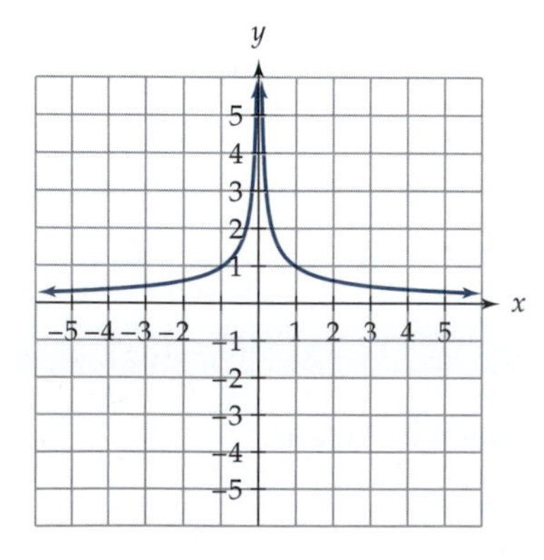

Figure 7 A graph of $g(x) = x^{-2/3}$.

Suppose $g(x) = x^{-2/3}$. This function can be written as $g(x) = \left(1/\sqrt[3]{x}\right)^2$. Because $\sqrt[3]{x}$ is an odd root, it is defined for negative inputs. Because the exponent, 2, is even, the function resembles a power function with a negative even integer exponent. A graph of g is shown in Figure 7.

[16] *Because we assume the exponent is a simplified fraction, both the numerator and denominator cannot be even.*

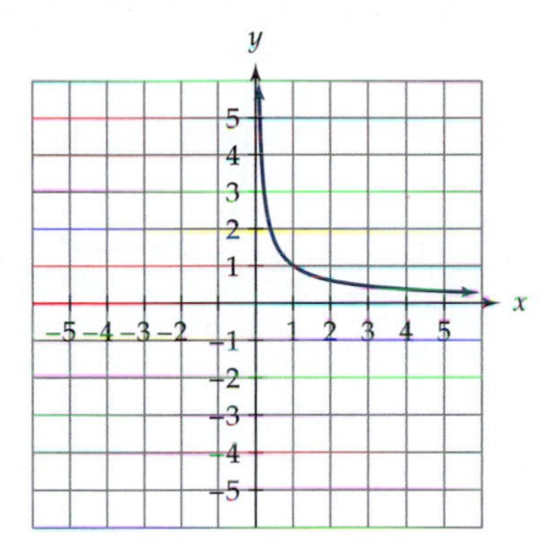

Figure 8 A graph of $h(x) = x^{-3/4}$.

Suppose $h(x) = x^{-3/4}$. This function can be written as $h(x) = \left(1/\sqrt[4]{x}\right)^3$. Because $\sqrt[4]{x}$ is an even root, it is not be defined for negative inputs. Therefore, this function is only defined for positive inputs. A graph of h is shown in Figure 8.

PROPERTIES

The basic shape of the graph of a power function with a negative integer exponent is dependent on whether the exponent is even or odd.

- If it is odd, then the graph exists in the first and third quadrants as shown in Figure 9(a).
- If it is even, then the graph exists in the first and second quadrants as shown in Figure 9(b).

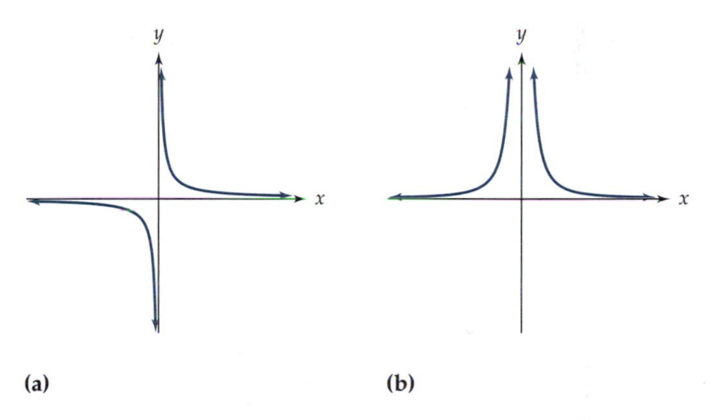

Figure 9 Graphs of power function with negative integer exponents. The graph in (a) has an odd exponent, and the graph in (b) has an even exponent.

PROPERTIES

The basic shapes of graphs of power functions of the form $f(x) = x^{-n/d}$, where n/d is a fraction in simplified form, depends on whether n and d are even or odd.

- If both n and d are odd, then the graph exists in the first and third quadrants as shown in Figure 10(a).
- If n is even and d is odd, then the graph exists in the first and second quadrants as shown in Figure 10(b).
- If n is odd and d is even, then the graph exists only in the first quadrant as shown in Figure 10(c).

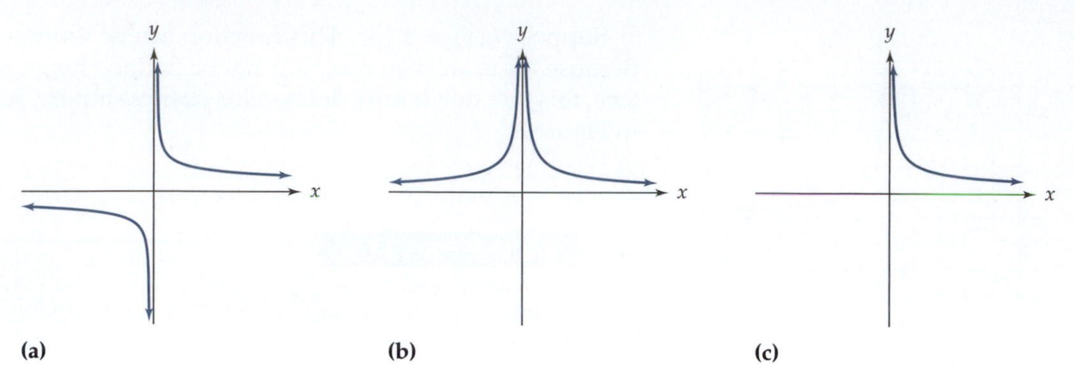

Figure 10 Graphs of power functions of the form $f(x) = x^{-n/d}$. In (a), both n and d are odd; in (b), n is even and d is odd; and in (c), n is odd and d is even.

READING QUESTIONS

7. Explain why the function $f(x) = x^{-5/6}$ is not defined for negative inputs.

8. Without using a calculator, sketch a graph of $g(x) = x^{-6/5}$.

INVERSES

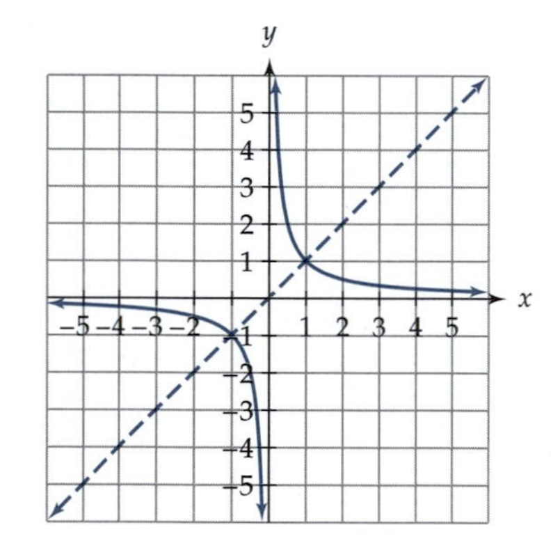

Figure 11 The graph of $f(x) = x^{-1}$ is symmetric about the line $y = x$ and is therefore its own inverse.

Inverses of power functions with negative exponents are interesting partly because the simplest of these functions, $f(x) = x^{-1}$, is its own inverse. This property can easily be verified by seeing that $f(f(x)) = f(x^{-1}) = (x^{-1})^{-1} = x$. It can also be seen by looking at the graph of $f(x) = x^{-1}$. See Figure 11. Because this graph is symmetric about the line $y = x$, it is its own inverse.

Remember that a function must be one-to-one to have an inverse.[17]. An easy method for determining if it is one-to-one is the horizontal line test. Using this test, you can easily see that power functions with negative even integer exponents (such as $f(x) = x^{-2}$) and those with rational exponents where the numerator is even (such as $f(x) = x^{2/3}$) do not pass the horizontal line test and therefore do not have inverses. As shown in section 3.5, if a power function, $f(x) = x^a$, has an inverse, then its inverse is $f^{-1}(x) = x^{1/a}$. If the power function has a negative exponent, then its inverse also has a negative exponent.

Example 2 Find the inverse of $f(x) = 8x^{-3}$.

Solution We first rewrite the function as $y = 8x^{-3}$ and solve for x. Multiply both sides by $\frac{1}{8}$ and then raise both sides to the $-\frac{1}{3}$ power:

$$\frac{1}{8}y = x^{-3}$$

$$\left(\tfrac{1}{8}y\right)^{-1/3} = (x^{-3})^{-1/3}$$

$$2y^{-1/3} = x.$$

[17] *A function is one-to-one if each output corresponds to exactly one input.*

By rewriting this equation in inverse function notation, we have $f^{-1}(x) = 2x^{-1/3}$.

Example 3 As you move farther above the surface of Earth, the effect of Earth's gravitational pull decreases. Thus, you weigh less the farther away you are from Earth's surface. The function that describes this relationship is $w = (1.58 \times 10^7)w_0 d^{-2}$, where w_0 is your weight on Earth's surface (in pounds) and d is the distance from the center of Earth (in miles). If you weigh 150 lb on the surface of Earth, then you can estimate your weight by

$$w = (2.37 \times 10^9)d^{-2}.$$

Find the inverse of this function.

Solution Writing this formula as $w = (2.37 \times 10^9)d^{-2}$, we need to solve for d. To do so, we divide both sides by 2.37×10^9 and then raise both sides to the $-\frac{1}{2}$ power:

$$d^{-2} = \frac{w}{2.37 \times 10^9}$$

$$(d^{-2})^{-1/2} = \left(\frac{w}{2.37 \times 10^9}\right)^{-1/2}$$

$$d \approx \frac{w^{-1/2}}{2.05 \times 10^{-5}}$$

$$d \approx 48{,}700w^{-1/2}.$$

Notice that the original function had an input of d, the distance, and an output of w, the weight. They were reversed in the inverse function.

Power functions with even exponents do not usually have inverses because they are not one-to-one. Functions that are not one-to-one, however, can have their domains restricted so that they are one-to-one. In the previous example, the function and its inverse are only defined for positive inputs because distance and weight have to be positive numbers. Therefore, we were able to find the inverse of a power function with an even exponent.

READING QUESTIONS

9. Why does $f(x) = 1/x^2$ not have an inverse?

10. Find the inverse of $f(x) = 2x^{-1/3}$.

SUMMARY

In this section, we explored what happens in power functions when the exponent is negative. These functions have horizontal and vertical asymptotes. A **horizontal asymptote** of a function, f, is the line $y = c$ where the outputs of f approach c as x approaches either positive or negative infinity. A **vertical asymptote** of a function, f, is the line $x = c$ where the outputs of f approach either positive or negative infinity as x approaches c.

For positive inputs, all power functions with negative exponents have the same type of behavior and the graphs have the same basic shape. For negative inputs, the behavior depends on whether the exponent is even or odd. The basic shapes of power functions with negative exponents are shown in Figures 9 and 10 in this section.

Power functions with negative even exponents or those with rational exponents with even numerators do not pass the horizontal line test and consequently do not have inverses. Functions with negative odd exponents, though, do have inverses, and the results can be very interesting. For example, the function $f(x) = x^{-1}$ is its own inverse.

EXERCISES

1. Describe the meaning of the following statements.
   (a) $f(x) \to +\infty$ as $x \to 0$.
   (b) $g(x) \to 0$ as $x \to 0^+$.
   (c) $h(x) \to 0$ as $x \to \pm\infty$.
2. Determine whether the following statements describe horizontal asymptotes, vertical asymptotes, or neither.
   (a) $f(x) \to +\infty$ as $x \to 0$.
   (b) $g(x) \to +\infty$ as $x \to +\infty$.
   (c) $h(x) \to 0$ as $x \to 0^+$.
   (d) $j(x) \to 0$ as $x \to \pm\infty$.
3. Describe the behavior in the first quadrant for power functions with negative exponents and explain why this behavior is the same for *all* power functions with negative exponents (and positive coefficients).
4. The following functions represent power functions with negative rational exponents where n/d is in simplified form. Match these functions with the following graphs.
   (a) $f(x) = x^{-n/d}$, where n is even and d is odd.
   (b) $g(x) = x^{-n/d}$, where n is odd and d is odd.
   (c) $h(x) = x^{-n/d}$, where n is odd and d is even.

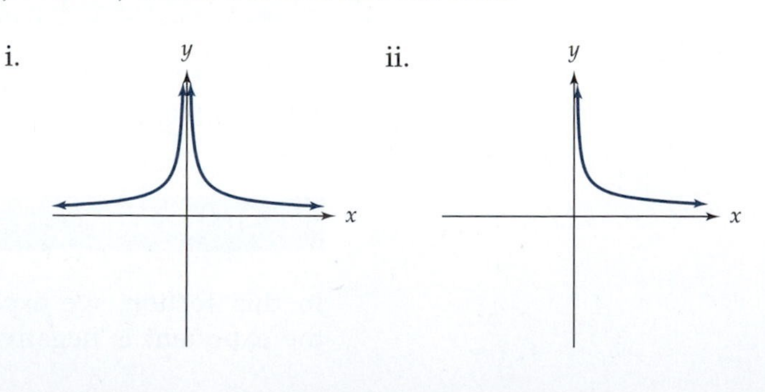

iii.

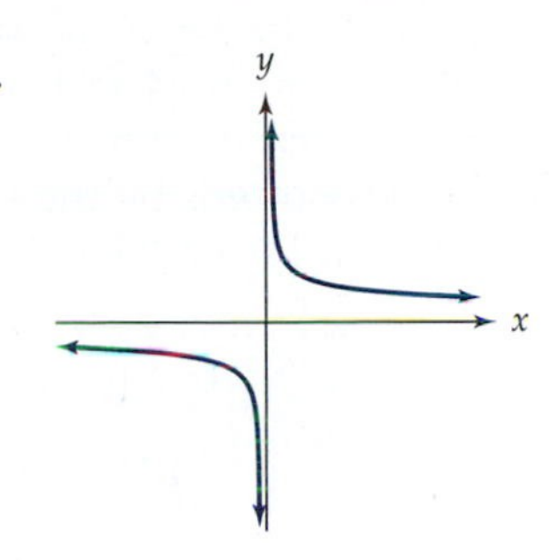

5. Describe the long-range behavior and the behavior close to the y-axis for the following functions.
   (a) $f(x) = 3x^{-2}$
   (b) $g(x) = 2x^{-7}$
   (c) $h(x) = 7x^{-3/4}$
   (d) $j(x) = 4x^{-4/3}$
6. Explain why $f(x) = 3x^{-9/4}$ is not defined for negative inputs.
7. Explain why $g(x) = 4x^{-6/5}$ only gives positive outputs.
8. Explain why $h(x) = 2x^{-4/3}$ does not have an inverse.
9. Find the inverses of the following functions.
   (a) $f(x) = 4x^{-7}$
   (b) $g(x) = 8x^{-5/3}$
   (c) $h(x) = 4x^{-1/2}$
10. Suppose you are taking a 200-mi trip on a freeway. For the first 100 mi of the trip, the speed limit is 70 mph, and for the last 100 mi, the speed limit is 55 mph.
    (a) If you travel the speed limit, what is your average speed for the entire trip?
    (b) Instead of traveling the speed limit of 55 mph for the last 100 mi, you decide to travel at 60 mph. How much time (in minutes) will you save?
    (c) Instead of traveling the speed limit of 55 mph for the last 100 mi, you decide to keep traveling at 70 mph. How much time (in minutes) will you save?
11. Let $f(x) = x^{-n/d}$, where n and d are positive and n/d is in simplified form.
    (a) What can you say about the behavior of $f(x)$ as $x \to +\infty$? Briefly justify your answer.
    (b) What can you say about the behavior of $f(x)$ as $x \to -\infty$? Briefly justify your answer.
    (c) What can you say about the behavior of $f(x)$ as $x \to 0^+$? Briefly justify your answer.
    (d) What can you say about the behavior of $f(x)$ as $x \to 0^-$? Briefly justify your answer.
    (e) What can you say about the domain and range of f?

12. As you move farther above Earth, the effect of Earth's gravitational pull decreases, which means you would weigh less. Someone who weighs w_0 pounds on the surface of Earth weighs $(1.58 \times 10^7)w_0d^{-2}$ pounds d miles from the center of Earth.
    (a) If you weigh 120 lb on the surface of Earth, how much will you weigh when you are 5000 mi from the center of Earth?
    (b) If you weigh 120 lb on the surface of Earth, how far from the center of Earth would you have to be to weigh 100 lb?
    (c) How far above the center of Earth would you have to be to weigh half as much as you do on the surface of Earth?
13. The measure of "brightness" of a light bulb is usually given in terms of lumens. For example, the output of a 60-W light bulb is about 840 lumens. The brightness of the light striking an object is dependent on the distance the object is away from the light source. This type of brightness is called illuminance and is measured in a unit called lux. To determine the amount of illuminance on an object r meters away from a light bulb emitting 840 lumens, we use the equation,

$$E = \frac{420}{\pi}r^{-2}.$$

    (a) What is the illuminance of an object 2 m away from an 840-lumen light bulb?
    (b) If the illuminance on an object from an 840-lumen light bulb is 10 lux, how far away from the source is the object?

INVESTIGATIONS

INVESTIGATION 1: SODA CANS

Cylindrical cans come in all sorts of shapes and sizes. Tuna cans are usually short and squat. Other cans, such as soda cans, are taller and more slender. It is interesting that tuna companies choose one shape and soft drink companies choose another. In this investigation, we look at the relationship between the radius and height of a 12-oz soda can.

1. Twelve fluid ounces is about 355 ml. Allowing for some air space, we assume a soda can is constructed to hold 370 ml. Using a volume of 370 ml, rewrite the volume formula for a cylinder, $V = \pi r^2 h$, so that the input is the radius, r, and the output is the height, h. (A volume of 370 ml is the same as a volume of 370 cm^3. Thus, we can use centimeters as our units for both the input and output.)
2. Using your answer from question 1, complete the second column of the following table by listing the height of a cylinder of volume 370 cm^3 for the given radius. The surface area formula for a cylinder is $SA = 2\pi r^2 + 2\pi rh$. Use this information to complete the third column of the table.

Radius (cm)	Height (cm)	Surface Area (cm^2)
1		
2		
3		
4		
5		

3. Measure a 12-oz soda can. Which radius and height from your table are closest to the actual radius and height of a 12-oz soda can? Do this radius and this height give the minimum surface area?
4. In the last question, you found the minimum surface area for a cylinder with integer values for the radius. To find the minimum surface area for any value for the radius, use your answer from question 1 and rewrite the surface area formula so that the only input is the radius, r, and the output is the surface area for a 370-cm^3 can, SA.
5. Graph your function from question 4 and estimate the minimum surface area for a 370-cm^3 can.
6. How does a 12-oz soda can with minimum surface area compare with an actual soda can?

INVESTIGATION 2: REPEATED COMPOSITIONS OF POWER FUNCTIONS

The notation $y = f^n(x)$ means to compose the function $y = f(x)$ with itself n times. For example, if $f(x) = 2x + 3$, then $f^2(x) = f(f(x)) = f(2x + 3) = 2(2x + 3) + 3 = 4x + 9$. In this investigation, we explore what happens when power functions with negative integer exponents are composed with themselves.

1. Let $f(x) = x^{-1}$.
   (a) What is $y = f^2(x)$?
   (b) What is $y = f^3(x)$?
   (c) What is $y = f^4(x)$?
   (d) What is $y = f^5(x)$?
   (e) What is $y = f^n(x)$ for even values of n?
   (f) What is $y = f^n(x)$ for odd values of n?
2. Let $g(x) = x^{-2}$.
   (a) What is $y = g^2(x)$?
   (b) What is $y = g^3(x)$?
   (c) What is $y = g^4(x)$?
   (d) What is $y = g^5(x)$?
   (e) What is $y = g^n(x)$?

3. Let $h(x) = x^{-3}$.
   (a) What is $y = h^2(x)$?
   (b) What is $y = h^3(x)$?
   (c) What is $y = h^4(x)$?
   (d) What is $y = h^5(x)$?
   (e) What is $y = h^n(x)$?
4. In general, if $p(x) = x^a$, where a is a negative integer, what is $y = p^n(x)$?
5. Does your answer to question 4 change if a is not an integer?

4.4 RATIONAL FUNCTIONS

In section 4.3, we discovered that power functions with negative exponents have asymptotes. Power functions are not the only functions that have asymptotes; rational functions can also have asymptotes. In this section, we define rational functions, determine their zeros and vertical asymptotes, examine their long-range behavior, and explore some applications.

DEFINITION OF RATIONAL FUNCTIONS

In section 1.4, we modeled the time a traffic light should remain yellow. This function was $t = 1 + v/20 + 70/v$, where v is the velocity of a car in feet per second and t is the time in seconds that a traffic light should remain yellow. Another way of writing this function is $t = (v^2 + 20v + 1400)/20v$. When written this way, we see that we have an example of a rational function. A **rational function** is a function that can be expressed as the quotient (or ratio) of two polynomial functions. Because a rational function involves a quotient, the domain must take into account that division by zero is undefined.

Example 1 Determine whether each of the following are rational functions.

(a) $f(x) = 2x^{-1}$

(b) $g(x) = \dfrac{\sqrt{x}}{x^{3/2}}$

(c) $h(x) = 3x^{-5/3}$

Solution

(a) The function f can be rewritten as $2/x$. Because both 2 and x are polynomials, it is a rational function.

(b) The function g can be rewritten as $1/x$. Since both 1 and x are polynomials, it is a rational function. Note, however, that the domain of g is restricted to $x > 0$ because of the way in which g was originally defined.[18]

[18] *It is possible to consider g as a nonrational function because of its restricted domain. For purposes of this book, we have chosen to define rational functions such that g fits the definition.*

(c) The function h is not a rational function because it cannot be rewritten as a quotient of polynomials.

Example 2 Let

$$f(x) = \frac{x^2 - x - 12}{x^2 + 2x + 1}.$$

Explain why f is a rational function and determine its domain.

Solution The function f is a rational function because it is the quotient of two polynomials. It can be rewritten as

$$f(x) = \frac{(x+3)(x-4)}{(x+1)^2}.$$

Because division by zero is undefined, the domain of f is all real numbers such that $x \neq -1$.

In section 4.3, we considered power functions with negative exponents. Some of these, such as $f(x) = 2x^{-1}$, can be written as a quotient of polynomials and therefore are also rational functions. So, we should expect some similarities between the behavior of rational functions and the behavior of power functions with negative exponents. Not all rational functions, however, are power functions, nor are all power functions rational functions. Recall that power functions with negative exponents are written symbolically as $f(x) = bx^a$, where $a < 0$. These functions always have the y-axis as a vertical asymptote and the x-axis as a horizontal asymptote, which is not true for all rational functions. For example, a graph of the function modeling the time needed for the yellow light, $t = (v^2 + 20v + 1400)/20v$ is shown in Figure 1. From the graph of this function, it appears that the y-axis is a vertical asymptote. The x-axis, however, is not a horizontal asymptote. In fact, this function does not appear to have a horizontal asymptote at all.

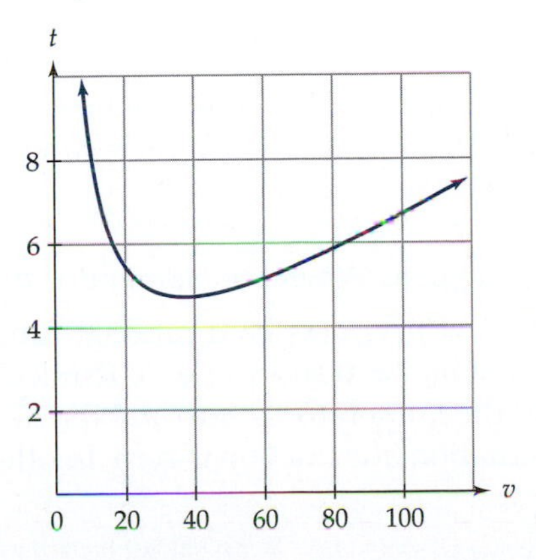

Figure 1 A graph of the time (in seconds) needed for a yellow light for given velocities (in feet per second) of cars.

Another difference between rational and power functions is that rational functions can have x-intercepts, whereas power functions with negative exponents cannot. A function of the form $f(x) = bx^a$, where $a < 0$, can never equal zero. The function $g(x) = (x^2 - x - 2)/x$, however, has zeros at $x = 2$ and $x = -1$. (See Figure 2.)

Finally, although power functions with negative exponents always have vertical asymptotes, it is not true of every rational function. For example, the outputs of the function $h(x) = 1/(x^2 + 1)$ never approach $+\infty$ or $-\infty$, so this function does not have a vertical asymptote. (See Figure 3.)

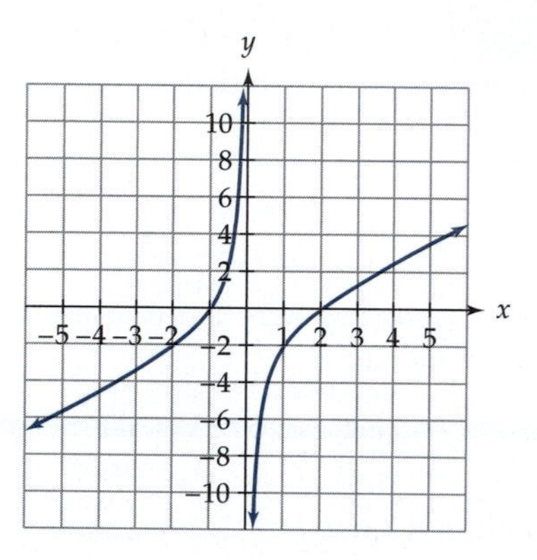

Figure 2 A graph of $g(x) = (x^2 - x - 2)/x$.

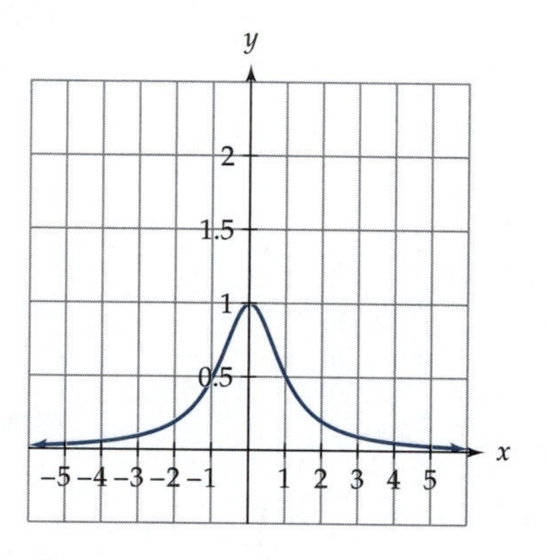

Figure 3 A graph of $h(x) = 1/(x^2 + 1)$.

READING QUESTIONS

1. Determine if each of the following is a rational function.

(a) $f(x) = 3x^{-2}$

(b) $g(x) = \dfrac{x}{\sqrt{x}}$

(c) $h(x) = \dfrac{1}{2x + 1}$

ZEROING IN ON RATIONAL FUNCTIONS: SHORT-TERM BEHAVIOR

Because a rational function can be expressed as the quotient of two polynomials, $f(x) = p(x)/q(x)$, interesting things happen where p, q, or both equal zero.

What Happens When the Numerator Is Zero

Let $f(x) = p(x)/q(x)$ be a rational function. Suppose there is an input c such that $p(c) = 0$ and $q(c) \neq 0$. Think about the result of having a fraction where the numerator is zero and the denominator is a nonzero number. In this situation, the fraction is zero. In other words, if $p(c) = 0$ (and $q(c) \neq 0$), then $f(c) = 0$. So, to find the x-intercepts for f, find the inputs that cause the numerator to equal zero. Because p is a polynomial, the inputs can be found by factoring p.

Example 3 Let

$$f(x) = \frac{x^2 + 5x - 4}{x - 2}.$$

Find the x-intercepts of f. Also give the domain of f.

Solution To determine the x-intercepts of f, we need to find where $x^2 + 5x - 4 = 0$. By using the quadratic formula, we find that

$$x = \frac{-5 \pm \sqrt{5^2 - 4(1)(-4)}}{2(1)} = \frac{-5 \pm \sqrt{41}}{2}.$$

Therefore, f has two zeros,

$$x = \frac{-5 + \sqrt{41}}{2} \approx 0.70 \quad \text{and} \quad x = \frac{-5 - \sqrt{41}}{2} \approx -5.70.$$

The domain is all real numbers such that $x \neq 2$ because division by zero is undefined. A graph of f is shown in Figure 4. Notice that the x-intercepts are just as we expected based on our analysis of the equation.

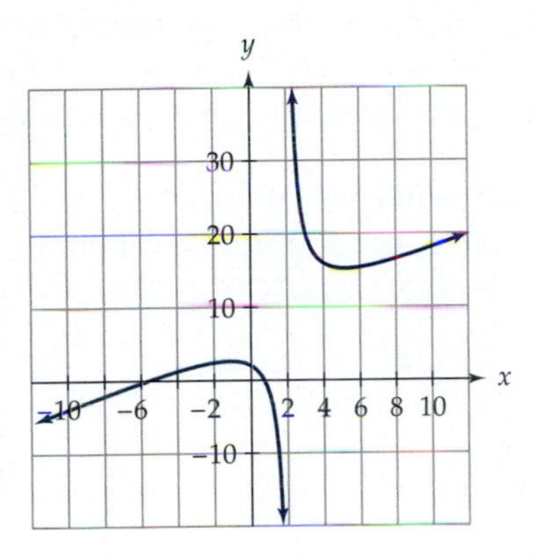

Figure 4

Notice that the graph in Figure 4 consists of two unconnected parts. A graph like this one on a calculator often shows these two parts connected. In section 1.3, on calculator graphics, we showed why this happens and how to fix it.

Not all rational functions have x-intercepts because not all polynomials have real roots, so there may be no value for which $p(c) = 0$. For example, the traffic light function mentioned earlier is

$$t = \frac{v^2 + 20v + 1400}{20v}.$$

If we use the quadratic formula to find values for which $p(v) = v^2 + 20v + 1400 = 0$, we find no solution because there is a negative number under the square root. This result also makes sense physically. In no situation should the traffic light remain yellow for 0 sec!

What Happens When the Denominator is Zero

Let $f(x) = p(x)/q(x)$ be a rational function. Suppose there is an input b such that $q(b) = 0$ but $p(b) \neq 0$. Think about what happens if you try to divide a nonzero number by zero. Such division is undefined. If you divide a nonzero number by a number close to zero (such as 0.00000001), however, then the absolute value of your answer is quite large. In symbols, as $x \to b$ (causing the denominator to approach zero), $|f(x)| \to +\infty$. For example, suppose

$$f(x) = \frac{3x^2 - 5x - 7}{x - 2}.$$

TABLE 1 Values of $f(x) = (3x^2 - 5x - 7)/(x - 2)$ for x Close to 2

x	1.9	1.99	1.999	2.001	2.01	2.1
$f(x)$	56.7	506.97	5006.997	−4992.997	−492.97	−42.7

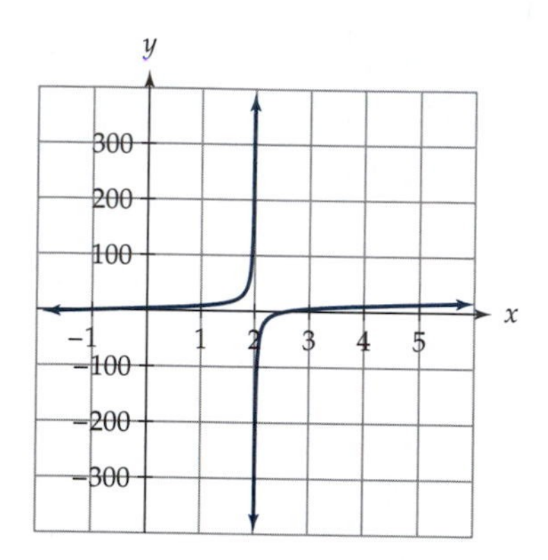

Figure 5 A graph of $f(x) = \dfrac{3x^2 - 5x - 7}{x - 2}$. Notice the vertical asymptote at $x = 2$.

Values for f near $x = 2$ are given in Table 1. Looking at Table 1 (or analyzing the function) shows us that $f(x) \to +\infty$ as $x \to 2^-$ and $f(x) \to -\infty$ as $x \to 2^+$.[19] A graph of f is shown in Figure 5. It is clear that f has a vertical asymptote at $x = 2$.

In general, if $f(x) = p(x)/q(x)$ and $q(b) = 0$ (while $p(b) \neq 0$), then f has a vertical asymptote at $x = b$. As with x-intercepts, not all rational functions have a vertical asymptote.

Example 4 Let

$$f(x) = \frac{3x + 3}{2x^2 + 5x - 3}.$$

Determine the vertical asymptote(s), x-intercept(s), and domain of f.

Solution To determine the vertical asymptotes and x-intercepts of f, we start by factoring the numerator and denominator, which gives us

$$f(x) = \frac{3x + 3}{2x^2 + 5x - 3} = \frac{3(x + 1)}{(2x - 1)(x + 3)}.$$

The numerator is zero when $x = -1$, and the denominator is zero when $x = \frac{1}{2}$ or $x = -3$. Thus, f has an x-intercept at $x = -1$ and vertical asymptotes at $x = \frac{1}{2}$ and $x = -3$. Because the denominator is zero at $x = \frac{1}{2}$ and $x = -3$, the domain of the function is all real numbers such that $x \neq \frac{1}{2}$ or $x \neq -3$. A graph of f is shown in Figure 6.

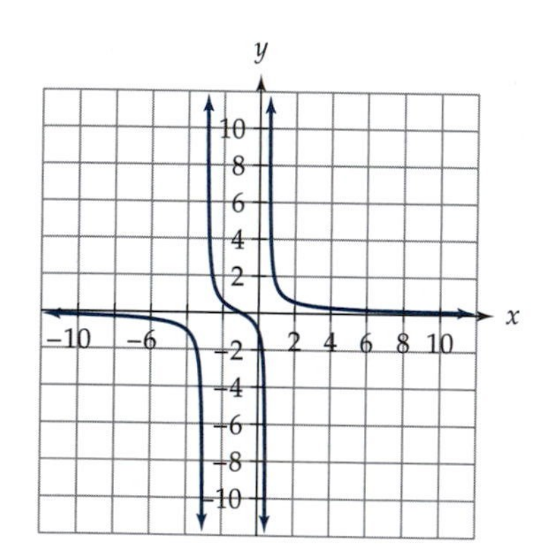

Figure 6

Our traffic light example, $t = (v^2 + 20v + 1400)/20v$, has a vertical asymptote at $v = 0$. Thus, as the velocity approaches zero, the amount of time the light should remain yellow becomes infinitely large.

What Happens When Both the Numerator and the Denominator Are Zero

Again, let $f(x) = p(x)/q(x)$. So far, it has been fairly easy to see that f has a zero (or x-intercept) at $x = c$ when $p(c) = 0$ and that f has a vertical asymptote at $x = b$ when $q(b) = 0$. What if there is an a such that *both* $p(a) = 0$ and $q(a) = 0$? It is impossible to have both an x-intercept and a vertical asymptote at the same place, so clearly this situation needs to be explored. First, note that f will not be defined at $x = a$ because division by

[19] *Recall that 2^- means that x approaches 2 from the left side and 2^+ means that x approaches 2 from the right side.*

zero is undefined. Just because a function is undefined, however, does not mean that it has a vertical asymptote. Recall that the definition of a vertical asymptote at $x = a$ is that the absolute value of the outputs get large as $x \to a$. Consider the function

$$g(x) = \frac{2x^2 + 11x + 15}{x + 3} = \frac{(x+3)(2x+5)}{x+3}.$$

Table 2 gives values of $g(x)$ for x-values close to -3. A graph of g is shown in Figure 7. By looking at either Table 2 or the graph in Figure 7, it is clear that g does *not* have a vertical asymptote at $x = -3$ even though g is undefined here. Instead, $g(x) \to -1$ as $x \to -3$, which can also be seen by looking at the equation for g. Because

$$\begin{aligned} g(x) &= \frac{2x^2 + 11x + 15}{x + 3} \\ &= \frac{(x+3)(2x+5)}{x+3} \\ &= 2x + 5 \qquad \text{for } x \neq -3, \end{aligned}$$

it is not surprising that g looks like a line with a hole at $x = -3$. In this case, when both the numerator and denominator were zero, the function had a "hole" at that point.

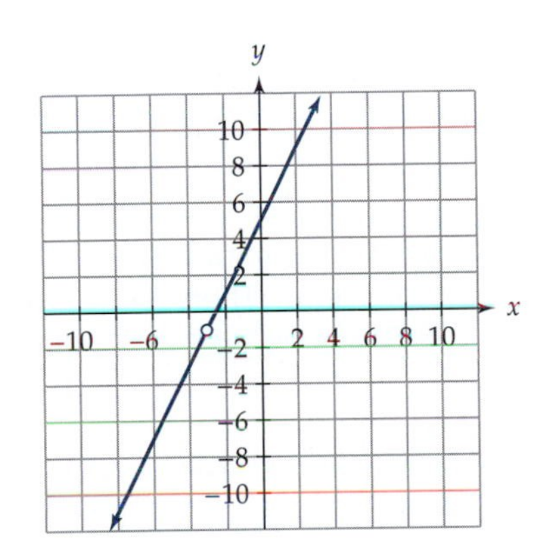

Figure 7 A graph of $g(x) = \dfrac{2x^2 + 11x + 15}{x + 3}$.

TABLE 2 Values of $g(x) = (2x^2 + 11x + 15)/(x + 3)$ for x Near -3

x	-3.1	-3.01	-3.001	-2.999	-2.99	-2.9
$g(x)$	-1.2	-1.02	-1.002	-0.998	-0.98	-0.8

This result does not always happen, though. Consider

$$h(x) = \frac{x^2 - 4}{x^2 + 4x + 4} = \frac{(x-2)(x+2)}{(x+2)^2}.$$

Table 3 gives values of h for x-values close to $x = -2$. A graph of h is shown in Figure 8. This time, when both the numerator and denominator equal zero at $x = -2$, the function has a vertical asymptote. Again, we can also see this asymptote by looking at the equation for h,

$$h(x) = \frac{x^2 - 4}{x^2 + 4x + 4}$$
$$= \frac{(x-2)(x+2)}{(x+2)^2}$$
$$= \frac{x-2}{x+2}.$$

In this case, the equation for h has a zero in the denominator when $x = -2$ even after simplifying, which causes the vertical asymptote. Even though $x = -2$ is a zero for both the numerator and denominator of h, a factor of $x + 2$ remains in the denominator after simplifying.

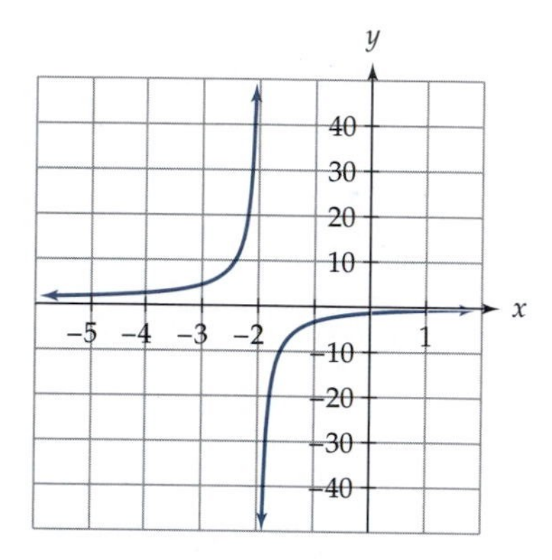

Figure 8 A graph of $h(x) = \dfrac{x^2 - 4}{x^2 + 4x + 4}$.

TABLE 3 Values of $h(x) = (x^2 - 4)/(x^2 + 4x + 4)$ Near $x = -2$

x	−2.1	−2.01	−2.001	−1.999	−1.99	−1.9
$h(x)$	41	401	4001	−3999	−399	−39

Again, let $f(x) = p(x)/q(x)$. As you know, if a is a root of a polynomial p, then $(x - a)$ is a factor of p. If a is a root for both p and q, then we know that $(x - a)$ is a factor for both the numerator and the denominator of our rational function. Whether a hole or a vertical asymptote is produced depends on the multiplicity of this factor for both p and q. Recall from section 4.2 that the *multiplicity* of a root, a, of a polynomial, p, is the number of times the factor $(x - a)$ appears in the factorization of the polynomial. The multiplicity of a root determines if any of those factors remain in a function after it is simplified. Let f be a rational function such that a is a root in both the numerator and the denominator. If the multiplicity of a in the denominator is greater than the multiplicity of a in the numerator, then, after simplification, there is still a factor of $(x - a)$ in the denominator and a vertical asymptote. If the multiplicity of the a in the numerator is equal to or greater than the multiplicity of a in the denominator, then there is a hole in the function.

Example 5 Determine the x-intercept(s), domain, and vertical asymptote(s) for

$$f(x) = \frac{x^3 - 3x^2 + 3x - 1}{(x+2)(x-1)^2}.$$

Solution To find the x-intercept(s), domain, and vertical asymptote(s), we start by factoring the numerator and denominator of f, which gives

$$f(x) = \frac{(x-1)^3}{(x+2)(x-1)^2}.$$

The domain of f is all real numbers suchthat $x \neq 1$ or $x \neq -2$. When we

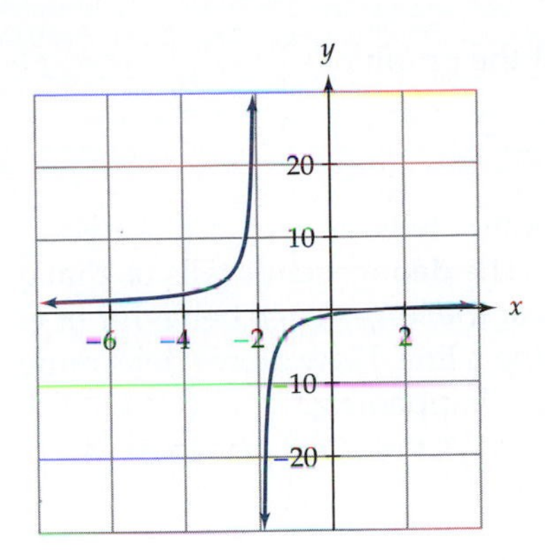

Figure 9

simplify this function, we get

$$f(x) = \frac{x-1}{x+2}.$$

Looking at the simplified f makes it clear that there is a vertical asymptote at $x = -2$ but not at $x = 1$; there must be a hole there instead. In summary, the domain of f is all real numbers such that $x \neq 1$ or $x \neq -2$; f has no x-intercepts; and f has a vertical asymptote at $x = -2$. A graph of f is shown in Figure 9.

PROPERTIES

If $f(x) = p(x)/q(x)$ is a rational function, then the following are true.

- If $p(a) = 0$ when $q(a) \neq 0$, then f has a x-intercept at $x = a$.
- If $q(a) = 0$ when $p(a) \neq 0$, then f has a vertical asymptote at $x = a$.
- If $p(a) = q(a) = 0$, then the behavior at $x = a$ depends on the multiplicity of $(x - a)$. If the multiplicity of $(x - a)$ in p is greater than or equal to the multiplicity of $(x - a)$ in q, then f has a hole at $x = a$. Otherwise, f has a vertical asymptote at $x = a$.

READING QUESTIONS

2. Let $f(x) = p(x)/q(x)$, where p and q are polynomials.

(a) Suppose $p(a) = 0$ and $q(a) \neq 0$. What do you know about f when $x = a$?

(b) Suppose $p(a) = 0$ and $q(a) = 0$. What do you know about f when $x = a$?

3. Explain why the rational function

$$f(x) = \frac{(x-2)^2}{x-2}$$

has a hole at the point $(2, 0)$.

4. Let

$$f(x) = \frac{x(x-1)(x+2)^2}{3(x-1)(x+3)}.$$

Determine the domain of f, its x-intercept(s), and its vertical asymptote(s), if there are any.

DIVIDE AND CONQUER: LONG-RANGE BEHAVIOR

Finding the zeros of the numerator and denominator of a rational function give us information about the x-intercepts, vertical asymptotes, and "holes" in the function, but it does not tell us everything about the behavior

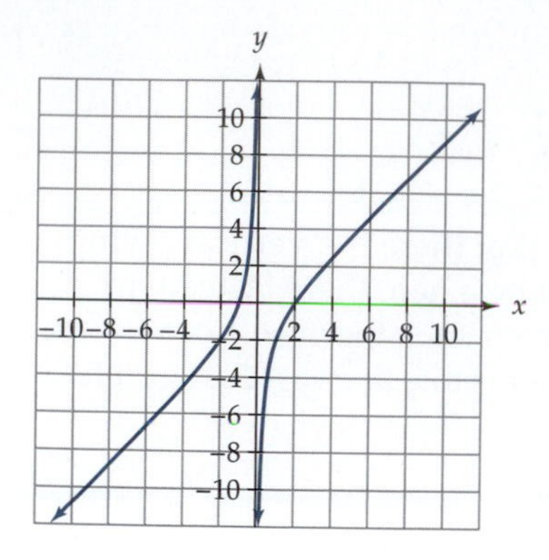

Figure 10 A graph of $g(x) = \dfrac{x^2 - x - 2}{x}$.

of rational functions. For example, look at the graph of

$$g(x) = \frac{x^2 - x - 2}{x}$$

shown in Figure 10. Factoring the numerator tells us that g has two x-intercepts, one at $x = 2$ and one at $x = -1$. The denominator tells us that g has a vertical asymptote at $x = 0$. Yet look at the long-range behavior of g. As $|x|$ gets large, g appears to be approaching a line. How do we determine that line? How do we know for sure what is happening?

The key to answering this question is to rewrite $g(x) = p(x)/q(x)$ as

$$g(x) = d(x) + \frac{r(x)}{q(x)},$$

where the degree of r is smaller than the degree of q. In other words, rather than factoring the numerator and denominator, the important algebraic step is dividing q into p and rewriting the rational function in terms of the quotient, d, and the remainder, r. Let's return to $g(x) = (x^2 - x - 2)/x$. We can rewrite this equation as $g(x) = x - 1 + (-2/x)$. When looking at the sum of two functions in section 3.3, we saw that if the outputs of one function are much larger than the outputs of the other, the former dominates that portion of the graph. If we compare $d(x) = x - 1$ and $r(x)/q(x) = -2/x$, we see that for large values of $|x|$, $d(x) = x - 1$ dominates because $-2/x \to 0$ as $x \to \pm\infty$. So, as $x \to +\infty$ and $x \to -\infty$, $2/x$ gets so small that $g(x)$ gets close to $d(x) = x - 1$. Table 4 compares values of $g(x)$ and values of $d(x) = x - 1$ for large values of x. Where appropriate, answers are rounded to three decimal places. Graphs of g and d are shown in Figure 11.

TABLE 4 Comparing $g(x) = x - 1 + (-2/x)$ and $d(x) = x - 1$

x	−1,000,000	−1000	−100	100	1000	1,000,000
$g(x) = x - 1 + \dfrac{-2}{x}$	−1,000,001.000	−1000.998	−100.98	98.98	998.998	999,999.000
$d(x) = x - 1$	−1,000,001	−1001	−101	99	999	999,999

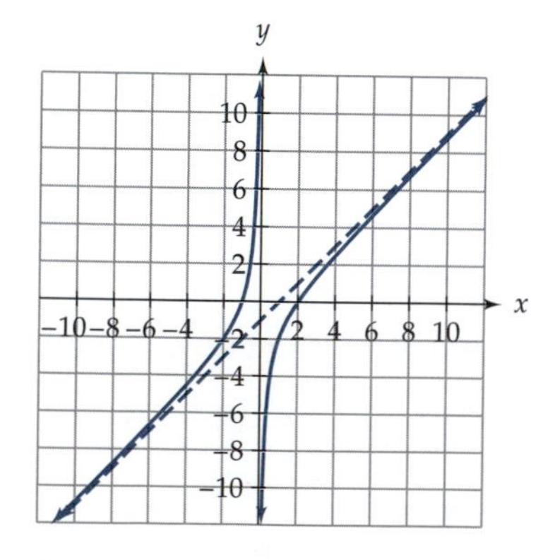

Figure 11 A graph of $g(x) = x - 1 + (-2/x)$ and $d(x) = x - 1$.

In general, long-range behavior is determined by the quotient obtained when the denominator is divided into the numerator, sometimes leading to asymptotic behavior and sometimes not. We consider several examples.

Example 6 Determine the long-range behavior of

$$f(x) = \frac{6x - 14}{2x + 4}.$$

Solution The first step in determining the long-range behavior is to divide $2x + 4$ into $6x - 14$.

$$\begin{array}{r} 3 \\ 2x + 4 \overline{)\, 6x - 14} \\ \underline{6x + 12} \\ -26 \end{array}$$

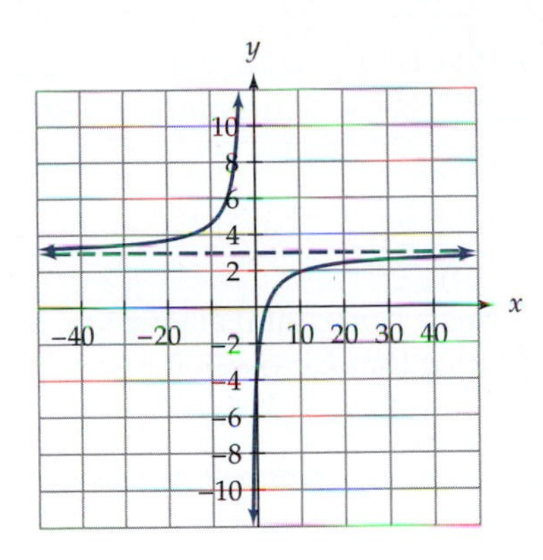

Figure 12

We can see that $g(x) = 3 - 26/(2x + 4)$. As $x \to \pm\infty$, $26/(2x + 4) \to 0$. Thus, as $x \to \pm\infty$, $g(x) = 3 - 26/(2x + 4)$ approaches 3. Therefore, the line $y = 3$ is a horizontal asymptote for $g(x) = (6x - 14)/(2x + 4)$. A graph of g is shown in Figure 12.

Example 7 Determine the long-range behavior of

$$f(x) = \frac{2x^2 + 6x + 9}{x + 4}.$$

Solution Again, the first step in determining the long-range behavior is to divide $x + 4$ into $2x^2 + 6x + 9$.

$$\begin{array}{r} 2x - 2 \\ x + 4 \overline{)\, 2x^2 + 6x + 9} \\ \underline{2x^2 + 8x} \\ -2x + 9 \\ \underline{-2x - 8} \\ 17 \end{array}$$

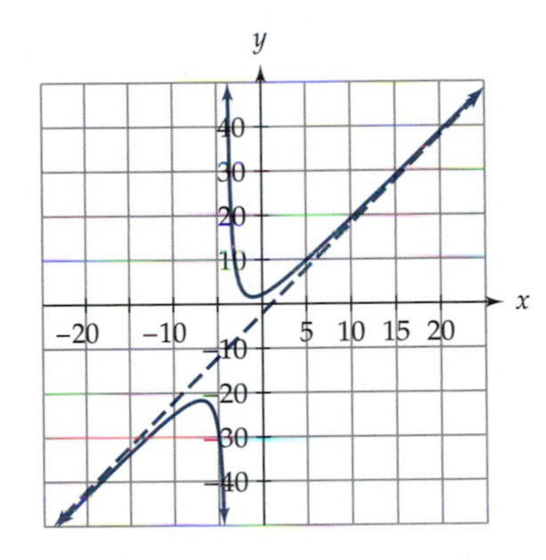

Figure 13

We see that $f(x) = 2x - 2 + 17/(x + 4)$. As $|x|$ gets large, $17/(x + 4)$ approaches zero. Therefore, the graph of f resembles the graph of $y = 2x - 2$ as $x \to \pm\infty$. Graphs of f and $y = 2x - 2$ are shown in Figure 13.

Example 8 Determine the long-range behavior of

$$g(x) = \frac{x^3 - 9}{x - 4}.$$

Solution To divide $x - 4$ into $x^3 - 9$, we rewrite $x^3 - 9$ as $x^3 + 0x^2 + 0x - 9$ to keep the columns lined up properly.

$$\begin{array}{r} x^2 + 4x + 16 \\ x - 4 \overline{)\, x^3 + 0x^2 + 0x - 9} \\ \underline{x^3 - 4x^2} \\ 4x^2 + 0x \\ \underline{4x^2 - 16x} \\ 16x - 9 \\ \underline{16x - 64} \\ 55 \end{array}$$

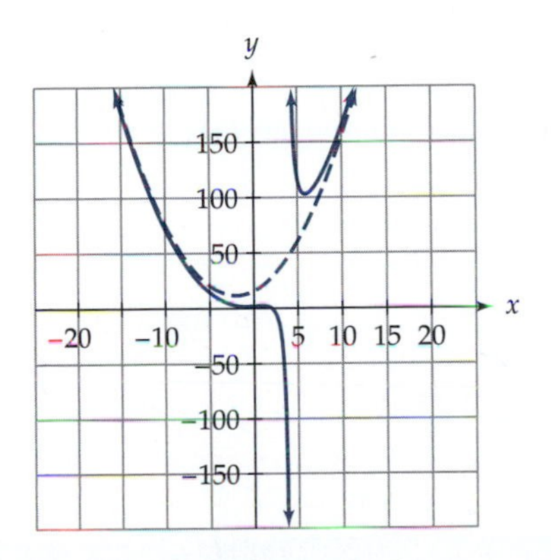

Figure 14

We see that $g(x) = x^2 + 4x + 16 + 55/(x - 4)$. As $|x|$ gets large, $55/(x - 4) \to 0$. Therefore, the graph of g resembles the graph of $y = x^2 + 4x + 16$ as $x \to \pm\infty$. Graphs of g and $y = x^2 + 4x + 16$ are shown in Figure 14.

Example 9 Determine the long-range behavior of

$$k(x) = \frac{x}{x^2 + 1}.$$

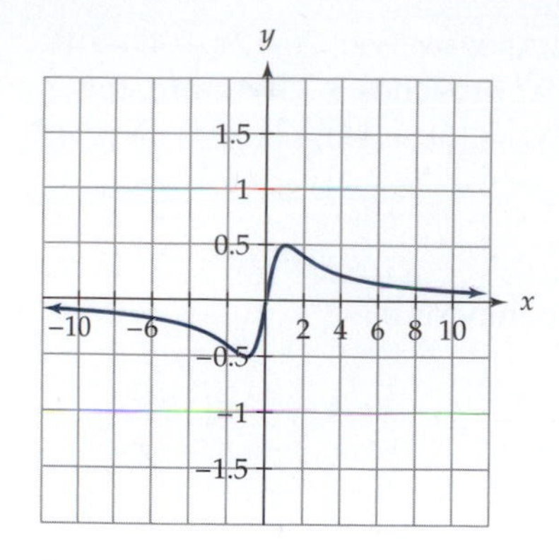

Figure 15

Solution In this case, there is no reason to divide because the degree of the numerator is already less than the degree of the denominator. We can rewrite this function as

$$k(x) = 0 + \frac{x}{x^2+1}.$$

As $|x|$ gets large, $k(x)$ approaches zero. So, k has a horizontal asymptote at $y = 0$. A graph of k is shown in Figure 15. Notice that k crosses its horizontal asymptote at the origin.

Determining the long-range behavior of a rational function is simple once you divide the denominator into the numerator to obtain $f(x) = p(x)/q(x) = d(x) + (r(x)/q(x))$. In the exercises, you are asked to prove that as $x \to \pm\infty$, $r(x)/q(x) \to 0$ because the degree of r is less than the degree of q. Therefore, as $x \to \pm\infty$, the graph of f resembles the graph of d. If the quotient is a constant, $d(x) = c$, then f has a horizontal asymptote at $y = c$. If the degree of the numerator, p, is smaller than the degree of the denominator, q, then there is no reason to divide. In this case, $d(x) = 0$, which means that the x-axis is the horizontal asymptote. If the quotient is a line, $d(x) = mx + b$, then f is said to have a *slant asymptote* because the long-range behavior of f resembles a line. If the degree of the quotient, $d(x)$, is 2 or greater, then f does not have any long-range asymptotic behavior. These properties are summarized below.

PROPERTIES

If f is a rational function such that $f(x) = p(x)/q(x)$, then the following are true.

- If the degree of the numerator, p, is smaller than the degree of the denominator, q, then the line $y = 0$ is a horizontal asymptote.
- If the degree of the numerator, p, is the same as the degree of the denominator, q, then the line $y = c$ is a horizontal asymptote. (The number c is the ratio of the leading coefficient of the numerator, p, and the leading coefficient of the denominator, q.
- If the degree of the numerator, p, is one larger than the degree of the denominator, q, then a slant asymptote exists.
- If the degree of the numerator, p, is more than one larger than the degree of the denominator, q, then no asymptote exists.

READING QUESTIONS

5. Determine the long-range behavior of

$$f(x) = \frac{x^2+1}{x^2+2x-2}.$$

6. Does every rational function exhibit long-range asymptotic behavior? Justify your answer.
7. When will a rational function have a slant asymptote?

FINDING FORMULAS, DETERMINING RANGE, AND AN APPLICATION

Finding a Formula for a Rational Function from a Graph

We considered how to determine the domain, x-intercepts, vertical asymptotes, and long-range asymptotic behavior, but not all rational functions exhibit all these behaviors. We already know that $k(x) = x/(x^2 + 1)$ does not have any vertical asymptotes (see Figure 15) and that $g(x) = (x^3 - 9)/(x - 4)$ does not have any long-range asymptotic behavior (see Figure 14). Thus, there is not a universal "shape" to the graphs of rational functions, nor are rational functions the only functions that exhibit asymptotic behavior. Earlier chapters showed that $y = \tan x$ has vertical asymptotes, whereas $y = 2^{-x}$ has the x-axis as a horizontal asymptote, but it is still often possible to find a plausible equation for the graph of a rational function.

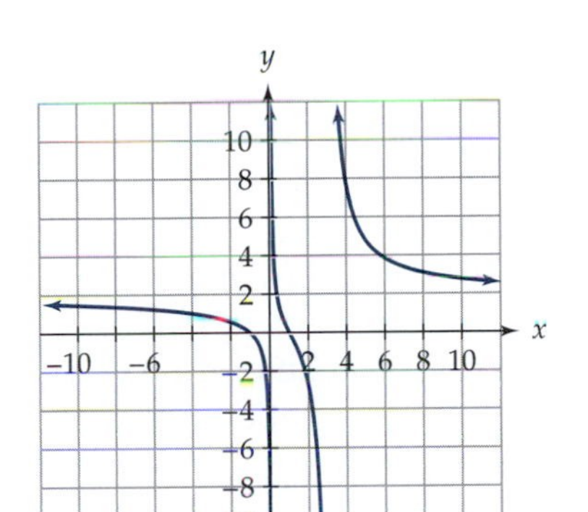

Figure 16

Example 10 Let f be the function shown in Figure 16. Assume f is rational. Determine a plausible equation for f.

Solution The graph of f appears to have x-intercepts at $x = 1$ and $x = -1$. There appear to be two vertical asymptotes, one at $x = 0$ and one at $x = 3$. So, one possibility is

$$f(x) = \frac{(x-1)(x+1)}{x(x-3)}.$$

This possibility, however, does not take into account the horizontal asymptote at $y = 2$. When we divide the denominator into the numerator, the quotient must be 2. Because the quotient is a constant, the degree of the numerator and denominator must be the same. One way of obtaining a quotient of 2 is if

$$f(x) = \frac{2(x-1)(x+1)}{x(x-3)}.$$

Determining the Range of a Rational Function

Going from a graph to an equation for a rational function is difficult; many different steps occur, and there is no universal shape for the graph of a rational function. It is also difficult to determine the range of a rational function. Determining the domain is easy as long as you avoid division by zero. Determining possible outputs, however, can be quite challenging. The best way to determine the range is to graph the function *and* algebraically analyze its equation. This is illustrated in the next two examples.

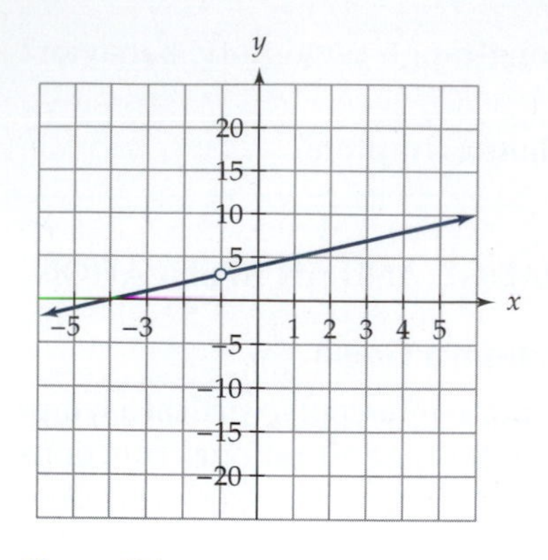

Figure 17

Example 11 Determine the domain, range, x-intercepts, vertical asymptotes, and long-range behavior of

$$g(x) = \frac{x^2 + 5x + 4}{x + 1}.$$

Solution Factoring gives

$$g(x) = \frac{(x + 4)(x + 1)}{x + 1}.$$

So, the domain of g is all real numbers such that $x \neq -1$. The x-intercept is at $x = -4$. When simplified, we see that $g(x) = x + 4$ for $x \neq -1$. So, g looks like a line with a "hole" at the point $(-1, 3)$. The range is all real numbers such that $y \neq 3$. A graph of g is shown in Figure 17.

Example 12 Determine the domain, range, x-intercepts, vertical asymptotes, and long-range behavior of

$$h(x) = \frac{10}{x^2 + 5}.$$

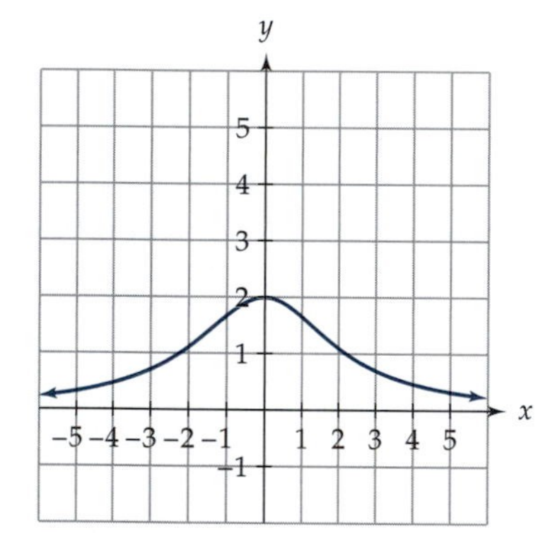

Figure 18

Solution The polynomial $q(x) = x^2 + 5$ has no real roots, so it cannot be factored. The domain is all real numbers, there are no x-intercepts, and there are no vertical asymptotes. Because the degree of the numerator ($p(x) = 10$) is less than the degree of the denominator ($q(x) = x^2 + 5$), the rational function g has a horizontal asymptote of $y = 0$. Finding the range is a bit trickier. Let's start with a graph of the function, shown in Figure 18. It appears that the range is $0 < y \leq 2$.

As we mentioned earlier, we are looking at only a small portion of the domain of the entire graph, so without analyzing the equation, we cannot be sure we know the range. Because the numerator is a positive constant and the denominator is also always positive regardless of the value of x, the outputs must always be positive. Thus, our range is, indeed, always greater than 0. There is also an upper bound for the outputs of h. To make the outputs of a fraction "large," we should make the denominator "small." In this case, the smallest value for the denominator is 5 (when $x = 0$). Thus, the largest value for $h(x)$ is

$$h(0) = \frac{10}{0^2 + 5} = 2.$$

Therefore, as we suspected, the range is $0 < y \leq 2$.

The Brightness of a Light

Let's explore the physical significance of the behavior of a rational function by considering an application. The brightness of light on an object (called illuminance) is inversely proportional to the square of the distance from the light source to the object. We represent illuminance by the formula

$$E = \frac{F}{2\pi r^2},$$

where F is the amount of luminous flux (measured in lumens) of the light source, r is the distance the object is away from the light source (measured in meters), and E is the amount of illuminance on the object (measured in lux).

Suppose we have a light on a 2.5-m ceiling and have an object on the floor, not necessarily directly under the light. Assuming the object is x meters away from the point directly under the light, we can model this situation as shown in Figure 19. By the Pythagorean theorem, the distance from the light to the object on the floor is $\sqrt{2.5^2 + x^2}$. The function that models this situation is

$$E = \frac{F}{2\pi(2.5^2 + x^2)}.$$

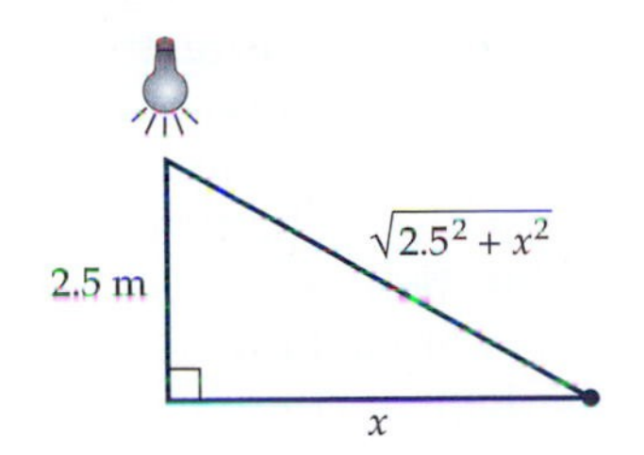

Figure 19 If a light is 2.5 m above the floor and an object is x meters away from the point directly below the light, then the object is $\sqrt{2.5^2 + x^2}$ meters away from the light.

The luminous flux (F) of a 60-W bulb is about 840 lumens. So, assuming a 60-W bulb, our function is now

$$E = \frac{840}{2\pi(2.5^2 + x^2)}.$$

A graph of this function is shown in Figure 20.

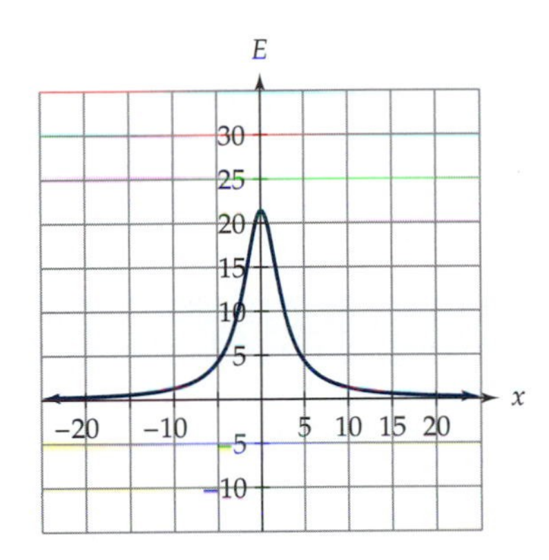

Figure 20 A graph of the function modeling the brightness of a light shining on an object x meters away from a point directly below a 60-W bulb hanging from a 2.5-m ceiling.

Let's look at the behavior of this rational function and its implications about the light. Notice that E has no vertical asymptotes. That there is no place you can place the object such that the brightness of the light is infinite makes sense because there is a bound on how bright a light can come from a 60-W bulb. In fact, the way to cause the light on the object to be as bright as possible is to place it directly under the light source. Maximum brightness occurs when $x = 0$, which is verified by noticing that the graph has a maximum when $x = 0$, corresponding to our physical knowledge of the model. There are no x-intercepts because there is no place (in this ideal room) where no light shines on our object. A horizontal asymptote at $y = 0$ suggests that the farther away you place the object, the dimmer the light shining on that object, again corresponding to our physical knowledge of the model.

READING QUESTIONS

8. Give an example (not in the text) of a rational function that does not have a vertical asymptote.

9. Determine the range for

$$f(x) = \frac{2x^2 + x - 1}{x^2 + 6x + 5}.$$

(*Hint:* Draw a graph.)

SUMMARY

A **rational function** is a function that can be expressed as a quotient of two polynomial functions.

Generally, when the numerator of a rational function is zero for a specific input, the function has a root at that input. When the denominator of a rational function is zero for a specific input, the function has a vertical asymptote at that input. If both the numerator and denominator are zero for the same input, then the behavior of the function depends on the multiplicity of the factors involved. If a is a root in both the numerator and denominator of a rational function, then there is a vertical asymptote if the multiplicity of a in the denominator is greater than the multiplicity of a in the numerator; otherwise, there is a hole in the function.

The long-range behavior of the function is determined by the relative degrees of the numerator and denominator. If the degree of the numerator is greater than the denominator, then the function approaches either positive or negative infinity as the input approaches positive or negative infinity. If the degrees are the same, then there is a horizontal asymptote. If the degree of the denominator is greater than that of the numerator, then the x-axis acts as a horizontal asymptote.

Finding a formula for a rational function based on its graph can be difficult. Vertical asymptotes usually indicate factors in the denominator, whereas horizontal asymptotes indicate the relative degrees of the numerator and denominator as well as the leading coefficients.

The domain of a rational function consists of all real numbers except those that cause the denominator to equal zero. Finding the range is challenging, though; usually the easiest way to do that is to look at the graph of the function.

EXERCISES

1. Determine if each of the following are rational functions.
   (a) $f(x) = \dfrac{x^{-2}}{x^3}$
   (b) $g(x) = \dfrac{\sqrt{x}}{x^2+1}$
   (c) $h(x) = \dfrac{3x}{x}$
   (d) $k(x) = \dfrac{x^{3/2}}{x^{5/2}+x^{3/2}}$
   (e) $m(x) = 2x^{-3}$
   (f) $n(x) = 2x^{-2/3}$
2. Determine whether each of the following are always true, sometimes true, or never true. Justify your answers.
   (a) If f is a power function with negative exponents, then it is also a rational function.
   (b) Let $f(x) = p(x)/q(x)$, where p and q are polynomials. If $q(a) = 0$, then f has a vertical asymptote at $x = a$.
   (c) If f is a rational function, then f has at least one x-intercept.
   (d) If f is a rational function, then f must have at least one vertical asymptote.

(e) Let $f(x)$ be a function that has a vertical asymptote. Then f is a rational function.

(f) Let $f(x) = p(x)/q(x)$, where p and q are polynomials. If $p(a) = q(a) = 0$, then f has a hole in the graph at $x = a$.

(g) Let $f(x) = p(x)/q(x)$, where p and q are polynomials. If $q(a) = 0$ when $p(a) \neq 0$, then f has a x-intercept at $x = a$.

(h) Let $f(x) = p(x)/q(x)$, where p and q are polynomials. If $p(a) = 0$ when $q(a) \neq 0$, then f has a zero at $x = a$.

3. For each of the following rational functions, determine the domain, x-intercepts, and vertical asymptotes. Draw a sketch of the graph.

(a) $y = \dfrac{2x+7}{3x-4}$

(b) $y = \dfrac{x^2+4x+4}{x-7}$

(c) $y = \dfrac{2x}{x^3-4x^2+3x}$

(d) $y = \dfrac{x^2}{x^2+1}$

(e) $y = \dfrac{(x-1)(x+5)^2}{x(x-2)^2(x+5)}$

(f) $y = \dfrac{x^2-4x+6}{3x^2+x+1}$

(g) $y = \dfrac{x^2+2x-8}{x^2-2x-24}$

(h) $y = \dfrac{(2x-6)(4x-5)}{6x^2-13}$

(i) $y = \dfrac{1}{4x^2-8}$

(j) $y = \dfrac{4x^3+6}{3x^2-2}$

4. Determine the long-range behavior of each of the following.

(a) $y = \dfrac{3x+4}{x}$

(b) $y = \dfrac{3x^2-9}{x^2+4}$

(c) $y = \dfrac{x^3-4x^2+1}{x^4}$

(d) $y = \dfrac{x^2-3x+1}{x-4}$

(e) $y = \dfrac{x^3-2x}{x^2+4x-1}$

(f) $y = \dfrac{x^3+3x^2-x}{x+1}$

5. In this exercise, you will prove that $r(x)/q(x) \to 0$ as $x \to \pm\infty$ if the degree of r is less than the degree of q.
   (a) Let
   $$g(x) = \frac{x^3 + 4x^2 - x + 10}{5x^4 - 3x^3 + x^2 - 5x + 1}.$$
      i. Multiply the numerator and denominator of g by $1/x^4$. Call it g_2.
      ii. Explain why $g_2(x) \to 0$ as $x \to \pm\infty$. Because $g_2(x) = g(x)$, the explanation also shows that $g(x) \to 0$ as $x \to \pm\infty$.
   (b) Let
   $$h(x) = \frac{r(x)}{q(x)} = \frac{a_n x^n + a_{n-1}x^{n-1} + \cdots + a_1 x + a_0}{b_m x^m + b_{m-1}x^{m-1} + \cdots + b_1 x + b_0}.$$
   Assume $a_n \neq 0$, $b_m \neq 0$, and the degree of r is less than the degree of q.
      i. Which is larger: n or m? Briefly justify your answer.
      ii. Multiply the numerator and denominator of h by $1/x^m$. Call the result h_2.
      iii. Explain why $h_2(x) \to 0$ as $x \to \pm\infty$. Because $h_2(x) = h(x)$, your explanation will also show that $h(x) \to 0$ as $x \to \pm\infty$.

6. Let $f(x) = p(x)/q(x)$ be a rational function where the degree of p is 2. For each of the following, what can you conclude (if anything) about the degree of q?
   (a) f has a horizontal asymptote at $y = 3$.
   (b) f has exactly one vertical asymptote.
   (c) f has no horizontal or slant asymptotes.
   (d) f has exactly two x-intercepts.
   (e) f is graphed in the accompanying figure.

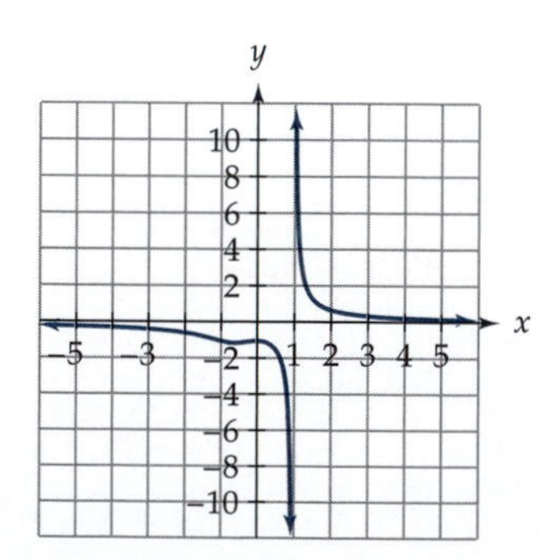

(f) f is graphed in the accompanying figure.

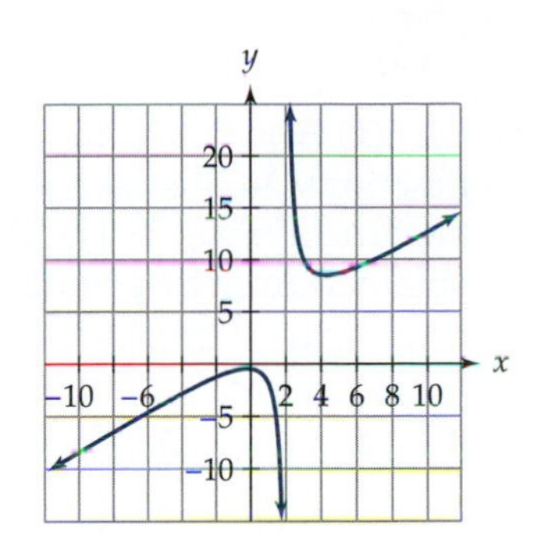

7. Let $f(x)$ be a rational function.
   (a) Suppose f has a horizontal asymptote. Is it possible for f to cross this asymptote? Why or why not?
   (b) Suppose f has a vertical asymptote. Is it possible for f to cross this asymptote? Why or why not?
8. Assume each of the following represents a rational function. Determine a plausible equation (more than one answer may be possible). The dotted lines indicate asymptotes.

(a)

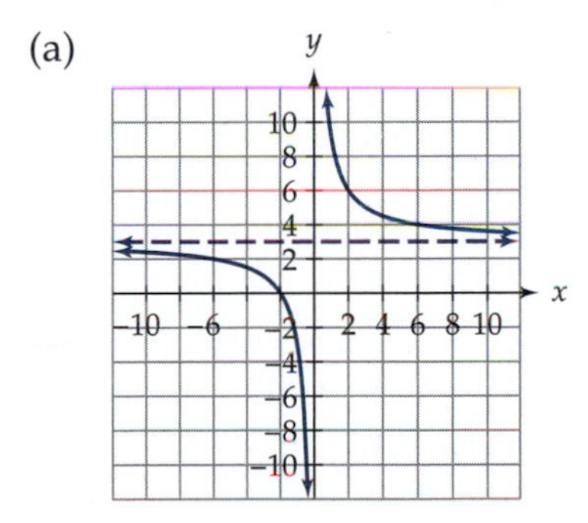

(b)

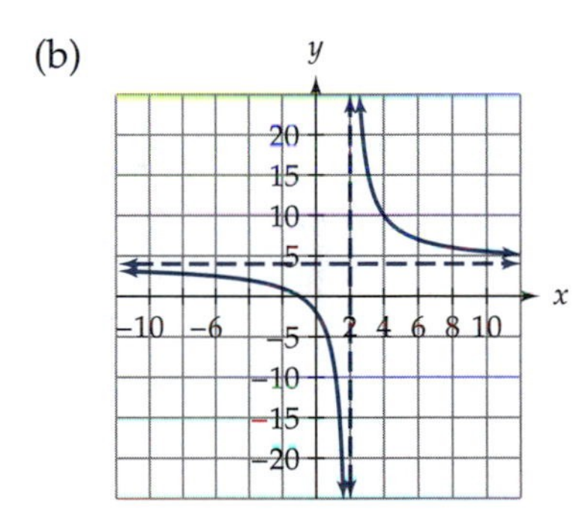

   (c) The function f has a vertical asymptote at $x = 2$, a hole at $(-1, 4)$, an x-intercept at $x = 3$, and a horizontal asymptote at $y = 3$.
   (d) The function g has a vertical asymptote at $x = 2$, x-intercepts at $x = \frac{1}{2}$ and $x = 3$, and a slant asymptote of $y = 2x - 3$.
9. Find the domain, range, x-intercepts, vertical asymptotes, and long-range behavior for each of the following. Draw a sketch of the graph.
   (a) $y = \dfrac{x^3 - 1}{x^3 - 2x^2 + x}$
   (b) $y = \dfrac{3}{x^2 - 1}$

(c) $y = \dfrac{3}{x^2+1}$

(d) $y = \dfrac{2}{x^3+1}$

10. Let $f(x) = p(x)/q(x)$ be a rational function where p and q are both linear functions. In this case, f can usually be written as a transformation of the function $v(x) = 1/x$. Recall that transformation means that f can be obtained from v by horizontally and vertically shifting and stretching.
    (a) Let
    $$g(x) = \frac{3x+2}{x-1}.$$
    Show that $g(x) = 5v(x-1) + 3$.
    (b) Rewrite
    $$h(x) = \frac{x+1}{4x-10}$$
    as a transformation of v.
    (c) Rewrite
    $$j(x) = \frac{ax+b}{cx+d}$$
    as a transformation of v. (Assume $a \neq 0$ and $c \neq 0$.)
    (d) It was stated that *most* rational functions composed of the ratio of two linear functions can be written as a transformation of the function $v(x) = 1/x$. When can such a function *not* be written as a transformation of v?
11. The function $w(d) = (1.58 \times 10^7)w_0 d^{-2}$ finds the weight, w (in pounds), for an object d miles from the center of Earth. The object's weight on Earth's surface (in pounds) is w_0.[20]
    (a) The radius of Earth is approximately 3960 mi. Rewrite the function w so that the input, s, is the distance (in miles) of the object above the surface of Earth. Your new function is $w(s)$.
    (b) The function $w(s)$ is a composition of two functions. Give these two functions.
    (c) Find the weight of a 200-lb person who is 300 mi above the surface of Earth.
    (d) Sketch a graph of the function $w(s)$ assuming $w_0 = 200$ lb.
    (e) What is the long-range behavior of $w(s)$?
    (f) Describe the physical meaning of the long-term behavior of $w(s)$.
12. The volume of a 12-oz soft drink can is 355 cm^3. Assume you want to change the dimensions of the can while keeping the volume constant.
    (a) Show that the function that gives the surface area of the new can is $A(r) = 2\pi r^2 + 710/r$, where $A(r)$ is the surface area (in square centimeters) and r is the radius (in centimeters).

[20] *This function was used in Example 3, section 4.3.*

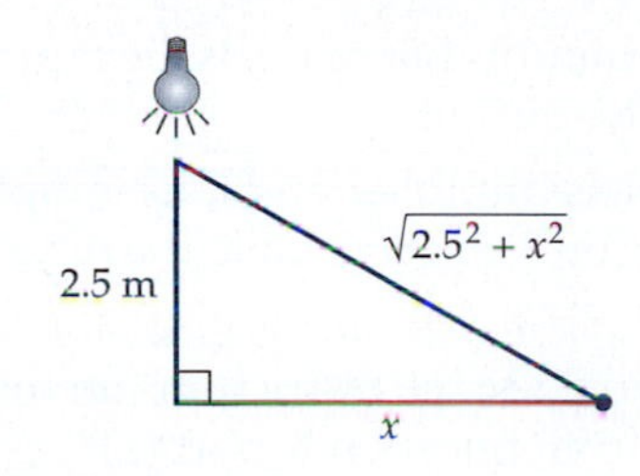

Figure 21

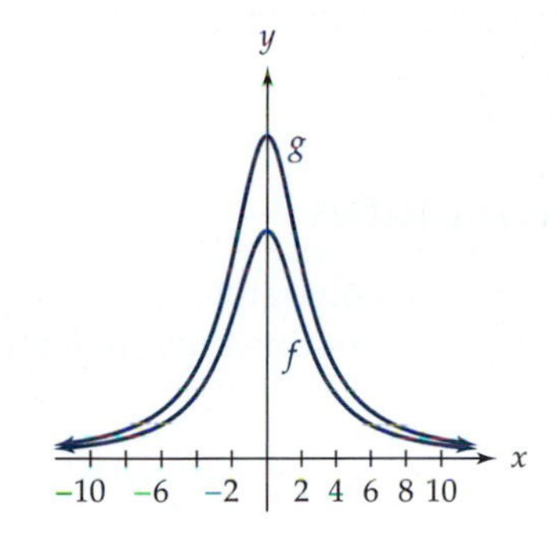

Figure 22

(b) Find the domain, zeros, vertical asymptotes, and long-range behavior for $A(r)$.

(c) Describe the physical meaning of the domain, zeros, vertical asymptotes, and long-range behavior.

13. In this section, we described a physical situation that found the brightness of light, E, on an object that was x meters from the point directly under the light (see Figure 21). Assume the height of the ceiling is 2.5 m. Figure 22 shows the graph of E for a 60-W bulb and the graph of E for a 75-W bulb.
    (a) Which graph is for the 75-W bulb? Briefly justify your answer.
    (b) Notice that as x gets large, the two graphs get closer to each other. What is the physical interpretation of this phenomenon?
    (c) Estimate how far an object should be from the point directly under the 75-W bulb for the brightness to be the same as the maximum possible brightness with a 60-W bulb.

INVESTIGATIONS

INVESTIGATION 1: CASHBACK BONUS AWARD

Discover, a credit card company, has a Cashback Bonus Award Program. In this program, each year a cardholder receives a check representing a portion of the money he or she charged to Discover during the year. The percentage that each cardholder receives depends on the amount of money that was charged, according to the following plan.

0.25%	of the first $1000 in purchases
0.50%	of the second $1000 in purchases
0.75%	of the third $1000 in purchases
1.00%	of the amount of purchases in excess of $3000

1. Write a piecewise function where the amount charged, a, is the input and the cashback bonus award, c, is the output.
2. If you charge $500 in a year, it is clear that you earn 0.25% as a cashback bonus rate. In general, if a is the amount charged, then your *cashback bonus rate* is r, where $r \cdot a =$ cashback bonus award. For any yearly charges less than $1000, notice that your cashback bonus rate is always 0.25%. If you charge above $1000 in a year, though, then your cashback bonus rate increases with each additional dollar charged.
    (a) What is your cashback bonus rate if you charged $1050 in a year?
    (b) What is your cashback bonus rate if you charged $1950 in a year?
    (c) What is the maximum possible cashback bonus rate if you charge up to $2000 in a year?
3. Assume someone charges over $3000 in a given year and, because of the restrictions on the credit card, is not allowed to charge over $2000 per month or $24,000 per year.

(a) Determine the function where the amount charged, a, is the input and the cashback bonus rate, r, is the output.

(b) Determine the x-intercepts, vertical asymptotes, and long-range behavior of the function from question 3, part (a). Give the physical interpretation of each.

(c) In view of the physical situation, what are the restrictions on the domain and range of the function from question 3, part (a)?

(d) Suppose you heard on a commercial, "You could earn up to 1% of the amount you charged as a cashback bonus award using your Discover Card." Would you agree or disagree with this statement? Explain.

INVESTIGATION 2: RANGE OF RATIONAL FUNCTIONS

Finding the range of a rational function can be quite challenging, as we saw in this section. In this investigation, we see how to determine the range for rational functions of the form

$$y = \frac{x^2 + bx + c}{x + d},$$

where $(x + d)$ is not a factor of the numerator.

1. Let

$$f(x) = \frac{x^2 - 4x + 7}{x - 3}.$$

   (a) Determine the domain, x-intercepts, vertical asymptotes, and long-range behavior of f.

   (b) Draw a sketch of the graph of f. Include dotted lines representing the vertical asymptote and the slant asymptote.

   (c) Looking at the graph, estimate the range of f. The range is of the form: "The range is all real numbers such that y is *not* between α and β."

   (d) What is the midpoint between α and β? What is the significance of this value in terms of your graph?

2. Let

$$f(x) = \frac{x^2 - 4x + 7}{x - 3}.$$

   From question 1, you should have estimated the range of f as all real numbers such that y is not between -2 and 6. We now prove that this estimation is correct.

   (a) If there was a value, x_0, such that $f(x_0) = 1$, then we should be able to find it by solving the equation

$$f(x_0) = 1$$

$$\frac{x_0^2 - 4x_0 + 7}{x_0 - 3} = 1$$

$$x_0^2 - 4x_0 + 7 = x_0 - 3$$
$$x_0^2 - 5x_0 + 10 = 0.$$

Show that this equation has no real solutions and hence that there is no value such that $f(x_0) = 1$.

(b) Now let y_0 represent a possible output, that is, $f(x_0) = y_0$. Then, repeating the procedure in part (a), show how you can get the equation

$$x_0^2 - (4 + y_0)x_0 + (7 + 3y_0) = 0.$$

(c) For what values of y_0 does this equation have no real solutions?

3. Let

$$g(x) = \frac{x^2 + bx + c}{x + d},$$

where $(x + d)$ is not a factor of the numerator. Using the same procedure as in question 2, part (b), show that the range of g is all real numbers such that y is *not* between $b - 2d - 2\sqrt{d^2 + c - bd}$ and $b - 2d + 2\sqrt{d^2 + c - bd}$.

THE AMAZING GOLF-O-METER

Professional golfers take great care in determining the distance from their golf ball to the green. Because they have a good idea how far they can hit the ball with each club, knowing the distance pays off. Golf courses help golfers determine these distances in a variety of ways. Some have elaborate books that give distances from the green to certain objects on the hole. Other golf courses have nothing at all, causing golfers to guess the distance they need to hit the ball. So, a device available in golf accessory catalogs claims to measure the distance to the flag stick quickly and easily. This product goes by different names, depending on the manufacturer, but we simply refer to it as a golf-o-meter. In this project, you learn how to make your own custom fit golf-o-meter and discover a flaw you may encounter in commercial ones.

A simple golf-o-meter consists of the graph in Figure 1 printed on a small piece of clear plastic. Looking toward the flag on the green, you hold this piece of plastic up at arm's length. You then move the graph so that the image of the bottom of the flag stick is on the horizontal axis and the top just touches the graph. The distance to the flag is then read on the horizontal axis.

1. To make a standard golf-o-meter, assume the distance from your eye to the end of your outstretched arm is 2 ft and the height of the flag stick on your golf course is 7 ft. When you look through your measuring device, the triangles shown in Figure 2 are formed.

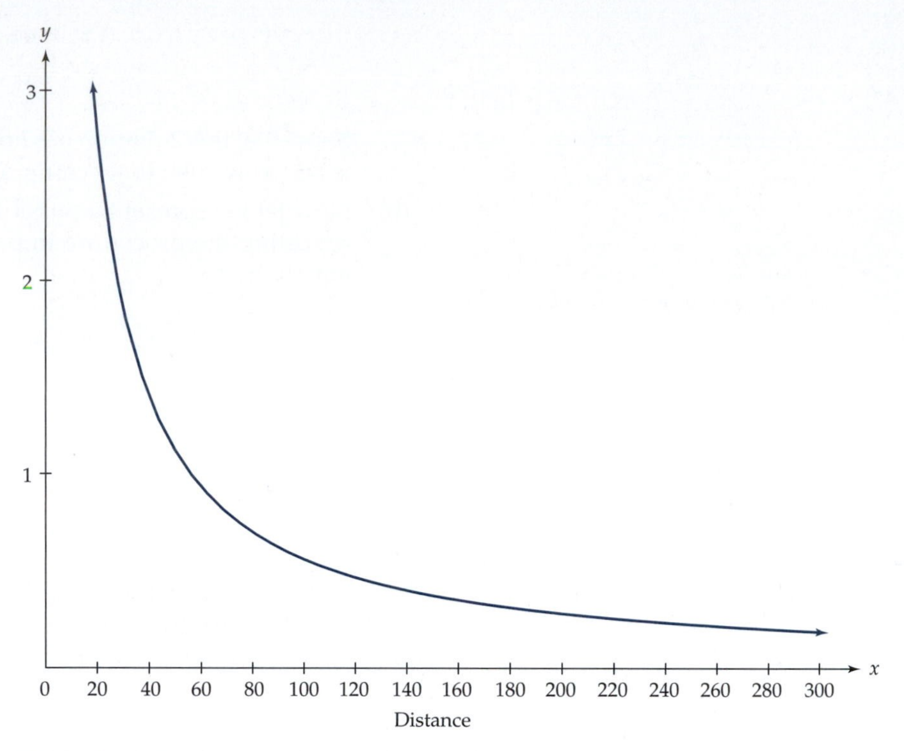

Figure 1

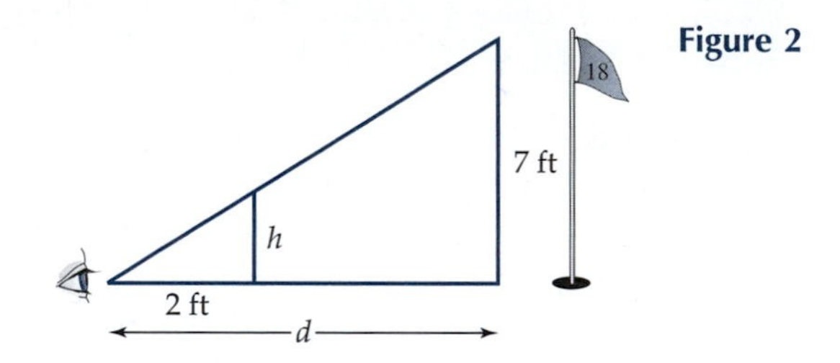

Figure 2

(a) Using Figure 2, find a function such that your input, h, is the apparent height of the flag stick as seen through your golf-o-meter and the output, d, is the distance you are away from the flag stick.

(b) Graph your function with a domain of $0 \text{ ft} \leq h \leq 0.5 \text{ ft}$ (reasonable answers for how tall the flag stick appears compared with the scale at the end of your outstretched arm) and with a range of $0 \text{ ft} \leq d \leq 900 \text{ ft}$.

(c) The way the graph is currently set up presents a problem with using your golf-o-meter. The input is the height of the flag stick, h, which is on the horizontal axis. This input, however, does not work on our graph; when using your golf-o-meter, the image of the flag stick is not horizontal, but vertical. Thus, the observed height of the flag stick needs to be on the vertical axis, even though we think of it as the input. There is nothing wrong with doing this rearrangement; it just runs counter to the way we normally think of graphs. By this change, the horizontal axis is now the distance to the green, d, and the vertical axis is now h. This arrangement is also handy because the image of the flag stick lines up with the numbers representing the distance, thus making the golf-o-meter easier to read. To switch things around, solve your equation for h, the observed height of the flag stick, and graph it with d on the horizontal axis and h on the vertical axis. (*Note:* Make sure you change the domain and range as well.)

(d) The distance of your outstretched arm and the height of the flag stick were both given in feet. Thus, the function you obtained has its input, the distance from the flag stick, as well as the output, the height of the curve, in feet. Golf course distances, however, are always given in yards, and because the height of the curve is quite small, feet is a cumbersome unit of measure. It would be more appropriate if the measure were in inches. Convert your function from question 1, part (c) so that the input is in yards and the output is in inches.

(e) Graph your modified function from question 1, part (d) with a domain of $0 \text{ yd} \leq d \leq 300 \text{ yd}$. This time, however, when you put it on paper, make sure that your vertical axis is scaled properly. For example, the point on the graph where the h-coordinate is 1 in. should be exactly 1 in. from the horizontal axis. Congratulations, you now have your own golf-o-meter! To use it, reproduce the graph on a clear piece of plastic.

2. Not all golf courses have 7-ft-high flag sticks. Our golf-o-meter, however, was constructed under this assumption, so our golf-o-meter will be inaccurate on courses with flag sticks that are not 7 ft.[21] Suppose you take your golf-o-meter to a course that has flag sticks that are 6.5 ft tall. Let's check the accuracy.

   (a) Derive the function for a golf-o-meter where the input, d, is in yards and the output, h, is in inches, for a 6.5-ft-tall flag stick.

   (b) Complete Table 1. Use your function from question 1, part (d) for the second column and your function from question 2, part (a) for the third column. Subtract your two answers for the fourth column. Because the flag stick is really 6.5 ft but the reading for the golf-o-meter is for a 7-ft flag stick, there is error. Subtracting the two columns gives you that error. (*Note:* The input for the table is h, not d.)

   (c) Using your numerical data, notice that this error is not constant but is dependent on your distance from the flag. In other words, the error is a function of distance.

      i. Use your numerical data to find a pattern between the distance read on the golf-o-meter (the 7-ft distance column)

[21] *The distance from your eye to the end of your outstretched arm can also vary and thus cause more inaccuracy.*

TABLE 1

h (in.)	7-ft Distance (yd)	6.5-ft Distance (yd)	Error (yd)
0.2			
0.4			
0.6			
0.8			
1.0			

and the error of this reading (the error column). Use this pattern to predict what the function will be when the input is the incorrect distance given by the 7-ft golf-o-meter and the output is the error.

ii. Manipulate your formulas symbolically to show that your predicted function is the actual function. To do so, remember that the error is the difference between the 7-ft and the 6.5-ft golf-o-meter readings. This function has to be written such that the input is d, not h.

iii. Should this error be added to or subtracted from the distance you read on your 7-ft golf-o-meter? Explain your answer.

(d) Let's determine if this error is significant. If you were a golfer, you are most apt to use a device like the golf-o-meter when you are between 50 and 250 yd away from the flag stick. Golf clubs with the lower numbers allow you to hit the ball farther. For each number lower, you can hit the ball about 15 yd farther. For example, if a seven-iron will give you about 150 yd, a six-iron will probably give you about 165 yd. Taking this all into account, do you think the error in using a 7-ft golf-o-meter on a course with 6.5-ft flag sticks is significant? Explain your answer.

REVIEW EXERCISES

1. Find the vertex of each of the following.
   (a) $f(x) = 2(x+4)^2 - \frac{3}{2}$
   (b) $g(x) = 2x^2 + 8x - 7$
   (c) $h(x) = -x^2 + 6x + 10$
2. Find the real roots of the following quadratic functions.
   (a) $f(x) = (x+3)^2 - 4$
   (b) $g(x) = x^2 + 10x + 24$
   (c) $h(x) = x^2 - 5$
   (d) $j(x) = 2x^2 + 3x - 8$

3. Find the equation for the quadratic function that contains the following.
   (a) Vertex at $(0, 0)$ and the point $(2, 5)$
   (b) Vertex at $(2, 5)$ and the point $(1, 3)$
   (c) The three points $(1, 0)$, $(4, 0)$ and $(0, 8)$
   (d) The three points $(2, 3)$, $(3, 4)$ and $(4, 6)$
4. Find the equations for the following quadratic functions.
   (a) The function $y = x^2$ is shifted one unit to the right and two units up and is vertically stretched by a factor of 2.
   (b) A quadratic function has a vertex at $(0, 5)$ and the leading coefficient is -1.
   (c) A quadratic function has roots at $x = -2$ and $x = 5$ and contains the point $(-1, -2)$.
5. Suppose a baseball is hit straight up in the air so that its height can be determined by the equation $h(t) = -16t^2 + 63t + 4$, where $h(t)$ is the height of the ball, in feet, after t seconds.
   (a) Find the maximum height of the ball.
   (b) Find the time it takes the ball to hit the ground.
6. Give the smallest possible degree for each of the following polynomial functions.
   (a) The polynomial function p has three real roots, changes direction two times, and changes concavity one time.
   (b) The polynomial function q has five real roots, changes direction six times, and changes concavity five times.
   (c) The polynomial function r has three real roots, changes direction two times, and changes concavity three times.
7. Assuming the graph does not change directions for x-values not shown, use each graph to give the smallest possible degree, determine if the leading coefficient is positive or negative, and classify the multiplicity of each root as 1, even, or an odd number greater than 1. Briefly justify your answers.

(a)

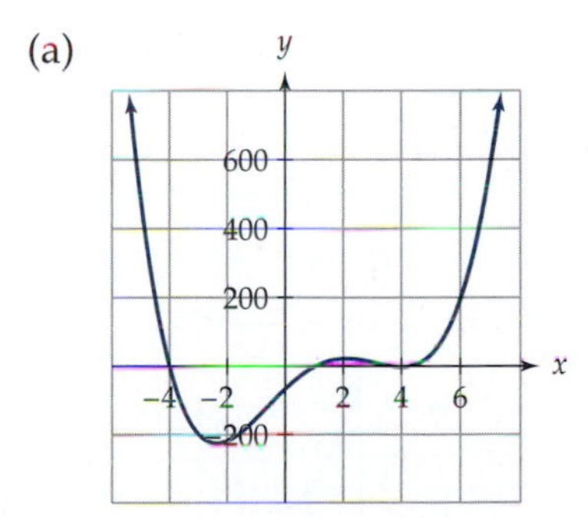

(b)

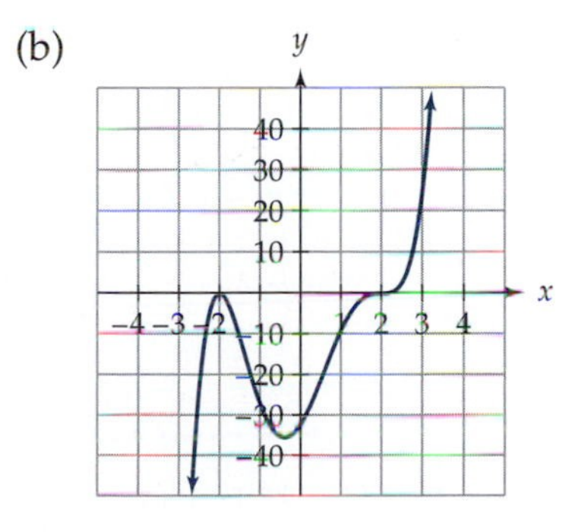

8. Match the functions to the graphs. Assume the graphs do not change direction outside the viewing window.
   (a) $y = x^6 - 10x^5$

(b) $y = -x^4 - 5x^3 - 4x^2 - x - 1$
(c) $y = -x^3 + 2x^2 - 10x$
(d) $y = 3x^5 + 25x^2 + 3$

i.

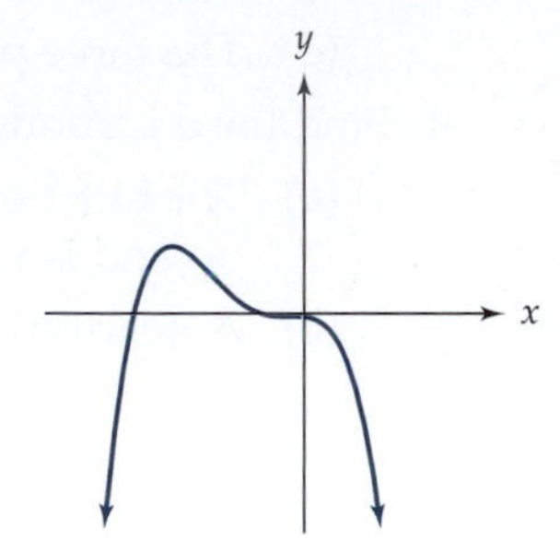

ii.

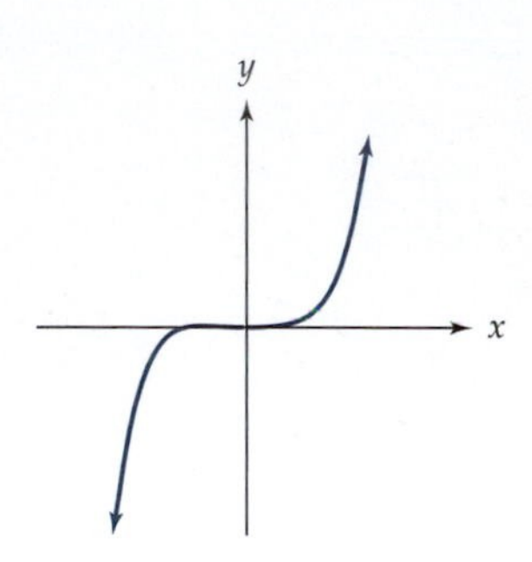

iii.

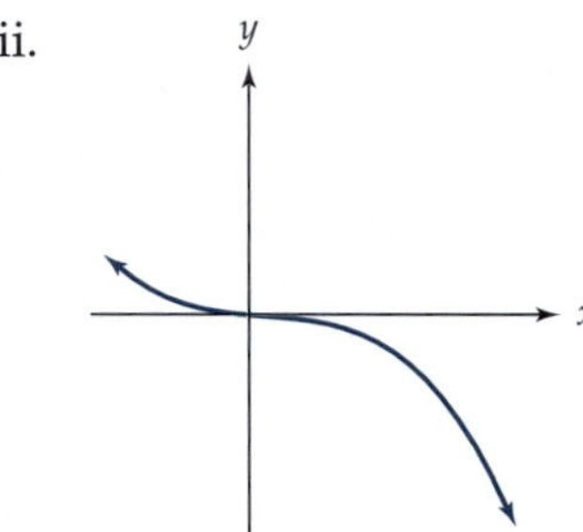

iv.

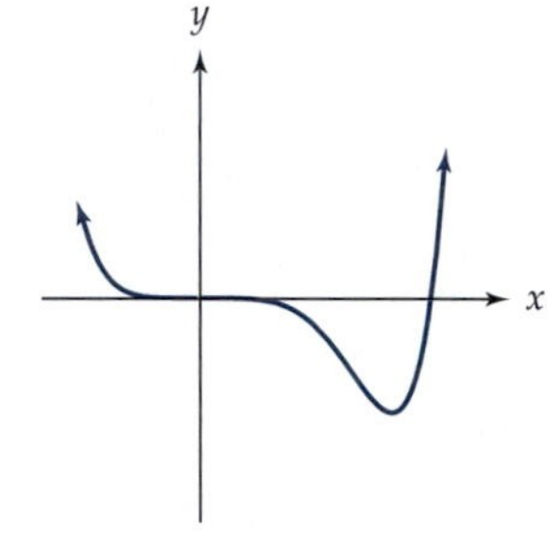

9. Find the exact values of the real roots for the following polynomial functions.
   (a) $y = x^3 - 9$
   (b) $y = 2(x - 2)(x + 3)(x - 5)^2$
   (c) $y = 2x^3 - 5x^2 + 2x$
10. Give a possible equation for the polynomial functions shown in the following graphs. You may assume that the absolute value of the leading coefficient is 1.

(a)

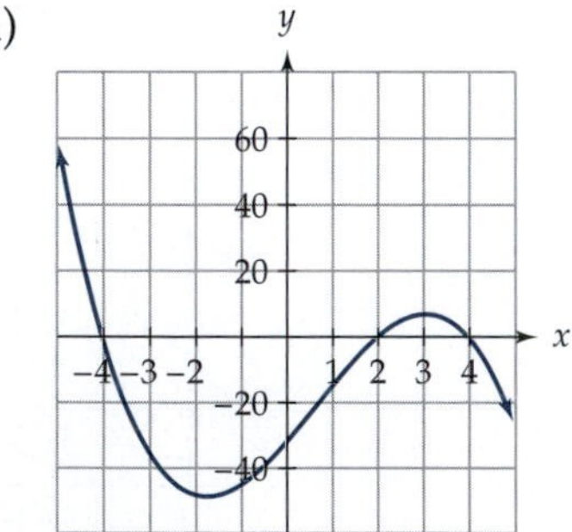

(b)

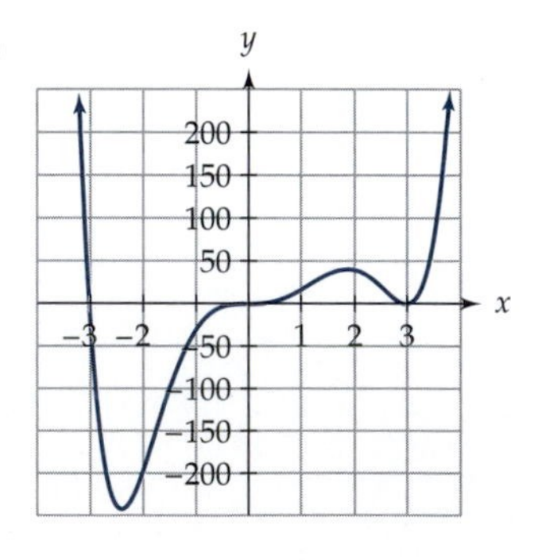

11. Determine whether the following statements describe horizontal asymptotes, vertical asymptotes, or neither.

(a) $h(x) \to 3$ as $x \to 3$.

(b) $j(x) \to 3$ as $x \to \infty$.

(c) $f(x) \to -\infty$ as $x \to 0$.

(d) $g(x) \to +\infty$ as $x \to -\infty$.

12. Describe the long-range behavior and the behavior close to the y-axis for the following functions.

    (a) $f(x) = 4x^{-3}$

    (b) $g(x) = -2x^{-3/2}$

    (c) $h(x) = 3x^{-4/3}$

13. Given the function $f(x) = x^{-8/3}$, explain the following.

    (a) Why f does not give negative outputs

    (b) Why f does not have an inverse

14. Given the function $g(x) = 2x^{-3/4}$, explain the following.

    (a) Why g is not defined for negative inputs

    (b) Why g only gives positive outputs

15. Suppose you are taking a 100-mi trip on a freeway. For 80 mi, the speed limit is 70 mph, and for 20 mi, the speed limit is 55 mph.

    (a) What is your average speed for the entire trip if you travel the speed limit?

    (b) Instead of traveling the speed limit, you decide to always travel 5 mph over the speed limit. How much time (in minutes and seconds) do you save?

16. For each of the following rational functions, determine the domain, x-intercepts, and vertical asymptotes. Draw a sketch of the graph.

    (a) $y = \dfrac{x+7}{x-1}$

    (b) $y = \dfrac{x^3 - x^2 - 6x}{x+2}$

    (c) $y = \dfrac{2x^2 - 24x + 70}{x^2 - 4x - 5}$

17. Determine the long-range behavior of each of the following rational functions.

    (a) $f(x) = \dfrac{2x+3}{5x-7}$

    (b) $g(x) = \dfrac{3x^3 + 7x^2 - 4x + 1}{x^2 + 2x + 1}$

    (c) $h(x) = \dfrac{5x^2 + 2x - 9}{x^3 - 2x^2 + 10x + 4}$

18. Assume each of the following represent a rational function. Determine a plausible equation. (More than one answer may be possible.)

    (a) The function f has a vertical asymptote at $x = -3$, a hole at $(-1, -6)$, an x-intercept at $x = 5$, and horizontal asymptote at $y = 3$.

(b) The function g has a vertical asymptote at $x = 4$, x-intercepts at $x = 5$ and $x = \frac{2}{3}$, and a slant asymptote of $y = 6x - 10$.

19. Suppose you were a single person with no dependents in 1998 and for tax purposes used the standard deduction. If you earned between \$7700 and \$32,300, you would have to pay 15% of the income you earned above \$7700 in federal income taxes.
    (a) Find a formula where the amount of income is the input and the proportion of income owed in federal income tax is the output.
    (b) What type of function is your answer to part (a)?
    (c) Find the domain and range for the function from part (a).

CHAPTER OUTLINE

CHAPTER OVERVIEW

- Definitions of sine, cosine and tangent functions
- Radian measure
- Arc length
- Areas of sectors and triangles
- Angular velocity
- Transformations of trigonometric functions
- Inverse trigonometric functions
- Trigonometric identities

TRIGONOMETRIC FUNCTIONS

In this chapter, the sine and cosine functions, first introduced in chapter 2, are reviewed and other trigonometric functions are introduced. By transforming these functions, we can use them in a variety of applications. The geometry of a circle, including arc length and area, is also explored. Although trigonometric identities are introduced throughout the chapter, they become the main focus in section 5.4.

5.1 TWO WAYS OF DEFINING TRIGONOMETRIC FUNCTIONS

In section 2.4, we introduced periodic functions, including an introduction to the two most commonly used periodic functions, the sine and the cosine. In this section, we look at these functions in more detail and also define the tangent function. We define these functions in two different ways, examine some of their properties, and look at ways to measure angles.

TWO DEFINITIONS OF SINE AND COSINE

The periodic functions sine and cosine were introduced in section 2.4. Our definitions for these involved the unit circle. Recall that we defined $\sin x$ as the vertical position of a point on the unit circle where x is the distance on the circle from the point $(1, 0)$ *counterclockwise* to the point. We also defined $\cos x$ as the horizontal position of a point on the unit circle. So, the coordinates of this point are $(\cos x, \sin x)$. (See Figure 1.)

The sine and cosine functions can also be defined in terms of triangles. If you have a right triangle with angle x, then the definitions of sine and cosine are $\sin x = \text{opposite}/\text{hypotenuse}$ and $\cos x = \text{adjacent}/\text{hypotenuse}$. (See Figure 2.) Obviously, the two ways of defining these functions must be related or they would not be called the same thing. To see this relationship, look at Figure 3. Using the right triangle definitions, the length of the vertical side of the triangle is equal to $\sin x$ (because the length of the hypotenuse is 1) and the length of the horizontal side of the triangle is equal to $\cos x$ (again, because the hypotenuse is 1). The lengths of these two lines also give the coordinates of the point P. So, the circle definitions and the right triangle definitions of sine and cosine seem to be the same. Are they?

Recall that when thinking of $\sin x$ and $\cos x$ in terms of right triangles, the input is the angle. When thinking of $\sin x$ and $\cos x$ in terms of the unit

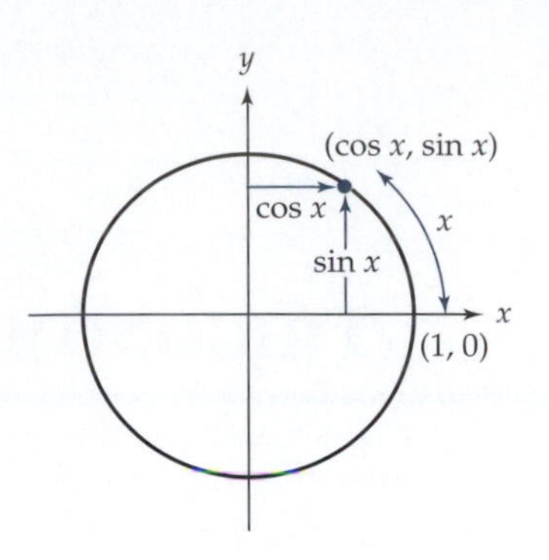

Figure 1 The coordinates of a point on the unit circle centered at the origin are $(\cos x, \sin x)$.

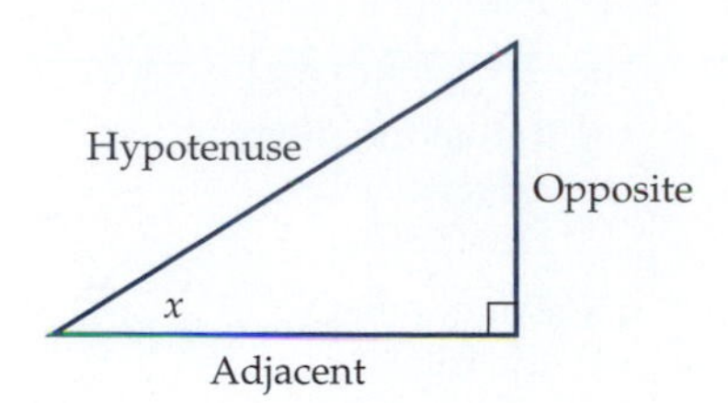

Figure 2 A right triangle showing $\sin x = \text{opposite/hypotenuse}$ and $\cos x = \text{adjacent/hypotenuse}$.

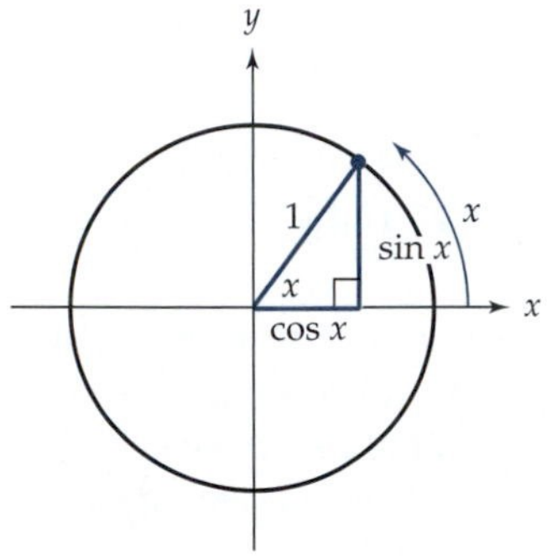

Figure 3 The sine and cosine functions for a right triangle in a unit circle.

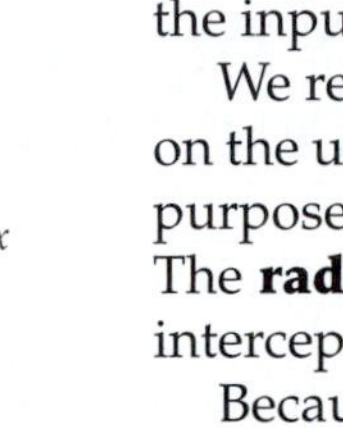

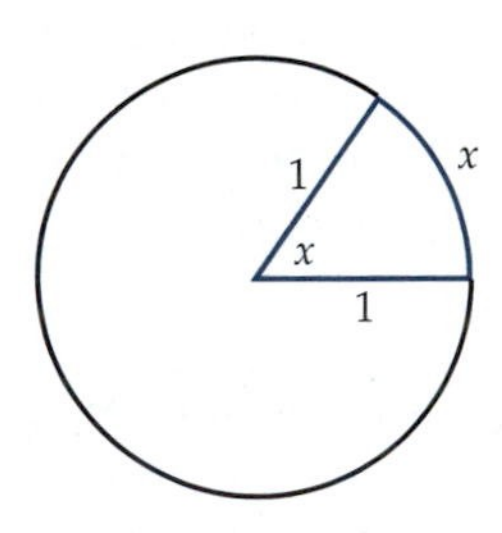

Figure 4 The measure of angle x is x radians.

circle, however, the input is the arc length. Figure 3 leads us to believe that the two ways of defining these functions are the same until we realize that the inputs do not correspond.

We resolve this discrepancy by having the angle equal to the arc length on the unit circle, thus forcing the inputs to be the same. This is exactly the purpose of using radians when working with the sine and cosine functions. The **radian** measure of an angle is the length of the arc on the unit circle intercepted by that angle. (See Figure 4.)

Because the arc length of the unit circle is $2\pi \cdot 1 = 2\pi$, the radian measure of the angle that intercepts the entire circle is also 2π. The radian measure of the angle that intercepts half of the circle must be π, the radian measure of the angle that intercepts a quarter of the circle must be $\pi/2$, and so forth. As long as you are using radians for your input, the right triangle definitions and the unit circle definitions of sine and cosine give the same answer for $0 < x < \pi/2$.

Why have two definitions for sine and cosine? The right triangle definitions are useful for solving problems involving right triangles. We do not have to think of these triangles as being in a circle, and we are not limited to having the length of the hypotenuse equal 1. The right triangle definitions, however, also have limitations. For these definitions, the input for sine or cosine must be one of the acute angles in the right triangle, which limits the input to being an angle between $0°$ and $90°$. In addition, this limitation does not allow you to see the periodic nature of these functions. In the unit circle definition, we do not have to limit ourselves to a domain of $0° < x < 90°$ or, equivalently, $0 < x < \pi/2$. In fact, we do not even have to limit ourselves to a domain of $0 \le x \le 2\pi$, which is one revolution around the circle. If x is greater than 2π, then the input just wraps around the circle more than once. If x is negative, then the input is the distance moving clockwise from $(1, 0)$ instead of counterclockwise. (See Figure 5.)

So, using the unit circle definitions, we see that the domain of the sine and cosine functions is any real number and that the outputs will repeat every 2π radians. Therefore, these functions are periodic with periods of 2π. The unit circle definitions also allow us to see that the range of the sine and cosine functions is -1 to 1. Because $\sin x$ and $\cos x$ are coordinates

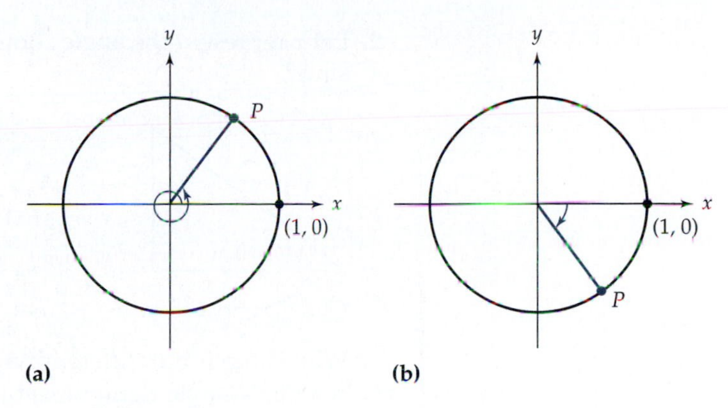

Figure 5 Angles can be greater than 2π, as shown in (a), and negative, as shown in (b).

of a point on the unit circle, $-1 \le \sin x \le 1$ and $-1 \le \cos x \le 1$. The domain, range, and the periodic behavior of these functions can be seen in the graphs in Figure 6.

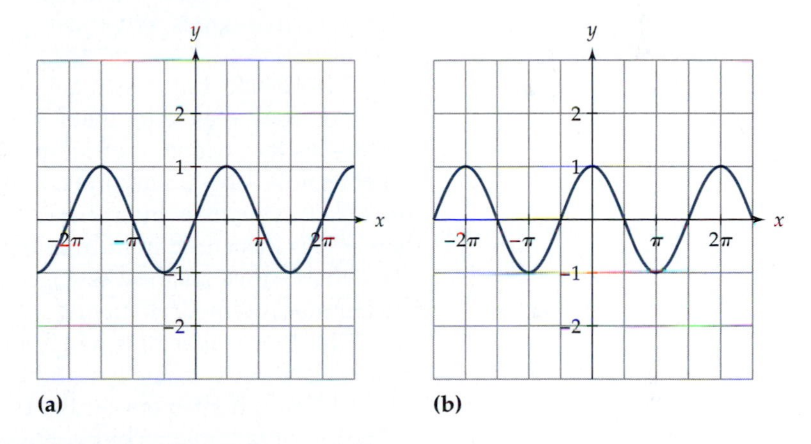

Figure 6 Graphs of (a) $y = \sin x$ and (b) $y = \cos x$.

READING QUESTIONS

1. The coordinates of point A in the accompanying figure are $(\frac{1}{2}, \sqrt{3}/2)$. What is $\cos x$?

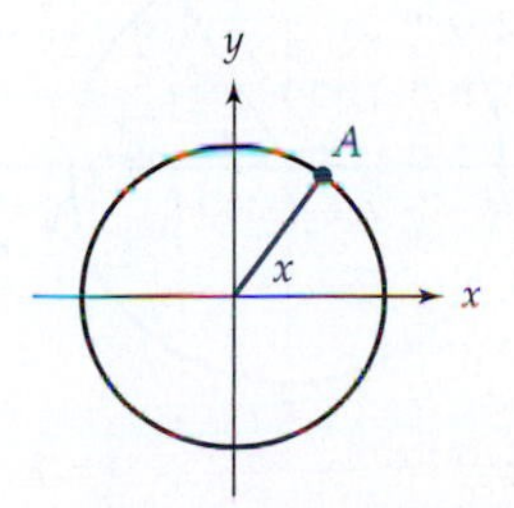

2. Let x represent the angle shown in the accompanying triangle. What is $\sin x$?

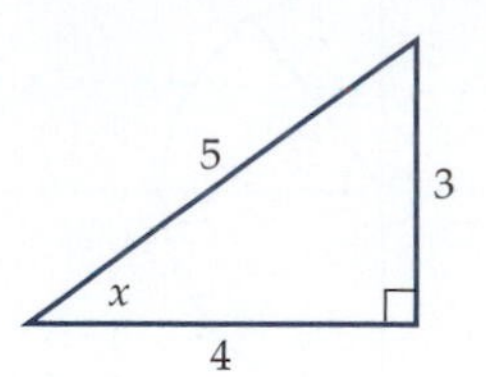

3. Why is the input different when considering the circle definitions versus the triangle definitions of sine and cosine? How is this difference reconciled?
4. Why is it impossible to find an angle x such that $\cos x = 5$?

RADIANS AND DEGREES

In addition to having two ways of defining trigonometric functions, there are two ways of measuring angles. When using triangles, the input is typically given in degrees. When using circles or thinking of sine and cosine as periodic functions, however, the input is typically given in radians. In both cases, because the input is an angle, it is common to denote that angle with the lowercase Greek letter theta, θ, instead of x. Writing $\cos \theta$ and $\sin \theta$ just reminds us that trigonometric functions are associated with angles.

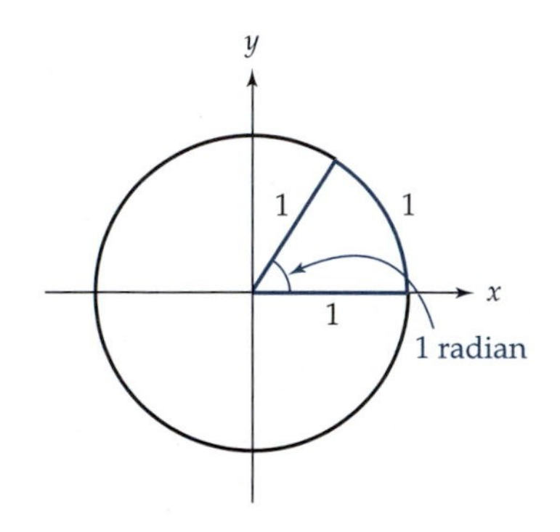

Figure 7 In a unit circle, 1 radian is the measure of an angle that intercepts an arc of length 1.

Let us now look at the connection between degrees and radians. Angles measured in degrees often have "nicer" numbers associated with them than angles measured in radians because there are 360° in a circle and 360 has many factors, such as 30, 45, 60, and 90. Radians, however, are a more natural measurement of an angle because they are based on properties of the circle. Remember that when an angle in a unit circle is measured in radians, it is equal to the length of the arc that it intercepts. Therefore, an angle of 1 radian intercepts an arc of length 1 on a unit circle. (See Figure 7.)

The use of radians can be expanded to include any circle. Figure 8 shows a 1-radian angle on a unit circle, a circle of radius 2, and a circle of radius r. Notice that as the radius of the circles increases, the length of the arc that is intercepted by an angle of 1 radian increases proportionally. On a circle of radius 2, an angle of 1 radian intercepts an arc that is two units long. On a

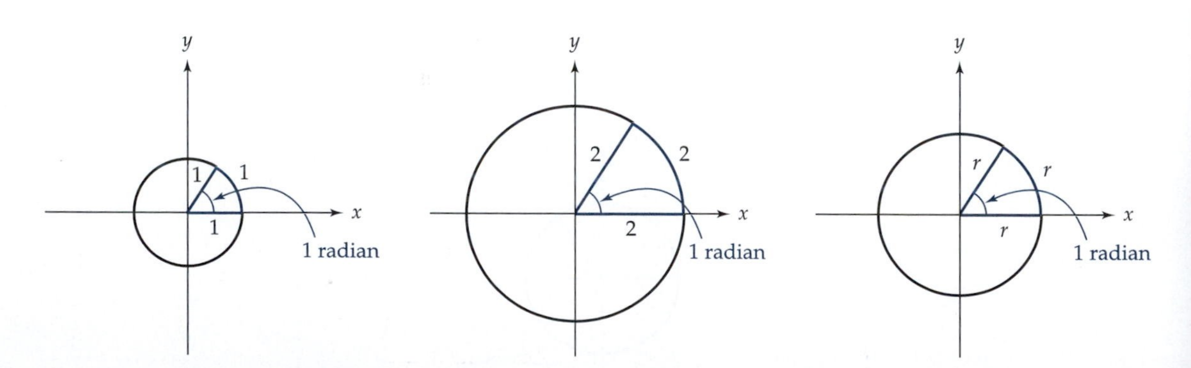

Figure 8 In a circle of radius r, a 1-radian angle intercepts an arc of length r.

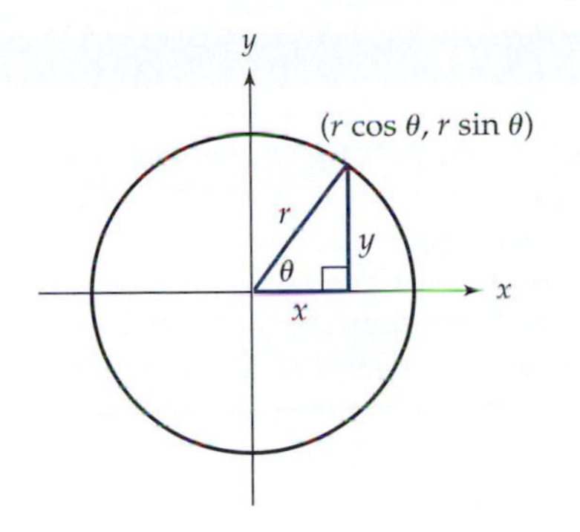

Figure 9 The coordinates of a point on a circle of radius r are $(r \cos \theta, r \sin \theta)$.

circle of radius r, an angle of 1 radian intercepts an arc that is r units long. In general, a 1-radian angle intercepts an arc that is the same length as the radius of the circle.

Just as we can use radians to measure more than angles in a unit circle, we can also find the values of $\cos \theta$ and $\sin \theta$ for points on a circle of radius r. Figure 9 shows a right triangle drawn inside a circle of radius r. Because $\cos \theta = \text{adjacent}/\text{hypotenuse}$, we have $\cos \theta = x/r$, which implies $x = r \cos \theta$. Similarly, we can show that $y = r \sin \theta$. So, the coordinates of a point on a circle centered at the origin of radius r are $(r \cos \theta, r \sin \theta)$.

Now let us look at how to convert from radians to degrees. There are 360° in a circle, and there are also 2π radians in a circle. Recall from section 2.1 that most conversions of units of measurement can be represented by a linear function. Changing from degrees to radians is simply converting units of measurement. For example, suppose you want to convert 90° to radians. Then

$$90° \times \frac{2\pi \text{ radians}}{360°} = 90° \times \frac{\pi \text{ radians}}{180°} = \frac{\pi}{2} \text{ radians.}$$

In general,

$$\theta° \times \frac{\pi \text{ radians}}{180°} = \text{measure of } \theta \text{ in radians}$$

$$\theta \text{ radians} \times \frac{180°}{\pi \text{ radians}} = \text{measure of } \theta \text{ in degrees.}$$

Example 1 Convert 50° to radians.

Solution To convert degrees to radians, we need to multiply by $\pi/180°$.

$$50° \cdot \frac{\pi}{180°} = \frac{50\pi}{180} = \frac{5\pi}{18} \text{ radians.}$$

Example 2 Convert $\pi/6$ radians to degrees.

Solution To convert radians to degrees, we need to multiply by $180°/\pi$.

$$\frac{\pi}{6} \cdot \frac{180°}{\pi} = \frac{180°}{6} = 30°.$$

Some angles, such as 30°, 45°, and 60°, frequently occur in applications. You should recognize these common angles in both degrees and radians. Figure 10 shows some of these common angles and both their degree measurement and their radian measurement.

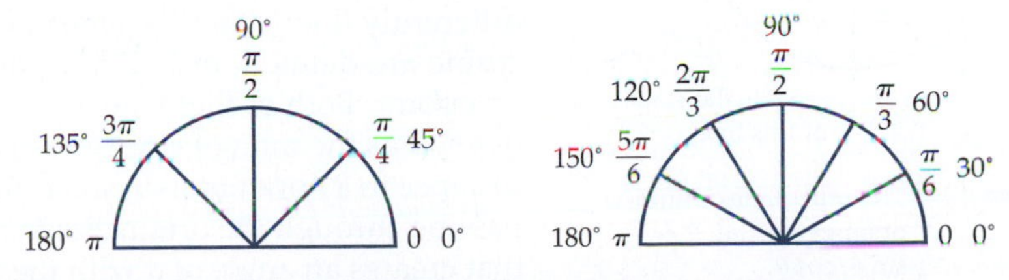

Figure 10 Angle measurements in degrees and radians for various angles.

TECHNOLOGY TIP

It is important to make sure your calculator is interpreting angles with the same measurement units you are. Many calculators use a menu accessed with a mode button to allow you to select either degrees or radians. Having your calculator set in the wrong mode almost always gives you inaccurate results. For example, if the calculator is set in radians, then $\cos \pi/2 = 0$. If the calculator is set in degrees, however, then $\cos (\pi/2)^\circ \approx 0.99962$. These graphs are of $y = \sin\theta$ for $-2\pi \le \theta \le 2\pi$. The left graph is when the calculator was in radian mode, and the right graph is when the calculator was in degree mode.

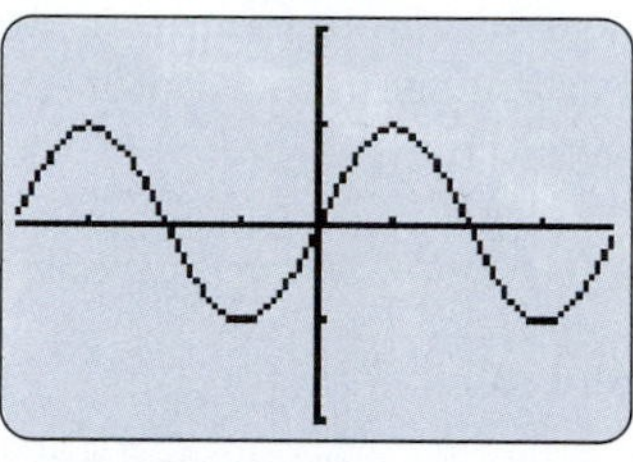

$y = \sin\theta$ for $-2\pi \le \theta \le 2\pi$

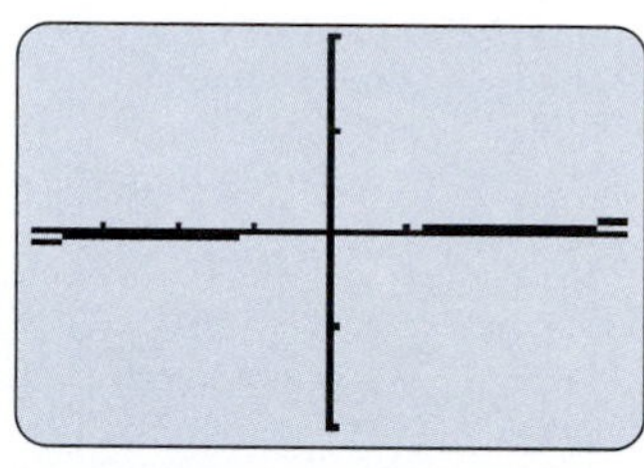

$y = \sin\theta$ for $-2\pi^\circ \le \theta \le 2\pi^\circ$

READING QUESTIONS

5. What is the length of the arc intercepted by a 1-radian angle in a circle of radius 5?
6. What is the x-coordinate of the point on a circle of radius 4 intercepted by an angle of π radians?
7. Convert 45° to radians.
8. Convert $6\pi/7$ radians to degrees.

THE TANGENT FUNCTION

Sine and cosine are not the only periodic functions associated with a triangle. Another commonly used trigonometric function is the tangent function. The **tangent** of θ, $\tan\theta$, is defined as

$$\tan\theta = \frac{\sin\theta}{\cos\theta}.$$

Although the tangent function is periodic, in many ways it behaves very differently than either the sine or cosine function. In a unit circle, sine and cosine are defined, respectively, as the y-coordinate and the x-coordinate of a point. Both of these are represented as *distances*. Tangent, however, is defined as the ratio of sine to cosine. Thus, tangent is the ratio of a vertical distance to a horizontal distance. Such a ratio, however, is the *slope* of a line passing through the origin. So, in a unit circle, $\tan\theta$ is the slope of the line that creates an angle of θ with the positive x-axis. (See Figure 11.)

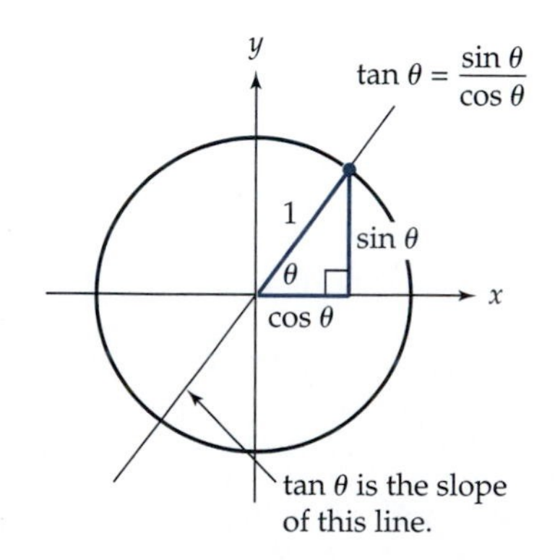

Figure 11 The unit circle definition of tangent is $\tan\theta = \sin\theta/\cos\theta$.

We can also find a right triangle definition of $\tan\theta$. Using our unit circle definition of tangent, our right triangle definitions of sine and cosine, and Figure 12, we have

$$\tan\theta = \frac{\sin\theta}{\cos\theta} = \frac{\text{opposite/hypotenuse}}{\text{adjacent/hypotenuse}} = \frac{\text{opposite}}{\text{adjacent}}.$$

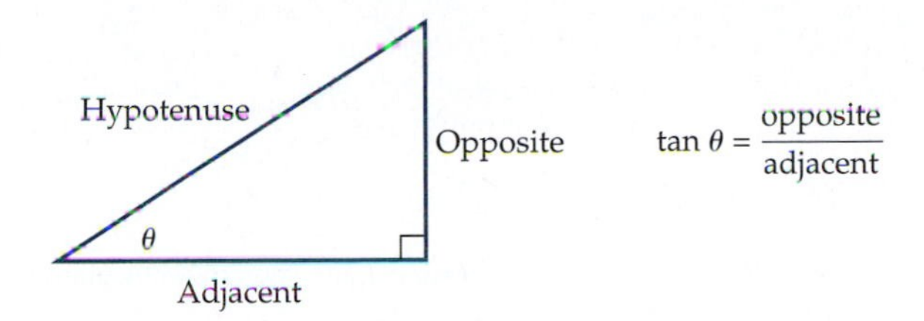

Figure 12 The right triangle definition of tangent is $\tan\theta = \text{opposite}/\text{adjacent}$.

Figure 13 shows a unit circle and a graph of $y = \tan\theta$. As the angle between the x-axis and a line passing through a point in the first quadrant increases, the slope of that line also increases. Starting at the x-axis, the slope is zero because you have a horizontal line. As the angle gets closer to $\pi/2$, the slope gets closer to infinity because the line is becoming more and more similar to a vertical line. Also, $\cos\pi/2 = 0$, so $\tan\pi/2$ is undefined. The slope of a line passing through a point on the unit circle in the second quadrant is negative because the line is decreasing. Therefore, the tangent function is negative in the second quadrant. As the angle gets closer to π, the slope is heading toward zero. Therefore, the tangent function will increase and at $\theta = \pi$, the slope will equal zero. As you can see in Figure 13, the lines in the third quadrant are the same as the lines in the first quadrant just as the lines in the fourth quadrant are the same as those in the second quadrant. Hence, the graph of $y = \tan\theta$ has a period of π. Every π units, the slopes repeat. The tangent function also has vertical asymptotes at $\pi/2$, $3\pi/2$, and every π units after that. Symbolically, the tangent function has a vertical asymptote when $\theta = \pi/2 + n\pi$, where n is an integer. Notice that, unlike the sine and cosine, the range of the tangent function is not restricted.

PROPERTIES

The definition and properties for the tangent function are:

- $\tan\theta = \sin\theta/\cos\theta$ or, equivalently, $\tan\theta$ = slope of the line that creates an angle of θ with the positive x-axis.
- $y = \tan\theta$ is periodic with period π.
- The domain of $y = \tan\theta$ is all real numbers except for $\pi/2 + n\pi$, where n is an integer.
- The range of $y = \tan\theta$ is all real numbers.

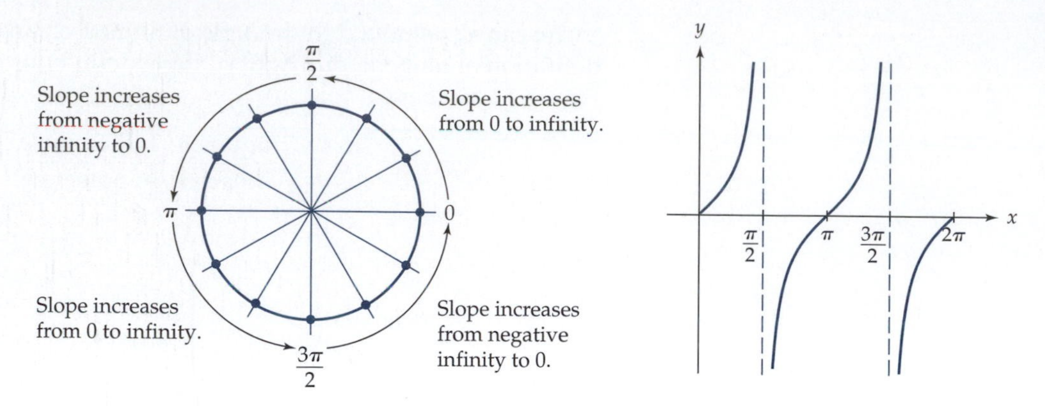

Figure 13 The tangent function is the slope of a line passing through the origin.

HISTORICAL NOTE

The word *trigonometry* comes from the Greek word *trigon,* meaning "triangle," and the Greek word *metron,* meaning "a measure." So, trigonometry means the measuring of triangles. The study of trigonometry goes back thousands of years. In 1858, a Scottish student of antiquities purchased a papyrus scroll that was written somewhere around 1800 B.C. This scroll, now known as the Rhind papyrus (named after the man who purchased it) contains evidence that the ancient Egyptians knew many of the basics of trigonometry.

Even though additional evidence shows that the Babylonians also knew some trigonometry, it is Hippocrates, a Greek who lived in the second century B.C., who is known as the "father of trigonometry" because he was the first to publish a table of the ratios of arcs to chords for a series of angles. Although Egyptian and Babylonian trigonometry seemed to be primarily related to triangles, the Greek tables were based on tables relating circles and their chords for use in astronomy.

The tangent function originated with the *gnomon.* A gnomon was a vertical rod that was used as a time-keeping device, similar to a sundial. The Greek gnomon and shadow functions became known in Latin as the *umbra versa* (turned shadow) and the *umbra recta* (straight shadow). In the late 1500s, these functions became known as the tangent and the cotangent.

The modern-day approach to trigonometry using the unit circle as well as triangles came into being with Leonhard Euler's work in the 1750s. Euler took an analytical approach to mathematics. He unified the study of many different mathematical topics by looking at the abstract study of functions. For example, he considered the sine function to be the y-coordinate of a point on a unit circle rather than the length of a line segment used in finding the distance from Earth to the moon. This analytical approach is used in the study of trigonometry today.

Sources: D. E. Smith, *History of Mathematics, Vol. II,* (New York: Dover 1953), pp. 600–622; Carl B. Boyer, *A History of Mathematics* (Princeton, NJ: Princeton University Press, 1968), pp. 485–86.

READING QUESTIONS

9. What is $\tan \pi/4$? (*Hint:* Recall that $\pi/4 = 45°$ and think of your answer in terms of slope.)
10. Let θ represent the angle shown in the triangle below. What is $\tan \theta$?

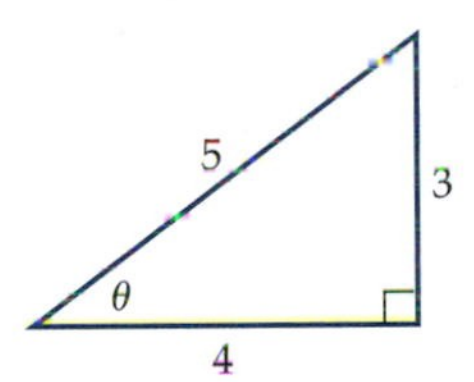

11. Why does $\tan \theta$ have a vertical asymptote when $\theta = 3\pi/2$?
12. Using a unit circle, draw a picture representing an angle θ such that $\tan \theta = 5$.

DETERMINING VALUES FOR SINE, COSINE, AND TANGENT

You can use your calculator to determine approximate values for sine, cosine, and tangent, but there are two things to remember when you do this. First, the answers will most likely be approximations. For example, $\sin \pi/3 = \sqrt{3}/2$. A calculator will give the decimal approximation as 0.8660254038. Second, the calculator must be in the correct mode when determining a trigonometric function of an angle. If asked to find sin 10, you need to know if you are looking for $\sin 10°$ or sin(10 radians). Because there is no degree symbol given, by default we assume you want 10 radians, which gives $\sin 10 \approx -0.5440211109$.

Example 3 Determine decimal approximations of $\cos 4$ and $\cos 4°$.

Solution Using a calculator in radian mode, we can determine that $\cos 4 \approx -0.6536$. Using a calculator in degree mode, we can determine that $\cos 4° \approx 0.9976$.

Integer Multiples of $\pi/2$

Most of the time, we use a calculator to find approximate values for sine, cosine, or tangent. We can, however, find a few values exactly. We start with the points that lie on either the x-axis or the y-axis. In the unit circle definitions of sine and cosine, sine is the vertical distance of a point on the unit circle and cosine is the horizontal distance. To be more precise, we describe this distance as *directed distance*. Direction is important if we think of the coordinates of a point on the unit circle as $(\cos \theta, \sin \theta)$. For example, $\cos 0 = 1$, whereas $\cos \pi = -1$. (See Figure 14.)

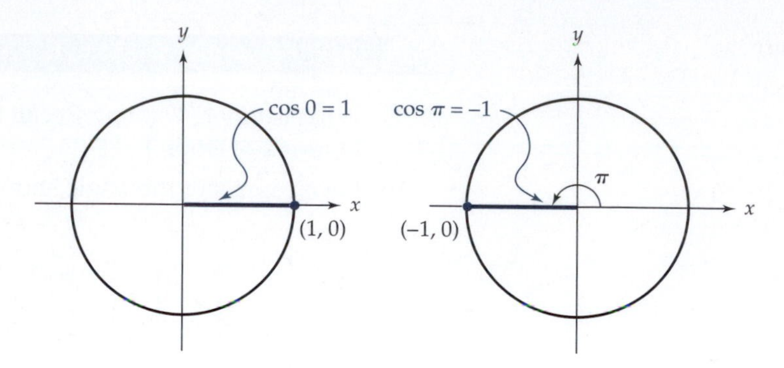

Figure 14 Directed distance is important, as illustrated by $\cos 0 = 1$ and $\cos \pi = -1$.

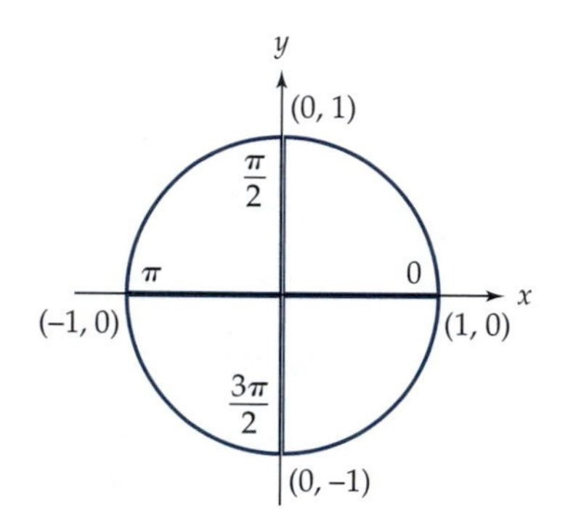

Figure 15

Example 4 Determine the values of sine and cosine for $\theta = 0, \pi/2, \pi,$ and $3\pi/2$.

Solution Using Figure 15 as a guide and knowing that cosine is the x-coordinate of a point on the unit circle and that sine is the y-coordinate, we can see that $\cos 0 = 1$, $\cos \pi/2 = 0$, $\cos \pi = -1$, and $\cos 3\pi/2 = 0$. We can also see that $\sin 0 = 0$, $\sin \pi/2 = 1$, $\sin \pi = 0$, and $\sin 3\pi/2 = -1$.

It is also easy to determine values for the tangent function for $\theta = 0, \pi/2, \pi$, and $3\pi/2$, because tangent is the ratio of sine and cosine or, equivalently, the slope of the line. Using our solutions to example 4 or Figure 15, we can see that $\tan 0 = 0/1 = 0$, $\tan \pi/2$ is undefined, $\tan \pi = 0/-1 = 0$, and $\tan 3\pi/2$ is undefined.

Integer Multiples of $\pi/4$ and $\pi/6$

There are a few other angles for which we can find the exact values of sine, cosine, and tangent. For these angles, it is possible to find the exact values of the sides of the triangles. In particular, because of their special properties, 45°–45°–90° and 30°–60°–90° triangles can be used to find the exact values of trigonometric functions for $\theta = 45°$ or $\pi/4$, $\theta = 30°$ or $\pi/6$, and $\theta = 60°$ or $\pi/3$.

For example, suppose we want to find $\sin \pi/4$. Knowing that sine is the y-coordinate of a point on the unit circle and that $\pi/4 = 45°$, we draw a 45°–45°–90° triangle inside the unit circle. (See Figure 16.) Using the Pythagorean theorem and Figure 16, we see that

$$a^2 + b^2 = 1^2.$$

Note, however, that $a = b$ because we have a 45° angle. Therefore, our equation is

$$b^2 + b^2 = 1.$$

Solving for b, we have

$$2b^2 = 1$$
$$b^2 = \frac{1}{2}$$

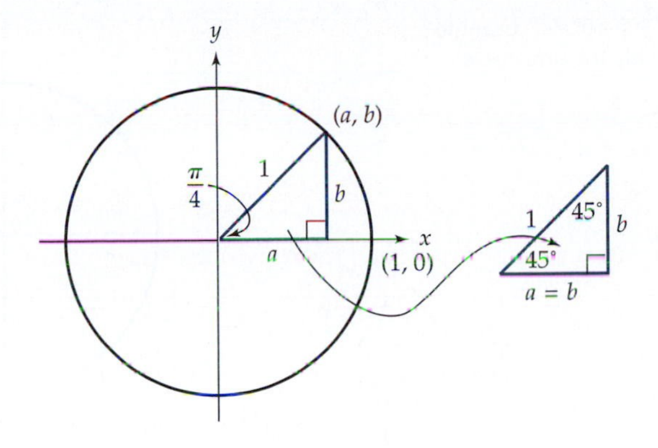

Figure 16 A 45°–45°–90° triangle inside the unit circle.

$$b = \pm\sqrt{\frac{1}{2}}$$
$$b = \pm\frac{\sqrt{2}}{2}.$$

Because b is positive, $b = \sqrt{2}/2$. Knowing that $\sin \pi/4$ is the y-coordinate of our point gives us

$$\sin\frac{\pi}{4} = \frac{\sqrt{2}}{2}.$$

The other triangle we can use to find exact values for our trigonometric functions is a 30°–60°–90° triangle. For example, suppose we want to determine the value of $\cos \pi/6$. Knowing that cosine is the x-coordinate of a point on the unit circle and that $\pi/6 = 30°$, we draw a 30°–60°–90° triangle inside the unit circle. If we flip this triangle over the x-axis, a 60°–60°–60° triangle is formed. (See Figure 17.) A 60°–60°–60° triangle is equilateral and, therefore, has sides of equal length. Because two of the sides are one unit, it is implied that the third side is also one unit. So, $2b = 1$ or $b = \frac{1}{2}$. Using the Pythagorean theorem, we have

$$a^2 + \left(\frac{1}{2}\right)^2 = 1^2$$
$$a^2 + \frac{1}{4} = 1$$
$$a^2 = \frac{3}{4}$$
$$a = \pm\frac{\sqrt{3}}{2}.$$

Because a is positive, we know that $a = \sqrt{3}/2$. Knowing that $\cos \pi/6$ is the x-coordinate of our point, we have

$$\cos\frac{\pi}{6} = \frac{\sqrt{3}}{2}.$$

Remembering that the sides of a 30°–60°–90° triangle in a unit circle are $\frac{1}{2}$, $\sqrt{3}/2$, and 1, whereas the sides of a 45°–45°–90° triangle in a unit circle

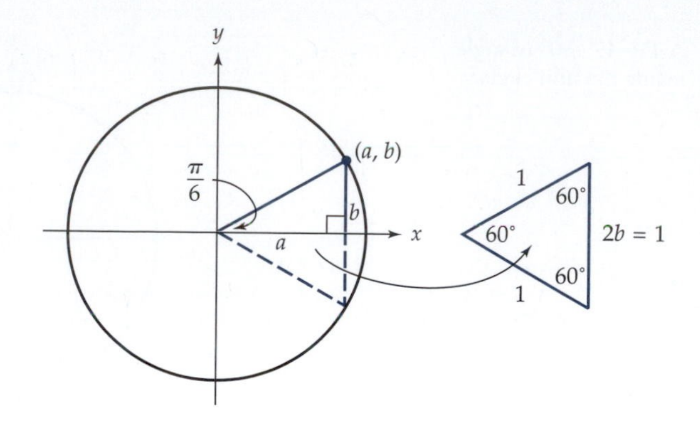

Figure 17 A 30°–60°–90° triangle inside the unit circle.

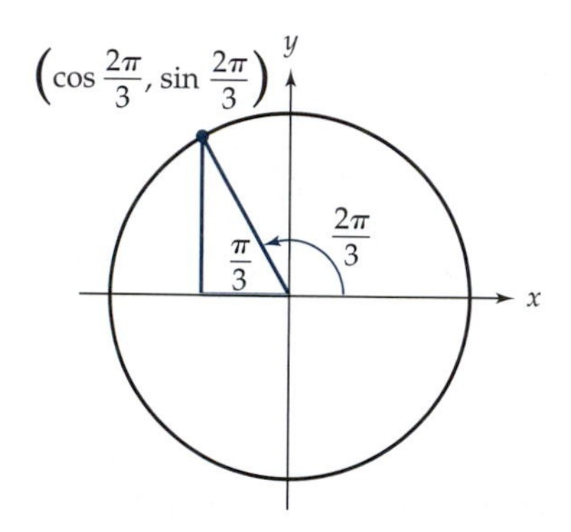

Figure 19 Finding $\cos 2\pi/3$ and $\sin 2\pi/3$ using the unit circle.

are $\sqrt{2}/2$, $\sqrt{2}/2$, and 1, can help determine the sine, cosine, and tangent of angles that are integer multiples of $\pi/6$ and $\pi/4$. (See Figure 18.) Not all these multiples, however, will be in the first quadrant. For example, Figure 19 shows an angle of $\theta = 2\pi/3$. To find the x-coordinate and the y-coordinate of the point on the unit circle (and hence $\cos 2\pi/3$ and $\sin 2\pi/3$), drop a perpendicular to the x-axis. Notice that this line forms a triangle with an angle of $\pi/3$. Knowing that $\cos \pi/3 = \frac{1}{2}$ and that $\sin \pi/3 = \sqrt{3}/2$ almost gives us the coordinates of our point. Because this point is in the second quadrant, however, the x-coordinate must be negative while the y-coordinate is positive. So, $\cos 2\pi/3 = -\frac{1}{2}$ and $\sin 2\pi/3 = \sqrt{3}/2$.

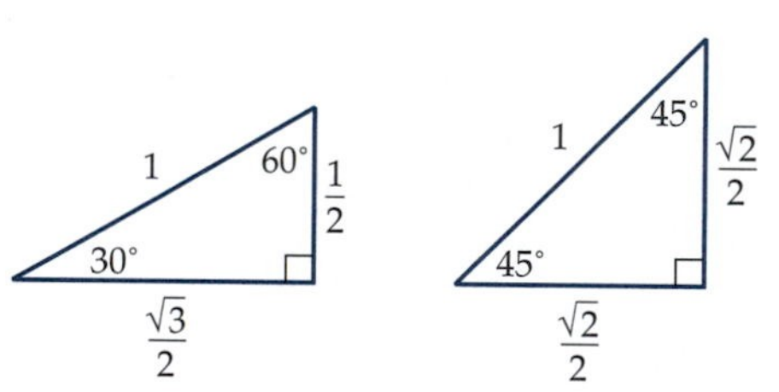

Figure 18 A 30°–60°–90° triangle and a 45°–45°–90° triangle.

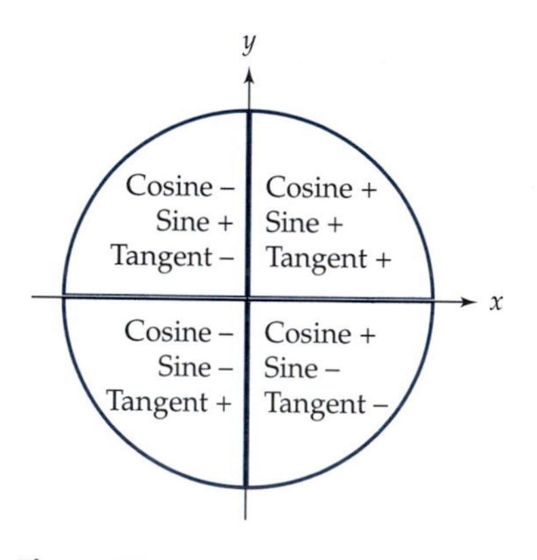

Figure 20 Summary of where the trigonometric functions are positive and negative.

Finding the exact values of the trigonometric functions for special angles involves not only knowing the triangles in Figure 18, but also remembering the signs of $\cos\theta$, $\sin\theta$, and $\tan\theta$ in each of the four quadrants. Because sine equals the y-coordinate, it will be positive in the first and second quadrants and negative in the other two. Because cosine equals the x-coordinate, it will be positive in the first and fourth quadrants and negative in the other two. Because tangent equals the slope (the ratio of sine and cosine), it will be positive in the first and third quadrants and negative in the other two. All this is summarized in Figure 20.

Example 5 Determine $\sin 5\pi/4$.

Solution Because $5\pi/4$ is $\pi/4$ more than π, a 45°–45°–90° triangle is formed. (See Figure 21.) Because the vertical leg of this triangle is below the x-axis, its directed distance is negative. Therefore, $\sin 5\pi/4 = -\sqrt{2}/2$.

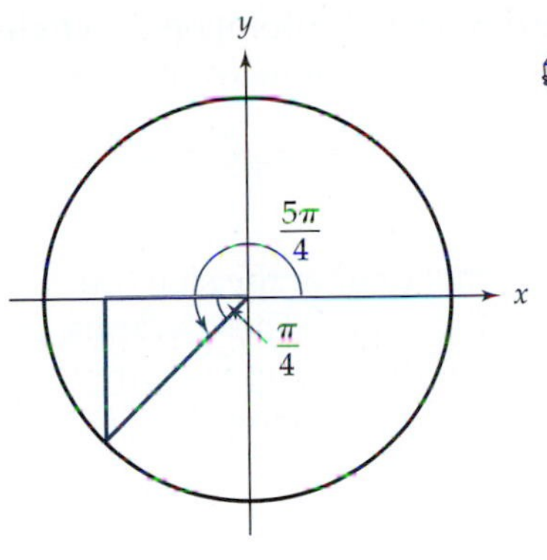

Figure 21

READING QUESTIONS

13. Determine a decimal approximation for cos 22° and cos 22.

14. Which trigonometric function is positive in both the first and the third quadrants?

15. Find each of the following.

(a) $\cos \frac{3\pi}{2}$

(b) $\sin \frac{7\pi}{4}$

(c) $\tan \frac{4\pi}{3}$

IDENTITIES DERIVED FROM THE GRAPHS OF TRIGONOMETRIC FUNCTIONS AND THE UNIT CIRCLE

We can derive some trigonometric identities just by looking at the graphs of trigonometric functions. Figure 22 shows the graphs of $y = \sin x, y = \cos x$, and $y = \tan x$. These graphs all have some sort of symmetry. In section 1.2, we categorized these types of symmetries as *symmetric about the y-axis* and *symmetric about the origin*. The graph of the cosine function is symmetric about the y-axis because the part of the graph to the left of the y-axis is a reflection of the part to the right.

We also learned in section 1.2 that if a function, f, is symmetric about the y-axis, then f has the property $f(x) = f(-x)$. A function that has this property is called an *even function*. We can see that $\cos x$ is an even function by looking at the diagram of the unit circle in Figure 23. The input of $f(x) = \cos x$ is the distance on the circle from $(1, 0)$ to the point. If the x is positive, then the input is the distance moving *counterclockwise* from the point $(1, 0)$. If x is negative, then the input is the distance moving *clockwise* from the point $(1, 0)$. Regardless of whether x is positive or negative, the

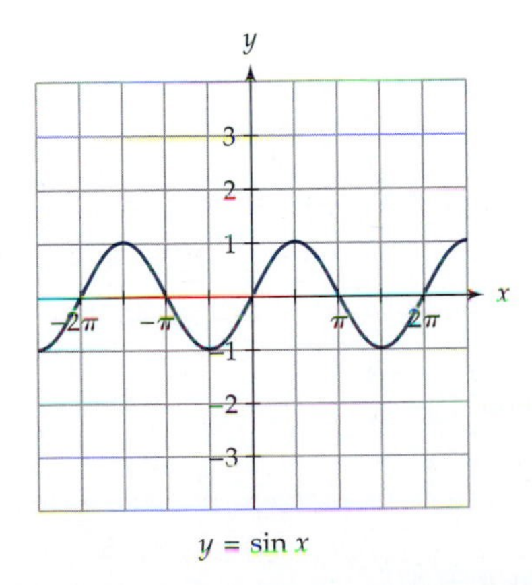

$y = \sin x$

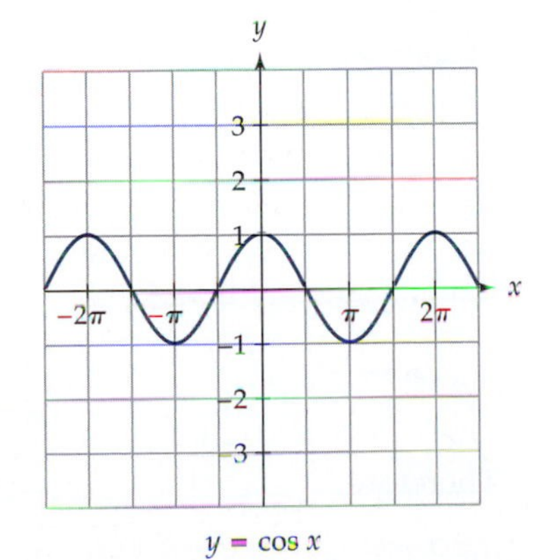

$y = \cos x$

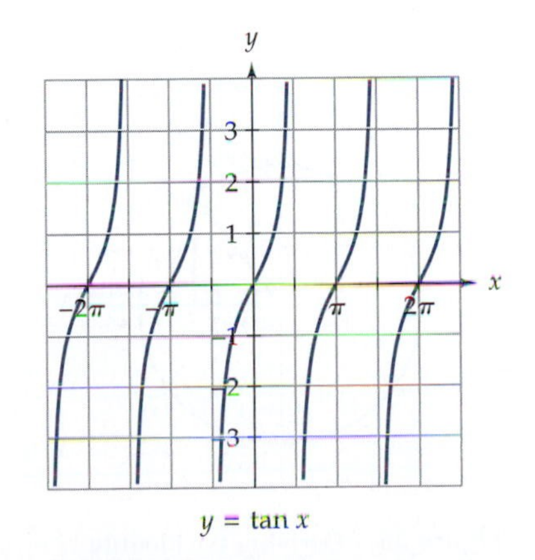

$y = \tan x$

Figure 22 Graphs of three trigonometric functions with a domain of $-2\pi \le x \le 2\pi$.

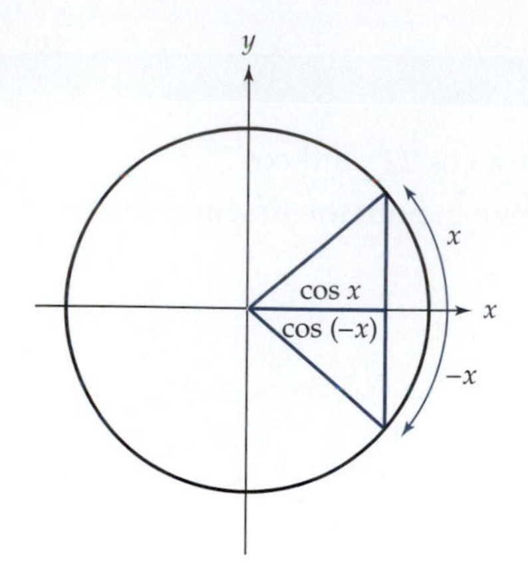

Figure 23 The same line segment represents $\cos x$ and $\cos(-x)$.

output of $f(x) = \cos x$ is the horizontal coordinate of the point on the circle. For a given value of x, this horizontal position is the same for both x and $-x$. (See Figure 23.) Therefore,

$$\cos x = \cos(-x).$$

Looking at the graphs in Figure 22, we see that neither the sine function nor the tangent function is symmetric about the y-axis. These graphs are, however, symmetric about the origin. Remember that a graph is symmetric about the origin if, when rotated 180° with the origin being the center of rotation, it looks the same as it did before it was rotated.

If a function, f, is symmetric about the origin, then f has the property $f(x) = -f(-x)$. A function that has this property is called an *odd function*. We can see that $\sin x$ is an odd function by looking at the diagram of the unit circle in Figure 24. The input of $f(x) = \sin x$ is the same as the input for $f(x) = \cos x$. The output, however, is the vertical position of the point on the circle. For a given value of x, this vertical position is the same distance away from the x-axis regardless of whether x is positive or negative. In one case, however, the output is positive, whereas in the other case, the output is negative. (See Figure 24.) Therefore,

$$\sin x = -\sin(-x).$$

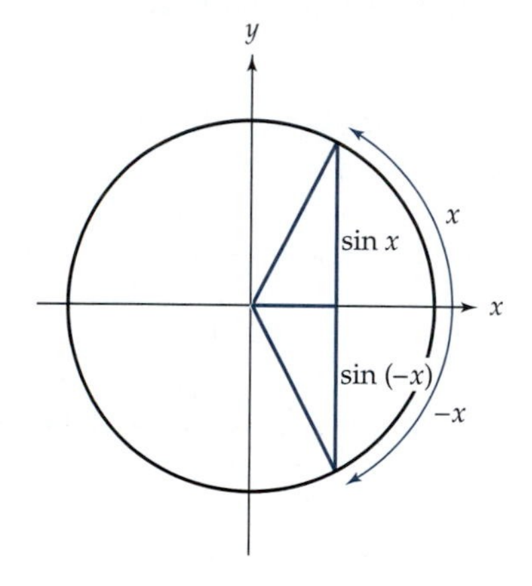

Figure 24 The line segment representing $\sin x$ is the same length as that representing $\sin(-x)$ but is below the x-axis, so it will have the opposite sign.

A similar situation occurs with the tangent function, showing that

$$\tan x = -\tan(-x).$$

The unit circle definitions of sine and cosine also lead us to the most widely used trigonometric identity. Recall that $\cos x$ is the horizontal position of a point on the unit circle and that $\sin x$ is the vertical position of a point on the unit circle. Using Figure 25 and the Pythagorean theorem, we see that

$$\cos^2 x + \sin^2 x = 1.$$

(*Note:* The notation $\cos^2 x$ means $(\cos\ x)^2$; that is, find the cosine of an angle and then square your answer.)

Example 6 If $\sin\theta = 2/3$, find $\cos\theta$.

Solution Using the identity $\cos^2\theta + \sin^2\theta = 1$ and substituting our value for $\sin\theta$, we have

$$\cos^2\theta + \left(\frac{2}{3}\right)^2 = 1.$$

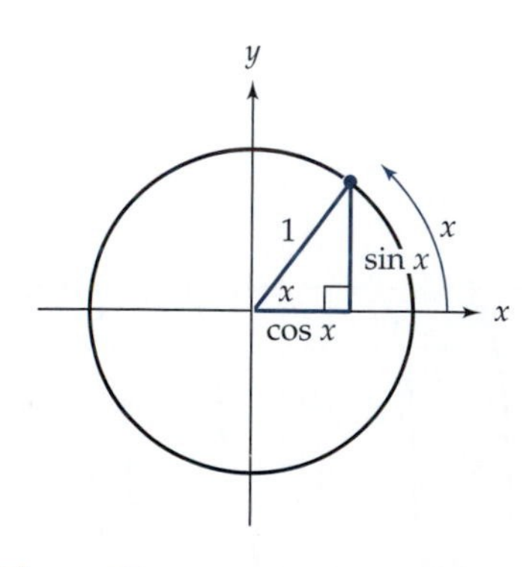

Figure 25 Deriving the identity $\cos^2 x + \sin^2 x = 1$.

Solving for $\cos^2\theta$ gives

$$\cos^2\theta = \frac{5}{9}.$$

Therefore,

$$\cos\theta = \pm\frac{\sqrt{5}}{3}.$$

Example 7 If $\tan\theta = 2$, find $\sin\theta$ and $\cos\theta$.

Solution Because $\tan\theta = \sin\theta/\cos\theta = 2$, then $\sin\theta = 2\cos\theta$. Using the identity $\cos^2\theta + \sin^2\theta = 1$ and substituting $2\cos\theta$ for $\sin\theta$, we have

$$\cos^2\theta + 4\cos^2\theta = 5\cos^2\theta = 1.$$

Solving for $\cos^2\theta$ gives

$$\cos^2\theta = \frac{1}{5}.$$

Therefore,

$$\cos\theta = \pm\sqrt{\frac{1}{5}} = \pm\frac{\sqrt{5}}{5}.$$

Because $\sin\ \theta = 2\cos\theta$,

$$\sin\theta = \pm\frac{2\sqrt{5}}{5}.$$

Because $\tan\theta$ is positive, both $\sin\theta$ and $\cos\theta$ are positive or both $\sin\theta$ and $\cos\theta$ are negative. So, either $\cos\theta = \sqrt{5}/5$ and $\sin\theta = 2\sqrt{5}/5$ or $\cos\theta = -\sqrt{5}/5$ and $\sin\theta = -2\sqrt{5}/5$.

READING QUESTIONS

16. If $\cos\theta = 0.34$, what is $\cos(-\theta)$?

17. Explain why $y = \tan x$ is an odd function.

18. Suppose $\sin\theta = \sqrt{15}/4$. Use the identity $\cos^2\theta + \sin^2\theta = 1$ to find $\cos\theta$.

SUMMARY

It is important to know and understand the connections between the unit circle definitions for the sine, cosine, and tangent functions and their right triangle definitions. In the unit circle definitions, $\sin x$ is defined as the vertical position of a point on the unit circle and $\cos x$ is the horizontal distance. The **tangent** function, or $\tan x$, is defined as $\tan x = \sin x/\cos x$, which can also be thought of as the slope of a line.

In terms of a right triangle, the sine, cosine, and tangent functions are defined as follows:

$$\sin x = \frac{\text{opposite}}{\text{hypotenuse}}, \qquad \cos x = \frac{\text{adjacent}}{\text{hypotenuse}}, \qquad \tan x = \frac{\text{opposite}}{\text{adjacent}}.$$

We often use **radians** to measure angles. An angle of 1 radian intercepts an arc of length 1 on a unit circle. To convert from degrees to radians, multiply by $\pi/180°$. To convert from radians to degrees, multiply by $180°/\pi$.

We can use our knowledge of special triangles to get exact values for the sine, cosine, and tangent functions where the input is a multiple of $\pi/4$ or $\pi/6$. Some basic identities of these functions are

- $\cos x = \cos(-x)$
- $\sin x = -\sin(-x)$
- $\tan x = -\tan(-x)$
- $\cos^2 x + \sin^2 x = 1$

EXERCISES

1. Why do we use radians rather than degrees when looking at the unit circle definitions of $\cos\theta$ and $\sin\theta$?
2. The letters in the accompanying figure represent the length of the arc beginning at $(1, 0)$ and ending at the points shown. Approximate the values of the sine, cosine, and tangent functions, for the arc lengths represented by each letter in the figure.

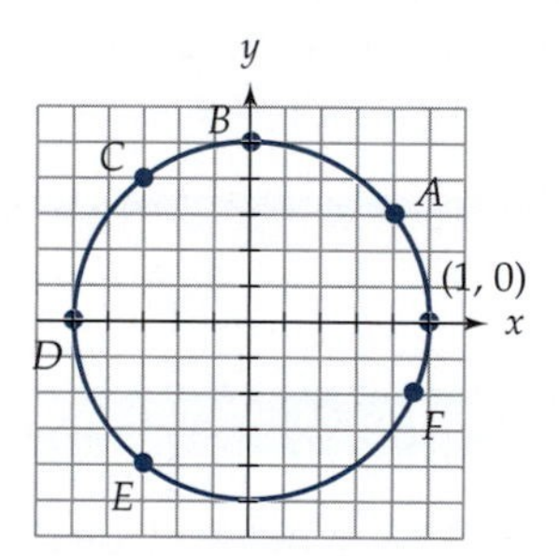

3. Convert the following angle measurements from degrees to radians.
   (a) $45°$
   (b) $36°$
   (c) $-60°$
   (d) $-95°$
   (e) $145°$
   (f) $315°$
4. Convert the following angle measurements from radians to degrees.
   (a) 3π
   (b) $\dfrac{5\pi}{6}$
   (c) $\dfrac{2\pi}{9}$
   (d) $\dfrac{4\pi}{5}$
   (e) 1.5
   (f) 6

5. Use your calculator to find decimal approximations for each of the following.
   (a) $\cos 42°$
   (b) $\sin \dfrac{\pi}{7}$
   (c) $\tan 132°$
   (d) $\tan 132$
6. With the calculator in degree mode rather than radian mode, we made the accompanying calculator graph of $y = \sin\theta$ for $-2\pi \leq \theta \leq 2\pi$. The graph increases very little in this viewing window. Explain why the calculator produced such a graph.

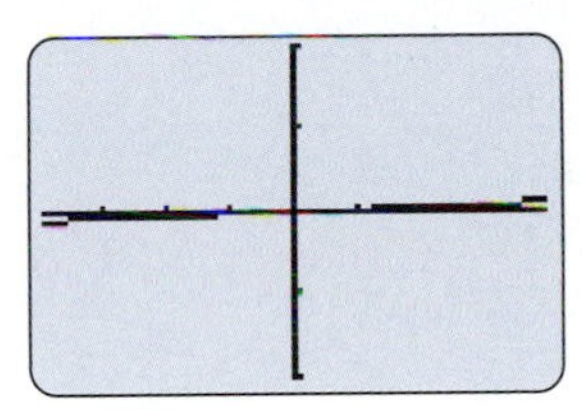

7. Using the right triangle definition, we say that $\tan\theta$ = opposite/adjacent. Explain why this wording is equivalent to saying that $\tan\theta = \sin\theta/\cos\theta$.
8. (a) Using the definition that $\tan\theta = \sin\theta/\cos\theta$, explain why $\tan\theta$ has a vertical asymptote for $\theta = \pi/2$.
   (b) Using the definition that $\tan\theta$ is the slope of the line that makes an angle of θ with the positive x-axis, explain why $\tan\theta$ has a vertical asymptote for $\theta = \pi/2$.
9. The accompanying figure shows graphs of $y = \cos\theta$ and $y = \sin\theta$.
   (a) Which graph is which?
   (b) Describe the points on the graphs where $\tan\theta = 1$.

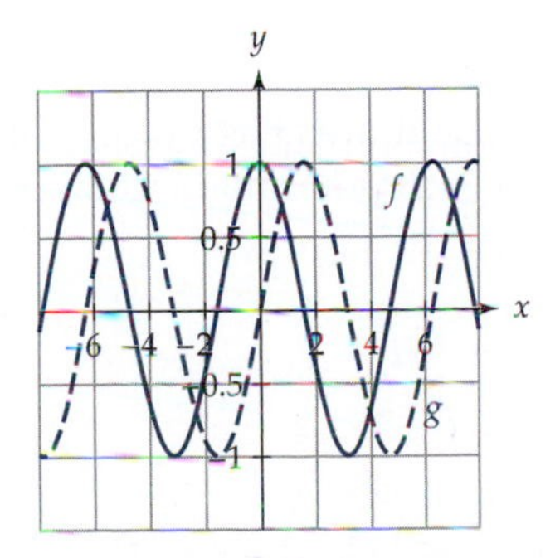

10. Complete the following table by filling in the missing entries.

θ (radians)	θ (degrees)	$\cos\theta$	$\sin\theta$	$\tan\theta$
$\frac{\pi}{6}$				
	$45°$			
$\frac{\pi}{3}$				
		$-\frac{1}{2}$	$\frac{\sqrt{3}}{2}$	
	$330°$			
			$\frac{\sqrt{2}}{2}$	-1

11. In the accompanying figure, the length of side AB is one unit. The measurement of $\angle CAB$ is $60°$, and the measurement of $\angle DAC$ is $45°$. Find the exact measurement of the length of side AD.

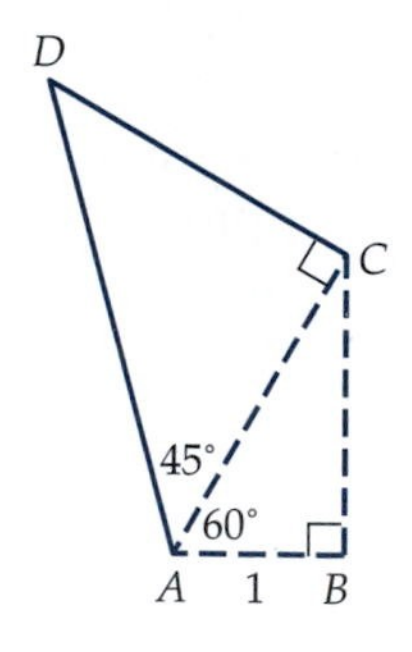

12. The circle shown has a radius of three units. Give the x- and y-coordinates of each point marked on the circle.

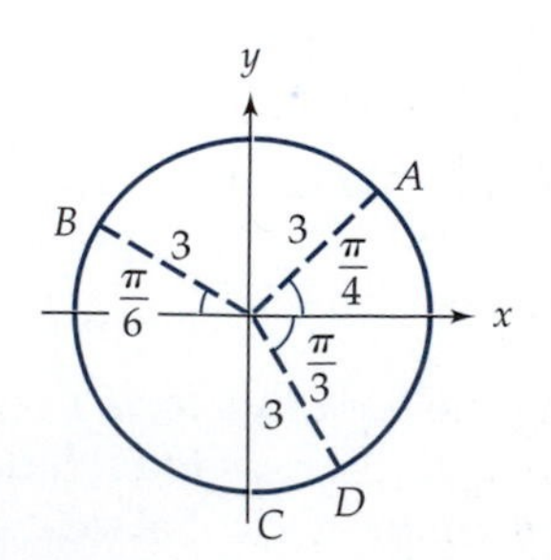

13. Let A be a point in the first quadrant such that the slope of the line connecting A to the origin is $\sqrt{3}$ and the distance from A to the origin is five units. Find the coordinates of point A.
14. Let B be a point in the second quadrant such that the slope of the line connecting B to the origin is -1 and the distance from B to the origin is four units. Find the coordinates of point B.
15. Let θ be an angle in the first quadrant such that $\sin\theta = \frac{4}{5}$. Find the exact values of each of the following.
    (a) $\cos\theta$
    (b) $\tan\theta$
    (c) $\sin(-\theta)$
16. Let θ be an angle in the first quadrant such that $\tan\theta = 3$. Find the exact values of each of the following.
    (a) $\cos\theta$
    (b) $\sin\theta$
    (c) $\tan(\theta+\pi)$
17. Let θ be an angle in the first quadrant such that $\cos\theta = \frac{1}{6}$. Find the exact values of each of the following.
    (a) $\sin\theta$
    (b) $\tan\theta$
    (c) $\cos(\theta+\pi)$
18. Let θ be an angle in the first quadrant such that $\cos\theta = b$.
    (a) What are the restrictions, if any, on the value of b?
    (b) Represent each of the following in terms of b.
        i. $\sin\theta$
        ii. $\tan\theta$
        iii. $\cos(-\theta)$
        iv. $\cos(\theta+2\pi)$
19. Let θ be an angle in the first quadrant such that $\tan\theta = c$.
    (a) What are the restrictions, if any, on the value of c?
    (b) Represent each of the following in terms of c.
        i. $\sin\theta$
        ii. $\cos\theta$
        iii. $\tan(-\theta)$
        iv. $\tan(\theta+\pi)$
20. (a) Explain, using the unit circle definition of the sine function, why $\sin\theta$ is an odd function.
    (b) Explain why you cannot use the right triangle definition of the sine function to show that $\sin\theta$ is an odd function.
21. What is wrong with the following logic?

    Using the equation for a circle, I know that the points on the circle, (x, y), satisfy the relationship $x^2 + y^2 = r^2$, where r is the radius of the circle. I also know that the coordinates on a circle are represented by trigonometric functions. So, the x-coordinate is equal to $\cos\theta$ and the y-coordinate is equal to $\sin\theta$. Using a circle of radius 3, I can conclude that $\cos^2\theta + \sin^2\theta = 9$.

22. Each of the following statements is false. Rewrite each so that it becomes a true statement.
    (a) The domain of $y = \sin\theta$ is $-1 \le \theta \le 1$.
    (b) The tangent function is defined for all real numbers.
    (c) $\tan\dfrac{\pi}{3} = 1$
    (d) $\sin\theta + \cos\theta = 1$
    (e) $y = \sin\theta$ is a periodic function with period π.
    (f) $y = \cos\theta$ is an odd function.
    (g) $\sin\theta = \sin(-\theta)$
23. What is the difference between $\sin^2\theta$ and $\sin\theta^2$?

INVESTIGATIONS

INVESTIGATION 1: A STITCH IN TIME

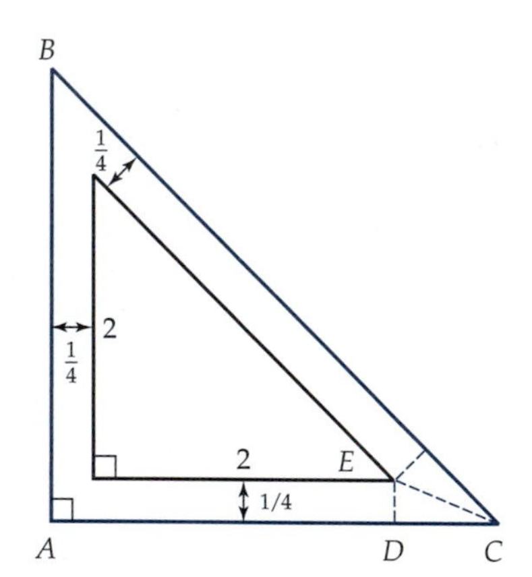

Figure 26

Part of the process of making a quilt involves piecing shapes together. One commonly used shape is a triangle. Often, different types of material are used to create intricate patterns. After the pattern is designed, the quilter must determine the sizes of the various pieces. To allow the pieces to be sewn together, a seam allowance must be added to each edge of each piece. In quilting, this seam allowance is usually $\frac{1}{4}$ in. The measurements must be very precise. If the piece is cut incorrectly, the quilt pieces will not fit together as they should.

1. Isosceles right triangles are frequently used by quilters. The first thing a quilter determines is the size of the finished shape. Then the quilter determines the actual size of the piece that needs to be cut, allowing for the seam. Figure 26 shows an isosceles right triangle with legs 2 in. long and a $\frac{1}{4}$-in. seam allowance added to all three sides.
    (a) Find AC by doing the following.
        i. Explain why $AC = \frac{1}{4} + 2 + DC$.
        ii. Use $\triangle DEC$ and an appropriate trigonometric function to approximate DC.
        iii. What is AC to the nearest $\frac{1}{16}$ in.?
    (b) Find the length of the legs of the triangle that should be cut out if the finished triangle were to have legs that measure 3 in. instead of 2 in. How does the amount added to each leg of this triangle compare with the amount added in question 1, part (a)?
    (c) Find the length of the legs of the triangle that should be cut out if your finished piece were n inches long. Justify your answer.

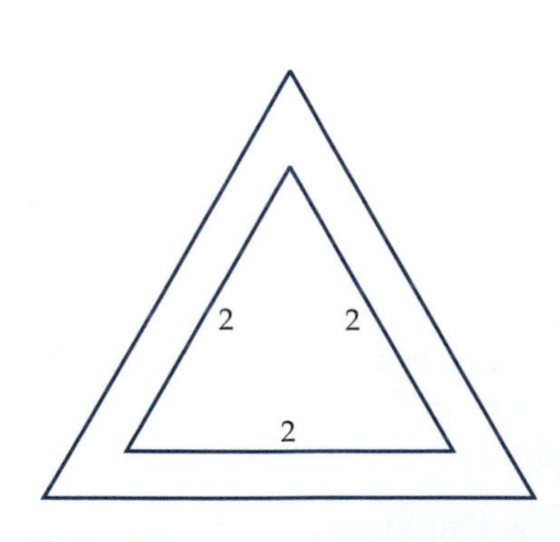

Figure 27

2. A different quilt design calls for equilateral triangles with 2-in. sides. (See Figure 27.) The quilter must still add a $\frac{1}{4}$-in. seam allowance. What length

should the sides of this triangle be? Can you use the same procedure that you used in question 1, part (a)? Why or why not?

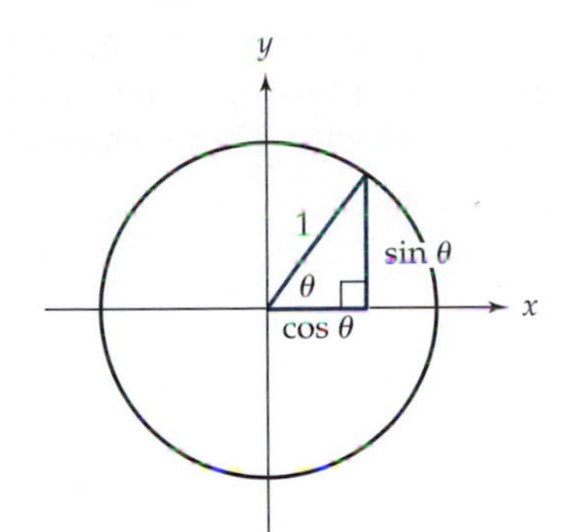

Figure 28

INVESTIGATION 2: TRIGONOMETRIC FUNCTIONS SHOWN AS LINE SEGMENTS WITH A UNIT CIRCLE

As you have seen, sine and cosine can be defined as lengths of line segments in a unit circle, as shown in Figure 28. In this investigation, we find ways to define other trigonometric functions as line segments.

1. In this section, we described the tangent function as the slope of a line passing through the origin. We can also, however, define the tangent function as the length of a line segment if the picture is drawn correctly. Figure 29 shows a right triangle in a unit circle. Explain why the length of segment AB is $\tan\theta$.
2. Three other trigonometric functions are commonly used: the cosecant function, written $\csc\theta$; the secant function, written $\sec\theta$; and the cotangent function, written $\cot\theta$. These three functions are defined as follows:

$$\csc\theta = \frac{1}{\sin\theta}, \qquad \sec\theta = \frac{1}{\cos\theta}, \qquad \cot\theta = \frac{1}{\tan\theta}.$$

   (a) Figure 30 shows the right triangle OCD in a unit circle. Explain why the length of segment OC is $\csc\theta$.
   (b) The key to finding a line segment that has the length of the desired trigonometric function is to construct a right triangle where the denominator in the defining ratio is 1. With that in mind, draw a right triangle in a unit circle in such a way that one of its lengths is $\sec\theta$. Explain which segment is $\sec\theta$ and why. (*Hint:* Your drawing should either look like Figure 29 or Figure 30.)
   (c) Draw a right triangle in a unit circle in such a way that one of its lengths is $\cot\theta$. Explain which segment is $\cot\theta$ and why.

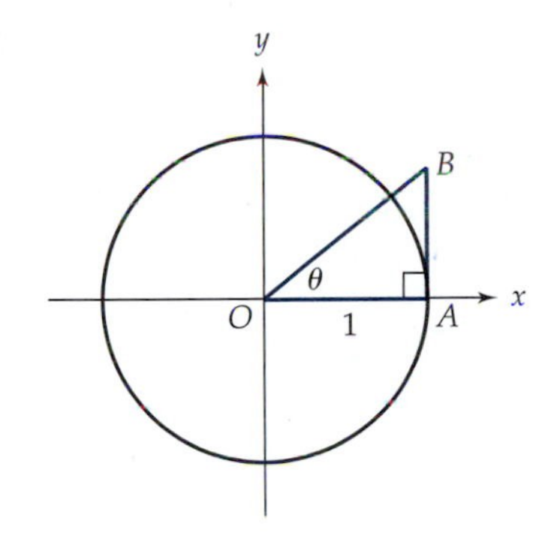

Figure 29

3. Three other trigonometric functions that are much less commonly used than the other six are the versed sine function, written versin θ; the coversine function, written cvs θ; and the external secant function, written exsec θ. These three functions are defined as follows:

$$\text{versin}\,\theta = 1 - \cos\theta, \qquad \text{cvs}\,\theta = 1 - \sin\theta, \qquad \text{exsec}\,\theta = \sec\theta - 1.$$

   (a) Make a drawing similar to Figure 28. Identify the line segment in your drawing whose length is versin θ. Explain why your segment is versin θ.
   (b) Make another drawing similar to Figure 28. By drawing an additional line segment in the figure, a line segment whose length is cvs θ is formed. Identify that segment and explain why it is cvs θ.
   (c) Draw a right triangle in a unit circle in such a way that a segment is formed whose length is exsec θ. Explain which segment is exsec θ and why.

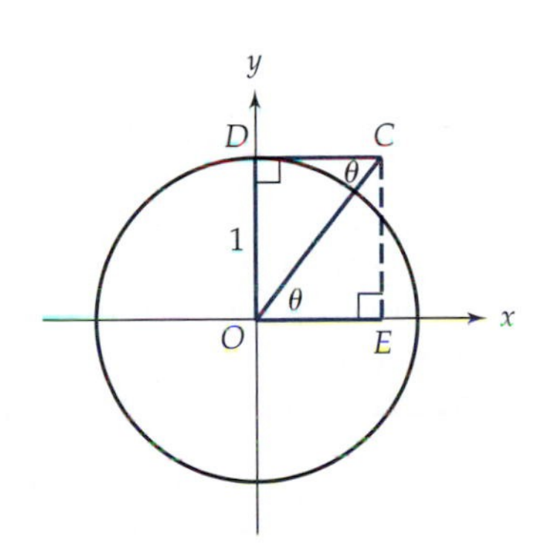

Figure 30

5.2 ARC LENGTH AND AREA

In section 5.1, we looked at periodic functions associated with circles and triangles: sine, cosine, and tangent. In this section, we look at circles and triangles. We learn how to calculate the arc length of a piece of a circle, find the area of a sector of a circle, and look at angular velocity. We also look at area of triangles and several applications.

ARC LENGTH

Suppose you wanted to construct a patio on the back of your house shaped like the one shown in Figure 1. The patio will be built of concrete and the perimeter will be edged with brick. To purchase the correct amount of both concrete and brick, you need to determine the area and the perimeter of the patio.

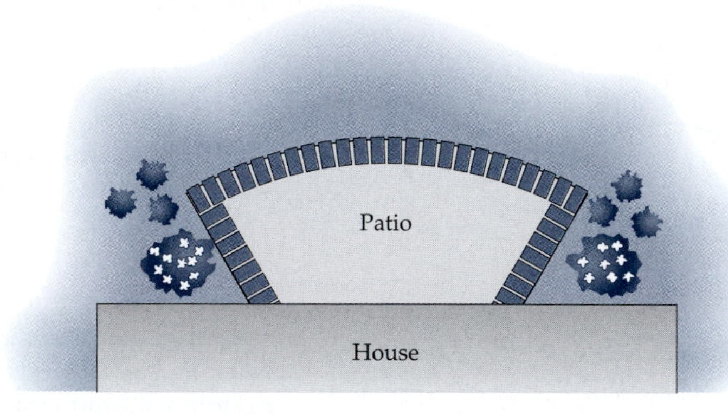

Figure 1 A patio built of concrete and edged with brick.

To start this process, a more detailed drawing, shown in Figure 2, is made. The patio is a portion of a sector of a circle. A **sector** of a circle is the region bounded by two radii and a portion of the circumference. The sector shown in Figure 2 has a central angle of 60°.

Our first task is to determine the perimeter of the patio. The lengths of the straight lines are shown as 10 ft, 20 ft, and 10 ft. We also need, however, to know the length of the arc. To do so, we first develop the formula for finding the arc length of a portion of the circumference of a circle.

The lowercase Greek letter pi, π, is defined to be the ratio of the circumference of any circle to its diameter. That is,

$$\pi = \frac{C}{d},$$

where C is the circumference and d is the diameter. The number $\pi \approx 3.14159$ is an irrational number, so its exact value cannot be represented by a fraction. From the definition of π, we see that $C = \pi d$, which gives us a formula for computing the circumference of a circle. Because $d = 2r$, where r is the radius of a circle, we also have the formula $C = 2\pi r$.

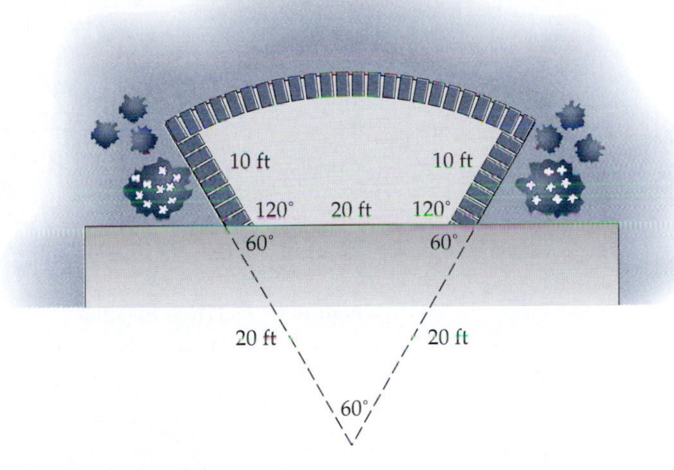

Figure 2 A detailed drawing of the patio showing how it is related to a sector with a central angle of 60°.

HISTORICAL NOTE

The number π has intrigued and fascinated people for centuries. William Jones first used the symbol π to represent the ratio of the circumference of a circle to its diameter in 1706, but the symbol was not generally accepted until Euler adopted it in 1737 and later used it in his book, *Analysis*. The estimation of the value of π, however, goes back much further. Around 1700 B.C., for example, the Egyptians used $\frac{256}{81} \approx 3.16$ as the value of π. Throughout the centuries, closer and closer approximations have been calculated. Today, computers have calculated π to more than two billion digits. According to the *Guinness Book of Records*, Hideaki Tomoyori from Yokdiang, Japan, recited the first 40,000 of those digits from memory in 1987. It took him 17 hr and 21 min. to complete this amazing feat.

Sources: W. W. Rouse Ball, and H. S. M. Coxeter, *Mathematical Recreations and Essays*, 13th ed. (New York: Dover, 1987), pp. 347–59; Donald McFarlan, ed., *The Guinness Book of Records 1991*, (Stanford, CT: Guinness Publishing, 1990), pp. 67, 78.

To find a formula for a portion of the circumference or arc length of the circle, recall from section 5.1 that a 1-radian angle intercepts an arc of length r in a circle of radius r. (See Figure 3.) Assume the radius of your circle is r. If a 1-radian angle intercepts an arc of length r, then a 2-radian angle intercepts an arc of length $r + r = 2r$, a 3-radian angle intercepts an arc of length $r + r + r = 3r$, and so forth, as illustrated in Figure 4. You can see that the length of the arc is the measure of the angle (in radians) times the radius of the circle.

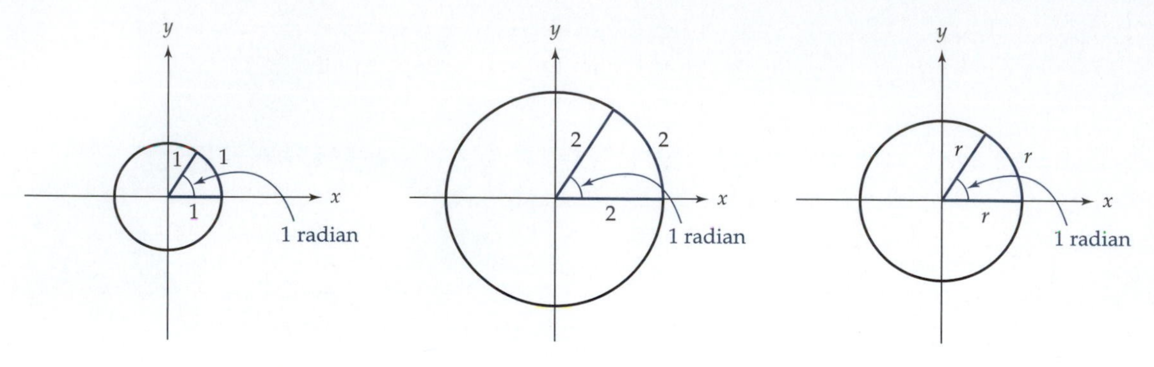

Figure 3 In a circle of radius r, a 1-radian angle intercepts an arc of length r.

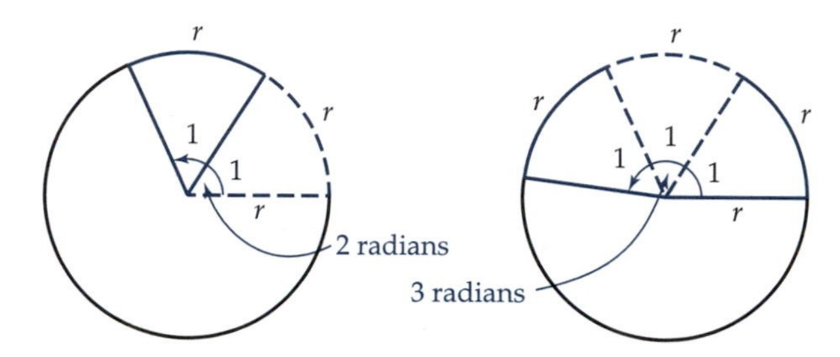

Figure 4 In a circle of radius r, a 2-radian angle intercepts an arc of length $2r$ and a 3-radian angle intercepts an arc of length $3r$.

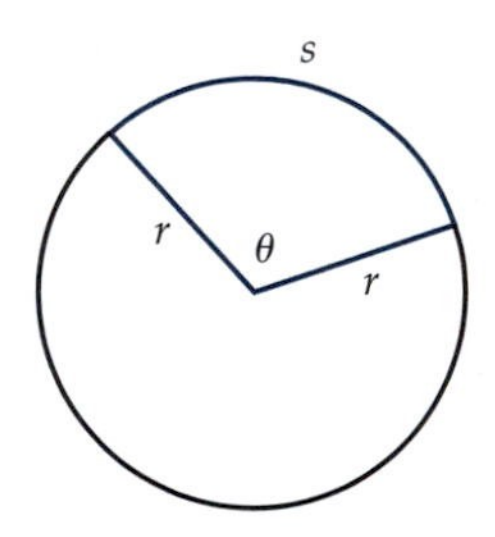

Figure 5 $s = r\theta$.

PROPERTIES

The length of an arc of a circle of radius r that is intercepted by a central angle of θ radians is $s = r\theta$. (See Figure 5.)

It is important that the measure of the angle be in *radians*. Radians define the angle in a unit circle in terms of the arc length intercepted by that angle. Because radians connect the measurement of the angle to the length of the arc, we can derive such a simple formula for the arc length of a circle. Degrees have no such connection. If given the measure of an angle in degrees, you must convert it to radians before finding the arc length. Notice also that the formula for arc length gives the same formula for the circumference of a circle that we derived earlier from the definition of π. Because the full circle angle is 2π, the circumference of a circle of radius r is $C = r \cdot 2\pi = 2\pi r$.

Now back to our patio example. Figure 2 shows that the curved side of the patio is part of the circumference of a circle. To determine the arc length, we need to know the radius of the circle and the measure of the central angle in radians. In the figure, the central angle is given as 60°. Converting 60° to radians gives $60° \cdot \pi/180° = \pi/3$ radians. We can see that the radius of the circle is $20 + 10 = 30$ ft. Therefore, the length of the arc in the patio is $30 \cdot \pi/3 = 10\pi \approx 31.4$ ft. So, the perimeter of the patio is approximately $10 + 20 + 10 + 31.4 = 71.4$ ft, which is about 71 ft 5 in.

With this information, we can determine how many bricks we need to go around the perimeter of our patio.

Figure 6

Example 1 A clock is constructed with a 6-in. second hand. Find the distance the tip of the second hand travels when it goes from the 1 on the face of the clock to the 5. (See Figure 6.)

Solution Going from 1 to 5 on the face of a clock is traveling $\frac{4}{12} = \frac{1}{3}$ of the way around. Therefore, the second hand will move through an angle of $\frac{1}{3} \cdot 2\pi = 2\pi/3$ radians. Because the hand is 6 in. long, its tip will travel $6 \cdot 2\pi/3 = 4\pi \approx 12.6$ in.

READING QUESTIONS

1. Suppose you have a circle of radius 5. Find the arc length intercepted by each of the following angles.
   (a) $\dfrac{\pi}{2}$
   (b) $\dfrac{2\pi}{3}$
   (c) $45°$
2. Why does the measure of your angle have to be in radians before you can find arc length?
3. Find the perimeter of the patio shown in the figure.

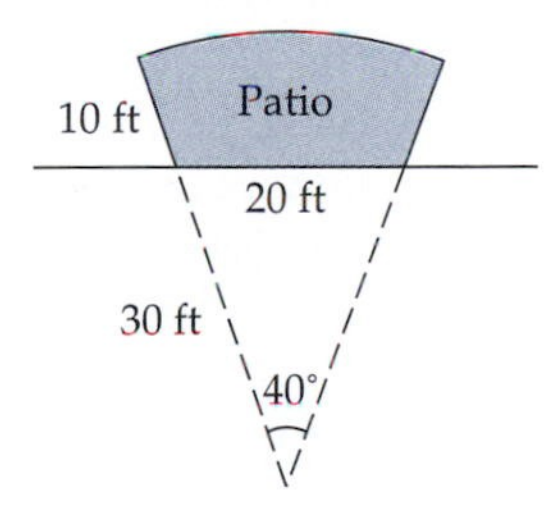

AREA OF A SECTOR

Let's return to building our patio. We found the perimeter so that we could determine the number of bricks we need. We still need to determine the area of the patio, however, to calculate how much concrete we need. To do so, we must find the area of the sector shown in Figure 2 and subtract the area of the equilateral triangle that is not part of the patio. We start by deriving the formula for the area of a sector.

To find the formula for a sector of a circle, look at Figure 7. If the central angle is $\theta = 2\pi$, then the shaded area is the entire circle. If the central angle is $\theta = \pi$ (half the full circle angle), then the shaded area is half the area of

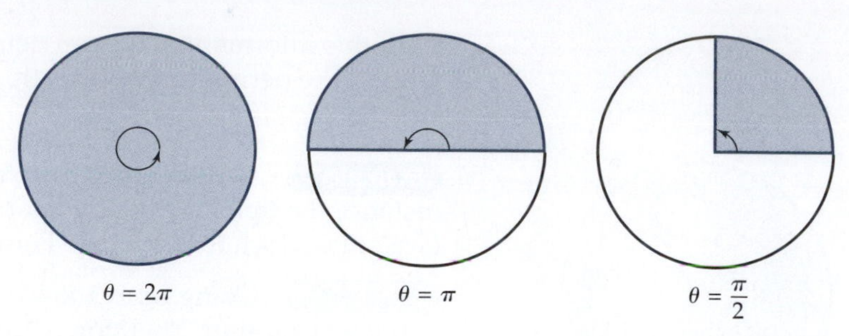

Figure 7 The ratio of the angle to 2π equals the ratio of the shaded area to the area of the entire circle.

the circle. If the central angle is $\theta = \pi/2$ (one-fourth the full circle angle), then the shaded area is one-fourth the area of the circle. In general, the ratio of the angle to the full circle angle of $\theta = 2\pi$ is equal to the ratio of the shaded area to the area of the entire circle. That is,

$$\frac{\theta}{2\pi} = \frac{A}{\pi r^2}.$$

Solving for A, we have the following.

PROPERTIES

The area of a sector of a circle of radius r that is intercepted by a central angle of θ radians is $A = \frac{1}{2}r^2\theta$. (See Figure 8.)

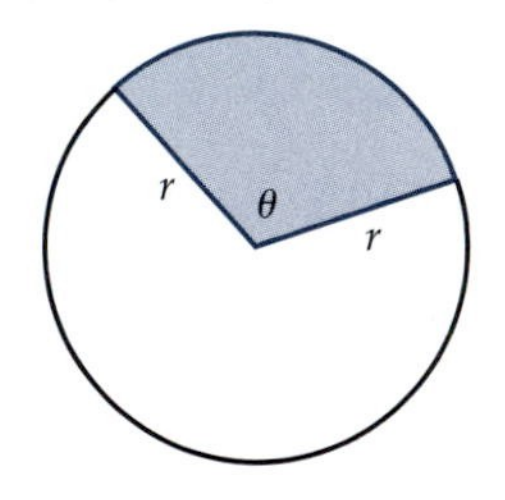

Figure 8 The area of a sector of a circle of radius r that is intercepted by a central angle of θ radians is $A = \frac{1}{2}r^2\theta$.

This formula, similar to the arc length formula, assumes the measure of your angle is in radians. If your angle is given in degrees, then be sure to convert it to radians before finding the area of the sector.

Now back to our patio example. Figure 9 shows that the patio is part of a sector of a circle. We need to find the area of this sector and subtract the area of the equilateral triangle.

To find the area of the sector, we convert the 60° central angle to $\pi/3$ radians. The radius is 30 ft, giving the area of the sector as $\frac{1}{2} \cdot 30^2 \cdot \pi/3 = 150\pi \approx 471$ ft^2. To determine the area of the equilateral triangle, we need to know its height, h. The segment labeled h in Figure 9 divides the equilateral triangle into two 30°–60°–90° triangles. In section 5.1, we saw that $\sin 60° = \sqrt{3}/2$. We also know, from Figure 9 and from the right triangle definition of sine, that $\sin 60° = h/20$. So, $h = \sin 60° \cdot 20 = 20\sqrt{3}/2 = 10\sqrt{3}$. The area of the equilateral triangle is $\frac{1}{2}bh = \frac{1}{2} \cdot 20 \cdot 10\sqrt{3} = 100\sqrt{3} \approx 173$ ft^2. Therefore, the area of the patio is approximately $471 - 173 = 298$ ft^2.

Example 2 In a circle of radius 6 in., find the area of a sector whose central angle is 100°.

Solution We first must convert 100° to radians: $100° \cdot \pi/180° = 5\pi/9$ radians. The area of the sector is $A = \frac{1}{2} \cdot 6^2 \cdot 5\pi/9 = 10\pi$ in.^{2}.

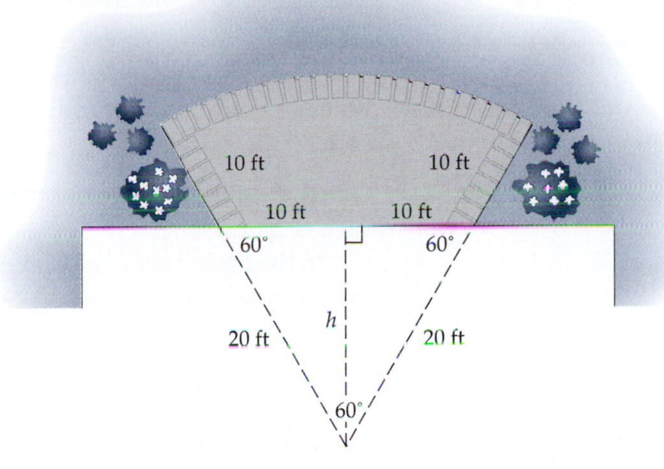

Figure 9 Finding the area of our patio involves finding the area of the sector and subtracting the area of the equilateral triangle.

READING QUESTIONS

4. Suppose you have a circle of radius 5. Find the area of the sector that is intercepted by each of the following angles.
   (a) $\dfrac{\pi}{2}$
   (b) $\dfrac{2\pi}{3}$
   (c) $45°$
5. What is the relationship between the area of a sector of a circle and the area of the entire circle?

AREA OF A TRIANGLE

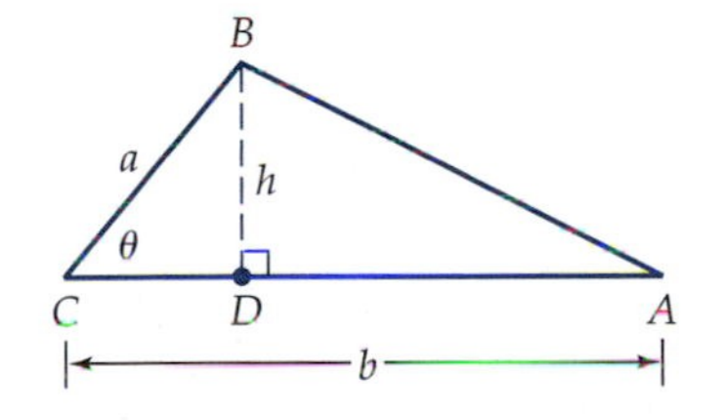

Figure 10 A triangle with the height, two sides, and one of the angles labeled.

The formula for the area of a triangle, $A = \frac{1}{2}bh$, where b is the base and h is the height, can be used to derive another formula for the area of the triangle in terms of two of the sides and one of the angles. Figure 10 shows a triangle with the height, two sides, and one of the angles labeled.

Notice that $\triangle BDC$ is a right triangle. Thus, using the right triangle definition of sine, $\sin\theta = h/a$. Solving for h gives $h = a\sin\theta$. Substituting into the standard formula for the area of a triangle leads us to $A = \frac{1}{2}b(a\sin\theta)$, which gives us the following formula for determining the area of a triangle.

PROPERTIES

The area of a triangle is $A = \frac{1}{2}ab\sin\theta$, where θ is the angle between the sides whose lengths are a and b.

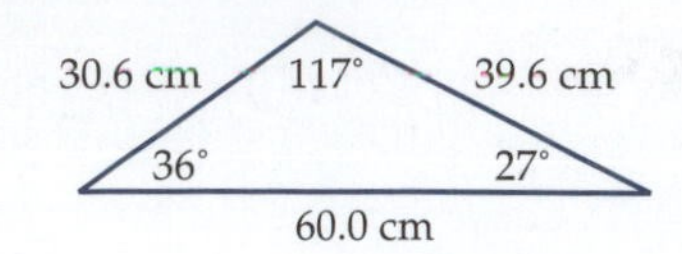

Figure 11

Example 3 Find the area of the triangle in Figure 11. Angles are measured to the nearest degree and sides to the nearest tenth of a centimeter.

Solution If we let $a = 30.6$ cm and $b = 60.0$ cm, then $\theta = 36°$. Using these values and the formula $A = \frac{1}{2}ab \sin\theta$, we have $A = 0.5 \cdot 30.6 \cdot 60.0 \cdot \sin 36° \approx 540\text{ cm}^2$.

In Example 3, we could have chosen *any* two sides as long as the angle we chose was between them. For example, if we let $a = 30.6$ cm and $b = 39.6$ cm, then $\theta = 117°$. In this case, $A = 0.5 \cdot 30.6 \cdot 39.6 \cdot \sin 117° \approx 540\text{ cm}^2$, which is the same answer we obtained when we used 36° as our angle. It does not matter which side you choose for a and which side you choose for b as long as θ is the angle between the two sides.

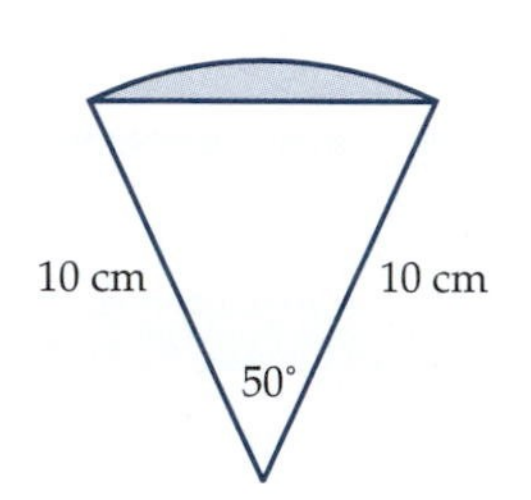

Figure 12

Example 4 Find the area of the shaded region in Figure 12 that is formed by the difference of the area of the sector and the triangle.

Solution To find the difference between the area of the sector and the area of the triangle, we need to subtract the two areas. Thus,

$$\text{area of shaded region } = \tfrac{1}{2}r^2\theta - \tfrac{1}{2}ab \sin\theta.$$

Because $a = b = r = 10$ and $\theta = 50° = 5\pi/18$ radians, we have

$$\begin{aligned}\text{area of shaded region } &= \frac{1}{2} \cdot 10^2 \cdot \frac{5\pi}{18} - \frac{1}{2} \cdot 10 \cdot 10 \cdot \sin\frac{5\pi}{18} \\ &\approx 5.33\text{ cm}^2.\end{aligned}$$

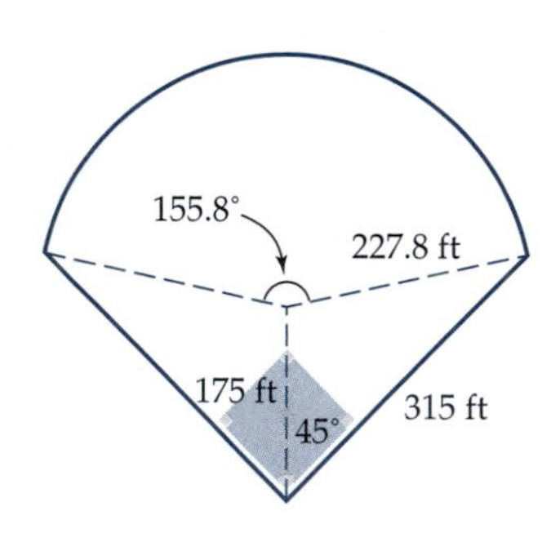

Figure 13

Example 5 The baseball field shown in Figure 13 consists of a sector and two triangles. Find the area of the baseball field.

Solution Since 155.8° is $155.8\pi/180$ radians, the area of the sector is

$$\frac{1}{2} \cdot \frac{155.8\pi}{180} \cdot 227.8^2 \approx 70{,}554\text{ ft}^2.$$

The area of each triangle is

$$\tfrac{1}{2} \cdot 315 \cdot 175 \cdot \sin 45° \approx 19{,}490\text{ ft}^2.$$

Therefore, the total area of the baseball field is approximately

$$70{,}554 + 2(19{,}490) = 109{,}534\text{ ft}^2.$$

READING QUESTIONS

6. Find the area of the triangle.

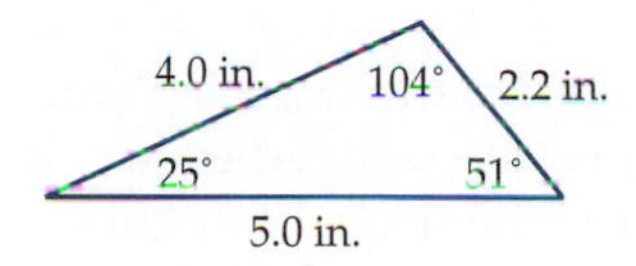

7. Find the area of the patio.

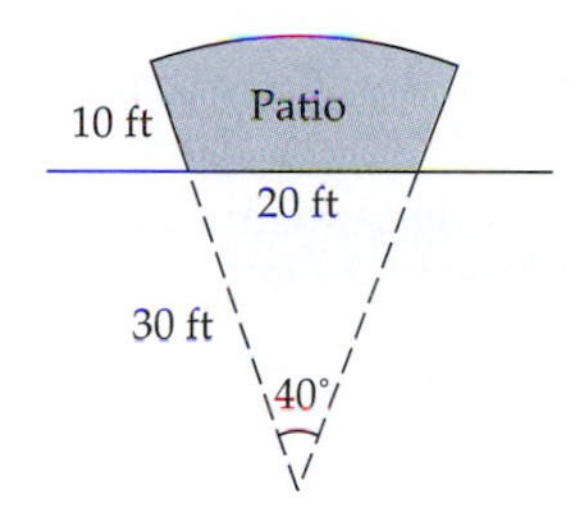

ANGULAR VELOCITY

Riding on a merry-go-round at a playground can feel very different for someone riding near the center compared with someone riding near the edge. Although both people are turning at the same rate and thus have the same angular velocity, both do not have the same linear velocity.

Suppose a merry-go-round is revolving at 6 revolutions per minute. This rate is the merry-go-round's angular velocity. **Angular velocity,** which we denote by the lowercase Greek letter omega, ω, is angular distance divided by time. **Angular distance** is some measure of the angle through which the radius of the circle moves. Angular distance can be given in degrees, radians, or revolutions. The most common unit used is revolutions because, most of the time, the angular distance will be more than one full circle. If a point is moving around the circumference of a circle at a constant angular velocity, then its angular velocity is given by

$$\omega = \frac{\theta}{t},$$

where θ is the angular measure through which the radius of the circle moved and t is the time required for the radius to move that distance. For example, if we have an angular velocity of 6 revolutions per minute, it means that the radius of the circle is moving six times completely around the circle in 1 min.

We can use the arc length formula described earlier to relate angular velocity with linear velocity. **Linear velocity** is linear distance divided by time. Linear velocity is what you typically think of when you think of velocity; it is described by the formula $r = d/t$, where d is distance and t is time. If a point is moving around the circumference of a circle at a constant velocity, then its linear velocity is given by

$$v = \frac{s}{t},$$

where s is the distance traveled around the circle (or arc length) and t is the time required to go that distance. Recall that the arc length formula tells us that $s = r\theta$. Because the angular measure must be in radians to use this formula, we assume that the angular velocity is in radians per unit time. Dividing both sides of $s = r\theta$ by t gives us

$$\frac{s}{t} = \frac{r\theta}{t}.$$

Because we know that $v = s/t$ and $\omega = \theta/t$, we obtain the following:

PROPERTIES

If a circle of radius r is rotating at ω radians per unit time, the linear velocity of a point on the circle is given by $v = r\omega$.

Let's return to our merry-go-round. Suppose there are two people on a merry-go-round that is spinning at 6 revolutions per minute. One person is 3 ft from the center and the other is 6 ft from the center. They are both revolving at the same angular velocity, but the one on the outside has a greater linear velocity. To find the linear velocity of each, we first convert the angular velocity from revolutions per minute to radians per minute. Because there are 2π radians in 1 revolution, there are $6 \cdot 2\pi = 12\pi$ radians in 6 revolutions. Therefore, the merry-go-round is revolving at a rate of 12π radians per minute. The person 3 ft from the center is traveling at a rate of

$$3 \cdot 12\pi = 36\pi \approx 113 \text{ ft/min}.$$

The person 6 ft from the center is traveling at a rate of

$$6 \cdot 12\pi = 72\pi \approx 226 \text{ ft/min}.$$

Note that the person 6 ft from the center is going twice as fast as the person 3 ft from the center.

Example 6 A 26-in.-diameter bicycle tire is rotating at 3 revolutions per second. Determine how fast the bicycle is traveling in miles per hour.

Solution The velocity of the bicycle is the same as the linear velocity of a point on the outside of the bicycle tire. To determine this velocity, we first need to convert 3 revolutions into radians and a 26-in.-diameter to a radius. Because 1 revolution is 2π radians, 3 revolutions is $3 \cdot 2\pi = 6\pi$ radians. The radius of a circle is half the diameter, so the radius is

$\frac{1}{2} \cdot 26 = 13$ in. Therefore, the linear velocity of the bicycle is $6\pi \cdot 13 = 78\pi$ in./sec. To convert 78π in./sec to miles per hour, we do the following.

$$\frac{78\pi \text{ in.}}{1 \text{ sec}} \cdot \frac{1 \text{ ft}}{12 \text{ in.}} \cdot \frac{1 \text{ mi}}{5280 \text{ ft}} \cdot \frac{3600 \text{ sec}}{1 \text{ h}} \approx 13.9 \text{ mph.}$$

Therefore, a bicycle that has 26-in.-diameter tires rotating at 3 revolutions per second will be traveling approximately 13.9 mph.

READING QUESTIONS

8. How is angular velocity different than linear velocity?
9. If a velocity is given as 30 radians per second, is it an angular velocity or a linear velocity?
10. A 24-in.-diameter bicycle tire is rotating at 2 revolutions per second. Determine how fast the bicycle is traveling in miles per hour.

SUMMARY

A **sector** of a circle is the region bounded by two radii and a portion of the circumference. The length of an arc of a circle of radius r that is intercepted by a central angle of θ radians is $s = r\theta$.

The area of a sector of a circle of radius r that is intercepted by a central angle of θ radians is $A = \frac{1}{2}r^2\theta$.

The area of a triangle is $A = \frac{1}{2}ab \sin\theta$, where θ is the measure of the angle between the sides whose lengths are a and b.

The **angular velocity** (ω) is angular distance divided by time. **Angular distance** is some measure of the angle through which the radius of the circle moves. If a point is moving around the circumference of a circle at a constant angular velocity, then its angular velocity is given by $\omega = \theta/t$.

Linear velocity is linear distance divided by time. If a point is moving around the circumference of a circle at a constant velocity, then its linear velocity is given by $v = s/t$.

If a circle of radius r is rotating at ω radians per unit time, the linear velocity of a point on the circle is given by $v = r\omega$.

EXERCISES

1. Each of the following represents either a linear measurement (such as a perimeter) or an area measurement for a common geometric figure. Determine if the formula is finding a linear measurement or an area measurement and then describe the geometric figure associated with the formula.

   (a) $2w + 2l$

(b) $\frac{1}{2}bh$
(c) s^2
(d) $2\pi r$
(e) $r\theta$
(f) $\frac{1}{2}ab \sin \theta$
(g) lw
(h) πr^2
(i) $4s$
(j) $\frac{1}{2}r^2\theta$

2. Complete the following table by filling in the missing answers associated with a sector of a circle.

	Sector *A*	**Sector *B***	**Sector *C***	**Sector *D***	**Sector *E***
Length of the Radius	2	7		$\frac{1}{4}$	
Measure of the Central Angle	$\frac{\pi}{3}$		2.5		45°
Length of the Intercepted Arc		$\frac{21\pi}{8}$	6.25	$\frac{\pi}{4}$	
Area of the Sector					$\frac{9\pi}{8}$

3. Find the perimeter and the area of the patio shown in the accompanying figure.

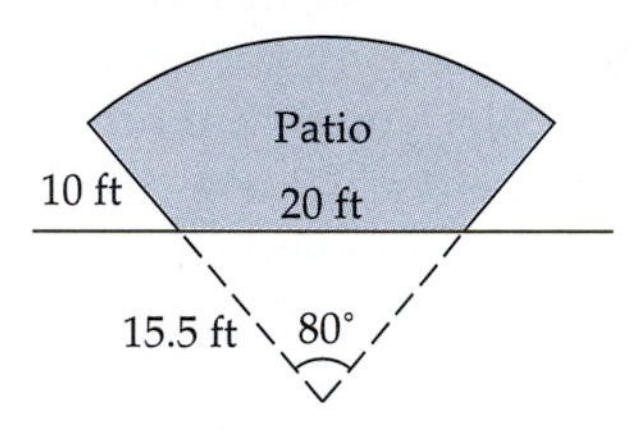

4. Glasgow, Scotland, is at a latitude of approximately 56° north and a longitude of approximately 4° west. Madrid, Spain, is at a latitude of approximately 40° north and a longitude of approximately 4° west. Thus, the distance from Glasgow to Madrid is the length an arc of a circle with a central angle of 16° and a radius of 3960 mi (the approximate radius of Earth). Approximately how far is it from Glasgow to Madrid?
5. Earth has a radius of about 3960 mi and has 24 time zones. What is the distance between time zones, at the equator, assuming they are equally spaced?

6. A nautical mile is 1 minute ($\frac{1}{60}$ of a degree) of the circumference of Earth. The radius of Earth is approximately 3960 (statute) miles. How far is 1 nautical mile in terms of (statute) miles?

7. A classic problem in mathematics is how Eratosthenes calculated the circumference of Earth in approximately 200 B.C. He knew that on a certain day of the year, the sun was directly overhead of Aswan, Egypt, because there was no shadow from the sun as it shown into a well. At noon on that same day, he measured the shadow of the sun in Alexandria, Egypt, which is 500 mi north of Aswan. From this, he concluded that the sun was 7.5° south of vertical in Alexandria. Using the accompanying figure, answer the following questions.

   (a) There are two angles labeled 7.5° in the figure. If one is 7.5°, why must the other also be 7.5°?

   (b) Calculate the radius and the circumference of Earth.

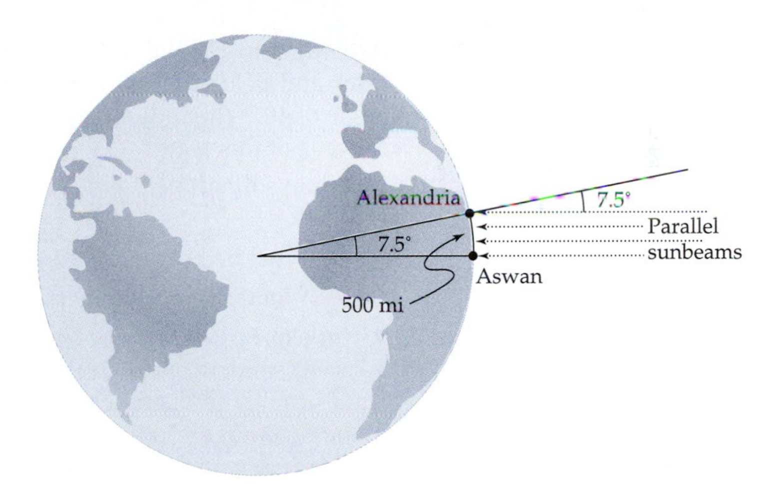

8. The accompanying figure shows a circular flower garden with a birdbath in the center. The radius of the circular base of the birdbath is 0.5 ft, and the distance from the base of the birdbath to the edge of the garden is 5 ft. Assume each plot is the same size. Determine the measurement of angle A and the area of each plot in the flower garden.

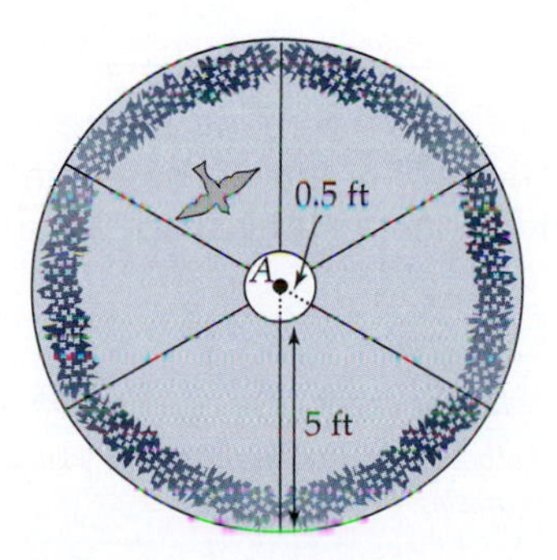

9. The back windshield wiper on a minivan is on a 22-in.-long arm. The wiper blade is 13.5 in. long and is located on the outer portion of the wiper arm. The arm moves through an angle of 100°. (See the accompanying figure.) What is the area of the back windshield that is wiped when the windshield wiper is operating?

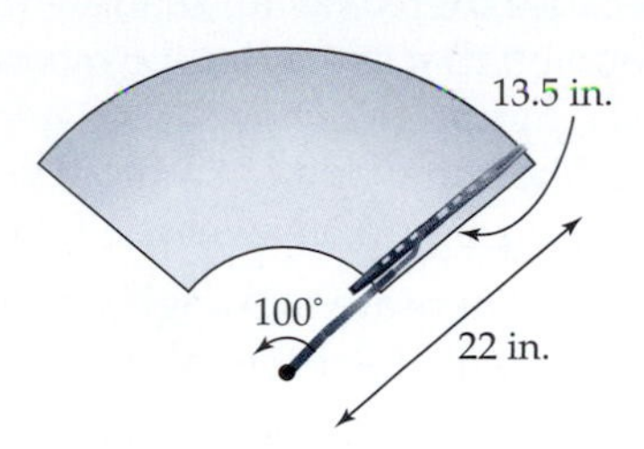

10. According to *The Rule Book*[1], the field on which a discus is thrown in a track and field competition must be a sector of a circle with a central angle of 40°. The radius can vary. Assume the radius is 250 ft. What is the area of this sector?
11. A certain softball field is a sector with a 90° central angle and a radius of 225 ft.
    (a) What is the area of the softball field?
    (b) What is the perimeter of the softball field?
12. An arbelos is the region bounded by a large semicircle and two smaller adjacent semicircles such that the sum of the two smaller diameters equals the large diameter. The placement of the semicircles is shown in the accompanying figure.[2] Archimedes showed that the area of the arbelos is $A = (\pi/4)d_1d_2$, where d_1 and d_2 are the diameters of the two smaller semicircles. Derive Archimedes' formula.

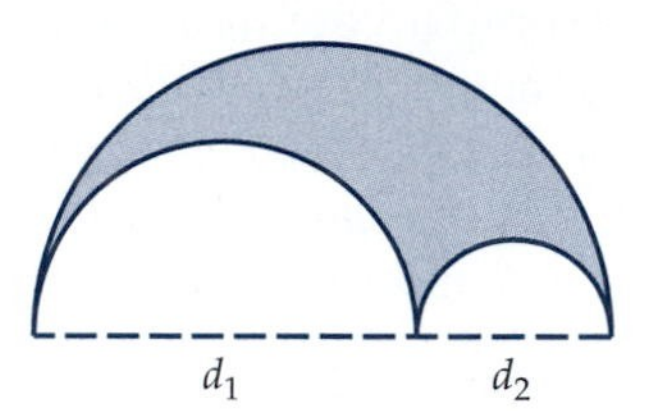

13. In the accompanying figure (shown on the next page), the arcs are portions of a circle whose center is one of the corners of the square. Find the area of the shaded region in terms of x.

[1] *Ruth Midgley, ed.,* The Rule Book *(New York: St. Martin's Press), p. 392.*

[2] *The word* arbelos *in Greek means cobbler's knife. Apparently, a cobbler's knife looks something like the shape of an arbelos.*

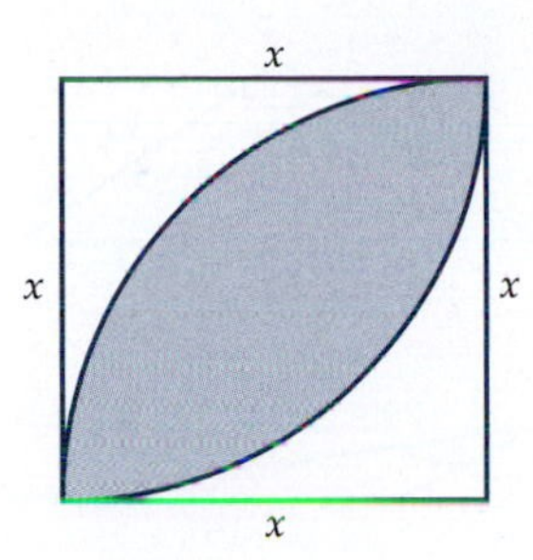

14. The accompanying drawing of the baseball field consists of a sector and two triangles. Find the perimeter and area of the baseball field.

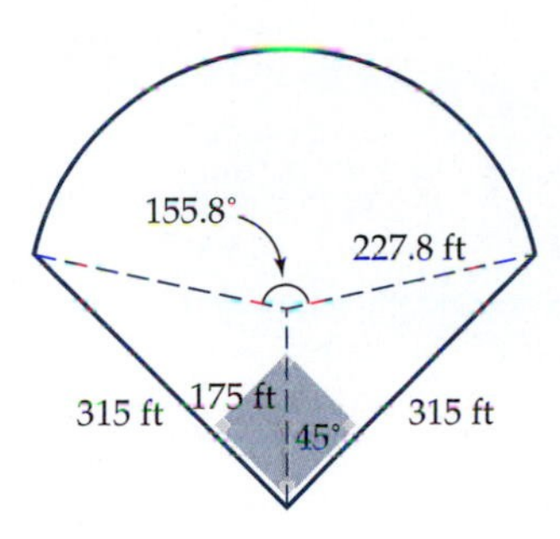

15. An arc is placed on top of a square with sides of length x. The arc is a portion of a circle whose center is also the center of the square. (See the accompanying figure.) Find the area of the entire region in terms of x.

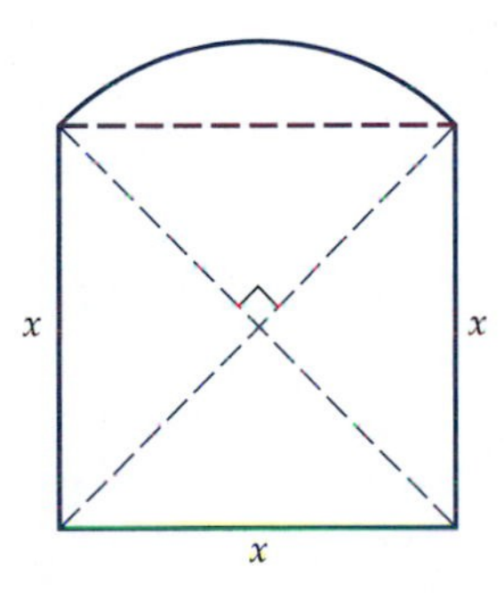

16. Find the area of the following triangles.

(a)

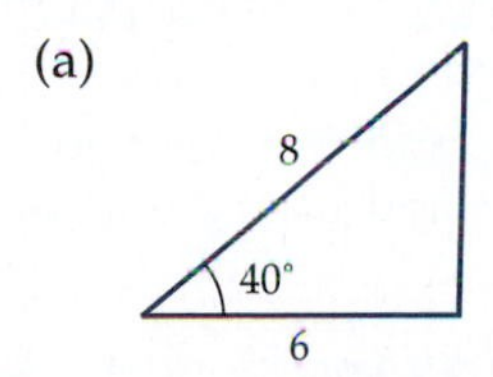

(b)

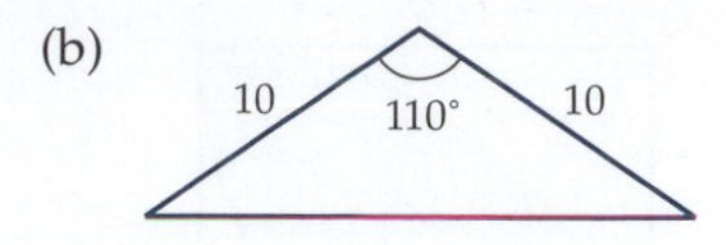

(c)

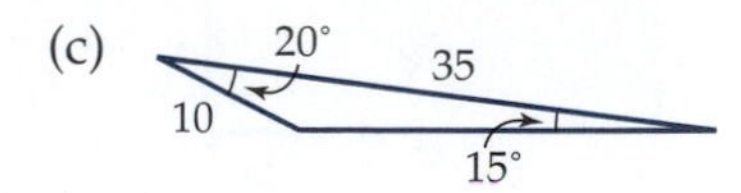

17. Find the area of the basketball backboard shown in the accompanying figure.

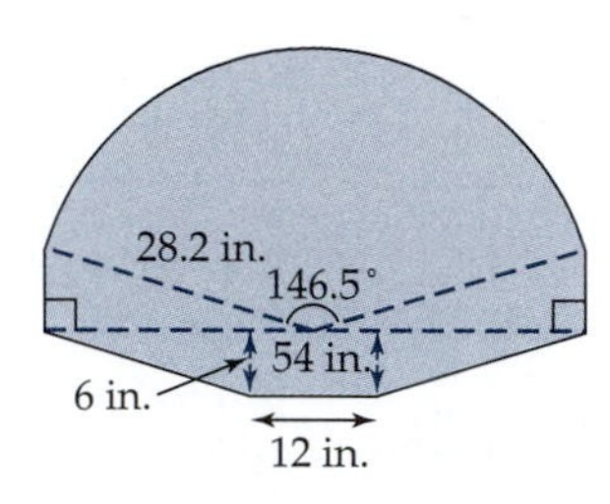

18. Two people are on a merry-go-round. Person A is sitting 1.5 ft from the center. How far from the center should person B sit to go three times as fast as person A?
19. One of the world's most famous clocks, commonly known as Big Ben[3], is in St. Stephen's Clock Tower at the House of Parliament in London, England. The minute hand of this clock is 14 ft long.
    (a) How far does the end of the minute hand travel in 10 min? Round your answer to the nearest inch.
    (b) What is the angular velocity of the minute hand in radians per second?
    (c) What is the linear velocity of a point on the end of the minute hand in inches per second?
    (d) What is the linear velocity of a point 7 ft from the end of the minute hand in inches per second?
20. A phonograph record is 12 in. in diameter. It revolves on the record player at $33\frac{1}{3}$ revolutions per minute.
    (a) What is the angular velocity of the record in radians per minute?
    (b) What is the linear velocity of a point on the outside edge of the record in inches per minute?
21. The world's tallest Ferris wheel, the London Eye, was built in London, England, and was open to the public in 2000. The wheel is 450 ft in diameter and has 32 enclosed capsules, each of which can carry up

[3]*Although the clock is commonly referred to as Big Ben, Big Ben is actually the name of the bell that chimes the hours.*

to 25 people. The wheel takes 30 min to make 1 revolution.[4] At what linear velocity, in feet per second, does each capsule on the London Eye move?

22. When a 10-speed bike is in first gear, the chain goes around a 7-in.-diameter sprocket and then back to a 4-in.-diameter sprocket. The outer diameter of the tire is 28 in. (See the accompanying figure.) Suppose the rider pedals the bike at a rate of 2 revolutions per second. Thus, the 7-in. sprocket will rotate at 2 revolutions per second.
    (a) The linear velocity of the chain is the same as the linear velocity of a point on the 7-in. sprocket. What is the linear velocity of the chain?
    (b) At what angular velocity will the 4-in. sprocket, and hence the wheel, be moving?
    (c) How fast will the bike be traveling?
    (d) When the same 10-speed bike is in its highest gear, the chain goes around an 8-in.-diameter sprocket and then back to a 2-in.-diameter sprocket. If the rider continues to pedal the bike at a rate of 2 revolutions per second, how fast will the bike be traveling?

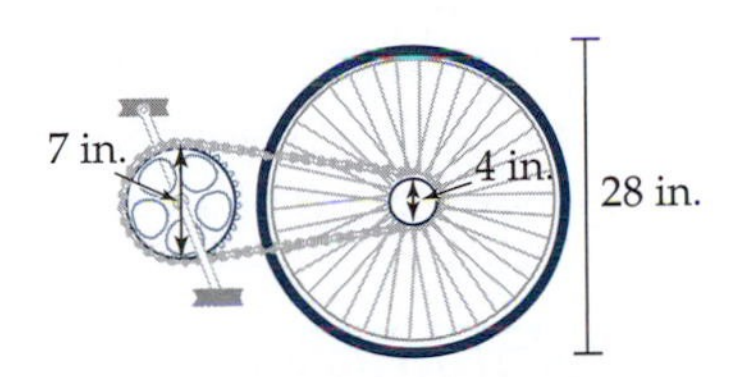

INVESTIGATIONS

INVESTIGATION 1: BROKEN WHEEL

If given an arc of a circle, you can determine both the radius of the circle and the radian measure of the angle forming the arc. Thus, given a piece of a circle, you can reconstruct the entire circle. Suppose an archeologist found the portion of the wheel shown in Figure 14 and is interested in reconstructing the dimensions of the wheel.

57.0 cm
6.7 cm
54.9 cm

Figure 14

1. We start by deriving the general formula for computing the dimensions of the circle given an arc. Figure 15(a) shows an arc that has a horizontal distance of x, a vertical distance of y, and an arc length of s. Notice that it would be easy to find measurements for x, y, and s if we had a physical object found by an archeologist. In Figure 15(b), we have taken the arc from Figure 15(a) and drawn in lines representing the radius, r, of the original circle. Notice that $\triangle ADQ$ and $\triangle BDQ$ are both right triangles.
    (a) Using the Pythagorean theorem, derive a function whose inputs are x and y and whose output is r.

[4] *British Airways*, London Eye–Statistics (*visited 20 December 1999*) ⟨*www.british-airways.com/millennium/docs/londoneye/stats.html*⟩.

(b) Using the arc length formula and your function from question 1, part (a), derive a function whose inputs are x, y, and s and whose output is θ.

Figure 15

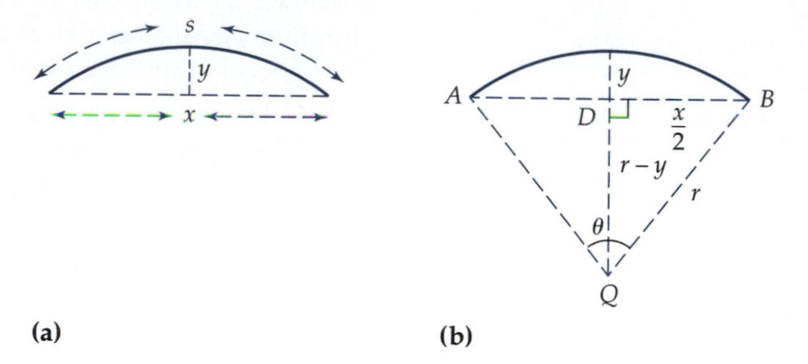

2. Using the portion of a wheel shown in Figure 14, determine the radius of the wheel. The archeologist is also interested in determining what portion of the wheel is represented by the arc. Describe this in terms of the angle θ formed by drawing two radii from the end points of the arc to the center of the wheel. Give your answer in degrees.
3. Write a short description of this arc and the reconstructed wheel from the viewpoint of an archeologist.

INVESTIGATION 2: AREA OF A SECTOR OF AN ANNULUS

An annulus, shown in Figure 16(a), is the region between two concentric circles. A sector of an annulus is much like a sector of a circle. Figure 16(b) shows a sector of an annulus. The formula for the area of a sector of an annulus is $A = \frac{1}{2}h(s_1 + s_2)$, where s_1 and s_2 are the two arc lengths and h is the width of the annulus. In this investigation, we derive the formula for the area of a sector of an annulus and compare it with the area formula for a trapezoid.

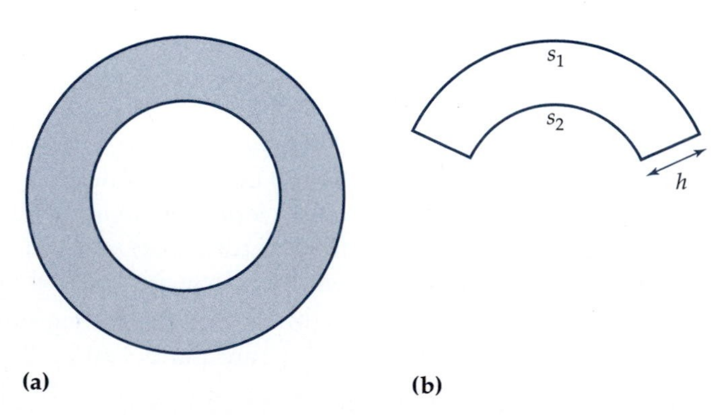

Figure 16 (a) Annulus; (b) section of an annulus.

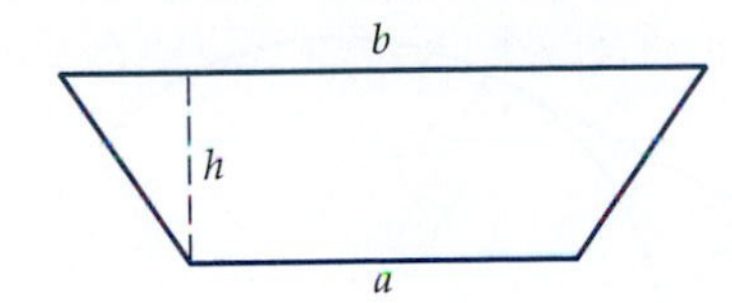

Figure 17

1. A sector of an annulus is similar to a "curved" trapezoid. Their area formulas are also similar. Figure 17 shows a trapezoid. Show that the area of a trapezoid is $A = \frac{1}{2}h(a+b)$. (*Hint:* Draw in appropriate lines to divide the trapezoid into either two triangles or two triangles and a rectangle.)

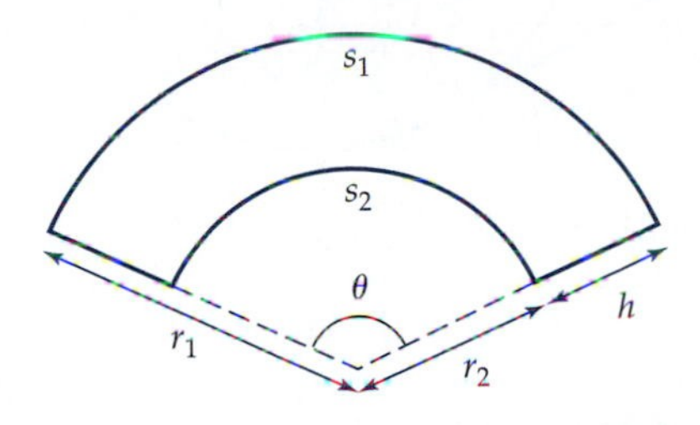

Figure 18

2. We want to derive the formula for the area of a sector of an annulus. Figure 18 shows such a sector, where s_1 is the arc length of the outer circle, s_2 is the arc length of the inner circle, r_1 is the radius of the outer circle, r_2 is the radius of the inner circle, and h is the width of the annulus. Note that $h = r_1 - r_2$.
   (a) Show that the area of a sector of an annulus can be written as $A = \frac{1}{2}\theta(r_1^2 - r_2^2)$.
   (b) Modifying your formula from part (a), show that $A = \frac{1}{2}h(s_1 + s_2)$. (*Hint:* Factor the difference of two squares.)

INVESTIGATION 3: FENCE POSTS

Suppose you have two circles "bound together" similar to two fence posts tied together with wire. In this investigation, we derive a formula for computing the length of the curve (or wire) "wrapped" around the two circles.

1. Figure 19 shows a cross-sectional view of a circular post of radius 9 in. and a circular post of radius 3 in. bound tightly together with wire. How long is the wire? (*Hint:* Begin by finding the lengths of the sides of $\triangle KOQ$.)

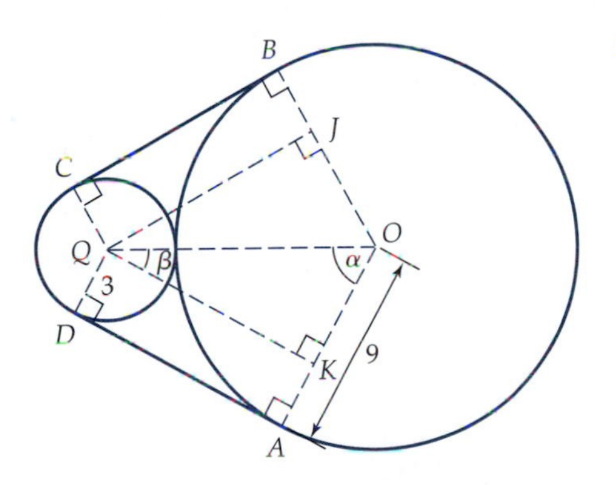

Figure 19

2. In general, show that $\gamma + \theta = 2\pi$. (See Figure 20.)
3. Find a formula for computing the length of the wire if the radius of the large circle is r_1 and the radius of the small circle is r_2. This function should contain an inverse cosine function. (See Figure 21).

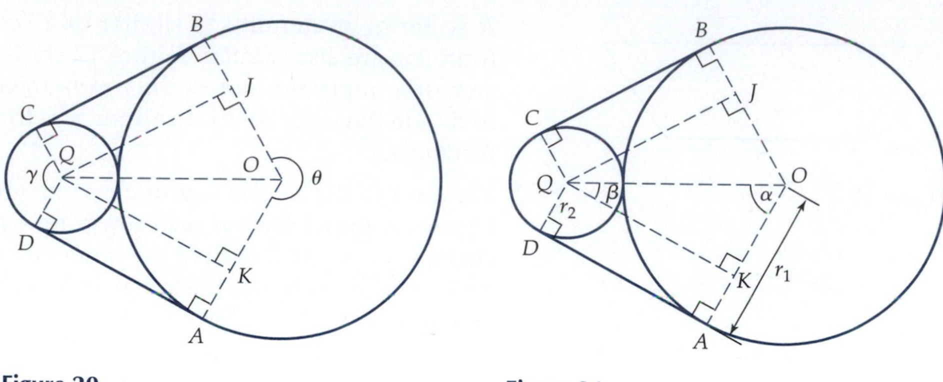

Figure 20

Figure 21

5.3 TRANSFORMATIONS OF TRIGONOMETRIC FUNCTIONS

In chapter 3, we looked at transformations of functions. In particular, in sections 3.1 and 3.2 we considered what happens when we change the input or output of a function by a constant. In this section, we revisit that material while concentrating on trigonometric functions. We discover how

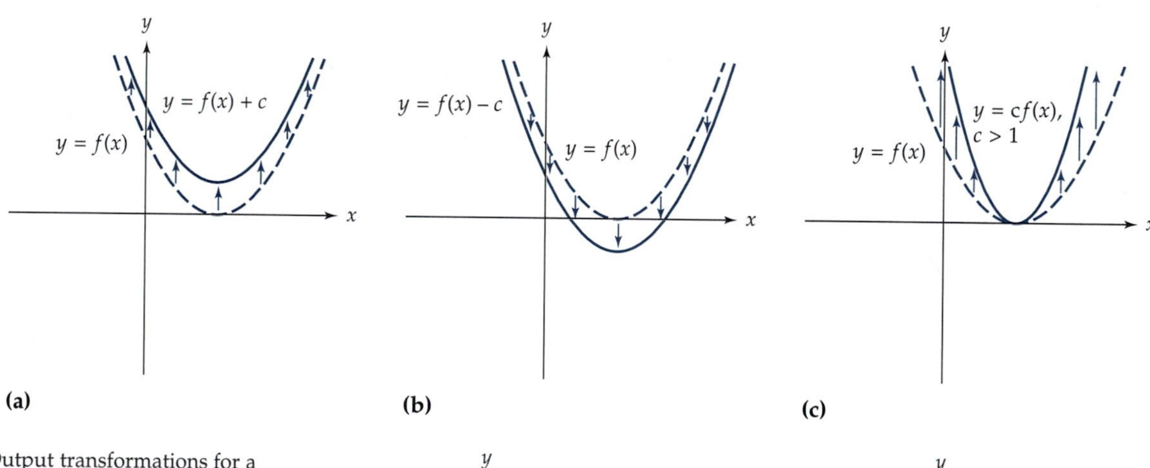

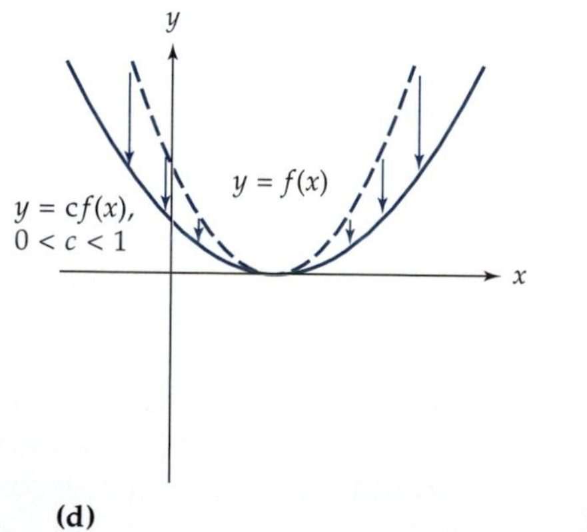

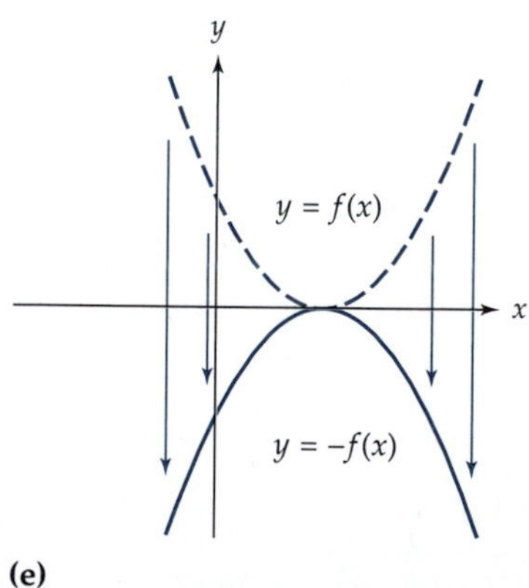

Figure 1 Output transformations for a function, $y = f(x)$, where $c > 0$. In (a) the graph of $y = f(x) + c$ is the graph of $y = f(x)$ shifted up c units. In (b) the graph of $y = f(x) - c$ is the graph of $y = f(x)$ shifted down c units. In (c) when $c > 1$, the graph of $y = c \cdot f(x)$ is the graph of $y = f(x)$ vertically stretched away from the x-axis by a factor of c. In (d) when $0 < c < 1$, the graph of $y = c \cdot f(x)$ is the graph of $y = f(x)$ vertically stretched towards the x-axis by a factor of c. (*Note:* This type of stretch is often referred to as a compression.) In (e) the graph of $y = -f(x)$ is a reflection of the graph of $y = f(x)$ across the x-axis.

to determine the amplitude, midline, period, and horizontal shift of either a sine or a cosine function. We also revisit the material in section 3.5 on inverse functions. In particular, we explore the inverse sine, inverse cosine, and inverse tangent functions and see how they can be used to solve trigonometric equations.

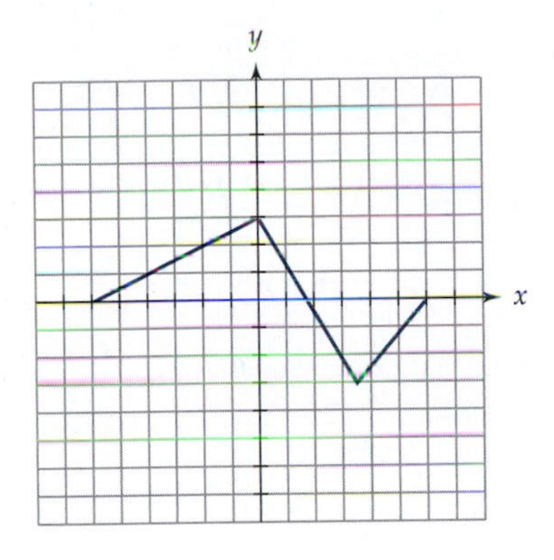

Figure 2

OUTPUT AND INPUT TRANSFORMATIONS

Output Changes

In section 3.1, we looked at the impact of changing the output of a function by a constant. Because output is represented on the vertical axis, changes in output cause vertical changes in the graph. These changes are summarized in Figure 1.

Example 1 The graph of $y = f(x)$ is shown in Figure 2. Sketch the graph of $y = -2f(x) - 1$.

Solution The graph of $y = 2f(x)$, shown in Figure 3(a), is the graph of $y = f(x)$ stretched away from the x-axis by a factor of 2. The graph of $y = -2f(x)$, shown in Figure 3(b), is the graph of $y = 2f(x)$ reflected across the x-axis. The graph of $y = -2f(x) - 1$, shown in Figure 3(c), is the graph of $y = -2f(x)$ shifted down 1 unit.

Input Changes

In section 3.2, we saw the impact of changing the input of a function by a constant. Because input is represented on the horizontal axis, changes in

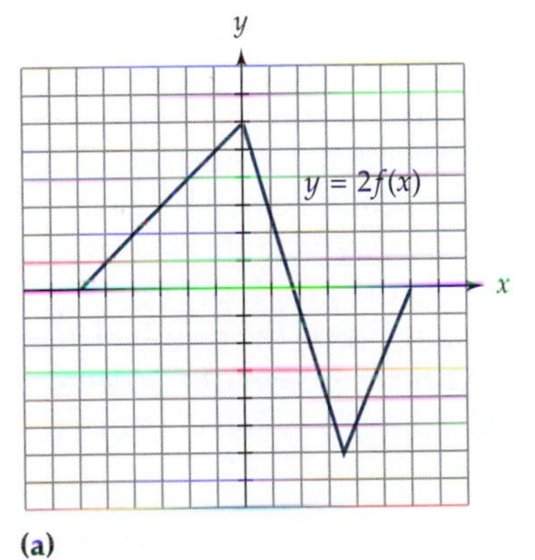

(a)

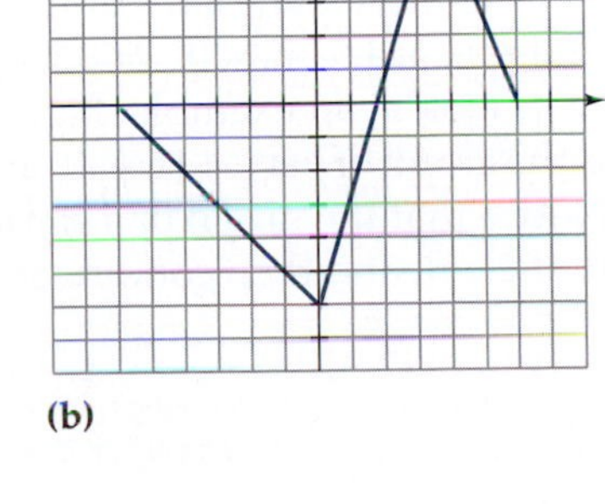

(b)

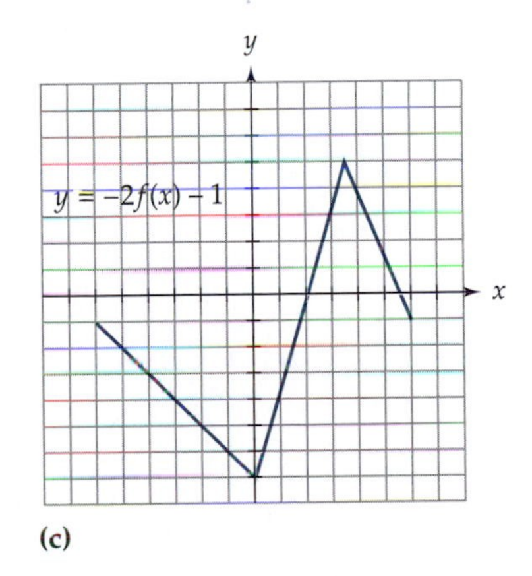

(c)

Figure 3

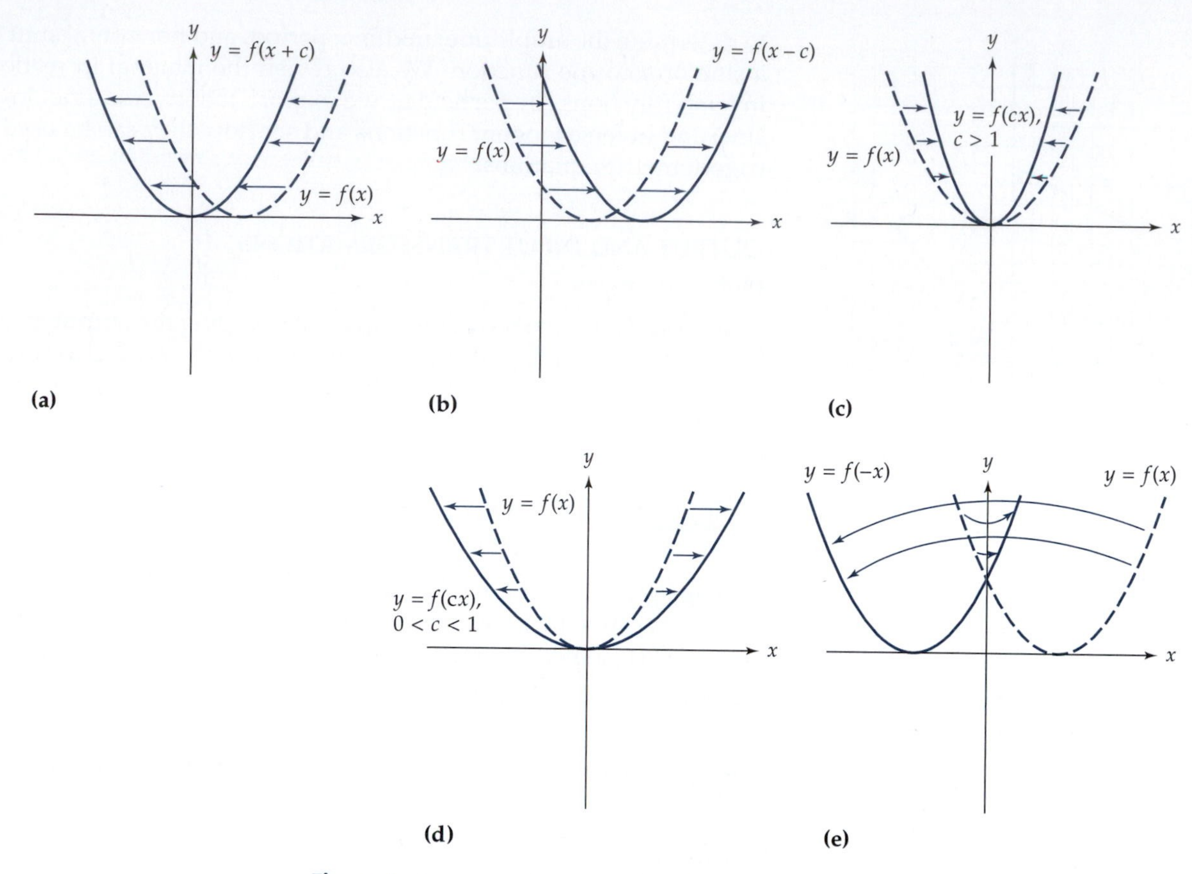

Figure 4 Input transformations for a function, f, where $c > 0$. In (a) the graph of $y = f(x + c)$ is the graph of $y = f(x)$ shifted to the left c units. In (b) the graph of $y = f(x - c)$ is the graph of $y = f(x)$ shifted to the right c units. In (c) when $c > 1$, the graph of $y = f(cx)$ is the graph of $y = f(x)$ horizontally compressed towards the y-axis by a factor of c. In (d) when $0 < c < 1$, the graph of $y = f(cx)$ is the graph of $y = f(x)$ horizontally compressed away from the y-axis by a factor of c. (*Note:* This type of compression is often referred to as a stretch.) In (e) the graph of $y = f(-x)$ is the reflection of the graph of $y = f(x)$ across the y-axis.

input cause horizontal changes in the graph. These changes are summarized in Figure 4.

In section 3.2, though, we did not see what happens if you both horizontally compress and horizontally shift a function. Given f, what is the graph of $g(x) = f(ax + b)$? The answer is that the graph of g is the graph of f first shifted to the left b units and then compressed by a factor of a. Notice that this order may be opposite to the way you thought it should work.[5] Let's look at an example. Figure 5(a) is the graph of $f(x) = |x|$, Figure 5(b) is the graph of $g(x) = |x + 4|$, and Figure 5(c) is the graph of $h(x) = |2x + 4|$. Notice that the graph of h started with the graph of f, shifted it four units to the left, and then compressed it horizontally by a factor of 2.

[5]*With inputs, many transformations are opposite to the way you may think they should be. For example, $y = f(x + a)$ may seem like it should shift to the right, but it really shifts to the left.*

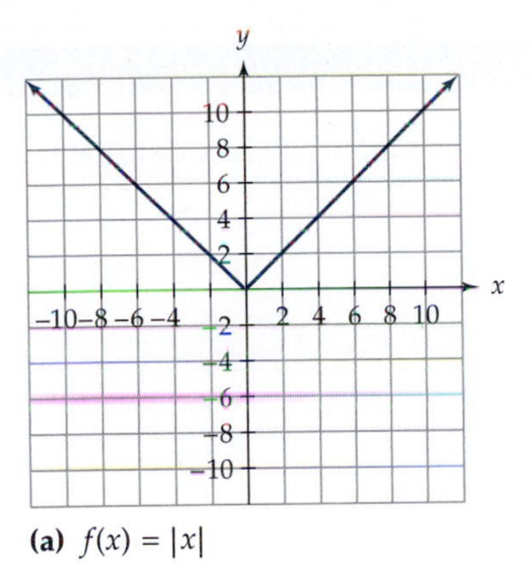

(a) $f(x) = |x|$

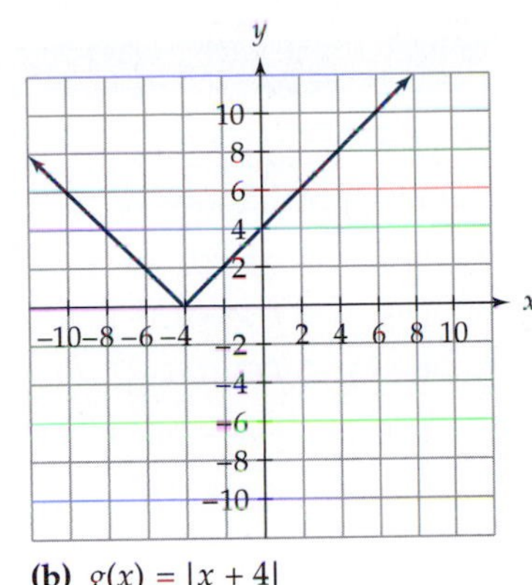

(b) $g(x) = |x + 4|$

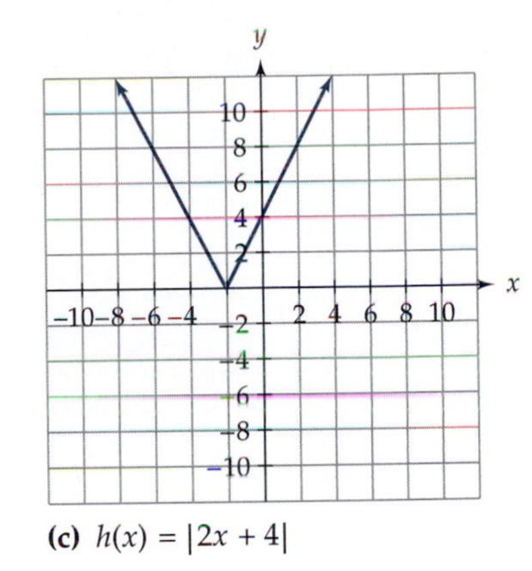

(c) $h(x) = |2x + 4|$

Figure 5 The graph in (b) is the graph in (a) shifted left four units. The graph of (c) is the graph in (a), first shifted four units and then horizontally compressed towards the y-axis by a factor of 2.

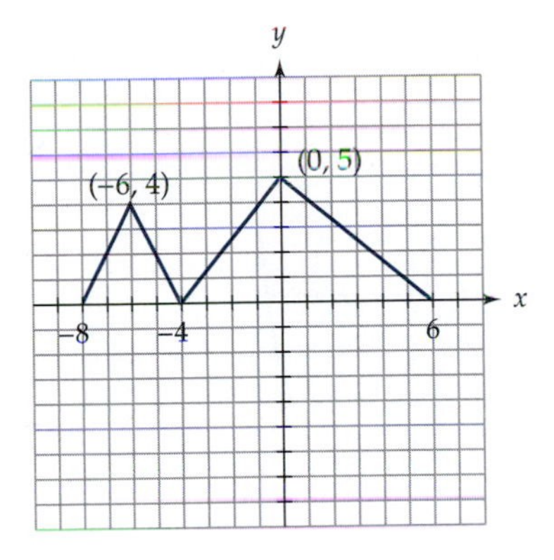

Figure 6

Example 2 The graph of $y = f(x)$ is shown in Figure 6. Sketch a graph of $y = f(-2x - 4)$.

Solution The graph of $y = f(x - 4)$, shown in Figure 7(a), is the graph of $y = f(x)$ shifted horizontally four units to the right. The graph of $y = f(2x + 4)$, shown in Figure 7(b), is the graph of $y = f(x - 4)$ horizontally compressed towards the y-axis by a factor of 2. The graph of $y = f(-2x - 4)$, shown in Figure 7(c), is the graph of $y = f(2x - 4)$ reflected across the y-axis.

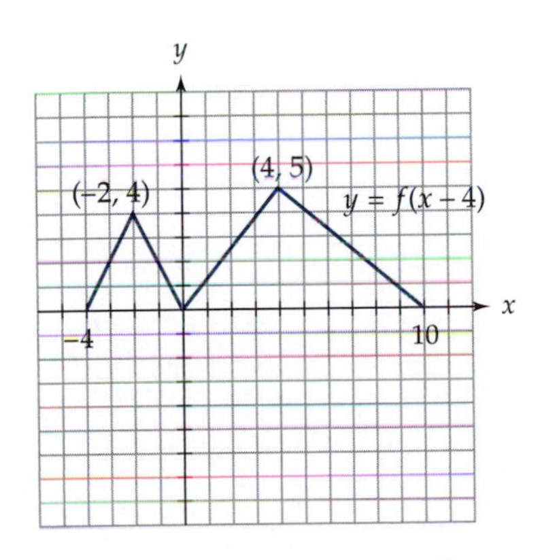

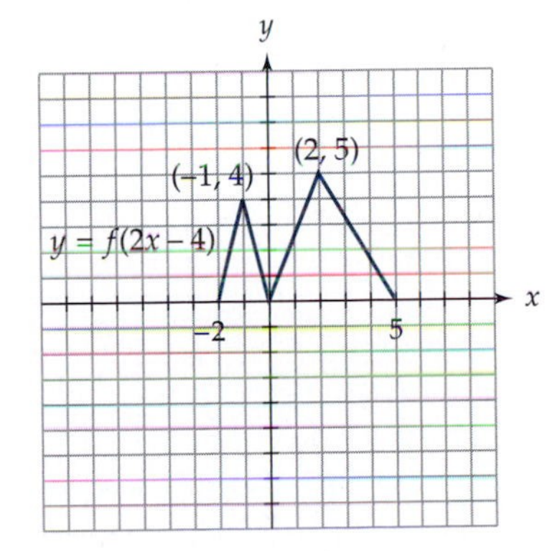

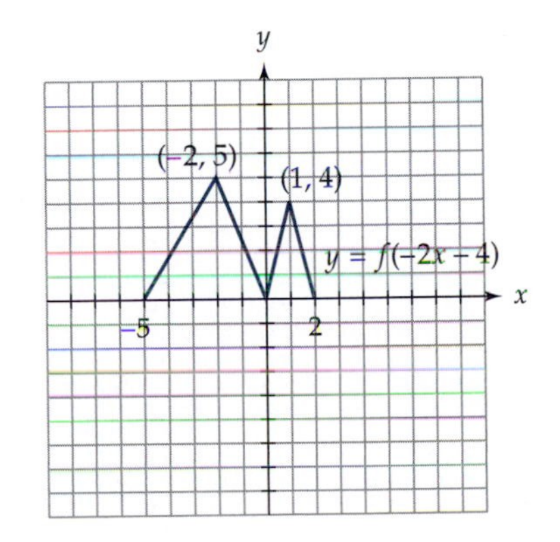

Figure 7

READING QUESTIONS

1. Use the accompanying graph of f to match these equations to the graphs of transformations of f.
   (a) $y = 3f(x)$
   (b) $y = f(x) + 3$
   (c) $y = f(3x)$
   (d) $y = f(x - 3)$
   (e) $y = f(x + 3)$

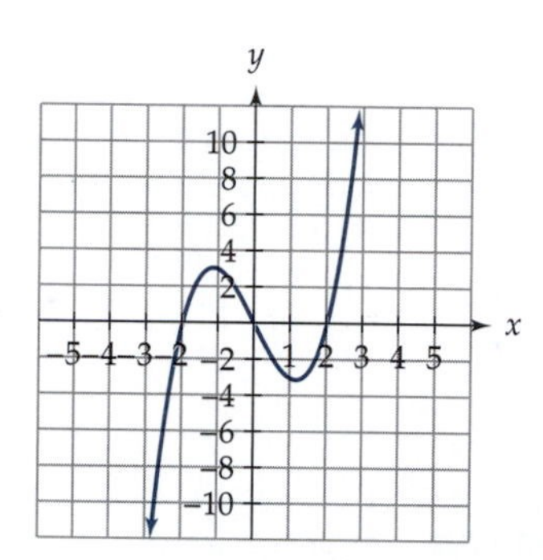

i.

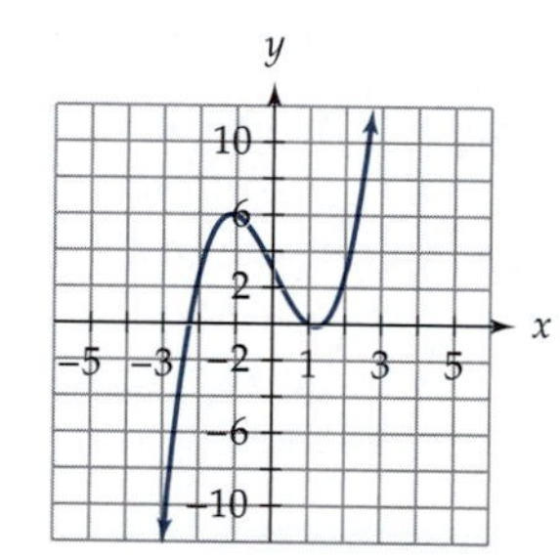

ii.

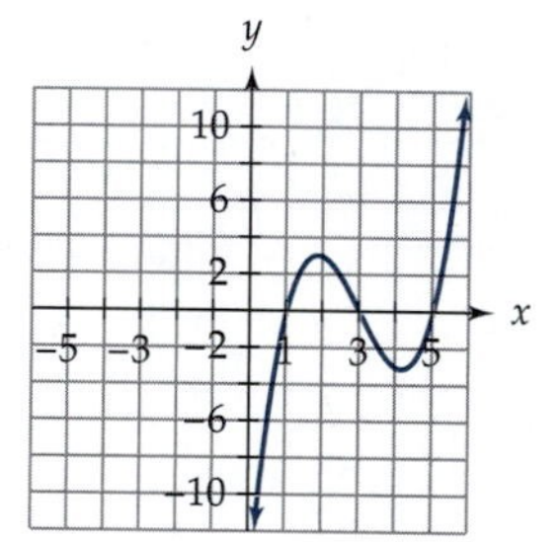

iii.

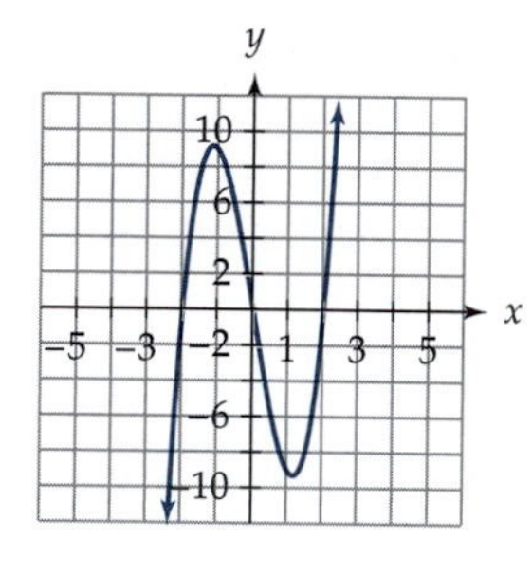

iv.

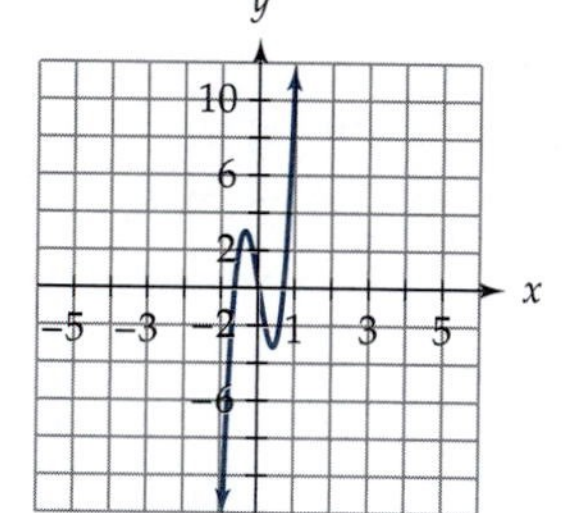

v.

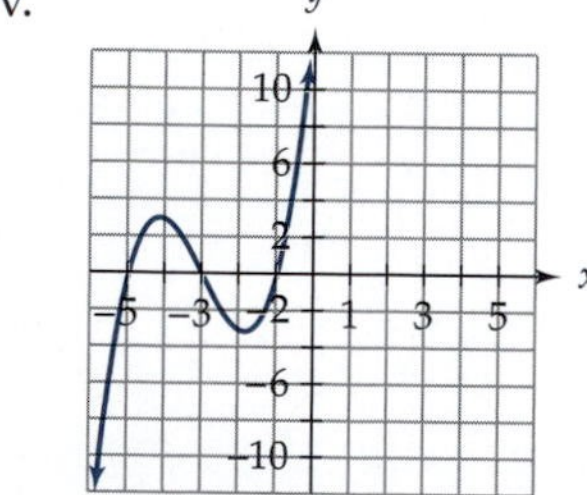

2. The accompanying figure shows the graphs of both $y = f(x)$ and $y = f(cx)$. Is $c > 1$, or is $0 < c < 1$? Briefly justify your answer.

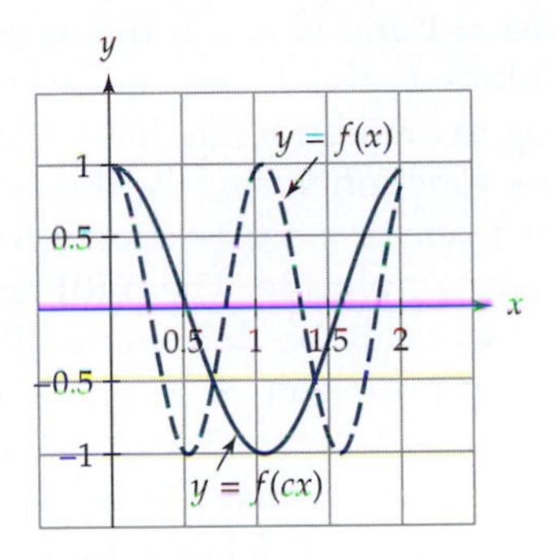

CHANGES IN OUTPUT FOR SINE AND COSINE FUNCTIONS: MIDLINE AND AMPLITUDE

Throughout this section, we always use radians as the input for the trigonometric functions because we are concentrating on graphical behavior. The familiar graphs of $y = \sin x$ and $y = \cos x$ are created from the unit circle definitions (rather than the triangle definitions). Typically, when using unit circle definitions, we assume the input is in radians.

Changing the output by a constant will either vertically shift or vertically stretch the function. For cosine and sine functions, these vertical changes impact the midline and the amplitude of the graph. The **midline** of a sine or a cosine function is the horizontal line that is halfway between the function's maximum and minimum outputs. If M is the maximum output and m is the minimum output, then the midline is $y = (M + m)/2$. For example, the function $f(x) = \sin x$ has a midline of $y = 0$ because the maximum output of f is 1 and the minimum output is -1 and $[1 + (-1)]/2 = 0$. See Figure 8(a). The **amplitude** of a sine or a cosine function is the distance from the midline to the maximum (or minimum) output, which is equal to half the difference between the maximum and minimum outputs. If M is the maximum output and m is the minimum output, then the amplitude is equal to $(M - m)/2$. The amplitude is always a positive number because $M > m$.

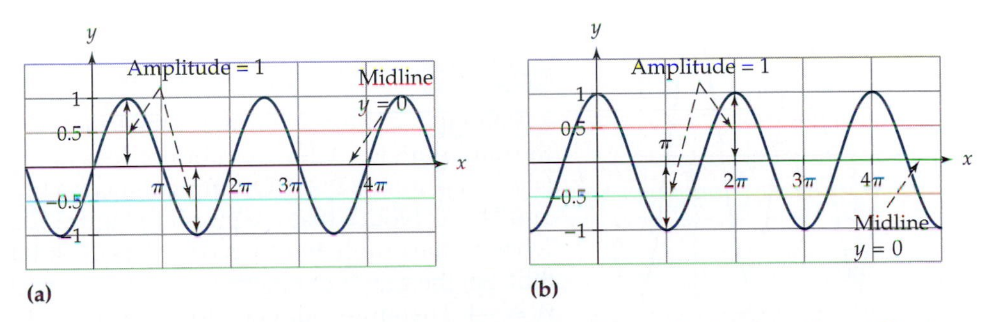

Figure 8 For (a) $f(x) = \sin x$ and (b) $g(x) = \cos x$, the amplitude is 1 and the midline is $y = 0$.

The function $f(x) = \sin x$ has an amplitude of 1 because the maximum output of f is 1 and the minimum output is -1 and $[1-(-1)]/2 = 1$. See Figure 8(a). Because the maximum and minimum outputs for $g(x) = \cos x$ are also $M = 1$ and $m = -1$, the midline for $g(x) = \cos x$ is also $y = 0$ and the amplitude is also 1. See Figure 8(b).

Adding to or subtracting from the output of a sine or a cosine function moves the function vertically, thus changing its midline. Multiplying the output of a sine or a cosine function stretches the function vertically, thus changing its amplitude. The midline and amplitude of a sine or a cosine function can easily be determined by looking at the symbolic form. For example, suppose $h(x) = 3\sin x + 2.5$. Because h is a vertical stretch of $f(x) = \sin x$ by a factor of 3 and f has an amplitude of 1, h has an amplitude of $3 \cdot 1 = 3$. Also, because h is shifted up 2.5 units from $f(x) = \sin x$ and f has a midline of $y = 0$, h has a midline of $y = 2.5 + 0 = 2.5$. The amplitude and midline of $h(x) = 3\sin x + 2.5$ can also be determined algebraically. We know that

$$-1 \leq \sin x \leq 1$$
$$-3 \leq 3\sin x \leq 3$$
$$-0.5 \leq 3\sin x + 2.5 \leq 5.5,$$

so the maximum value for $h(x)$ is $M = 5.5$, whereas the minimum value is $m = -0.5$. Therefore, the amplitude is $[M-m]/2 = [5.5-(-0.5)]/2 = 6/2 = 3$, whereas the midline is $[M+m]/2 = [5.5+(-0.5)]/2 = 5/2 = 2.5$. A graph of this function is shown in Figure 9. Notice that the amplitude, 3, and the midline, $y = 2.5$, are both numbers appearing in the equation for $h(x) = 3\sin x + 2.5$. The amplitude is the vertical stretch and the midline is the vertical shift, which is always true for any sine or cosine function.

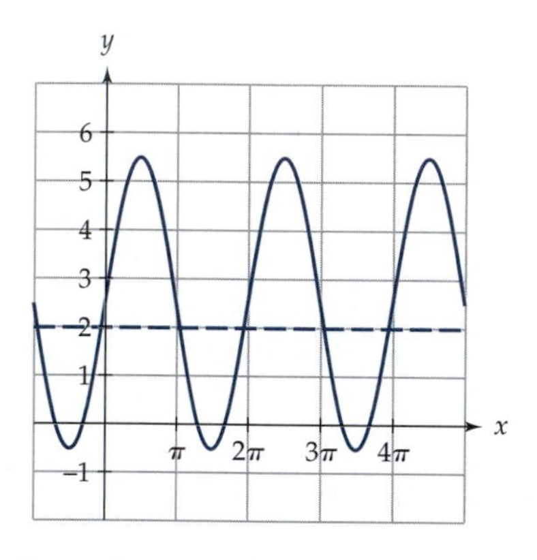

Figure 9 A graph of $h(x) = 3\sin x + 2.5$.

PROPERTIES

In general, for $f(x) = A\sin x + D$ or $g(x) = A\cos x + D$:

- The amplitude is $|A|$. If A is negative, then the graph will be reflected over the x-axis.
- The midline is $y = D$.

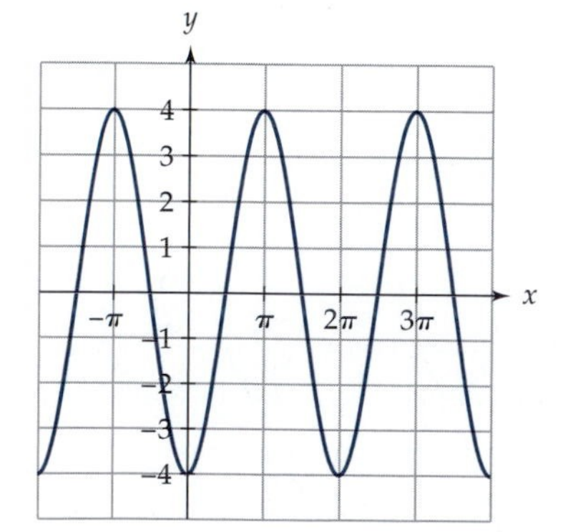

Figure 10

Example 3 Find an equation for the cosine function shown in Figure 10.

Solution The maximum for the function shown in Figure 10 is 4 and the minimum is -4. So, the amplitude is $[4-(-4)]/2 = 4$ and the midline is $y = [4+(-4)]/2 = 0$. Our function is $h(x) = A\cos x + D$, where $|A| = 4$ and $D = 0$. Notice that the graph in Figure 10 has a y-intercept of $(0, -4)$, whereas the graph of $g(x) = \cos x$ has a y-intercept of $(0, 1)$. See Figure 8(b). So, the graph in Figure 10 has been reflected over the x-axis. Hence, $A = -4$. Therefore, our equation is $h(x) = -4\cos x + 0 = -4\cos x$.

READING QUESTIONS

3. In the accompanying graph of $y = A \sin x + D$, is A positive or negative? Briefly justify your answer.

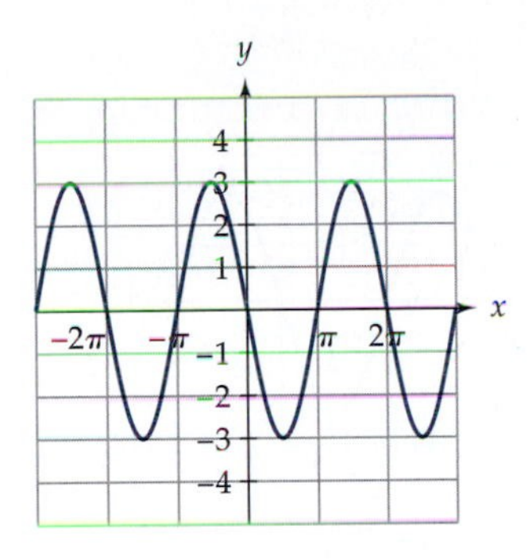

4. Let $f(x) = 2 \cos x + \pi$. What is the amplitude of f? What is the midline of f?

5. Find an equation for the sine function, $y = A \sin x + D$, shown in the accompanying graph.

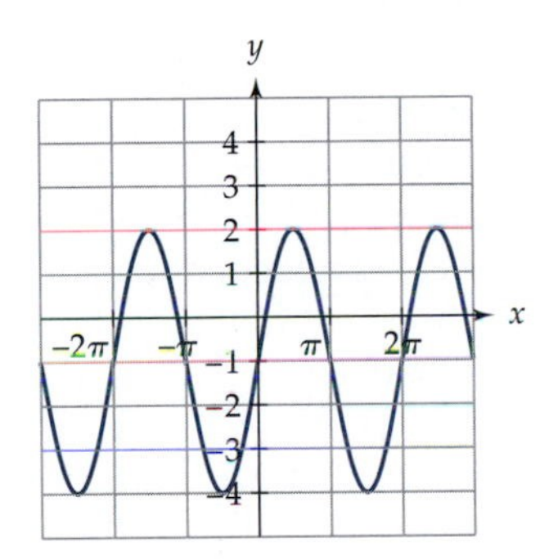

CHANGES IN INPUT FOR SINE AND COSINE FUNCTIONS: PERIOD AND HORIZONTAL SHIFT

Period

Recall from section 2.4 that a **periodic function** is one that gives the same output for inputs a fixed distance apart. Symbolically, if f is a periodic function, then for some constant $p, f(x) = f(x + p)$ for all x. The **period** of this function is the smallest value of p for which this relationship is true. The functions $y = \sin x$ and $y = \cos x$ have a period of 2π. (See Figure 11.) Thus, $\sin x = \sin(x + 2\pi)$ and $\cos x = \cos(x + 2\pi)$ for all x. For $y = \sin x$ and $y = \cos x$, inputs that are 2π units apart will have the same output. Notice that for p to be the period, the relationship $f(x) = f(x + p)$ has to work for *all* x. For example, $f(x) = \sin x$ is zero when x is a multiple of π. Looking at Figure 11(a), we see that $\sin 0 = \sin \pi = \sin(2\pi) = \sin(4\pi) = 0$.

At first glance, you may erroneously conclude that the period of $f(x) = \sin x$ is π, but $\sin(x+\pi) \neq \sin x$ for *all* values of x. In particular, $1 = \sin(\pi/2) \neq \sin(\pi/2+\pi) = \sin(3\pi/2) = -1$. So, be careful when finding the period of a function by looking at points with the same output.

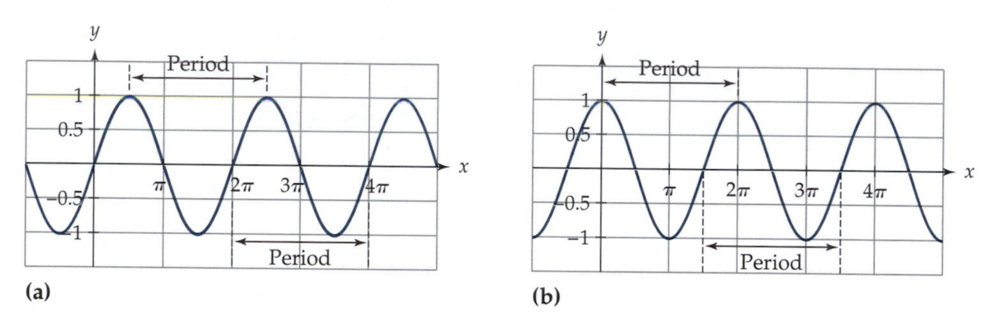

Figure 11 The period of (a) $y = \sin x$ and (b) $y = \cos x$ is 2π.

Because multiplying the input of a function by a constant compresses it horizontally toward the y-axis, multiplying the input of a trigonometric function by a constant will change the period. If you compress a function horizontally, then you are changing the frequency at which the outputs will repeat. We can see this horizontal compression symbolically. For example, if $g(x) = \cos x$, then the period of g is 2π. Thus, the graph of g will complete one period between $x = 0$ and $x = 2\pi$. Now suppose $h(x) = \cos(3x)$. Because h is compressed horizontally by a factor of 3, the outputs will repeat three times as often. Thus, the graph of $h(x) = \cos(3x)$ will complete three periods between $x = 0$ and $x = 2\pi$. So, the period of h is one-third of 2π, or $2\pi/3$. The behavior of the sine function is similar. In general, the period of $f(x) = \sin(Bx)$ and $g(x) = \cos(Bx)$ is $2\pi/B$. Notice that the horizontal compression, B, is equal to 2π/period.

Example 4 Determine the amplitude, midline, and period for $f(x) = 3\sin(\pi x)$.

Solution Starting with the function $y = \sin x$, the function $f(x) = 3\sin(\pi x)$ has been vertically stretched by a factor of 3 and horizontally compressed by a factor of π. It has not been shifted either vertically or horizontally. So, the amplitude of f is three times the amplitude of $y = \sin x$, and the period of h is $1/\pi$ times the period of $y = \sin x$. Thus, the amplitude of $f(x) = 3\sin(\pi x)$ is 3, and the period is $2\pi/\pi = 2$. Because there is no vertical shift, the midline of f is the same as the midline for $y = \sin x$, which is $y = 0$.

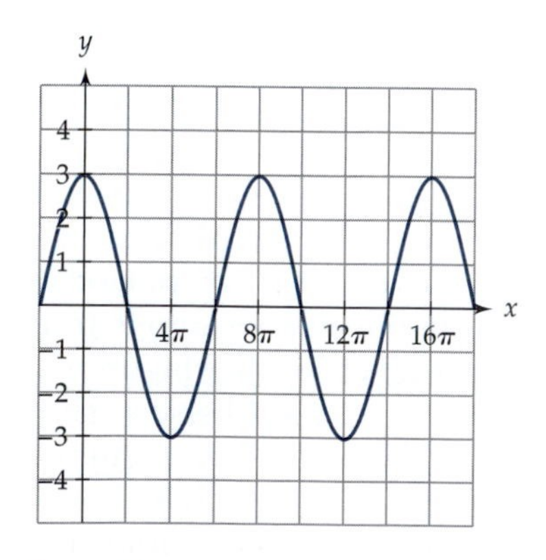

Figure 12

Example 5 Figure 12 is a graph of a cosine function that has been vertically stretched and horizontally compressed. Using the amplitude and period, find the equation of the function.

Solution The amplitude is 3, and the period is 8π. Thus, there has been a vertical stretch of 3 and a horizontal compression of $\frac{1}{4}$ (because $2\pi/8\pi = \frac{1}{4}$). Therefore, the equation is $f(x) = 3\cos(x/4)$.

Horizontal Shift

Now that we looked extensively at multiplying the input of $y = \sin x$ or $y = \cos x$ by a constant, let's see what happens when we *add* a constant to the input. Recall that adding a constant to the input horizontally shifts the graph of the function. Look back at Figure 11. Notice that the graphs of $y = \sin x$ and $y = \cos x$ will be identical if we appropriately shift them horizontally. If the graph of $y = \sin x$ were shifted to the left $\pi/2$ units, it would be the same as the graph of $y = \cos x$. Also, if the graph of $y = \cos x$ were shifted to the right $\pi/2$ units, it would be the same as the graph of $y = \sin x$. The observations give us the following two identities:

$$\sin\left(x + \frac{\pi}{2}\right) = \cos x$$
$$\cos\left(x - \frac{\pi}{2}\right) = \sin x.$$

When determining a horizontal shift given the graph of a sine or cosine function, focus on "familiar" points. For example, you know that the maximum values of $y = \sin x$ occur for $x = \pi/2, 5\pi/2, 9\pi/2, \ldots$. You know that the maximum values of $y = \cos x$ occur at $x = 0, 2\pi, 4\pi, \ldots$. Be careful about concentrating on points where a sine or cosine function is zero. At some zeros, the function is changing from positive to negative while at other zeroes the function is changing from negative to positive. When using zeros to determine a horizontal shift, be sure to match them appropriately.

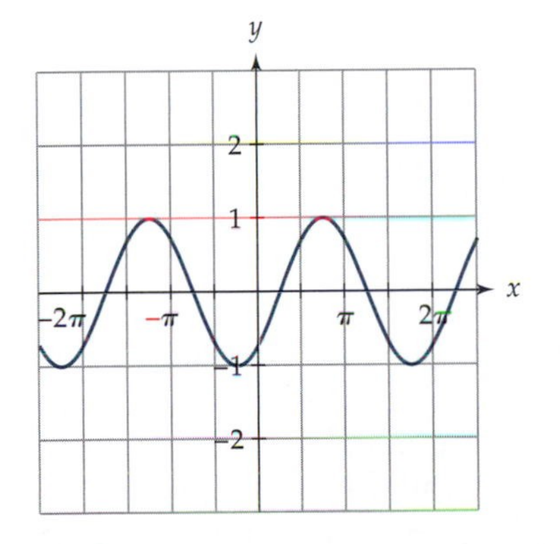

Figure 13

Example 6 Figure 13 is the graph of a sine function that has been shifted horizontally. Find the equation of the function.

Solution The graph in Figure 13 has a zero changing from negative to positive when $x = \pi/4$. The function $y = \sin x$ has a zero changing from negative to positive when $x = 0$. Thus, $y = \sin x$ has been shifted $\pi/4$ units to the right. Alternatively, we can notice that the graph in Figure 13 appears to have a maximum value when $x = 3\pi/4$, whereas the graph of $y = \sin x$ has a maximum when $x = \pi/2$. Thus, the graph in Figure 13 has been shifted $\pi/4$ units to the right. So, the equation of the function shown in Figure 13 is $f(x) = \sin(x - \pi/4)$.

The equation $f(x) = \sin(x - \pi/4)$ is not the only equation that represents the graph in Figure 13. Because a sine function is periodic, we could have also used $f(x) = \sin(x - \pi/4 + 2\pi) = \sin(x + 7\pi/4)$ as the equation. If Example 6 had not stated that we were looking for the equation of a sine function, we could have written the equation for Figure 13 as a cosine function. Because $\cos x = \sin(x - \pi/2)$, we know that $f(x) = \sin(x - \pi/4) =$

$\cos(x - \pi/2 - \pi/4) = \cos(x - 3\pi/4)$. When finding equations for periodic functions, remember that answers are not unique!

Combining Changes in the Period with a Horizontal Shift

Combining a change in the period with a horizontal shift can be tricky. Earlier we mentioned that the graph of $f(x) = f(ax + b)$ is the graph of f, first shifted b units horizontally and then horizontally compressed by a factor of a. Consider the two functions $f(x) = \sin(2x - \pi/2)$ and $g(x) = \sin[2(x - \pi/2)]$. For $f(x) = \sin(2x - \pi/2)$, the graph of $y = \sin x$ has first been shifted to the right $\pi/2$ units and then horizontally compressed by a factor of 2. For $g(x) = \sin[2(x - \pi/2)]$, the graph of $y = \sin x$ has first been horizontally compressed by a factor of 2 and then shifted to the right $\pi/2$ units. Both f and g are sine functions whose period has been changed to π and whose graph has been shifted to the right $\pi/2$ units, but these two transformations have been done in the opposite order. The graphs of f and g are shown in Figure 14.

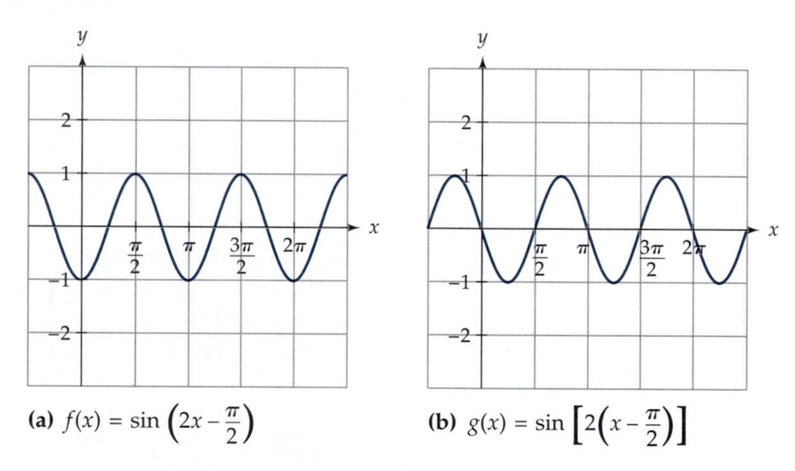

Figure 14 Two sine functions that have been shifted $\pi/2$ units to the right and have a period of π. (a) The function with the shift first and the change in period second; (b) the function with the change in period first and the shift second.

Let's take a closer look at these two functions. In particular, notice that $g(x) = \sin[2(x - \pi/2)] = \sin(2x - \pi)$. We can think of g as compressing the period by a factor of 2 and then shifting $\pi/2$ units to the right (as we did earlier), or we can think of g as shifting π units to the right and then compressing the period by a factor of 2. Similarly, notice that $f(x) = \sin(2x - \pi/2) = \sin[2(x - \pi/4)]$. We can think of f as shifting $\pi/2$ units to the right and then compressing the period by a factor of 2 (as we did earlier), or we can think of f as compressing the period by a factor of 2 and then shifting $\pi/4$ units to the right. Which is the better way to think about this type of transformation? Should we typically write functions as $y = \sin(Bx + E)$ or as $y = \sin[B(x + C)]$? Look carefully at the graphs in Figure 14. Using Figure 14(a), it is fairly easy to see the period is π and there is a shift of $\pi/4$ units to the right.[6] Using Figure 14(b), it is fairly easy to see that the period is π and there is a shift of $\pi/2$ units to the right. The numbers 2 and $\pi/4$ for f and the numbers 2 and $\pi/2$ for g occur when the formulas are in the form $y = \sin[B(x + C)]$. This form gives a clearer connection between the graph and the equation. So, even though algebraically it is simpler to multiply through and remove the parenthesis, we typically write sine and cosine functions in the form $y = \sin[B(x + C)]$ or $y = \cos[B(x + C)]$.

PROPERTIES

In general, for $f(x) = \sin[B(x + C)]$ or $g(x) = \cos[B(x + C)]$:

- The period is $2\pi/B$. Also, $B = 2\pi/\text{period}$.
- The horizontal shift is C units to the left if C is positive.
- The horizontal shift is C units to the right if C is negative.

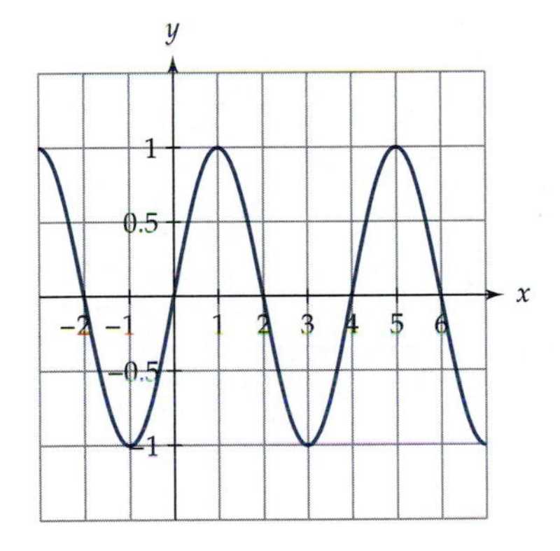

Figure 15

Example 7 Determine the equation of the function shown in Figure 15. Write your solution two ways, first as $f(x) = \sin[B(x + C)]$ and then as $f(x) = \cos[B(x + C)]$.

Solution From Figure 15, we can see that the function has a period of 4. Therefore, $B = 2\pi/4 = \pi/2$. The function $y = \sin x$ has a zero at $x = 0$, where the function is changing from negative to positive. The graph in Figure 15 does the same. Therefore, this function has no horizontal shift from $y = \sin x$, so $C = 0$. The equation of this function is $f(x) = \sin(\pi x/2)$.

For the cosine function, we again have a period of 4, which means that $B = \pi/2$. The function $y = \cos x$ has a maximum at $x = 0$, whereas the graph in Figure 15 has a maximum at $x = 1$. Therefore, we have a cosine graph that has been shifted one unit to the right, so $C = -1$. The equation of this function is $g(x) = \cos[\pi/2(x - 1)]$. Notice once again that the equation of a periodic function is not unique!

READING QUESTIONS

6. Match the following equations to the graphs.

(a) $y = \sin(\pi x)$

(b) $y = \sin(x - \pi)$

[6] *When looking for a horizontal shift combined with a horizontal compression, it is best to use points that use to be at $x = 0$ rather than concentrating on maximums or minimums because a horizontal compression changes the location of all points except for $x = 0$.*

(c) $y = \cos(\pi x)$

(d) $y = \cos(x - \pi)$

i.

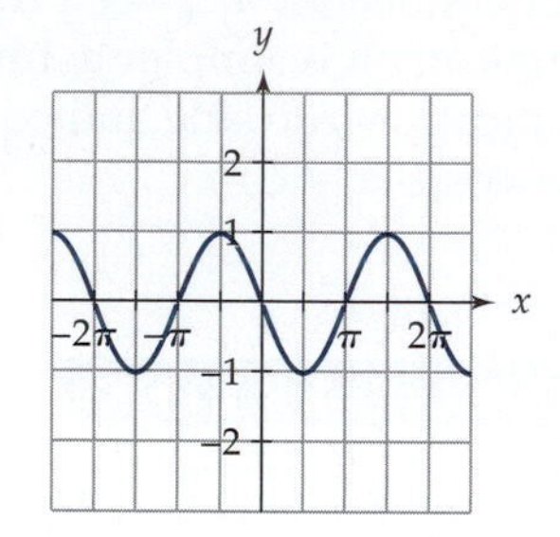

ii.

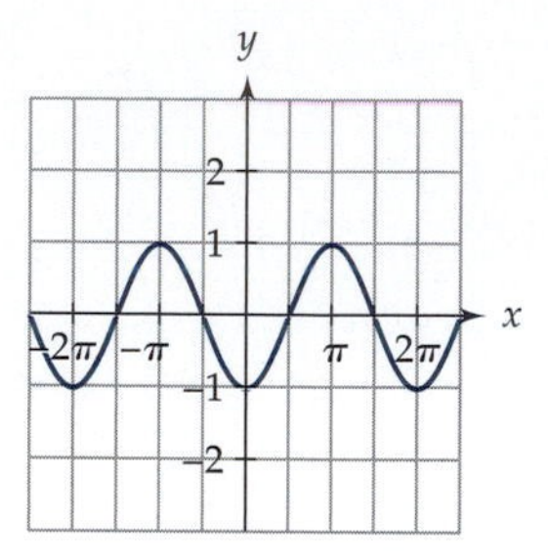

iii.

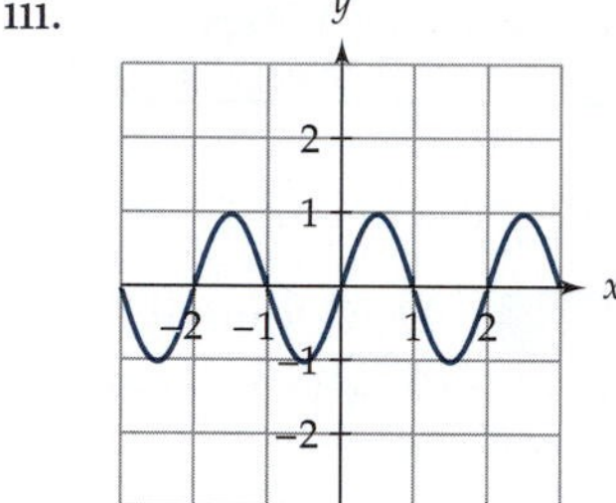

iv.

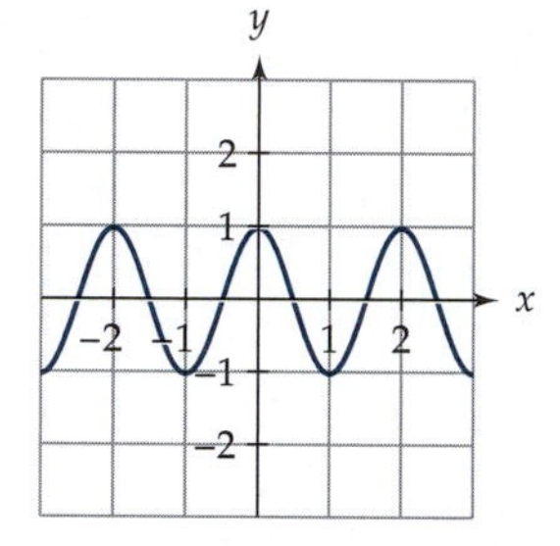

7. If a cosine function has been horizontally compressed by a factor of 10, what is its period? What is its equation?

8. True or false:

   (a) $\sin(x + \pi/2) = \cos x$

   (b) $\cos(x + \pi/2) = \sin x$

   (c) The period of $y = \sin(2x)$ is 2.

   (d) The horizontal shift of $y = \cos(x - 3)$ is three units to the right.

9. The equation for the accompanying graph can be written in the form $f(x) = \cos[B(x + C)]$. Find the value of B and C.

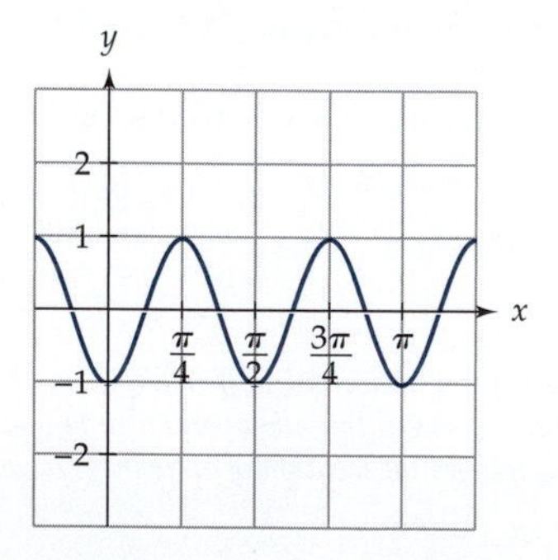

PUTTING IT ALL TOGETHER

PROPERTIES

The general formulas for the sine and cosine functions are:

$$f(x) = A\sin[B(x + C)] + D \quad \text{and} \quad g(x) = A\cos[B(x + C)] + D,$$

where

- The amplitude is $|A|$.
- The period is $2\pi/B$.
- The horizontal shift is C.
- The equation of the midline is $y = D$.

By combining input and output changes, we can find the equation for any sine or cosine function. More important is that we can find equations to model real-world periodic behavior.

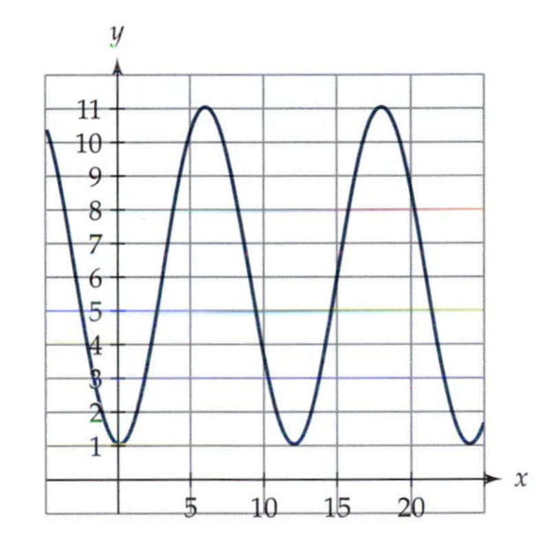

Figure 16

Example 8 The graph of f is shown in Figure 16. Determine an equation for f.

Solution We can use either a sine or cosine function for the equation for f. We choose a cosine function because this graph is similar to $y = \cos x$, reflected over its midline, with no horizontal shift. The maximum value is 11 and the minimum value is 1, so the amplitude is $(11 - 1)/2 = 5$. The equation of the midline is $y = (11 + 1)/2 = 6$. Because, for $x = 0$, the output is below the midline, we know that $A = -5$ and $D = 6$. We chose a cosine function so that we could ignore a horizontal shift. Therefore, $C = 0$. Finally, to find B, we need to look at the period. According to the graph, the period is 12, so $B = 2\pi/12 = \pi/6$. Therefore, the equation of the function shown in Figure 16 is $f(x) = -5\cos[(\pi/6)x] + 6$.

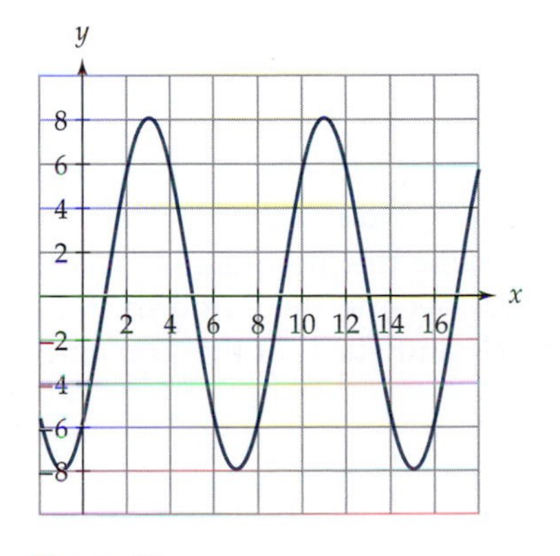

Figure 17

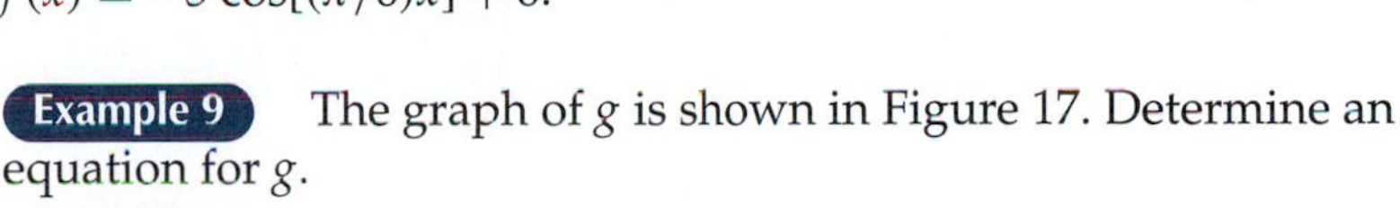

Example 9 The graph of g is shown in Figure 17. Determine an equation for g.

Solution The graph shown in Figure 17 can be modeled by either a sine or a cosine function. Both require a horizontal shift. We choose a sine function. The maximum value is 8 and the minimum value is -8, so the amplitude $[8 - (-8)]/2 = 8$ and the midline is $y = [8 + (-8)]/2 = 0$. Because we are assuming we have a sine function that has been shifted to the right, the graph of g is not reflected over its midline. So, $A = 8$ and $D = 0$. To find B, we determine that the period is also 8. So, $B = 2\pi/8 = \pi/4$. The graph of $y = \sin x$ crosses its midline and is increasing when $x = 0$. The graph in Figure 17 crosses its midline ($y = 0$) and is increasing when $x = 1$. So, g is a sine function shifted one unit to the right. Therefore, $C = -1$. The equation of a function that fits this graph is $g(x) = 8\sin[(\pi/4)(x - 1)]$.

Example 10 The world's tallest Ferris wheel is the London Eye. The wheel is 450 ft in diameter and has 32 enclosed capsules, each of which can carry up to 25 people. The wheel takes 30 min to make 1 revolution.[7] Assume the capsule starts at the bottom of the Ferris wheel at height 0. Find a function whose input is time (in minutes) and whose output is height of the capsule (in feet).

Solution Because we have a periodic function where the minimum height occurs at $t = 0$, a cosine function reflected over the x-axis is a good choice for our modeling function. The maximum height of the wheel is 500 ft, and the minimum height is 0. Therefore, the amplitude is $(450 - 0)/2 = 225$, and the midline is $y = (450 + 0)/2 = 225$. The function is reflected over its midline (because it starts at the minimum rather than the maximum value), so $A = -225$ and $D = 225$. To find the period, we know that it takes 30 min to make 1 revolution. So, $B = 2\pi/30 = \pi/15$. We do not need a horizontal shift because the function starts at the minimum value. Therefore, a function that models the height of the capsule, in feet, is $h(t) = -225\cos(\pi t/15) + 225$, where t is the time in minutes.

READING QUESTIONS

10. Determine an equation for the function shown.

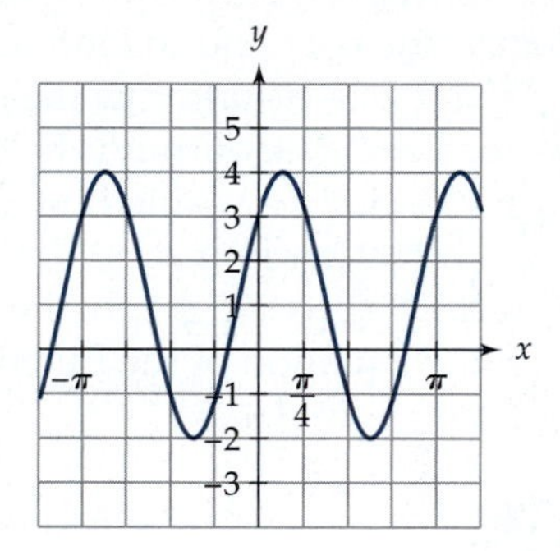

11. Suppose you have a kitchen clock with a 4-in. minute hand. Find an equation to model the height of the tip of the minute hand as it travels around the clock. Your input should be time (in minutes), and your output should be height from the line connecting the 3 and the 9 on the clock (in inches). Assume the minute hand starts at the 12 o'clock position.

[7] *British Airways,* London Eye—Statistics *(visited 20 December 1999) ⟨www.british-airways.com/millennium/docs/londoneye/stats.html⟩.*

TRANSFORMATIONS OF THE TANGENT FUNCTION

The tangent function is a periodic function, like sine and cosine, but it is different in many ways. Sine and cosine functions have maximum and minimum outputs, but the tangent function does not. Thus, amplitude is not defined for the tangent function. We can still vertically stretch tangent functions (as we can for any function), but without maximums and minimums, it is difficult to see this stretch when looking at the graph. A midline is also not defined for the tangent function. We can, however, still vertically shift tangent functions. Because we cannot focus on a maximum or minimum point when determining the vertical shift, we instead focus on the point where the tangent function changes concavity. For $f(x) = \tan x$, the graph changes from concave down to concave up at the points where $y = 0$. See Figure 18(a). In the graph of $g(x) = \tan x + 2$, g changes from concave down to concave up at the points where $y = 2$. See Figure 18(b).

A horizontal shift can be seen using these same points. For $f(x) = \tan x$, the graph changes concavity when $x = n\pi$, where n is an integer. See Figure 18(a). For $h(x) = \tan(x - 1)$, the graph changes concavity when $x = n\pi + 1$, where n is an integer. In Figure 19, this graph has been shifted one unit to the right.

A horizontal compression can be seen by observing the change in period. The period of $f(x) = \tan x$ is π. See Figure 18(a). The period of $k(x) = \tan(2x)$ is $\pi/2$ because of compression by a factor of 2. The easiest way to see the period is to view the distance between vertical asymptotes (or the distance between points where the graph changes concavity).

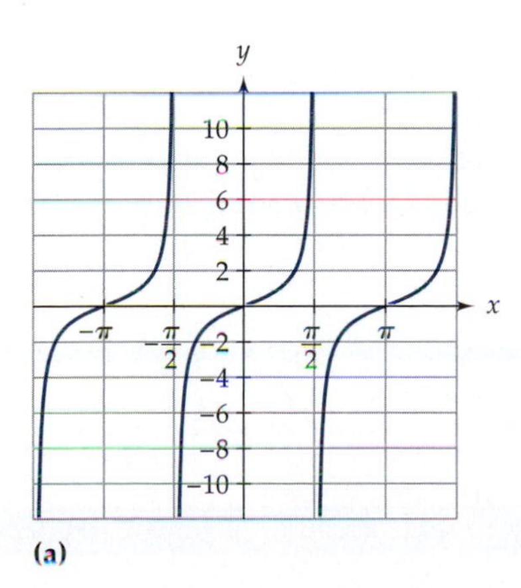

(a)

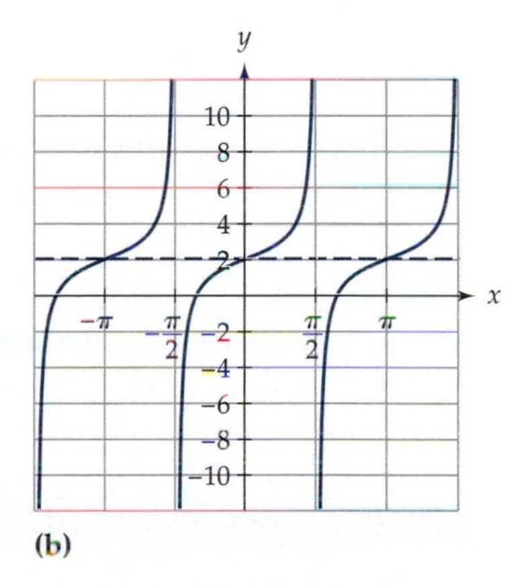

(b)

Figure 18 (a) The graph of $f(x) = \tan x$ changes concavity at points where $y = 0$. (b) The graph of $y = \tan x + 2$ changes concavity at the points where $y = 2$.

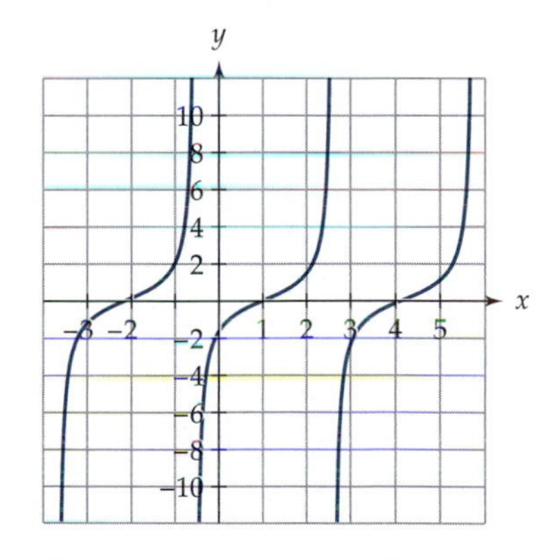

Figure 19 The graph of $h(x) = \tan(x - 1)$ changes concavity at the point $(1, 0)$. It is a tangent function that has been shifted to the right one unit.

PROPERTIES

In general, for $y = A\tan[B(x + C)] + D$:

- The vertical stretch is $|A|$. If the graph changes from concave down to concave up between the vertical asymptotes, then A is positive. If the graph changes from concave up to concave down between the vertical asymptotes, then A is negative. (The value of A is difficult to determine from a graph.)
- The vertical shift is D. The function will change concavity at the points where the output is D.
- The period is π/B. Notice that $B = \pi/\text{period}$.
- The horizontal shift is C. One of the points where the function changes concavity is (C, D).

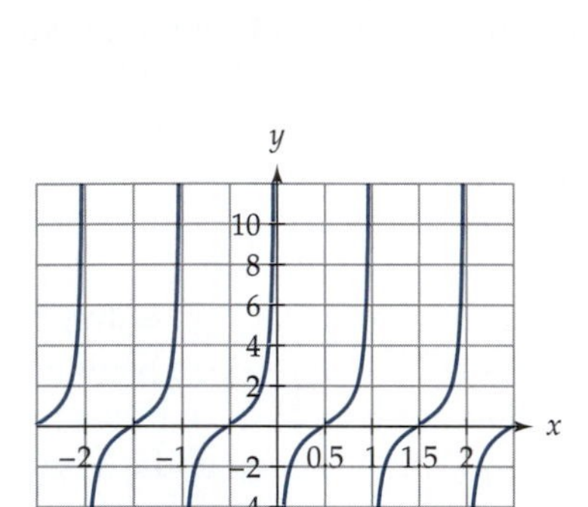

Figure 20

Example 11 Determine the equation of the tangent function shown in Figure 20.

Solution The graph shown in Figure 20 is similar to the graph of $y = \tan x$. The points where the graph changes concavity are at $y = 0$, so the function has not been shifted vertically and $D = 0$. One of the points where the concavity changes is $(0.5, 0)$, so the function has been shifted to the right 0.5 unit and $C = -0.5$. The period has also changed. The period of the function shown in Figure 20 is 1, so $B = \pi/1 = \pi$. Therefore, the equation of this function is $f(x) = \tan[\pi(x - 0.5)]$.

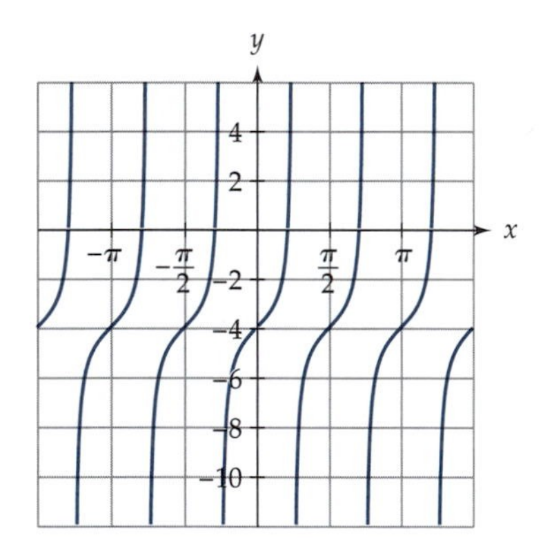

Figure 21

Example 12 Determine the equation of the tangent function shown in Figure 21.

Solution The graph shown in Figure 21 is similar to $y = \tan x$. The points where it changes concavity, however, are at $y = -4$. So, there is a vertical shift of -4 units, meaning that $D = -4$. One point where it changes concavity is $(0, -4)$. So, there is no horizontal shift, meaning that $C = 0$. The period of this function is $\pi/2$, so $B = \pi/(\pi/2) = 2$. Therefore, the equation is $f(x) = \tan(2x) - 4$.

READING QUESTIONS

12. Why doesn't the tangent function have an amplitude or a midline?

13. For $y = A\tan[B(x + C)] + D$, how do you tell, by looking at the graph, if A is positive or negative?

14. Determine the equation of the tangent function shown in the accompanying figure.

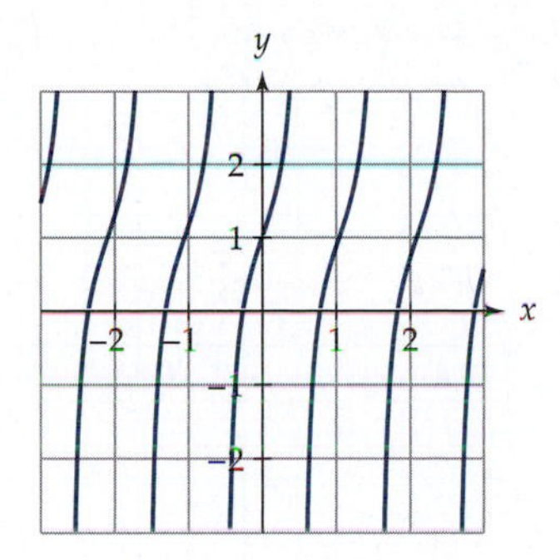

INVERSE TRIGONOMETRIC FUNCTIONS

Inverse trigonometric functions are frequently used to solve trigonometric equations. Recall from section 3.5 that inverse trigonometric functions have a restricted domain because, for the inverse of a function to exist, the function must pass the "horizontal line test" and not have repeated outputs. By the very nature of being periodic, trigonometric functions do not satisfy this condition unless the domain is restricted to avoid repeated outputs.

PROPERTIES

We define the inverse of the sine and cosine functions as follows:

$$y = \sin^{-1} x \quad \text{if and only if} \quad x = \sin y \text{ and } \frac{-\pi}{2} \leq y \leq \frac{\pi}{2}$$

$$y = \cos^{-1} x \quad \text{if and only if} \quad x = \cos y \text{ and } 0 \leq y \leq \pi.$$

The inverse tangent function is defined with a similar restricted domain as the sine function.

PROPERTIES

The inverse tangent function is defined as follows:

$$y = \tan^{-1} x \quad \text{if and only if} \quad x = \tan y \text{ and } \frac{-\pi}{2} < y < \frac{\pi}{2}.$$

Figure 22 shows the restricted domains of $y = \sin x$, $y = \cos x$, and $y = \tan x$ needed so that the inverse functions can be defined.

We illustrate how inverse trigonometric functions can be used to solve equations by finding a solution to

$$2 = 3 \cos\left(\frac{\pi}{2}x\right).$$

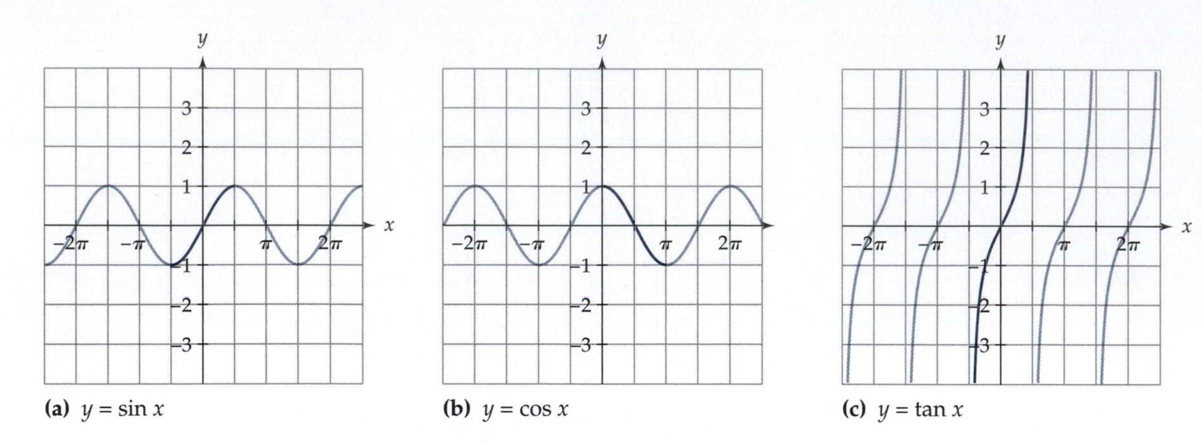

Figure 22 The bold portions of these graphs show the restricted domains of the sine, cosine, and tangent functions so their inverse functions can be defined.

To begin, we divide both sides of this equation by 3 to obtain

$$\frac{2}{3} = \cos\left(\frac{\pi}{2}x\right).$$

Applying the inverse cosine function we obtain

$$\cos^{-1}\left(\frac{2}{3}\right) = \frac{\pi}{2}x.$$

Solving for x gives

$$\frac{2}{\pi}\cos^{-1}\left(\frac{2}{3}\right) = x.$$

Using a calculator to find a decimal approximation, we see that $x \approx 0.535$. (*Note:* There are an infinite number of solutions to $2 = 3\cos[(\pi/2)x]$, but the way inverse cosine is defined, it gives us only one solution. We find other solutions in Chapter 6.)

Example 13 Solve $1 = 2\tan[\pi(x-2)] + 3$ for x.

Solution We must first solve $1 = 2\tan[\pi(x-2)] + 3$ for $\tan[\pi(x-2)]$:

$$\begin{aligned} 1 &= 2\tan[\pi(x-2)] + 3 \\ -2 &= 2\tan[\pi(x-2)] \\ -1 &= \tan[\pi(x-2)]. \end{aligned}$$

Now we apply the inverse tangent function and solve for x:

$$\begin{aligned} -1 &= \tan[\pi(x-2)] \\ \tan^{-1}(-1) &= \pi(x-2) \\ \frac{\tan^{-1}(-1)}{\pi} &= x-2 \\ \frac{\tan^{-1}(-1)}{\pi} + 2 &= x. \end{aligned}$$

In section 5.1, we used a 45°–45°–90° triangle to find that $\tan 45° = \tan(\pi/4) = 1$. Because $\tan x$ is an odd function, $\tan(-\pi/4) = -1$. So, $\tan^{-1}(-1) = -\pi/4$. Thus,

$$x = \frac{-(\pi/4)}{\pi} + 2 = -\frac{1}{4} + 2 = 1.75.$$

Example 14 Solve $3 = \sin[\pi(x+1)]$ for x.

Solution This equation does not have a solution because the maximum output of $y = \sin[\pi(x+1)]$ is 1, so $\sin[\pi(x+1)]$ will never equal 3. If we did not notice this fact and simply begin to solve this equation, we would encounter a problem. To try to solve this, we apply the inverse sine function and solve for x:

$$\sin^{-1}(3) = \pi(x+1)$$

$$\frac{\sin^{-1}(3)}{\pi} = x+1$$

$$\frac{\sin^{-1}(3)}{\pi} - 1 = x.$$

When trying to find a decimal approximation, we get an error message on our calculator because the domain of the inverse sine function is the range of the sine function, $-1 \le x \le 1$, and thus $\sin^{-1}(3)$ is not defined. Therefore, our equation has no solution.

READING QUESTIONS

15. Why must the domains of trigonometric functions be limited for their inverse functions to be defined?

16. Find a solution to $3 = \cos(x+1) + 4$.

SUMMARY

Transformations of trigonometric functions follow the same patterns established earlier. They are summarized at the beginning of this section.

The **midline** of a sine or a cosine function is the horizontal line halfway between the function's maximum and minimum outputs. The **amplitude** of a sine or a cosine function is the distance from the midline to the maximum (or minimum) output, which is equal to half the difference between the maximum and minimum outputs.

The **period** of a **periodic function** is the smallest value p such that $f(x+p) = f(x)$ for all x. We found the following two identities:

- $\sin\left(x + \frac{\pi}{2}\right) = \cos x$
- $\cos\left(x - \frac{\pi}{2}\right) = \sin x$

The general formulas for the sine and cosine functions are $f(x) = A\sin[B(x+C)]+D$ or $g(x) = A\cos[B(x+C)]+D$, where

- The amplitude is $|A|$.
- The period is $2\pi/B$.
- The horizontal shift is C.
- The midline is $y = D$.

In general, for $y = A\tan[B(x+C)]+D$,

- The vertical stretch is $|A|$. If the graph changes from concave down to concave up between the vertical asymptotes, then A is positive. If the graph changes from concave up to concave down between the vertical asymptotes, then A is negative.
- The vertical shift is D. The function will change concavity at the points where the output is D.
- The period is π/B, and $B = \pi/\text{period}$.
- The horizontal shift is C. One of the points where the function changes concavity is (C, D).

We need to restrict the domain of these trigonometric functions to define inverse functions. We define the inverse functions as follows:

- $y = \sin^{-1} x$ if and only if $x = \sin y$ and $-\pi/2 \le y \le \pi/2$.
- $y = \cos^{-1} x$ if and only if $x = \cos y$ and $0 \le y \le \pi$.
- $y = \tan^{-1} x$ if and only if $x = \tan y$ and $-\pi/2 < y < \pi/2$.

EXERCISES

1. Indicate whether each of the following statements is true or false. If false, correct the statement so that it becomes a true statement.
   (a) The graph of $y = f(x+3)$ is the graph of $y = f(x)$ shifted to the right three units.
   (b) The graph of $y = f(2x+3)$ is the graph of $y = f(x)$ first horizontally compressed toward the y-axis by two units and then shifted to the left three units.
   (c) The period of $y = \sin x$ is π units.
   (d) The amplitude of $y = 3\cos(2x) - 1$ is 3.
   (e) The midline of $y = \sin(2x) - 1$ is $y = 2$.
   (f) The period of $y = 3\sin(2x)$ is 2.
   (g) The function $y = \tan x$ is undefined for $x = n\pi$, where n is an integer.
   (h) The domain of $y = \sin^{-1} x$ is $-1 \le x \le 1$.
2. Use the accompanying graph of $y = f(x)$ to match the following functions to their graphs.
   (a) $y = f(x-3)$

(b) $y = f(x + 3)$

(c) $y = 3f(x)$

(d) $y = \frac{1}{3}f(x)$

(e) $y = f(3x)$

(f) $y = f(\frac{1}{3}x)$

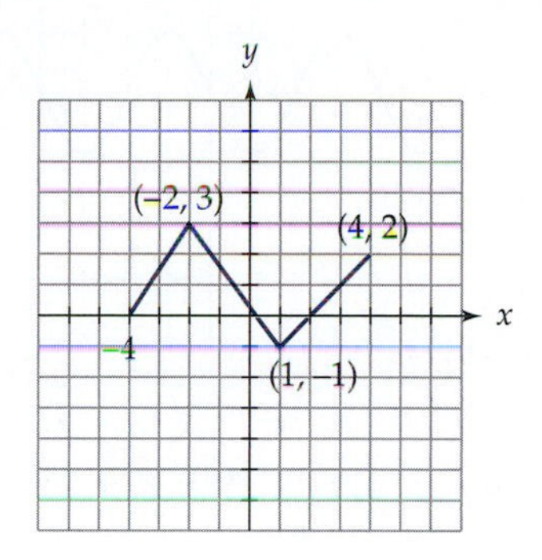

i.

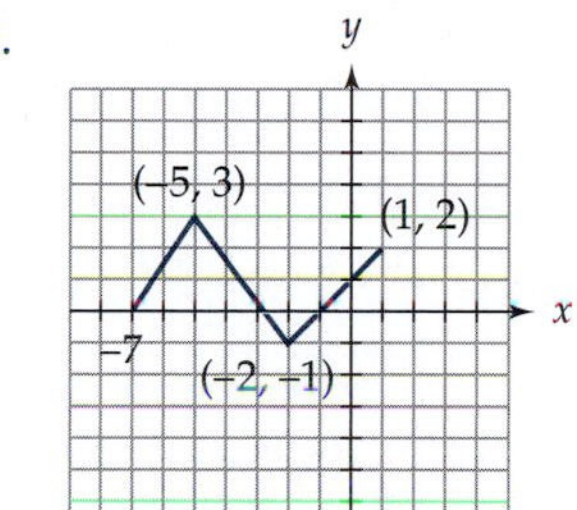

ii.

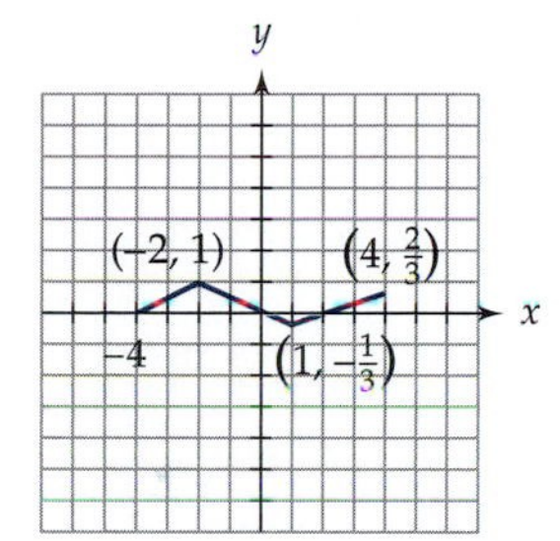

iii.

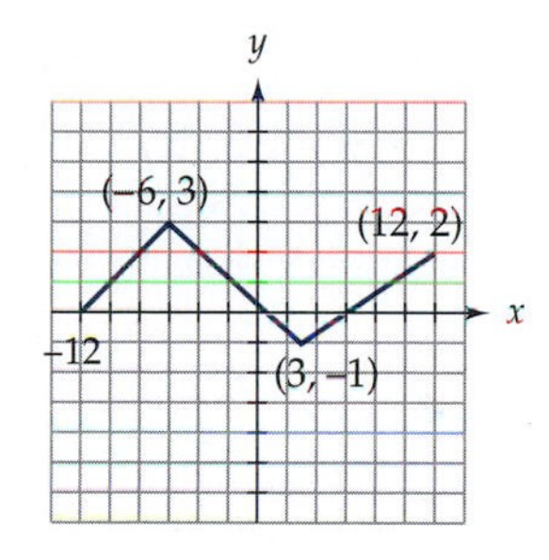

iv.

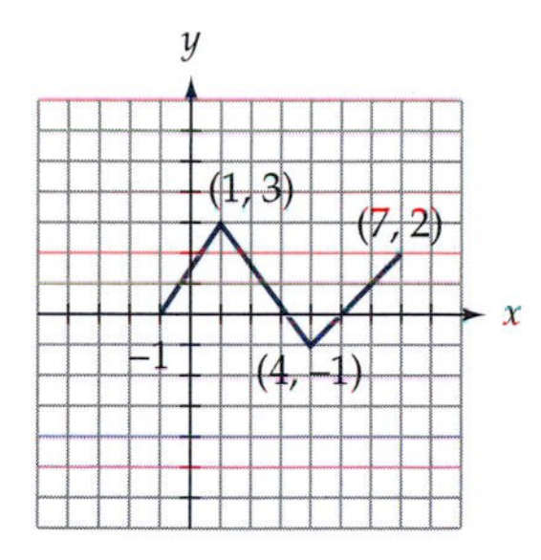

v.

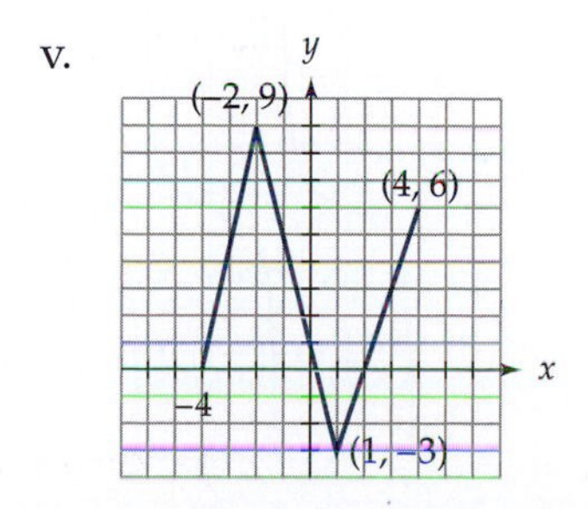

vi.

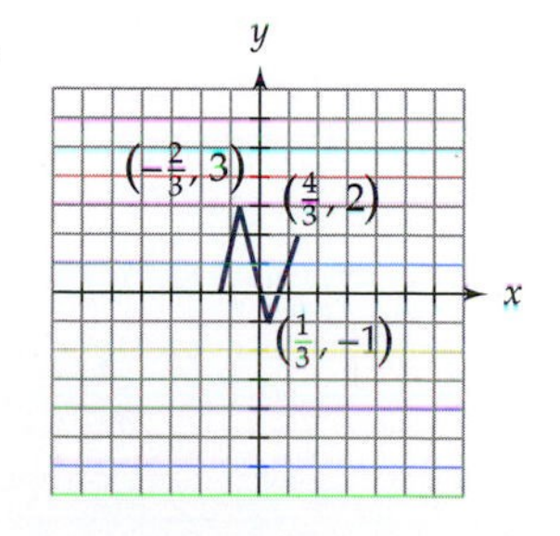

3. Give the amplitude and the midline for each of the following functions.
   (a) $y = 3\sin(2x) + 4$
   (b) $y = -3\cos x - \pi$

   (c)

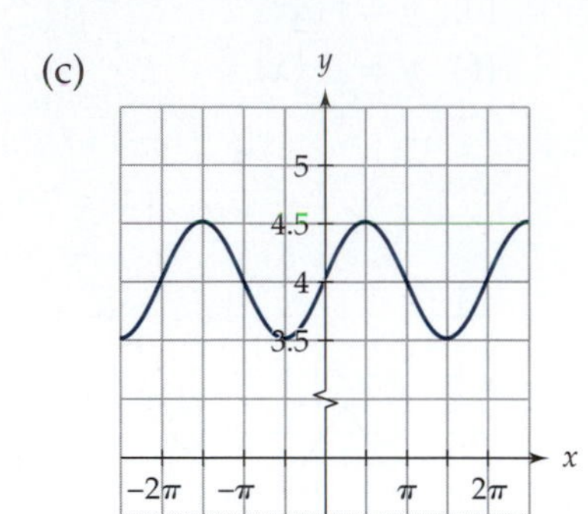

   (d)

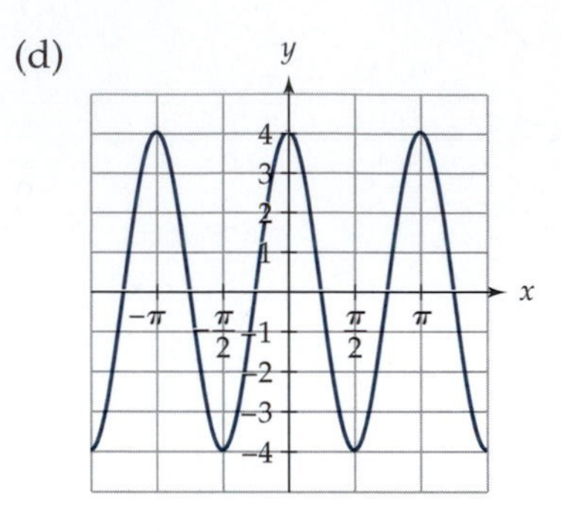

   (e) A cosine function whose maximum is 10 and whose midline is $y = -2$
   (f) A sine function vertically stretched by a factor of π and then shifted down one unit
4. Give the period for each of the following functions.
   (a) $y = 2\cos(x + \pi) - 1$
   (b) $y = -\sin(3x)$
   (c) $y = 3\tan\left[4\left(x - \frac{\pi}{4}\right)\right]$

   (d)

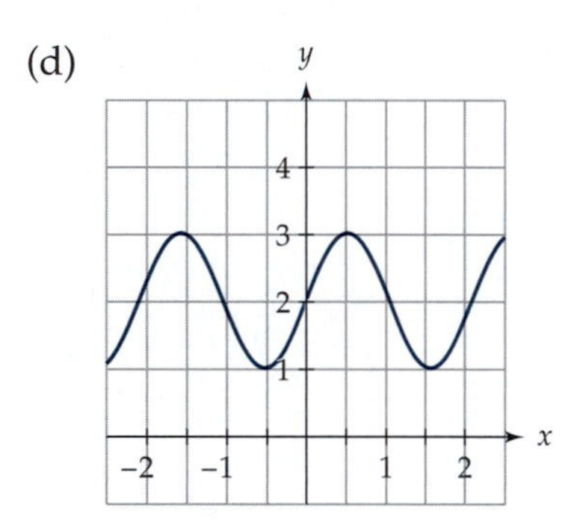

   (e)

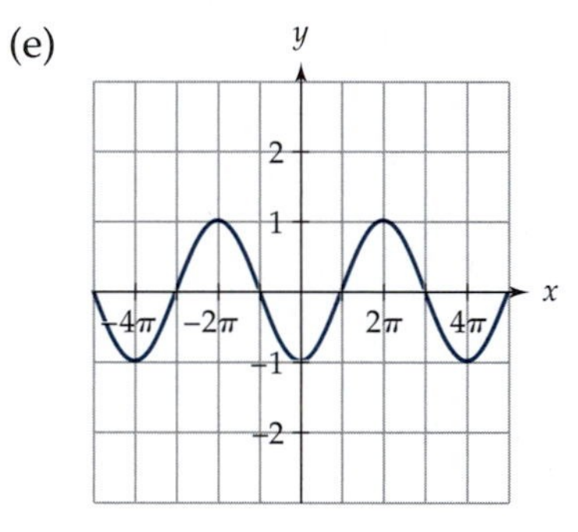

   (f)

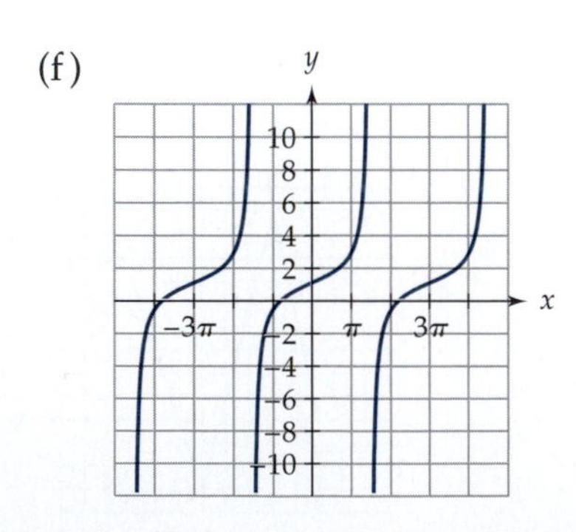

(g) A sine function that has been vertically stretched by a factor of 4, horizontally compressed by a factor of $\frac{1}{2}$, and then reflected over the y-axis.

(h) A cosine function that has a maximum at $x = 0$ and at $x = 3$ and no maximums for $0 < x < 3$.

(i) A tangent function that changes concavity at $(1, 1)$. The nearest point to the right where it has a vertical asymptote is $x = 2$.

5. (a) Write the equation of a cosine function with a period of $\pi/2$ that has been shifted one unit to the left.

   (b) Rewrite your answer to part (a) as a sine function.

6. (a) Rewrite $y = 2\cos(3x - 1) + 5$ as a sine function.

   (b) Rewrite $y = A\cos[B(x + C)] + D$ as a sine function.

   (c) Rewrite $y = -3\sin(\pi x + 2) - \pi/4$ as a cosine function.

   (d) Rewrite $y = A'\sin(B'x + C') + D'$ as a cosine function.

7. Match the following functions to their graphs.

   (a) $y = \sin x$

   (b) $y = \sin(2x)$

   (c) $y = 2\sin x$

   (d) $y = \sin(x + 2)$

   (e) $y = \sin(x - 2)$

   (f) $y = \sin x + 2$

i.

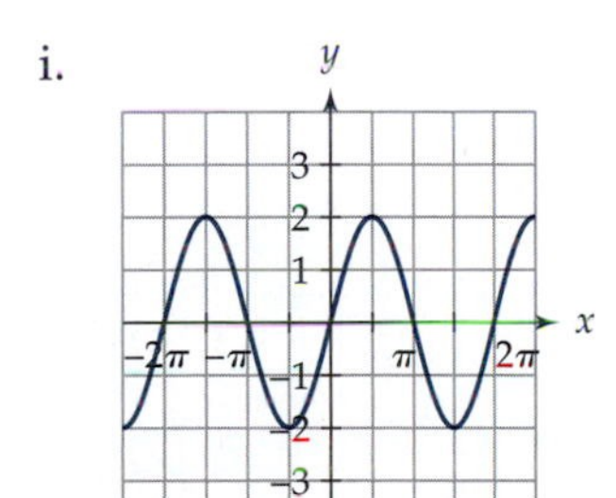

ii.

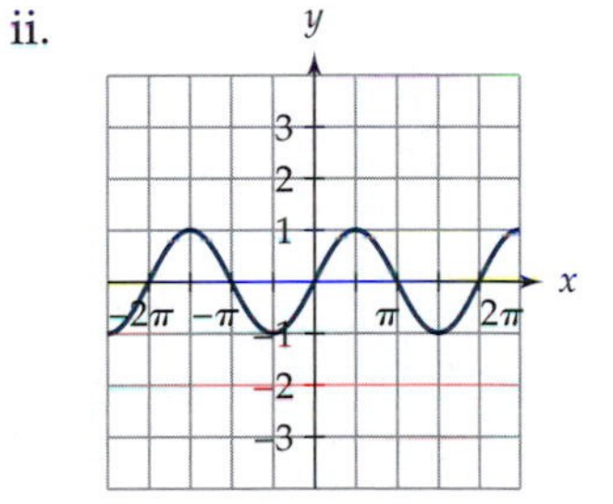

iii.

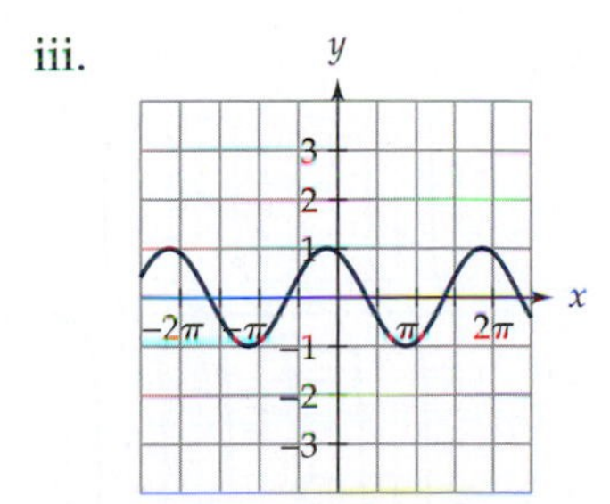

iv.

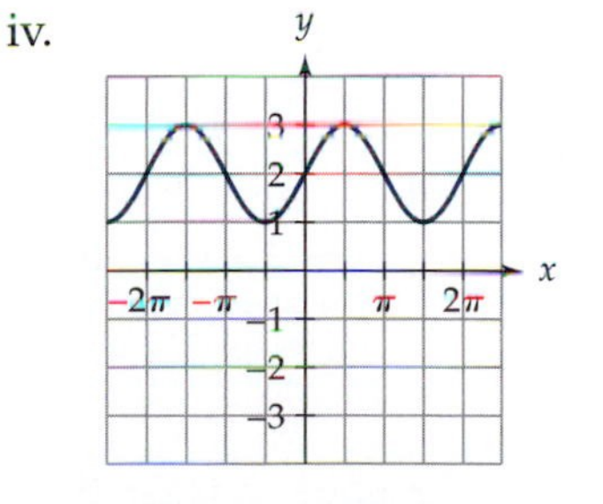

v.

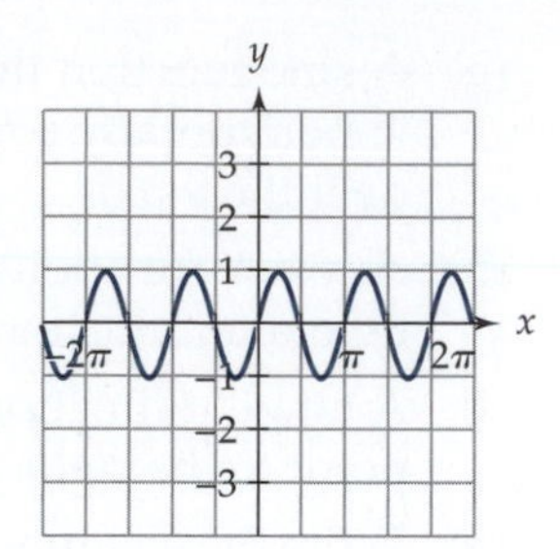

vi.

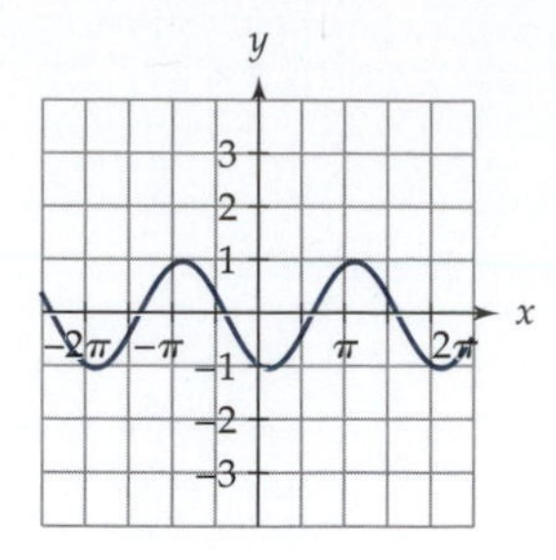

8. Match the following functions to their graphs.

(a) $y = 3\cos x$

(b) $y = \cos(3x)$

(c) $y = \cos(\frac{1}{3}x)$

(d) $y = \tan x + 2$

(e) $y = \tan(2x)$

(f) $y = \tan(0.5x)$

i.

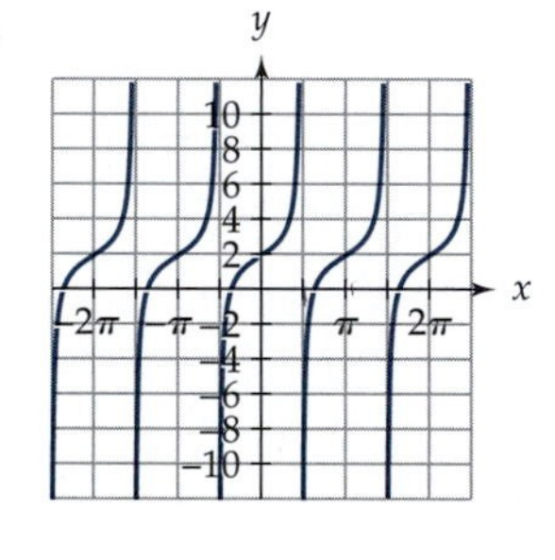

ii.

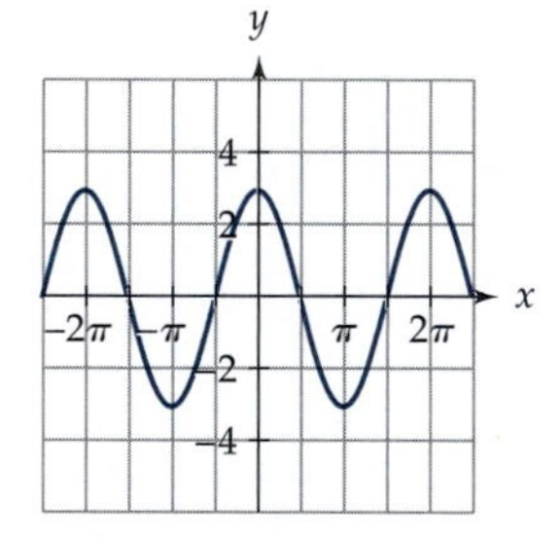

iii.

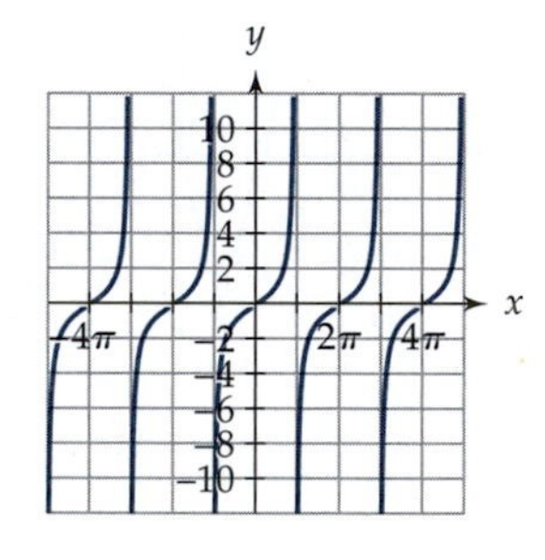

iv.

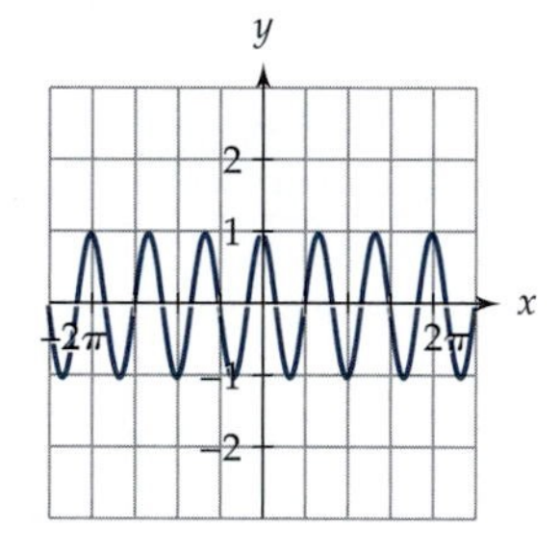

v.

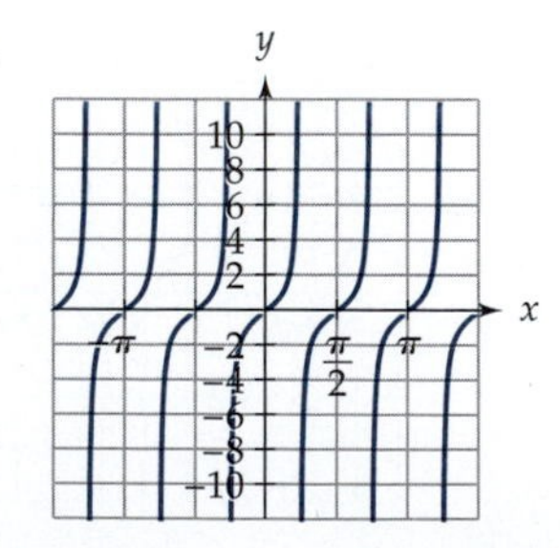

vi.

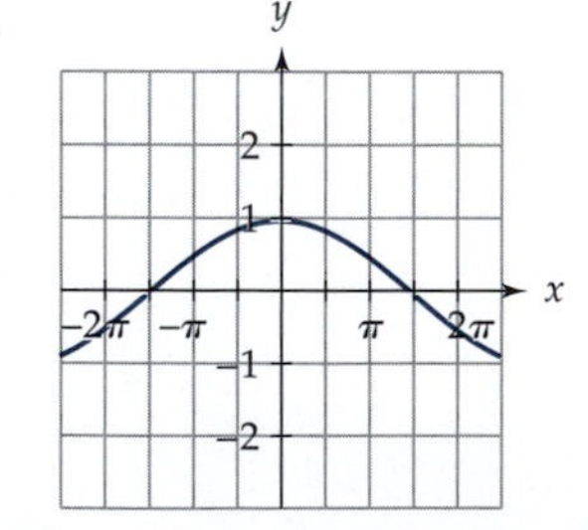

9. The following graphs are functions of the form $y = A \tan x + D$. Determine if A and D are positive or negative. Justify your answer.

(a)

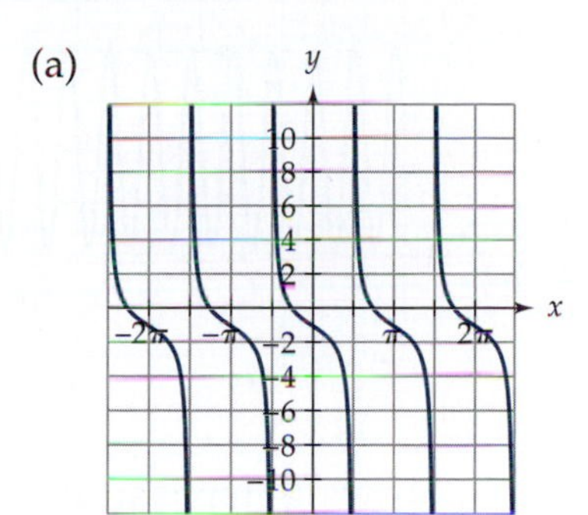

(b)

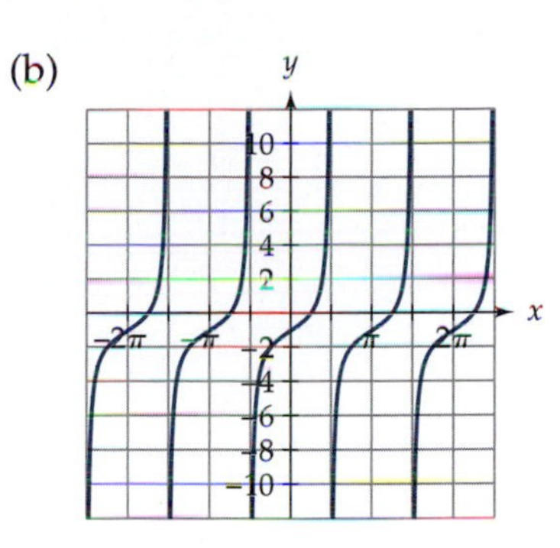

(c)

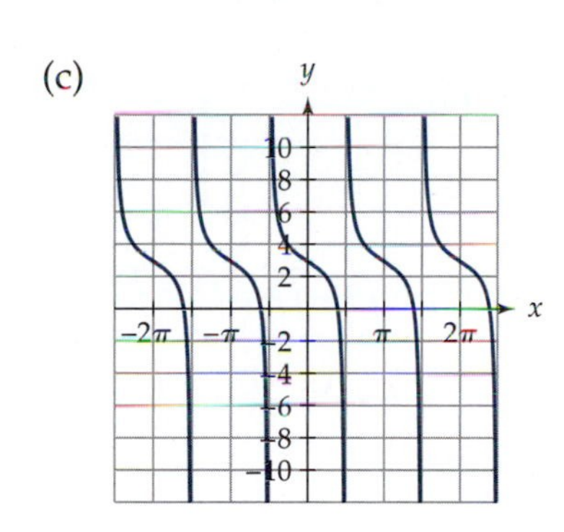

10. The following graphs are functions of the form $y = A \sin[B(x + C)] + D$. Find A, B, C, and D.

(a)

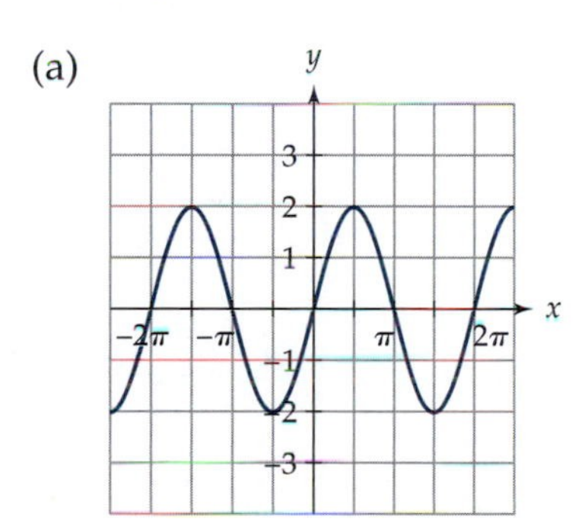

(b)

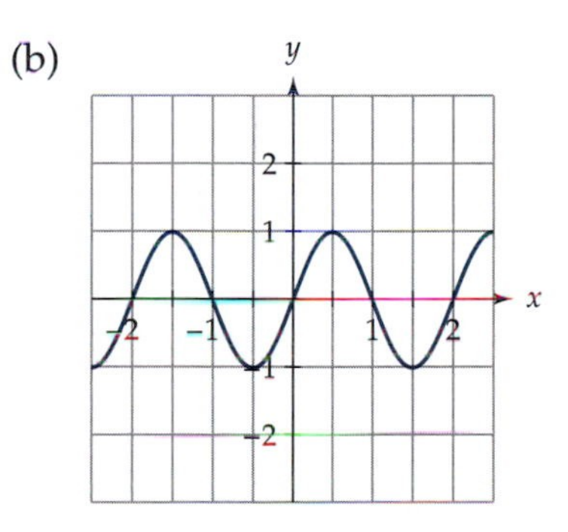

(c)

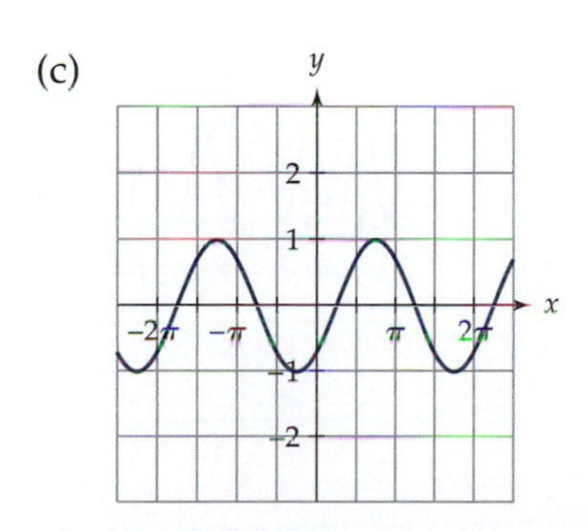

(d)

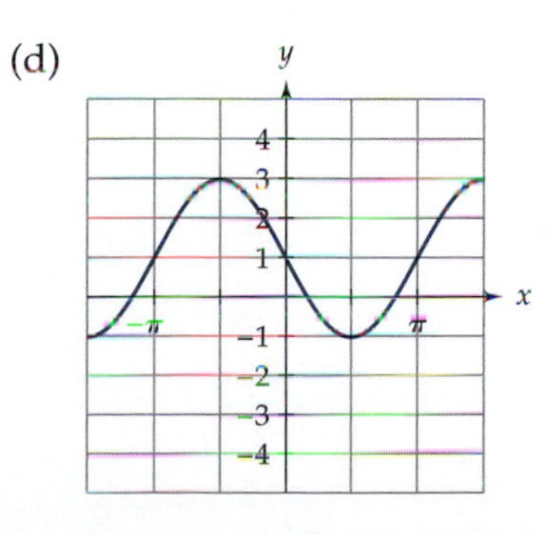

(e)

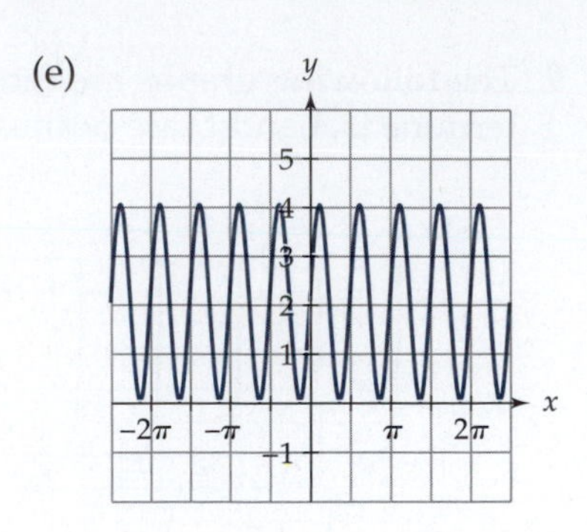

(f)

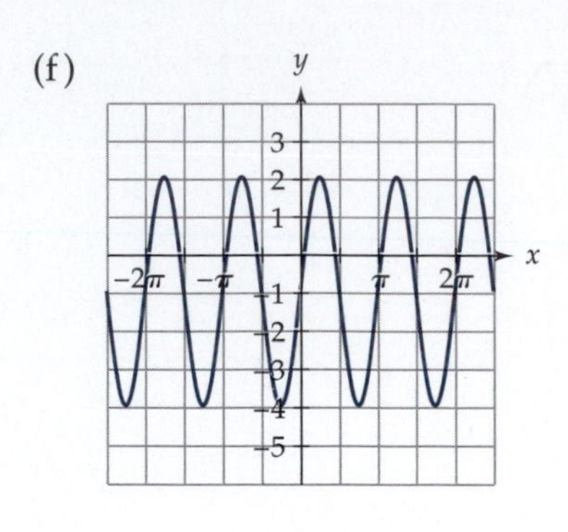

11. The following graphs are functions of the form $y = A\cos[B(x + C)] + D$. Find A, B, C, and D.

(a)

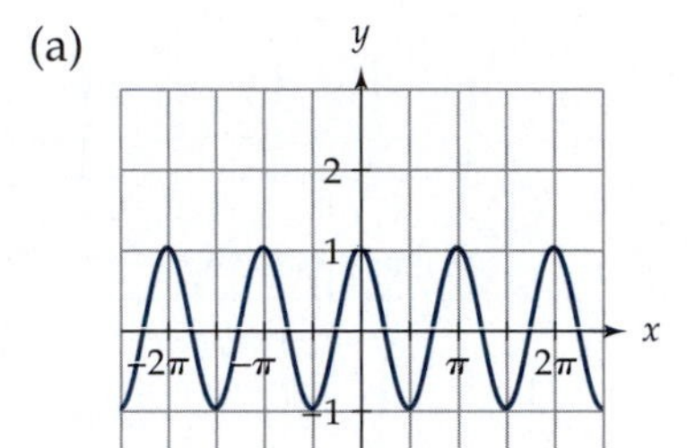

(b)

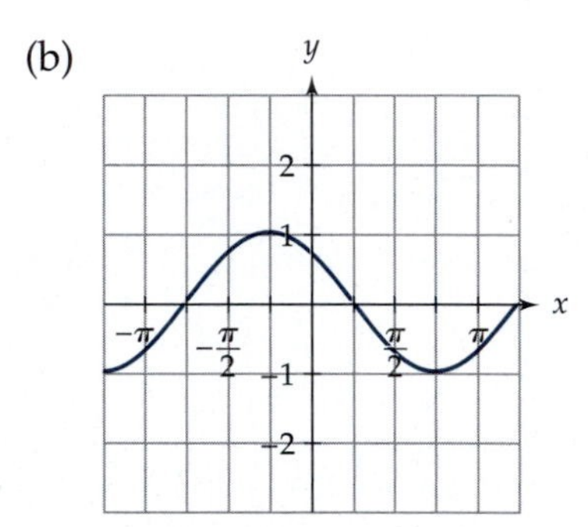

(c)

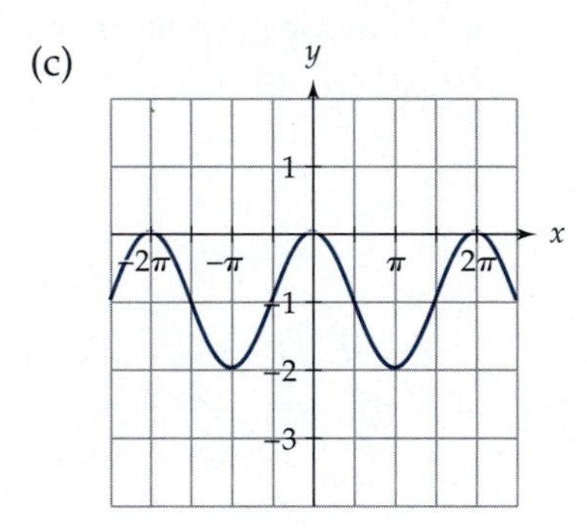

(d)

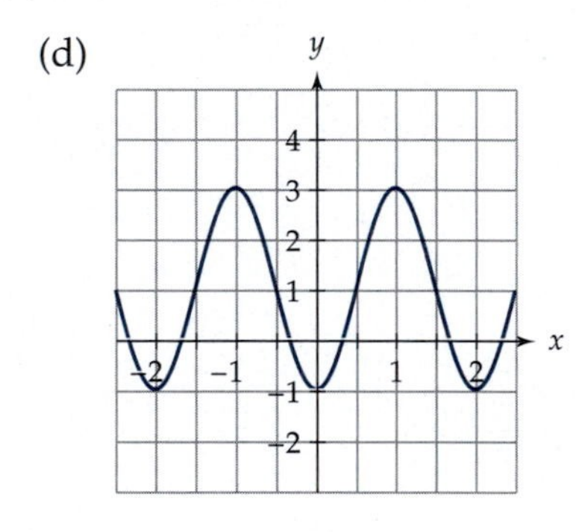

(e)

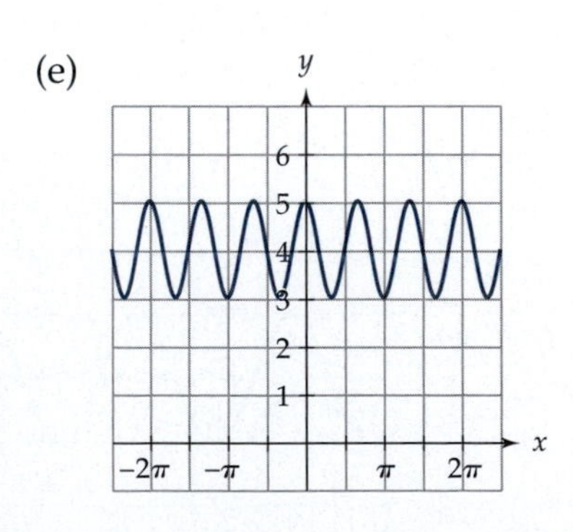

(f)

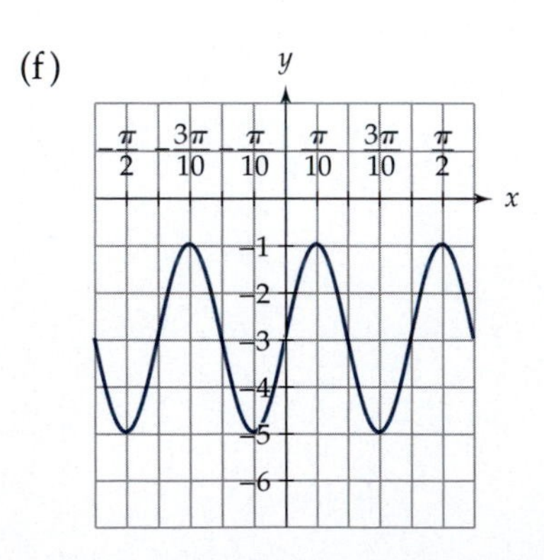

12. Find the equations for the functions given in the following graphs.

(a)

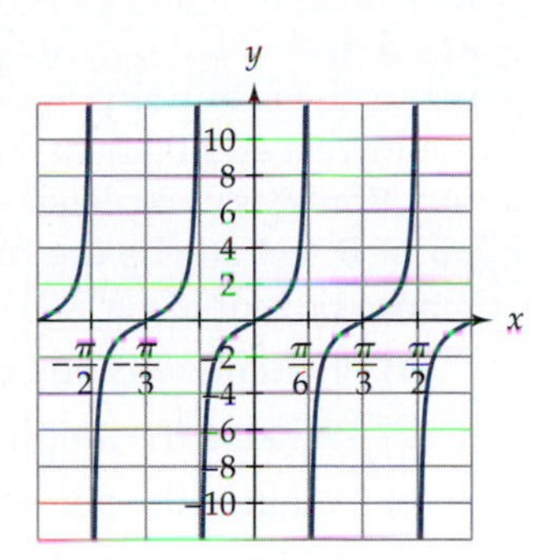

(b)

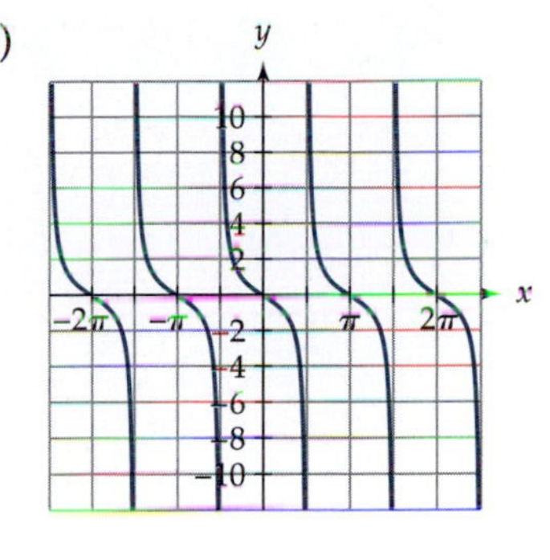

(c)

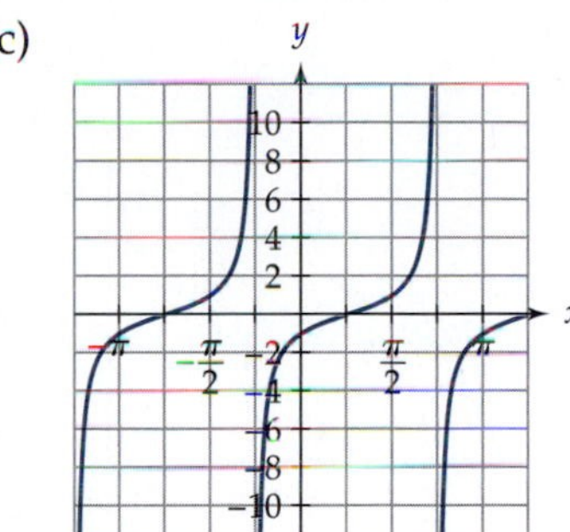

(d)

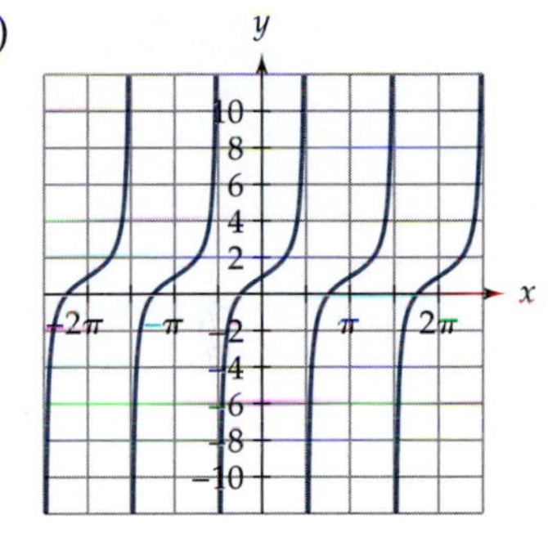

13. Correct each of the following statements.

(a) The domain of $y = \sin^{-1} x$ is $-1 \le x \le 1$ and the range is all real numbers.

(b) The range of $y = \tan^{-1} x$ is $0 \le y \le 2\pi$.

(c) The range of $y = \cos^{-1} x$ is $-\pi \le x \le \pi$.

(d) $\cos^{-1} \dfrac{\pi}{4} = \dfrac{1}{\sqrt{2}}$

(e) $\sin^{-1}(3)$ is positive.

(f) $\tan^{-1} 1 = \dfrac{\pi}{2}$

14. Solve the following equations. Give exact answers where possible. Otherwise, round answers to two decimal places.

(a) $\dfrac{1}{2} = \sin\left(\dfrac{\pi}{4}x\right)$

(b) $4 = \tan\left(\dfrac{x}{3}\right)$

(c) $1 = 3\cos(4x - 3)$

(d) $1 = \cos\left(\dfrac{\pi}{3}x\right)$

(e) $0 = \sin[\pi(x - 6)]$

(f) $5 = \tan\left(x - \dfrac{\pi}{4}\right) - 4$

15. The accompanying figure shows a child's swing. As the child swings forward, she starts 1 m above the ground (the highest position), then is at 0.5 m above the ground (the lowest position), then is back up to 1 m when she is at point B. Because she went from the highest position to the lowest and back to the highest, one period for the vertical position function was completed. It takes 1 sec for her to swing from point A to point B. Let $s(t)$ be the function whose input is time (in seconds) and whose output is vertical distance above the ground (in meters).
    (a) Explain why $s(t)$ can be modeled by a cosine function.
    (b) What is the amplitude for s?
    (c) What is the period for s?
    (d) Write an equation for s. Check your answer by making sure it gives you the points $(0, 1)$, $(0.5, 0.5)$, and $(1, 1)$.

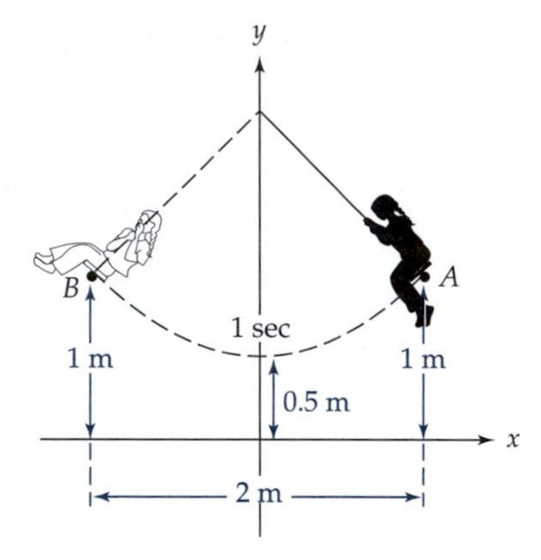

16. Common household electricity is known as alternating current. In the United States, this current continually alternates between a maximum value of $110\sqrt{2}$ V and a minimum value of $-110\sqrt{2}$ V, at a rate of 60 times per second. Write the symbolic representation for a cosine function that models common household alternating current, where input is time in seconds.
17. Sound waves are periodic. A "pure" tone can be represented by a sine or cosine function. The period of the function is the reciprocal of the frequency of the note; that is, frequency $= 1/\text{period}$. The following table shows the frequency of the notes in hertz (or cycles per second) from middle C to the C one octave higher. When you play the same note one octave lower, the frequency is half as much. When you play the same note one octave higher, the frequency is twice as much. Using this information and the table, determine what note is represented by the following functions.
    (a) $y = \sin(880\pi x)$
    (b) $y = \cos(466\pi x)$
    (c) $y = \sin(494\pi x)$
    (d) $y = \cos(1568\pi x)$

Note	C	C♯	D	E♭	E	F	F♯	G	A♭	A	B♭	B	C
Frequency	262	277	294	311	330	349	370	392	415	440	466	494	524

18. At one time, the Ferris wheel in Osaka, Japan, held the title of the world's tallest. It measures 112.5 m from the ground to its apex, and the wheel has a diameter of 100 m (another Ferris wheel, in Otsu, has the same diameter, but the wheel in Osaka is 4 m higher). One revolution of this wheel takes about 15 min.[8] Assuming a point on the rim of the wheel starts at its minimum height, write a symbolic representation for the height of that point where time is the input (in minutes) and height above the ground is the output (in meters).

INVESTIGATIONS

INVESTIGATION 1: ALIASING

Using calculators or computers is an easy way to see a graph of a function quickly. The picture on the screen, however, is never completely accurate. Most of the time, the graph conveys enough information to analyze the function correctly, but not always. For example, when graphing periodic functions and the calculator or computer is not choosing a good set of inputs, you can misinterpret the period of the function and therefore think you have one periodic function when the graph actually represents a different periodic function. In physics, this situation is known as "aliasing." One place aliasing causes problems is when the graphs are representing sound waves.

1. To explore the effect of aliasing, we begin by using the function $f(x) = \cos x$.
   (a) Complete Table 1 for the function $f(x) = \cos x$. Be sure your calculator is set in radians.

TABLE 1

x	0	6	12	18	24	30	36	42	48	54	60
$\cos x$											

   (b) On graph paper (not on a calculator), sketch the graph you get by plotting the points on Table 1. Make each square on your graph paper equal to one unit so that there are six horizontal squares between each point you plot. Connect the points with a smooth curve.

[8] *Japan Information Network*, Monthly News, Aug 1997: World's Tallest Ferris Wheel in Osaka (visited 27 May 1998) (http://www.jinjapan.org/kidsweb/news/97-8/wheel.html).

(c) What do you know about the period of $f(x) = \cos x$? What does the graph from part (b) imply about the period of $f(x) = \cos x$?

(d) Complete Table 2 and plot the points on the graph you used in part (b). Connect the points with a smooth curve. How do the two graphs compare?

TABLE 2

x	0	1	2	3	4	5	6	7	8	9	10	11
$\cos x$												

(e) Why do the points in Table 1 give an inaccurate graph of $f(x) = \cos x$?

2. To show the effect of aliasing on a calculator, we look at the function $f(x) = \cos x$, using various dimensions for the viewing window.
   (a) Have your calculator graph f for $0 \le x \le 10\pi$. Sketch the graph you see on your calculator screen. By looking at this graph, estimate the period of f.
   (b) Have your calculator graph f for $0 \le x \le 50\pi$. Sketch the graph you see on your calculator screen. By looking at this graph, estimate the period of g.
   (c) Have your calculator graph f for $0 \le x \le 100\pi$. Sketch the graph you see on your calculator screen. By looking at this graph, estimate the period of f.
   (d) Have your calculator graph f for $0 \le x \le 200\pi$. Sketch the graph you see on your calculator screen. By looking at this graph, estimate the period of f.
3. We should also see the effect of aliasing on a calculator when we change the period of a function. For each of the following sine functions, do the following.
   - Determine the period.
   - Graph on your calculator with a window where x has a minimum value of -2π and a maximum value of 2π.
   - Using your graph, approximate the period of the function.
   - Note any discrepancies between what the period should be and what it appears to be on the calculator and explain why these discrepancies occurred.

   (a) $y = \sin(\pi x)$
   (b) $y = \sin(5\pi x)$
   (c) $y = \sin(10\pi x)$
   (d) $y = \sin(15\pi x)$
   (e) $y = \sin(20\pi x)$
   (f) $y = \sin(25\pi x)$
   (g) $y = \sin(30\pi x)$

Aliasing has applications in digital recording. A digital recorder samples sound at a certain rate (over 40,000 times each second) and records a picture of the sound wave in much the same way that your calculator pictures a function. The highest sound most people can hear is about 20,000 Hz, or a sound wave that has about 20,000 periods per second. By sampling the sound at more than twice that rate, the aliasing risk is eliminated for sounds that we can hear. If the recorder samples the sound too slowly, it records a wave much different from the one it was trying to reproduce, much like the graph you made based on Table 1. The result is a wave with a relatively long period and a low frequency that emits a low rumble when you would expect to hear the high sound made by the wave with the much higher frequency. Aliasing can be such a problem, in fact, that filters are used to eliminate any sounds above 20,000 Hz before they are digitized, thus making sure that every sound the digital recorder tries to sample has a frequency that is less than half of the sampling rate.

INVESTIGATION 2: DAYS OF OUR LIVES

The amount of daylight on a given day is an important part of our lives. As the seasons change, so does the amount of daylight. Newspapers and weather reports often give the times for the sunrise and sunset. In this investigation, you derive a function that gives the amount of daylight in Grand Rapids, Michigan, for any day throughout the year. Table 3 represents the amount of daylight for January 1 and every 10th day after that for Grand Rapids.

Graph the points from the table with the x-axis representing the day of the year (numbered 1 through 365) and the y-axis representing the number

TABLE 3

Day	Hour	Minute	Total Minutes	Day	Hour	Minute	Total Minutes
1	9	5	545	191	15	11	911
11	9	16	556	201	14	56	896
21	9	33	573	211	14	37	877
31	9	55	595	221	14	14	854
41	10	20	620	231	13	49	829
51	10	47	647	241	13	23	803
61	11	13	673	251	12	54	774
71	11	42	702	261	12	26	746
81	12	11	731	271	11	57	717
91	12	40	760	281	11	28	688
101	13	9	789	291	10	59	659
111	13	36	816	301	10	32	632
121	14	3	843	311	10	6	606
131	14	27	867	321	9	43	583
141	14	49	889	331	9	24	564
151	15	5	905	341	9	9	549
161	15	16	916	351	9	2	542
171	15	21	921	361	9	2	542
181	15	19	919				

of minutes of daylight.[9] Notice that this data can be modeled by a sine function.

1. What is the vertical shift of the function that fits the data? What does this number mean in terms of minutes of daylight?
2. What is the amplitude of the function that fits the data?
3. What is the period of the function that fits the data?
4. What is the horizontal shift of a sine function that fits the data? What does this number mean in terms of number of days?
5. Find the equation of a sine function that fits the data.
6. Graph your function along with the data. If you used a calculator, then draw a sketch of your graph.

5.4 TRIGONOMETRIC IDENTITIES

Often, there is more than one way to write a mathematical expression. Sometimes, it is obvious that two mathematical expressions are the same, such as x and $\frac{1}{2} \cdot 2x$. Other times, it takes a bit more work to see that two expressions, such as $(x-4)(x+3)$ and $x^2 - x - 12$, are equivalent. In this section, we derive several trigonometric identities and learn to tell if two expressions are equivalent. A summary of trigonometric identities completes the section.[10]

REVIEW OF SOME TRIGONOMETRIC IDENTITIES

Equivalent trigonometric expressions are called **trigonometric identities.** Earlier in this chapter we saw several identities. In Section 5.1, we described three identities that involved the symmetry of the graphs of $y = \sin x$, $y = \cos x$, and $y = \tan x$. (See Figure 1.) Because $y = \cos x$ is symmetric about the y-axis and $y = \sin x$ and $y = \tan x$ are symmetric about the origin, we have the following identities.

PROPERTIES

Symmetry Identities

$\sin x = -\sin(-x)$.
$\cos x = \cos(-x)$.
$\tan x = -\tan(-x)$.

[9] *If you prefer, you could use data for your own location. One source of data is the U.S. Naval Observatory. As of March 16, 2000, their web site address for this information was ⟨http://aa.usno.navy.mil/AA/data/docs/RS_OneYear.html⟩.*

[10] *In this section, we assume the input of the trigonometric function is measured in radians. These identities are also true, however, if the input is measured in degrees.*

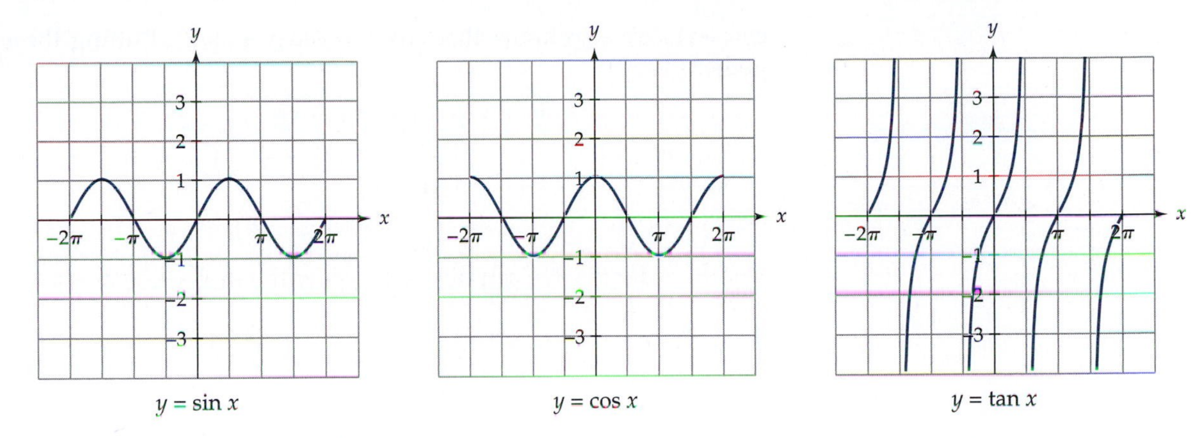

Figure 1 Graphs of three trigonometric functions with a domain of $-2\pi \leq x \leq 2\pi$.

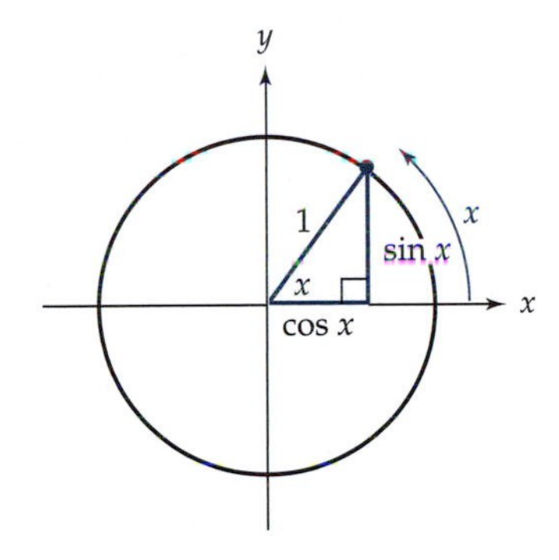

Figure 2 You can use a triangle in the unit circle to derive the identity $\cos^2 x + \sin^2 x = 1$.

Another identity mentioned earlier was derived from the Pythagorean theorem and the unit circle definitions of $y = \sin x$ and $y = \cos x$. Using Figure 2, we can see that the following is true.

PROPERTIES

Pythagorean Identity

$\cos^2 x + \sin^2 x = 1.$

In Section 5.3, we saw that by using a horizontal shift, we could find a relationship between the graph of $y = \sin x$ and the graph of $y = \cos x$. Looking back at Figure 1, notice that the graph of $y = \cos x$ looks like the graph of $y = \sin x$ shifted to the left $\pi/2$ units. Similarly, the graph of $y = \sin x$ looks like the graph of $y = \cos x$ shifted to the right $\pi/2$ units. These similarities give us the following two identities.[11]

PROPERTIES

Horizontal Shift Identities

$$\sin x = \cos\left(x - \frac{\pi}{2}\right).$$

$$\cos x = \sin\left(x + \frac{\pi}{2}\right).$$

Let's see how we can use the identities described so far to create new identities. We know that $y = \cos x$ is symmetric about the y-axis, so $\cos x =$

[11] *These identities assume x is measured in radians. If x is measured in degrees, substitute 90° for $\pi/2$.*

$\cos(-x)$. We also know that $\cos x = \sin(x + \pi/2)$. Putting these two together gives

$$\begin{aligned}\cos x &= \cos(-x)\\ &= \sin\left(-x + \frac{\pi}{2}\right)\\ &= \sin\left(\frac{\pi}{2} - x\right).\end{aligned}$$

Another identity we can derive starts with $\sin x = \cos(x - \pi/2)$ and that $\cos x = \cos(-x)$.

$$\begin{aligned}\sin x &= \cos\left(x - \frac{\pi}{2}\right)\\ &= \cos\left(-\left[x - \frac{\pi}{2}\right]\right)\\ &= \cos\left(-x + \frac{\pi}{2}\right)\\ &= \cos\left(\frac{\pi}{2} - x\right).\end{aligned}$$

The two new identities we have derived are called complement identities because angles with measures of x and $\pi/2 - x$ are complements of each other.[12]

PROPERTIES

Complement Identities

$$\sin x = \cos\left(\frac{\pi}{2} - x\right).$$

$$\cos x = \sin\left(\frac{\pi}{2} - x\right).$$

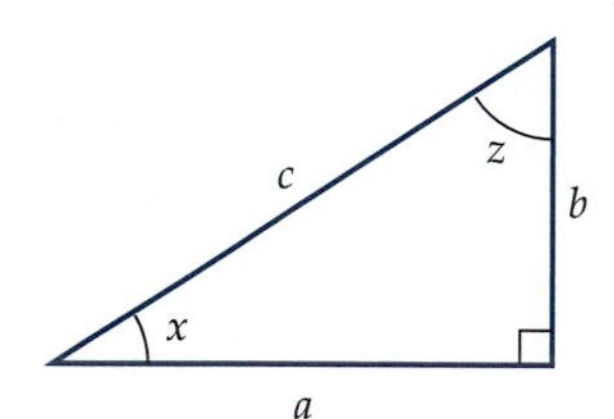

Figure 3 A right triangle can be used to show a relationship between $y = \sin x$ and $y = \cos x$.

We could also have derived these two identities using the right triangle definitions of $y = \sin x$ and $y = \cos x$. Figure 3 shows a right triangle with angles x and z. Because the two angles of a triangle that are not right angles add to 90° (or $\pi/2$ radians), we know that $z = \pi/2 - x$. From the right triangle definitions of $y = \sin x$ and $y = \cos x$, we see that $\sin x = b/c$ and $\cos z = b/c$. Figure 3 clearly shows that what we proved algebraically, $\sin x = \cos(\pi/2 - x)$, can also be derived using right triangles. Similarly, we can use right triangles to show that $a/c = \cos x = \sin z = \sin(\pi/2 - x)$.[13] Usually, there is more than one way to derive a trigonometric identity. Throughout this section, we derive identities algebraically, geometrically, using right triangles, using unit circles, or some combination of these methods. There is no "right way" when it comes to deriving trigonometric identities, but rather an array of approaches that you should become comfortable using.

[12] *These identities assume x is measured in radians. If x is measured in degrees, substitute 90° for $\pi/2$. Angles that sum to be 90° are called complementary angles.*

[13] *When using right triangles, there is the assumption that $0 < x < \pi/2$. You can use right triangles to prove identities for other values of x by drawing appropriate triangles in the second, third, or fourth quadrants of the xy-plane.*

READING QUESTIONS

1. Use the definition of $\tan x = \sin x/\cos x$ to show that $\tan x = -\tan(-x)$.
2. Explain, using Figure 2, why $\cos^2 x + \sin^2 x = 1$.
3. What property of the cosine function allows us to write $\cos x = \cos(-x)$?
4. Another identity mentioned in Section 5.3 is $\sin x = \sin(x + 2\pi)$. What are the equivalent identities for $y = \cos x$ and $y = \tan x$?

COSECANT, SECANT, AND COTANGENT

So far, we have concentrated on the three most commonly used trigonometric functions: sine, cosine, and tangent. Three other trigonometric functions are not used as frequently as these: cosecant (abbreviated as $\csc x$), secant ($\sec x$), and cotangent ($\cot x$). These functions can be defined in terms of the other trigonometric functions:

$$\csc x = \frac{1}{\sin x}, \qquad \sec x = \frac{1}{\cos x}, \qquad \cot x = \frac{1}{\tan x} = \frac{\cos x}{\sin x}.$$

Notice that $y = \csc x$, $y = \sec x$, and $y = \cot x$ are not defined for all real numbers because of division by zero. The domain for $y = \csc x$ is all real numbers except $x = \ldots, -2\pi, -\pi, 0, \pi, 2\pi, \ldots$. The domain can also be written $x \neq n\pi$, where n is an integer, because $y = \sin x$ is zero when x is a multiple of π. To find the range of $y = \csc x$, remember that the range of $y = \sin x$ is $-1 \leq \sin x \leq 1$. Because you are taking the reciprocal of numbers between -1 and 1, the range of $y = \csc x$ is $y \leq -1$ or $y \geq 1$. Similarly, the domain of $y = \sec x$ is all real numbers except $x = \ldots, -3\pi/2, -\pi/2, \pi/2, 3\pi/2, \ldots$ (or $x \neq \pi/2 + n\pi$, where n is an integer). The range is $y \leq -1$ or $y \geq 1$. The domain for $y = \cot x$ is all real numbers except $x = \ldots, -2\pi, -\pi, 0, \pi, 2\pi, \ldots$ (or $x \neq n\pi$, where n is an integer). To find the range of $y = \cot x$, remember that the range of $y = \tan x$ is all real numbers. Because $y = \cot x$ is the reciprocal of $y = \tan x$, its range will also be all real numbers. The definitions, domain, and range of these three functions are summarized below. Their graphs are shown in Figure 4.

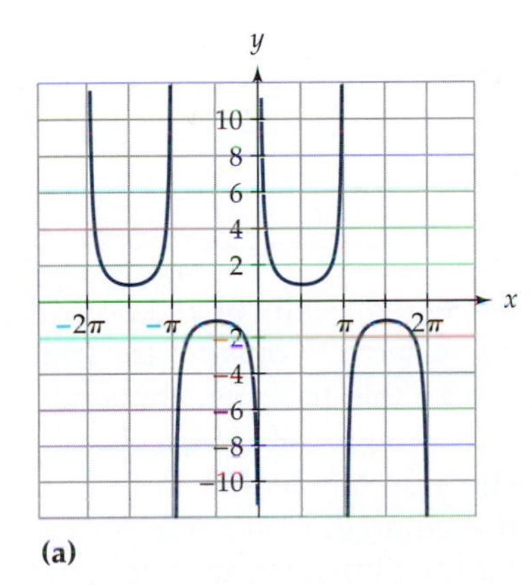

(a)

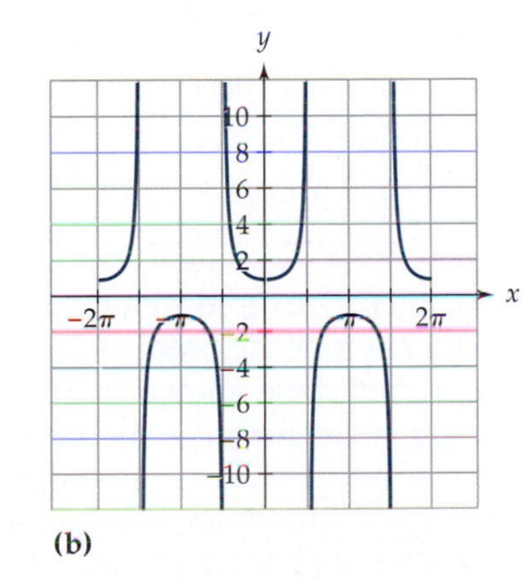

(b)

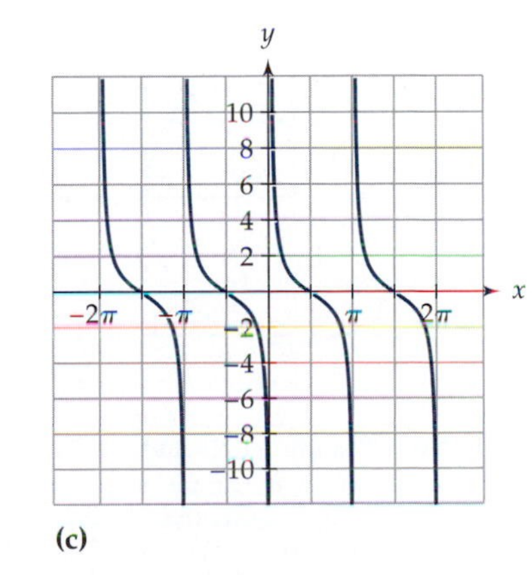

(c)

Figure 4 Graphs of (a) $y = \csc x$, (b) $y = \sec x$, and (c) $y = \cot x$ for $-2\pi \leq x \leq 2\pi$.

PROPERTIES

Definitions and Properties of Cosecant, Secant, and Cotangent

$y = \csc x = \dfrac{1}{\sin x}$:

- Domain is all real numbers except for $x = n\pi$, where n is an integer.
- Range is $y \leq -1$ or $y \geq 1$.

$y = \sec x = \dfrac{1}{\cos x}$:

- Domain is all real numbers except for $x = \pi/2 + n\pi$, where n is an integer.
- Range is $y \leq -1$ or $y \geq 1$.

$y = \cot x = \dfrac{1}{\tan x} = \dfrac{\cos x}{\sin x}$:

- Domain is all real numbers except for $x = n\pi$, where n is an integer.
- Range is all real numbers.

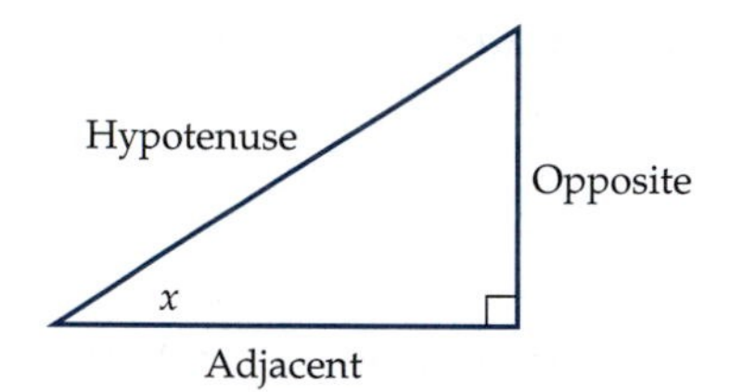

Figure 5 Using a right triangle to find definitions of the secant, cosecant, and cotangent functions.

Like the other trigonometric functions, $y = \csc x$, $y = \sec x$, and $y = \cot x$ can also be defined in terms of the lengths of the sides of a right triangle. Using Figure 5 as a guide, the right triangle definitions of cosecant, secant, and cotangent are as follows:

$$\csc\theta = \frac{\text{hypotenuse}}{\text{opposite}}, \qquad \sec\theta = \frac{\text{hypotenuse}}{\text{adjacent}}, \qquad \cot\theta = \frac{\text{adjacent}}{\text{opposite}}.$$

Notice that these are consistent with the definitions we gave earlier. For example,

$$\csc x = \frac{1}{\sin x} = \frac{1}{(\text{opposite}/\text{hypotenuse})} = \frac{\text{hypotenuse}}{\text{opposite}}.$$

Two identities involving these three functions are the following.

PROPERTIES

Pythagorean Identities

$1 + \tan^2 x = \sec^2 x.$
$1 + \cot^2 x = \csc^2 x.$

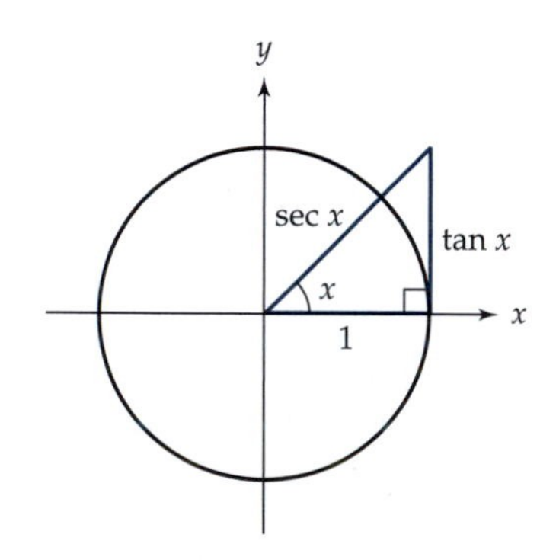

Figure 6 The unit circle and a triangle can be used to show that $1 + \tan^2 x = \sec^2 x$.

We can prove these identities either geometrically or algebraically. We start with a geometric proof of $1 + \tan^2 x = \sec^2 x$. Consider the triangle drawn in a unit circle shown in Figure 6. In this triangle, the side adjacent to angle x has length 1. Because $\tan x = \text{opposite}/\text{adjacent}$ and length of the adjacent side is 1, the opposite side has length $\tan x$. Also, because $\sec x = \text{hypotenuse}/\text{adjacent}$ and the length of the adjacent side is 1, the hypotenuse has length $\sec x$. Using the Pythagorean theorem, we see that

$$1^2 + \tan^2 x = \sec^2 x \qquad \text{or} \qquad 1 + \tan^2 x = \sec^2 x,$$

which proves our identity for $0 < x < \pi/2$. Similar drawings in other quadrants can be used to prove this identity for other values of x where $y = \tan x$ and $y = \sec x$ are defined. We can also prove this identity algebraically, as shown in Example 1.

Example 1 Algebraically prove that $1 + \tan^2 x = \sec^2 x$.

Solution One of the identities mentioned earlier is $\cos^2 x + \sin^2 x = 1$. If we divide both sides of this equation by $\cos^2 x$, we get

$$\frac{\cos^2 x}{\cos^2 x} + \frac{\sin^2 x}{\cos^2 x} = \frac{1}{\cos^2 x},$$

which simplifies to

$$1 + \tan^2 x = \sec^2 x.$$

Notice that our proof does not work if $\cos^2 x = 0$.

In the exercises, you are asked to prove the other identity, $1 + \cot^2 x = \csc^2 x$, both geometrically and algebraically.

READING QUESTIONS

5. Show that the right triangle definition of $y = \cot x$ is equivalent to the definition $\cot x = \cos x/\sin x$.
6. What is the relationship between the range of $y = \sin x$ and the range of $y = \csc x$?

DOUBLE-ANGLE IDENTITIES

A Double-Angle Identity for Sine

Sometimes, it is useful to have trigonometric identities for double angles. We start by looking at a double-angle identity for $\sin(2x)$. One way to derive this identity geometrically is to start with a triangle drawn in a unit circle. You want to draw a triangle where it is easy to identify an angle of x radians and another angle of $2x$ radians so that you can find a relationship between $\sin(2x)$ and trigonometric functions involving x. Two such triangles are shown in Figure 7. Notice that the triangle in Figure 7(b), $\triangle OCD$, is a rotated version of the triangle in Figure 7(a), $\triangle OAB$.

Because we are trying to prove something about $\sin(2x)$, it makes sense to identify a line segment of length $\sin(2x)$ in Figure 7. Using Figure 7(b) and the unit circle definition of $y = \sin x$, we see that $\sin(2x)$ is the y-coordinate of point C. In other words, $\sin(2x)$ is the length of CF. Also notice that $\overline{CF}$ is the height of $\triangle OCD$. Using that the area of a triangle

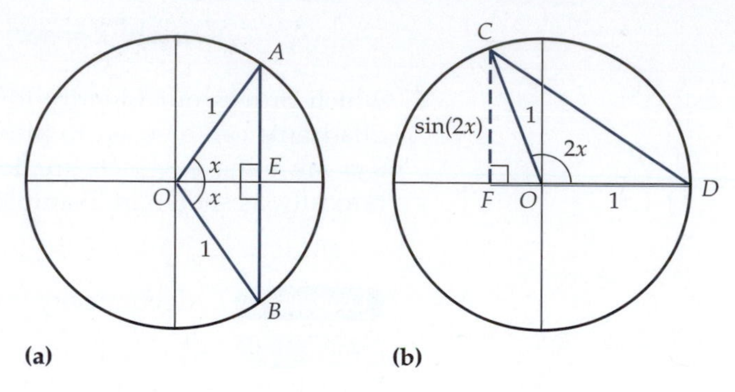

Figure 7 We can use these two triangles to derive an identity for $\sin(2x)$ geometrically.

is one-half the base times the height, we see that the area of $\triangle OCD$ is $\frac{1}{2} \cdot CF \cdot OD = \frac{1}{2} \cdot \sin 2x \cdot 1 = \frac{1}{2}\sin(2x)$.[14]

Now let's look at $\triangle OAB$ in Figure 7(a). The area of this triangle is $\frac{1}{2} \cdot OE \cdot AB$. Using the unit circle definition of $y = \cos x$, we see that $\cos x$ is the x-coordinate of point A. In other words, $\cos x$ is the length of OE. To find the length of AB, we use the unit circle definition of $y = \sin x$ to see that $\sin x$ is the y-coordinate of point A and $\sin(-x)$ is the y-coordinate of point B. Because $\sin(-x) = -\sin x$, the y-coordinate of B can also be written as $-\sin x$. Therefore, the length of AB is $\sin x - (-\sin x) = 2\sin x$. Putting this information all together tells us that the area of $\triangle OAB = \frac{1}{2} \cdot OE \cdot AB = \frac{1}{2} \cdot \cos x \cdot 2\sin x = \cos x \sin x$.

As mentioned earlier, $\triangle OCD$ is a rotated version of $\triangle OAB$, which we can also see by observing that these are congruent triangles because $\triangle OAB \cong \triangle OCD$ by *side–angle–side*. We know that congruent triangles have the same area. Therefore,

$$\begin{aligned} \text{area of } \triangle OAB &= \text{area of } \triangle OCD \\ \cos x \sin x &= \tfrac{1}{2}\sin(2x) \\ 2\cos x \sin x &= \sin(2x). \end{aligned}$$

We now have the following identity.

PROPERTIES

Double Angle Identity for Sine

$\sin(2x) = 2\cos x \sin x.$

Double-Angle Identities for Cosine

Three commonly used double-angle identities for cosine are as follows.

[14] *An alternate way of finding the area of $\triangle OCD$ is to use the formula developed in Section 5.2:* Area $= \frac{1}{2}ab\sin\theta$.

PROPERTIES

Double Angle Identities for Cosine

$\cos 2x = \cos^2 x - \sin^2 x.$
$\cos 2x = 1 - 2\sin^2 x.$
$\cos 2x = 2\cos^2 x - 1.$

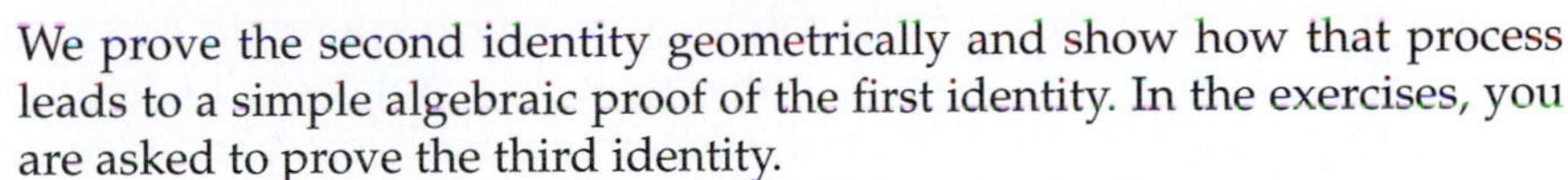

We prove the second identity geometrically and show how that process leads to a simple algebraic proof of the first identity. In the exercises, you are asked to prove the third identity.

Once again, to start a geometric proof, the goal is to set up a figure where it is easy to identify a line segment of length $\cos(2x)$. We use the same figure we used to prove the identity for $\sin(2x)$. Figure 8 shows the same triangle as Figure 7(b).[15] We saw earlier that the length of CF is $\sin(2x)$ and the length of CD is $2\sin x$. Using the unit circle definitions of $y = \cos x$, notice that $\cos(2x)$ is the x-coordinate of point C. Because point C is in the second quadrant, we know that $\cos(2x)$ is negative. Therefore, the length of OF is $-\cos(2x)$. By the Pythagorean theorem (using $\triangle FCD$), $\sin^2(2x) + (-\cos(2x) + 1)^2 = (2\sin x)^2$. Simplifying gives

Figure 8 We can use this triangle to find an identity for $\cos(2x)$.

$$\begin{aligned} \sin^2(2x) + (-\cos(2x) + 1)^2 &= (2\sin x)^2 \\ \sin^2(2x) + \cos^2(2x) - 2\cos(2x) + 1 &= 4\sin^2 x. \end{aligned}$$

Because $\sin^2(2x) + \cos^2(2x) = 1$,

$$\begin{aligned} (\sin^2(2x) + \cos^2(2x)) - 2\cos(2x) + 1 &= 4\sin^2 x \\ 2 - 2\cos(2x) &= 4\sin^2 x \\ 1 - \cos(2x) &= 2\sin^2 x \\ 1 - 2\sin^2 x &= \cos(2x). \end{aligned}$$

From this identity, it is easy to derive the other two using $\cos^2 x + \sin^2 x = 1$. We demonstrate by algebraically deriving the identity $\cos(2x) = \cos^2 x - \sin^2 x$.

$$\begin{aligned} \cos(2x) &= 1 - 2\sin^2 x \\ &= (\cos^2 x + \sin^2 x) - 2\sin^2 x \\ &= \cos^2 x - \sin^2 x. \end{aligned}$$

A Double-Angle Identity for Tangent

There is also a double-angle identity for the tangent function. To derive it, we start with the definition of $\tan x = \sin x/\cos x$, which means that $\tan(2x) = \sin(2x)/\cos(2x)$. Using the identity we just found for $\sin(2x)$ and

[15] *In the exercises, you are asked to prove this identity using a triangle where* $2x < \pi/2$.

one of the identities for $\cos(2x)$, we get

$$\begin{aligned}\tan(2x) &= \frac{\sin(2x)}{\cos(2x)}\\ &= \frac{2\sin x\cos x}{\cos^2 x - \sin^2 x}.\end{aligned}$$

We want to rewrite the right-hand side of this identity in terms of $\tan x$ rather than $\sin x$ and $\cos x$. To do this, we divide both the numerator and the denominator by $\cos^2 x$ and then simplify.

$$\begin{aligned}\frac{2\sin x\cos x}{\cos^2 x - \sin^2 x} &= \frac{(2\sin x\cos x)/\cos^2 x}{(\cos^2 x - \sin^2 x)/\cos^2 x}\\ &= \frac{2\cdot\ \sin x/\cos x}{\cos^2 x/\cos^2 x - \ \sin^2 x/\cos^2 x}\\ &= \frac{2\cdot\ \sin x/\cos x}{1 - \ \sin^2 x/\cos^2 x}\\ &= \frac{2\tan x}{1-\tan^2 x}.\end{aligned}$$

This result gives us the following identity.

PROPERTIES

Double Angle Identity for Tangent

$$\tan(2x) = \frac{2\tan x}{1-\tan^2 x}.$$

READING QUESTIONS

7. In Figure 7, suppose $x < \pi/4$. How would Figure 7(b) look in this case? Explain why this change in the figure would not change the proof of $\sin(2x) = 2\cos x\sin x$.
8. In Figure 8, explain why the length of OF is $-\cos(2x)$ rather than $\cos(2x)$.
9. Let a be an angle such that $\tan a = \frac{12}{7}$. What is $\tan(2a)$?

SUM AND DIFFERENCE IDENTITIES

Other identities are those involving sums and differences of angles. Some of these are as follows.

PROPERTIES

Sum and Difference Identities

$\sin(x+y) = \sin x \cos y + \cos x \sin y.$
$\sin(x-y) = \sin x \cos y - \cos x \sin y.$
$\cos(x+y) = \cos x \cos y - \sin x \sin y.$
$\cos(x-y) = \cos x \cos y + \sin x \sin y.$
$\tan(x+y) = \dfrac{\tan x + \tan y}{1 - \tan x \tan y}.$
$\tan(x-y) = \dfrac{\tan x - \tan y}{1 + \tan x \tan y}.$

We give proofs for the identities involving $\cos(x+y)$ and $\sin(x-y)$. In the exercises, you are asked to prove the other identities.

Proving the Sum Identity for the Cosine Function

We begin by proving that $\cos(x+y) = \cos x \cos y - \sin x \sin y$. Figure 9 is a unit circle with five points of interest. We are assuming $y > x$. The characteristics of these points are as follows.

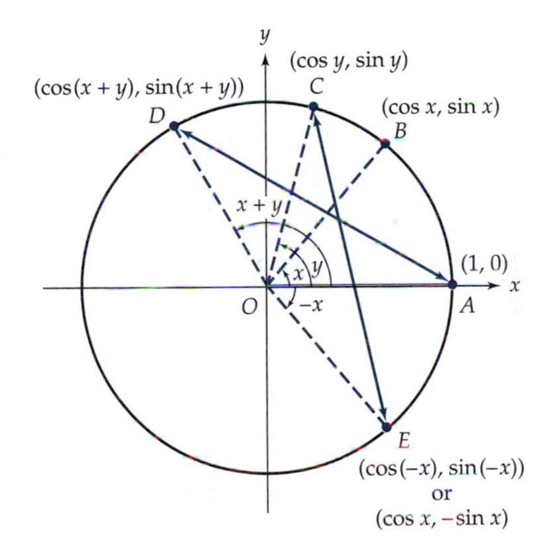

Figure 9 A figure used to prove the identity $\cos(x+y) = \cos x \cos y - \sin x \sin y$.

Point A is on the x-axis.

Point B is formed by the angle whose measure is x.

Point C is formed by the angle whose measure is y.

Point D is formed by the angle whose measure is $x+y$.

Point E is formed by the angle whose measure is $-x$.

The coordinates of each point are given. Note that because $\cos(-x) = \cos x$ and $\sin(-x) = -\sin x$, the coordinates of point E are also given as

$(\cos x, -\sin x)$. The measure of arc $\overset{\frown}{AD}$ is $x+y$. The measure of the arc $\overset{\frown}{EC}$, however, is also $x+y$. Because these two arcs have the same measure, $AD = EC$.

Using the distance formula[16] and setting these two distances equal to each other gives

$$\sqrt{(\cos(x+y)-1)^2+(\sin(x+y)-0)^2}$$
$$= \sqrt{(\cos y - \cos x)^2 + (\sin y - (-\sin x))^2}.$$

Squaring both sides and expanding the expressions gives

$$(\cos(x+y)-1)^2+(\sin(x+y)-0)^2$$
$$= (\cos y - \cos x)^2 + (\sin y + \sin x))^2$$
$$\cos^2(x+y) - 2\cos(x+y) + 1 + \sin^2(x+y)$$
$$= \cos^2 y - 2\cos x \cos y + \cos^2 x + \sin^2 y + 2\sin x \sin y + \sin^2 x$$
$$[\cos^2(x+y) + \sin^2(x+y)] - 2\cos(x+y) + 1$$
$$= [\cos^2 y + \sin^2 y] + [\cos^2 x + \sin^2 x] - 2\cos x \cos y + 2\sin x \sin y.$$

Using the identity $\sin^2 x + \cos^2 x = 1$, we can replace the expressions in the square brackets with 1. So, our equation simplifies to

$$-2\cos(x+y) + 2 = -2\cos x \cos y + 2\sin x \sin y + 2.$$

By subtracting 2 from both sides and dividing both sides by -2, we derive our identity:

$$-2\cos(x+y) = -2\cos x \cos y + 2\sin x \sin y$$
$$\cos(x+y) = \cos x \cos y - \sin x \sin y.$$

Proving the Difference Identity for the Sine Function

To prove that $\sin(x-y) = \sin x \cos y - \cos x \sin y$, we begin by using the identity $\sin x = \cos(\pi/2 - x)$ to convert $\sin(x-y)$ to a cosine function.

$$\sin(x-y) = \cos\left[\frac{\pi}{2} - (x-y)\right]$$
$$= \cos\left[\left(\frac{\pi}{2} - x\right) + y\right].$$

Using the sum identity for cosine that we just proved gives

$$\cos\left[\left(\frac{\pi}{2} - x\right) + y\right] = \cos\left(\frac{\pi}{2} - x\right)\cos y - \sin\left(\frac{\pi}{2} - x\right)\sin y.$$

By using the identities $\sin x = \cos(\pi/2 - x)$ and $\cos x = \sin(\pi/2 - x)$, we derive our result:

$$\cos\left[\left(\frac{\pi}{2} - x\right) + y\right] = \cos\left(\frac{\pi}{2} - x\right)\cos y - \sin\left(\frac{\pi}{2} - x\right)\sin y$$
$$= \sin x \cos y - \cos x \sin y.$$

[16] *The distance between any two points (x_1, y_1) and (x_2, y_2) on the xy-plane is* $d = \sqrt{(x_2 - x_1)^2 + (y_2 - y_1)^2}$.

Using the Difference Identities for the Tangent Function to Find Angles Formed by Intersecting Lines

There is an interesting consequence of the difference identity for the tangent function,

$$\tan(x - y) = \frac{\tan x - \tan y}{1 + \tan x \tan y},$$

that allows us to determine the angle formed by the intersection of two lines. You may recall that the unit circle definition of $y = \tan\theta$ is the slope of the line that creates an angle of θ with the x-axis. Using this definition, we can see from Figure 10 that $\tan\theta_1 = m_1$ and $\tan\theta_2 = m_2$.

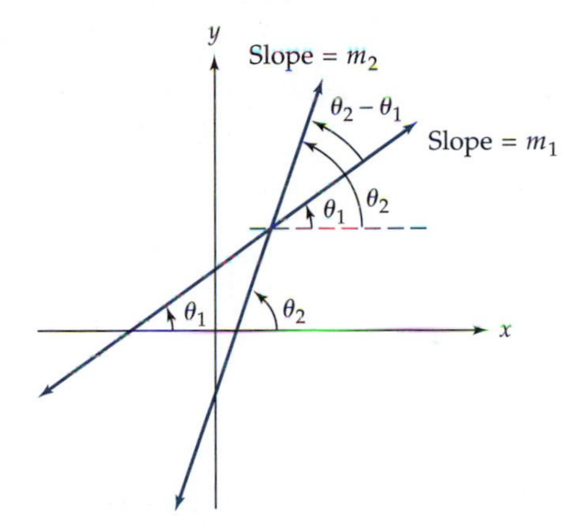

Figure 10 Knowing the slopes of two lines allows us to find the angle between them.

The acute angle formed by the intersection of the two lines in Figure 10 has measure $\theta_2 - \theta_1$. Using the identity for the tangent of the difference of two angles and substituting slopes where appropriate gives

$$\begin{aligned}\tan(\theta_2 - \theta_1) &= \frac{\tan\theta_2 - \tan\theta_1}{1 + \tan\theta_2 \tan\theta_1}\\ &= \frac{m_2 - m_1}{1 + m_2 m_1}.\end{aligned}$$

Therefore, if two lines intersect to form an angle of θ, then

$$\tan\theta = \frac{m_2 - m_1}{1 + m_2 m_1},$$

where m_1 and m_2 are the slopes of the two lines. By taking the inverse tangent of both sides, we can find the angle of intersection. Because the range of the inverse tangent is $-\pi/2 < y < \pi/2$, we will always get an acute angle.[17]

[17] *The sign of the angle simply indicates direction. In this case, it makes no difference if our angle is measured counterclockwise or clockwise.*

PROPERTIES

Let L_1 be a line with slope m_1 and let L_2 be a line with slope m_2. The acute angle, θ, formed by the intersection of the two lines can be found as follows.

$$\theta = \tan^{-1}\left(\frac{m_2 - m_1}{1 + m_2 m_1}\right)$$

Example 2 Determine the angle formed by the intersection of $y = 2x + 3$ and $y = \frac{1}{3}x - 4$.

Solution We let $m_2 = 2$ and $m_1 = \frac{1}{3}$. Using the formula $\tan\theta = (m_2 - m_1)/(1 + m_2 m_1)$ gives

$$\begin{aligned} \tan\theta &= \frac{m_2 - m_1}{1 + m_2 m_1} \\ &= \frac{2 - \frac{1}{3}}{1 + 2 \cdot \frac{1}{3}} \\ &= \frac{\frac{5}{3}}{\frac{5}{3}} \\ &= 1. \end{aligned}$$

So, $\theta = \tan^{-1} 1 = \pi/4 = 45°$.

READING QUESTIONS

10. Show that the sum identity for cosine gives the same result as the double-angle identity for cosine when $x = y$.

11. Use the sum identity for sine to prove the difference identity for sine by replacing y with $-y$.

12. The formula for finding the angle between two lines is $\tan\theta = (m_2 - m_1)/(1 + m_2 m_1)$. What happens if the lines have the same slope (and thus are parallel)?

PROVING TRIGONOMETRIC IDENTITIES

It is helpful to be able to prove trigonometric identities. First, these type of proofs give you practice in algebraic and geometric skills and reinforce the identities already proved in this section. Second, at times two answers involving trigonometric functions look different and yet are actually the same. Using trigonometric identities appropriately allows you to determine whether two expressions are truly different or are actually the same. As seen in the proofs done so far, there is no general algorithm for proving trigonometric identities. Certain steps, however, are frequently useful. When doing a geometric proof, try drawing a triangle involving a unit circle. Ideally, find a side whose length involves one of the trigonometric functions in your identity. When doing an algebraic proof, try to use the identities we have already shown. It is often helpful to convert all the functions to sine or cosine functions. Start with one side of the identity (typically the more complicated expression) and simplify it until it is obvious that it equals the expression on the other side of the identity. We work several examples to illustrate ways to prove identities algebraically.

Example 3 Prove that $\tan x \sin x + \cos x = \sec x$.

Solution Because the left side of the equation is more complicated than the right side, we try simplifying it until it is obvious that it equals $\sec x$. We begin by converting the left side to an expression involving only sine and cosine functions, which gives

$$\begin{aligned}\tan x \sin x + \cos x &= \frac{\sin x}{\cos x}\sin x + \cos x\\ &= \frac{\sin^2 x}{\cos x} + \cos x.\end{aligned}$$

Finding a common denominator, adding, and using the identity $\cos^2 x + \sin^2 x = 1$ gives

$$\begin{aligned}\frac{\sin^2 x}{\cos x} + \cos x &= \frac{\sin^2 x}{\cos x} + \frac{\cos^2 x}{\cos x}\\ &= \frac{\sin^2 x + \cos^2 x}{\cos x}\\ &= \frac{1}{\cos x}.\end{aligned}$$

Because we know that

$$\frac{1}{\cos x} = \sec x,$$

we have proved that $\tan x \sin x + \cos x = \sec x$.

Example 4 Prove that

$$\frac{1 - \cos^4 x}{1 + \cos^2 x} = \sin^2 x.$$

Solution Starting with the left side of the equation, we factor the numerator as a difference of two squares and simplify to get

$$\begin{aligned}\frac{1 - \cos^4 x}{1 + \cos^2 x} &= \frac{(1 - \cos^2 x)(1 + \cos^2 x)}{1 + \cos^2 x}\\ &= 1 - \cos^2 x.\end{aligned}$$

Because $\cos^2 x + \sin^2 x = 1$, $1 - \cos^2 x = \sin^2 x$, which proves that $(1 - \cos^4 x)/(1 + \cos^2 x) = \sin^2 x$.

As mentioned earlier, you may have a solution to a problem that does not match someone else's solution. One way to see if they are actually the same is by graphing both solutions. If the graphs look different, then the two solutions are obviously different. If the graphs look the same, then the two solutions *may* be the same. To make sure they are the same, you must provide a proof.

Example 5 Graph each of the following equations. If the two graphs appear to be the same, try to prove that the two equations are the same.

$$f(x) = (\sin x + \cos x)^2$$
$$g(x) = 1.$$

Solution The graph of $f(x) = (\sin x + \cos x)^2$ along with the graph of $g(x) = 1$ is shown in Figure 11. Because these graphs are clearly different, $(\sin x + \cos x)^2 \neq 1$. Therefore, this statement is not an identity.

Figure 11 These two graphs are clearly different.

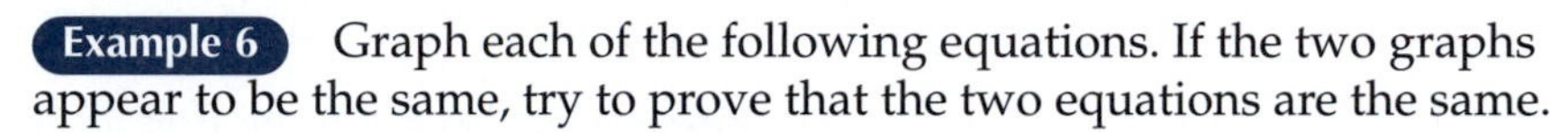

Example 6 Graph each of the following equations. If the two graphs appear to be the same, try to prove that the two equations are the same.

$$h(x) = \sec x$$
$$k(x) = \csc x \tan x.$$

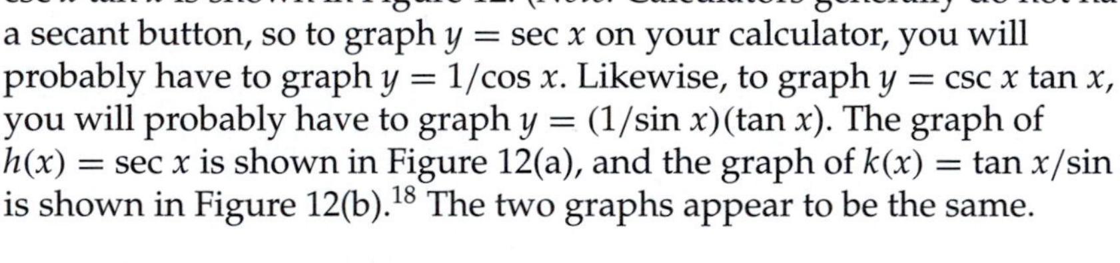

Solution The graph of $h(x) = \sec x$ along with the graph of $k(x) = \csc x \tan x$ is shown in Figure 12. (*Note:* Calculators generally do not have a secant button, so to graph $y = \sec x$ on your calculator, you will probably have to graph $y = 1/\cos x$. Likewise, to graph $y = \csc x \tan x$, you will probably have to graph $y = (1/\sin x)(\tan x)$. The graph of $h(x) = \sec x$ is shown in Figure 12(a), and the graph of $k(x) = \tan x/\sin x$ is shown in Figure 12(b).[18] The two graphs appear to be the same.

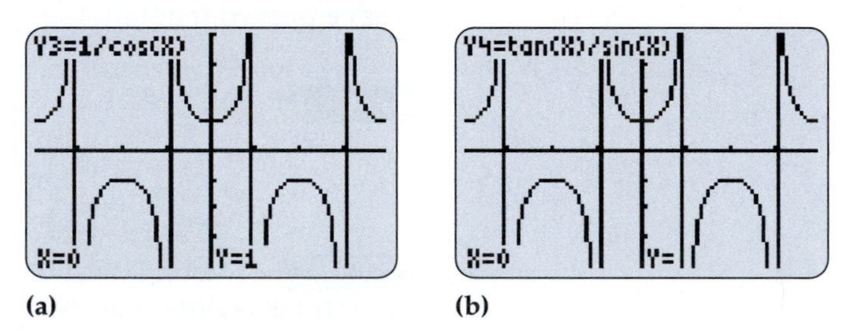

Figure 12 These graphs, (a) $y = \sec x$ and (b) $y = \csc x \tan x$, appear to be the same.

To prove that these two expressions are the same and thus show that we have an identity, we simplify the right side of the equation to get

[18] *The calculator graphs show vertical lines where there are actually vertical asymptotes.*

$$\begin{aligned}\csc x \tan x &= \frac{1}{\sin x} \cdot \frac{\sin x}{\cos x} \\ &= \frac{1}{\cos x} \\ &= \sec x.\end{aligned}$$

So, it appears that $\sec x = \csc x \tan x$. Yet there is a subtle difference. The domain of $h(x) = \sec x$ is all real numbers x such that $x \neq \ldots, -3\pi/2, -\pi/2, \pi/2, 3\pi/2, \ldots$ because you must avoid x-values where $\cos x = 0$. The domain of $\csc x \tan x$, however, is all real numbers x such that $x \neq \ldots, -3\pi/2, -\pi, -\pi/2, 0, \pi/2, \pi, 3\pi/2, \ldots$ because you must avoid x-values where either $\sin x = 0$ or $\cos x = 0$. Therefore, $\sec x = \csc x \tan x$ only *for values where both sides of the equation are defined*.

Example 6 illustrates an important yet subtle point. It is easy to simplify an expression algebraically to prove an identity yet ignore issues involving the domain. This error is similar to saying that $[(x-3)(x+2)]/(x+2) = x-3$ and ignoring that it is only true when $x \neq -2$. Often, when dealing with fractions or square roots, there is a restriction on the values for which the identity is true.

HISTORICAL NOTE

Although trigonometric identities are most often used in mathematical proofs, they have also been used for mathematical computation. Before the invention of logarithms by John Napier in 1614, trigonometric identities were used to aid in the multiplication of large numbers. Like using logarithms, appropriate trigonometric identities could convert a multiplication problem to an addition or subtraction problem. This method was known as *prosthaphaeresis*, a Greek word meaning addition and subtraction. This method of multiplying was used by astronomers such as Tycho Brahe (1546–1601) in Denmark. Learning of Brahe's work with prosthaphaeresis prompted John Napier to redouble his efforts to develop logarithms, which became an easier method of doing such calculations.

To illustrate the method of prosthaphaeresis, suppose you were an ancient astronomer who needed to multiply 94,562 by 3253. You could use the identity $\cos x \cos y = [\cos(x+y) + \cos(x-y)]/2$ and let $\cos x = 0.94562$ and $\cos y = 0.3253$. (You had to convert your numbers to decimals less than 1 because of the range of the cosine function. To convert back to the original product, multiply by 10^9.) Using the trigonometric tables available at the time, you would find that $x \approx 0.331300886$ and $y \approx 1.239467361$. By using the identity and the trigonometric tables, you calculate

$$\begin{aligned}&94{,}562 \cdot 3253 \\ &\quad = 10^9 \cdot 0.94562 \cdot 0.3253 \\ &\quad \approx 10^9 \cdot \frac{\cos(0.331300886 + 1.239467361) + \cos(0.331300886 - 1.239467361)}{2}\end{aligned}$$

continued on page 428

$$= 10^9 \cdot \frac{\cos(1.570768247) + \cos(-0.908166475)}{2}$$
$$\approx 10^9 \cdot \frac{0.000028080 + 0.615192292}{2}$$
$$\approx 10^9 \cdot 0.307610186$$
$$= 307{,}610{,}186.$$

In this example, we used enough decimal places so that our final answer is exact rather than an approximation. As you can see, this method of multiplying was not a great labor-saving device. For large numbers, however, it apparently did save some time. The use of logarithms soon made this method of multiplying obsolete. Now, through the use of calculators, using logarithms (or slide rules) to multiply large numbers also seems like an obsolete method.

Source: Carl B. Boyer, *A History of Mathematics,* (Princeton, NJ: Princeton University Press, 1968), pp. 339–43.

READING QUESTIONS

13. Why isn't it sufficient to prove an identity by showing that the graphs of the two expressions are the same?

14. (a) Prove that $\cot x \sec x = \csc x$.

(b) For what values of x is this identity true?

SUMMARY

In this section, we have reviewed or introduced the following **trigonometric identities.**

Symmetry Identities

- $\sin x = -\sin(-x)$
- $\cos x = \cos(-x)$
- $\tan x = -\tan(-x)$

Horizontal Shift Identities

- $\sin x = \cos\left(x - \frac{\pi}{2}\right)$
- $\cos x = \sin\left(x + \frac{\pi}{2}\right)$

Complement Identities

- $\sin x = \cos\left(\frac{\pi}{2} - x\right)$
- $\cos x = \sin\left(\frac{\pi}{2} - x\right)$

Definition Identities

- $\tan x = \frac{\sin x}{\cos x}$
- $\csc x = \frac{1}{\sin x}$
- $\sec x = \frac{1}{\cos x}$
- $\cot x = \frac{1}{\tan x}$

Pythagorean Identities

- $\cos^2 x + \sin^2 x = 1$
- $1 + \tan^2 x = \sec^2 x$
- $1 + \cot^2 x = \csc^2 x$

Double-Angle Identities

- $\sin(2x) = 2 \sin x \cos x$
- $\cos(2x) = \cos^2 x - \sin^2 x$
- $\cos(2x) = 1 - 2 \sin^2 x$
- $\cos(2x) = 2 \cos^2 x - 1$
- $\tan 2x = \frac{2 \tan x}{1 - \tan^2 x}$

Sum and Difference Identities

- $\sin(x + y) = \sin x \cos y + \cos x \sin y$
- $\cos(x + y) = \cos x \cos y - \sin x \sin y$
- $\tan(x + y) = \frac{\tan x + \tan y}{1 - \tan x \tan y}$
- $\sin(x - y) = \sin x \cos y - \cos x \sin y$
- $\cos(x - y) = \cos x \cos y + \sin x \sin y$
- $\tan(x - y) = \frac{\tan x - \tan y}{1 + \tan x \tan y}$

In addition to these identities, we found that if two lines intersect to form an angle of θ, then $\tan\theta = (m_2 - m_1)/(1 + m_2 m_1)$, where m_1 and m_2 are the slopes of the two lines.

EXERCISES

1. Fill out the following table. All answers are to be exact.

x (radians)	x (degrees)	$\sec x$	$\csc x$	$\cot x$
$\frac{\pi}{6}$				
$\frac{\pi}{3}$				
$\frac{\pi}{4}$				

2. Each of the following statements is false. Explain why.
   (a) $1 + \tan^2 x = \sec^2 x$ for all values of x.
   (b) The function $y = \cot x$ is undefined for $x = \pi/2$.
   (c) It is possible to find a value of x such that $\csc x = \frac{1}{4}$.
   (d) The function $y = \cot x$ is equal to the slope of the line containing the points $(\cos x, \sin x)$ and $(0, 0)$.
   (e) Because $y = \cos x$ is symmetric about the origin, I know that $\cos x = \cos(-x)$.
   (f) $\sin x = \sin(\pi/2 - x)$.
3. To prove $1 + \cot^2 x = \csc^2 x$ geometrically, consider the triangle drawn in a unit circle shown. In this triangle, the side opposite angle x has length 1.
   (a) Explain why the length of $OA = \csc x$.
   (b) Explain why the length of $OB = \cot x$.
   (c) Prove that $1 + \cot^2 x = \csc^2 x$ for $0 < x < \pi/2$.
   (d) Draw a similar triangle in the second quadrant to prove that $1 + \cot^2 x = \csc^2 x$ for $\pi/2 < x < \pi$.

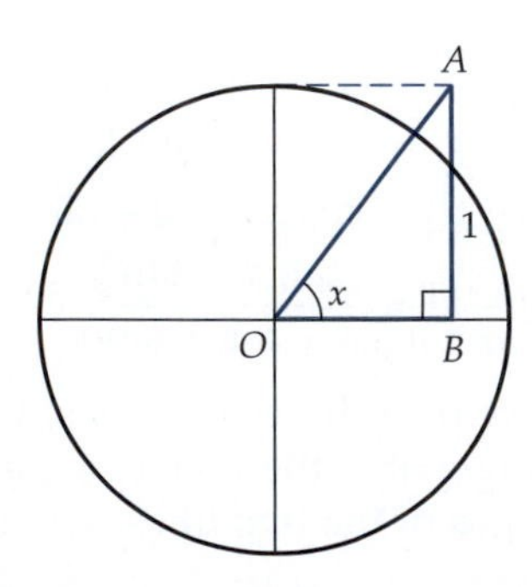

4. Prove algebraically that $1 + \cot^2 x = \csc^2 x$.
5. Use the identity $\cos(2x) = 1 - 2\sin^2 x$ to prove $\cos(2x) = 2\cos^2 x - 1$.
6. Let a be an angle in the first quadrant such that $\sin a = \frac{2}{3}$. Find each of the following.
   (a) $\cos a$
   (b) $\tan a$
   (c) $\sin 2a$
   (d) $\cos 2a$
   (e) $\tan 2a$
7. Let b be an angle in the second quadrant such that $\sin b = \frac{2}{3}$. Find each of the following.
   (a) $\cos b$
   (b) $\tan b$
   (c) $\sin 2b$
   (d) $\cos 2b$
   (e) $\tan 2b$
8. Let c be an angle in the first quadrant such that $\sin c = r$.
   (a) What are the restrictions on r?
   (b) Find each of the following.
      i. $\cos c$
      ii. $\tan c$
      iii. $\sin 2c$
      iv. $\cos 2c$
      v. $\tan 2c$
9. (a) Explain why the identity $\tan(2x) = (2\tan x)/(1 - \tan^2 x)$ does not work for $x = \pi/4$.
   (b) For what other values of x does the identity $\tan(2x) = (2\tan x)/(1 - \tan^2 x)$ not work?
10. (a) Draw a figure similar to Figure 8 in this section, except assume $x < \pi/4$. This assumption means that $2x < \pi/2$, so your triangle will be contained in the first quadrant.
   (b) Using your figure from part (a), geometrically prove that the identity $\cos(2x) = 1 - 2\sin^2 x$ for $0 < x < \pi/4$.
11. The value of π can be approximated by finding the perimeter of regular polygons inscribed in a circle of radius $\frac{1}{2}$. As the number of sides of the polygon increases, its perimeter becomes closer and closer to the circumference of the circle, which is π. One method of calculating the perimeter of such a polygon with n sides leads to the function

$$q(n) = n \sin\left(\frac{180^\circ}{n}\right).$$

Another method of calculating the same perimeter leads to the function

$$p(n) = \frac{n/2 \sin(360°/n)}{\sin(90° - 180°/n)}.$$

Show that these two functions, which look different, are actually the same.

12. In this section, we proved the sum identity for cosine and the difference identity for sine. Use them to prove the following identities.
    (a) $\cos(x - y) = \cos x \cos y + \sin x \sin y$
    (b) $\sin(x + y) = \sin x \cos y + \cos x \sin y$
13. Prove the following identities for the tangent function.
    (a) $\tan(x + y) = \dfrac{\tan x + \tan y}{1 - \tan x \tan y}$
    (b) $\tan(x - y) = \dfrac{\tan x - \tan y}{1 + \tan x \tan y}$
14. Find the measure of the acute angle formed by the following pairs of lines. Give your answer in degrees and round to the nearest tenth of a degree.
    (a) $y = 4x - 7$ and $y = -2x + 9$
    (b) $y = \frac{1}{4}x + 3$ and $y = 4x - \frac{1}{2}$
    (c) $y = -\frac{3}{5}x - 4$ and $y = \frac{5}{3}x + 19$
15. We can find exact values for trigonometric functions when $x = \pi/6, \pi/3$, or $\pi/4$. Use these values and the appropriate sum or difference formulas to complete the following table. All answers are to be exact.

x (radians)	**x (degrees)**	**cos x**	**sin x**	**tan x**
$\frac{5\pi}{12}$				
$\frac{7\pi}{12}$				
$\frac{11\pi}{12}$				
$\frac{\pi}{12}$				

16. (a) Graph $f(x) = \cos^2 x - \sin^2 x$.
    (b) Using the transformation techniques described in section 5.3, find the equation of a cosine function whose graph looks like the graph of f.
    (c) Using the appropriate sum identity, prove that your cosine function from part (b) is equal to f.

17. (a) Graph $g(x) = \cos x + \sin x$.
    (b) Using the transformation techniques described in section 5.3, find the equation of a sine function whose graph looks like the graph of g.
    (c) Using the appropriate sum identity, prove that your sine function from part (b) is equal to g.

18. Prove the following identities.
    (a) $\sin^2 x = \dfrac{1 - \cos 2x}{2}$
    (b) $\cos^2 x = \dfrac{1 + \cos 2x}{2}$
    (c) $\tan^2 x = \dfrac{1 - \cos 2x}{1 + \cos 2x}$

19. Prove the following trigonometric identities. Indicate for which values of x the identity is true.
    (a) $\dfrac{\cos x}{\sec x - \tan x} = 1 + \sin x$
    (b) $\tan x = \dfrac{1 - \cos 2x}{\sin 2x}$
    (c) $\tan x(1 - \sin^2 x) = \frac{1}{2} \sin 2x$
    (d) $\sec^2 x \cot x - \cot x = \tan x$

20. Simplify each of the following expressions completely. Your answer, in each case, will be a number.
    (a) $\cos^4 x + \sin^4 x + \frac{1}{2} \sin^2(2x)$
    (b) $\sin^2 x - \sin^4 x - \cos^2 x \sin^2 x$
    (c) $1 + \cos(2x) + 2 \sin^2 x$

21. Prove the following identities.
    (a) $\cos(3x) = \cos^3 x - 3 \cos x \sin^2 x$
    (b) $\sin(3x) = 3 \cos^2 x \sin x - \sin^3 x$

22. Prove the following identities.
    (a) $\cos x \cos y = \frac{1}{2}(\cos(x + y) + \cos(x - y))$
    (b) $\sin x \sin y = \frac{1}{2}(\cos(x - y) - \cos(x + y))$

23. (a) Algebraically "prove" that $\sin x = \sqrt{1 - \cos^2 x}$.
    (b) Sketch a graph $f(x) = \sin x$ and $g(x) = \sqrt{1 - \cos^2 x}$ for $-2\pi < x < 2\pi$.
    (c) The graphs of f and g from part (b) are different, yet your "proof" in part (a) says that the graphs should be the same. Explain this discrepancy.

24. Graph to determine if each of the following pairs of functions are the same. If they appear to be the same, then prove that they are.
    (a) $f(x) = \dfrac{1}{\cos x}, g(x) = \dfrac{\csc(-x)}{\cot(-x)}$
    (b) $f(x) = \cos(2x) + \sin(2x), g(x) = (\cos x - \sin x)^2$

(c) $f(x) = \sqrt{1 + \cot^2 x}, g(x) = \csc x$

(d) $f(x) = 2 \sin x + \tan x, g(x) = \tan x(2 \cos x + 1)$

25. The accompanying figure consists of two triangles where the measure of $\angle CAD = x$, the measure of $\angle EAB = y$, the measure of $\angle GAE = (y - x)/2$, and the measure of $\angle GEF = (x + y)/2$. Right angles are labeled. The length of $\overline{AE} = 1$, and the measure of $\overline{AD} = 1$. Use this figure to prove the following identities geometrically.[19]

(a) Use

$$\sin\left(\frac{x+y}{2}\right) = \frac{FD}{ED}$$

to prove that

$$\cos x - \cos y = 2 \sin\left(\frac{y-x}{2}\right) \sin\left(\frac{x+y}{2}\right).$$

(b) Use

$$\cos\left(\frac{x+y}{2}\right) = \frac{EF}{ED}$$

to prove that

$$\sin y - \sin x = 2 \sin\left(\frac{y-x}{2}\right) \cos\left(\frac{x+y}{2}\right).$$

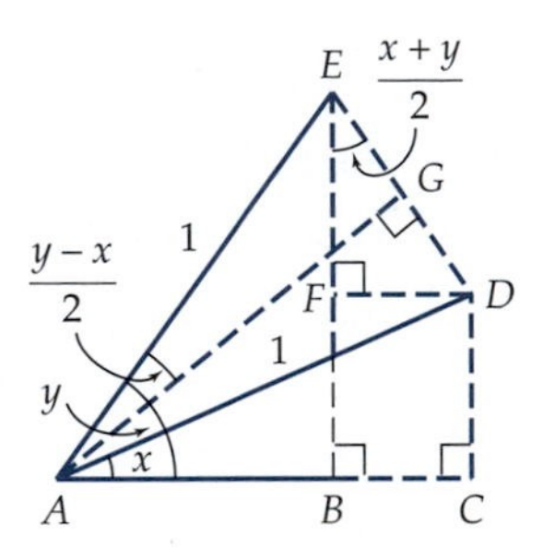

26. The figure on the following page shows a rectangle inscribed in a circle of radius 1. Let θ be the measure of the angle formed by a diagonal of the rectangle and the horizontal line bisecting the rectangle.

(a) Let A be the area of the rectangle. Show that $A = 2 \sin 2\theta$.

(b) Notice that $0° \leq \theta \leq 90°$. What value of θ will give a rectangle with the largest area? What are the dimensions of this rectangle?

[19] *The proof for these identities was shown by Yukio Kobayashi as a proof without words in March 1998 ("Trigonometric Identity: The Difference of Two Sines or Two Cosines,"* The College Mathematics Journal, *29, (2): p. 133).*

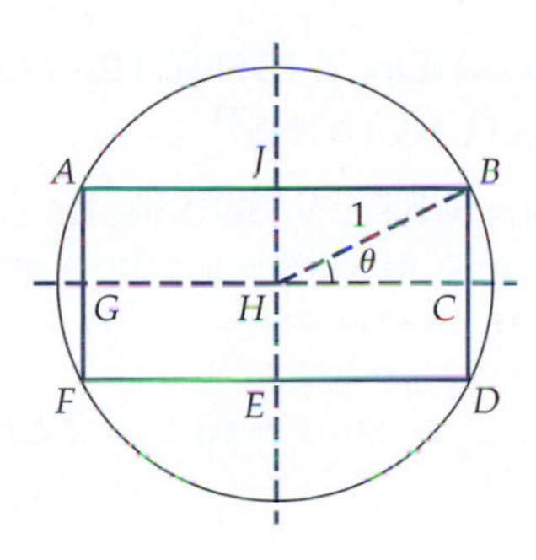

27. In the accompanying figure, $AHGB$, $BGFC$, and $CFED$ are squares with sides of length x.
    (a) Notice that $\tan\gamma = 1$. Show that $\tan(\alpha+\beta)$ also equals 1. Explain why this step shows that $\alpha+\beta=\gamma$.
    (b) Geometrically prove that $\alpha+\beta=\gamma$ without using trigonometry.[20] (*Hint:* Show that $\triangle BCE$ is similar to $\triangle ECA$ by showing that the appropriate sides are proportional.)

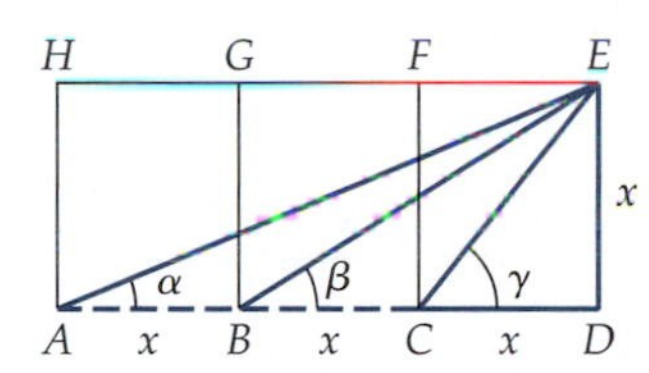

INVESTIGATIONS

INVESTIGATION 1: GEOMETRIC PROOFS OF SUM IDENTITIES

In the text, we gave a geometric proof for the identity $\cos(x+y) = \cos x \cos y - \sin x \sin y$. In this investigation, you work an alternate geometric proof for $\cos(x+y)$ and also prove that $\sin(x+y)$. These proofs use the right triangles shown in Figure 13. The measure of $\angle BOD = x$, and the measure of $\angle DOE = y$. Right angles are labeled. Notice that these triangles are *not* in a unit circle.

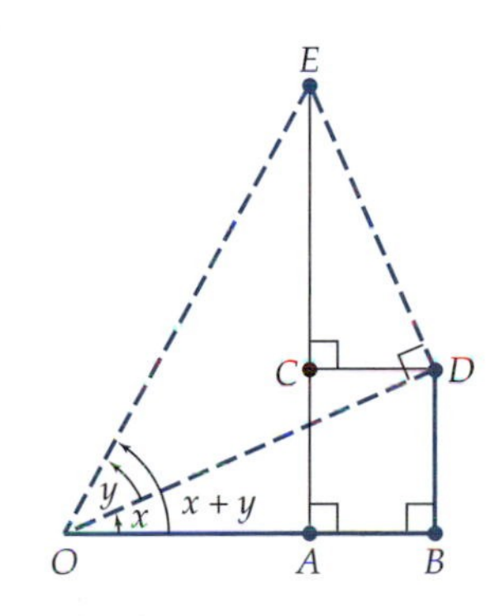

Figure 13

1. Show that the measure of $\angle ODC = x$. Use this fact to show that the measure of $\angle CED = x$.
2. Using the right triangle definitions of sine and cosine as well as Figure 13, show that
$$\sin(x+y) = \sin x \cos y + \cos x \sin y.$$
3. Using the right triangle definitions of sine and cosine as well as Figure 13, show that
$$\cos(x+y) = \cos x \cos y - \sin x \sin y.$$
4. Figure 13 assumes what restrictions on x and y?

[20] *Charles Trigg demonstrated 54 proofs of $\alpha+\beta=\gamma$ in "A Three Square Geometry Problem,"* Journal of Recreational Mathematics, *4, (2): pp. 90–9.*

INVESTIGATION 2: IDENTITIES FOR THE SUM OF SINES OR COSINES FUNCTIONS[21]

In addition to having identities for sine and cosine of the sum of two angles, there are also identities for the sum of two sine functions and the sum of two cosine functions:

$$\sin x + \sin y = 2\cos\left(\frac{y-x}{2}\right)\sin\left(\frac{x+y}{2}\right)$$

and

$$\cos y + \cos x = 2\cos\left(\frac{y-x}{2}\right)\cos\left(\frac{x+y}{2}\right).$$

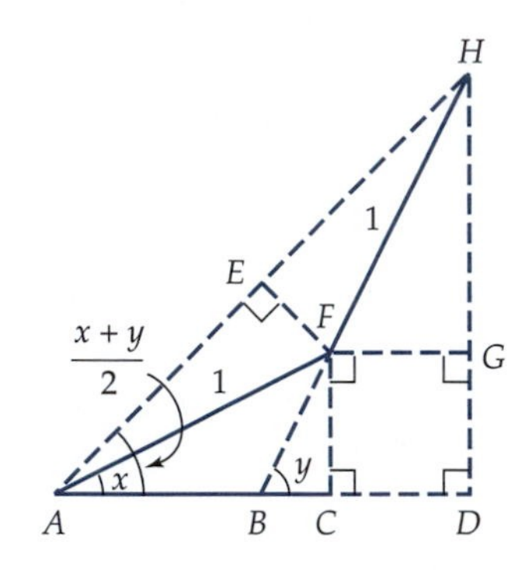

Figure 14

In this investigation, we geometrically prove these identities using Figure 14. The measure of $\angle BAF = x$, the measure of $\angle CBF = y$, and the measure of $\angle EAB = (x+y)/2$. Right angles are labeled. The length of $\overline{AF} = 1$, and the length of $\overline{FH} = 1$.

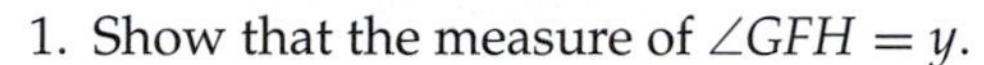

1. Show that the measure of $\angle GFH = y$.
2. Show that the measure of $\angle EAF = (y-x)/2$.
3. Use

$$\cos\left(\frac{x+y}{2}\right) = \frac{AD}{AH}$$

to prove that

$$\cos x + \cos y = 2\cos\left(\frac{y-x}{2}\right)\cos\left(\frac{x+y}{2}\right).$$

4. Use

$$\sin\left(\frac{x+y}{2}\right) = \frac{DH}{AH}$$

to prove that

$$\sin x + \sin y = 2\cos\left(\frac{y-x}{2}\right)\sin\left(\frac{x+y}{2}\right).$$

INVESTIGATION 3: PROJECTILE MOTION

Suppose a projectile was fired into the air at an angle of θ. The vertical distance of the projectile (in feet) is given by $v(t) = -16t^2 + v_0 t \sin\theta$, where t is time (in seconds) and v_0 is the initial velocity (in feet per second). The horizontal distance of the projectile (in feet) is given by $h(t) = v_0 t \cos\theta$. We are interested in determining the total horizontal distance (known as *range*) traveled by the projectile.

[21] *The proof for these identities was shown by Yukio Kobayashi as a proof without words in March 1998 ("Trigonometric Identity: The Sum of Two Sines or Two Cosines,"* The College Mathematics Journal, *29, (2): p. 157).*

1. Let r be the range of the projectile. Show that $r = \frac{1}{32}v_0^2 \sin 2\theta$. (*Hint:* First use the vertical distance to find an expression for the time when the projectile has landed.)
2. Calculate the range of a projectile shot at an angle of 60° that has an initial velocity of 120 ft/sec.
3. Calculate the range of a projectile shot at an angle of 30° that has an initial velocity of 120 ft/sec.
4. Suppose the first projectile is shot at an angle of θ and a second projectile is shot at an angle of $90° - \theta$. Assume the initial velocity is the same for both projectiles. Which one travels farther? Justify your answer.
5. Suppose, given a fixed initial velocity, you want the projectile to travel as far as possible. At what angle should you fire the projectile? Justify your answer.
6. Suppose you have two projectiles. The first is shot at an angle of α, and the second is shot an an angle of β. Assume the initial velocity is the same for both projectiles. If you want the first projectile to travel the farthest, what do you know about the relationship between α and β?

PROJECTS 5.5 LOOKING OUT TO SEA

The Chicago-to-Mackinac sailboat race is held every summer on Lake Michigan. The residents along the shore of the lake have an opportunity to view the boats as they make their way from Chicago, on the southern end of the lake, to Mackinac Island, on the northern end. One year, the winds died down and stranded many of the boats. If you stood on the shore, you could see about 20 boats stalled out in the lake. If you climbed one of the tall bluffs near the shore, however, your vision was greatly enhanced and almost 100 boats came into view. Going just a little bit higher dramatically increased how far you can see. Exactly how much farther? That is the question answered in this project.

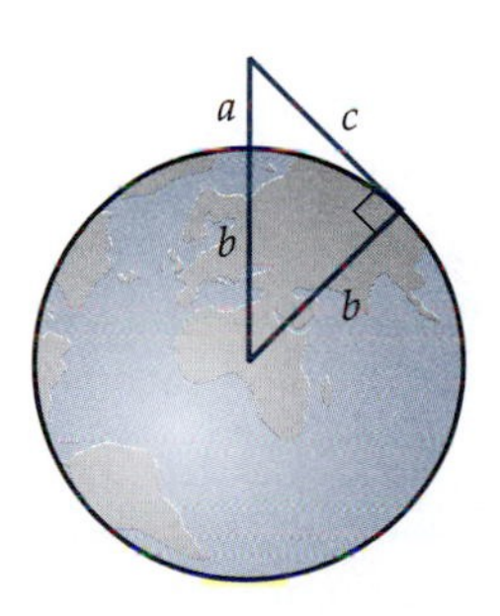

Figure 1

1. Study Figure 1. The circle represents the circumference of Earth. The line segment a is the distance one is above the surface of Earth. The line segments b are the radii of Earth, and c is the line of sight distance. Because c is tangent to the circle and b is a radius, the angle where they intersect is a right angle.
   (a) The radius of Earth is about 3950 mi. Use this fact to find a function whose input is a and whose output is c.
   (b) In the function found in part (a), a and c are in terms of miles, which is a bit cumbersome because we usually do not talk about being so many miles off the surface of Earth. Rewrite your function so that c, your output, is in terms of miles and a, your input, is in terms of feet.[22]

[22] 5280 ft = 1 mi *and* 5280^2 ft^2 = 1 mi^2.

2. Let's explore the properties of the function from question 1, part (b).
   (a) Complete Table 1. The first column is how high you are above Earth. The second column is your line of sight distance. The third column is the difference between your current and previous output divided by the difference between your current and previous input, that is, an approximate "slope" of the function at that point.

TABLE 1

Height (ft)	Distance (mi)	Difference in Output/Difference in Input (mi/ft)
5	2.7352	—
10		
15		
20		

   (b) Using the information in Table 1, could the graph of this function be a line? Why or why not?
   (c) Graph the function you found in question 1, part (b). What is the relationship between the general shape of the graph and the information given by the numbers in the third column of the chart in question 2, part (a)?
   (d) How high above Earth would you have to be to see 20 mi out?
3. The distance we have found so far is the distance from the observer to the farthest point in a straight line, that is, the line of sight distance. Often, however, when talking about how far one can see, people think of the distance along the horizon of Earth, that is, the ground distance. This distance is the arc length, s, in Figure 2. Find a formula to express this arc length in terms of a where the input is in miles. Also, find another version of this formula where the input is in feet.

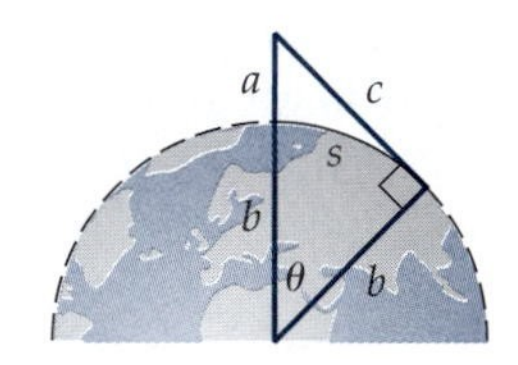

Figure 2

4. We want to compare the line of sight distance with the ground distance along the horizon.
   (a) Compare the ground distance to the line of sight distance when a is 10 ft, 1000 ft, 10,000 ft, and 200 mi by completing Table 2. What do you notice about the difference between the arc length and the line of sight distance as a gets larger? (*Note:* The formula commonly used for arc length assumes the angle is measured in *radians*, not degrees. Be sure to set your calculator appropriately.)
   (b) Using the functions where the input is in miles, graph both the line of sight distance and the ground distance on the same set of axes using a domain of 0 to 500 mi. Then graph using a domain of 0 to 50,000 mi.
   (c) We are interested in exploring what happens to both the line of sight distance and the ground distance as a gets larger. Some

TABLE 2

Height	Straight Line Distance (mi)	Arc Length (mi)
10 ft		
1000 ft		
10,000 ft		
200 mi		

functions increase without bound; that is, they keep getting bigger and bigger. Other functions have a limit or upper bound, that is, a limit as to how large they will get.

i. Think of the physical situation. (Refer to Figure 2.) The line of sight distance will increase without bound, and the ground distance will have a limit. Explain why. What is the limit for the ground distance?

ii. Look at the second graph for question 4, part (b). Explain how this graph is compatible with your answer to question 4, part (c)i.

iii. Look at the behavior of your symbolic formulas as a gets large. Explain how this behavior is compatible with your answer to question 4, part (c)i.

5. The space shuttle orbits approximately 200 mi above the surface of Earth. Is it possible for the someone in the space shuttle to view the entire contiguous United States? Explain your answer. Use the included map of the United States as a guide as shown in Figure 3.

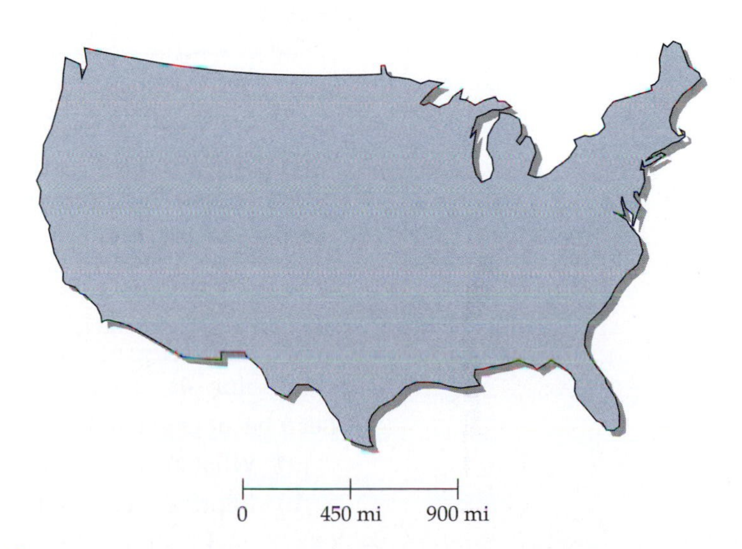

Figure 3

REVIEW EXERCISES

1. Convert the following angle measurements from degrees to radians.
   (a) 135°
   (b) 270°
   (c) 72°
   (d) −120°
2. Convert the following angle measurements from radians to degrees.
   (a) $\frac{5\pi}{4}$
   (b) $\frac{\pi}{9}$
   (c) $\frac{5\pi}{6}$
   (d) 2
3. Complete the following table by filling in the missing entries.

θ (radians)	θ (degrees)	$\cos\theta$	$\sin\theta$	$\tan\theta$
$\frac{3\pi}{4}$				
	60°			
		$\frac{1}{2}$	$-\frac{\sqrt{3}}{2}$	
			$-\frac{\sqrt{2}}{2}$	1

4. Let P be a point in the first quadrant such that the slope of the line connecting P to the origin is $\sqrt{2}$ and the distance from P to the origin is three units. Find the coordinates of point P.
5. Let θ be an angle in the first quadrant such that $\sin\theta = \frac{3}{5}$. Find the exact values of each of the following.
   (a) $\cos\theta$
   (b) $\tan\theta$
   (c) $\sin(-\theta)$
6. Let θ be an angle in the first quadrant such that $\cos\theta = c$.
   (a) What are the restrictions, if any, on the value of c?
   (b) Represent each of the following in terms of c.
      i. $\sin\theta$
      ii. $\tan\theta$

iii. $\cos(-\theta)$
iv. $\cos(\theta + 2\pi)$

7. Complete the following table by filling in the missing answers associated with a sector of a circle.

	Sector A	Sector B	Sector C
Length of the Radius	3	5	
Measure of the Central Angle	2		$\frac{\pi}{6}$
Length of the Intercepted Arc		30	
Area of the Sector			3π

8. The drawing of the baseball field shown consists of a sector and two triangles. Find the perimeter and area of the baseball field.

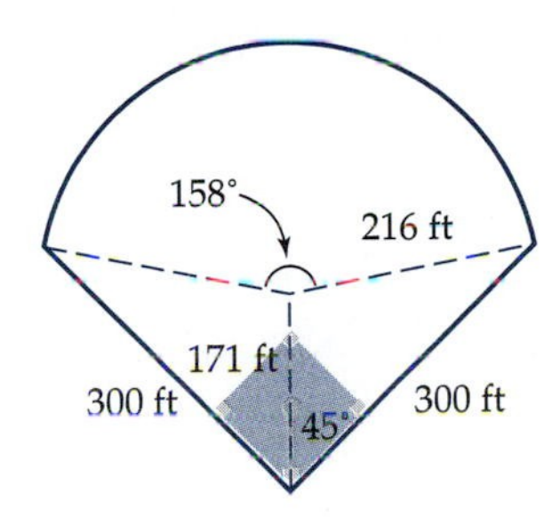

9. Find the area of the following triangles.

(a)

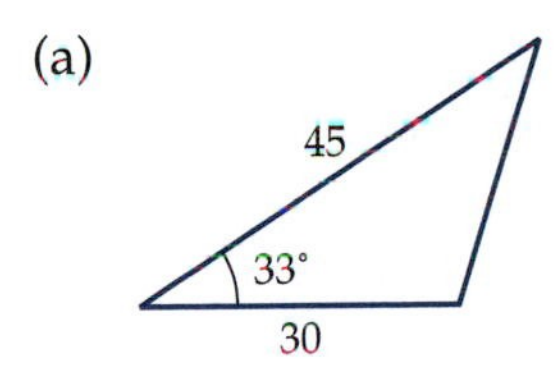

(b)

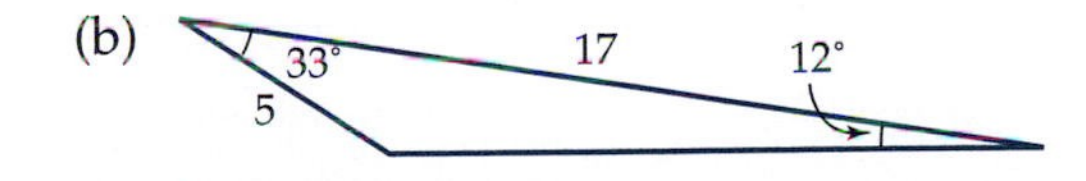

10. Two people are on a merry-go-round. Person A is sitting 2 ft from the center. How far from the center should person B sit if person B wants to go twice as fast as person A?
11. A bicycle wheel is 16 in. in diameter. It is revolving at 2 revolutions per second as the bicycle is traveling down a road.
    (a) What is the angular velocity of the wheel in radians per second?

(b) What is the linear velocity of a point on the outside edge of the wheel in inches per second?

(c) How fast is the bicycle traveling in miles per hour?

12. A lampshade is to be made from the portion of a sector that is shaded in the accompanying figure. What is the area of the lampshade?

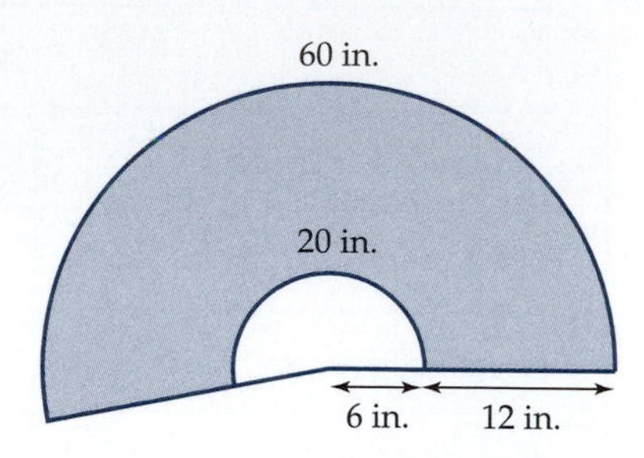

13. If the surface of a cone (not including the base) is opened up, it will form a sector. (See the accompanying figures.) The slant height of the cone is l, and the radius is r. We are interested in finding a formula for the surface area of a cone.

(a) In the figure showing the sector, explain why $\theta = 2\pi r/l$.

(b) Explain why the area of the sector is $A = \pi rl$.

(c) Explain why the surface area of the cone is $A = \pi rl + \pi r^2$.

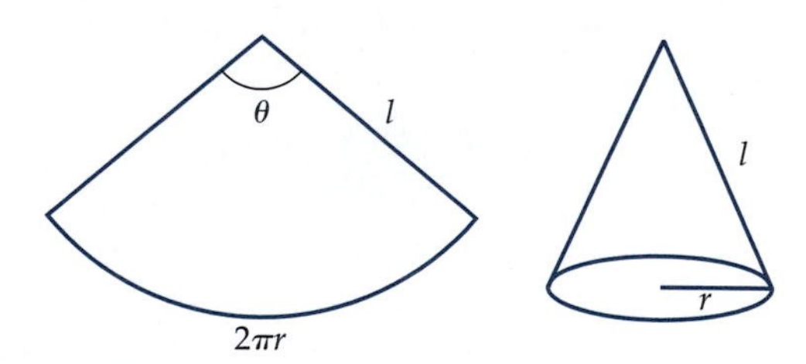

14. Give the amplitude, midline, and period for each of the following sine and cosine functions.

(a) $y = 5\sin(2x) - 7$

(b) $y = -2\cos(3x - 6) + 3$

(c)

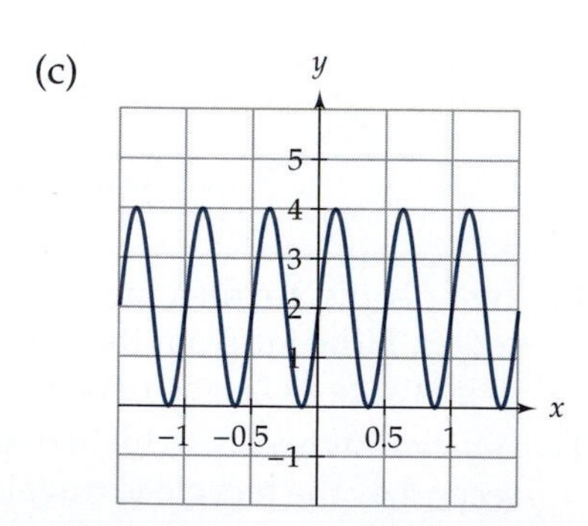

(d)

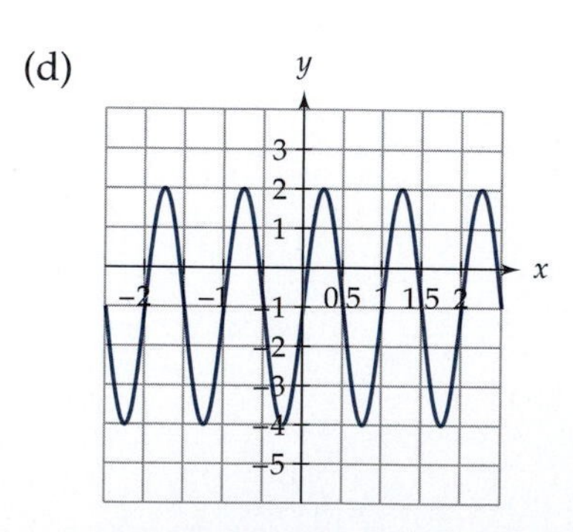

15. The following graphs are functions of the form $y = A\sin[B(x + C)] + D$. Find A, B, C, and D.

(a)

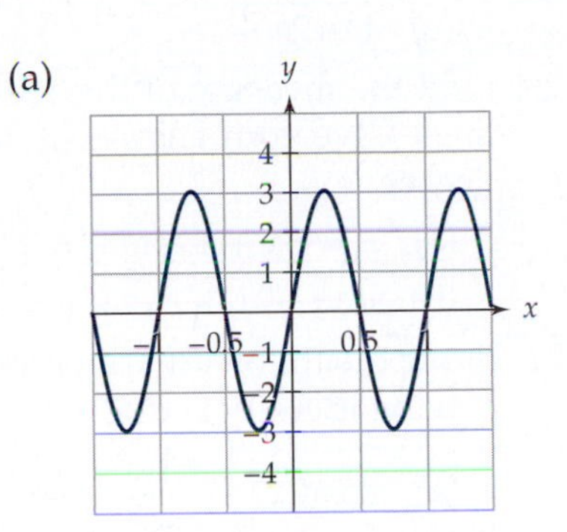

(b)

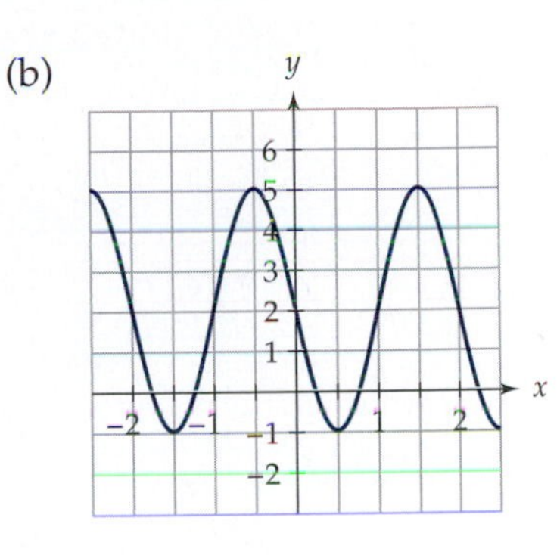

16. The following graphs are functions of the form $y = A\cos[B(x + C)] + D$. Find A, B, C, and D.

(a)

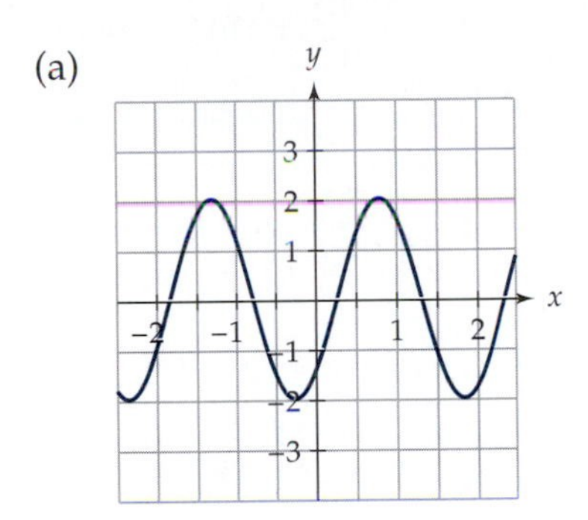

(b)

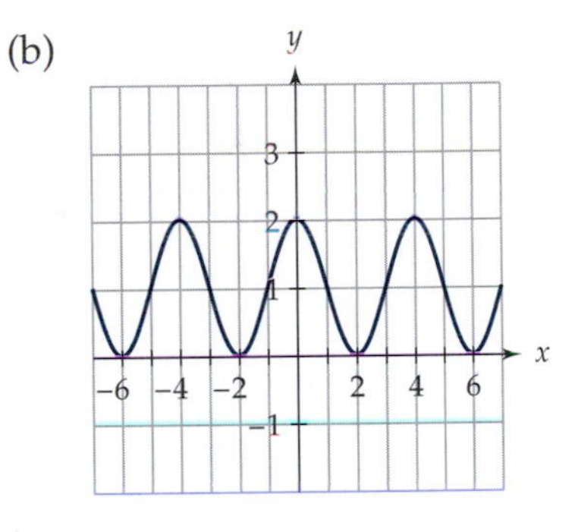

17. Find one solution for each of the following equations. Give exact answers if possible.
    (a) $1 = 2\cos(\pi x)$
    (b) $\sqrt{2} = 2\sin[2(x - 3)]$
    (c) $0 = 5\cos(x - \pi)$
18. The London Eye, the world's tallest Ferris wheel, is located along the bank of the River Thames. The wheel has a diameter of 450 ft. There are 32 capsules, each capable of holding 25 people. One revolution of the wheel takes 30 min.[23] Assuming a point on the rim of the wheel starts at ground level, write a symbolic representation for the height of that point where time is the input (in minutes) and height above the ground is the output (in feet).
19. Let a be an angle in the first quadrant such that $\cos a = \frac{1}{3}$. Find each of the following.
    (a) $\sin a$
    (b) $\tan a$

[23] *British Airways,* London Eye—Statistics.

(c) $\sin 2a$
(d) $\cos 2a$
(e) $\tan 2a$

20. Find the measure of the acute angle formed by the following pairs of lines. Give your answer in degrees and round to the nearest tenth of a degree.
    (a) $y = 3x + 1$ and $y = \frac{1}{3}x + 4$
    (b) $y = -2x - 7$ and $y = x + 3$

21. Prove the following trigonometric identities. Indicate for which values of x the identity is true.
    (a) $\cos(2x) + \dfrac{\sin x}{\csc x} + \sin^2 x = 1$
    (b) $\dfrac{\sin x}{\csc x - \cot x} = 1 + \cos x$
    (c) $1 + \sin(2x) = (\sin x + \cos x)^2$
    (d) $\dfrac{\cos x}{\tan x} + \sin x = \csc x$

CHAPTER OUTLINE

CHAPTER OVERVIEW

- Proofs of the Pythagorean theorem
- Using right triangle definitions to solve two- and three-dimensional applications
- The law of sines and the law of cosines
- Using the sine and cosine functions to model periodic behavior
- Sums of sine and cosine functions
- Combining periodic functions with functions that are not periodic

APPLICATIONS OF TRIGONOMETRIC FUNCTIONS

Using trigonometric functions to model situations in the world is the focus of this chapter. We begin by looking at problems involving triangles. We then combine the periodic functions of sine and cosine with similar functions as well as with nonperiod functions, which allows us to expand the areas in which we can use trigonometric functions to model applications.

6.1 RIGHT TRIANGLE APPLICATIONS

As mentioned in section 5.1, the right triangle approach to trigonometry is much older than the unit circle approach. In fact, the word *trigonometry* comes from the Greek words *trigon*, meaning triangle, and *metron*, meaning measure. Therefore, trigonometry literally means "the measuring of triangles." Right triangle applications of trigonometry are numerous. In this section, we derive the Pythagorean theorem and use it—as well as trigonometric functions—to solve right triangle applications.

THE PYTHAGOREAN THEOREM

One of the most well known theorems in all mathematics is the Pythagorean theorem. This theorem states that the square of the length of the hypotenuse of a right triangle is equal to the sum of the squares of the lengths of the two legs. Figure 1 illustrates the Pythagorean theorem. In this figure, the area of the square on the hypotenuse is equal to the sum of the areas of the other two squares. The Pythagorean theorem is often used to find the third side of a right triangle when given the lengths of two sides.

The Pythagorean Theorem has many interesting proofs.[1] Many of these proofs are based on the square of a length being thought of as an area. For example, the two squares shown in Figure 2 each have sides of length $(a + b)$. Square *LEFT* has been subdivided into four right triangles and a square. Square *RITE* has been subdivided into four right triangles and two squares. Square *LEFT* has area

$$4 \cdot \tfrac{1}{2}ab + c^2.$$

[1] *In 1927, Elisha Scott Loomis published* The Pythagorean Proposition, *in which he describes 370 different proofs for this famous theorem.*

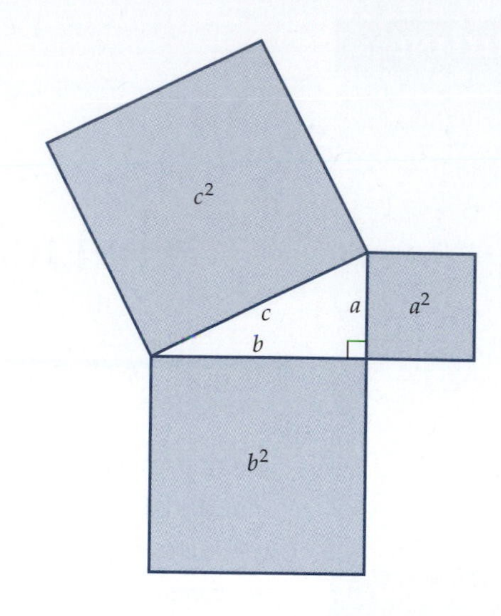

Figure 1 Using this figure, the Pythagorean theorem can be stated as $a^2 + b^2 = c^2$.

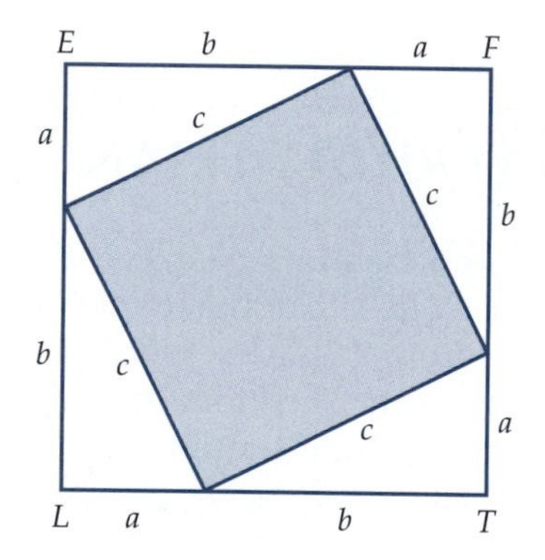

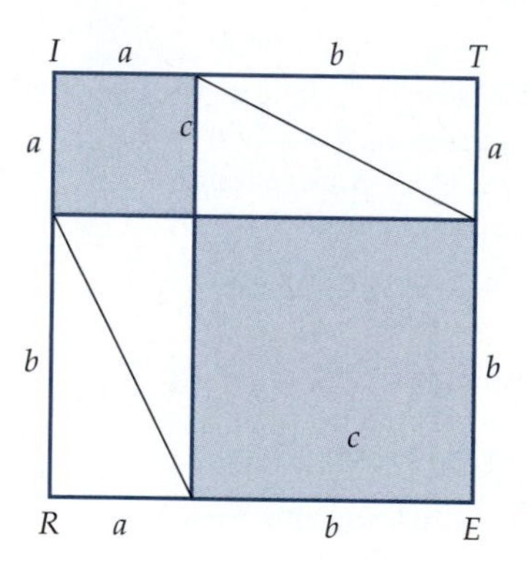

Figure 2 These two figures can be used to show that $c^2 = a^2 + b^2$.

Square *RITE* has area

$$4 \cdot \tfrac{1}{2}ab + a^2 + b^2.$$

Because square *LEFT* and square *RITE* have the same area, we know that

$$4 \cdot \tfrac{1}{2}ab + c^2 = 4 \cdot \tfrac{1}{2}ab + a^2 + b^2$$
$$c^2 = a^2 + b^2.$$

This result is a proof of the Pythagorean theorem because c is the length of the hypotenuse and a and b are the lengths of the legs of the right triangle. Other geometric proofs of the Pythagorean theorem are included in the exercises.

If three positive integers represent the sides of a right triangle, then they are known as a **Pythagorean triple.** Pythagorean triples are the set of positive integer solutions to the equation $a^2 + b^2 = c^2$. A *coprime Pythagorean triple* is a Pythagorean triple that has no common divisor other than 1. The smallest Pythagorean triple is $\{3, 4, 5\}$, which is also a coprime Pythagorean triple because the only positive integer that divides 3, 4, and 5 is 1. One way of generating Pythagorean triples is to use the set of equations

$$a = m^2 - n^2$$
$$b = 2mn$$
$$c = m^2 + n^2,$$

where m and n are positive integers such that $m > n$. It can be verified that these equations work by substituting them into $a^2 + b^2 = c^2$. Using these equations allows us to find all possible coprime Pythagorean triples.[2] For example, if $m = 2$ and $n = 1$, then we find that $a = 3$, $b = 4$, and $c = 5$. Because any multiple of a Pythagorean triple is again a Pythagorean triple, using these equations to find all the coprime Pythagorean triples allows us to find every Pythagorean triple. For example, because we found the triple $\{3, 4, 5\}$, we also know that $\{6, 8, 10\}$ and $\{9, 12, 15\}$ are Pythagorean triples.

There are an infinite number of positive integer solutions for $a^2 + b^2 = c^2$. This is not true for $a^n + b^n = c^n$, where n is an integer larger than 2. Pierre de Fermat (1601–1665) proved that $a^3 + b^3 = c^3$ has no positive integer solutions and wrote in the margin of a book that he had a proof showing that there were no solutions for $n > 3$. He stated, however, that his proof was too large to fit in the margin, and he never wrote it elsewhere. The theorem that no positive integer solutions exist for $a^n + b^n = c^n$, where n is an integer larger than 2, became know as Fermat's last theorem. For over 300 years, this theorem was one of the most famous unsolved problems in mathematics. In 1993, Andrew Wiles, a mathematics professor at Princeton University, announced that he had proved Fermat's last theorem. Although his original proof contained some flaws, these flaws have been fixed, and it is now generally accepted that Wiles's paper, more than 100 pages in length, proves Fermat's last theorem.

HISTORICAL NOTE

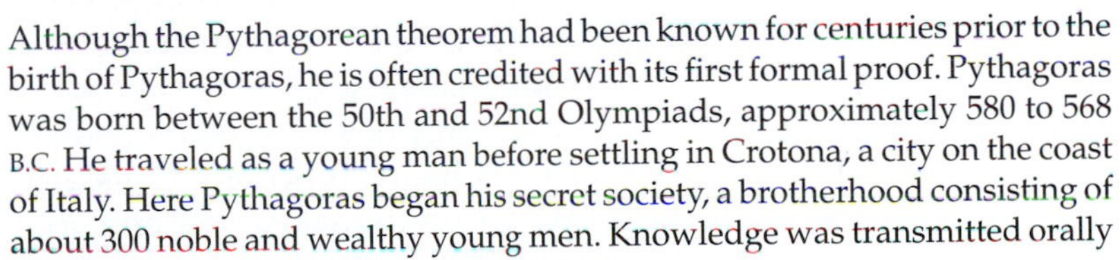

Although the Pythagorean theorem had been known for centuries prior to the birth of Pythagoras, he is often credited with its first formal proof. Pythagoras was born between the 50th and 52nd Olympiads, approximately 580 to 568 B.C. He traveled as a young man before settling in Crotona, a city on the coast of Italy. Here Pythagoras began his secret society, a brotherhood consisting of about 300 noble and wealthy young men. Knowledge was transmitted orally because writing materials were quite scarce.

The Pythagorean theorem also appears in ancient Chinese documents. Establishing the original date of these documents is difficult because copies were made over the course of hundreds of years. It is also difficult to date ancient Chinese documents because, in 213 B.C., Emperor Shi Huang-ti ordered all books burned. The oldest existing classic Chinese mathematics text is the *Chou-Pei*. The date of this book is estimated to be somewhere between 1200 B.C. and 200 B.C. The person or people responsible for writing this text were familiar with what we now call the Pythagorean theorem and had a geometrical argument to justify it. The difficulty in dating this text, however, makes it unclear whether the Chinese had a proof before or after the Greeks. The *Chou-Pei* contains the figure on the following page, which can be

continued on page 448

[2] *This statement is a result of the derivation of the three formulas, something you are asked to do in Investigation 3.*

used to prove the Pythagorean theorem. In the exercises, you are asked to use this figure to prove the Pythagorean theorem.

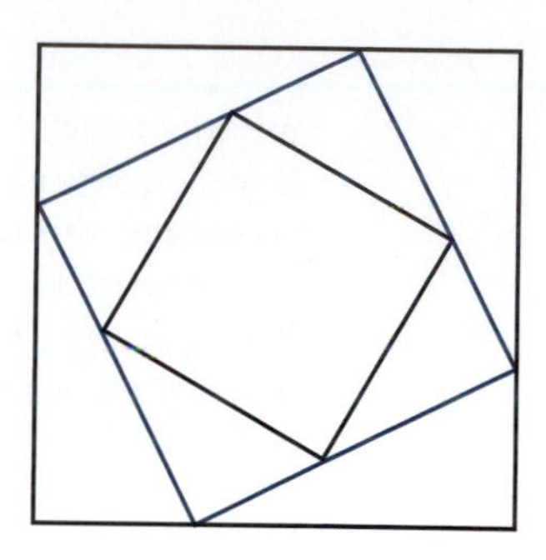

Sources: D. E. Smith, *History of Mathematics, Vol. I* (New York: Dover, 1951), pp. 29–30, 69–77; Carl B. Boyer, *A History of Mathematics* (Princeton, NJ: Princeton University Press, 1968), pp. 217–18.

READING QUESTIONS

1. Find the length of the hypotenuse of a right triangle given that the legs have lengths 16 and 30.
2. Use $a = m^2 - n^2$, $b = 2mn$, and $c = m^2 + n^2$ to find a Pythagorean triple where $m = 20$ and $n = 9$.

RIGHT TRIANGLE TRIGONOMETRY

Simple Right Triangle Applications

Although the Pythagorean theorem is useful for finding the third side of a right triangle when given two sides, it is not helpful if we have information about only one of the sides and an acute angle. In these cases, we can use trigonometric functions. Consider, for example, $\triangle ABC$ shown in Figure 3. We are given that $AC = 2$ and $m\angle B = 60°$. Let's say that we are interested in finding x, the length of $\overline{BC}$. In this case, we cannot use the Pythagorean theorem because we do not have the lengths of two sides. We do know, however, that $\sin B = $ opposite/hypotenuse. So, we have

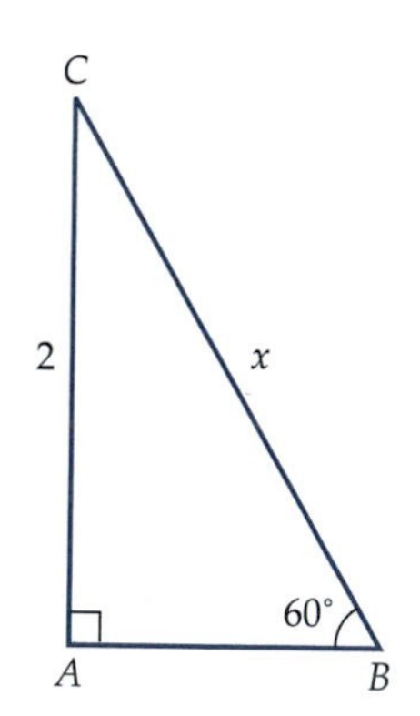

Figure 3 We can use the sine function to find x.

$$\sin 60° = \frac{2}{x}$$
$$\frac{\sqrt{3}}{2} = \frac{2}{x}$$
$$x = 2 \cdot \frac{2}{\sqrt{3}}$$
$$x = \frac{4}{\sqrt{3}}.$$

The length of $\overline{BC}$ is $4/\sqrt{3} \approx 2.3$ units. If we want the length of $\overline{AB}$ instead, we use the tangent function. Using the sine, cosine, and tangent functions as well as the Pythagorean theorem allows us to find missing lengths or angles of a right triangle when we are given the measurement of any two sides or one side and an acute angle.

One common application of right triangle trigonometry is finding heights of objects. It is relatively easy to measure a shadow because it is on the ground. It is also possible to find the angle of inclination using a clinometer.[3] Given these two pieces of information, it is possible to find the height of the object, as illustrated in Example 1.

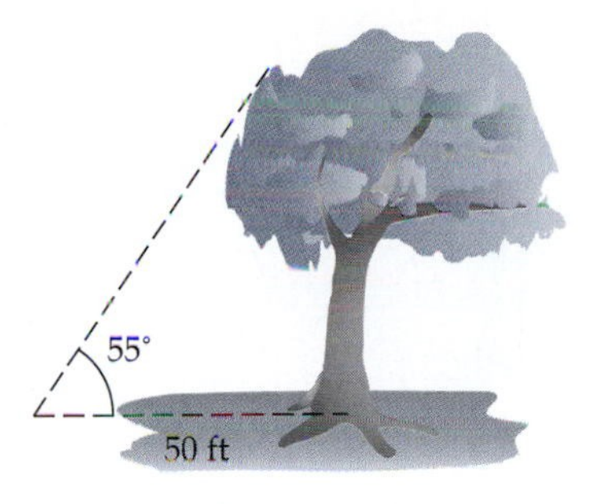

Figure 4

Example 1 There is a large oak tree outside the office where the authors are working. We found that the top of this tree was 55° above horizontal from a spot 50 ft away from the base of the tree. (See Figure 4.) Estimate the height of the tree.

Solution Let t represent the height of the tree. Then $\tan 55° = t/50$, which implies that $t = 50 \tan 55° \approx 71$. So, the height of the tree is about 71 ft.

HISTORICAL NOTE

Issues such as determining the location of objects, the placement of property lines, and the height of things have long interested people. Many ancient cultures, such as the Babylonians and the Chinese, developed methods for surveying land. As early as 1400 B.C., the Egyptians used surveying techniques to divide land into plots so that they could levy taxes. The ancient Greeks are credited with standardizing these methods. These ancient surveyors generally measured distances with ropes or wooden rods. Right angles were laid off by using what is now known as a carpenter's square or by using the 3-4-5 Pythagorean relationship applied to a stretched cord.

In addition to measuring distances, ancient civilizations also developed a variety of instruments for measuring angles. Examples are the astrolabe, the quadrant, and the planisphere. These instruments are all similar to what we now call a clinometer. To use a clinometer such as the one shown here, an object is sighted through holes in the two projections on the upper right edge. The angle of elevation is then indicated by the plumb line on the arc.

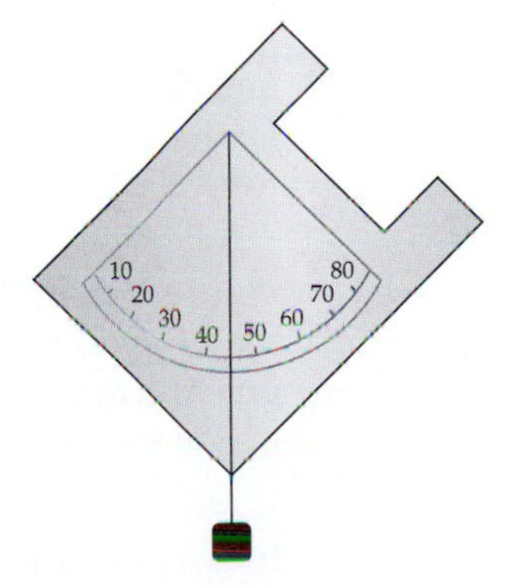

continued on page 450

[3] *A clinometer is a tool that can be made with a protractor, a string, and a weight.*

Using a speculum is an interesting method of finding the height of an object. This device involves a horizontal mirror laid on the ground. Through the use of similar triangles, the height of an object is calculated, as demonstrated in the accompanying figure. This device was used extensively in the sixteenth and seventheenth centuries.

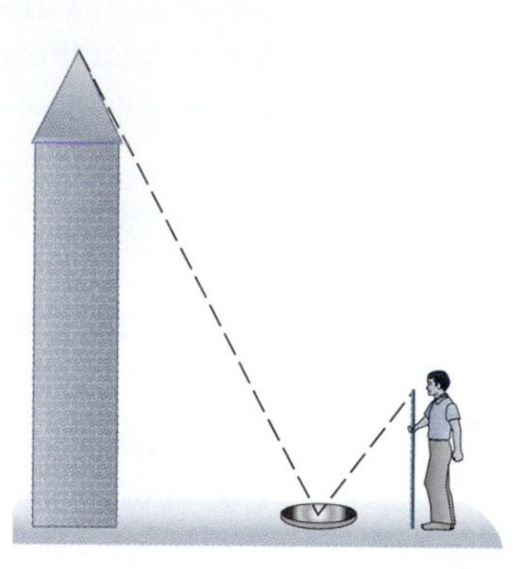

A transit is a common surveying tool used in our era. Surveyors out on the roads often look through a transit. The main function of a transit is to measure horizontal and vertical angles, but it can also be used for measuring distances. Transits consist of a telescope that can be swiveled through angles that are measured by scales on the transit. These devices are quite accurate and can measure angles to the nearest minute. Surveying equipment has also gone high tech with devices that use laser beams to measure distances and with global positioning systems to determine your position on Earth.

Sources: David Dux, *The History of Surveying* (visited 10 July 1998) ⟨http://pasture.ecn.purdue.edu/agen215/history.html⟩; Michael W. Smirnoff, *Measurements for Engineering and Other Surveys* (Englewood Cliffs, NJ: Prentice-Hall, 1961), p. 23; D. E. Smith, *History of Mathematics, Vol. II* (New York: Dover, 1951), pp. 344–63.

Estimating a Value for π

Trigonometry can be used in places where you might not expect it. For example, trigonometry can help us approximate the value of π. The number π, defined as the ratio of the circumference of a circle to its diameter, has intrigued and fascinated people for centuries. Over the years, closer and closer approximations for π have been calculated. Currently, computers have calculated π to 51,539,607,552 digits.[4] We use the circumference formula, $C = 2\pi r$, and inscribed polygons to estimate the value of π.

We choose a circle of radius $\frac{1}{2}$ because its circumference is exactly π. Figure 5 shows a regular octagon inscribed in this circle. The perimeter of the octagon is close to the perimeter of the circle, so finding its perimeter gives an approximation for π. Notice that because the octagon is inscribed inside the circle, its perimeter is less than π.

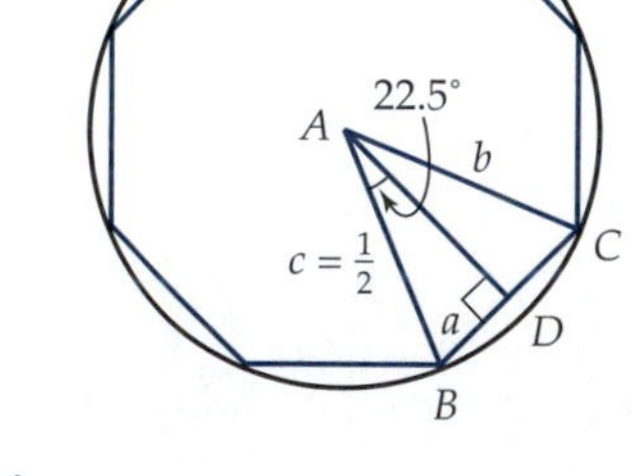

Figure 5 The perimeter of an octagon inscribed in a circle of radius $\frac{1}{2}$ can be used to approximate π.

[4] *Guinness Media, Inc.,* The Guinness Book of World Records 1998 *(Stanford, CT: Guinness Publishing, 1997), p. 68.*

Because this octagon is a regular octagon, $m\angle BAC$ must be one-eighth of 360°. So, $m\angle BAC = 360°/8 = 45°$. The sides b and c are both radii of the circle, so their length is $\frac{1}{2}$. Because $\triangle ABC$ is isosceles, $\overline{AD}$ is a perpendicular bisector of $\overline{BC}$ and a bisector of $\angle BAC$. Thus, in the right triangle $\triangle ABD$, the measure of $\angle BAD = 45°/2 = 22.5°$. Now that we know one side and one acute angle, we can use trigonometric functions to find the other sides. We are interested in the perimeter of the octagon, so we are most interested in finding the length of $\overline{BD}$. Let a represent this length. Using the sine function, we have

$$\sin 22.5° = \frac{a}{1/2}$$

$$\frac{1}{2}\sin 22.5° = a.$$

Because $\triangle ABC$ is isosceles, $BC = 2a$. Therefore, the perimeter of the octagon is $8 \cdot 2a = 8 \cdot 2(\frac{1}{2}\sin 22.5°) = 8\sin 22.5° \approx 3.0615$. This result is a reasonable underestimate of π, which is approximately 3.1416.

As we increase the number of sides of the inscribed polygon, it looks more like a circle and the perimeter more closely approximates the value for π. This time, we find the perimeter of a regular polygon with n sides inscribed in a circle of radius $\frac{1}{2}$. To find a in terms of n, we use the same procedure as with the octagon except that $m\angle BAC = 360°/n$ instead of 45°. (See Figure 6.) This time, we have

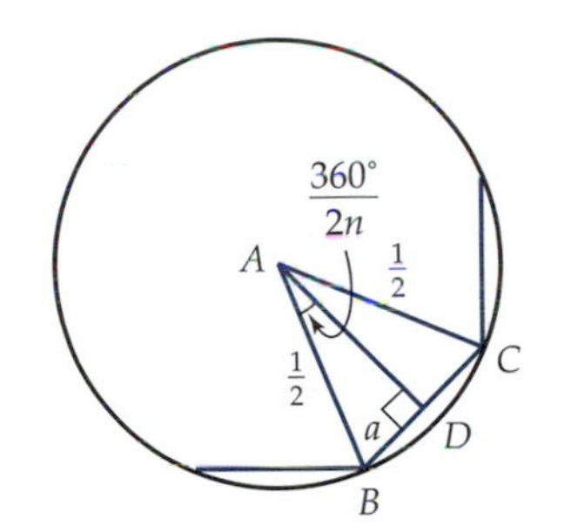

Figure 6 The perimeter of a regular polygon with n sides inscribed in a circle of radius $\frac{1}{2}$ can be used to approximate π.

$$\sin\left(\frac{360°}{2n}\right) = \frac{a}{1/2}$$

$$\frac{1}{2}\sin\left(\frac{180°}{n}\right) = a.$$

Therefore the perimeter, $p(n)$, of the n-sided polygon is $p(n) = n \cdot 2a = n \cdot 2 \cdot \frac{1}{2}\sin(180°/n) = n\sin(180°/n)$.

Because this polygon is inscribed in the circle, the perimeter is always less than the circumference of the circle. As the number of sides of the polygon increases, however, the perimeter gets closer and closer to the circumference of the circle, which is π. Table 1 gives the perimeter for various polygons where n is the number of sides and $p(n)$ is the perimeter. Answers are rounded to three decimal places. You can see from Table 1 that $p(n)$ is getting closer to π as n increases, which can also be seen in the graph of $y = p(n)$. (See Figure 7.) Notice that the values of $p(n)$ are always less than π.

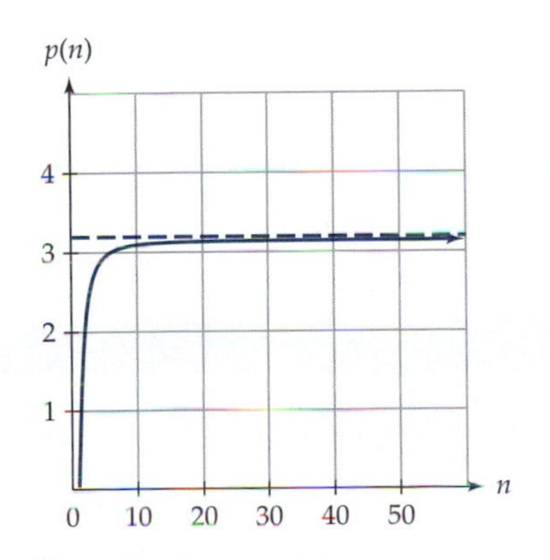

Figure 7 As n increases, $p(n) = n\sin(180°/n)$ approaches π.

TABLE 1 As the Number of Sides of a Regular Polygon Inscribed in a Circle of Radius $\frac{1}{2}$ Increases, its Perimeter Approaches π

n	10	20	30	40	50
$p(n) = n\sin\left(\frac{180°}{n}\right)$	3.090	3.129	3.136	3.138	3.140

Three-Dimensional Objects

Although solving problems where triangles can be drawn on a piece of paper is useful, we live in a three-dimensional world. Solving three-dimensional problems using right triangles, the Pythagorean theorem, and trigonometry often takes multiple steps and some creativity. Let's look at an application involving sports. When kicking a field goal in football, the distance from the uprights affects the difficulty of the kick not only because the player must kick the ball further, but because the "angle of success"[5] is smaller. This situation is illustrated in Example 2.

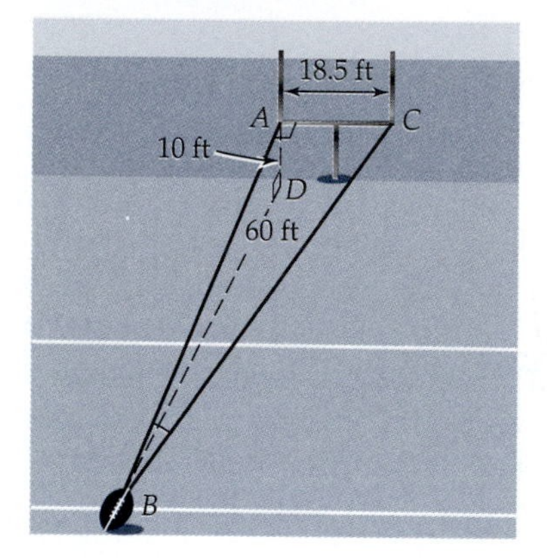

Figure 8

Example 2 Suppose you were to kick a field goal from the position shown in Figure 8. The ball is 60 ft away from the point on the ground directly below the left goal post. The goal posts are 18.5 ft apart, and the crossbar connecting them is 10 ft off the ground.[6] Determine the "angle of success," that is, the angle whose vertex is the ball's position and whose rays connect the ball to the two goal posts. Remember that the goal posts are 10 ft off the ground.

Solution We use the Pythagorean theorem with $\triangle ABD$ to find the length of $\overline{AB}$.

$$\begin{aligned} AB^2 &= 60^2 + 10^2 \\ AB^2 &= 3700 \\ AB &= \sqrt{3700}. \end{aligned}$$

We use the tangent function with $\triangle ABC$ to find the measure of $\angle ABC$.

$$\begin{aligned} \tan(m\angle ABC) &= \frac{18.5}{\sqrt{3700}} \\ m\angle ABC &= \tan^{-1}\left(\frac{18.5}{\sqrt{3700}}\right) \\ m\angle ABC &\approx 16.9°. \end{aligned}$$

Therefore, the football must be kicked within an angle of approximately 16.9° to be successful.

READING QUESTIONS

3. In $\triangle ABC$ (shown in the following figure), the length of side $\overline{AC}$ is 4 and the measure of $\angle B$ is 30°.

[5] *The "angle of success" is the angle formed by the ball's position on the ground and the two goal posts.*

[6] *These dimensions of the goal posts are used by the National Football League.*

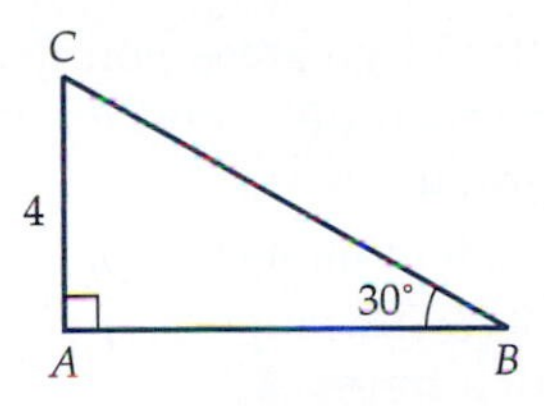

(a) What trigonometric function would you use to find the length of side $\overline{AB}$?

(b) Find the length of $\overline{AB}$.

4. Suppose you inscribe a regular polygon in a circle of radius $\frac{1}{2}$.

(a) As the number of sides of the polygon increases, the perimeter of the polygon approaches what value?

(b) Let a represent the perimeter of an inscribed 10-sided polygon and b represent the perimeter of an inscribed 20-sided polygon. Which is larger, a or b? Briefly justify your answer.

SUMMARY

A proof of the Pythagorean theorem, one of the most well known theorems in mathematics, was given in this section. The theorem can be stated as $a^2 + b^2 = c^2$, given that a and b represent the lengths of the legs of a right triangle and c represents the length of the hypotenuse.

If three positive integers represent the sides of a right triangle, then they are known as a **Pythagorean triple.** One way of generating a set of Pythagorean triples is to use the set of equations: $a = m^2 - n^2$, $b = 2mn$, and $c = m^2 + n^2$, where m and n are positive integers such that $m > n$.

We can use our right triangle definitions of trigonometric functions to find the missing measures of any right triangle. This method is especially useful in applications. By using these functions and a polygon inscribed inside a circle, we are able to estimate π. By using repeated applications of right triangles, we are able to solve problems involving three-dimensional figures.

EXERCISES

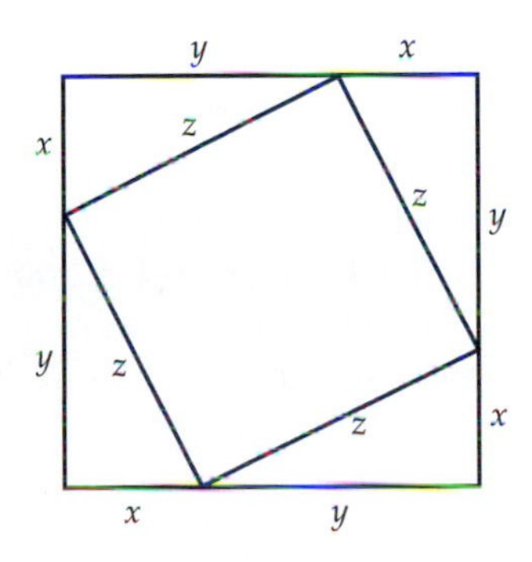

1. The accompanying figure composed of four congruent right triangles and a square is similar to that found in the *Chou Pei,* an ancient Chinese mathematics text.[7]

(a) Find the area of the larger square by using the lengths of the sides of the larger square.

(b) Find the area of the larger square by finding the sum of the areas of the four right triangles and the smaller square.

[7] *See the Historical Note in this section about the Pythagorean theorem.*

(c) Set the two areas you found in parts (a) and (b) equal to each other, simplify, and explain why this proves the Pythagorean theorem.

2. When President James Garfield was a member of the House of Representatives, he developed a proof of the Pythagorean theorem using a diagram of a trapezoid.

(a) Find the area of the accompanying trapezoid.

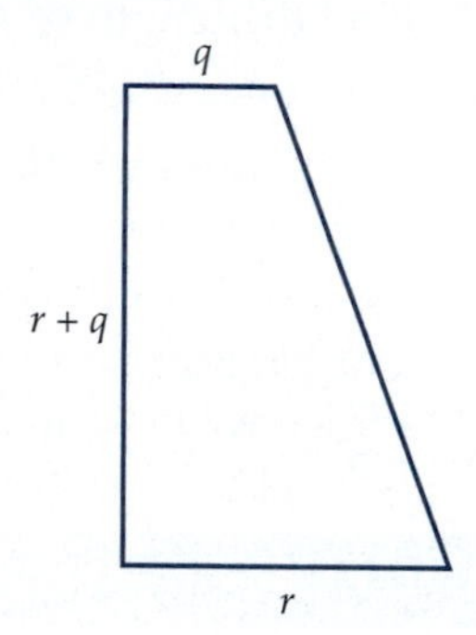

(b) Find the area of the same trapezoid by finding the sum of the areas of the three right triangles shown.

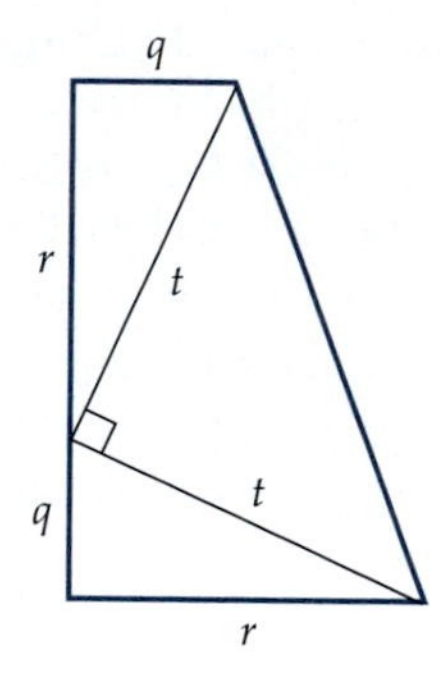

(c) Set the two areas you found in parts (a) and (b) equal to each other, simplify, and explain why this process proves the Pythagorean theorem.

3. The twelfth-century Hindu mathematician Bhaskara used the accompanying figure, consisting of four congruent triangles and a small square within a larger square, to prove the Pythagorean theorem. He just included one word as his proof: "Behold!"[8] We ask you to fill in the details of his proof.

[8] *Stephanie J. Morris,* The Pythagorean Theorem *(visited 31 March 2000) ⟨http://jwilson.coe.uga.edu/emt669/Student.Folders/Morris.Stephanie/EMT.669/Essay.1/Pythagorean.html⟩.*

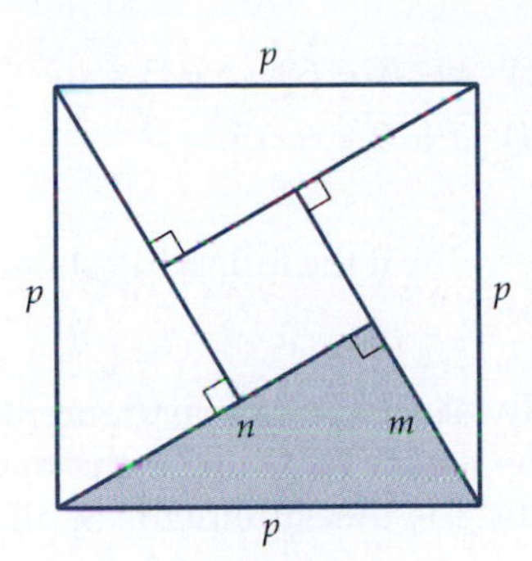

(a) Determine the length of the side of the interior square in terms of m and n.

(b) Find the area of the large square by finding the sum of the areas of the four right triangles and the smaller square.

(c) Set your area from part (b) equal to the area of the large square, p^2, simplify, and explain why this process proves the Pythagorean theorem.

4. Let $a = m^2 - n^2$, $b = 2mn$, and $c = m^2 + n^2$, where $m > n$. Show that $a^2 + b^2 = c^2$.

5. One way of generating Pythagorean triples is to use the equations $a = m^2 - n^2$, $b = 2mn$, and $c = m^2 + n^2$, where m and n are positive integers such that $m > n$.

   (a) If three numbers (such as 3, 4, and 5) form a Pythagorean triple, then multiples of these three numbers (such as 6, 8, and 10) also form a Pythagorean triple. Explain why this fact is true in general by assuming r, s, and t form a Pythagorean triple and showing that qr, qs, and qt, where q is a positive integer, also form a Pythagorean triple.

   (b) Find all the Pythagorean triples such that the length of the hypotenuse is less than or equal to 30. When doing so, do not forget to include appropriate multiples of Pythagorean triples found by using the three equations given earlier in this problem.

6. Given each of the following measurements for a right triangle whose right angle is $\angle C$, find the missing measurements for the remaining angles and sides. Where appropriate, round answers to one decimal place.

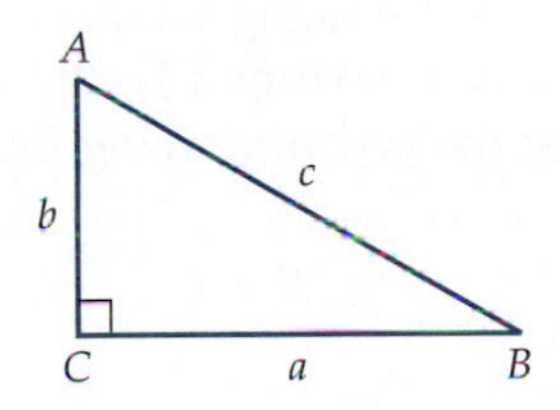

   (a) $a = 5, b = 12$

   (b) $m\angle B = 40°, b = 6$

(c) $m\angle B = 60°, c = 5$
(d) $a = 9, c = 15$

7. Determine if the following statement is true or false. Briefly justify your answer.

   Given any two measurements of a right triangle consisting of the length of a side or the measure of an acute angle, you can find the measurements of all three angles and all three sides.

8. Suppose you are standing in Liberty State Park in New Jersey at a spot 2000 ft from the Statue of Liberty. You look up at the torch of the statue, which is 305 ft above the ground. What is the angle of elevation from your eyes? (Assume your eyes are at ground level for the Statue of Liberty.)

9. Along large bodies of water, houses are often built on top of bluffs overlooking the water. One way to determine the height of the bluff is to run a string from a person at the top of the bluff to someone standing at the bottom and measure both the string and the angle of depression (indicated in the accompanying figure by θ). Express the height of the bluff in terms of the length of string, s, and θ.

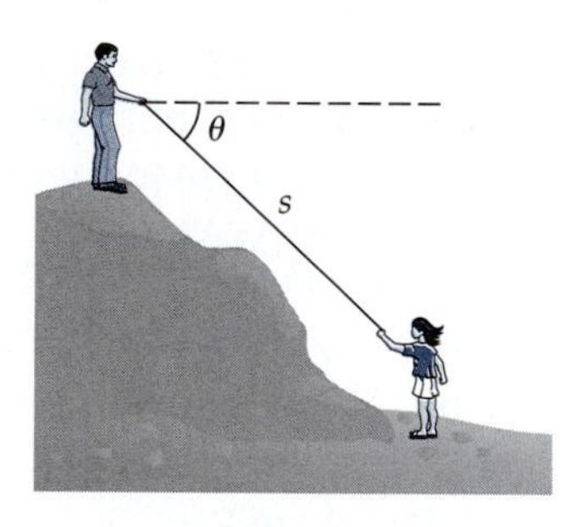

10. Eratosthenes estimated the circumference of Earth by knowing that, at noon on the summer solstice, the sun was 7.5° south of vertical.[9] At Hope College (in Holland, Michigan), we placed a 50-cm stick vertically in a south-facing window and found that, at solar noon[10] on the summer solstice (June 21), the stick cast a shadow that was 17.9 cm long. (See the accompanying figure.)

[9] *See section 5.2, Exercise 7.*

[10] *In western Michigan, solar noon, the time when the sun is as high in the sky as it gets on a given day, occurs at about 1:30 P.M. on June 21.*

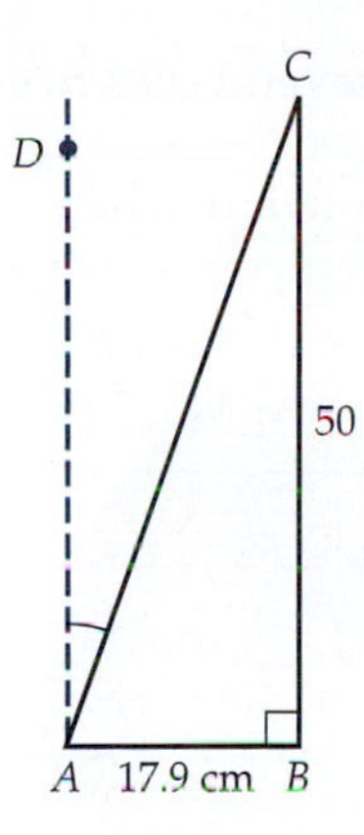

(a) Determine how far the sun is south of vertical to the nearest tenth of a degree. (This measurement is that of $\angle DAC$.)

(b) On the summer solstice, the sun is directly above the Tropic of Cancer, which is located at a latitude of approximately 23.5° north. Adding the measure of $\angle DAC$ (the amount that the sun is south of vertical on the summer solstice) to 23.5° gives the latitude of the location where the measurement was taken. What is the latitude of Holland, Michigan?

11. The measurement guidelines set by the South Dakota Register of Big Trees[11] says that one way of measuring the height of a tree is to do the following:

> Use a stick whose length is the same as the length of your arm. Hold the stick vertically at arm's length and walk backwards away from the tree until the stick above your hand appears to be the same length as the tree. Measure how far you have backed away from the tree. This measurement is approximately equal to the height of the tree.

Explain why this method works.

12. Assume building A and building B are 25 ft apart. By looking out the third-floor window of building A, and using a clinometer, you find that the angle of elevation to the top of the building is approximately 62°. You also measure the angle to the bottom of the building and find that it is approximately 43°. Estimate the height of building B.

13. As you are approaching a mountain in your car you notice a sign that states, "CAUTION: Steep Slope Ahead—10% Grade." The 10% grade means that the road has a slope of 0.10. Every time the horizontal distance changes by 10 ft, the vertical distance changes by 1 ft.

(a) As you start driving up the mountain, what is your angle of elevation?

(b) If, according to your odometer, you have driven up the mountain for 2 mi (10,560 ft), what is your increase in elevation?

[11] Division of Forestry: Big Tree Register *(visited 6 July 1998)* ⟨*http://people.enternet.com.au/wothersp/home/gtc/bigtree1.htm*⟩.

14. Your eyes use several cues to estimate the distance to an object. One of these is *convergence*, a muscular cue that indicates the extent to which the eyes turn inward when looking at an object.[12] Assume that the distance between your eyes is 7.1 cm. (See the accompanying figure.)

(a) While looking at a coffee cup, your eyes turn inward at an angle of 87°. Let AD represent the distance from you to the coffee cup. Is the cup within arm's reach (about 60 cm)?

(b) While sitting in class, you look up at the clock, which is about 7 m away. What is the angle of convergence of your eyes?

(c) If you look at your finger on the tip of your nose (about 5 cm away), what is the angle of convergence of your eyes?

15. While building a floor-to-ceiling bookcase, you carefully measure and discover that the distance from the floor to ceiling is 8 ft. In your workshop, you build a bookcase that is 8 ft high and 1 ft wide.

(a) When you bring the bookcase into the house, you have to tilt it on its side to get it through the door. When you try to stand it up, you discover that it will not fit. Why not? (See the accompanying figure.)

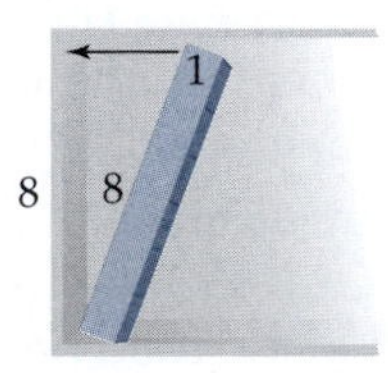

(b) What is the tallest bookcase that you can build in your workshop, bring into the house, and successfully stand up?

16. Suppose you decide to go tobogganing on a hill that is 60 ft high and whose angle of depression is 30°. The toboggan runs on an ice-covered track. (See the accompanying figure.)

[12] *David Myers,* Psychology, *2d ed. (New York: Worth Publishers, 1989), p. 174.*

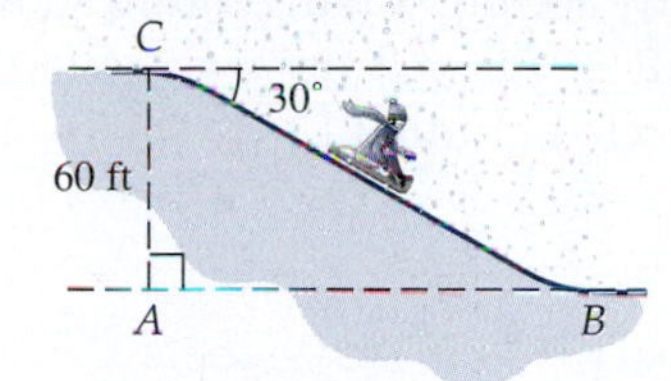

(a) How far is it from the top of the hill to the bottom of the hill (i.e., what is the length of $\overline{BC}$)?

(b) The equation

$$gh = \tfrac{1}{2}(v_f^2 - v_i^2), \qquad (1)$$

where g is the acceleration due to gravity, h is the height of the hill, v_f is the velocity when the toboggan reaches the bottom of the hill, and v_i is the velocity of the toboggan at the top of the hill, can be used to find the velocity of the toboggan. Assume $v_i = 0$ and use $g = 32$ ft/sec^2. Find the approximate velocity (in miles per hour) of the toboggan when it reaches the bottom of the hill.

(c) Notice that the equation used for finding the final velocity is dependent on the *height* of the hill but is independent of the steepness of the hill. Thus, if there are two toboggans, one on a hill with a very gentle slope and another on a hill of the same height with a very steep slope, then both toboggans have the same velocity when they reach the bottom of the hill (assuming the toboggans are on solid ice where friction can be ignored). The "thrill" of going down these two hills, however, is not the same because of the acceleration. We are interested in finding the acceleration of the toboggan on our hill, which is 60 ft high and has an angle of depression of 30°. Continue to assume $v_i = 0$ and use $g = 32$ ft/sec^2.

i. If a, the acceleration, is constant (as it is in this case), then $a = (v_f - v_i)/t$. Use equation (1) to find an expression for a in terms of h and t.

ii. Find an expression for d, the distance the toboggan travels, in terms of h, the height of the hill, and θ, the angle of depression.

iii. The average velocity, $\overline{v}$, is given by $\overline{v} = d/t$. In this case, because acceleration is constant and $v_i = 0$, we also know that $\overline{v} = v_f/2$. Use these two equations to find an expression for t in terms of h and θ.

iv. Using your answers to part (i) and part (iii), find an expression for a in terms of θ.

v. What is the acceleration of the toboggan on the hill whose angle of depression is 30°?

vi. What is the acceleration of the toboggan on the hill whose angle of depression is 10°?

17. The accompanying figure shows $\triangle ABC$, an isosceles triangle such that $\overline{AB} \cong \overline{BC}$. All three sides of the triangle are tangent to the inscribed circle of radius 1.

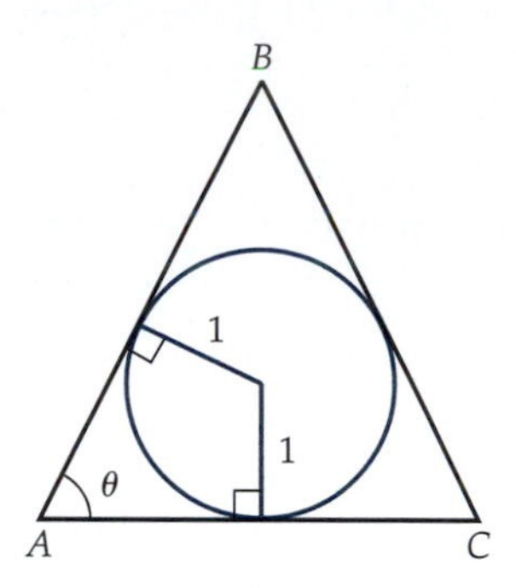

 (a) Express the area of the triangle in terms of θ.
 (b) Graph the area function from part (a) and use this graph to estimate the angle that gives the triangle with the smallest area.

18. One of the authors had a room added onto his house. During this process, the contractor needed to get sheets of drywall down to a bedroom but was concerned about getting them around the corner in a hallway. (See the accompanying figure.)

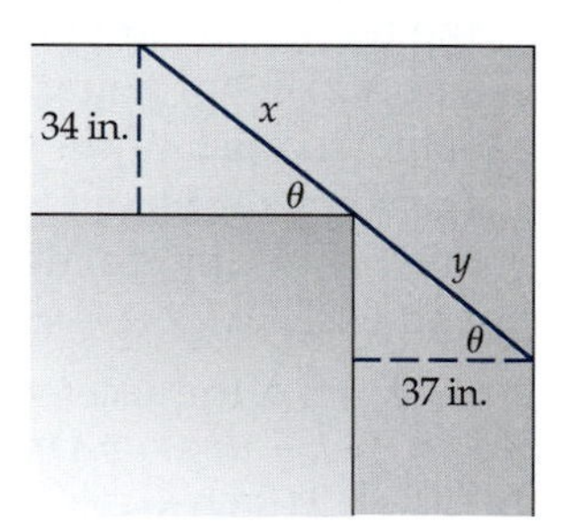

 (a) Explain why the two angles labeled θ are equal.
 (b) Find a function that gives the length, $x + y$, in terms of θ.
 (c) Graph your function from part (b). Drywall comes in sheets that are 8 ft and sheets that are 12 ft in length. Which size, if either, will fit around the corner?

19. In this section, we used a polygon inscribed in a circle to estimate π. In this case, the estimate was always an underestimate. Suppose, instead, you drew a *circumscribed* regular polygon around a circle of radius $\frac{1}{2}$ and used this polygon to estimate π. The accompanying figure shows one segment of a circumscribed regular n-gon.

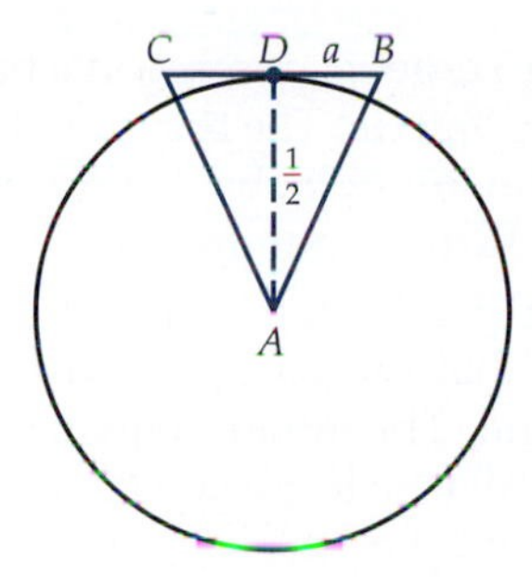

(a) Find the perimeter of a circumscribed regular octagon (eight-sided polygon).

(b) If r is the perimeter of a circumscribed regular 50-gon and s is the perimeter of a circumscribed regular 60-gon, which is larger, r or s? Justify your answer.

(c) Find the formula for the perimeter of a regular circumscribed n-gon.

(d) Use a circumscribed regular 100-gon to estimate π. Is this estimate an overestimate or an underestimate? Why?

20. The Great Pyramid in Egypt has a square base where each side is approximately 230.4 m in length. Its height is approximately 142 m. (See the accompanying figure.)

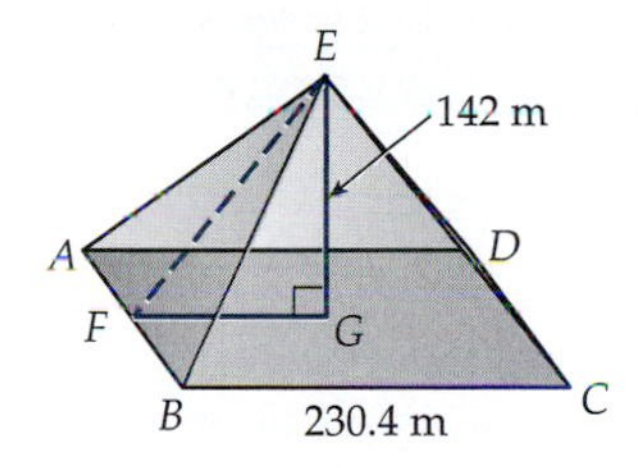

(a) It has been said that the ratio of the perimeter of the Great Pyramid to its height is approximately 2π. Do you agree or disagree with this statement? Briefly justify your answer.

(b) Find the angle formed by a side of the Great Pyramid and its base ($\angle EFG$).

(c) Find the slant height of the Great Pyramid (length of $\overline{EF}$).

21. An ancient Chinese text, *Chui-chang suan-shu*, or *Nine Chapters on the Mathematical Art*, includes many interesting problems. The last chapter includes problems involving right triangles, in which one is similar to the following. A square pond that is 10 ft on each side has a reed growing in the center and extending 1 ft above the water. When drawn to the edge of the pond, the reed just reaches the surface of the pond. Find the depth of the pond.[13]

[13] *Boyer,* A History of Mathematics, *p. 218.*

22. Aunt Veronica purchased a baseball bat as a Christmas present for her niece, Valerie. The box she plans to use to mail the Christmas presents is 22 in. by 22 in. by 20 in. The bat is $33\frac{1}{2}$ in. long. Will the bat fit in the box? Justify your answer.
23. The rules for the National Collegiate Athletic Association (NCAA)[14] state that the goal posts for football are $23\frac{1}{3}$ ft wide and 10 ft off the ground. The inbound line is 15 ft from the edge of the goal post. If the football is to be kicked 60 ft from the goal line, determine the "angle of success" (i.e., the angle whose vertex is the ball's position and whose rays connect the ball to the two goal posts). See the accompanying figure.

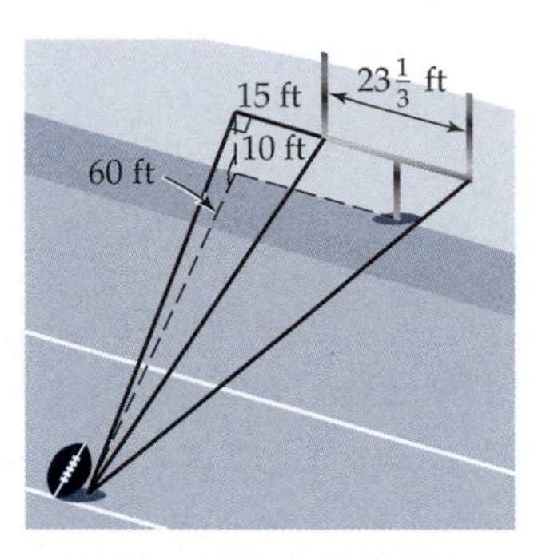

INVESTIGATIONS

INVESTIGATION 1: SORTING STRIPS OF VENEER

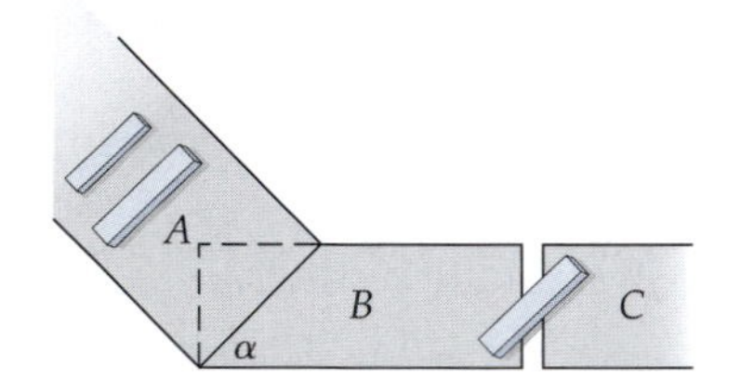

Figure 9

On the television show *Bob Vila's Home Again*, the host often takes his viewers on a tour of various factories that make items used in the construction of homes. One episode showed Bob touring a factory that made engineered beams, large wooden beams made by gluing strips of wood veneer together. These beams are used in construction because of their size and because they are stronger than a beam made from a solid piece of wood. The veneer comes to the manufacturer in large sheets. The manufacturer does not need to use perfect sheets of veneer because the wood is cut into narrow strips before it is used in the beam. Therefore, even a poor sheet of veneer can produce many usable strips. After the veneer has been dried and cut, strips that are too short to be used must be sorted out. On this episode of *Bob Vila's Home Again*, the method used by the manufacturer to sort the strips was a three-conveyor system similar to the one shown in Figure 9.

Assume the strips of wood have uniform thickness and do not overlap as they move down conveyor A. They move perpendicular to the sides of conveyor A and then drop onto conveyor B. The angle between conveyors A and B causes the strips to go down conveyor B at an angle α. The short strips fall through the gap between conveyors B and C.

[14] *The Diagram Group,* The Rule Book *(New York: St. Martin's, 1990), pp. 112–23.*

Assume the manufacturer can adjust both α and the gap between conveyors B and C. The manufacturer wants to eliminate all strips less than 16 in. long. Assume that strips must have at least 60% of their length off the conveyor before they fall through.[15]

1. Give the best angle, α, for each of the gap settings listed in the following table.

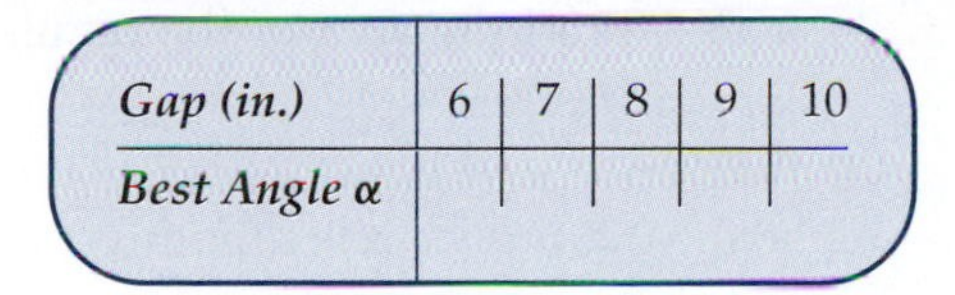

Gap (in.)	6	7	8	9	10
Best Angle α					

2. Determine a function in which the gap, g, is the input and the best angle, α, is the output. Graph your function using a domain of 6 to 10 in.
3. Consider what happens to the accuracy of the process as the gap and the angle are modified. Do you think one of the five gap/angle settings in Table 1 is more likely to do a better job of sorting the strips than the other settings? Justify your answer. (*Hint:* Use the numerical data from question 1 and the graph from question 2 to help in your analysis.)

INVESTIGATION 2: USING THE AREA OF A CIRCLE TO ESTIMATE π

In this section and in the exercises, we used the perimeter of inscribed or circumscribed polygons to approximate π. Another method of estimating π is to use the area of inscribed polygons because we know the area of a circle is πr^2.

1. This time, we use a circle of radius 1 rather than a circle of radius $\frac{1}{2}$. Explain why.
2. Start with an octagon inscribed inside a circle of radius one. (See the accompanying figure.)

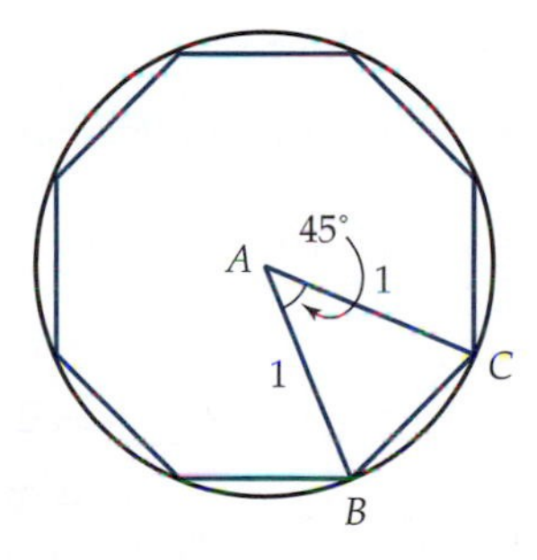

   (a) Will the area of the octagon be less than or greater than π? Briefly justify your answer.

[15] *The wood has some momentum as it is moving down the conveyor and therefore must have more than 50% of its length off the conveyor before it falls through.*

(b) Using the formula $A = \frac{1}{2}ab\sin\theta$,[16] find the area of $\triangle ABC$.

(c) What is the area of the octagon?

3. As the number of sides of the inscribed regular polygon increases, its area becomes closer to π. This time, use a regular n-gon instead of an octagon.
   (a) Find the area of one triangle in terms of n, the number of sides of a regular polygon.
   (b) Find the area of the regular polygon (a lower bound for π) in terms of n.
4. Find an upper bound for π using the area of a regular n-gon circumscribed about a circle of radius one. (See the accompanying figure.)

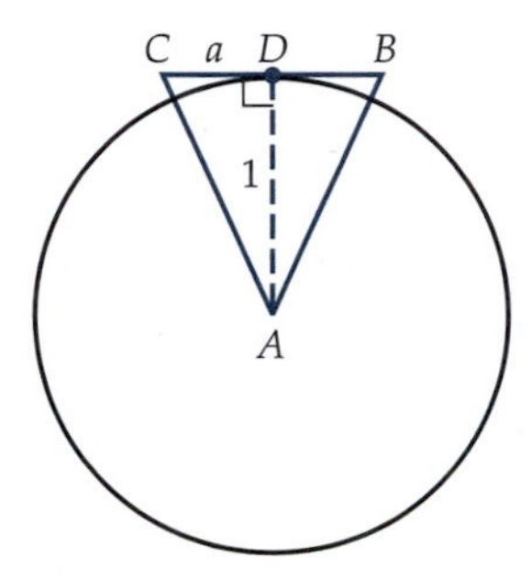

5. Graph your function from question 3, part (b) and your function from question 4 on the same set of axes using a domain of $3 \le n \le 60$ and a range of $1 \le y \le 5$. Estimate the smallest value of n where the two graphs are so close that they appear to touch.
6. Find the smallest positive integer n such that the lower bound estimate for π and the upper bound estimate for π agree when rounded to four decimal places.

INVESTIGATION 3: DERIVING THE EQUATIONS FOR GENERATING PYTHAGOREAN TRIPLES

In this section, we defined a Pythagorean triple as a set of three positive integers that satisfy $a^2 + b^2 = c^2$. We also mentioned that one way of finding all coprime triples is to use the equations

$$a = m^2 - n^2$$
$$b = 2mn$$
$$c = m^2 + n^2,$$

where m and n are positive integers such that $m > n$. Recall that coprime Pythagorean triples have a greatest common factor of 1. For example, 3, 4, and 5 are coprime, whereas 6, 8, and 10 are not because they have a

[16] *This formula was derived in section 5.2; θ is the angle between the sides whose lengths are a and b.*

common factor of 2. Once you find a coprime Pythagorean triple, you can find other triples by multiplying by a common factor.[17]

In this investigation, we show that if $\{a, b, c\}$ forms a Pythagorean triple, then there exist numbers m and n such that $a = m^2 - n^2$, $b = 2mn$, and $c = m^2 + n^2$. If $\{a, b, c\}$ forms a coprime Pythagorean triple, then we can assume that m and n are positive integers such that $m > n$.

Throughout this investigation, we assume that $a^2 + b^2 = c^2$.

1. Often, when writing a mathematical proof, we try to rephrase the problem in a simpler context. We do so by changing to a problem involving a unit circle centered at the origin. Now we not only have a simpler context for our question, but we can also use both geometry and algebra. Mathematical reasoning is often strengthened if we can exploit the interplay between algebra and geometry.
   (a) What is the equation for the unit circle centered at the origin? (*Note:* A *unit circle* has a radius of 1.)
   (b) Let $x = a/c$ and $y = b/c$. Because $a^2 + b^2 = c^2$, explain why (x, y) is a point on the unit circle. (See the accompanying figure.)

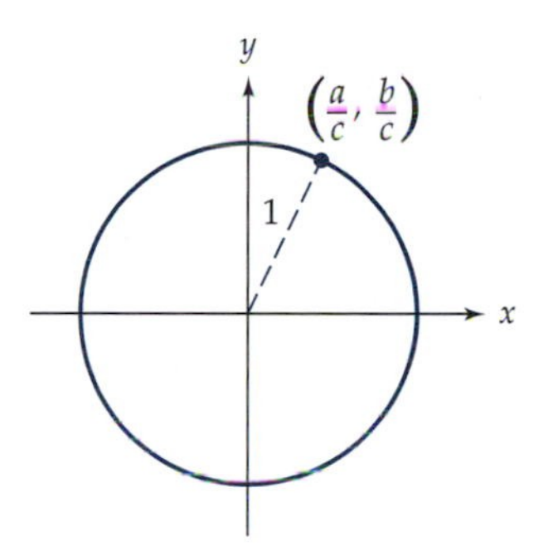

2. Introducing the unit circle makes the problem more concrete and allows us to use geometry. To make the problem even simpler, however, we want to find a way to describe a generic point on the unit circle with one variable, t, instead of two variables, x and y.[18] One way to do so (which you have seen before) is to let $x = \cos\theta$ and $y = \sin\theta$. This time, however, we want to use another method.
   (a) Consider the line that connects the point $(-1, 0)$ to a point on the unit circle, (x, y). Explain why the y-intercept of this line is equal to the slope of this line. Call this number t. (See the figure on the next page.)

[17] *See Exercise 5.*

[18] *This process is known as* parameterizing a curve.

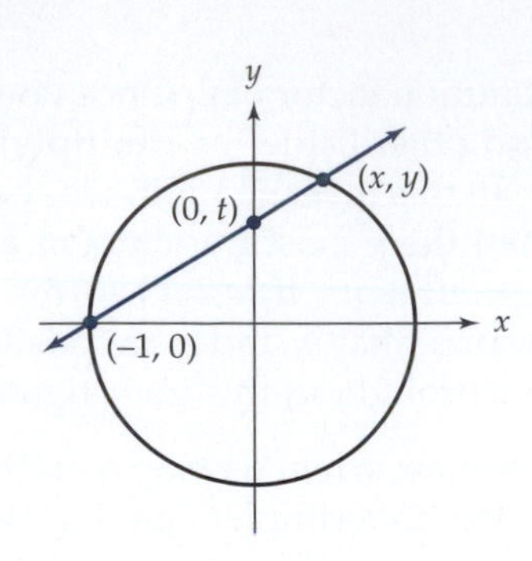

(b) Because the y-intercept equals the slope, the equation of this line is $y = tx + t$. Assume the point (x, y) is in the first quadrant (so that both x and y are positive). Explain why the point being in the first quadrant implies that $0 < t < 1$.

3. We now know two things about our point (x, y).

   - The point (x, y) is on the unit circle.
   - The point (x, y) is on the line $y = tx + t$.

   Whenever two equations are both true for the same point, we can solve the equations simultaneously. In this case, our two equations are $x^2 + y^2 = 1$ and $y = tx + t$.

   (a) Because we know that $y = tx + t$, substitute this expression for y into the equation $x^2 + y^2 = 1$.

   (b) Simplify and show that your result can be written as

   $$(x + 1)[(1 + t^2)x + (t^2 - 1)] = 0.$$

   (c) We are interested in finding values of t and x that cause the left side of this equation to equal zero. If the product of two factors is zero, then one of the factors must itself be zero. To make the first factor zero, we let $x = -1$, which gives us the point $(-1, 0)$. We are interested in making the second factor zero. Show that, in this case, $x = (1 - t^2)/(1 + t^2)$, $y = 2t/(1 + t^2)$.

4. Assume $t = n/m$ is a rational number.[19] There are many different ways to write a rational number. For example, if $t = \frac{1}{2}$, then we also know that $t = \frac{3}{6} = \frac{5}{10} = \sqrt{2}/(2\sqrt{2})$. Let us assume t is written in lowest terms. In other words, m and n are relatively prime positive integers. By making this assumption, we can only guarantee that we can find values of m and n for coprime Pythagorean triples. Without the assumption that t is written in lowest terms, we can find values of m and n for all Pythagorean triples (however, we can no longer assume that m and n are integers).

   (a) Earlier, you explained why $0 < t < 1$. What limits does this restriction place on m and n?

   (b) Show that $x = (m^2 - n^2)/(m^2 + n^2)$ and $y = 2mn/(m^2 + n^2)$.

[19] *Rational numbers are those that can be written as the quotient of two integers.*

(c) Substitute these values for x and y into the equation $x^2 + y^2 = 1$. By algebraically manipulating this equation and knowing that $a^2 + b^2 = c^2$, show that you can derive the equations

$$a = m^2 - n^2$$
$$b = 2mn$$
$$c = m^2 + n^2.$$

6.2 LAW OF SINES AND LAW OF COSINES

Section 6.1 concentrated on applications involving right triangles. Many applications however, involve other types of triangles. In this section, we introduce the law of sines and the law of cosines that, when given sufficient information, allow us to find the dimensions of triangles that are not right triangles. We look at applications using these laws and derive Heron's formula for finding the area of a triangle.

LAW OF SINES

In a geometry course, you probably learned that two triangles are congruent if you know that they have two sides and an included angle in common (called SAS for *side–angle–side*). You also learned that two triangles are congruent if three sides of one have the same lengths as the three sides of the other (SSS). Another way to prove that two triangles are congruent is if the measures of two angles and the length of one side of one triangle are the same as those of the other (ASA or AAS). Because knowing that these combinations create a unique triangle, we should be able to determine the dimensions of any triangle where we know side–angle–side, side–side–side, angle–side–angle, or angle–angle–side.

Finding the Dimensions of a Triangle Using Angle–Angle–Side

Look at the triangle in Figure 1. In $\triangle ABC$, we are given that $m\angle A = 50°$, $m\angle C = 60°$, and $AB = 10$. We have drawn $\overline{BD}$ perpendicular to $\overline{AC}$. Let h represent the length of $\overline{BD}$ and a represent the length of $\overline{BC}$. Using the right triangles $\triangle BAD$ and $\triangle BCD$, we notice that

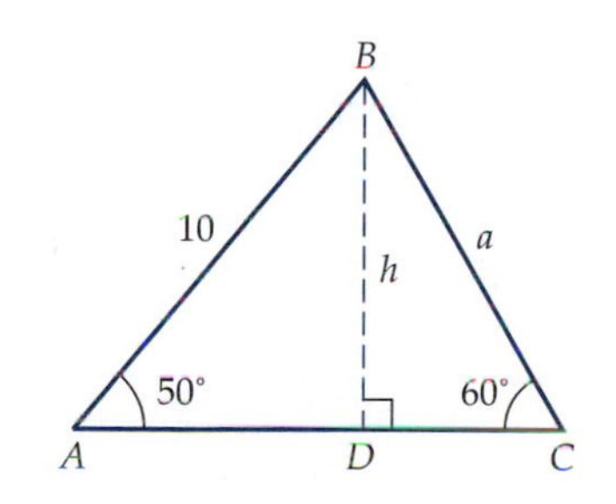

Figure 1 Finding the dimensions of a triangle where we know angle–angle–side.

$$\sin 50° = \frac{h}{10} \qquad \text{and} \qquad \sin 60° = \frac{h}{a}.$$

Solving each for h and setting them equal to each other gives

$$10 \sin 50° = h \qquad \text{and} \qquad a \sin 60° = h$$
$$10 \sin 50° = a \sin 60°.$$

Now we can easily solve for a.

$$a = \frac{10 \sin 50°}{\sin 60°}$$
$$\approx 8.8.$$

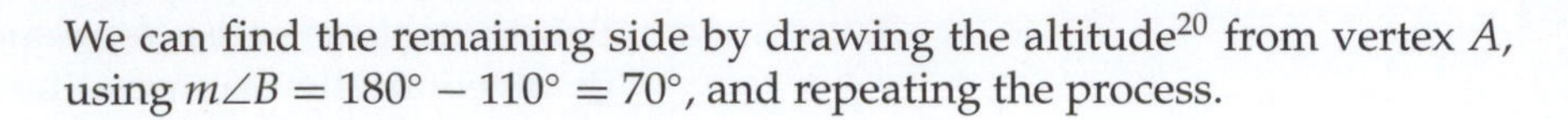

We can find the remaining side by drawing the altitude[20] from vertex A, using $m\angle B = 180° - 110° = 70°$, and repeating the process.

Law of Sines Using Acute Triangles

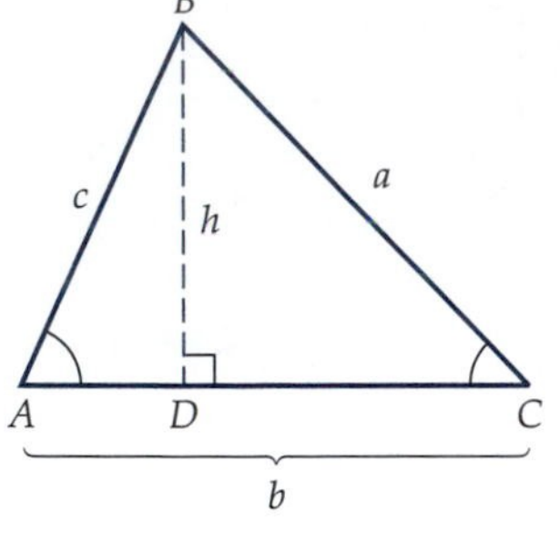
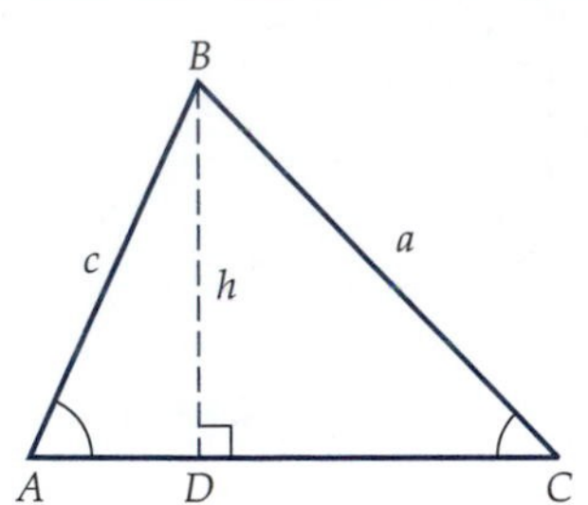

Figure 2 Finding the dimensions of a generalized triangle.

The procedure used in finding the dimension of $\triangle ABC$ in Figure 1 can be generalized to any acute triangle where we are given the measurement of two angles and a side. We use $\triangle ABC$ in Figure 2 to illustrate this process.

Assume $\overline{BD} \perp \overline{AC}$. Let a represent the length of $\overline{BC}$, b represent the length of $\overline{AC}$, and c represent the length of $\overline{AB}$. We see that

$$\sin A = \frac{h}{c} \qquad \text{and} \qquad \sin C = \frac{h}{a}.$$

Solving both equations for h and setting them equal to each other gives us

$$h = c\sin A \qquad \text{and} \qquad h = a\sin C$$
$$c\sin A = a\sin C.$$

Dividing both sides by ac gives

$$\frac{\sin A}{a} = \frac{\sin C}{c},$$

where a is the side opposite $\angle A$ and c is the side opposite $\angle C$. By repeating the procedure using the altitude from vertex A, we see that $\sin B/b = \sin C/c$. So, in general, we have

$$\frac{\sin A}{a} = \frac{\sin B}{b} = \frac{\sin C}{c}, \tag{1}$$

where a is the side opposite $\angle A$, b is the side opposite $\angle B$, and c is the side opposite $\angle C$. The relationship given in equation (1) is known as the **law of sines.**

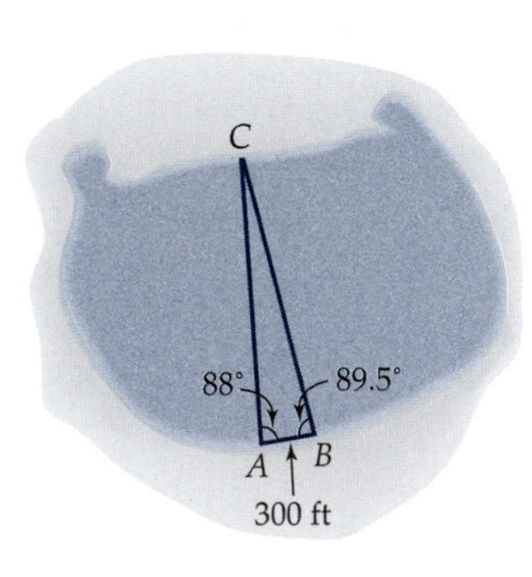

Figure 3 Using the law of sines to find the distance across a lake.

When he was a high school student one of the authors used this method to find the distance across Bear Lake in Michigan. His class walked down to the lake and, using a surveyor's transit, sighted the same point across the lake from two different spots that were 300 ft apart. (See Figure 3.) They measured the angles between their surveying points and point C on the other side of the lake. They discovered that $m\angle A \approx 88°$ and $m\angle B \approx 89.5°$. They then used the law of sines to estimate the distance across Bear Lake. Because they knew two of the three angles in the triangle, they subtracted to find the third angle.

$$\begin{aligned} m\angle C &= 180° - 88° - 89.5° \\ &= 2.5°. \end{aligned}$$

[20] *An altitude of a triangle is a perpendicular line drawn from one side of a triangle to the opposite vertex.*

They assumed BC represented the distance across the lake because $\angle B$ is almost a right angle.[21] To find BC, we use the law of sines.

$$\frac{\sin 88^\circ}{BC} = \frac{\sin 2.5^\circ}{300}$$
$$BC \sin 2.5^\circ = 300 \sin 88^\circ$$
$$BC = \frac{300 \sin 88^\circ}{\sin 2.5^\circ}$$
$$\approx 6873.$$

According to this calculation, the distance across Bear Lake is about 6873 ft. Be careful when you are calculating distances, however, let's say that the students made an error of 0.5° when measuring $\angle A$ and that actually $m\angle A = 87.5^\circ$. In this case, $m\angle C = 3^\circ$ and we have $BC \approx (300 \sin 87.5^\circ)/\sin 3^\circ \approx 5726$, a difference of over 1000 ft from our original calculation! Small errors in measurement can lead to large differences in the results, particularly if you are dividing by small numbers (as in this case, because $\sin 3^\circ \approx 0.0523$). As a result, surveying equipment is extremely accurate, and surveyors are taught to check their work by making several different measurements.

Law of Sines Using Obtuse Triangles

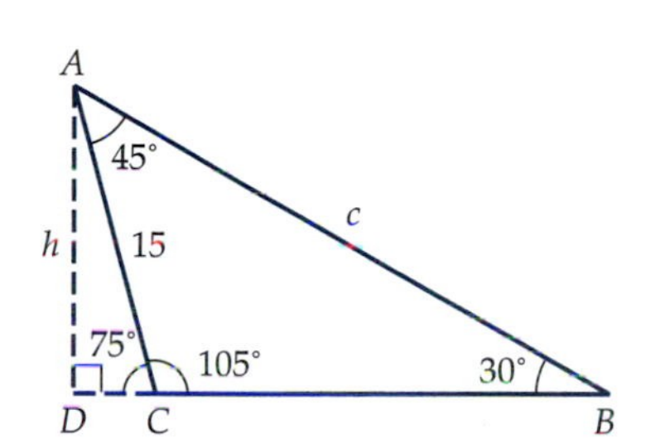

Figure 4 Finding the dimensions of an obtuse triangle.

So far, our examples and the derivation of the law of sines have all used acute triangles. Does the law of sines work for obtuse triangles as well? Look at $\triangle ABC$ in Figure 4. In this case, $m\angle C = 105^\circ$ and, to draw the altitude from vertex A, we need to extend side $\overline{BC}$. Let c represent the length of $\overline{AB}$. We find that

$$\sin 30^\circ = \frac{h}{c} \quad \text{and} \quad \sin 75^\circ = \frac{h}{15}.$$

Notice that instead of using the obtuse angle, 105°, we use the supplement of that angle, 75°. Solving each equation for h, setting them equal to each other, and rewriting gives

$$\frac{\sin 75^\circ}{c} = \frac{\sin 30^\circ}{15}.$$

Using the law of sines, we would expect to get

$$\frac{\sin 105^\circ}{c} = \frac{\sin 30^\circ}{15}.$$

How do $\sin 75^\circ$ and $\sin 105^\circ$ compare? Recall from section 5.4 that $\sin(x - y) = \sin x \cos y - \cos x \sin y$. Because 75° and 105° are supplementary angles, $\sin 105^\circ = \sin(180^\circ - 75^\circ) = \sin 180^\circ \cos 75^\circ - \cos 180^\circ \sin 75^\circ = 0 \cdot \cos 75^\circ - (-1) \sin 75^\circ = \sin 75^\circ$. In fact, the sine of every angle and its supplement (180° − the angle) are the same. (See Figure 5.) Because the sine function is defined as the vertical position of the point on the unit circle, you can see that the vertical position of the point is the same for an angle of

[21] *We could also have decided to use $\overline{AC}$ to represent the distance across the lake because $m\angle A$ is also close to 90°. The shore of a lake is not actually a straight line, so there is not "one" distance across a lake.*

measure θ as it is for an angle of measure $180° - \theta$. Because of this equality, we can still write the law of sines for obtuse angles as

$$\frac{\sin A}{a} = \frac{\sin B}{b} = \frac{\sin C}{c},$$

even though we actually use the supplement of the obtuse angle to geometrically find the dimension of the missing side. So, the law of sines is true for every triangle (although, with a right triangle, the law of sines reduces to the right triangle definition of sine).

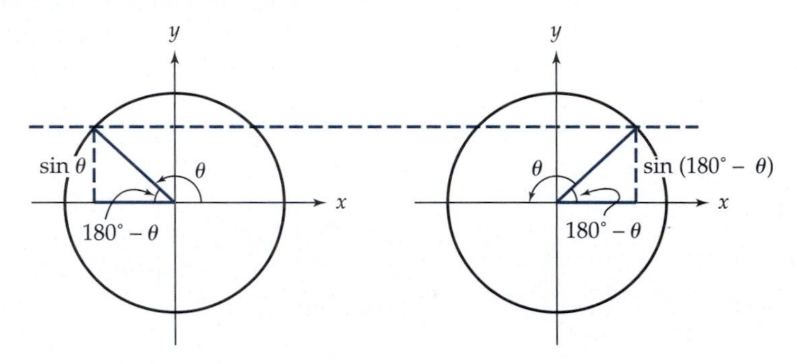

Figure 5 The sine of an angle and its supplement are the same.

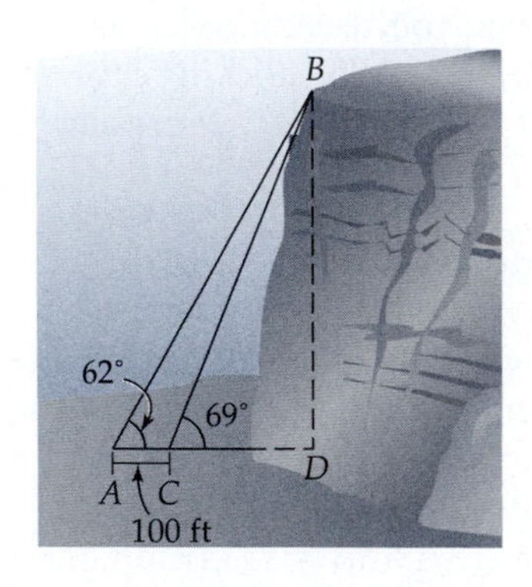

Figure 6

Example 1 Suppose you want to find the height of a cliff where you were able to determine the angle of elevation from two different points 100 ft apart. (See Figure 6.) The angle of elevation from point C is 69° and the angle of elevation from point A is 62°. Using this information, estimate the height of the cliff, BD.

Solution Before we can determine the length of $\overline{BD}$, we need to find the length of $\overline{AB}$ using the law of sines. To do so, determine $m\angle ABC$ and $m\angle ACB$. We know that $m\angle ACB = 180° - 69° = 111°$. So, $m\angle ABC = 180° - (62° + 111°) = 7°$. Using the law of sines, we have

$$\frac{\sin 111°}{AB} = \frac{\sin 7°}{100}.$$

Solving for AB gives

$$AB = \frac{100 \sin 111°}{\sin 7°} \approx 766 \text{ ft}.$$

Now using the right triangle $\triangle ADB$, we can find BD.

$$\sin 62° \approx \frac{BD}{766}$$
$$BD \approx 766 \sin 62°$$
$$\approx 676.$$

Therefore, the cliff is approximately 676 ft tall.

READING QUESTIONS

1. Use the law of sines to find x in the following triangle.

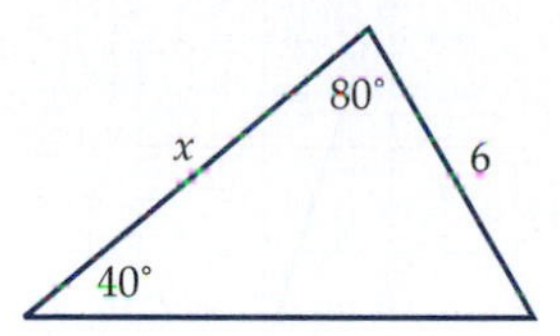

2. Explain what is incorrect in the statement $b = (6 \sin 45°)/\sin 30°$ using the accompanying triangle.

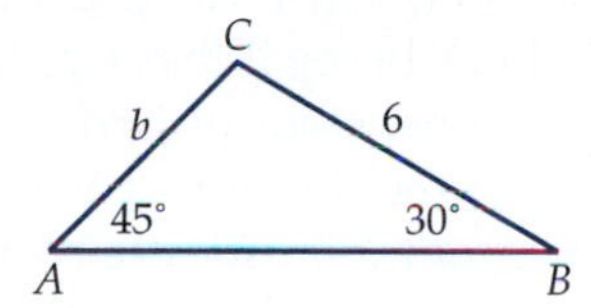

THE AMBIGUOUS CASE

In our previous examples, we used the law of sines to determine the length of a side of a triangle when given the measure of two angles and one side. You can also use the law of sines to find the measure of an angle when given the measure of two sides and a nonincluded angle. Using the law of sines to solve for the measure of the angle, however, can often lead to ambiguities because you are solving for the sine of the angle rather than the measure of the angle directly. Thus, you have to use the inverse sine function. Recall from section 3.5 that the range of $\theta = \sin^{-1} x$ is $-90° \leq \theta \leq 90°$. Hence, when you use the inverse sine function on your calculator, your answer will *always* be an acute angle (either positive or negative).[22] You need to be aware, however, that sometimes the correct solution may be an obtuse angle.

How do we determine if we want an obtuse angle instead of an acute angle? If we decide the angle is actually obtuse, how do we find it? Let's answer the second question first. Except for 90°, there are always two angles between 0° and 180° for which the sine is the same. For example, $\sin 45° = \sin 135° = \sqrt{2}/2$. See the horizontal line in Figure 7.

Recall from Figure 5 that the sine of an angle and its supplement are always the same. So, the two angles where a horizontal line crosses the function $y = \sin\theta$ are θ and $180° - \theta$. Therefore, for your answer to be an obtuse angle, determine the supplement of the acute angle found using the inverse sine function.

How do we determine if we want an obtuse angle instead of an acute angle? Sometimes it is clear from the physical context of the problem. More

[22] *When finding the dimensions of a triangle, the angle will always be positive.*

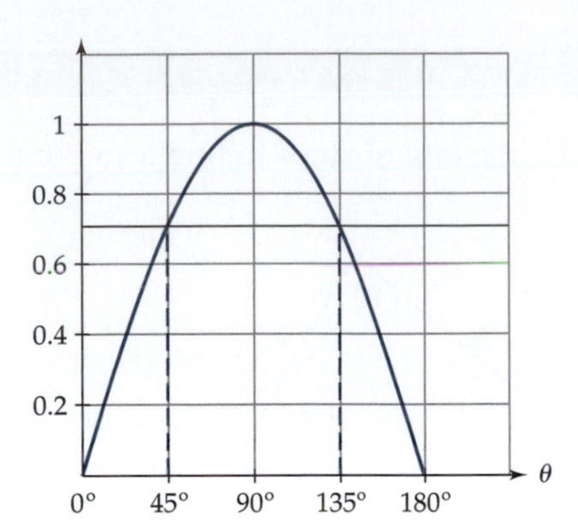

Figure 7 A graph of $\sin\theta$ for $0° \leq \theta \leq 180°$. Notice that $\sin 45° = \sin 135°$.

often, it is a matter of determining if an obtuse angle forms a triangle of the desired dimensions. Once you use the law of sines to find the measure of an acute angle, check to see if the supplement is also a solution. We illustrate as we find the dimensions of $\triangle ABC$.

In $\triangle ABC$, assume the side opposite a given vertex is referred to by the same letter. For example, side a is opposite $\angle A$. Assume $m\angle A = 40°, a = 10$, and $b = 12$. We want to find $m\angle B$. Using the law of sines, we set up the equation

$$\frac{\sin B}{12} = \frac{\sin 40°}{10}.$$

Solving for $\sin B$ gives

$$\sin B = \frac{12 \sin 40°}{10} \approx 0.7713.$$

Using the inverse sine function, we find that $\sin^{-1}(0.7713) \approx 50.5°$. Because $\sin\theta = \sin(180° - \theta)$, however, $\sin 50.5° = \sin(180° - 50.5°) = \sin 129.5°$. Thus, $m\angle B$ could be either 50.5° or 129.5°, so there are two solutions for this problem. Figure 8 shows a drawing of the two triangles that fit our original data.

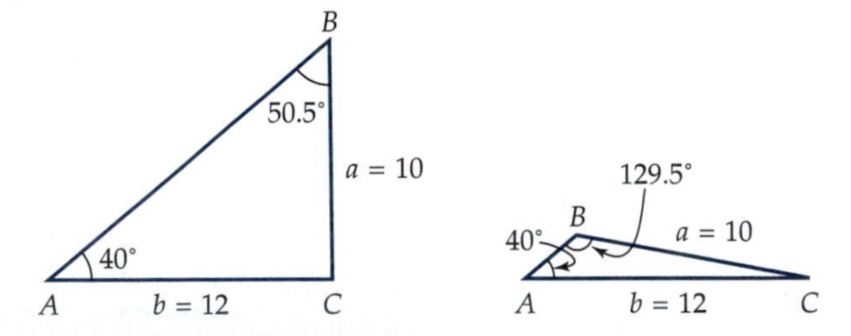

Figure 8 Two triangles where $m\angle A = 40°$, $a = 10$, and $b = 12$. When solving for $m\angle B$, we find that B can either be an acute angle (50.5°) or an obtuse angle (129.5°).

If you are finding the measure of an angle when given two sides and a nonincluded angle, you need to be aware that there will be two angles with the same sine, giving you two possible angles. There will not necessarily be two triangles with the required dimensions, however. A theorem from geometry states that if two sides of a triangle are unequal, then the angles opposite them are unequal and the larger angle is opposite the longer side. In our last example, you can see that both of the solutions for $m\angle B$ are greater than $m\angle A$, which is correct because $b > a$. If b were less than a, however, the only solution would have been the acute angle. This idea is illustrated in the next example.

Example 2 Find $m\angle B$ in $\triangle ABC$ if $m\angle A = 80°$, $a = 14$, and $b = 10$.

Solution To find $m\angle B$, we can use the law of sines to obtain

$$\frac{\sin B}{10} = \frac{\sin 80°}{14}.$$

Solving for $\sin B$ gives

$$\sin B = \frac{10 \sin 80°}{14} \approx 0.7034.$$

By using the inverse sine function, we find that $B \approx \sin^{-1}(0.7713) \approx 44.7°$. Because $\sin 44.7° = \sin(180° - 44.7°) = \sin 135.3°$, we might think that $B \approx 135.3$ is another possible solution, but that is not the case. Because $a > b$, $m\angle A$ must be greater than $m\angle B$. The only solution for B, therefore, is $44.7°$. (See Figure 9.)

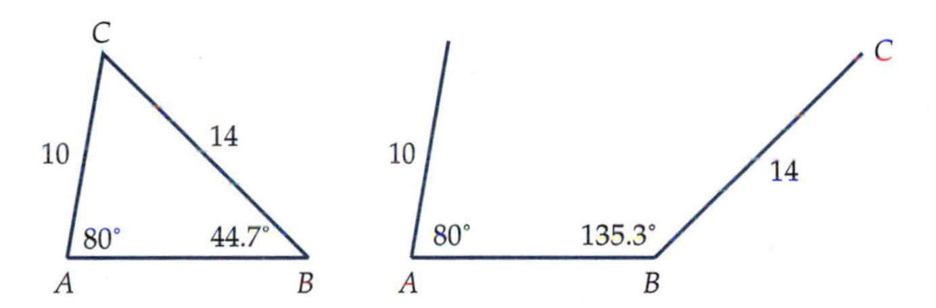

Figure 9

This ambiguous case helps explain why you were unable to show in geometry that two triangles were congruent using SSA.[23] Given the measurements of two sides and an nonincluded angle, there may be two triangles with these dimensions instead of just one. If we know two sides and an *included* angle (SAS), then we have enough information to determine one triangle. In this case, we need to use the law of cosines.

READING QUESTIONS

3. For each of the following, determine if there are 1 or 2 solutions for $m\angle B$.

(a) $m\angle A = 45°, a = 5, b = 4$

(b) $m\angle A = 45°, a = 5, b = 6$

[23] *Some students preferred to list these in reverse order for comic relief.*

4. Use the law of sines to find θ in the triangle shown.

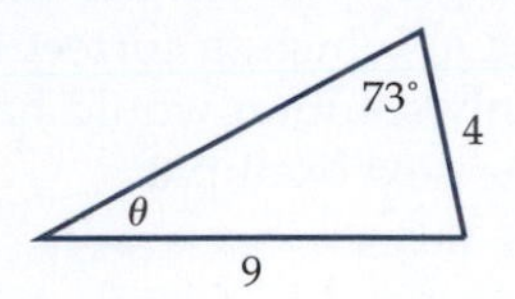

LAW OF COSINES

So far, we have looked at how to determine the dimensions of a triangle when given ASA or AAS. We have also considered what happens if you are given two sides and a nonincluded angle. In geometry, however, you also learned that two triangles are congruent if you have three sides congruent (SSS) or two sides and an included angle (SAS). The law of sines cannot be used for either of these two cases. Yet because a unique triangle is determined, there must be some method of finding the missing dimensions. This method, which we derive, is the law of cosines.

Suppose we do not know the measurement of any of the angles of a triangle but we do know the length of the three sides. Let's start with $\triangle ABC$ in Figure 10. Let h represent the length of the altitude from vertex B and x represent the length of side $\overline{DC}$. Use of the Pythagorean theorem with $\triangle ABD$ gives

$$c^2 = h^2 + (b - x)^2.$$

Expanding gives

$$c^2 = h^2 + b^2 - 2bx + x^2.$$

By using the Pythagorean theorem with $\triangle BCD$, we can see that $a^2 = h^2 + x^2$. Substituting a^2 for $h^2 + x^2$ in our equation and rearranging gives

$$c^2 = a^2 + b^2 - 2bx.$$

By using $\triangle BCD$, we can see that $\cos C = x/a$ or $x = a\cos C$. Making this substitution in our equation gives

$$c^2 = a^2 + b^2 - 2ab \cos C. \tag{2}$$

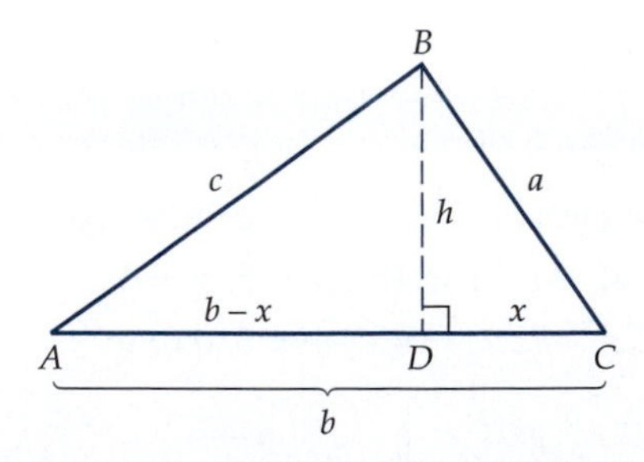

Figure 10 Determining the dimensions of a triangle when given three sides.

Equation 2 is known as the **law of cosines.**[24] Notice that c is the length of the side opposite $\angle C$. It does not matter what notation you use to represent the three sides as long as the side on the left side of the equation is the one opposite your angle. Alternatively, you can write the law of cosines as

$$a^2 = b^2 + c^2 - 2bc \cos A \quad \text{or} \quad b^2 = a^2 + c^2 - 2ac \cos B.$$

Notice this relationship between the lengths of the sides is similar to the Pythagorean theorem. In fact, if the angle is a right angle, then the law of cosines simplifies to the Pythagorean theorem. Thus, the law of cosines can be thought of as a general version of the Pythagorean theorem.

The law of cosines can also be used to find the measurements of a triangle where you are given two sides and an included angle (SAS), as illustrated in the following example.

Example 3 On a baseball field, the distance from home plate to the pitcher's mound is 60 ft 6 in. and the distance from home plate to first base is 90 ft. (See Figure 11.) Find the distance from the pitcher's mound to first base.

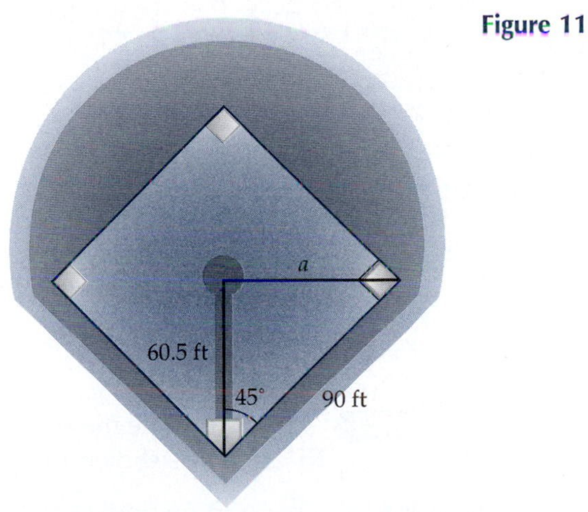

Figure 11

Solution Let a denote the distance from the pitcher's mound to first base. Using the law of cosines gives

$$\begin{aligned} a^2 &= 60.5^2 + 90^2 - 2 \cdot 60.5 \cdot 90 \cdot \cos 45^\circ \\ &= 11{,}760.25 - 10{,}890 \cos 45^\circ \\ &\approx 4059.85 \\ a &\approx 63.7. \end{aligned}$$

[24] *In our proof, we assumed that $\angle C$ was an acute angle. In the exercises, you are asked to prove this theorem when $\angle C$ is an obtuse angle.*

Therefore, the distance from the pitcher's mound to first base is approximately 63.7 ft.

If we are given the measurements of three sides of a triangle, then we can use the law of cosines to solve for the cosine of any angle. To find the measurement of the angle, we need to use the inverse cosine function. Recall that in the law of sines, using the inverse sine led to two cases, depending on whether the angle should be obtuse or acute. Does the same thing happen here? To answer this question, recall the graph of the cosine function as shown in Figure 12.

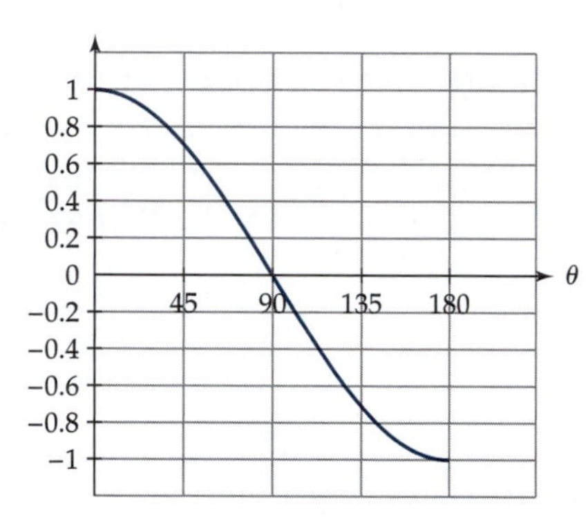

Figure 12 A graph of $y = \cos\theta$ for $0° \leq \theta \leq 180°$.

Notice that $y = \cos\theta$ is one-to-one on the interval $0° < \theta < 180°$. In other words, each output has only one corresponding input. So, unlike the inverse sine function, there is no ambiguity in using the inverse cosine function to find the measure of any angle between 0° and 180°.

Figure 13

Example 4 Find the angle measures of the triangle whose sides measure 12, 15, and 20 as shown in Figure 13.

Solution We start by finding $m\angle A$. This angle will be the largest because it is opposite the largest side. Using the law of cosines gives

$$\begin{aligned} 20^2 &= 12^2 + 15^2 - 2 \cdot 12 \cdot 15 \cdot \cos A \\ 400 &= 369 - 360 \cdot \cos A \\ 31 &= -360 \cdot \cos A \\ -\frac{31}{360} &= \cos A. \end{aligned}$$

So $A = \cos^{-1}\left(-\frac{31}{360}\right) \approx 94.9°$.

To find the next angle, we could use either the law of cosines or the law of sines (because we now have two sides and an angle). Because there are fewer computations with the law of sines than with the law of cosines and because the other two angles, being smaller than $\angle A$, are acute (so that there is no ambiguity with the law of sines), we use the law of sines.

Using this law to solve for $m\angle B$ gives

$$\begin{aligned}\frac{\sin B}{15} &= \frac{\sin 94.9^\circ}{20}\\ \sin B &= \frac{15 \sin 94.9^\circ}{20}\\ &\approx 0.7472.\end{aligned}$$

So, $B \approx \sin^{-1} 0.7472 \approx 48.3^\circ$.

The sum of the three angles of a triangle is 180°, so the last angle is approximately

$$180^\circ - 94.9^\circ - 48.3^\circ \approx 36.8.$$

Therefore, the measures of the three angles in our triangle are approximately 94.9°, 48.3°, and 36.8°.

READING QUESTIONS

5. Use the accompanying triangle to answer the following.

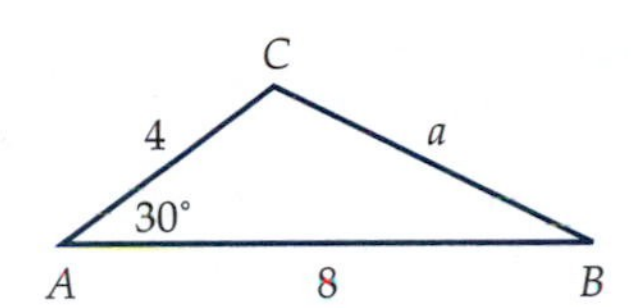

(a) Why can't you use the law of sines to find a?

(b) Using the law of cosines, approximate a.

6. Why is there no ambiguity when you use the law of cosines to find the measurement of an angle?

HERON'S FORMULA

Earlier, we used two formulas for finding the area of a triangle:

- Area $= \frac{1}{2}$ base × height
- Area $= \frac{1}{2}ab \sin C$, where C is the measure of the angle in between the sides whose lengths are a and b[25]

Notice that the first formula requires you to find an altitude of the triangle and the second formula requires you to know the measurement of an angle. At times, however, neither formula is applicable, yet you have a unique triangle and therefore a unique area. One of these cases is side–side–side. Suppose you have a triangle where you are given the lengths of the three sides, a, b, and c. Let C be the measure of the angle opposite side c. (See Figure 14.)

Figure 14 Finding the area of a triangle when given three sides.

[25] *This formula was derived in section 5.2.*

According to our second area formula, the area of the triangle is $A = \frac{1}{2}ab\sin C$. Unfortunately, we do not know the measure of $\angle C$. By using the law of cosines, however, it is possible to find $\cos C$. To change $\sin C$ to an expression involving $\cos C$, recall that $\cos^2 C + \sin^2 C = 1$. Therefore, $\sin C = \sqrt{1 - \cos^2 C}$ when $0° \leq C \leq 180°$.[26] Substituting $\sqrt{1 - \cos^2 C}$ for $\sin C$ gives

$$A = \tfrac{1}{2}ab\sqrt{1 - \cos^2 C}.$$

From the law of cosines, we know that $c^2 = a^2 + b^2 - 2ab\cos C$, which implies that $\cos C = (a^2 + b^2 - c^2)/2ab$. Our new formula for finding the area of a triangle becomes

$$A = \frac{1}{2}ab\sqrt{1 - \left(\frac{a^2 + b^2 - c^2}{2ab}\right)^2}.$$

This formula is for the area of a triangle that only depends on the lengths of the sides of the triangle.

Algebraically simplifying this formula (something you are asked to do in the investigations) leads to a more well known version. This version, commonly referred to as **Heron's formula,** is stated as follows.

PROPERTIES

The area of a triangle, A, is given by $A = \sqrt{s(s-a)(s-b)(s-c)}$, where a, b, and c are the lengths of the sides of the triangle and s is the semiperimeter (half of the perimeter); that is, $s = (a + b + c)/2$.

Example 5 Sometimes buildings or rooms are designed in the shape of triangles. Such is the case with the East Wing of the National Gallery of Art in Washington, D.C. Parts of this building are triangular in shape, as are some of its rooms. Suppose you have a triangular room with sides of 20, 24, and 30 ft. Find the area of the floor of the room.

Solution The length of the semiperimeter is $s = (20 + 24 + 30)/2 = 37$. Using Heron's formula gives

$$\begin{aligned} \text{area} &= \sqrt{s(s-a)(s-b)(s-c)} \\ &= \sqrt{37(37-20)(37-24)(37-30)} \\ &= \sqrt{37(17)(13)(7)} \\ &= \sqrt{57{,}239}. \end{aligned}$$

So, the area of the floor of the room is $\sqrt{57{,}239} \approx 239.2$ ft^2.

[26] *This restriction is necessary because we are using the positive square root.*

HISTORICAL NOTE

Heron[27] lived around 50 B.C. in the city of Alexandria. This Greek city, founded by Alexander the Great in 331 B.C., was located at the mouth of the Nile. Before Heron's time, Alexandria was the home of the great geometer Euclid. In this city, he wrote his famous geometry book *Elements*, which, after the Bible, is considered the world's best-selling book. Alexandria was also where Eratosthenes measured the shadow of the midday sun to determine the circumference of Earth. (See section 5.2, Exercise 7.) Cleopatra, who lived during the time of Heron, was also from Alexandria.

Much of classical Greek mathematics, such as that in Euclid's *Elements*, was abstract and nonnumerical. Heron, however, followed the more traditional Egyptian route and wrote about the "nonclassical" type of mathematics in his book *Metrica*, where he used numerical data to calculate, for instance, the area of a triangle. *Metrica* contains the first proof of what we now know as Heron's formula. Even though the formula was likely known and proved 300 years earlier by Archimedes, we know it by Heron's name because he was the first one to publish it.

In addition to his expertise as a mathematician, Heron was also famous as an inventor. His inventions include a primitive steam engine, a fire engine that pumped water, an altar fire that lit as soon as the temple doors opened, and a wind organ.

Sources: Boyer, *A History of Mathematics*, pp. 190–93; Theoni Pappas, *The Joy of Mathematics* (San Carlos, CA: Wide World Publishing/Tetra, 1989), pp. 62; Edward M. Forester, *Alexandria: A History and a Guide* (Woodstock, NY: Overlook Press, 1974).

[27] *Heron is sometimes referred to as Hero.*

READING QUESTIONS

7. Find the area of a triangle whose sides are 3, 10, and 11.

8. Explain why $\sin C = \sqrt{1 - \cos^2 C}$ is only true when $0° \leq C \leq 180°$.

SUMMARY

We can find the missing lengths and angle measures of any triangles (not just right triangles) by using the law of sines or the law of cosines. If you know two angles and any side of a triangle (ASA or AAS), then you can use the law of sines to find the other side. The **law of sines** is summarized by

$$\frac{\sin A}{a} = \frac{\sin B}{b} = \frac{\sin C}{c}.$$

When using the law of sines to solve for an angle, you are really solving for the sine of that angle. To find the angle, you must remember that $\sin^{-1} x$

has two positive values between 0° and 180°. So, if your calculator gives you $\sin^{-1} x = y$, then $180° - y$ may also be a solution.

Because the inverse sine has more than one possible solution, we sometimes have a triangle in which there is more than one possible solution. This situation is known as the ambiguous case. Often, only one of these solutions makes sense because either potential solutions make the sum of the angles of the triangle greater than 180° or because the solution gives the greatest angle in the triangle opposite something other than the longest side. Sometimes, though, there are two sets of solutions.

If you know two sides and an included angle (SAS) or all three sides (SSS), then you can use the law of cosines to find another side or angle. The **law of cosines** can be expressed as

$$c^2 = a^2 + b^2 - 2ab \cos C.$$

Because the inverse cosine function has only one positive solution between 0° and 180°, there is no ambiguous case for the law of cosines.

Heron's formula for the area of a triangle, A, is given by

$$A = \sqrt{s(s-a)(s-b)(s-c)},$$

where a, b, and c are the lengths of the sides of a triangle and s is the semiperimeter (half the perimeter), or $(a+b+c)/2$.

EXERCISES

In questions 1 through 3, assume the sides of $\triangle ABC$ are labeled as shown in the accompanying figure.

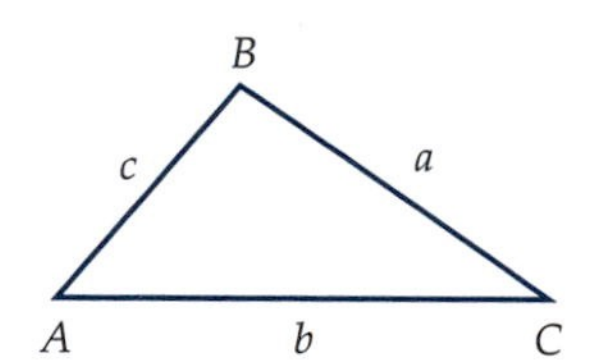

1. For each of the following, you are given three pieces of information about a triangle and are asked to determine a third. To do so, determine if you would use the law of sines, if you would use the law of cosines, or if this information cannot be determined.
   (a) Given $\angle A$, $\angle B$, and a, find side c.
   (b) Given $\angle A$, a, and b, find side c.
   (c) Given a, b, and c, find $\angle A$.
   (d) Given $\angle A$, $\angle B$, and b, find side c.
   (e) Given $\angle A$, $\angle B$, and $\angle C$, find side b.
   (f) Given $\angle A$, b, and c, find a.
   (g) Given $\angle A$, a, and b, find $\angle B$.

2. Find the measurement of all three angles and all three sides for each of the following. Round answers to two decimal places.
   (a) $m\angle A = 40°,\ m\angle C = 80°,\ a = 12$
   (b) $m\angle A = 20°,\ a = 13,\ b = 18$
   (c) $m\angle A = 40°,\ m\angle C = 110°,\ a = 12$
   (d) $m\angle A = 63°,\ c = 14,\ b = 15$
   (e) $a = 10,\ b = 6,\ c = 9$
3. (a) Explain why each of the following dimensions will not produce a triangle.
      i. $a = 14, b = 6, m\angle A = 30°, m\angle C = 15°$
      ii. $a = 6, b = 2, m\angle A = 30°, m\angle B = 50°$
      iii. $a = 3.7, b = 2.1, m\angle A = 33°, m\angle B = 162°$
      iv. $a = 3, b = 5, m\angle A = 45°, m\angle B = 60°$
      v. $a = 3, b = 4, c = 5, m\angle C = 80°$
      vi. $a = 4, b = 5, c = 4.87, m\angle C = 30°$
   (b) What happens if you try to use the law of sines or the law of cosines with dimensions that do not actually produce a triangle?
4. In this section, we used the accompanying figure to prove that $(\sin A)/a = (\sin C)/c$. Using a similar method, prove that $(\sin A)/a = (\sin B)/b$.

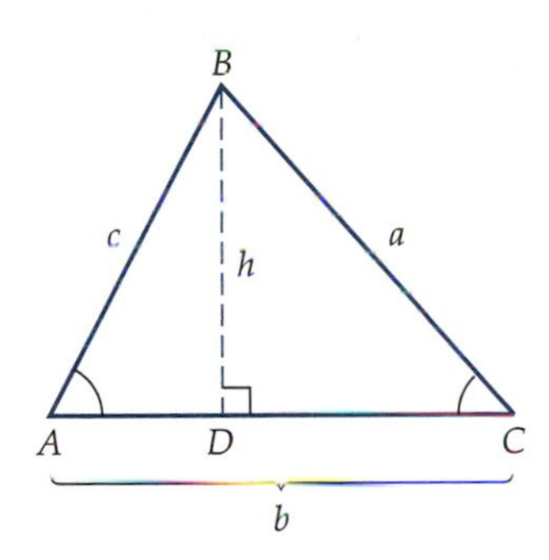

5. Use the law of sines to prove the following: The base angles of an isosceles triangle are congruent.
6. Show that if $\triangle ABC$ is a right triangle, then the law of sines reduces to the definition of $\sin\theta$ = opposite/hypotenuse.
7. Show that if $\triangle ABC$ is a right triangle, then the law of cosines reduces to the Pythagorean theorem.
8. In this section, we discussed what can happen when you use the law of sines to find the dimensions of a triangle when given the measurement of two sides and a nonincluded angle (SSA). This exercise explores this concept further.
   (a) Explain why, given SSA, there may be zero, one, or two triangles with the given dimensions.
   (b) Suppose you are given the length of $\overline{AC}$ and $m\angle A$ such that $\angle A$ is an acute angle. (See the accompanying figure.) The length of

side a (the side opposite $\angle A$) is also given and is represented by the distance from B to the arc drawn by a compass. In the figure, a triangle cannot be drawn because all the sides do not connect. Assume the length of $\overline{AC}$ and $m\angle A$ are fixed and you can change the length of a.

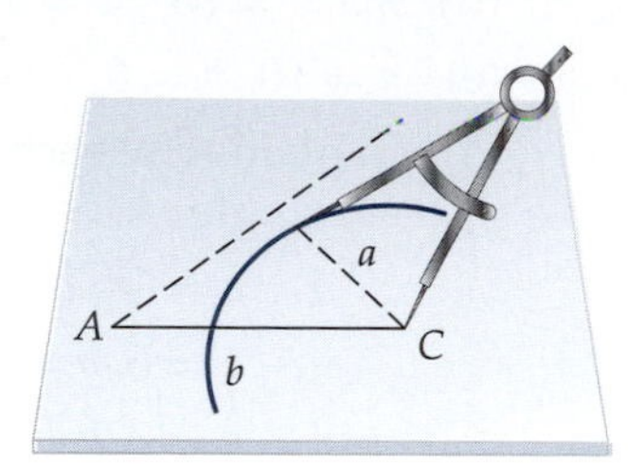

i. What must be true about the length of a if no triangle is formed (as in the picture)?

ii. Suppose you make a longer. At what length will there be exactly one triangle formed?

iii. What length of a would cause two triangles to be formed?

iv. We have already seen that if a is too short, then no triangle can be formed. Can a ever be so large that no triangle can be formed?

(c) The accompanying figure is similar to the one in part (b) except that $\angle A$ is obtuse. What lengths of a will form zero, one, or two triangles? Is this case any different than when $\angle A$ is an acute angle?

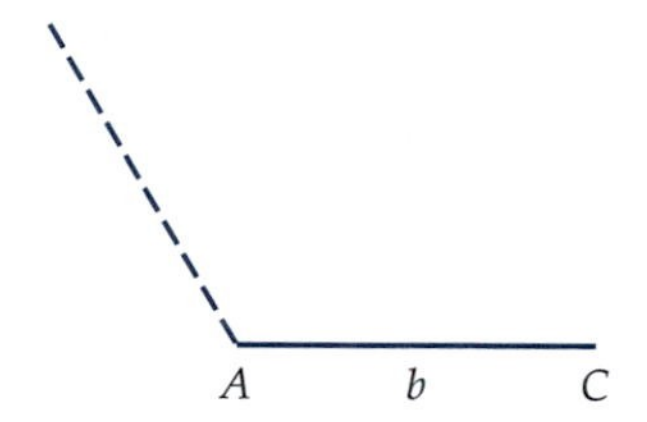

9. Derive the law of cosines for obtuse triangles by using the accompanying figure. Use $\angle C$, the obtuse angle.

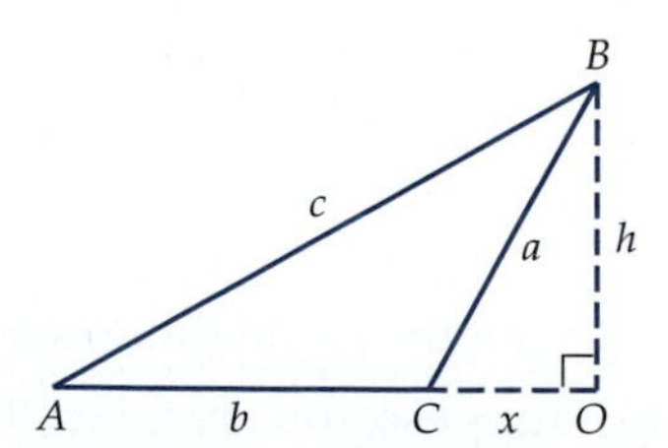

10. The world's tallest building is the Petronas Tower in Kuala Lumpur, Malaysia.[28] Suppose you look at the top of the building and find that the angle of elevation is 48°. (See the accompanying figure.) You move 40 m closer to the building and find the angle of elevation is now 50.9°. Estimate the height of the building.

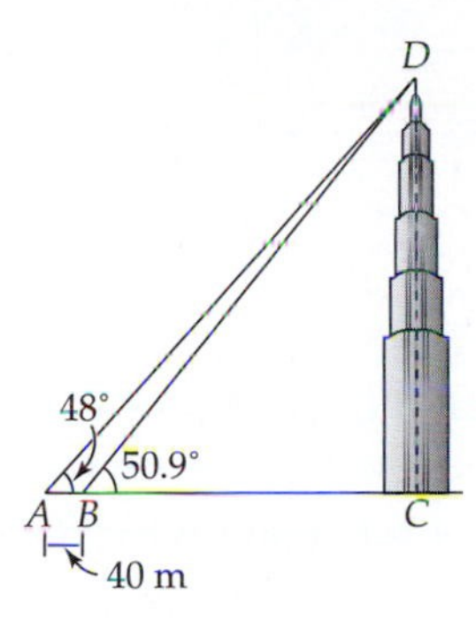

11. The accompanying figure shows the old Tiger Stadium in Detroit, Michigan. The distance from home plate to the center field wall is 440 ft. If the distance from home plate to third base is 90 ft, how far would a player have to throw the ball if he catches a fly ball at the wall in center field and tries to throw out a runner heading for third base? That is, what is the length of $\overline{BC}$?

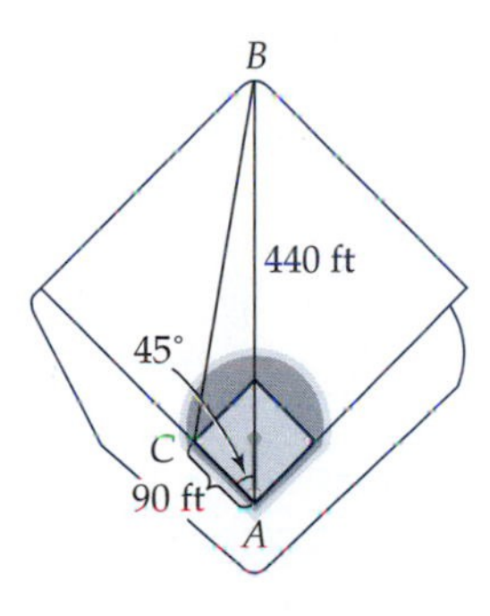

12. The length of the Leaning Tower of Pisa, from the base to the top, is 56 m.[29] Suppose you are at a point 50 m from the base and find that the angle of elevation is 51.2°. (See the accompanying figure.)

[28] *There are actually two towers, Petronas Tower 1 and Petronas Tower 2, that are the same height. The highest floor of the Sears Tower in Chicago is 200 ft higher than that of the highest floor at the Petronas Tower. When determining the height of the tallest building, however, the height of the spire of the Petronas Tower is included because it is an "architectural feature," whereas the antennae on top of the Sears Tower are not considered "architectural features" and so are not included when calculating the height.* The World's Tallest Building Page: Technical. *(visited 16 June 1998) ⟨http://www.dcircle.com/wtb/93.html⟩.*

[29] *Britannica Online,* Leaning Tower of Pisa *(visited 16 June 1998) ⟨http://www.eb.com:180/cgi-bin/g?DocF=micro/342/14.html⟩.*

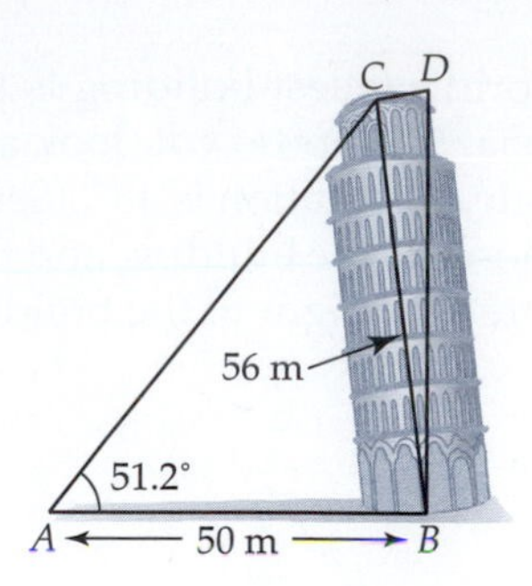

(a) We are interested in finding how far from vertical is the tower leaning, that is, the measure of $\angle CBD$ where $\overline{BD}$ is the height of the tower if it were vertical.

  i. Explain why, when using the law of sines, there is only one possible answer for $m\angle CBD$.

  ii. Find $m\angle CBD$.

(b) How far away from being vertical is the top of the tower? That is, find the length of $\overline{CD}$.

13. The 210-feet-high Coit Tower is at the top of Telegraph Hill in San Francisco.[30] Suppose, while standing on the hill, you measure the angle of elevation to the bottom of the tower and find that it is 10°. You also measure the angle to the top of the tower and find that it is 32°. How far are you from the tower? That is, what is the length of $\overline{AB}$? (See the accompanying figure.)

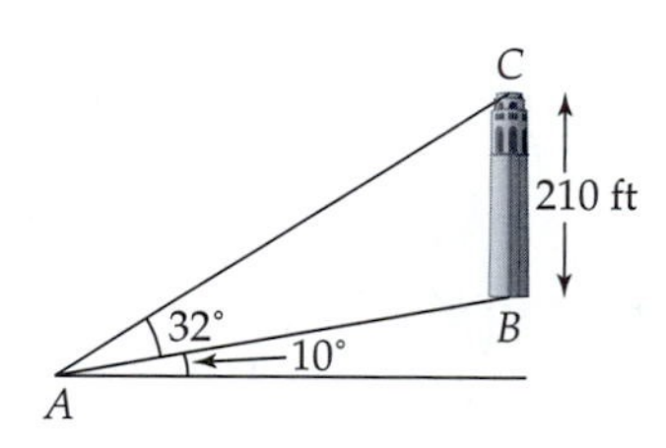

14. Wiffle ball is a game similar to baseball except that a plastic bat and ball are used. There are organized adult Wiffle ball leagues with official rules. According to the *North American Wiffle Ball Championship Rule Book*,[31] the field is designed as shown in the accompanying diagram. Determine the measure of $\angle BAC$.

[30]Coit Tower, San Francisco Online *(visited 17 June 1998)* *⟨http://www.sanfranciscoonline.com/bcg/City_sights/Coit_Tower/Coit_Tower.html⟩*.

[31]North American Wiffle Ball Championship Rule Book *(visited 21 July 1998)* *⟨http://www.wiffleball.com⟩*.

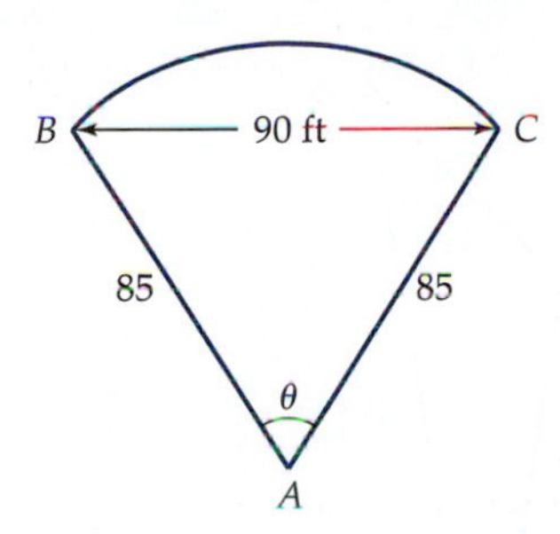

15. An airplane is traveling at 100 mph with its course set at 20° south of east as shown by $\overline{AB}$ in the accompanying figure. There is a 20 mph wind coming from the north as shown by $\overline{BC}$. Because of the wind, the plane actually flies in the direction of $\overline{AC}$. Its speed relative to the ground is the length of $\overline{AC}$.

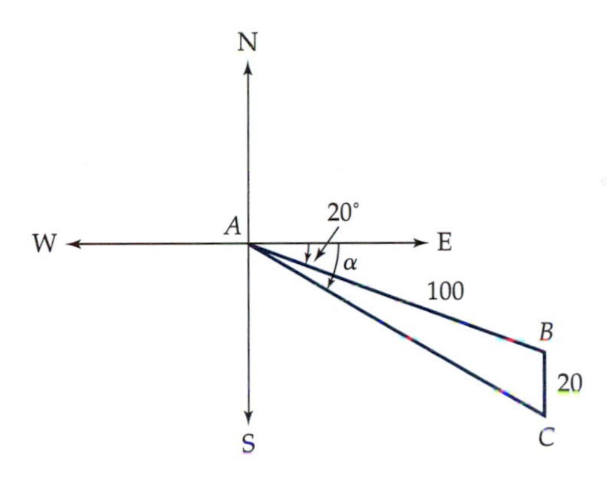

(a) Explain why $m\angle B = 110°$.
(b) Find the ground speed of the plane by finding the length of $\overline{AC}$.
(c) The direction the plane will actually be flying is given by α. Find the measure of this angle.

16. A calculator-based ranger (CBR) connected to a graphing calculator was used to graph the path of a slow moving toy truck as it passed through the path of the CBR beam. The CBR emits a cone-shaped sonic beam at an angle of 20°. The time from when the ultrasonic pulses are emitted and the first echo is returned is used to calculate the distance an object is away from the CBR. The calculator gives the time when the truck entered the beam and its distance from the CBR as well as the time when the truck left the beam and its distance from the CBR. Using the data collected on the calculator, you are asked to calculate the speed of the truck. In the accompanying figure, side $\overline{AB}$ indicates that the truck was 18.0 in. from the CBR when it entered the beam. Side $\overline{AC}$ indicates that the truck was 25.3 in. from the CBR when it left the beam. Side $\overline{BC}$ represents the distance the truck traveled.

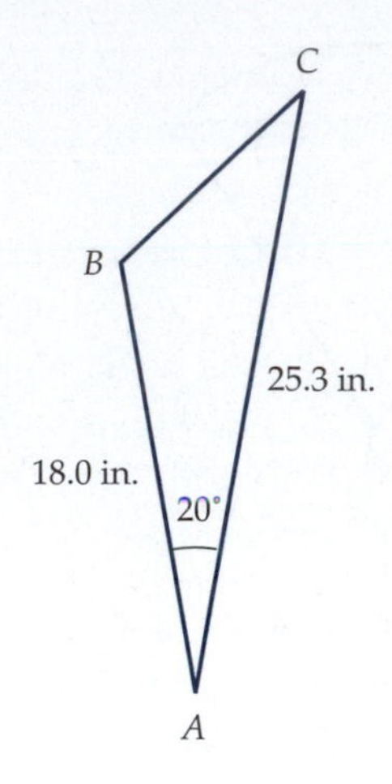

(a) Find the length of $\overline{BC}$.

(b) The truck entered the beam after the program was running for 17.4 sec and exited the beam after the program was running for 49.0 sec. Using this information, find the speed of the truck in inches per second.

17. For each of the following, find the area of a triangle shown with the given dimensions.

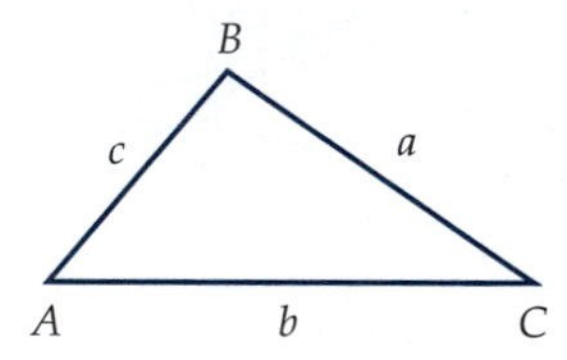

(a) $a = 6, b = 10, c = 15$ (b) $a = 3, b = 5, m\angle C = 30°$

(c) $a = 4, b = 2, m\angle B = 15°$ (d) $a = 5, m\angle A = 50°, m\angle B = 60°$

18. Most property lots are rectangular, which makes finding the area of the lot simple. Some, like the one owned by one of the authors, are not rectangular in shape, however, which makes finding the area of the lot more difficult. Find the area of the lot shown in the accompanying figure.

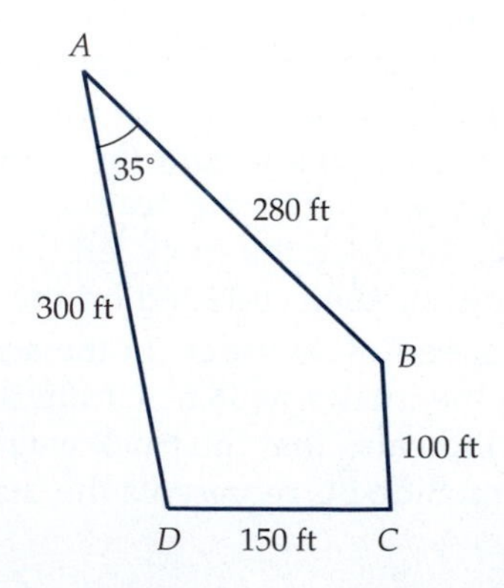

19. In this problem, we investigate what happens to the area of a triangle when the lengths of the sides are doubled, tripled, and multiplied by a factor of n.
    (a) Find the area of a triangle with sides of length 3, 5, and 7.
    (b) Find the area of a triangle with sides of length 6, 10, and 14. How is this area related to the area of the triangle from part (a)?
    (c) Find a relationship between a triangle whose sides have length $a, b,$ and c and a triangle that has sides of length $2a, 2b,$ and $2c$. In general, what happens to the area of a triangle when the length of each side is doubled?
    (d) In general, what happens to the area of a triangle when the length of each side is tripled?
    (e) In general, what happens to the area of a triangle when the length of each side is multiplied by a factor of n?
20. The accompanying figure is a pentagram. Ratios of the lengths of certain sides of this figure form the golden ratio,[32] which is $(1+\sqrt{5})/2$.

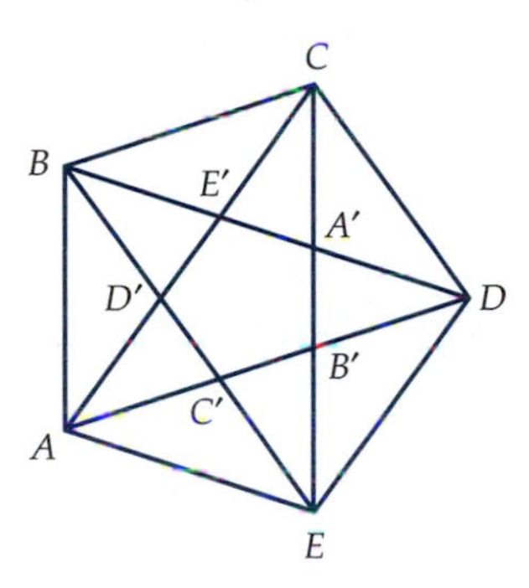

    (a) Find the sum of the angles of a pentagon. (*Hint:* Inserting appropriate diagonals in the pentagon will allow you to use triangles as a tool for finding the sum of the angles.)
    (b) The figure $A'B'C'D'E'$ is a regular pentagon, which means that each angle has the same measure (e.g., $m\angle A'B'C' = m\angle B'C'D' = m\angle C'D'E'$) and each side has the same length. Find $m\angle D'E'A'$.
    (c) Find the measures of the three angles in $\triangle A'E'C$.
    (d) Let $A'C = s$. Approximate $A'C/A'E'$. Compare your approximation with the golden ratio.
21. In addition to the law of sines and law of cosines, other formulas can be used to find the missing dimensions of a triangle. One such formula is

$$\tan A = \frac{a \sin C}{b - a\cos C}.$$

[32] *The golden ratio was explored in Investigation 2 in section 4.1.*

(a) Use the accompanying diagram to show that this formula is true when $\angle C$ is acute.

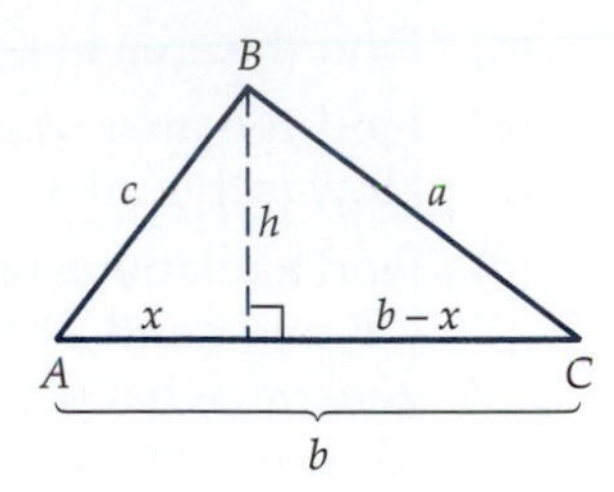

(b) Show that this formula is also true when $\angle C$ is obtuse. (*Hint:* Modify the picture appropriately.)

INVESTIGATIONS

INVESTIGATION 1: A PROOF FOR HERON'S FORMULA

In this section, we obtained the formula

$$A = \frac{1}{2}ab\sqrt{1 - \left(\frac{b^2 + c^2 - a^2}{2ab}\right)^2},$$

where a, b, and c are lengths of the sides of a triangle and A is the area of the triangle. We stated that this formula is equivalent to Heron's formula,

$$A = \sqrt{s(s-a)(s-b)(s-c)},$$

where a, b, and c are the lengths of the three sides of a triangle, s is the semiperimeter $s = (a + b + c)/2$, and A is the area of the triangle. In this investigation, you show that these two formulas are equivalent.

1. Recall that $\sqrt{x^2} = |x|$. Explain why we ignore the absolute value signs throughout this investigation.
2. Show that
$$A = \frac{1}{2}ab\sqrt{1 - \left(\frac{a^2 + b^2 - c^2}{2ab}\right)^2}$$
can be rewritten as $A = \frac{1}{4}\sqrt{4a^2b^2 - (a^2 + b^2 - c^2)^2}$.
3. Show that
$$A = \sqrt{s(s-a)(s-b)(s-c)}$$
can be rewritten as $A = \frac{1}{4}\sqrt{[2ab + (a^2 + b^2 - c^2)][2ab - (a^2 + b^2 - c^2]}$ (*Hint:* Multiply s by $(s - c)$ and $(s - a)$ by $(s - b)$).
4. Explain why
$$\frac{1}{4}\sqrt{4a^2b^2 - (a^2 + b^2 - c^2)^2} = \frac{1}{4}\sqrt{[2ab + (a^2 + b^2 - c^2)][2ab - (a^2 + b^2 - c^2)]}.$$
Conclude that the two area formulas are equivalent.

INVESTIGATION 2: ANOTHER FORMULA FOR THE AREA OF A TRIANGLE

Geometry teaches that a unique triangle is formed when you have side–side–side, side–angle–side, angle–side–angle or angle–angle–side. Therefore, given any of these four combinations of information, you should be able to find the missing dimensions of a triangle, as we did using the law of sines or the law of cosines. But what about area? Because any of these combinations give a unique triangle, we should be able to determine its area. In this section, we found that Heron's formula works for side–side–side. Earlier, we showed that you can use the formula $A = \frac{1}{2}ab \sin C$ if you are given side–angle–side. In this investigation, we derive an area formula if you are given angle–side–angle. Notice that formula also works for angle–angle–side because, if you know the measure of two angles in a triangle, you also know the third.

1. In the accompanying figure, assume you know $m\angle A$, $m\angle B$, and the length of $\overline{AB}$. We have drawn a perpendicular to AB and labeled its length as h. We have also labeled the length of AD as x, which implies that the length of DB is $c - x$.

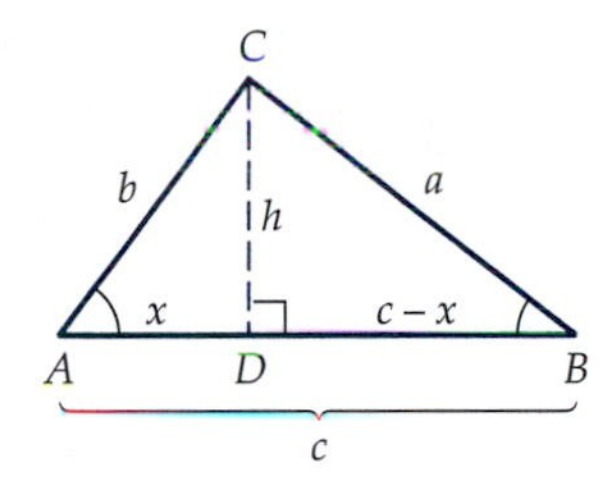

   (a) Find equations for $\tan A$ and $\tan B$ in terms of c, x, and h.

   (b) Solve both equations for h and set them equal to each other.

   (c) Solve this equation for x.

   (d) i. Explain why $h = x \tan A$.

   ii. Using $h = x \tan A$, substitute the expression for x you obtained in part (c) to get an expression for h in terms of $\angle A$, $\angle B$, and c.

   (e) Using the area formula, area $= \frac{1}{2}$ base · height, show that

$$\text{area} = \frac{1}{2}c^2 \frac{\tan A \tan B}{\tan A + \tan B}. \tag{3}$$

2. In mathematical books of formulas, you can often find the area of a triangle given as

$$\text{area} = \frac{1}{2}c^2 \frac{\sin A \sin B}{\sin C}. \tag{4}$$

   We want to show that this equation is equivalent to equation (3) in question 1, part (e).

   (a) Rewrite equation (3) in terms of the cosine and sine functions. Simplify your answer.

(b) Explain why $\sin(A + B) = \sin C$. Explain why it implies that $\sin A \cos B + \sin B \cos A = \sin C$.

(c) Explain why the answer to part (b) shows that equation (3) is equivalent to equation (4).

3. We started this investigation by saying that we wanted an area formula we could use if we were given angle–side–angle. Yet equation (4) requires knowing three angles and one side. Reconcile this seeming discrepancy.

INVESTIGATION 3: HOW ACCURATE ARE THE RESULTS?

One common way to find the height of a building or other object is to measure the angle of elevation from one position, walk in a straight line to a second position, and again measure the angle of elevation. Using the law of sines appropriately, you can then find the height of the object. (See Figure 15.) The accuracy of your calculation of the height depends on more than how accurately you take your measurements. Where you take your measurements is also important. In this investigation, we use the Sears Tower to illustrate that the distance between your measurements has a significant effect on accuracy.

Figure 15

1. The height of the Sears Tower is 1454 ft. Suppose you were to stand 200 ft from the building and find the angle of elevation, α. Moving 100 ft farther away, you take a second measurement of the angle of elevation, β. (See the accompanying figure.) By using α and β as well as knowing that the measurements are 100 ft apart, you can estimate the height of the Sears Tower.

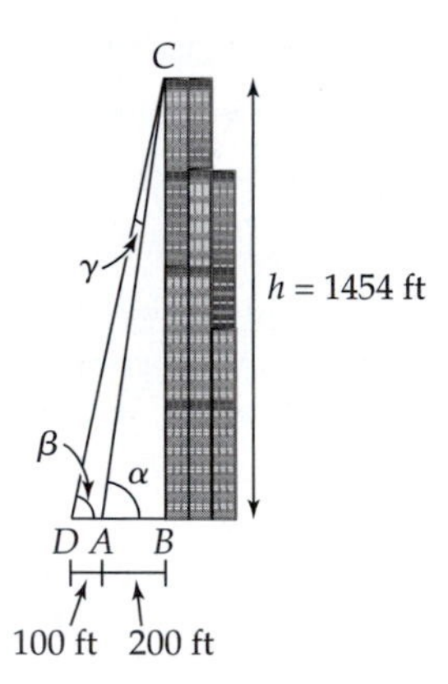

(a) Explain why $\alpha - \beta = \gamma$.

(b) Using the actual height of the Sears Tower, calculate the measurements of α, β, and γ to the nearest tenth of a degree. This calculation gives the measurements you would obtain from a clinometer or other surveying device, assuming there was no instrument or human error (and assuming you were measuring to the nearest tenth of a degree).

(c) Let h represent the height of the Sears Tower. Show that

$$h = \frac{100 \sin \alpha \sin \beta}{\sin \gamma}.$$

(d) Using your values for α, β, and γ as well as the formula from part (c), approximate the height of the Sears Tower to the nearest foot. How accurate is your result?

(e) Suppose the measurement for α was off by a tenth of a degree but the measurement for β was accurate. Increase your value of α by one-tenth of a degree (which will also increase your value of γ by one-tenth of a degree) and approximate the height of the Sears Tower. How does this compare with your answer to part (d)? How does it compare with the true height of the building?

(f) In general, would you recommend that surveyors estimate the height of an object by taking angle measurements 100 ft apart? Why or why not?

2. We are going to again approximate the height of the Sears Tower using two angular measurements. This time, however, the two measurements will be 1000 ft apart. (See the accompanying figure.)

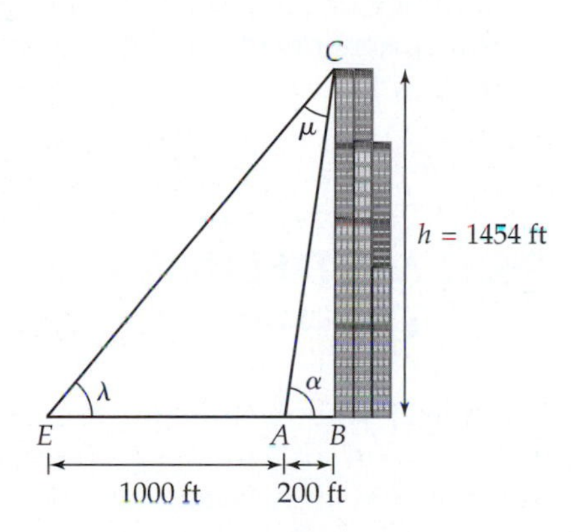

(a) Using the actual height of the Sears Tower, calculate the measurements of α, λ, and μ to the nearest tenth of a degree. This calculation gives the measurements you would obtain from a clinometer or other surveying device, assuming there was no instrument or human error (and assuming you were measuring to the nearest tenth of a degree).

(b) Show that the height of the Sears Tower, h, can be expressed as

$$h = \frac{1000 \sin \alpha \sin \lambda}{\sin \mu}.$$

(c) Using your values for α, λ, and μ as well as the formula from part (b), approximate the height of the Sears Tower. How accurate is your result?

(d) Suppose your measurement for α was off by a tenth of a degree but your measurement for λ was accurate. Increase your value of

α by one-tenth of a degree (which will also increase your value of μ by one-tenth of a degree) and approximate the height of the Sears Tower. How does this compare with your answer for part (c)? How does this compare with the true height of the building?

3. You should have discovered that the distance between the two measurements has a significant impact on the accuracy of the result. In general, measurements that are farther apart produce more accurate results. This "rule of thumb," however, can also lead to problems.
   (a) Describe a situation in which it might be better to take two measurements that are 100 ft apart rather than 1000 ft apart.
   (b) What would you recommend as an optimal distance between the two measurements? Why?
4. What is causing the accuracy to change so dramatically when you increase the distance between the two measurements? (*Hint:* Look carefully at the formulas for h given in question 1, part (c) and question 2, part (b). Think about the behavior of fractions and the values of the various pieces in this formula.)
5. Suppose you are finding the height of an object by measuring the angle of elevation from two positions d feet apart. List five things that affect the accuracy of your result. Do not include either the distance between the measurements or error in measurements (either human or instrument) in your list.

6.3 MODELING BEHAVIOR WITH SUMS OF SINE AND COSINE

Finding a symbolic function to model data obtained from a physical situation is useful for a number of reasons. For instance, it allows us to predict future behavior and estimate values between the data points. A formula also gives a great deal of information in a compact form. The periodic functions $y = \cos x$ and $y = \sin x$ can often be used to model periodic behavior. In this section, we look at what types of applications can be modeled by these two periodic functions. We also see what types of applications can be modeled with sums of sine and cosine functions.

USING SINE AND COSINE TO MODEL PERIODIC BEHAVIOR

Figure 1 The general shape of a cosine (or sine) function.

In section 5.3, we saw that the general shape of a function whose graph resembles Figure 1 is $f(x) = A\cos[B(x + C)] + D$. The amplitude of such a function is $|A|$, the period is $2\pi/B$, the horizontal shift is C, and the vertical shift is D. For example, $f(x) = 4\cos[2(x - \pi/2)] + 2$ has an amplitude of 4, a period of $2\pi/2 = \pi$, a vertical shift of two units upward, and a horizontal shift of $\pi/2$ units to the right. (See Figure 2.) The sine function, $g(x) = A\sin[B(x + C)] + D$, is similar. Multiplying, dividing, or adding a constant to the input or the output of a cosine or sine function stretches, compresses, or shifts the graph of $y = \cos x$ or $y = \sin x$. By choosing

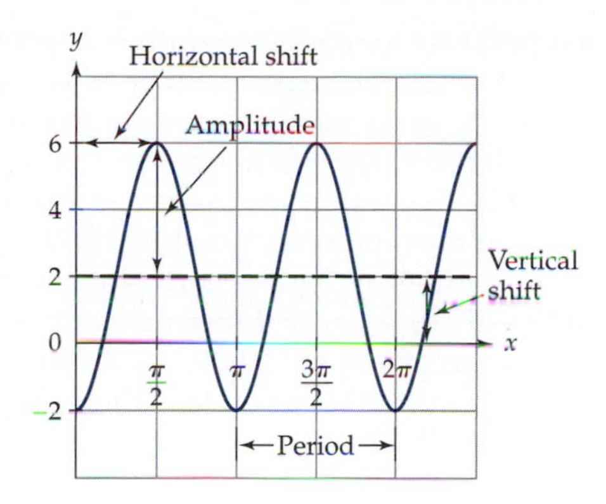

Figure 2 A graph of $f(x) = 4\cos[2(x - \pi/2)] + 2$.

TABLE 1 Average Monthly Maximum Temperatures for Grand Rapids, Michigan

Month	Temp. (°F)
January	29.0°
February	31.6°
March	42.8°
April	56.6°
May	69.3°
June	78.7°
July	82.8°
August	80.5°
September	72.0°
October	59.8°
November	45.8°
December	33.5°

appropriate constants, we can use $f(x) = A\cos[B(x + C)] + D$ or $g(x) = A\sin[B(x + C)] + D$ to model some types of periodic behavior.

One data set that can be modeled with a cosine or sine function is the average monthly temperature for a given location. In the Northern Hemisphere, the temperatures are relatively cool in the winter and warm in the summer. The average maximum temperature (using data from 1961 to 1990) for Grand Rapids, Michigan, is given in Table 1.[33]

A plot of the data in Table 1 is shown in Figure 3. For this plot, we have used the number 1 for January, 2 for February, and so on. Looking at Figure 3, we notice that a cosine or sine function will be a good model for the temperature function because the shape of the graph in Figure 3 is similar to the shape of the graph in Figure 1. We now need to determine reasonable values of A, B, C, and D in the generic formula[34] $y = A\cos[B(x + C)] + D$. The period of this function is 12 because the average monthly maximum

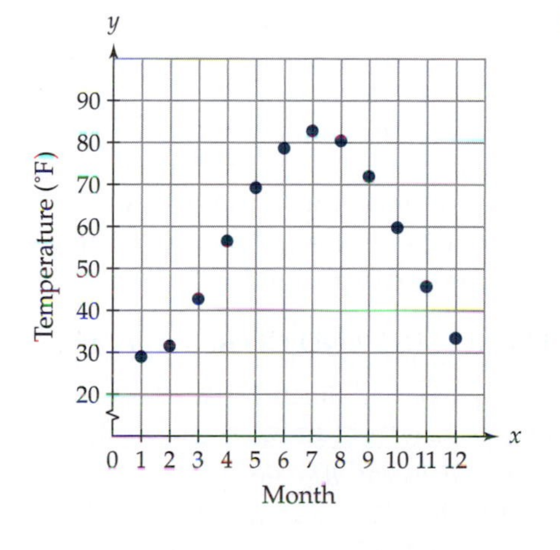

Figure 3 A plot of the average monthly maximum air temperature for Grand Rapids, Michigan.

[33]Temperature Summary, Grand Rapids Weather Service Office Airport *(visited 2 November 1998)* ⟨*http://mcc.sws.uiuc.edu/Summary/Data/203333.txt*⟩.

[34]*It is often easier to use the cosine function rather than the sine function to model this type of periodic data because $y = \cos x$ has a maximum at $x = 0$, which makes it easy to determine the horizontal shift.*

temperature for a given month is approximately the same year after year. By knowing that the period is $2\pi/B$, we find that $12 = 2\pi/B$ or $B = \pi/6$. The highest average monthly maximum temperature occurs in July (month 7) when the average maximum temperature is 82.8°F, and the lowest average monthly maximum temperature occurs in January (month 1) when the average maximum temperature is 29.0°F. Subtracting these two numbers and dividing by 2 gives a reasonable value for the amplitude. So, $A = (82.8 - 29.0)/2 = 26.9$. Because our maximum is at month 7 while the maximum of $y = \cos x$ is at $x = 0$, this graph can be modeled by a cosine function shifted seven units to the right, meaning that $C = -7$. Finally, the vertical shift gives us the midline, $y = D$, which can be approximated by finding the average of the largest and smallest maximum temperatures. So, $D = (82.8 + 29.0)/2 = 55.9$. Our equation is now $T(m) = 26.9\cos[\pi/6(m - 7)] + 55.9$, where m is the month and $T(m)$ is the average monthly maximum temperature. Figure 4 shows the graph of this equation along with the data. You can see that although the graph does not go through every data point, it is a good model for the temperature data.

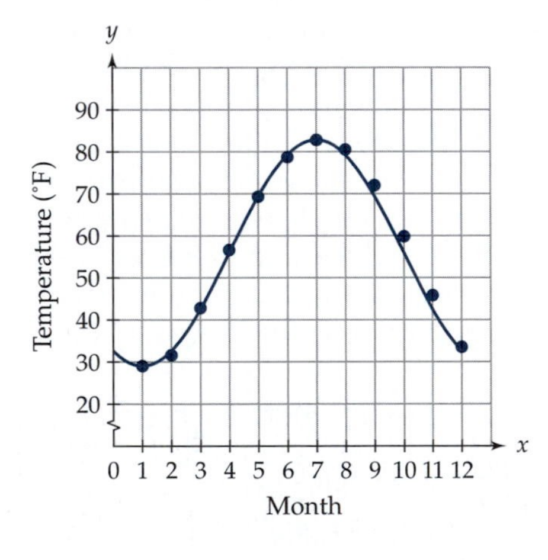

Figure 4 A plot of the average monthly maximum temperature for Grand Rapids, Michigan, along with the function $T(m) = 26.9\cos[\pi/6(m - 7)] + 55.9$.

Sound waves are another example of periodic behavior that can be modeled by cosine or sine functions. The amplitude of the sound wave is related to the intensity or loudness of the sound. The period of the sound wave is related to the frequency of the sound. In fact, the frequency is the reciprocal of the period; that is, $f = 1/p$. Thus, if you have a graph of a sound wave, then you can determine the frequency of the note by looking at the period. One way of obtaining a graph of a sound wave is to use a microphone probe attached to a calculator-based laboratory. Such a graph is shown in Figure 5.

We are interested in finding the equation of this graph. From its shape, we know that using a cosine or sine function will work well. Again, we choose to use a cosine function, $y = A\cos[B(x + C)] + D$. Four periods are completed from $x = 0$ to $x = 0.02$, so the period of the function is

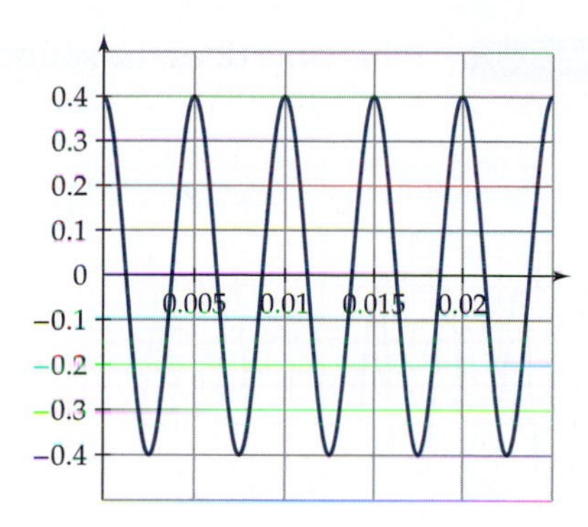

Figure 5 The graph of a sound wave.

$0.02/4 = 0.005$. Thus, $B = 2\pi/0.005 = 400\pi$. There is no vertical shift, so the amplitude is equal to the maximum output, which is $A = 0.4$. Because $y = \cos x$ has a maximum at $x = 0$ as does our function, there is also no horizontal shift. So, the equation is $S(x) = 0.4\cos(400\pi x)$.

We mentioned earlier that the period of a sound wave, p, is related to the frequency of the note played, f, by the equation $f = 1/p$. For the graph in Figure 5, the frequency is $f = 1/0.005 = 200$. The frequencies are given in the standard unit of frequency, the **hertz,** abbreviated **Hz,** which equals one period per second. Table 2 contains the names of the notes and their frequencies for the octave starting at middle C. This table can be used to find the frequency of any note because every time you go up an octave, the frequency doubles. So, the note that was used to produce Figure 5 must have been close to a G above middle C (because $200 \times 2 = 400 \approx 392$).

TABLE 2 Notes and Their Frequencies for the Octave Starting with Middle C

Note	Frequency (Hz)
C	262
C♯ or D♭	277
D	294
D♯ or E♭	311
E	330
F	349
F♯ or G♭	370
G	392
G♯ or A♭	415
A	440
A♯ or B♭	466
B	494
C (next octave)	524

Example 1 What note does the sound wave in Figure 6 represent?

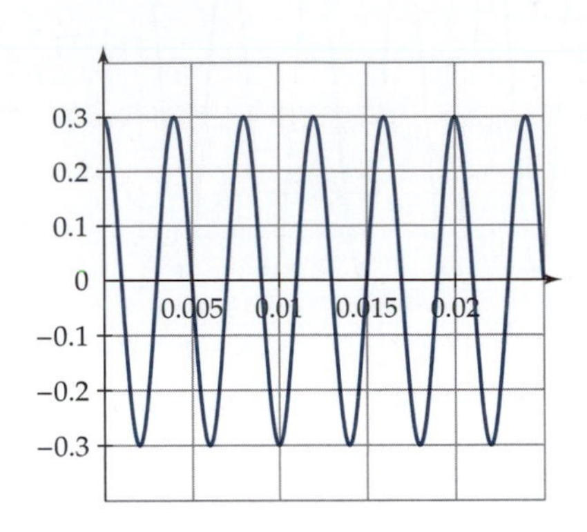

Figure 6

Solution This function completes five periods from $x = 0$ to $x = 0.02$, so the period is $0.02/5 = 0.004$. Thus, the frequency of the note is $1/0.004 = 250$. Using Table 2, this note is close to a B above middle C because $2 \times 250 = 500$ is closest to 494.

READING QUESTIONS

1. Name something other than temperature and sound waves that can be modeled by a sine or cosine function.
2. The following table contains the average low temperature (from 1961 to 1990) for Milwaukee, Wisconsin.[35]

Month	January	February	March	April	May	June
Temp. (°F)	11.6°	15.9°	26.2°	35.8°	44.8°	55.0°
Month	July	August	September	October	November	December
Temp. (°F)	62.0°	60.8°	52.8°	41.8°	30.7°	17.5°

   (a) Why can these data be modeled by a sine or cosine function?
   (b) Find the amplitude of a sine or cosine function that could be used to model the data.
3. What note will produce a sound wave whose equation is $S(x) = 0.35 \cos(880\pi x)$?

[35]Temperature Summary, Milwaukee Weather Service Office *(visited 19 November 1998)* ⟨*http://mcc.sws.uiuc.edu/Summary/Data/4754.79.txt*⟩.

SUMS OF SINE AND COSINE FUNCTIONS

So far, we have looked at periodic behavior that can be modeled by a single sine or cosine function. Using a single function, however, leads to restrictions such as constant amplitude. We can model more complex behavior if we use sums of sine or cosine functions.

The graphs of $f(x) = \cos(4\pi x)$ and $g(x) = \cos[(8\pi/3)x]$ are shown in Figure 7. Notice that these two functions have the same amplitude and have no horizontal or vertical shifts. The only difference is the period. Let $h(x) = f(x) + g(x) = \cos(4\pi x) + \cos[(8\pi/3)x]$. The graph of h is shown in Figure 8. Recall that adding functions means adding the outputs for a given input. The maximum value for $h(x)$ is 2, which occurs when the maximum values for $f(x)$ and $g(x)$ coincide at $x = 0$ and again at $x = 1.5$. The minimum values for $f(x)$ and $g(x)$ never coincide, so the minimum value for $h(x)$ is larger than -2. Although $h(x) = f(x) + g(x)$ is no longer a cosine function, it is still periodic. Recall that a periodic function is one that gives the same output for inputs a fixed distance apart. Thus, there exists a constant p such that $f(x + p) = f(x)$ for all x. The period is the smallest value of p for which this relationship is always true. The function h appears to have a period of 1.5. Let's see why.

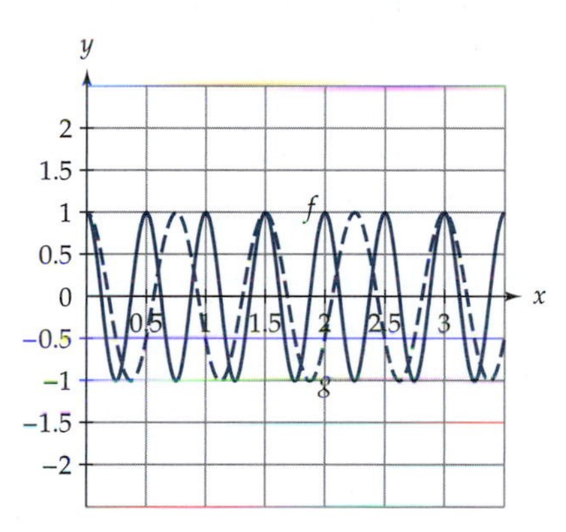

Figure 7 A graph of $f(x) = \cos(4\pi x)$ and $g(x) = \cos[(8\pi/3)x]$.

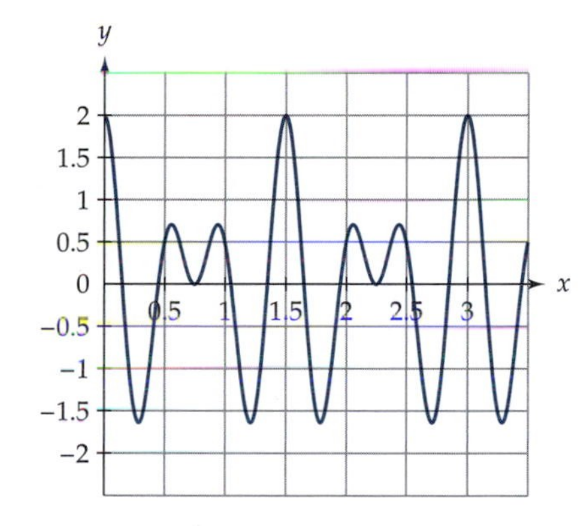

Figure 8 A graph of $h(x) = \cos(4\pi x) + \cos[(8\pi/3)x]$.

The period of $f(x) = \cos(4\pi x)$ is $2\pi/4\pi = \frac{1}{2}$. If we start by looking at $x = 0$, then $f(0) = f(\frac{1}{2}) = f(1) = f(\frac{3}{2}) = f(n/2)$, where n is an integer. The function, f, has the same output for every integer multiple of $\frac{1}{2}$. The period of $g(x) = \cos[(8\pi/3)x]$ is $2\pi/(8\pi/3) = \frac{3}{4}$ so, similarly, g has the same output for every integer multiple of $\frac{3}{4}$. When the two functions are added together, the output is the same (in this case, a maximum of 2) for every *common* integer multiple of $\frac{1}{2}$ and $\frac{3}{4}$. That is, h has the same output when both f and g have the same output. The period of h is the smallest positive common integer multiple of $\frac{1}{2}$ and $\frac{3}{4}$.

We can find the smallest positive common integer multiple of two rational[36] numbers by listing the integer multiples of both numbers until the

[36] *Recall that a rational number is one that can be written as the quotient of two integers.*

same number appears in both lists. With rational numbers, however, we can also use the following algorithm.

1. Rewrite both of the rational numbers so that they share a common denominator.
2. Find the least common multiple of the two numerators.
3. Divide the least common multiple of the numerators by the common denominator.

In this case, our two periods are $\frac{1}{2} = \frac{2}{4}$ and $\frac{3}{4}$. The least common multiple of 2 and 3 is 6. So, the smallest positive common integer multiple of $\frac{1}{2}$ and $\frac{3}{4}$ is $\frac{6}{4} = \frac{3}{2} = 1.5$, the answer we predicted by looking at the graph.

If we are adding two cosine functions, two sine functions, or a cosine function and a sine function, then the sum is always periodic if the periods of the two functions are both rational because, using the algorithm described earlier, we can always find a smallest positive common integer multiple of the two rational numbers. If one period is rational and one is not, however, then the resulting sum is not periodic. If both periods are irrational,[37] the sum might be periodic.

Let's look at some more examples. Suppose $f(x) = \cos x$ and $g(x) = \cos(\pi x)$. Let $h(x) = f(x) + g(x)$. The graphs of f and g are shown in Figure 9(a), and the graph of h is shown in Figure 9(b).

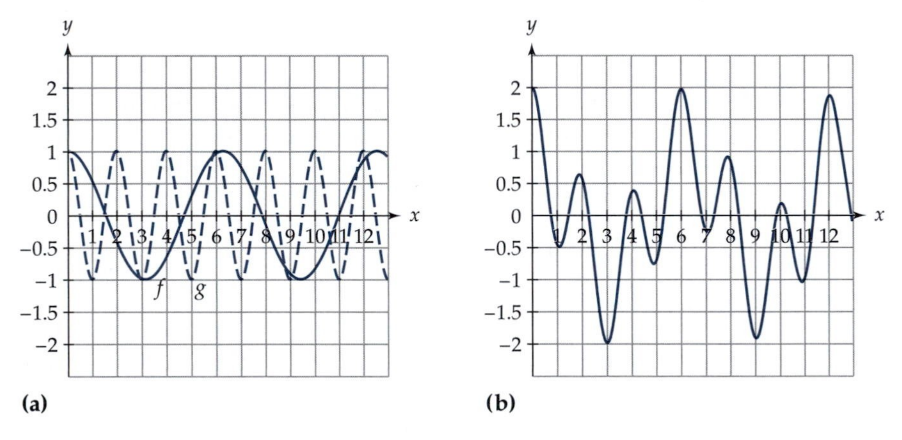

Figure 9 The graphs of (a) $f(x) = \cos x$ and $g(x) = \cos(\pi x)$ and (b) the sum $h(x) = f(x) + g(x)$.

The period of f is 2π, and the period of g is $2\pi/\pi = 2$. The output of f repeats for every integer multiple of 2π, and the output of g repeats for every integer multiple of 2. If we list the integer multiples of 2π and also list the integer multiples of 2, we never see the same number in both lists because all the numbers in the first list will be irrational and all the numbers in the second list will be rational. Thus, $h(x) = \cos x + \cos(\pi x)$ is not a periodic function even though the graph looks periodic, which is

[37] *Recall that an irrational number is one that is not rational.*

always true when the period of one of the functions is rational while the other is irrational. If both periods are irrational, then the sum may or may not be periodic. This situation is illustrated in Example 2.

Example 2 Determine if each of the following are periodic. If so, find the period.

(a) $f(x) = \cos(5x/2) + \cos(3x)$

(b) $g(x) = \sin(\sqrt{2}x) + \cos(2x)$

Solution

(a) The period of $y = \cos(5x/2)$ is $2\pi/(5/2) = 4\pi/5$, whereas the period of $y = \cos(3x)$ is $2\pi/3$. Both of these periods are irrational. Because both are rational multiples of π, however, we can find a smallest common integer multiple of $\frac{4}{5}$ and $\frac{2}{3}$. Because $\frac{4}{5} = \frac{12}{15}$ and $\frac{2}{3} = \frac{10}{15}$, we need to find the least common multiple of 12 and 10, which is 60. So, the smallest common integer multiple of $\frac{4}{5}$ and $\frac{2}{3}$ is $\frac{60}{15} = 4$. Therefore, f is periodic with a period of 4π.

(b) The period of $y = \sin(\sqrt{2}x)$ is $2\pi/\sqrt{2}$, whereas the period of $y = \cos(2x)$ is $2\pi/2 = \pi$. In this case, g is not periodic because there is no common integer multiple of $2\pi/\sqrt{2}$ and π.

READING QUESTIONS

4. For each of the following, determine if the function is periodic. If so, find the period. If not, explain why not.
   (a) $f(x) = \cos(2\pi x) + \sin(\pi x)$
   (b) $g(x) = \sin(3x) + \sin(\pi x)$
   (c) $h(x) = \cos(3x) + \cos(4x)$

AN APPLICATION: THE BEAT GOES ON

Now that we have explored what happens when you add sine and cosine functions, let's see how this information can be applied to sound waves. Earlier in this section, we looked at the sound waves generated by a single tone. What happens if you play two notes at the same time? If you play two identical notes on an instrument (such as a guitar), then the sound is the same as playing a single note, only louder. When you play two different notes, you usually hear two different sounds. Two or more notes together create harmonies in music. If the frequencies of the two notes are close together, however, then you often hear "beats" rather than two distinct notes. If you have ever tuned an instrument, then you are probably familiar with this phenomenon. We perceive the two notes as a single sound whose volume oscillates (hence, the term *beats* is used). As the frequencies of the two notes get closer together, the rate at which the volume oscillates becomes slower, until you finally hear a single note with a constant volume. At this point, you know that your instrument is in tune.

Let's explore why beats occur. Consider an A note with frequency 440 Hz. Because the period is the reciprocal of the frequency, the sound wave of this note has a period of $\frac{1}{440}$ seconds and can be represented by the function $f(t) = \cos(880\pi t)$. If we played the same note on a different tuning fork or instrument at the same time with the same intensity (loudness), then the sound is represented by $g(t) = \cos(880\pi t) + \cos(880\pi t) = 2\cos(880\pi t)$. Figure 10 shows both f and g on the same graph for $0 \le t \le 0.005$. Notice that g, representing two identical sounds added together, is simply the graph of f stretched vertically by a factor of 2. Thus, g represents a sound wave that is louder than f (because the amplitude of the graph corresponds to the loudness of the sound) but that still corresponds to an A note.

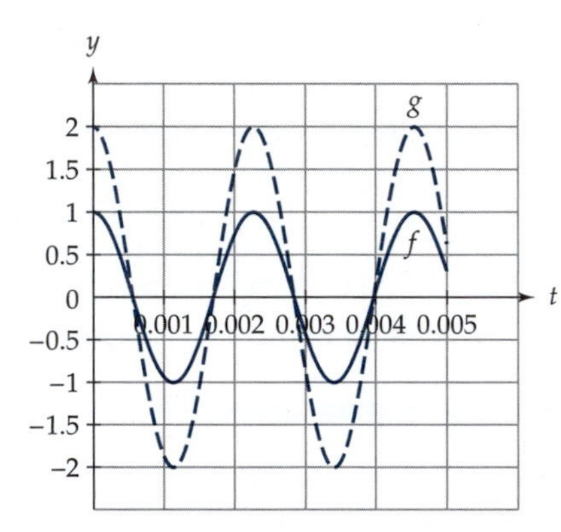

Figure 10 A graph of $f(t) = \cos(880\pi t)$ and $g(t) = 2\cos(880\pi t)$.

Let's now play two different notes with the same intensity (loudness). When we do so, the result is a new wave that is the sum of the two individual sound waves. If we assume the period of a sound wave is always a rational number, then the sum of the two waves is also a periodic function. If we play an A at 440 Hz and a G at 400 Hz, then the resulting sound is modeled by the function $h(t) = \cos(880\pi t) + \cos(800\pi t)$. Because the period of $y = \cos(880\pi t)$ is $\frac{1}{440} = \frac{11}{4840}$ and the period of $y = \cos(800\pi t)$ is $\frac{1}{400} = \frac{10}{4840}$, the period of h is $\frac{110}{4840} = \frac{1}{40}$. Figure 11 is a graph of h using the domain $0 \le t \le \frac{1}{40}$. Notice that the amplitude of this sound wave is getting smaller and smaller until about 0.12 sec, when it starts to increase again. This variation of amplitude corresponds to a variation in loudness that produces beats. The **beat frequency** is equal to the difference in the individual frequencies of the two waves;[38] in this case, $440 - 400 = 40$. Thus, this sound has a maximum intensity 40 times per second, or, in other words, there is $\frac{1}{40}$ of a second between beats. Forty times per second is too fast for our ear to distinguish, so we hear these two notes as two distinct sounds. In fact, the fastest fluctuation in loudness that our ear can detect is 15 to 20 beats per second.[39]

[38] *You are asked to show this fact in Investigation 1.*

[39] Paul A. Tipler, Physics for Scientists and Engineers—Extended Version, *3d ed. (New York: Worth Publishers, 1991), pp. 452–53.*

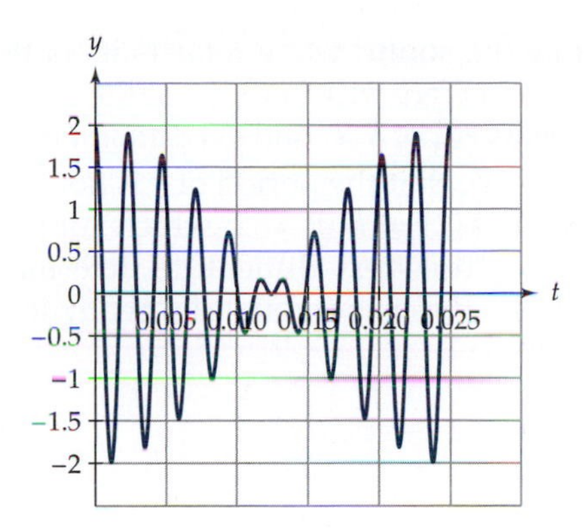

Figure 11 A graph of $h(t) = \cos(880\pi t) + \cos(800\pi t)$ for $0 \le t \le 0.025$.

Let $j(t) = \cos(880\pi t) + \cos(870\pi t)$. For j, the beat frequency is $440 - 435 = 5$ Hz. Thus, this sound has a maximum intensity five times per second or, in other words, there is $\frac{1}{5}$ of a second between beats. In this case, we hear a single note that is varying in volume. Figure 12 shows a graph of j for $0 \le t \le 1$. The graph corresponds to our calculation that the maximum intensity of sound occurs five times per second. Hearing these beats is useful for tuning instruments such as guitars because, if beats can be heard, then you know that the instruments are not in tune. The closer the frequencies of the notes, the slower the beats. Because the period of $y = \cos(880\pi t)$ is $1/440 = 87/38{,}280$ and the period of $y = \cos(870\pi t)$ is $1/435 = 88/38{,}280$, the period of j is $7656/38{,}280 = \frac{1}{5}$. In this case, the period of the function is the same as the time between beats, but that is not always true.

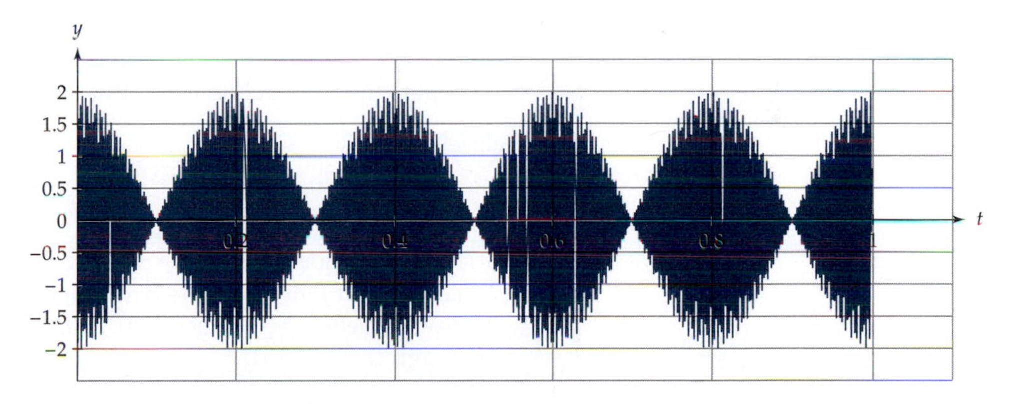

Figure 12 A graph of $j(t) = \cos(880\pi t) + \cos(870\pi t)$ for $0 \le t \le 1$. Notice that there are five beats per second.

Consider the function $k(t) = \cos(881\pi t) + \cos(871\pi t)$. The beat frequency for k is $440.5 - 435.5 = 5$ Hz, the same as the beat frequency for $j(t) = \cos(880\pi t) + \cos(870\pi t)$. The period of k, however, is different. The period of $y = \cos(881\pi t)$ is $2/881 = 1742/767{,}351$, and the period of $y = \cos(871\pi t)$ is $2/871 = 1762/767{,}351$. So, the period of k is $1{,}534{,}702/767{,}351 = 2$. Although the maximum intensity occurs five times per second, it is 2 sec before the sound wave repeats its pattern. A graph of k for $0 \leq t \leq 1$ is shown in Figure 13. Looking at the graph, it is easy to see that the period of the beats is $\frac{1}{5}$ of a second yet difficult to see that the period is 2. Numerically, we can see that the period of k is not $\frac{1}{5} = 0.2$ because $k(0) = 2$ but $k(0.2) \approx 1.62$. Yet $k(2) = k(0) = 2$, which supports our calculation that the period of k is 2. It is very difficult to determine the period (or even if a sum of cosines/sines is periodic) simply by looking at a graph.

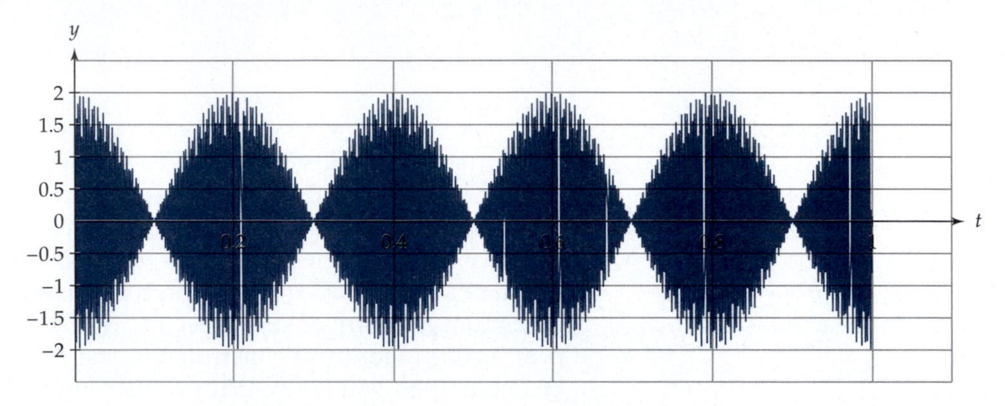

Figure 13 A graph of $k(t) = \cos(881\pi t) + \cos(871\pi t)$ for $0 \leq t \leq 1$. Notice that there are five beats per second.

Example 3 A note is played at a frequency of 300 Hz and another is played at a frequency of 302 Hz.

(a) Give a function that represents both notes played at the same time at the same intensity (or loudness).

(b) What is the beat frequency of this sound? Can you hear the beats?

Solution

(a) $h(x) = \cos(600\pi x) + \cos(604\pi x)$ represents the sound wave because the period is 1/frequency.

(b) The beat frequency is $302 - 300 = 2$ Hz, which means that the maximum intensity occurs two times each second (or that the time between beats is $\frac{1}{2}$ of a second). This beat frequency is slow enough to be heard.

READING QUESTIONS

5. Explain why the sum of two sound waves is always periodic.
6. Suppose two notes are played, one at a frequency of 400 Hz and another at a frequency of 420 Hz. What is the length of time between beats? Are you able to hear the beats?

SUMMARY

In this section, we were reminded that we can use $f(x) = A\cos[B(x + C)] + D$ or $g(x) = A\sin[B(x + C)] + D$ to model periodic behavior where the amplitude is $|A|$, the period is $2\pi/B$, the horizontal shift is C, and the vertical shift is D.

If we are adding two sine or cosine functions, then the sum is always periodic if the periods of the two functions are both rational. We can find the period of this new function by using the following algorithm.

1. Rewrite both of the rational numbers so that they share a common denominator.
2. Find the least common multiple of the two numerators.
3. Divide the least common multiple of the numerators by the common denominator.

The sum of two sine or cosine functions may be periodic even if the periods are not rational. As long as there is a common integer multiple of the two periods, the sum is periodic.

Sound waves can be modeled with sine or cosine functions. In doing so, information is often given about the wave's frequency. The standard unit of frequency is the **hertz (Hz),** which equals one period per second. Sound waves often result from the sum of tones with different frequencies. As a result, a beat is often heard. The **beat frequency** is equal to the difference in the individual frequencies of the two waves.

EXERCISES

1. The following function can be used to model the minutes of daylight for a given day, d, in Grand Rapids, Michigan. (*Note:* The domain for this function is $1 \leq d \leq 365$.)

$$L(d) = 732 + 190\sin\left[2\pi\left(\frac{d-81}{365}\right)\right].$$

   (a) How many minutes of daylight can be expected in Grand Rapids, Michigan, on January 15?
   (b) How many minutes of daylight can be expected in Grand Rapids, Michigan, on March 15?
   (c) What is the amplitude? What is its physical meaning?
   (d) What is the period? What is its physical meaning?
   (e) What is the horizontal shift? What is its physical meaning?
   (f) What is the vertical shift? What is its physical meaning?

2. The following table contains the average low temperature (from 1961 to 1990) for Milwaukee, Wisconsin.[40]

[40]Temperature Summary, Milwaukee Weather Service Office.

Month	January	February	March	April	May	June
Temp. (°F)	11.6°	15.9°	26.2°	35.8°	44.8°	55.0°
Month	July	August	September	October	November	December
Temp. (°F)	62.0°	60.8°	52.8°	41.8°	30.7°	17.5°

(a) Find a function, L, that models the data where m indicates the month (January = 1, February = 2, etc.) and $L(m)$ indicates the average low temperature.

(b) Suppose that the average high temperature was typically 20°F higher than the average low temperature. Find a function, H, that models the average high temperature.

(c) Suppose you wanted a function whose output was in degrees Celsius rather than degrees Fahrenheit. Find a function, C, that models the average low temperature in degrees Celsius. (*Note:* The formula that is used to convert from Fahrenheit to Celsius is $C = \frac{5}{9}(F - 32)$.)

3. Electric service is provided in the form of alternating current (AC) at 110 V and 60 Hz. Thus, the voltage cycles between −110 V and +110 V, and 60 cycles occur each second. Using a calculator-based laboratory, a light probe was held up to a fluorescent light and a time versus light intensity graph was created on the calculator. The function that best modeled the time versus light intensity graph was

$$L(t) = 0.2254445\ \sin(791.7320196x) + 0.521225.$$

Use the function L to answer the following questions.

(a) Why are the data obtained from this experiment best modeled by a periodic function?

(b) What is the period of this function?

(c) The period represents seconds per cycle (the time required for one complete on–off cycle). The frequency represents cycles per second (the number of cycles that occur in 1 sec). Find the frequency.

(d) What is the connection between the frequency of the graph and the information you were given about AC?

4. What note will produce a sound wave whose equation is $S(x) = 0.35 \cos(784\pi x)$? Use Table 2 in this section to help determine your answer.
5. Determine a reasonable equation for a sound wave that is produced by a tuning fork labeled E (above middle C). Use Table 2 in this section to help determine your answer. (*Note:* More than one answer is possible.)
6. A calculator-based laboratory and a microphone probe were used to model the sound wave produced by a tuning fork. A graph similar to the one in the accompanying figure was obtained.

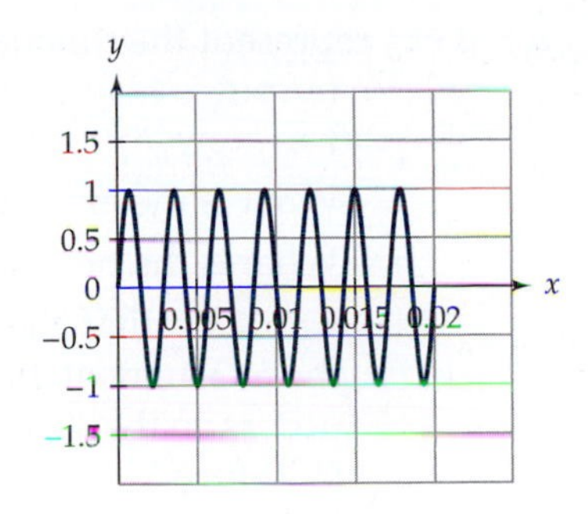

(a) Find a function that models this graph.

(b) What note produced this graph? Use Table 2 in this section to help determine your answer.

7. The height of the sun in the sky at solar noon (when the sun is the highest in the sky) depends on the latitude and on the day of the year. The maximum height of the sun varies by about 47° from the first day of summer to the first day of winter; Earth's axis is tilted about 23.5°, so the difference between being tilted toward or away from the sun is a total of $23.5° + 23.5° = 47°$. In the Northern Hemisphere, the maximum angular height of the sun on the first day of summer (the summer solstice) is $A = 90° + 23.5° - L$, where L is the latitude. The maximum angular height on the first day of winter (the winter solstice) is 47° less than that, or $A = 90° - 23.5° - L$.

   (a) Find the height of the sun at solar noon on both the summer solstice and the winter solstice for the following cities in the Northern Hemisphere. The latitude of each city is given.

      i. Winnipeg, Manitoba, Canada; 49°

      ii. Karachi, Pakistan; 24°

      iii. New York, New York; 40°

      iv. Tegucigalpa, Honduras; 14°

      v. Nome, Alaska; 64°

   (b) Let's compare the height of the sun at solar noon for the summer and winter solstice.

      i. Explain why the height of the sun at solar noon on the summer solstice is the highest it will be for that year.

      ii. Explain why the height of the sun at solar noon on the winter solstice is the smallest it will be for that year.

      iii. What does it mean if the height of the sun at solar noon on the summer solstice is more than 90°? Can this happen?

      iv. What does it mean if the height of the sun at solar noon on the winter solstice is zero? Can this happen?

      v. What does it mean if the height of the sun at solar noon on the winter solstice is negative? Can this happen?

(c) Let f represent the function whose input is day of the year and whose output is height of the sun at solar noon (in degrees) on that day in New York City. We use a cosine function to model f.

i. What is the amplitude of f?

ii. What is the period of f?

iii. Knowing that the summer solstice typically occurs on the 172d day of the year and the winter solstice typically occurs on the 356th day of the year, what is the horizontal shift for f?

iv. What is the vertical shift for f? What does that number indicate?

v. Find a formula for f.

8. The accompanying graph[41] shows the mean temperatures in degrees Celsius for Kalamazoo, Michigan.

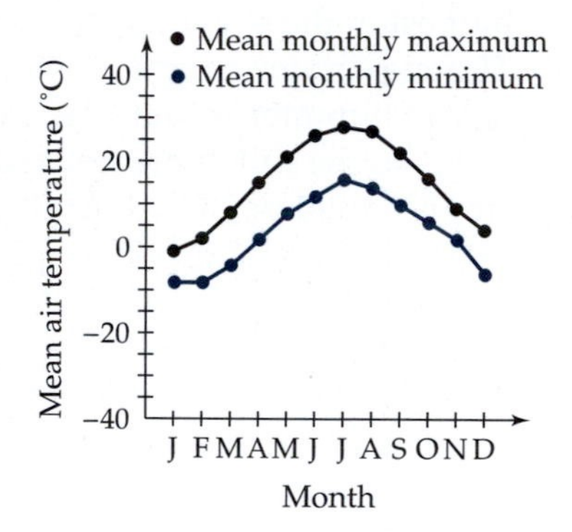

(a) What should you expect for the high temperature in September? Give your answer in degrees Fahrenheit ($F = 32 + \frac{9}{5}C$).

(b) What should you expect for the low temperature in September? Give your answer in degrees Fahrenheit ($F = 32 + \frac{9}{5}C$).

(c) Find a function that models the high temperature in degrees Celsius.

(d) Find a function that models the high temperature in degrees Fahrenheit.

9. The table on the next page gives the cost of a homeowner's gas bill for a 24-month period. In this home, gas is used primarily to run the furnace and hot water heater.

(a) Does this person most likely live in Michigan or in Florida? Why?

(b) Sketch a graph of these data with time on the horizontal axis and cost on the vertical axis.

[41] Mean Air Temperature *(visited 8 July 1998)* ⟨*http://lter.kbs.msu.edu/Weather/Summary%20Figs/temperature.html*⟩.

Year 1		Year 2	
Month	Cost of Gas	Month	Cost of Gas
January	$75.87	January	$58.15
February	$75.29	February	$59.62
March	$72.73	March	$53.95
April	$45.44	April	$43.60
May	$27.73	May	$32.15
June	$19.73	June	$16.48
July	$11.30	July	$12.92
August	$11.76	August	$12.42
September	$12.81	September	$12.92
October	$23.96	October	$15.49
November	$34.16	November	$29.34
December	$50.85	December	$57.57

(c) Find the equation of a periodic function that models the data for year 1. For the input, let 1 = January, 2 = February, and so forth.

(d) What is the period and the vertical shift of your function? What do these numbers mean in terms of the gas bill?

(e) Use your function from part (c) to predict the cost of gas for May of year 2. How well did your function predict the actual value? Why?

10. The table gives the amount of electricity furnished by a large Midwestern power plant. The hour is based on a 24-hr clock (for example, 16 = 4 P.M.).

Weekday	Date	Hour	Average Megawatts
Tuesday	14 July 1998	4	3910
Tuesday	14 July 1998	8	4955
Tuesday	14 July 1998	12	6327
Tuesday	14 July 1998	16	6833
Tuesday	14 July 1998	20	6405
Tuesday	14 July 1998	24	5193
Wednesday	15 July 1998	4	4112
Wednesday	15 July 1998	8	5191
Wednesday	15 July 1998	12	6519
Wednesday	15 July 1998	16	6901
Wednesday	15 July 1998	20	6467
Wednesday	15 July 1998	24	5366
Thursday	16 July 1998	4	4293
Thursday	16 July 1998	8	5214
Thursday	16 July 1998	12	6456
Thursday	16 July 1998	16	6631
Thursday	16 July 1998	20	6120
Thursday	16 July 1998	24	4897

(a) Average the data for each given hour from the 3 days and fill out the following table.

Average Electricity Use on a Week Day in July

Hour	Average Megawatts
4	
8	
12	
16	
20	
24	

(b) Find the equation for a periodic function that models the data given in the table from part (a).

(c) What is the period and vertical shift of your function? What do these numbers mean in terms of the electrical output of the power plant?

(d) Use your function from part (b) to predict the electricity used at 4 P.M. on Tuesday, 14 July 1998. Would you recommend that the power company use your function to estimate the amount of electricity it needs to produce? Why or why not?

(e) At what time of day is the maximum amount of electricity used? Why?

(f) Would you expect different usage patterns on the weekend? Why or why not?

11. Let $f(x) = \cos Ax + \cos Bx$.

(a) Explain why f will not necessarily be periodic.

(b) Give a condition for A and B that guarantees that f will be a periodic function.

12. Let $g(x) = \cos Ax + \cos Bx$. Assume g is periodic.

(a) Let P_1 be the period of $y = \cos Ax$ and P be the period of g. Is it possible for $P \leq P_1$? Explain.

(b) Let M be the maximum value of $g(x)$.

i. Is it possible for $M > 2$? Explain.

ii. Is it possible for $M = 2$? Explain.

iii. Is it possible for $M < 2$? Explain.

(c) Let m be the minimum value of $g(x)$.

i. Is it possible for $m > -2$? Explain.

ii. Is it possible for $m = -2$? Explain.

iii. Is it possible for $m < -2$? Explain.

13. Determine if each of the following are periodic. If so, find the period. If not, explain why not.
    (a) $f(x) = \sin(\frac{3}{4}x) + \sin(\frac{4}{5}x)$
    (b) $f(x) = \cos(2\pi x) + \sin(\frac{4}{5}x)$
    (c) $f(x) = \cos(\frac{5}{4}x) + \cos(5x)$
    (d) $f(x) = \sin(\frac{1}{4}x) + \sin(\frac{3}{7}x)$
    (e) $f(x) = \sin(\sqrt{2}\pi x) + \cos(2\pi x)$
    (f) $f(x) = \cos(2x) + \sin(\frac{1}{5}x)$
    (g) $f(x) = \sin(\pi x) + \sin(3\pi x)$
14. Let $h(x) = \sin^2(x) + \cos^2(x)$.
    (a) Is $y = \sin^2(x)$ periodic? If so, what is its period?
    (b) Is $y = \cos^2(x)$ periodic? If so, what is its period?
    (c) Is h periodic? If so, what it its period?
15. Let f be a periodic function with period p. Thus, $f(x+p) = f(x)$ for all values of x. If n is an integer, explain why $f(np) = f(0)$.
16. The Touch-Tone phone system uses pairs of tones (high and low) to represent the various keys. The low tones vary according to the horizontal row of the tone button and the high tones correspond to the vertical column of the tone button. For example, when the 4 button is pressed, the 770-Hz and 1209-Hz tones are sent together. The telephone central office then decodes the number from this pair of tones.
    (a) Give a function that models the sound heard when the 4 button on a phone is pressed, assuming the two tones are equally loud.
    (b) What is the period of the sound wave representing this sound?
17. Some people believe that our behavior is partially governed by biorhythms. Biorhythms are three cycles (physical, emotional, and intellectual) that begin at birth. The physical cycle has a period of 23 days, the emotional cycle has a period of 28 days, and the intellectual cycle has a period of 33 days. According to proponents of this theory, when our cycles are positive, we feel better physically, are more cheerful, and think more quickly. Because we assume all three cycles are zero at birth, we can model biorhythms with a sine function with no horizontal or vertical shift.
    (a) For each biorhythm, find a sine function that models it. The input is *days since you were born*, and the output is between 1 and -1.
    (b) To use the biorhythm functions, you must first compute your age in terms of days. Remembering that leap years, which usually occur in years that are divisible by 4 (such as 1980), have 366 days and that other years have 365 days, compute your current age in terms of days since you were born.
    (c) Find your current value for each of the three biorhythms. Do these values seem to correspond to the way you are feeling physically, emotionally, and intellectually today?
    (d) Graph the three biorhythms for the next 4 weeks of your life. Sketch your graph, being sure that everything is clearly labeled.

(e) According to proponents of biorhythms, you are most vulnerable on days when your biorhythm is zero because your levels are changing the most rapidly at this point. For example, if your physical biorhythm is zero, then you are more accident prone.[42] A day when at least one of your biorhythms is zero is called a "critical day."

i. How often is your physical biorhythm zero? Answer the same question for the other two biorhythms. (*Note:* The functions will be zero when the biorhythm is changing from positive to negative as well as when it is changing from negative to positive.)

ii. Assume both your physical and emotional biorhythms were zero at the same moment. How long will it be before they are both at zero again?

iii. One source suggests that all three cycles are zero approximately every 7 years.[43] Is that source correct?

18. Suppose a homeowner's monthly gas bill is modeled by the function $g(m) = 38\cos[(\pi/6)(m-1)] + 50$, where t represents the month (January $= 1$, etc.). The same homeowner's monthly electricity bill is modeled by the function $e(m) = 8\cos[(\pi/6)(m-8)] + 32$. Let $h(m) = g(m) + e(m)$.

(a) What does $h(m)$ represent in terms of the homeowner?

(b) Is h a periodic function? If so, what is its period? If not, explain why.

(c) i. What is the maximum value of $g(m)$? When does it occur?

ii. What is the maximum value of $e(m)$? When does it occur?

iii. What is the maximum value of $h(m)$? When does it occur?

iv. Could you have predicted the maximum value of $h(m)$ and when it occurred by looking at the maximum values and their locations of $g(m)$ and $e(m)$? Justify your answer.

(d) i. What is the minimum value of $g(m)$? When does it occur?

ii. What is the minimum value of $e(m)$? When does it occur?

iii. What is the minimum value of $h(m)$? When does it occur?

iv. Could you have predicted the minimum value of $h(m)$ and when it occurred by looking at the minimum values and their locations of $g(m)$ and $e(m)$? What if you also knew the maximum value of $h(m)$ and where it occurred? Justify your answer.

19. A note is played at a frequency of 440 Hz and another is played at a frequency of 448 Hz.

(a) Let S be the function representing the sound wave produced when both notes are played at the same time at the same intensity (or loudness). Find an equation for S.

[42]Biorhythms *(visited 8 July 1998)* ⟨*http://www.value.net/ esoteric/biorhythm/biorhyth.html*⟩.

[43]BioRythms [sic] *(visited 30 June 1998)* ⟨*http://freespace.virgin.net/anthony.edey/biorythms.html*⟩.

(b) Is S periodic? Justify your answer.

(c) What is the beat frequency of this sound? Can you hear the beats?

20. Tone A is played at a frequency of 400 Hz and tone B is played at a frequency of 410 Hz. Assume both are played at the same intensity (or loudness).

    (a) Determine a reasonable modeling function for tone A. Do the same for tone B. (*Note:* More than one answer is possible.)

    (b) Let h represent the function modeling the sound wave produced when both tones are played at the same time. Graph h and give its equation.

    (c) Is h periodic? If so, what is its period?

    (d) When both tones are played simultaneously, will you hear beats? If so, what is the time between beats? If not, why not?

21. Suppose a 220-Hz (A) tuning fork is struck simultaneously with the playing of the A note on a guitar and you hear five beats per second. Assuming the tuning fork is accurate, what do you know about the note played on the guitar?

22. One reason musical instruments do not all sound the same is that the shape and material used to construct the instrument determine the harmonics produced (other factors, such as the way the tone begins, the vibrato or tremolo, and the rate at which the harmonics build up, are at play as well). The harmonics are multiples of the fundamental frequency. For example, if a note is played at a frequency of 200 Hz, then its harmonic is called the fundamental frequency (or the first harmonic). Its second harmonic has a frequency of 400 Hz, its third harmonic has a frequency of 600 Hz, and so on. The resultant sound is the sum of the harmonics.[44]

    (a) The approximate relative intensity of the harmonics for a cornet are given in the following table.

Harmonic	1	2	3	4	5
Relative Intensity	3.2	4.5	7.8	3.0	1.0

    i. Write an equation for the function that describes the sound of a cornet playing an A at a frequency of 440 Hz. Assume the relative intensity of a note is the same as the relative amplitude.

    ii. Graph your function from part (i).

[44] *Paul A. Tipler,* College Physics *(New York: Worth Publishers, 1987), pp. 416–17.*

(b) The harmonics for a clarinet are given in the following table.

Harmonic	1	2	3	4	5	6	7
Relative Intensity	6.0	1.6	6.8	2.0	4.5	2.0	3.4

i. Write an equation for the function that describes the sound of a clarinet playing an A at a frequency of 440 Hz. Assume the relative intensity of a note is the same as the relative amplitude.

ii. Graph your function from part (i).

23. In the exercises for section 5.4, question 22, you proved the following identities:

$$\cos x \cos y = \tfrac{1}{2}(\cos(x+y) + \cos(x-y))$$

$$\sin x \sin y = \tfrac{1}{2}(\cos(x-y) - \cos(x+y)).$$

(a) Use the first identity to show that $y = \cos(2\pi x)\cos(4\pi x)$ is periodic. Find the period.

(b) Use the second identity to show that $y = \sin(4x)\sin(9x)$ is periodic. Find the period.

INVESTIGATIONS

INVESTIGATION 1: THE LENGTH OF TIME BETWEEN BEATS[45]

In this section, we stated that the beat frequency is the difference in the frequencies between two almost identical tones and that the length of time between beats is the reciprocal of the beat frequency. In this investigation, you show why.

1. Suppose you have two similar tones, one with a frequency of n_1 and the other with a frequency of n_2, that are played with the same loudness at the same time. Give a function, f, using the sum of sine functions, that models the sound wave that would be produced. Assume the loudness of both sounds at time zero is zero.

2. In Investigation 2 in section 5.4, you derived the identity

$$\sin x + \sin y = 2\cos\left(\frac{y-x}{2}\right)\sin\left(\frac{x+y}{2}\right).$$

Use this identity to rewrite your function from question 1 as a product of a cosine and a sine function.

3. When looking at the representation of f given in question 2, we can think of f as a sine function whose amplitude is given by a cosine function. In

[45] *Paul A. Tipler,* Physics for Scientists and Engineers—Extended Version *(New York: Worth Publishers, 1991), pp. 452–53.*

other words, we can think of it as a sine function with a variable amplitude. The sound is loud whenever the amplitude is either at a maximum or a minimum. By looking at the cosine in the product representation of f, determine how often a maximum or a minimum occurs.

4. Your answer to question 3 gives you the *period* of the beats (i.e., the number of seconds between beats). The beat *frequency* is the number of beats *per second*. Use the beat period to find the beat frequency. (*Note:* The beat frequency should be the difference in the frequencies of the original two tones.)
5. When two almost identical tones are played together, the tone that you hear is given by the sine function in the representation of f from question 2. What can you say about the relationship between the original two tones and the sound that you hear when these tones are played simultaneously at the same loudness?

INVESTIGATION 2: DISTANCE BETWEEN PLANETS

The distance between Earth and another planet is a complex relationship because the planets are moving in different orbits and at different rates. We begin this investigation by finding the distance between a stationary point and a moving point on a unit circle. We expand the calculation to find the distance between two points revolving on two different circles. Finally, we find a function that models the distance between Earth and Mars, assuming the orbits are circular.

1. We first find the distance between a stationary point and a point moving on a unit circle. See the accompanying figure.

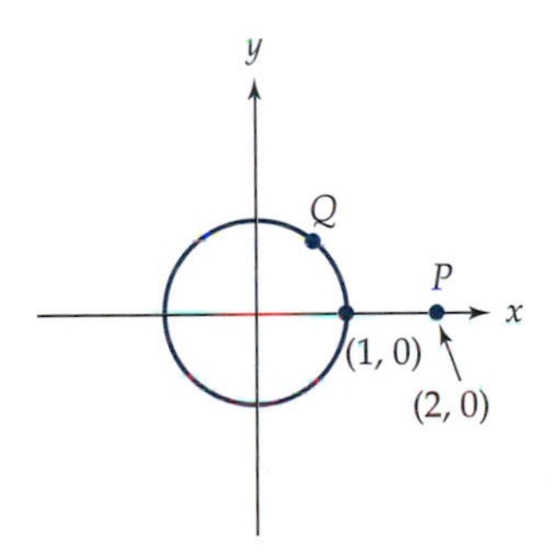

(a) Explain why the coordinates of point Q can be expressed as $(\cos\theta, \sin\theta)$.

(b) Show that the distance between the stationary point, P, at $(2, 0)$ and point Q is $d(\theta) = \sqrt{5 - 4\cos\theta}$.

(c) Graph your function from part (b). Where is point Q when the two points are the closest? Where is point Q when the points are the farthest apart?

(d) Now suppose Q is moving around the circle at a rate of 2 revolutions per second and, at $t = 0$, is at the point $(1, 0)$. Let $h(t) = \theta$ be the function whose input is time in seconds and whose output is the angle, θ, such that the coordinates of Q are $(\cos\theta, \sin\theta)$. Let $f = d \circ h$, the composition of d and h.

i. Find the formula for f. It will be the function whose input is time in seconds and whose output is the distance between Q and $(2, 0)$.

ii. When is point Q closest to the point $(2, 0)$? When is it farthest from the point $(2, 0)$?

2. We now assume both points are moving. Let Q be a point on the unit circle centered at the origin that is moving at the rate of 2 revolutions per second. Let P be a point on a circle of radius 2 centered at the origin that is moving at the rate of 1 revolution per minute (or $\frac{1}{60}$ of a revolution per second). Assume at time zero that Q is located at $(1, 0)$ and P is located at $(2, 0)$. (See the accompanying figure).

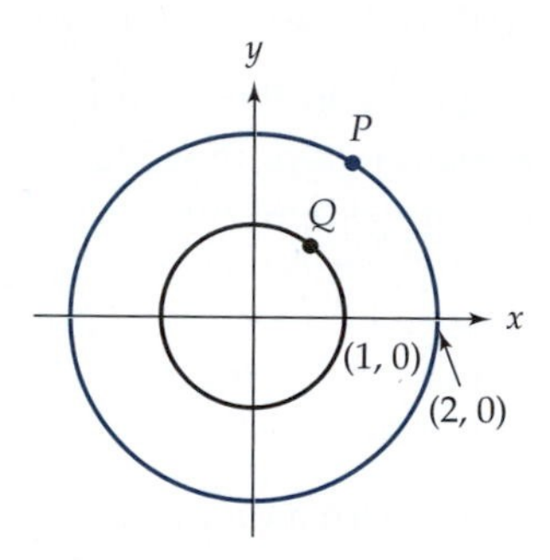

(a) What is the ordered pair that gives the position of P in terms of t, time in seconds?

(b) Write a function for the distance between P and Q in terms of t. Simplify your answer as much as possible. (Do not forget that both P and Q are moving.)

(c) Graph your answer from part (b). Use the graph to answer the following questions.

i. After $t = 0$, when are P and Q the closest? Where are the two points when that happens?

ii. When are P and Q farthest apart? Where are the two points when that happens?

3. Planets orbit in ellipses, not circles. They also orbit in different planes. Most orbital paths, however, are almost circular, and most orbital planes are close together. Assume Earth and Mars have circular orbits that are in the same plane.

(a) The radius of Earth's orbit is about 150 million km, and Earth completes 1 revolution in 1 year. The radius of Mars's orbit is about 228 million km, and Mars completes 1 revolution in 1.88 Earth years. Assume the two planets are as close together as possible at $t = 0$. Find a function that gives the distance between the two planets, in millions of kilometers, where the input is time, in Earth years.

(b) Let r_1 (150 million km) represent the radius of Earth's orbit and r_2 (228 million km) represent the radius of Mars's orbit.

i. Show that the minimum distance between the two planets is $r_2 - r_1$.

ii. Show that the maximum distance between the two planets is $r_2 + r_1$.

(c) According to NASA,[46] the minimum distance between Earth and Mars is 56 million km and the maximum distance between the two planets is 399 million km.

i. The percentage of error for an estimate is defined as (actual – estimate)/actual. Find the percentage error in the estimate of the minimum distance given in question 3, part (b)(i).

ii. Find the percentage error in the estimate of the maximum distance given in question 3, part (b)(ii).

iii. Assuming circular orbits makes calculations easier yet less accurate than other assumptions. Looking at the percentage errors, when (if ever) do you think assuming circular orbits is appropriate for calculating the distance between Earth and Mars?

INVESTIGATION 3: DOUBLE FERRIS WHEEL

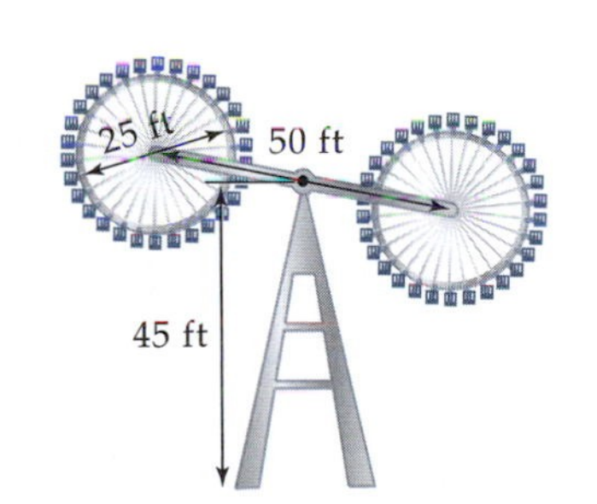

Figure 14

Most amusement parks have a Ferris wheel. A few amusement parks, such as Pacific Ocean Park in California, however, took the concept one step further by having a double Ferris wheel, which consists of two Ferris wheels on a long rotating arm. (See Figure 14.) Assume each small wheel has a diameter of 25 ft, the larger rotating arm has a length of 50 ft, and the center of the large rotating arm is 45 ft off the ground. Assume the small wheels rotate at 5 revolutions per minute (rpm) and the large arm rotates at 2 rpm. A wheel at its lowest point is 7.5 ft off the ground, and both wheels as well as the arm rotate clockwise. Let $t = 0$ represent when a passenger is boarding the right-hand Ferris wheel when it is at its lowest point.

1. We are interested in finding a function that models the position of a person on this double Ferris wheel.
   (a) Find a function, f, modeling the height of the point at the center of the right wheel where the input, t, is time in minutes.
   (b) Assume the arm is stationary and the right wheel is in its lowest position. Find a function, g, modeling the height of a point on the revolving right wheel relative to the hub where the input, t, is time in minutes.

[46] *NASA,* Mars Lithograph *(visited 6 April 1999) (http://spacelink.msfc.nasa.gov/Instructional.Materials/NASA.Educational.Products/Solar.System.Lithograph.Set/Mars/Mars.Lithograph.pdf).*

(c) Let $h(t) = f(t) + g(t)$.

i. Sketch a graph of h for $0 \leq t \leq 15$.

ii. What is the maximum height for h? What is the minimum height for h?

iii. What is the period of h?

2. You should have found that the maximum for h is less than the highest possible height.

(a) In terms of the position of a person on the double Ferris wheel, what must be true if the maximum value of $h(t)$, the function describing the height, is the maximum possible height?

(b) In terms of the revolution of the small wheel and the revolution of the large arm, what will make the maximum value of $h(t)$ the maximum possible height?

(c) Modify the description of the double Ferris wheel given at the beginning of this investigation so that the maximum value of $h(t)$ is equal to the maximum possible height.

3. Let m be the movement constant for a double Ferris wheel. We define m as the number of times a person on the double Ferris wheel is going up during one period. For example, for a single Ferris wheel, $m = 1$ because you are going up once. For the double Ferris wheel described at the beginning of this investigation, $m = 5$ because you are going up five times before the end of one period. (See your graph from question 1, part (c)(i)).

(a) We are interested in building a double Ferris wheel that satisfies the following conditions.

The double Ferris wheel is physically possible to construct and (relatively) safe to ride.

A person riding the double Ferris wheel will achieve the maximum possible height at some point during the ride.

The length of the ride is no less than 5 min and no more than 20 min. (Assume you get on and get off when the small wheel is at the lowest possible point.)

The movement constant is greater than 8.

i. Give the dimensions for a double Ferris wheel that satisfies the above conditions.

ii. Find the function, h, that models the position of a person on the double Ferris wheel you just described.

iii. Graph your function, h, for one period.

iv. Verify that your double Ferris wheel meets each of the stated conditions.

(b) We are now interested in building a double Ferris wheel that satisfies the conditions given in part (a) except we want the movement constant to be less than 3.

i. Give the dimensions for a double Ferris wheel that satisfies the above conditions.

ii. Find the function, h, that models the position of a person on the double Ferris wheel you just described.

iii. Graph your function, h, for one period.

iv. Verify that your double Ferris wheel meets each of the stated conditions.

6.4 OTHER APPLICATIONS FOR TRIGONOMETRIC FUNCTIONS

In the previous section, we used sums of sine and cosine functions to model some types of periodic behavior. In this section, we are interested in modeling behavior that, although it has a cyclic component, is not periodic. For example, the monthly cost of heating a house varies seasonally yet gradually increases over time due to the rising cost of natural gas. To model this situation, we need a function that oscillates (such as sine or cosine) yet also increases (such as $y = x$ or $y = 1.5^x$). We look at situations, such as the cost of heating a home, that can be modeled with sums of functions where one of the terms is a sine or a cosine function. We also consider applications that can be modeled by a product of a periodic function and a nonperiodic function.

SUMS OF PERIODIC AND NONPERIODIC FUNCTIONS

Recall that $y = A\cos[B(x + C)] + D$ is a periodic function where the amplitude is $|A|$, the period is $2\pi/B$, the horizontal shift is C, and the vertical shift is D. What about a function of the form $y = A\cos[B(x + C)] + f(x)$ such as $g(x) = \cos(2\pi x) + x$? Figure 1 shows a graph of g and a graph of $y = x$ for $0 \leq x \leq 5$. Notice that g oscillates (similar to a cosine function), but its vertical shift is along the line $y = x$. Thus, g is not periodic because it does not repeat the same outputs at equal intervals.

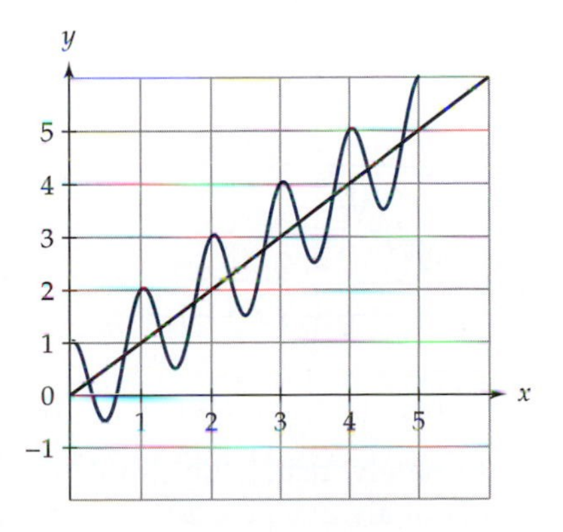

Figure 1 A graph of $g(x) = \cos(2\pi x) + x$ and $y = x$ for $0 \leq x \leq 5$.

Figure 2 shows a graph of g for $0 \le x \le 120$. By "zooming out," we can see that the graph of g is similar to the line $y = x$. It is difficult to see the oscillation produced by the cosine graph because $y = \cos x$ varies from 1 to -1 and our window is 120 units wide. On such a large window, it is difficult to see such a small difference. The function g exhibits characteristics of both $y = \cos x$ (e.g., oscillating) and $y = x$ (e.g., it looks like $y = x$ when we zoom out).

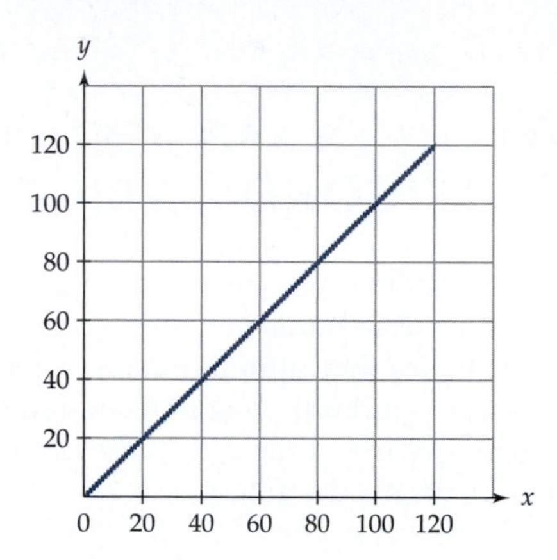

Figure 2 A graph of $g(x) = \cos(2\pi x) + x$ for $0 \le x \le 120$.

A graph of h is given in Figure 3. The shape of the graph suggests that h might be the sum of a periodic function and a nonperiodic function because it oscillates but has a variable vertical shift. The pattern of the vertical shift appears to be exponential, so h is probably the sum of a cosine (or sine) function and an exponential function. In fact, the formula used to produce the graph for h is $h(x) = \cos(2\pi x) + 1.16^x$. Although it is difficult or impossible to determine the exact formula for h by looking at the graph, we were able to determine the general form of the answer.

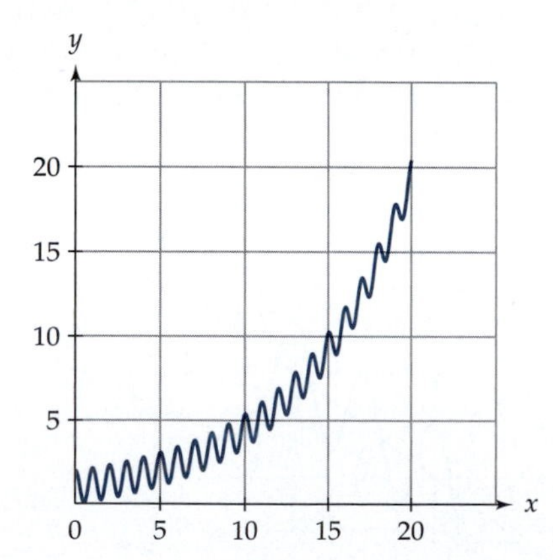

Figure 3 A graph of h for $0 \le x \le 20$.

Unfortunately, using the graph to determine the general form of the answer is not always possible. The graph of a sum of a periodic function

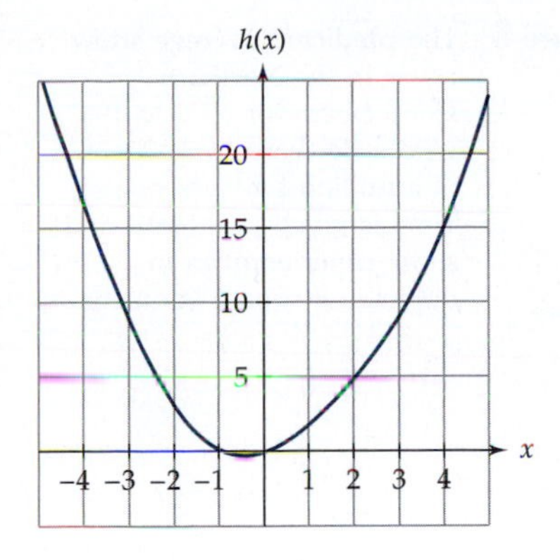

Figure 4 A graph of $h(x) = \sin x + x^2$.

and a nonperiodic function does not necessarily exhibit the characteristics of both terms. For example, Figure 4 shows the graph of $h(x) = \sin x + x^2$. This graph looks like a distorted version of $y = x^2$, but there is no visible oscillation because the output of $y = \sin x$ is too small and the oscillation is too infrequent to be visible. In this case, you can guess the form of only one of the terms by looking at the graph.

Let's consider an application that can be modeled by the sum of a periodic function and a nonperiodic function. Consider the amount of snow cover in the Northern Hemisphere. According to a paper by D. A. Robinson,[47] the snow cover in the Northern Hemisphere in 1972 varied from a high of 48.5 million km^2 in February to a low of 4.8 million km^2 in August. Using this information, the snow cover can be modeled by the function $s(t) = 21.85\cos[(\pi/6)(t-2)] + 26.65$, where t represents time in months and $s(t)$ represents snow cover in millions of square kilometers. Let $t = 1$ represent January 1972. The graph of s for 1972 is shown in Figure 5.

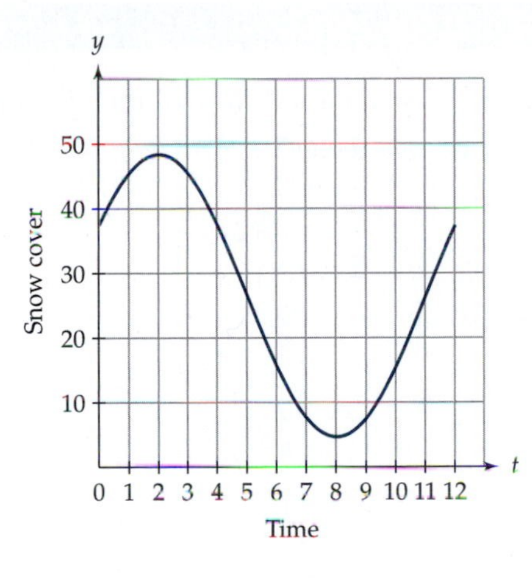

Figure 5 The graph of a function modeling snow cover in the Northern Hemisphere for 1972. Time is given in months and snow cover is given in millions of square kilometers.

Many people believe that the amount of snow cover is decreasing from year to year because of global warming. In fact, the annual snow cover in the Northern Hemisphere has decreased by about 0.000628 million km^2 per month.[48] A function that models the snow cover in the Northern Hemisphere including this decrease is $g(t) = 21.85\cos[(\pi/6)(t-2)] + 26.65 - 0.000628t$, where t represents time in months and $g(t)$ represents snow cover in millions of square kilometers. This function is no longer a periodic function because there is now a variable vertical shift given by $y = 26.65 - 0.000628t$. A graph of g and the line $y = 26.65 - 0.000628t$ for 1972 to 1982 is given in Figure 6. This model incorporates the oscillation of the snow throughout the year along with the gradual decrease of snow cover from year to year.

[47] *D. A. Robinson, "Recent Trends in Northern Hemisphere Snow Cover,"* NSIDC: Northern Hemisphere Snow Charts *(visited 10 May 1999) (http://www.nsidc.colorado.edu/NSIDC/EDUCATION/SNOW/snow_Robinson.html).*

[48] *The slope of the "line of best fit" for the annual snow cover from 1972 to 1997 is 0.00754. This number divided by 12 is 0.000628.*

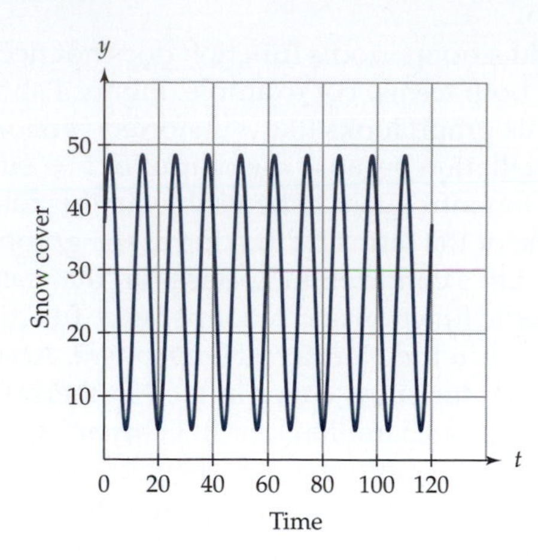

Figure 6 The predicted average snow cover in the Northern Hemisphere for 1972 to 1982 with a decrease of 0.000628 of a million km^2 per month. Time is given in months and snow cover is given in millions of square kilometers.

READING QUESTIONS

1. Match each equation with the appropriate graph.
   (a) $y = 3\cos(10x) - x^2$
   (b) $y = 0.5\sin\left(\dfrac{x}{2}\right) + \sqrt{x}$
   (c) $y = 2\cos(9x) + 2^x$
   (d) $y = \cos(20x) + x^3$

i.

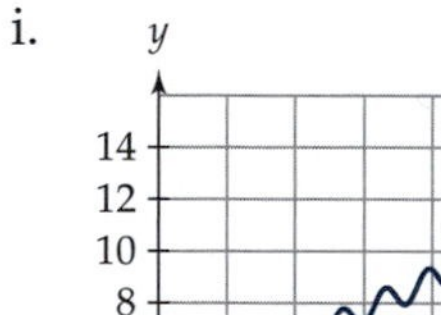
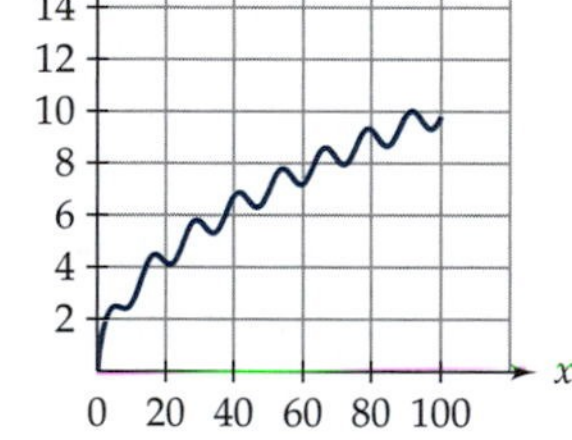

ii.

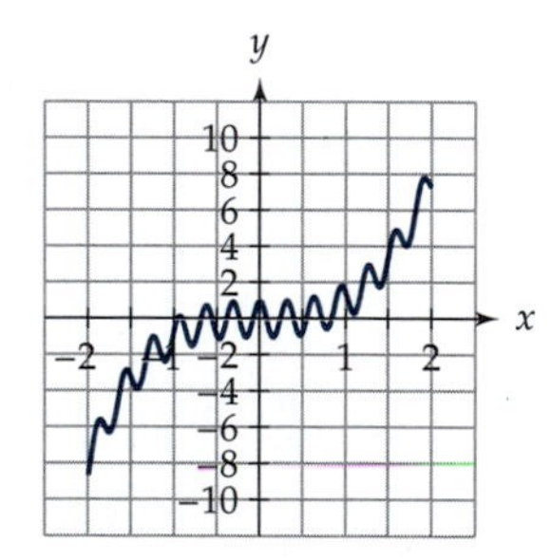

iii.

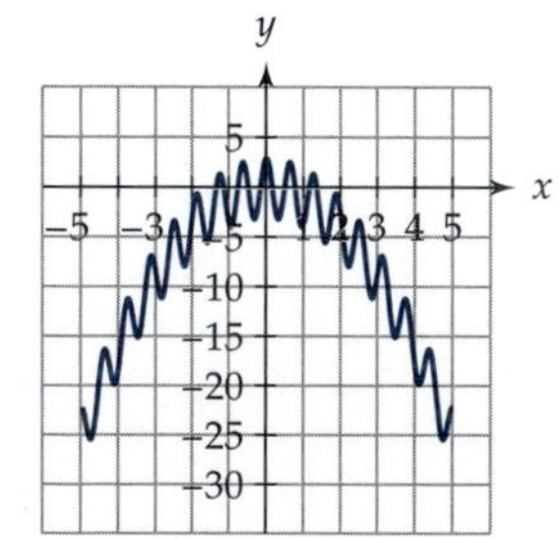

iv.

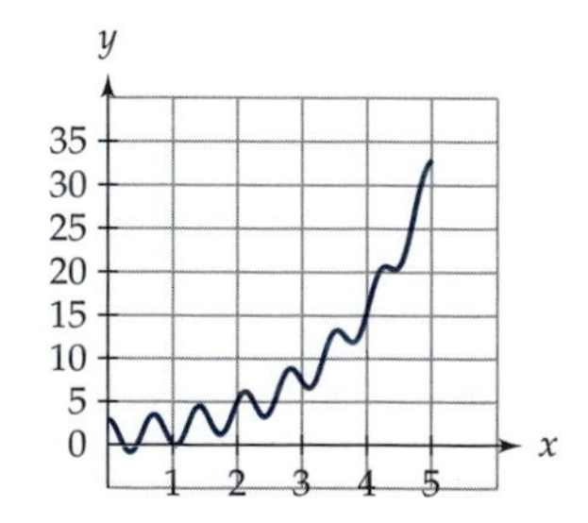

2. Give an example, other than snow cover, that can be modeled by the sum of a periodic function and a nonperiodic function.

PRODUCTS OF PERIODIC AND NONPERIODIC FUNCTIONS

We have seen that the sum of a sine or cosine function and a nonperiodic function can be thought of as an oscillating function with a "variable" vertical shift. What happens if we multiply instead of adding? Recall that for the function $y = A\cos[B(x + C)] + D$, the amplitude is given by $|A|$. If we are looking at functions of the form $y = f(x)\cos x$, then the values of $f(x)$ vary the amplitude. For example, let $h(x) = x\cos x$. The graph of h for $0 \leq x \leq 25$ is given in Figure 7. Notice that the graph is oscillating similar to a cosine or sine function yet the amplitude is increasing as x increases. This behavior is different from what happened with the sum; in that case, the vertical shift changed but the height of the oscillations remained the same. Notice that as with the sum, this function is no longer periodic. Figure 8 shows the graph of h, $y = x$, and $y = -x$. The cosine behavior is indicated by the oscillation in the graph, and the $y = x$ behavior is indicated by the heights of the oscillation. Because $-1 \leq \cos x \leq 1$ for all values of x, then $-x \leq x\cos x \leq x$ for all positive values of x. In other words, the outputs of $h(x) = x\cos x$ are "trapped" between $y = x$ and $y = -x$.

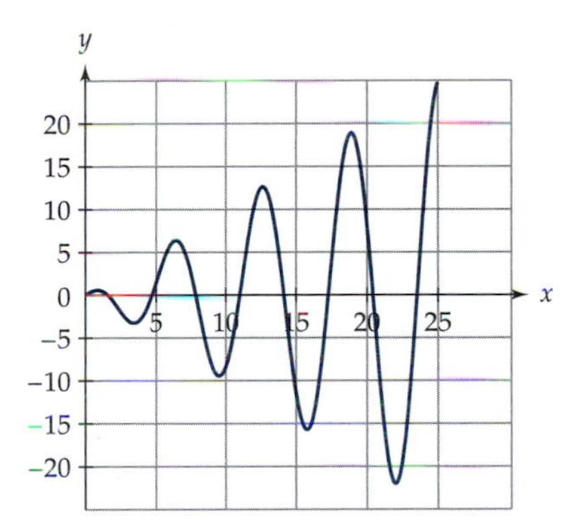

Figure 7 A graph of $h(x) = x\cos x$.

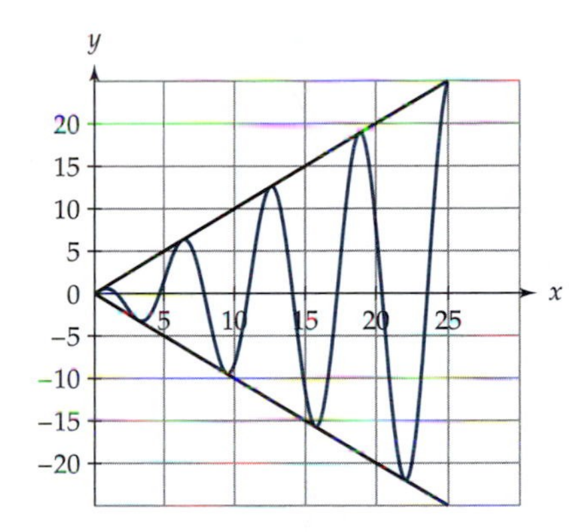

Figure 8 A graph of $h(x) = x\cos x$, $y = x$, and $y = -x$. Notice that the amplitude of h is "trapped" by $y = x$ and $y = -x$.

The graph of f is given in Figure 9. It appears to be the product of a periodic function and a nonperiodic function because it oscillates but the amplitude changes. The curve that "connects" the peaks looks exponential, therefore f seems to be the product of a sine or a cosine function and an exponential function. In fact, the function used to produce the graph in Figure 9 is $f(x) = 2^x\sin(25x)$. As with the sum, it is usually impossible to find the equation of the function by looking at the graph, although you can often determine the general form of the answer.

Just as with the sum, however, the graph of a product of a periodic function and a nonperiodic function can be misleading. For example, Figure 10(a) is a graph of $g(x) = (x^2 - 1)\cos 5x$, and Figure 10(b) is a graph of $k(x) = x^2 - 1$. Although the graph in Figure 10(a) oscillates, it also

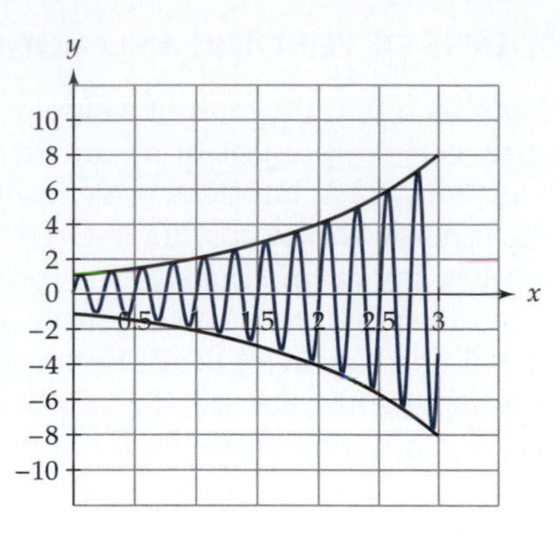

Figure 9 A graph of a product of a periodic function and a nonperiodic function.

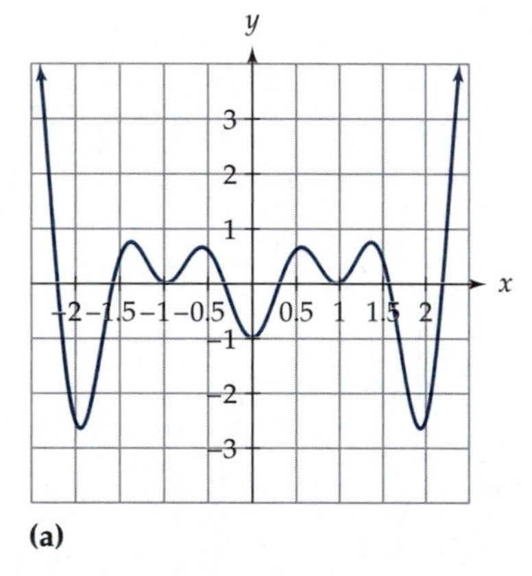

(a)

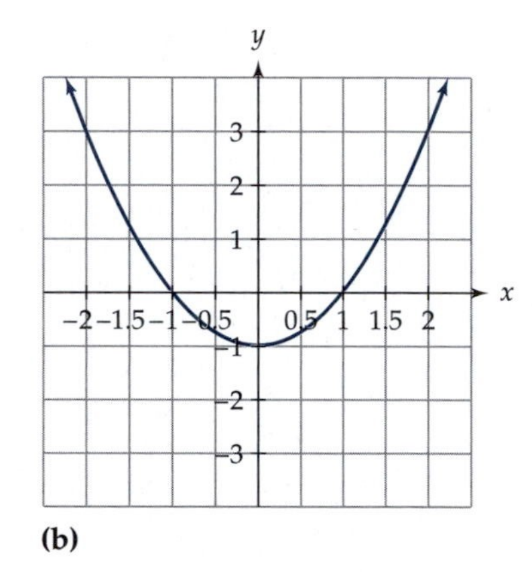

(b)

Figure 10 (a) A graph of $g(x) = (x^2 - 1)\cos 5x$; (b) a graph of $k(x) = x^2 - 1$.

resembles the graph of a polynomial. If, by zooming out and seeing that the graph continues to oscillate, you decide that it is the product of a sine or a cosine function and a nonperiodic function, it is still difficult to deduce information about the nonperiodic function. The reason for this is partly because the nonperiodic function, $y = x^2 - 1$, changes from positive to negative, so some of the "high" points need to be connected to some of the "low" points to determine the shape of the nonperiodic function. It is often difficult to deduce useful information about the equation of a function simply by looking at the graph.

Let's look at an application that can be modeled by the product of a periodic function and a nonperiodic function. Let p represent the motion of a pendulum that is slowing down due to air resistance. This type of pendulum is known as a *damped* pendulum. The input, t, is time in seconds, and the output, $p(t)$, is the directed distance from the end of the pendulum to the center position (i.e., where it would hang when it is at rest). Let $p(t) = 0.9^t \cos(2\pi t)$ represent a pendulum where the initial swing is one unit from the center (at rest) position and the height of each swing is 90% of the height of the previous swing (due to air resistance). Also, each swing takes 1 sec. A graph of p for $0 \le t \le 10$ is shown in Figure 11.

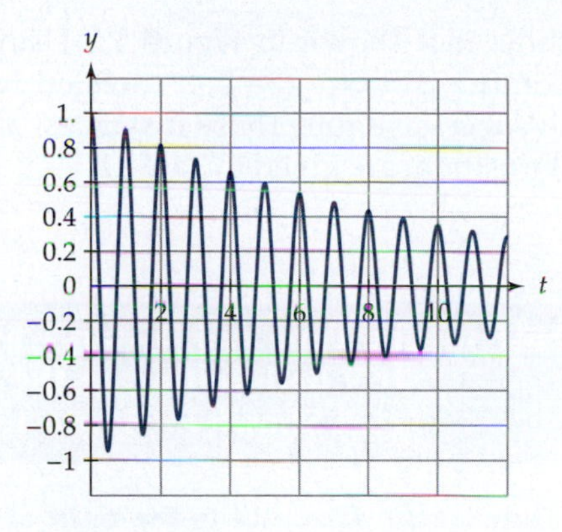

Figure 11 A graph of the position of a damped pendulum.

Example 1 Find a function that appropriately models a damped pendulum where the intial swing is 1.5 ft from the center (at rest) position, the height of each swing is 94% of the height of the previous swing, and each swing takes 0.6 sec.

Solution Because the height of the swing is decaying, it is best modeled with an exponential function where the growth factor is less than 1.[49] Because the initial swing is 1.5 ft and the height of each swing is 94% of the height of the previous swing, the appropriate exponential function is $y = 1.5 \cdot 0.94^t$. This function, however, only models the heights and does not account for the oscillation of the pendulum. To do that, we need to include a cosine or a sine function. A cosine function with period 0.6 is $y = \cos(2\pi t/0.6)$. A function that models the damped pendulum is $f(x) = 1.5 \cdot 0.94^t \cos(2\pi t/0.6)$. A graph of this function and the graphs of $y = 1.5 \cdot 0.94^t$ and $y = -1.5 \cdot 0.94^t$ are shown in Figure 12.

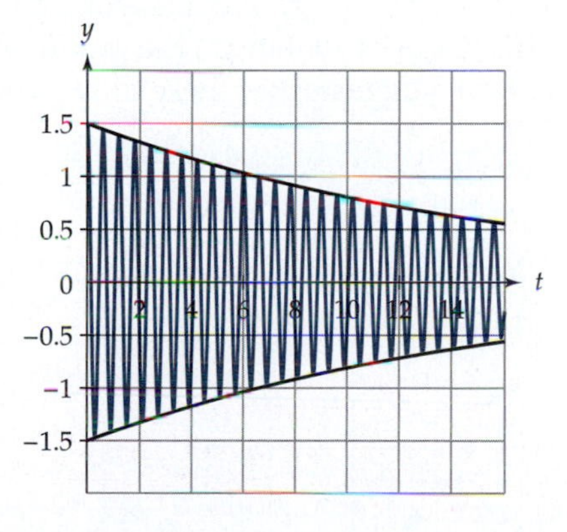

Figure 12

Let's see how to construct a modeling function starting with data. A situation similar to a damped pendulum is rolling a ball on a curved track

[49] *See section 2.2.*

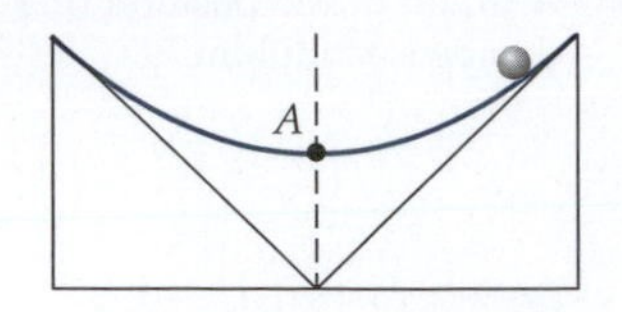

Figure 13 Rolling a steel ball on a curved track is similar to the motion of a damped pendulum.

such as that shown in Figure 13. Using such a track, we measured the maximum distance the ball traveled to the right of its resting point for each successive roll. These distances, along with the time when it was at that position, are given in Table 1.

TABLE 1 The Time and Distance of a Steel Ball Rolling on the Curved Track Shown in Figure 13

Time (sec)	Distance to the right of A (cm)
0	31.0
1.55	24.8
3.10	20.0
4.65	17.1
6.20	14.3
7.75	12.4
9.30	10.5
10.85	9.2
12.40	7.9
13.95	7.0
15.50	5.7
17.05	4.4
18.60	3.2

We want to find a function that models the distance the ball is away from its resting point with time, in seconds, as the input and directed distance, in centimeters, as output. A positive number indicates that the ball is to the right of A, and a negative number indicates that the ball is to the left of A.

Let's begin by looking at the graph of the data in Table 1. (See Figure 14.) The graph suggests that these data points can be modeled by an exponen-

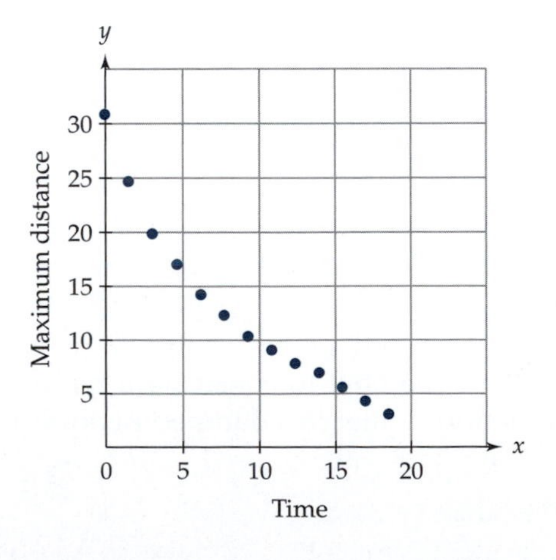

Figure 14 A graph of the data in Table 1 for the steel ball rolling on the curved track.

tial function of the form $y = b \cdot a^x$. Using an exponential function is also reasonable in light of the physical situation involving decay of distance due to friction. Because b is the y-intercept or starting distance, $b = 31.0$. To find a value for a, we substitute our value for b as well as one of the data points and solve for a. We decided to use the last data point in Table 1, which is (18.60, 3.2).

$$\begin{aligned} y &= b \cdot a^x \\ 3.2 &= 31.0 \cdot a^{18.60} \\ \frac{3.2}{31.0} &= a^{18.60} \\ \left(\frac{3.2}{31.0}\right)^{1/18.60} &= a \\ 0.885 &\approx a. \end{aligned}$$

So, a function modeling the maximum distances is $y = 31.0 \cdot 0.885^x$. Figure 15 shows the graph of the data points and the function we found.

The function $y = 31.0 \cdot 0.885^x$ does not take into account that the ball is oscillating from side to side as it rolls. To show this oscillation, we need to include a cosine or a sine function. From Table 1, you can see that the period of the oscillation is about 1.55 sec. Because the first maximum occurs at 0 sec, a cosine function is a good choice for modeling the oscillation. Therefore, a function that describes the directed distance the ball is away from the starting point t seconds after it begins to roll is $d(t) = 31.0 \cdot 0.885^t \cdot \cos(2\pi t/1.55)$. A graph of $d(t)$ is shown in Figure 16.

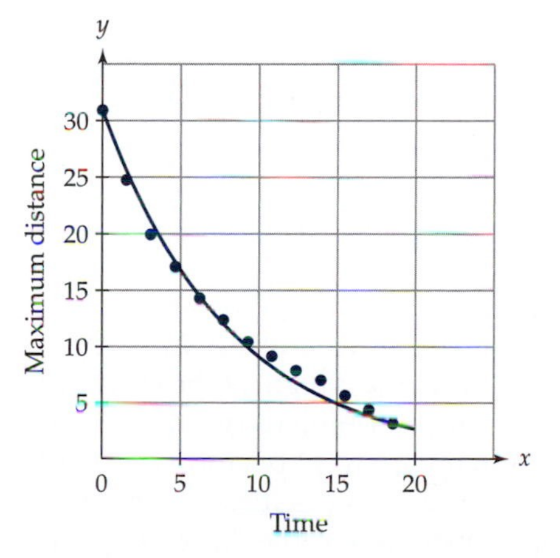

Figure 15 A graph of the data points from Table 1 and the function $y = 31.0 \cdot 0.885^x$.

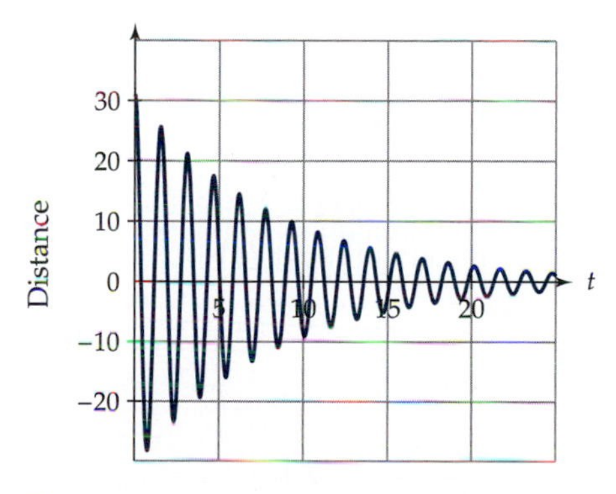

Figure 16 A graph of $d(t) = 31.0 \cdot 0.885^t \cdot \cos(2\pi t/1.55)$, a function that models a ball rolling on a curved track.

READING QUESTIONS

3. How is the behavior of $f(x) = 3^x \cos x$ different than that of $g(x) = 3^x + \cos x$?
4. Match each equation with the appropriate graph.
   (a) $y = 0.5x^2 \cos(20x)$
   (b) $y = 2x \sin(20x)$
   (c) $y = 0.3^x \cos(20x)$
   (d) $y = 2^x \sin(30x)$

i.

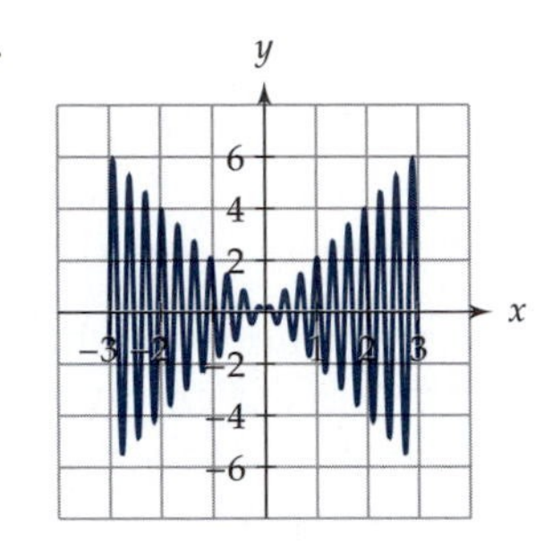

ii.

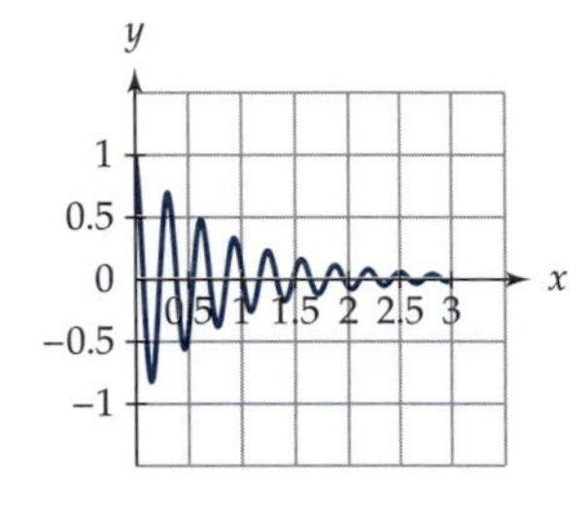

iii.

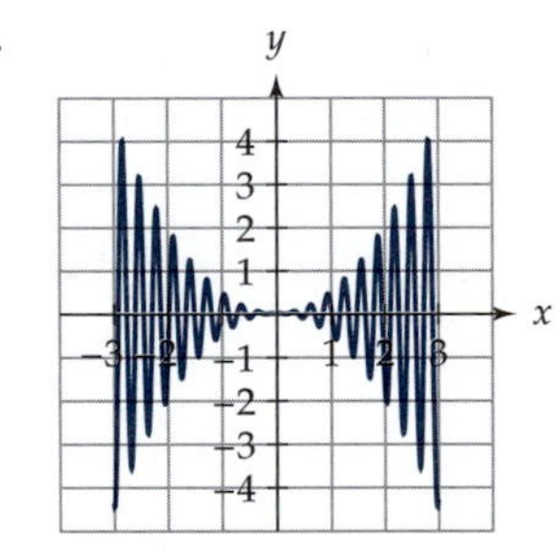

iv.

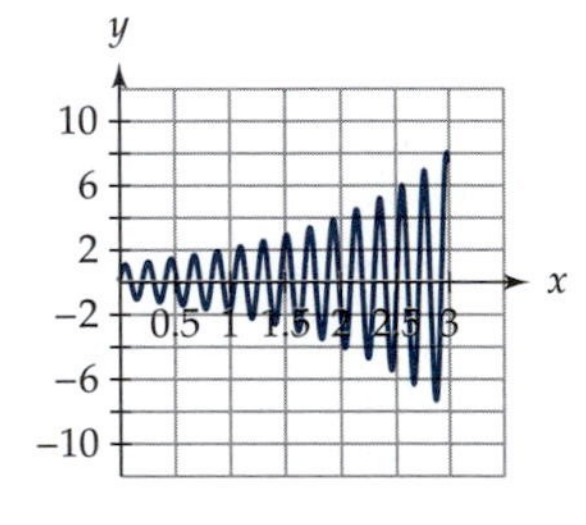

5. Find a function that models a damped pendulum where the intial swing is 1 ft from the center (at rest) position, the height of each swing is 96% of the previous swing, and each swing takes 0.8 sec.

COMBINING SUMS AND PRODUCTS OF PERIODIC AND NONPERIODIC FUNCTIONS

One of the authors made a pendulum and measured the distance, while swinging, from the weight at the end of the pendulum to a motion detector. The function describing the motion of this pendulum if the distance is measured from the "at rest" position is $y = 1.53 \cdot 0.97^t \cos(2\pi t/2.54)$, where t is time in seconds and y is distance in meters. The motion detector, however, was not positioned at the "at rest" position but rather was 3.38 m away. In this case, the function that best models the motion of the pendulum as recorded by the motion detector is $d(t) = 1.53 \cdot 0.97^t \cos(2\pi t/2.54) + 3.38$. A graph of d for $0 < t < 50$ is shown in Figure 17. The amplitude decreases

Figure 17 A graph of the position of a damped pendulum measured by a motion detector 3.38 m away.

because we have a damped pendulum, but there is also a vertical shift because the motion detector is 3.38 m away.

Our function $d(t) = 1.53 \cdot 0.97^t \cos(2\pi t/2.54) + 3.38$ combines a variable vertical stretch with a constant vertical shift. In certain situations, however, both the stretch and the shift are variable. For example, suppose the natural gas bill for a residential home reaches a high of \$120 in January and a low of \$40 in July. The monthly cost of natural gas could be modeled with the function $f(t) = 40\cos[(\pi/6)(t-1)] + 80$, where $t = 1$ represents January, $t = 2$ represents February, and so on. The graph of f is shown in Figure 18.

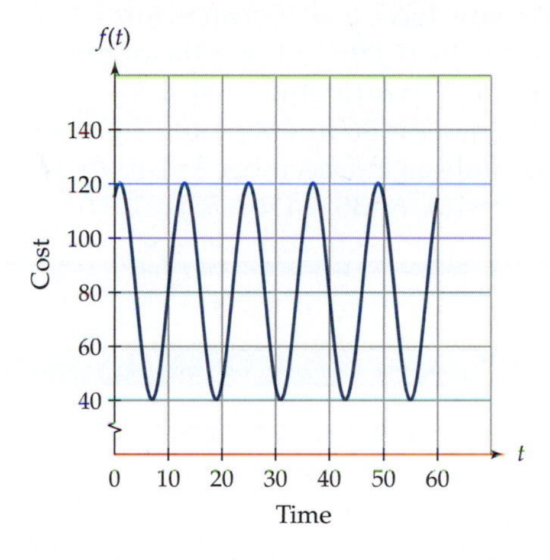

Figure 18 The graph of $f(t) = 40\cos[(\pi/6)(t-1)] + 80$ for $1 \le t \le 60$. This function models the monthly cost of natural gas for 5 years.

Now suppose there is a 6% annual increase in the cost of gas. We cannot express this function simply as $g(t) = f(t) \cdot (1.06)^t$ because the input for f is time in *months* and the input for $(1.06)^t$ is time in *years*. This problem can be easily solved by writing $(1.06)^t$ as $(1.06)^{t/12}$. The $t/12$ converts months to years. Thus, the monthly cost of natural gas, including the increase in the cost per year, can be modeled by the function $g(t) = [40\cos[(\pi/6)(t-1)] + 80] \cdot (1.06)^{t/12}$. A graph of g is shown in Figure 19. Notice that both the vertical shift and the amplitude are changing.

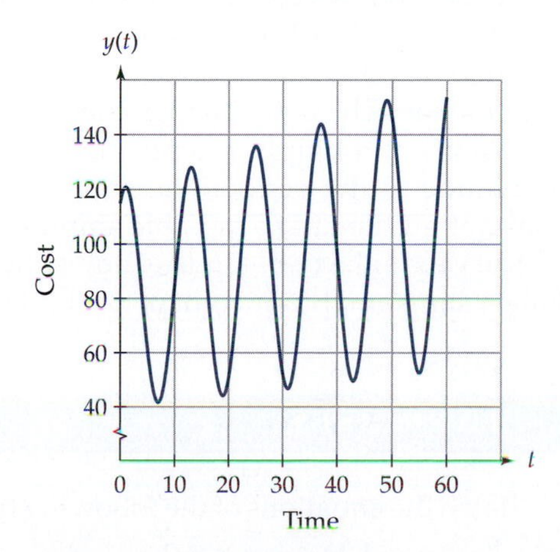

Figure 19 A graph of $g(t) = [40\cos[(\pi/6)(t-1)] + 80] \cdot (1.06)^{t/12}$ for $1 \le t \le 60$. This function models the monthly cost of natural gas for 5 years assuming a 6% annual increase in the cost.

Example 2 Find a function that models the monthly cost for natural gas where, before inflation, the cost is at a high of \$90 in January and a low of \$20 in July. Assume the rate of inflation is 9% annually. Use this function to predict the cost of natural gas three years from now in December. (Assume the starting point is January represented by $t = 1$.)

Solution Without considering inflation, the function that models the monthly cost of natural gas is of the form $y = A\cos[B(x + C)] + D$. The amplitude is $A = (90 - 20)/2 = 35$. The vertical shift is $D = (90 + 20)/2 = 55$. The period is 12, so $B = 2\pi/12 = \pi/6$, and the horizontal shift is $C = -1$ because we are starting with January $= 1$ instead of January $= 0$. We now have $y = 35\cos[(\pi/6)(t - 1)] + 55$. Inflation causes the monthly cost to be $(1.09)^{1/12}$ times the original cost. So, our function is $f(t) = [35\cos[(\pi/6)(t - 1)] + 55] \cdot (1.09)^{t/12}$. Three years from now in December is 36 months later. So, our function predicts the cost of natural gas in December 3 years from now to be $f(36) = [35\cos[(\pi/6)(36 - 1)] + 55] \cdot (1.09)^{36/12} \approx 110.48$.

READING QUESTIONS

6. Suppose $d(t) = 1.1 \cdot 0.98^t \cos(2\pi t/2.54) + 2.45$ models the distance (in feet) from a motion detector to the weight at the end of a swinging pendulum. How far is the motion detector from the weight when the pendulum is in the at rest position?
7. How is the behavior of $f(x) = 3^x \sin x + 5$ different from that of $g(x) = 3^x(\sin x + 5)$?
8. In this section, we used the function $g(t) = f(t) \cdot (1.06)^{t/12}$ to model the cost of natural gas.
   (a) Why is our exponent in this equation $t/12$ and not just t?
   (b) Why is the base of the exponential function 1.06 and not 0.06?

SUMMARY

In this section, we considered what happens when you combine periodic functions with functions that are not periodic. When you add a nonperiodic function to a periodic function, you change the vertical shift or midline of the function. The new function then varies around this changing midline.

When you multiply a nonperiodic function by a periodic function, the amplitude of the periodic function is changed. The new function has a constant midline but a variable amplitude.

You can make both kinds of adjustments to periodic functions by combining sums and products of periodic functions and nonperiodic functions.

EXERCISES

1. Match the equations of the following functions to the appropriate graph.
   (a) $y = 1.1^x + \cos x + 2$

(b) $y = 1.1x\cos x + 2$

(c) $y = 1.1^x(\cos x + 2)$

(d) $y = 1.1^x \cos x + 2$

(e) $y = x^2 \cos x + 2$

(f) $y = 1.1x + \cos x + 2$

i.

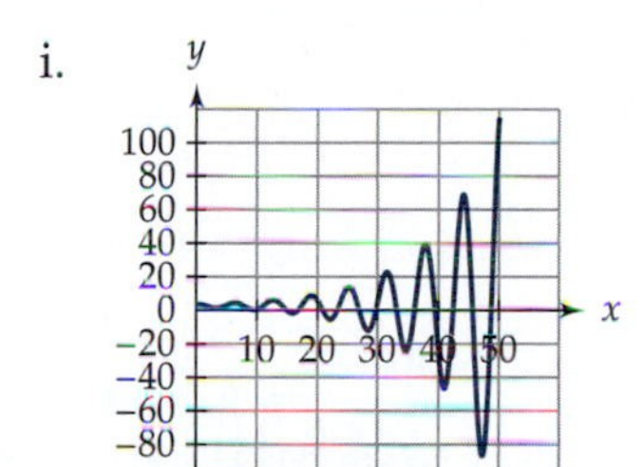

ii.

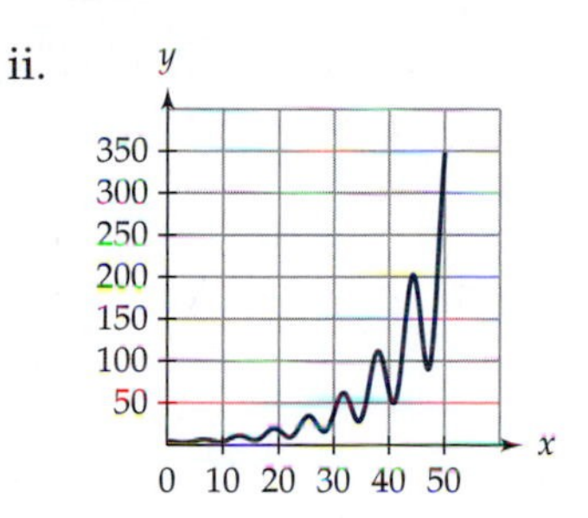

iii.

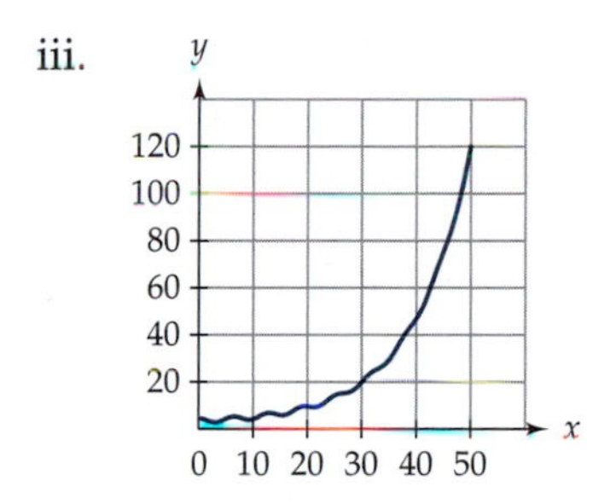

iv.

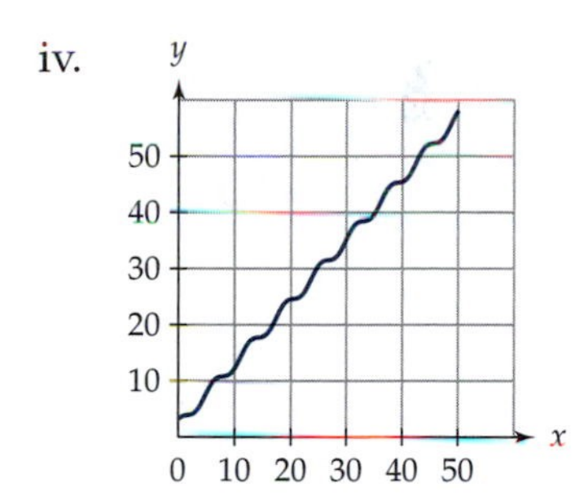

v.

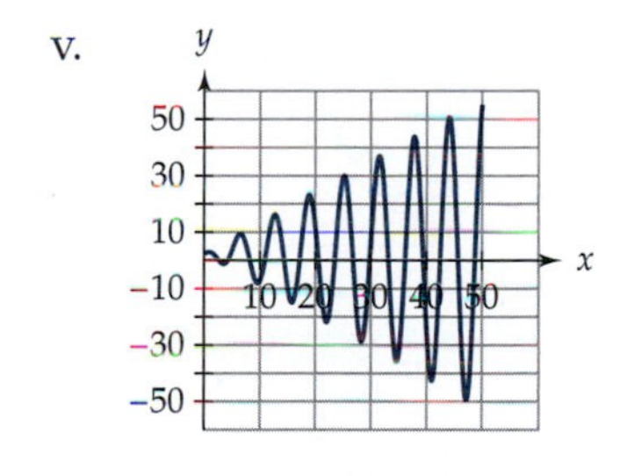

vi.

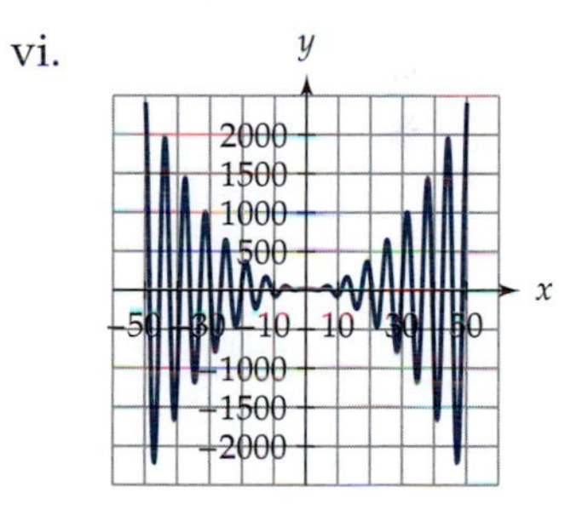

2. For each of the following pairs of functions, describe how the behavior differs.

   (a) $f(x) = 2^x \cos x$ and $g(x) = 2^x + \cos x$

   (b) $f(x) = 2^x + \cos x$ and $g(x) = 2x + \cos x$

   (c) $f(x) = 2^x \cos x$ and $g(x) = 2x \cos x$

   (d) $f(x) = 2^x \cos x + 3$ and $g(x) = 2^x(\cos x + 3)$

3. Each of the following graphs represents either the function $f(x) = a^x \cos(2\pi x)$ or $g(x) = mx\cos(2\pi x)$. Determine which function it is and estimate the value of a or m.

(a)

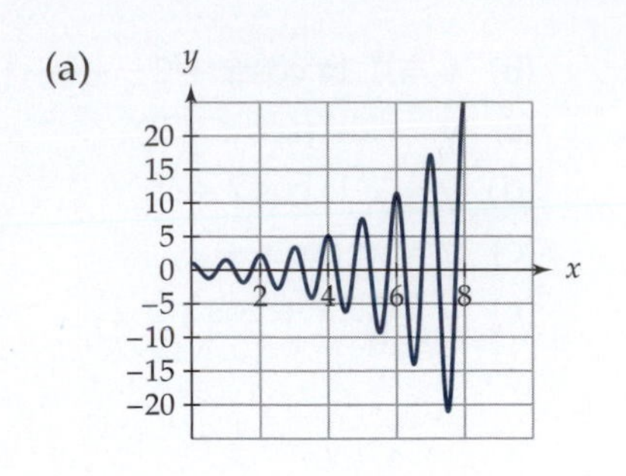

(b)

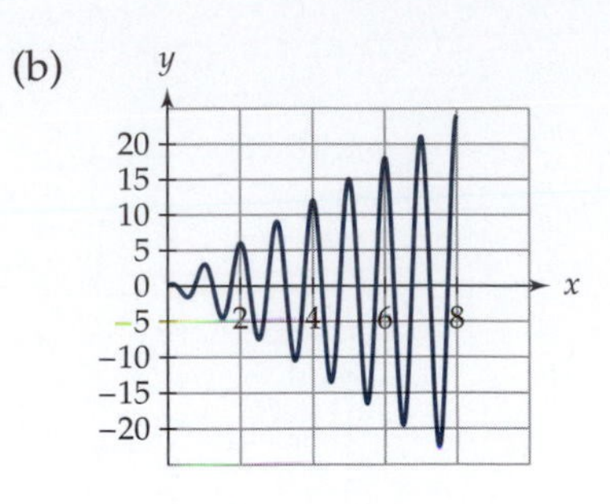

(c)

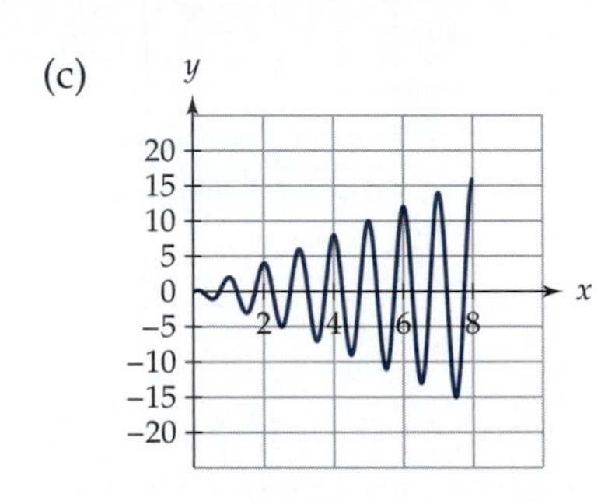

4. Each of the following graphs represents either the function $f(x) = a^x + \cos(2\pi x)$ or $g(x) = mx + \cos(2\pi x)$. Determine which function it is and estimate the value of a or m.

(a)

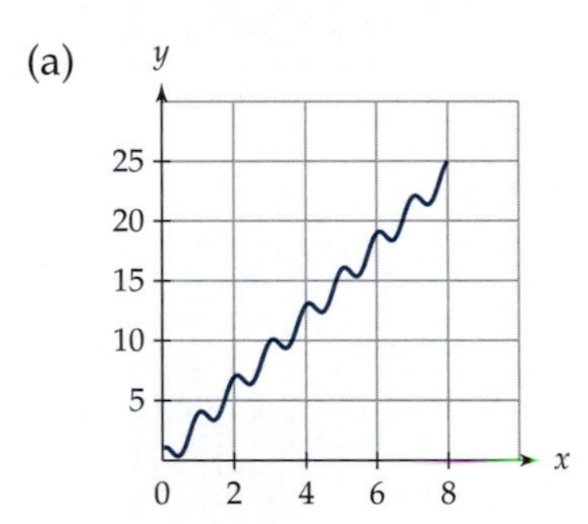

(b)

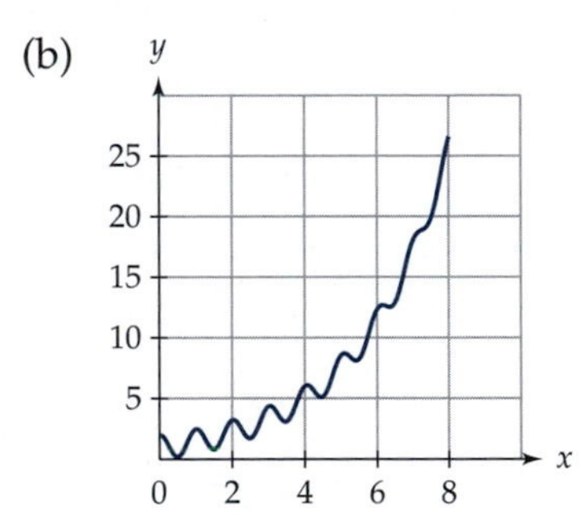

(c)

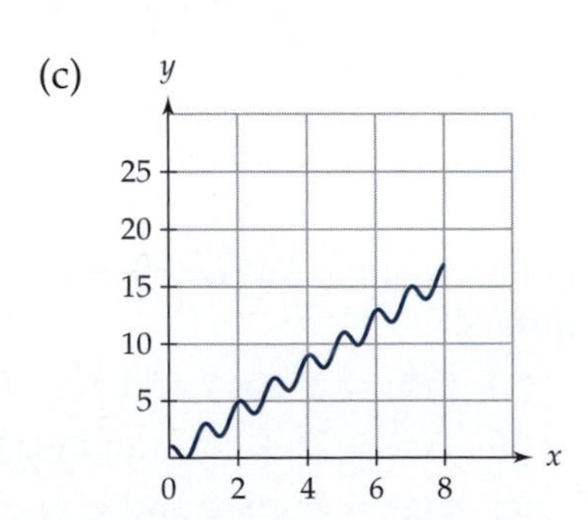

5. A pendulum consisting of a large plastic ball and fishing line was constructed. We measured the distance from a CBR to the ball as it swung.

Some of the data collected are shown in the table and the graph. Using these data and assuming that they can be modeled by the product of an exponential function and a sine or a cosine function, find an equation for a function that models the distance from the CBR to the ball.

x	0	0.3	0.5	1.2	1.4	2.1	2.5	3.2	3.5	4.1	4.5	5.2
y	1.43	1.53	1.40	0.78	0.83	1.46	1.31	0.82	0.96	1.42	1.25	0.85
x	5.5	6.1	6.4	7.1	7.5	8	8.4	9.1	9.4	10	10.5	11
y	1.02	1.39	1.28	0.87	1.06	1.37	1.24	0.89	1.03	1.36	1.13	0.91
x	11.6	11.9	12.5	13	13.9	14.6	15	15.3	15.8	16.6		
y	1.20	1.33	1.10	0.92	1.32	1.01	0.94	1.06	1.30	0.99		

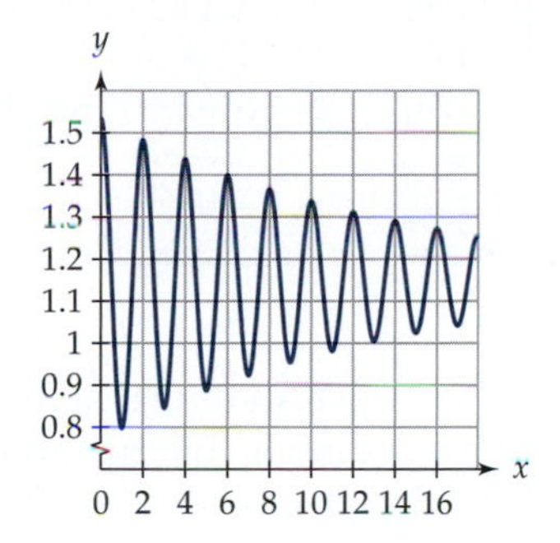

6. The concentration of CO_2 in the atmosphere is continuously changing on a yearly cycle, primarily because of the change in plants. The amount of CO_2 decreases in spring and summer with photosynthesis and rises in the fall and winter when many plants are without their leaves. One series of samples of CO_2 taken in Hawaii since 1958 shows that the concentration of CO_2 in the atmosphere (measured in parts per million, or ppm) varies by about 6 ppm each year.[50]
   (a) The highest concentration of CO_2 in 1958 was 317 ppm (in March), and the lowest concentration of CO_2 in 1958 was 311 ppm (in September). Find an equation that models this yearly variation in CO_2 concentration where time, in months, is the input, and CO_2 concentration, in parts per million, is the output. Let $t = 0$ by January 1958.
   (b) The burning of fossil fuels and, to a lesser extent, the production of cement have caused additional CO_2 to be released into the atmosphere over the years so that the average level of CO_2 has risen since at least 1958. The average amount for each year is

[50] *E. Berner and R. Berner,* Global Environment: Water, Air and Geochemical Cycles *(Upper Saddle River, NJ: Prentice Hall, 1996), pp. 28–30.*

given in the table. A graph of these data indicates that the data might be modeled by either a linear or an exponential function.

Year	CO_2 Concentration	Year	CO_2 Concentration
1958	314.60	1974	330.67
1959	315.75	1975	331.25
1960	316.91	1976	332.17
1961	317.17	1977	335.21
1962	317.61	1978	335.72
1963	318.72	1979	337.40
1964	318.92	1980	338.81
1965	319.60	1981	340.25
1966	321.33	1982	341.00
1967	321.64	1983	342.89
1968	322.94	1984	345.45
1969	324.40	1985	345.60
1970	325.06	1986	347.27
1971	326.28	1987	349.19
1972	327.14	1988	352.06
1973	330.13	1989	353.14

i. If we assume that the average level of CO_2 concentration is going up by the same amount each year, then the function that models this increase will be linear. Based on these data, give an equation in which the input is the year since 1958 and the output is the average concentration of CO_2 in the atmosphere above Hawaii.

ii. Now combine the two equations so that the yearly cycle and the linear equation representing the rise in average level are both included.

iii. Graph the equation.

(c) Suppose the average level of CO_2 concentration is not a linear increase but an exponential increase.

i. Based on these data, give an equation in which the input is the year and the output is the average concentration of CO_2 in the atmosphere above Hawaii.

ii. Now combine the two equations so that the yearly cycle and the exponential equation representing the rise in average level are both included.

iii. Graph the equation.

(d) It has been suggested[51] that a value of 550 ppm, twice the preindustrial level of CO_2, would raise the average temperature of

[51] *Minnesota Daily Online,* Urgency of Global Warming Action Hotly Debated *(visited 29 July 1998) ⟨http://www.daily.umn.edu/daily/1997/11/20/world_nation/wn4.ap/⟩.*

Earth by 2° to 6°F. When would this rise occur if the level of CO_2 continued to rise at a rate similar to what it has in our data? Give two answers, one based on the linear model and one based on the exponential model.

7. A weight attached to a spring was allowed to oscillate, producing a graph on a calculator attached to a CBR similar to the graph shown here. The input for this graph is time in seconds, and the output is height above the floor in meters. Find a function that models the graph.

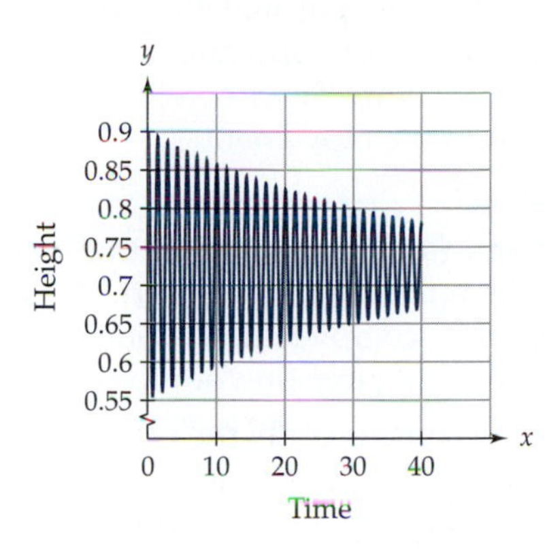

8. When cars speed by you, the frequency of the sound emitted by them seems to change. The frequency does not change, but observed frequency does. This phenomenon is known as the Doppler effect. As the car is traveling toward you, the length of the sound waves decreases, causing the frequency to increase. As the car is traveling away from you, the length of the sound waves increases, causing the frequency to decrease. You notice the sudden change in frequency as the car passes. The same effect occurs if the source of the sound is stationary and the person hearing it is moving.[52] If the receiver is moving away from the source of the sound at v feet per second, then the perceived frequency, f', of a tone that has a frequency of f_0 is

$$f' = f_0\left(1 - \frac{v}{1090}\right).$$

   (a) Suppose you are traveling at 73 ft/sec (about 50 mph) away from a tone that has a frequency of 220 Hz. What is the perceived frequency of that tone?

   (b) Suppose a tone is being emitted from a stationary source at a frequency of 220 Hz. Also suppose tone can be modeled by the function $f(t) = \cos(440\pi t)$, where t is time in seconds. Now suppose you are in a car that is accelerating at 10 ft/sec^2 away from that tone.

      i. Find the function that models the perceived sound wave.

[52] *Tiper,* College Physics, *pp. 384–86.*

ii. Graph the function in a window of $0 \le t \le 0.01$. Also graph the function in a window of $10 \le t \le 10.01$. Describe the difference you see in the two windows.

INVESTIGATIONS

INVESTIGATION 1: CYCLOID

If we take a wheel and put a spot of paint somewhere on the wheel, what path will the spot make as the wheel rolls along? It turns out that this question is more difficult to answer than you might first think. In fact, it is so difficult to describe with one equation that we use two equations to represent the movement.

1. Sketch a circle of radius 1 such that the lowest point on the circle is the point $(0, 0)$. Put a dot on that point. Let θ represent the angle through which the circle has rolled. What is the value of θ when the circle has made a complete revolution? What ordered pair will give the location of the dot after 1 revolution?
2. Let's consider only the vertical position of the dot. Give an equation for the vertical position of the dot as the circle rolls where the input is θ and the output, p_v, is the vertical position of the dot.
3. Let's consider only the horizontal position of the dot.
   (a) Give an equation for the horizontal position of the dot if the circle is merely turning and not rolling.
   (b) Now, take into account that the circle is rolling. Give a function that represents the horizontal position of the center of the circle as it rolls.
   (c) Combine the two equations you just made so that the input is θ and the output, p_h, is the horizontal position of the dot.
4. Sometimes combining two equations, like those for the horizontal and vertical positions, into one equation is very difficult or impossible. This case is like that. So, we simply leave our results as two equations.
   (a) To graph these two equations, fill in the table.[53]

θ	0	$\frac{\pi}{6}$	$\frac{\pi}{3}$	$\frac{\pi}{2}$	$\frac{2\pi}{3}$	$\frac{5\pi}{6}$	π	$\frac{7\pi}{6}$	$\frac{4\pi}{3}$	$\frac{3\pi}{2}$	$\frac{5\pi}{3}$	$\frac{11\pi}{6}$	2π
p_h													
p_v													

   (b) Make a graph of the data in your table. Connect the points with a smooth curve.
   (c) Is the function you graphed periodic? If so, what is the period?

[53] *Another way to graph these two equations is to use your graphing calculator in parametric mode. We discuss parametric equations in chapter 9.*

5. Imagine that you see a bicycle with a reflector on its wheel traveling down the road. The tire has a 26-in. diameter, and the reflector is 10 in. from the center of the wheel. Give the horizontal and vertical equations for the position of this reflector.

INVESTIGATION 2: BOUNCING BALL

An inflatable ball was dropped and was allowed to bounce. A calculator connected to a CBR was used to record the ball's height above the floor. Some of the data collected and the resulting graph are shown in Table 2 and Figure 20. The input for the graph and the table is time in seconds, and the output is height above the floor in centimeters. Looking at the graph, it does not appear that the ball touched the floor each time it bounced, but that is only because the calculator samples a finite number of points and does not always record the times when the ball is on the floor. The ball did, in fact, land on the floor as it bounced.

TABLE 2

Time (sec)	0.00	0.09	0.17	0.22	0.34	0.47	0.60	0.69	0.82	0.95	1.08	1.16	1.24
Height (cm)	25.3	19.8	8.2	0.00	15.0	21.4	13.8	1.1	13.7	19.2	10.3	1.3	12.4
Time (sec)	1.33	1.46	1.54	1.63	1.76	1.84	1.93	2.02	2.10	2.23	2.27	2.36	2.45
Height (cm)	17.1	12.4	1.2	9.7	15.6	11.6	1.6	10.5	14.3	8.2	3.1	9.1	13.2
Time (sec)	2.58	2.62	2.79	2.92	3.10	3.22	3.35	3.48	3.61	3.74	3.87	4.00	
Height (cm)	6.9	2.1	11.4	2.5	10.3	1.9	9.2	2.6	8.6	2.5	8.2	3.7	

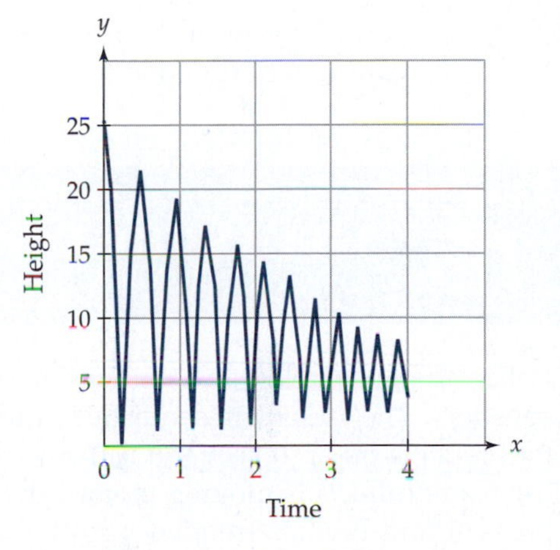

Figure 20

Our goal in this investigation is to find a function that models these data points. The basic function we use to start this process is the absolute value of a cosine function. Because the local maximum heights of the ball decreases with time, we have to find a function for the "amplitude." Because the length of time between bounces also decreases, we have to find a function for the "period." We use exponential functions for both.

1. Explain why the graph of the absolute value of a cosine function is similar to the graph shown in Figure 20.
2. Plot the bouncing ball data from Table 2 on your calculator. Along with these data, graph the function $y = |25.3 \cos x|$. This function should not fit the data, but it is a place to start. What is the period of $y = |25.3 \cos x|$? (*Hint:* The answer is not 2π.)
3. Now we transform our function to take into account a diminishing "amplitude."
    (a) Starting with the first point (0, 25.3), list all the other local maximum points. (There should be 12.) Find an exponential function that fits these points.
    (b) The exponential function you found in part (a) will act as the "amplitude" for our absolute value cosine function. Using this function as the "amplitude," rewrite the function from question 2. Graph this new height function along with the data.
4. The last thing we need to do is adjust the period.
    (a) Looking at the times for the local maximum heights, find the difference in times for each pair of adjacent local maximum heights. In doing so, make a collection of ordered pairs. For example, the second difference is $0.95 - 0.47 = 0.48$, which gives the ordered pair (2, 0.48). Find an exponential function that fits these ordered pairs.
    (b) Using the exponential function in part (a) as the "period," rewrite your absolute value cosine function from question 3, part (b). Graph this new height function along with the data. (*Note:* Because of a limited number of points that were sampled when determining the height of the ball, a function will not fit these data points all that well. Therefore, when you graph your function along with the data, make any necessary adjustments in your height function to get it to fit the data as best you can.)

PROJECTS 6.5 LIFE IN THE FAST LANE

Police often use radar (an acronym for *ra*dio *d*etection *a*nd *r*anging) to catch speeders. The radar unit sends out a radio wave at a designated frequency that reflects off an object and returns to the unit. If the object from which the wave reflects is moving toward the radar unit, then the wave is compressed and thus its frequency is changed. This phenomenon, known as the Doppler effect, also causes a train's sound to change as it passes you. An analysis of this change in frequency gives the speed of the approaching object.

To measure speed accurately, police officers who use radar should be directly in front of the moving object. Obviously, that is not very practical, unless you want to have a very short career in law enforcement! In this project, we explore some factors that impact the accuracy of using radar to measure the speed of oncoming vehicles.

1. Suppose an officer is in a car 15 ft off the side of a road (point B in Figure 1). A vehicle approaches traveling 70 mph (point A in Figure 1). We want to calculate the speed of the car reported by the radar unit when the car is 100 ft away.

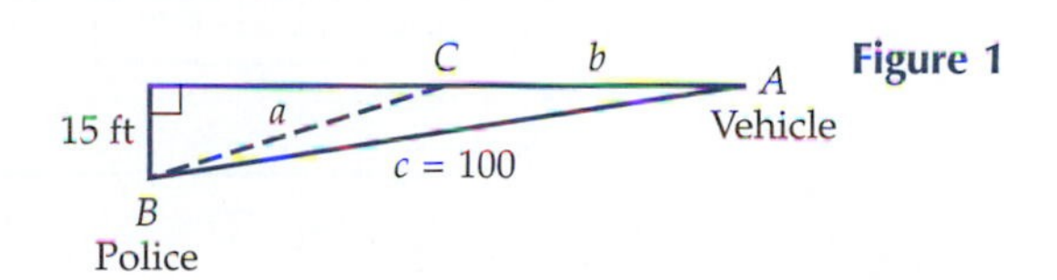

Figure 1

(a) We need to compute the speed of the car by measuring the difference in the length of $\overline{BA}$ and the length of $\overline{BC}$. Generally, the length of $\overline{AC}$ (the true distance traveled by the car in t seconds) is not the same as the difference in the lengths of $\overline{BA}$ and $\overline{BC}$ (the distance the radar gun uses to compute the speed of the car). Where would the police officer need to be positioned for the distance the car travels and the distance the radar gun measures to be exactly the same?

(b) We need to compute the length of $\overline{BC}$ as the first step in obtaining a function for computing the speed reported by the radar gun. To do so, we use the formula

$$a = \sqrt{b^2 + c^2 - 2bc \cos A}, \tag{1}$$

where a is the length of $\overline{BC}$, b is the length of $\overline{AC}$, and c is the length of $\overline{BA}$.

i. Explain why equation (1) is valid.

ii. Find $\cos A$.

iii. Determine an expression for b in terms of rate and time with the rate in feet per second and time as a variable in seconds. Remember that the car is traveling at 70 mph.

iv. Substitute your expressions for b and $\cos A$ into equation (1). You should now have a function where your input, t, is in seconds and your output, a, is in feet.

(c) The speed reported by the radar gun is

$$r_{\text{rep}} = \frac{100 - a}{t}.$$

Find a function where your input, t, is in seconds and your output, r_{rep}, is in feet per second.

(d) Use the list of times given in the following table to compute various values of r_{rep}. Report your answers in both feet per second and miles per hour. Round to four decimal places.

Time (sec)	r_{rep} **(ft/sec)**	r_{rep} **(mph)**
0.1		
0.01		
0.0001		
0.000001		

2. In the previous table, the time becomes progressively smaller. Let's now assume the time is instantaneous. In this case, it can be shown (using calculus) that $r_{rep} = r_{car} \cos A$. Use this formula to compute r_{rep} for question 1. How does this answer compare with the answers you obtained in part (d)?

For the rest of this project, use the formula $r_{rep} = r_{car} \cos A$ to compute the speed reported by the radar gun.

3. An officer is in a car 15 ft off the side of a road. A vehicle approaches traveling 70 mph. What will the radar read as the car's speed when the car is 50 ft away?
4. Compare your answers in questions 2 and 3. Notice that the accuracy of the measurement (i.e., $|r_{car} - r_{rep}|$) changed as the vehicle came closer. Let's determine how the accuracy of the radar is affected by changing its distance from the vehicle, the speed of the vehicle, and the distance of the police officer from the side of the road.
   (a) First, let's look at this problem graphically.
      i. Holding speed of the car constant at 70 mph and distance from the police car to the side of the road constant at 25 ft, find a function where the input is the distance from the police car to the speeding car and the output is the reported speed. Graph this function.
      ii. Holding distance from the speeding car constant at 100 ft and distance from the side of the road constant at 25 ft, find a function where the input is the speed of the car and the output is the speed reported by the radar unit. Graph this function.
      iii. Holding distance from the car constant at 100 ft and speed of the car constant at 70 mph, find a function where the input is the distance from the side of the road and the output is the speed reported by the radar unit. Graph this function.

(b) Using your graphs from part (a) as a guide, determine what happens to the accuracy of the radar reading as the vehicle comes closer. Determine what happens to the accuracy of the radar reading as the target vehicle goes faster. Determine what happens to the accuracy of the radar reading as the police officer sits farther off the side of the road. Justify each of your three explanations with either a symbolic or a geometric argument.

5. In a realistic situation, people often first decide on how much error they are willing to have and then change the other variables accordingly. For example, the police officer may want the reading of the radar unit to be within 1 mph of the actual speed of the car. Assume the police officers are sitting in their car 10 ft off the side of an interstate highway (speed limit 65 mph). Suppose they would like the reading of the radar unit to be within 1 mph of the actual speed of an approaching car traveling 75 mph. How far away should the car be when the police officers use their radar unit to ensure that their reading is within the desired accuracy range?

6. A person who received a speeding ticket for going 57 mph in a 45 mph zone contests the ticket on the grounds that, because of the angle effect, his speed was actually less than what the radar indicated. As the court's expert witness (on the law of cosines), write a response to the judge indicating why you think the ticketed driver should or should not be fined for speeding.

REVIEW EXERCISES

1. Given each of the following measurements for a right triangle whose right angle is $\angle C$, as shown, find the missing measurements for the remaining angles and sides. Where appropriate, round answers to one decimal place.

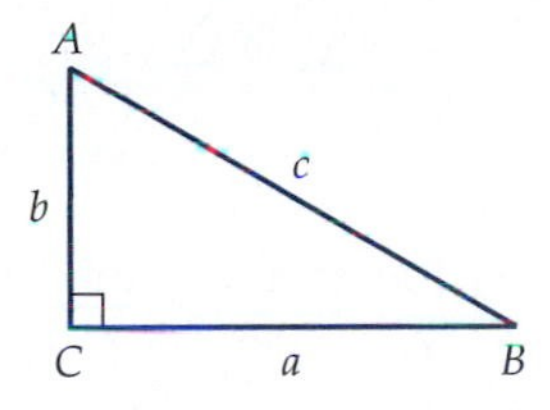

   (a) $a = 9, b = 12$
   (b) $m\angle B = 50°, a = 4$
   (c) $b = 5, c = 10$

2. Suppose you are standing in Paris, France, at a spot 500 m from the Eiffel Tower. You look up to the top of the tower, which is 300 m above the ground. What is the angle of elevation from your eyes to the top of the tower? (Assume your eyes are at ground level for the tower.)

3. As you are approaching a mountain in your car you notice a sign that states, "Caution: Steep Slope Ahead—15% Grade." The 15% grade means that the road has a slope of 0.15. Every time the horizontal distance changes by 100 ft, the vertical distance changes by 15 ft.
   (a) As you start driving up the mountain, what is your angle of elevation?
   (b) If, according to your odometer, you have driven up the mountain for 1 mi (5280 ft), what is your increase in elevation?
4. Suppose you are building a bookcase that is 9 ft high and 18 in. deep in your workshop. When you bring the bookcase into the house, you have to tilt it on its side to get it through the door. When you try to stand it up, you discover that it will not fit in a room with ceilings that are 9 ft high. (See the accompanying figure.) What is the tallest bookcase that you can build in your workshop, bring into the house, and successfully stand up?

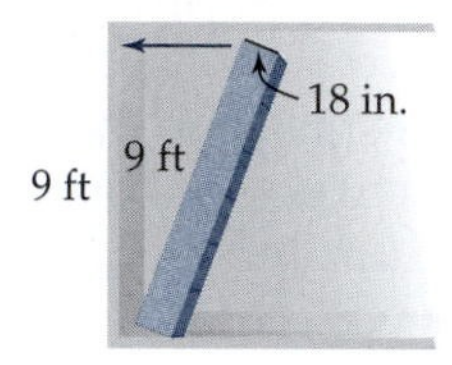

5. The smallest of the three pyramids in Giza, Egypt, has a square base where each side is approximately 109 m in length. Its height is approximately 66 m. (See the accompanying figure.)

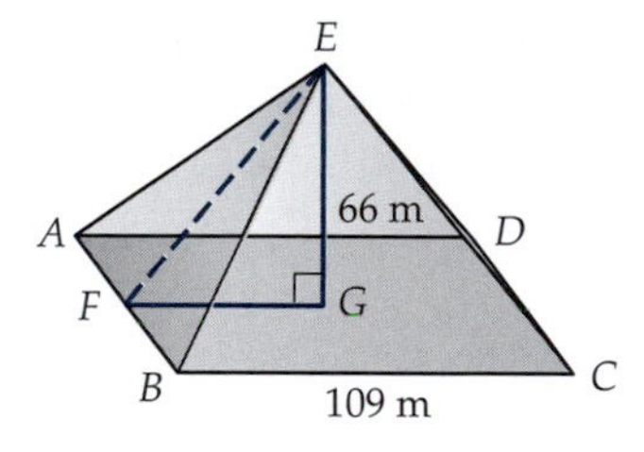

   (a) Find the angle formed by a side of the pyramid and its base ($\angle EFG$).
   (b) Find the slant height of the pyramid (length of $\overline{EF}$).

In questions 6 and 7, assume the sides of $\triangle ABC$ are labeled as shown in the figure.

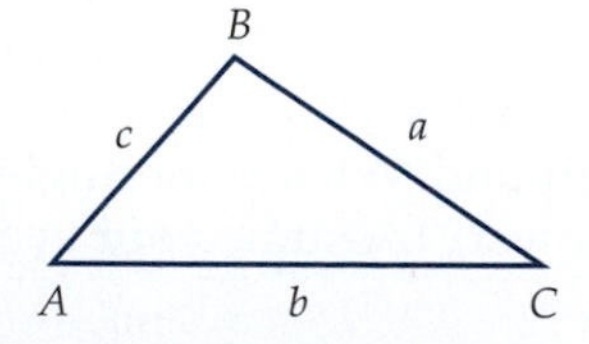

6. For each of the following, you are given three pieces of information about a triangle and are asked to determine a third. To do so, determine if you would use the law of sines, if you would use the law of cosines, or if this information cannot be determined.
   (a) Given $\angle A$, $\angle B$, and $\angle C$, find side c.
   (b) Given $\angle A$, $\angle B$, and c, find side a.
   (c) Given a, b, and c, find $\angle C$.
   (d) Given $\angle A$, b, and c, find side a.
7. Find the measurement of all three angles and all three sides for each of the following. Round answers to two decimal places.
   (a) $m\angle A = 30°, m\angle B = 40°, a = 5$ (b) $m\angle A = 35°, b = 10, c = 12$
   (c) $m\angle A = 35°, b = 10, a = 6$ (d) $a = 12, b = 8, c = 10$
8. The world's tallest self-supporting structure is the CN Tower in Toronto, Canada. Suppose you look at the top of the tower and find that the angle of elevation is 56.6°. You move 200 ft closer to the tower and find that the angle of elevation is now 61.2°. Estimate the height of the tower.
9. For each of the following, find the area of a triangle shown with the given dimensions.

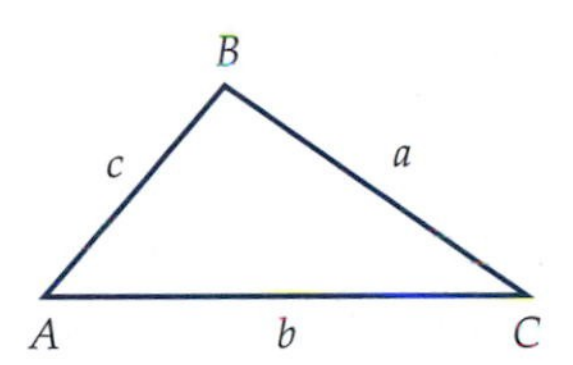

   (a) $a = 3, b = 4, c = 6$ (b) $a = 3, b = 4, m\angle C = 100°$
   (c) $a = 4, b = 5, m\angle B = 80°$
10. In Seattle, Washington, the 356th day of the year has the least amount of daylight, about 8.5 hr. The 173d day of the year has the most daylight, about 16 hr. Given that a year has 365 days, use a sine function to model the number of hours of daylight, h, for a given day of the year, d, in Seattle.
11. The following table contains the average high temperature in degrees Fahrenheit for Champaign-Urbana, Illinois.[54]

Month	January	February	March	April	May	June
Temp. (°F)	31.7°	36.3°	48.8°	62.4°	73.6°	82.9°
Month	July	August	September	October	November	December
Temp. (°F)	85.3°	83.0°	77.7°	65.1°	50.3°	36.5°

[54] *Illinois State Climatologist Office,* Climate Data for Champaign-Urbana, *(visited 4 January 2000) (mcc.sws.uiuc.edu/atmos/statecli/Champ-Urb/cu-averages.htm).*

(a) Find a function, H, that models the data where m indicates the month (January = 1, February = 2, etc.) and $H(m)$ indicates the average high temperature.

(b) Suppose you wanted a function whose output was in degrees Celsius rather than degrees Fahrenheit. Find a function, C, that models the average high temperature in degrees Celsius. (*Note:* The formula to convert from degrees Fahrenheit to degrees Celsius is $C = \frac{5}{9}(F - 32)$.)

12. Determine if each of the following is periodic. If so, find the period. If not, explain why not.

(a) $f(x) = \sin(\frac{2}{3}x) + \cos(\frac{5}{6}x)$

(b) $f(x) = \cos(2\pi x) + \sin(\frac{2}{3}x)$

(c) $f(x) = \cos(\frac{5}{2}x) + \cos(3x)$

(d) $f(x) = \sin(3\pi x) + \cos(2\pi x)$

13. The Touch-Tone phone system uses pairs of tones (high and low) to represent the various keys. The low tones vary according to the horizontal row of the tone button, and the high tones correspond to the vertical column of the tone button. For example, when the 5 button is pressed, the 770-Hz and 1336-Hz tones are sent together. The telephone central office then decodes the number from this pair of tones.

(a) Give a function that models the sound heard when the 5 button on a phone is pressed, assuming the two tones are equally loud.

(b) What is the period of the sound wave representing this sound?

14. A note is played at a frequency of 500 Hz and another is played at a frequency of 510 Hz.

(a) Let T be the function representing the sound wave produced when both notes are played at the same time at the same intensity (or loudness). Find an equation for T.

(b) Is T periodic? Justify your answer.

(c) What is the beat frequency of this sound? Can you hear the beats?

15. Match the equations of the following functions to the appropriate graph.

(a) $y = x^2 + \sin x + 2$

(b) $y = x^2 \sin x + 2$

(c) $y = x^2(\sin x + 2)$

(d) $y = 2x(\sin x + 2)$

i.

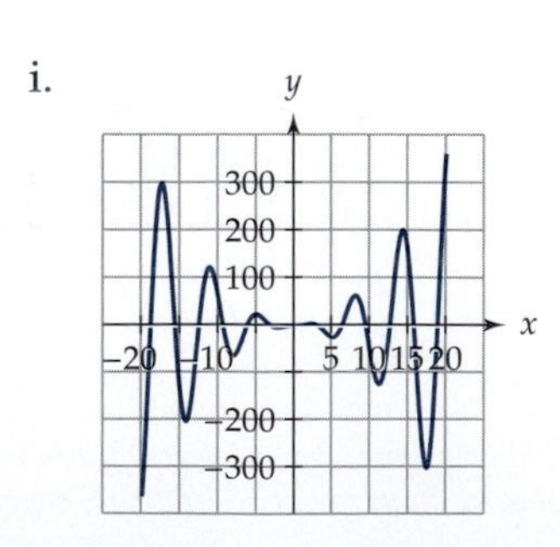

ii.

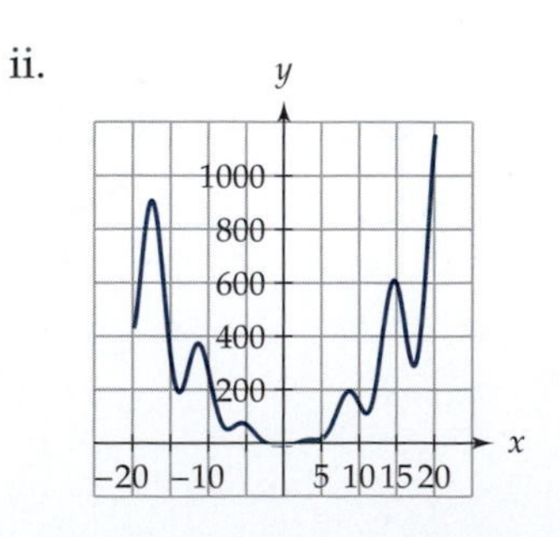

iii.

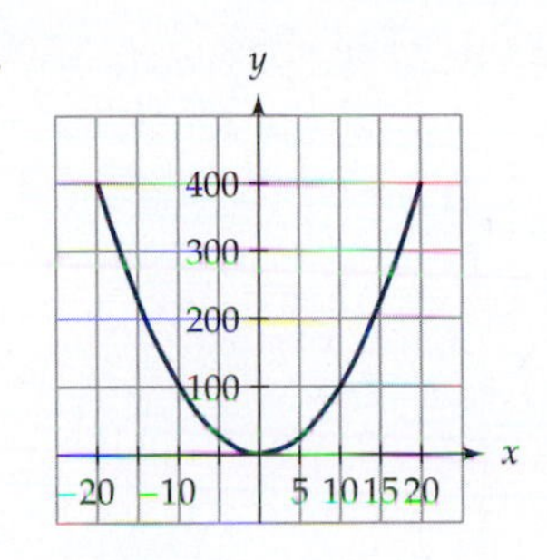

iv.

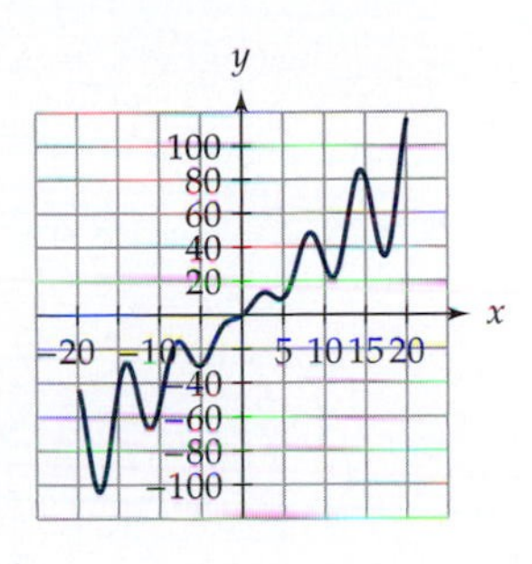

16. Each of the following graphs represents either the function $f(x) = a^x \sin(2\pi x)$ or $g(x) = mx \sin(2\pi x)$. Determine which function it is and estimate the value of a or m.

(a)

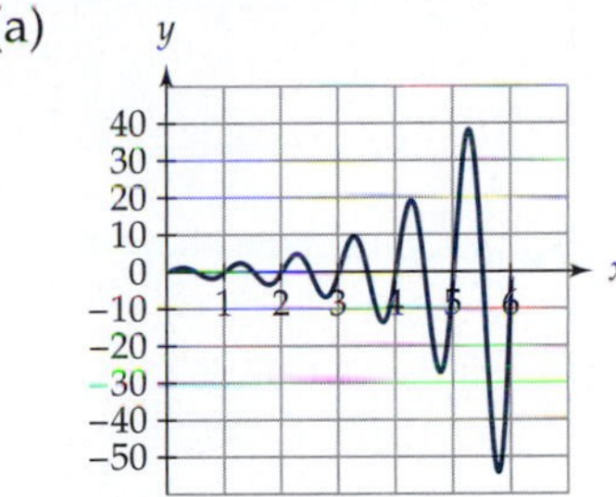

(b)

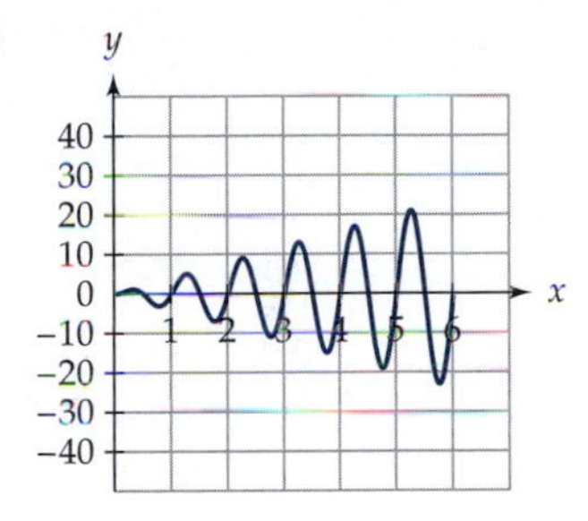

17. Each of the following graphs represents either the function $f(x) = a^x + \sin(2\pi x)$ or $g(x) = mx + \sin(2\pi x)$. Determine which function it is and estimate the value of a or m.

(a)

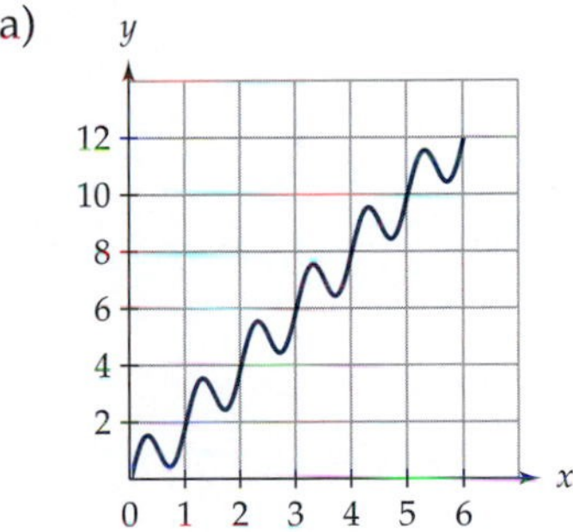

(b)

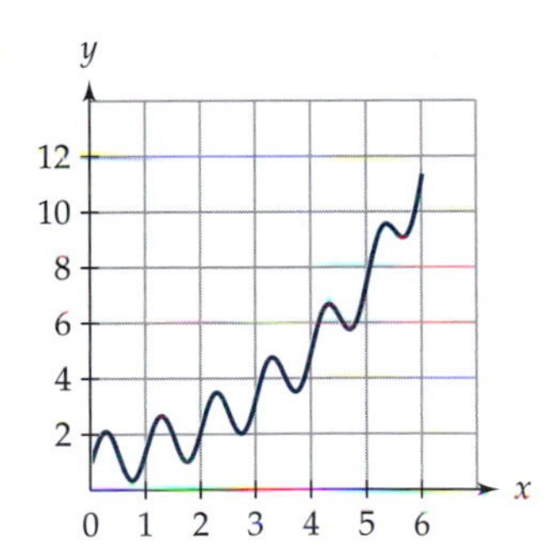

CHAPTER OUTLINE

CHAPTER OVERVIEW

- Solving equations symbolically and graphically
- Solving inequalities symbolically and graphically
- Solving exponential equations using logarithms
- Solving trigonometric equations
- Using trigonometric identities to solve equations
- Least squares regression
- Correlation
- Transforming exponential and power functions into linear functions
- Exponential and power regression

SOLVING EQUATIONS AND FITTING FUNCTIONS TO DATA

Solving equations and inequalities is the focus of this chapter. We begin with a review of the basic equation solving algorithms and then proceed to solve more difficult equations. We also study fitting equations to data and least squares regression lines. When fitting power and exponential functions to data, we see how these functions can be transformed to linear functions by using logarithms.

7.1 INTRODUCTION TO SOLVING EQUATIONS AND INEQUALITIES

Throughout this text, you have had to solve equations to solve specific problems. In this chapter, we solve equations in general both to get more practice at doing so and to see the similarities and differences between solving equations involving different types of functions. In this section, we solve equations and inequalities involving linear, power, polynomial, and rational functions and see how they are used in applications. In particular, we show how graphs can be an aid when solving equations.

SOLVING EQUATIONS

To solve an equation, you need to find *every* value for the variable that makes the equation true. The solution sometimes has more than one value. The primary concept in solving equations in their symbolic form is to isolate the variable on one side of the equation. When determining what the solution is or how many solutions exist, it is often advantageous to graph the functions involved first. Therefore, throughout this section we use graphs to help us see the solutions.

Linear and Power Functions

In solving linear equations like $3x - 5 = 13$, you merely add 5 to both sides of the equation and then divide by 3 to find that $x = 6$. An alternative method of solving this equation is to graph both sides independently. To do so, you graph $y = 3x - 5$ and $y = 13$. The x-coordinate of the point or points where the two lines intersect represents solutions. Because these equations represent two different lines, they can intersect at most once, so there is at most one solution. Figure 1 shows that the lines intersect

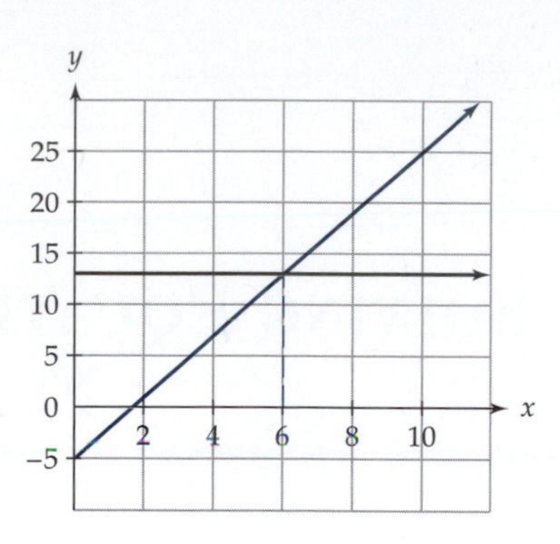

Figure 1 Graphing $y = 3x - 5$ and $y = 13$ shows that the two lines intersect when $x = 6$, the solution to $3x - 5 = 13$.

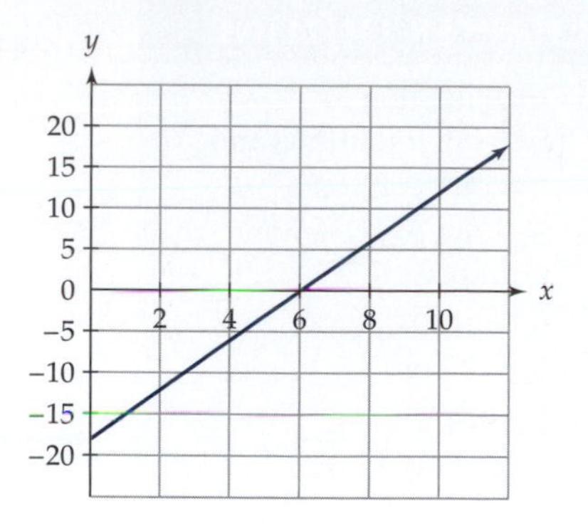

Figure 2 Finding the zero of $y = 3x - 18$ also gives the solution to $3x - 5 = 13$.

when $x = 6$, the same solution we found when we solved the equation symbolically. It is sometimes easier to transform the original equation so that one side of the equation is equal to zero. Then all you need to do to find the solution is graph the nonzero side and find where it crosses the x-axis (where $y = 0$). To solve $3x - 5 = 13$ this way, we subtract 13 from each side and rewrite it as $3x - 18 = 0$. Now by graphing $y = 3x - 18$, we can see that the zero is $x = 6$. (See Figure 2.)

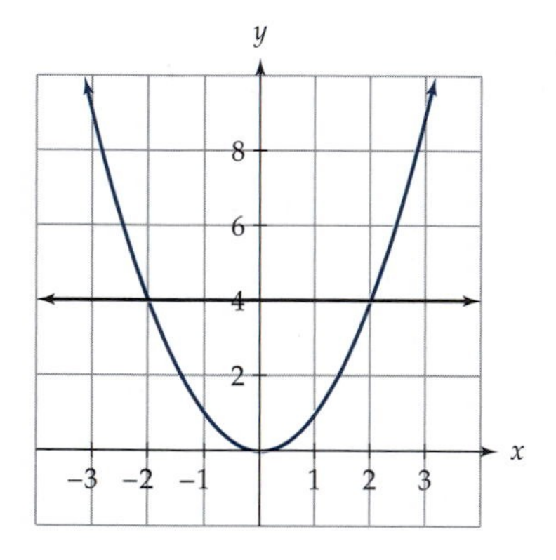

Figure 3 A graph of $y = x^2$ and $y = 4$ shows two solutions, -2 and 2, to the equation $x^2 = 4$.

To solve power functions symbolically, you must be a little more careful because more than one solution could exist. For example, to find a solution to $x^2 = 4$, we take the square root of both sides (or raise both sides to the $\frac{1}{2}$ power), which gives us $x = 4^{1/2} = 2$. This solution is not the only solution to $x^2 = 4$. A solution of -2 also exists because $(-2)^2 = 4$. We can easily see this by graphing $y = x^2$ and $y = 4$ as in Figure 3.

We do not always get two solutions when solving equations involving a power function. For example, to solve $x^3 = 64$, we get $x = 64^{1/3} = 4$. Because $(-4)^3 = -64$, we only have one solution, 4. How can you tell if there will be one solution or two? To answer this question, we need only remember some of the properties of power functions from section 2.5, when we first introduced them. There, we noted that a power function of the form $y = x^{n/d}$ is symmetric about the y-axis when d is odd and n is even. Remember also that if a function, f, is symmetric about the y-axis then $f(x) = f(-x)$. In other words, two different inputs can give the same output. Hence, we can get two solutions when solving equations involving power functions. (We could get no solution when solving this type of equation as well.) So, when solving an equation of the form $x^{n/d} = c$ where n/d is a rational exponent in simplified form and c is a constant, two solutions may exist if n is even.

When a power function has a coefficient other than 1, we must first divide both sides of the equation by the coefficient before we start raising both sides to the appropriate exponent. In other words, instead of solving for x as in a linear equation, we first solve for $x^{n/d}$. This process is illustrated in the following example.

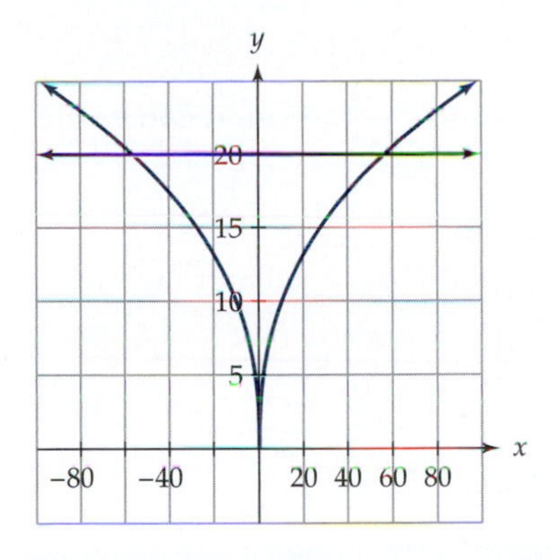

Figure 4

Example 1 Solve $4x^{2/5} = 20$.

Solution Because the numerator of the exponent is even, there is a chance that there are two solutions to this equation. Before we raise both sides of this equation to the $\frac{5}{2}$ power, however, we divide both sides by 4.

$$4x^{2/5} = 20$$
$$x^{2/5} = 5$$
$$x = \pm 5^{5/2}$$
$$x \approx \pm 55.9.$$

Graphing $y = 4x^{2/5}$ and $y = 20$ shows us that there are in fact two solutions. (See Figure 4.)

Just as you sometimes run the risk of losing solutions, there is a risk of finding solutions that do not really exist when you raise both sides of an equation to some power. For example, suppose you are asked to solve $3x^{3/2} = -12$. If you did so symbolically, your process might be

$$3x^{3/2} = -12$$
$$x^{3/2} = -4$$
$$x = (-4)^{2/3}$$
$$x \approx 2.5.$$

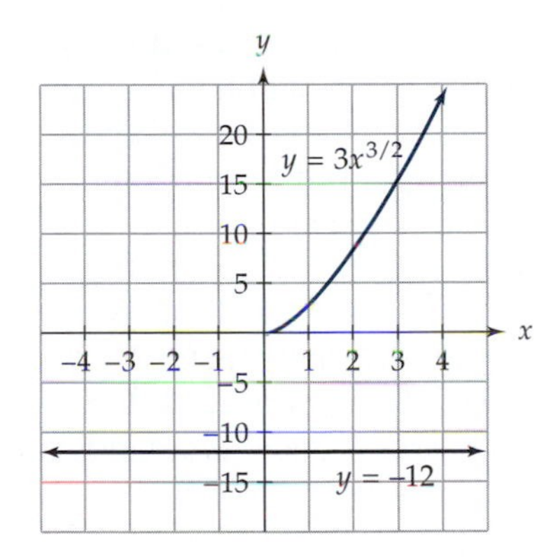

Figure 5 Because the graphs of $y = 3x^{3/2}$ and $y = -12$ do not intersect, there are no solutions to the equation $3x^{3/2} = -12$.

This result, however, is not a solution to the equation $3x^{3/2} = -12$. In fact, there are no solutions to this equation, which we can see by looking at a graph of $y = 3x^{3/2}$ and $y = -12$ as shown in Figure 5.

A false solution like this one is sometimes called an **extraneous solution.** Extraneous solutions can be obtained when both sides of an equation involving power functions are raised to some power. The easiest way to handle this situation is simply to check your solutions. Go back to the original equation and substitute your potential solutions to make sure they work. Graphing also helps determine if your potential solutions are in fact solutions.

Polynomial and Rational Functions

Solving polynomial equations presents additional challenges because often we can no longer isolate the variable by simple means. In section 4.1, we showed methods for finding the roots of quadratic functions. These methods are, of course, very useful in solving quadratic equations. Although the quadratic formula,

$$x = \frac{-b \pm \sqrt{b^2 - 4ac}}{2a},$$

can be used to solve second-degree equations, formulas for solving higher-degree equations are so cumbersome that they are rarely used. So, to solve these kinds of equations, we use other methods. One such method is factoring. An equation can sometimes be easily factored, as that given in Example 2, to allow us to find the solutions symbolically.

Example 2 Solve $x^3 - 5x^2 = -6x$.

Solution Even though this equation is a cubic equation, it can easily be factored.

$$\begin{aligned} x^3 - 5x^2 &= -6x \\ x^3 - 5x^2 + 6x &= 0 \\ x(x^2 - 5x + 6) &= 0 \\ x(x-2)(x-3) &= 0. \end{aligned}$$

Our solutions are therefore $x = 0$, $x = 2$, and $x = 3$.

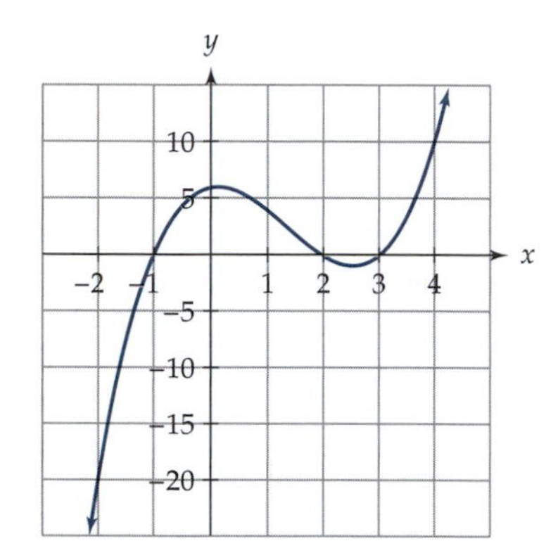

Figure 6 The solutions to $x^3 - 4x^2 + x + 6 = 0$ can be found by finding the zeros of $y = x^3 - 4x^2 + x + 6$.

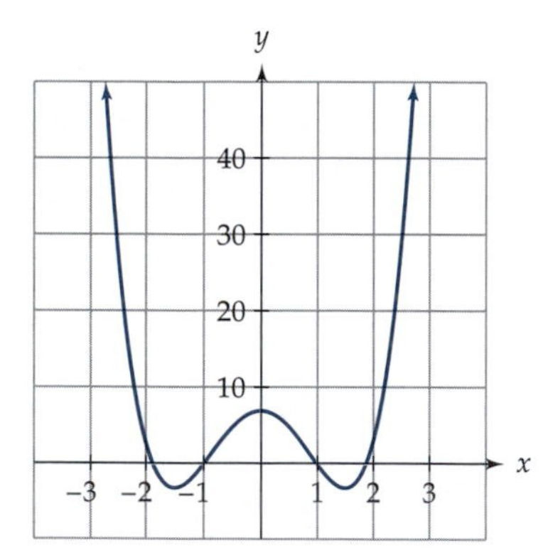

Figure 7 The graph of $y = 2x^4 - 9x^2 + 7$ shows four zeros.

When equations cannot easily be factored or cannot be factored at all, graphing can be helpful. By doing so, you can get approximate solutions and may even get enough information to allow you to find exact solutions as well. For example, to solve $x^3 - 4x^2 + x + 6 = 0$, we use the graph of $y = x^3 - 4x^2 + x + 6$, shown in Figure 6. We can see that -1, 2, and 3 appear to be zeros of the function and therefore solutions to our equation. To check, we substitute these values for x in the original equation and see that they are solutions. For example, $(2)^3 - 4(2)^2 + (2) + 6 = 8 - 16 + 2 + 6 = 0$ and is therefore an exact solution. We also know, based on our knowledge of third-degree polynomials, that the graph will have at most three zeros. Because we see all three zeros, we know that we have them all.

To solve the equation $2x^4 - 9x^2 + 7 = 0$, we graph the function, $y = 2x^4 - 9x^2 + 7$, as shown in Figure 7. There appear to be four solutions, one at $x = -1$, one at $x = 1$, and two that appear to be close to, but not equal to, $+2$ and -2. To determine the other two solutions, we can approximate using the trace feature in a graphing calculator (or the intermediate value theorem as discussed in section 4.2), or we can try to get exact solutions by dividing. We should first check to make sure that 1 and -1 are indeed solutions by substituting those values into the original equation.

$$\begin{aligned} 2(1)^4 - 9(1)^2 + 7 &= 2 - 9 + 7 = 0 \\ 2(-1)^4 - 9(-1)^2 + 7 &= 2 - 9 + 7 = 0. \end{aligned}$$

Because 1 and -1 are solutions, $(x - 1)$ and $(x + 1)$ must be factors. We divide $2x^4 - 9x^2 + 7$ by $(x - 1)(x + 1)$ or $x^2 - 1$ to find the other factors. By dividing,

$$\begin{array}{r} 2x^2 - 7 \\ x^2 - 1 \overline{)\, 2x^4 - 9x^2 + 7} \\ \underline{2x^4 - 2x^2} \\ -7x^2 + 7 \\ \underline{-7x^2 + 7} \\ 0 \end{array}$$

you can see that $2x^4 - 9x^2 + 7 = (x - 1)(x + 1)(2x^2 - 7)$. Thus, by solving $2x^2 - 7 = 0$, we can find the other two solutions to the original equation. By adding, dividing, and finding the square root, we find that the values of $x = \pm\sqrt{7/2} \approx \pm 1.87$ are solutions to the equation. Although you cannot always find exact solutions to higher-degree equations, it is sometimes possible to find other solutions if you find one or more.

Example 3 Solve $x^5 - 16x^3 = 2x^4 - 22x^2 - 55x + 60$.

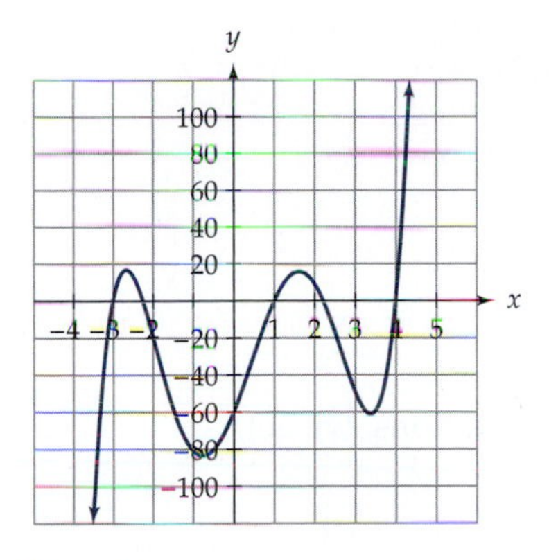

Figure 8

Solution We first write the equation as $x^5 - 2x^4 - 16x^3 + 22x^2 + 55x - 60 = 0$, which shows us that we are looking for the roots of $y = x^5 - 2x^4 - 16x^3 + 22x^2 + 55x - 60$. A graph of this function is shown in Figure 8. The graph suggests that there are five solutions and that -3, 1, and 4 may be exact solutions. (Because a fifth-degree polynomial can have at most five zeros, we can see them all in our graph.) After checking, we can see that they all are indeed solutions.

$$
\begin{aligned}
&(-3)^5 - 2(-3)^4 - 16(-3)^3 + 22(-3)^2 + 55(-3) - 60 \\
&= -243 - 162 + 432 + 198 - 165 - 60 = 0 \\
&(1)^5 - 2(1)^4 - 16(1)^3 + 22(1)^2 + 55(1) - 60 \\
&= 1 - 2 - 16 + 22 + 55 - 60 = 0 \\
&(4)^5 - 2(4)^4 - 16(4)^3 + 22(4)^2 + 55(4) - 60 \\
&= 1024 - 512 - 1024 + 352 + 220 - 60 = 0
\end{aligned}
$$

Because -3, 1, and 4 are solutions, $(x + 3)$, $(x - 1)$, and $(x - 4)$ must be factors of the function. By dividing all these solutions into our function using long division, we find that $(x^2 - 5)$ is also a factor. Hence, the equation can be written as $y = (x - 4)(x + 3)(x - 1)(x^2 - 5)$. Therefore, the only solutions we do not know are the solutions to $x^2 - 5 = 0$. We easily see that these solutions are $x = \pm\sqrt{5}$. All the solutions to this equation are therefore -3, 1, 4, and $\pm\sqrt{5}$.

As you learned in section 4.4, rational functions are made up of quotients of polynomial functions. Therefore, we can easily transform rational equations into polynomials by clearing the fractions. When we do so, however, we again run the risk of getting extraneous or false solutions. Let's do an example.

Example 4 Solve

$$\frac{2x - 1}{3x + 2} = 4x - 7.$$

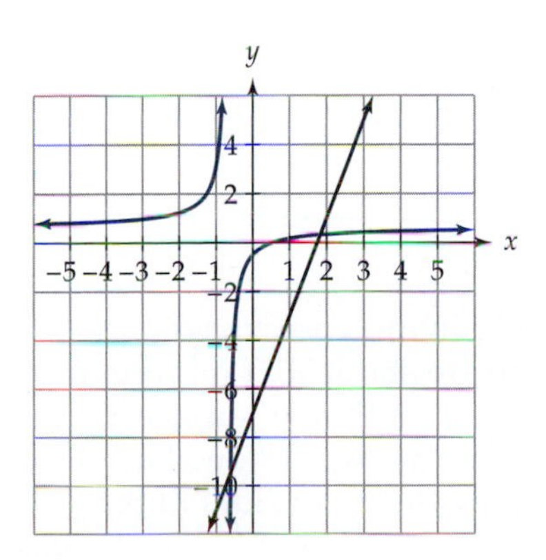

Figure 9

Solution Graphing

$$y = \frac{2x - 1}{3x + 2} \quad \text{and} \quad y = 4x - 7$$

shows how many solutions there are and their approximate value. A graph of both these functions is shown in Figure 9. This figure shows that the two graphs intersect at two points, one when the x-coordinate is a little more than -1 and one where the x-coordinate is a little less than 2.

Let's find these solutions symbolically. To do so, we multiply both sides of the equation by $3x + 2$, which gives us

$$\frac{2x-1}{3x+2} = 4x - 7$$
$$2x - 1 = (4x-7)(3x+2),$$

which then expands and simplifies to

$$2x - 1 = 12x^2 - 13x - 14$$
$$0 = 12x^2 - 15x - 13.$$

By using the quadratic formula to find the solutions, we get

$$x = \frac{15+\sqrt{849}}{24} \approx 1.839 \quad \text{and} \quad x = \frac{15-\sqrt{849}}{24} \approx -0.589.$$

These solutions correspond to the approximate solutions we saw from the graphs in Figure 9.

Example 5 In the exercises in section 4.4, we saw that the formula $w = (1.58 \times 10^7)w_0 d^{-2}$ gives the weight, w, in pounds for an object d miles from the center of Earth. The object's weight on Earth's surface is w_0. How far away from the center of Earth should you be so that your weight is one-half your weight on Earth's surface?

Solution In this case, we want $w = \frac{1}{2}w_0$, so we substitute $\frac{1}{2}w_0$ for w to make our equation $\frac{1}{2}w_0 = (1.58 \times 10^7)w_0 d^{-2}$. Dividing both sides by w_0 gives $\frac{1}{2} = (1.58 \times 10^7)d^{-2}$. Rewriting and solving for d gives

$$\frac{1}{2} = (1.58 \times 10^7)d^{-2}$$
$$\frac{1}{2}d^2 = (1.58 \times 10^7)$$
$$d^2 = 2 \cdot (1.58 \times 10^7)$$
$$d^2 = (3.16 \times 10^7)$$
$$d = \pm\sqrt{3.16 \times 10^7}$$
$$d \approx \pm 5621.$$

Because negative numbers do not make sense in this context, we can ignore the negative answer. Therefore, if you are about 5621 mi from the center of Earth, your weight is half your weight on Earth's surface. (*Note:* When we divided both sides of our equation by w_0, we assumed $w_0 \neq 0$. If w_0 could equal 0, then we could not have divided both sides of the equation by 0.)

READING QUESTIONS

1. Find solutions to the following.
   (a) $2x + 7 = 3x - 4$
   (b) $3x^3 + 5 = 15$
2. Explain why there are two solutions to $2x^{2/3} = 16$.
3. Explain why there are no solutions to $3x^{1/2} = -6$.

SOLVING INEQUALITIES

Inequalities—expressions that use $>$, $\geq$, $<$, $\leq$ or $\neq$—can also be solved symbolically and graphically in much the same way as equations. You must be careful when solving these expressions symbolically because the direction of the inequality is reversed when multiplying or dividing by a negative number. We start by considering a couple of linear examples.

Example 6 Solve $-4x + 6 < 14$ both symbolically and graphically.

Solution To solve this inequality symbolically, we need to subtract 6 from both sides and then divide both sides by -4.

$$-4x + 6 < 14$$
$$-4x < 8$$
$$x > -2.$$

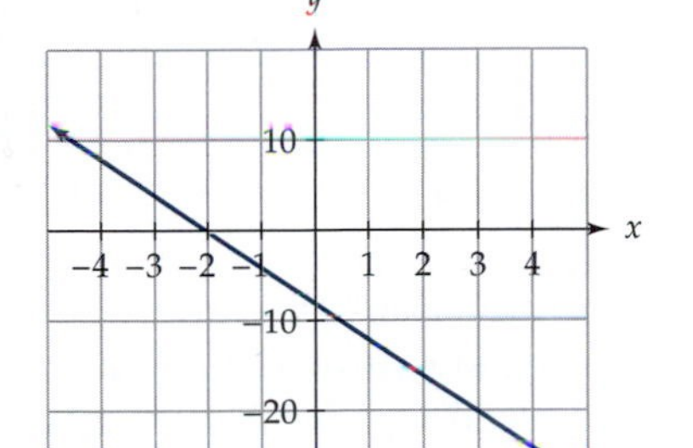

Figure 10 A graph of $y = -4x - 8$ shows that $-4x - 8 < 0$ when $x > -2$.

Note in the last step that because we divided both sides of the inequality by -4, we reversed the direction of the inequality sign. The solution indicates that *all* values of x greater than -2 make the original inequality true. We can use a graph to find this solution as well. By rewriting the inequality as $-4x - 8 < 0$, we can graph $y = -4x - 8$ as shown in Figure 10. The solution set consists of all values of x for which the output of the function is less than zero. In this case, just as with the symbolic solution, the solution is $x > -2$.

In some instances, you need to simplify the inequality and gather all like terms together before you can solve. The concept is still the same, however. You isolate the variable, and the solution is on the other side. Let's consider a common example of using inequalities when comparing prices.

Cellular phones are available with a large variety of payment plans. Thumb Cellular serves a portion of Michigan with cellular phone service and offers the services outlined in Table 1.[1] Based on Table 1, how many minutes of service should you use before you upgrade from one service plan to the next? This question is complicated because cost is a piecewise function for many of the plans. For example, in the basic plan, when your time is 15 min or less, your cost is just \$19.95 and begins changing after that.

[1] *Account Executives*, Airtime Plans *(visited 19 May 1999)* ⟨*http://www.thumbcellular.com*⟩.

TABLE 1 The Monthly Charges and Costs for Extra Minutes of Cellular Phone Service for Thumb Cellular

Plan	Monthly Charge	Cost over Free Minutes	Free Time per Month
Security	\$12.95	\$0.75/min	None
Basic	\$19.95	\$0.40/min	15 min
Value	\$35.00	\$0.35/min	60 min
Budget	\$49.00	\$0.30/min	100 min

Let's compare the security plan to the basic plan and leave the others as an exercise. The question we are asking is, When is the cost for the basic plan less than the cost for the security plan? If t is the number of minutes used, then we can express the cost for the security plan as

$$C_s = 12.95 + 0.75t$$

and the cost of the basic plan as

$$C_b = \begin{cases} 19.95, & \text{if } 0 \le t \le 15 \\ 19.95 + 0.4(t - 15), & \text{if } t > 15. \end{cases}$$

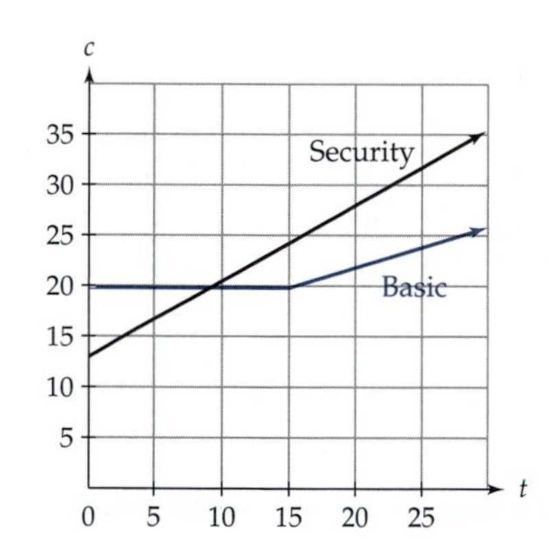

Figure 11 Graphs representing the cost of two cellular phone service plans.

(*Note:* To be more precise, the inputs for these functions should include only integer values such that fractions of a minute are always rounded up to the next minute. For example, if you make a call that is 6.5 min long, you are charged for 7 min. We adjust our final answer to accommodate this problem.) The inequality we want to solve is $C_b < C_s$. Let's first look at a graph of these two functions as shown in Figure 11. From the graph, we see that the basic plan and the security plan cost approximately the same when $t = 10$ and when $t > 10$ the basic plan is cheaper. Thus, to solve this problem symbolically, we need only consider the piece of the basic plan function when t is less than 15. Doing so gives

$$\begin{aligned} C_b &< C_s \\ 19.95 &< 12.95 + 0.75t \\ 7 &< 0.75t \\ \frac{7}{0.75} &< t \\ \frac{28}{3} &< t. \end{aligned}$$

Because $\frac{28}{3} \approx 9.33$, it is cheaper to use the basic plan if you intend to make more than 9 min of calls a month. For 9 min, the basic plan would cost \$19.95 and the security plan would cost $\$12.95 + \$0.75(9) = \$19.70$. For slightly more than 9 min (but not more than 10), the basic plan would still cost \$19.95 and the security plan would cost $\$12.95 + \$0.75(10) = \$20.45$.

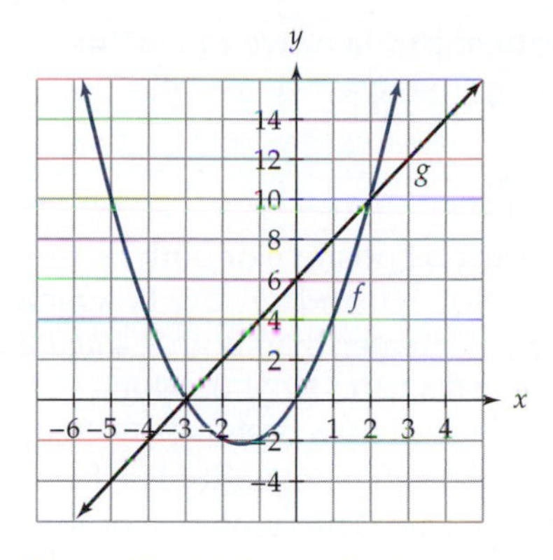

Figure 12 With a graph of $f(x) = x^2 + 3x$ and $g(x) = 2x + 6$, we see that $f(x) \leq g(x)$ when $-3 \leq x \leq 2$.

Solving inequalities is a bit more difficult when more than just lines are involved. For example, to find where $x^2 + 3x \leq 2x + 6$, we can first look at a graph of both the left and right sides to see where the left side is less than or equal to the right side. Let's call $f(x) = x^2 + 3x$ and $g(x) = 2x + 6$. Figure 12 shows the two functions. By looking at the graph, we see that the outputs for f are less than or equal to the outputs for g when $-3 \leq x \leq 2$, which can also be seen numerically as shown in Table 2. From the table, we also see that $f(x) \leq g(x)$ when $-3 \leq x \leq 2$.

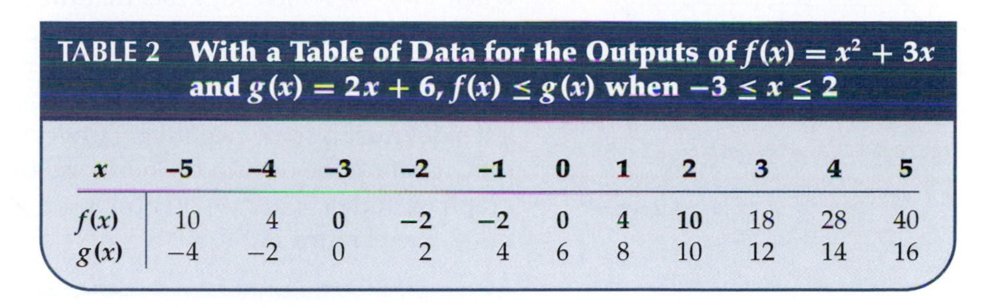

TABLE 2 With a Table of Data for the Outputs of $f(x) = x^2 + 3x$ and $g(x) = 2x + 6$, $f(x) \leq g(x)$ when $-3 \leq x \leq 2$

x	−5	−4	−3	−2	−1	0	1	2	3	4	5
$f(x)$	10	4	0	−2	−2	0	4	10	18	28	40
$g(x)$	−4	−2	0	2	4	6	8	10	12	14	16

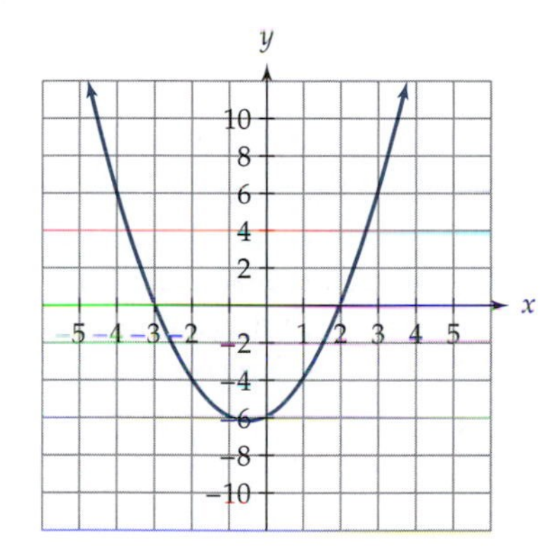

Figure 13 From a graph of $y = x^2 + x - 6$, we see that $x^2 + x - 6 \leq 0$ when $-3 \leq x \leq 2$.

This solution can also be seen, perhaps more clearly, if we rewrite the inequality $x^2 + 3x \leq 2x + 6$ as $x^2 + x - 6 \leq 0$. Here we want to find when the outputs of $y = x^2 + x - 6$ are less than or equal to zero. A graph is shown in Figure 13; you can see that the outputs of the function are less than or equal to 0 when $-3 \leq x \leq 2$.

This inequality can also be solved symbolically. First we factor.

$$x^2 + x - 6 \leq 0$$
$$(x - 2)(x + 3) \leq 0$$

The product of factors, $(x - 2)(x + 3)$, will be less than or equal to zero when the two factors, $(x - 2)$ and $(x + 3)$, have opposite signs or when at least one of them is zero. Because $(x + 3)$ is always greater than $(x - 2)$ for a given value of x, $(x + 3)$ cannot be negative when $(x - 2)$ is positive. Therefore, our solution set occurs when $(x - 2) \leq 0$ and $(x + 3) \geq 0$, which happens when $x \leq 2$ and $x \geq -3$. Another way to write this solution is $-3 \leq x \leq 2$, which is the same solution we saw with our graphs.

Example 7 Solve $\sqrt{2x + 12} < x + 2$.

Solution We begin our solution by squaring both sides of the inequality and simplifying.

$$\sqrt{2x + 12} < x + 2$$
$$2x + 12 < (x + 2)^2$$
$$2x + 12 < x^2 + 4x + 4$$
$$0 < x^2 + 2x - 8.$$

At this point, we can either use the quadratic formula or we can factor. Because a factored solution is fairly simple in this case, we use that method.[2]

$$0 < (x+4)(x-2).$$

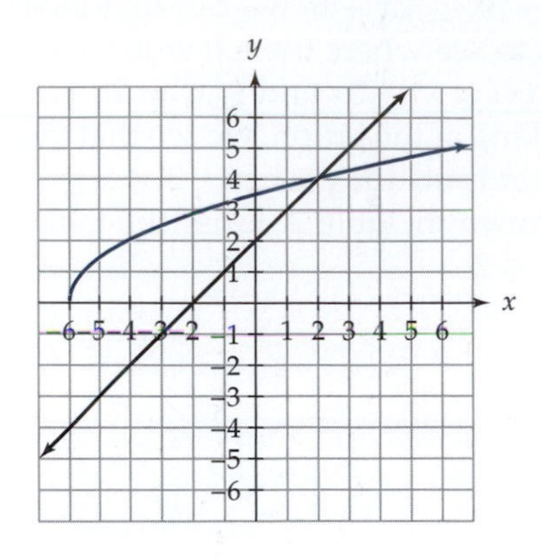

Figure 14 A graph of $y = \sqrt{2x+12}$ and $y = x+2$ shows that $\sqrt{2x+12} < x+2$ when $x > 2$.

For this inequality to be true, both factors must be positive or both factors must be negative. If both are positive, then $x+4 > 0$ and $x-2 > 0$, which means that $x > -4$ and $x > 2$. The variable x is greater than both -4 *and* 2 when $x > 2$. If both are negative, then $x+4 < 0$ and $x-2 < 0$, which means that $x < -4$ and $x < 2$. The variable x is less than both -4 *and* 2 when $x < -4$, which suggests that the solution to the inequality is when $x > 2$ or $x < -4$. This solution, however, is not true. The right side of the original inequality, $\sqrt{2x+12} < x+2$, is negative when $x < -4$. If the right side is negative and the left side is less than the right side, then the left side must also be negative. However, because the left side is a square root, it cannot be negative. Therefore, the only solution is when $x > 2$. A graph of both $y = \sqrt{2x+12}$ and $y = x+2$ shows that the solution is just $x > 2$. (See Figure 14.)

Solving rational inequalities can also be easily done with graphs or with a combination of symbolic methods and graphing. These methods can be seen in the following example.

Example 8 In section 1.4, we introduced a formula that can be used to determine the time (in seconds) needed for a yellow traffic light given the speed limit (in feet per second) for vehicles driving on the street where the light is located. This formula is $t = 1 + v/20 + 70/v$. This function is a rational function because it can be written as a ratio of two polynomials, or

$$t = \frac{20v + v^2 + 1400}{20v}.$$

Find the speed limits required for yellow lights that are less than 5 sec in duration.

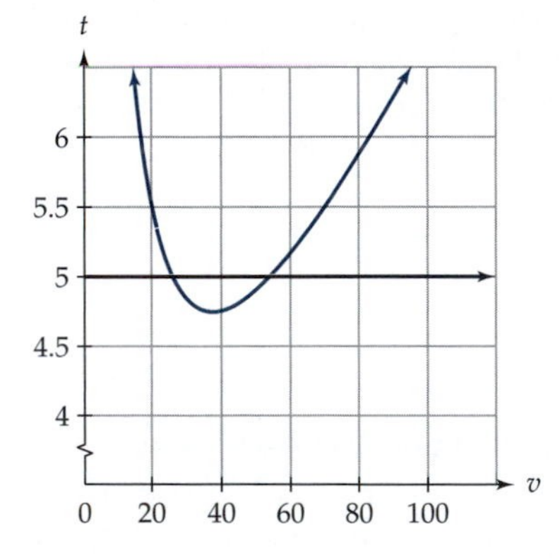

Figure 15

Solution We need to solve the inequality

$$\frac{20v + v^2 + 1400}{20v} < 5.$$

To approximate the solution set, we graph

$$y = \frac{20v + v^2 + 1400}{20v} \quad \text{and} \quad y = 5$$

as shown in Figure 15. From this graph we see that when the speed limit is between approximately 25 ft/sec and 55 ft/sec, the time needed for a yellow light is less than 5 sec.

[2] *Often, a quick scan of possible factors of a quadratic yields a factored solution and saves you time. If it does, as in this case, then it is quicker and easier to use than other methods. If no factored solution is clear to you quickly, then use the quadratic formula.*

To determine the solution set more precisely, we solve this inequality symbolically. Multiplying both sides of $(20v + v^2 + 1400)/20v < 5$ by $20v$ gives

$$20v + v^2 + 1400 < 100v.$$

Subtracting $100v$ from both sides and rearranging the terms gives

$$v^2 - 80v + 1400 < 0.$$

Using the quadratic formula, we find that the left side equals zero when $v = 40 \pm 10\sqrt{2}$, or approximately 25.9 and 54.1. From our graph, we know that for the left side to be less than 5, the velocity must be between these two numbers, or

$$40 - 10\sqrt{2} < v < 40 + 10\sqrt{2}.$$

READING QUESTIONS

4. Find solutions for the following.

(a) $-2(x-3) \leq 4x + 2$

(b) $x^2 + 2x \geq 8$

5. Explain why there cannot be any solutions for x in $\sqrt{x+10} \leq 2x + 4$ that are less than -2.

SUMMARY

Solving linear and power equations was reviewed. When solving power equations, it is important to know that there is often more than one solution. First solve for the variable raised to the given power and then raise both sides of the equation to the exponent's reciprocal. This process sometimes leads to **extraneous solutions,** answers that are not actual solutions to the equations. It is therefore important to check all solutions, either by substituting your answers in the original equation or by graphing.

In solving quadratic equations, the quadratic formula,

$$x = \frac{-b \pm \sqrt{b^2 - 4ac}}{2a},$$

can be used to find all solutions, but no such convenient formula exists for higher-degree polynomials. One way to solve these equations is through factoring, which works some of the time. In more difficult problems, you can rewrite the equation so that one side equals zero; then a graph of the other side will allow you to approximate the solutions. You can check these solutions to determine if they are exact solutions. Even if you cannot find all the exact solutions to a higher-degree polynomial in this manner, this process sometimes allows you to find some solutions. You can then use polynomial division to rewrite the equation in factored form and use other techniques (such as the quadratic formula) to finish the problem. Sometimes, the best you can do is approximate solutions with a calculator.

When solving inequalities, if you multiply (or divide) both sides of an inequality by a negative, then you must reverse the direction of the inequality. With inequalities that are not linear, the process can be more complicated. Often, a graph or a table helps to determine which intervals are solutions to the inequalities.

EXERCISES

1. Solve the following equations. Give exact solutions.
   (a) $5x^2 = 42$
   (b) $x^{4/3} = 12$
   (c) $x^{-3/5} = 2$
2. Solve the following equations. Give exact solutions.
   (a) $x^2 + 3x = 15$
   (b) $x^3 + 3x^2 - 10x = 24$
   (c) $x^4 - 2x^3 = -2x^2 + x$
3. Solve the following equations. Give exact solutions.
   (a) $\dfrac{x}{x^2+1} = \dfrac{1}{3}$
   (b) $\dfrac{x^2-9}{x-4} = 150$
   (c) $\dfrac{2x^2+6x+9}{x+4} = -10$
   (d) $x - 1 - \dfrac{2}{x} = 5$
4. Solve the following inequalities. Give exact solutions.
   (a) $x^2 - 2x < 6$
   (b) $4x^{2/5} \geq 8$
   (c) $\sqrt{2x-6} \leq 4$
   (d) $x^3 - 10x > -3x^2$
5. Suppose $f(x) = c$, where f is some function and c is some constant. Determine the least possible number of solutions and the most possible number of solutions if f is a
   (a) linear function.
   (b) power function.
   (c) exponential function.
   (d) sine function.
   (e) quadratic function.
   (f) cubic function.

6. For each of the following equations, first determine how many solutions exist and then find the solutions.
   (a) $\sqrt{x} = 2x - 1$
   (b) $\sqrt{x} = 2x$
   (c) $\sqrt{x} = 2x + 1$
7. Solve each of the following formulas for the indicated variable.
   (a) Solve $F = (Gm_1m_2)/R^2$ for G.
   (b) Solve $p = sf/(s - f)$ for s.
   (c) Solve $A = \pi r(r + \sqrt{r^2 + h^2})$ for h.
   (d) Solve $A = 2\pi rh + 2\pi r$ for r.
8. If $x = 1$ and $y = 1$, explain what is wrong with the following.

$$\begin{aligned} x &= y \\ x^2 &= xy \\ x^2 - y^2 &= xy - y^2 \\ (x - y)(x + y) &= y(x - y) \\ x + y &= y \\ 2 &= 1 \end{aligned}$$

9. In the text, we compared the security plan with the basic plan for cellular phone service for a company in Michigan. In this exercise, you similarly compare other plans shown in the following table.

Plan	**Monthly Charge**	**Cost over Free Minutes**	**Free Time per Month**
Security	\$12.95	\$0.75/min	None
Basic	\$19.95	\$0.40/min	15 min
Value	\$35.00	\$0.35/min	60 min
Budget	\$49.00	\$0.30/min	100 min

   (a) How many minutes would you have to use your cellular phone service for the value plan to be cheaper than the basic plan?
   (b) How many minutes would you have to use your cellular phone service for the budget plan to be cheaper than the value plan?
10. Kepler's third law of planetary motion is stated as follows.

    The square of the length of time it takes any planet to orbit the sun is proportional to the cube of the planet's mean distance from the sun.

   (a) Given that Earth is about 150,000,000 km away from the sun and orbits the sun in approximately 365 days, use Kepler's third law to determine a function where a planet's distance from the sun (in kilometers) is the input and the time it takes to orbit the sun (in Earth days) is the output.

(b) Mars is approximately 230,000,000 km away from the sun. Use your function from part (a) to determine the length of time it takes Mars to orbit the sun.

(c) Suppose a planet existed that took twice as long as Earth to orbit the sun. How far from the sun would this planet be?

11. The formula for the height of an object in free fall is $h(t) = -16t^2 + v_0 t + h_0$, where $h(t)$ is the height of the object (in feet), t is the time since it was launched (in seconds), v_0 is the initial velocity (in feet per second), and h_0 is the initial height (in feet). Assume the object is launched upward from the ground ($h_0 = 0$) at 90 ft/sec ($v_0 = 90$).

    (a) What time or times is the object 100 ft above ground?

    (b) Find the maximum height of the object by finding the vertex of the quadratic function. (Methods for finding the vertex of a quadratic function were shown in section 4.1.)

    (c) How long will it take for the object to fall back to the ground after it is launched?

12. Suppose a golf ball is hit off the ground at an angle of 45° with an initial velocity of 120 ft/sec. To simplify, the height of the golf ball (in feet) can be modeled with the equation $h = -16t^2 + \left(120/\sqrt{2}\right)t$, where t is the time in seconds since it was hit. The horizontal distance of the golf ball along the ground (in feet) can be modeled with the equation $d = (120/\sqrt{2})t$, where t is the time in seconds since it was hit.

    (a) Find the function where the horizontal distance of the golf ball, d, is the input and the height of the golf ball, h, is the output.

    (b) What horizontal distance will the golf ball have traveled when it hits the ground?

    (c) What is the maximum height of the golf ball?

    (d) Would the golf ball be able to go over the top of an 80-ft tree that is 350 ft away from where it was hit? Explain your answer.

13. In section 4.2, we looked at the regulations for mailing packages. For the U.S. Postal Service, the regulations state that the girth plus the length cannot exceed 108 in. In that section, we set the sum of girth and length at 108, giving us a box with dimensions as large as possible. The formula for such a box is $V(w) = w^2(108 - 4w)$, where w is the width of the square base as shown in the accompanying figure. Find the width, w, of a box with a volume of 4500 in.^{3}.

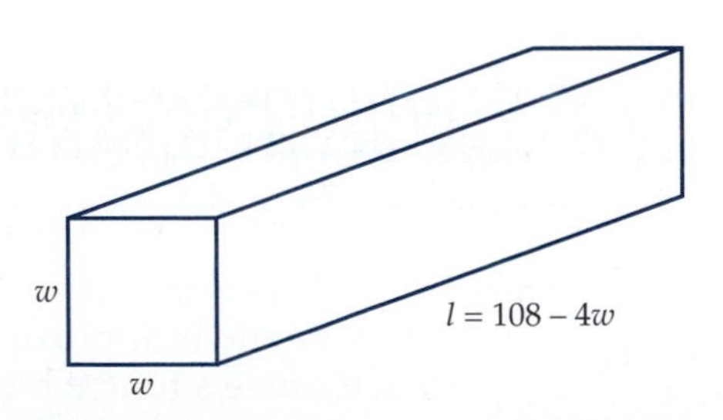

14. In section 1.4, you first worked with the formula for determining the length of a yellow light in a traffic intersection based on the speed of the vehicles. In that section, you were given the formula $t = 1 + v/20 + 70/v$, where v is the speed of the car and t is the time needed for a yellow light.
    (a) Using symbolic methods and the quadratic formula, find the speed (or speeds) that cars should be traveling if the yellow light were set for 6 sec.
    (b) We want to find the minimum time for a yellow light.
        i. Show that the original formula, $t = 1 + v/20 + 70/v$, can be written as $v^2 + (20 - 20t)v + 1400 = 0$.
        ii. Use the quadratic formula to solve $v^2 + (20 - 20t)v + 1400 = 0$ for v. Leave your solution in terms of t.
        iii. For a solution to exist, the expression under the square root from your solution from part (ii) must be nonnegative. Explain why the smallest value of t occurs when the expression under the square root sign equals 0.
        iv. Using your solution from part (ii), set the expression under the square root equal to zero and solve for t. The answer should be the minimum time needed for a yellow light.

15. In section 4.4 we discussed the brightness of light coming from a source like a light bulb. In the accompanying figure, you see a light bulb that is 2.5 m above the floor. For this particular bulb (60 W with a luminous flux of about 840 lumens), the function that models the brightness is given by

$$E = \frac{840}{2\pi(2.5^2 + x^2)},$$

where E is the amount of illuminance (given in lux). Find the distance away from directly underneath the bulb where the illuminance is greater than 10 lux.

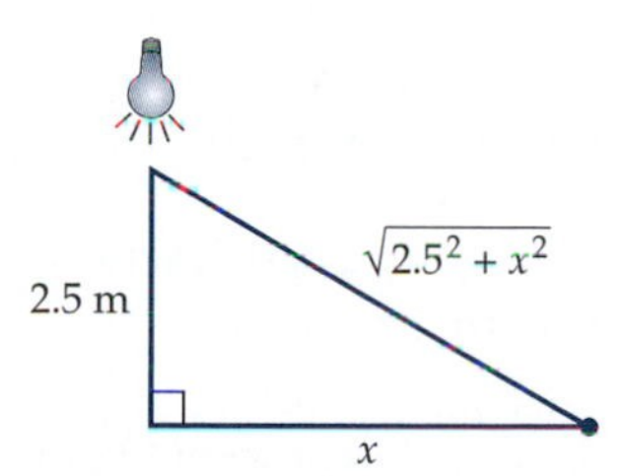

INVESTIGATIONS

INVESTIGATION 1: PARTIAL FRACTIONS

A graph of

$$h(x) = \frac{2x}{x^2 - 4}$$

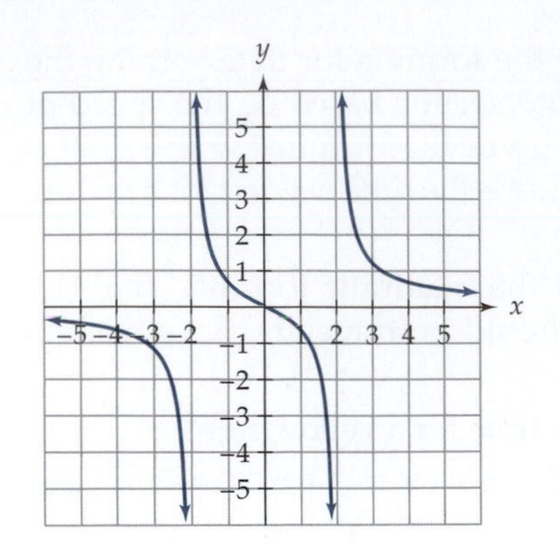

Figure 16

is shown in Figure 16. This rational function has a horizontal asymptote of $y = 0$ and vertical asymptotes of $x = 2$ and $x = -2$. These asymptotes can be seen in the graph, but they are also evident from the equation. Because the denominator has zeros of $x = 2$ and $x = -2$ (and the numerator does not have zeros there), the function has vertical asymptotes at these places. Because the rational function consists of a linear function divided by a quadratic function (something of higher degree than the numerator), it has a horizontal asymptote of $y = 0$. Although these asymptotes are evident from the equation, just how the function fits between these asymptotes may not be. Another way to examine this function symbolically is to rewrite it as

$$h(x) = \frac{1}{x+2} + \frac{1}{x-2}.$$

You can think of h as the sum of the two functions $f(x) = 1/(x+2)$ and $g(x) = 1/(x-2)$. When the functions f, g, and h are shown together, as in Figure 17, it is easy to see why h looks the way it does. In this figure, f and g are shown with dashed lines and h is shown with a thick solid curve.

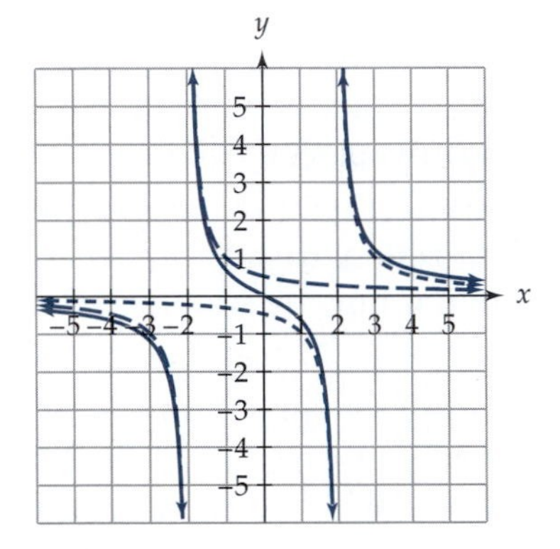

Figure 17

In this investigation, you see how to rewrite some rational functions into sums of simpler rational functions in the same way h was rewritten as the sum of f and g. These simpler rational functions are known as partial fractions. Rewriting a rational function as the sum of partial fractions can be an important technique for some types of problems in calculus. We use it as simply another way to view a function.

1. Verify that $f(x) = 1/(x+2) + 1/(x-2)$ is the same as $f(x) = 2x/(x^2-4)$ by showing

$$\frac{1}{x+2} + \frac{1}{x-2} = \frac{2x}{x^2-4}.$$

2. Although it is quite straightforward to write

$$f(x) = \frac{1}{x+2} + \frac{1}{x-2}$$

as a polynomial divided by a polynomial, it is not straightforward to rewrite

$$f(x) = \frac{2x}{x^2-4}$$

as the sum of partial fractions. To do so, we first need to factor the denominator,

$$f(x) = \frac{2x}{(x+2)(x-2)}.$$

Now writing

$$f(x) = \frac{2x}{(x+2)(x-2)} = \frac{A}{x+2} + \frac{B}{x-2},$$

we need to solve this for A and B. Solve

$$\frac{2x}{(x+2)(x-2)} = \frac{A}{x+2} + \frac{B}{x-2}$$

by doing the following.

(a) Multiply the equation by $(x+2)(x-2)$, the common denominator, which should clear out the denominators of both sides.

(b) Show that the equation can be written as $2x = (A+B)x + (-2A+2B)$.

(c) Explain why this rewriting means that $A+B=2$ and $-2A+2B=0$.

(d) Solve your system of equations for A and B.

3. Use a similar procedure as in question 2 to find values for A and B so that

$$\frac{3x+1}{(x+3)(x-1)} = \frac{A}{x+3} + \frac{B}{x-1}.$$

4. Any rational function $p(x)/q(x)$ can be written as the sum of partial fractions. Partial fractions are rational functions whose denominators are irreducible factors of $q(x)$ or powers of the irreducible factors of $q(x)$. These irreducible factors are either linear or quadratic. Furthermore, proper rational functions—those that have numerators with lower degrees than denominators—can be written as the sum of proper partial fractions. Use these ideas to show that

$$f(x) = \frac{2x-1}{x^2+3x+2}$$

can be written as the sum of two rational functions whose denominators are linear functions.

INVESTIGATION 2: CONJUGATE ROOTS THEOREM

So far, when we look at roots of polynomials, we look at those roots that are real numbers, which are the roots that we can see graphically as a function crosses or touches the x-axis. Some polynomial functions also have complex roots, or roots that are not real numbers, but complex numbers. If we were to find the roots of $y = x^2 + 4$, then we would have to solve the equation $x^2 + 4 = 0$. In doing so, we could subtract 4 from both sides of the equation to get $x^2 = -4$. Because we cannot take the square root of a negative number in the real number system, we would conclude that there were no solutions to this equation. If we extend this to the complex number system, however, then there are solutions to this equation.

You can think of the complex number system as an enlargement of the real number system. In other words, the set of all real numbers is a subset of the set of all complex numbers. You have seen this idea before. The set of all integers, for example, is a subset of the set of all rational numbers, and the set of all rational numbers is a subset of the set of all real numbers.

Complex numbers can be written in the form $a + bi$, where a and b are real numbers and $i = \sqrt{-1}$. Thus, complex numbers allow for square roots of negative numbers. We can take the square root of -4 in the complex number system. To do so, we get $\sqrt{-4} = 2\sqrt{-1} = 2i$. To write this equation in the standard form of $a + bi$, we could write $0 + 2i$. The complex number $2i$ is not the only solution to our original equation of $x^2 + 4 = 0$ because $-2i$ is also a solution, just as $x = \pm 2$ are the two solutions to the equation $x^2 = 4$. Both the positive and negative solutions are more evident if we solve these equations using the quadratic formula.

Because we stated earlier that the set of all real numbers is a subset of the set of all complex numbers, we must be able to write any real number in the form $a + bi$. We can do so by having $b = 0$. For example, 7 written in the standard complex form is $7 + 0i$.

In this investigation, we want to prove the conjugate root theorem. This theorem states that if f is a polynomial with real coefficients and if a complex number is a root of f, then its conjugate is also a root of f. You may remember that the conjugate of the complex number $a + bi$ is $a - bi$. Likewise, the conjugate of $a - bi$ is $a + bi$.

Before we prove the conjugate root theorem, let's first see how complex numbers are added and multiplied. The usual rules for adding and multiplying hold for complex numbers. For example, $(3 + 4i) + (9 + 2i) = 3 + 9 + 4i + 2i = 12 + 6i$. When two complex numbers are multiplied, we need to remember that because $i = \sqrt{-1}$, $i^2 = -1$. Therefore, $(3 + 4i)(9 + 2i) = 27 + 6i + 36i + 8i^2 = 27 + 42i - 8 = 19 + 42i$.

1. Perform the indicated operations on the following complex numbers.
   (a) $(4 + 5i) + (2 - 8i)$
   (b) $(2 - 10i) - (6 + 9i)$
   (c) $(2 + 5i)(1 + 3i)$
   (d) $(3 + 2i)(1 - 6i)$
   (e) $(2 + 3i)(2 - 3i)$
2. Show that the product of a complex number and its conjugate is a real number. In particular, show that $(a + bi)(a - bi) = a^2 + b^2$.
3. The real number, a, written in standard complex notation is $a + 0i$. Explain why the conjugate of the real number a is a.
4. One way of denoting a conjugate is to put a bar over the top of the number. For example, the conjugate of $a + bi$ can be written as $\overline{a + bi}$. Let $s = a + bi$ and $t = c + di$. Prove the following.
   (a) $\overline{s + t} = \bar{s} + \bar{t}$. (*Hint:* For the left side of the equation, add the numbers together and then find the conjugate of the sum. For the right side of the equation, first find the conjugates of each number and then add them.)
   (b) $\overline{s \cdot t} = \bar{s} \cdot \bar{t}$
5. Now we prove the conjugate roots theorem.
   (a) Suppose $f(x) = ax^3 + bx^2 + cx + d$ is a polynomial with real coefficients and t is a complex root of f. We want to show that $\overline{f(t)} = f(\bar{t})$. To do so, use your results from the previous questions to show that $\overline{at^3 + bt^2 + ct + d} = a\bar{t}^3 + b\bar{t}^2 + c\bar{t} + d$. Remember that because t is a number, these results are not functions, but numbers.
   (b) Explain why if t is a root of f then $\bar{t}$ is a root of f. (*Hint:* Remember that $0 = \bar{0}$.)
   (c) In the previous two questions, you showed that if t is a complex root of a third-degree polynomial, then $\bar{t}$ is a root of the polynomial. Explain why you can extend this concept to all polynomials with real coefficients.

7.2 SOLVING EXPONENTIAL EQUATIONS

In section 2.2, we first looked at exponential functions. We defined what they were, looked at some of their properties, and used them in a few applications. In this section, we take another look at exponential functions. We also see how logarithms can be used to solve equations involving exponential functions. In addition to using the common logarithm, as in section 2.3, we also introduce and use the natural logarithm function.

REVIEW OF EXPONENTIAL FUNCTIONS AND THEIR APPLICATIONS

In section 2.2, we defined an **exponential function** as a function where the growth factor is constant. Symbolically, an exponential function is

$$f(x) = ba^x,$$

where $a > 0$ and $a \neq 1$. The y-intercept (or starting value) is b. The **growth factor** is a, which is also $f(x+1)/f(x)$.

Exponential functions are often used to model growth. For example, suppose a colony of bacteria is growing exponentially such that the colony starts with 1000 organisms and that these organisms double every hour. This population can be modeled by an exponential function that has a starting value of 1000 and a growth factor of 2. The equation that models this population is

$$p(t) = 1000 \cdot 2^t,$$

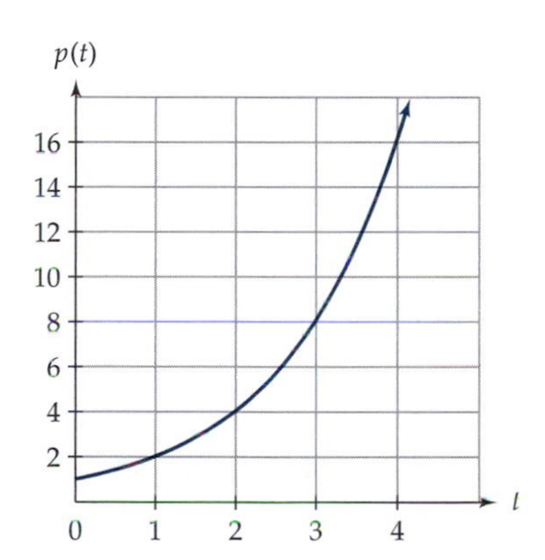

Figure 1 A graph of $p(t) = 1000 \cdot 2^t$. The output is given in thousands of bacteria.

where $p(t)$ is the number of bacteria after t hours. A graph of this function is shown in Figure 1, where the output is the number of bacteria. This graph is the classic shape of an exponential function where the growth factor is greater than 1 because it is increasing and concave up.

Determining a formula for the bacteria example was straightforward because we were given both the growth factor and the starting value. We are not always as fortunate, as the following situation illustrates. According to a newspaper article,[3] the average tuition for public universities in Michigan increased 114.1% from 1986 to 1996. The article also stated that the average yearly tuition in 1996 was \$3816. Suppose an exponential function is a good model for describing the average tuition for public universities in Michigan over this period. To determine the equation to model this situation, we need two things, a starting value (the cost of tuition in 1986), b, and the annual growth factor, a. Putting those two values into the equation $c(t) = ba^t$, where $c(t)$ is the average cost of tuition and t is the time in years since 1986, gives us the exponential function we seek.

We first determine the growth factor. Simply converting the 114.1% to a decimal, 1.141, does not give us the growth factor. The 114.1% given was the amount of *increase*. Suppose the tuition in 1986 was \$1000. If there is a 114.1% increase, then the new tuition is

[3] *"Tuition at State Universities Continues to Outpace Inflation,"* Holland Sentinel, *30 September 1996, p. D1.*

$$\begin{aligned} 1000 + 1000 \cdot 1.141 &= 1000(1 + 1.141) \\ &= 1000(2.141) \\ &= 2141. \end{aligned}$$

The growth factor over the 10-year period of time is therefore 2.141. The 114.1% = 1.141 is called a growth rate. A **growth rate** is always 1 less than the growth factor. (We could also describe the growth factor as 1 more than the growth rate.) In our situation, we want to convert the growth factor over a 10-year period to an annual growth factor. If this function is exponential and a is the annual growth factor, then $a^{10} = 2.141$. Solving for a gives $a = 2.141^{1/10} \approx 1.079$. When determining a value for b, the cost of tuition in 1986, we know that the tuition in 1996 is 2.141 times larger than that in 1986. If b is the cost of tuition in 1986, then $2.141b = 3816$. Thus, b is $3816/2.141 \approx 1782$. Therefore, our exponential function is $c(t) = 1782 \cdot 1.079^t$.

Example 1 Suppose the value of a car depreciates exponentially. The value of a certain 3-year-old car is $10,000, and the value of that same car when it is 10 years old is $2000. Find the equation where the output is the value of the car and the input is the age of the car in years.

Solution If the value of the car depreciates exponentially, then we can use the equation $v(t) = ba^t$, where v is the value of the car and t is the age of the car. When the car was 3 years old, it was worth $10,000. Substituting these two numbers into our equation gives

$$10{,}000 = ba^3.$$

When the car was 10 years old, it was worth $2000. Substituting these two numbers into our equation gives

$$2000 = ba^{10}.$$

We then have two equations with two unknowns. If we solve the first equation for b, then we get

$$b = \frac{10{,}000}{a^3}.$$

Substituting $10{,}000/a^3$ for b in the second equation gives

$$2000 = \left(\frac{10{,}000}{a^3}\right) a^{10}.$$

Solving for a gives

$$\begin{aligned} 2000 &= \left(\frac{10{,}000}{a^3}\right) a^{10} \\ 2000 &= 10{,}000a^7 \\ \frac{2000}{10{,}000} &= a^7 \end{aligned}$$

$$\left(\frac{2000}{10{,}000}\right)^{1/7} = a$$

$$0.7946 \approx a$$

Therefore, the growth (or decay) factor is approximately 0.7946. Thus, the value of the car decreases by about 20% every year. (We could also say that it retains about 80% of is value.) To find the starting value, b, we simply substitute 0.7946 into one of our original equations and solve for b.

$$10{,}000 = ba^3$$

$$10{,}000 = b(0.7946^3)$$

$$\frac{10{,}000}{0.7946^3} = b$$

$$19{,}932 \approx b.$$

Therefore, the value of a new car is about \$19,932 (assuming the same type of exponential depreciation occurred since the car was new). The equation that models this type of depreciation is therefore

$$v(t) = 19{,}932(0.7946)^t.$$

READING QUESTIONS

1. How does a growth rate differ from a growth factor?
2. Suppose a college's tuition in 1990 was \$4000 and grew by 5% each of the following years.
   (a) How much was the tuition in 1995?
   (b) How much would the tuition be in t years after 1990?

SOLVING EXPONENTIAL EQUATIONS USING LOGARITHMS

We have done quite a bit of solving with exponential equations, but we have not yet solved for the exponent. In Example 1, we found the equation that models the value of a car over time, $v(t) = 19{,}932(0.7946)^t$, where v is the value of the car and t is the age in years. Suppose we want to determine the age of the car when its value is \$5000. To do so, we need to solve the equation $5000 = 19{,}932(0.7946)^t$ for t. Adding, multiplying, or raising both sides of the equation to some exponent will not help us solve for t because the variable stubbornly remains in the exponent. We need a way to get the variable out of the exponent where we can work with it, something that will "undo" what the exponential function did. We need an inverse. The logarithm function is the inverse of the exponential function. In section 2.3, you learned that $\log x = y$ is equivalent to $10^y = x$, which means, for example, that if $10^y = 4$, then $\log 4 = y$. The equation we want to solve is a bit more complicated than this one, but using logarithms is still the solution we need.

To solve $5000 = 19{,}932(0.7946)^t$, we first solve for $(0.7946)^t$. Doing so gives

$$\frac{5000}{19{,}932} = (0.7946)^t.$$

Now by taking the log of both sides we get

$$\log\left(\frac{5000}{19{,}932}\right) = \log(0.7946)^t.$$

You might think that this equation does not look any better, or perhaps worse, than it did before. However, let's remember the laws of logarithms that we learned in section 2.3. Specifically, we learned that $\log a^b = b \log a$, which allows us to rewrite our equation as

$$\log\left(\frac{5000}{19{,}932}\right) = t\log(0.7946).$$

Now the variable is not in the exponent and we can just divide both sides by log(0.7946) to find our solution.

$$\begin{aligned}\frac{\log\left(\frac{5000}{19{,}932}\right)}{\log(0.7946)} &= t \\ 6.015 &\approx t.\end{aligned}$$

Thus, when the car is approximately 6 years old, its value is \$5000.

Example 2 Solve $16^{x+6} = 9$.

Solution We first take the log of both sides of this equation and then use the laws of logarithms to move the exponent and solve for x.

$$\begin{aligned}16^{x+6} &= 9 \\ \log(16^{x+6}) &= \log 9 \\ (x+6)\log 16 &= \log 9 \\ x + 6 &= \frac{\log 9}{\log 16} \\ x &= \frac{\log 9}{\log 16} - 6.\end{aligned}$$

This answer is the exact solution to the equation. By using a calculator to get a decimal approximation, we get $x \approx -5.208$.

Example 3 Solve $5^x = 3 \cdot 8^{3x}$.

Solution We first take the log of both sides of this equation and then use the laws of logarithms and algebra to solve for x.

$$\begin{aligned}5^x &= 3 \cdot 8^{3x} \\ \log(5^x) &= \log(3 \cdot 8^{3x})\end{aligned}$$

$$\begin{aligned} \log(5^x) &= \log 3 + \log(8^{3x}) \\ x \log 5 &= \log 3 + 3x \log 8 \\ x \log 5 - 3x \log 8 &= \log 3 \\ x(\log 5 - 3 \log 8) &= \log 3 \\ x &= \frac{\log 3}{\log 5 - 3 \log 8}. \end{aligned}$$

The decimal approximation to this solution is $x \approx -0.2373$.

It is very important to remember that you can always check your answer by plugging it in to the original equation when solving equations. For example, in Example 3's last solution, $x \approx -0.2373$, the right side of the original equation is $5^{-0.2373} \approx 0.6825$. The left side of the equation, however, is $3 \cdot 8^{3(-0.2373)} \approx 0.6827$. Because of rounding, the two sides are not exactly the same, but this check does confirm that we have the approximate solution.

READING QUESTIONS

3. Solve $3^x = 6$ for x.

4. Why would you not want to use logarithms to solve $5x^3 = 6$?

CONTINUOUS GROWTH RATES AND THE NUMBER e

Suppose you have \$1000 invested in an account that has an annual interest rate of 8% per year. The amount of money you have in the account after the first year is $A = \$1000 + \$1000(0.08) = \$1000(1 + 0.08) = \$1080$, an increase of \$80. The amount of money that you have at the end of the second year is $A = \$1080 + \$1080(0.08) = \$1080(1 + 0.08) = 1080(1.08) = \$1166.40$, an increase of \$86.40 over the previous year. The amount of money earned during the second year is more than the amount earned during the first year because the interest is being compounded. That is, the interest earned in the first year is earning interest during the second. Note that the amount of money in the account at the end of the second year could have been found by multiplying the original amount, \$1000, by (1.08) twice. In other words, the amount of money in the account at the end of the second year is $A = 1000(1.08)^2$. In general, the amount of money in the account after t years can be expressed as $A = 1000(1.08)^t$. In this case, 1.08 is the growth factor and 0.08 is the growth rate. If we let P be the amount of money put into an account, or the principal, and r be the annual interest rate, then the amount of money in an account after t years can be expressed as

$$A = P(1 + r)^t.$$

Now let's suppose that this account is compounded quarterly. In other words, interest based on one-fourth of the yearly interest rate is added to the account four times over the course of the year. Thus, if the annual interest rate is 8%, then the quarterly interest rate is $8\%/4 = 2\%$. Let's look at the amount in an account where $1000 is invested at 8% compounded quarterly. The amount at the end of the first quarter is $A = 1000(1 + 0.08/4) =$ $1020. To find the second quarter, we do exactly the same thing except now our value for P is $1020 instead of $1000. The amounts earned for all four quarters are shown in Table 1.

As you can see, by the time 1 year is up, the account has $2.43 more than it would have without compounding. As the number of compounding periods increases, the amount of money in the account after 1 year also increases. Fortunately, we do not have to make a table every time we want to figure compound interest. We can express the amount in the account at any time by using the formula

$$A = P\left(1 + \frac{r}{n}\right)^{nt},$$

where r is the annual interest rate, n is the number of compounding periods in a year, and t is the time in years. Thus, r/n is the interest rate for the specific compounding period and nt is the number of times the interest is compounded during the time the money is in the account.

Let's see what happens as we let the number of compounding periods per year increase. Again, let $P = 1000$, $r = 0.08$, and $t = 1$, which makes our equation $A = 1000(1 + 0.08/n)^n$. Some selected values for A are given in Table 2.

The amount keeps going up, but the rate at which it increases seems to be slowing down. Let's look at another table, Table 3, with some very large values for n.

As you can see from Tables 2 and 3, the amount in the account continues to grow as the number of compounding periods increases, but the rate at which it grows seems to be slowing as n gets larger. In fact, we can let n be as large as we like and we will never reach $1084. Let's see why. To do so, we first need to look at the function $g(k) = (1+1/k)^k$, where k is an integer. In Table 4, we have given values for $g(k)$ for larger and larger inputs. These

TABLE 1 Interest and the Total Amount in an Account Where $1000 Is Compounded Quarterly at an Annual Interest Rate of 8%

Quarter	Interest	Total
0	$0.00	$1000.00
1	$20.00	$1020.00
2	$20.40	$1040.40
3	$20.81	$1061.21
4	$21.22	$1082.43

TABLE 2 Total Amount in an Account after 1 Year Where $1000 Is Compounded *n* Times a Year at an Annual Interest Rate of 8%

n	$A = 1000\left(1 + \frac{0.08}{n}\right)^n$
10	$1082.94
20	$1083.11
30	$1083.17
40	$1083.20
50	$1083.22
60	$1083.23

TABLE 3 For Very Large Values of n, the Total Amount in an Account after 1 Year Where \$1000 Is Compounded n Times a Year at an Annual Interest Rate of 8% Levels Off

n	$A = 1000\left(1 + \frac{0.08}{n}\right)^n$
1000	\$1083.284
2000	\$1083.285
3000	\$1083.286
4000	\$1083.286
5000	\$1083.286
6000	\$1083.286

TABLE 4 As k Increases, $g(k) = (1 + 1/k)^k$ Gets Closer to Some Fixed Number

k	$g(k) = \left(1 + \frac{1}{k}\right)^k$
1	2
10	2.5937
100	2.7048
1,000	2.7169
10,000	2.7181
100,000	2.7183

values are rounded to four decimal places. Notice what happens to $g(k)$ as k gets large.

Notice that as k gets larger, $(1 + 1/k)^k$ increases. The value of $g(k)$ does not increase without bound, but instead gets closer to a fixed number called e, an irrational number *approximately* equal to 2.7182818. Mathematically speaking, $(1 + 1/k)^k \to e$ as $k \to \infty$.

Now let's return to our compound interest formula and see how $A = P(1 + r/n)^{nt}$ changes when n approaches infinity. To do so, we let $k = n/r$ and $n = kr$. Thus,

$$P\left(1 + \frac{r}{n}\right)^{nt} = P\left(1 + \frac{1}{k}\right)^{krt} = P\left[\left(1 + \frac{1}{k}\right)^{k}\right]^{rt}.$$

As n approaches infinity, k does so also because $k = n/r$ and r is fixed. So, knowing that $(1 + 1/k)^k \to e$ as $k \to \infty$, we can conclude that

$$P\left(1 + \frac{r}{n}\right)^{nt} = P\left[\left(1 + \frac{1}{k}\right)^{k}\right]^{rt} \to Pe^{rt}$$

as $n \to \infty$. So, when we assume that our money is compounded continuously, the formula becomes $A = Pe^{rt}$.

Now let's see what happens to our \$1000 when it is compounded continuously at 8% for 1 year. Using the formula $A = Pe^{rt}$, we find that

$$A = 1000e^{0.08 \cdot 1} \approx \$1083.29.$$

Example 4 Suppose you have some money that you want to put into a bank account. One account offers you 5% annual interest compounded monthly. Another account offers only 4.9% interest, but it compounds the interest continuously. Which account gives a greater return?

Solution We look at how much money we have in each account after 1 year. If we invested P dollars at 5% compounded monthly for 1 year, then we would have

$$\begin{aligned} A &= P\left(1+\frac{0.05}{12}\right)^{12\cdot 1} \\ &\approx P(1.0512). \end{aligned}$$

If we invested P dollars at 4.9% compounded continuously for 1 year, we would have

$$\begin{aligned} A &= Pe^{0.049\cdot 1} \\ &\approx P(1.0502). \end{aligned}$$

Because $1.0512P > 1.0502P$ for positive values of P, putting our money in the account that pays 5% compounded monthly gives a greater return.

Other things besides money in a bank account grow continuously. For example, populations of people or animals can have continuous exponential growth. The formula for such growth is the same as it is for money. If we use the same formula, $A = Pe^{rt}$, then A represents the population at time t, P represents the initial population, and r represents the growth rate. For example, from 1990 to 1995, the United Nations[4] estimated that the population of the United States is growing continuously at a rate of 0.71%. They also estimated that the population of the United States in 1990 was 249,224,000. Assuming these facts to be true, the function that gives the population of the United States for t years since 1990 is

$$A = 249{,}224{,}000e^{0.0071t}.$$

An estimate of the population in 1995 is then

$$A = 249{,}224{,}000e^{0.0071\cdot 5} \approx 258{,}230{,}000.$$

HISTORICAL NOTE

The number e is often said to have received its name in honor of Leonhard Euler, a Swiss mathematician in the eighteenth century. Euler (pronounced "oil-er") was the first to use e as the designation for this number. He may have chosen e to represent this number because it is the first letter in the word *exponential*. Because e is coincidently the first letter in Euler's last name, it is now used to also honor him. Both e and π are examples of *transcendental*

continued on page 571

[4] *United Nations Population Fund,* Population and the Environment: The Challenges Ahead *(London: Banson Productions, 1991), p. 43.*

numbers. When we hear of something transcending something else, we think of going above and beyond, and that is what the definition of transcendental numbers is about. Transcendental numbers are those that cannot be the roots of a polynomial equation with integral coefficients. They "go beyond" algebraic numbers to form a different set of numbers. It is easy to see that any rational number can be the root of a polynomial. For example, if we want to see how $\frac{4}{19}$ could be the root of a polynomial, we simply write $y = 19x - 4$. Most irrational numbers can also be written as the roots of polynomials. For example, $\sqrt{23}$ is a root of $y = x^2 - 23$. In much the same way, nearly every number that you can think of can be the solution to a polynomial equation with integer coefficients. The numbers e and π, however, are different.

Stating that e and π are transcendental and proving it are two different matters. Just proving that *any* transcendental number existed was a major step by Joseph Liouville in 1844. He then tried to prove that e was transcendental, but he was only able to prove that neither e nor e^2 were roots of any quadratic. Not until almost thirty years later, in 1873, did Charles Hermite prove that e was, indeed, transcendental. Proving it for π turned out to be even more challenging. Not until C. L. F. Lindemann published a proof for π in 1882 was the matter finally settled.

Lindemann's proof also settled one of the classic problems of Greek geometry, squaring the circle. If given a circle, can you construct (using only compass and straight edge) a square with the same area? Lindemann proved that π cannot be the root of a polynomial and that the length of a square clearly must be the root of a polynomial. Mathematicians had been trying to square the circle since about 200 B.C. Finally putting this quest to rest by proving it could not be done was accomplished in the late 1800s.

Sources: S. Schwartzman, *The Words of Mathematics* (Washington, DC: Mathematical Association of America, 1994), pp. 79–80; Carl B. Boyer, *A History of Mathematics* (Princeton, NJ: Princeton University Press, 1968), pp. 602–3; E. Sander, *Squaring the Circle Part One* (visited 11 August 1998) ⟨http://www.forum.swarthmore.edu/news.archives/geometry.college/article106.html⟩.

READING QUESTIONS

5. In the formula $A = P(1 + r)^t$, is the $(1 + r)$ a growth factor or a growth rate? Is the r a growth factor or a growth rate?
6. Suppose \$1000 is put into a bank account that earns 6% annual interest.
   (a) How much is in the account after 2 years if the interest is compounded monthly?
   (b) How much is in the account after 2 years if the interest is compounded continuously?

THE NATURAL LOGARITHM

In our last example, we gave a function that describes the population of the United States since 1990:

$$A = 249{,}224{,}000e^{0.0071t}.$$

We might be interested in finding the time it takes the population to double. To do so, we do not even need to know the size of the population; all we need to know is the growth rate. If our population is P, then our function is then $A = Pe^{0.0071t}$, and we would want to find the time it takes until our population reaches $2P$. In other words, we want to solve for t in the equation

$$2P = Pe^{0.0071t}.$$

To do so, we first need to divide both sides of the equation by P, which gives us

$$2 = e^{0.0071t}.$$

Now we need to take the log of both sides. We could use the common logarithm as we did in the past, but, there is an easier method.

The common logarithm has a base of 10, which can be seen because of its relationship with the exponential function that had a base 10, or $\log x = y$, equivalent to $10^y = x$. Logarithms can have other bases, however. Another quite useful logarithm is the natural logarithm. The **natural logarithm** has a base of e and is abbreviated ln. Thus, $\ln x = y$ is equivalent to $e^y = x$. The properties of the common logarithm seen in section 2.3 are still true for the natural logarithm.

PROPERTIES

$\ln(ab) = \ln a + \ln b$

$\ln a^n = n \ln a$

Now let's return to the problem of solving $2 = e^{0.0071t}$ for t. If we use the natural logarithm, then we can transform this equation to

$$\ln 2 = \ln(e^{0.0071t}),$$

which is simply

$$\ln 2 = 0.0071t.$$

Dividing both sides by 0.0071 gives

$$\frac{\ln 2}{0.0071} = t.$$

Using a calculator to get the decimal approximation gives $t \approx 98$ years.

Example 5 The United Nations estimated that the population of China was growing continuously at a rate of 1.42% from 1990 to 1995. Assuming the population kept growing at this rate, how long would it take the population to triple?

Solution With a continuous growth rate of 1.42%, the population of China can be described with the formula

$$A = Pe^{0.0142t},$$

where P is the initial population and A is the population after time t. To find the time it takes to triple, we need to let $A = 3P$. Doing so gives

$$3P = Pe^{0.0142t}.$$

To solve for t, we first divide both sides by P,

$$3 = e^{0.0142t},$$

and then take the natural logarithm of both sides,

$$\ln 3 = 0.0142t.$$

Now dividing both sides by 0.0142 gives

$$\frac{\ln 3}{0.0142} = t,$$

which gives a decimal approximation of $t \approx 77$ years.

READING QUESTIONS

7. Solve $5 = e^t$ for t. Give both the exact solution and a decimal approximation rounded to three decimal places.
8. Can the common logarithm be used to find a decimal approximation for x in $5e^x = 8$? If it can, then find an approximation for x to three decimal places. If it cannot, then explain why.

SUMMARY

In this section, we reviewed exponential functions. Remember that **exponential functions** have a constant growth factor and can be written as $f(x) = ba^x$. The **growth factor,** which is a, is also $f(x+1)/f(x)$. A **growth rate** is always 1 less than the growth factor.

Taking the logarithm of both sides of an exponential equation can help you find a solution to the equation. By using the laws of logarithms, you can rewrite the side with the variable in such a way that the variable becomes a factor rather than an exponent. Then you can simply solve for the variable.

By looking at continuous growth rates, we determined that the value of $g(k) = (1 + 1/k)^k$ approaches the number known as e (approximately 2.7183) as k approaches infinity. The number e comes up in situations involving continuous growth. We also determined a formula for continuous population growth, $A = Pe^{rt}$, where A is the population at time t, P is the initial population, and r is the growth rate.

The **natural logarithm** has a base of e and is abbreviated ln.

EXERCISES

1. Solve the following equations. Give exact answers.
   (a) $10^x = 50$

(b) $4 \cdot 10^x = 100$
(c) $e^x = 5$
(d) $8 \cdot e^x = 40$

2. Solve the following equations. Round answers to three decimal places.
   (a) $10^x = 42$
   (b) $1.05^x = 2$
   (c) $14^{3x} + 16 = 938$
   (d) $e^{2x} = 14$
   (e) $4^x + 13 = 0$
3. Solve the following equations. Round answers to three decimal places.
   (a) $10^x = 5^{x+2}$
   (b) $8^{(x+7)} = 29$
   (c) $e^{3x} = 10^x$
   (d) $5^{2x+1} = 3^{x-4}$
4. Find the equation of an exponential function that passes through each of the following pairs of points.
   (a) $(0, 3)$ and $(1, 15)$
   (b) $(0, 2)$ and $(4, 162)$
   (c) $(2, 20)$ and $(3, 5)$
   (d) $(-1, 2)$ and $(3, 512)$
5. In 1980, there were approximately 1,200,000 African elephants. In 1990, there were approximately 700,000 African elephants.[5] Assume the population growth factor is constant.
   (a) Find the function that models these data where the input is the time since 1980 and the output is the number of African elephants.
   (b) Use your function to estimate the size of the population of African elephants in 2010.
   (c) Use your function to estimate the year the African elephant population will be half the size it was in 1980.
   (d) When will the African elephant population be half the size it was in 1990?
6. Suppose a ball is dropped from a height of 10 ft. Each time it bounces, it rises to 80% of the height from which it last fell.
   (a) Find the maximum height of the ball after its third bounce.
   (b) Find the maximum height of the ball after its nth bounce.
   (c) After how many bounces will the ball no longer rise above 6 in.?
7. An investment grows by 70% in 7 years.
   (a) What is the investment's annual growth rate?

[5]*John A. Burton, ed.,* The Atlas of Endangered Species *(New York: Macmillan Publishing, 1991), p. 138.*

(b) What is the investment's continuous annual growth rate?

8. Moore's law, created by Intel cofounder Gordon Moore, states that microprocessors in computers will get twice as powerful for the same price every 18 months.[6]
   (a) If Moore's law is true, then what is the annual growth rate in the power of microprocessors?
   (b) If Moore's law is true, then what is the continuous annual growth rate in the power of microprocessors?
9. Living organisms contain a constant ratio of carbon 14 to carbon 12. When the organism dies, the carbon 14 decays into nitrogen 14. The change in the carbon 14 to carbon 12 ratio is the basis for carbon 14 dating. Half of the carbon 14 will decay into nitrogen 14 in 5730 years. This 5730 years is known as the half-life of carbon 14.
   (a) If the carbon 14 decays continuously, what is the continuous annual decay rate?
   (b) It is determined that an object contains 60% of the carbon 14 from when it was alive. How long ago did the object die?
   (c) Find the function that converts the percent of carbon 14 remaining in an object to the age of the object.
10. In 1989, the suggested retail price for a new Ford Escort was $7680. The suggested retail price for that same car 10 years later was $1600.[7] Assume the value of the car depreciated exponentially in those 10 years.
   (a) What proportion of the value of 1989 Ford Escorts was lost each year?
   (b) Assume the rate at which the 1989 Ford Escort depreciated in value is the same as that for a 1999 Ford Escort. The suggested retail price for a 1999 Ford Escort was $11,455. Determine a function where the input is the number of years since 1999 and the output is the suggested retail value of a used 1999 Ford Escort.
11. In 1990, the United Nations estimated the population of Guyana to be 796,000 and growing at a continuous annual growth rate of 0.81%. It also estimated the population of Suriname to be 422,000 and growing at a continuous annual growth rate of 1.76%. If the populations of these two countries continue to grow at their respective rates, then in what year will the two populations be the same?
12. In 1990, the United Nations estimated the population of India to be 853,094,000 and growing at a continuous annual growth rate of 2.08%. Assume the population of India continues to grow at this rate.
   (a) In what year will the population reach 1 billion?
   (b) How long will it take the population of India to double?
13. If you invest $10,000 into a money market account that pays 6.5% compounded quarterly, how much will the account be worth after 5 years?

[6] *"Computer Chip May Double Memory,"* Holland Sentinel, *18 September 1997, p. A6.*

[7] Auto World *(visited 13 November 1998) (http://www.autoworld.com).*

14. Based on a Bureau of Justice Statistics report, a newspaper headline read, "U.S. Prison Population Doubled in 12 Years."[8]
    (a) Assuming exponential growth, what was the annual growth rate for the prison population during those 12 years?
    (b) If the prison population continues to grow exponentially at the same rate, then how long will it take the population to triple? To quadruple?
15. Suppose you have two bank accounts. One has a 6% annual interest rate compounded monthly, and the other has 5.9% annual interest rate compounded continuously. Which bank account gives you the most return for your money?
16. A cup of hot cappuccino is placed on a kitchen table. According to Newton's law of cooling, the difference between the temperature of the cappuccino and the room can be modeled with an exponential function. Suppose the initial temperature of the cappuccino is 90°C, the room temperature is 20°C, and the difference in temperature between the cappuccino and the room is decreasing by 1% per minute.
    (a) Find a function where the input is the time, in minutes, since the cappuccino was put on the kitchen table and the output is the difference in temperature between the cappuccino and the room. (*Hint:* The initial difference in temperature, 70°C.)
    (b) How long will it take until the difference in temperature between the cappuccino and the room is 30°C?
    (c) Find a function where the input is the time, in minutes, since the cappuccino was put on the kitchen table and the output is the temperature of the cappuccino.
    (d) How long will it take until the temperature of the cappuccino is 60°C?

INVESTIGATIONS

INVESTIGATION 1: RULE OF 72

With exponential equations, we can determine the amount of time it takes for an investment to reach a certain point. One interesting way of expressing growth is to consider the amount of time it takes for an investment to double in size. In this investigation, we look at a simple rule sometimes used to determine the time it takes something that is growing exponentially to double in size.

1. Let's consider some problems where interest on a bank account is compounded annually.
    (a) Suppose you have an account that earns 5% compounded annually. How long will it take the money in that account to double?
    (b) Suppose you have an account that earns R% compounded annually. Develop a formula where the input is R and the output is

[8] *"U.S. Prison Population Doubled in 12 Years,"* Holland Sentinel, *15 March 1999, p. A1.*

the time it takes the money in the account to double. (*Note:* The variable R is given as a percent. Thus, if the interest rate is 5%, then R is 5, not 0.05.)

(c) In the following table, the interest rate written as a percent, R, is given. Complete the table by including the time needed for the account to double, and $R\times$ (the doubling time).

R	2	4	6	8	10	12
Doubling Time						
$R\times$ *(Doubling Time)*						

(d) The rule of 72 is a well-known shortcut for determining the amount of time it takes for an investment to double. According to the rule, the number 72 divided by the interest rate (in percent) gives the time it takes an investment to double. For example, an investment at 5% doubles in about $72/5 = 14.4$ years. Does this shortcut seem to be accurate based on the last row of your table in part (c)? Explain.

2. Let's now look at the same series of questions as in question 1 except with the interest on a bank account compounded monthly.

(a) Suppose you have an account that earns 5% compounded monthly. How long will it take the money in that account to double?

(b) Suppose you have an account that earns R% compounded monthly. Develop a formula where the input is R and the output is the time it takes the money in the account to double. (*Note:* The variable R is given as a percent. Thus, if the interest rate is 5%, then R is 5, not 0.05.)

(c) In the following table, the interest rate written as a percent, R, is given. Complete the table by including the time needed for the account to double and $R\times$ (the doubling time).

R	2	4	6	8	10	12
Doubling Time						
$R\times$ *(Doubling Time)*						

(d) How does the rule of 72 seem to work given that the interest is now compounded monthly? Explain.

3. Let's now look at the same series of questions as in question 1 except with the interest on a bank account compounded continuously.

(a) Suppose you have an account that earns 5% compounded continuously. How long will it take the money in that account to double?

(b) Suppose you have an account that earns R% compounded continuously. Develop a formula where the input is R and the output is the time it takes the money in the account to double. (*Note:* The variable R is given as a percent. Thus, if the interest rate is 5%, then R is 5, not 0.05.)

(c) In the following table, the interest rate written as a percent, R, is given. Complete the table by including the time needed for the account to double and $R\times$ (the doubling time).

R	2	4	6	8	10	12
Doubling Time						
$R\times$ *(Doubling Time)*						

(d) How does the rule of 72 seem to work given that the interest is now compounded continuously? Does there seem to be a better number to use than 72? Explain.

4. Develop a similar rule as the rule of 72 that could be used as a shortcut to determine the time needed for a bank account to triple in value.

INVESTIGATION 2: FRACTALS

A fractal is a geometric shape that commonly exhibits the property of self-similarity.[9] A self-similar object is one whose component parts resemble the whole. Each part of the object, when magnified, looks much like the object as a whole. Thus, a self-similar object looks the same even when viewed with a different scale. When we put a fractal under a microscope, its appearance does not change. Although not all fractals are exactly self-similar, most have this property.

A Koch snowflake is a common fractal. The beginning of the construction of one side of the Koch snowflake is shown in Figure 2. At each step, the middle third of every segment is replaced by a triangular "roof." The length of each segment of each roof is a third of the length of the segment from the previous step. If the roof had a "floor," then it would be an equilateral triangle.

[9] *The term* fractal, *coined by Benoit B. Mandelbrot, is derived from the Latin word* fractus *(meaning "broken"). The concept of a fractal has given rise to a new system of geometry that has had a significant impact on mathematics recently.*

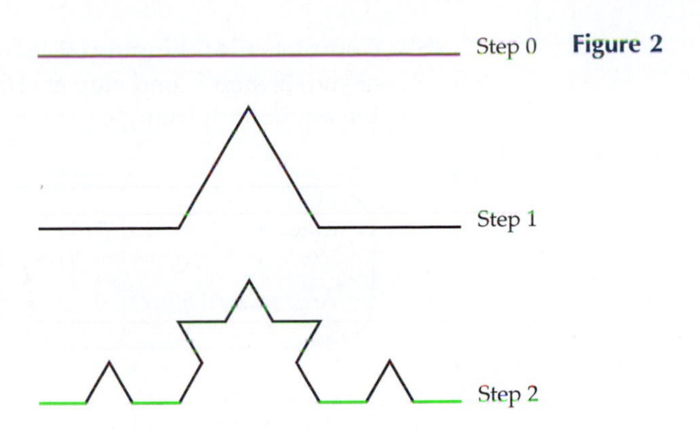

Figure 2

In this investigation, we explore the number of segments, the length of the sum of the segments, and the area under the figure at each step of growth, which all use exponential equations.

1. Complete the following table by finding the number of segments at step 3, step 4, and step n.

Step	0	1	2	3	4	n
Number of Segments	1	4	16			

2. Assuming the segment at step 0 has length 1, complete the following table by finding the sum of the lengths of the segments at step 2, step 3, step 4, and step n. As $n \to \infty$, what does the sum of the lengths of the segments approach?

Step	0	1	2	3	4	n
Sum of the Segments	1	$\frac{4}{3}$				

3. As the number of steps increases, the sum of the lengths of the segments increases and approaches infinity. Although the area under the figure increases, it does not approach infinity. At step 0, the area under the line segment is considered to be 0. At step 1, the area under the roof can be considered the same as the area of an equilateral triangle whose sides are length $\frac{1}{3}$.
   (a) Show that the area of an equilateral triangle can be written as $A = (\sqrt{3}/4)s^2$, where the length of the side of the triangle is s.

(b) Complete the following table by finding the area under the figure at step 2, step 3, and step 4. (*Hint:* Find the number and areas of the equilateral triangles of each size and add them together.)

Step	0	1	2	3
Area under Figure	0	$\frac{\sqrt{3}}{36}$		

(c) To find the area under the figure at step n, we first need to find the sum of n-terms beginning with the area under the figure at step 1.

i. Show that this sum can be written in the form

$$S = a + ar + ar^2 + \cdots + ar^{n-1},$$

where a is the area under the figure at step 1 and r is a rational number. This type of sum is called a geometric series. What are your values for a and r?

ii. Show that in general (not with your specific values for a and r) $S - rS = a - ar^n$. Then, solve for S to obtain

$$S = \frac{a(1 - r^n)}{1 - r}.$$

iii. Use your values for a and r to find the area under the figure at step n.

iv. As $n \to \infty$, what does the area under the figure approach?

7.3 SOLVING TRIGONOMETRIC EQUATIONS

In the last two sections, we solved equations involving polynomial, rational, power, and exponential functions. In this section, we look at how to solve equations involving trigonometric functions. Solving trigonometric equations can result in an infinite number of solutions. We show how these solutions can be obtained and some situations where these types of equations model real-world applications. We begin the section, however, by looking at how extraneous solutions can occur.

DOMAIN AND RANGE

When solving equations, we must remember to check the domain and range for the functions initially involved as well as the functions involved when the equation is transformed. For example, if we try to solve $\sqrt{4x+2} = -4$ symbolically, we might do the following:

$$\sqrt{4x+2} = -4$$
$$(\sqrt{4x+2})^2 = (-4)^2$$

$$\begin{aligned} 4x + 2 &= 16 \\ 4x &= 14 \\ x &= 3.5. \end{aligned}$$

When we go back to the original equation and check the answer, however, we see that $\sqrt{4(3.5)+2} = \sqrt{16} = 4$. Why did our "solution" turn out to be false? Did we make a mistake in our algebra? The answer is no, although there are things we could have done to prevent the problem.

The problem is one of range. The range of $y = \sqrt{4x+2}$ is $y \geq 0$, so the output could never be -4. Therefore, there are no values for x that would make $\sqrt{4x+2} = -4$. The reason we found a "solution" that was not really a solution is that we squared both sides of the equation. When we squared both sides, this square root function changed to the linear function $y = 4x + 2$. The range of this function is all real numbers, so we could find a solution to $4x + 2 = 16$. That change in the nature of the equation can result in what are sometimes called extraneous solutions, as considered in section 7.1.

By taking a simple equation like $x - 4 = 0$, it is easy to see how these extra solutions can occur. Even though the solution to this equation is very simple, we multiply both sides of the equation by x to make our point. This step gives us $x^2 - 4x = 0$. Our new equation now has two solutions, 4 and 0, where our original equation only had 4 as a solution. By multiplying both sides by the variable, you can see that we introduced an additional solution into the equation.

This process is tricky because multiplying both sides of an equation by a variable does not always produce extraneous solutions. For example, when we solve

$$\frac{3}{x+1} = 7,$$

we need to multiply both sides of the equation by $x + 1$, which gives

$$\begin{aligned} 3 &= 7x + 7 \\ -4 &= 7x \\ -\tfrac{4}{7} &= x. \end{aligned}$$

Checking our solution shows that $-\frac{4}{7}$ is, indeed, a solution to the equation. Even though we multiplied both sides of the equation by a function containing a variable, we did not get an extraneous solution. When you multiply both sides of an equation by a variable or when you raise both sides of an equation to a power, extraneous solutions are a possibility. Because of this, and because it is the prudent thing to do, you should check all your answers when solving equations.

READING QUESTIONS

1. Why are there no solutions to $10^x = -8$?
2. In your own words, what are extraneous solutions?
3. Why is the equation $x^2 = 2x$ not equivalent to $x = 2$?

SOLVING TRIGONOMETRIC EQUATIONS

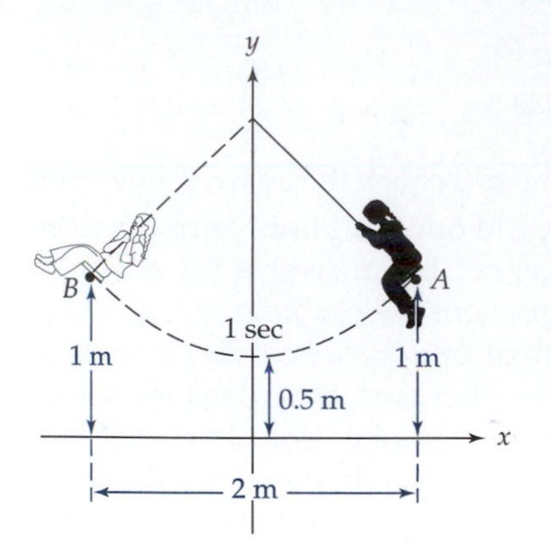

Figure 1 As a child swings from point A to point B, the swing goes from a high of 1 m to a low of 0.5 m and back to a high of 1 m.

Thinking about the domain and range are important in solving trigonometric functions too. In section 2.4, we looked at a situation involving a child's swing. (See Figure 1.) In this example, the swing started at its highest point, 1 m above the ground, at point A. As it swung forward, it went past its lowest point, 0.5 m above the ground. At point B, it was again at its highest point above the ground, 1 m. It took 1 sec to go from point A to point B. The function that describes the height of a swing in meters is $h(t) = 0.25\cos(2\pi t) + 0.75$, where t is time in seconds. Suppose we want to find the times when the height of the swing is 0.8 m. To do so, we need to solve the equation $0.8 = 0.25\cos(2\pi t) + 0.75$. Solving this equation graphically is fairly easy. Simply graph $y = 0.25\cos(2\pi x) + 0.75$ and $y = 0.8$ on the same graph and find all the points of intersection, as shown in Figure 2. Notice that there are many solutions.

Solving the equation $0.25\cos(2\pi t) + 0.75 = 0.8$ symbolically requires a bit of thought. Our first steps involve solving for $\cos(2\pi t)$.

$$\begin{aligned} 0.25\cos(2\pi t) + 0.75 &= 0.8 \\ 0.25\cos(2\pi t) &= 0.05 \\ \cos(2\pi t) &= 0.2. \end{aligned}$$

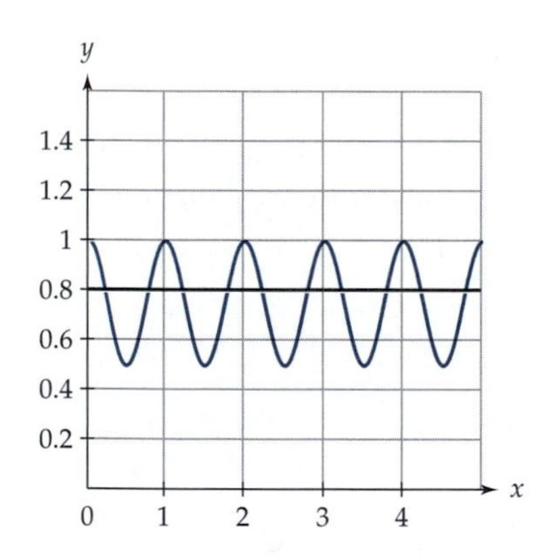

Figure 2 The x-coordinates of the points of intersection for $y = 0.25\cos(2\pi x) + 0.75$ and $y = 0.8$ are the solutions to the equation $0.25\cos(2\pi x) + 0.75 = 0.8$.

Now, to find when the cosine function has a value of 0.2, we take the inverse cosine of both sides of the equation and get

$$2\pi t = \cos^{-1}(0.2).$$

A calculator gives an approximate value of 1.369 for the inverse cosine of 0.2. We now write

$$2\pi t \approx 1.369.$$

Thus, $t \approx 1.369/2\pi \approx 0.218$. At about 0.218 sec, the swing is at a height of 0.8 m. Comparing our solution with the graph, however, reminds us that we only have one of many solutions.

Remember that the range of the inverse cosine function was specifically limited so that it would actually be a function with only one output for each input. The inverse sine and tangent functions also had limited ranges. These functions are defined as follows.

$y = \sin^{-1} x$ if and only if $x = \sin y$ and $-\pi/2 \leq y \leq \pi/2$.

$y = \cos^{-1} x$ if and only if $x = \cos y$ and $0 \leq y \leq \pi$.

$y = \tan^{-1} x$ if and only if $x = \tan y$ and $-\pi/2 < y < \pi/2$.

Thus, when we find the inverse cosine of any value, we get only one solution, the one between zero and π. We need to find a way to get the other solutions. By looking at the graph and thinking about the physical situation of the swing, we can easily see that because the period of the function is 1 sec, the swing is at a height of 0.8 m 0.218 sec after each high point. We also know that the high points occur, in this example, every second. Thus, we also have a height of 0.8 m at 1.218 sec, at 2.218 sec, and so forth. In fact, if this swing were to go on at this rate endlessly, then we could say that the solutions to this problem are at $n + 0.218$ sec for all whole number values of n.

Even though we now have an infinite number of solutions for this problem, we do not have them all. Look again at the graph in Figure 2. Not only is the swing at a height of 0.8 m 0.218 sec after each high point, but it is also at the height 0.218 sec *before* each high point. Thus, another series of solutions occurs at $1 - 0.281 = 0.782$, $2 - 0.281 = 1.782$, and so on. We can write these solutions as $n - 0.218$ sec for all whole number values of n greater than or equal to 1. By comparing these solutions with the graph and the physical situation, it is now clear that we have all the solutions.

In this case, we easily see that there is a second set of solutions and how to arrive at them. In other problems, these solutions are not as obvious. Finding all the solutions to a trigonometric equation involves a couple of steps, just as finding the solutions to our swing example did. First, due to the periodic nature of trigonometric functions, if a is a solution, then $a +$ (the period of the function) is also a solution. In the swing example, because 0.218 was a solution and the period was 1 sec, then $0.218 + 1$ was also a solution. To generate additional solutions, we can add the period to each new solution as well. In fact, adding every multiple of the period to a solution gives you another solution. In other words, if a is a solution, then $a +$ (k times the period of the function) is also a solution for all integer values of k.

Just as in our swing example, though, there is still a set of solutions not yet found. This process is sometimes challenging but always necessary. Look again at the graph in Figure 2. Notice that the two solutions in each period are equidistant from the high points, which is always the case. In other words, because the swing was at a height of 0.8 m 0.218 sec after each high point, it will also be at the same height 0.218 sec before each high point. Graphing these equations can be extremely helpful in understanding where additional solutions are found. Once we know one solution from the second set, we simply add or subtract the period an integral number of times to find other solutions.

In the swing example, we used the periodic nature of the function to find one set of solutions. We then used the symmetry of the graph to get the first value of the other set. We then again used the periodic nature of the function to get all the other solutions. We performed the steps in this order because we had a nice cosine function with no horizontal shift. If a trigonometric function is shifted horizontally, then it may be easier to look at the periodic nature of the function *before* you solve the equation completely for the first solution. We illustrate in the following example.

Example 1 Find all the solutions for $6\sin(2x - 5) = 3$.

Solution We first solve for $(2x - 5)$.

$$6\sin(2x - 5) = 3$$
$$\sin(2x - 5) = \tfrac{1}{2}$$
$$2x - 5 = \sin^{-1}(\tfrac{1}{2}),$$

which means that $2x - 5 = \pi/6$. Because the period of the sine function is

2π, one set of solutions is $\pi/6 + 2\pi k$ for all integer values of k. Thus, we can now go back to our solution and write

$$\begin{aligned} 2x - 5 &= \frac{\pi}{6} + 2\pi k \\ 2x &= 5 + \frac{\pi}{6} + 2\pi k \\ x &= \frac{5}{2} + \frac{\pi}{12} + \pi k. \end{aligned}$$

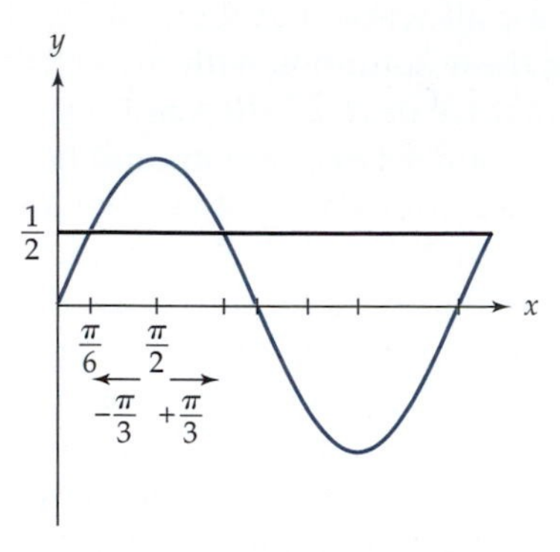

Figure 3 The x-coordinate of the points of intersection for $y = \sin x$ and $y = \frac{1}{2}$ shows the many solutions to the equation $\sin x = \frac{1}{2}$.

We now have one complete set of solutions. To find the other set of solutions, we need to go back to $\sin^{-1}(\frac{1}{2})$. Let's look at a graph of $y = \sin x$ and $y = \frac{1}{2}$, as shown in Figure 3. You can see that $y = \sin x$ has a maximum output at $\pi/2$. The first solution we found was $\pi/6$, which is $\pi/2 - \pi/6 = \pi/3$ units to the *left* of the maximum output. There is another solution at $\pi/3$ units to the *right* of the maximum output, which is at $\pi/2 + \pi/3 = 5\pi/6$. Because this function is a periodic function with a period of 2π, there are solutions at $5\pi/6 + 2\pi k$. We can now go back and solve for x.

$$\begin{aligned} 2x - 5 &= \frac{5\pi}{6} + 2\pi k \\ 2x &= \frac{5\pi}{6} + 5 + 2\pi k \\ x &= \frac{5\pi}{12} + \frac{5}{2} + \pi k. \end{aligned}$$

Our solutions, therefore, are $\pi/12 + 5/2 + \pi k$ and $5\pi/12 + 5/2 + \pi k$.

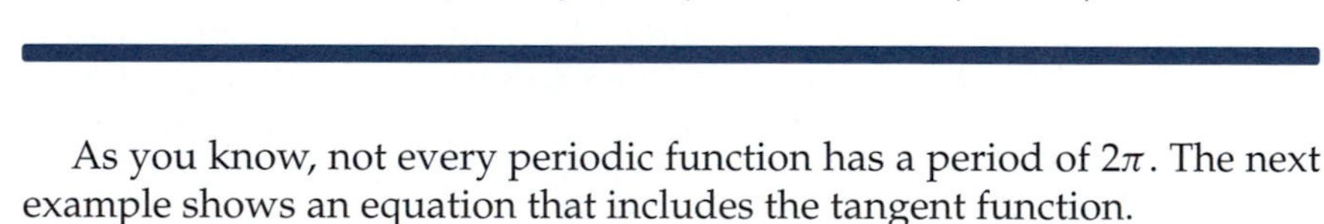

As you know, not every periodic function has a period of 2π. The next example shows an equation that includes the tangent function.

Example 2 In section 5.3, we solved $1 = 2\tan[\pi(x-2)] + 3$, but only found one solution. Let's go back to that problem and find all the solutions.

Solution

$$\begin{aligned} 2\tan[\pi(x-2)] + 3 &= 1 \\ 2\tan[\pi(x-2)] &= -2 \\ \tan[\pi(x-2)] &= -1 \\ \pi(x-2) &= \tan^{-1}(-1) \\ \pi(x-2) &= -\frac{\pi}{4}. \end{aligned}$$

Solving for x in $\pi(x-2) = -\pi/4$ gives one solution. Because the period of the tangent function is π, we use that information to finish our solution. If k is some integer, then

$$\pi(x-2) = -\frac{\pi}{4} + \pi k$$
$$x - 2 = -\frac{1}{4} + k$$
$$x = \frac{7}{4} + k.$$

Because the tangent function is unlike the sine and cosine functions in that its outputs equal a given value only once per period, we have all the solutions.

Example 3 The number of hours of daylight in Grand Rapids, Michigan, can be modeled with the function, $d(t) = 190\sin[2\pi/365(t - 81)] + 731$, where t is the number of the day of the year (i.e., January 1 = 1, January 2 = 2, ..., December 31 = 365). According to this model, which days of the years will have fewer than 10 hr (600 min) of daylight?

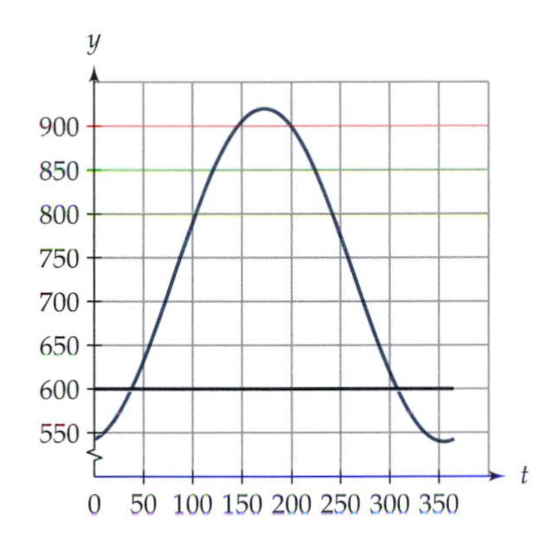

Figure 4

Solution We need to solve the inequality $190\sin[2\pi/365(t-81)] + 731 < 600$. To help see our solutions, we first graph $d(t) = 190\sin[2\pi/365(t-81)] + 731$ and $y = 600$ on the same set of axes, as shown in Figure 4. The domain of d is $1 \le t \le 365$. Therefore, we do not have an infinite number of solutions as we had in previous examples. From Figure 4, we can roughly see that the graph of the length of the daylight is below 600 when $1 < t < 40$ or $300 < t < 365$. To solve this inequality symbolically, we find the two times when the length of daylight is 600 h.

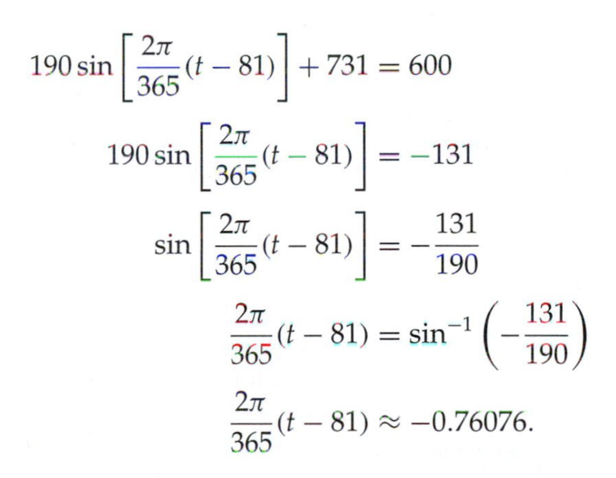

$$190\sin\left[\frac{2\pi}{365}(t-81)\right] + 731 = 600$$
$$190\sin\left[\frac{2\pi}{365}(t-81)\right] = -131$$
$$\sin\left[\frac{2\pi}{365}(t-81)\right] = -\frac{131}{190}$$
$$\frac{2\pi}{365}(t-81) = \sin^{-1}\left(-\frac{131}{190}\right)$$
$$\frac{2\pi}{365}(t-81) \approx -0.76076.$$

Solving for t gives $t \approx 37$. So, according to this model, there will be fewer than 600 min of daylight for days 1 through 37. The day with the most daylight will occur when $2\pi/365(t-81) = \pi/2$. Solving for t gives $t \approx 172$. Because a day with approximately 600 min of daylight occurred $172 - 37 = 135$ days before the day with the most daylight, another day with approximately 600 min will occur 135 days after the day with the most daylight. This day is $172 + 135 = 307$. Therefore, days that have fewer than 600 min of daylight are days 1 through 37 and days 307 through 365.

READING QUESTIONS

4. Find all the solutions for $x = \sin^{-1}(0.5)$.
5. Why did we add $2\pi k$ to the solutions in Example 1 but only πk to the solutions in Example 2?

USING TRIGONOMETRIC IDENTITIES

Occasionally, you may need to use trigonometric identities to solve equations. For example, trying to use simple techniques to solve $8 \sin x \cos x + 6 = 8$ leaves us with a problem: We cannot isolate the variable. We could still graph $y = 8 \sin x \cos x + 6$ and we might be able to estimate the period so that we could generate a set of solutions, but it would be better if we could get the solution symbolically. Fortunately, we can. One option is to rewrite the equation so that it contains a single trigonometric function. Although this method will not always work, it sometimes does. The identity $2 \sin x \cos x = \sin 2x$ can be used to help us rewrite this equation as follows.

$$\begin{aligned} 8 \sin x \cos x + 6 &= 8 \\ 4(2 \sin x \cos x) + 6 &= 8 \\ 4(\sin(2x)) + 6 &= 8. \end{aligned}$$

We can now complete this solution just as we did in the previous examples.

$$\begin{aligned} 4 \sin(2x) + 6 &= 8 \\ 4 \sin(2x) &= 2 \\ \sin(2x) &= \frac{1}{2} \\ 2x &= \sin^{-1}\left(\frac{1}{2}\right) \\ 2x &= \frac{\pi}{6}. \end{aligned}$$

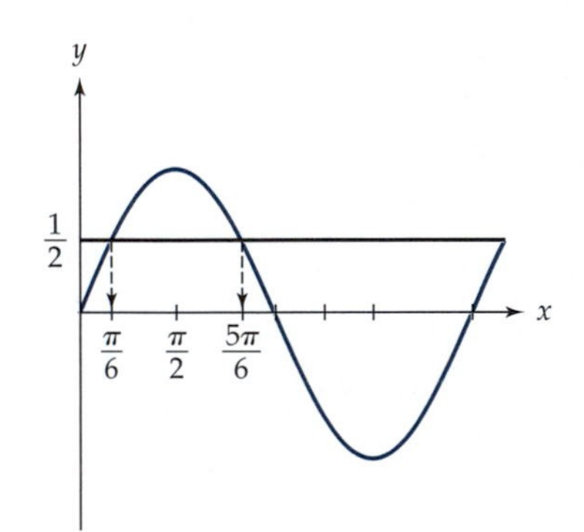

Figure 5 The intersection points of the graph of $y = \sin x$ and $y = \frac{1}{2}$ shows that there are many solutions to the equation $\sin x = \frac{1}{2}$. In particular, it shows that $\pi/6$ and $5\pi/6$ are solutions.

This result gives one solution for $2x$. Figure 5, showing $y = \sin x$ and $y = \frac{1}{2}$, reminds us of another solution, $\sin 5\pi/6 = \frac{1}{2}$. Knowing that the period of the sine function is 2π, we add integer multiples of 2π to our solutions. We now have

$$2x = \frac{\pi}{6} + 2\pi k \quad \text{or} \quad 2x = \frac{5\pi}{6} + 2\pi k$$

for all integral values of k. Solving each of these equations for x gives

$$x = \frac{\pi}{12} + \pi k \quad \text{or} \quad x = \frac{5\pi}{12} + \pi k,$$

where k is an integer.

Example 4 Find all the solutions to $\tan x + 2 \sin x = 0$, where $0 \le x \le 2\pi$.

Solution A graph of $y = \tan x + 2 \sin x$, shown in Figure 6, shows that there are five zeroes in the interval $0 \le x \le 2\pi$ for this function. It is

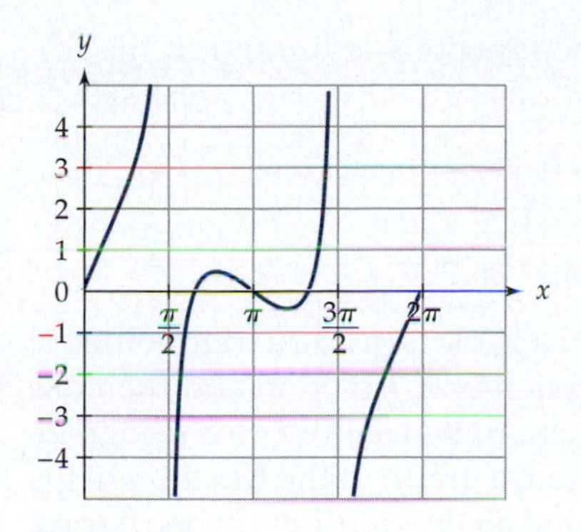

Figure 6

not immediately obvious how to solve this equation symbolically. A good first step in solving equations like $\tan x + 2\sin x = 0$ is to put all the trigonometric functions in terms of sine and cosine. By doing so, the equation becomes

$$\frac{\sin x}{\cos x} + 2\sin x = 0.$$

This equation now has $\sin x$ in both terms. We can factor $\sin x$ out of each term to obtain

$$\sin x\left(\frac{1}{\cos x} + 2\right) = 0.$$

We can now set each of these factors equal to zero and solve them separately. First,

$$\sin x = 0$$

when $x = 0, \pi$, or 2π. Setting the other factor equal to zero and solving for $\cos x$ gives

$$\begin{aligned} \frac{1}{\cos x} + 2 &= 0 \\ \frac{1}{\cos x} &= -2 \\ \cos x &= -\frac{1}{2}. \end{aligned}$$

Because $\cos x = -\frac{1}{2}$ when $x = 2\pi/3$ or $x = 4\pi/3$, the five solutions to our original equation are

$$x = 0, \frac{2\pi}{3}, \pi, \frac{4\pi}{3} \text{ or } 2\pi.$$

READING QUESTIONS

6. Which trigonometric identity would you use to solve $\sin x + \sin^2 x + \cos^2 x = 0.5$?
7. Find all the solutions to $4\sin x\cos x = 2$, where $0 \le x \le \pi$.

SUMMARY

This section began with a reminder to check the range and domain of functions when solving certain equations. Extraneous solutions sometimes result when transforming the original equation.

The range of the inverse trigonometric functions is limited in the following ways:

- $y = \sin^{-1} x$ if and only if $x = \sin y$ and $-\pi/2 \le y \le \pi/2$.
- $y = \cos^{-1} x$ if and only if $x = \cos y$ and $0 \le y \le \pi$.
- $y = \tan^{-1} x$ if and only if $x = \tan y$ and $-\pi/2 < y < \pi/2$.

When solving an equation in which the variable is inside a trigonometric function, you first need to isolate the trigonometric function and then use the appropriate inverse trigonometric function. You should also recognize that more than one solution may exist. Often, a graph of the function helps to find other solutions. When you have found all the solutions in one period of the function, other solutions may exist at your solution plus integer multiples of the period.

Occasionally, you may need to use trigonometric identities to solve equations. Rewriting functions so that they use a single trigonometric function often simplifies an equation enough to solve it.

EXERCISES

1. Find all the solutions to each of the following.
   (a) $4x - 1 = \sqrt{x}$
   (b) $\sqrt{x} = 3x - 4$
   (c) $\sqrt{x} = 2x$
   (d) $\sqrt{3x - 4} = -8$
   (e) $\dfrac{2x+3}{x} = \dfrac{3}{x}$
2. Given the equation $2x - 1 = \sqrt{x}$, do the following.
   (a) Show that by solving the equation symbolically, you can get "solutions" of $x = \frac{1}{4}$ and $x = 1$.
   (b) Explain why if you square both sides of $2x - 1 = \sqrt{x}$ and if $x = 1$ is a solution, both sides are multiplied by 1.
   (c) Explain why if you square both sides of $2x - 1 = \sqrt{x}$ and if $x = \frac{1}{4}$ is a solution, one side is multiplied by $-\frac{1}{2}$ and the other by $\frac{1}{2}$.
   (d) Explain why $x = \frac{1}{4}$ is an extraneous solution.
3. Find all the exact solutions to the following equations.
   (a) $\sin x = 0.5$
   (b) $\cos x = 0.5$
   (c) $\tan x = 1$
   (d) $\cos x = 1$
4. Find all solutions for x such that $0 \le x \le 2\pi$. Round your answers to three decimal places.
   (a) $\sin x < 0.2$
   (b) $\cos x > 0.3$
   (c) $3 \tan x < 2$

5. Find all solutions for x such that $0 \leq x \leq \pi$. Round your answers to three decimal places.
   (a) $5\sin(2x) = 4.3$
   (b) $2\tan(3x) = 8.2$
   (c) $-4\cos(6x) = 2.5$
6. Find all solutions for x such that $0 \leq x \leq 4$. Round your answers to three decimal places.
   (a) $4\cos(\pi - 3x) + 6 = 4$
   (b) $2\sin(2x - \pi) + 7 = 2$
   (c) $3\tan(\pi + 4x) + 1 = 3$
7. Find all solutions for x such that $0 \leq x \leq 2\pi$.
   (a) $4\sin x\cos x = \sin x$
   (b) $\tan x\csc x - 2\cos x = 0$
   (c) $4\sin x\cos x + 3 = 5$
   (d) $\tan x + 4\sin x = 0$
8. In this section, we looked at a function that modeled the number of minutes of daylight for Grand Rapids, Michigan. This function was given as

$$d(t) = 190\sin\left[\frac{2\pi}{365}(t - 81)\right] + 731,$$

   where t is the number of the day of the year (i.e., January 1 = 1, January 2 = 2, ..., December 31 = 365).
   (a) Find two days when there are approximately 900 min of daylight.
   (b) Are there any days when there are 1000 min of daylight? How can you tell without actually solving the equation?
   (c) What days have more than 800 min of daylight?
9. The range, r, of a projectile fired at an angle of θ with an initial velocity of 120 ft/sec can be expressed with the formula[10]

$$r = 900\sin\theta\cos\theta.$$

   (a) Explain why this function can be written as $r = 450\sin(2\theta)$.
   (b) What angles of θ give a range of 200 ft?
   (c) What angles of θ give a range of 500 ft?
   (d) What value of θ gives a maximum range for this projectile?
10. Suppose the average high temperature for a location in Michigan can be modeled with the formula

$$T(n) = 27\cos\left[\frac{2\pi}{365}(n - 230)\right] + 56,$$

   where $T(n)$ is the average high temperature (in degrees Fahrenheit) on the nth day of the year. Which days of the year have average high temperatures that are above 80°F?

[10] *A formula similar to this one was developed in Investigation 3 in section 5.4.*

11. One of the world's tallest Ferris wheels is in Osaka, Japan. It measures 112.5 m from the ground to its apex, and the wheel has a diameter of 100 m (another Ferris wheel in Otsu has the same diameter, but the wheel in Osaka is 4 m higher). A complete revolution of this wheel takes about 15 min.[11]
    (a) Assuming a point on the rim of the wheel starts at its minimum height, write a symbolic representation for the height of that point where time is the input (in minutes) and height above the ground is the output (in meters).
    (b) Find the times between 0 and 60 min when the point on the wheel is at a height of 100 m.
12. Let a be some real number.
    (a) For what values of a are there no solutions for $a \sin x = 8$?
    (b) For what values of a are there no solutions for $a \cos x = 8$?
    (c) For what values of a are there no solutions for $a \tan x = 8$?
13. Label each of the following as always true, sometimes true, or never true. Briefly justify your answers.
    (a) For $-1 \le x \le 1$, $\cos(\cos^{-1} x) = x$.
    (b) For all real numbers, $\cos^{-1}(\cos x) = x$.
    (c) If $x = y$, then $\sin x = \sin y$.
    (d) If $\sin x = \sin y$, then $x = y$.
    (e) If $\sin x = \sin y$, then $\cos x = \cos y$.
    (f) If $\sin x = \sin y$, then $\tan x = \tan y$.
14. Find all exact solutions greater than 0.1 for $\sin(1/x) = 0$.
15. Graph the function $f(x) = \cos^{-1}(\cos x)$ on a domain of $-2\pi \le x \le 2\pi$. You should see a periodic function made of line segments.
    (a) What are the amplitude and the period of f?
    (b) What linear function describes f on a domain of $0 \le x \le \pi$?
    (c) What linear function describes f on a domain of $\pi \le x \le 2\pi$?

INVESTIGATIONS

INVESTIGATION: USING IDENTITIES TO SOLVE "SIMPLE" TRIGONOMETRIC EQUATIONS

Solutions are easily seen when we solve equations like $\sin x = \tan x$ graphically. Because the solutions to this equation are "nice," they are also easy to verify. In this investigation, however, we forgo graphical solutions. Instead, we focus on symbolic solutions to these types of equations. In doing so, we need not use only algebraic techniques, but also trigonometric identities. We also must be aware of the domains and ranges for trigonometric functions. We start by finding solutions to $\sin x = \tan x$ and then solve some other equations of the form $\sin x =$ ____.

[11] *Japan Information Network, "World's Tallest Ferris Wheel in Osaka,"* Monthly News, August 1997 *(visited 27 May 1998) ⟨http://www.jinjapan.org/kidsweb/news/97-8/wheel.html⟩.*

1. We want to find all the solutions to $\sin x = \tan x$.
   (a) Show that $\sin x = \tan x$ can be written as $\sin x - (\sin x/\cos x) = 0$.
   (b) Write the difference, $\sin x - (\sin x/\cos x)$, as a product by factoring out $\sin x$. Set each factor to zero to find all the solutions to our original equation.
2. We want to find all the solutions to $\sin x = \sec x$.
   (a) Show that $\sin x = \sec x$ can be written as $\sin x \cos x = 1$.
   (b) Using a double-angle identity, find all the solutions to our original equation.
3. Symbolically, find all the solutions to $\sin x = \cos x$.
4. Symbolically, find all the solutions to $\sin x = \csc x$.

7.4 REGRESSION AND CORRELATION

We started chapter 2 with a look at linear functions. In section 2.1, we stated that linear functions are the simplest type of and most commonly used function. One reason they are used so often is because in many situations, data naturally fall into a linear pattern. The most common method of fitting a linear function to data is through the method of least squares. In this section, we learn to describe a linear relationship in a scatterplot by using a least squares regression line and a correlation number.

LINEAR RELATIONSHIPS

Robert Pershing Wadlow was born in Alton, Illinois, on February 22, 1918. He weighed a normal 8 lb 6 oz. Due to an overactive pituitary gland, however, he grew at an astounding rate. He continued to grow his entire life. When he died at age 22, he had reached a height of 8 ft 11.1 in., which qualified him as the tallest person in history as recorded in the *Guinness Book of World Records*. Table 1 shows Wadlow's heights for various ages as reported there.[12]

TABLE 1 Heights for Robert Wadlow, the Tallest Person in History

Age (years)	5	8	9	10	11	12	13	14
Height (in.)	64	72	74.5	77	79	82.25	85.75	89
Age (years)	15	16	17	18	19	20	21	22.4
Height (in.)	92	94.5	96.5	99.5	101.5	102.75	104.25	107.1

[12] *Norris McWhirter, ed.,* 1977 Guinness Book of World Records *(New York: Sterling Publishing, 1976), pp. 12–13.*

A graph of Wadlow's age on the horizontal axis and his height in inches on the vertical axis is shown in Figure 1. By looking at this graph, it is evident that a linear function will do a reasonably good job of fitting these data.

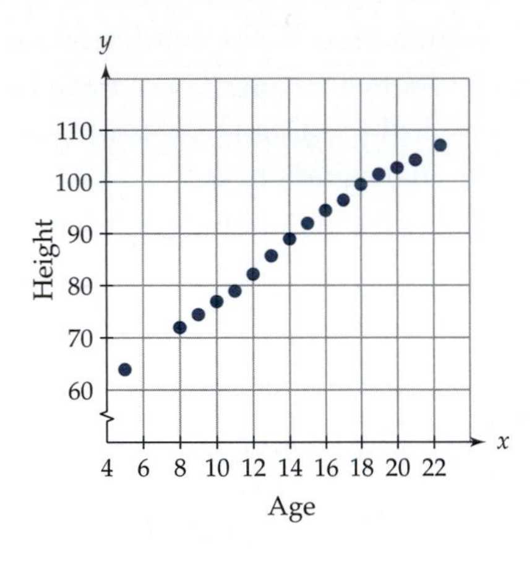

Figure 1 Heights (in inches) for Robert Wadlow, the tallest person in history.

The linear equation that fits the height data given in Table 1 is

$$\text{height} = 2.57 \times \text{age} + 51.8.$$

This equation can also be written as

$$y = 2.57x + 51.8.$$

We can see how the linear equation closely approximates our data by drawing the line in the same graph as the scatterplot. (See Figure 2.)

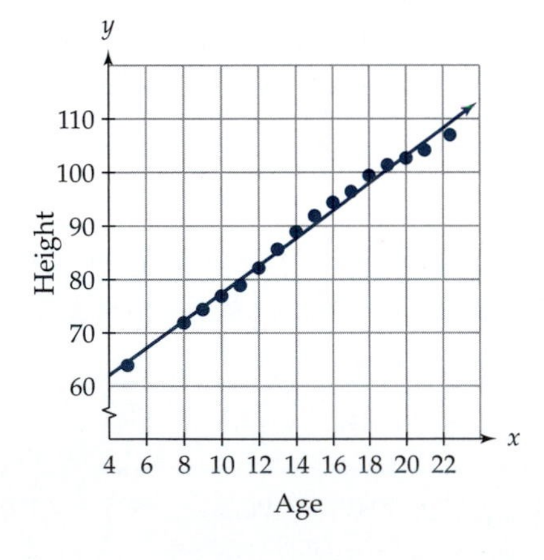

Figure 2 Heights (in inches) for Robert Wadlow along with a sketch of the line that fits the data.

The slope of the line, 2.57, means that Wadlow grew about 2.57 in./year from age 5 until his death. It is clear that the equation is not always valid outside of the ages we were given in Table 1. If it were, then the y-intercept of 51.8 would mean that Wadlow was 51.8 in. tall at birth. Because he was of normal weight at birth, we can probably safely assume he was not anywhere near 51.8 in. in height.

Now that we have an equation describing the relationship between age and the height of Wadlow, we can use it to predict his height given his age. For example, to predict his height at age 7, we just substitute 7 into our equation to find

$$\text{height} = 2.57 \times 7 + 51.8 \approx 70 \text{ in.}$$

Because the data points in Figure 2 lie close to the line, we know that 70 in. is a fairly good estimate for his height when he was 7 years old. A word of warning is necessary, however. Although our equation may do a good job of predicting Wadlow's height from age 7 to 22, it does not necessarily do a good job of predicting his height at any age outside that range. In general, using the equation for the line of best fit to make predictions *only* works well when your input is close to the inputs of your data points.

READING QUESTIONS

1. Suppose a baby boy's birth weight was 7 lb and he gained 2.5 lb each month for the first few months of his life. The function describing his weight is a line and could be written as

$$y = 2.5x + 7,$$

where x is his age in months and y is his weight in pounds. In this equation, the 2.5 is the slope and the 7 is the y-intercept.

(a) Determine the baby's weight after 4 months.

(b) Determine the child's weight after 48 months. Is your answer reasonable? Explain.

(c) If the baby's birth weight was 8 lb instead of 7 lb, how would the function change?

(d) What would it mean if the slope of the function were 3?

LEAST SQUARES REGRESSION

In section 1.2, we looked at a scatterplot of students' heights and arm spans. Because we were trying to determine if a person's arm span is about the same as his or her height, we also included the line $y = x$ in the graph of our scatterplot. That scatterplot and the line $y = x$ are repeated in Figure 3. The line $y = x$ seems to fit the data, but is there a line that fits better? The answer to this question is yes. In fact, using the method of least squares, we can find the equation of the line that best fits the data.

The most common method of determining the equation of the line of best fit is the *method of least squares*. This method involves finding the difference between the predicted output from the line and the actual output from the data points and then finding the equation that minimizes a combination of these differences. The line that does so is referred to as the least squares regression line, or simply the regression line. A **least squares regression**

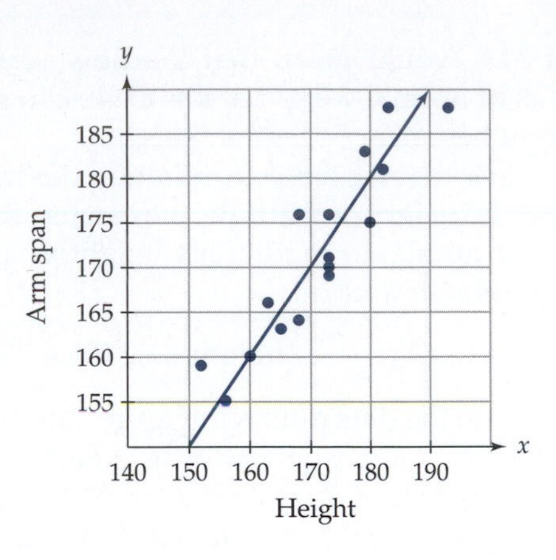

Figure 3 A scatterplot of students' heights (in centimeters) and arm spans (in centimeters) with the line $y = x$.

TABLE 2 A Subset of the Students' Heights and Arms Spans

Height (cm)	152	160	165	168	173	173	180	183
Actual Arm Span (cm)	159	160	163	164	170	176	175	188
Predicted Arm Span Using the Line (cm)	154.29	161.29	165.66	168.28	172.66	172.66	178.78	181.40
Error	4.71	−1.29	−2.66	−4.28	−2.66	3.34	−3.78	6.60

line is a line that best fits two-variable data by minimizing the sum of the squares of the vertical distances between the points in the scatterplot and the line. To illustrate, look at Table 2, which includes part of the height and arm span data in Figure 3. Using a calculator, the least squares regression line for these data is found to be

$$\text{arm span} = 0.8746 \times \text{height} + 21.35.$$

The last row of Table 2 gives the error of our line. This error is expressed as the difference between the actual output (using the data points) and the predicted output (using the equation of the line). The sum of the squares of all these errors for the least squares regression line will be smaller than the sum obtained using any other line. Thus, the least squares regression line is often called the line of best fit.

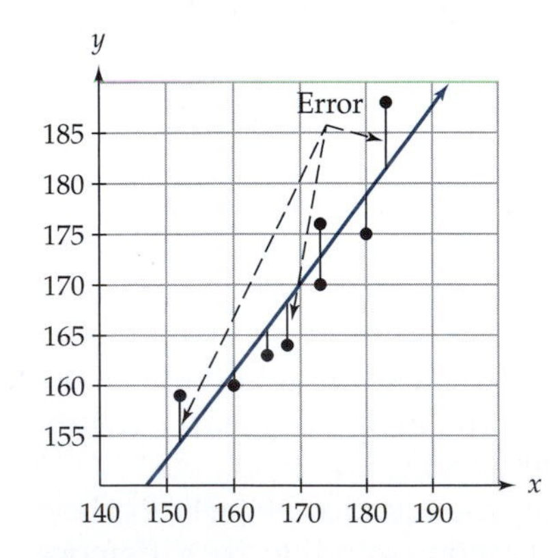

Figure 4 The vertical distance from a point to a line that fits the data is the error. For a regression line, this error is minimized.

The error between actual outputs (using the data points) and predicted outputs (using the equation of the line) is illustrated in Figure 4. This figure shows the points for the heights and the actual arm spans from Table 2 (represented by dots) along with the least squares regression line, $y = 0.8746x + 21.35$. In this figure, the lengths of the vertical lines between the points and the regression line represent the error. When fitting a line to the data, our goal is to make these vertical lengths as small as possible. The method of least squares does so by minimizing the sum of the squares of these distances.

The equation for the least squares regression line can be found using rather complicated formulas, but we use calculators instead. Calculator instructions for a TI-83 are given at the end of this section.

READING QUESTIONS

2. Suppose you have a scatterplot and you draw in a line that you think nicely fits the data. How is a regression line different from the line you drew?
3. Can you ever have a least squares regression line where all the data points are below the line? Explain.

CORRELATION

The two examples of regression lines considered earlier did a reasonably good job of predicting. Both the infant weight equation and the height and arm span equation had predicted outputs that were very close to the actual outputs. Some regression equations, however, do not give such accurate results. For example, in a statistics class, students were asked to give their mother's and father's heights. We wanted to see if the father's height could be predicted given the mother's height. A scatterplot of these results along with the least squares regression line are shown in Figure 5. The regression equation does not give accurate predictions because many of the data points are not close to the line; the data do not have a linear pattern. To measure how well a regression line fits a set of data, we use a number called correlation. **Correlation** measures both the strength and the direction of a linear relationship between two numerical variables. The height and arm span data in Figure 3 are said to be highly correlated, whereas the mother and father height data in Figure 5 are not.

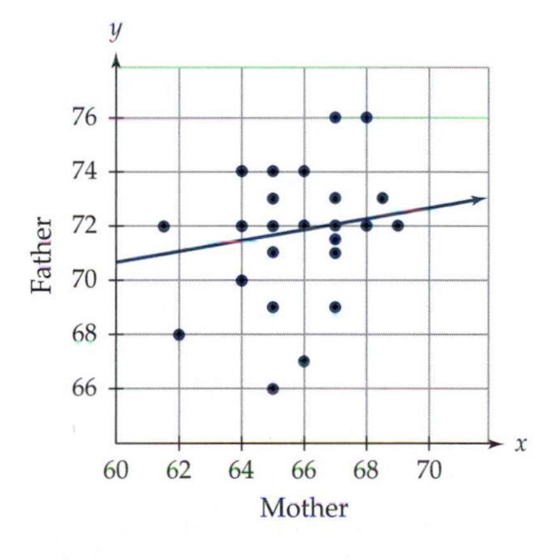

Figure 5 A scatterplot of the heights of students' mothers and fathers along with the regression line.

Correlation is usually denoted by the letter r such that $-1 \leq r \leq 1$. If the correlation is close to -1 or 1, then the data points are highly correlated. In fact, if $r = 1$ or $r = -1$, the data all fall in a straight line and the regression line perfectly predicts the outcomes. If the correlation is close to 0, then the data points have little correlation. If the correlation is positive, then there is a positive association between the two variables and the regression line has a positive slope. Therefore, as one variable increases, the other also increases (such as height and arm span). If the correlation is negative, then there is a negative association between the two variables and the regression line has a negative slope. Therefore, as one variable increases, the other decreases. (An example is the relationship between the age of a car and its value.)

The correlation of the height and arm span data shown in Table 2 and Figure 3 is 0.90, whereas the correlation of the father and mother height data in Figure 5 is 0.17. Therefore, the height and arm span data are highly correlated and have a positive association: people who are taller have longer arm spans than people who are not as tall. The father and mother height data have a very low correlation. Therefore, knowing one parent's height is not a good predictor of the height of the other parent.

You can often approximate the correlation by looking at a scatterplot of the data. Figure 6 shows a number of different scatterplots with corresponding correlations. Like the least squares regression equation, there are rather complicated formulas to determine the correlation, but we again rely on a calculator to compute these numbers. Calculator instructions on how to compute a correlation on a TI-83 are given at the end of this section.

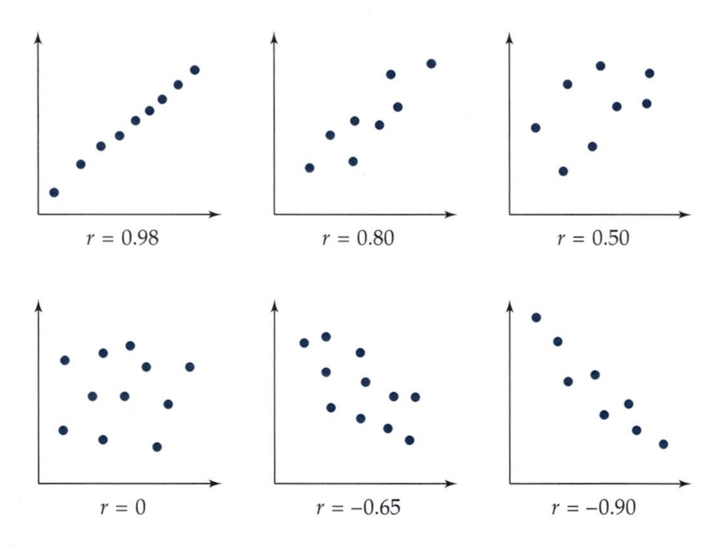

Figure 6 Examples of various scatterplots and their corresponding correlation.

Correlation is often confused with causation, but they are not the same. Two variables can be highly correlated without one necessarily causing the other. For example, if you were studying the mathematical achievement of elementary school students, you would probably find a high positive correlation between a child's shoe size and his or her achievement in mathematics. This correlation, however, does not mean that the growth of a child's foot causes the child to become better at mathematics. Instead, it means that as children grow older, their feet are growing longer and at the same time they are learning more mathematics in school. If two variables are highly correlated, then it simply means that there is a relationship between them; it does not mean that one causes the other.

HISTORICAL NOTE

In the past, both regression and correlation were not defined as they are today, but concepts behind these statistical tools have an interesting history. Sir Francis Galton (1822–1911) developed the ideas of regression and correlation. As a young man, he studied medicine at Cambridge. After receiving his inheritance, however, he abandoned his medical career and set out to explore the world and live the life of a gentleman. From 1850 to 1852, he explored Africa, and in 1853, he received a gold medal from the Royal Geographical Society for his achievements. In the 1860s he began to study weather, and he created elaborate weather maps of Europe. Although he studied and published in many other fields, from 1865 on he became mostly concerned with the study of heredity, perhaps because of the works of Charles Darwin, Galton's more famous cousin. Darwin published *Origin of the Species* in 1859.

In Galton's study of heredity, he compared the heights of children with the heights of their parents. He examined the heights of parents and their grown children in an attempt to understand to what degree height is an inherited characteristic. In doing so, he multiplied the women's heights by 1.08 to make them comparable with the men's heights. He then defined the parents' height as the average of the two parents (which he called midparent's height).

Galton found that the heights of the children tended to be more moderate than the heights of their parents. For example, very tall parents had children that tended to be tall, but shorter than their parents. Very short parents had children that tended to be short, but taller than their parents. He called this discovery "regression to the mean," with the word *regression* meaning "to go back to."

Much of Galton's work on regression appeared in his book *Natural Inheritance*. After the publication of this book, Galton presented a short paper titled, "Co-relations and their measurement, chiefly from anthropometric data." Within a year, the term *co-relation* had changed its spelling to *correlation*.

Although Galton introduced the ideas of correlation and regression, Karl Pearson (1857–1936), who was also a Cambridge graduate, provided the mathematical framework for these methods that are familiar to us today.

Source: Stephen M. Stigler, *The History of Statistics: The Measurement of Uncertainty before 1900* (Cambridge, MA: Belknap Press of Harvard University Press, 1986).

READING QUESTIONS

4. What does correlation measure?

5. Match the following correlations with the corresponding scatterplot.

(a) $r = 0$ (b) $r = 0.85$ (c) $r = 0.98$

(d) $r = -0.6$ (e) $r = -0.85$ (f) $r = 0.6$

i. 

ii. 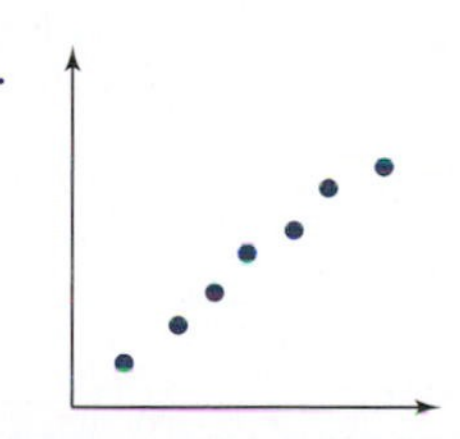

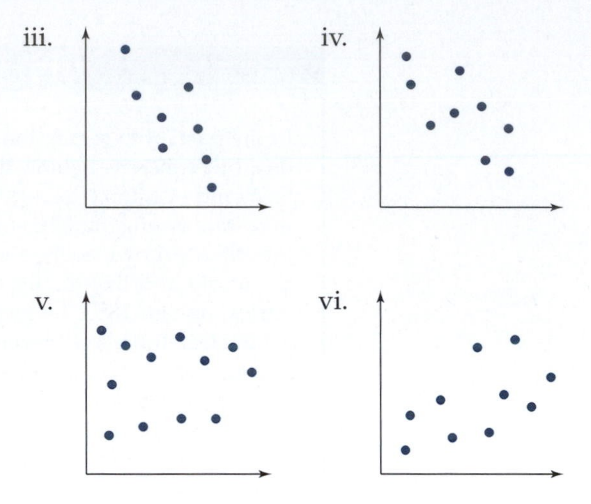

6. Give an example of two variables that you think would have a negative correlation.

7. If a person's arm span was *always* 2 in. shorter than his or her height, then what would the correlation between arm span and height be? Explain.

TECHNOLOGY TIP

USING A CALCULATOR TO COMPUTE THE REGRESSION EQUATION AND THE CORRELATION

To use your TI-83 to compute the linear regression equation and the correlation, you first need to enter your data into two lists. We demonstrate with the data from Table 2.

Height	152	160	165	168	173	173	180	183
Arm Span	159	160	163	164	170	176	175	188

1. Before you enter the data in the calculator, make sure it is set in the right mode to calculate correlation.[13] You must turn the "diagnostic" on. To do so, press `2nd` `0`, which is the catalog, a list of calculator commands in alphabetical order. Toggle down to **DiagnosticOn** by holding down `▽`. Then press `ENTER` `ENTER`, and the calculator is now in the proper mode to calculate correlation.

[13] *This step is specifically for the TI-83. If you are using a TI-82, then skip to number 2.*

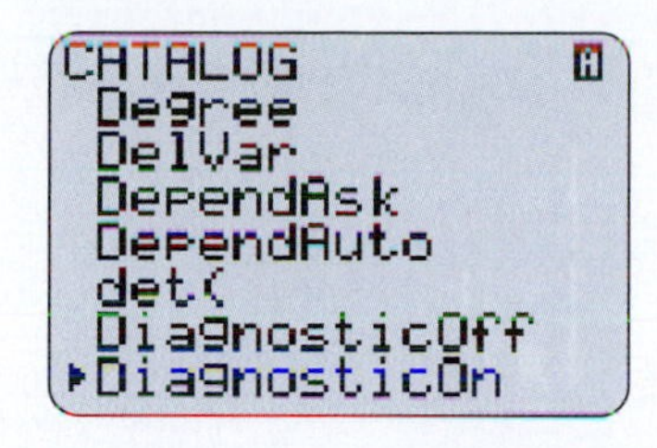

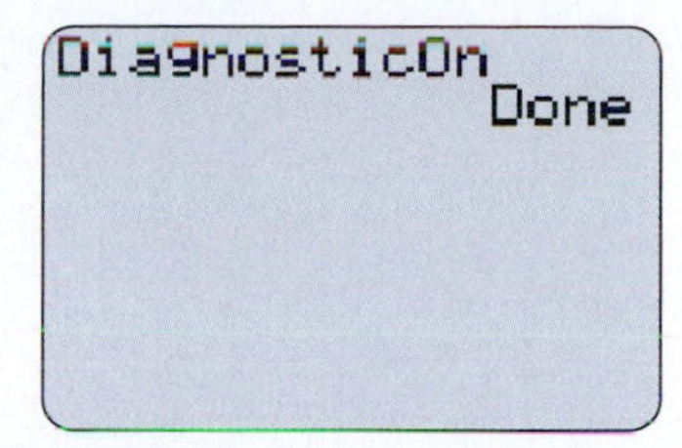

2. To open up the statistical editor, press STAT and select **1:Edit** from the menu by pressing ENTER. Input the height data in **L1** and the arm span data in **L2.** If necessary, first clear lists **L1** and **L2** by using the arrow keys to scroll the cursor to the top of the list and then press CLEAR ENTER.

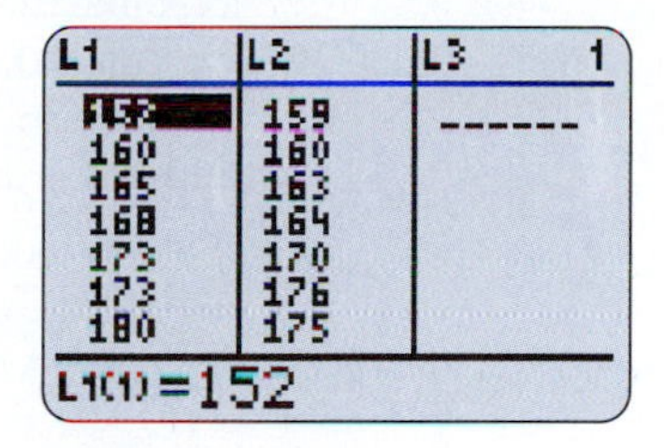

3. Once the data are entered, press STAT and move the cursor over to the **CALC** menu by pressing ▷. Now press ▽ three times so that the cursor is on **4:LinReg(ax+b)**.

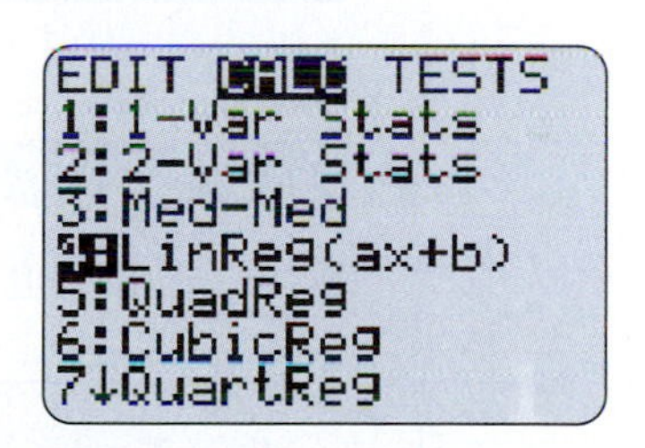

4. Press ENTER to get back to the home screen. To get the calculator to compute the regression equation and correlation for **L1** and **L2**, input **L1,L2** after **LinReg(ax+b).**[14] By having the lists in the order **L1,L2,** your calculator makes the **L1** list (height) the input or independent variable and the **L2** list (arm span) the output or dependent variable when it calculates the regression equation. To calculate the regression equation, press 2nd 1 , 2nd 2. Pressing ENTER gives the slope and y-intercept for the regression equation and the correlation. The regression equation for this set of data can be written as $y = 0.874575119x + 21.35316111$. The correlation is 0.9043440664.

[14] *The calculator's default list is* **L1** *and* **L2,** *so if your list of numbers is in* **L1** *and* **L2,** *then you do not really need to enter* **L1,L2.** *If it is any other lists, however, then you must identify those particular lists.*

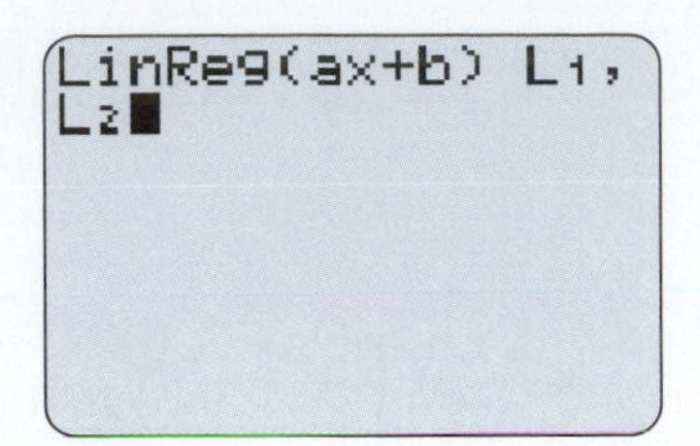

5. To graph your regression line along with a scatterplot of the data, type in the equation manually or use the shortcut. The shortcut allows you to have the regression equation automatically stored as a function so that you can graph it. To do so, go back to the previous step after you have **LinReg(ax+b) L1,L2** on your screen. You now need to insert **,Y1** after **LinReg(ax+b) L1,L2.** To do so, press [,] [VARS] [▷] [ENTER] [ENTER]. Your screen should now look like the left side of the accompanying picture. Now when you press [ENTER], the calculator calculates the regression equation and puts this equation in **Y1** so that it can be graphed. This line can be graphed along with the scatterplot.

LinReg(ax+b) L1,
L2,Y1

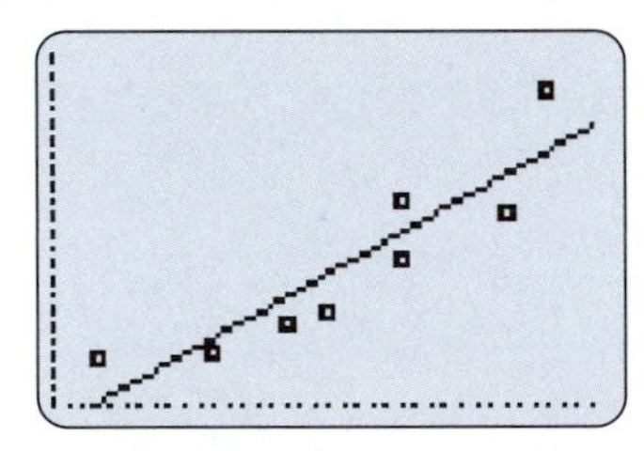

READING QUESTIONS

8. Input the following data into your calculator.

x	1	2	3	4	5	6	7	8
y	19	17	15	13	11	9	7	5

(a) Determine the regression equation and correlation for the data.

(b) What does the correlation tell you about the data and the regression equation?

SUMMARY

A **least squares regression line** is a line that best fits two-variable data by minimizing the sum of the squares of the vertical distances between the points in the scatterplot and the line. Although the equation for this line can be found by using a complicated formula, it is more easily found with a graphing calculator.

Correlation measures both the strength and the direction of a linear relationship between two numerical variables. Correlation is usually denoted by the letter r, and values for correlations are between -1 and 1 inclusive. A correlation close to 1 means that there is a strong positive relationship between the two variables. In other words, as one variable gets larger, so does the other. A correlation close to -1 means that there is a strong negative relationship. In other words, as one variable gets larger, the other gets smaller. A correlation close to zero indicates that there is no relationship between the variables.

EXERCISES

1. The following table represents the height and length of arm span for 16 college students. Enter the data in your calculator to answer the following questions.

Height (cm)	152	156	165	163	165	168	168	173
Arm span (cm)	159	155	160	166	163	176	164	171
Height (cm)	173	173	173	179	180	182	183	193
Arm span (cm)	170	169	176	183	175	181	188	188

   (a) Determine the linear regression equation for these data with height as the input and arm span as the output.

   (b) What is the correlation? What does this number tell you about the relationship between a person's height and the length of his or her arm span?

2. Suppose you started a job at age 25 and earned $30,000 per year. For each of the next 20 years after that you receive a $2000 raise. Suppose your salary for your first year on the job is represented by the point $(0, 30)$, the salary for the following year as $(1, 32)$, and so on.

   (a) What is the equation for the least squares regression line for the collection of points?

   (b) What is the correlation for this collection of points?

3. Throughout history, different measures of length have been linked to human anatomy such as the foot. Leonardo da Vinci (1452–1519) was very interested in the proportions of the human body. He wrote about the views of the Roman architect and engineer Vitruvius Pollio (first century B.C.):

   > Vitruvius declares that Nature has thus arranged the measurements of a man: four fingers make one palm; four palms make one foot; six palms make one cubit; four cubits make once a man's height; four cubits make a pace, and twenty-four palms make a man's height.[15]

[15] *H. Arthur Klein,* The World of Measurements *(New York: Simon and Schuster, 1974), p. 68.*

If Vitruvius is correct that four palms make one foot and 24 palms make one's height, then six times the length of someone's foot should be about the same as that person's height. The following table represents the foot length and height of 10 college students.

Foot Length (cm)	21.5	22.0	21.5	24.5	23.7	22.5	25.0	24.0	22.5	24.5
Height (cm)	152.4	156.2	160.0	162.6	165.1	167.6	172.7	172.4	172.7	172.9

To determine how accurate Vitruvius suggestion that six times the length of someone's foot is about the same as this person's height, do the following.

(a) Construct a scatterplot for the data given in the table. Put the foot length on the horizontal axis and the height on the vertical axis.

(b) What linear equation represents the statement, "Six feet make one's height"?

(c) Graph your equation from part (b) on your scatterplot. Does the line seem to fit the data? Does this statement seem accurate?

(d) Find the regression equation for your data. Describe what the slope means in terms of foot length and height.

(e) How accurately does your equation estimate someone's height given a foot length?

4. The following table represents the winning times for the 100-m dash in the Olympic Games from 1928 to 1992.[16]

Year	**Men's Winning Time (sec)**	**Women's Winning Time (sec)**
1928	10.8	12.2
1932	10.3	11.9
1936	10.3	11.5
1948	10.3	11.9
1952	10.4	11.5
1956	10.5	11.5
1960	10.2	11.0
1964	10.0	11.4
1968	9.95	11.0
1972	10.14	11.07
1976	10.06	11.08
1980	10.25	11.6
1984	9.99	10.97
1988	9.92	10.54
1992	9.96	10.82

[16]The World Almanac and Book of Facts 1996 *(Mahwah, NJ: Funk and Wagnalls, 1995), pp. 855, 858.*

(a) Determine the regression equation where the year is the input and the men's winning time is the output. Do the same for the women's winning times.

(b) Why does it make sense that the slope of each regression line is negative?

(c) Using each of your regression equations, determine the predicted winning times for the men's and the women's 100-m dashes in the 1952 Olympic Games. Are these numbers close to the actual winning times?

(d) Suppose the Olympic Games were held in the year 0. Using each of your regression equations, determine the predicted winning times for the men's and the women's 100-m dashes. Do these numbers seem reasonable for winning times? Explain.

(e) Using each of your regression equations, determine the predicted winning times for the men's and the women's 100-m dashes for the Olympic Games in 2100. Do these numbers seem reasonable for winning times? Explain.

(f) Based on your answers to the previous questions, when do you think regression equations work well in predicting outcomes?

5. The following table lists years and speed for running 1 mi. These speeds are the world record for that year. A scatterplot of these data is also shown. The years on the scatterplot are given as number of years since 1800.

Year	Speed (mph)	Year	Speed (mph)
1865	13.02	1943	14.84
1868	13.39	1944	14.90
1874	13.53	1945	14.91
1875	13.61	1954	15.13
1880	13.68	1957	15.18
1882	13.77	1958	15.35
1884	13.93	1962	15.36
1894	13.94	1964	15.38
1895	14.08	1965	15.41
1911	14.10	1966	15.56
1913	14.15	1967	15.58
1915	14.25	1975	15.69
1923	14.38	1979	15.72
1931	14.50	1980	15.73
1933	14.54	1981	15.84
1934	14.59	1985	15.91
1937	14.61	1993	16.04
1942	14.62		

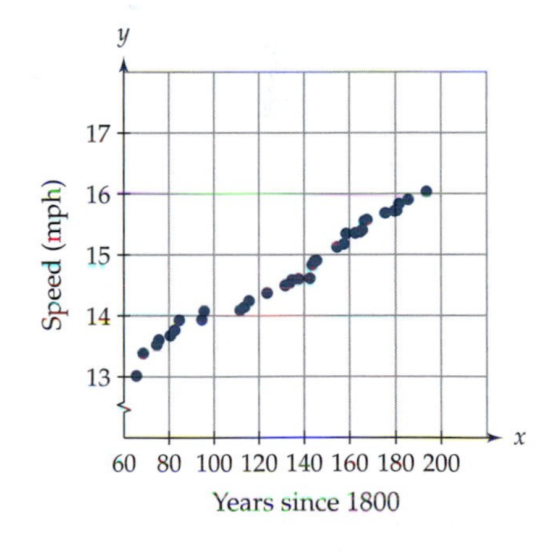

(a) Find the regression line that fits these data. Your input should be number of years since 1800, and your output should be speed in miles per hour.

i. What is the physical meaning of the slope of the regression line?

ii. What is the physical meaning of the y-intercept of the regression line?

iii. Is the regression line a good fit for your data? Justify your answer.

(b) World records for the mile are typically given as time rather than as speed.

i. What was the fastest time (in minutes and seconds) for running the mile in 1865?

ii. What was the fastest time (in minutes and seconds) for running the mile in 1993?

iii. The regression line you found in part (a) iii gave the speed for that year. How can you convert this equation so that the input is time (since 1800) and the output is time (in minutes)?

(c) Roger Bannister was the first person to run 1 mi in less than 4 min, making headlines around the world. Use the data to determine the year this occurred.

6. The following table gives carbon emissions from fossil fuel energy consumption for the commercial sector in the United States.[17]

Year	1980	1990	1991	1992	1993	1994	1995	1996	1997
Carbon Emissions (million metric tons)	178	207	206	205	212	214	218	226	236

(a) Find the equation of the line that best fits these data. Denote the year 1980 by 0.

(b) What is the correlation? What does it tell you about your line?

(c) What is the y-intercept of your line? What is the physical interpretation of the y-intercept?

(d) What is the slope of your line? What is the physical interpretation of the slope?

(e) Use your line to predict carbon emmissions in the year 2000.

7. Life expectancy at birth has increased during the last 100 years. Since 1920, female life expectancy has been greater than male life expectancy, as can be seen in the following table.[18]

[17] U.S. Carbon Emissions for Fossil Energy Consumption by End-Use Sector *(visited 6 July 1998)* ⟨*http://www.eia.doe.gov/oiaf/1605/flash/sld003.html*⟩.

[18] The World Almanac and Book of Facts 1996, *p. 974.*

Year	1920	1930	1940	1950	1960	1970	1980	1990
Male	53.6	58.1	60.8	65.6	66.6	67.1	70.0	71.8
Female	54.6	61.6	65.2	71.1	73.1	74.7	77.5	78.8

(a) Make a scatterplot with the male life expectancy on the horizontal axis and the female life expectancy on the vertical axis. Do these data fall in a linear pattern?

(b) Find the regression equation where the input is male life expectancy and the output is female life expectancy.

(c) Based on your regression equation, determine if the gap between male life expectancy and female life expectancy is increasing or decreasing. How can you tell simply from the equation?

8. Explain what is wrong with the following statements about correlation and regression.

   (a) The correlation between the number of years of education and yearly income is 1.23.

   (b) The regression equation that describes the relationship between the age of a used Ford Mustang and its value is $y = -1000x + 10,000$, where the input is the age of the automobile (in years) and the output is its value (in dollars). The correlation describing this relationship is 0.83.

9. Suppose a friend of yours is telling you about a study reported in a popular magazine. The friend said that the study reported a correlation of zero between the amount of coffee a person drank as a child and his or her height as an adult. Your friend interprets this correlation to mean that the more coffee a child drinks, the shorter that person will be as an adult. In other words, drinking coffee stunts one's growth. Explain what is wrong with your friend's interpretation and explain what this type of correlation means.

10. Just because two variables are highly correlated does not mean that one necessarily causes the other. Something else could be causing one or both events. For each of the following statements, explain why the causation listed may or may not be true.

    (a) There is a high correlation between a child's height and his or her achievement in mathematics. Does this correlation mean that a child's physical growth is causing the increased achievement in math?

    (b) There is a high negative correlation between the outside temperature and the number of colds people get. Does this correlation mean the number of colds people get causes the temperature to decrease?

    (c) There is a high correlation between the number of drinks someone has at a bar and incidents of lung cancer. Does this correlation mean that drinking causes lung cancer?

INVESTIGATIONS

INVESTIGATION 1: MEDIAN-MEDIAN LINE

A least squares regression line is not the only way to fit a line to data. Another method is through the use of a median-median line. A least squares regression line uses all the data points to get an equation, but a median-median line does not. In producing a median-median line, the data are divided into three parts, summary points are obtained for each part, and only these three summary points are used to produce the median-median line.

1. A summary point is the median of all the x-values and the median of all the y-values in that part of the data set. We call the three summary points (x_1, y_1), (x_2, y_2), and (x_3, y_3). We use these three points to develop the formulas for the slope and y-intercept of the median-median line.
   (a) The slope of the median-median line is the same as the slope of the line that goes through (x_1, y_1) and (x_3, y_3). If m is the slope of the median-median line, then find a formula for m.
   (b) The y-intercept for the median-median line, b, is the average (mean) of y-intercepts of the three lines going through the three summary points with the same slope as that of the median-median line, m. Explain why an equation to find b can be given as

$$b = \frac{y_1 + y_2 + y_3 - m(x_1 + x_2 + x_3)}{3}.$$

2. Given the data in the table, do the following.

x	1	2	3	4	5	6	7	8	9
y	5	8	10	12	14	18	20	24	25

   (a) Find the three summary points for the data. The data are already divided into three parts by the vertical lines.
   (b) Find the median-median line for the data given in the table.
   (c) Find the least squares regression line for the data given in the table. Is this equation similar to the one for the median-median line?

3. The data shown in the following table are the same as those in question 2 except the y-value for the last entry has been changed from 25 to 50.

x	1	2	3	4	5	6	7	8	9
y	5	8	10	12	14	18	20	24	50

(a) Find the three summary points for the data. The data are already divided into three parts by the vertical lines.

(b) Find the median-median line for the data given in the table. Is your equation for this median-median line much different than the median-median line you obtained in question 2?

(c) Find the least-squares regression line for the data given in the table. Is your equation for this regression line much different than the regression line you obtained in question 2?

(d) With the exception of the last data point, the data from the table fall in a pattern that is fairly linear. Does the median-median line or the least squares regression line do a better job of showing this linear pattern? Explain.

INVESTIGATION 2: FORMULAS FOR REGRESSION AND CORRELATION

The equation of a least squares regression line and correlation can be calculated with formulas as well as on a calculator. Formulas are given in a variety of forms. In this investigation, we show a couple of formulas for determining the slope of a least squares regression line, reconcile these formulas, and then use the formulas in an example. The following are formulas for correlation, r, the slope of the regression line, m, and the y-intercept for the regression line, b.

$$r = \frac{\sum xy - n\overline{xy}}{\sqrt{\sum x^2 - n\overline{x}^2}\sqrt{\sum y^2 - n\overline{y}^2}}$$

$$m = \frac{\sum xy - n\overline{xy}}{\sum x^2 - n\overline{x}^2} \qquad m = r\frac{s_y}{s_x}$$

$$b = \overline{y} - m\overline{x}$$

The symbol Σ is an oversized uppercase Greek letter sigma. It is used to indicate addition or summation. For example, to find the mean of a set of numbers, you need to find the sum of the numbers and then divide by the number of numbers you have. Symbolically, this sum can be indicated by

$$\overline{x} = \frac{\sum x}{n}.$$

The $\Sigma\, x$ simply means that you need to add the numbers associated with the variable x. The notation $\overline{x}$ represents the mean of this set. It is often referred to as x-bar.

1. Our first task is to show that the two formulas given for m are the same. To do so, we first need to define s_x and s_y.

   (a) The notation s_x and s_y are used to signify the standard deviations of the x-coordinates and the y-coordinates, respectively. Standard deviation is a measure of the spread of a set of numbers. The formulas are

$$s_x = \sqrt{\frac{\sum x^2 - \left(\sum x\right)^2/n}{n-1}} \quad \text{and} \quad s_y = \sqrt{\frac{\sum y^2 - \left(\sum y\right)^2/n}{n-1}},$$

where x represents the x-coordinates, y represents the y-coordinates and n represents the number of data points you have. Use these formulas to determine s_x and s_y for the following set of ordered pairs: $(2, 3)$, $(3, 6)$, $(4, 10)$, and $(5, 12)$. (*Hint:* $\Sigma\, x^2$ means that you need to first square the numbers and then add them, whereas $(\Sigma\, x)^2$ means that you first add the numbers and then square the sum.)

(b) Knowing that $\overline{x} = \Sigma\, x/n$ and $\overline{y} = \Sigma\, y/n$, show that the two formulas for s_x and s_y can be written as

$$s_x = \sqrt{\frac{\sum x^2 - n\overline{x}^2}{n-1}} \quad \text{and} \quad s_y = \sqrt{\frac{\sum y^2 - n\overline{y}^2}{n-1}}.$$

(c) Show that $m = r(s_y/s_x)$ is equivalent to

$$m = \frac{\sum xy - n\overline{x}\overline{y}}{\sum x^2 - n\overline{x}^2}.$$

2. Now we use the formulas to determine the correlation and regression line for a set of data.

   (a) Explain why the formula $b = \overline{y} - m\overline{x}$ shows that the regression line must contain the point $(\overline{x}, \overline{y})$.

   (b) Using the formulas, determine the correlation and regression line for the points $(2, 3)$, $(3, 6)$, $(4, 10)$, and $(5, 12)$. (*Hint:* The notation $\Sigma\, xy$ means to find the products of an x-coordinate and the corresponding y-coordinate and then find the sum of the products.)

7.5 FITTING EXPONENTIAL AND POWER FUNCTIONS TO DATA

In the last section, we saw how least squares regression can be used to fit a line to data that fall in a linear pattern. Not all data, of course, fit a linear pattern. When data fit the pattern of an exponential function or a power function, techniques can be used to transform the data to fit a linear pattern. Scientists looking for relationships in data often transform data so it looks more linear. In this section, we see how exponential functions and power functions can be transformed into linear functions. By doing so, we see how least squares regression can be used to fit exponential and power functions to data.

TRANSFORMING EXPONENTIAL FUNCTIONS INTO LINEAR FUNCTIONS

The national debt of the United States has been increasing quite rapidly for many years and is now in the trillions of dollars. One way to get a better sense of what this huge amount of debt represents is to think of it as debt per person or debt per capita. If the public debt of the U.S. government were spread out among all the citizens of the United States in 1994, each person would owe $18,026. Let's look at a table of this per capita debt since 1975.[19] (See Table 1.)

TABLE 1 National Debt of the U.S. Government per Capita

Year	1975	1976	1977	1978	1979	1980	1981	1982	1983	1984
Debt per Capita ($)	2475	2852	3170	3463	3669	3985	4338	4913	5870	6640
Year	1985	1986	1987	1988	1989	1990	1991	1992	1993	1994
Debt per Capita ($)	7598	8774	9615	10,534	11,545	13,000	14,436	15,846	16,871	18,026

From Table 1, you can see that the amount of the national debt per capita has increased every year from 1975 to 1994. You can also see that it has not increased linearly. The differences in the amount of debt in the earlier years is much smaller than the increase in the later years. We can see these differences better with a graph. Figure 1 shows a scatterplot of this debt where the input is years since 1900 and the output is the debt per capita in thousands of dollars. From this graph, it appears that an exponential function might model these data points well.

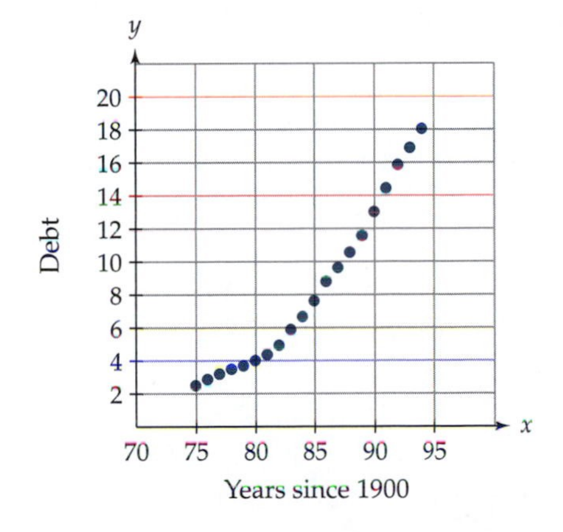

Figure 1 The national debt of the U.S. government per capita (in thousands of dollars) given the number of years since 1900.

Often, scientists transform data that is nonlinear into linear data. One way to do so is to graph the data so that one or both of the axes use a

[19]The World Almanac and Book of Facts 1996, *p. 112.*

logarithmic scale instead of a linear scale. If we graph the same data from Figure 1 on a graph whose vertical scale is a common logarithm scale, then we get the graph shown in Figure 2. On this graph, the powers of 10 are evenly spaced. For example, notice that the 5, 10, and 20 are evenly spaced. This even spacing is because $10^{0.7} \approx 5$, $10^1 = 10$, and $10^{1.3} \approx 20$, and the difference between 0.7 and 1 is the same as the difference between 1 and 1.3. A graph in which the vertical axis uses a log scale, commonly called a log graph, is used to make an exponential function look like a linear function. Another way to transform data that look exponential into data that look linear is to take the logarithm of the output. In Table 2, we took the common logarithm of the debt per capita in Table 1. Graphing the data in Table 2 gives a graph that has the same shape as the graph using a log scale. (See Figure 3.)

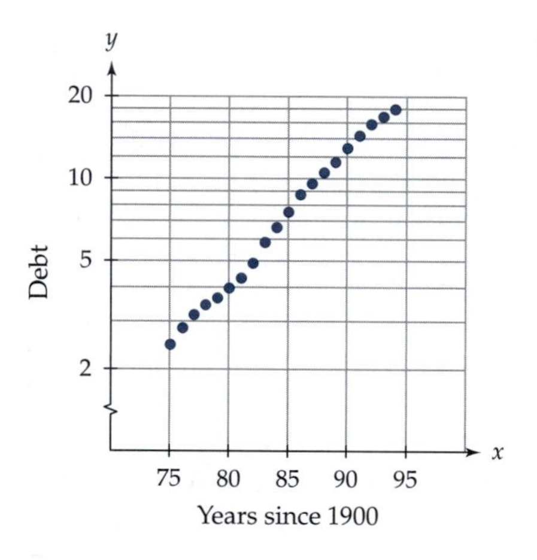

Figure 2 The national debt of the U.S. government per capita (in thousands of dollars) given the number of years since 1900 on a log graph.

TABLE 2 Common Logarithm of the National Debt of the U.S. Government per Capita

Year	1975	1976	1977	1978	1979	1980	1981	1982	1983	1984
log(*debt per capita*)	3.39	3.46	3.50	3.54	3.56	3.60	3.64	3.69	3.77	3.82
Year	1985	1986	1987	1988	1989	1990	1991	1992	1993	1994
log(*debt per capita*)	3.88	3.94	3.98	4.02	4.06	4.11	4.16	4.20	4.23	4.26

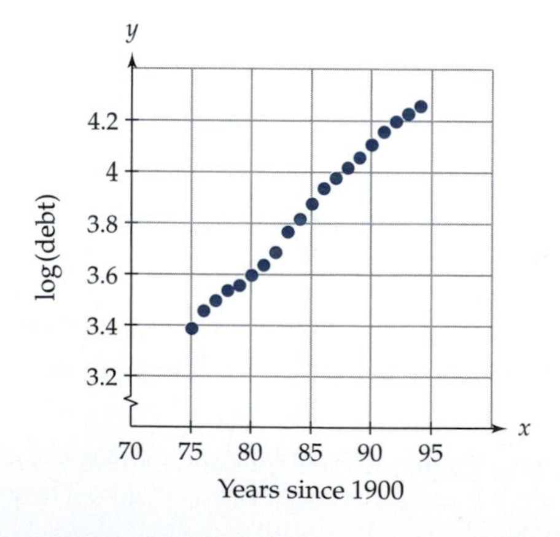

Figure 3 The common logarithm of the national debt of the U.S. government per capita given the number of years since 1900.

How does taking the logarithm of the output of an exponential function transform it into a linear function? To answer this question, remember that an exponential function can be written in the form $y = b \cdot a^x$. By taking the common logarithm of each side of this equation, the equation is transformed to $\log y = \log(b \cdot a^x)$. By using some properties of logarithms, we can write this transformed equation as

$$\begin{aligned}\log y &= \log(b \cdot a^x)\\ &= \log b + \log a^x\\ &= \log b + x \cdot \log a\end{aligned}$$

This equation, $\log y = \log b + x \cdot \log a$, is a linear function with an input of x, an output of $\log y$, a y-intercept of $\log b$, and a slope of $\log a$.

Using least squares regression, we can find a linear equation that will fit the data shown in Table 2 and Figure 3. This equation is $y = 0.0474x - 0.1653$. Now we need to find the exponential equation of the form $y = b \cdot a^x$ that fits the data before it was transformed. Because the slope of our linear function is $\log a$, $\log a = 0.0474$, which means that $a = 10^{0.0474}$ or $a \approx 1.115$. Because the y-intercept of our linear function is $\log b$, $\log b = -0.1653$, which means that $b = 10^{-0.1653}$ or $b \approx 0.683$. Thus, the equation $y = 0.683 \cdot 1.115^x$ should reasonably fit our original data from Table 1. These data along with the graph of $y = 0.683 \cdot 1.115^x$ are shown in Figure 4.

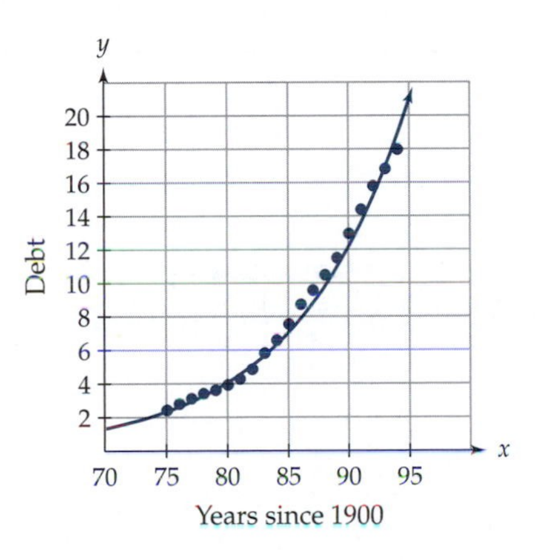

Figure 4 The national debt of the U.S. government per capita (in thousands of dollars) given the number of years since 1900 along with the equation $y = 0.683 \cdot 1.115^x$.

Summary

Data that fit an exponential equation can be transformed into data that fit a linear equation by taking the common logarithm of the output. Linear regression can then be used on the transformed data, and a least squares regression line can be obtained. By using the slope and the y-intercept of the linear regression line, an exponential function of the form $y = b \cdot a^x$ that fits the original data can be obtained by the following.

- $\log a = \text{slope}$ or $a = 10^{\text{slope}}$
- $\log b = y\text{-intercept}$ or $b = 10^{y\text{-intercept}}$

Example 1 Audio compact discs (or CDs) are currently the most popular format of recorded music. They increased in popularity during the late 1980s and early 1990s. Table 3 shows the number of CDs sold in the U.S. from 1987 to 1994.[20] Find an exponential function to fit the data.

TABLE 3

Years Since 1987	0	1	2	3	4	5	6	7
Millions of CDs	102.1	149.7	207.2	286.5	333.3	407.5	495.4	662.1

Solution First we transform the data in Table 3 by taking the common logarithm of the output. This step is shown in Table 4. The data in Table 4 are plotted in Figure 5. Because these data are fairly linear, a linear function will fit these data and an exponential function will fit the original data.

TABLE 4

Years Since 1987	0	1	2	3	4	5	6	7
log(*millions of CDs*)	2.009	2.175	2.316	2.457	2.523	2.610	2.695	2.821

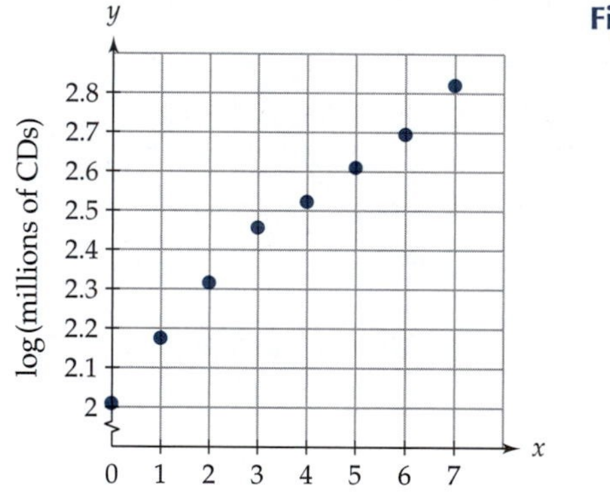

Figure 5

Using least squares regression, we can find a linear equation that will fit the data shown in Table 4 and Figure 5. This equation is $y = 0.1099x + 2.0661$.

Now we need to find the exponential equation of the form $y = b \cdot a^x$ that fits the data before it was transformed. Because the slope of our linear function is $\log a$, $\log a = 0.1099$, so $a = 10^{0.1099}$ or $a \approx 1.288$. Because the y-intercept of our linear function is $\log b$, $\log b = 2.0661$, so $b = 10^{2.0661}$ or $b \approx 116.439$. Thus, the equation $y = 116.439 \cdot 1.288^x$ should reasonably

[20]The World Almanac and Book of Facts 1996, *p. 258.*

fit our original data from Table 3. These data along with the graph of $y = 116.439 \cdot 1.288^x$ are shown in Figure 6.

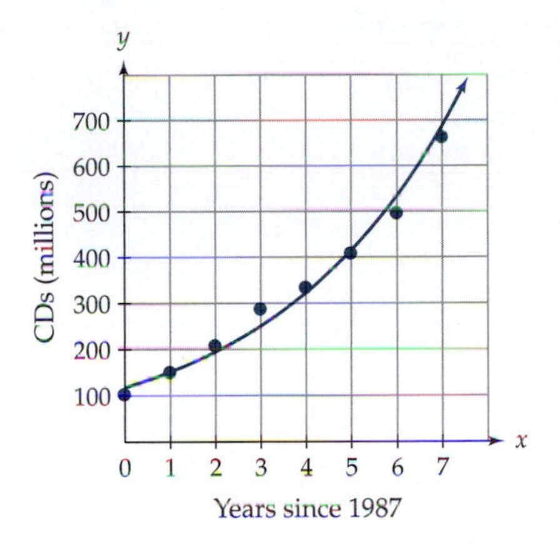

Figure 6

READING QUESTIONS

1. Show that the function $y = 5.2 \cdot 8.7^x$ can be transformed to a linear function by taking the logarithm of both sides of the equation.
   (a) What are the input and the output of the linear function?
   (b) What are the slope and the y-intercept of the linear function?
2. Suppose data that fit an exponential function have been transformed by taking the common logarithm of the output. The linear regression line $y = 7.8x + 3.0$ has been obtained using the transformed data. What exponential function fits the original data?

TRANSFORMING POWER FUNCTIONS INTO LINEAR FUNCTIONS

The wings of different species of birds perform different functions. For example, the wings of a hummingbird are well adapted for hovering, and the wings of a hawk are well adapted for soaring. Whatever the purpose, in general the wings of all flying birds need to support the weight of the bird. Let's examine the relationship between the surface area of birds' wings and their weights. Table 5 gives the weight, in grams, and the wing area, in square centimeters, for six different species of birds. These data are plotted in Figure 7.

We might suspect that relating the weight of a bird to the surface area of its wings is similar to relating the volume of an object to its surface area. These types of functions tend to be power functions. For example, the surface area of a cube is given by the formula $A = 6s^2$, where s is the length of one side of the cube. The volume of a cube is given by the formula $V = s^3$. We can combine these two equations to have a formula where volume is the input and surface area is the output. To do so, we first solve

TABLE 5 Wing Area of a Bird Given Its Weight

Species	Weight (g)	Wing Area (cm^2)
House wren	11.0	48.4
Chimney swift	17.3	118.5
Song sparrow	22.0	86.5
Purple martin	43.0	185.5
Red-winged blackbird	70.0	245.0
Mourning dove	130.0	357.0

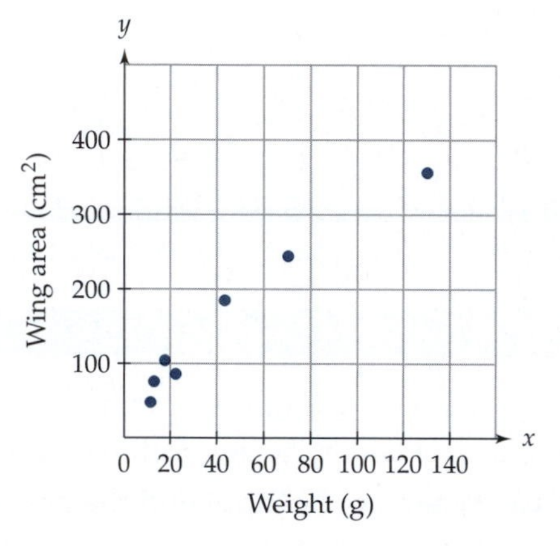

Figure 7 The wing area of a bird (in square centimeters) given its weight (in grams).

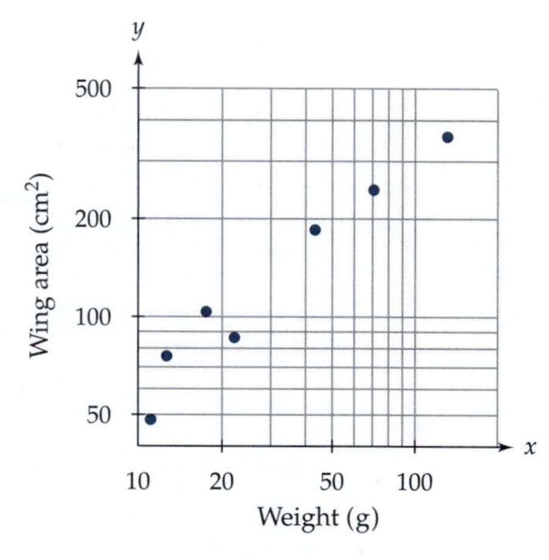

Figure 8 A log-log plot of the wing area of a bird given its wing area.

the volume formula for s to get $s = V^{1/3}$. We then substitute $V^{1/3}$ for s in the area formula to get $A = 6V^{2/3}$. This result is a power function. This same exponent, $\frac{2}{3}$, is used in what is called the wing-load index for birds. The wing-load index is $W^{2/3}/A$, where W is the weight of a bird and A is the surface area of a bird's wing.[21]

Let's assume that a power function would fit our data. Data that have the form of a power function can be transformed to look like a linear function if both axes are log scales instead of linear scales. These types of graphs are commonly called log-log plots. Figure 8 shows this transformation. The data in Figure 8 should look more linear than those in Figure 7.

Another way to transform the data that have the form of a power function into data that have the form of a linear function is to take the logarithm of the input and plot it against the logarithm of the output. We did this step in Table 6 and plotted these data in Figure 9.

[21] *Donald S. Farner and James R. King, eds.,* Avian Biology Vol. IV *(New York: Academic Press, 1974), p. 419.*

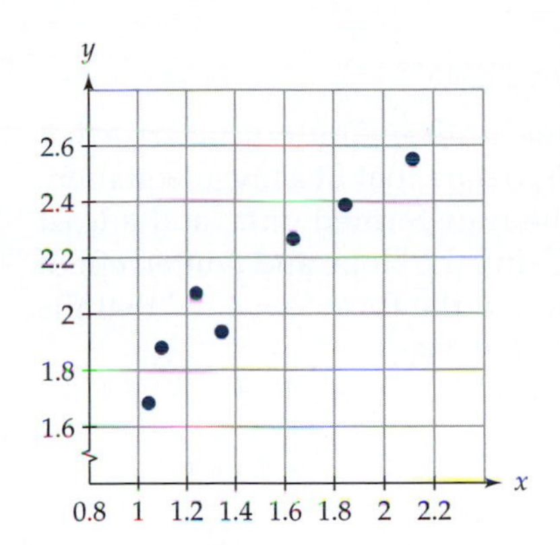

Figure 9 A plot of the common logarithms of the wing area of a bird given the common logarithm of its weight.

TABLE 6 Common Logarithm of the Wing Area of a Bird Given the Common Logarithm of Its Weight

Species	log(weight)	log(wing area)
House wren	1.041	1.685
Chimney swift	1.238	2.074
Song sparrow	1.342	1.937
Purple martin	1.633	2.268
Red-winged blackbird	1.845	2.389
Mourning dove	2.114	2.553

Let's see why taking the logarithm of both sides of a power function transforms it into a linear function. A power function has the form $y = b \cdot x^a$. By taking the common logarithm of both sides, the equation is transformed to $\log y = \log(b \cdot x^a)$. By using some properties of logarithms, we can write this transformed equation as follows.

$$\begin{aligned} \log y &= \log(b \cdot x^a) \\ &= \log b + \log x^a \\ &= \log b + a \log x \end{aligned}$$

This equation, $\log y = \log b + a \log x$, is a linear equation with an input of $\log x$, an output of $\log y$, a y-intercept of $\log b$, and a slope of a.

Using least squares regression, we can find a linear equation that will fit the data shown in Figure 9. This equation is $y = 0.756x + 0.991$. Now we need to transform this back to a power function of the form $y = b \cdot x^a$. The slope of our linear function is a, so $a = 0.756$. The y-intercept of our linear function is $\log b$, so $\log b = 0.991$. Therefore, $b = 10^{0.991} \approx 9.795$. The power equation $y = 9.795 \cdot x^{0.756}$ should reasonably fit our original bird data. These data along with the graph of $y = 9.795 \cdot x^{0.756}$ are shown in Figure 10.

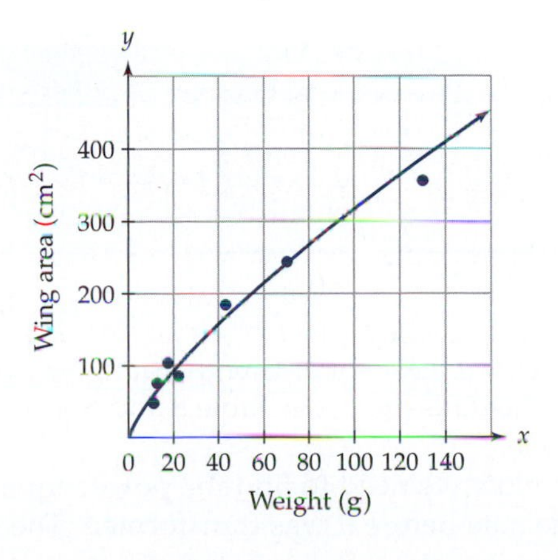

Figure 10 The wing area of a bird (in square centimeters) given the weight of the bird (in grams) along with the graph of $y = 9.795 \cdot x^{0.756}$.

Summary

By taking the common logarithm of both the input and the output, data that fit a power equation can be transformed into data that fit a linear equation. Linear regression can then be used on the transformed data, and a least squares regression line can be obtained. Using the slope and y-intercept of the linear regression line, a power function of the form $y = b \cdot x^a$ that fits the original data can be obtained by the following.

- $a = \text{slope}$
- $\log b = y\text{-intercept}$ or $b = 10^{y\text{-intercept}}$

Example 2 As babies get older, their weights naturally increase. This increase is not usually linear, however. Babies usually gain a lot of weight during the first few months of life, then the weight gain slows down. Table 7 shows the average weight of infants, according to the National Center for Health Statistics, given their age in months. Find a power function that fits the data in Table 7 where the input is a baby's age, in months, and the output is the average weight, in pounds.

TABLE 7

Age (months)	3	6	9	12	15	18	21	24
Weight (lb)	13.0	17.2	20.3	22.2	24.0	25.3	26.6	27.8

Solution First we transform the data in Table 7 by taking the common logarithm of the input and the output. This step is shown in Table 8. The data in Table 8 are plotted in Figure 11. Because these data are fairly linear, a linear function will fit these data and a power function will fit the original data.

TABLE 8

log(*age*)	0.477	0.778	0.954	1.079	1.176	1.255	1.322	1.380
log(*weight*)	1.114	1.236	1.307	1.346	1.380	1.403	1.425	1.444

Using least squares regression, we can find a linear equation that will fit the data shown in Table 8 and Figure 11. This equation is $y = 0.362x + 0.950$.

Now we need to find the power equation of the form $y = b \cdot x^a$ that fits the data before it was transformed. The slope of our linear function is a;

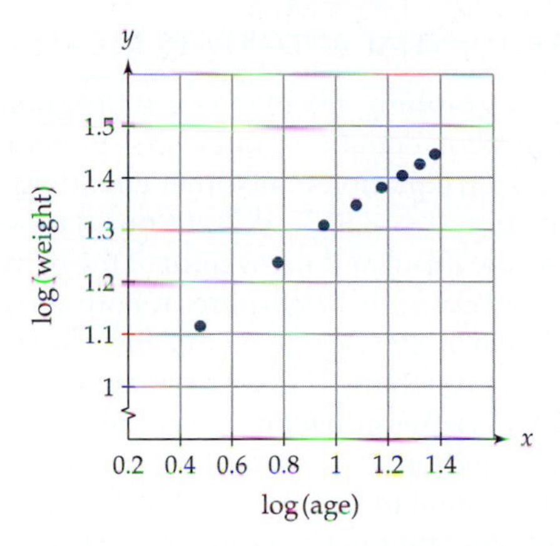

Figure 11

therefore, $a = 0.362$. Because the y-intercept of our linear function is $\log b$, $\log b = 0.950$, so $b = 10^{0.950}$ or $b \approx 8.913$. Therefore, the equation $y = 8.913 \cdot x^{0.362}$ should reasonably fit our original data from Table 7. These data along with the graph of $y = 8.913 \cdot x^{0.362}$ are shown in Figure 12.

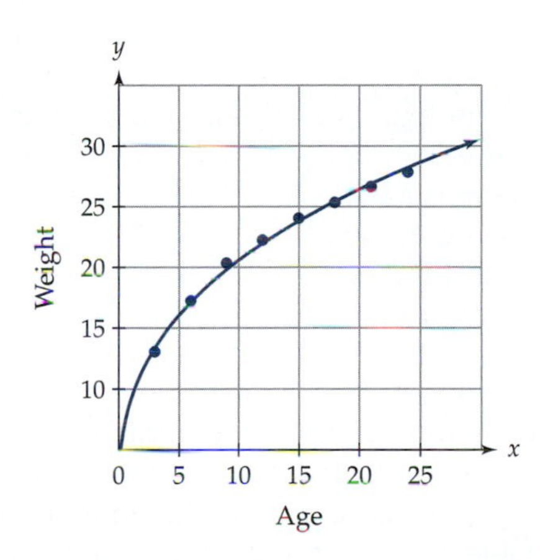

Figure 12

READING QUESTIONS

3. Show that the function $y = 4.5x^{7.3}$ can be transformed to a linear function by taking the logarithm of both sides of the equation.
   (a) What are the input and the output of this linear function?
   (b) What are the slope and the y-intercept of this linear function?
4. Suppose data that fit a power function have been transformed by taking the common logarithm of the input and the output. The linear regression line $y = 3.5x - 4.2$ has been obtained using the transformed data. What power function fits the original data?

EXPONENTIAL AND POWER REGRESSION

Most graphing calculators are programmed to determine not only linear regression equations, but also exponential regression equations, power regression equations, and other equations. In Example 2, we found the power function $y = 8.913 \cdot x^{0.362}$ fit our data where the age of a baby in months was the input and the weight of the baby in pounds was the output. We did so by taking the logarithm of both the input and the output, determining a linear regression line, and then transforming the linear function into a power function. This process can be done much more easily by using the power regression feature on a calculator. When using a TI-83, we obtained the equation $y = 8.916 \cdot x^{0.363}$ (rounding each number to three places after the decimal point). Notice that this result is very similar to our original result. We obtained a slightly different equation because of all the rounding we did when transforming our original data using logarithms. If we had not done as much rounding, our two equations would be the same.

Because power regression is the easier method, why go through the process of transforming power equations into linear equations to find the regression equation? We do it for a number of reasons. For one, it shows some nice applications for logarithms. It is also the way many people (particularly scientists) still use regression. Calculators also go through a similar process when determining exponential and power regression equations. Perhaps a more important reason for understanding the process of how power or exponential regression equations are obtained is to understand the difference between these types of regression and linear regression.

There is a big difference between exponential or power regression and linear regression. Remember that linear regression is based on minimizing the differences between the actual outcomes and the outcomes predicted by the regression equation by minimizing the sum of the squares of these differences, hence the name least squares regression. Exponential and power regression do not do so. Even though the sums of the squares of these differences are minimized when these functions are transformed to linear equations, it does not hold true when they are transformed back to exponential or power equations. To demonstrate this, let's use the data in Table 9.

TABLE 9 Data That Fit an Exponential Function Nicely

x	1	2	3	4	5
y	50	25	15	8	7

Using a calculator, the exponential regression function was used on the data in Table 9 and the equation $\hat{y} = 73.285 \cdot 0.602^x$ was obtained. (*Note:* The standard statistical notation $\hat{y}$, pronounced "y hat," is used to indicate the predicted outputs rather than the actual outputs, y.) The differences between the actual outputs and the predicted outputs (as well as the squares of these differences) are shown in Table 10. By looking at the graph on our calculators and adjusting the numbers on a trial and error basis, we also fitted an exponential equation to the data and obtained $\hat{y} = 90 \cdot 0.55^x$. The differences between the actual outputs and the predicted outputs (as well as the squares of these differences) are shown in Table 11 using this equation.

Which equation is a better fit? Using the regression equation, the sum of the squared differences is 42.10. Using our self-fitted equation, the sum of the squared differences is 11.38. Therefore, the self-fitted equation actually fits better. So, although linear regression does produce the line of best fit,

TABLE 10 Differences between the Actual Outputs and the Predicted Outputs Using the Exponential Regression Equation, $\hat{y} = 73.285 \cdot 0.602^x$

x	1	2	3	4	5
y	50	25	15	8	7
$\hat{y}$	44.12	26.56	15.99	9.63	5.79
$y - \hat{y}$	5.88	−1.56	−0.99	−1.63	1.21
$(y - \hat{y})^2$	34.57	2.43	0.98	2.66	1.46

TABLE 11 Differences between the Actual Outputs and the Predicted Outputs Using Our Self-Fitted Equation, $\hat{y} = 90 \cdot 0.55^x$

x	1	2	3	4	5
y	50	25	15	8	7
$\hat{y}$	49.5	27.23	14.97	8.24	4.53
$y - \hat{y}$	0.5	−2.23	0.03	−0.24	2.47
$(y - \hat{y})^2$	0.25	4.97	0.00	0.06	6.10

exponential and power regression do not necessarily produce exponential or power functions of best fit.

READING QUESTIONS

5. Using the data in the following table, two exponential equations were obtained. The first equation, $\hat{y} = 0.50 \cdot 2.658^x$, was obtained using exponential regression. The second equation, $\hat{y} = 0.57 \cdot 2.5^x$, was obtained by trial and error.

x	1	2	3	4
y	1	5	11	20

(a) By graphing the data along with each equation, determine which equation appears as though it is a better fit.

(b) By finding the sum of the squared differences, $(\hat{y} - y)^2$, for each equation, determine which equation is a better fit.

SUMMARY

Data that fit an exponential equation can be transformed into data that fit a linear equation by taking the common logarithm of the output. Linear regression can then be used on the transformed data, and a least squares regression line can be obtained. Using the slope and y-intercept of the linear regression line, an exponential function of the form $y = b \cdot a^x$ that fits the original data can be obtained by the following.

- $\log a = \text{slope}$ or $a = 10^{\text{slope}}$
- $\log b = y\text{-intercept}$ or $b = 10^{y\text{-intercept}}$

Data that fit a power equation can be transformed into data that fit a linear equation by taking the common logarithm of both the input and the output. Linear regression can then be used on the transformed data, and a least squares regression line can be obtained. By using the slope and y-intercept of the linear regression line, a power function of the form $y = b \cdot x^a$ that fits the original data can be obtained by the following.

- $a = \text{slope}$
- $\log b = y\text{-intercept}$ or $b = 10^{y\text{-intercept}}$

Although linear regression gives a line of best fit, exponential and power regression do not necessarily give an exponential or power function of best fit. You may be able to determine a better equation yourself. You can check by looking at the differences between the predicted outcomes and the actual outcomes.

EXERCISES

1. The following table gives outputs for $f(x) = 2^x$.

x	1	2	3	4	5
$f(x) = 2^x$	2	4	8	16	32
$g(x) = \log(2^x)$					
$h(x) = \ln(2^x)$					

(a) Complete the table by filling in the data for $g(x) = \log(2^x)$ and $h(x) = \ln(2^x)$.

(b) Explain, by using your data from the table, why the functions $g(x) = \log(2^x)$ and $h(x) = \ln(2^x)$ are linear functions.

(c) Explain, by using only the formulas $g(x) = \log(2^x)$ and $h(x) = \ln(2^x)$, why g and h are linear functions.

2. The following table gives outputs for $f(x) = x^2$.

x	1	2	3	4	5
$f(x) = x^2$	1	4	9	16	25
$\log x$					
$\log(x^2)$					

(a) Complete the table by filling in the data for $\log x$ and $\log(x^2)$.

(b) Explain, by using your data from the table, why the function where $\log x$ is the input and $\log(x^2)$ is the output is a linear function.

3. The following table contains three functions. One function is best fit with a linear equation, one with an exponential equation, and one with a power equation.

x	1	2	3	4	5	6
$f(x)$	30	33	35	37	38	39
$g(x)$	32	38	43	48	54	60
$h(x)$	24	28	33	39	49	60

(a) Make scatterplots of x versus $f(x)$, x versus $\log(f(x))$, and $\log x$ versus $\log(f(x))$. Which scatterplot looks the most linear? Which type of function will best fit the data for x versus $f(x)$? Find an equation that fits the data for f.

(b) Make scatterplots of x versus $g(x)$, x versus $\log(g(x))$, and $\log x$ versus $\log(g(x))$. Which scatterplot looks the most linear? Which type of function will best fit the data for x versus $g(x)$? Find an equation that fits the data for g.

(c) Make scatterplots of x versus $h(x)$, x versus $\log(h(x))$, and $\log x$ versus $\log(h(x))$. Which scatterplot looks the most linear? Which type of function will best fit the data for x versus $h(x)$? Find an equation that fits the data for h.

4. The number of farms and the number of farmers declined during much of the twentieth century. The following table shows the percent of the workforce in the United States employed on farms.[22]

Year	1940	1950	1960	1970	1980	1990
Percent of Workforce Employed on Farms	17.4	11.6	6.1	3.6	2.7	2.4

[22]The World Almanac and Book of Facts 1996, *p. 137.*

(a) Find the logarithm of the percent of workforce employed on farms for each year.

(b) Find the linear regression equation where the years since 1900 is the input and the logarithm of the percent of workforce employed on farms is the output.

(c) Transform your linear equation into an exponential equation where years since 1900 is the input and the percent of workforce employed on farms is the output. How well does your exponential equation fit the data?

(d) Using your exponential equation, determine the percent of the workforce employed on farms in 1900. The actual percent of the workforce employed on farms in 1900 was 37.5%. Is this number close to your estimate using the equation? Explain any discrepancy.

(e) Using your exponential equation, determine the percent of the workforce working on farms in 1890. Does this number make sense? Explain.

5. While CD sales increased greatly in the late 1980s and early 1990s, record sales decreased greatly. The following table gives the number of single record (45's) sales in millions.[23]

Year	1985	1987	1988	1989	1990	1991	1992	1993	1994
Single Record Sales (in millions)	120.7	82.0	65.6	36.6	27.6	22.0	19.8	15.1	11.7

(a) Find the logarithm of the record sales for each year given.

(b) Find the linear regression equation where the time since 1980 is the input and the logarithm of the record sales is the output.

(c) Transform your linear equation into an exponential equation where time since 1980 is the input and the record sales is the output. How well does your exponential equation fit the data?

6. Two hundred thumbtacks were tossed onto a table. The ones that landed point up were removed. The remaining were again tossed onto the table, and again the ones landing point up were removed. This process continued until all the thumbtacks were removed from the table. The results are shown in the following table.

Tosses	0	1	2	3	4	5	6	7	8
Number of Thumbtacks Remaining	200	76	28	12	10	1	1	1	0

[23]The World Almanac and Book of Facts 1996, *p. 258.*

(a) Why does it make sense that an exponential function would fit a situation like this?

(b) Why does the last point, $(8, 0)$, make it impossible to use exponential regression on the data?

(c) Eliminate the last point and find an exponential equation where the input is the toss number and the output is the number of thumbtacks remaining.

(d) How would you think your exponential function would change if instead of removing the thumbtacks that landed point up, the ones not landing point up were removed?

7. The following table gives the percent of the U.S. population that lives in rural areas, as defined by the Census Bureau, for 1900 through 1990.[24]

Year	1900	1910	1920	1930	1940	1950	1960	1970	1980	1990
Percent Rural	60.4	54.6	48.8	43.9	43.5	36.0	30.1	26.4	26.3	24.8

(a) Find a linear equation that fits the data where the number of years after 1900 is the input and the percent of the population that lives in rural areas is the output.

(b) Find an exponential equation that fits the data where the number of years after 1900 is the input and the percent of the population that lives in rural areas is the output.

(c) Which equation fits the data better? Explain.

(d) People are defined as living in either a rural area or an urban area according to the Census Bureau. Using your linear equation, find an equation where the input is years since 1900 and the output is the percent of the population that live in urban areas. Using your exponential equation, find an equation where the input is years after 1900 and the output is the percent of the population that live in urban areas.

8. Suppose you wanted to find both an exponential function and a power function to fit the following data.

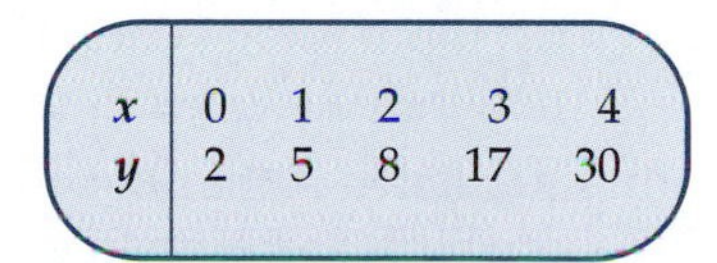

x	0	1	2	3	4
y	2	5	8	17	30

(a) Use the exponential regression and the power regression features on your calculator to get two functions to fit the data.

[24] *U.S. Census Bureau,* Selected Historical Census Data *(visited 1 March 1999)* ⟨*http://www.census.gov/population/www/censusdata/hiscendata.html*⟩.

(b) You should have gotten an error in finding one of the functions. Which function gave you an error, and why?

9. Variation in gasoline mileage among different types of cars is influenced by a number of factors. One factor is that of the horsepower of the engine. The following table gives the make and model of 10 different cars along with their horsepower and mileage based on city driving. All the cars are 1999 models with automatic transmissions.[25]

Make and Model	Horsepower	Miles per Gallon
Suzuki Swift	79	30
Hyundai Accent	92	27
Saturn SCI	100	27
Ford Escort	110	25
Toyota Celica	130	22
Chevrolet Lumina	160	20
Ford Mustang	190	20
Lexus ES300	200	19
Cadillac Seville	275	17
Jaguar XJ8	290	17

(a) For each car, find the common logarithm of the horsepower and the common logarithm of the miles per gallon.

(b) Find the linear regression equation that fits the logarithm of the horsepower as the input and the logarithm of the miles per gallon as the output.

(c) Transform your linear regression equation into a power equation with the horsepower as an input and the miles per gallon as the output.

10. In a book about the biology of birds, the equation

$$\log M = \log 89 + 0.64 \log W$$

is given. In this equation, W is the weight of a bird in kilograms and M is the bird's metabolic rate in kilocalories per day.[26]

(a) The given equation can be considered a linear equation where the input is $\log W$ and the output is $\log M$. Transform this equation into one where the input is W and the output is M.

(b) It is stated in the book that a twofold increase in the body weight of a bird is accompanied by less than a doubling of the metabolic rate. Using your transformed equation from part (a), explain why.

[25] Edmund's Automobile Buyers Guide, *(visisted 18 February 1999) ⟨http://www.edmunds.com⟩.*

[26] *A. J. Marshall,* Biology and Comparative Physiology of Birds, *Vol. II (New York: Academic Press, 1961), pp. 227–29.*

11. The following table gives the average weight and length for babies according to the National Center for Health Statistics.

Age (months)	3	6	9	12	15	18	21	24
Weight (lb)	13.0	17.2	20.3	22.2	24.0	25.3	26.6	27.8
Length (in.)	24.0	26.7	28.6	30.0	31.4	32.5	33.6	34.5

(a) Make graphs of both age versus weight and age versus length. (Both graphs should have age on the horizontal axis.) Explain why power functions should do a better job of fitting each of these plots than exponential functions.

(b) Find a power function that fits the data where age is the input and weight is the output.

(c) Find a power function that fits the data where age is the input and length is the output.

(d) Combine your two power equations in such a way that length is the input and weight is the output. Is this function also a power function?

12. A pendulum made with fishing line and a small metal weight was constructed. The length of the pendulum was changed and the number of periods per minute were counted for various lengths. The results are shown in the following table.

Length of String (in.)	1	3	5	10	15	20	25
Number of Periods (per minute)	165	101	80	57	47	41	37

(a) Decide if a power equation or an exponential equation better fits the data. Find the appropriate equation that fits the data where the input is the length of the string and the output is the number of periods per minute.

(b) Find the length of a pendulum (in inches) that would swing at a rate of one period per second.

13. Prison populations have increased in recent years. The following table gives the total number of sentenced prisoners in state and federal prisons from 1985 to 1995. It also gives the number of sentenced prisoners per 100,000 residents.[27]

[27] *U.S. Department of Justice, Office of Justice Programs, Bureau of Justice Statistics,* Correctional Populations in the United States, 1995 *(Washington, DC: GPO, June 1997).*

Year	Number of Sentenced Prisoners	Rate (per 100,000)
1985	480,568	202
1986	522,084	217
1987	560,812	231
1988	603,732	247
1989	680,907	276
1990	739,980	297
1991	789,610	313
1992	846,277	332
1993	932,074	359
1994	1,016,691	389
1995	1,085,363	411

(a) Find an exponential equation where the input is the years since 1985 and the output is the total number of prisoners. What is the growth factor in your equation?

(b) Find an exponential equation where the input is the years since 1985 and the output is the rate of prisoners per 100,000 residents. What is the growth factor in your equation?

(c) The growth factor for the rate per 100,000 residents should be lower than the growth factor for the total number of prisoners. Explain why.

14. The value of cars depreciate over time. A similar sort of "depreciation" can be seen when comparing current values of the same model of car for different model years. For example, the following table shows the suggested retail values for different years of the Chevrolet Camaro. The first entry in the table, $16,065, is for a 1-year-old Camaro.[28] (Because these data were collected in 1999, this information is for the 1998 model year.)

Age	Retail Price ($)
1	16,065
2	14,275
3	12,285
4	10,265
5	8,575
6	7,435
7	6,025
8	5,085
9	4,340
10	3,385
11	3,250
12	2,940

[28] *Autoweb.com (visited 23 February 1999) (http://www.autoweb.com).*

(a) Determine if a power function or an exponential function better models these data and find the appropriate equation that fits the data.

(b) Using the equation you determined in part (a), estimate the suggested retail price of a brand new Camaro. The suggested retail price for a new Camaro in 1999 was $20,870. Is this price close to your answer? If not, explain why you think there may be a difference.

15. The Fibonacci sequence begins with two 1s. To get the third term, the two 1's are added together to get 2. To get the fourth term, the second 1 and the 2 are added together to get 3, and so on. The first few terms of the Fibonacci sequence are given the following table.

Term	1	2	3	4	5	6	7	8	9	10
Fibonacci Number	1	1	2	3	5					

(a) Complete the table by finding the Fibonacci numbers for terms 6 through 10.

(b) Find an exponential function that fits the Fibonacci sequence where the term number is the input and the corresponding Fibonacci number is the output.

(c) As the Fibonacci numbers increase, the ratio of adjacent Fibonacci numbers approaches the golden ratio, $(1 + \sqrt{5})/2$. Your exponential equation should contain a number similar to this one. What number in your equation is similar to the golden ratio?

16. Two hundred pennies were tossed. If the penny landed tail side up, it was removed from the group. The remaining pennies were tossed, and again the tails were removed from the group. This process was repeated until all the pennies were removed. The following table gives the number of coins remaining after each toss.

Number of Tosses	0	1	2	3	4	5	6	7	8	9
Number of Coins Remaining	200	95	48	23	10	5	4	3	1	0

(a) If the probability of a coin landing tail side up is 0.5 when tossed and we start with 200 coins, then what exponential function should model the situation where the input is the number of tosses and the output is the number of coins remaining after all the tails are removed?

(b) Using regression, find the exponential function that fits the data given in the table.

(c) Which function fits the data better, the theoretical model from part (a) or the regression model from part (b)?

17. Twenty-five pennies were balanced so that they stood on their edges. The table on which they were balanced was jarred, and the pennies fell. If the penny landed tail side up, it was removed from the group. This process was repeated. The remaining pennies were again balanced on edge, the table jarred, and the tails were removed from the group. The following table gives the number of pennies remaining after each trial.

Number of Trials	0	1	2	3	4	5	6	7	8	9	10	11	12	13	14	15
Number of Coins Remaining	25	19	17	12	9	6	5	5	4	3	2	1	1	1	1	1

(a) Using regression, find the exponential function that fits the data given in the table.

(b) Based on your regression equation, what is the probability that a penny will land tail side up if it is bumped when balanced on edge?

(c) Find an exponential equation that fits the data better than your exponential regression equation.

18. Many movies are advertised heavily just before they are released into theaters, sending moviegoers to theaters soon after a movie's release. The number of people seeing the movie then slows down after a few weeks. One such movie is *Antz*, which opened on 2 October 1998. The following table shows the cumulative amount of money the movie earned (in millions of dollars) in the United States for the first eight weeks after it was released.[29] A plot of these data shows a function that is increasing and concave down. A power function does a reasonable job of fitting the data, but a logarithmic function of the form $y = b + m\log x$ does better. Note that this function is a linear function where $\log x$ is the input.

Time Since Opening (weeks)	1	2	3	4	5	6	7	8
Total Box Office (millions of dollars)	17.2	35.6	51.4	61.7	67.8	75.0	81.1	84.2

(a) Convert the numbers in the table so that the input is log(time since opening) and the output remains total box office. Make a scatterplot of the transformed data.

(b) Find a linear function that fits the transformed data.

[29] The Internet Movie Database *(visited 17 February 1999)* ⟨*http://us.imdb.com*⟩.

(c) Using your slope, m, and your y-intercept, b, from your linear equation, graph the original data along with the function $y = b + m \log x$. Does this equation fit the data?

(d) Find a function of the form $y = b + m \log x$ that fits the cumulative U.S. box office dollars for the movie *Saving Private Ryan*. The data for the first 10 weeks of this movie's release are given in the following table.

Time Since Opening (weeks)	Total Box Office (millions of dollars)
1	30.6
2	73.4
3	103.8
4	126.0
5	142.7
6	155.3
7	167.1
8	173.1
9	178.1
10	181.8

INVESTIGATIONS

INVESTIGATION: HOT WATER

When a cup of hot water is set on a kitchen counter, the difference between the temperature of the hot water and the air temperature can be modeled with an exponential function. This exponential decay is known as Newton's law of cooling. We performed a similar activity with a temperature probe placed in a cup of hot water. After the temperature probe adjusted to the temperature of the hot water, it was removed and allowed to cool to room temperature. Every 10 sec, for the first 100 sec after the probe was removed from the water, the temperature was recorded. These temperatures, in degrees Celsius, are shown in Table 12. The room temperature during this process was 20°C.

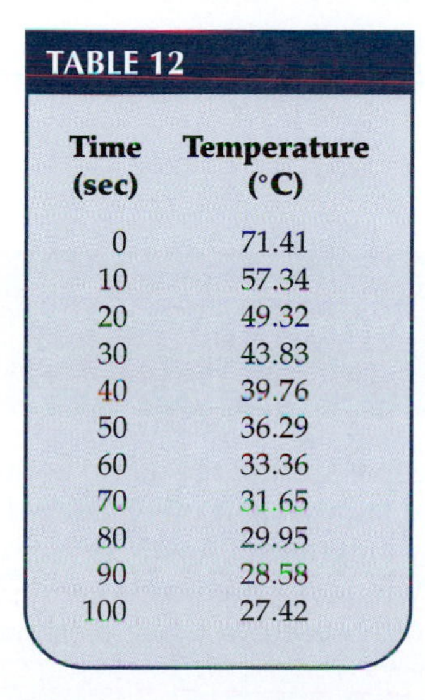
TABLE 12

Time (sec)	Temperature (°C)
0	71.41
10	57.34
20	49.32
30	43.83
40	39.76
50	36.29
60	33.36
70	31.65
80	29.95
90	28.58
100	27.42

1. Let's first look at why an exponential function will not be a good fit for the data given in Table 12.
   (a) If the time of this experiment were extended indefinitely, what would the temperature of the probe eventually become?
   (b) Graph the data in the table. Explain what attributes of your plot make it look like an exponential function. Also explain what attributes of the data (and the data if this experiment were extended indefinitely) make it so that an exponential function would not be a good fit.
   (c) Construct a graph where time is the input and the logarithm of the temperature is the output. Based on this graph, explain why

an exponential function will not be a good fit for the data from the table.

2. Although an exponential function will not be a good fit for our original data, it will be a good fit if we transform the data by subtracting the room temperature from our output.
   (a) Complete the following table.

Time	Temperature	Temperature − Room Temperature	log(temperature − room temperature)
0	71.41		
10	57.34		
20	49.32		
30	43.83		
40	39.76		
50	36.29		
60	33.36		
70	31.65		
80	29.95		
90	28.58		
100	27.42		

   (b) Graph the time versus log(temperature − room temperature) data. Using this graph, explain why the data where time is the input and (temperature − room temperature) is the output would fit an exponential function.
   (c) Find a linear equation to fit the time versus log(temperature − room temperature) data.
   (d) Using your linear equation, find an exponential equation that will fit the time versus (temperature − room temperature) data.
   (e) Determine the equation that can be used to model time versus temperature. Graph this function along with the original time versus temperature data.

PROJECTS 7.6 THE POPULATION PROBLEM

Population growth is a concern of many people around the world because of its impact on the environment, the food supply, and other limited resources. Although this growth, particularly long-term growth, is very difficult to predict, it is nonetheless necessary to make predictions and plan for the future. Mathematical models are often used to make reasonable predictions. These models can be quite complicated because population growth patterns are dependent on many variable factors such as birth rate, death rate, and the age distribution of the population. Relatively accurate information, however, can be generated from simple mathematical models, particularly when their use is restricted to a short time period. The goal of this project is to help you understand simple mathematical models for population growth and what a growth rate number (something you might see in a newspaper article) represents.

In this project, we simplify the situation by assuming that the world growth rate (birth rate − death rate) is constant within a calendar year.

1. A table produced by the U.S. Census Bureau gives the world population in 1996 as 5,771,938,438. It also states that the growth rate that year was 1.38% and that the increase in population in 1996 was 80,272,274.[30] We are interested in the relationship between the growth rate of 1.38% and the increase in population of 80,272,274. One possibility (perhaps the one that occurs to most people) is that the annual growth rate means that the world's population will increase by 1.38% of the existing population. What is 1.38% of 5,771,938,438? What is the difference between the answer you obtained and the increase given by the Census Bureau?
2. You should have found that your calculation from question 1 produced a significant underestimate. By calculating increase in this way, you are assuming that all the increase takes place at the same time. This ignores the fact that as the population grows, there are more women of child-bearing age at the end of a given year than at the beginning; hence, there are more children born at the end of that year.
   (a) Let's try again, assuming the population increases twice during the year. The semiannual growth rate is 0.0138/2, or 0.69%. We begin by finding a formula for the yearly increase in population.
      i. Write a formula for the total population (original population + the population increase) after only a half year of growth. Give your answer in factored form.
      ii. Building on your formula from part (i), write a formula for the total population after a full year of growth. Again, give your answer in factored form.
      iii. Subtract the original population to get a formula for the population increase.
   (b) Using the formula from part (a)(iii), determine the yearly increase in population if we assume it increases twice a year. What is the difference between the answer you obtained and the increase given by the Census Bureau?
3. We are getting closer, but we still have not discovered how to get the increase given by the Census Bureau. Extending the idea developed in question 2, let's look at a model where population increases monthly.
   (a) What is the monthly growth rate?

[30] *U.S. Bureau of Census, "Total Midyear Population for the World: 1950–2050,"* International Data Base *(visited 2 June 1997) ⟨http://www.census.gov/ftp/pub/ipc/www/worldpop.html⟩.*

(b) Write a formula similar to that in question 2, part (a)(iii) for the yearly increase in population, assuming population increases monthly.

(c) Using your formula from part (b), compute the yearly increase in population if we assume population increases monthly. Again, compare it with the increase given by the Census Bureau. What is the difference?

4. By compounding monthly, we obtain an answer closer to the increase given by the Census Bureau. Our answer is still not quite right because our population actually changes almost continuously. This continuous change, believe it or not, will make this problem a bit easier.

   (a) Consider the function $f(n) = (1 + 1/n)^n$, where n is an integer. Explore what happens when n gets large. Do so by completing the following table, rounding your answers to four decimal places.

n	$\left(1+\frac{1}{n}\right)^n$
1	2
10	
100	
1,000	
10,000	
100,000	
1,000,000	

Notice that as n gets larger, $(1 + 1/n)^n$ increases. The value of $f(n)$ does not increase without bound but instead gets closer to e, an irrational number *approximately* equal to 2.7182818. Mathematically speaking, we say that the limit, as n goes to infinity, of $(1 + 1/n)^n$ is e. We sometimes write limits as

$$\lim_{n\to\infty}\left(1+\frac{1}{n}\right)^n = e.$$

We are interested in this limit because it can be used in our population problem. To answer question 2, part (b) and question 3, part (c), you should have used a formula similar to

$$P_i = P_0\left(1+\frac{r}{n}\right)^{nt} - P_0,$$

where P_i is the population increase, P_0 is the initial population, r is the annual growth rate, t is the time (1 year in this case), and n is the number of times that the growth is compounded per year.

(b) Let's figure out what happens to this formula as n, the number of times the population increases each year, goes towards infinity. We want to show that $\lim_{n\to\infty} P_0(1 + r/n)^{nt} = P_0e^{rt}$. Let $k = n/r$ and give reasons for both of the steps in the argument below.

$$P_0\left(1+\frac{r}{n}\right)^{nt} = P_0\left(1+\frac{1}{k}\right)^{krt}$$
$$= P_0\left[\left(1+\frac{1}{k}\right)^{k}\right]^{rt}$$

As n increases to infinity, k does also because $k = n/r$ and r is fixed. Therefore,

$$\lim_{n\to\infty} P_0\left(1+\frac{r}{n}\right)^{nt} = \lim_{k\to\infty} P_0\left[\left(1+\frac{1}{k}\right)^{k}\right]^{rt} = P_0e^{rt}.$$

So, when we assume our population grows continuously, the formula becomes $P_i = P_0e^{rt} - P_0$.

(c) Use the formula for continuous population growth to compute the population increase and compare it with that given in the census data. Compute the difference.

(d) Note that your answer still does not exactly match the census data even though it is significantly better than when we first started. The reason is because the growth rate number has been rounded to the nearest hundredth of a percent. In some cases, even a little rounding can make a big difference. Because 1.38% is rounded to the nearest hundredth of a percent, the yearly growth rate given could actually be anywhere from 1.375% to 1.385%. Using 1.375% and the final formula from part (b), determine the effect of rounding on the accuracy of our approximation of the population increase.[31] (*Hint:* To do so, subtract the value you get for P_i using 1.375% from the value you got using 1.38%.) Is your answer from part (c) within this range?

5. Now that we have seen a simple mathematical model used to find short-term increases in population, let's try to gain a better understanding of the effects of growth rate on population. To do so, we look at another way to describe population growth, the time it takes the population to double.

   (a) If the world population grows continuously at an annual rate of 1.38%, then how long will it take to double in size?

   (b) Write a function where growth rate is the input and the time needed to double in size (known as the doubling time) is the output.

6. Instead of looking at world population, we now compare population growth rates from three different countries. In doing so, we focus not on the increase in population, but on the size of the total population after a

[31] *We chose the smaller number because our result was an underestimate.*

given period. A document produced by the United Nations Population Fund gives an annual growth rate of approximately 1% for Ireland, a growth rate of approximately 2% for Mexico, and a growth rate of approximately 3% for Ethiopia from 1990 to 1995. (For this period, the United States had a growth rate of 0.71%.)[32] For comparison purposes, assume each of these countries had an initial population of 1,000,000.

(a) Assuming the growth rates of Ireland, Mexico, and Ethiopia stay constant at the rates mentioned above, complete the following table.[33]

Country	Growth Rate	Population after 10 Years	Population after 100 Years	Time to Double
Ireland	1%			
Mexico	2%			
Ethiopia	3%			

(b) Graph the population for these three countries, on the same set of axes, with time on the horizontal axis and population on the vertical axis. Use a domain of 0 to 10 years.

(c) Graph the population for these three countries with a domain of 0 to 100 years.

(d) Describe the differences and similarities between the graphs in the 10-year window and those in the 100-year window. How are the shapes of the graphs in these two windows related to the resulting differences in populations of the three countries after 10 years and after 100 years?

(e) The populations of each of these countries in 1990 was not 1,000,000, but was actually much larger. In 1990, Mexico had a population of about 90,000,000 and Ethiopia had a population of about 50,000,000. Determine the year that Ethiopia's population exceeds that of Mexico if the growth rates of these two countries stay constant. Find the answer both graphically (include the graph as part of your answer) and symbolically.

7. If the growth rate of one country is double that of another country, describe what "double" means. Does it mean that the population of the second country is always double that of the first? Does it mean that the increase in population of the second is always double that of the first? Exactly what is doubled? Use examples from previous questions or make up other examples to support your conclusions.

[32] *United Nations Population Fund,* Population and the Environment: The Challenges Ahead *(London: Banson Productions, 1991), pp. 39–43.*

[33] *In reality, population growth rates do not stay constant, but for simplification, we assume they do.*

REVIEW EXERCISES

1. Solve the following equations. Give exact solutions.
   (a) $4x^2 = 33$
   (b) $x^{2/3} = 5$
   (c) $2x^2 + 3x = 10$
   (d) $6x^3 + 15x^2 = -6x$
   (e) $\dfrac{3x^2 + 5x - 8}{x + 2} = 22$
   (f) $3x - 12 - \dfrac{16}{x} = 10$
2. Solve the following inequalities. Give exact solutions.
   (a) $2x^{3/5} \geq 10$
   (b) $\sqrt{x - 4} < 4$
   (c) $2x^3 + 3x^2 > 9x$
3. Solve each of the following formulas for the indicated variable.
   (a) Solve $f = (1/2L)\sqrt{T/\rho}$ for L.
   (b) Solve $f = (1/2L)\sqrt{T/\rho}$ for T.
   (c) Solve $s = \sqrt{r^2 + h^2}$ for h.
   (d) Solve $(x^2/a^2) + (y^2/b^2) = 1$ for y.
4. Solve the following equations. Give exact answers.
   (a) $10^x = 35$
   (b) $3 \cdot 10^x = 300$
   (c) $5 \cdot e^x = 75$
5. Solve the following equations. Round answers to three decimal places.
   (a) $1.06^x = 2$
   (b) $3^{5x} + 10 = 900$
   (c) $e^{6x} = 38$
   (d) $5^{(x+5)} = 75$
   (e) $3^{2x+3} = 4^{x+2}$
6. Find the equation of an exponential function that passes through each of the following pairs of points.
   (a) $(0, 2)$ and $(1, 10)$
   (b) $(0, 8)$ and $(4, 40.5)$
   (c) $(2, 4)$ and $(4, \frac{16}{9})$
7. The element radium decays into radon exponentially. If you have 1 million atoms of radium, half of those will change into radon in about 1600 years. Thus, 1600 years is known as the half-life for radium. Half-lives continue exponentially. Therefore, after 3200 years, 25% of the original 1 million atoms of radium will remain.
   (a) If radium decays continuously, then what is the continuous annual decay rate?

(b) What portion of a quantity of radium will be remaining after 500 years? What portion will remain after 800 years?

(c) How long will it take until 30% of the initial quantity of radium remains?

8. In 1995, the suggested retail price for a new Saturn SCI Coupe was $11,895. The suggested retail price for that same car 5 years later was $4800.[34] Assume the value of the car depreciated exponentially in those 5 years.

   (a) Find an exponential function where the input is years since 1995 and the output is the retail price for a 1995 Saturn SCI Coupe.

   (b) Assuming the car depreciates in value into the future, find the value of the car in 2005.

9. In 1990, the United Nations estimated the population of Switzerland to be 6,609,000 and growing at a continuous annual growth rate of 0.22%. It also estimated the population of Austria to be 7,583,000 and growing at a continuous annual growth rate of 0.05%.

   (a) If the populations of these two countries continue to grow at their respective rates, how can you tell that Switzerland will eventually surpass Austria in population?

   (b) If the populations of these two countries continue to grow at their respective rates, how many years will it take for the two populations to be the same?

10. Suppose you have two bank accounts. One has a 4% annual interest rate compounded monthly, and the other has a 3.9% annual interest rate compounded continuously. Which bank account gives you the most return for your money?

11. Find all the solutions to each of the following.

    (a) $\sqrt{x} = 2x - 1$

    (b) $\sqrt{x+2} = -4$

    (c) $\dfrac{x+1}{x} = \dfrac{1}{x}$

12. Find all the exact solutions to the following equations.

    (a) $\cos x = \dfrac{\sqrt{3}}{2}$

    (b) $\cos x = 0$

    (c) $\sin^2 x = 1$

13. Find all solutions for x between 0 and 2π. Round your answers to three decimal places.

    (a) $10 \sin x > 1$

    (b) $2 \cos x < \frac{1}{2}$

    (c) $\tan x < 3$

14. Find all solutions for x between 0 and 2. Round your answers to three decimal places.

[34] Auto World *(visited 12 January 2000).*

(a) $-3\tan(2x) = 2$

(b) $-2\cos\left(\frac{x}{2}\right) = 1$

(c) $4\sin[2\pi(x-1)] = 1$

(d) $2\tan[\pi(x+2)] = 5$

15. Find all solutions for x between 0 and 2π.

(a) $(\cos x \cdot \sin x)^2 = 0$

(b) $4\sin x\cos x + 1 = 2$

16. A function that models the number of minutes of daylight for Seattle, Washington, is

$$d(t) = 228\sin\left[\frac{2\pi}{365}(t-81)\right] + 733,$$

where t is the number of the day of the year (i.e., January $1 = 1$, January $2 = 2, \ldots$, December $31 = 365$).

(a) Find two days when there are approximately 900 min of daylight.

(b) What days have fewer than 600 min of daylight?

17. Automobiles decrease in value as time goes on. The following table gives the age of a Saturn SL Sedan along with its retail value.[35]

Age	0	1	2	3	4	5	6	7	8	9
Retail Value	$12,085	$11,480	$10,680	$9600	$8300	$6475	$5590	$4625	$3985	$3370

(a) Determine a linear regression equation for the data where age is the input and retail value of the Saturn is the output.

(b) What does the slope of your regression line mean in terms of age and retail value? Be specific.

(c) What does the y-intercept of your regression line mean in terms of age and retail value? Be specific.

(d) What is the correlation? What does this number tell you about the relationship between the age and the retail value?

18. The more a person drinks alcohol, the higher the level of blood alcohol concentration in that person's body. A number of other factors, however, also affect someone's blood alcohol concentration. One factor is a person's weight. The following table gives the approximate blood alcohol concentration for people who have three drinks given their weight.

[35] Kelly Blue Book *(visited 15 January 2000)* ⟨*http://www.kbb.com*⟩.

Weight (lb)	100	120	140	160	180	200	220	240
Blood Alcohol Concentration	0.113	0.094	0.080	0.070	0.063	0.056	0.051	0.047

(a) By taking the appropriate logarithms of the input and the output, determine whether an exponential regression equation or a power regression equation will have a better fit. Explain how you made your choice.

(b) Find the regression equation (exponential or power) that fits best.

19. The following table contains three functions. One function is best fit with a linear equation, one with an exponential equation, and one with a power equation.

x	1	2	3	4	5	6
$f(x)$	14	19	24	31	41	53
$g(x)$	16	21	27	32	38	43
$h(x)$	16	23	28	32	36	39

(a) Make scatterplots of x versus $f(x)$, x versus $\log(f(x))$, and $\log x$ versus $\log(f(x))$. Which scatterplot looks the most linear? Which type of function will best fit the data for x versus $f(x)$? Find an equation that fits the data for f.

(b) Make scatterplots of x versus $g(x)$, x versus $\log(g(x))$, and $\log x$ versus $\log(g(x))$. Which scatterplot looks the most linear? Which type of function will best fit the data for x versus $g(x)$? Find an equation that fits the data for g.

(c) Make scatterplots of x versus $h(x)$, x versus $\log(h(x))$, and $\log x$ versus $\log(h(x))$. Which scatterplot looks the most linear? Which type of function will best fit the data for x versus $h(x)$? Find an equation that fits the data for h.

20. Fifty regular six-sided dice were tossed. If a die lands so that it shows a 1, it was removed from the group. The remaining dice were tossed and again the 1s were removed from the group. This process was repeated until all the dice were removed. The following table gives the number of dice remaining after each toss.

Number of Tosses	0	1	2	3	4	5	6	7	8	9	10	11	12
Number of Dice Remaining	50	42	38	31	29	22	16	14	11	9	7	5	3
Number of Tosses	13	14	15	16	17	18	19	20	21	22	23	24	25
Number of Dice Remaining	2	2	2	2	2	2	2	2	2	2	1	1	0

(a) If the probability of a die not showing a one is $\frac{5}{6}$ when tossed and we start with 50 dice, then what exponential function should model the situation where the input is the number of tosses and the output is the number of dice remaining after all the 1s are removed?

(b) Using regression, find the exponential function that fits the data given in the table.

(c) Which function fits the data better, the theoretical model from part (a) or the regression model from part (b)?

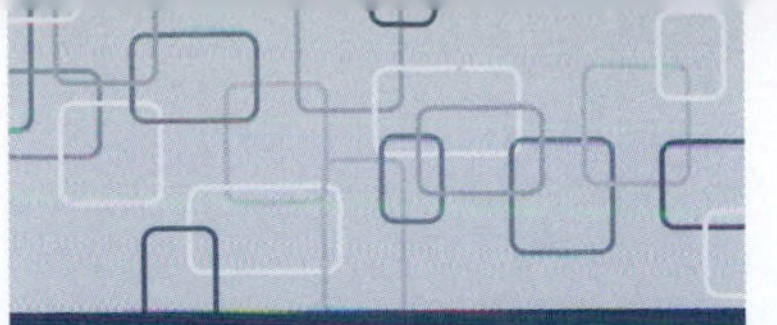

CHAPTER OUTLINE

CHAPTER OVERVIEW

- An informal definition of the limit
- One-sided limits and vertical asymptotes
- The number e
- Continuity
- The secant line
- The derivative
- Sequences and series
- The area under a curve
- Riemann sums and the integral

GETTING READY FOR CALCULUS

The focus of this chapter is introductory calculus concepts. We begin with limits and then move on to derivatives and integrals. These topics prepare you for your study of calculus but serve as introductions only.

8.1 LIMITS

In this chapter, we turn our attention to some topics that help you prepare for the study of calculus. An important topic in calculus is that of a limit, a concept in mathematics that allows us to consider the behavior of a function *near* a value instead of *at* a value. In this section, we informally define a limit, consider one-sided limits, see how we have used limits in past sections of this book, and look at how continuity is defined.

AN INFORMAL DEFINITION OF A LIMIT

In many parts of this book, especially in section 4.1, we looked at quadratic functions that gave the height of an object tossed up in the air. In doing so, we gave the formula

$$h(t) = -16t^2 + v_0 t + h_0,$$

where $h(t)$ is the height of the object in feet after t seconds, v_0 is the initial velocity, and h_0 is the initial height of the object. Let's take a closer look at the initial velocity.

Suppose a small toy rocket was launched from the ground and had an initial velocity of 100 ft/sec. The equation for the height of the rocket is then

$$h(t) = -16t^2 + 100t.$$

Let's see if this equation does in fact give an initial velocity of 100 ft/sec. To do so, remember that average velocity is change in position (or height) divided by the change in time. Then, we can find the average velocity for any period of time we want. Let's find the average velocity for the first second the rocket is in the air. To do so, we need to find the change in height, $h(1) - h(0)$, and divide that number by the change in time, 1 sec. We get

$$\begin{aligned}\frac{h(1) - h(0)}{1} &= \frac{(-16 \cdot 1^2 + 100 \cdot 1) - (-16 \cdot 0^2 + 100 \cdot 0)}{1} \\ &= 84 - 0 \\ &= 84.\end{aligned}$$

According to the equation, the average velocity of the rocket for the first second is 84 ft/sec. The initial velocity, however, is the velocity of the rocket at the moment it is launched, not for the first second. We can approximate the initial velocity by making the time interval used to find the average velocity smaller and smaller. This approximation is shown in Table 1.

TABLE 1 As the Time Interval Gets Smaller, the Average Velocity Approaches 100

Time Interval	Average Velocity
$0 \leq t \leq 1$	84
$0 \leq t \leq 0.1$	98.4
$0 \leq t \leq 0.01$	99.84
$0 \leq t \leq 0.001$	99.984
$0 \leq t \leq 0.0001$	99.9984

From Table 1, we can see that as the time interval gets smaller, the average velocity gets closer to 100. Why not simply let the time interval equal zero to find the initial velocity? We cannot do that because the denominator of our fraction would be zero. What we can say, though, is that the average velocity approaches 100 as the time interval approaches zero. This concept is the idea of a limit.[1]

The initial velocity can also be described as the *instantaneous* velocity when $t = 0$. Instantaneous velocity (or speed) is what you observe when you are looking at your speedometer in a car. We use this same idea to find the instantaneous velocity of the rocket at any time during its flight.

Example 1 Suppose a small toy rocket is launched from the ground and has an initial velocity of 100 ft/sec. The equation for the height of the rocket is then

$$h(t) = -16t^2 + 100t.$$

Find the instantaneous velocity for the rocket 2 sec after it is launched.

Solution We find the average velocity for time intervals near 2 sec. Let's find the average velocity for $2 \leq t \leq 2.1$. To do so, we need to find the change in height, $h(2.1) - h(2)$, and divide that by the change in time, 0.1 sec. Doing so gives

$$\begin{aligned} \frac{h(2.1) - h(2)}{0.1} &= \frac{(-16(2.1^2) + 100(2.1)) - (-16(2^2) + 100(2))}{0.1} \\ &= \frac{139.44 - 136}{0.1} \end{aligned}$$

[1] *To be more exact, we are simply looking at this limit as the time is approaching zero from the positive side. We look at one-sided limits like this one later in this section.*

$$= \frac{3.44}{0.1}$$
$$= 34.4.$$

Thus, the average velocity of the rocket for $2 \le t \le 2.1$ is 34.4 ft/sec. We have progressively reduced this interval in Table 2 to get an idea of the instantaneous velocity of the rocket after 2 sec.

TABLE 2

Time Interval	Average Velocity
$2 \le t \le 2.01$	$\frac{h(2.01) - h(2)}{2.01 - 2} = \frac{136.3584 - 136}{0.01} = 35.84$
$2 \le t \le 2.001$	$\frac{h(2.001) - h(2)}{2.001 - 2} = \frac{136.035984 - 136}{0.001} = 35.984$
$2 \le t \le 2.0001$	$\frac{h(2.0001) - h(2)}{2.0001 - 2} = \frac{136.00359984 - 136}{0.0001} = 35.9984$
$2 \le t \le 2.00001$	$\frac{h(2.00001) - h(2)}{2.00001 - 2} = \frac{136.0003599984 - 136}{0.00001} = 35.99984$

From Table 2, we can see that as the time interval gets smaller, the average velocity gets closer to 36 ft/sec.[2] Therefore, it appears that the instantaneous velocity of the rocket at 2 seconds is 36 ft/sec.

In finding the average velocity using a time interval starting at 2 and by defining this function as v, we can express our average velocity as

$$\begin{aligned} v(x) &= \frac{h(x) - h(2)}{x - 2} \\ &= \frac{(-16x^2 + 100x) - (136)}{x - 2} \\ &= \frac{-16x^2 + 100x - 136}{x - 2} \\ &= \frac{(x - 2)(-16x + 68)}{x - 2}. \end{aligned}$$

After writing the function this way, it is easy to see that we cannot simply substitute 2 for x to find the initial velocity. We can, however, say that $v(x)$ approaches 36 as x approaches 2. (Note that v is simply the linear function

[2] *Notice that we did not round the numbers we used in finding our approximations. Because we are dealing with such small intervals, rounding these numbers do not allow you to find a reasonable limit.*

$y = -16x + 68$ with a hole at $x = 2$.) In chapter 4, we used arrows to represent the word *approaches*. Using this notation, we write

$$v(x) \to 36 \quad \text{as} \quad x \to 2.$$

We could also use limit notation to describe this situation. We write

$$\lim_{x \to 2} v(x) = 36,$$

or, in words, "the limit of $v(x)$, as x approaches 2, equals 36." In general, an informal definition of a limit is as follows.

DEFINITIONS

If we can make the values of $f(x)$ arbitrarily close to L by making x sufficiently close to a but not equal to a, then we say that

$$\lim_{x \to a} f(x) = L.$$

In words,

"The limit of $f(x)$, as x approaches a, equals L."

Notice that we are not saying that $f(a) = L$. We are also not saying that $f(a) \approx L$. Instead, we are saying that the *limit* of $f(x) = L$ as x approaches a. That wording may seem like nitpicking, but these distinctions are important when using these concepts. Actually, we are still not being as formal as we could, but for our purposes, this informal definition of limit suffices.

Example 2 For

$$f(x) = \frac{x^2 + 5x + 6}{x^2 - 2x - 8},$$

find

$$\lim_{x \to -2} f(x).$$

Solution If we try to find $f(-2)$, we see that the denominator equals zero and the function is not defined at that point. Let's make a table of values of $f(x)$ for points close to -2 to help us determine what the limit is. These values are shown in Table 3, where the outputs are rounded to four decimal places. From this table, we see that $\lim_{x \to -2} f(x)$ is probably somewhere between -0.1669 and -0.1665. To be more accurate, let's look at the function more carefully.

We can rewrite our function by factoring both the numerator and denominator:

$$f(x) = \frac{(x+3)(x+2)}{(x+2)(x-4)}.$$

TABLE 3

x	-2.1	-2.01	-2.001	-2.0	-1.999	-1.99	-1.9
$f(x)$	-0.1475	-0.1647	-0.1665	Undefined	-0.1669	-0.1686	-0.1864

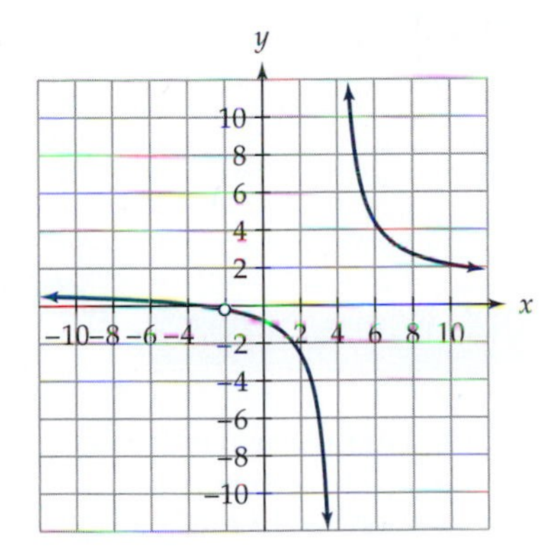

Figure 1

It is clear that f is not defined for $x = -2$ or $x = 4$. If we eliminate the $(x+2)$ factor in both the numerator and denominator, then we get the function

$$g(x) = \frac{x+3}{x-4}.$$

This function is the same as f except that it is defined when $x = -2$. Using this function, we can see that

$$g(-2) = \frac{-2+3}{-2-4} = -\frac{1}{6}.$$

Notice that $-\frac{1}{6}$ is between -0.1669 and -0.1665. In fact, $\lim_{x \to -2} f(x) = -\frac{1}{6}$. A graph of f is shown in Figure 1. From this figure, you can see that f has a hole at the point $(-2, -\frac{1}{6})$.

Notice again that in our previous example, we did not say that $f(-2) = -\frac{1}{6}$ or even that $f(-2) \approx -\frac{1}{6}$. Remember that $f(-2)$ is undefined. What the limit tells us is that $f(x)$ gets close to $-\frac{1}{6}$ as x gets close to -2. A function does not have to be undefined at a point for us to consider the limit at that point. Look at $\lim_{x \to 5} x^2 + 1$. In this case, the limit is simply $5^2 + 1 = 26$. We do not dwell on these types of limits, but instead look mostly at those limits where the function is not defined at the particular point.

Where a function is defined or not may not have any effect on the limit. The graphs of the functions shown in Figure 2 all have a limit of 2 as x approaches 0. The function in Figure 2(a) has an output of 2 when $x = 0$,

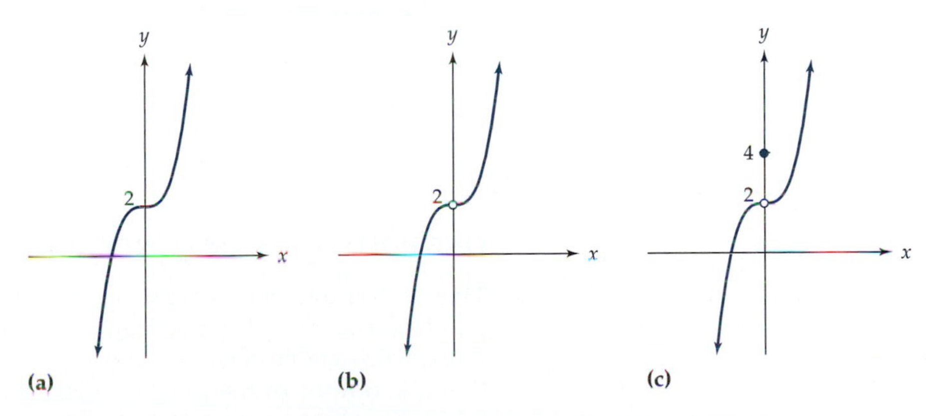

Figure 2 For each of these functions, the limit, as x approaches 0, equals 2.

the function in Figure 2(b) is not defined when $x = 0$, and the function in Figure 2(c) has an output of 4 when $x = 0$. For inputs close to 0, however, all these functions have outputs that are close to 2. For this reason, all these functions have a limit of 2 as x approaches 0.

READING QUESTIONS

1. How is instantaneous velocity different than average velocity?
2. How do you say in words that $\lim_{x\to 0} f(x) = 3$?
3. What is the value of

$$\lim_{x\to 2} \frac{(x+1)(x-2)}{(x-2)}?$$

HISTORICAL NOTE

The idea of the limit might seem trivial, but defining a limit turned out to be a major step in the development of calculus. When Isaac Newton first worked with this idea, he used the word *vanish*. His descriptions were full of ideas and language that were not as carefully defined as mathematicians require today, which made people suspicious that Newton was involved in some sort of mathematical slight of hand.

Newton used the word *fluxion* to define what we now call a derivative. Finding instantaneous velocity is similar to finding a derivative of a function at a point. Newton's description of a fluxion was also disputed. One of the biggest critics of Newton's ideas was Bishop George Berkeley, who took Newton to task in *The Analyst: a Discourse addressed to an Infidel Mathematician*. Here we read Berkeley's response to Newton's idea, one of Berkeley's most famous quotes: "And what are these fluxions? The velocities of evanescent increments. And what are these same evanescent increments? They are neither finite quantities nor quantities infinitely small, nor yet nothing. May we not call them the ghosts of departed quantities?"

These fluxions were often seen as little more than ghosts until around 1820, about 150 years after Newton, when Augustin-Louis Cauchy (1789–1846) gave a sufficiently rigorous definition of the limit that satisfied mathematicians. Newton was on the right track, but it took a long time for someone to make his ideas rigorous.

Source: Carl B. Boyer, *A History of Mathematics* (Princeton, NJ: Princeton University Press, 1968), p. 470.

ONE-SIDED LIMITS AND VERTICAL ASYMPTOTES

There are a number of ways you can find the limit of a function at a certain point. You can find it symbolically by factoring, you can look at a graph, or you can use a table of values. The important issue is that if $\lim_{x\to a} f(x) = L$, then the output of the function must approach L from *both directions* as $x \to a$. That is, the output of the function must approach L as x approaches a

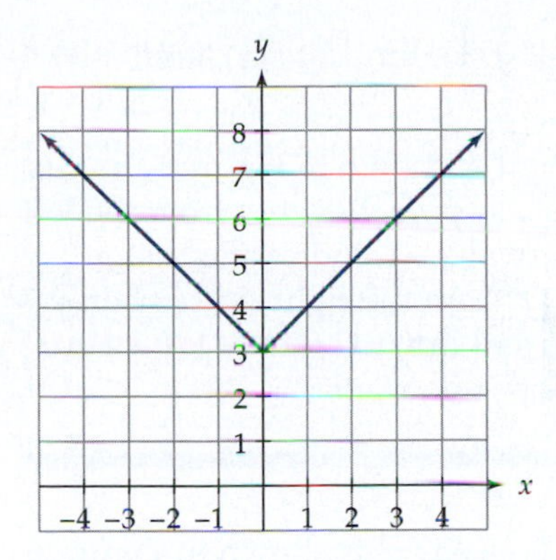

Figure 3 As x approaches 0 from both the left and the right, $f(x)$ approaches 3.

TABLE 4

x	$f(x)$
−0.1	3.1
−0.01	3.01
−0.001	3.001
0	Undefined
0.001	3.001
0.01	3.01
0.1	3.1

from both the left side and the right side. So far, we have looked at functions that approached a specific value at the points in question. Consider the piecewise function

$$f(x) = \begin{cases} -x+3, & \text{if } x < 0 \\ x+3, & \text{if } x > 0. \end{cases}$$

The graph of f is shown in Figure 3. Suppose we want to find $\lim_{x\to 0} f(x)$. In this case, as in previous cases, the function does not exist at the point in question because f is not defined when $x = 0$. As we look at the graph and at Table 4, however, we see that as x gets closer to zero from the right or when x is positive, $f(x)$ approaches 3. This is also true as x approaches zero from the left or when x is negative. Because the function behaves the same no matter from which direction we approach zero, we say that $\lim_{x\to 0} f(x) = 3$.

Based on the examples you have seen so far, you might be tempted to think that every function has a limit at every point, but that is not true. In the previous example, we specified that the behavior of this function had to be the same as it approached our target point from either side. If the behavior is different, then the limit does not exist. Look at the graph of

$$g(x) = \begin{cases} -x+2, & \text{if } x < 0 \\ x+3, & \text{if } x > 0, \end{cases}$$

which is shown in Figure 4. In this function, $\lim_{x\to 0} g(x)$ does not exist because from the right side $g(x)$ is approaching 3, but from the left side $g(x)$ is approaching 2. For a limit to exist, the two sides have to agree.

Even though both sides must equal the same thing for limits to exist, we refer to one-sided limits. For the function whose graph is shown in Figure 4, we could write $\lim_{x\to 0^+} g(x) = 3$ and $\lim_{x\to 0^-} g(x) = 2$ and say "the limit of $g(x)$, as x approaches 0 from the right, equals 3" and "the limit of $g(x)$, as x approaches 0 from the left, equals 2."

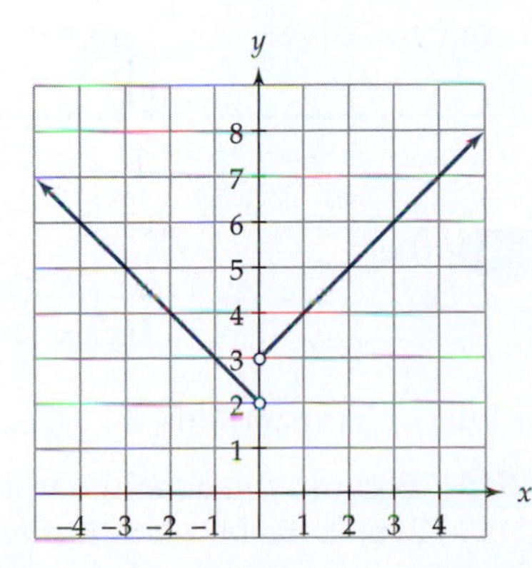

Figure 4 In

$$g(x) = \begin{cases} -x+2, & \text{if } x < 0 \\ x+3, & \text{if } x > 0, \end{cases}$$

$\lim_{x\to 0} g(x)$ does not exist.

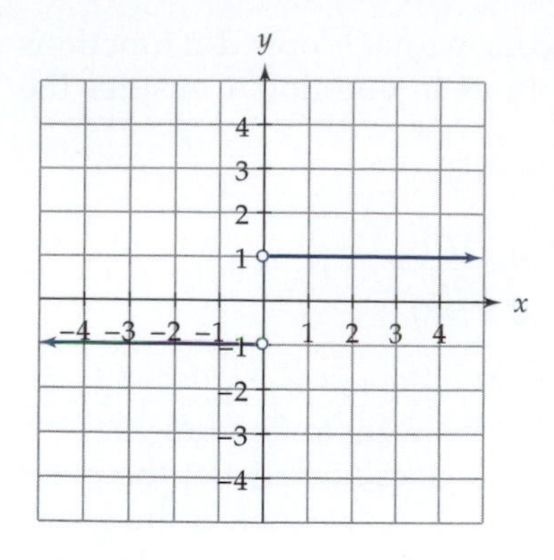

Figure 5

Example 3 Let $f(x) = |x|/x$. Find $\lim_{x\to0^-} f(x)$, $\lim_{x\to0^+} f(x)$, and $\lim_{x\to0} f(x)$ if they exist.

Solution Look at the graph of $f(x) = |x|/x$ shown in Figure 5. In this figure, if we consider the values of $f(x)$ as x approaches zero from the left side (that is, if x is negative), then $f(x)$ is always -1, indicating that $\lim_{x\to0^-} f(x) = -1$. If, however, we look at f from the right side (when x is positive), then $f(x)$ is always 1, indicating that $\lim_{x\to0^+} f(x) = 1$. Because $\lim_{x\to0^-} f(x) \neq \lim_{x\to0^+} f(x)$, $\lim_{x\to0} f(x)$ does not exist.

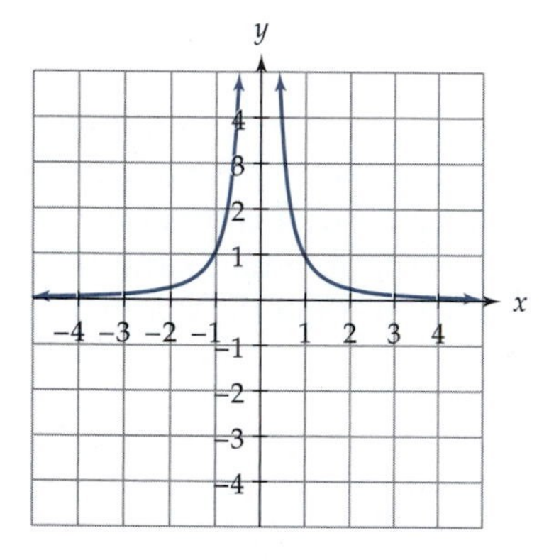

Figure 6 As x approaches 0 from both the left and the right, $f(x)$ approaches infinity.

Consider $\lim_{x\to0}(1/x^2)$. The graph of $f(x) = 1/x^2$ is shown in Figure 6. In this case, as in previous cases, the function does not exist at the point in question. As we look at the graph and at Table 5, however, we see that even though the function is not approaching a specific value, the function is behaving in the same way from either direction. That is, as we look at x getting closer to zero from the right or when x is positive, $f(x)$ approaches infinity. The same is true as x approaches zero from the left or when x is negative. Because the function does not approach a specific value, the limit does not exist. To indicate this type of behavior, however, we can still write $\lim_{x\to0}(1/x^2) = \infty$. This notation does not indicate that the limit exists, but it describes the specific way the limit does not exist. In particular, this notation describes a vertical asymptote.

TABLE 5

x	$f(x)$
−0.1	100
−0.01	10,000
−0.001	1,000,000
0	Undefined
0.001	1,000,000
0.01	10,000
0.1	100

In section 4.3, when we looked at power functions with negative exponents, we introduced the concept of a vertical asymptote. As you have just seen, this concept can be discussed using limits. We defined a vertical asymptote of a function, f, as the line $x = c$, where the outputs of f approach either positive or negative infinity as x approaches c. This asymptote can be redefined using the concept of limits. The definition using limits is more involved, but it is more precise.

DEFINITIONS

The line $x = c$ is a **vertical asymptote** of a function, f, if at least one of the following statements is true.

$$\lim_{x\to c} f(x) = \infty \qquad \lim_{x\to c^+} f(x) = \infty \qquad \lim_{x\to c^-} f(x) = \infty$$

$$\lim_{x\to c} f(x) = -\infty \qquad \lim_{x\to c^+} f(x) = -\infty \qquad \lim_{x\to c^-} f(x) = -\infty$$

Example 4 Let

$$f(x) = \frac{x+3}{x-2}.$$

Describe $\lim_{x\to2^-} f(x)$ and $\lim_{x\to2^+} f(x)$.

Solution Because the denominator of this rational function has a factor of $(x - 2)$ and the numerator does not, the function has a vertical

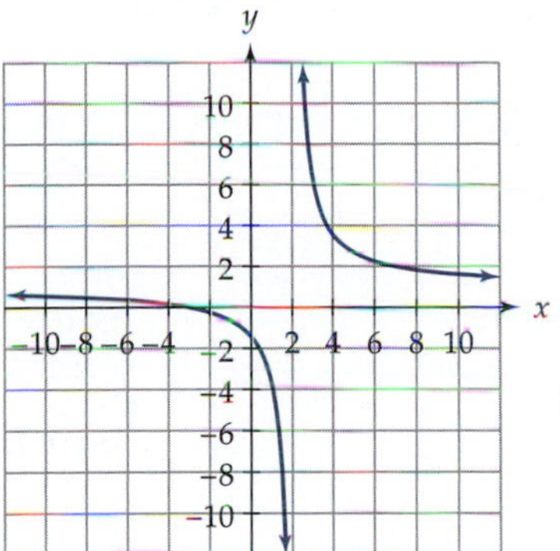

Figure 7

asymptote at $x = 2$. We can see by the graph of this function in Figure 7 that the function approaches negative infinity as x approaches 2 from the left (or $\lim_{x\to 2^-} f(x) = -\infty$) and positive infinity as x approaches 2 from the right (or $\lim_{x\to 2^+} f(x) = \infty$).

READING QUESTIONS

4. If $\lim_{x\to a^+} f(x) \neq \lim_{x\to a^-} f(x)$, what can we say about $\lim_{x\to a} f(x)$?
5. Suppose $\lim_{x\to a} f(x) = 4$. Does that mean that $f(a) = 4$?
6. Suppose a function, f, has the property $\lim_{x\to 2} f(x) = -\infty$. What does that tell us about f?

LONG-RANGE BEHAVIOR AND THE NUMBER e

In addition to introducing the concept of vertical asymptotes in section 4.3 when we looked at power functions with negative exponents, we also introduced the concept of horizontal asymptotes. These asymptotes can also be discussed using limits. We defined a horizontal asymptote of a function, f, as the line $y = c$, where the outputs of f approach c as x approaches either negative or positive infinity. This asymptote can be redefined using the concept of limits.

DEFINITIONS

The line $y = c$ is a **horizontal asymptote** of a function, f, if at least one of the following statements is true.

$$\lim_{x\to\infty} f(x) = c \qquad \lim_{x\to -\infty} f(x) = c$$

Example 5 Find

$$\lim_{x\to -\infty} \frac{x^2-4}{3x^2-2x+1}.$$

Solution As in section 4.4 on rational functions, the first step in determining the long-range behavior is to divide $x^2 - 4$ by $3x^2 - 2x + 1$.

$$\begin{array}{r} \frac{1}{3} \\ 3x^2-2x+1 \overline{) x^2 - 0x - 4} \\ \underline{x^2 - \frac{2}{3}x + \frac{1}{3}} \\ \frac{2}{3}x - \frac{13}{3} \end{array}$$

From this step we can see that

$$\frac{x^2-4}{3x^2-2x+1} = \frac{1}{3} + \frac{(2/3)x - 13/3}{3x^2-2x+1}.$$

As $x \to -\infty$,

$$\frac{(2/3)x - 13/3}{3x^2 - 2x + 1} \to 0.$$

Therefore,

$$\frac{1}{3} + \frac{(2/3)x - 13/3}{3x^2 - 2x + 1} \to \frac{1}{3}.$$

Thus,

$$\lim_{x \to -\infty} \frac{x^2 - 4}{3x^2 - 2x + 1} = \frac{1}{3}.$$

Notice in this case that it does not matter whether x is approaching positive or negative infinity because

$$\lim_{x \to \infty} \frac{x^2 - 4}{3x^2 - 2x + 1} = \frac{1}{3}$$

as well.

In section 7.2, we introduced the number e. This number is an irrational number approximately equal to 2.7183. The number is used when looking at continuous exponential growth. Even though we did not use the word *limit* when we described e, we actually described it in terms of a limit. We said

$$\left(1 + \frac{1}{k}\right)^k \to e \quad \text{as} \quad k \to \infty,$$

which can be rewritten as

$$\lim_{k \to \infty} \left(1 + \frac{1}{k}\right)^k = e.$$

The equation $y = e$ is actually a horizontal asymptote to the function $f(x) = (1 + 1/x)^x$.

By inputing larger and larger values for k, we see that

$$\lim_{k \to \infty} \left(1 + \frac{1}{k}\right)^k \approx 2.7183.$$

Table 6 shows this result where the outputs are rounded to four decimal places. As k increases, the value of $(1 + 1/k)^k$ approaches e. A calculator,

TABLE 6 $\lim_{k \to \infty} \left(1 + \frac{1}{k}\right)^k = e \approx 2.7183$

k	100	1000	10,000	100,000
$\left(1 + \frac{1}{k}\right)^k$	2.7048	2.7169	2.7181	2.7183

however, does not always show this result. As k increases, the value of $1/k$ approaches 0, which means that $(1 + 1/k)$ approaches 1. We are then raising a number close to 1 to a very large exponent. Because of rounding, a calculator will eventually recognize $(1+1/k)$ as 1, and when 1 is raised to a very large power, the result is 1. For example, let $k = 100{,}000{,}000{,}000{,}000$ in $(1+1/k)$ on a TI-83; the calculator gives a result of 1. Therefore, be careful; although much of the time you can input large numbers in a calculator to approximate a limit, at times the numbers get so large that rounding errors occur and your result is not correct. Such errors are not only true for this example, but for many others as well.

READING QUESTIONS

7. Is it always true that $\lim_{x\to\infty} f(x) = \lim_{x\to-\infty} f(x)$ for any function f?
8. Suppose a function, f, has the property $\lim_{x\to\infty} f(x) = -3$. What does that tell us about f?
9. Find

$$\lim_{x\to\infty} \frac{2x+3}{x-2}.$$

CONTINUITY

In past sections, we mentioned that some functions are "connected" like the one in Figure 8(a) and some have skips or jumps in them like the one in Figure 8(b). The idea that some functions have jumps or skips in them is not new. We have considered functions that have vertical asymptotes and piecewise functions since very early in this book. Here, though, we define this concept more carefully.

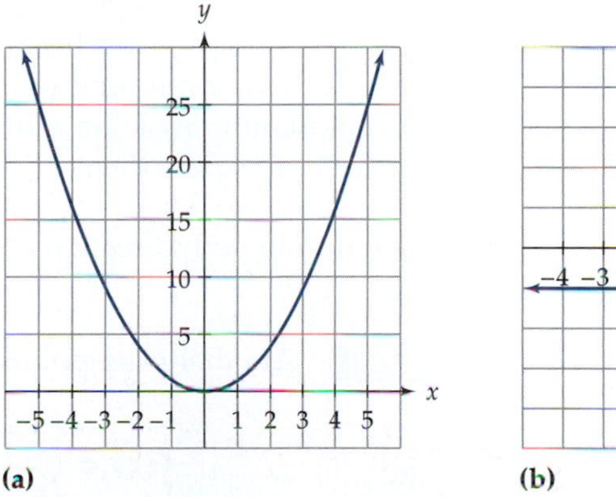

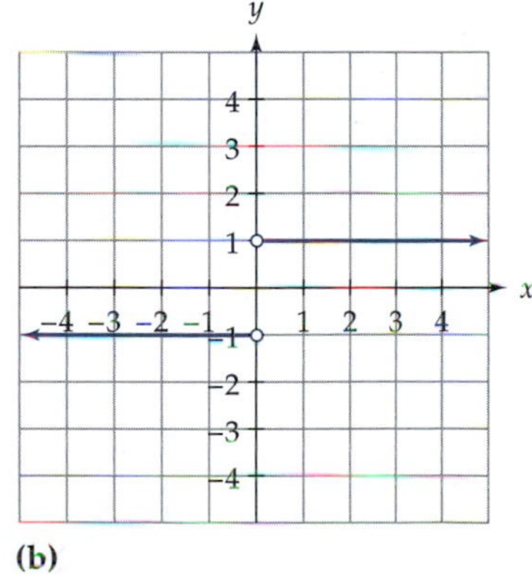

Figure 8 The graph in (a) is "connected," but the graph in (b) is not.

For most elementary functions, it is sufficient to define a *continuous* function as one that is "connected" or one that you can draw without lifting

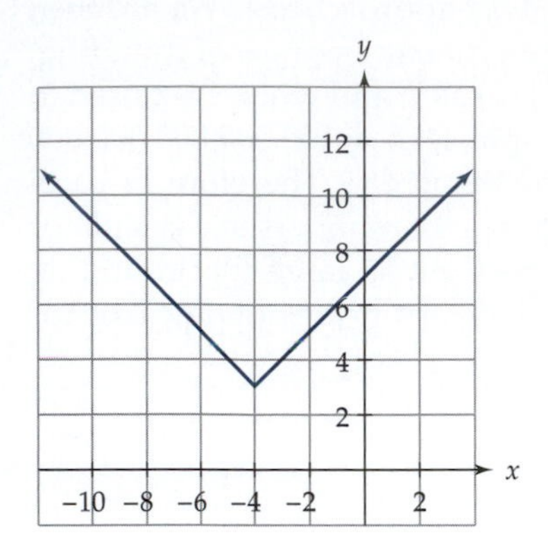

Figure 9 Is this function continuous at the point where $x = -4$?

your pencil from the page. With this informal definition, you see that many functions, such as linear, polynomial, or exponential functions, are all continuous. They all have no places where the function is undefined or where the function jumps from one value to another. In fact, continuity is not unusual. Most functions we have studied are continuous. With functions that were not continuous, we spent extra time wtih those points of discontinuity, such as where vertical asymptotes occur, because they can be the most interesting. Although some functions were obviously continuous, others were obviously not continuous. For example, rational functions with vertical asymptotes or holes are not continuous.

For some functions, though, such as $f(x) = |x+4|+3$, shown in Figure 9, continuity may not be quite as clear. The informal definition of a continuous function—having no skips or jumps and being able to sketch the graph without lifting your pencil—seems to work, but without a formal definition, we are less convinced. We need to be more precise about what it means for a function to be continuous.

DEFINITIONS

A function, f, is **continuous** at a point a if

$$\lim_{x\to a} f(x) = f(a).$$

A function is **continuous on an interval** if it is continuous at every point on the interval.

By using these definitions and looking back to the function $f(x) = |x+4|+3$ in Figure 8, we can determine if f is continuous. It is clear, and was clear even before we made our more careful definition, that the function is continuous everywhere, except perhaps when $x = -4$. At that point, the function abruptly went from decreasing to increasing. Yet $\lim_{x\to -4} |x+4| + 3 = 3$ and $f(-4) = 3$. Because $\lim_{x\to -4} f(x) = f(-4)$, the function is continuous at $x = -4$, and because it is clearly continuous everywhere else, it is therefore continuous over all real numbers.

From our definition of continuity, we see that continuous functions behave just as we expect them to. There are no surprises in a continuous function. Even the "kink" in $f(x) = |x+4|+3$ is not a surprise if you approach it from both directions, as you must do with limits.

Example 6 Are there any points of discontinuity in

$$f(x) = \begin{cases} x+4, & \text{if } x < 4 \\ x-3, & \text{if } x \geq 4? \end{cases}$$

Solution The only possible point of discontinuity is when $x = 4$, so let's look at that point. We know that $f(4) = 1$, but $\lim_{x\to 4^-} f(x) = 8$, whereas $\lim_{x\to 4^+} f(x) = 1$. Because the limit from the left does not equal the limit from the right, a limit does not exist here. Therefore, this function is not continuous at $x = 4$. By looking at a graph of this function, as shown in Figure 10, it is also clear that this function is not continuous.

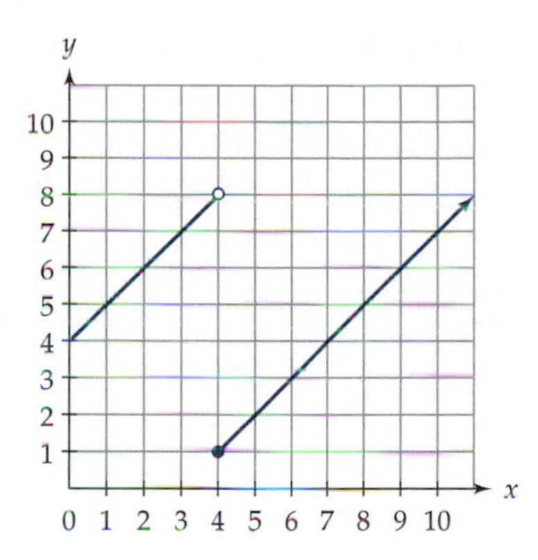

Figure 10

When a function is not continuous at a point, it happens in specific ways. The function in the previous example was not continuous at $x = 4$ because $\lim_{x\to 4} f(x)$ did not exist due to a "jump" in the function. A function, f, can also be not continuous at a specific point, a, if $f(a)$ does not exist due to a "skip" in a function, like a function with a hole. Another way for a function to be not continuous is if $f(a)$ exists but does not equal $\lim_{x\to a} f(x)$. Because there are specific ways that functions can be discontinuous, it can be relatively easy to find points of discontinuity. Usually, as you analyze a function, you look at those specific points anyway.

Because continuity can be defined over an interval, even functions that have points of discontinuity can be continuous over some interval. For example, $y = 1/x$ is continuous over the interval $1 \le x \le 10$. A similar situation occurs in the next example.

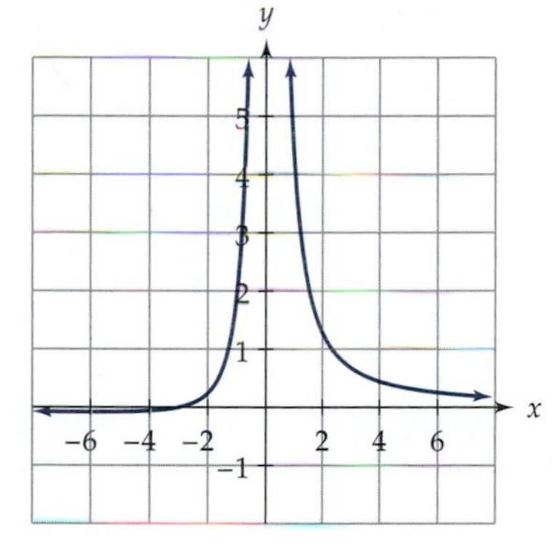

Figure 11

Example 7 Are there any points of discontinuity in $f(x) = (3 + x)/x^2$? Give an interval over which f is continuous.

Solution Because the function is not defined for $x = 0$, the function is not continuous over any interval containing that point. The function is continuous over any interval not containing that point—for example, over the interval $1 \le x \le 3$—so the function is continuous. The graph is shown in Figure 11.

READING QUESTIONS

10. List three types of functions that are always continuous over all real numbers.

11. List two types of functions that are often not continuous.

12. Find an interval over which $g(x) = (3x - 7)/(x + 4)$ is not continuous.

SUMMARY

In this section, we informally defined limit in the following manner. If we can make the values of $f(x)$ arbitrarily close to L by making x sufficiently close to a but not equal to a, then we say that $\lim_{x\to a} f(x) = L$. In words, "the limit of $f(x)$, as x approaches a, equals L."

We can sometimes factor rational functions and eliminate some factors from both the numerator and denominator to find the limit of functions. Other times we rely on graphs or tables.

For a limit to exist, it must exist in both directions. We can refer to one-sided limits, however. A function can have a limit as x approaches a certain number from the right and a different limit as x approaches the same number from the left. This type of function has different one-sided limits at that point and therefore does not have a limit defined at that point.

The line $x = c$ is a **vertical asymptote** of a function, f, if at least one of the following statements is true.

$$\lim_{x\to c} f(x) = \infty \qquad \lim_{x\to c^+} f(x) = \infty \qquad \lim_{x\to c^-} f(x) = \infty$$

$$\lim_{x\to c} f(x) = -\infty \qquad \lim_{x\to c^+} f(x) = -\infty \qquad \lim_{x\to c^-} f(x) = -\infty$$

The line $y = c$ is a **horizontal asymptote** of a function, f, if at least one of the following statements is true.

$$\lim_{x\to\infty} f(x) = c \qquad \lim_{x\to-\infty} f(x) = c$$

To find the long-range behavior of a rational function in which the degree of the numerator and denominator are the same, we use polynomial division to rewrite the function as a rational number plus a new rational function. The limit as the variable approaches infinity is equal to the rational number.

Finally, we defined continuity. A function, f, is **continuous** at a point a if $\lim_{x\to a} f(x) = f(a)$. A function is **continuous on an interval** if it is continuous at every point on the interval.

EXERCISES

1. Suppose a ball is thrown into the air so that its height can be determined by the equation $h(t) = -16t^2 + 50t$, where $h(t)$ is the height of the ball, in feet, after t seconds.
   (a) Find the average velocity of the ball between 0 and 1 sec.
   (b) Find the average velocity of the ball between 0.5 and 1 sec.
   (c) Find the average velocity of the ball between 0.9 and 1 sec.
   (d) What is the instantaneous velocity of the ball at 1 sec?
   (e) What is the initial velocity of the ball?
2. Suppose a baseball is hit straight into the air so that its height can be determined by the equation $h(t) = -16t^2 + 64t + 4$, where $h(t)$ is the height of the ball, in feet, after t seconds.
   (a) Find the average velocity of the ball between 1 and 2 sec.
   (b) Find the average velocity of the ball between 1.9 and 2 sec.
   (c) Find the average velocity of the ball between 1.99 and 2 sec.
   (d) What is the instantaneous velocity of the ball at 2 sec?
   (e) Is the ball moving upward or downward at 2 sec? Explain.
3. *Car and Driver* magazine road tested a Chevrolet Impala Police Package.[3] One of the tests performed is that of acceleration. The accompa-

[3] *Car and Driver Online,* Road Test: Chevrolet Impala Police Package *(visited 23 July 1999)* ⟨*http://www.caranddriver.com*⟩.

nying table gives the time needed to accelerate to the speeds shown. Speeds are given in both miles per hour and feet per second.

Time (sec)	**0**	**3.0**	**4.4**	**6.4**	**9.0**	**11.8**	**15.5**	**21.2**
Speed (mph)	0	30	40	50	60	70	80	90
Speed (ft/sec)	0	44	58.7	73.3	88	102.7	117.3	132

Average acceleration is change in speed divided by change in time. For parts (a) through (d), use the time and the speed in feet per second to find the acceleration for the car between the following times.

(a) 0 and 21.2 sec

(b) 4.4 and 21.2 sec

(c) 9.0 and 21.2 sec

(d) 15.5 and 21.2 sec

(e) Does the acceleration of the car seem to be increasing or decreasing as time passes?

4. *Car and Driver* magazine road tested a Jaguar S-type 4.0 V8.[4] One of the tests performed is that of acceleration. The accompanying table gives the time needed to accelerate to the speeds shown. Speeds are given in both miles per hour and feet per second.

Time (sec)	**0**	**2.5**	**3.7**	**5.2**	**7.0**	**9.6**	**11.9**	**15.1**	**19.2**	**24.0**	**30.1**	**40.3**
Speed (mph)	0	30	40	50	60	70	80	90	100	110	120	130
Speed (ft/sec)	0	44	58.7	73.3	88	102.7	117.3	132	146.7	161.3	176	190.7

Average acceleration is change in speed divided by change in time. For parts (a) through (e), use the time and the speed in feet per second to find the acceleration for the car between the following times.

(a) 0 and 40.3 sec

(b) 3.7 and 40.3 sec

(c) 9.6 and 40.3 sec

(d) 19.2 and 40.3 sec

(e) 30.1 and 40.3 sec

(f) Does the acceleration of the car seem to be increasing or decreasing as time passes?

[4] *Car and Driver Online,* Road Test: Jaguar S-type 4.0 V8 *(visited 23 July 1999)* (*http://www.caranddriver.com*).

5. Find limits for the following.
   (a) $\lim_{x \to 5} \dfrac{x^2 - 3x - 10}{x^2 - 2x - 15}$
   (b) $\lim_{x \to 0} \dfrac{x^2 + 4x}{x^2 - 6x}$
   (c) $\lim_{x \to -1} \dfrac{x^2 - 2x - 3}{x^2 + 6x + 5}$
6. Write a symbolic representation for functions that have the following properties.
   (a) $\lim_{x \to 2} f(x) \neq f(2)$
   (b) $\lim_{x \to 4} f(x)$ does not exist.
   (c) $\lim_{x \to 2} f(x) = 4$
7. Because of air resistance, skydivers reach a limit in their velocity when they jump out of a plane. This limit is called terminal velocity. Suppose the velocity of a skydiver can be determined using the equation $v(t) = 160(1 - e^{-0.1t})$, where $v(t)$ is the velocity of the skydiver, in feet per second, after t seconds. Find the terminal velocity of the skydiver.
8. Consider the function

$$f(x) = \frac{\ln(x+1)}{x}.$$

   (a) Using a graph of f, determine $\lim_{x \to 0} f(x)$.
   (b) Using numerical methods, determine $\lim_{x \to 0} f(x)$.
   (c) Using a calculator, find $[\ln(1 \times 10^{-15} + 1)]/(1 \times 10^{-15})$.
   (d) Are there any discrepancies in your answers to parts (a) through (c)? If so, explain why.
9. Sketch graphs of functions that have the following properties.
   (a) $\lim_{x \to 2^+} f(x) = 4$ and $\lim_{x \to 2^-} f(x) = 6$
   (b) $\lim_{x \to 2} f(x) = 4$ and $f(2) \neq 4$
   (c) $\lim_{x \to 2^+} f(x) = 4$, $\lim_{x \to 2^-} f(x) = 4$, and $f(2) = 4$
10. Find the following limits for the accompanying graph of f.

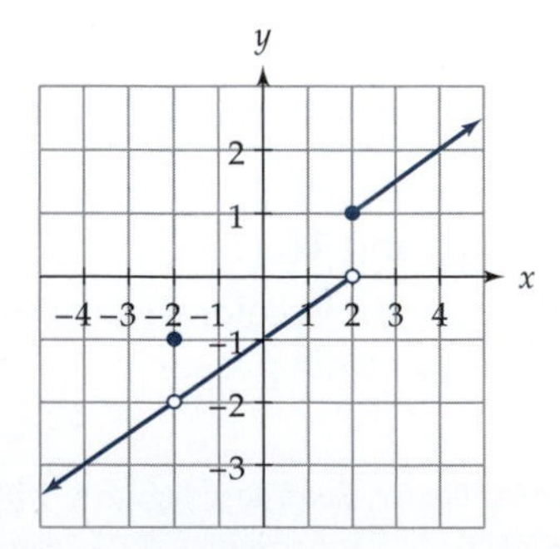

(a) $\lim_{x \to -2} f(x)$

(b) $\lim_{x \to 2^+} f(x)$

(c) $\lim_{x \to 2^-} f(x)$

11. Find the following limits for the accompanying graph of g.

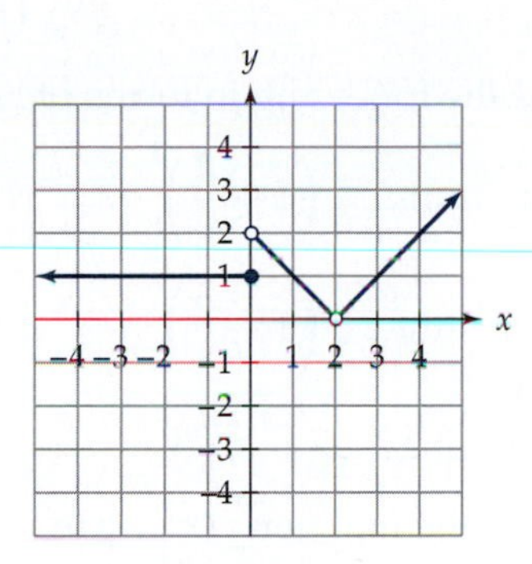

(a) $\lim_{x \to 0^-} g(x)$

(b) $\lim_{x \to 0^+} g(x)$

(c) $\lim_{x \to 2^-} g(x)$

(d) $\lim_{x \to 2^+} g(x)$

12. Let

$$g(x) = \begin{cases} x+4, & \text{if } x \leq 8 \\ x-2, & \text{if } x > 8. \end{cases}$$

(a) Find $\lim_{x \to 8^-} g(x)$.

(b) Find $\lim_{x \to 8^+} g(x)$.

(c) Find $\lim_{x \to 8} g(x)$.

13. Let

$$g(x) = \begin{cases} x+4, & \text{if } x \leq 2 \\ 3x-2, & \text{if } 2 < x \leq 6 \\ 16, & \text{if } x > 6. \end{cases}$$

(a) Find $\lim_{x \to 2^-} g(x)$.

(b) Find $\lim_{x \to 2^+} g(x)$.

(c) Find $\lim_{x \to 2} g(x)$.

(d) Find $\lim_{x \to 6^-} g(x)$.

(e) Find $\lim_{x \to 6^+} g(x)$.

(f) Find $\lim_{x \to 6} g(x)$.

14. Find limits for the following.

(a) $\lim_{x \to \infty} \dfrac{2x-3}{8x-15}$

(b) $\lim_{x\to\infty} \frac{6x^2+4x}{2x^2-6x}$

(c) $\lim_{x\to\infty} \frac{5}{1+6e^{-x}}$

15. In this section, it was mentioned that

$$\lim_{k\to\infty} \left(1+\frac{1}{k}\right)^k = e.$$

Find the following in terms of e.

(a) $\lim_{k\to\infty} \left(1+\frac{2}{k}\right)^k$

(b) $\lim_{k\to\infty} \left(1+\frac{3}{k}\right)^k$

(c) $\lim_{k\to\infty} \left(1+\frac{n}{k}\right)^k$

16. Given the function $f(x) = \sin x/x$, find the following.

(a) $\lim_{x\to 0} f(x)$

(b) $\lim_{x\to\infty} f(x)$

17. Given the function $g(x) = \tan x/x$, find the following.

(a) $\lim_{x\to 0} g(x)$

(b) $\lim_{x\to\infty} g(x)$

18. Given the function $h(x) = \cos x/x$, find the following.

(a) $\lim_{x\to 0} h(x)$

(b) $\lim_{x\to\infty} h(x)$

19. Consider the following two methods for determining income tax.

A. The tax rate is 15% of the first \$10,000 earned and 20% on the amount earned above \$10,000.

B. The tax rate is 15% if your income is \$10,000 or less and 20% if your income is above \$10,000.

(a) Which income tax method describes a continuous function, and which describes a function that is not continuous?

(b) Which tax method is less controversial for those earning close to \$10,000?

20. Are there any points of discontinuity in

$$f(x) = \frac{x^2+2x+1}{x+1}?$$

If so, what are they? Give an interval over which f is continuous.

21. Let

$$g(x) = \begin{cases} x+4, & \text{if } x \le 2 \\ 3x-2, & \text{if } 2 < x \le 6 \\ 16, & \text{if } x > 6. \end{cases}$$

Are there any points of discontinuity in g? If so, what are they? Give an interval over which g is continuous.

INVESTIGATIONS

INVESTIGATION 1: THE SQUEEZE THEOREM

A graph of $g(x) = x/\sin x$, shown in Figure 12, indicates that

$$\lim_{x\to 0} \frac{x}{\sin x} = 1.$$

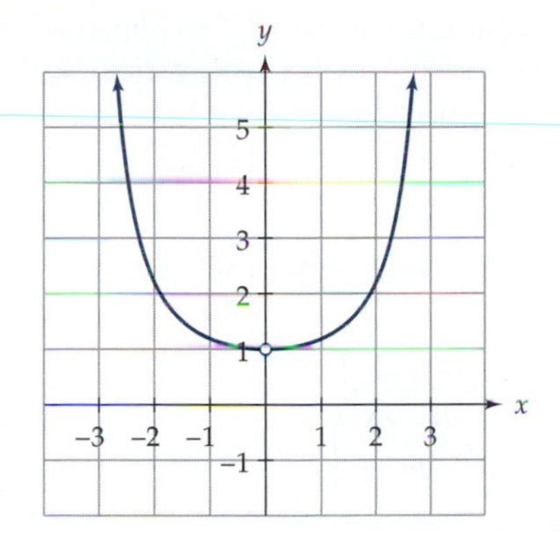

Figure 12

In this investigation, we want to prove symbolically that $\lim_{x\to 0} x/\sin x = 1$. We cannot find this limit directly, but we can find two functions—one whose output is less than that of $g(x)$ near 0 and one whose output is greater than that of $g(x)$ near 0—such that both have limits of 1 when $x \to 0$. Therefore, you can think of $g(x)$ being "squeezed" between $f(x)$ and $h(x)$, and because both $\lim_{x\to 0} f(x) = 1$ and $\lim_{x\to 0} h(x) = 1$, $\lim_{x\to 0} g(x)$ must also equal 1. This idea uses what is called the *squeeze theorem*.

The squeeze theorem states if

$$f(x) \le g(x) \le h(x)$$

when x is near a and

$$\lim_{x\to a} f(x) = \lim_{x\to a} h(x) = L,$$

then

$$\lim_{x\to a} g(x) = L.$$

We begin this proof geometrically with Figure 13. In this figure, an angle with measure x is shown in a unit circle.

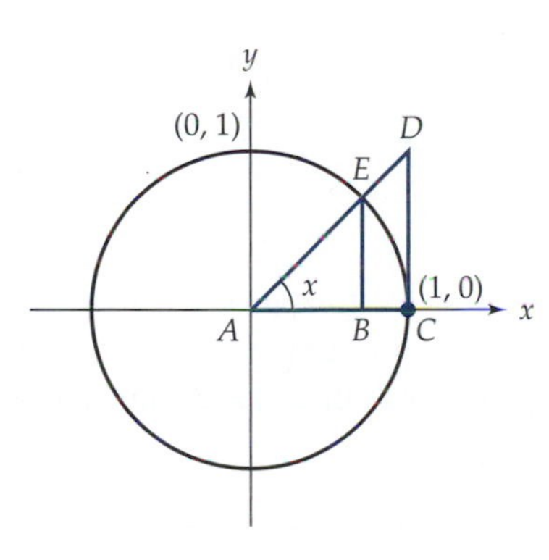

Figure 13

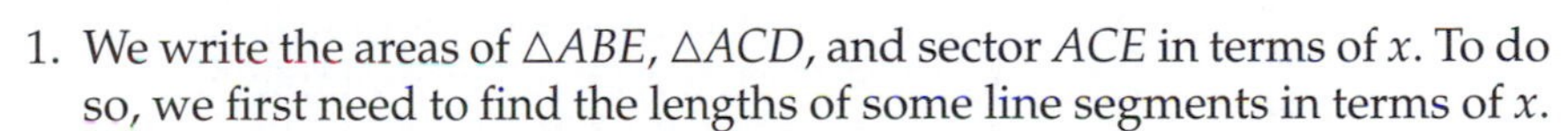

1. We write the areas of $\triangle ABE$, $\triangle ACD$, and sector ACE in terms of x. To do so, we first need to find the lengths of some line segments in terms of x.
   (a) Write an expression for AB in terms of x.
   (b) Write an expression for BE in terms of x.
   (c) Write an expression for CD in terms of x.
2. Now, using the lengths of the line segments you just found, we find the areas of the triangles and the sector.
   (a) Write an expression for the area of $\triangle ABE$ in terms of x.
   (b) Write an expression for the area of $\triangle ACD$ in terms of x.
   (c) Write an expression for the area of sector ACE in terms of x.
3. It is clear from Figure 13 that area $\triangle ABE \le$ area of sector $ACE \le$ area $\triangle ACD$. Rewrite this inequality using the areas you found in the previous questions.
4. Multiply each part of the inequality you got from question 3 by $2/\sin x$ and simplify.
5. Using the squeeze theorem, explain why your results prove that $\lim_{x\to 0} x/\sin x = 1$.

INVESTIGATION 2: ANOTHER LOOK AT e

There are a number of different ways to find a value for e. One way mentioned in the text is

$$e = \lim_{n\to\infty}\left(1+\frac{1}{n}\right)^n.$$

Another method uses factorial notation, $e = 1 + 1/1! + 1/2! + 1/3! + \cdots$. When we see the exclamation points, we read them "factorial," so 3! is read "3 factorial." When we use factorial notation, we multiply the number given by every positive integer less than that number. Thus, $3! = 3\cdot 2\cdot 1 = 6$. As you can see, factorial numbers get large very quickly; $4! = 4\cdot 3\cdot 2\cdot 1 = 24$ and $5! = 120$.

1. Using the factorial approximation for e, complete the following table.

Number of Terms	1	2	3	4	5	6	7
Approximation for e	1	2	2.5				

2. As you can see, this method of approximating e gives a good approximation very quickly. This approximation appears to be the same as the limit definition for e, so we should be able to show that they are equal. That is our task in this investigation. Let's start with $(1+1/n)^n$. Expanding binomials like this one can be long and tedious. Just think about what it would take to expand $(1+x)^{14}$. All the multiplying would take a very long time. Fortunately, the binomial theorem is a shorter way. This theorem, discovered by Isaac Newton, makes use of combinations.

   A combination is a way of counting how many different ways you can choose items from a list if the order in which you choose them does not matter. For example, if we have five items, A, B, C, D and E, and wish to choose two of them, then we say we have a combination of five things taken two at a time and write it as $\binom{5}{2}$. In this case, we have 10 ways of choosing two things from this list of five items: AB, AC, AD, AE, BC, BD, BE, CD, CE, and DE. Remember, that because order is not important, BA is the same as AB. We can, in general, find the value of a combination by using the formula,

$$\binom{n}{r} = \frac{n!}{(n-r)!r!}.$$

   In the case of $\binom{5}{2}$, we obtain

$$\frac{5!}{(5-2)!2!} = \frac{5\cdot 4\cdot 3\cdot 2\cdot 1}{(3\cdot 2\cdot 1)(2\cdot 1)} = 10.$$

   Let's try a few of these combinations.

   (a) Find $\binom{10}{3}$.

   (b) Find $\binom{10}{7}$.

   (c) Find $\binom{4}{4}$. (*Hint:* $0! = 1$.)

3. We can use combinations to present the binomial theorem:

$$(a+b)^n = \binom{n}{0}a^n \cdot b^0 + \binom{n}{1}a^{n-1} \cdot b^1 + \binom{n}{2}a^{n-2} \cdot b^2 + \binom{n}{3}a^{n-3} \cdot b^3 + \cdots + \binom{n}{n-1}a^1 \cdot b^{n-1} + \binom{n}{n}a^0 \cdot b^n.$$

   (a) Using the binomial theorem, expand $(1+x)^3$.
   (b) Using the binomial theorem, expand $(a+x)^6$.

4. Now we are ready to look at the expansion of $(1+1/n)^n$. We look at only the first five terms of the expansion.
   (a) Write the first five terms of the expansion of $(1+1/n)^n$.
   (b) Rewrite and show how you can get

$$1 + \frac{1}{1!} + \frac{1}{2!} \cdot \frac{(n-1)}{n} + \frac{1}{3!} \cdot \frac{(n-1)(n-2)}{n^2} + \frac{1}{4!} \cdot \frac{(n-1)(n-2)(n-3)}{n^3}.$$

   (c) Now we simply take the limit as $n \to \infty$. One important property of limits that you need to use is to recognize that the limit of a sum is equal to the sum of the limits. In other words, as we take the limit of this expansion, we can take the limit of each term separately. Find

$$\lim_{n\to\infty}\left(1 + \frac{1}{1!} + \frac{1}{2!} \cdot \frac{(n-1)}{n} + \frac{1}{3!} \cdot \frac{(n-1)(n-2)}{n^2} + \frac{1}{4!} \cdot \frac{(n-1)(n-2)(n-3)}{n^3}\right).$$

   (d) Explain why what you have done now seems to show that

$$\lim_{n\to\infty}\left(1+\frac{1}{n}\right)^n = 1 + \frac{1}{1!} + \frac{1}{2!} + \frac{1}{3!} + \frac{1}{4!} + \cdots.$$

8.2 SLOPES OF SECANT LINES AND THE DERIVATIVE

One of the important concepts of calculus allows us to find functions for rates of change. You have already worked with the concept of rate of change when you learned about slopes of linear functions. It was easy to study rate of change (or slope) for these functions because it was constant. As we worked with nonlinear functions, however, it became less possible to carefully consider the rate of change. We could determine if the rate of change was increasing or decreasing, but little else. In this section, you learn how to find the rate of change of some nonlinear functions and learn about the derivative.

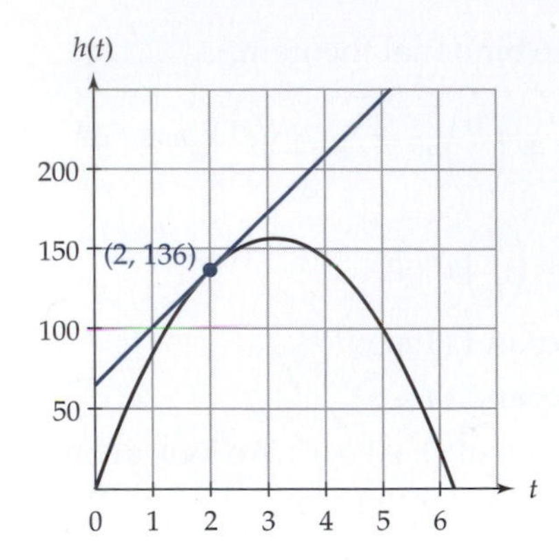

Figure 1 The instantaneous velocity at $t = 2$ is the same as the slope of the line tangent to the curve at $t = 2$.

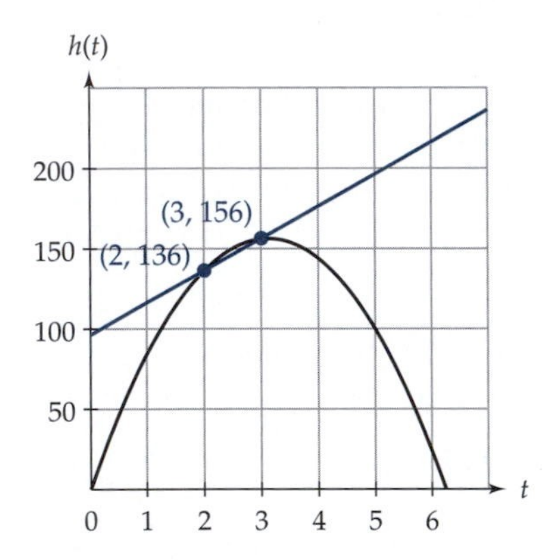

Figure 2 To approximate the slope of the tangent line, we can find the slope of a secant line.

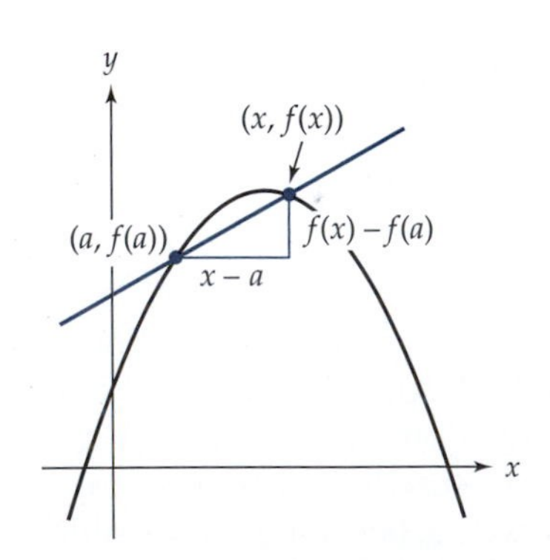

Figure 3 In general, the slope of a secant line is $m = f(x) - f(a)/x - a$.

THE SECANT LINE

In the last section, on limits, we began with an equation for the height of a toy rocket. This equation was $h(t) = -16t^2 + 100t$, where $h(t)$ is the height of the rocket, in feet, after t seconds. Using this equation and some numerical methods, we found that the instantaneous velocity of the rocket after 2 sec was 36 ft/sec. We want to repeat the same example here, but we focus on the problem more graphically.

Because velocity is the rate of change of the position, it can be considered the "slope" of a position function like h. Because h is not a linear function, we know that the rate of change is not constant. That is, the rate of change itself is continually changing. Therefore, we cannot find one number for the slope of the entire function like we can for a linear function. We can, however, find the slope at a point. In this case, it is the same as the instantaneous velocity for a specific time. To find the slope graphically, we need to find the slope of a line that is *tangent* to the curve at the point $(2, 136)$. (See Figure 1.)

A tangent line touches the curve at our point of interest. To find the slope of a line, however, we need to have two points, not just one. We can use the same procedures we used in the last section to find the slope of a *secant* line. Our secant line crosses the curve at the point $(2, 136)$ and at a point near $(2, 136)$. If we make the point *near* $(2, 136)$ closer and closer to $(2, 136)$, the slope of the secant line gets closer and closer to the slope of the tangent line.

Figure 2 shows a graph of h along with a line that passes through two points on h, $(2, 136)$ and $(3, 156)$. Our first estimate of the slope of the tangent line is the slope of this secant line. This secant line is an approximation, not necessarily a good one, for the slope of the tangent line we seek. To find the slope of that line, remember that slope is defined as change in output divided by change in input. In this case,

$$m = \frac{h(3) - h(2)}{3 - 2} = \frac{156 - 136}{3 - 2} = \frac{20}{1} = 20.$$

Therefore, our first approximation for the slope of $h(t) = -16t^2 + 100t$ at the point $(2, 136)$ is 20.

As mentioned, this first approximation is not a very good one. By finding slopes of other secant lines where the interval between $(2, 136)$ and our second point on the curve gets smaller and smaller, we can find the slope of the tangent line. We have done so in Table 1, which is a repeat of a table from the last section. From the table, we can see that as the time interval gets smaller, the slopes of the secant lines get closer to 36 ft/sec. We can say

$$\lim_{t \to 2} \frac{h(t) - h(2)}{t - 2} = 36.$$

Therefore, the slope of the tangent line is 36, which means that the instantaneous velocity of our function at $t = 2$ is 36 ft/sec.

Let's generalize this procedure before we move to another example. Look at Figure 3. A function, f, is shown with a secant line. As you can see,

TABLE 1 As the Time Interval Gets Smaller, the Average Velocity Approaches 36

Time Interval	Average Velocity
$2 \le t \le 2.01$	$\dfrac{h(2.01)-h(2)}{2.01-2} = \dfrac{136.3584-136}{0.01} = 35.84$
$2 \le t \le 2.001$	$\dfrac{h(2.001)-h(2)}{2.001-2} = \dfrac{136.035984-136}{0.001} = 35.984$
$2 \le t \le 2.0001$	$\dfrac{h(2.0001)-h(2)}{2.0001-2} = \dfrac{136.00359984-136}{0.0001} = 35.9984$
$2 \le t \le 2.00001$	$\dfrac{h(2.00001)-h(2)}{2.00001-2} = \dfrac{136.0003599984-136}{0.00001} = 35.99984$

the slope of the secant line is

$$m = \frac{f(x)-f(a)}{x-a}.$$

Therefore, in general, the slope of a line tangent to the graph of a function f is

$$m = \lim_{x \to a} \frac{f(x)-f(a)}{x-a}.$$

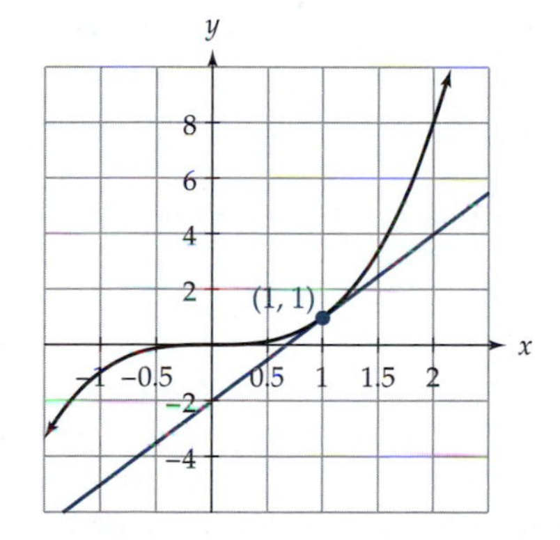

Figure 4

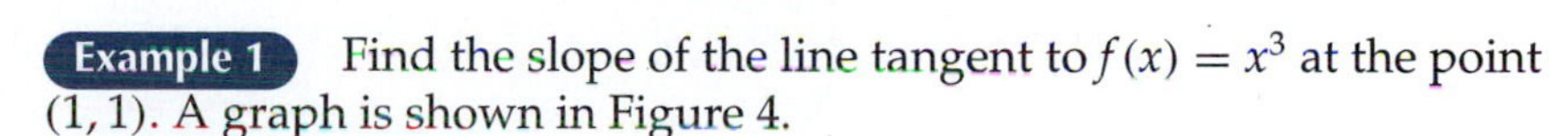

Example 1 Find the slope of the line tangent to $f(x) = x^3$ at the point $(1, 1)$. A graph is shown in Figure 4.

Solution To find this slope, we start by finding the slope of the secant line that goes through the points $(1, f(1))$ and $(1.1, f(1.1))$ and gradually get these two points closer and closer.

$$\frac{f(1.1)-f(1)}{1.1-1} = \frac{1.331-1}{1.1-1} = 3.31$$

$$\frac{f(1.01)-f(1)}{1.01-1} = \frac{1.030301-1}{1.01-1} = 3.0301$$

$$\frac{f(1.001)-f(1)}{1.001-1} = \frac{1.003003001-1}{1.001-1} = 3.003001$$

$$\frac{f(1.0001)-f(1)}{1.0001-1} = \frac{1.000300030001-1}{1.0001-1} = 3.00030001.$$

It appears that

$$\lim_{x \to 1} \frac{f(x)-f(1)}{x-1} = 3.$$

Therefore, the slope of the line tangent to $f(x) = x^3$ at the point $(1, 1)$ is 3.

READING QUESTIONS

1. What is the difference between

$$m = \frac{f(x) - f(a)}{x - a}$$

and

$$m = \lim_{x \to a} \frac{f(x) - f(a)}{x - a}?$$

2. What is the slope of a line tangent to $f(x) = x^3$ at the point $(2, 8)$?

THE DERIVATIVE

One of the big ideas of calculus is to find the slopes of tangent lines to curves. In doing so, we find the "slope" of the curve at a specific point, which allows us to find, for example, instantaneous velocity. Finding these slopes is the same as finding the *derivative* at a point. The derivative of f is the rate of change of f at x, or the instantaneous rate of change of f. The derivative of a function at a point is noted by writing $f'(x)$, pronounced "f prime of x." One definition for this is as follows.

DEFINITIONS

The **derivative of a function, f, at a point** $(a, f(a))$ is

$$f'(a) = \lim_{x \to a} \frac{f(x) - f(a)}{x - a}.$$

This limit can be written differently, in a way that often makes computations a bit easier, by substituting $a + h$ for x. Because our limit had $x \to a$, we now have $a + h \to a$, which means the same as $h \to 0$. The denominator of our function, which was $x - a$, now becomes $a + h - a$ or just h. Therefore, we can rewrite our definition of a derivative at a point as follows.

DEFINITIONS

The **derivative of a function, f, at a point** $(a, f(a))$ is

$$f'(a) = \lim_{h \to 0} \frac{f(a + h) - f(a)}{h}.$$

In some cases, as in our next example, you can find a specific value for the slope of the function at a given point.

Example 2 Find the slope of the line tangent to $f(x) = 3x^2 - 4x + 2$ at $x = 0$.

Solution By using $f'(0) = \lim_{h\to 0}(f(a+h) - f(a))/h$, and evaluating at $(0+h)$ and 0, we get

$$
\begin{aligned}
f'(a) &= \lim_{h\to 0}\frac{f(0+h) - f(0)}{h} \\
&= \lim_{h\to 0}\frac{(3h^2 - 4h + 2) - 2}{h} \\
&= \lim_{h\to 0}\frac{3h^2 - 4h}{h} \\
&= \lim_{h\to 0} 3h - 4 \\
&= -4.
\end{aligned}
$$

The slope of the line tangent to f at $x = 0$ is -4.

In fact, with *any* polynomial function, we can use this technique to evaluate the slope at any point. We do not even have to determine ahead of time the point at which we want to find the slope. As we see in the next example, we can often find f' directly.

Example 3 Find the derivative of $f(x) = 4x^3 - 2x$.

Solution

$$
\begin{aligned}
f'(x) &= \lim_{h\to 0}\frac{f(x+h) - f(x)}{h} \\
&= \lim_{h\to 0}\frac{(4(x+h)^3 - 2(x+h)) - (4x^3 - 2x)}{h} \\
&= \lim_{h\to 0}\frac{(4(x^3 + 3x^2h + 3xh^2 + h^3) - 2(x+h)) - (4x^3 - 2x)}{h} \\
&= \lim_{h\to 0}\frac{4x^3 + 12x^2h + 12xh^2 + 4h^3 - 2x - 2h - 4x^3 + 2x}{h} \\
&= \lim_{h\to 0}\frac{12x^2h + 12xh^2 + 4h^3 - 2h}{h} \\
&= \lim_{h\to 0}(12x^2 + 12xh + 4h^2 - 2) \\
&= 12x^2 + 0 + 0 - 2 \\
&= 12x^2 - 2.
\end{aligned}
$$

Therefore, the derivative of $f(x) = 4x^3 - 2x$ for all values of x is $12x^2 - 2$.

Finding a derivative at a point is not always as easy as we have shown in our examples. We have purposely chosen functions and points such that the derivative at that point is a nice number. Obviously, though, derivatives are not always nice numbers, and some can be very difficult to approximate. In fact, derivatives at some points in the domain of a function do not even exist. Remember that a derivative is defined in terms of a limit and that limits of some functions at certain points do not exist.

One intuitive way to see if a function has a derivative at a certain point is to graph the function on a calculator and then zoom in. Functions that have a derivative will look like linear functions if you zoom in enough. For example, the graph of $y = x^2 + 3$ is shown in Figure 5(a). In this window, it is clear that the graph does not look linear. The same function is shown in Figure 5(b) in a different window. In this window, the function looks linear.

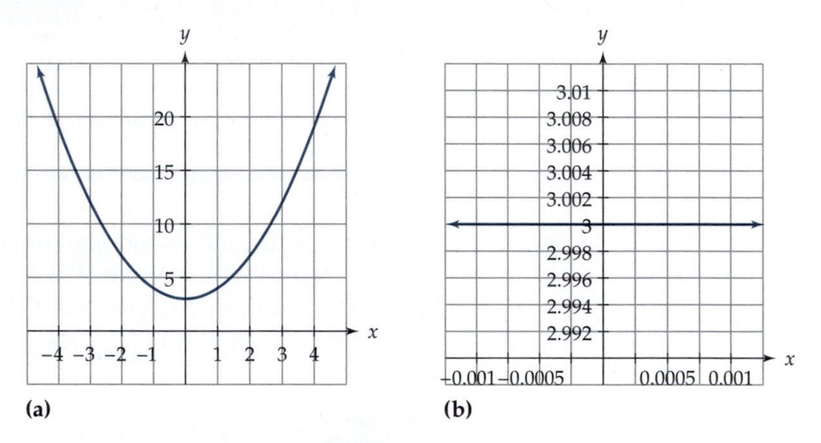

Figure 5 (a) A graph of $y = x^2 + 3$. (b) This same function after "zooming in" on the point $(0, 3)$. In this case, the graph looks like a horizontal line.

In contrast, the graph of $y = |x|$ is shown in Figure 6(a). If we zoom in on this function at the point $(0, 0)$, then the function does not look any different. It will never look like a line at this point no matter how much you zoom in. For example, the same function is shown in Figure 6(b) in a much smaller window. This graph intuitively shows that the function does not have a derivative at that point. In the exercises, you are asked to use symbolic methods to show why $y = |x|$ does not have a derivative when $x = 0$.

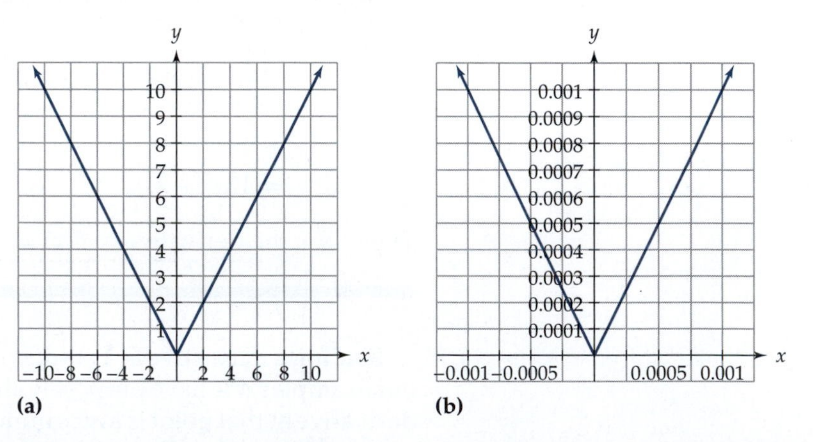

Figure 6 (a) A graph of $y = |x|$. (b) This same function after "zooming in" on the point $(0, 0)$. In this case, the graph looks the same as it did in the larger window.

HISTORICAL NOTE

Calculus was discovered by Isaac Newton and Gottfried Leibniz who, independently, came up with very similar ideas around the same time. Sir Isaac Newton was an amazing man who was responsible for a number of great discoveries. In fact, there was a period in 1665–1666 in which Trinity College at Cambridge University, where young Isaac received his A.B. degree, was closed due to the plague. During this time, Newton went home to live and think. Apparently, he thought quite a bit, because during this short period he made four of his most important discoveries: the binomial theorem, calculus, the law of gravitation, and the nature of color.

In 1669, Newton was appointed to the prestigious Lucasian professorship at Cambridge and was later appointed to represent the university at Parliament. Even though politics held little interest for him, he was reelected in 1701 and was knighted by the queen in 1705. Newton had many eccentricities and was so absorbed in his work that he often lost sight of the world around him. One story has it that while hosting a dinner party, he left the table to get a bottle of wine. He forgot why he left the table, put on his church garments, and ended up in the chapel.

Although perhaps not as colorful as Newton, Gottfried Willhelm Leibniz was also a person whose accomplishments went way beyond mathematics. A child prodigy, he entered the University of Leipzig as a law student at age 15. By age 20, he was prepared for his doctor's degree in law, but it was refused him. Officially, he was refused his degree because of his young age, but some historians have suggested that it was actually jealousy on the part of the faculty because Leibniz knew more law than the faculty did. Outraged, he left the university and went to work in the law for a time, even becoming a diplomat. At age 26, he met mathematician Christian Huygens in Paris. At this point, his mathematics education really began. Within 3 years, if we accept his own account, he had worked out some of the elementary formulas of calculus.

In addition to his work in mathematics and law, Leibniz was a master of many other fields: religion, statecraft, history, literature, logic, and metaphysics. He made noteworthy contributions to many of these areas, and his work in them may have secured his place in history books even without his work in mathematics.

Looking back, it seems that Newton discovered the ideas of calculus before Leibniz, but Leibniz published them first (a result of Newton's general reluctance to publish). There was a bitter quarrel between the two men about who actually discovered these concepts. It is now generally accepted that they both arrived at similar conclusions independently, and modern mathematicians use the ideas of both men in the study of calculus.

Sources: Carl B. Boyer, *A History of Mathematics,* (Princeton, NJ: Princeton University Press, 1968), pp. 429–52; D. E. Smith, *History of Mathematics, Vol. I* (New York: Dover, 1951), p. 404; E. T. Bell, *Men of Mathematics* (New York: Simon and Schuster, 1937) p. 117–30.

READING QUESTIONS

3. What does the derivative at a point tell you about a function?

4. What is the relationship between

$$f'(a) = \lim_{x \to a} \frac{f(x) - f(a)}{x - a}$$

and

$$f'(a) = \lim_{h \to 0} \frac{f(a+h) - f(a)}{h}?$$

5. What is the slope of a line tangent to $f(x) = 4x^3 - 2x$ at $x = 3$?

SKETCHING THE GRAPH OF THE DERIVATIVE

Now that we realize that the derivative of a function can also be written as a function, it is natural to consider the graph of the derivative and see how it relates to the graph of the original function. In doing so, we need to remember that the graph of the derivative of a function is the graph of its rate of change. Graphing the derivative of a linear function is simple. Because a linear function has a constant rate of change, the graph of its derivative is simply a horizontal line. If the line has a positive slope, then the graph of the derivative is a horizontal line above the x-axis. If the line has a negative slope, then the graph of the derivative is a horizontal line below the x-axis. Let's look at a more complicated example. Consider the graph shown in Figure 7.

Because we can think of the derivative as the rate of change, or slope, we need to think about the way the *slope* changes as x changes. To help, we have labeled four points, A, B, C, and D, on the graph in Figure 7. We label four corresponding points, A', B', C', and D', on the graph of the derivative.

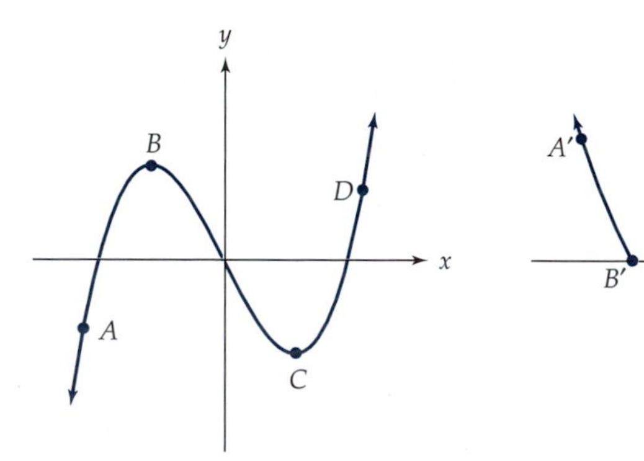

Figure 7 We want to graph the derivative of the function shown here.

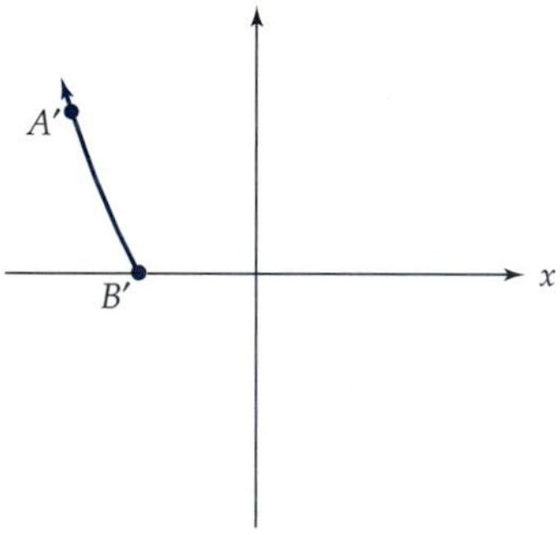

Figure 8 Because the slope of our function is positive from A to B, the graph of the derivative has positive outputs from A' to B'.

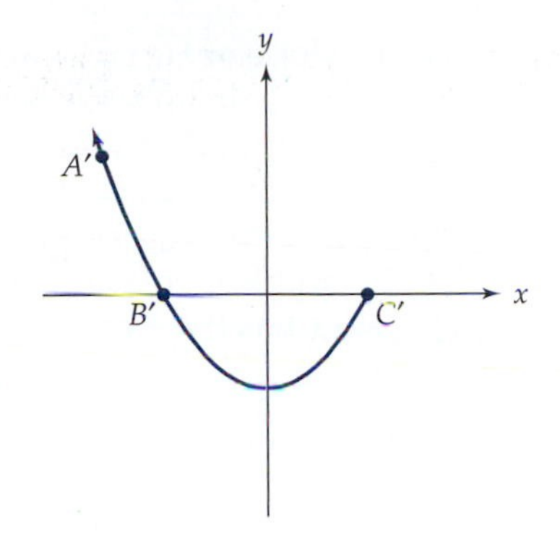

Figure 9 Because the slope of our function is negative from B to C, the graph of the derivative has negative outputs from B' to C'.

On the graph in Figure 7, the slope is positive at point A. Between points A and B, the slope is still positive but is decreasing until point B, where the slope looks like zero. Thus, the graph of the derivative has positive outputs between A' and B' but decreases until B' where the output will be zero. (See Figure 8.)

Between points B and C, in Figure 7, the slope is negative. It starts out close to zero. About half way between B and C, it is the steepest (or most negative). Close to C it is near zero again, and at C it looks like it equals zero. Thus, the graph of the derivative has negative outputs between B' and C', is the most negative half way between B' and C', and is back at zero at C'. (See Figure 9.)

Between points C and D, in Figure 7, the slope is positive. It starts out close to zero and gradually gets steeper. Thus, the graph of the derivative has positive outputs between C' and D', and is increasing. (See Figure 10.)

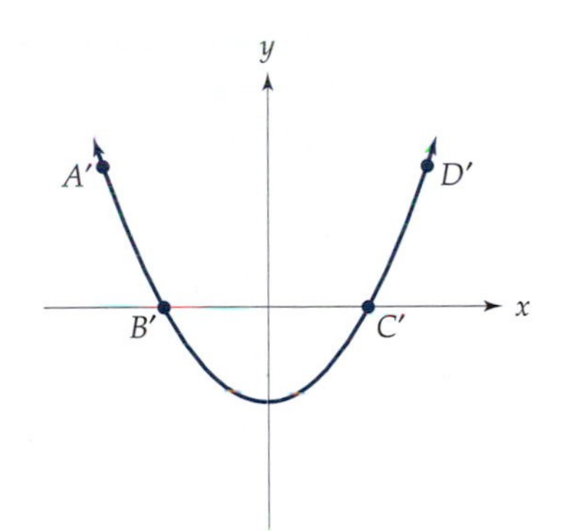

Figure 10 Because the slope of our function is positive from C to D, the graph of the derivative has positive outputs from C' to D'.

Example 4 Sketch the derivative of the function whose graph is shown in Figure 11.

Solution The slope of the graph to the left of the y-axis in Figure 11 is positive, the slope looks like it is zero at $x = 0$, and the slope is negative to the right of the y-axis. Thus, the output for the graph of the derivative is positive for negative inputs, goes through the origin, and is negative for positive inputs. Hence, there is no change in direction for the graph of the derivative, and it could look simply like a line with a negative slope that passes through the origin. A graph like that is shown in Figure 12.

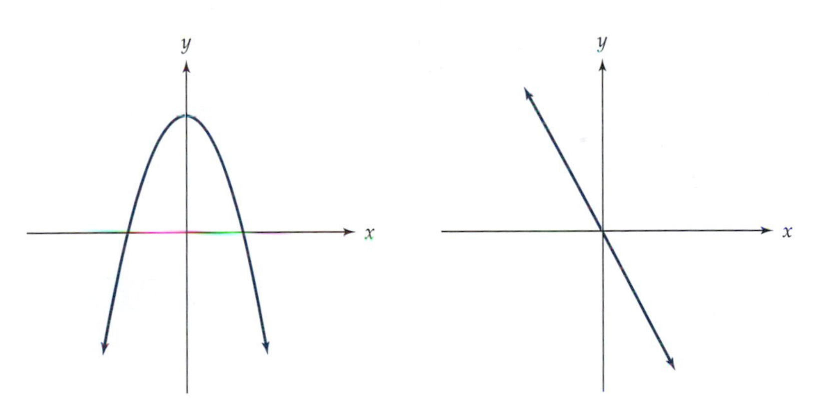

Figure 11 **Figure 12**

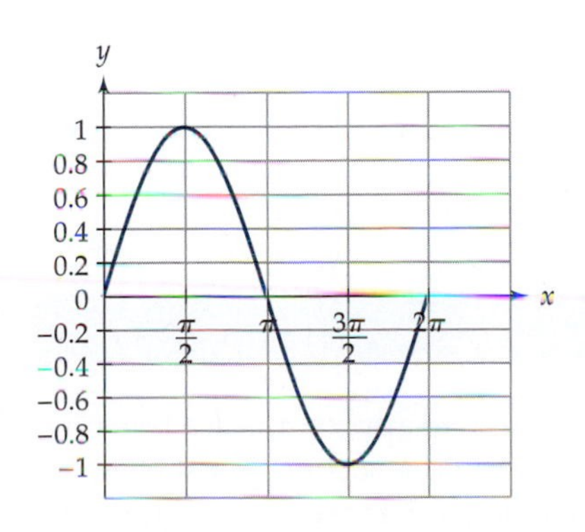

Figure 13

Example 5 Figure 13 shows a graph of $f(x) = \sin x$ on the interval $0 \le x \le 2\pi$. Use this graph to draw a graph of the derivative of $f(x) = \sin x$.

Solution On the interval $0 \le x \le \pi/2$, the graph of $f(x) = \sin x$ has a positive slope and appears to be decreasing until at $\pi/2$ the slope is zero. Thus, the graph of the derivative has positive outputs and is decreasing, where at $\pi/2$ the output is zero. From $\pi/2$ to $3\pi/2$, the slope of $f(x) = \sin x$ is negative and is the steepest around π. At $3\pi/2$, it again appears to be zero. Thus, the graph of the derivative has negative outputs between $\pi/2$ and $3\pi/2$ and is the most negative at π. From $3\pi/2$ to 2π, the graph of $f(x) = \sin x$ has a positive slope and is getting steeper. Thus, the graph of the derivative has positive outputs from $3\pi/2$ to 2π and is increasing. To summarize, the graph of the derivative of $f(x) = \sin x$ has the following properties.

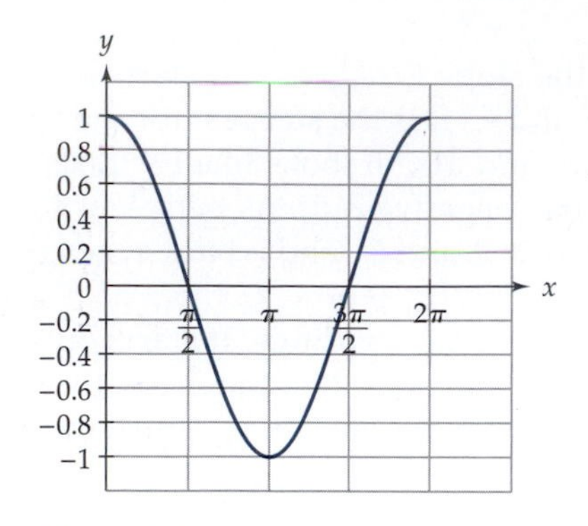

Figure 14

- The graph has positive outputs and is decreasing on the interval $0 < x < \pi/2$, and at $\pi/2$, the output will be zero.
- The graph has negative outputs on the interval $\pi/2 < x < 3\pi/2$ and is the most negative at π.
- The graph has positive outputs on the interval $3\pi/2 < x < 2\pi$ and is increasing.

A graph is shown in Figure 14.

The graph of the derivative of $f(x) = \sin x$, shown in Figure 14, looks like the graph of the cosine function, and indeed it is. (Because the sine function is periodic, the graph of its derivative is also periodic.) Later, in an investigation, we consider the derivative of the sine function from a symbolic perspective.

Most graphing calculators have some sort of program that gives a derivative (or an approximate derivative) at a point. Such a program can be used to graph the derivative of a function as well. One word of caution though. When dealing with technology, as always, the answers you get may not be accurate. In fact, sometimes the answers are just wrong. For example, on the TI-83, the numerical derivative of $f(x) = |x|$ when $x = 0$ is given as 0, when in fact, as we pointed out earlier, the derivative of that function at that point does not exist.

READING QUESTIONS

6. What does the graph of the derivative of a line look like?
7. If the rate of change of a function is always positive on a certain interval, what can you say about the graph of the derivative of that function?
8. How will the graph of the derivative of a parabola that opens down differ from one that opens up?

SUMMARY

In this section, we began with using a secant line to approximate a line tangent to a function at a particular point. We picked a second point close to the point in question and made a secant line with those two points. We then used that secant line to approximate the tangent line.

We defined the **derivative of a function, *f*, at a point** $(a, f(a))$ as

$$f'(a) = \lim_{x \to a} \frac{f(x) - f(a)}{x - a},$$

which we rewrote as

$$f'(a) = \lim_{h \to 0} \frac{f(a+h) - f(a)}{h}.$$

To find the derivative of a particular function, we substitute that function for $f(a)$ in the definition and simplify. This substitution often (but not always) simplifies nicely and gives us the new function for the derivative.

In sketching the graph of a derivative, you need to pay special attention to points at which the function has a derivative of zero and points at which the slope of the original function is steepest. Connecting these points can help you sketch a graph of the derivative of a function.

EXERCISES

1. The first discussion in this section was of a toy rocket whose height could be found by using the function $h(t) = -16t^2 + 100t$. Find the instantaneous velocity of the rocket 3 sec into its flight.
2. Suppose a ball is tossed up in the air such that the equation giving its height in feet is $h(t) = -16t^2 + 40t + 5$, where t is time in seconds since the ball was released.
   (a) Find values for $h(2)$ and $h'(2)$.
   (b) What is the meaning, in terms of the height of the ball and time, of $h(2)$ and $h'(2)$?
3. Suppose the bacteria count on fresh chicken (per square centimeter in a 40°F refrigerator) can be modeled by the function $B(t) = 360 \cdot 16^t$, where t is time in days since the chicken was put in the refrigerator.
   (a) Find a value for $B(2)$ and estimate a value for $B'(2)$.
   (b) What is the meaning, in terms of the number of bacteria on the chicken and the time, of $B(2)$ and $B'(2)$?
4. What is the derivative of any function, $f(x) = c$, where c is some constant number? Explain how you determined your answer.
5. Find the slope of the tangent line for each function at the given point.
   (a) $f(x) = x^2 - 4$ at the point $(3, 5)$
   (b) $g(x) = x^3 - 3$ at the point $(-1, -4)$
   (c) $j(x) = x^3 - x^2$ at the point $(2, 4)$
   (d) $k(x) = x^2 - 4x + 1$ at the point $(-2, 13)$
   (e) $f(x) = x^4 - 1$ at the point $(2, 15)$
6. Estimate the slope of the tangent line for each function at the given point. Do so by finding the slope of a secant line and taking smaller and smaller intervals.
   (a) $f(x) = \sqrt{x+3}$ at the point $(1, 4)$

(b) $g(x) = 2^x$ at the point $(2, 4)$

(c) $h(x) = \log x$ at the point $(10, 1)$

7. Find the derivative of each function.

(a) $f(x) = 3x - 5$

(b) $g(x) = x^2 + 6x - 4$

(c) $f(x) = x^3 - 4$

(d) $f(x) = 2x^2 - 6x$

8. Find the equation of the tangent line for each function at the given point.

(a) $f(x) = x^3 - 4$ at the point $(1, -3)$

(b) $f(x) = x^3 - 4$ at the point $(-1, -5)$

(c) $g(x) = x^2 + 8x$ at the point $(-4, -16)$

(d) $g(x) = x^2 + 8x$ at the point $(2, 20)$

9. Four points are shown on the accompanying graph. Put the derivatives of this function at these points in order from smallest to largest.

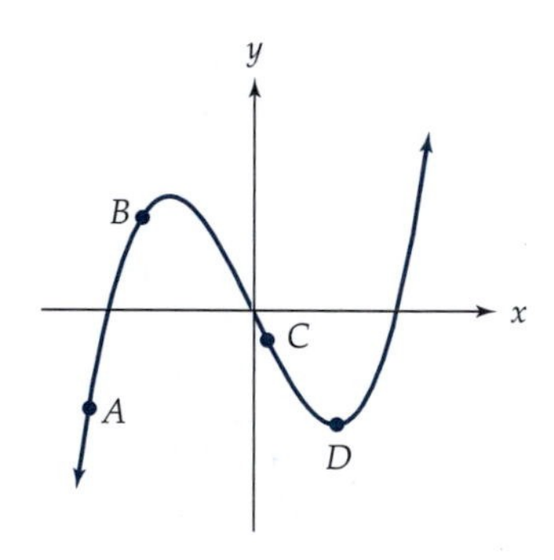

10. The following table gives the population data for walleye in Mille Lacs Lake, Minnesota.[5]

Year	1985	1986	1987	1988	1989	1990	1991	1992	1993	1994	1995
Population	7809	6455	9205	8958	5400	5510	8500	8430	4376	4113	4688

(a) Plot the data from the table with the year on the horizontal axis and the population on the vertical axis. Connect the points with a smooth curve.

(b) We want to construct a table of the derivatives of this function. Because we have discrete points, we cannot find the exact derivatives, so we will have to estimate. To estimate the derivative for 1986, for example, find the average rate of change from

[5]*NERC Centre for Population Biology, Imperial College,* The Global Population Dynamics Database *(visited 29 July 1999)* ⟨*http://cpbnts1.bio.ic.ac.uk/gpdd*⟩.

1985 to 1986 and the average rate of change from 1986 to 1987. Then use the average of these two average rates of change as an estimate for the derivative for 1986. Using this method, estimate the derivatives for years 1986 to 1994.

(c) Plot the derivatives calculated in part (b). Connect the points with a smooth curve.

(d) What do positive values for the derivative tell you about the number of walleye? What do the negative values tell you?

11. The accompanying graph represents the height of an elevator over time. Use the graph to answer the following.

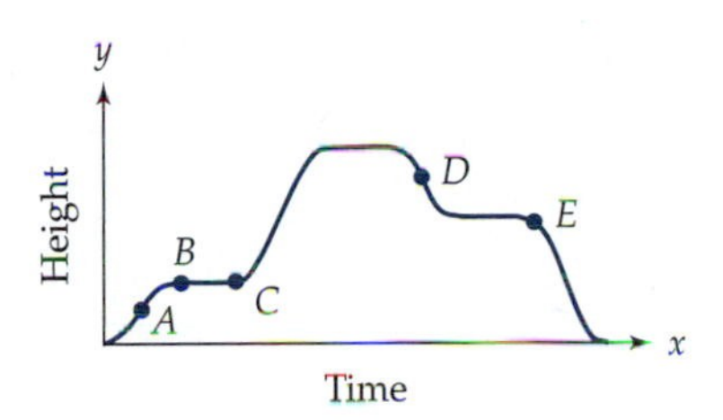

(a) Describe what is happening to the movement of the elevator at point A.

(b) Describe what is happening to the movement of the elevator between points B and C.

(c) Describe what is happening to the movement of the elevator at point D.

(d) Describe what is happening to the movement of the elevator at point E.

(e) How many stops did the elevator make at floors above point A?

12. Match the following graphs, labeled (a) through (d), with the graphs of their derivatives, labeled (i) through (iv).

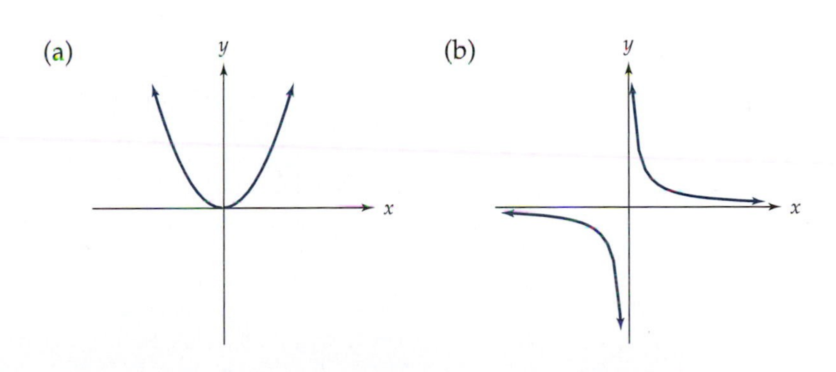

(c)

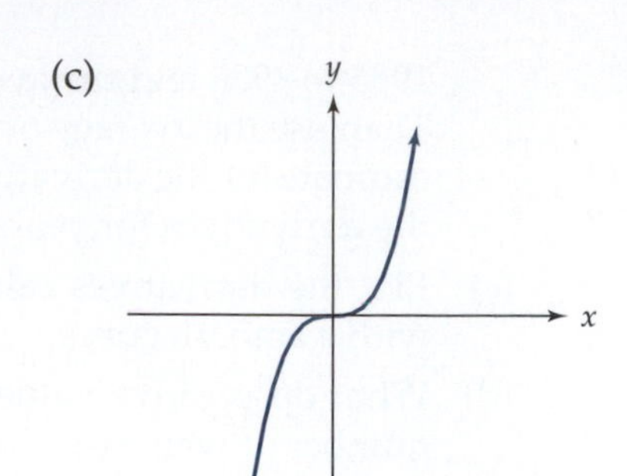

(d)

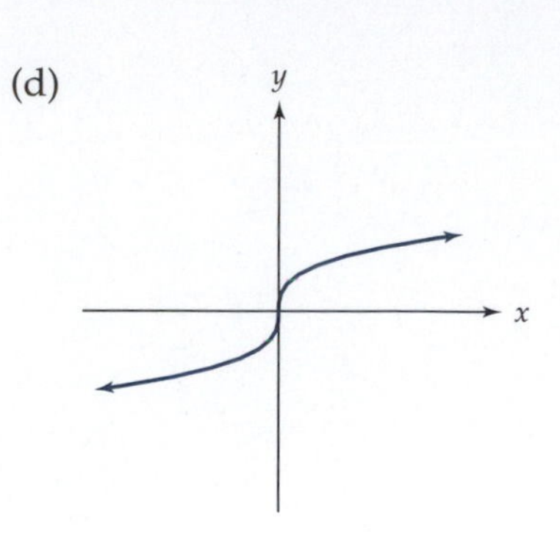

i.

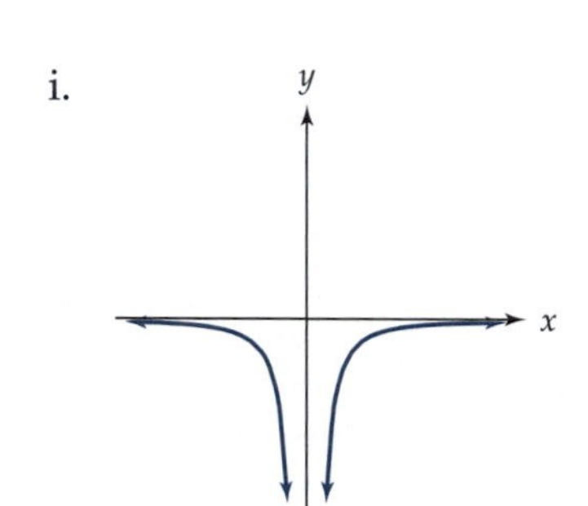

ii.

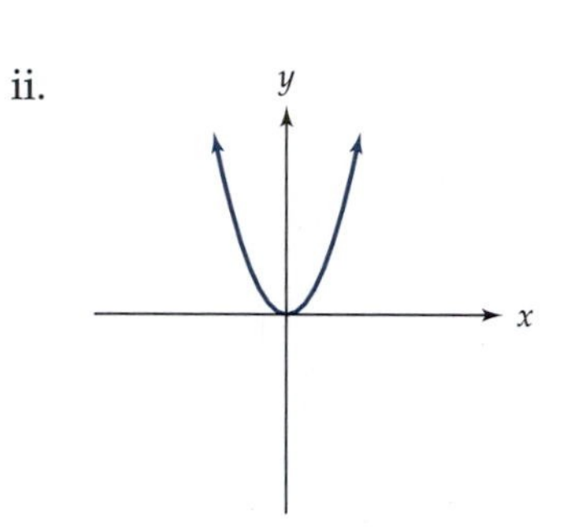

iii.

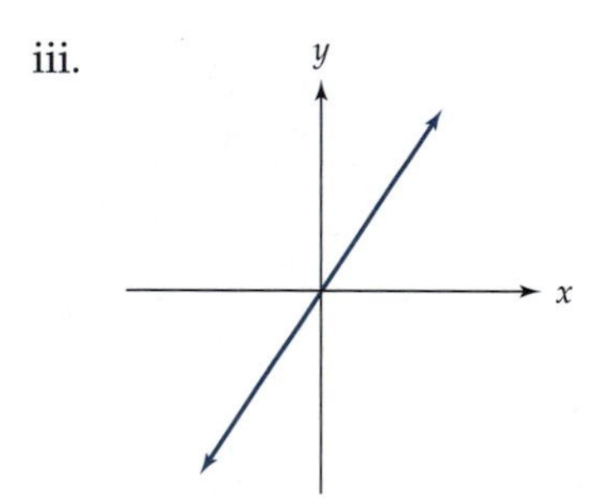

iv.

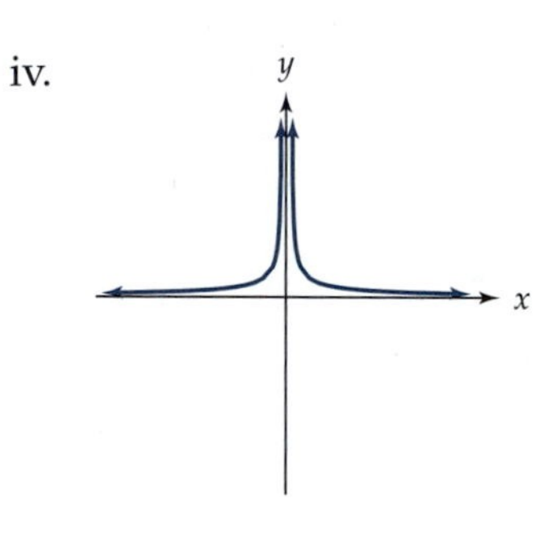

13. The accompanying graph gives the average weight for young children, in pounds, given their age in months. Use the graph to answer the following questions.

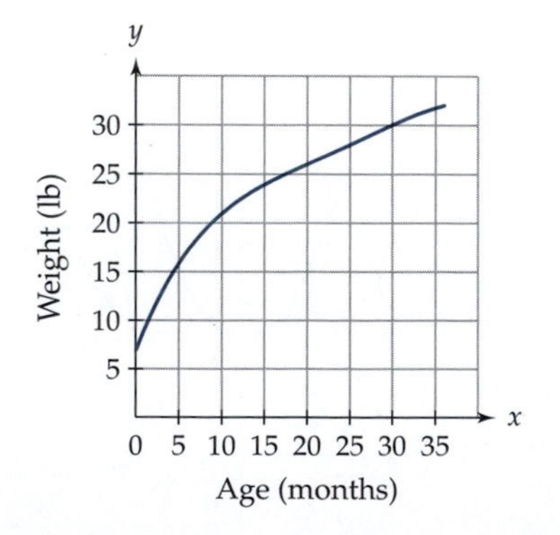

(a) Estimate the rate of growth, in pounds per month, for the average baby at 6 months.

(b) Estimate the rate of growth, in pounds per month, for the average baby at 18 months.

(c) During what ages is the baby gaining the most weight?

14. In this section, we stated that the derivative of $f(x) = |x|$ does not exist at $x = 0$. If the derivative did exist at this point, then we would want to find

$$\lim_{x \to 0} \frac{|x| - |0|}{x - 0}.$$

This limit, however, does not exist. Show this fact by answering the following questions. (*Hint:* Remember that $|x| = x$ when x is positive and $|x| = -x$ when x is negative.)

(a) Find

$$\lim_{x \to 0^-} \frac{|x| - |0|}{x - 0}.$$

(b) Find

$$\lim_{x \to 0^+} \frac{|x| - |0|}{x - 0}.$$

(c) Based on your answers to the previous questions, explain why the $f(x) = |x|$ does not have a derivative at $x = 0$.

15. Sketch a graph of the derivative for each of the functions shown.

(a)

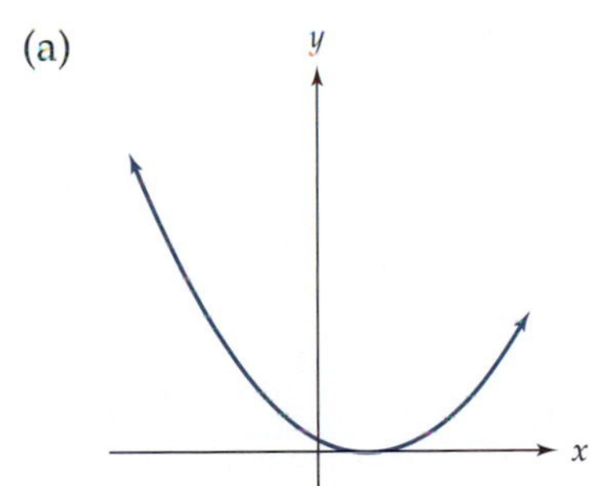

(b)

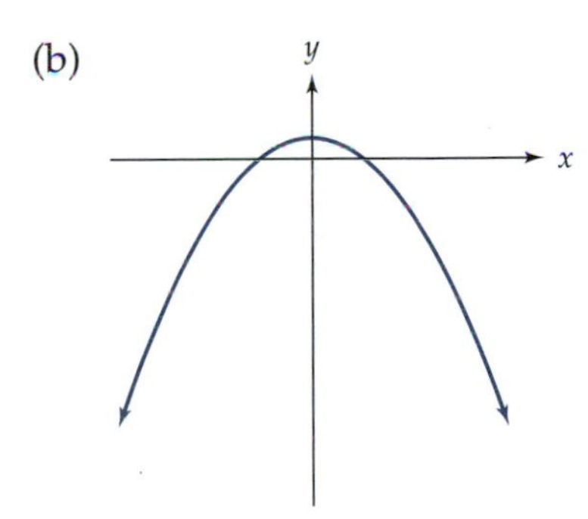

(c)

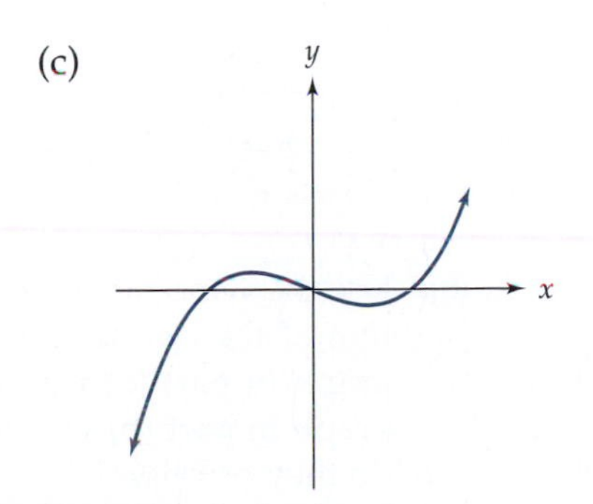

(d)

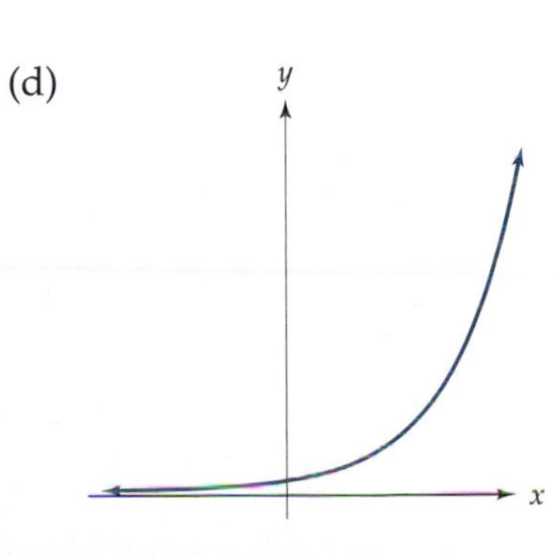

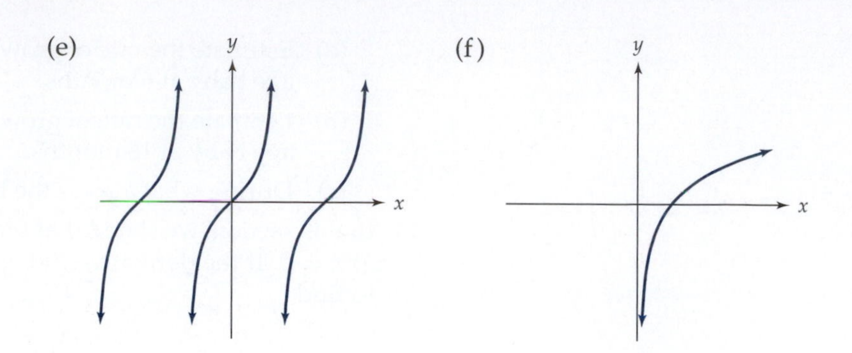

16. In this problem, we investigate the relationship between the derivatives of two functions that differ only by a constant.
    (a) Sketch the graph of the derivative of the function shown.

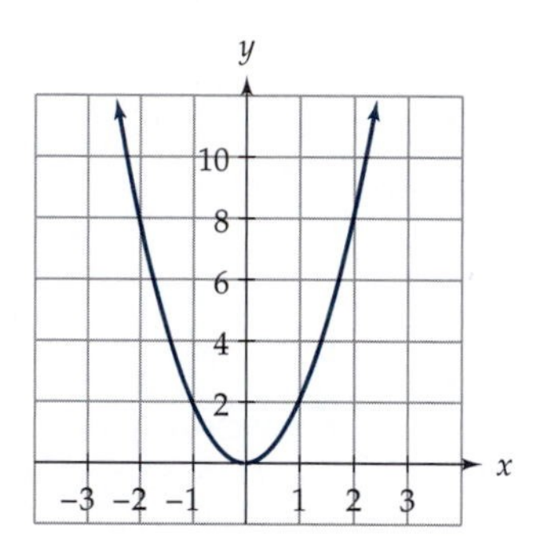

    (b) The function shown in the accompanying figure is the same function as that from part (a), except this one is shifted up four units. Sketch the graph of the derivative of this function.

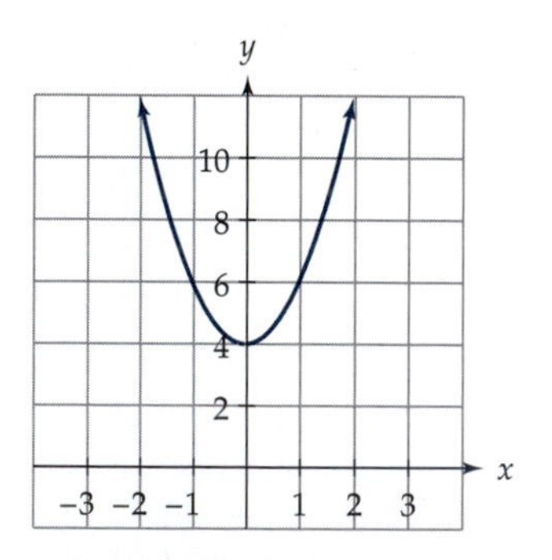

    (c) Use the symbolic definition of the derivative to find the derivative of the two functions shown in the previous figures. The graph in part (a) was made by graphing $f(x) = 2x^2$, and the graph in part (b) was made by graphing $f(x) = 2x^2 + 4$. How do they compare?

(d) What can you say about the derivatives of two functions that differ only by a constant? Explain why you think your statement is true.

17. What happens to the derivative of a function when that function is multiplied by a constant? Support your answer with two arguments, one based on the graph of a function and one based on the symbolic representation.

18. A graph of the cosine function is shown. Determine which of the following functions is a graph of the derivative of the cosine function. For the two graphs that are *not* graphs of the derivative, explain why they are not.

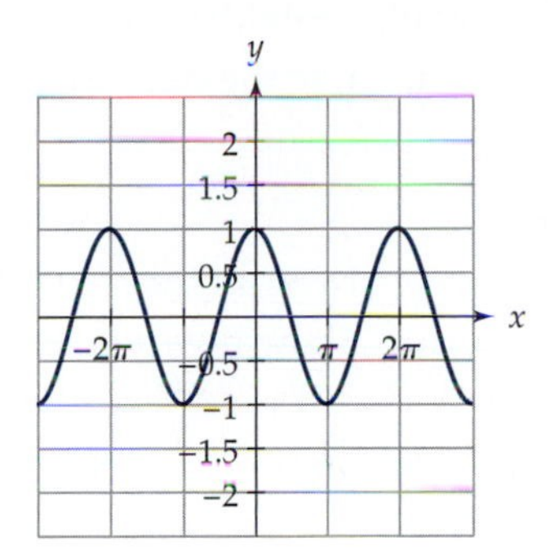

(a)

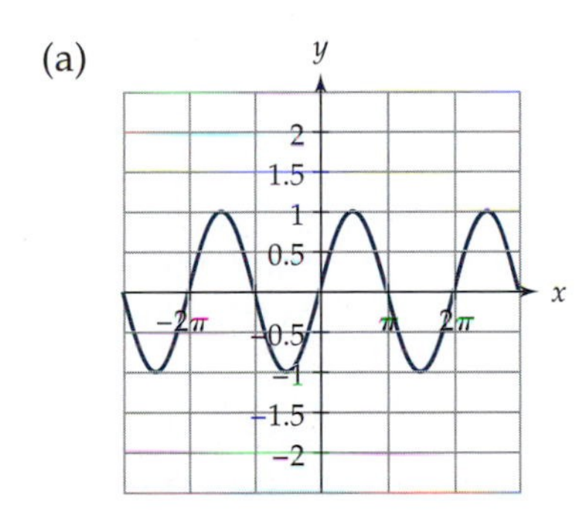

(b)

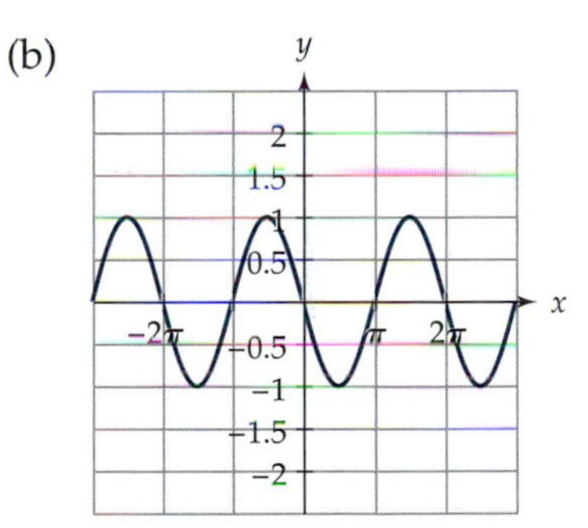

(c)

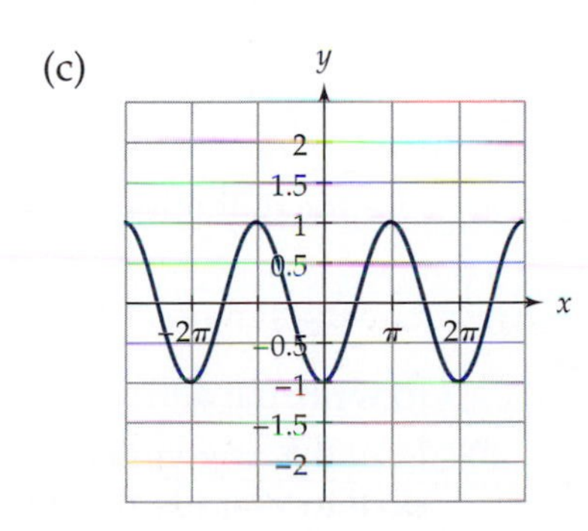

INVESTIGATIONS

INVESTIGATION 1: FINDING A FORMULA FOR THE DERIVATIVE OF EXPONENTIAL FUNCTIONS AND THE SINE FUNCTION

In this section we looked at how to find the derivative of polynomials symbolically, but we did not look at the symbolic derivatives of any other types of functions. We did that for a very good reason, because finding the derivative of other types of functions can be very difficult and requires techniques that you will not learn until you take calculus. We can explore some of these functions, though.

1. Let's begin by finding the derivative of some exponential functions.
   (a) We first want to approximate the derivative of $f(x) = 2^x$.
      i. Use the definition
      $$f'(x) = \lim_{h \to 0} \frac{f(x+h) - f(x)}{h},$$
      to write the derivative of f as a limit.
      ii. Show how you can simplify your limit to get
      $$f'(x) = \lim_{h \to 0} \left[(2^x) \frac{(2^h - 1)}{h} \right].$$
      iii. In working with limits, it is true that $\lim_{x \to a} c \cdot f(x) = c \cdot \lim_{x \to a} f(x)$. In other words, the limit of a constant times a function is the same as the constant times the limit of a function. Explain how this property allows us to write
      $$f'(x) = (2^x) \lim_{h \to 0} \frac{(2^h - 1)}{h}.$$
      iv. If we can find a value for $\lim_{h \to 0} (2^h - 1)/h$, then we can find a function for f'. Approximate this limit by making a table or a graph to approximate $\lim_{h \to 0} (2^h - 1)/h$.
      v. Substitute your numerical value for $\lim_{h \to 0} (2^h - 1)/h$ to get an approximation for the derivative of $f(x) = 2^x$.
   (b) Use the same process as in part (a) to approximate the derivative of $f(x) = 3^x$.
   (c) Again, use the same procedure as in part (a) to find the derivative of $f(x) = e^x$.
2. Now let's look at some functions that are not exponential. Follow the steps listed to find the derivative of $f(x) = \sin x$.
   (a) Use the definition
   $$f'(x) = \lim_{h \to 0} \frac{f(x+h) - f(x)}{h}$$
   to write the derivative of f as a limit.
   (b) Use the trigonometric identity $\sin(x + y) = \sin x \cos y + \cos x \sin y$ to simplify the function.

(c) Show how you can factor and simplify to get

$$f'(x) = \lim_{h\to 0}\left[(\sin x)\frac{(\cos h - 1)}{h} + (\cos x)\frac{\sin h}{h}\right].$$

(d) Another property of limits that we can use is $\lim_{x\to a}[f(x) + g(x)] = \lim_{x\to a} f(x) + \lim_{x\to a} g(x)$. Use this property and others to show how you can get

$$f'(x) = (\sin x)\lim_{h\to 0}\frac{(\cos h - 1)}{h} + (\cos x)\lim_{h\to 0}\frac{\sin h}{h}.$$

(e) Now you only need to determine the values of

$$\lim_{h\to 0}\frac{(\cos h - 1)}{h} \quad \text{and} \quad \lim_{h\to 0}\frac{\sin h}{h}.$$

Determine these values in the same way as you did before, by approximating with tables.

(f) Substitute these values in your equation for the derivative to find an approximation for f'. Does your approximation match what we thought the derivative of $f(x) = \sin x$ was in the text?

INVESTIGATION 2: SECOND DERIVATIVE

In this section, we looked at the derivative as the slope or the rate of change of a function. In the case of our toy rocket, this rate of change was the velocity of the rocket. The velocity of our rocket was continually changing, which showed up in our derivative when our result was a linear equation. What if we took the derivative of our derivative? Because the derivative function gives the rate of change of the position, the derivative of the derivative (called the *second derivative*) gives the rate of change of the velocity. This rate of change is known, of course, as acceleration.

1. Our original function is $p(t) = -16t^2 + 100t$.
   (a) Find a formula for $p'(t)$ (which is the function for the velocity.)
   (b) Find the derivative of the function you obtained in part (a).
   (c) What type of function did you get for the second derivative?
   (d) Why does the type of function you obtained for the second derivative make sense if you consider what you know about falling objects?
2. Now that we have looked at the second derivative as acceleration, let's look at it from a more general perspective. Just as the first derivative is noted by using f', the second derivative has a similar notation, f''. Using the accompanying graph, we want to see what the graph of the first and second derivatives looks like.

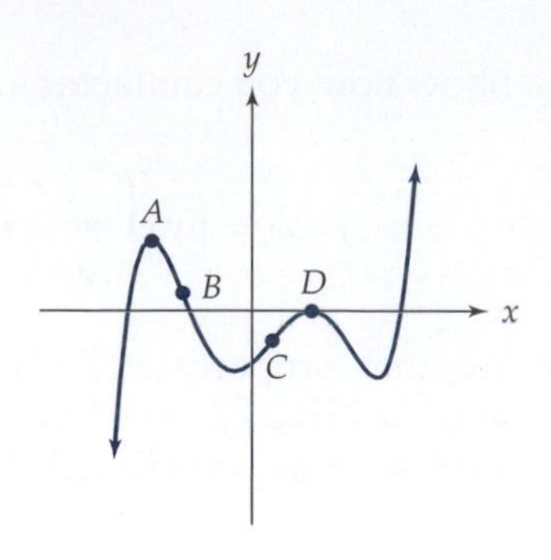

(a) Construct a graph of the derivative of the function shown in the graph. Label points corresponding to the derivative of the function at points A, B, C, and D.

(b) Construct a graph of the second derivative of the function shown in the graph. Do so by making a graph of the derivative of your answer from part (a). Label points corresponding to the second derivative of the original function at points A, B, C, and D.

(c) When the first derivative is zero at a point, what do you know about the original function?

(d) When the second derivative is zero at a point, what do you know about the original function?

8.3 SEQUENCES AND SERIES

As we continue in our preparation for calculus, we turn our attention to sequences and series. In this section, we look at the words and notation associated with sequences and series. In particular, we take a close look at arithmetic and geometric sequences.

INTRODUCTION TO SEQUENCES

At some point in your study of mathematics, you probably made some sort of sequence of numbers and then tried to figure out the "rule" that would allow you to generate that sequence. For example, the sequence $\{1, 3, 6, 10, \ldots\}$ is sometimes called the sequence of *triangular numbers* because the numbers can be found by counting the number of dots used when you make progressively larger equilateral triangles, as shown in Figure 1.

Figure 1 The first four terms in the sequence of triangular numbers are 1, 3, 6, and 10.

In general, a **sequence** is a list of numbers written in a definite order. The numbers in a sequence are referred to as **terms.** Terms in a sequence are often referred to by a variable with a subscript. For example, in the sequence of triangular numbers, the first term is $a_1 = 1$, the second term is $a_2 = 3$, and so on. The entire sequence itself can be referred to as $\{a_n\}$.

Suppose we want to find the eighth term, a_8, in our sequence of triangular numbers. There are a couple of approaches we can use. In one approach, we work our way out to the eighth term by using the previous terms. In the sequence $\{1, 3, 6, 10, \ldots\}$, the second term is 2 more than the first, the third is 3 more than the second, and the fourth is 4 more than the third. See Figure 1. In that figure, another row is added to each progressively larger triangle. Thus, the fifth term is $10 + 5 = 15$, the sixth term is $15 + 6 = 21$, the seventh term is $21 + 7 = 28$, and so on. Using this method, we are finding each term *recursively*. More generally, a term in this sequence can be found by adding the previous term to the number of the term we are trying to find. We write this term symbolically as $a_n = a_{n-1} + n$, where a_n is the nth term, a_{n-1} is the previous term, and n is the number of the term we are trying to find. We then define this sequence by the *recurrence relation*

$$a_1 = 1 \qquad \text{and} \qquad a_n = a_{n-1} + n.$$

This recurrence relation gives the first term and instructions on how to find each next term. With that information, the entire sequence can be determined.

One of the most famous sequences is the Fibonacci sequence. This sequence is most often defined recursively. The Fibonacci sequence was given as the solution to a problem about rabbits. Leonardo of Pisa (ca. 1180–1250), better known as Fibonacci, wrote a book about algebraic methods called *Liber abaci* in which he proposed the following problem: "How many pairs of rabbits will be produced in a year, beginning with a single pair, if in every month each pair bears a new pair which becomes productive from the second month on?"[6] The solution to this problem generates the Fibonacci sequence $\{1, 1, 2, 3, 5, 8, \ldots\}$ in which each term is the sum of the previous two terms. Expressed as a recurrence relation, the Fibonacci sequence can be defined as

$$f_1 = 1, \qquad f_2 = 1, \qquad \text{and} \qquad f_n = f_{n-2} + f_{n-1}.$$

Notice in this definition that both the first and second terms need to be defined because the two previous terms are being used to find the next.

Example 1 Use the following recurrence relation to find the first five terms of the sequence $\{a_n\}$.

$$a_1 = 1 \qquad \text{and} \qquad a_n = a_{n-1} + 3n - 2.$$

Solution By knowing that the first term is 1, we know that the second term is $a_2 = 1 + 3(2) - 2 = 5$, the third term is $a_3 = 5 + 3(3) - 2 = 12$, the fourth term is $a_4 = 12 + 3(4) - 2 = 22$, and the fifth term is $a_5 = 22 + 3(5) - 2 = 35$. Therefore, the first five terms of the sequence are $\{1, 5, 12, 22, 35\}$.

Determining a sequence recursively works well if you are only trying to find a small number of terms. Suppose we wanted to find just the 100th term. Going through this recursive process is time consuming and tedious. Another approach is to find a *general formula* for the sequence. Because

[6] *Boyer,* A History of Mathematics, *pp. 280–81.*

finding the general formula for our triangular numbers is rather difficult unless you have worked at finding formulas like these before, we supply it here. The general formula for triangular number sequence, is

$$a_n = \frac{n(n+1)}{2}.$$

From this formula, we see that $a_1 = 1(2)/2 = 1$, $a_2 = 2(3)/2 = 3$, $a_3 = 3(4)/2 = 6$, and so on. Using this method, we can easily find any term in the sequence without having to find lots of others. For example, the 100th term is $a_{100} = 100(101)/2 = 5050$.

Example 2 Use the general formula

$$a_n = \frac{3n(n-1)+2}{2}$$

to find the first five terms of the sequence $\{a_n\}$.

Solution To find the first five terms of the sequence, we simply use the general formula five times.

$$a_1 = \frac{3 \cdot 1(1-1)+2}{2} = 1$$

$$a_2 = \frac{3 \cdot 2(2-1)+2}{2} = 4$$

$$a_3 = \frac{3 \cdot 3(3-1)+2}{2} = 10$$

$$a_4 = \frac{3 \cdot 4(4-1)+2}{2} = 19$$

$$a_5 = \frac{3 \cdot 5(5-1)+2}{2} = 31.$$

Therefore, the first five terms of the sequence are $\{1, 4, 10, 19, 31\}$.

READING QUESTIONS

1. Extend the Fibonacci sequence five terms beyond $\{1, 1, 2, 3, 5, 8\}$.
2. Use the recurrence relation

$$a_1 = 1 \quad \text{and} \quad a_n = a_{n-1} + 2n$$

to find the first three terms of the sequence $\{a_n\}$.

3. Use the general formula

$$a_n = \frac{n(n-1)}{2}$$

to find the 10th term of the sequence $\{a_n\}$.

ARITHMETIC AND GEOMETRIC SEQUENCES

Two common sequences are the arithmetic and geometric sequences. Each term in an **arithmetic sequence** is formed by adding a constant to each preceding term. For example, $\{4, 7, 10, 13, \ldots\}$ is an arithmetic sequence formed by starting with 4 and adding 3 to each term. In this sequence, the 3 is referred to as the *common difference* because 3 is the difference between a term and its preceding term. If you know a sequence starts with 4 and has a common difference of 3, then you can find any term in that sequence quite easily. For example, the second term is $4 + 3 = 7$, the third term is $4 + 3 + 3 = 4 + 3(2) = 10$, the fourth term is $4 + 3 + 3 + 3 = 4 + 3(3) = 13$, and the fifth term is $4 + 3 + 3 + 3 + 3 = 4 + 3(4) = 16$. Can you see a pattern here? Because repeated addition is multiplication, the nth term is $4 + 3(n - 1)$. In general, the nth term of any arithmetic sequence is $a_n = a_1 + d(n - 1)$, where a_1 is the first term and d is the common difference.

Example 3 Find the 20th term of an arithmetic sequence in which the first term is -4 and the common difference is 3.

Solution The general formula for the nth term of an arithmetic sequence is $a_n = a_1 + d(n - 1)$. In this case, $a_1 = -4$, $d = 3$, and $n = 20$. The 20th term, then, is $a_{20} = -4 + 3 \cdot (20 - 1) = 53$.

Notice the similarities between the arithmetic sequence and a linear function. In fact, one way of thinking about linear functions is in terms of a sequence where the y-intercept is the first term and the slope is the common difference.

Another common sequence is the geometric sequence. Instead of adding a constant, each term in a **geometric sequence** is formed by multiplying each preceding term by a constant. For example, $\{3, 6, 12, 24, \ldots\}$ is a geometric sequence. This sequence is formed by starting with 3 and multiplying by 2 to get the next term. The 2 is referred to as the *common ratio* because 2 is the ratio of a term and its preceding term. As with an arithmetic sequence, if you know that a geometric sequence starts with 3 and has a common ratio of 2, then you can find any term in the sequence quite easily. For example, the second term is $3 \cdot 2 = 6$, the third term is $3 \cdot 2 \cdot 2 = 3 \cdot 2^2 = 12$, the fourth term is $3 \cdot 2 \cdot 2 \cdot 2 = 3 \cdot 2^3 = 24$, and the fifth term is $3 \cdot 2 \cdot 2 \cdot 2 \cdot 2 = 3 \cdot 2^4 = 48$, again you should see a pattern. Because repeated multiplication is raising a number to an exponent, the nth term is $3 \cdot 2^{n-1}$. In general, the nth term of any geometric sequence is $a_n = a_1 \cdot r^{n-1}$, where a_1 is the first term and r is the common ratio.

Example 4 Find the eighth term in a geometric sequence in which the first term is 20 and the common ratio is 0.5.

Solution Because this sequence is geometric, our solution is $a_n = a_1 \cdot r^{n-1}$, where $a_1 = 20$, $r = 0.5$, and $n = 8$. Thus, the eighth term is $a_8 = 20 \cdot (0.5)^{(8-1)} = 0.15625$.

If the common ratio is greater than 1, then the geometric sequence grows very quickly; if the common ratio is between 0 and 1, then it decreases like an exponential function. In fact, the relationship between the geometric sequence and the exponential function is similar to the relationship between the arithmetic sequence and the linear function. In a geometric sequence, the first term is like the y-intercept or starting value in an exponential function and the common ratio is like the growth factor.

The following is a summary of the general formulas for arithmetic and geometric sequences.

PROPERTIES

- The nth term for an arithmetic sequence is given by $a_n = a_1 + d(n - 1)$, where d is the common difference.
- The nth term for a geometric sequence is given by $a_n = a_1 \cdot r^{n-1}$, where r is the common ratio.

READING QUESTIONS

4. If you were to write the first 100 even positive integers as an arithmetic sequence, then what would be the values for a_1 and d?

5. Find the 10th term in an arithmetic sequence in which the first term is 2 and the common difference is −4.

6. Find the 10th term in a geometric sequence in which the first term is 5 and the common ratio is 2.

FINDING THE SUMS OF ARITHMETIC AND GEOMETRIC SEQUENCES

One interesting property of sequences, especially as we look forward to our study of calculus, is what happens when we add the terms in a sequence. This concept is certainly not new. A frequently told story about Carl Friedrich Gauss, a famous mathematician from the late eighteenth and early nineteenth centuries, concerns finding the sum of a sequence. As the story goes, Gauss was a young boy when his teacher, looking for ways to occupy the class, asked his students to add all the integers from 1 to 100. The teacher was astonished when young Carl produced the correct answer almost immediately.[7] Let's see how he did it.

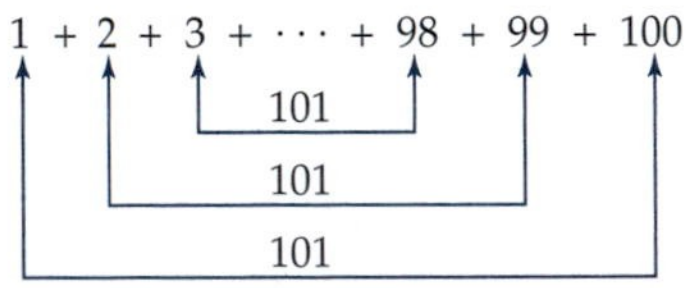

$1 + 2 + 3 + \cdots + 98 + 99 + 100$

101

101

101

Figure 2 Finding the sum of the first 100 positive integers can be done easily if you notice that you have 50 pairs of numbers whose sum is 101.

Gauss apparently noticed that if you add the first and last terms in the sequence, then you get a sum of 101. If you add the second and second-to-last terms, then you get the same sum, 101. You will continue to get the same sum if you continue to take pairs of terms in the same manner. (See Figure 2.) In fact, there are 50 of those pairs that add up to 101. To find the answer, you only need to multiply 50 by 101 to get 5050. Because the first

[7] *Boyer,* A History of Mathematics, *p. 544.*

100 integers are simply an arithmetic sequence, let's see if we can extend this idea to work with *any* arithmetic sequence.

Before we generalize, let's look at some new notation. The oversized uppercase Greek letter sigma, $\sum$, is used to designate sums. The sum, written as $\sum_{i=1}^{100} i$, is pronounced "the sum of i as i goes from 1 to 100." This notation represents the sum of the first 100 positive integers, or $\sum_{i=1}^{100} i = 1 + 2 + 3 + \cdots + 100$.

Now that we have a convenient way of expressing this summation, let's look at another arithmetic sequence and see if we can find a quick way to find its sum. Let's find the sum of the first 20 terms of the arithmetic sequence where the first term is 6 and the common difference is 4. To do so, we first need to find the 20th term. Because we know the first term and the common difference, the 20th term is $a_{20} = 6 + 4(20 - 1) = 82$. Our sequence can then be written as $\{6, 10, 14, \ldots, 78, 82\}$. Because we are interested in finding the sum, we can write this sequence with summation notation. We write

$$\sum_{i=1}^{20}(6 + 4(i - 1)) = \sum_{i=1}^{20} 4i + 2.$$

Using Gauss's reasoning, we see that adding the first and last terms of this sum gives 88. Adding the second and second-to-last terms also gives 88. Because there are 20 terms, there are 10 pairs of numbers whose sum is 88. Because $10 \cdot 88 = 880$, we can write

$$\sum_{i=1}^{20}(4i + 2) = 880.$$

This pattern works for any arithmetic sequence. The sum of the first and last terms is the same as the sum of the second and the second-to-last terms and so on. Because you are pairing numbers of a sequence with n numbers, there are $n/2$ pairs. Therefore, to find the sum of the terms in any arithmetic sequence, you need to add the first and the last and then multiply this sum by the number of terms divided by two. Symbolically, we write

$$S = \frac{n}{2}(a_1 + a_n),$$

where S is the sum of the first n terms of an arithmetic sequence, a_1 is the first term, and a_n is the nth term.

Example 5 Find the sum of the first 30 terms of $\{20, 25, 30, 35, \ldots\}$.

Solution To find the sum of these terms, we need to find the 30th term. Because the first term is 20 and the common difference is 5, the 30th term is $a_{30} = 20 + 5(30 - 1) = 165$. By using the formula $S = (n/2)(a_1 + a_n)$, we get,

$$\tfrac{30}{2}(20 + 165) = 2775.$$

Therefore, the sum of the first 30 terms is 2775.

Example 6 Find $\sum_{i=1}^{11}(224 + 9(i - 1))$.

Solution Here we are asked to find the sum of the first 11 terms in an arithmetic sequence whose first term is 224 and whose common difference is 9. To do so, we first need to find the 11th term. The 11th term is $a_{11} = 224 + 9(11 - 1) = 314$. By using the formula $S = (n/2)(a_1 + a_n)$, we get

$$\tfrac{11}{2}(224 + 314) = 2959.$$

Therefore, the sum of the first 11 terms is 2959.

Notice in Example 6 that the sum of the first and last terms is $224 + 314 = 538$. Because there are 11 terms, we have five pairs of numbers whose sum is 538 with one term left over. Not coincidently, the term that is left over is equal to half of 538, or 269. Hence, we have $5\frac{1}{2}$, or $\frac{11}{2}$, pairs of 538.

Finding the sum of the terms in a geometric sequence requires a different process than that for an arithmetic sequence. Let's start out with a simple geometric sequence and refer to its sum as S. Let

$$S = 1 + 2 + 4 + 8 + 16 + 32 + 64 + 128 + 256 + 512.$$

If we multiply both sides of this sum by the common ratio, 2, then we get another sum with many of the same terms,

$$2S = 2 + 4 + 8 + 16 + 32 + 64 + 128 + 256 + 512 + 1024.$$

If we now combine our two equations by subtracting, then most of the terms are eliminated.

$$\begin{array}{rcl} 2S &=& \phantom{-(1+{}} 2 + 4 + 8 + 16 + 32 + 64 + 128 + 256 + 512 + 1024 \\ -\ S &=& -\ (1 + 2 + 4 + 8 + 16 + 32 + 64 + 128 + 256 + 512) \\ \hline S &=& -\ 1 \hfill +\ 1024. \end{array}$$

Because $S = -1 + 1024 = 1023$, the sum is 1023. This process finds the sum of the terms for any geometric sequence. Let's do another example before we generalize.

Example 7 Find the sum of the first eight terms of the geometric sequence where the first term is 2 and the common ratio is 3.

Solution The sum of the terms of the geometric sequence is

$$S = 2 + 6 + 18 + 54 + 162 + 486 + 1458 + 4374.$$

Multiplying both sides by 3 gives

$$3S = 6 + 18 + 54 + 162 + 486 + 1458 + 4374 + 13{,}122.$$

Subtracting the first equation from the second gives

$$2S = 13{,}122 - 2$$
$$2S = 13{,}120$$
$$S = 6560.$$

Therefore, the sum is 6560.

We can now find a formula for the sum of the terms in any geometric sequence. We can express the sum of the first n terms of any geometric sequence as

$$S = a_1 + a_1r + a_1r^2 + a_1r^3 + \cdots + a_1r^{n-2} + a_1r^{n-1},$$

where a_1 is the first term and r is the common ratio. When we multiply both sides of the equation by r, we get

$$Sr = a_1r + a_1r^2 + a_1r^3 + \cdots + a_1r^{n-2} + a_1r^{n-1} + a_1r^n.$$

Subtracting S from Sr gives

$$Sr - S = a_1r^n - a_1,$$

which simplifies to

$$S(r - 1) = a_1(r^n - 1).$$

We can then solve for S by dividing both sides by $(r - 1)$. Finally, we get

$$S = \frac{a_1(r^n - 1)}{r - 1}.$$

Example 8 Find the sum of the first 15 terms of a geometric sequence that has a first term of 5 and a common ratio of 3.

Solution With $a_1 = 5$, $r = 3$, and $n = 15$, we can use $S = a_1(r^n - 1)/(r - 1)$ to get

$$S = \frac{5(3^{15} - 1)}{3 - 1}$$
$$= \frac{5(3^{15} - 1)}{2}$$
$$= \frac{5(14{,}348{,}906)}{2}$$
$$= 35{,}872{,}265.$$

So, the sum of these terms is 35,872,265.

In summary, the formulas for the sums of the terms for both arithmetic and geometric sequences are as follows.

PROPERTIES

- The sum of the first n terms of an arithmetic sequence is

$$S = \frac{n}{2}(a_1 + a_n),$$

where a_1 is the first term and a_n is the nth term.

- The sum of the first n terms of a geometric sequence is

$$S = \frac{a_1(r^n - 1)}{r - 1},$$

where a_1 is the first term and r is the common ratio.

READING QUESTIONS

7. What is the sum of the first 100 even integers?

8. Find $\sum_{i=1}^{10}(2 + 4(i - 1))$.

9. Find $\sum_{i=1}^{5} 3 \cdot 4^{i-1}$.

INFINITE SERIES

Suppose we want to find the sum of the first 10 terms of a geometric sequence where the first term is 1 and the common ratio is $\frac{1}{2}$. This sum can be found by using the formula for the sum of the terms of a geometric sequence:

$$\begin{aligned} S &= \frac{a_1(r^n - 1)}{r - 1} \\ &= \frac{1(\frac{1}{2}^{10} - 1)}{\frac{1}{2} - 1} \\ &\approx 1.998. \end{aligned}$$

Even after adding these 10 terms together, we still have a sum less than 2. What if we extended this sequence to find the sum of first 20 terms? Then

$$\begin{aligned} S &= \frac{a_1(r^n - 1)}{r - 1} \\ &= \frac{1(\frac{1}{2}^{20} - 1)}{\frac{1}{2} - 1} \\ &\approx 1.999998. \end{aligned}$$

The solution is, remarkably, still slightly less than 2.

How far can we go with this sequence and still get a sum less than 2? Amazingly, no matter how many terms we add, it will *never* exceed 2. Let's look at the formula for n terms to see why. Starting with $S = a_1(r^n - 1)/(r - 1)$, we have $a_1 = 1$ and $r = \frac{1}{2}$. So,

$$\begin{aligned} S &= \frac{1\left(\left(\frac{1}{2}\right)^n - 1\right)}{\frac{1}{2} - 1} \\ &= \frac{\left(\frac{1}{2}\right)^n - 1}{-\frac{1}{2}} \\ &= \frac{\left(\frac{1}{2}\right)^n}{-\frac{1}{2}} + \frac{1}{\frac{1}{2}} \\ &= -\left(\frac{1}{2}\right)^{n-1} + 2 \\ &= 2 - \left(\frac{1}{2}\right)^{n-1}. \end{aligned}$$

Because $\left(\frac{1}{2}\right)^{n-1}$ is always positive, the final expression, $2 - \left(\frac{1}{2}\right)^{n-1}$, is less than 2 for all finite values of n. Consequently, the sum of the terms in the sequence never exceeds 2. We can say, however, that

$$\lim_{n\to\infty}\left[2 - \left(\frac{1}{2}\right)^{n-1}\right] = 2.$$

By so doing, we are looking at the sum of an infinite number of terms of our geometric sequence. This sum can be written as

$$\sum_{i=1}^{\infty}\left(\frac{1}{2}\right)^{i-1} = 2.$$

What we are doing here is rather remarkable. We are looking at the sum of an infinite number of terms and getting a finite number.

The sum of an infinite number of terms of a sequence is called an **infinite series,** or just a **series.** The sum of a finite number of terms in a sequence is referred to as a **partial sum.** When the sum of a series gives a finite answer, as above, we say that the series *converges*. For our previous sequence, we can say that the series converges to 2.

There is nothing special about the particular geometric series that we looked at except that the common ratio was less than 1. Let's look at a general geometric series and see if we can generalize our steps. The formula we used to find the sum of the terms in a geometric sequence was

$$\frac{a_1(r^n - 1)}{r - 1}.$$

If we take the limit of this sequence, then, as n approaches infinity, we get

$$\lim_{n\to\infty}\frac{a_1(r^n - 1)}{r - 1}.$$

The only part of this expression with n in it is r^n. If we specify that $-1 < r < 1$, then $\lim_{n\to\infty} r^n = 0$. Therefore, for all r, where $-1 < r < 1$,

$$\lim_{n\to\infty} \frac{a_1(r^n - 1)}{r - 1} = \frac{a_1(0-1)}{r-1} = \frac{-a_1}{r-1} = \frac{a_1}{1-r}.$$

This gives us the property

$$S = \frac{a_1}{1-r}$$

where S is the sum of an infinite geometric series, a_1 is the first term, and r is the common ratio such that $-1 < r < 1$.

Example 9 Find the sum of an infinite geometric series whose first term is 3 and whose common ratio is $\frac{3}{4}$.

Solution Because the common ratio of this geometric series is between -1 and 1, we can find the sum by using the formula $S = a_1/(1-r)$. Doing so gives

$$S = \frac{3}{1-\frac{3}{4}} = \frac{3}{\frac{1}{4}} = 12.$$

To give us more confidence in this answer, let's look at values for this series as n increases, shown in Table 1. As you can see in the table, the sum seems to approach 12 without exceeding it.

TABLE 1

i	5	10	15	20	25	30	35
$\sum_{i=1}^{n} 3\cdot\left(\frac{3}{4}\right)^{i-1}$	9.152	11.324	11.840	11.962	11.991	11.998	11.999

Example 10 Find

$$\sum_{i=1}^{\infty} \frac{1}{3^i}.$$

Solution If we write out the first few terms of the series, then we can see that it is a geometric series:

$$\sum_{i=1}^{\infty} \frac{1}{3^i} = \frac{1}{3} + \frac{1}{9} + \frac{1}{27} + \cdots.$$

We can also see that the first term is $a_1 = \frac{1}{3}$ and the common ratio is $r = \frac{1}{3}$. Because $-1 < r < 1$, we can use the formula $S = a_1/(1-r)$ to find the sum. Doing so gives

$$S = \frac{\frac{1}{3}}{1 - \frac{1}{3}} = \frac{\frac{1}{3}}{\frac{2}{3}} = \frac{1}{2}.$$

Therefore,

$$\sum_{i=1}^{\infty} \frac{1}{3^i} = \frac{1}{2}.$$

So far, we have only dealt with geometric sequences and series where the common ratio is positive, but that does not have to be the case. For a geometric series to converge, it must have a common ratio between -1 and 1. In the next example, we see a geometric series with a common ratio that is negative.

Example 11 Find the sum of $2 - \frac{4}{3} + \frac{8}{9} - \frac{16}{27} + \frac{32}{81} + \cdots$.

Solution The series given is geometric with a first term of $a_1 = 2$ and a common ratio of $r = -\frac{2}{3}$. Because $-1 < r < 1$, we can use the formula $S = a_1/(1-r)$ to find the sum. Doing so gives

$$S = \frac{2}{1 - (-\frac{2}{3})} = \frac{2}{\frac{5}{3}} = \frac{6}{5}.$$

Therefore, the sum of the series is $\frac{6}{5}$.

READING QUESTIONS

10. If an infinite geometric series converges, what do you know about the value of r?

11. Find the sum of an infinite geometric series whose first term is 5 and whose common ratio is $\frac{1}{4}$.

SUMMARY

A **sequence** is a list of numbers written in a definite order. The numbers in a sequence are referred to as **terms.** Sometimes sequences are written in terms of a recurrence relation in which each term is defined by changing the previous term.

Each term in an **arithmetic sequence** is formed by adding a constant to each preceding term. The nth term for an arithmetic sequence is given by $a_n = a_1 + d(n-1)$, where d is the common difference. The sum of the first n terms of an arithmetic sequence can be found by using $S = (n/2)(a_1 + a_n)$, where S is the sum of the first n terms, a_1 is the first term, and a_n is the nth term.

Each term in a **geometric sequence** is formed by multiplying each preceding term by a constant. The nth term for a geometric sequence is given by $a_n = a_1 \cdot r^{n-1}$, where r is the common ratio. The sum of the first n terms of a geometric sequence can be found by using $S = a_1(r^n - 1)/(r - 1)$, where S is the sum of the first n terms, a_1 is the first term, and r is the common ratio.

The sum of an infinite number of terms of a sequence is called an **infinite series,** or just a **series.** The sum of a finite number of terms of a sequence is referred to as a **partial sum.** When the sum of a series gives a finite answer, we say the series *converges*.

If the common ratio in an infinite geometric series is between -1 and 1, the sum of the series is $S = a_1/(1 - r)$.

EXERCISES

1. Find the first five terms of the sequence defined by each of the following recurrence relations.
   (a) $a_1 = 1, a_n = a_{n-1} + 4$
   (b) $a_1 = 1, a_n = a_{n-1} \cdot 3$
   (c) $a_1 = 2, a_n = a_{n-1} + n$
   (d) $a_1 = 5, a_n = a_{n-1} - 2n$
   (e) $a_1 = 2, a_2 = 3, a_n = a_{n-1} + a_{n-2}$
2. Find the 20th term of the sequence defined by each of the following general formulas.
   (a) $a_n = -6 + 7n$
   (b) $a_n = 4 - 3(n - 1)$
   (c) $a_n = 4 \cdot 3^{n-1}$
   (d) $a_n = 2 \cdot \left(\frac{6}{5}\right)^n$
   (e) $a_n = \dfrac{n(n+1)(2n+1)}{6}$
3. Write a formula for the nth term of each arithmetic sequence.
   (a) $\{4, 8, 12, 16, \ldots\}$
   (b) $\{-3, 7, 17, 27, \ldots\}$
   (c) $\{-2, -16, -30, -44, \ldots\}$
4. Write a formula for the nth term of each geometric sequence.
   (a) $\{4, 8, 16, 32, \ldots\}$
   (b) $\{-9, -27, -81, -243, \ldots\}$
   (c) $\{-2, 16, -128, 1024, \ldots\}$
5. Find the following sums.
   (a) $4 + 7 + 10 + \cdots + 82 + 85 + 88$

(b) $(-12)+(-10)+(-8)+\cdots+22+24+26$
(c) $16+11+6+\cdots+(-9)+(-14)+(-19)$

6. Find the following sums.
   (a) $\sum_{i=1}^{10}(3i+1)$
   (b) $\sum_{i=7}^{14}(7+4i)$
   (c) $\sum_{i=3}^{100}(4+6i)$
7. Find the sum of the 16th through the 24th terms of the arithmetic sequence whose first term is 10 and whose common difference is 4.
8. Find the sum of the 6th through the 32nd terms of the arithmetic sequence whose first term is 2 and whose common difference is 7.
9. Find the sum of each of the following series.
   (a) $2+\frac{2}{3}+\frac{2}{9}+\frac{2}{27}+\cdots$
   (b) $\frac{1}{4}+\frac{1}{2}+1+2+\cdots$
   (c) $3+\frac{3}{5}+\frac{3}{25}+\frac{3}{125}+\cdots$
10. Find the following partial sums.
   (a) $\sum_{i=1}^{10}2\cdot 3^i$
   (b) $\sum_{i=1}^{14}7\cdot\left(\frac{1}{5}\right)^i$
   (c) $\sum_{i=1}^{8}5\cdot\left(\frac{2}{3}\right)^i$
11. Find the sum of the 19th through the 24th terms of the geometric sequence whose first term is 10 and whose common ratio is $\frac{3}{5}$.
12. Find the sum of the 6th through the 12th terms of the geometric sequence whose first term is 2 and whose common difference is $\frac{1}{7}$.
13. Find the sum of the infinite geometric series that has a first term of 4 and a common ratio of $\frac{3}{8}$.
14. Find the sum of the infinite geometric series that has a first term of 3 and a common ratio of $\frac{7}{9}$.
15. Find the sum of each series.
   (a) $\sum_{i=1}^{\infty}12\cdot\left(\frac{8}{11}\right)^i$
   (b) $\sum_{i=1}^{\infty}7\cdot\left(\frac{2}{5}\right)^i$
   (c) $\sum_{i=1}^{\infty}2\cdot\left(\frac{2}{7}\right)^i$

16. In this section, we gave a general formula, $a_n = n(n+1)/2$, to find the nth triangular number. Explain why this formula is the same as that used for finding the sum of the first n positive integers.
17. The repeating decimal $0.\overline{3} = 0.3333\ldots$ can be written as the infinite geometric series $0.3 + 0.03 + 0.003 + \cdots$. This series has a first term of 0.3 and a common ratio of 0.1. Thus, the sum of the series is $S = a_1/(1-r) = 0.3/(1-0.1) = 0.3/0.9 = \frac{1}{3}$. As you can see, this process can be used to convert a repeating decimal into a fraction. Using this process, convert the following repeating decimals to fractions.
    (a) $0.\overline{5}$
    (b) $0.\overline{9}$
    (c) $0.\overline{25}$
    (d) $2.\overline{73}$
18. Consider the series defined by

$$\sum_{n=0}^{\infty} \frac{1}{n!}.$$

Remember that the symbol ! is a factorial symbol and that $n! = n(n-1)(n-2)\cdots 2 \cdot 1$. Also remember that 0! is defined to be 1.
    (a) Find the partial sum $\sum_{n=0}^{6}(1/n!)$.
    (b) The infinite series $\sum_{n=0}^{\infty}(1/n!)$ is equal to a commonly used mathematical number. Which number is it? (*Hint:* Your answer to part (a) should be a good approximation for this number.)
19. Chain letters have been sent out for years and often promise great things to you if you do not break the chain. For example, one chain letter might say: "Send $1 to the top name on this list, cross out that name, put your name at the bottom of the list of people, and send copies of this letter to 10 of your friends. Soon you will receive thousands of dollars!" This instruction is often followed by the names of people who got rich simply by sending $1! The idea is simple. If you send the letter to 10 friends and they each send it to 10 friends who in turn do the same, 1000 people have the letter and are sending $1 to the person whose name is on the top of the list.
    (a) Suppose the first person sends this letter out to 10 people. Express the number of recipients of the letter as a sequence where each term represents another mailing of the letter for five rounds (where a round is what happens when everyone receives the letter and then mails it out again).
    (b) Suppose there are five names above yours on the list. How much money can you expect to receive if no one breaks the chain?
    (c) Suppose there are nine names above yours on the list. How much money can you expect to receive if no one breaks the chain?
    (d) The population of the United States in 1999 was approximately 273,000,000.[8] If the letter has nine names above yours, why

[8] U.S. Census Bureau Home Page *(visited 6 July 1999)* ⟨*http://www.census.gov*⟩.

is it unlikely that you will see the amount of money you are promised in the letter?

20. Consider the following first four steps in a sequence of dot patterns.

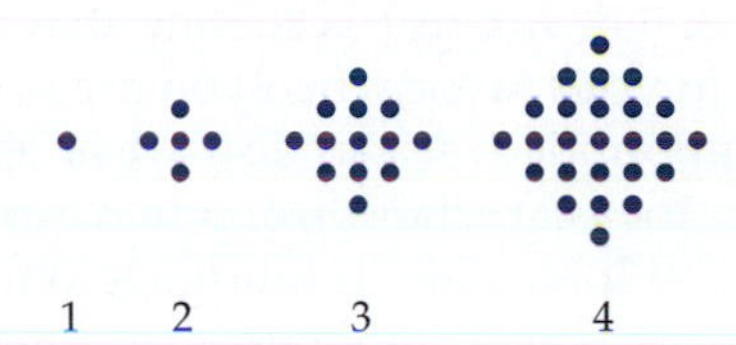

(a) How many dots will there be in the fifth step?

(b) How many dots will there be in the nth step?

(c) Find a recurrence relation for the sequence of the number of dots in each step.

21. Consider the following first four steps in a sequence of dot patterns. The number of dots in these patterns represents the sequence of *pentagonal numbers*.

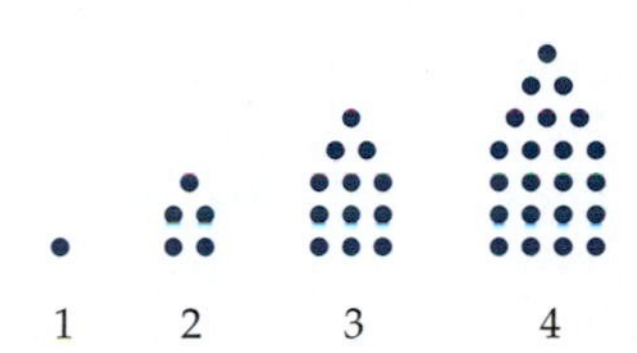

(a) What is the fifth pentagonal number?

(b) What is the nth pentagonal number?

(c) Find a recurrence relation that describes the sequence of pentagonal numbers.

22. The general formula for the Fibonacci sequence is

$$f_n = \frac{1}{\sqrt{5}}\left(\frac{1+\sqrt{5}}{2}\right)^{n+1} - \frac{1}{\sqrt{5}}\left(\frac{1-\sqrt{5}}{2}\right)^{n+1}.$$

Use this formula to verify the first six terms of the Fibonacci sequence.

23. In an old story, the person that developed the game of chess presented his new game to his king. The king was so pleased with the game that he told the inventor that he could have anything he desired. After much thought, the inventor requested that he be given grains of wheat in the following way. "One grain for the first square on the chess board, two grains for the second square, four grains for the third, and so on by doubling the number of grains of wheat for each additional square on the board." There are 64 squares on a chess board. How many grains of wheat did the inventor request?

INVESTIGATIONS

INVESTIGATION 1: BUY NOW, PAY LATER

The power that time has on money can be quite surprising when interest is involved. Interest is money that is paid for the use of money. It can work in your favor when you are saving money and it earns interest that is compounded. It can also work against you when you are borrowing money for a car, a house, or purchases bought using a credit card. Payments on many loans, like a mortgage on a house, are done in equal monthly payments. The formula used to compute such payments is

$$M = \frac{iP}{1-(1+i)^{-n}}, \tag{1}$$

where M is the periodic payment (usually monthly), P is the principal,[9] i is the annual interest rate divided by the number of times compounded per year (usually monthly), and n is the total number of payments you will make (the number of payments per year multiplied by the number of years). To help understand equation (1), let's look at a simple case. Assume you borrowed \$1000 for 1 year at 12% compounded quarterly (quarterly means interest is added to your account four times during the year, or every 3 months). What actually happens is that at the end of the first quarter, you are charged 3% on the amount you owe, in this case \$30. Your payment must cover the interest payment plus some portion of the principal (the amount you borrowed in the first place). According to equation (1), that payment is \$269.03. After you make that first payment, you now owe \$1030 − \$269.03 = \$760.97. When it is time to make your second payment, you are charged interest on the amount left, that is, on \$760.97. So, your new interest charge for the second quarter is 0.03(\$760.97) = \$22.83. This process happens twice more. Each time, you are charged interest on the money you still owe until, on your final payment, you pay off all the money. Table 2 summarizes what happens.

TABLE 2

Quarter	Interest Charged	Amount Owed	Payment	Amount Owed after Payment
First	\$30	\$1030.00	\$269.03	\$760.97
Second	\$22.83	\$783.80	\$269.03	\$514.77
Third	\$15.44	\$530.21	\$269.03	\$261.18
Fourth	\$7.84	\$269.02	\$269.02	\$0

1. Use equation (1) to determine the payments on \$3000 at 8% compounded quarterly for 1 year. Show how much interest and principal has been

[9] *The initial amount of money you deposit or borrow is called the principal.*

paid in each of the four payments by completing Table 3. A schedule, such as Table 3, listing each of the four payments including the amount of interest and principal is called an *amortization* schedule.[10]

TABLE 3

Quarter	Interest Charged	Amount Owed	Payment	Amount Owed after Payment
First				
Second				
Third				
Fourth				

2. Formulas such as the one given for finding payments do not simply drop out of the sky. Before continuing to use this formula, let's derive it, starting with a simple case. Let P be the amount borrowed, r be the annual interest rate, and M be the monthly payment. Assume we make quarterly payments and plan to pay back the loan by the end of the year. The interest rate, i, we use is actually $r/4$, because we are making the payments quarterly. At the end of the first quarter, we owe $P + iP = (1+i)P$. We then make our payment, M, leaving us with a balance of $(1+i)P - M$. At the end of the second quarter, we owe

$$\begin{aligned}[(1+i)P - M] + i[(1+i)P - M] &= (1+i)[(1+i)P - M] \\ &= (1+i)^2P - (1+i)M.\end{aligned} \tag{2}$$

   (a) Continuing this process, show the steps to get

$$(1+i)^4P - (1+i)^3M - (1+i)^2M - (1+i)M - M \tag{3}$$

   at the end of the fourth quarter.

   (b) Because this process paid off the whole loan, we can set it equal to zero.

$$(1+i)^4P - (1+i)^3M - (1+i)^2M - (1+i)M - M = 0. \tag{4}$$

   Show the steps needed to get from equation (4) to

$$P = \left(\frac{1}{1+i} + \frac{1}{(1+i)^2} + \frac{1}{(1+i)^3} + \frac{1}{(1+i)^4}\right)M. \tag{5}$$

   (c) Equation (5) would look nicer if we could simplify the part in the large parentheses. Let's call that whole piece X, that is

$$X = \frac{1}{1+i} + \frac{1}{(1+i)^2} + \frac{1}{(1+i)^3} + \frac{1}{(1+i)^4}. \tag{6}$$

[10] *The word* amortization *contains the root word* mort, *the same root used in* mortician *and* mortal, *referring to death. An amortization schedule tells you how you "kill off" your loan.*

At this point, we use a procedure previously used when finding the sum of a geometric series. Multiply both sides of equation (6) by $(1+i)$ and subtract equation (6) from the resulting equation. Lots of terms should disappear! Now solve this equation for X and show that you get

$$X = \frac{1}{i}\left(1 - \frac{1}{(1+i)^4}\right).$$

(d) Place this result into equation (5) and solve for M. How does this result compare with the equation given for M at the beginning of this investigation? Explain why there is a difference.

3. Explain why it is necessary that how often you make your payment and how often the interest is compounded be the same for this formula to be valid.

4. Suppose you are planning to buy a house and determine that you can afford about a \$600 monthly house payment. Assume the current annual interest rate is 9%. You have \$3000 saved for the down payment.
   (a) Show that if you take out a 15-year loan, then you should plan to buy a home that costs about \$62,000. Show that if you take out a 30-year loan, then you should plan to buy a home that costs about \$77,500.
   (b) Suppose you purchase a \$62,000 home with a 15-year loan. What is the total price you will have paid for this house at the end of that 15 years? How much of this money was interest?
   (c) Suppose you purchase a \$77,500 home with a 30-year loan. What is the total price you will have paid for this house at the end of that 30 years? How much of this money was interest?
   (d) What are your advantages in taking out a 15-year loan? What are your advantages of taking out a 30-year loan?

INVESTIGATION 2: MATHEMATICAL INDUCTION

Mathematical induction is a powerful technique that can be used to prove many statements in mathematics. We use it to prove that a general formula for different types of sums is correct.

To prove a statement using mathematical induction, you must show that the statement is true for the first step. You then assume that it is true for some arbitrary step and show that it must be true for the next step. This method is much like the "domino effect." If you know that the first domino will fall when you push it over and if you know that if one domino falls then the next in line will, then you can conclude that all the dominoes will fall.

Let's use mathematical induction to prove that

$$1+2+3+\cdots+n = \frac{n(n+1)}{2}.$$

To do so, we first need to show that this statement is true for the first step, (or $n = 1$). Because

$$\frac{1(1+1)}{2} = \frac{1(2)}{2} = 1,$$

we know that this statement is true for $n = 1$.

Next we assume this statement is true for some arbitrary step. We call our arbitrary step k. So, we need to assume the statement is true for $n = k$ and use that assumption to show that it is true for the next step (or $n = k + 1$). Assuming it is true for $n = k$ means that

$$1 + 2 + 3 + \cdots + k = \frac{k(k+1)}{2}.$$

From this result, we need to show that the statement is true for $n = k + 1$, or

$$1 + 2 + 3 + \cdots + k + (k+1) = \frac{(k+1)(k+2)}{2}.$$

Starting with the left side of this equation, we substitute $k(k+1)/2$ for $1 + 2 + 3 + \cdots + k$ and simplify our result to prove that this statement is true for $n = k + 1$. So,

$$\begin{aligned} 1 + 2 + 3 + \cdots + k + (k+1) &= \frac{k(k+1)}{2} + (k+1) \\ &= \frac{k(k+1)}{2} + \frac{2(k+1)}{2} \\ &= \frac{k(k+1) + 2(k+1)}{2} \\ &= \frac{(k+1)(k+2)}{2}. \end{aligned}$$

We now have proved, through induction, that

$$1 + 2 + 3 + \cdots + n = \frac{n(n+1)}{2}.$$

1. We want to prove by using mathematical induction that the following is true.

$$1 \cdot 2 + 2 \cdot 3 + 3 \cdot 4 + \cdots + n(n+1) = \frac{n(n+1)(n+2)}{3}.$$

   (a) Show that this statement is true for $n = 1$.

   (b) Assume the statement is true for $n = k$. In other words, assume

$$1 \cdot 2 + 2 \cdot 3 + 3 \cdot 4 + \cdots + k(k+1) = \frac{k(k+1)(k+2)}{3}.$$

   Using this statement, show that our original statement is true for $n = k + 1$. In other words, show that

$$\begin{aligned} &1 \cdot 2 + 2 \cdot 3 + 3 \cdot 4 + \cdots + k(k+1) + (k+1)(k+2) \\ &= \frac{(k+1)(k+2)(k+3)}{3}. \end{aligned}$$

2. Prove by mathematical induction that the following statements are true.

   (a) $\dfrac{1}{1\cdot 2}+\dfrac{1}{2\cdot 3}+\dfrac{1}{3\cdot 4}+\cdots+\dfrac{1}{n\cdot(n+1)}=\dfrac{n}{n+1}$

   (b) $1+3+5+7+\cdots+(2n-1)=n^2$

   (c) $1^2+2^2+3^2+4^2+\cdots+n^2=\dfrac{n(n+1)(2n+1)}{6}$

8.4 AREA AND THE INTEGRAL

When we considered the derivative, we looked at one of the two major concepts addressed by the study of calculus: the instantaneous rate of change of a function. The other concept, that of finding the area under a curve, is answered by the integral. In this section, you learn how to approximate the area under a curve, what an integral is, and how integrals can be applied in some applications.

AREA UNDER A CURVE

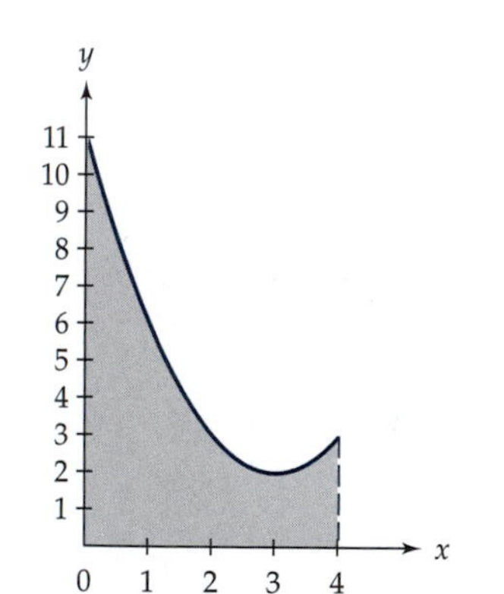

Figure 1 We want to determine the area of the shaded region under the parabola.

When studying geometry, even back in elementary school, you found areas of regions that were composed of straight sides, like rectangles and triangles. The only region without straight sides in which students typically find the area is that of a circle or a part of a circle. Suppose we are interested in finding the area under a parabola. In particular, suppose we want to determine the area bounded by $f(x)=(x-3)^2+2$ on the top, the x-axis on the bottom, and the vertical lines $x=0$ and $x=4$ on the left and the right. This region is shown in Figure 1.

This problem is often referred to as *finding the area under the curve from 0 to 4*. Although the region is bounded by lines on the bottom, left, and right, what makes this problem a challenge is that the region is not bounded by lines (or by a circle) on the top. If it were, then we would simply use one of the formulas that we have already learned. What we can do, however, is approximate the area.

One way to approximate the area is to use rectangles where the height of the left edge of the rectangle corresponds to the height of the curve.[11] By splitting the region into two sections, from $x=0$ to $x=2$ and from $x=2$ to $x=4$, we can make two rectangles. Because $f(0)=11$ and $f(2)=3$, our left rectangle has a height of 11 and the right rectangle has a height of 3. See Figure 2(a). Thus, the approximation for the area under the curve is the sum of the areas of the two rectangles,

$$L_2=11\cdot 2+3\cdot 2=28.$$

This approximation is fair, but as you can probably guess, we can do better by making more rectangles. Figure 2(b) shows four rectangles, each with

[11] *There is nothing particularly special about using the left edge except for convenience. We could also use the right edge or the center of each rectangle, which would give different approximations, but the concept is the same.*

a width of 1. The heights are $f(0) = 11, f(1) = 6, f(2) = 3$, and $f(3) = 2$. Together, the areas of these four rectangles sum to be

$$L_4 = 11 \cdot 1 + 6 \cdot 1 + 3 \cdot 1 + 2 \cdot 1 = 22.$$

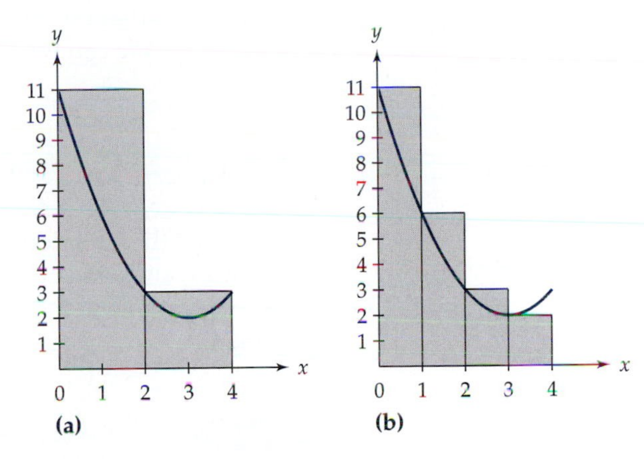

Figure 2 Using four rectangles to approximate the area under the parabola gives a better approximation than using two.

We can continue to increase the number of rectangles in this way and continue to improve our approximation. Let's increase the number of rectangles once more and generalize the process along the way. Each time we approximate the area, we get a series of rectangles of equal width. We are trying to find the area under the curve from $x = 0$ to $x = 4$, so if we want n rectangles, then each rectangle has width $(4 - 0)/n = 4/n$. In general, if our interval goes from a to b, then $b - a$ is the width of the interval and $(b - a)/n$ is the width of each rectangle. We refer to this latter width as Δx. If we divide this area into eight rectangles, the width is then $\Delta x = (4 - 0)/8 = \frac{1}{2}$. (See Figure 3.)

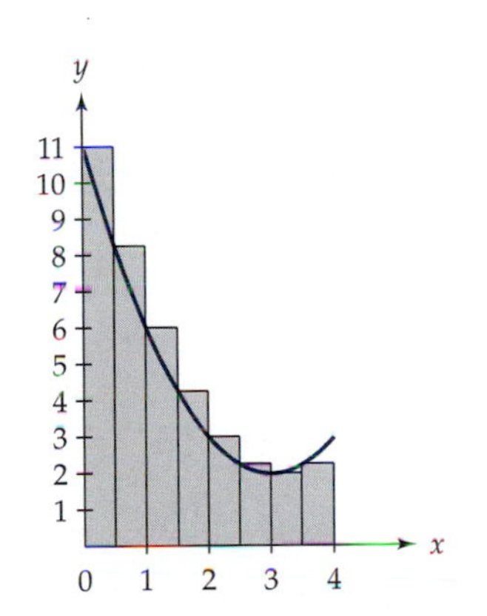

Figure 3 Using eight rectangles gives a better approximation than using four.

To approximate the area, we need to find the sum of the areas of the eight rectangles. Each rectangle has a width of $\frac{1}{2}$. We again let the height of the left edge of the rectangle correspond to the height of the curve at that point. Because we are using the left edge, we use the letter L to indicate the sum, and because we are summing eight rectangles, we use a subscript 8. If we had used the right edge, then we would have called our sum R_8. If we had used the center (or middle), then we would have called our sum M_8. We consider these different ways of approximating the area under the curve later. The sum of the eight rectangles is

$$\begin{aligned} L_8 &= \tfrac{1}{2} \cdot f(0) + \tfrac{1}{2} \cdot f(0.5) + \tfrac{1}{2} \cdot f(1) + \tfrac{1}{2} \cdot f(1.5) + \tfrac{1}{2} \cdot f(2) + \tfrac{1}{2} \cdot f(2.5) \\ &\quad + \tfrac{1}{2} \cdot f(3) + \tfrac{1}{2} \cdot f(3.5) \\ &= \tfrac{1}{2}(f(0) + f(0.5) + f(1) + f(1.5) + f(2) + f(2.5) + f(3) + f(3.5)) \\ &= \tfrac{1}{2}(11 + 8.25 + 6 + 4.25 + 3 + 2.25 + 2 + 2.25) \\ &= 19.5. \end{aligned}$$

This approximation of 19.5 is an even better approximation than our earlier ones.

In general, we can use the summation notation learned in the last section to write these types of sums more compactly. Suppose we wanted to approximate the area under a curve from a to b by finding the sum of n rectangles such that the left edge of the rectangle corresponds to the height of the curve, f. We can write this sum as

$$L_n = \sum_{i=1}^{n} f(x_i)\Delta x,$$

where $x_1 = a, x_2 = a + \Delta x, x_3 = a + 2\Delta x, \ldots, x_n = a + (n-1)\Delta x, f(x_i)$ is the output of the function at the left endpoints of the n different equally spaced intervals, and $\Delta x = (b-a)/n$ is the width of each interval. As the number of rectangles increases, the closer our approximation gets to the true area.

Example 1 Approximate the area under the curve $g(x) = (x-2)^3 + 8$ from 0 to 2 by subdividing the interval into eight equally spaced rectangles and letting the left edge of the rectangle correspond to the height of the curve.

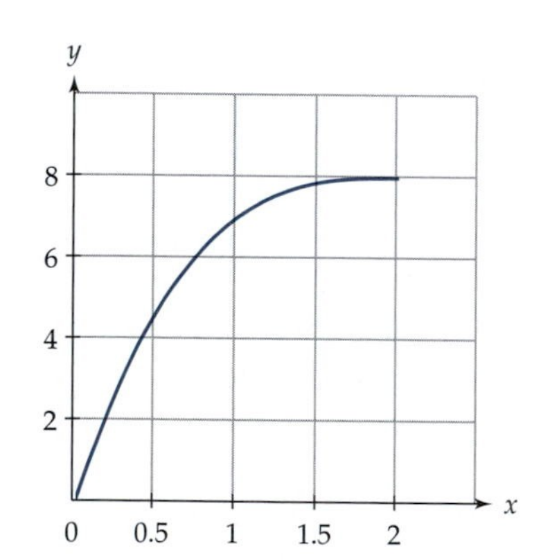

Figure 4

Solution A graph of the function on this interval is shown in Figure 4. From this graph, you see that the maximum height of the function on the interval $0 \le x \le 2$ is 8. Because the width of the entire interval is 2, the area under the curve must be less than $2 \times 8 = 16$.

Because the width of our entire interval is 2 and we need eight subintervals, the width of each rectangle is $\frac{2}{8} = \frac{1}{4} = 0.25$. The height of the rectangles correspond to $g(x_i)$ for values of x_i that run from 0 to 1.75 in increments of 0.25. Thus,

$$\begin{aligned} L_8 &= \tfrac{1}{4}\cdot g(0) + \tfrac{1}{4}\cdot g(0.25) + \tfrac{1}{4}\cdot g(0.5) + \tfrac{1}{4}\cdot g(0.75) + \tfrac{1}{4}\cdot g(1) \\ &\quad + \tfrac{1}{4}\cdot g(1.25) + \tfrac{1}{4}\cdot g(1.5) + \tfrac{1}{4}\cdot g(1.75) \\ &= \tfrac{1}{4}\left(g(0) + g(0.25) + g(0.5) + g(0.75) + g(1) + g(1.25) \right. \\ &\quad \left. + g(1.5) + g(1.75)\right) \\ &\approx \tfrac{1}{4}(0 + 2.641 + 4.625 + 6.047 + 7 + 7.578 + 7.875 + 7.984) \\ &= \tfrac{1}{4}(43.75) \\ &= 10.9375. \end{aligned}$$

We thus have an approximate area of 10.9375.

READING QUESTIONS

1. When using the sum of the areas of rectangles to approximate the area under a curve, how can you improve the approximation?

2. For the area under the curve in Example 1, we obtained an answer of 10.9375. Is this an underestimate or an overestimate? How do you know?

RIEMANN SUMS AND THE INTEGRAL

So far, all our examples have been done with the left edge of each rectangle corresponding to the height of the curve. As mentioned earlier, this setup need not be the one we use. Let's again look at the function from Example 1. There we used eight rectangles whose left edges each corresponded to the height of the function to approximate the area. A drawing is shown in Figure 5(a). We could have used eight rectangles whose right edges each corresponded to the height of the function. A drawing of that is shown in Figure 5(b).

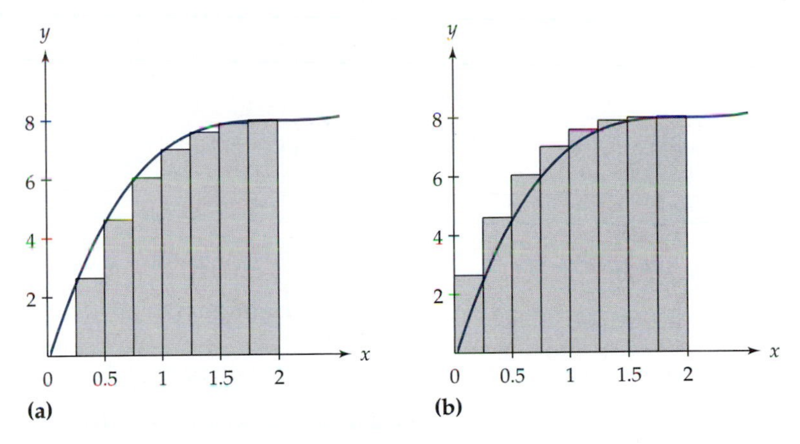

Figure 5 When approximating the area under the curve shown here, using left-edge rectangles gives an underestimate and using right-edge rectangles gives an overestimate.

By looking at Figure 5(a), you can see that when using the left-edge rectangles, we get an underestimate of the area in this situation. When using the right-edge rectangles, we get an overestimate of the area. Let's calculate the area using the right-edge rectangles. To do so, remember that our function is $g(x) = (x-2)^3 + 8$. The width of our eight rectangles is the same as before, $\frac{1}{4}$. The only difference is the points at which we evaluate the function. Instead of starting at 0, we start at $\frac{1}{4} = 0.25$. Doing so gives

$$\begin{aligned}
R_8 &= \tfrac{1}{4} \cdot g(0.25) + \tfrac{1}{4} \cdot g(0.5) + \tfrac{1}{4} \cdot g(0.75) + \tfrac{1}{4} \cdot g(1) + \tfrac{1}{4} \cdot g(1.25) + \tfrac{1}{4} \cdot g(1.5) \\
&\quad + \tfrac{1}{4} \cdot g(1.75) + \tfrac{1}{4} \cdot g(2) \\
&= \tfrac{1}{4}\big(g(0.25) + g(0.5) + g(0.75) + g(1) + g(1.25) + g(1.5) \\
&\quad + g(1.75) + g(2)\big) \\
&\approx \tfrac{1}{4}(2.641 + 4.625 + 6.047 + 7 + 7.578 + 7.875 + 7.984 + 8) \\
&= \tfrac{1}{4}(51.75) \\
&= 12.9375.
\end{aligned}$$

By using the right-edge rectangles, we have an approximate area of 12.9375. Because this value is an overestimate and our earlier estimate of 10.9375 was an underestimate, we know that the area under the curve must be between 10.9375 and 12.9375.

Example 2 Approximate the area under the curve $f(x) = \left(\frac{1}{2}\right)^x$ from -2 to 1 by subdividing the interval into six equally spaced rectangles and letting the right edge of the rectangle correspond to the height of the curve. Is the area you find an underestimate or an overestimate?

Solution Because the width of our entire interval is 3 and we need six subintervals, the width of each rectangle is $\frac{3}{6} = \frac{1}{2}$. The height of the rectangles correspond to $f(x_i)$ for values of x_i that run from -1.5 to 1 in increments of $\frac{1}{2}$. Thus,

$$\begin{aligned} R_8 &= f(-1.5)\cdot\tfrac{1}{2} + f(-1)\cdot\tfrac{1}{2} + f(-0.5)\cdot\tfrac{1}{2} + f(0)\cdot\tfrac{1}{2} + f(0.5)\cdot\tfrac{1}{2} + f(1)\cdot\tfrac{1}{2} \\ &= \tfrac{1}{2}(f(-1.5) + f(-1) + f(-0.5) + f(0) + f(0.5) + f(1)) \\ &\approx \tfrac{1}{2}(2.828 + 2 + 1.414 + 1 + 0.707 + 0.5) \\ &= \tfrac{1}{2}(8.449) \\ &= 4.2245. \end{aligned}$$

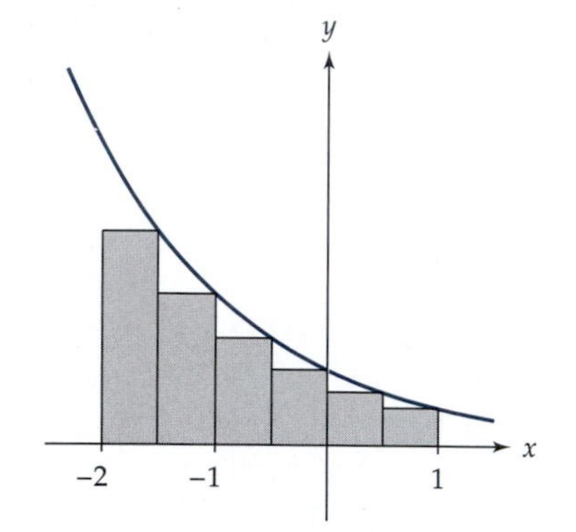

Figure 6 For this function, using right-edge rectangles gives an underestimate.

Using the right-edge rectangles, we have an approximate area of 4.2245. By looking at a graph of this area with rectangles included, as shown in Figure 6, we can see that this result is an underestimate.

All of the estimates of areas under curves that we have done were found using special types of *Riemann sums*, named after German mathematician G. F. B. Riemann. Simply put, a Riemann sum is a method of approximating the area under curves using the sum of the areas of rectangles. It does not matter if the left edge or the right edge is used to correspond to the height of the curve. It also does not matter if any other portion of the rectangle corresponds to the height of the curve. The area of these rectangles are all considered Riemann sums. In fact, the rectangles need not even be of the same width.

A Riemann sum can be written similarly to how we wrote sums of rectangles earlier in this section. In general, if we have a function f defined on an interval from a to b, if we divide that interval into n subintervals such that

$$a = x_0 < x_1 < x_2 < \cdots < x_n = b,$$

and if we let $\Delta x_i = x_i - x_{i-1}$ be the ith subinterval such that c_i is any point in that interval, then the sum

$$\sum_{i=0}^{n} f(c_i)\Delta x_i$$

is a Riemann sum with n subdivisions.

HISTORICAL NOTE

Georg Friedrich Bernhard Riemann was born in Germany in 1826. Even though he was raised in modest circumstances, he managed to get a good education and make a name for himself in mathematics. Riemann submitted his doctoral dissertation to the already legendary mathematician Gauss for his consideration. Gauss, who was not known for giving praise, said about Riemann's dissertation, "The dissertation submitted by Herr Riemann offers convincing evidence of the author's thorough and penetrating investigations in those parts of the subject treated in the dissertation, of a creative, active, truly mathematical mind, and of a gloriously fertile originality."

After completing his doctoral work, Riemann became *Privatdozent*, or private lecturer, at the University of Göttingen. The custom there is for the *Privatdozent* to deliver a special lecture or a professorial dissertation for the faculty. The lecture turned out to be a very important one in which Riemann presented a broad view of the entire field of geometry. Riemann's work, along with that of his student, Felix Klein, is sometimes referred to as the high point of the golden age of modern geometry.

This interest in geometry and in finding the distance between two points that are infinitesimally close together brought Riemann to the definition of the definite integral presented in this section. Even though other definitions have been suggested, the work of Riemann is still presented in most undergraduate calculus courses. Riemann went on to do many things in mathematics and physics in the fields of number theory and, within 5 years of arriving at the University of Göttingen, held the same chair as Gauss.

Felix Klein also achieved a certain amount of notoriety and is perhaps best known for the Klein bottle. The Klein bottle, named for him after his death in 1925, is a bottle whose neck is turned back into the bottle in such a way that there is only one surface, the inside and outside flowing seamlessly into one another. He was such an inspiring lecturer and took such a great interest in mathematics pedagogy—that is, the teaching of mathematics—that people flocked from around the world to the University of Göttingen to study mathematics with him and his colleagues.

Sources: E. T. Bell, *Men of Mathematics,* (New York: Simon and Schuster, 1937), p. 495; Carl B. Boyer, *A History of Mathematics,* (Princeton, NJ: Princeton University Press, 1968), p. 358, 588–94.

As the number of rectangles increases in a Riemann sum, the approximation of the area gets closer to the true area. In fact, if we find

$$\lim_{n\to\infty} \sum_{i=0}^{n} f(c_i)\Delta x_i,$$

then we have the exact area under the curve. This limit is called an *integral*.

In Example 2, we were finding the approximate area under the curve $f(x) = \left(\frac{1}{2}\right)^x$ from -2 to 1. We can express the exact area using the integral sign, $\int dx$, as

$$\int_{-2}^{1} \left(\frac{1}{2}\right)^x dx.$$

The dx indicates that the variable x is the variable in the function we are using. If we were using a different variable, then we would simply replace the x with that letter, but the d is always there. You can see that the lower and upper bounds of the area in question are noted on the integral sign (which is simply an elongated letter s, reminding us that it is a sum). We read this notation as "the integral from -2 to 1 of $f(x) = \left(\frac{1}{2}\right)^x$." In this text we find integrals (or areas under curves) only for areas in which we can easily find a geometric formula (as in our next example). For all others, we use a Riemann sum to approximate the area.

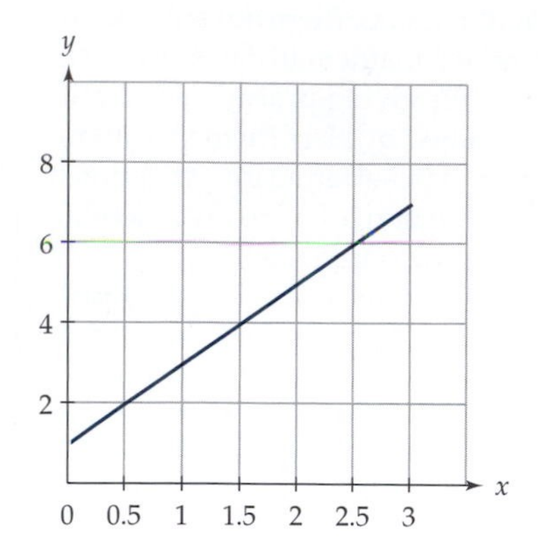

Figure 7

Example 3 A graph of the line $f(x) = 2x + 1$ on the interval from $x = 0$ to 3 is shown in Figure 7. Find $\int_0^3 (2x + 1)\,dx$.

Solution The area under this line from $x = 0$ to $x = 3$ forms a trapezoid. The area of a trapezoid is $A = \frac{1}{2}h(a + b)$, where h is the height and a and b are the lengths of the two parallel sides. Because our parallel sides are vertical, the "height" of this trapezoid is actually the horizontal width of the figure. Because this distance is $3 - 0 = 3$, $h = 3$. The length of the parallel sides are $a = f(0) = 1$ and $b = f(3) = 7$. Therefore, the area of this trapezoid is $A = \frac{1}{2} \cdot 3(1 + 7) = 12$. Thus, $\int_0^3 (2x + 1)\,dx = 12$.

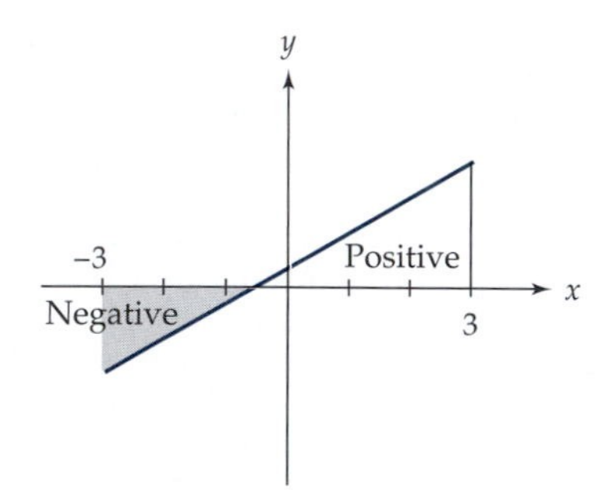

Figure 8 The area above the x-axis and below the graph of the function is positive, whereas the area below the x-axis and above the graph of the function is negative.

Although we usually think of area as always being a positive number, finding an integral does not always yield a positive result. In finding an integral, you are actually finding *signed area*. All area below the x-axis is defined to be negative, whereas all area above (like every example we have shown so far) is defined to be positive. For example, suppose we wanted to find $\int_{-3}^3 (2x + 1)\,dx$. Where the output for this function is positive, the area between the graph of the function and the x-axis is positive, and where the output for this function is negative, the area between the graph of the function and the x-axis is negative. (See Figure 8.)

In Figure 8, the shaded region above the x-axis, from $x = -\frac{1}{2}$ to $x = 3$, has a signed area of $+12.25$, and the shaded region below the x-axis, from $x = -3$ to $-\frac{1}{2}$, has a signed area of -6.25. Thus, the total signed area is $12.25 - 6.25 = 6$, or $\int_{-3}^3 (2x + 1)\,dx = 6$.

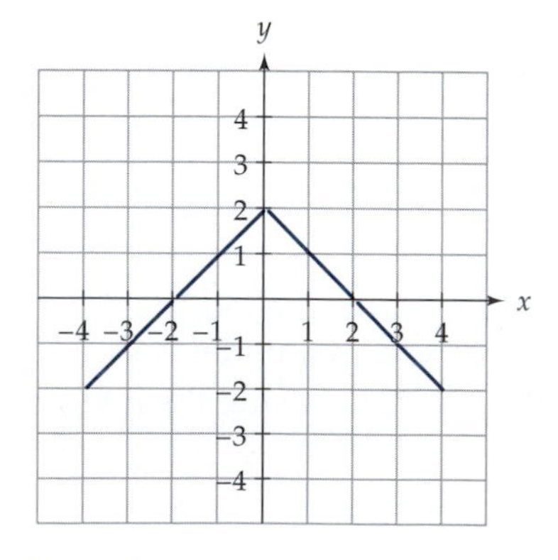

Figure 9

Example 4 Find $\int_{-4}^4 (2 - |x|)\,dx$. A graph of $f(x) = 2 - |x|$ from -4 to 4 is shown in Figure 9.

Solution The area between the graph of the function and the x-axis is positive between $x = -2$ and $x = 2$. This region is a triangle with a base of 4 and a height of 2. Therefore, this area is $\frac{1}{2} \cdot 4 \cdot 2 = 4$. The areas from $x = 2$ to $x = 4$ and from $x = -4$ to $x = -2$ are negative. Each of these regions is a triangle with a base of 2 and a height of -2. Therefore, each area is $\frac{1}{2} \cdot 2 \cdot -2 = -2$, for a total of -4 for both. Therefore, $\int_{-4}^4 (2 - |x|)\,dx = 4 - 4 = 0$.

READING QUESTIONS

3. How are Riemann sums and integrals related?
4. Describe the type of situations that can give a negative area when calculating an integral.
5. Find $\int_0^4 2x\,dx$.

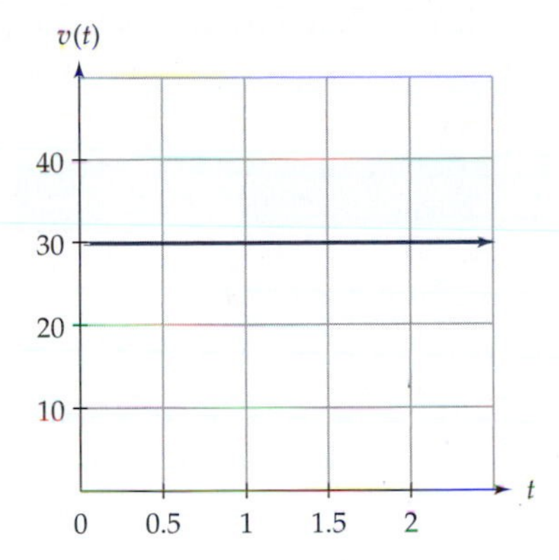

Figure 10 A graph of the velocity of a car going at a constant 30 mph.

APPLYING THE INTEGRAL

Suppose you are traveling in a car at a constant velocity of 30 mph for 2 hours. Because the speed is constant, the graph of the velocity is simply the horizontal line $v(t) = 30$. The graph of the velocity of the car as a function of time is shown in Figure 10. The area under the curve (or in this case, under the line) from $t = 0$ to $t = 2$ is $2 \times 30 = 60$. What does this number represent? Let's look at it a little more closely. To find the area of this rectangle, we multiply 2 *hours* by 30 *miles per hour,* which means that the result is 60 *miles*. This 60 mi represents the total distance that you traveled during the 2 hours, which should make sense because velocity × time = distance. In the following example, we consider this scenario again but with a slightly more complicated driving pattern.

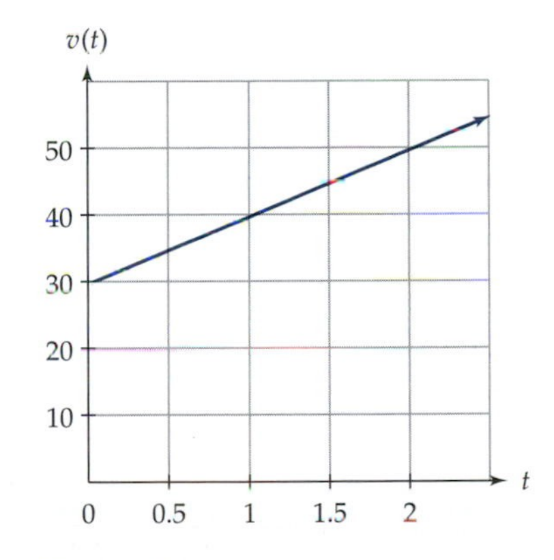

Figure 11

Example 5 Suppose at $t = 0$ a car is going 30 mph and is accelerating at a constant rate such that after 1 hr ($t = 1$) the car is going 40 mph and after 2 hr ($t = 2$) the car is going 50 mph. The function describing the velocity is $v(t) = 10t + 30$, as seen in Figure 11. Find $\int_0^2 v(t)\,dt$ and explain what your answer means.

Solution We need to find the area of the region bounded vertically by the lines $x = 0$ and $x = 2$ and by the x-axis on the bottom and the function $v(t) = 10t + 30$ on the top. This region is that of a trapezoid. Its area is simply $\frac{1}{2} \cdot 2(30 + 50) = 80$. Therefore, $\int_0^2 v(t)\,dt = 80$, which means that the car traveled 80 mi from $t = 0$ to $t = 2$ hr.

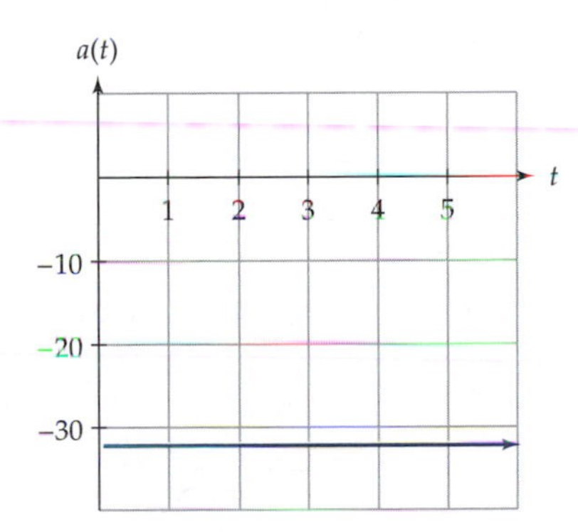

Figure 12

Example 6 Consider an object in free fall after being dropped off an airplane. Ignoring wind resistance, the object has a constant acceleration of -32 ft/sec^2, or $a(t) = -32$. Find $\int_0^5 a(t)\,dt$ and explain what this integral represents.

Solution This "curve" is actually a horizontal line. The area between the line and the horizontal axis is simply a rectangle. (See Figure 12.)

The area can be found by multiplying the acceleration (in *feet per second squared*) by the time (in *seconds*). The result is *ft/sec*, a measure of velocity. In this case, for 5 sec, the result is $5(-32) = -160$. Therefore, the velocity of the object after 5 sec has elapsed is -160 ft/sec. The negative velocity indicates that the object is traveling downward.

READING QUESTIONS

6. The integral of an acceleration function gives what type of answer?
7. The integral of a velocity function gives what type of answer?
8. Suppose W is a function that gives the rate at which water is flowing into a tank in liters per second at any given time. What would $\int_0^4 W(t)\,dt$ represent?

SUMMARY

To approximate the area under a graph of a function, we can form a series of rectangles with a common width and a height that is the value of the function at either the right edge, the left edge, or the center of the rectangle. We write this sum as

$$L_n = \sum_{i=0}^{n} f(x_i)\Delta x,$$

where $x_1 = a$, $x_2 = a + \Delta x$, $x_3 = a + 2\Delta x, \ldots, x_n = a + (n-1)\Delta x$, $f(x_i)$ is the output of the function at the left endpoints of the n different equally spaced intervals, and $\Delta x = (b-a)/n$ is the width of each interval. As the number of rectangles increases, the closer our approximation gets to the true area.

The estimates for areas under curves like this are all special types of *Riemann sums*. As the number of rectangles increases in a Riemann sum, the approximation of the area gets closer to the true area. In fact, if we find

$$\lim_{n\to\infty} \sum_{i=0}^{n} f(c_i)\Delta x_i,$$

then we will have the exact area under the curve. This limit is called an *integral*, and we use integral notation to represent it.

EXERCISES

1. Approximate the area under the curve in the accompanying figure by finding the sum of the areas of the rectangles shown.

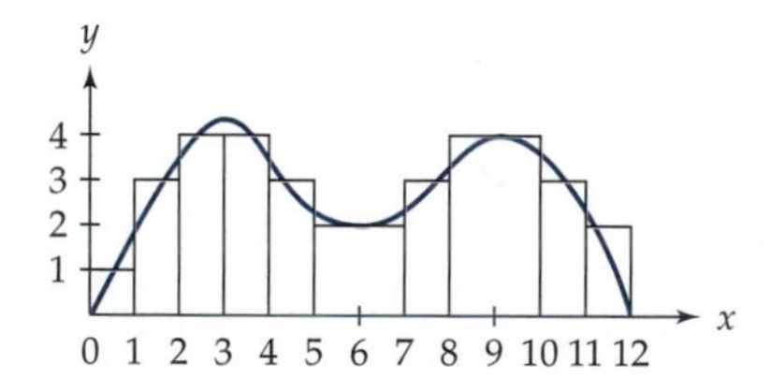

2. Find the following sums.

(a) $\sum_{n=1}^{4} (2n-1) \cdot \frac{1}{2}$

(b) $\sum_{n=1}^{6} n^2 \cdot \frac{1}{3}$

(c) $\sum_{n=1}^{8} 2^{n-1} \cdot \frac{1}{4}$

3. Approximate the area under the following curves by using six rectangles each of whose left edge corresponds to the height of the curve.
   (a) $f(x) = -x^2 - 2x + 6$ from $x = -3$ to $x = 1$
   (b) $g(x) = 2 \cdot 3^x$ from $x = 0$ to $x = 3$
   (c) $h(x) = \sqrt{x}$ from $x = 0$ to $x = 6$
4. In this section, we estimated the area under the function $f(x) = (x - 3)^2 + 2$ by using rectangles whose heights were based on the value of the function at the left edge of each rectangle.
   (a) Approximate the area under f from $x = 0$ to $x = 4$ using eight rectangles whose heights are based on the value of the function at the right edge of each rectangle.
   (b) Approximate the area under f from $x = 0$ to $x = 4$ using eight rectangles whose heights are based on the value of the function at the center of each rectangle.
5. Given the function $g(x) = \sqrt{x + 2}$, answer the following.
   (a) Approximate the area under g from $x = 0$ to $x = 3$ using six rectangles whose heights are based on the value of the function at the left edge of each rectangle. Does this approximation give an underestimate or overestimate?
   (b) Approximate the area under g from $x = 0$ to $x = 3$ using six rectangles whose heights are based on the value of the function at the right edge of each rectangle. Does this approximation give an underestimate or overestimate?
   (c) Approximate the area under g from $x = 0$ to $x = 3$ using six rectangles whose heights are based on the value of the function at the center of each rectangle. How does this approximation compare with your first two?
6. Find the area between the x-axis and each of the functions shown in the following graphs from $x = -4$ to $x = 4$.

(a)

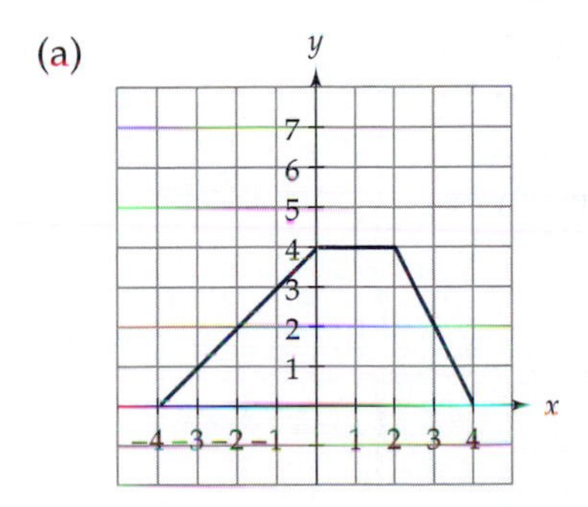

(b)

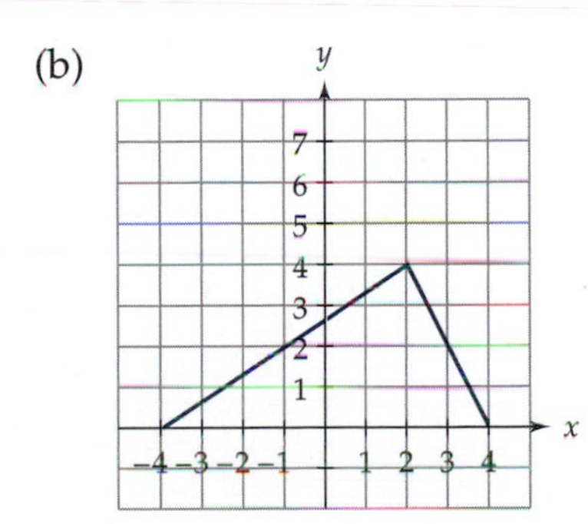

7. Find the signed area between the x-axis and each of the functions shown in the following graphs.

(a)

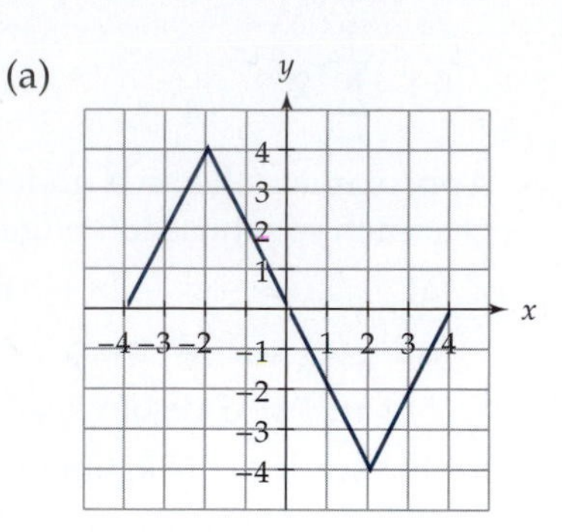

(b)

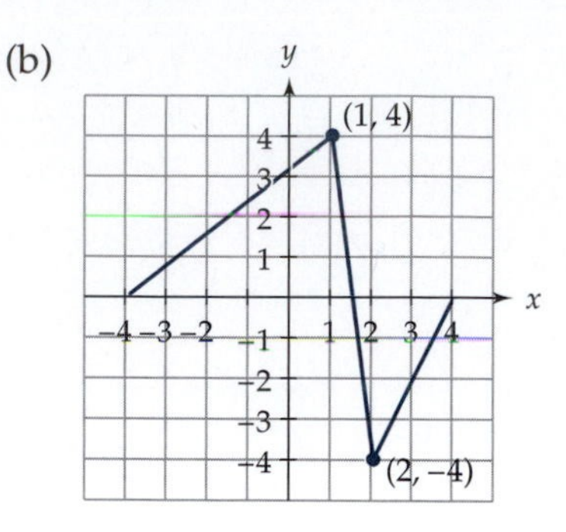

8. By looking at graphs of the functions, match each of the following values, (a)–(d) to their corresponding integrals, (i)–(iv).

(a) $\int_{-2}^{2} x^2\,dx$

(b) $\int_{-2}^{2} x^3\,dx$

(c) $\int_{0}^{1} x^{3/5}\,dx$

(d) $\int_{0}^{1} x^{5/3}\,dx$

i. $\frac{3}{8}$

ii. $5\frac{1}{3}$

iii. $\frac{5}{8}$

iv. 0

9. By calculating areas, find the following integrals.

(a) $\int_{0}^{8} (2x+1)\,dx$

(b) $\int_{-2}^{2} |x|\,dx$

(c) $\int_{-2}^{2} \sqrt{4-x^2}\,dx$

10. By calculating areas, both positive and negative, find the value of each of the following integrals.

(a) $\int_{0}^{7} (-x+5)\,dx$

(b) $\int_{-4}^{10} (4-|x|)\,dx$

(c) $\int_{-2}^{2} (x-2)\,dx$

11. Suppose there is a function, f, such that $\int_{-2}^{0} f(x)\,dx = 2$, $\int_{0}^{2} f(x)\,dx = -5$, and $\int_{2}^{4} f(x)\,dx = 3$. Find the following integrals.

    (a) $\displaystyle\int_{0}^{4} f(x)\,dx$

    (b) $\displaystyle\int_{-2}^{4} f(x)\,dx$

    (c) $\displaystyle\int_{-2}^{2} f(x)\,dx$

12. In this question, we look at the integral of the sine function on various intervals.
    (a) By looking at a graph of the $y = \sin x$, determine $\int_{0}^{2\pi} \sin x\,dx$.
    (b) By using a Riemann sum, estimate $\int_{0}^{\pi} \sin x\,dx$.
    (c) Using your answer from part (b), estimate $\int_{0}^{\pi/2} \sin x\,dx$.
    (d) Using your answer from part (b), estimate $\int_{-\pi/2}^{\pi/2} \cos x\,dx$.

13. Given that $\int_{0}^{\pi/2} \cos x\,dx = 1$, find the following integrals by looking at the graphs of the functions.

    (a) $\displaystyle\int_{-\pi/2}^{\pi/2} \cos x\,dx$

    (b) $\displaystyle\int_{0}^{2\pi} \cos x\,dx$

    (c) $\displaystyle\int_{0}^{2\pi} |\cos x|\,dx$

    (d) $\displaystyle\int_{0}^{\pi} \sin x\,dx$

14. Suppose water is flowing through a pipe at a constant rate of 10 gal/min. The function describing this rate is r with an input of t, time in minutes.
    (a) Find a formula for r.
    (b) Find $\int_{0}^{10} r(t)\,dt$.
    (c) Give the meaning of the value of the integral from part (b).
15. Suppose a car is traveling in such a way that it starts ($t = 0$) at a speed of 30 mph and its speed is increasing by 5 mph each hour.
    (a) Give a linear function, v, that describes the velocity of the car where time, in hours, is the input.
    (b) Find $\int_{0}^{5} v(t)\,dt$.
    (c) Give the meaning of the value of the integral from part (b).
16. Suppose a car is slowing down in such a way that its velocity can be described with the function $v(t) = -10t + 40$, where t is time in seconds and $v(t)$ is velocity in feet per second.
    (a) At what time will the car have a velocity of 0?
    (b) How far will the car travel before its velocity is 0?

INVESTIGATIONS

INVESTIGATION 1: TRAPEZOIDAL APPROXIMATIONS

Using rectangles to approximate the area under a curve can often give a fair amount of error because the rectangles may not fit well. Another way to estimate area under a curve is to use trapezoids, which gives a better approximation. For example, Figure 13 shows a curve from the text with rectangles used to approximate the area and the same curve using trapezoids to approximate the area. As you can see, the trapezoidal approximation is much better than that done with rectangles.

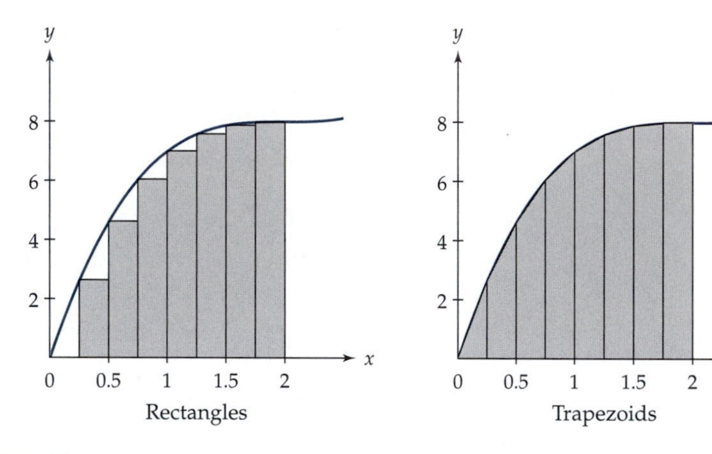

Figure 13

The trapezoids that we use are like those in the figure. The bottom is perpendicular to the two vertical sides and only the top side slants. To find the area of such a trapezoid, you need to find the sum of the two vertical sides and multiply this sum by one-half the width. This process can be expressed symbolically as the sum

$$\sum_{i=1}^{n} \frac{\Delta x}{2}(f(x_i) + f(x_{i+1})).$$

1. We use various techniques to estimate the integral $\int_0^2 (x^2 + 4)\, dx$. For each method, find the estimate, draw a picture of the curve along with your rectangles, and decide if your method is an underestimate or overestimate.
   (a) Approximate $\int_0^2 (x^2 + 4)\, dx$ using four rectangles such that the left edge of each rectangle corresponds to the height of the curve.
   (b) Approximate $\int_0^2 (x^2 + 4)\, dx$ using four rectangles such that the right edge of each rectangle corresponds to the height of the curve.
   (c) Approximate $\int_0^2 (x^2 + 4)\, dx$ using four trapezoids.

2. How does the trapezoidal method compare with the two rectangular methods?
3. Show that the trapezoidal method is always the mean of the left rectangular method and the right rectangular method.

INVESTIGATION 2: FINDING INTEGRALS USING LIMITS

In this section, we found exact integrals only through geometry. Therefore, our examples were rather simplistic. In calculus, you will learn how to evaluate all types of integrals. We already have much of the background needed to find exact integrals for some areas that do not involve straight lines or circles, however. For example, in Investigation 2 in section 8.3, you were asked to show that

$$\sum_{i=1}^{n} i^2 = 1^2 + 2^2 + 3^2 + \cdots + n^2 = \frac{n(n+1)(2n+1)}{6}.$$

We use this sum to determine a value for $\int_0^1 x^2\,dx$ symbolically. Using similar methods, we also determine a value for $\int_0^1 x^3\,dx$ symbolically.

1. We first find $\int_0^1 x^2\,dx$.
   (a) Use four rectangles of equal width such that the right edge of each rectangle corresponds to the height of the curve to approximate $\int_0^1 x^2\,dx$.
   (b) Use eight rectangles of equal width such that the right edge of each rectangle corresponds to the height of the curve to approximate $\int_0^1 x^2\,dx$.
   (c) Explain why

$$\sum_{i=1}^{n} \left(\frac{i}{n}\right)^2 \frac{1}{n},$$

   for some positive integer value of n, can be used to approximate $\int_0^1 x^2\,dx$.
   (d) To find the exact value of $\int_0^1 x^2\,dx$, we need to find

$$\lim_{n\to\infty} \sum_{i=1}^{n} \left(\frac{i}{n}\right)^2 \frac{1}{n}.$$

      i. Show that this limit can be written as

$$\lim_{n\to\infty} \frac{1}{n^3} \sum_{i=1}^{n} i^2.$$

      ii. Because

$$\sum_{i=1}^{n} i^2 = \frac{n(n+1)(2n+1)}{6},$$

      make this substitution and take the limit to find an exact value. This limit should then be the exact value of $\int_0^1 x^2\,dx$.

2. Using similar methods as the previous questions and that

$$\sum_{i=1}^{n} i^3 = \left(\frac{n(n+1)}{2}\right)^2,$$

find an exact value for $\int_0^1 x^3\,dx$.

ZERO TO SIXTY

Car magazines often do road tests and publish performance data on new cars. One item often included is the time it takes for a car to go from 0 mph to 60 mph, usually called "0 to 60." In the July 1994 issue of *Car and Driver*, the BMW M3 was tested. This car, then selling for $38,760, goes from 0 to 60 in 5.6 seconds.[12] In this project, we will analyze the velocity and the distance traveled for this car when accelerating from 0 to 60.

1. For the first part of this project, we assume the acceleration of the car is constant. When testing cars, the drivers try to make the car accelerate as quickly as possible, and it is unlikely that the result is constant acceleration over the whole 5.6 sec. Using this assumption, however, makes the problem more manageable. Later we look at a more accurate model.
   (a) Assuming this acceleration is constant, what would it be? Give your answer in feet per second squared. (*Note:* Acceleration = change in velocity/time.)
   (b) Because the acceleration is assumed to be constant, the velocity changes at a constant rate with respect to time, so $v = kt$ for some constant k. Find k. We now have a formula whose input is time, in seconds, and whose output is the velocity, in feet per second. Graph this function.
   (c) We wish to estimate the distance the car traveled in those 5.6 sec. It is clear that the car went less than 60 mph (or 88 ft/sec) during most of the 5.6 sec. Therefore, we can safely say that the distance traveled is less than 492.8 ft (88 ft/sec · 5.6 sec). We can also say that the car went faster than 0 mph (or 0 ft/sec) for almost all the 5.6 sec, so the distance must be greater than 0 ft. We now have upper and lower bounds for the distance traveled. Therefore, the actual distance traveled is somewhere between those two values. These bounds are not very useful, though, because they are very crude estimates; after all, the car obviously went farther than 0 ft! Let's see if we can improve on them.

   Suppose we divide the time into two equal intervals, the first interval from 0 sec to 2.8 sec and the second interval from 2.8 sec to 5.6 sec. We then say that during the first interval, the car traveled somewhere between 0 ft and 123.2 ft (because its top

[12] *Barry Winfield, "BMW M3,"* Car and Driver, *July 1994, pp. 44–48.*

speed is 44 ft/sec for 2.8 sec). Find the upper and lower bounds for the distance the car traveled during the second time interval (2.8 sec to 5.6 sec). Add the two lower bounds together to find an underestimate for the total distance traveled in 5.6 sec. Similarly, add the upper bounds together to find an overestimate for the total distance traveled in 5.6 sec.

2. We can improve on these estimates even more.

   (a) Use this same process again. This time, split the 5.6 sec into four intervals and find lower and upper estimates for the total distance traveled. Complete the table. What are your lower and upper estimates for the cumulative distance after 5.6 sec?

Time (sec)	Velocity at Start of the Interval	Distance (Lower Estimate)	Velocity at End of the Interval	Distance (Upper Estimate)	Cumulative Distance (Lower Estimate)	Cumulative Distance (Upper Estimate)
1.4						
2.8						
4.2						
5.6						

   (b) Now repeat the process for eight intervals by completing the next table. What are your lower and upper estimates for the cumulative distance after 5.6 sec?

Time (sec)	Velocity at Start of the Interval	Distance (Lower Estimate)	Velocity at End of the Interval	Distance (Upper Estimate)	Cumulative Distance (Lower Estimate)	Cumulative Distance (Upper Estimate)
0.7						
1.4						
2.1						
2.8						
3.5						
4.2						
4.9						
5.6						

   (c) What is happening to your cumulative distance estimates as you split the time into more and more subintervals?

   (d) Based on these results, how far do you think the car went during those 5.6 sec?

3. Let's consider the graph of what happened to the distance as the car continued to accelerate.
   (a) Sketch time versus cumulative distance, using your *lower estimates* from question 2, part (b). *On the same graph,* sketch time versus cumulative distance using your *upper estimates* from question 2, part (b). Assuming the approximation for distance traveled is half way between your lower and upper limits, draw a graph for that approximation on the same set of axes as your upper and lower bounds. Because it is clear that when $t = 0$ the distance must be zero, this approximation should pass through the origin.
   (b) Explain why the formula is *not* of the form $y = mx + b$.
   (c) By using other methods (upon which we do not elaborate here), it can be shown that the graph is a parabola whose vertex is at the origin. Thus, the equation of the parabola is of the form $d = kt^2$, where k is some constant. Approximate a point on your graph of the average of the lower and upper estimates. Use this point to determine a reasonable equation for the function whose input is t and whose output is total distance.
4. Let's compare our scenario, in which we assumed the acceleration was constant, to what really happened when the car was tested.
   (a) Assuming your estimates and graphs are accurate, how long do you think it would take for the car to go one-quarter of a mile?
   (b) What would its speed be by the time it reached the quarter-mile mark?
   (c) The actual car, as reported by *Car and Driver*, took 14.3 sec and was traveling at 98 mph at the end of one-quarter of a mile. What may be causing the discrepancy between the actual test data and your prediction based on your graphs and formulas?
5. Here is a list of data taken from the July 1994 issue of *Car and Driver* for the acceleration time of the BMW M3.

Time (sec)	**Velocity (mph)**	**Velocity (ft/sec)**
1.9	30	
3.1	40	
4.3	50	
5.6	60	
7.5	70	
9.5	80	
12.0	90	
14.9	100	
18.4	110	
23.5	120	
29.8	130	

(a) Based on this table, do you think acceleration was constant? Why or why not? What happened to the acceleration?

(b) Complete the table by converting velocity to feet per second.

(c) Graph the function you found in question 1, part (b) again with a domain of $0 < t < 5.6$. Plot the data points from the actual car as given in the table on the same graph. Looking at your graph, how do you think the distance traveled by the actual car compares with the distance you estimated in question 2, part (d)? Explain your answer.

(d) Again, graph the function from question 1, part (b), but this time with a domain of $0 < t < 29.8$. Plot the data points from the table on that graph. Based on the two graphs, under what circumstances is the function you found in question 1, part (b) a good model for the velocity of the car?

REVIEW EXERCISES

1. Suppose a ball is hit into the air so that its height can be determined by the equation $h(t) = -16t^2 + 70t + 4$, where $h(t)$ is the height of the ball, in feet, after t seconds.
   (a) Find the average velocity of the ball between 1.5 and 2 sec.
   (b) Find the average velocity of the ball between 1.9 and 2 sec.
   (c) Find the average velocity of the ball between 1.99 and 2 sec.
   (d) What is the instantaneous velocity of the ball at 2 sec?
2. Find limits for the following.
   (a) $\lim_{x \to 0} \dfrac{3x^2 + 7x}{5x^2 - 4x}$
   (b) $\lim_{x \to -2} \dfrac{3x^2 + 2x - 8}{2x^2 + 7x + 6}$
   (c) $\lim_{x \to \infty} \dfrac{3x^2 + 7x}{5x^2 - 4x}$
   (d) $\lim_{x \to \infty} \dfrac{4}{2 \cdot 6^x}$
3. Find the following limits for the accompanying graph of f.

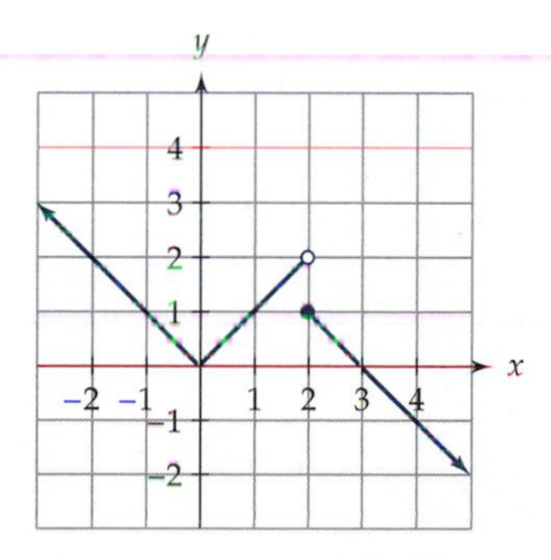

(a) $\lim_{x\to 0} f(x)$

(b) $\lim_{x\to 2^+} f(x)$

(c) $\lim_{x\to 2^-} f(x)$

4. Let

$$g(x) = \begin{cases} x+2, & \text{if } x \le 4 \\ 2x-1, & \text{if } x > 4. \end{cases}$$

(a) Find $\lim_{x\to 4^-} g(x)$.

(b) Find $\lim_{x\to 4^+} g(x)$.

(c) Find $\lim_{x\to 4} g(x)$.

5. Let

$$g(x) = \begin{cases} x+2, & \text{if } x \le 2 \\ x^2, & \text{if } 2 < x \le 4 \\ 2^x, & \text{if } x > 4. \end{cases}$$

Are there any points of discontinuity in g? If so, what are they?

6. Suppose a car is traveling away from you such that its position (in feet) is given by the equation $p(t) = 66t + 20$, where t is time in seconds.
   (a) Find values for $p(10)$ and $p'(10)$.
   (b) What is the meaning, in terms of the position of the car and time, of $p(10)$ and $p'(10)$?
7. Find the slope of the tangent line for each function at the given point.
   (a) $f(x) = x^2 + 2$ at the point $(1, 3)$
   (b) $g(x) = x^3 - 4x$ at the point $(-1, 3)$
   (c) $j(x) = x^3 + 5x^2$ at the point $(2, 28)$
8. Estimate the slope of the tangent line for each function at the given point. Do so by finding the slope of a secant line and taking smaller and smaller intervals.
   (a) $f(x) = \sqrt{x-2}$ at the point $(3, 1)$
   (b) $g(x) = e^x$ at the point $(2, e^2)$
9. Find the derivative of each function.
   (a) $f(x) = 5x + 2$
   (b) $g(x) = 2x^2 + 8$
   (c) $h(x) = x^3 - 5x$
10. Sketch a graph of the derivative for each of the functions shown.

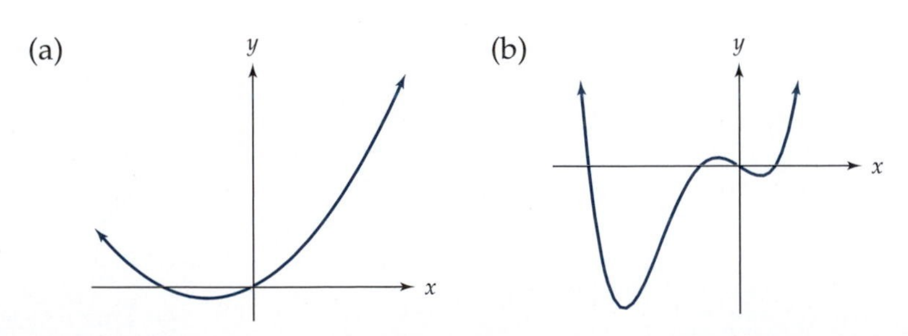

(c)

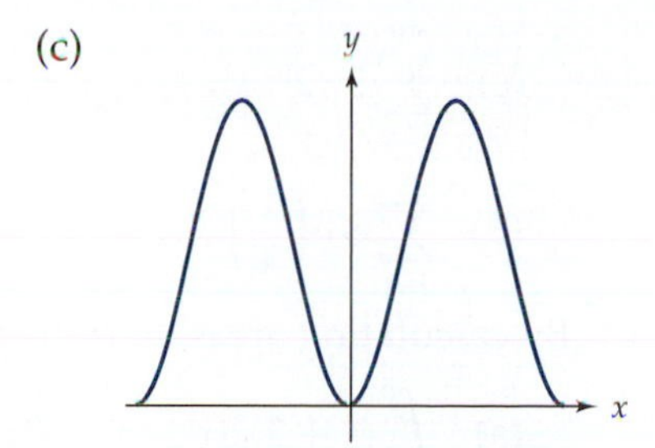

(d)

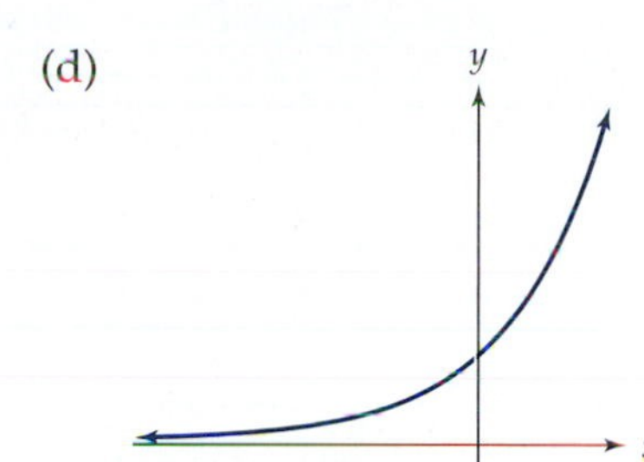

11. Find the equation of the tangent line for each function at the given point.
    (a) $f(x) = x^3 + 8$ at the point $(1, 9)$
    (b) $g(x) = x^2 - 4x$ at the point $(-2, 12)$
12. Find the first five terms of the sequence defined by each of the following recurrence relations.
    (a) $a_1 = 1, a_n = a_{n-1} + 5$
    (b) $a_1 = 1, a_n = a_{n-1} \cdot 2$
    (c) $a_1 = 2, a_n = a_{n-1} + 2n$
13. Find the 20th term of the sequence defined by each of the following general formulas.
    (a) $a_n = 4 + 2(n - 1)$
    (b) $a_n = 5 \cdot 2^{n-1}$
    (c) $a_n = \dfrac{n(n+1)(n+2)}{6}$
14. Write a formula for the nth term for each of the following arithmetic or geometric sequences.
    (a) $\{2, 10, 18, 26, \ldots\}$
    (b) $\{16, 8, 4, 2, \ldots\}$
    (c) $\{-8, 12, -18, 27, \ldots\}$
    (d) $\{16, 4, -8, -20 \ldots\}$
15. Find the following sums.
    (a) $3 + 6 + 9 + \cdots + 81 + 84 + 87$
    (b) $12 + 13.5 + 15 + \cdots + 90 + 91.5 + 93$
    (c) $4 + 3 + \frac{9}{4} + \frac{27}{16} + \cdots$
    (d) $3 + \frac{3}{2} + \frac{3}{4} + \frac{3}{8} + \cdots$
16. Find the following sums.
    (a) $\displaystyle\sum_{i=0}^{10} (2i + 1)$
    (b) $\displaystyle\sum_{i=8}^{28} (5i + 2)$

(c) $\sum_{i=0}^{10} 3 \cdot 2^i$

(d) $\sum_{i=1}^{\infty} 8 \cdot \left(\frac{1}{4}\right)^i$

17. By calculating areas, find the following integrals.

(a) $\int_0^5 (x+2)\,dx$

(b) $\int_{-4}^6 (8-|x|)\,dx$

(c) $\int_0^3 \sqrt{9-x^2}\,dx$

18. Approximate the area under the following curves by using six rectangles whose left edge corresponds to the height of the curve.

(a) $f(x) = -x^2 + 9$ from $x = -3$ to $x = 3$

(b) $g(x) = 3 \cdot 2^x$ from $x = 0$ to $x = 3$

(c) $h(x) = \sqrt{x}$ from $x = 6$ to $x = 12$

19. By calculating areas, both positive and negative, find the following integrals.

(a) $\int_{-2}^2 (2x-1)\,dx$

(b) $\int_{-4}^4 (|x|-2)\,dx$

20. Suppose there was a function, f, such that $\int_{-2}^0 f(x)\,dx = 3$, $\int_0^2 f(x)\,dx = 7$, and $\int_2^4 f(x)\,dx = -5$. Find the following integrals.

(a) $\int_{-2}^2 f(x)\,dx$

(b) $\int_0^4 f(x)\,dx$

(c) $\int_{-2}^4 f(x)\,dx$

21. Suppose water is flowing through a pipe at a constant rate of 5 gal/min. The function describing this rate is r with an input of t, time in minutes.

(a) Find a formula for r.

(b) Find $\int_0^{10} r(t)\,dt$.

(c) Give the meaning of the value of the integral from part (b).

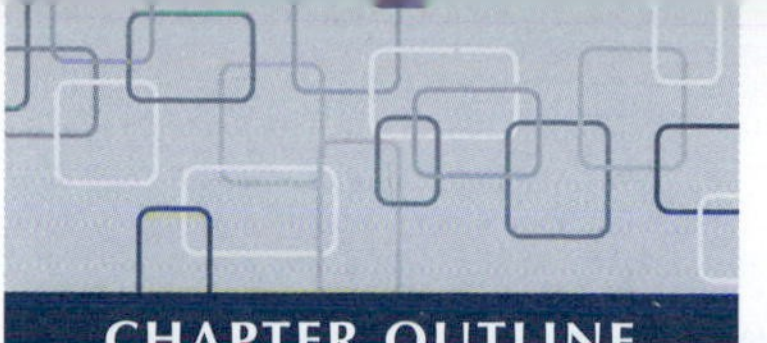

CHAPTER OUTLINE

CHAPTER OVERVIEW

- An introduction to parametric equations
- Using parametric equations to model trajectories
- Cycloids and other curves
- An introduction to vectors
- Adding vectors
- Using vectors to model forces
- An introduction to multivariable functions
- Using contour curves to graph multivariable functions

ADDITIONAL TOPICS

In this chapter, supplemental topics to a precalculus course are covered. Although these topics are not vital for a preparation of the study of calculus, they are of interest because many applications use these mathematical concepts.

9.1 PARAMETRIC EQUATIONS

A function cannot have more than one output for any given input. Therefore, something as simple as a circle cannot be graphed using a single function. By using parametric equations, however, circles and much more interesting curves can be graphed. In this section, we see how parametric equations can be used to describe the position of objects in space: swings, hands on a clock, a baseball after it is hit, and a point on a moving bicycle tire, for example.

INTRODUCTION

Imagine a point traveling in typical "mathematical" manner around the unit circle shown in Figure 1. In other words, it starts at the point (1, 0) and travels counterclockwise. Using Figure 1, we see that the x-coordinate of the point is $x = \cos\theta$. We also see that the y-coordinate of the point is $y = \sin\theta$.

Now let's put these two functions together. Table 1 gives the horizontal ($x = \cos\theta$) and vertical ($y = \sin\theta$) positions, rounded to two decimal places, for certain values of θ.

Plotting these points shows, not surprisingly, that they are on the unit circle. (See Figure 2.) The two functions, $x = \cos\theta$ and $y = \sin\theta$, both have the same input (θ), but the output for one is horizontal position (x) and the output for the other is vertical position (y). These two equations are called parametric equations. **Parametric equations** are equations in which the coordinates of a point are separately expressed by functions with a common input. This common input is referred to as the **parameter.** The parameter in this case is θ. In our set of parametric equations, each value of θ corresponds to a point $(x, y) = (\cos\theta, \sin\theta)$. In other words, our input is θ and our output is the coordinates of a point. By using these parametric equations, we can express a circle in terms of two functions.

The parametric equations $x = \cos\theta$, $y = \sin\theta$ for $0 \le \theta \le 2\pi$ describe a graph of a unit circle centered at the origin. The domain, $0 \le \theta \le 2\pi$, shows that 1 revolution of the circle is being drawn. This domain need not

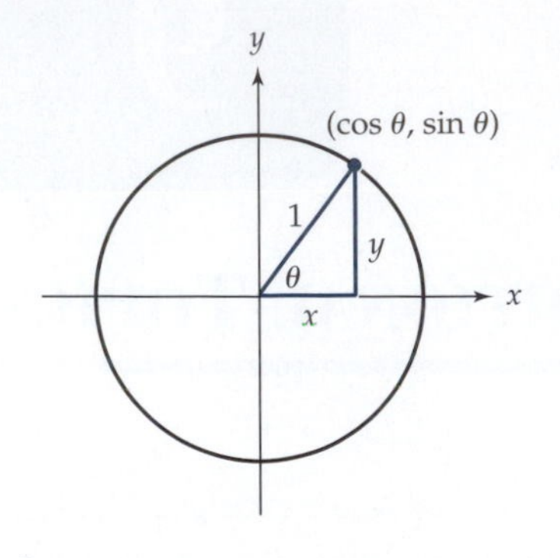

Figure 1 The x-coordinate of a point on the unit circle is $x = \cos\theta$, and the y-coordinate is $y = \sin\theta$.

TABLE 1 Decimal Approximations for $x = \cos\theta$ and $y = \sin\theta$ for Certain Values of θ

θ	0	$\frac{\pi}{4}$	$\frac{\pi}{2}$	$\frac{3\pi}{4}$	π	$\frac{5\pi}{4}$	$\frac{3\pi}{2}$	$\frac{7\pi}{4}$	2π
$x = \cos\theta$	1	0.71	0	−0.71	−1	−0.71	0	0.71	1
$y = \sin\theta$	0	0.71	1	0.71	0	−0.71	−1	−0.71	0

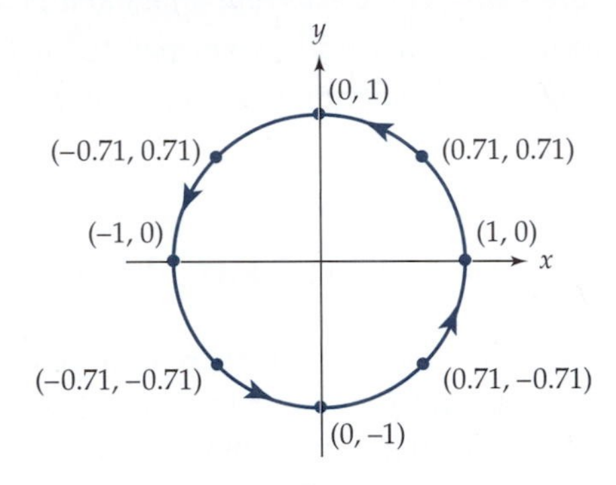

Figure 2 The points (x, y), where $x = \cos\theta$ and $y = \sin\theta$, are located on the unit circle.

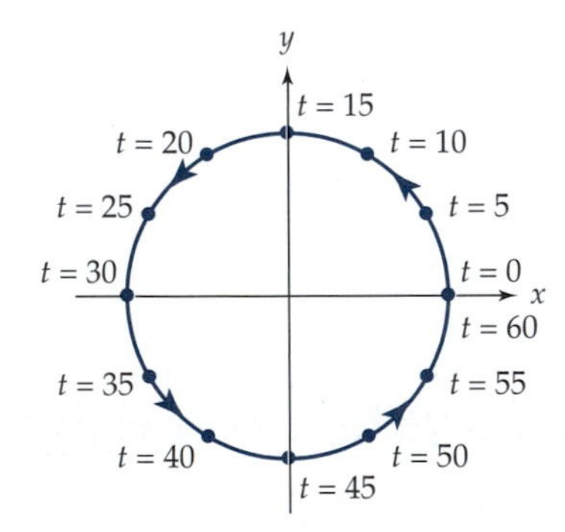

Figure 3 A graph of the set of parametric equations $x = \cos((\pi/30)t)$, $y = \sin((\pi/30)t)$ with a domain of $0 \le t \le 60$. Some outputs are shown as points along with the corresponding input.

be $0 \le \theta \le 2\pi$. For a domain of $0 \le \theta \le \pi$, only half the circle is described. Our input also does not have to be an angle. It can be many different things. Suppose we want time to be the input for a point on the unit circle that is revolving around the circle at a rate of 1 revolution every 60 sec. Because the period is 60 sec, we need our input to be $(2\pi/60)t = (\pi/30)t$. Our equations are now $x = \cos((\pi/30)t)$, $y = \sin((\pi/30)t)$. If we want the domain of these equations to include 1 revolution, then it will be $0 \le t \le 60$ sec. A graph is shown in Figure 3, with some outputs shown as points along with the corresponding input.

Parametric equations allow us to do many things that cannot be done with regular functions. Parametric equations allow us to graph many interesting curves that cannot be represented by regular functions. Parametric equations can also be used to describe any type of function that we have studied so far. Example 1 shows how parametric equations can be used to describe a line.

Example 1 Describe the path drawn out by the parametric equations

$$x = 2t + 2, \qquad y = 4t - 2 \qquad \text{for } 0 \le t \le 6$$

by determining the coordinates of the points determined by the equations using integer values of $0 \le t \le 6$.

Solution For $t = 0$, $x = 2(0) + 2 = 2$ and $y = 4(0) - 2 = -2$, which gives a point of $(2, -2)$. For $t = 1$, $x = 2(1) + 2 = 4$ and $y = 4(1) - 2 = 2$, which gives a point of $(4, 2)$. As we continue, for $t = 2$ the output is the point $(6, 6)$, for $t = 3$ the output is the point $(8, 10)$, for $t = 4$ the output is

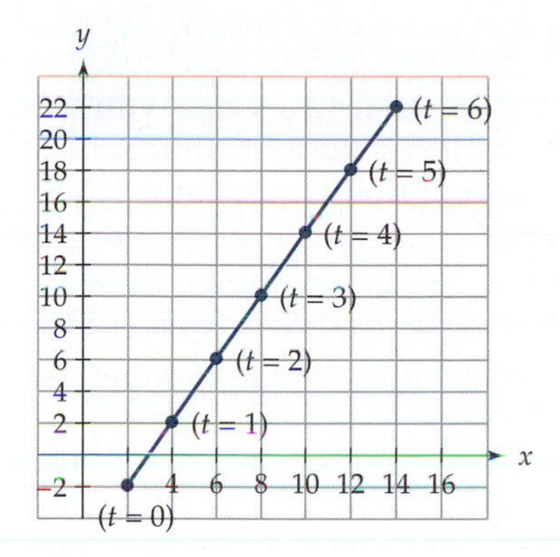

Figure 4

the point of $(10, 14)$, for $t = 5$ the output is the point $(12, 18)$, and for $t = 6$ the output is the point of $(14, 22)$. These points and the line connecting them are shown in Figure 4.

In the previous example, we saw by looking at the outputs that a linear function is given by the set of parametric equations. As the function goes from one point to another, there is an increase in 4 for the y-coordinate and an increase of 2 for the x-coordinate, which gives a line with a slope of $\frac{4}{2} = 2$. This increase can also be seen if we combine the parametric equations into a function that has x as an input and y as an output. This process is often referred to as "eliminating the parameter." For this step, all we need to do is solve $x = 2t + 2$ for t and substitute this expression for t into $y = 4t - 2$. Solving $x = 2t + 2$ for t gives $t = \frac{1}{2}x - 1$. Substituting $\frac{1}{2}x - 1$ for t in $y = 4t - 2$ gives $y = 4(\frac{1}{2}x - 1) - 2$, which simplifies to $y = 2x - 6$. As we can see from our points in the solution to Example 1, the domain of the equation $y = 2x - 6$ is $2 \leq x \leq 14$.

Graphing parametric equations on a calculator can be done quite easily. You first need to change the mode of the calculator to parametric, which on the TI-83 can be done with the mode button and changing the menu from "Func" to "Par". After inputting your pair of equations, you need to set up a proper viewing window. This process is similar to setting up a viewing window for a graph of a function with the addition of three items, T_{min}, T_{max}, and T_{step}. The items T_{min} and T_{max} define the endpoints of your domain, and the T_{step} describes how often points are plotted. Setting the T_{step} at too large a value causes your graph to look distorted because the calculator is not plotting enough points. Setting the T_{step} at too small a value causes your calculator to graph very slowly.

READING QUESTIONS

1. What is the parameter in $x = 3t - 7$ and $y = 4t + 2$ for $0 \leq t \leq 10$?
2. What sort of figure is made using the set of parametric equations $x = \cos\theta$ and $y = \sin\theta$ for $0 \leq \theta \leq \pi/2$?

TRAJECTORIES

A number of different types of applications can be best modeled using parametric equations. Such modeling is often done when describing the movement of an object in a plane. By using parametric equations, we can describe the horizontal position and the vertical position separately. By combining them, we can then see a path of the object with the horizontal position on one axis and the vertical position on the other. This path is referred to as the trajectory of the object. We consider two types of applications involving trajectories, ones considered previously using other methods. We examine the trajectories of swings and a second hand on a clock, and then we look at the trajectory of a projectile.

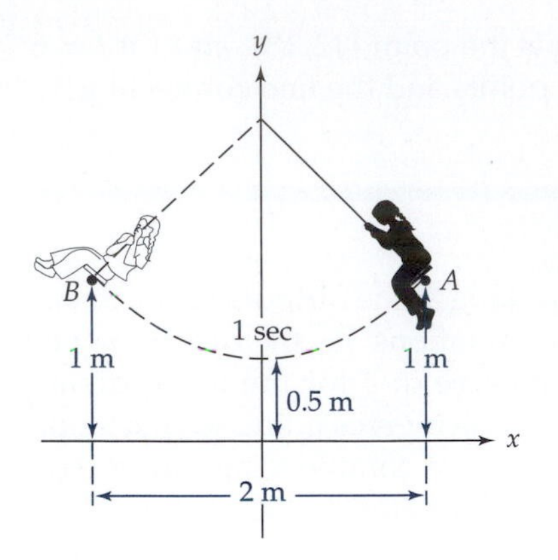

Figure 5 A child's swing.

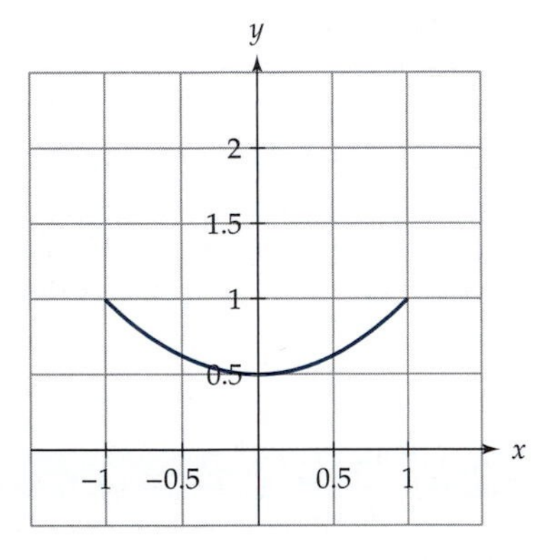

Figure 6 A graph of the path of the child's swing.

Swings and Clocks

A child swinging back and forth in a swing is similar to a pendulum of a clock swinging back and forth. When we introduced periodic functions in section 2.4, we gave a picture of a child swinging that is similar to Figure 5. This figure illustrates a child starting at point A, 1 m above the ground and 1 m to the right of the vertical line given by the resting point of the swing. In 1 sec, the child swings forward, passing by the lowest point (0.5 m above the ground) and then swinging up to the highest point (1 m above the ground) at point B.

We first find equations for the vertical position and the horizontal position where the input is time. Because the vertical position starts at the maximum height, we can use a cosine function without any horizontal shift. Because the maximum height is 1 m and the minimum height is 0.5 m, the amplitude is $(1 - 0.5)/2 = 0.25$ and the midline is $y = (1 + 0.5)/2 = 0.75$. Putting all of this information together, along with the period of 1 sec, gives the equation that describes the vertical position of the swing as $y = 0.25\cos(2\pi t) + 0.75$.

Although the output for the vertical position describes the height above ground level, the output for the horizontal position describes the distance right (positive) or left (negative) the swing is from its resting position. Because the swing is starting at the point farthest right, a cosine function without any horizontal shift is used again. There are differences in this function, however. Because the resting position is in the center, there is no vertical shift for this function. Because the swing moves horizontally a maximum of 1 m away from the resting position, the amplitude is 1. The period is also different from that of the vertical position. One period consists of the swing moving from the far right, to the far left, and then back to the right. It takes 2 sec to complete this period. Putting all this information together gives the equation that describes the horizontal position as $x = \cos(\pi t)$.

Graphing the equations for the vertical and horizontal positions separately gives two typical cosine functions. If we graph them as two parts of a set of parametric equations, however, then we see a graph of the position of the swing as it swings back and forth.[1] In Figure 6, we have a graph of $x = \cos(\pi t)$ and $y = 0.25\cos(2\pi t) + 0.75$ for $0 \leq t \leq 2$.

Example 2 Assuming the center of a clock is at the origin, find a pair of parametric equations that describes the trajectory of a point at the end of a 6-in. second hand where time, in seconds, is the input. Assume the hand starts on the 12 and moves for 60 sec.

Solution Both the vertical and horizontal components can be modeled by sine or cosine functions. Both need to have periods of 60 sec and amplitudes of 6 in. Thus, the amplitude is 6 and the input is $(2\pi/60)t = (\pi/30)t$. The horizontal component starts in the middle of the clock, and as the second hand moves away from the 12, the horizontal component increases because it moves to the right. Becaue the sine

[1] *The swing's position is best seen on a calculator because you can see the swing's movement as the calculator is making the graph.*

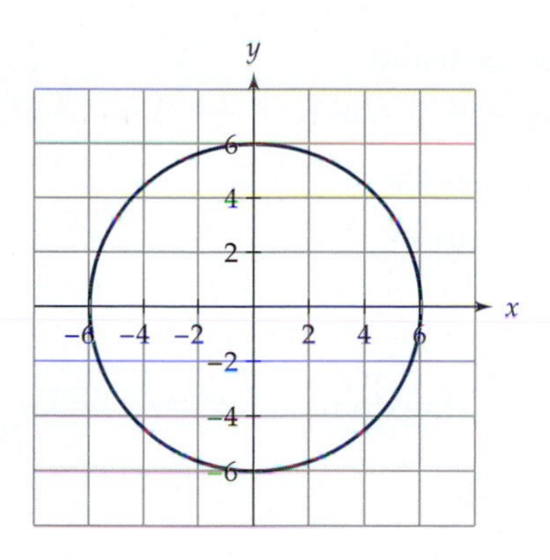

Figure 7

function passes through the origin while increasing, it is the better function to use (it requires no vertical or horizontal shifts). The only transformations required deal with the period and the amplitude. So, the equation $x = 6\sin((\pi/30)t)$ describes the horizontal component.

The vertical component starts at the highest point and decreases. Therefore, the cosine function can be used with no horizontal or vertical transformations. Again, the only transformations required deal with the period and the amplitude. So, the equation $y = 6\cos((\pi/30)t)$ describes the vertical component. The set of parametric equations that describes the trajectory of a point on the end of a 6-in. second hand is

$$x = 6\sin\left(\frac{\pi}{30}t\right) \quad \text{and} \quad y = 6\cos\left(\frac{\pi}{30}t\right) \quad \text{for } 0 \le t \le 60.$$

A graph is shown in Figure 7.

Projectiles

Just as in our previous examples with a swing and a second hand on a clock, we can find the trajectory of a projectile by looking at its horizontal and vertical components separately with parametric equations. In doing so, we simplify the process by not considering the effects of air (e.g., air resistance or lift obtained by a spinning ball).

Suppose a golf ball is hit from the ground, at an angle of 55°, in such a way that the initial velocity is 100 ft/sec. In a world without gravity, the ball would shoot off in a straight line and continue to travel at 100 ft/sec. Let's look at this situation before we bring gravity into the picture. To break apart the vertical and horizontal components, we need to first write the distance traveled by the golf ball as a function of time. Because rate × time = distance, the distance the ball travels on an angle is $100t$. We can use a right triangle, as illustrated in Figure 8, to find the horizontal and vertical components of the golf ball. Because $\sin 55° = \text{opposite}/(100t)$, the opposite side (or the vertical component) is $100t\sin 55°$. By a similar method, we find that the horizontal component is $100t\cos 55°$. Because gravity is a downward force, it does not affect the horizontal component, but it does affect the vertical component. We can model this effect simply by adding $-16t^2$ to the vertical component.[2] Therefore, the vertical component becomes $100t\sin 55° - 16t^2$. Thus, the set of parametric equations that describe the trajectory of the golf ball is $x = 100t\cos 55°$ and $y = 100t\sin 55° - 16t^2$.

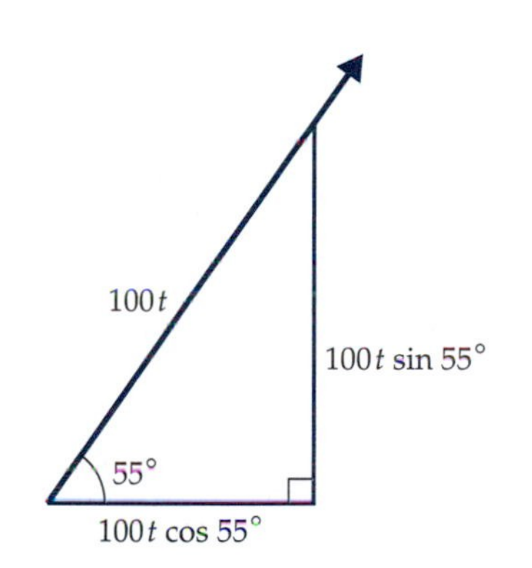

Figure 8 A right triangle can be used to find the vertical and horizontal components of the path of the golf ball.

Let's also define a domain for these equations. We know that the time cannot be negative, so we only need to find the upper bound to the interval. We do this by setting the vertical component to 0 (to represent the time at which the ball hits the ground) and factoring to get

$$100t\sin 55° - 16t^2 = 0$$
$$t(100\sin 55° - 16t) = 0.$$

[2] *The addition of this term is based on an acceleration of gravity of 32 ft/sec². This term is always a part of the vertical component for a projectile as long as the input is in seconds and the output is distance in feet.*

Now by setting each factor to zero, we get $t = 0$ and

$$\begin{aligned} 100 \sin 55° - 16t &= 0 \\ 16t &= 100 \sin 55° \\ t &= \frac{100 \sin 55°}{16} \\ t &\approx 5.12. \end{aligned}$$

Since the projectile hits the ground at 5.12 sec, the domain for these equations is approximately $0 \le t \le 5.12$.

A graph of the trajectory of the golf ball using

$$x = 100t \cos 55° \quad \text{and} \quad y = 100t \sin 55° - 16t^2 \quad \text{for } 0 \le t \le 5.12$$

is shown in Figure 9.

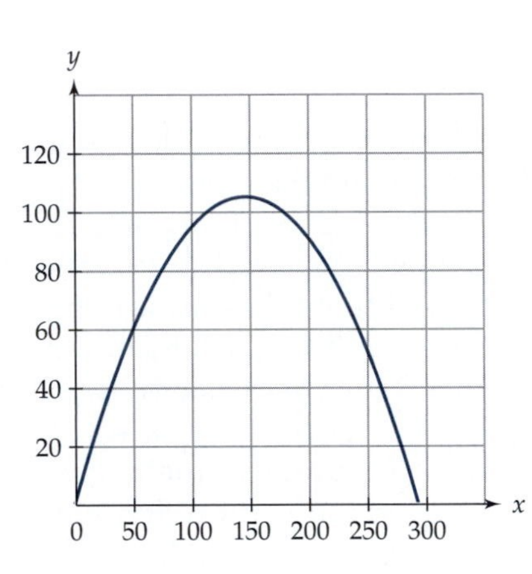

Figure 9 A graph of the trajectory of a golf ball hit at a 55° angle with an initial velocity of 100 ft/sec. The numbers on the x- and y-axes represent distance in feet.

Example 3 Find the parametric equations that describe the trajectory of a baseball hit 3 ft off the ground at an angle of 35°. The initial velocity of the ball is 120 ft/sec. Also find the time it takes the ball to hit the ground.

Solution We again simplify by not including gravity and by assuming the ball is hit off the ground. Just as with the previous golf ball example, the vertical and horizontal components can be found by assuming the angular distance traveled is $120t$, and the horizontal and vertical components can by found using right triangle trigonometry. (See Figure 10.) Thus, the horizontal component is $x = 120t \cos 35°$ and the vertical component is $y = 120t \sin 35°$. By including the effects of gravity and that the ball is being hit from 3 ft off the ground, only the vertical component changes. Including gravity adds $-16t^2$ to the vertical component. Including that the ball was hit 3 ft off the ground adds 3 to the vertical component. Therefore, the equations that describe the trajectory of the ball are

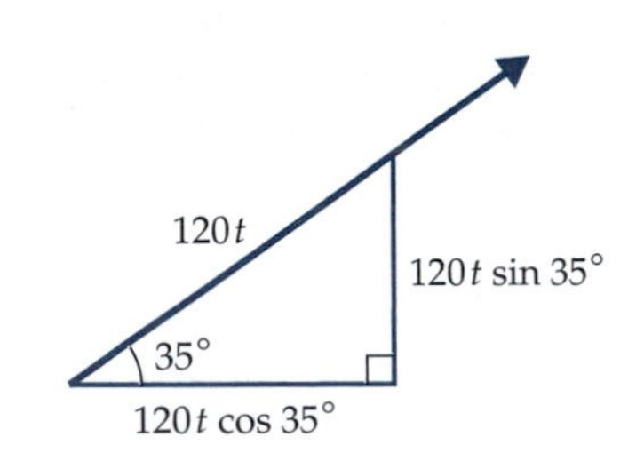

Figure 10

$$x = 120t \cos 35° \quad \text{and} \quad y = 120t \sin 35° - 16t^2 + 3.$$

To find the time it takes the ball to hit the ground, we need to find the time that the height, y, is 0. We therefore need to solve $120t \sin 35° -$

$16t^2 + 3 = 0$ for t. To do so, we need to use the quadratic formula, with $a = -16$, $b = 120 \sin 35°$, and $c = 3$.

$$t = \frac{-120 \sin 35° \pm \sqrt{(120 \sin 35°)^2 - 4(-16)(3)}}{2(-16)}$$

$$t \approx -0.043 \quad \text{and} \quad t \approx 4.345.$$

The positive time, 4.345 sec, is the time it takes the ball to hit the ground. It also means that the domain for the equations is approximately $0 \le t \le 4.345$. A graph of $x = 120t \cos 35°$ and $y = 120t \sin 35° - 16t^2 + 3$ for $0 \le t \le 4.345$ is shown in Figure 11.

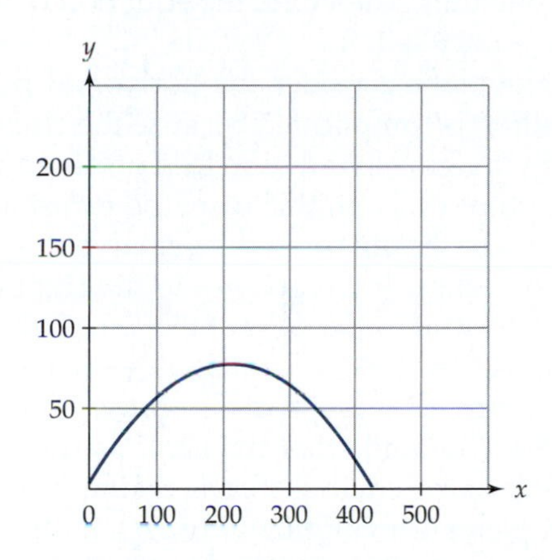

Figure 11

READING QUESTIONS

3. Why is the $-16t^2$ included in the equation describing the vertical component of the trajectory of a golf ball but not in the equation describing the horizontal component?
4. The parametric equations describing the movement at the tip of a 6-in. second hand on a clock were given in this section. Give the parametric equations that describe a point on the tip of a 6-in. second hand that is rotating counterclockwise.

CYCLOIDS AND OTHER INTERESTING CURVES

Imagine watching the movement of a spot on a bicycle wheel as the wheel rolls down the road. The curve that is traced by this spot is known as a cycloid. More generally, a **cycloid** is the curve traced by a point on the circumference of a circle that rolls along a straight line. (See Figure 12.)

Like our previous examples with trajectory, we again want to determine the parametric equations that give the horizontal and vertical positions as outputs. To determine the set of equations that describe a cycloid, we start with a circle of radius 1 such that the lowest point on the circle is the point

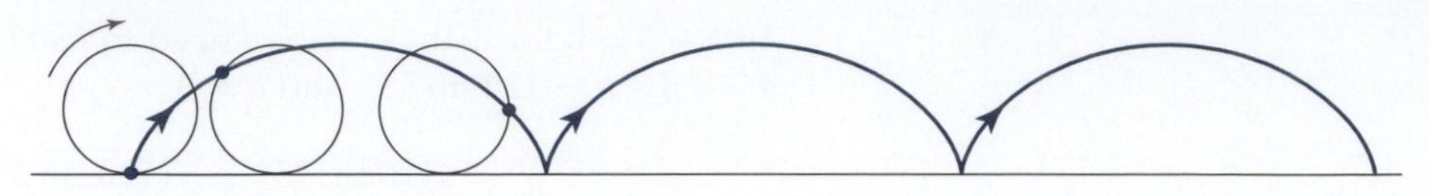

Figure 12 A cycloid is the curve traced by a point on the circumference of a circle that rolls along a straight line

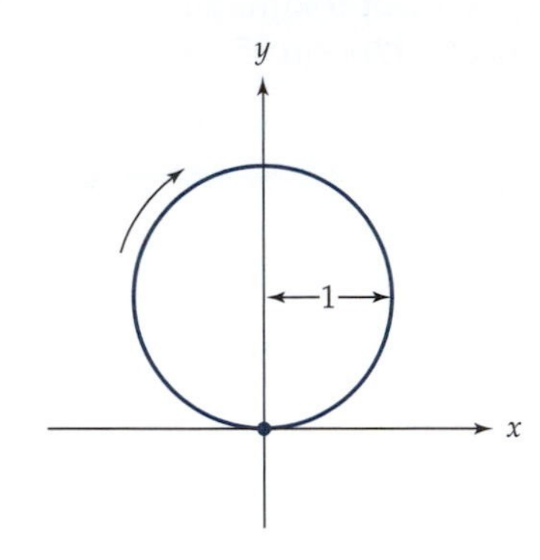

Figure 13 To find the equations describing a cycloid, we start with a point at the origin on a circle of radius 1.

$(0, 0)$. Imagine a dot located on the circle at that point initially. (See Figure 13.) Let's first consider the vertical position of the dot. As the circle rolls to the right, the dot increases in height until the circle has rolled through π radians; then it decreases in height until the circle has rolled through 2π radians. This graph is simply the graph of a negative cosine function shifted up one unit. Therefore, the equation describing the vertical component is $y = -\cos\theta + 1$.

Now let's consider the horizontal position of the dot. The horizontal position is complicated because the circle is traveling horizontally. To simplify, imagine that the circle is turning but that the center of it is staying at the point $(0, 1)$. In this case, the dot starts at the origin on the y-axis and moves to the left (or in the negative direction) until it is one unit to the left of the y-axis. It then moves to the right (or in the positive direction) until it is one unit to the right of the y-axis, then it moves back to the y-axis. The graph of the this horizontal movement starts at the origin and goes down. This graph is a graph of a negative sine function, or $x = -\sin\theta$.

Now imagine that the circle is rolling. The horizontal position of the center of the circle as it rolls is simply $x = \theta$ because as the circle rolls and completes 1 revolution ($\theta = 2\pi$), it also travels the distance of its circumference ($x = 2\pi$). Combining these two equations, $x = -\sin\theta$ and $x = \theta$, together gives $x = -\sin\theta + \theta$, which represents the horizontal position of a point on the circumference of the circle as it rolls. Therefore, the parametric equations that describe a cycloid are

$$x = -\sin\theta + \theta \quad \text{and} \quad y = -\cos\theta + 1.$$

For a cycloid that is traced out by a circle of radius r, the equations are

$$x = r(-\sin\theta + \theta) \quad \text{and} \quad y = r(-\cos\theta + 1).$$

Example 4 Imagine that you see a bicycle traveling down the road with a white dot painted on its 26-in.-diameter tire. Suppose the bicycle wheel is revolving at 2 revolutions per second. Find the parametric equations that describe the position of the dot such that time, in seconds, is the input for both equations.

Solution A transformed cycloid models this situation. Because the tire has a diameter of 26 in., the radius is 13 in. Thus, the pair of equations describing this cycloid are $x = 13(-\sin\theta + \theta)$ and $y = 13(-\cos\theta + 1)$. If the wheel revolves at 2 revolutions per second, then it is completing 1 revolution (or one period) in 0.5 sec. Therefore, instead of an input of θ, we need an input of $(2\pi/0.5)t = 4\pi t$. With an input of time (in seconds), the parametric equations describing the position of the dot are $x = 13(-\sin(4\pi t) + 4\pi t)$ and $y = 13(-\cos(4\pi t) + 1)$. A graph of these parametric equations for 1 sec is shown in Figure 14.

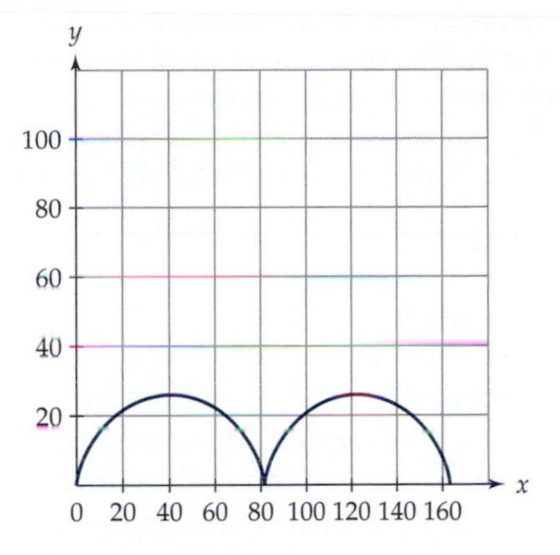

Figure 14

HISTORICAL NOTE

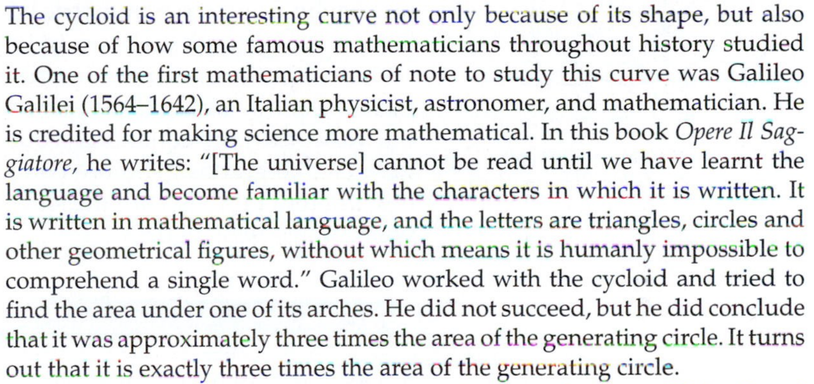

The cycloid is an interesting curve not only because of its shape, but also because of how some famous mathematicians throughout history studied it. One of the first mathematicians of note to study this curve was Galileo Galilei (1564–1642), an Italian physicist, astronomer, and mathematician. He is credited for making science more mathematical. In this book *Opere Il Saggiatore,* he writes: "[The universe] cannot be read until we have learnt the language and become familiar with the characters in which it is written. It is written in mathematical language, and the letters are triangles, circles and other geometrical figures, without which means it is humanly impossible to comprehend a single word." Galileo worked with the cycloid and tried to find the area under one of its arches. He did not succeed, but he did conclude that it was approximately three times the area of the generating circle. It turns out that it is exactly three times the area of the generating circle.

The next mathematician of note to study the cycloid was Blaise Pascal (1623–1662), a French mathematician, physicist, and theologian. He is best known in mathematics as one of the founders of the science of probability. One night in 1654, Pascal experienced a religious vision that caused him to abandon the study of science and mathematics for that of theology. He only studied mathematics for a brief period in 1658 and 1659 because a toothache or illness prevented him from falling asleep. To distract him from the pain, he started studying the cycloid. Miraculously, the pain went away, and Pascal took that as a sign that the study of mathematics was not displeasing to God. Having answered some of the questions about the cycloid himself, Pascal proposed a number of other questions about the cycloid and offered prizes for their solutions. The solutions to the questions posed by Pascal contained some errors in computation, and no prizes were ever awarded.

Other interesting players in the history of the cycloid were two Swiss brothers, Jacque Bernoulli (1654–1705) and Jean Bernoulli (1667–1748). The

continued on page 730

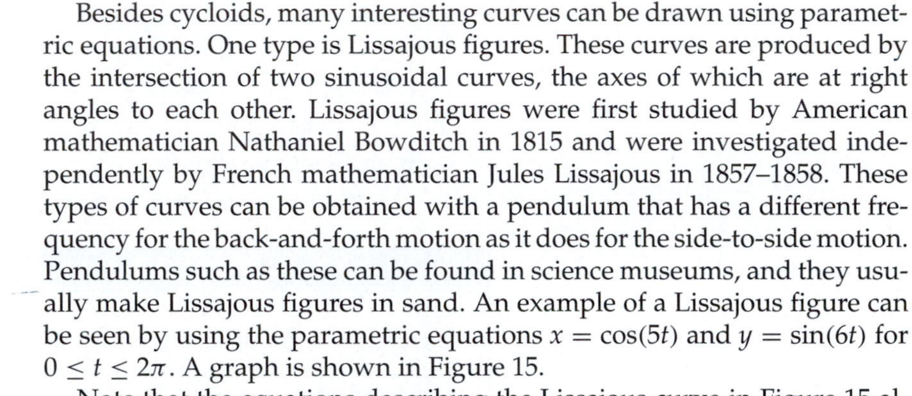
Bernoulli family produced many famous mathematicians for over two centuries. These particular two Bernoulli brothers were trying to find the curve along which a particle will slide in the shortest time from one point to a second lower point not directly beneath the first. This problem amounts to finding the shape of a playground slide that will give the quickest ride. Jean found an incorrect proof that this curve was an inverted cycloid. He then challenged his brother to find a correct proof. When Jacque correctly proved that the inverted cycloid was indeed the correct curve, Jean tried to substitute his brother's proof for his own.

Source: Boyer, *A History of Mathematics*, pp. 359, 400, 457.

Besides cycloids, many interesting curves can be drawn using parametric equations. One type is Lissajous figures. These curves are produced by the intersection of two sinusoidal curves, the axes of which are at right angles to each other. Lissajous figures were first studied by American mathematician Nathaniel Bowditch in 1815 and were investigated independently by French mathematician Jules Lissajous in 1857–1858. These types of curves can be obtained with a pendulum that has a different frequency for the back-and-forth motion as it does for the side-to-side motion. Pendulums such as these can be found in science museums, and they usually make Lissajous figures in sand. An example of a Lissajous figure can be seen by using the parametric equations $x = \cos(5t)$ and $y = \sin(6t)$ for $0 \leq t \leq 2\pi$. A graph is shown in Figure 15.

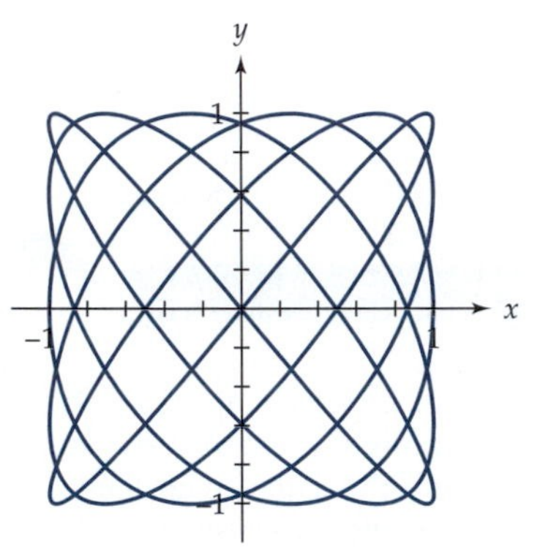

Figure 15 A graph of a Lissajous curve using the equations $x = \cos(5t)$ and $y = \sin(6t)$ for $0 \leq t \leq 2\pi$.

Note that the equations describing the Lissajous curve in Figure 15 almost describe a circle. If the periods of the cosine and sine functions were the same, then the equations would describe a circle. By having different periods, the graphs get more complex. It is interesting to experiment using different periods with the sine and cosine functions to create different types of Lissajous figures.

READING QUESTIONS

5. Given the equations for a basic cycloid, $x = -\sin\theta + \theta$ and $y = -\cos\theta + 1$, explain the following.
   (a) Why is the θ added to the $-\sin\theta$ in the equation describing the horizontal component?
   (b) Why is the 1 added to the $-\cos\theta$ in the equation describing the vertical component?

SUMMARY

Parametric equations are equations in which the coordinates of a point are separately expressed by functions with a common input. This common input is referred to as the **parameter.**

When modeling the trajectories of objects, we can use parametric equations, having one equation represent the horizontal position of the object and another represent the vertical position.

A **cycloid** is the curve traced by a point on the circumference of a circle that rolls along a straight line. The parametric equations that represent a cycloid traced out by a circle of radius r are $x = r(-\sin\theta + \theta)$ and $y = r(-\cos\theta + 1)$.

EXERCISES

1. For each of the following, plot the curve described by the parametric equations. Along with the curve, label the points that correspond to $t = 0$, $t = 1$, $t = 2$, and $t = 3$.
   (a) $x = 2t + 1$ and $y = t - 5$ for $0 \le t \le 3$
   (b) $x = \sqrt{t+2}$ and $y = t^2$ for $0 \le t \le 3$
   (c) $x = \sin t$ and $y = \cos t$ for $0 \le t \le 3$
2. Graph the following sets of parametric equations. Also eliminate the parameter and give a single function, with an appropriate domain, that gives the same graph.
   (a) $x = 8t - 4$ and $y = 2t + 6$ for $0 \le t \le 10$
   (b) $x = t^3$ and $y = t^6 + 3$ for $-2 \le t \le 2$
   (c) $x = 1/t$ and $y = 2/t$ for $0.1 \le t \le 5$
3. Each of the following two sets of parametric equations give graphs that look identical, but they are traced out differently. Describe these differences.
   (a) $x_1 = t^3$ and $y_1 = t^6 + 6$ for $-3 \le t \le 3$, and $x_2 = t$ and $y_2 = t^2 + 6$ for $-27 \le t \le 27$
   (b) $x_1 = \sin t$ and $y_1 = \cos t$ for $0 \le t \le 2\pi$, and $x_2 = \cos t$ and $y_2 = \sin t$ for $0 \le t \le 2\pi$
   (c) $x_1 = 1/t$ and $y_1 = 2/t$ for $0.25 \le t \le 5$, and $x_2 = t + 3$ and $y_2 = 2t + 6$ for $-2.8 \le t \le 1$
4. In the text, we showed that the parametric equations $x = \cos\theta$ and $y = \sin\theta$ for $0 \le \theta \le 2\pi$ describe the graph of a unit circle centered at the origin. Other circles can be described similarly. Consider the set of parametric equations

$$x = a + r\cos\theta \quad \text{and} \quad y = b + r\sin\theta \quad \text{for } 0 \le \theta \le 2\pi.$$

   (a) Graph the set of parametric equations above for $a = 2$, $b = 3$, and $r = 9$. What point is at the center of the circle, and what is the length of the circle's radius?
   (b) Find the set of parametric equations that describes a circle centered at $(4, -3)$ with a radius of 5.
   (c) The equation $(x - a)^2 + (y - b)^2 = r^2$ also describes a circle centered at (a, b) with a radius of r. Show that $(x - a)^2 + (y - b)^2 = r^2$ by substituting $a + r\cos\theta$ for x and $b + r\sin\theta$ for y.
5. The equations $x = 5\cos\theta$ and $y = 5\sin\theta$ for $0 \le \theta \le 2\pi$ describe a circle of radius 5.
   (a) What sort of shape do you think is described by the equations $x = 8\cos\theta$ and $y = 5\sin\theta$ for $0 \le \theta \le 2\pi$?

(b) The equation

$$\frac{x^2}{8^2} + \frac{y^2}{5^2} = 1$$

describes an ellipse. Show that this equation is true when $8\cos\theta$ is substituted for x and $5\sin\theta$ is substituted for y.

6. The equations $x = 5\cos\theta$ and $y = 5\sin\theta$ for $0 \le \theta \le 2\pi$ plot a circle of radius 5. If we think of our calculator screen as a face on a clock, then this circle is traced by starting at the 3 and moving counterclockwise. Change these equations so that our "clock" starts on the 12 and moves clockwise.
7. Assuming the center of a clock is at the origin, find a pair of parametric equations that describe the trajectory of a point at the end of a 10-in. second hand where time, in seconds, is the input. Assume the hand starts on the 3 and moves for 60 sec.
8. Find the set of parametric equations that describe the trajectory of a golf ball that is hit at an initial velocity of 140 ft/sec at an angle of 30°. How far does the golf ball travel before it hits the ground?
9. Find the parametric equations that describe the trajectory of a baseball that is hit 3 ft off the ground at an angle of 40°. The initial velocity of the ball is 100 ft/sec. How far does the baseball travel before it hits the ground?
10. The accompanying figure shows a child's swing. The swing takes 2.2 sec to travel from point A to point B (which are 20 ft apart). The swing varies from a low of 2 ft above the ground to a high of 6 ft above the ground. Find the set of parametric equations that describe the trajectory of the swing. Time, in seconds, should be the input for your equations, and at $t = 0$, the swing is at point A.

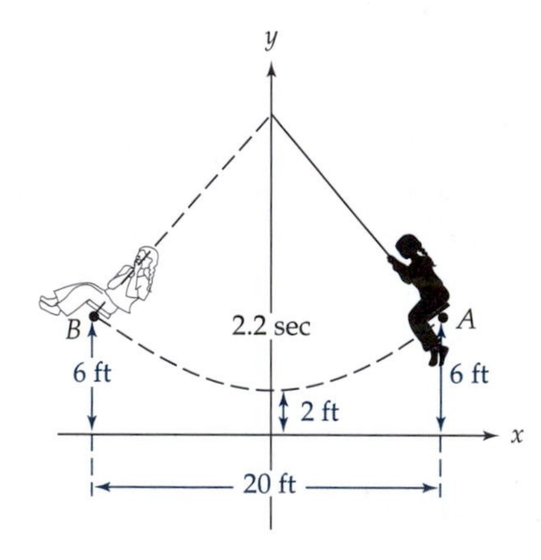

11. Imagine you see a bicycle traveling down a road with a white dot painted on its 24-in.-diameter tire. Suppose the bicycle wheel is revolving at 3 revolutions per second.
    (a) Find a set of parametric equations that describe the position of the dot such that the input is the angle through which the wheel has turned, θ.

(b) Find a set of parametric equations that describe the position of the dot such that the input is time, in seconds.

(c) If we want the parametric equations you found in part (b) to include 4 revolutions of the bicycle wheel, what should be the domain for the equations?

12. In the text, we look at the Lissajous figure described by the equations $x = \cos(5t)$ and $y = \sin(6t)$ for $0 \le t \le 2\pi$. Let's look for patterns in these types of figures. Let $x = \cos(at)$ and $y = \sin(bt)$ for $0 \le t \le 2\pi$. We vary the values of a and b and see how the graphs change.

    (a) Make graphs for Lissajous figures given the following constants.

    i. $a = 1$ and $b = 1$

    ii. $a = 1$ and $b = 2$

    iii. $a = 1$ and $b = 3$

    (b) Predict what the Lissajous figure for $a = 1$ and $b = 4$ will look like.

    (c) Make graphs for Lissajous figures given $a = 2$ and $b = 4$. Also make a Lissajous figure given $a = 3$ and $b = 6$. How are your two figures the same, and how are they different?

13. The parametric equations $x = \theta \cos\theta$ and $y = \theta \sin\theta$ give a graph known as Archimedes' spiral.

    (a) Graph these equations with $0 \le \theta \le 4\pi$.

    (b) Explain why the graph spirals outward.

14. Consider the following two sets of parametric equations.

$$x_1 = \cos\theta \quad \text{and} \quad y_1 = \sin\theta$$
$$x_2 = \theta \quad \text{and} \quad y_2 = \sin\theta$$

The first set will graph a unit circle and the second will make a graph of the sine function. Graph these two sets of equations on a graphing calculator for $0 \le \theta \le 2\pi$ in a window that goes from -1.5 to 7 on the x-axis and -3 to 3 on the y-axis. Do so in such a way that the calculator will draw the two graphs simultaneously. (On a TI-83, you need to change the "MODE" to "Simul".)

    (a) Make a sketch of your final graph.

    (b) Explain why what you saw; when the graphs were drawn on your calculator screen, gives a good demonstration for a graph of the sine function based on the definition of the sine function.

15. Suppose a ball is thrown at an angle of θ with the horizontal at an initial velocity of 30 ft/sec. The parametric equations

$$x_1 = 30t\cos\theta \quad \text{and} \quad y_1 = 30t\sin\theta - 16t^2$$

describe the position of this ball. Also suppose another ball is thrown directly upward, 15 ft from where the first ball was thrown, at an initial velocity of 25 ft/sec. Parametric equations that describe the position of this ball are

$$x_2 = 15 \quad \text{and} \quad y_2 = 25t - 16t^2.$$

(a) Assume θ, the angle at which the first ball is thrown, is 45°. Graph both sets of parametric equations in such a way that the calculator draws the two graphs simultaneously. (On a TI-83, you need to change the "MODE" to "Simul".) Graph with a window of $0 \leq X \leq 30$ and $0 \leq Y \leq 20$. Do the balls appear to hit each other?

(b) Find the proper angle where the balls would hit. To do so, set the equations for y_1 and y_2 equal to each other and solve for θ.

(c) Using your angle for θ found in part (b), graph both sets of parametric equations, except now make $x_2 = 6$. Do the balls still hit? Do the balls always hit whatever the value of x_2?

16. The horizontal component for a projectile shot at ground level at an angle of θ is $x = vt\cos\theta$, where v is the initial velocity and t is the time since launch. The vertical component for the same projectile is $y = vt\sin\theta - 16t^2$. We are interested in finding the angle, θ, that makes the projectile travel the farthest.

    (a) Using the vertical component, find the time it takes the projectile to land in terms of θ and v.

    (b) Using your answer from part (a), show that the range of the projectile is given by $x = \frac{1}{32}v^2\sin(2\theta)$.

    (c) Determine the angle, θ, that gives the maximum range for the projectile.

17. Graph a circle using the equations $x = 5\cos T$ and $y = 5\sin T$, where T is in degrees, not radians. Do so with a domain of $0° \leq T \leq 360°$ and $T_{step} = 3$. Your viewing window should have horizontal components of $-10 \leq x \leq 10$ and vertical components of $-7 \leq y \leq 7$.

    (a) Change your graph so that $T_{step} = 60$. You should see a hexagon graphed instead of a circle. Explain why a hexagon is graphed.

    (b) By changing the T_{step} different regular polygons can be drawn.

        i. What should the T_{step} be so that a triangle is drawn?

        ii. What should the T_{step} be so that a square is drawn?

        iii. What should the T_{step} be so that a pentagon is drawn?

        iv. What should the T_{step} be so that an n-gon is drawn?

INVESTIGATIONS

INVESTIGATION 1: SPIROGRAPH

The cycloids described in this section are similar to the curve drawn out by a point on a bicycle tire as the bicycle travels horizontally. Imagine (if you can) a bicycle traveling around a circle. The curve drawn by a point on the tire of this bicycle is called an *epicycloid*. You can think of it as a cycloid around a circle. An epicycloid is one type of figure that can be drawn by the children's toy Spirograph. In this investigation, we determine the equations for epicycloids and other similar figures.

The parametric equations given in this section for a cycloid of radius r are

$$x = r(-\sin\theta + \theta) \qquad \text{and} \qquad y = r(-\cos\theta + 1).$$

We can rewrite these equations as

$$x = -r\sin\theta + r\theta \qquad \text{and} \qquad y = -r\cos\theta + r.$$

The addition of $r\theta$ in the horizontal component describes how the circle is moving horizontally at a constant rate. The addition of r in the vertical component is just a vertical shift upward of r units. Both $r\theta$ and r need to be changed because the circle is not traveling in a straight line horizontally, but is traveling around a circle.

1. Instead of traveling in a straight line, our circle of radius r travels around a circle of radius R. (See the accompanying figure.) We use $x = -r\sin\theta$ and $y = -r\cos\theta$ to describe the movement of a point on the outer circle if it were revolving while staying in one place. To define the movement of the outer circle as in goes around the inner circle, we need to describe the position of the center of the outer circle in terms of θ.

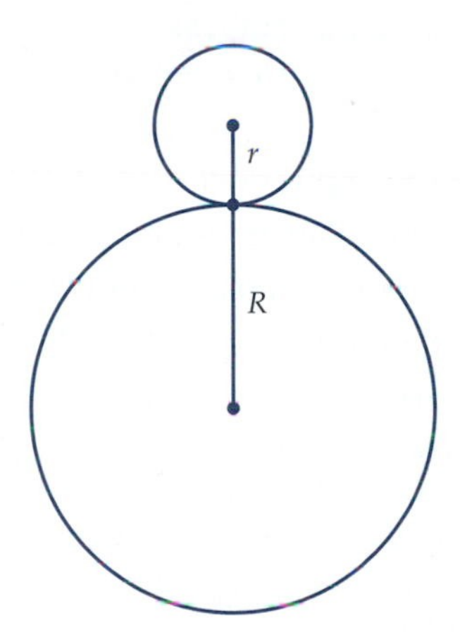

   (a) Imagine the *center* of the outer circle traveling around the inner circle. As it revolves, the point traces out a circle. What is the radius of this circle?

   (b) If the outer circle starts directly on top of the inner circle, as shown, explain why the sine function works well to describe the horizontal component of the position of the center of the outer circle and the cosine function works well to describe the vertical component of the position of the center of the outer circle.

   (c) When finding the formula for a cycloid, we used the number of radians that a point on the circle had turned as the input. In this case, there are two circles, an inner circle and an outer circle. Our input is the number of radians that a point on the outer circle turns as it goes around the inner circle. You can see that this problem becomes more complicated because, depending on the relative sizes of the two circles, the outer circle may have to make a large number of trips around the inner circle for the point that we are tracking to end up in the same spot as it did when it started. The number of trips depends on the ratio of the circumference of the circle formed by the center of the outer circle as it makes its way around the inner circle to the circumference of the outer circle. Find the ratio of these circumferences.

(d) Assuming the period of the revolving outer circle is 2π, explain why the period of the center of the outer circle as it travels around the inner circle is $2\pi \cdot (R+r)/r$.

(e) Look at the equations for a cycloid given in this section. Notice that each equation has two parts, one that gives the position of the center of the moving circle and another that gives the position of the point on that circle. Use your answers from parts (a) through (d) and your knowledge of the equations for cycloids to explain why the graph of an epicycloid can be described using the equations

$$x = -r\sin\theta + (R+r)\sin\left(\frac{r\theta}{R+r}\right) \qquad \text{and}$$

$$y = -r\cos\theta + (R+r)\cos\left(\frac{r\theta}{R+r}\right).$$

2. Now let's graph some epicycloids.
   (a) Graph an epicycloid where $R = 4$ and $r = 2$ for $0 \le \theta \le 6\pi$.
   (b) Graph an epicycloid where $R = 5$ and $r = 2$ for $0 \le \theta \le 14\pi$.
   (c) Graph an epicycloid where $R = 7$ and $r = 10$ for $0 \le \theta \le 34\pi$.
3. When using a Spirograph, you do not always draw the curve using the outermost position on the outer circle. For example, if you use the center of the circle, then the drawing is simply a circle. If you use any other position, then you get some sort of epicycloid. We can make these types of figures simply by changing our equations slightly. The set of parametric equations $x = -p\sin\theta + (R+r)\sin(r\theta/(R+r))$ and $y = -p\cos\theta + (R+r)\cos(r\theta/(R+r))$ makes a "spiro-graph," where R is the radius of the inner circle, r is the radius of the outer circle, and p is the distance the point (or pen) is from the center of the outer circle.
   (a) Graph an epicycloid such that $R = 7, r = 10$, and $p = 7$ in a domain of $0 \le \theta \le 34\pi$.
   (b) One thing you can do with the graphs of these equations that you cannot do with a Spirograph is make graphs where the distance from the point to the center of the outer circle, p, is greater than the radius of the outer circle, r. Make this kind of graph where $R = 7, r = 10$, and $p = 15$ for $0 \le \theta \le 34\pi$.

INVESTIGATION 2: POLAR COORDINATES

When graphing a function or a point with rectangular coordinates, the input is the horizontal component and the output is the vertical component. Using parametric equations allows the input to be something else (like

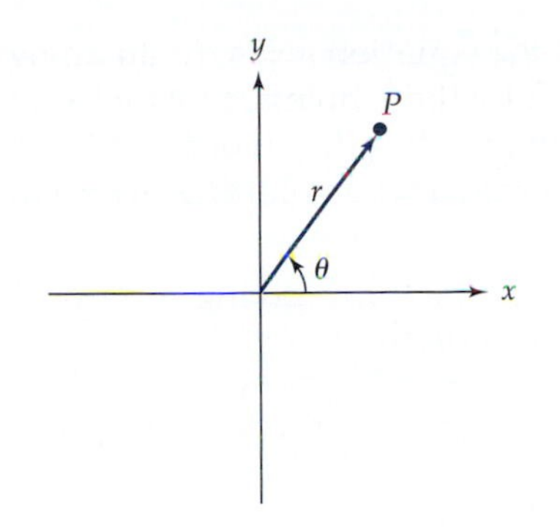

Figure 16

time), and one equation gives the horizontal component and the other gives the vertical component. This method still uses the rectangular coordinate system, but differently. Graphing an equation or a point using polar coordinates is a completely different method. To describe a point in polar coordinates, the point is given a distance from the origin, r, and an angle, θ, that is measured counterclockwise from the positive x-axis to a line connecting the point and the origin. (See Figure 16.)

To convert the point $(4,4)$ in the rectangular coordinate system to the polar coordinate system, we have to find r and θ shown in Figure 17.

By using the Pythagorean theorem, we can find r:

$$r^2 = 4^2 + 4^2$$
$$r^2 = 32$$
$$r = \sqrt{32} = 4\sqrt{2}.$$

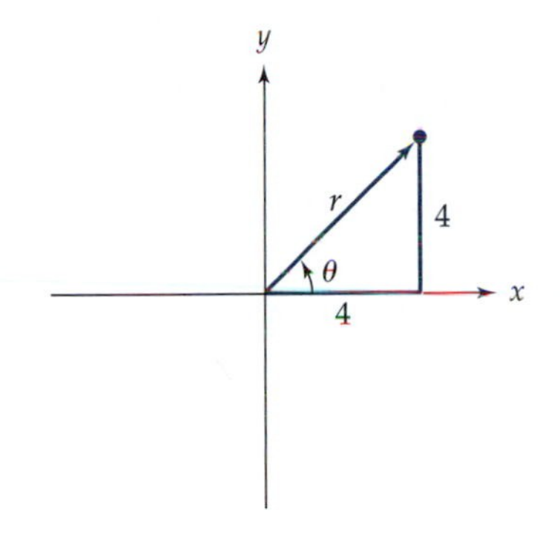

Figure 17

To find θ, we can use $\tan\theta = \frac{4}{4}$; thus, $\theta = \pi/4$. Therefore, the polar coordinates of the point $(4,4)$ are $(4\sqrt{2}, \pi/4)$. Polar coordinates are usually given as an ordered pair, (r, θ).

1. Convert the following points given in rectangular coordinates to those of polar coordinates. Give decimal approximations for the angle θ to two places after the decimal point.
   (a) $(3,4)$
   (b) $(-5,8)$
   (c) $(-6,-12)$

   Some curves can be easily defined using polar coordinates. For example, because circles have a constant radius, defining them with polar coordinates is very simple. Graphs of polar coordinates are written with the angle, θ, as the input and the distance, r, as the output. Because a circle has constant radius, the graph of $r = 5$ gives a circle with radius 5 centered at the origin. To graph this circle on a calculator, make sure it is polar mode and then input the equation. To see the entire circle, make sure that it graphs from $0 \le \theta \le 2\pi$ and that the viewing window is set appropriately.

2. The curve called Archimedes' spiral can be graphed using the polar equation $r = \theta$. We want to graph this curve by plotting points by hand.
   (a) Given that $r = \theta$, complete the table by giving a decimal approximation for r for the given values of θ.

θ	0	$\frac{\pi}{4}$	$\frac{\pi}{2}$	$\frac{3\pi}{4}$	π	$\frac{5\pi}{4}$	$\frac{3\pi}{2}$	$\frac{7\pi}{4}$	2π
$r = \theta$									

   (b) Plot the points from the table and connect them with a smooth curve.
   (c) Graph $r = \theta$ on a calculator. How does the calculator graph compare with the one you drew by hand?

3. We want to plot points to get a graph of the equation $r = \sin\theta$. In doing so, we get some negative values for r. To plot these points, plot the point in the opposite direction (180° or π radians). In other words, plotting the point where $r = -1$ and $\theta = \pi/4$ is the same as plotting the point where $r = 1$ and $\theta = 5\pi/4$.
   (a) Given that $r = \sin\theta$, complete the table by giving a decimal approximation for r for the given values of θ.

θ	0	$\frac{\pi}{4}$	$\frac{\pi}{2}$	$\frac{3\pi}{4}$	π	$\frac{5\pi}{4}$	$\frac{3\pi}{2}$	$\frac{7\pi}{4}$	2π
$r = \sin\theta$									

   (b) Plot the points from the table and connect them with a smooth curve.
   (c) Graph $r = \sin\theta$ on a calculator. How does the calculator graph compare with the one you drew by hand?
4. Using a calculator, we want to plot $r = \sin(n\theta)$ for various values of n when $0 \le \theta \le 2\pi$. These graphs are called roses. The number of petals vary, depending on the value of n.
   (a) Graph $r = \sin(2\theta)$. How many petals does it have?
   (b) Graph $r = \sin(3\theta)$. How many petals does it have?
   (c) Graph $r = \sin(4\theta)$. How many petals does it have?
   (d) Graph $r = \sin(5\theta)$. How many petals does it have?
   (e) How many petals, in terms of n, does $r = \sin(n\theta)$ have?

9.2 VECTORS

When taking a trip in a car, we often use maps to tell us which roads to take. By following the map, we can be fairly assured that our route will take us where the map says it will, at least most of the time. If you are a pilot of a small plane, you do not have that luxury because you do not travel on roads. In this case, you not only have to be sure what course you need to follow to reach your destination, but also how the wind will affect the course you choose. In this section, we look at vectors. Using vectors is one way to understand how a plane's planned course and the wind combine to give a plane's actual course. We learn what vectors are, how to add them, and how to use them in applications.

INTRODUCTION AND NOTATION

Magnitude and Direction

To consider what happens to a plane when the wind pushes it, we need to consider not only the amount of force on the plane from its engines and the wind, but also the direction of those forces. In mathematics, we use vectors to describe such forces. A **vector** is a unit of measure that consists of both a *magnitude* and a *direction*. If a plane is traveling at 100 mph directly east, then the magnitude is 100 mph and the direction is east. We depict a vector geometrically as a directed line segment or arrow. The length of the directed line segment should correspond to the magnitude of the vector. If a vector starts at point A and goes to point B, then the vector is denoted as **AB**. Vectors can also be represented by a single letter, such as **v**. The length or magnitude of the vector is written with two vertical lines on either side of the name of the vector.[3] For example, we write the magnitude of **AB** as $\|\mathbf{AB}\|$ and the magnitude of **v** as $\|\mathbf{v}.\|$

Let's look at a simple example to see how vectors and their notation can be used. Suppose you are the pilot in a plane. You point your plane due east and begin flying with an airspeed (the speed you can fly if there is no wind) of 100 mph.[4] A wind with a speed of 25 mph pushes you due north. It is helpful to show these forces using directed line segments. The directed line segments **AB** and **BC** represent the plane's airspeed and wind speed, respectively. (See Figure 1.)

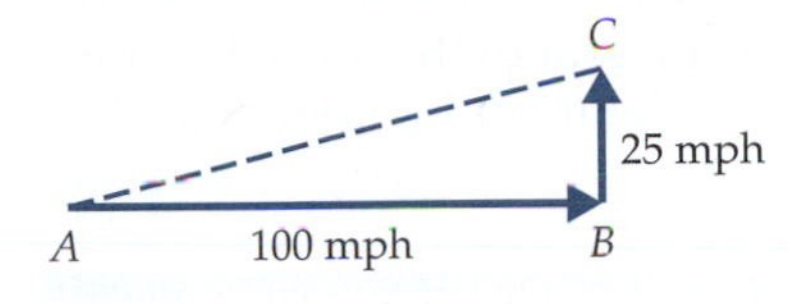

Figure 1 The vector **AC** is the resultant vector of a plane flying with an air speed of 100 mph due east in a 25 mph wind blowing due north.

Notice in Figure 1 that if there were no wind, then you start at point A and end at point B. The wind, however, pushes you north of point B to point C. You can see that these two vectors have both magnitude (or length) as well as direction. You can also see that by placing the "head" of one on the "tail" of the other, we can show what happens when both of these vectors are combined. In other words, the vector **AC** is the result of adding the other two vectors. Let's find both the magnitude and direction of **AC**.

We use the Pythagorean theorem to find the $\|\mathbf{AC}\|$:

$$\begin{aligned}\|\mathbf{AC}\|^2 &= \|\mathbf{AB}\|^2 + \|\mathbf{BC}\|^2\\ \|\mathbf{AC}\| &= \sqrt{\|\mathbf{AB}\|^2 + \|\mathbf{BC}\|^2}\\ &= \sqrt{100^2 + 25^2}\\ &\approx 103.08.\end{aligned}$$

We also find the direction of **AC** by using the tangent function.

$$\begin{aligned}\tan(m\angle A) &= \tfrac{25}{100}\\ m\angle A &= \tan^{-1}\left(\tfrac{25}{100}\right)\\ &\approx 14.04^\circ.\end{aligned}$$

Thus, the plane ends up traveling at a ground speed of about 103 mph in a direction of 14° north of due east.

The Components of a Vector

The magnitude and direction of vector **AC** above were determined by knowing the lengths of the two vectors we depicted as horizontal and

[3] *This notation is similar to that of absolute value because, in both cases, we consider amount without regard to direction.*

[4] *Actually, pilots use knots, or nautical miles per hour. One knot equals about 1.15 mph. One nautical mile is 1 minute (i.e., $\frac{1}{60}$ of a degree) of Earth's circumference. To simplify for this problem, we use miles.*

vertical directed line segments, **AB** and **BC.** These two vectors can be considered the component vectors of **AC.** The magnitude and direction of any vector can be determined by knowing the vector's components. When vectors are defined in a plane, as we do throughout this section, they consist of two components. We use the standard notation **i** to represent the horizontal component of magnitude 1 and **j** to represent the vertical component of magnitude 1.

Because we represented **AB** horizontally in Figure 1 and **AB** had magnitude 100, it can be represented as 100**i**. Because we represented **BC** vertically and **BC** had magnitude 25, it can be represented as 25**j**. The resultant vector **AC** can be represented as $100\mathbf{i} + 25\mathbf{j}$.

Example 1 Suppose a person walks 2 mi east and then turns and walks 3 mi north. Express the displacement of this person in terms of **i** and **j**.

Solution We let a 1-mi displacement of this person to the east be represented by **i** and a 1-mi displacement of this person to the north be represented by **j**. A displacement of 2 mi east and 3 mi north can then be represented by $2\mathbf{i} + 3\mathbf{j}$.

In Example 1, a person was walking east and north. On a set of axes, we typically place east to the right and north on the top. Because we consider "right" and "top" to be positive, the components we found were positive. If, on the other hand, a person was walking west 2 mi and south 3 mi, then this displacement vector would be represented as $-2\mathbf{i} - 3\mathbf{j}$.

READING QUESTIONS

1. What two units of measure make up a vector?
2. Suppose $\mathbf{v} = 2\mathbf{i} + 4\mathbf{j}$.
   (a) What is the direction of **v**?
   (b) What is the magnitude of **v**?

Figure 2 A walk of 2 mi east followed by 3 mi north followed by 3 mi east followed by 1 mi north results in a walk of 5 mi east and 4 mi north.

ADDING VECTORS

In Example 1, we showed that we could write the displacement of a person who walked 2 mi east and 3 mi north in vector notation as $2\mathbf{i} + 3\mathbf{j}$. Suppose that person then walked another 3 mi east and then 1 mi more north. (See Figure 2.) How can we write the displacement of this walk?

It is easy to see that the second portion of this walk can be expressed by $3\mathbf{i} + \mathbf{j}$. To find the displacement of the entire walk, we need to find the sum of $2\mathbf{i} + 3\mathbf{j}$ and $3\mathbf{i} + \mathbf{j}$. Because our walker walked a total of 5 mi east and 4 mi north, the displacement of the entire walk can be described as $5\mathbf{i} + 4\mathbf{j}$. We also say that

$$(2\mathbf{i} + 3\mathbf{j}) + (3\mathbf{i} + \mathbf{j}) = 5\mathbf{i} + 4\mathbf{j},$$

which implies that to add vectors together, we only need to add their respective components.

PROPERTIES

In general, we say if $\mathbf{v} = a\mathbf{i} + b\mathbf{j}$ and $\mathbf{u} = c\mathbf{i} + d\mathbf{j}$, then $\mathbf{u} + \mathbf{v}$ is expressed as

$$(a\mathbf{i} + b\mathbf{j}) + (c\mathbf{i} + d\mathbf{j}) = (a + c)\mathbf{i} + (b + d)\mathbf{j}.$$

By using right triangle trigonometry, we easily determine the magnitude and direction of a vector when given its components.

PROPERTIES

If $\mathbf{v} = a\mathbf{i} + b\mathbf{j}$, then the following are true.

- The magnitude of $\mathbf{v}$ is $\|\mathbf{v}\| = \sqrt{a^2 + b^2}$.
- If θ is the direction of $\mathbf{v}$, then $\tan\theta = b/a$.

To determine θ, we use the inverse tangent function, but we may need to make an adjustment depending in which quadrant our vector lies. For example, if the coefficients on **i** and **j** are both positive, then the vector lies in the first quadrant, but if the coefficient on **i** is negative and the coefficient on **j** is positive, then the vector lies in the second quadrant. Depending on the signs of these coefficients, you can determine in which quadrant the vector lies and make the appropriate adjustment to the angle.

Let's look at a more complicated example. Let's consider a pilot who flies 20° north of due east at 100 mph for 1 hr on a windless day. The pilot then makes a course correction, turns an additional 20° north, and flies in that direction for 2 hrs. (See Figure 3.) We want to find the magnitude and direction of the resultant vector, **AC**. In order to do so, it is helpful to break **AB** and **BC** into their component parts and then add the components. (See Figure 4.)

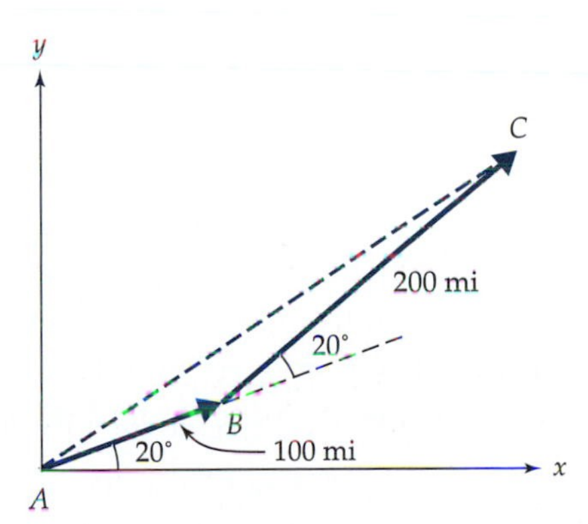

Figure 3 We need to find the magnitude and direction of **AC**.

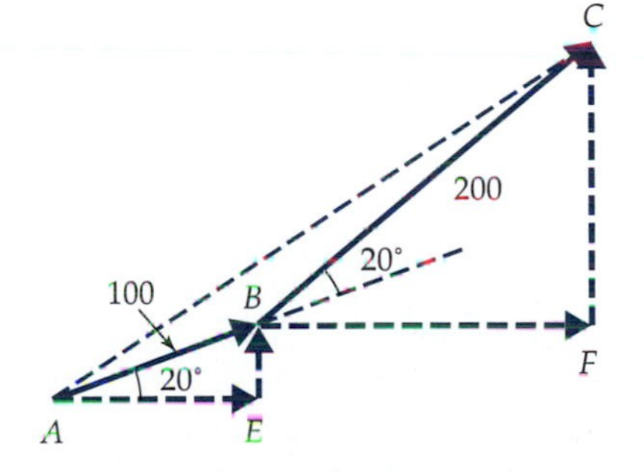

Figure 4 Finding the components of **AB** and **BC** will help find the direction and magnitude of **AC**.

Let's first consider **AB.** This vector can be expressed as the sum of vectors **AE** and **EB.** By using trigonometry, we can find the magnitudes of these vectors.

$$\begin{aligned}\|\mathbf{AE}\| &= 100\cos 20^\circ \\ &\approx 93.97 \text{ mi}\end{aligned}$$

and

$$\begin{aligned}\|\mathbf{EB}\| &= 100\sin 20^\circ \\ &\approx 34.20 \text{ mi.}\end{aligned}$$

Thus, **AB** can be expressed as $93.97\mathbf{i} + 34.20\mathbf{j}$. We do the same with **BC.**

$$\begin{aligned}\|\mathbf{BF}\| &= 200\cos 40^\circ \\ &\approx 153.21 \text{ mi}\end{aligned}$$

and

$$\begin{aligned}\|\mathbf{FC}\| &= 200\sin 40^\circ \\ &\approx 128.56 \text{ mi.}\end{aligned}$$

Thus, **BC** can be expressed as $153.21\mathbf{i} + 128.56\mathbf{j}$. Therefore, **AC** can be expressed as

$$\begin{aligned}\mathbf{AC} &= \mathbf{AB} + \mathbf{BC} \\ &= (93.97\mathbf{i} + 34.20\mathbf{j}) + (153.21\mathbf{i} + 128.56\mathbf{j}) \\ &= 247.18\mathbf{i} + 162.76\mathbf{j}.\end{aligned}$$

The magnitude of this vector is

$$\begin{aligned}\|\mathbf{AC}\| &\approx \sqrt{247.18^2 + 162.76^2} \\ &\approx 295.95.\end{aligned}$$

This plane will have then flown 295.95 mi at an angle of

$$\begin{aligned}\tan\theta &\approx \frac{162.76}{247.18} \\ \theta &\approx \tan^{-1}\left(\frac{162.76}{247.18}\right) \\ &\approx 33.36^\circ \text{ north of due east.}\end{aligned}$$

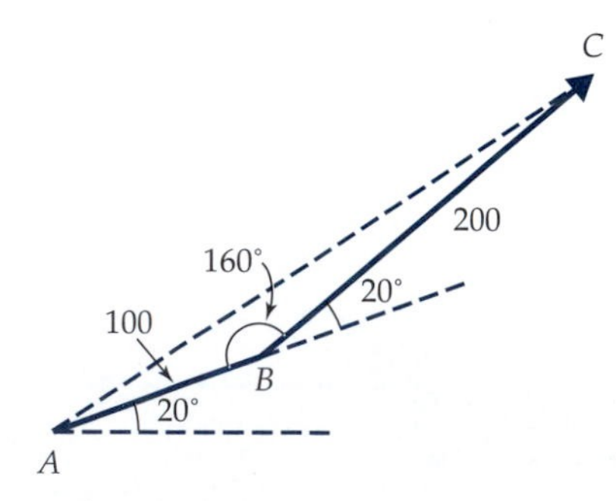

Figure 5 The direction and magnitude of **AC** can be found using the law of cosines and the law of sines.

If all we wanted to find was the distance the plane flew and the direction, then we could have used the law of cosines and the law of sines. By using Figure 5, we can see that, in $\triangle ABC$, the measure of $\angle B$ is 160° $(180^\circ - 20^\circ)$ and the lengths of sides $\overline{AB}$ and $\overline{BC}$ are 100 and 200, respectively.

Using the law of cosines to find the length of $\overline{AC}$ gives

$$\begin{aligned}(AC)^2 &= 100^2 + 200^2 - 2 \cdot 100 \cdot 200\cos 160^\circ \\ &\approx 87{,}587.7 \\ AC &\approx 295.95 \text{ mi.}\end{aligned}$$

To find the direction in which the plane flew, we use the law of sines to find $m\angle A$:

$$\frac{\sin A}{200} = \frac{\sin 160^\circ}{295.95}$$
$$\sin A = \frac{200 \sin 160^\circ}{295.95}$$
$$A = \sin^{-1}\left(\frac{200 \sin 160^\circ}{295.95}\right)$$
$$A \approx 13.36^\circ.$$

By adding 13.36° to the 20° that the plane originally was flying, we find that the direction is 33.36° north of east. Let's look at another example.

Example 2 Suppose you are in a plane with a heading of 10° west of due north traveling 140 mph. If there is a wind pushing you 20° north of due east at 10 mph, then what is your final position relative to your starting point?

Solution Let's look at the vectors associated with the plane and those associated with the wind separately. (See Figure 6.) We find the components of each vector and then add the components to find the magnitude and direction of the plane.

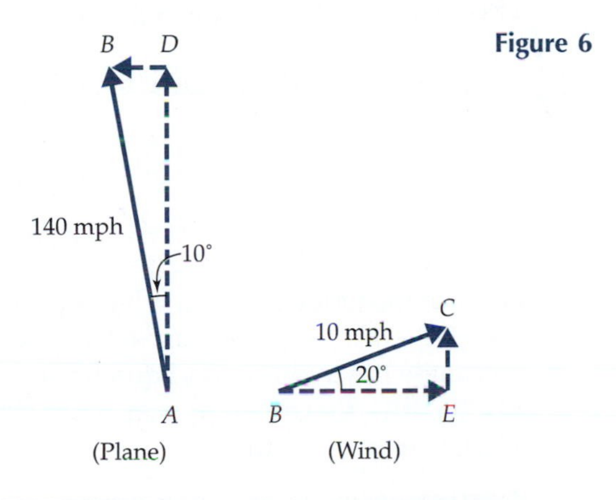

Figure 6

First, we write **AB** as component parts.

$$\|\mathbf{AD}\| = 140 \cos 10^\circ$$
$$\approx 137.87$$

and

$$\|\mathbf{DB}\| = 140 \sin 10^\circ$$
$$\approx 24.31.$$

So, **AB** has a western component of 24.31 and a northern component of 137.87. We write this western component as an eastern component of −24.31. Thus, **AB** can be described as $-24.31\mathbf{i} + 137.87\mathbf{j}$.

To write the wind vector, **BC**, in its component parts, we do the following.

$$\begin{aligned} \|\mathbf{BE}\| &= 10\cos 20^\circ \\ &\approx 9.40 \end{aligned}$$

and

$$\begin{aligned} \|\mathbf{EC}\| &= 10\sin 20^\circ \\ &\approx 3.42. \end{aligned}$$

The vector **BC** therefore has an eastern component of 9.40 and a northern component of 3.42. Thus, **BC** can be described as $9.40\mathbf{i} + 3.42\mathbf{j}$. Therefore, **AC** can be expressed as

$$\begin{aligned} \mathbf{AC} &= \mathbf{AB} + \mathbf{BC} \\ &= (-24.31\mathbf{i} + 137.87\mathbf{j}) + (9.40\mathbf{i} + 3.42\mathbf{j}) \\ &= -14.91\mathbf{i} + 141.29\mathbf{j}. \end{aligned}$$

The magnitude of this vector is

$$\begin{aligned} \|\mathbf{AC}\| &\approx \sqrt{(-14.91)^2 + 141.29^2} \\ &\approx 142.07. \end{aligned}$$

This plane will be flying at a ground speed of 142.07 mph at an angle of

$$\begin{aligned} \tan\theta &\approx \frac{141.29}{-14.91} \\ \theta &\approx \tan^{-1}\left(\frac{141.29}{-14.91}\right) \\ &\approx -83.98^\circ. \end{aligned}$$

Because the tangent function has a period of 180° and because $-83.98^\circ + 180^\circ = 96.02^\circ$, $\tan 96.02^\circ$ also equals $141.29/-14.91$. Because this angle lies in the second quadrant, as does the vector $-14.91\mathbf{i} + 141.29\mathbf{j}$, this direction is appropriate. This direction can also be described as 6.02° west of north.

READING QUESTIONS

3. Why is it sometimes useful to consider the components of a vector when doing problems that require vector addition?
4. Simplify $(4\mathbf{i} - 7\mathbf{j}) + (5\mathbf{i} + 37\mathbf{j})$.
5. Explain why a vector with a western component of 20 is the same as a vector with an eastern component of −20.

MAY THE FORCES BE WITH YOU

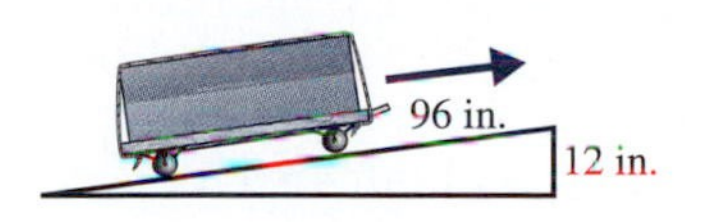

Figure 7 The use of a ramp makes it easier to move a 400-lb piano up 12 in.

We have already used vectors to find the solutions to problems involving the force of wind on airplanes. Vectors can also be used to find the solution to other problems involving forces. One of the authors bought a 400-lb piano and needed to get it up two steps to get it into his living room. Rather than lift the piano up the stairs, he used a ramp to make the process easier. The piano only had to go up 12 in., but he used an 8-ft (96-in.) ramp. (See Figure 7.)

As soon as he started pushing the piano up the ramp, it was obvious that he was not lifting 400 lb. It helped that he had a good piano dolly, but clearly the ramp was a huge help. What weight did it feel like he was pushing by using this ramp?

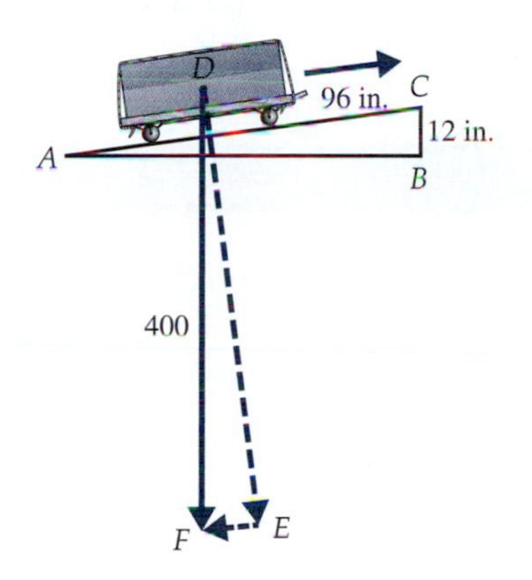

Figure 8 The magnitude of **EF** is the force needed to push the piano up the ramp.

Using the ramp meant that the force of gravity, which pulls straight down, was split into two components, one perpendicular to the ramp, **DE**, and one parallel to the ramp, **EF**. The magnitude of **EF** is the force needed to push the piano up the ramp. (See Figure 8.) Let's find the magnitude of **EF**. To do so, we need to find the measures of one of the angles in $\triangle DEF$. We use that $\triangle ABC$ is similar to $\triangle DEF$ because the measures of the corresponding angles are the same. Because the triangles are similar, their sides are proportional. In particular,

$$\frac{EF}{BC} = \frac{DF}{AC}$$

$$\frac{EF}{12} = \frac{400}{96}$$

$$EF = \frac{400 \cdot 12}{96}$$

$$EF = 50.$$

So, the force needed to push the piano up the ramp (if you ignore the friction of the wheels on the ramp) is only about 50 lb.

When doing problems such as this one, the force of gravity can always be expressed as a vector pointing straight down. We can think of the components of this vector as those perpendicular to and parallel to the ramp. The perpendicular force is exerted on the ramp while the parallel force is exerted on the person pushing the piano up the ramp.

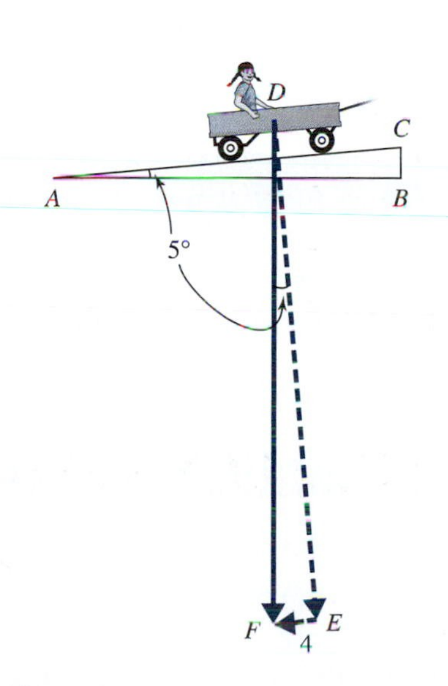

Figure 9

Example 3 Suppose a child in a wagon is being pushed up a ramp going into a building. The ramp is at an angle of 5°, and a force of 4 lb is being used to push the wagon up the ramp. Ignoring any friction, determine the combined weight of the child and the wagon.

Solution A drawing of this situation is shown in Figure 9. The measure of $\angle EDF$ is the same as the measure of $\angle A$. Knowing that $\|\mathbf{EF}\| = 4$, we want to determine the magnitude of **DF**.

$$\sin 5^\circ = \frac{\|\mathbf{EF}\|}{\|\mathbf{DF}\|}$$

$$\sin 5^\circ = \frac{4}{\|\mathbf{DF}\|}$$

$$\|\mathbf{DF}\| = \frac{4}{\sin 5^\circ}$$
$$\|\mathbf{DF}\| \approx 45.9 \text{ lb.}$$

The combined weight of the wagon and the child is about 45.9 lb.

READING QUESTIONS

6. Suppose a child in a wagon is being pushed up a ramp. The ramp is at an angle of 10°, and the combined weight of the child and the wagon is 50 lb. Ignoring any friction, determine the force needed to push the wagon up the ramp.

SUMMARY

A **vector** is a unit of measure that consists of both a *magnitude* and a *direction*. We depict a vector as a directed line segment or an arrow. When vectors are defined in a plane, they consist of two components.

We use the standard notation $\mathbf{i}$ to represent the horizontal component of magnitude 1 and $\mathbf{j}$ to represent the vertical component of magnitude 1. When adding vectors written in this standard notation, you simply add the separate components. In general, we say if $\mathbf{v} = a\mathbf{i} + b\mathbf{j}$ and $\mathbf{u} = c\mathbf{i} + d\mathbf{j}$, then $\mathbf{u} + \mathbf{v}$ can be expressed as $(a\mathbf{i} + b\mathbf{j}) + (c\mathbf{i} + d\mathbf{j}) = (a + c)\mathbf{i} + (b + d)\mathbf{j}$.

By using right triangle trigonometry, we can easily determine the magnitude and direction of a vector when given its components. If $\mathbf{v} = a\mathbf{i} + b\mathbf{j}$, then the magnitude of $\mathbf{v}$ is $\|\mathbf{v}\| = \sqrt{a^2 + b^2}$. If θ is the direction of $\mathbf{v}$, then $\tan\theta = b/a$.

When vectors are used to solve problems involving forces, it is helpful to draw a diagram of the forces. A triangle often results from this diagram, and the appropriate use of trigonometry can be applied to help you solve the problem.

EXERCISES

1. For each of the following pairs of points, assume the first point is the intial point of a vector and the second point is the terminal point of a vector. Write each vector in terms of **i** and **j**.
   (a) $(0, 0)$ and $(7, 10)$
   (b) $(0, 0)$ and $(-8, -5)$
   (c) $(2, 3)$ and $(6, -4)$
   (d) $(-3, -6)$ and $(4, 2)$
2. For each of the following, express the displacement of the person making the walk in terms of **i** and **j**.
   (a) A person walks 3 mi east and 2 mi north.
   (b) A person walks 3 mi west and 2 mi south.

(c) A person walks 2 mi east and 1 mi north and 3 mi west.

(d) A person walks 1 mi west and 2 mi north and 3 mi east and 4 mi south.

3. Determine in which quadrant each of the following vectors lie.

   (a) $-4\mathbf{i} + 5\mathbf{j}$

   (b) $3\mathbf{i} + 3\mathbf{j}$

   (c) $-2\mathbf{i} - 3\mathbf{j}$

   (d) $\mathbf{i} - 2\mathbf{j}$

4. For each of the following vectors, determine both its magnitude and direction.

   (a) $3\mathbf{i} + 7\mathbf{j}$

   (b) $-2\mathbf{i} + 4\mathbf{j}$

   (c) $5\mathbf{i} - 2\mathbf{j}$

   (d) $-3\mathbf{i} - 8\mathbf{j}$

5. Given a vector's magnitude and direction, write each of the following in terms of **i** and **j**.

   (a) $\|\mathbf{v}\| = 4$ and $\theta = 50°$

   (b) $\|\mathbf{v}\| = 6$ and $\theta = 120°$

   (c) $\|\mathbf{v}\| = 3$ and $\theta = 200°$

   (d) $\|\mathbf{v}\| = 2$ and $\theta = 340°$

6. Suppose an object is acted upon by two forces with magnitudes $\|\mathbf{u}\|$ and $\|\mathbf{v}\|$ and directions θ_u and θ_v, respectively. Determine the magnitude and direction of the resultant force.

   (a) $\|\mathbf{u}\| = 4$ lb, $\theta_u = 20°$; $\|\mathbf{v}\| = 4$ lb, $\theta_v = 40°$

   (b) $\|\mathbf{u}\| = 4$ lb, $\theta_u = 20°$; $\|\mathbf{v}\| = 8$ lb, $\theta_v = 40°$

   (c) $\|\mathbf{u}\| = 4$ lb, $\theta_u = 20°$; $\|\mathbf{v}\| = 8$ lb, $\theta_v = 110°$

   (d) $\|\mathbf{u}\| = 4$ lb, $\theta_u = 20°$; $\|\mathbf{v}\| = 4$ lb, $\theta_v = 200°$

7. Find the sum of the following vectors, **u** and **v**, for each of the following situations.

   (a) $\mathbf{u} = 6\mathbf{i} + 3\mathbf{j}$ and $\mathbf{v} = -2\mathbf{i} + 4\mathbf{j}$

   (b) $\mathbf{u} = -5\mathbf{i} - 2\mathbf{j}$ and $\mathbf{v} = 9\mathbf{i} - 14\mathbf{j}$

   (c) **u** has a magnitude of 4 and a direction of 40° and **v** has a magnitude of 6 and has a direction of 60°

   (d) **u** has a magnitude of 7 and has a direction of 120° and **v** has a magnitude of 7 and has a direction of 60°

8. In the text, we gave formulas used to determine the magnitude and direction of a vector. In this exercise, you are asked to show why each of them is true. For each, assume $\mathbf{v} = a\mathbf{i} + b\mathbf{j}$.

   (a) Explain why $\|\mathbf{v}\| = \sqrt{a^2 + b^2}$.

   (b) Explain why if θ is the direction of **v**, then $\tan\theta = b/a$.

9. A plane is flying 20° west of due south at an airspeed of 120 mph. A wind is pushing the plane 10° south of due west at 20 mph. What is the ground speed of the plane, and in which direction is it flying?

10. A plane is flying at 100 knots at 40° east of north for 1 hr and then adjusts its course to go 50° east of north for 2 hr. Assuming there is no wind, find the plane's position relative to takeoff after 3 hr.
11. A plane is flying at 140 knots at 42° west of north for 1 hr and then adjusts its course to go 47° west of north for half an hour. During the entire flight, the wind is 35 knots, pushing the plane 10° south of west. Find the plane's position relative to takeoff after $1\frac{1}{2}$ hr.
12. Gravity causes objects in free fall to fall with an acceleration of 32 ft/sec^2. If instead of falling straight down, you are on a sled sliding down a hill, that acceleration is changed into two vectors, one parallel to the hill and one perpendicular to the hill. Suppose you are on a hill with a 20° slope as shown in the accompanying figure.

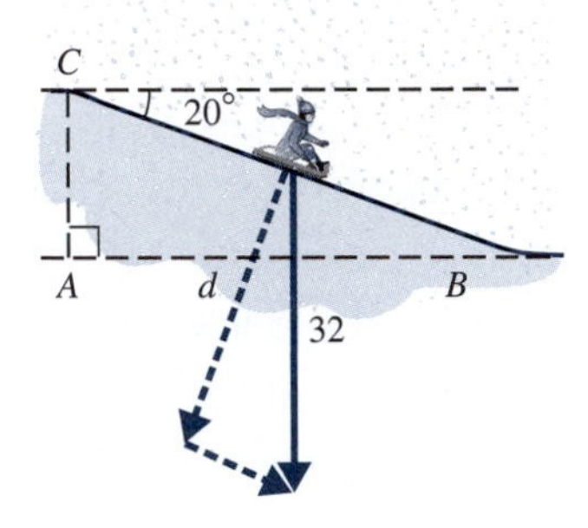

    (a) Ignoring any friction, what is the rate of acceleration of the sled?
    (b) How fast will the sled be traveling after 4 sec? (*Hint:* For a constant acceleration, velocity = acceleration × time.)
13. As mentioned in the text, you combine forces by adding vectors together. Assume we have two forces, a force of 6 lb being applied in the direction of 90° and a force of 2 lb whose direction is variable.
    (a) If you want to maximize the sum of the two forces, what should the direction of the second vector be?
    (b) What is the smallest force you can get by combining the forces of the two vectors? What should the direction of the second vector be?
    (c) If the 2-lb force is perpendicular to the 6-lb force, what are the magnitude and the direction of the resultant force? Is there more than one answer? If so, give another answer.
    (d) What direction should you apply the second force for the combined forces to have a resultant force of 6 lb? Is there more than one answer? If so, give another answer. (*Hint:* Be careful when taking the square root of both sides of an equation. Remember that if $x^2 = 4$, then $x = \pm 2$.)
14. Two ropes are tied to a fallen tree as it is being dragged through the woods. The ropes are tied to the same point on the tree. The angle

between the ropes is 20°. (See the accompanying figure.) One person is pulling with a force of 30 lb, and the other is pulling with a force of 40 lb. What is the total force required to pull the tree?

15. A force of 20 units is being applied to an object. Another force of 30 units is also going to be applied to this object. Can you arrange this second force so that the resultant force is zero? If you can, give the angle between the forces. If not, explain why not.
16. You have to move a 600-lb piano up a short flight of stairs (3 ft) by yourself. You have plenty of room for a ramp and an excellent frictionless piano dolly. You would like to reduce the weight that you have to push up the ramp to about 75 lb if possible. To do so, how long should your ramp be?
17. Suppose a 100-lb weight is suspended from two ropes as shown in the accompanying figure. Because of the weight, suppose the force on the rope on the right has magnitude m and the force on the rope on the left has magnitude n.

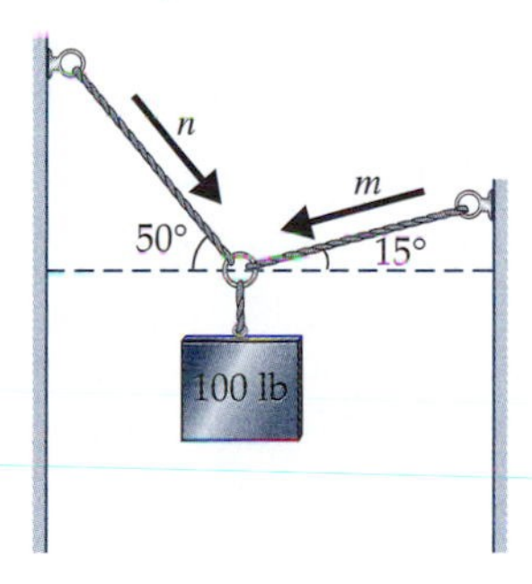

(a) Explain why the vector describing the force of the rope on the right can be described as $m \cos 15°\mathbf{i} + m \sin 15°\mathbf{j}$.
(b) Explain why the vector describing the force of the rope on the left can be described as $n \cos 130°\mathbf{i} + n \sin 130°\mathbf{j}$.
(c) Explain why the sum of the vectors, $(m \cos 15°\mathbf{i} + m \sin 15°\mathbf{j}) + (n \cos 130°\mathbf{i} + n \sin 130°\mathbf{j}) = 0\mathbf{i} + 100\mathbf{j}$.
(d) Find the magnitudes m and n.

18. Suppose a 100-lb weight is suspended with two ropes of equal length as shown in the accompanying figure. Because of the weight, suppose the force on each rope has magnitude m.

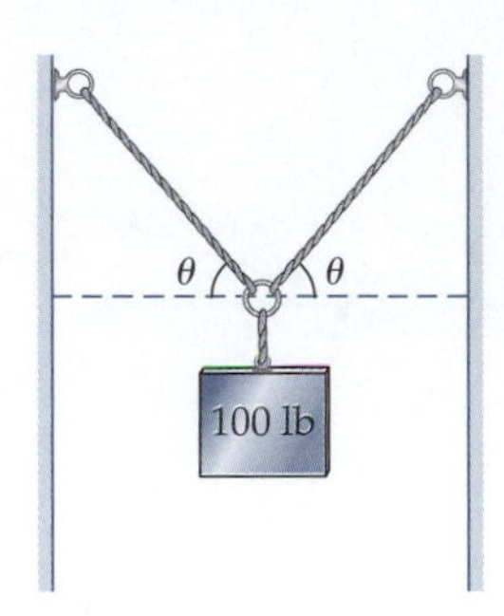

(a) Explain why the vector describing the force of the rope on the right can be described as $m\cos\theta\mathbf{i} + m\sin\theta\mathbf{j}$.

(b) Explain why the vector describing the force of the rope on the left can be described as $-m\cos\theta\mathbf{i} + m\sin\theta\mathbf{j}$.

(c) Explain why $(m\cos\theta\mathbf{i} + m\sin\theta\mathbf{j}) + (-m\cos\theta\mathbf{i} + m\sin\theta\mathbf{j}) = 0\mathbf{i} + 100\mathbf{j}$.

(d) Find a function where m is the output and θ is the input.

(e) If $\theta = 20°$, what is the value of m?

(f) If $\theta = 80°$, what is the value of m?

(g) If $m = 100$ lb, what is the value of θ?

(h) Can θ ever equal 0? Explain.

19. A cable is hanging from two poles as shown in the accompanying figure. If a force of 45 lb is being exerted on each end, what is the weight of the cable?

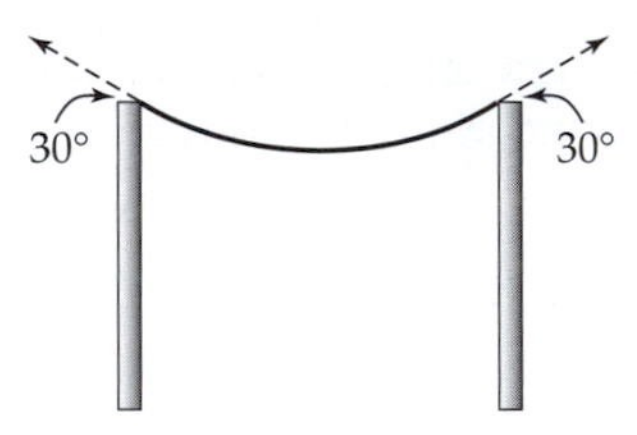

20. A piece of exercise equipment called the Total Gym is advertised on an infomercial. A person lies on a sliding platform on an incline plane to complete a workout on this machine. The machine then uses a person's body weight as resistance. In the infomercial, a statement is made that at the lowest incline, a person is lifting only 4% of his or her body weight. At the highest incline, a person is lifting 60% of his or her body weight. (See the accompanying figure.) Assuming the sliding platform

is frictionless, what is the range of the angle of inclination for the Total Gym?

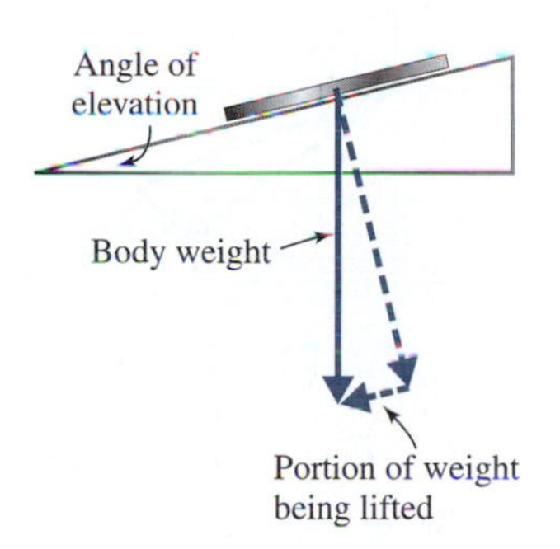

INVESTIGATIONS

INVESTIGATION: WINGING IT

In this section, we had a number of examples from aviation. In all our problems so far, we have considered what the wind does to a plane flying in certain directions. As a pilot, though, it is not enough to know where the wind will take you. You need to be able to compensate for that wind by pointing the plane in a slightly different direction from where you want to go. That way, the wind will blow you on course instead of off course.

Let's begin with some elementary aviation terms. A heading of 310° means that you are going in a direction 310° clockwise from due north. (See Figure 10.) The wind direction signifies the direction *from* which the wind is blowing. A wind with direction 90°, for example, is blowing from *east to west*. Other aviation terms include the following.

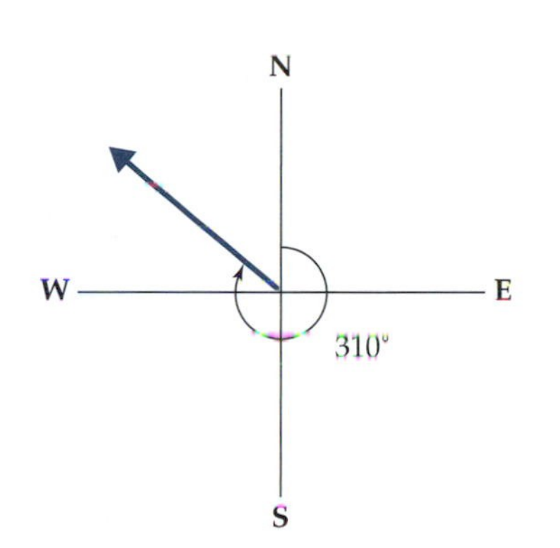

Figure 10

- Ground speed: the speed of a plane relative to the ground
- True airspeed: the speed a plane can travel if there is no wind
- True heading: the actual heading needed to get from starting point to destination with no wind

1. A pilot's true heading is 125° with a true airspeed of 110 knots.[5] The wind is 20 knots at 40°. (Remember that a wind at 40° indicates the direction from which it is blowing. It actually is blowing toward a direction of 220°.) If the pilot heads directly at 125°, then the plane will be blown off course. We wish to find the heading the pilot should use as well as the plane's ground speed. In the accompanying figure on the next page, **AC** represents the true heading, **BC** represents the wind, and **AB** represents the heading the pilot must use to compensate for the wind.

[5] A knot *is a measure of speed. One knot equals 1 nautical mile per hour, or approximately 1.15 mph. One nautical mile is 1 minute (i.e.,* $\frac{1}{60}$ *of a degree) of Earth's circumference.*

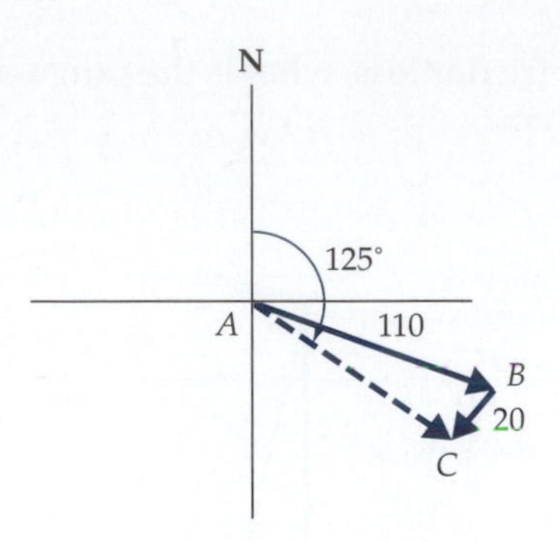

(a) Show that $m\angle BCA = 95°$.

(b) Find $m\angle CAB$ to the nearest tenth of a degree. This is the measure of your wind correction angle. (*Hint:* The law of sines may be helpful in this question.)

(c) What heading should the pilot use?

(d) Find the ground speed of the plane. (*Hint:* The ground speed is the magnitude (or length) of **AC**.)

2. A plane is preparing to go from Holland, Michigan, to Sheboygan, Wisconsin. This trip requires a true heading of 313°. The plane is capable of a true airspeed of 125 knots. The winds aloft forecast reports that the winds are 25 knots at 50°.

   (a) What is the wind correction angle?

   (b) To what heading should the pilot direct the plane?

   (c) What is the actual ground speed?

9.3 MULTIVARIABLE FUNCTIONS

In this book, we looked at functions and graphs of functions and used a calculator to explore functions. While doing so, we assumed that each function has one input and one output. Often, however, functions can be more complicated. Sometimes a function has more than one input. For example, consider the process of computing a car loan. The monthly payment depends on the price of the car, the interest rate, and the length of the loan. Actually, three inputs (price, interest rate, and length of the loan) are used to determine the output (car payment). In this section, we study functions with more than one input and look at graphs associated with these functions.

INTRODUCTION AND NOTATION

Suppose you are planning to take out a 5-year loan to buy a new car. In determining your monthly payment, the price of the car is important, but so is the interest rate. In fact, both are used as variables in the standard formula that computes the monthly payment, M, for a 5-year loan. The formula for the monthly payment, M, is

$$M = \frac{(r/12)P}{1-(1+r/12)^{-60}}, \tag{1}$$

where r is the yearly interest rate and P is the amount of the loan (or the principal).[6] This function is an example of a multivariable function. A **multivariable function** is a function that has more than one input. Note that the car loan formula is still a function. That is, for any pair of inputs (i.e., for any fixed loan amount and fixed interest rate), there is only one possible output.

Let's look at the car payment example again, changing the notation slightly. By using function notation, we can write it as

$$M(r,P) = \frac{(r/12)P}{1-(1+r/12)^{-60}}. \tag{2}$$

This notation is similar to other function notation. In this case, the notation $M(r,P)$ reminds us that M, the monthly payment, is the output and that r, the annual interest rate, and P, the amount of the loan, are the inputs.

Example 1 If you take out a 5-year loan for $12,000 with a yearly interest rate of 10%, what is your monthly payment?

Solution To use formula (2) to compute the monthly payment, we need to have the interest rate, 10%, written as a decimal, 0.10. Then, by using $P = 12{,}000$ and $r = 0.1$, we get

$$M(0.1, 12000) = \frac{(0.10/12)\cdot 12000}{1-(1+0.10/12)^{-60}} \approx \frac{100}{1-0.60779} \approx 254.96.$$

Our monthly car payment is $254.96.

Earlier in this text, we considered two-variable functions by holding one of the variables constant. In this case, if we hold the interest rate at 10%, then we can substitute 0.10 for r and can get a new function,

$$M(0.10, P) = \frac{(0.10/12)P}{1-(1+0.10/12)^{-60}},$$

which represents the monthly payment for a 5-year loan of amount P assuming an annual interest rate of 10%. This function may look complicated, but when simplified, you see that it is just a linear function. When simplified, we get

$$M(0.10, P) \approx 0.021246P.$$

Because the y-intercept of this function is zero, when the amount borrowed doubles, so does the payment. For example, a loan of $5000 has a payment

[6] *The exponent 60 in this equation comes from making 60 payments, one payment per month for 5 years.*

of \$106.23, whereas a loan of \$10,000 has a payment of \$212.46. This function can also be written as

$$M(0.10, P) \approx \tfrac{1}{47}P,$$

which shows that after about 47 payments, the borrower has paid off the entire principal. Because there are a total of 60 payments, 13 payments are needed solely to pay off the interest.

We can also hold the amount of the loan constant and let the interest rate vary to create a different one-variable function. Suppose we are looking to borrow \$12,000. The function for the monthly payment where the interest rate is the input is

$$M(r,\ 12{,}000) = \frac{(r/12)\cdot 12000}{1-(1+r/12)^{-60}} = \frac{1000r}{1-(1+r/12)^{-60}}.$$

This function is not a linear function, but it is almost linear for interest rates in which you expect to pay for a car loan. The graph is seen in Figure 1. This graph is slightly concave up. For example, for an interest rate of 5%, the payment is \$226.45; for an interest rate of 10%, the payment is \$254.96; and for an interest rate of 15%, the payment is \$285.48. From 5% to 10%, the payment went up \$28.51. From 10% to 15%, the payment went up \$30.52. Thus, the graph is not linear, but is slightly concave up.

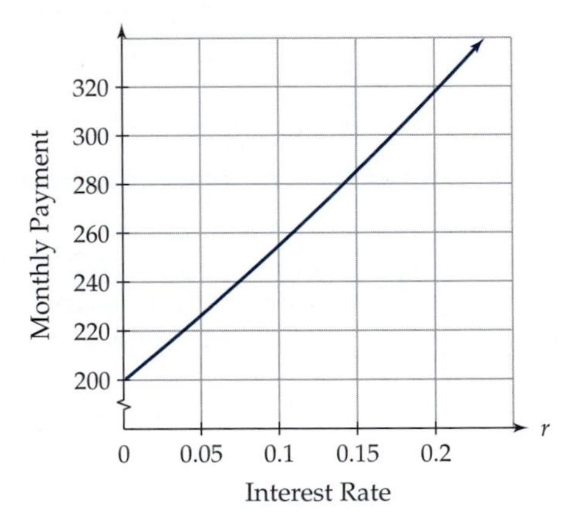

Figure 1 A graph of

$$M(r, 12{,}000) = \frac{1000r}{1-(1+r/12)^{-60}}.$$

Let's consider another two-variable function. Table 1 represents the two-variable function for computing the windchill equivalent temperature, a measure of how cold it feels on a cold windy day. This function has two inputs (wind speed and air temperature) and one output (windchill). We can represent this function by $W(s, t)$, where $W(s, t)$ is the windchill equivalent temperature in degrees Fahrenheit, s is the wind speed in miles per hour, and t is the temperature in degrees Fahrenheit.

TABLE 1 Windchill Equivalent Temperature

Wind Speed (mph)	Air Temperature (°F)												
	35	30	25	20	15	10	5	0	−5	−10	−15	−20	−25
5	32	27	22	16	11	6	0	−5	−10	−15	−21	−26	−31
10	22	16	10	3	−3	−9	−15	−22	−27	−34	−40	−46	−52
15	16	9	2	−5	−11	−18	−25	−31	−38	−45	−51	−58	−65
20	12	4	−3	−10	−17	−24	−31	−39	−46	−53	−60	−67	−74
25	8	1	−7	−15	−22	−29	−36	−44	−51	−59	−66	−74	−81
30	6	−2	−10	−18	−25	−33	−41	−49	−56	−64	−71	−79	−86
35	4	−4	−12	−20	−27	−35	−43	−52	−58	−67	−74	−82	−89
40	3	−5	−13	−21	−29	−37	−45	−53	−60	−69	−76	−84	−92

Example 2 Find $W(20, 10)$.

Solution Use the left column of Table 1 to find 20 for the wind speed and use the top row to find 10° for the air temperature. At their intersection, we see that the windchill equivalent temperature is −24°F.

When evaluating a multivariable function, be consistent in the notation. If the function is given as $W(s, t)$, then the number substituted for the first input, s, must *always* be wind speed and the number substituted for the second input, t, must be air temperature.

Example 3 Explain the meaning of $z = W(s, 5)$.

Solution The function $z = W(s, 5)$ is the one-variable function whose input is wind speed and whose output is the windchill equivalent temperature when the air temperature is 5°F.

READING QUESTIONS

1. Give an example of a multivariable function that is different from those mentioned in the text. Define the input and output variables.
2. Find the monthly car payment on a 5-year loan of $19,000 at 6% annual interest.
3. Give the formula for $z = M(r, 12{,}500)$ and explain what it means.
4. Using the windchill table, determine how fast the wind needs to blow to make it feel like −22°F if the air temperature is 15°F.

CONTOUR CURVES

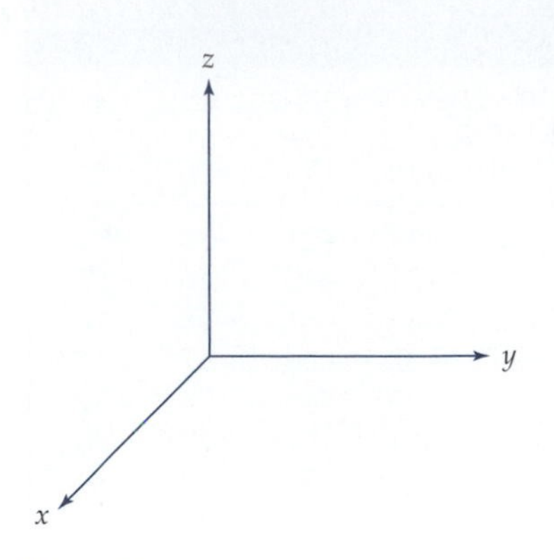

Figure 2 The axes for the three-dimensional coordinate system.

The typical Cartesian plane has an x-axis and a y-axis at right angles to each other. We can add a third axis, called the z-axis, at a right angle to both. Just as we can represent any point in a plane with an ordered pair, we can now represent any point in space with an ordered triple. The point $(1, 2, 3)$, for example, lines up with 1 on the x-axis, 2 on the y-axis, and 3 on the z-axis. Sometimes we draw the three axes on a page as shown in Figure 2. Using function notation, the outputs $f(x, y)$ are graphed using the z-axis. So, a graph of $f(x, y) = x + y$, for example, is the same as $z = x + y$.

Consider the function $f(x,y) = 5\left(\frac{1}{2}\right)^{x^2+y^2}$. This function is a multivariable function with two inputs, x and y, and one output, $f(x, y)$. This function cannot be graphed with most calculators because they only graph functions with one input and one output (i.e., two variables total). This function should be graphed in three-dimensions because there are two inputs and one output (a total of three variables). Two different views of the graph of this function are given in Figure 3.

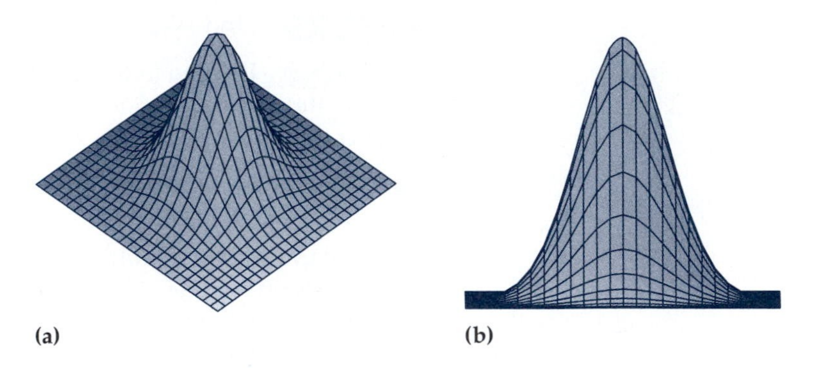

Figure 3 Two views of the graph of $f(x,y) = 5\left(\frac{1}{2}\right)^{x^2+y^2}$. The view in (a) is as though you are looking down on the xy-plane while in (b) the view is as though you were standing in the xy-plane.

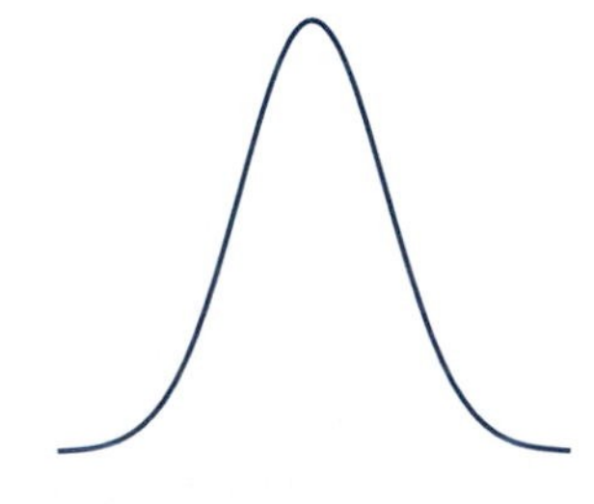

Figure 4 A graph of $f(x,0) = 5\left(\frac{1}{2}\right)^{x^2}$.

Notice that the graph of a two-variable function is a surface (like a molded sheet of plastic). Because it is difficult to draw surfaces on a flat sheet of paper, we often look at alternative types of graphs. One possibility is to create a one-variable function by fixing one of the variables. For example, let's look at the graph of $f(x, 0)$. Because $f(x,y) = 5\left(\frac{1}{2}\right)^{x^2+y^2}$, then $f(x,0) = 5\left(\frac{1}{2}\right)^{x^2+0} = 5\left(\frac{1}{2}\right)^{x^2}$. Figure 4 is the graph of $f(x,0) = 5\left(\frac{1}{2}\right)^{x^2}$.

Notice that this graph looks similar to the highest curve on the graph in Figure 3(b), which is no accident. Fixing one of the inputs is the same as cutting a vertical cross section of the surface.[7]

It is also valuable to look at horizontal cross sections of a surface. The horizontal cross sections of a surface are called **contour curves.** Symbolically, a contour curve is the collection of input pairs that gives a fixed output. To

[7] *Fixing the other input cuts a vertical cross section that is perpendicular to this one.*

create a horizontal cross section, the output (in this case, height) is fixed. Figure 5 shows two graphs of contour curves for $f(x, y) = 5\left(\frac{1}{2}\right)^{x^2+y^2}$. Figure 5(a) shows the contour curves at various heights, and Figure 5(b) is the same set of contour curves "flattened" onto a single plane (like a piece of paper). Equivalently, Figure 5(b) is what you see when looking at the stack of contour curves from directly above.

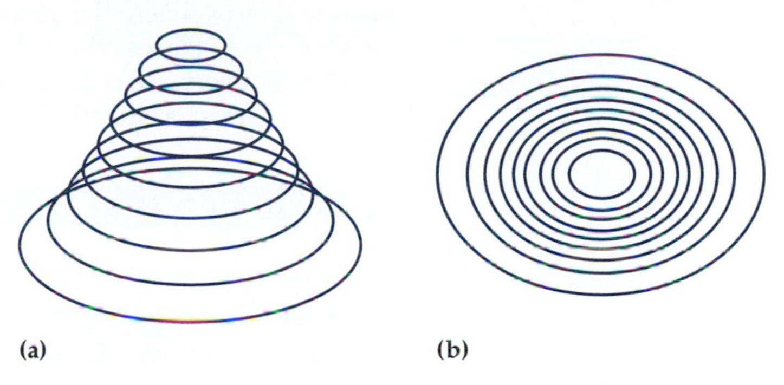

(a) (b)

Figure 5 Two views of the contour curves for $f(x, y) = 5\left(\frac{1}{2}\right)^{x^2+y^2}$. The view in (a) is from slightly above the xy-plane and the view in (b) is from directly above the center of the graph.

Making contour curves does not necessarily require a computer. One way is to plot a series of graphs on the same axes, as shown in Example 4.

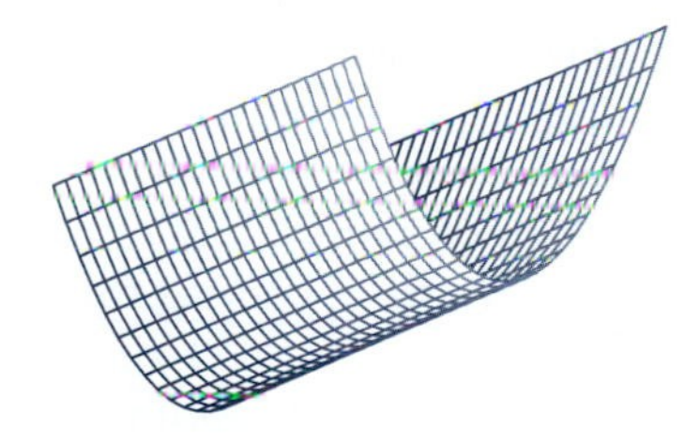

Figure 6

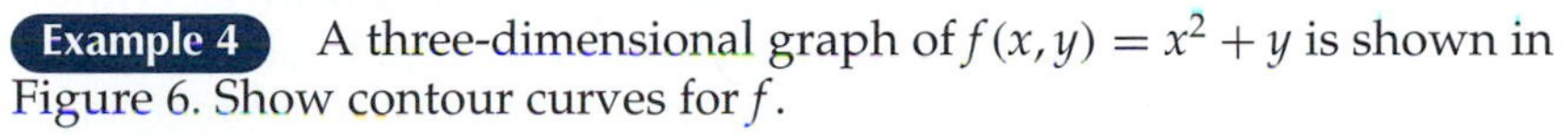

Example 4 A three-dimensional graph of $f(x, y) = x^2 + y$ is shown in Figure 6. Show contour curves for f.

Solution We set $f(x, y)$ equal to the integers $-3, -2, \ldots, 3$ to obtain seven equations. We then solve each of these equations for y to obtain seven functions.

$$-3 = x^2 + y \text{ or } y = -x^2 - 3$$
$$-2 = x^2 + y \text{ or } y = -x^2 - 2$$
$$-1 = x^2 + y \text{ or } y = -x^2 - 1$$
$$0 = x^2 + y \text{ or } y = -x^2$$
$$1 = x^2 + y \text{ or } y = -x^2 + 1$$
$$2 = x^2 + y \text{ or } y = -x^2 + 2$$
$$3 = x^2 + y \text{ or } y = -x^2 + 3$$

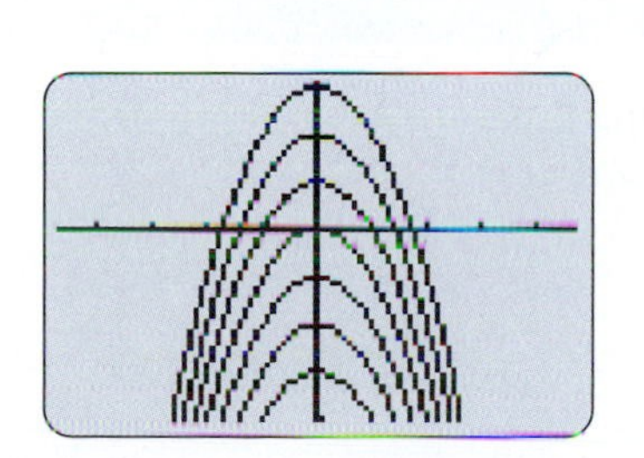

Figure 7

These seven functions are plotted on the same set of axes in a calculator graph shown in Figure 7.

Our orignal function, $f(x, y) = x^2 + y$, can be thought of as a sum of a quadratic function and a linear function, which is shown in the contour graphs. The quadratic part can be seen in that each graph is a parabola. The linear part can be seen in that the graphs are one unit apart vertically.

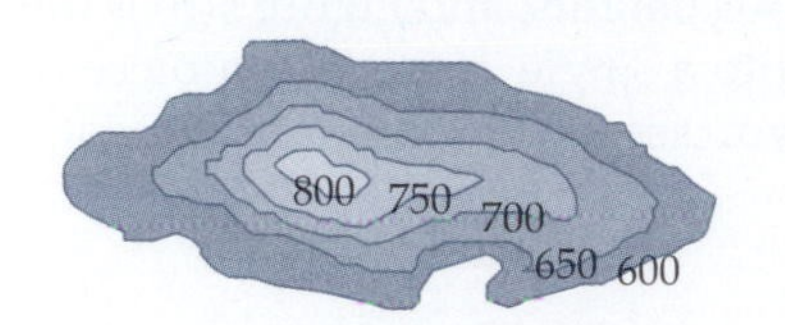

Figure 8 A topographical map of a hill.

Two common examples using contour curves are topographical maps (commonly used by geologists and hikers) and weather maps (commonly given in newspapers). In a typical topographical map, we define f as the function whose input is the latitude, x, and longitude, y, of the position where you are located and whose output, $f(x,y)$, is the elevation or height of that point above sea level. Figure 8 shows a topographical map of a hill. The numbers on the contour curves describe the height (output) for the function. Places where the curves are close together indicate areas where the hill is steep. Places where the curves are far apart indicate areas where the hill is not very steep. In this way, contour curves help you understand a three-dimensional function displayed in only two dimensions.

READING QUESTIONS

5. Describe the process of constructing contour curves for a function with two inputs.

6. In Figure 5(b), we showed contour curves of the function $f(x,y) = 5\left(\frac{1}{2}\right)^{x^2+y^2}$ that look like concentric circles. Show that simplifying $1 = 5\left(\frac{1}{2}\right)^{x^2+y^2}$ gives the equation for a circle. (*Hint:* The equation for a circle centered at the origin is $x^2 + y^2 = r^2$, where r is the radius.)

SUMMARY

A **multivariable function** is a function that has more than one input. In this section, we considered a number of examples of functions with more than one input, such as a function that gives the payment on a car loan. The notation for these functions is similar to the function notation for a single-variable function, but it allows for more than one input.

We can show some multivariable functions graphically by using three-dimensional graphs. We also sometimes look at the horizontal cross sections of a surface, called **contour curves.** Contour curves are commonly used, for example, as topographical maps. To create contour curves, you need to fix the output of the function and then make a graph of the resulting equation. Repeat this process a number of times by fixing the output at different values and graphing the different equations.

EXERCISES

1. Given that $f(x,y) = x^2 + 2y$, evaluate the following.
   (a) $f(3,5)$
   (b) $f(5,3)$
   (c) $f(3,y)$
   (d) $f(x,3)$
2. Use Table 1 from this section on the windchill equivalent temperature to answer the following. Assume the function can be represented by

$W(s,t)$, where s is the wind speed in miles per hour and t is the temperature in degrees Fahrenheit.

(a) Find $W(15,10)$.

(b) Find $W(10,15)$.

(c) Explain the meaning of $W(s,10)$.

(d) Explain the meaning of $W(10,t)$.

3. Table 1 from this section gave windchill equivalent temperatures. This function can be represented by

$$W(s,t) = 91.4 - \frac{(10.45 + 10\sqrt{0.45s} - 0.45s)(91.4 - t)}{22.04},$$

where s is the wind speed in miles per hour and t is the temperature in degrees Fahrenheit.

(a) Using the formula, find $W(15,10)$. How does this number compare with that using Table 1?

(b) Find a formula for $z = W(10,t)$.

(c) What type of function is $W(10,t)$? How can you see that by only using Table 1?

4. The body mass index (BMI) is used to compare someone's weight with their height. "In 1995 the National Institutes of Health and the American Health Foundation issued new guidelines that define healthy weight as a BMI below 25. According to a recent report by the Institute of Medicine, 59 percent of American adults exceed that threshold."[8] The formula for finding a person's BMI is $B = M/H^2$, where B is the BMI, M is a person's weight (in kilograms), and H is a person's height in meters.

(a) Convert the BMI formula so that the inputs are weight in pounds and height in inches. (There are about 2.2 lb in 1 kg and 39.4 in. in 1 m.)

(b) Determine the BMI for someone who is 68 in. tall and weighs 140 lb.

(c) Determine a formula for the BMI for people who are 68 in. tall. (Your only input should be weight in pounds.)

(d) Suppose a person is 68 in. tall and trying to lose weight. How much weight does this person have to lose to reduce the BMI by one unit?

5. Relative humidity can have a great impact on how comfortable you feel. Just as windchill helps us understand the effect of wind on our perception of cold temperatures, the heat index can help us understand how hot it feels when we factor in the humidity. One function that approximates the heat index[9] is given as $I(t,h) = -42.379 + 2.04901523t + 10.14333127h - 0.22475541th - (6.83783 \times 10^{-3})t^2 - (5.481717 \times$

[8] *W. Wayt Gibbs, "A Shifting Scale,"* Scientific American, *(visited 6 July 1999) ⟨http://www.sciam.com/0896issue/0896trends_box1.html⟩.*

[9] *"Heat Index Formula,"* News 10 Online *(visited 6 July 1999) ⟨http://www2.sacbee.com/kxtv/weather/heatindx.htm⟩.*

$10^{-2})h^2 + (1.2287 \times 10^{-3})t^2h + (8.5282 \times 10^{-4})th^2 - (1.99 \times 10^{-6})t^2h^2$, where t is the temperature in degrees Fahrenheit and h is the relative humidity written as a whole number (i.e., 70% relative humidity is 70, not 0.70).

(a) Determine a formula for $I(90, h)$. What type of function is it?

(b) Determine a value for $I(90, 80)$.

6. The function $f(x, y) = 2^x + y$ can be thought of as the sum of a linear function and an exponential. You should be able to see both an exponential function and a linear function in a graph of some contour curves for f. Make contour curves where $f(x, y) = -3, -2, -1, 0, 1, 2,$ and 3. Describe how these curves show the linear function and how they show the exponential function.

7. Make contour curves of the following. In doing so, let $f(x, y) = 1, 2, 3,$ and 4.

(a) $f(x, y) = x^3 + 3y$

(b) $f(x, y) = 0.5^{y^2 - x^2}$

(c) $f(x, y) = 2x + y^2$

(d) $f(x, y) = x \cdot 2^y$

8. Sometimes we can get insights into fairly common functions by looking at contour curves. The well-known formula $t = d/r$ gives time as a function of both distance and rate.

(a) Make contour curves with distance on the horizontal axis and speed on the vertical axis. In doing so, use times of 1, 2, 3, 4, and 5. Label each line you obtain with the time it represents.

(b) Using your contour curves, fix the distance at a specific number and describe what you notice about the change in time as you change the speed.

9. A warm summer day can feel warmer than the temperature indicates if it is also quite humid. The heat index can help us understand how hot it feels when we factor in the humidity. The heat index is usually determined by using a table. The following is a portion of that table.

Temperature (°F)	Relative Humidity																
	10	15	20	25	30	35	40	45	50	55	60	65	70	75	80	85	90
80°								80	81	81	82	83	84	85	86	87	88
85°	80	81	82	83	84	85	86	87	88	89							
90°	84	85	86	87	88												

Create contour curves for the heat index by plotting the relative humidity on the horizontal axis and temperature on the vertical axis. Plot the points of like temperature and connect them with lines. For example, the humidity and temperatures that give a heat index of 81 are (50, 80), (55, 80), and (15, 85). These points should be plotted and connected by two lines.

10. The accompanying contour curves represent the volume of a pyramid with a square base. The horizontal axis shows the length of one side of the base, s, and the vertical axis shows the height, h. The output of the function, the volume, is represented by the curves. Volumes of 10, 20, 30, 40, 50, and 0 are labeled. Use the contour curves to answer the following.

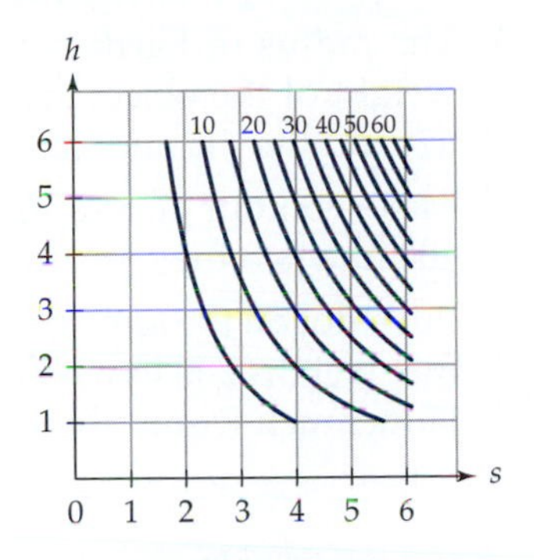

   (a) Find the height of a pyramid with a volume of 10 and a side of length 4.
   (b) Find the length of the side of a pyramid with a volume of 20 and a height of 2.
   (c) Find the volume of a pyramid with a side of length 5 and a height of 3.

11. In the text, the formula for a monthly payment for a 5-year loan was given as

$$M(r, P) = \frac{(r/12)P}{1 - (1 + r/12)^{-60}},$$

where r is the interest rate (written as a decimal) and P is the principal (or amount borrowed).

   (a) Make contour curves for this function by letting $M(r, P)$ equal 100, 150, 200, 250, and 300. (*Hint:* Solve your equations for P.)
   (b) Using your contour curves, what is the most you could borrow and have a payment no larger than \$200 with an interest rate no lower than 6%?

12. In the text, we used a formula to find the monthly payment for a 5-year loan. If we let the length of the loan vary, then the function becomes

$$M(r, p, t) = \frac{(r/12)P}{1 - (1 + r/12)^{-12t}},$$

where r is the interest rate (written as a decimal), P is the principal (or amount borrowed), and t is the length of the loan in years. Suppose you were borrowing \$100,000 at an annual interest rate of 7.5% to buy a house.

(a) Find a formula for $M(0.075, 100000, t)$. Graph this function where time varies from 10 to 30 years.

(b) Find the monthly payments for loans with a period of 15, 20, 25, and 30 years.

(c) Determine the total amount paid for borrowing \$100,000 for 15, 20, 25, and 30 years.

13. The function $w(w_0, d) = (1.58\times 10^7)w_0 d^{-2}$ can be used to find the weight of a small object d miles from the center of Earth given that the object weighs w_0 pounds on the surface of Earth.

(a) The radius of Earth is about 3960 mi. Find a function for the weight of the object given the weight on the surface of Earth, w_0, and the distance above the surface of Earth, D.

(b) Make contour curves for objects that weigh 100, 200, 300, and 400 lb when they are out in space.

(c) Put a dot on the appropriate contour curve that shows an object that weighs 200 lb when it is 100 mi above the surface of Earth. How much would that object weigh on the surface of Earth?

INVESTIGATIONS

INVESTIGATION 1: A SINE OF THE TIMES

One of the authors was in a class with a number of mathematics teachers when another teacher took a hollow cylinder (like a rolled-up sheet of paper) and cut the top off at a 45° angle. When the sheet was unrolled, it appeared that the top of the paper had been cut in the shape formed by the graph of a sine or cosine function. But was it? Sometimes appearances can be deceiving.

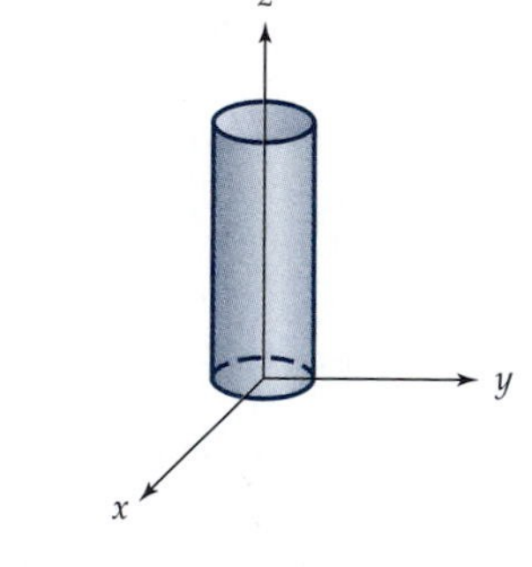

Figure 9

Let's look at a cylinder on a three-dimensional coordinate system as shown in Figure 9. The cylinder is placed in such a way that the center of the circular base is at the origin of the xy-plane and the z-axis goes right up the middle of the cylinder. To make our life a little easier, we let the cylinder have a radius of 1. We can generalize later.

Now we need to cut this cylinder at a 45° angle as in Figure 10, which shows both a front view and a side view. As you can see, from the side view this cut looks like a line.

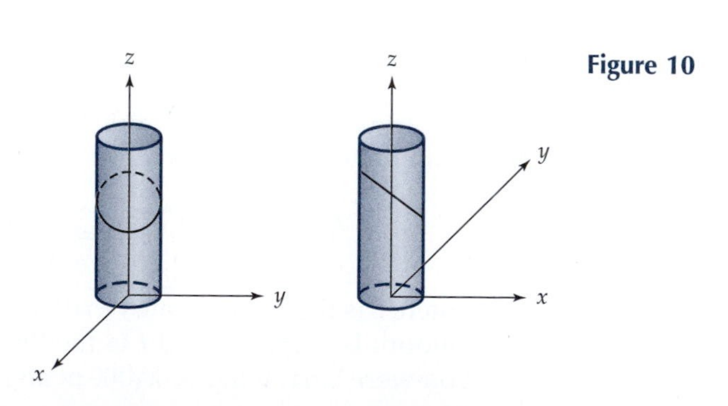

Figure 10

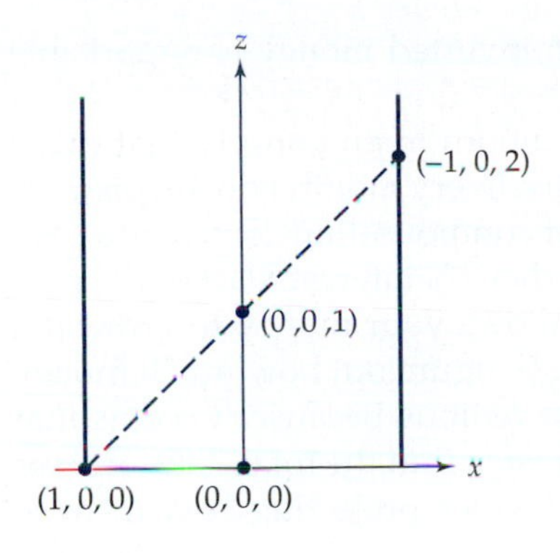

Figure 11

Again, to make our problem easier, we let this cut go through the point (1, 0, 0) so that it looks like Figure 11.

1. The equation of the plane formed by the cut can be written just in terms of x and z (much like a line). What is the equation of the plane shown in Figure 11?
2. Now that we have the equation for the plane, we need to find a way to express its intersection with the edge of the cylinder such that it gives the height of the intersection as we go around the cylinder. Imagine looking down on the cylinder from above, that is, looking down on the xy-plane. You should see a circle. We want to write the height change (z) as we travel around that circle or as the angle θ changes. Assuming we start at the point (1,0) in the xy-plane and go counterclockwise, find the equation where θ is the input and the x-coordinate is the output. (See the figure.)

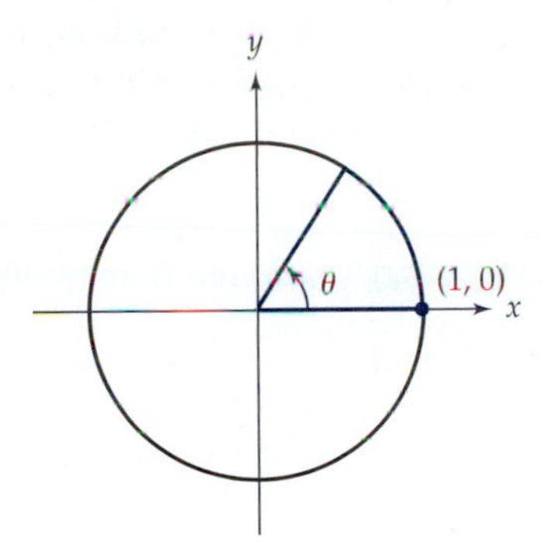

3. Compose the two functions from questions 1 and 2 so that θ is the input and z is the output to give an equation expressing the height of the cut cylinder in terms of how far around the cylinder you are. A graph of this equation is equivalent to the shape of the top edge of the paper after the cylinder is cut lengthwise and unrolled, the very thing we wanted to find out.
4. Earlier, we made a couple of assumptions to make the process easier. Rewrite the function such that the radius of the cylinder does not have to be 1 but can instead be any radius, r.
5. If the cylinder had been cut at another angle, then the result would be different. Find a function to give the height of the cut cylinder if it had been cut at a 30° angle.
6. Develop a multivariable function for the height of a cut cylinder of radius r and cut at angle α.

INVESTIGATION 2: HOW TO BE A MILLIONAIRE

Retirement might not sound very appealing to you right now, especially because you may not have even started your career yet. At some time, however, you will want to think about how you will support yourself after you quit working. If you are still young, then you have a powerful tool at your disposal: time. The effect of time on invested money is remarkable and important to know.

Suppose you are able to put $50 each month in an annuity that earns 12% interest per year compounded monthly. Every month you are paid 1% interest on your investment. The power of compounding comes into play in the next month, when you are paid another 1% interest. Part of the new principal is last month's interest, which allows your money to grow at a rate faster than it would otherwise. Trying to figure out how much money you would have at a certain point can be tedious because it seems that you have to go through each month's interest to find the total. This process would involve much calculation, especially when projecting 30 years or so to retirement!

Fortunately, we can tackle the problem with a more symbolic approach. Start by considering your first month. After your initial $50 investment at the beginning of the month, your money grows to a whopping $50.50 at the end of the month. At the beginning of month 2, you add another $50, giving a total of $100.50 in your account. This approach certainly does not seem like a get rich quick scheme, but then again, you did not start with much. Table 2 shows the growth of your account.

TABLE 2

Month	Amount at Beginning of the Month	Interest Earned	Amount at End of the Month
1	$50.00	$0.50	$50.50
2	$100.50	$1.01	$101.51
3	$151.51	$1.52	$153.03
4	$203.03	$2.03	$205.06
5	$255.06	$2.55	$257.61
6	$307.61	$3.08	$310.69
7	$360.69	$3.61	$364.30
8	$414.30	$4.14	$418.44
9	$468.44	$4.68	$473.12
10	$523.12	$5.23	$528.35
11	$578.35	$5.78	$584.13
12	$634.13	$6.34	$640.47

Without interest, you would have merely $600 in this account. Compounding gives you the opportunity to begin earning interest right away and for that interest to become part of the new principal and earn additional interest.

To try to anticipate how much money you would have if you kept investing at this rate, we could continue to make our chart for 30 years (360 chart entries) or we can develop a formula to help us. One advantage to writing the formula is that it should work for any amount, not only at these specific amounts. Otherwise, we need to make a table like Table 2 for each individual scenario.

1. Let's begin at the end of the first month. Your monthly deposit (let's call it M) has been sitting in the account for a month now and is ready to get its first dose of interest. This step is like multiplying the money currently sitting in the account by $1+R$, where R is the monthly rate. (Note that this rate is the yearly rate divided by 12). This equation is

$$T = M(1+R). \tag{1}$$

The same process happens at the end of the next month as well; we add M, and our total gets multiplied by $(1+R)$, giving

$$T = (M(1+R)+M)(1+R). \tag{2}$$

This process continues for as long as you put money in your account.

(a) Extend equation (2) and multiply to show that, for n deposits,

$$T = M(1+R)^n + M(1+R)^{n-1} + M(1+R)^{n-2} + \cdots + M(1+R). \tag{3}$$

(b) The very long equation (3) may seem unwieldy, but it is actually quite simple to deal with because it is merely a geometric series. Multiply both sides of equation (3) by $(1+R)$.

(c) Now subtract equation (3) from your result after the multiplication. (On the left side you should have RT, and most of the terms on the right should drop out.)

(d) Divide both sides of your new equation by R. You now have a general equation for T.

2. Now that you have your equation and a better idea of how the variables relate, let's use the equation on a couple of scenarios.

(a) Assume you have 45 years to retirement and you can invest your money in an account at 8% that is compounded monthly. How much should you invest every month to have $1,000,000 at retirement? What if you have only 35 years? 25 years?

(b) Frank goes to work right after high school and is able to put $100 a month in an account that earns 12% interest per year compounded monthly. He does this for 10 years and then stops, letting his money receive interest for the next 20 years. Sheryl goes to college, pays off her student loans, and 10 years after Frank started begins investing $200 every month for 20 years. At retirement (30 years after Frank started and 20 years after Sheryl started), who has more money saved up? If the time to retirement had been longer or shorter, would it have made a difference?

INVESTIGATION 3: HOW COLD IS IT?

If you have ever been outside on a cold, windy day, you have experienced the chilling effect of the wind. We looked at windchill briefly in this section but in this investigation, we take a closer look. The windchill equivalent temperature, commonly known as windchill, is a number that represents the effect of the wind on your perception of cold. Equations for windchill were first published in 1939 by Paul Siple and Charles Passel based on experiments done in Antarctica.[10] Table 3 gives the windchill equivalent temperature for air temperatures given in degrees Celsius and wind speeds given in meters per second.[11]

TABLE 3 Windchill Equivalent Temperature

Wind Speed (m/sec)	Air Temperature (°C)												
	6	**3**	**0**	**−3**	**−6**	**−9**	**−12**	**−15**	**−18**	**−21**	**−24**	**−27**	**−30**
3	3	−1	−4	−7	−11	−14	−18	−21	−24	−28	−31	−34	−38
6	−2	−6	−10	−14	−18	−22	−26	−30	−34	−38	−42	−46	−50
9	−6	−10	−14	−18	−23	−27	−31	−35	−40	−44	−48	−53	−57
12	−8	−12	−17	−21	−26	−30	−35	−39	−44	−48	−53	−57	−62
15	−9	−14	−18	−23	−27	−32	−37	−41	−46	−51	−55	−60	−65
18	−10	−14	−19	−24	−29	−33	−38	−43	−48	−52	−57	−62	−67
21	−10	−15	−20	−25	−29	−34	−39	−44	−49	−53	−58	−63	−68
24	−10	−15	−20	−25	−30	−35	−39	−44	−49	−54	−59	−63	−68

Source: Joseph M. Moran and Michael D. Morgan, *Meteorology, The Atmosphere, and the Science of Weather*, 2d ed. (London: Macmillan, 1989), p. 73.

1. (a) Keeping the temperature constant at −15°C (5°F), make a graph by plotting pairs of numbers with wind speed on the horizontal axis and windchill on the vertical axis. Connect your points with a reasonable curve. Is your graph linear? If not, what is its shape?

   (b) Keeping the wind speed constant at 15 m/sec, make a graph with air temperature on the horizontal axis and windchill on the vertical axis. Connect your points with a reasonable curve. Is your graph linear? If not, what is its shape? Does a 15 m/sec wind cause a greater difference between air temperature and windchill when temperatures are warmer or colder? (*Hint:* Insert the line $y = x$ on your graph.)

   (c) What does the shape of the graph from part (a) tell you about the effect of wind speed on windchill? What does the shape of the graph from part (b) tell you about the effect of air temperature on windchill?

2. The equation for windchill is

$$W = 33 - \frac{(10.45 + 10\sqrt{V} - V)(33 - T)}{22.04}, \qquad (1)$$

where W is the windchill temperature in degrees Celsius, V is the speed of the wind in meters per second, and T is the air temperature in de-

[10] *John E. Oliver,* The Encyclopedia of Climatology *(New York: Van Nostrand Reinhold, 1987), pp. 928–29.*

[11] *This table is similar to the one in the text except that the temperature here is given in degrees Celsius and the wind speed is given in meters per second. The table in the text is given using English units.*

grees Celsius. Let $T = -15°C$ and $V = 15$ m/sec. Use equation (1) to compute the windchill for these values. Does your answer agree with the windchill given in the chart?

3. (a) In equation (1), substitute $T = -15$ and simplify. Graph the resulting function and compare it with your graph for question 1, part (a). Are they similar?
   (b) In equation (1), substitute $V = 15$ and simplify.
      i. Graph the resulting function and compare it with your graph for question 1, part (b). Are they similar?
      ii. Justify symbolically that this equation is the equation of a line. What is its slope? What does the slope mean in terms of air temperature and windchill?
      iii. What is the y-intercept? What does the y-intercept mean in terms of air temperature and windchill?

4. We have looked at the effect of wind speed on windchill and the effect of air temperature on windchill. We now hold the windchill constant and compare wind speed with air temperature.
   (a) Using Table 3, find three different temperature and wind speed pairs that give a windchill of $-30°C$.
   (b) Set equation (1) equal to $-30°C$. You now have a function with two variables, T and V. Solve for T. This result gives a function whose input is V and whose output is T. Are the three different temperature and wind speed pairs you found in part (a) solutions to your equation?
   (c) Make contour graphs for windchill temperatures $0°$, $-10°$, $-20°$, $-30°$, and $-40°C$.
   (d) By looking at the contour graphs, how can you tell that if you have a constant temperature, then the windchill temperature decreases as the velocity increases?

PROJECTS 9.4 ELLIPTIPOOL

One of the authors enjoyed visiting his uncle for a number of reasons, one of which was an interesting pool table called Elliptipool. Instead of being constructed in the shape of a rectangle, like most pool tables, this one was in the shape of an ellipse. It had only one hole (or pocket) located somewhere near the middle right of the table and one dot located near the middle left of the table.[12] These two spots, the hole and the dot, were not chosen randomly or chosen because of aesthetics, but rather were chosen because the spots had an important function. It seemed any ball that passed over the dot (without any extra spin or "English") would bounce off the cushion

[12] *These locations, of course, depend on where you are standing.*

and go directly to the hole.[13] These two points on the pool table, the hole and the dot, were located at the foci of the ellipse.

In this project, you theoretically prove that any ball passing over the dot and bouncing off the cushion goes directly to the hole. First, though, let's review some of the properties of an ellipse. To draw an ellipse, all you need to do is tack down the two ends of a piece of string, giving the string plenty of slack. Putting a pencil against the string, move the pencil while keeping the string taut. (See Figure 1.) The two places where the string is tacked down are called the foci of the ellipse.

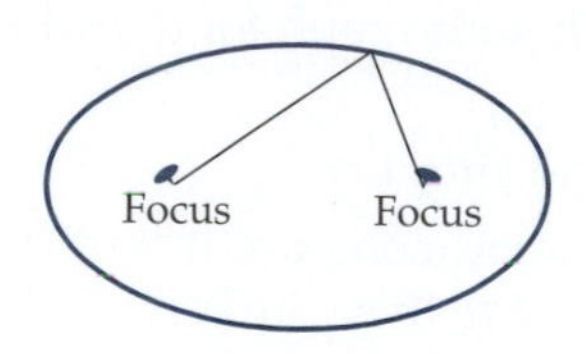

Figure 1 An ellipse can be easily drawn with string and a couple of tacks.

An ellipse is said to be in standard position if its center lies at the origin with its two axes located on the x-axis and y-axis. (See Figure 2.) The equation for such an ellipse is

$$\frac{x^2}{a^2} + \frac{y^2}{b^2} = 1,$$

where the vertices are the points $(-a, 0)$, $(a, 0)$, $(0, -b)$, and $(0, b)$. We can assume, without loss of generality, the ellipse is positioned such that $a > b$. The foci are the two points $(-c, 0)$ and $(c, 0)$. There is a nice relationship between a, b, and c: $c^2 = a^2 - b^2$.

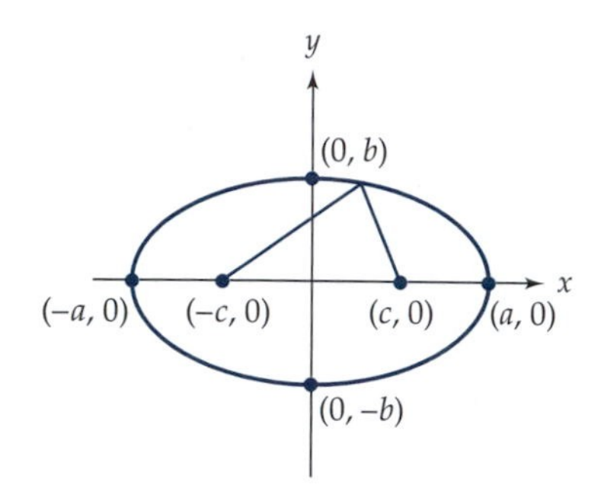

Figure 2 The equation for an ellipse centered at the origin is $(x^2/a^2) + (y^2/b^2) = 1$. The points $(-c, 0)$ and $(c, 0)$ are the foci.

In calculus, you will study how to find the equation for a line tangent to a curve, something we need to do this problem. When you study calculus, you will be able to determine this equation for yourself, but for now, we give the equation for the slope of the tangent line for the "top half"[14] of the ellipse, which is

$$m = \frac{-bx}{a^2\sqrt{1 - (x^2/a^2)}}. \tag{1}$$

You can also use a graphing calculator to determine the slope of the tangent line at various points on the ellipse.

Now let's return to the elliptical pool table. In physics, one learns that the angle of incidence (the angle made when the ball hits the side) is equal to the angle of reflection (the angle made when the ball rebounds). So, if the ball indeed follows the path noted from one focus to the other, then the measure of the angle of incidence, α, should equal the measure of the angle of reflection, β. (See Figure 3.) We want to show that this relationship is true.

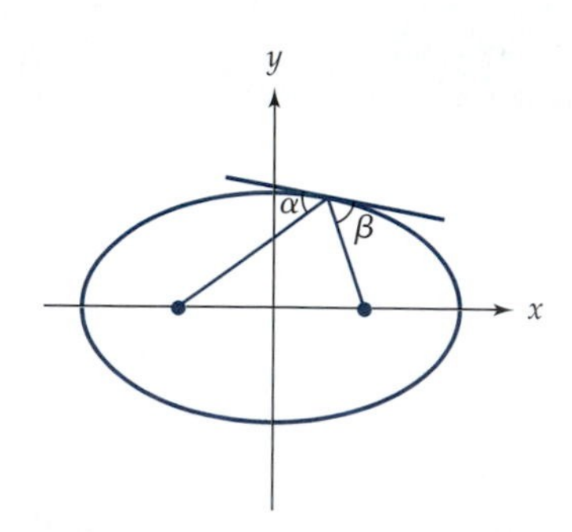

Figure 3 We want to show that the angle of incidence, α, is the same as the angle of reflection, β.

1. First, we prove a couple of properties of an ellipse.
   (a) The distance from one focus of an ellipse to any point on the ellipse and back to the other focus is a constant, which is why drawing an ellipse with a piece of string and two pins works. Therefore, one way to define an ellipse is as the set of all points such that this distance is constant. Choose a specific point to show that this constant is $2a$. (*Hint:* You should choose your point wisely.)

[13] *Actually hitting the ball in the hole while playing a game was still not all that easy. You could not hit the ball too hard or too soft, and usually there were other balls in the way.*

[14] *The "bottom half" is similar, differing only in sign.*

(b) Using the accompanying figure, explain why the property mentioned earlier, $c^2 = a^2 - b^2$, is true.

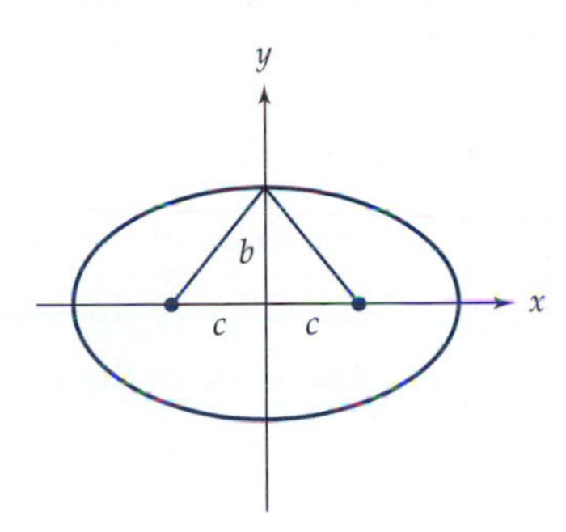

2. Solving our problem with a general ellipse and a general point is somewhat complicated. Before we do so, let's consider a specific ellipse and a specific point on that ellipse. We use the ellipse $(x^2/5^2) + (y^2/3^2) = 1$ and the point $(3, 2.4)$.[15] (See the accompanying figure.)

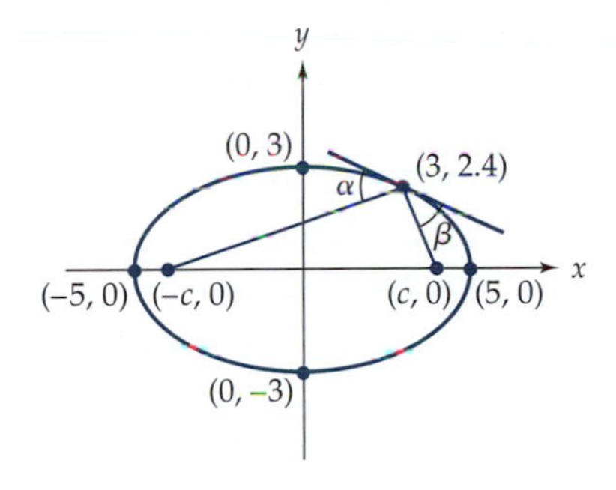

(a) Determine the location of the foci for the ellipse in the figure.

(b) Determine the slopes of the line segments that connect each focus to the point $(3, 2.4)$. Leave your answer in exact form; do not round.

(c) Using equation (1), determine the slope of the tangent line that passes through the point $(3, 2.4)$ for our ellipse.

(d) Recall from section 5.4 on trigonometric identities that

$$\tan\theta = \frac{m_2 - m_1}{1 + m_1 m_2},$$

where θ is the angle formed by the intersection of two lines with slopes m_1 and m_2. (This property is derived from the trigonometric identity $\tan(x - y) = (\tan x - \tan y)/(1 + \tan x \tan y)$.) For consistency in this problem, we assume $m_2 > m_1$. Use this property to determine the measures of angles α and β in the figure.

[15] *There is nothing particularly special about this ellipse. We picked it and the point because they made some of the calculations easier.*

(e) Exactly what can we conclude from knowing that α and β are the same for this particular ellipse and this particular point?

3. To see that this property holds true for any ellipse, we must generalize. A general point on the "top half" of the ellipse $(x^2/a^2) + (y^2/b^2) = 1$ is given by $(x, b\sqrt{1-(x^2/a^2)})$. We label the two foci as $(c, 0)$ and $(-c, 0)$. (See the accompanying figure.) Let's define some variables:

$m_1 =$ the slope of the line that contains $(x, b\sqrt{1-(x^2/a^2)})$ and $(c, 0)$
$m_2 =$ the slope of the line that contains $(x, b\sqrt{1-(x^2/a^2)})$ and $(-c, 0)$
$m_3 =$ the slope of the tangent line using equation (1).

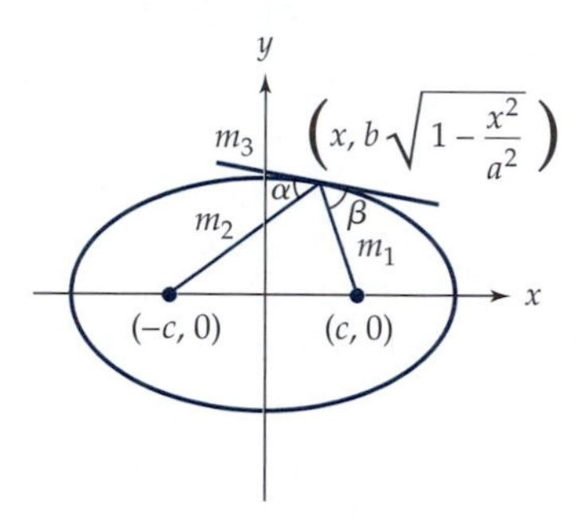

(a) Find m_1 and m_2.

(b) Now, for the hard part; take a deep breath and *don't panic*. Use the formulas $\tan\alpha = (m_2 - m_3)/(1 + m_2 m_3)$ and $\tan\beta = (m_3 - m_1)/(1 + m_1 m_3)$ to show that

$$\tan\alpha = \frac{a^2 b + bcx}{(a^2 x + a^2 c - b^2 x)\sqrt{1-(x^2/a^2)}}$$

$$\tan\beta = \frac{bcx - a^2 b}{(a^2 x - a^2 c - b^2 x)\sqrt{1-(x^2/a^2)}}$$

(c) To show that $\tan\alpha = \tan\beta$, set the right-hand parts of the preceding equations equal to each other and simplify until you reduce it to a true statement. (*Hint:* Do not forget that $c^2 = a^2 - b^2$.)

(d) Now that you have shown that $\tan\alpha = \tan\beta$, explain why that implies that $\alpha = \beta$.

4. Suppose you are the supervisor in a company that is thinking about making a pool table much like the Elliptipool. As the highly paid and very busy boss, you ask one of your less well paid but still very busy underlings to prove that any ball passing over the dot always goes into the hole. (The legal department wants proof so that you will not get sued for misrepresentation during your media blitz.) Your underling sends you a report a few days later with an example of a specific ellipse and 12 chosen points on the ellipse along with demonstrations that the angle of incidence is indeed equal to the angle of reflection in all 12 cases. Exasperated, you fire back a memo explaining the difference between giving examples and giving a proof. Write the memo.

REVIEW EXERCISES

1. Find the set of parametric equations that describe the trajectory of a golf ball that is hit at an initial velocity of 120 ft/sec at an angle of 45°. How far does the golf ball travel before it hits the ground?
2. Graph the following sets of parametric equations. Also eliminate the parameter and give a single function, with an appropriate domain, that gives the same graph.
   (a) $x = 2t - 3$ and $y = 4t + 3$ for $0 \leq t \leq 10$
   (b) $x = e^t$ and $y = t + 2$ for $0 \leq t \leq 10$
   (c) $x = t^3$ and $y = t^2$ for $0 \leq t \leq 10$
3. Suppose a Ferris wheel rotates at 2 revolutions per minute clockwise. It has a diameter of 50 ft, and at its lowest point, it is 5 ft off the ground.
   (a) Find the parametric equations that describe the position of a point on the wheel. Time, in seconds, should be the input for your equations, and at $t = 0$, the point on the wheel should be at its lowest position.
   (b) How would your equations change if the wheel were turning counterclockwise?
4. Imagine you see a bicycle traveling down a road with a white dot painted on its 26-in.-diameter tire. Suppose the bicycle wheel is revolving at 2 revolutions per second.
   (a) Find a parametric equation that describes the position of a dot such that the input is the angle through which the wheel has turned, θ.
   (b) Find a parametric equation that describes the position of the dot such that the input is time, in seconds.
5. For each of the following pairs of points, assume the first point is the intial point of a vector and the second point is the terminal point of a vector. Write each vector in terms of **i** and **j**.
   (a) $(0, 0)$ and $(-2, 10)$
   (b) $(-1, 5)$ and $(6, 3)$
   (c) $(-2, -3)$ and $(-4, 6)$
6. Determine the magnitude and direction for each vector.
   (a) $-2\mathbf{i} + 8\mathbf{j}$
   (b) $3\mathbf{i} + 4\mathbf{j}$
   (c) $5\mathbf{i} - 2\mathbf{j}$
7. Given a vector's magnitude and direction, write each of the following in terms of **i** and **j**. (Give exact answers.)
   (a) $\|\mathbf{v}\| = 3$ and $\theta = 45°$
   (b) $\|\mathbf{v}\| = 8$ and $\theta = 120°$
   (c) $\|\mathbf{v}\| = 4$ and $\theta = 300°$
8. Suppose an object is acted upon by two forces with magnitudes $\|\mathbf{u}\|$ and $\|\mathbf{v}\|$ and directions θ_u and θ_v, respectively. Determine the magnitude

and direction of the resultant force. (Round answers to two decimal places.)

(a) $\|\mathbf{u}\| = 5$ lb, $\theta_u = 30°$; $\|\mathbf{v}\| = 5$ lb, $\theta_v = 60°$

(b) $\|\mathbf{u}\| = 5$ lb, $\theta_u = 60°$; $\|\mathbf{v}\| = 10$ lb, $\theta_v = 30°$

9. A plane is flying 10° west of due south at an airspeed of 100 mph. A wind is pushing the plane 10° south of due west at 20 mph. What is the ground speed of the plane and in which direction is it flying?

10. A plane is flying at 100 knots at 10° east of north for 2 hr and then adjusts its course to go 20° east of north for 1 hr. Assuming there is no wind, find the plane's position relative to takeoff after 3 hr.

11. You have to move a 500-lb piano up a ramp that is 8 ft long. You set your ramp so that it rises 3 ft. You, of course, have your excellent frictionless piano dolly. If you were pushing the piano up this ramp by yourself, how much weight would you be pushing up the ramp?

12. Given that $f(x, y) = 2x^3 + y^2$, evaluate the following.

    (a) $f(2, 4)$

    (b) $f(4, 2)$

    (c) $f(2, y)$

    (d) $f(x, 2)$

13. The formula for the volume of a cylinder is $V(r, h) = \pi r^2 h$, where r is the radius and h is the height.

    (a) Find $V(3, 6)$ and explain what it means in terms of the volume of the cylinder.

    (b) Find $V(3, h)$ and explain what it means in terms of the volume of the cylinder.

    (c) Find $V(r, 6)$ and explain what it means in terms of the volume of the cylinder.

14. Make contour curves of the following. In doing so, let $f(x, y) = 1, 2, 3$, and 4.

    (a) $f(x, y) = 2x^3 + y^2$

    (b) $f(x, y) = 2^{x+y}$

    (c) $f(x, y) = \dfrac{x}{y}$

APPENDIX

ANSWERS TO ODD-NUMBERED EXERCISES

Section 1.1

1. The r in $A(r)$ is the input.
3. The rule is a function because for every input (numbers of gallons of gasoline), there is at most one output (the cost of the gasoline).
5. (a) Domain: all real numbers. Range: all real numbers.
   (b) Domain: $r > 0$. Range: $V > 0$.
   (c) The difference is the context of the function in part (b). The radius and volume must be positive.
7. (a) $f(10) = \frac{1}{10}; f(\frac{1}{10}) = 10$
   (b) $h(\pi) = 3\pi - 4; h(\pi + 1) = 3(\pi + 1) - 4 = 3\pi - 1$
   (c) $g(2) = 3; g(x^2) = x^2 + 1$
   (d) $f(20) = 68; f(40) = 104$
   (e) $f(-5) = 2(-5) + 3 = -7; f(5) = 5^2 = 25$
9. (a) $n(x + 2)$ is the number of hamburgers you can make from $x + 2$ pounds of ground beef.
   (b) $n(x) + 2$ is 2 more hamburgers than you can make from x pounds of ground beef.

Section 1.2

1. (a) Independent variable: time. Dependent variable: temperature.
   (b) Independent variable: number of gallons of gasoline. Dependent variable: total cost of gasoline.
   (c) Independent variable: date. Dependent variable: amount of daylight.
   (d) Independent variable: length of race. Dependent variable: time taken to finish the race.
   (e) Independent variable: month of the year. Dependent variable: cost of heating your home.
   (f) Independent variable: age of a used car. Dependent variable: price of a used car.
3. Answers will vary. The following domains and ranges are examples of appropriate answers.
   (a) Input: number of guests. Domain: $0 \leq x \leq 500$. Output: total meal cost. Range: $\$0 \leq y \leq \$10{,}000$.
   (b) Input: latitude in the Northern Hemisphere. Domain: $0° \leq x \leq 90°$. Output: amount of daylight. Range: $0 \text{ h} \leq y \leq 12 \text{ h}$.
   (c) Input: day of the year. Domain: $1 \leq x \leq 365$. Output: average high daily temperature in Chicago. Range: $-25°\text{F} \leq y \leq 80°\text{F}$.
   (d) Input: distance you live from work. Domain: 1 mi $\leq x \leq$ 50 mi. Output: time it takes to drive. Range: 1 min $\leq y \leq$ 80 min.
   (e) Input: time. Domain: 0 sec $\leq x \leq$ 120 sec. Output: speed of car. Range: 45 mph $\leq y \leq$ 70 mph.
5. A function cannot have more than one y-intercept because then it would have more than one output for an input of 0.
7. (a) If f is symmetric about the origin, then $f(0) = 0$.
   (b) If f is symmetric about the y-axis, then $f(0)$ could be anything.
9. (a) The person starts walking toward the motion detector at a constant rate and slows down and goes back to the starting point at a constant rate. This process is repeated again.
   (b) The person walks toward the motion detector slowing down, then goes back, slowly at first, and then speeds up.
   (c) The person does not move for a while. Then the person walks away from the motion detector at a constant rate, then walks very slowly for a while and then continues to walk away quickly.
   (d) The person walks away from the motion detector at a constant rate, stops for a while and then walks back towards the motion detector at a constant rate.
11. The impact would be less on a graph with a vertical scale that goes from 0 to 600.

13. (a)

(b) A team below this line is highly payed compared with their percentage of games won. The team above this line has a high percentage of games won compared with their pay.

(c) The Boston Celtics were highly paid compared with their percentage of games won.

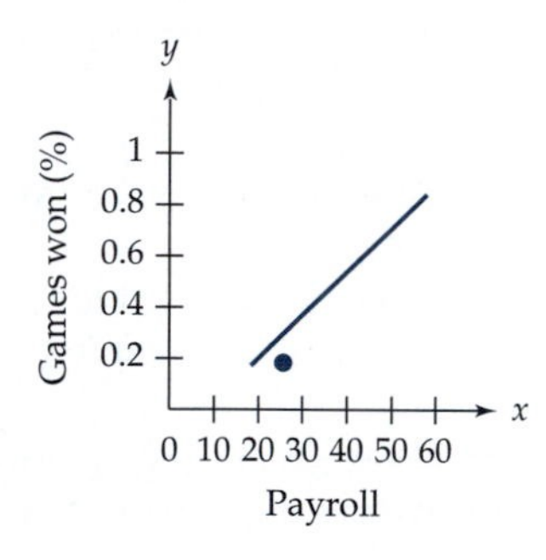

(d) The Utah Jazz have a high percentage of games won compared with their pay.

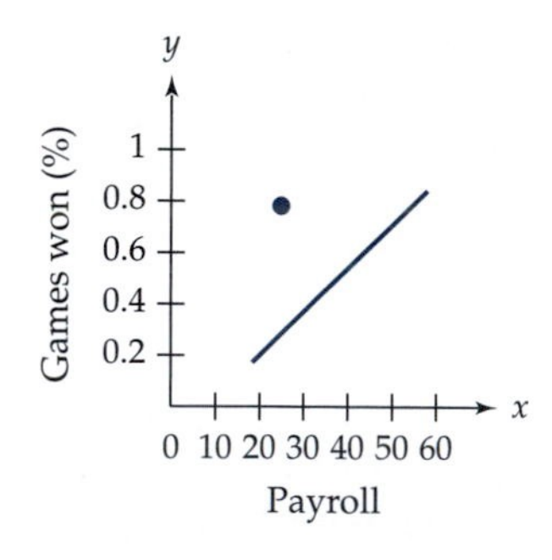

(e) Most teams would be close to this line because most are paid according to their ability, and ability would show in their percentage of games won.

Section 1.3

1. (a) The graph of f is concave down, increasing between -10 and 0 and decreasing between 0 and 10.

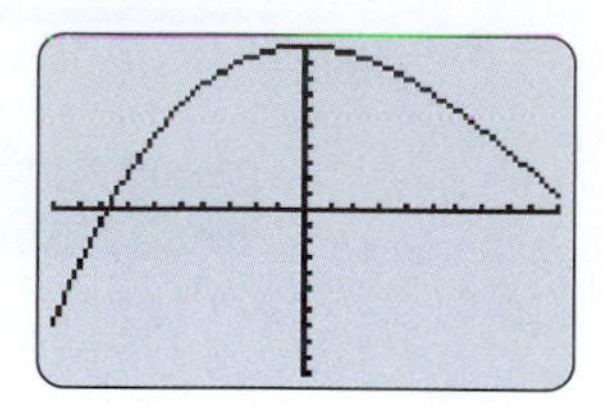

(b) The graph of f is concave down between -10 and 10 and concave up between 10 and 32. It is increasing between -10 and 0, decreasing between 0 and 21, and increasing between 21 and 32.

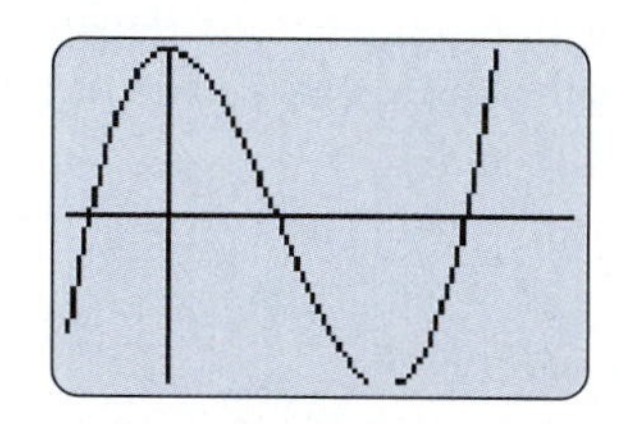

(c) The description in part (b) best describes the shape of the graph because the domain of $-10 \leq x \leq 40$ shows some characteristics of the graph not seen in the other domain.

3. (a) The function $f(x) = 2/(x-3)$ is disconnected because it is undefined at $x = 3$.

(b)

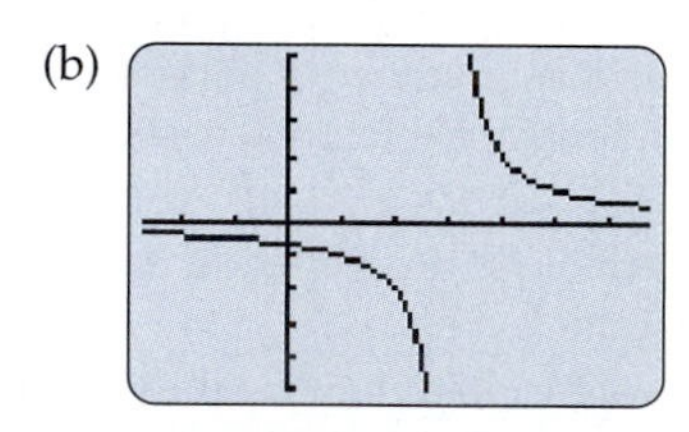

(c) The dimensions of the viewing window on our TI-83 are $-2.7 \leq x \leq 6.7$ and $-5 \leq y \leq 5$. There are 95 tenths between -2.7 and 6.7, when both -2.7 and 6.7 are included. There are also 95 pixels on the screen horizontally.

5. (a) The lines do not look perpendicular because the screen is wider than it is tall and the distance between the tick marks on the x-axis are farther apart than those on the y-axis.

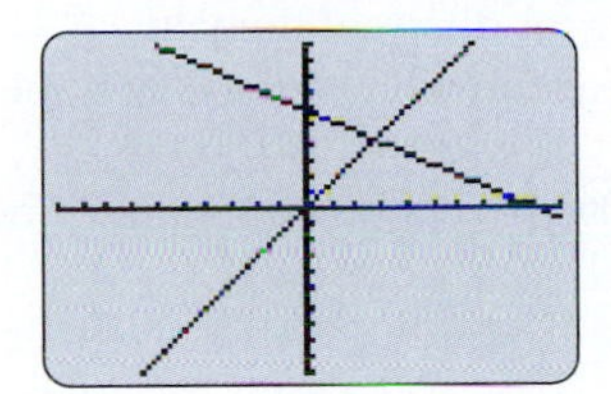

(b) The dimensions of the viewing window on our TI-83 are $-4.7 \le x \le 4.7$ and $0 \le y \le 6.2$. A TI-83 has a screen that is 95 pixels wide and 63 pixels tall. The relationship between the dimensions and the pixels is $\frac{95}{63} = (4.7 - (-4.7) + 0.1)/(6.2 - 0 + 0.1)$. To get the total number of pixels, subtract the extremes and add 1.

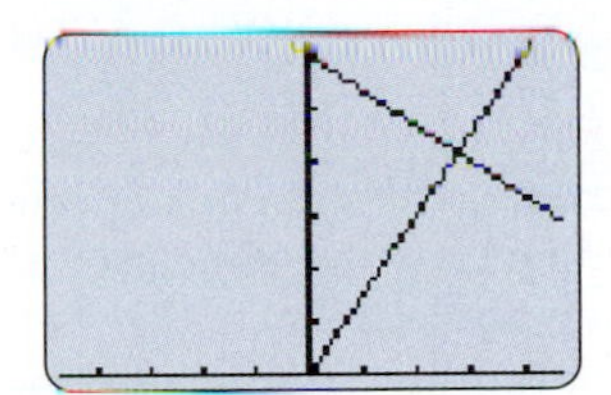

7. The zeros are $x \approx -3.83$ and $x \approx 1.83$. They are both approximations because $f(-3.83) \neq 0$ and $f(1.83) \neq 0$.

9. The point of intersection of f and g is $(-1.1, -2, 4)$. It is an approximation because $f(-1.1) \neq g(-1.1)$.

Section 1.4

1. (a) $C = \frac{5}{9}(F - 32)$

(b) $h = \dfrac{A - 2\pi r^2}{2\pi r}$

(c) $r = \sqrt{\dfrac{V}{\pi h}}$

(d) $a = \dfrac{2\sqrt{A}}{\sqrt[4]{3}}$

3. (a) The area of the circle is $\pi A/2$.

(b) The area of the square is $4A/\pi$.

(c) The difference in area of a square of area A and the circumscribed circle is greater.

(d) It is the same because the circle inscribed in a square fits better in both cases.

5. The height above Earth must be given in miles because the radius of Earth was given in miles.

7. Approximately 0.0792 mi, or 418.17 ft, tall

9. $t = 1 \text{ sec} + \dfrac{V}{20} \dfrac{\text{ft/sec}}{\text{ft/sec}^2} + \dfrac{70 \text{ ft}}{V \text{ ft/sec}}$

$= 1 \text{ sec} + \dfrac{V}{20} \text{ sec} + \dfrac{70}{V} \text{ sec}$

Review Exercises

1. (a) $\dfrac{25\sqrt{3}}{4}$

(b) -3

(c) $g(2) = 2, g(4) = 3$

(d) 492

(e) \$0.99, \$1.99

3. (a) $\frac{1}{25}, 25$

(b) $5, x^2 + 2x + 2$

(c) $-1, 2x^2 - 5$

(d) $20, -40$

(e) $-12, 26$

5. (a) There is an x-intercept at 1; it is increasing and concave down.

(b) There is a y-intercept at $-\frac{1}{2}$, decreasing over the entire domain, concave down when $x < 2$, and concave up when $x > 2$.

(c) There is a y-intercept at -4, x-intercepts at 2 and -2, symmetric about the y-axis, and concave up.

7. (a)

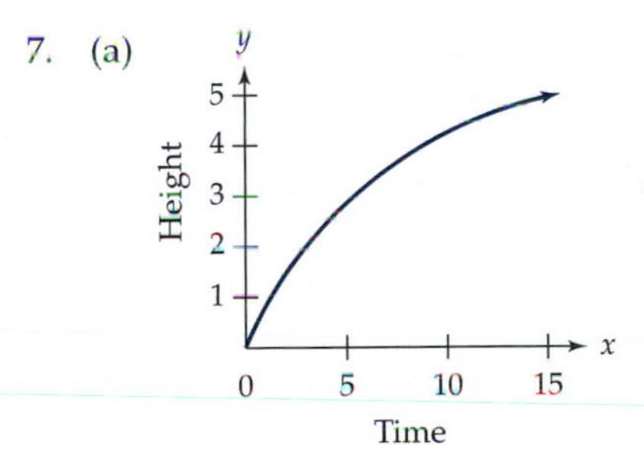

(b)

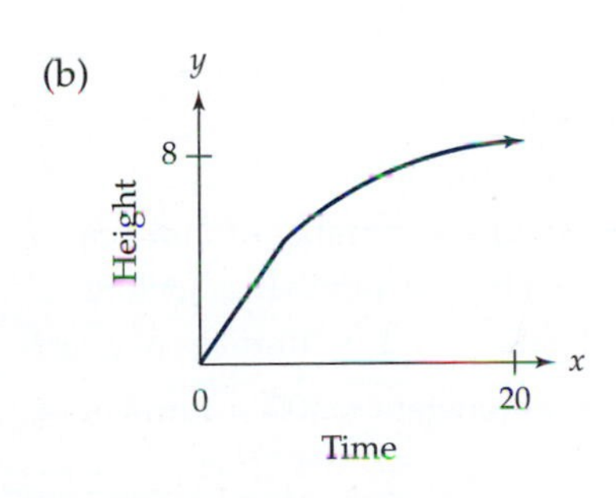

9. Answers will vary.
11. The exact point of intersection is $(-1, 1)$.
13. (a) $A = \left(4 + \frac{\pi}{2}\right) r^2$
    (b) $(6 + \pi)r$
15. (a) $0 \le t \le 10$
    (b) $0 \le V(t) \le 1000$
    (c) 250 gal
    (d) ≈ 2.93 min

Section 2.1

1. Yes, f is a linear function. The slope of f is $\frac{3}{2}$.
3. The slope of the linear function f is negative because it is decreasing. The y-intercept of the linear function f is positive because it intersects the y-axis above the x-axis.
5. (a) i. The domain is all real numbers and the range is all real numbers.
       ii. The domain is all real numbers and the range is all real numbers.
       iii. The domain is all real numbers and the range is 7.
   (b) In general, the domain and the range of a linear function are all real numbers. When the slope of the linear function is 0, the range is a constant.
7. (a) $y = \frac{1}{2}x$
   (b) $y = 2x - 6$
   (c) $y = -2x + 3$
   (d) $y = 3x - 4$
   (e) $y = 6$
9. (a) It is a linear function. The slope is $\frac{3}{2}$ and y-intercept is $\frac{7}{2}$.
   (b) It is a linear function. The slope is $\frac{5}{2}$ and y-intercept is $\frac{-39}{2}$.
   (c) It is a linear function. The slope is 3 and y-intercept is 14.
   (d) It is a linear function. The slope is 3 and y-intercept is -2.
   (e) It is not a linear function.
   (f) It is not a linear function.
   (g) It is a linear function. The slope is 0 and y-intercept is 8.
   (h) It is not a linear function.
11. (a) Total cost = \$0.15 $\times$ number of minutes
    (b) Total price = \$1.18 $\times$ number of gallons
    (c) Number of gallons = $\frac{1}{4}$ $\times$ number of quarts
    (d) Number of centimeters = 100 $\times$ number of meters
    (e) Number of inches = $\frac{100}{254}$ $\times$ number of centimeters
    (f) Number of U.S. dollars = $\frac{100}{483}$ $\times$ number of French francs
13. (a) c is a piecewise function because there are different rules for different domains.
    (b) i. c_1 is a linear function because its slope is constant.
        ii. $c_1 = 0.090906$ $\times$ number of kilowatt-hours

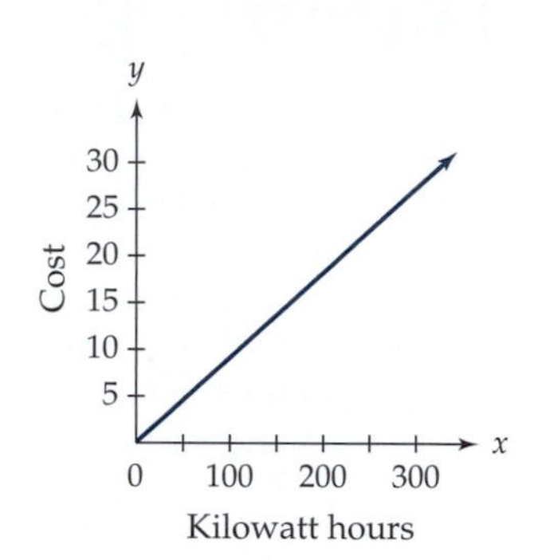

        iii. The slope of c_1 is the cost per kilowatt-hour consumed and the y-intercept is the cost of using no electricity.
    (c) i. c_2 is a linear function because its slope is constant.
        ii. $c_2 = 0.076206$ $\times$ (number of kilowatt-hours $-$ 300) + 27.27, where number of kilowatt-hours > 300.

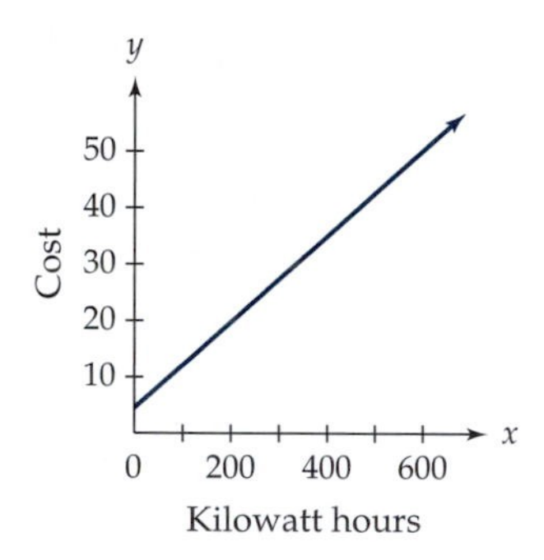

        iii. The slope of c_2 is the cost per kilowatt-hour consumed (between 300 and 600) and the y-intercept is the cost for using 300 kWh.
    (d) i. c_3 is a linear function because its slope is constant.

ii. $c_3 = 0.090906 \times$ (number of kilowatt-hours $-$ 600) + 50.13, where number of kilowatt-hours ≥ 600

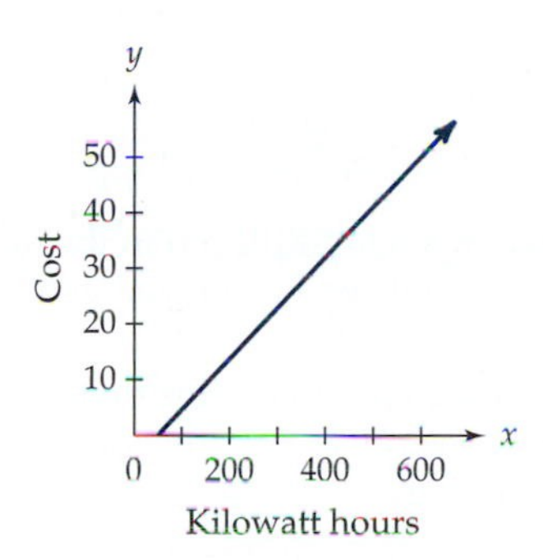

iii. The slope of $c_3(e)$ is the cost per kilowatt-hour consumed (between 600 and 649) and the y-intercept is the cost for using 600 kWh.

15. (a) The conversion from cups to jacks is linear because the rate of conversion from cups to jacks is constant.
(b) Two cups is approximately 8 jacks.
(c) Ten jacks is approximately 2.5 cups.
(d) The equation that converts from cups into jacks is $j = 4c$.

17. (a) The slope of the function is 3. This number means that if the length of a person's foot increases by 1 in., the shoe size increases by 3.
(b) The y-intercept of this function is -22. This number means that when the length of a person's foot in inches is 0, the shoe size is -22, which is not realistic because 0 is not in the domain of the function.
(c) The equation for this function is $s = 3l - 22$, where s is the shoe size and l is the length of a person's foot in inches.
(d) It is similar to the function in Example 6. In fact, if you take the function from Example 6 and solve for s, you get this function.

19. (a) The function that determines the temperature beneath the surface of Earth is linear because the temperature increases at a constant rate of 1° per 80 ft.
(b) $T(d) = \frac{1}{80}d + 50$
(c) $T(d) = \frac{1}{80}(6200) + 50 = 127.5°\text{F}$

21. (a) These data are not a function because some inputs (for instance, 22) have more than one output.
(b)

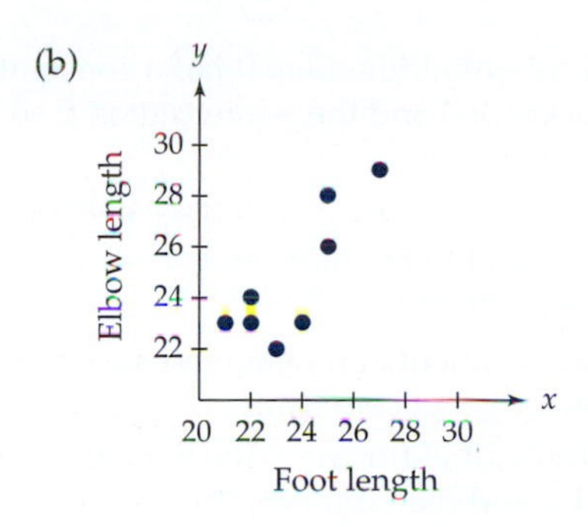

(c) The points are not perfectly linear because there is no fixed relationship between foot length and wrist-to-elbow length.
(d) Answers will vary.

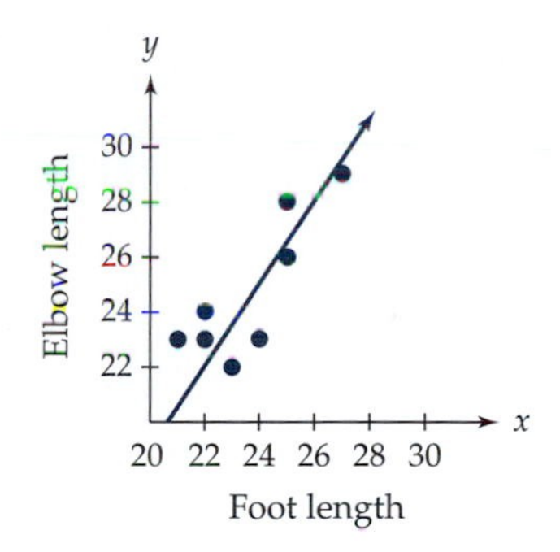

(e) Answers will vary.
(g) No
(h) Answers will vary.

Section 2.2

1. (a) $h(-2) = \frac{1}{4}$
(b) $h(-1) = \frac{1}{2}$
(c) $h(0) = 1$
(d) $h(\frac{1}{2}) = \sqrt{2}$
(e) $h(10) = 1024$

3. (a) It is an exponential function. The growth factor is 3 and the starting value is 1.
(b) It is not an exponential function.
(c) It is an exponential function. The growth factor is $\sqrt{2}$ and the starting value is $\frac{1}{3}$.
(d) It is an exponential function. The growth factor is π and the starting value is 1.
(e) It is not an exponential function.

5. (a) It is an exponential function. It has a constant growth factor of 3 and the y-intercept is 1, so the function is $y = 3^x$.
   (b) It is an exponential function. It has a constant factor of $\sqrt{2}$ and the y-intercept is 1, so the function is $y = (\sqrt{2})^x$.
   (c) It is not an exponential function because the growth factor is not constant.
   (d) It is an exponential function. It has a constant factor of 1.2 and the y-intercept is 1, so the function is $y = (1.2)^x$.
   (e) It is not an exponential function because the y-intercept is 0.
7. (a) $f(x) = 2^x$
   $f(-3) = 2^{-3} = \frac{1}{8}$
   $f(-2) = 2^{-2} = \frac{1}{4}$
   $f(-1) = 2^{-1} = \frac{1}{2}$
   $f(1) = 2^1 = 2$
   $f(2) = 2^2 = 4$
   $f(3) = 2^3 = 8$
   $f(x)$ is a reciprocal of $f(-x)$ for this function and these inputs.
   (b) $f(x) = a^x$
   $f(-3) = a^{-3} = \dfrac{1}{a^3}$
   $f(-2) = a^{-2} = \dfrac{1}{a^2}$
   $f(-1) = a^{-1} = \dfrac{1}{a^1}$
   $f(1) = a^1 = a$
   $f(2) = a^2 = a^2$
   $f(3) = a^3 = a^3$
   $f(x)$ is a reciprocal of $f(-x)$ for this function.
   (c) The property of exponents that causes this behavior is $a^{-b} = 1/a^b$.
   (d) $f(x) = a^x$ is neither symmetric about the y-axis nor about the origin.
9. (a) He first earned more than $1,000,000 on his 20th game.
   (b) He earned half of $536,870,912, or $268,435,456, for playing the second to the last game.
   (c) $P(n) = (1.5)^n$ represents his pay for the nth game. He would have earned $57.66 for the 10th game. He would have earned $127,834.04 for the last game.
   (d) 1.61
11. (a) $B(t)$ is a linear function because the rate of change ($1000) is constant.
   (b) $B(t) = 1000t + 33{,}500$
   (c) $C(t)$ is an exponential function because the growth rate (2.7%) is constant.
   (d) $C(t) = 33{,}500 \cdot (1.027)^t$
   (e) The $1000 annual bonus is the better choice.
   (f) The 2.7% cost of living increase is the better choice.
13. (a) $L(5)$
   (b) $E(20)$
   (c) $L(c)$
   (d) $E(d)$
15. (a) Three rounds will be played in the first strategy and four rounds will be played in the second strategy.
   (b) A linear function models the first strategy, whereas an exponential function models the second strategy.
   (c) $T(x) = 16 - 5x$, where $x = 0, 1, 2, 3$
   (d) $y = 16 \cdot \left(\frac{1}{2}\right)^x$, where $x = 0, 1, 2, 3, 4$
17. (a) 6 h
   (b) After 66 h (6 h before the third day)

Section 2.3

1. (a) This table does not represent a linear function because the rate of change is not constant.
   (b) This table does not represent an exponential function because the growth rate is not constant.
5. (a) $\log 50 \approx 1.69897$
   (b) $\log 5000 \approx 3.69897$
   (c) $\log 25 \approx 1.39794$
   (d) $\log 125 \approx 2.09691$
7. (a) $\log(10a) = 1.4$
   (b) $\log(100a) = 2.4$
   (c) $\log(1000a) = 3.4$
   (d) $\log(a^2) = 0.8$
   (e) $\log(a^{1/2}) = 0.2$
   (f) $\log(a^3) = 1.2$
9. $\log \dfrac{a}{b} = \log\left(a \cdot \dfrac{1}{b}\right) = \log a + \log \dfrac{1}{b} =$
   $\log a + \log b^{-1} = \log a - \log b$

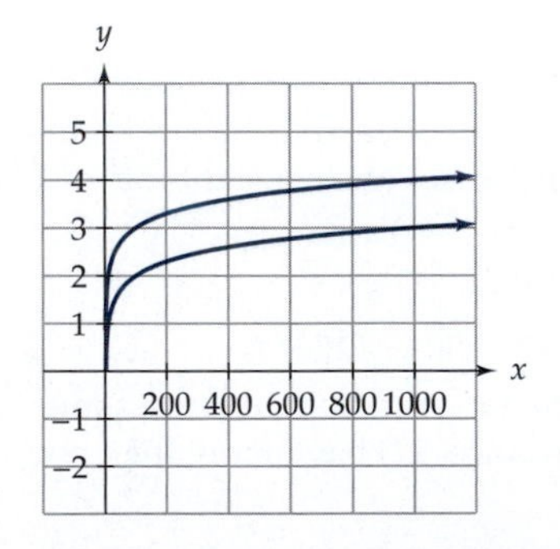

11. (a) The graph of $g(x) = \log 10x$ is a vertical shift upward of the graph $f(x) = \log x$ by one unit.
    (b) The graph of $h(x) = \log 100x$ would be a vertical upward shift of the graph $f(x) = \log x$ by two units.
    (c) The graph of $j(x) = \log 0.1x$ would be a vertical downward shift of the graph $f(x) = \log x$ by one unit.
13. The decibel level at 500 is approximately 27. So, when sound level goes from 500 to 700, $(1.26)^{27+2} \approx 814$. Therefore, the decibel level increases by less than 2.

Section 2.4

1. (a) It is not a periodic function.
   (b) It is a periodic function with period 8 and amplitude 5.
   (c) It is a periodic function with period 4 and amplitude 3.
   (d) It is not a periodic function.
3. (a) $f(1) = 0.5$
   (b) $f(5) = 0.5$
   (c) $f(10) = 1$
   (d) $f(20) = 0$
   (e) $f(-3) = 0.5$
5. They would remain the same.
7. The period is 12 h and the amplitude is 2 in.
9. (a) The period is 1.5 sec and the amplitude is 1.5 ft.

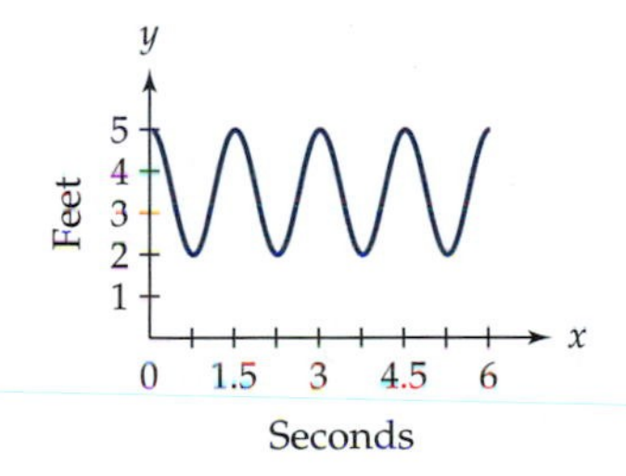

   (b) The period is 3 sec and the amplitude is 5 ft.

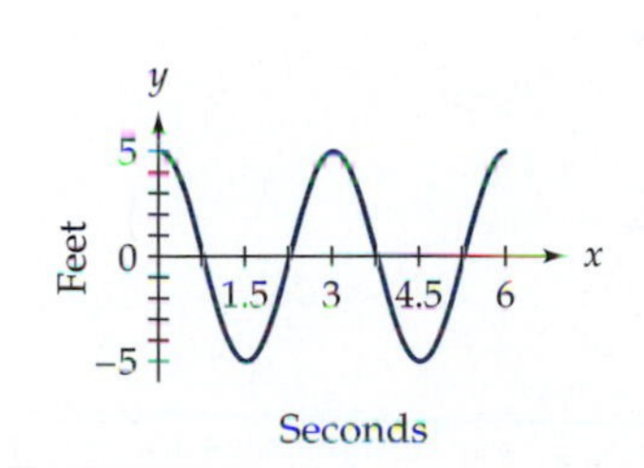

11. (a) $\cos A = 0.6$
    (b) $\sin B = 1$
    (c) $\cos C = -0.8$
    (d) $\sin D = 0$
    (e) $\sin E = -0.8$
    (f) $\cos F = 0.8$

13. (a)

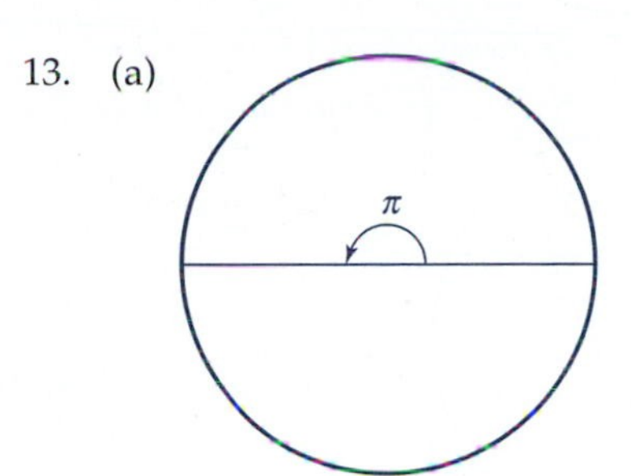

(b)

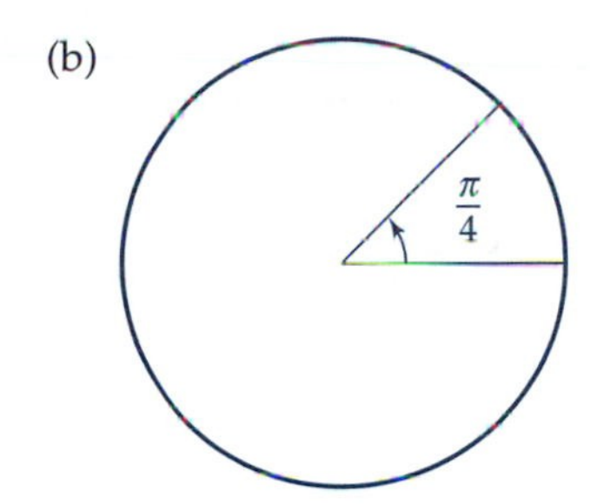

(c)

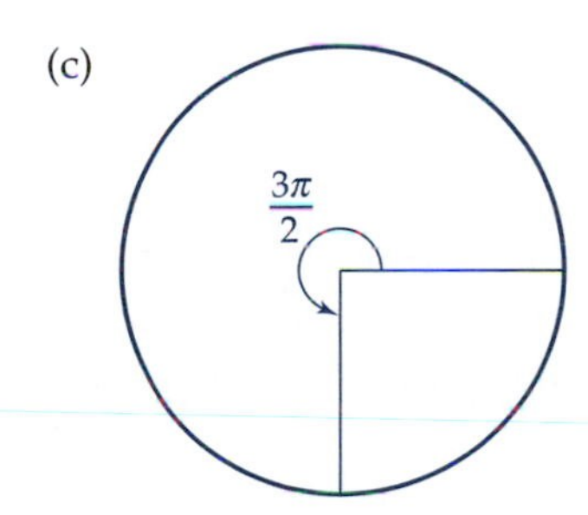

(d)

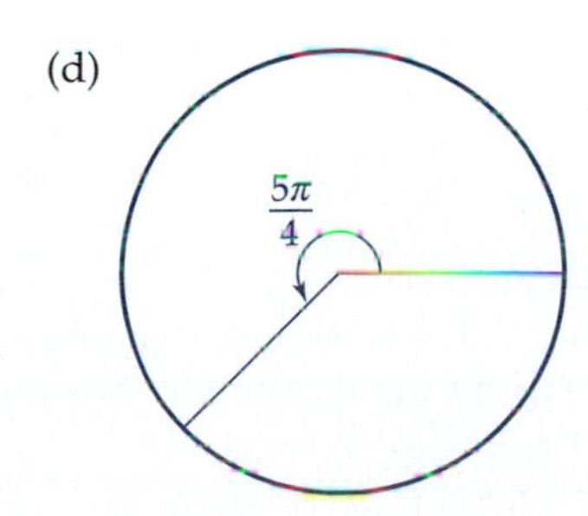

(e)

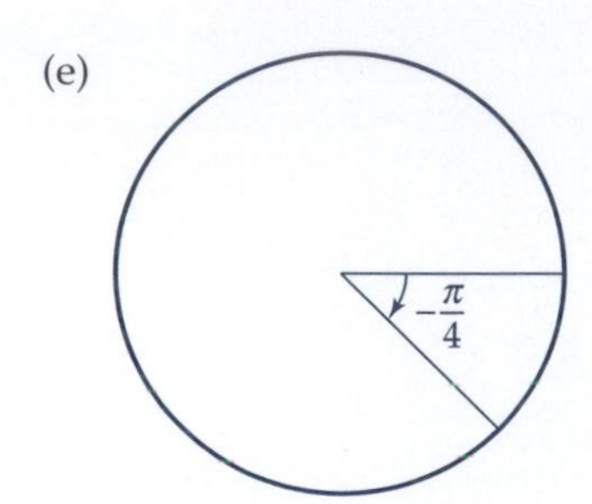

(f)

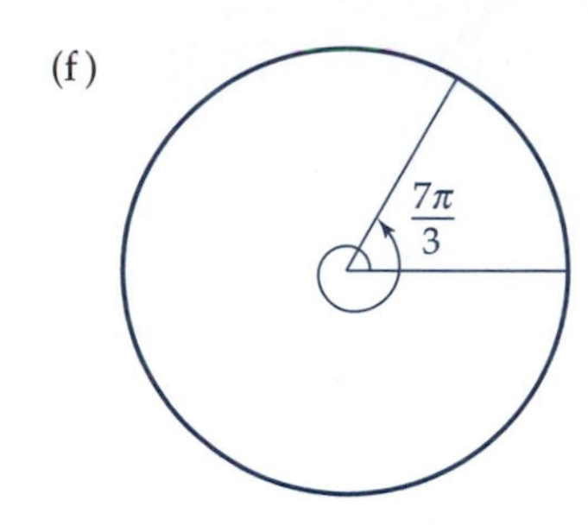

15. (a) $\sin\frac{\pi}{2} = 1$
    (b) $\cos 2\pi = 1$
    (c) $\sin 0 = 0$
    (d) $\cos\frac{3\pi}{2} = 0$
    (e) $\cos\pi = -1$

Section 2.5

1. (a) It is not a linear function because the rate of change (slope) is not constant.
   (b) $a < 1$ because the function is concave down.
3. (a) vi
   (b) ii
   (c) viii
   (d) v
   (e) iii
   (f) vii
   (g) i
   (h) iv
5. (a) Sometimes true
   (b) Sometimes true
   (c) Always true
   (d) Sometimes true
   (e) Sometimes true
7. (a) When $n = 2$, $f(x) = f(-x)$ because squaring a positive or a negative input results in the same positive output.
   (b) When n is even, $f(x) = f(-x)$ because raising a positive or a negative input to an even power results in the same positive output. Therefore, f is symmetric about the y-axis.
9. (a) The domain is all real numbers and the range is $f(x) \geq 0$.
   (b) The domain is all real numbers and the range is all real numbers.
   (c) The domain is $x \geq 0$ and the range $f(x) \geq 0$.
   (d) The domain is all real numbers and the range is all real numbers.
   (e) The domain is $x \geq 0$ and the range is $f(x) \geq 0$.
   (f) The domain is all real numbers and the range is $f(x) \geq 0$.
11. (a) It is a power function; $f(x) = x^6$.
    (b) It is a power function; $f(x) = 3x^2$.
    (c) It is a power function; $f(x) = x^{3/5}$.
    (d) It is not a power function because $f(0) \neq 0$.
    (e) It is a power function; $f(x) = 2x^{1/4}$.
    (f) It is a power function; $f(x) = x^{8/3}$.
    (g) It is a power function; $f(x) = 5x^{11/2}$.
13. $f(x) = 2.5x^3$, $g(x) = 4 \cdot 1.5^x$, $h(x) = 7.5x + 3$
15. $b = 1$ and $a > 1$ such that a must have an odd numerator and denominator.
17. (a) $t = [d/(1.496 \times 10^8)]^{3/2}$.
    (b) In the new formula, $b = [1/(1.496 \times 10^8)]^{3/2}$. In the formula we found using astronomical units, $b = 1$.

Review Exercises

1. $f(3) = 11, f(10) = 39$
3. (a) $g = \frac{1}{231}i$
   (b) $f = 5280m$
   (c) $c = 0.35m$
   (d) $p = 0.15f$
5. (a) 16
   (b) $2^{1/3} \approx 1.26$
   (c) $2^{48} \approx 2.8 \times 10^{14}$
7. (a) $y = 5 \cdot 3^x$
   (b) $y = 5 \cdot 3^{x/5}$
   (c) $y = \frac{5}{4} \cdot 2^x$
9. (a) $\frac{7}{4}$
   (b) $\frac{11}{4}$
   (c) $\frac{3}{2}$
   (d) 1
11. (a) 60 dB

(b) 90 dB
(c) 10,000

13.

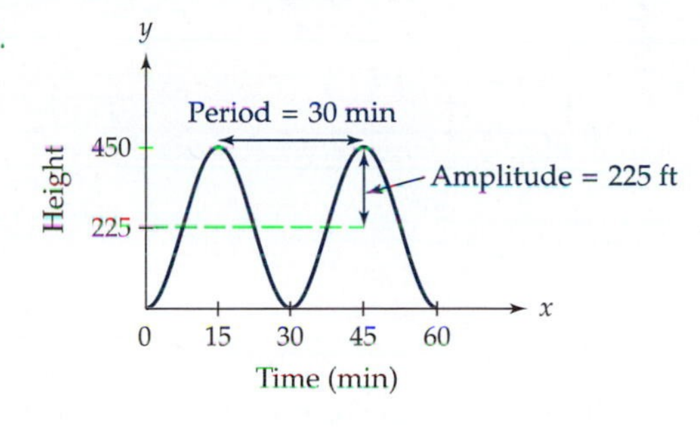

15. (a)

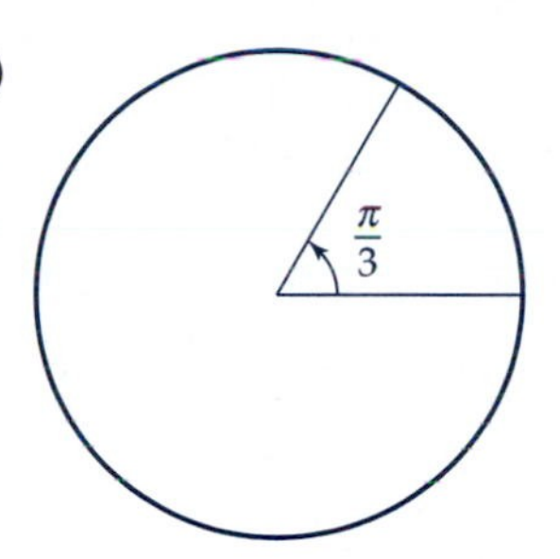

(b)

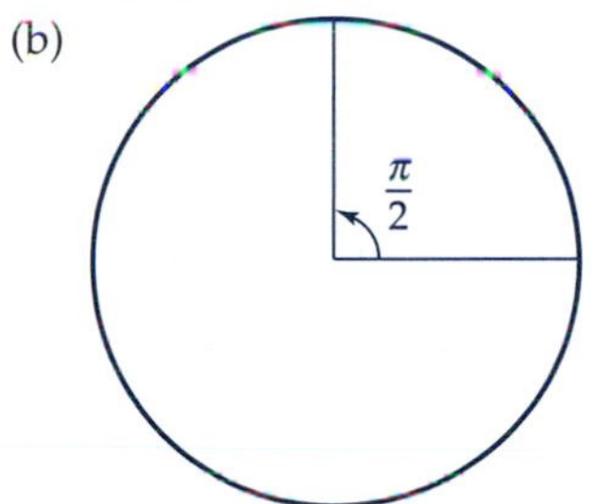

(c)

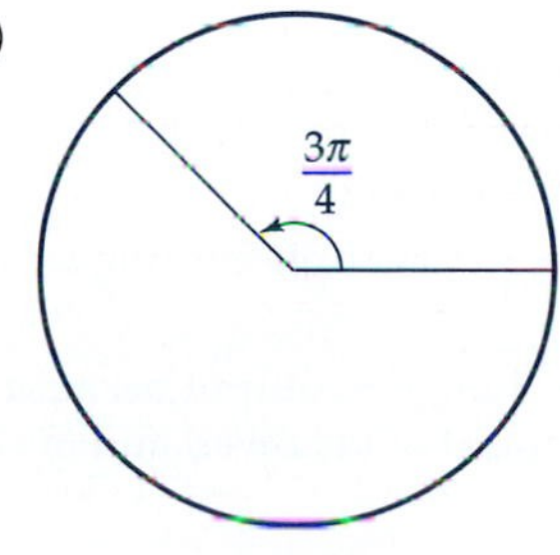

(d)

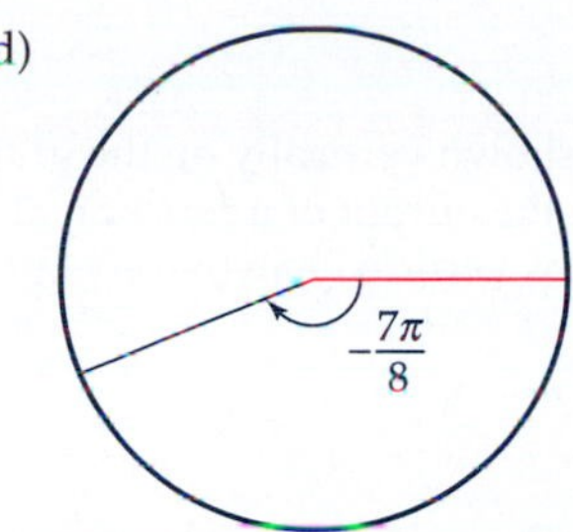

(e)

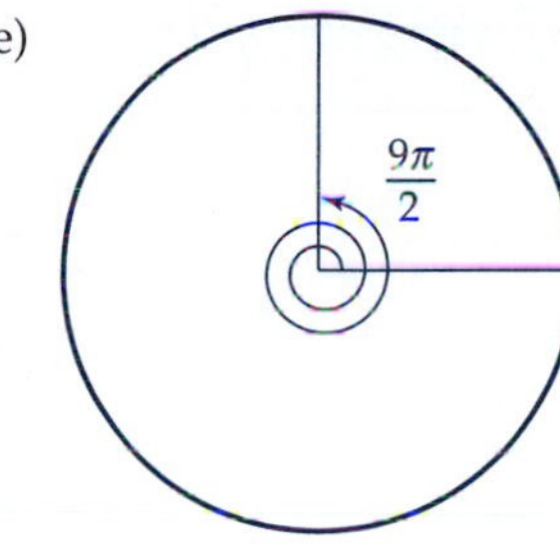

(f)

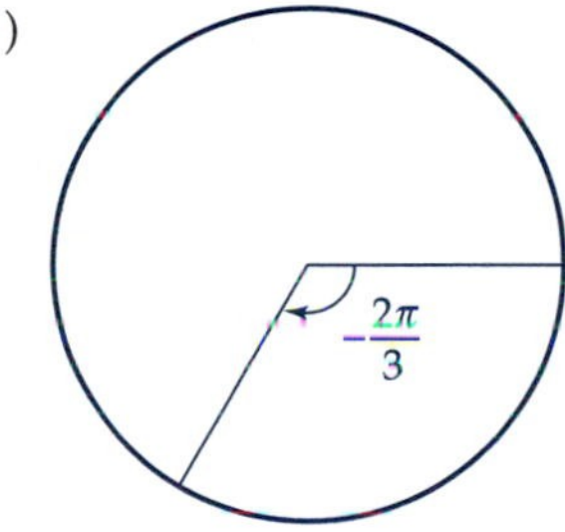

17. (a) iv
(b) i
(c) v
(d) iii
(e) ii
(f) vi

19. (a) $\sqrt[4]{x}$
(b) $4\sqrt[11]{x^7}$
(c) $\sqrt[4]{x^{11}}$
(d) $15\sqrt[9]{x^{14}}$

21. $f(x) = 0.15x^2$, $h(x) = 20 \cdot 0.4^x$, $j(x) = 0.15 + 0.4x$

Section 3.1

1. Because output is shown vertically on the graph, transformations in the output of a function always shows themselves as vertical changes on a graph.

3.

x	1	2	3	4	5
$f(x)$	3	−4	5	−3	1
$f(x)+5$	8	1	10	2	6
$2f(x)-7$	−1	−15	3	−13	−5

5. (a)

(b)

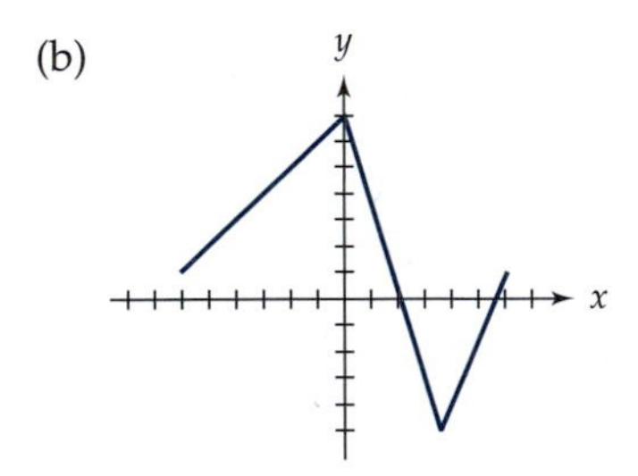

(c)

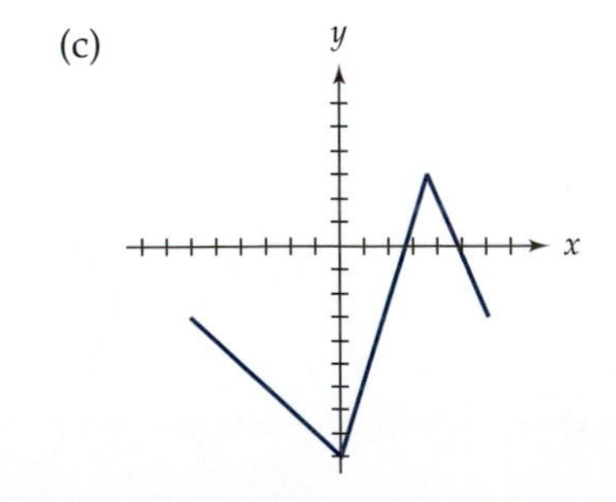

7. (a)

(b) i. It is a change in the output because the output (the temperature) has been changed.

ii.

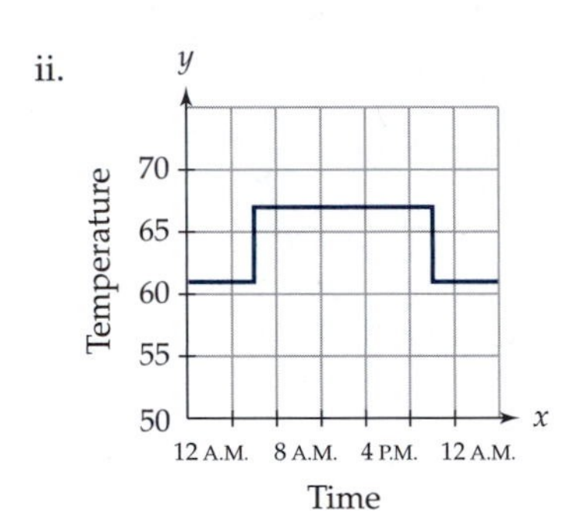

iii. The person could change the timing of the thermostat setting.

9. (a) g is a stretch of f.
   (b) $f(3) = 0$

11. (a) $0 < a < 1$
    (b) $a > 1$
    (c) $a > 0$
    (d) $a < 0$

13. $g(x) = f(x) - 4$, $h(x) = 2f(x) + 3$.

15. (a) $(i - 99{,}600)$ represents how much money over \$99,600 you have earned.
    (b) 22,532 is the tax for a couple earning exactly \$99,600.
    (c) This credit is a change in output because tax credit is the amount of tax owed and tax owed is the output.
    (d) $t(i) = 17{,}532 + 0.31(i - 99{,}600)$

Section 3.2

1. Transformation of the input results in horizontal change because the input is on the horizontal x-axis.

3.

x	−4	−3	−2	−1	0	1	2	3	4
$f(x)$	−2	5	7	−3	10	−1	6	0	8
$f(2x)$			−2	7	10	6	8		
$f(x+3)$	−3	10	−1	6	0	8			
$4f(-x)+1$	33	1	25	−3	41	−11	29	21	−7

5. (a)

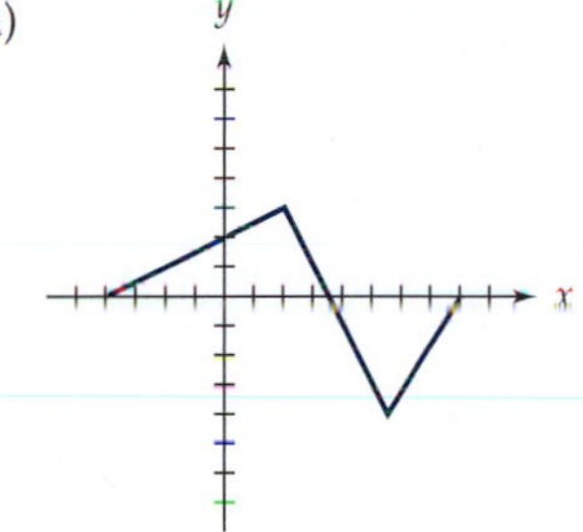

(b)

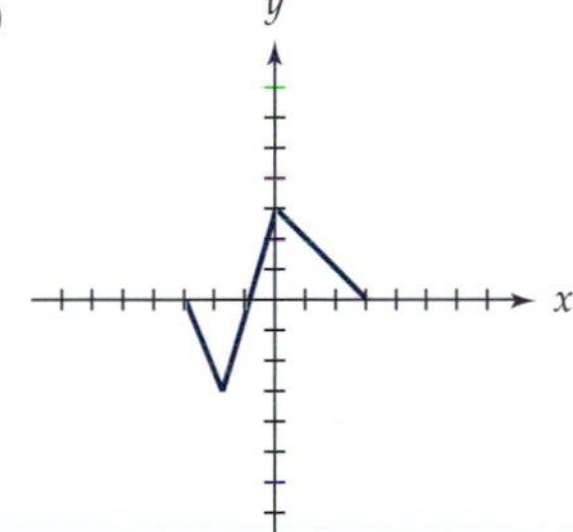

7. Situation B
9. (a) vii
   (b) iv
   (c) i
   (d) ii
11. (a) The y-intercept will change if $m \neq 0$.
    (b) The slope will change.
13. (a) $N_{70}(4) = 100 \cdot 2^4 = 1600$
    (b) $N_{70}(t) = 100(2^t)$
    (c) Bacteria are doubling every hour.

Section 3.3

1. Answers will vary.
3. Make the first \$13,300 taxable.
5. 9.58 sec
7. The graph of g has the same shape as that of f, but it is shifted to the right two units.

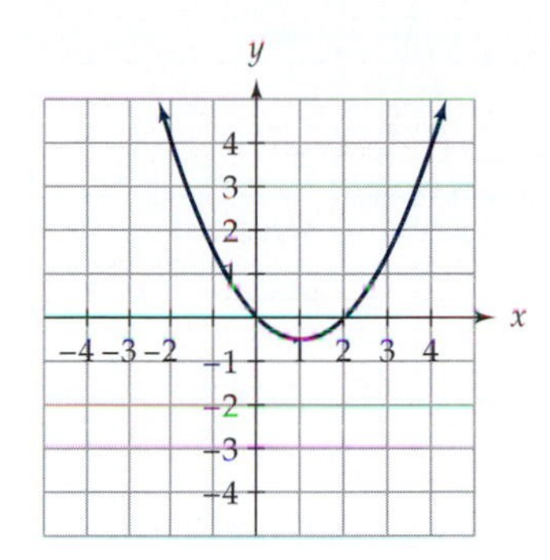

9. (a)

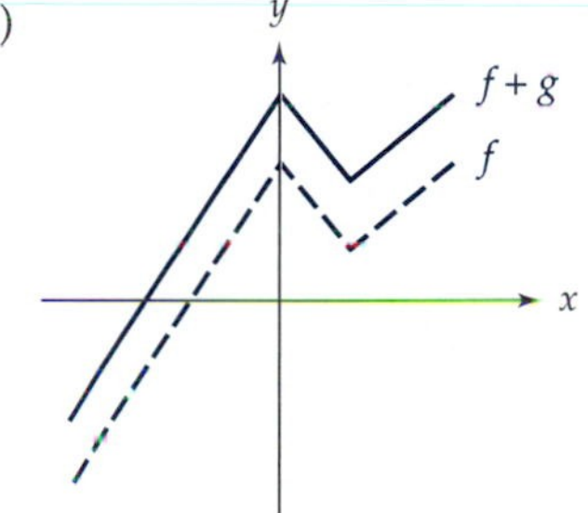

(b)

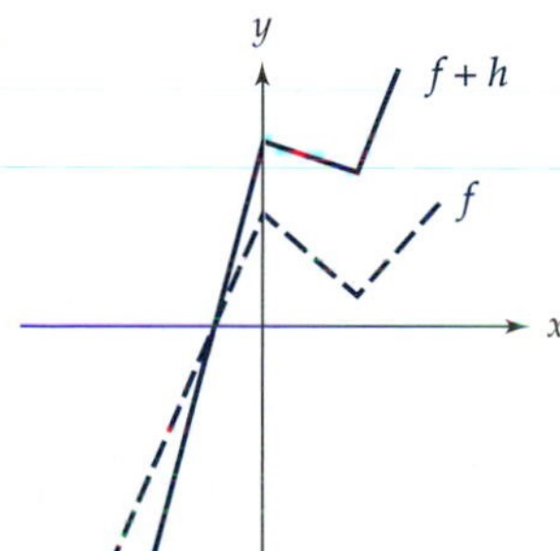

11. $y = f(x)$ will appear similar to $y = 3^{-x}$ when x is negative; $y = f(x)$ will appear similar to $y = x$ when x is positive.

13. (a) When $x > 0$, the graph looks like the graph of $y = \sin x$, but the amplitude is getting larger as x gets farther from zero. The portion of the graph to the left side of the y-axis is a reflection of the portion of the graph to the right of the y-axis.

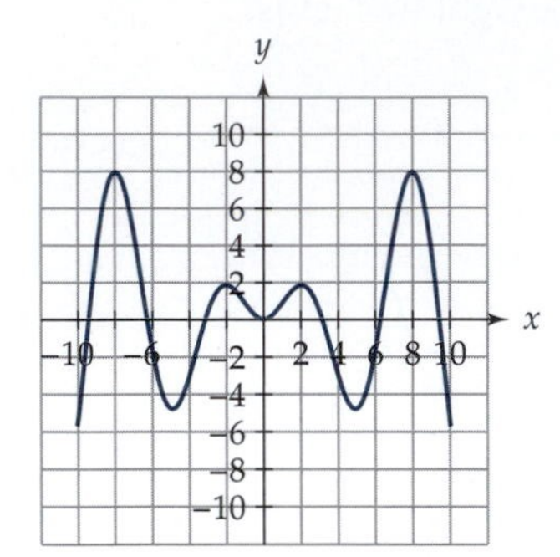

(b) The graph looks like $y = \sin x$, with an amplitude that gets larger as x gets farther from zero.

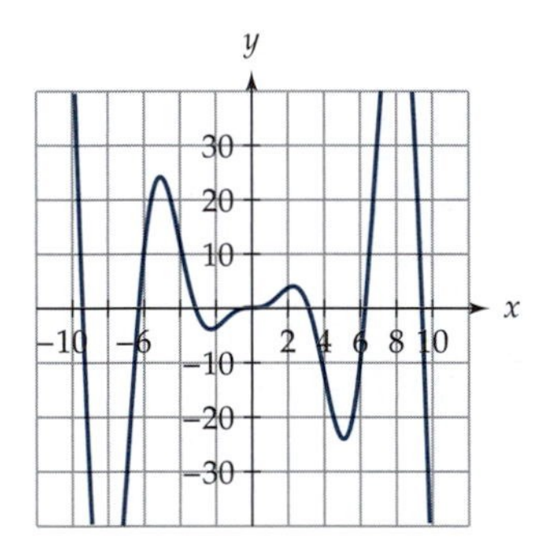

(c) The graph looks like $y = \sin x$, but it has an amplitude that gets smaller as x increases and larger for negative values of x.

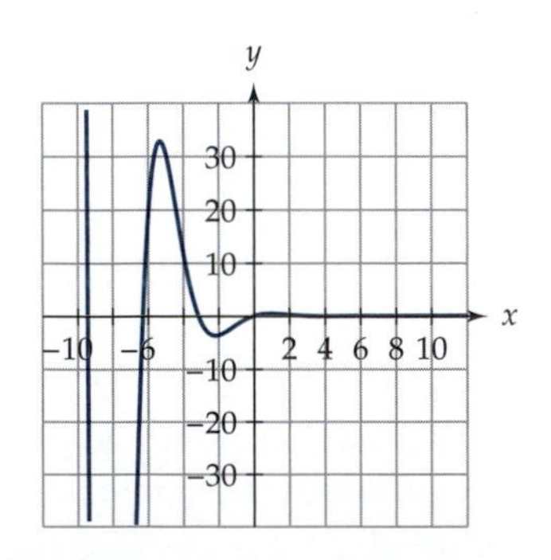

15. (a) i. $g(m) = 1.05(m/22)$
ii. $l(m) = 22.5(m/3000)$
iii. $r(m) = 200(m/25{,}000)$
iv. $t(m) = 4173(m/66{,}000) \approx 0.0632m$
v. 6.32 cents per mile
(b) \$1300

Section 3.4

1. (a) $S \circ A$ represents writing the address on the envelope followed by the process of placing a letter in an envelope and sealing it.
(b) Yes, it is a valid process. $A \circ S$ represents placing the letter in an envelope and sealing it followed by writing the address on the envelope.

3. (a) $g(f(0)) = 7$
(b) $(g \cdot f)(0) = 4$

5.

x	−3	−2	−1	0	1	2	3
$f(x)$	4	−1	0	2	3	−2	1
$g(x)$	−2	0	2	4	2	1	−1
$g(f(x))$		2	4	1	−1	0	2

7. (a) $g(f(x)) = 3(2^x) - 4$
(b) $f(h(x)) = 2^{\sqrt{x}}$
(c) $h(g(x)) = \sqrt{3x - 4}$
(d) $h(f(g(x))) = \sqrt{2^{3x-4}}$

9. (a) $F(x) = x + 38$
(b) The slope is the change in temperature (degrees Fahrenheit) for each increase of one chirp in 15 sec. The y-intercept is the temperature (degrees Fahrenheit) at which there are no chirps in 15 sec.
(c) You would have a composition because the output of one function is the input of the other.
(d) $C = \frac{5}{9}(x + 6)$
(e) The slope is the change in temperature (degrees Celsius) for each increase of one chirp in 15 sec. The y-intercept is the temperature (degrees Celsius) at which there are no chirps in 15 sec.

11. (a) $k(f(0)) = -1$
(b) $k(k(f(1))) = 4$
(c) $f(k(g(0))) = 1$
(d) $g(g(2)) = -4$
(e) $f(k(g(1))) = 2$

13. The domain is $\{0, 2, 3, 4, 5, 6\}$ and the range is $\{-2, -1, 0, 1, 2\}$.
15. (a) $g(f(x)) = x$
    (b) $g(f(x)) = x$
    (c) $g(f(x)) = x$
    (d) $g(f(x)) = x$
    (e) $g(f(x)) = x$
    (f) Each of these compositions has its output the same as its input.

Section 3.5

1. You take your left shoe off, walk forward five paces, put your right shoe on, walk backward five paces, and put your left shoe back on.
3. (a) $g^{-1}(x) = 2x + 10$
   (b) $h^{-1}(x) = 3x - 1$
   (c) $j^{-1}(x) = 5x + 12$
5. (a) No, the function is not one-to-one because the graph of any quadratic function has two inputs for all but one output. Possible restricted domains are $x \leq 0$ and $x \geq 0$.

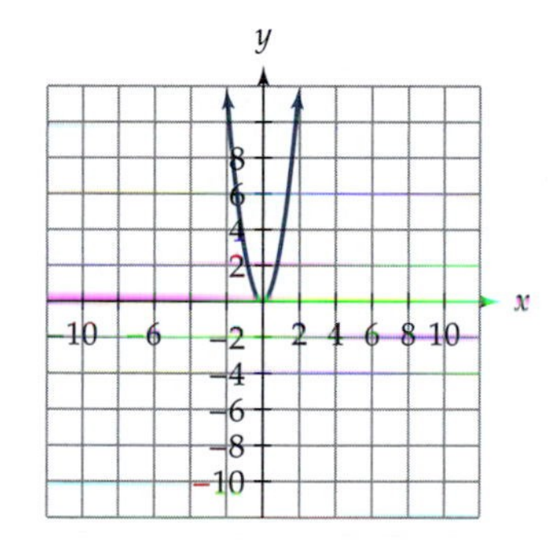

(b) Yes

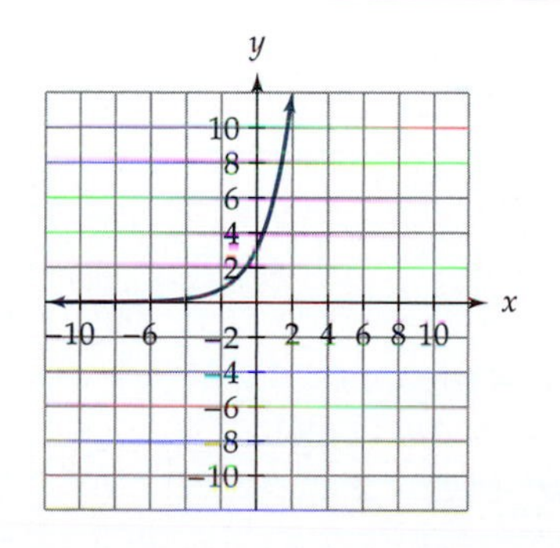

(c) Yes

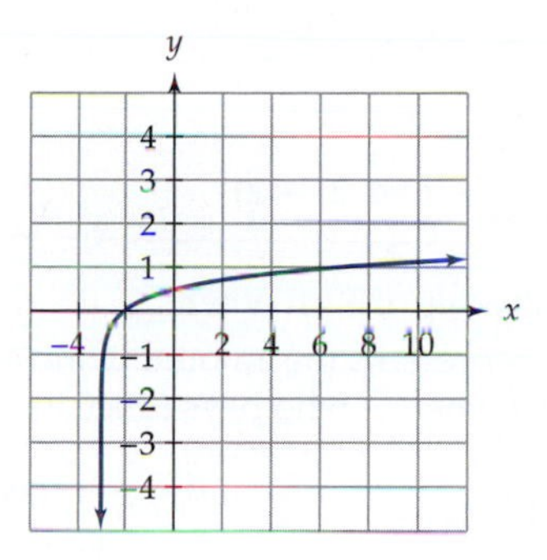

(d) No. One possible restricted domain is $-\pi/4 \leq x \leq \pi/4$.

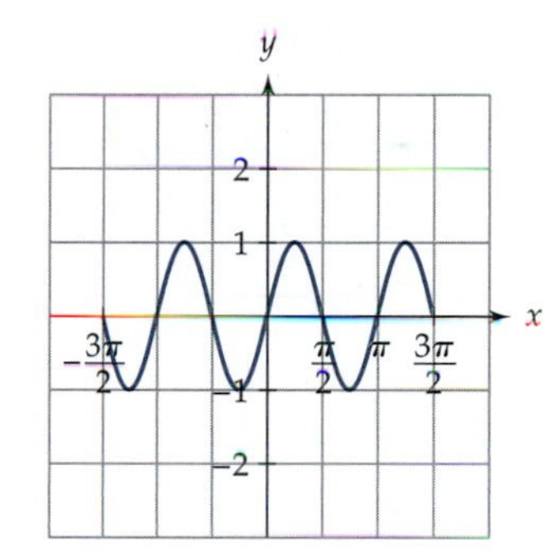

(e) Yes
(f) No. Possible restricted domains are $x < -1$ and $x > 0$.

7.

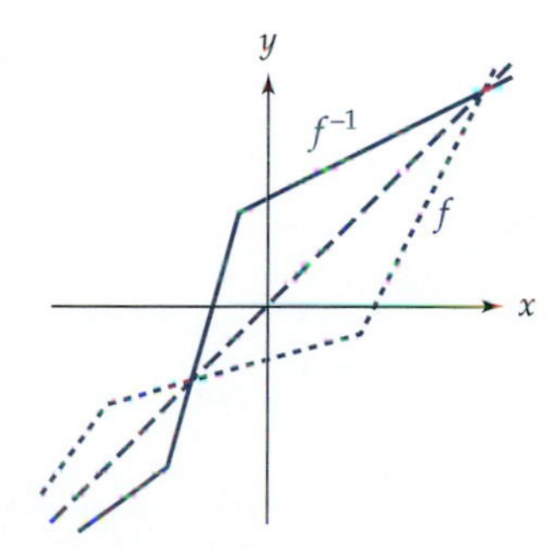

9. $f(g(x)) = f\left(\frac{x-2}{3}+6\right)$

$$= 3\left[\left(\frac{x-2}{3}+6\right)-6\right]+2 = x,$$

$g(f(x)) = g(3(x-6)+2)$

$$= \frac{[(3(x-6)+2)-2]}{3}+6 = x$$

11. They are reciprocals of each other.

13. (a) $f^{-1}(2000)$ represents the amount of income earned by someone who pays \$2000 in taxes.

(b) $f^{-1}(x) = \begin{cases} 10x & \text{if } 0 \le x \le 1000 \\ 5(x+1000) & \text{if } x > 1000 \end{cases}$

Review Exercises

1.

x	−4	−3	−2	−1	0	1	2	3	4
$f(x)$	−2	5	7	−3	10	−1	6	0	8
$f(-2x)$			8	6	10	7	−2		
$f(x+4)$	10	−1	6	0	8				
$f(\frac{x}{2})+1$	8		−2		11		0		7

3. (a)

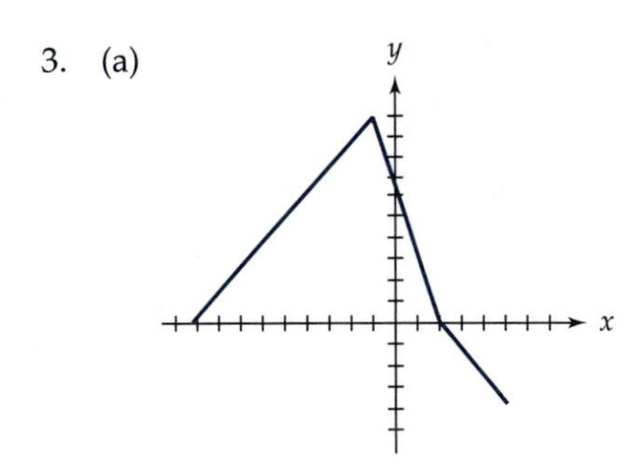

(b)

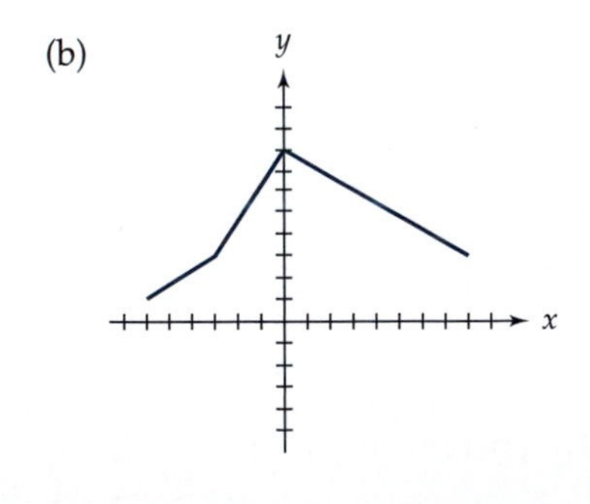

5. (a) The y-intercept will change.
(b) The growth factor will change.

7.

x	0	1	2	3	4	5	6
$f(x)$	2	4	6	1	3	5	0
$g(x)$	0	6	1	5	2	4	3
$(f+g)(x)$	2	10	7	6	5	9	3
$(f \cdot g)(x)$	0	24	6	5	6	20	0
$f(g(x))$	2	0	4	5	6	3	1
$g(f(x))$	1	2	3	6	5	4	0

9. (a) The domain of f is $x \ge -2$. The domain of g is all $x \ne 0$.
(b) The domain of h is $-2 \le x < 0$ or $x > 0$. The domain of k is $-2 < x < 0$ or $x > 0$.

11. (a) $(g \circ f)(x) = (x^2+3x)^{3/2}$
(b) $(f \circ g)(x) = x^3 + 3x^{3/2}$
(c) $(h \circ g)(x) = 5x^{3/2} - 8$
(d) $(h \circ f \circ g)(x) = 5x^3 + 15x^{3/2} - 8$

13. (a) This function is one-to-one.
(b) This function is not one-to-one. We can restrict the domain to $0 \le x \le \pi/2$.
(c) This function is not one-to-one. We can restrict the domain to $x \ge 0$.
(d) This function is one-to-one.

15.

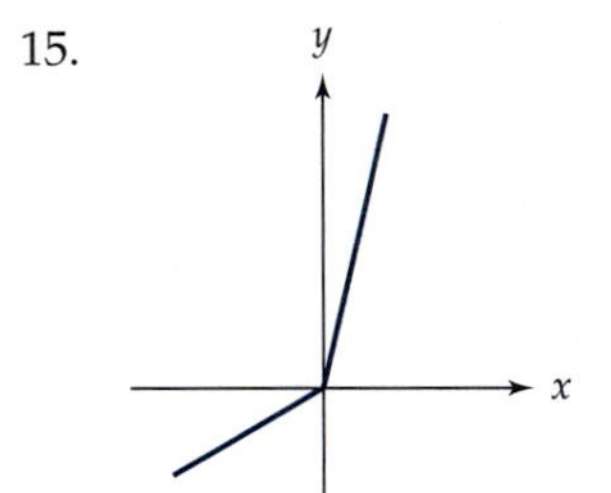

19. (a) $D^{-1}(x)x = 10^{(x/10)-16}$
(b) 10^{-11}

Section 4.1

1. (a) $(5, -4)$
(b) $(2, 3)$
(c) $(-3, -4)$

3. (a) $f(x) = 4x^2 - 16x + 19$
(b) $g(x) = -3x^2 - 36x - 109$
(c) $h(x) = 6x^2 + 12x + 2$

7. $x = 2$
11. (a) $x = -6$ and $x = 2$
    (b) $x = -3$ and $x = 1$
    (c) $x = 3$ and $x = 6$
    (d) $x = 4$
    (e) $x = -2$ and $x = 2$
    (f) No real roots
    (g) No real roots
13. (a) $f(x) \geq 2$
    (b) $g(x) \geq -13.25$
    (c) $h(x) \geq -5.125$
15. (a) $y = -x^2 - 4x - 5$
    (b) $y = -2x^2 - 4$
    (c) $y = -8x^2 + 8x$
    (d) $y = -\frac{1}{4}x^2 + \frac{1}{2}x + 2$
17. (a) Making k negative
    (b) Making $k = 0$
    (c) Making k positive
19. $\sqrt{5} \approx 2.2$ sec.

Section 4.2

1. (a) i. $V = -4w^3 + 130w^2$
       ii. $0 < w < 32.5$
       iii. $w = 21.66$ in. and $L = 43.33$ in.
   (b) i. $V = -4w^3 + 165w^2$
       ii. $0 < w < 41.25$
       iii. $w = 27.5$ in. and $L = 55$ in.
3. (a) The polynomial is degree 3.
   (b) The polynomial is degree 5.
   (c) The polynomial is degree 5.
   (d) The polynomial is degree 4.
   (e) The polynomial is degree 4 or greater.
   (f) The polynomial is degree 4 or greater.
5. (a) This function is at least degree 4. The leading coefficient is positive. The roots of multiplicity 1 are $x = 0$ and $x = 0.4$. -0.4 is a root with even multiplicity.
   (b) This function is at least degree 5. The leading coefficient is negative. The roots of multiplicity 1 are $x = -1$, $x = 0$, and $x = 2$. The root $x = 1$ has even multiplicity.
   (c) This function is at least degree 3. The leading coefficient is positive. The root $x = -1$ has multiplicity 1. The root $x = 1$ has even multiplicity.
   (d) This function is at least degree 3. The leading coefficient is negative. The root $x = 0$ has odd multiplicity greater than 1.
7. Answers will vary.
9. Because a polynomial of even degree approaches ∞ as $x \to \pm\infty$ or approaches $-\infty$ as $x \to \pm\infty$, it needs to change direction an odd number of times.
11. (a) i. The graph changes direction.
        ii. The degree is 3 or greater and the roots are $x = -1$, $x = 0$, and $x = 2$.
    (b) The degree is 5 or greater.
13. n is odd.
15. Answers will vary.
17. An input of -1 produces a negative output, whereas an input of 0 produces a positive output.
19. The function is never negative.
21. A periodic function changes direction infinite times, so the degree of the polynomial would be infinite.
23. (a) A fourth-degree polynomial can change concavity up to two times.
    (b) No. Lines (or curves) of best fit often do not work well in predicting outputs for inputs outside the range of those given.

Section 4.3

1. (a) $f(x)$ approaches positive infinity as x approaches zero.
   (b) $g(x)$ approaches zero as x approaches zero from the right.
   (c) $h(x)$ approaches zero as x approaches positive or negative infinity.
3. Every power function with a negative exponent approaches zero as x approaches positive infinity and approaches positive infinity as x approaches zero.
5. (a) $f(x) \to \infty$ as $x \to 0, f(x) \to 0$ as $x \to \pm\infty$.
   (b) $g(x) \to \infty$ as $x \to 0^+, g(x) \to -\infty$ as $x \to 0^-$, $g(x) \to 0$ as $x \to \pm\infty$.
   (c) $h(x) \to \infty$ as $x \to 0^+, h(x) \to 0$ as $x \to \infty$.
   (d) $j(x) \to \infty$ as $x \to 0, j(x) \to 0$ as $x \to \pm\infty$.
7. The 6 in the numerator of the exponent is equivalent to raising x to the sixth power, and any even power gives a positive output for both negative and positive inputs.
9. (a) $f^{-1}(x) = \left(\frac{x}{4}\right)^{-1/7}$
   (b) $g^{-1}(x) = \left(\frac{x}{8}\right)^{-3/5}$
   (c) $h^{-1}(x) = 16x^{-2}$
11. (a) $f(x) \to 0$
    (b) If d is odd, then $f(x) \to 0$. If d is even, however, then f is not defined for $x < 0$.

(c) $f(x) \to \infty$

(d) If d is odd and n is even, then $f(x) \to \infty$. If d is odd and n is odd, then $f(x) \to -\infty$. If d is even, then f is not defined for $x < 0$.

(e) If d is odd, then the domain of x is all real numbers not equal to zero. If d is even, then the domain is all positive real numbers. The range is all real numbers not equal to zero unless either n or d is even. If so, the negative real numbers are excluded from the domain.

13. (a) $\frac{105}{\pi} \approx 33.42$ lux

(b) $\sqrt{\frac{42}{\pi}} \approx 3.66$ m

Section 4.4

1. (a) Rational
   (b) Not rational
   (c) Rational
   (d) Rational
   (e) Rational
   (f) Not rational

3. (a) The domain of the function is all real numbers such that $x \neq \frac{4}{3}$, the x-intercept is at $x = -\frac{7}{2}$, and the vertical asymptote is at $x = \frac{4}{3}$.

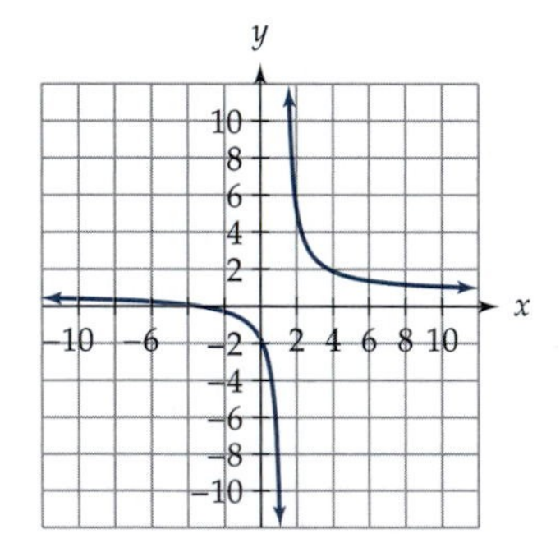

(b) The domain of the function is all real numbers such that $x \neq 7$, the x-intercept is at $x = -2$, and the vertical asymptote is at $x = 7$.

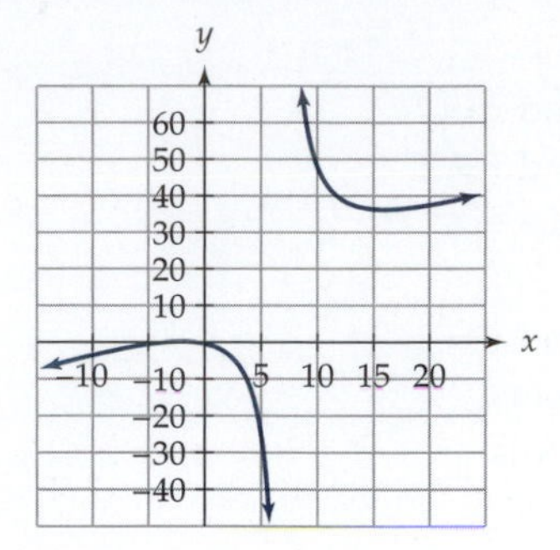

(c) The domain of the function is all real numbers such that $x \neq 0$, $x \neq 1$, or $x \neq 3$; there are no x-intercepts; and the vertical asymptotes are at $x = 1$ and $x = 3$. There is a hole at $x = 0$.

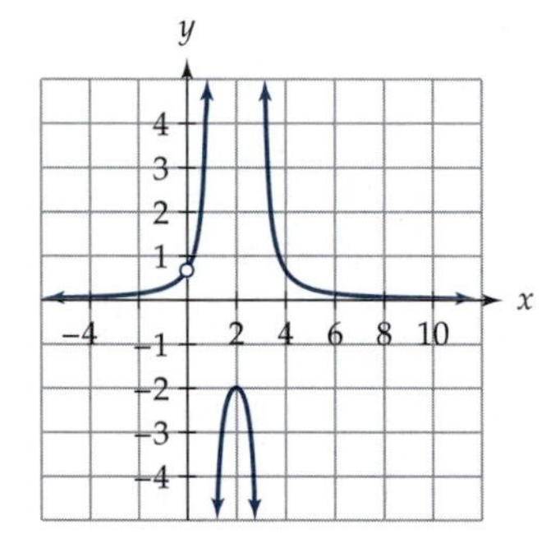

(d) The domain of the function is all real numbers, the x-intercept is at $x = 0$, and there are no vertical asymptotes.

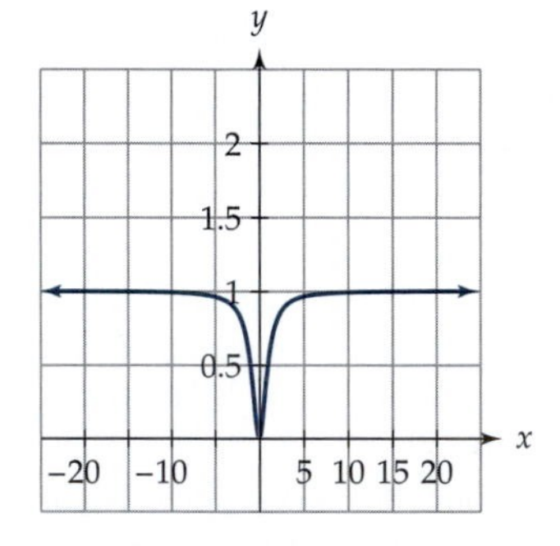

(e) The domain of the function is all real numbers such that $x \neq -5$, $x \neq 0$, or $x \neq 2$; the x-intercept is at $x = 1$; and the vertical asymptotes are at $x = 0$ and $x = 2$. There is a hole at $x = -5$.

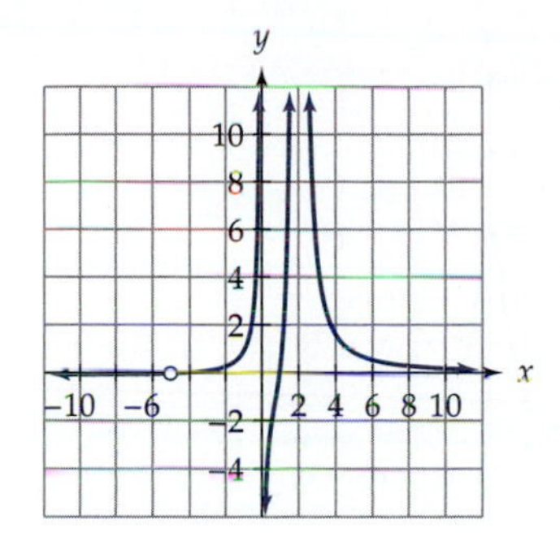

(f) The domain of the function is all real numbers, there are no x-intercepts, and there are no vertical asymptotes.

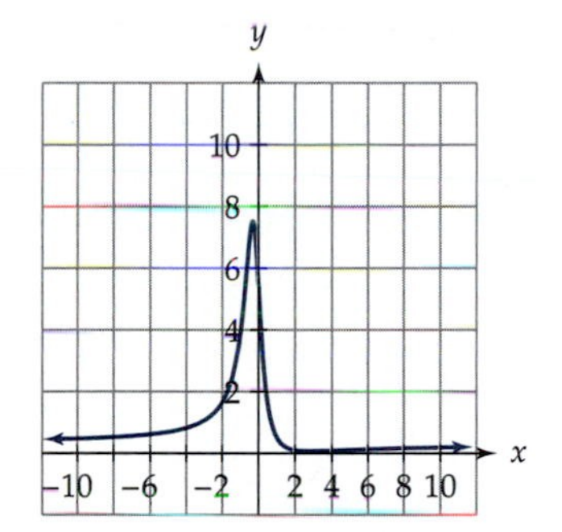

(g) The domain of the function is all real numbers such that $x \neq -4$ or $x \neq 6$, the x-intercept is at $x = 2$, and the vertical asymptote is at $x = 6$. There is a hole at $x = -4$.

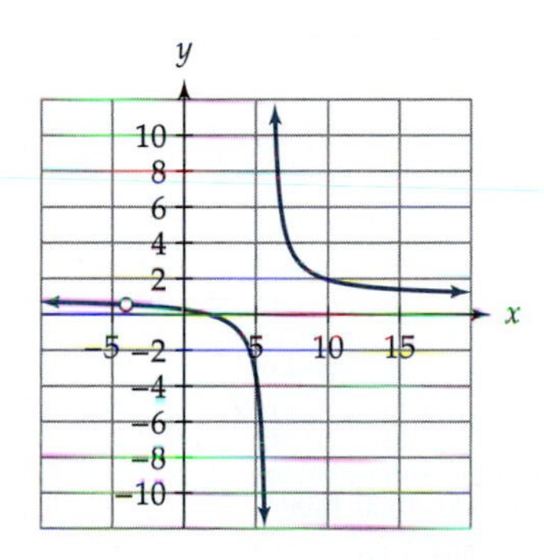

(h) The domain of the function is all real numbers such that $x \neq \sqrt{\frac{13}{6}}$ or $x \neq -\sqrt{\frac{13}{6}}$, the x-intercepts are at $x = 3$ and $x = \frac{5}{4}$, and the vertical asymptotes are at $x = \sqrt{\frac{13}{6}}$ and $x = -\sqrt{\frac{13}{6}}$.

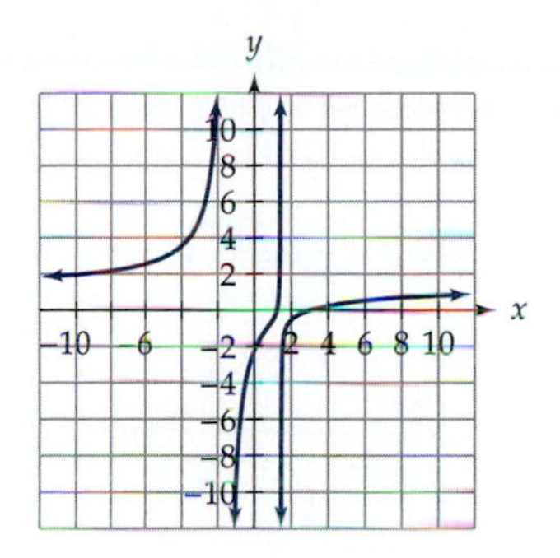

(i) The domain of the function is all real numbers such that $x \neq \sqrt{2}$ or $x \neq -\sqrt{2}$, there are no x-intercepts, and the vertical asymptotes are at $x = \sqrt{2}$ and $x = -\sqrt{2}$.

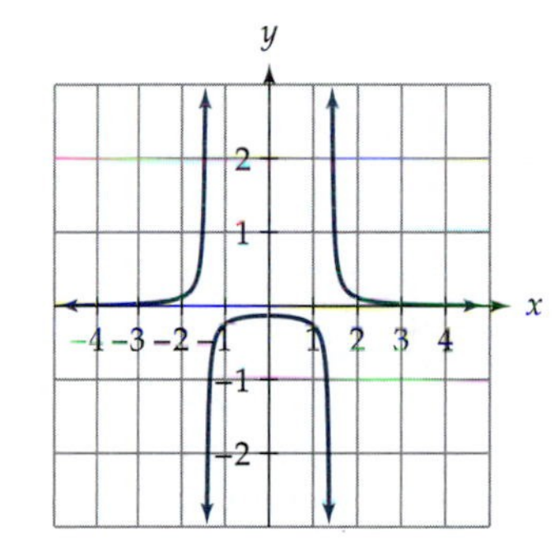

(j) The domain of the function is all real numbers such that $x \neq \sqrt{\frac{2}{3}}$ or $x \neq -\sqrt{\frac{2}{3}}$, the x-intercept is at $x = (-\frac{3}{2})^{1/3}$, and the vertical asymptotes are at $x = \sqrt{\frac{2}{3}}$ and $x = -\sqrt{\frac{2}{3}}$.

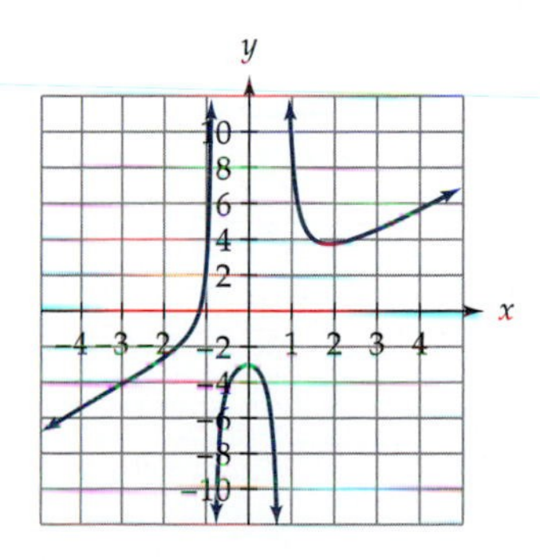

7. (a) Yes
   (b) No

9. (a) The domain of the function is all real numbers such that $x \neq 0$ or $x \neq 1$, there are no x-intercepts, and the vertical asymptotes are $x = 0$ and $x = 1$. As $x \to \pm\infty$, $y \to 1$. The range is approximately $y < -6.5$ or $y > 0.5$.

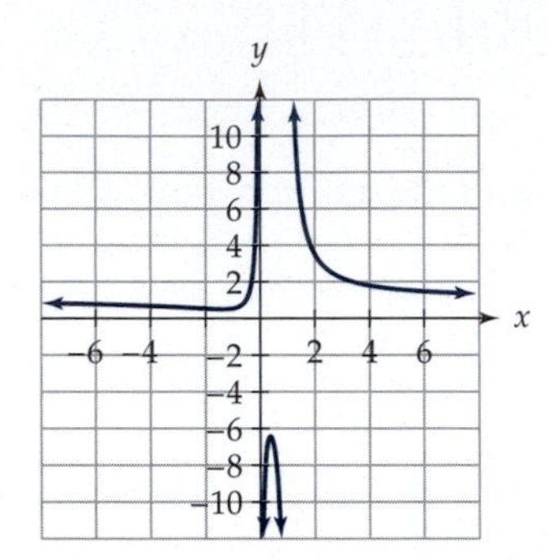

(b) The domain of the function is all real numbers such that $x \neq -1$ or $x \neq 1$, the range is $y \leq -3$ or $y > 0$, and there are no x-intercepts. The vertical asymptotes are $x = -1$ and $x = 1$. As $x \to \pm\infty$, $y \to 0$.

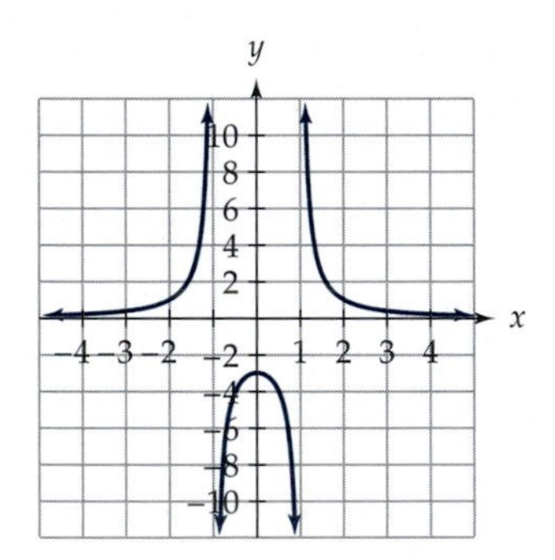

(c) The domain of the function is all real numbers, the range is $0 < y \leq 3$, and there are no x-intercepts or vertical asymptotes. As $x \to \pm\infty$, $y \to 0$.

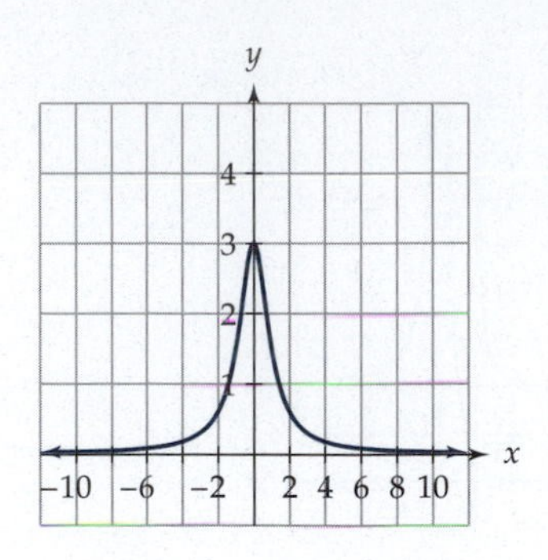

(d) The domain of the function is all real numbers such that $x \neq -1$, the range is all real numbers such that $y \neq 0$, and there are no x-intercepts. The vertical asymptote is $x = -1$. As $x \to \pm\infty$, $y \to 0$.

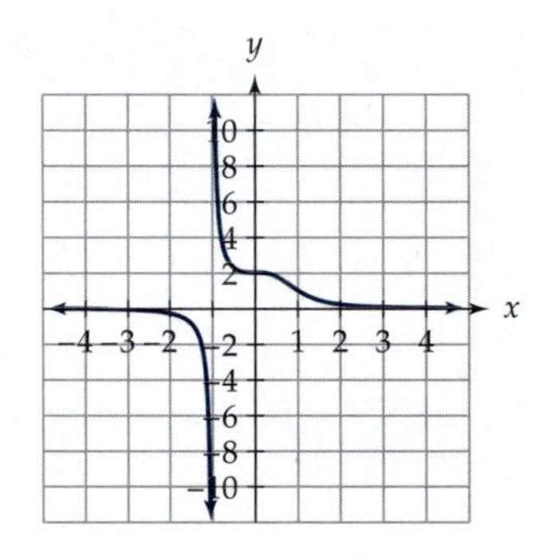

11. (a) $w(s) = (1.58 \times 10^7)w_0(s + 3960)^{-2}$

(b) The two functions are $w(d) = (1.58 \times 10^7)w_0(d)^{-2}$ and $d = s + 3960$.

(c) $\dfrac{200(1.58 \times 10^7)}{(3960 + 300)^2} \approx 174$ lb

(d)

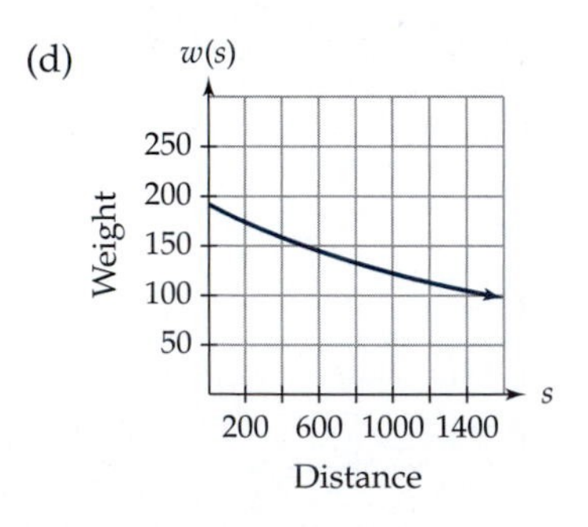

(e) As $s \to \infty$, $w(s) \to 0$.

13. (a) The graph of g represents the 75-W bulb.

(b) Physically, if the distance between the object and the bulbs is very large, then the brightness of light on the object is about the same for each bulb.

(c) It appears that the object should be approximately 2 m from directly under the 75-W bulb.

Review Exercises

1. (a) $(-4, -\frac{3}{2})$
   (b) $(-2, -15)$
   (c) $(3, 19)$
3. (a) $y = \frac{5}{4}x^2$
   (b) $y = -2x^2 + 8x - 3$
   (c) $y = 2x^2 - 10x + 8$
   (d) $y = \frac{1}{2}x^2 - \frac{3}{2}x + 4$
5. (a) $66\frac{1}{64}$ ft
   (b) 4 sec
7. (a) The polynomial is at least degree 4 with a positive leading coefficient, the roots at -4 and 1 are of multiplicity 1, and the root at 4 is of even multiplicity.
   (b) The polynomial is at least degree 5 with a positive leading coefficient, the root at -2 is of even multiplicity, and the root at 2 is of odd multiplicity greater than 1.
9. (a) $\sqrt[3]{9}$
   (b) 2, -3, and 5
   (c) 0, $\frac{1}{2}$, and 2
11. (a) Neither
   (b) Horizontal asymptote
   (c) Vertical asymptote
   (d) Neither
15. (a) 66.38 mph
   (b) 6 min and 23 sec
17. (a) $f(x) \to \frac{2}{5}$ as $x \to \pm\infty$.
   (b) $g(x) \to \infty$ as $x \to \infty$ and $g(x) \to -\infty$ as $x \to -\infty$.
   (c) $h(x) \to 0$ as $x \to \pm\infty$.
19. (a) $T(i) = \dfrac{0.15(i - 7700)}{i}$.
   (b) A rational function
   (c) The domain is $\$7700 < i < \$32{,}300$ and the range is $0 < t < 0.11424$.

Section 5.1

1. We use radians because we are defining the cosine and sine in terms of arc length and an angle of x radians will intercept an arc of length x.
3. (a) $\dfrac{\pi}{4}$
   (b) $\dfrac{\pi}{5}$
   (c) $-\dfrac{\pi}{3}$
   (d) $-\dfrac{19\pi}{36}$
   (e) $\dfrac{29\pi}{36}$
   (f) $\dfrac{7\pi}{4}$
5. (a) 0.7431
   (b) 0.4338
   (c) -1.111
   (d) 0.0532
9. (a) $g(x) = \sin\theta$ and $f(x) = \cos\theta$
   (b) $\tan\theta = 1$ where $\sin x = \cos x$ and the graphs intersect
11. $2\sqrt{2}$ units
13. $(\frac{5}{2}, \frac{5}{2}\sqrt{3})$
15. (a) $\frac{3}{5}$
   (b) $\frac{4}{3}$
   (c) $-\frac{4}{5}$
17. (a) $\dfrac{\sqrt{35}}{6}$
   (b) $\sqrt{35}$
   (c) $-\frac{1}{6}$
19. (a) $c \geq 0$
   (b) i. $\dfrac{c}{\sqrt{1+c^2}}$
      ii. $\dfrac{1}{\sqrt{1+c^2}}$
      iii. $-c$
      iv. c
21. The coordinates of a *unit* circle represent the trigonometric functions. The circle in this statement is not a unit circle.
23. $\sin^2\theta = (\sin\theta)\cdot(\sin\theta)$, whereas $\sin\theta^2 = \sin(\theta\cdot\theta)$. In $\sin^2\theta$, the output is squared, whereas in $\sin\theta^2$, the input, θ, is squared.

Section 5.2

1. (a) $2w + 2l$ is a linear measurement for the perimeter of any rectangle with length l and width w.
   (b) $\frac{1}{2}bh$ is an area measurement for any triangle with base b and height h.

(c) s^2 is an area measurement for any square with sides of length s.
(d) $2\pi r$ is a linear measurement for the circumference of a circle with radius r.
(e) $r\theta$ is a linear measurement for the length of an arc of angle θ on a circle with radius r.
(f) $\frac{1}{2}ab \sin\theta$ is an area measurement for any triangle with adjacent sides a and b and an included angle of measure θ.
(g) lw is an area measurement for any rectangle with length l and width w.
(h) πr^2 is an area measurement for a circle with radius r.
(i) $4s$ is a linear measurement for the perimeter of a square with sides of length s.
(j) $\frac{1}{2}r^2\theta$ is an area measurement for the area of a sector of angle θ on a circle of radius r.

3. Perimeter ≈ 75.6 ft, area ≈ 336 ft^2
5. $330\pi \approx 1036.7$ mi
7. (a) The two angles are corresponding angles of parallel lines cut by a transversal and therefore have the same measure.
   (b) 24,000 mi
9. $114.375\pi \approx 359.3$ in.^{2}
11. (a) $\dfrac{50,625\pi}{4} \approx 39,760.7$ ft^2
    (b) $450 + 112.5\pi \approx 803.4$ ft
13. $x^2\left(\dfrac{\pi}{2} - 1\right)$
15. $\frac{3}{4}x^2 + \frac{\pi}{8}x^2$
17. 1434.1 in.^{2}
19. (a) 176 in.
    (b) 0.00175 radians/sec
    (c) 0.293 in./sec
    (d) 0.147 in./sec
21. 0.79 ft/sec

Section 5.3

1. (a) False
   (b) False
   (c) False
   (d) True
   (e) False
   (f) False
   (g) False
   (h) True
3. (a) The amplitude is 3 and the midline is $y = 4$.
   (b) The amplitude is 3 and the midline is $y = -\pi$.
   (c) The amplitude is $\frac{1}{2}$ and the midline is $y = 4$.
   (d) The amplitude is 4 and the midline is $y = 0$.
   (e) The amplitude is 12 and the midline is $y = -2$.
   (f) The amplitude is π and the midline is $y = -1$.
5. (a) $y = \cos(4(x + 1))$.
   (b) $y = \sin\left(4(x + 1) + \dfrac{\pi}{2}\right)$.
7. (a) ii
   (b) v
   (c) i
   (d) iii
   (e) vi
   (f) iv
9. (a) A is negative and D is negative.
   (b) A is positive and D is negative.
   (c) A is negative and D is positive.
11. (a) $A = 1, B = \dfrac{2\pi}{\pi} = 2, C = 0, D = 0$
    (b) $A = 1, B = 1, C = \dfrac{\pi}{4}, D = 0$
    (c) $A = 1, B = 1, C = 0, D = -1$
    (d) $A = -2, B = \pi, C = 0, D = 1$
    (e) $A = 1, B = 3, C = 0, D = 4$
    (f) $A = 2, B = 5, C = -\dfrac{\pi}{10}, D = 3$
13. (a) The domain of $y = \sin^{-1} x$ is $-1 \le x \le 1$ and the range is $-\pi/2 \le y \le \pi/2$.
    (b) The range of $y = \tan^{-1} x$ is $-\pi/2 < y < \pi/2$.
    (c) The range of $y = \cos^{-1} x$ is $0 \le y \le \pi$.
    (d) $\cos^{-1}(1/\sqrt{2}) = \pi/4$
    (e) $\sin^{-1}(3)$ does not exist.
    (f) $\tan^{-1}(1) = \pi/4$
15. (a) $s(t)$ can be modeled by a cosine function because the vertical distance starts at a high point and goes down and up in a periodic manner.
    (b) $\frac{1}{4}$ m
    (c) 1 sec
    (d) $s(t) = \frac{1}{4}\cos(2\pi x) + \frac{3}{4}$
17. (a) A
    (b) B♭
    (c) B
    (d) G

Section 5.4

1.

x (radians)	x (degrees)	sec x	csc x	cot x
$\frac{\pi}{6}$	30°	$\frac{2\sqrt{3}}{3}$	2	$\sqrt{3}$
$\frac{\pi}{3}$	60°	2	$\frac{2\sqrt{3}}{3}$	$\frac{\sqrt{3}}{3}$
$\frac{\pi}{4}$	45°	$\sqrt{2}$	$\sqrt{2}$	1

7. (a) $-\frac{\sqrt{5}}{3}$

(b) $\frac{-2\sqrt{5}}{5}$

(c) $-\frac{4\sqrt{5}}{9}$

(d) $\frac{1}{9}$

(e) $-4\sqrt{5}$

9. (a) When $x = \pi/4$, both sides of the equation are undefined.

(b) The identity will not work whenever $x = \pi/4 + n(\pi/2)$, where n is an integer.

15.

x (radians)	x (degrees)	cos x	sin x	tan x
$\frac{5\pi}{12}$	75°	$\frac{\sqrt{3}-1}{2\sqrt{2}}$	$\frac{\sqrt{3}+1}{2\sqrt{2}}$	$\frac{\sqrt{3}+1}{\sqrt{3}-1}$
$\frac{7\pi}{12}$	105°	$\frac{1-\sqrt{3}}{2\sqrt{2}}$	$\frac{\sqrt{3}+1}{2\sqrt{2}}$	$\frac{1+\sqrt{3}}{1-\sqrt{3}}$
$\frac{11\pi}{12}$	165°	$\frac{-\sqrt{3}-1}{2\sqrt{2}}$	$\frac{\sqrt{3}-1}{2\sqrt{2}}$	$\frac{\sqrt{3}-1}{-\sqrt{3}-1}$
$\frac{\pi}{12}$	15°	$\frac{\sqrt{3}+1}{2\sqrt{2}}$	$\frac{\sqrt{3}-1}{2\sqrt{2}}$	$\frac{\sqrt{3}-1}{\sqrt{3}+1}$

17. (a)

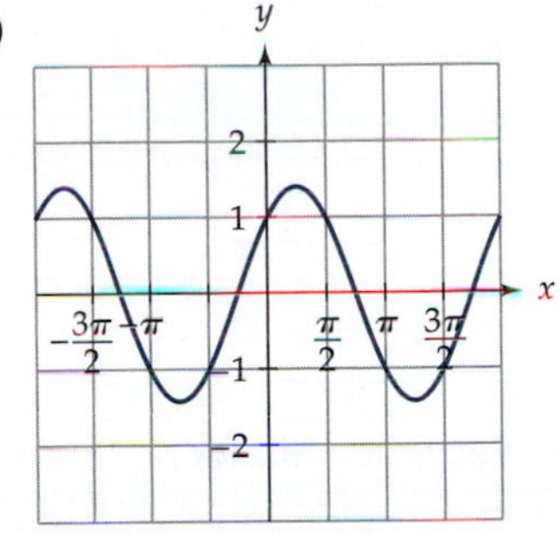

(b) $y = \sqrt{2}\sin\left(x + \frac{\pi}{4}\right)$

23. (a) $\sin^2 x + \cos^2 x = 1$
$\sin^2 x = 1 - \cos^2 x$
$\sin x = \sqrt{1 - \cos^2 x}$

(c) In the proof, we were not careful to make $\sqrt{\sin^2 x} = |\sin x|$. If we had been, then both the graphs would be the same.

Review Exercises

1. (a) $\frac{3\pi}{4}$

(b) $\frac{3\pi}{2}$

(c) $\frac{2\pi}{5}$

(d) $-\frac{2\pi}{3}$

3.

θ (radians)	θ (degrees)	cos θ	sin θ	tan θ
$\frac{3\pi}{4}$	135°	$-\frac{\sqrt{2}}{2}$	$\frac{\sqrt{2}}{2}$	-1
$\frac{\pi}{3}$	60°	$\frac{1}{2}$	$\frac{\sqrt{3}}{2}$	$\sqrt{3}$
$-\frac{\pi}{3}$	$-60°$	$\frac{1}{2}$	$-\frac{\sqrt{3}}{2}$	$-\sqrt{3}$
$\frac{5\pi}{4}$	225°	$-\frac{\sqrt{2}}{2}$	$-\frac{\sqrt{2}}{2}$	1

5. (a) $\frac{4}{5}$
   (b) $\frac{3}{4}$
   (c) $-\frac{3}{5}$

7.

	Sector A	Sector B	Sector C
Length of the Radius	3	5	6
Measure of the Central Angle	2	6	$\frac{\pi}{6}$
Length of the Intercepted Arc	6	30	π
Area of the Sector	9	75	3π

9. (a) 367.6 square units
   (b) 23.1 square units
11. (a) 4π radians/sec
   (b) $32\pi \approx 100.5$ in./sec
   (c) 5.7 mph
15. (a) $A = 3, B = 2\pi, C = 0, D = 0$
   (b) $A = 3, B = \pi, C = -1, D = 2$
17. (a) $\frac{1}{3}$
   (b) $3 + \frac{\pi}{8}$
   (c) $\frac{3\pi}{2}$
19. (a) $\frac{2\sqrt{2}}{3}$
   (b) $2\sqrt{2}$
   (c) $\frac{4\sqrt{2}}{9}$
   (d) $-\frac{7}{9}$
   (e) $-\frac{4\sqrt{2}}{7}$

Section 6.1

1. (a) Area $= x^2 + 2xy + y^2$
   (b) Area $= z^2 + 2xy$
3. (a) $s = n - m$
   (b) Area $= n^2 + m^2$
5. (b) (3, 4, 5), (6, 8, 10), (5, 12, 13), (9, 12, 15), (8, 15, 17), (12, 16, 20), (7, 24, 25), (15, 20, 25), (10, 24, 26), (20, 21, 29), (18, 24, 30)
7. The statement is false.
9. $h = s \cdot \sin\theta$
13. (a) 5.7°
   (b) 1049 ft
15. (b) 7 ft $11\frac{1}{4}$ in.
17. (a) Area $= \dfrac{\tan\theta}{\tan^2(\theta/2)}$
   (b) The smallest area occurs when $\theta = 60°$.
19. (a) 3.314
   (b) $r > s$
   (c) $P = n \tan\left(\dfrac{180°}{n}\right)$
   (d) 3.142. This estimate is an overestimate.
21. 12 ft
23. 18.5°

Section 6.2

1. (a) Law of sines
   (b) Law of sines
   (c) Law of cosines
   (d) Law of sines
   (e) We cannot determine one side if we are only given the three angles.
   (f) Law of cosines
   (g) Law of sines
11. 381.70 ft
13. 294.5 ft
15. (b) 108.5 mph
   (c) 30° south of east
17. (a) 20.1 square units
   (b) 3.75 square units
   (c) 2.88 square units
   (d) 13.3 square units
19. (a) 6.5 square units
   (b) 26 square units, four times the area from part (a).
   (c) When the length of each side is doubled, the area of the triangle quadruples.
   (d) When the length of each side is tripled, the area is nine times as large.
   (e) The area of a triangle is n^2 times as large when the length of each side is multiplied by a factor of n.

Section 6.3

1. (a) 559.67 min
   (b) 709.16 min
   (c) 190 min
   (d) 365 days

(e) 81 days

(f) 732 min

3. (a) The data obtained from this experiment are best modeled by a periodic function because the voltage is cyclic.

(b) 0.007936

(c) 126

(d) The frequency is approximately twice the number of times the voltage cycles between −110 V and +110 V.

5. The equation $S(x) = \cos(660\pi x)$ will model the note E.

7. (a) i. 64.5° and 17.5°

ii. 89.5° and 42.5°

iii. 73.5° and 26.5°

iv. 99.5° and 52.5°

v. 49.5° and 2.5°

(c) i. 23.5°

ii. 365 days

iii. 172 days to the right

iv. 50°

v. $f(t) = 23.5 \cos\left[\frac{2\pi}{365}(t - 172)\right] + 50$

9. (a) Michigan

(b)

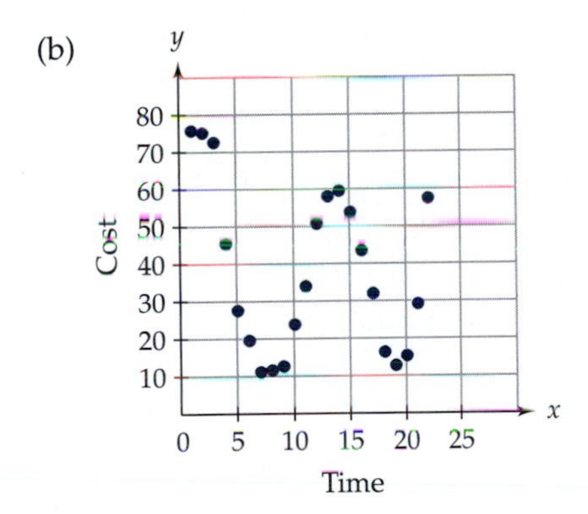

(c) $C(m) = 32.3 \cos\left[\frac{\pi}{6}(m - 1)\right] + 43.6$

(d) The period is 12 months and the vertical shift is 43.6.

(e) $C(5) = 32.3 \cos\left[\frac{\pi}{6}(5 - 1)\right] + 43.6 = \$27.45$. Our estimate is under by \$4.70.

11. (a) For example, f will not be periodic if the period of $y = \cos Ax$ is irrational and $y = \cos Bx$ is rational.

(b) If the periods for both $y = \cos Ax$ and $y = \cos Bx$ were rational numbers, then f would be periodic.

13. (a) Periodic with a period of 40π

(b) Not periodic

(c) Periodic with a period of 1.6π

(d) Periodic with a period of 56π

(e) Not periodic

(f) Periodic with a period of 10π

(g) Periodic with a period of 2

15. If n is an integer, then $f(np) = f(p) = f(0 + p) = f(0)$.

17. (a) The function for the physical cycle is $p(d) = \sin[(2\pi/23)d]$, the function for the emotional cycle is $e(d) = \sin[(2\pi/28)d]$, and the function for the intellectual cycle is $i(d) = \sin[(2\pi/33)d]$.

(b) Answers will vary.

(c) Answers will vary.

(d) Answers will vary.

(e) i. The physical biorhythm is zero every 11.5 days, emotional every 14 days, and intellectual every 16.5 days.

ii. Both functions are zero at the same moment every 322 days.

iii. No

19. (a) $S(x) = \cos(880\pi x) + \cos(896\pi x)$

(b) Yes, h is periodic.

(c) 8 beats per second; yes

21. You know that the note is either five cycles per second high (225 Hz) or five cycles low (215 Hz).

Section 6.4

1. (a) iii

(b) v

(c) ii

(d) i

(e) vi

(f) iv

3. (a) The function is f and it appears that $a \approx 1.5$.

(b) The function is g and it appears that $m \approx 3$.

(c) The function is g and it appears that $m \approx 2$.

5. $f(x) = 0.38(0.93)^x \cos(\pi x) + 1.15$

7. $f(t) = 0.175(0.972)^t \cos(4.40t) + 0.725$

Review Exercises

1. (a) $c = 15, A \approx 36.9°, B \approx 53.1°$

(b) $A = 40°, b \approx 4.7, c \approx 6.2$

(c) $a \approx 8.7, A = 60°, B = 30°$

3. (a) 8.53°

(b) 783 ft

5. (a) 50.45°
   (b) 85.6°
7. (a) $C = 110^\circ, b \approx 6.43, c \approx 9.40$
   (b) $a \approx 6.89, B \approx 56.35^\circ, C \approx 88.65^\circ$
   (c) $B \approx 72.93^\circ, C \approx 72.07^\circ, c \approx 9.95$ or $B \approx 107.07^\circ, C \approx 37.93^\circ, c \approx 6.43$
   (d) $A \approx 82.82^\circ, B \approx 41.41^\circ, C = 55.77^\circ$
9. (a) 5.33
   (b) 5.91
   (c) 7.43
11. (a) $H(m) = 26.8 \cos\left[\frac{\pi}{6}(m-7)\right] + 58.5$
   (b) $C(m) = 14.89 \cos\left[\frac{\pi}{6}(m-7)\right] + 14.72$
13. (a) $y = \cos(1540\pi x) + \cos(2672\pi x)$
   (b) 1 sec
15. (a) iii
   (b) i
   (c) ii
   (d) iv
17. (a) $g(x) = 2x + \sin(2\pi x)$
   (b) $f(x) = 1.5^x + \sin(2\pi x)$

Section 7.1

1. (a) $x = \pm\sqrt{\frac{42}{5}}$
   (b) $x = \pm 12^{3/4}$
   (c) $x = 2^{-5/3}$
3. (a) $x = \frac{3}{2} \pm \frac{1}{2}\sqrt{5}$
   (b) $x = 75 \pm \sqrt{5034}$
   (c) There are no real solutions.
   (d) $3 \pm \sqrt{11}$.
5. (a) If f is a linear function, then $f(x) = c$ has at most one solution and may have no solutions.
   (b) If f is a power function of the form $f(x) = bx^a$, then $f(x) = c$ has at most two solutions and may have no solutions.
   (c) If f is an exponential function, then $f(x) = c$ has either one solution or no solutions.
   (d) If f is a sine function, there is either no solution or an infinite number of solutions.
   (e) If f is a quadratic function, there are as many as two solutions but as few as none.
   (f) If f is a cubic function, there are as many as three solutions but as few as one.
7. (a) $G = \frac{FR^2}{m_1 m_2}$
   (b) $s = \frac{pf}{p-f}$
   (c) $h = \pm\sqrt{\left(\frac{A}{\pi r} - r\right)^2 - r^2}$
   (d) $r = \frac{A}{(2\pi h + 2\pi)}$
9. (a) more than 52 min.
   (b) more than 100 min.
11. (a) 1.52 and 4.10 sec
   (b) 126.56 ft
   (c) 5.625 sec
13. ≈ 7.62 or ≈ 25.23 in.
15. Less than 2.67 m

Section 7.2

1. (a) $\log 50$
   (b) $\log 25$
   (c) $\ln 5$
   (d) $\ln 5$
3. (a) 4.644
   (b) −5.381
   (c) 0
   (d) −2.832
5. (a) $p(t) = 1{,}200{,}000 \cdot \left(\frac{7}{12}\right)^{0.1t}$
   (b) About 240,000
   (c) 1993
   (d) 2003
7. (a) 7.875%
   (b) 7.58%
9. (a) 0.00012
   (b) About 4257 years ago
   (c) $t = \frac{\ln 0.01\,N}{0.00012}$
11. 2057
13. $13,804.20
15. 6% compounded monthly

Section 7.3

1. (a) $\frac{9+\sqrt{17}}{32}$
   (b) $\frac{16}{9}$
   (c) 0 or $\frac{1}{4}$
   (d) No solution
   (e) No solution
3. (a) $\frac{\pi}{6} + 2\pi k$ or $\frac{5\pi}{6} + 2\pi k$ for integer values of k

(b) $\frac{\pi}{3} + 2\pi k$ or $\frac{5\pi}{3} + 2\pi k$ for integer values of k

(c) $\frac{\pi}{4} + \pi k$ for integer values of k

(d) $2\pi k$ for integer values of k

5. (a) 0.518 or 1.053
   (b) 0.444, 1.491, or 2.538
   (c) 0.374, 0.673, 1.421, 1.720, 2.468, or 2.768

7. (a) 0, 1.318, π, 4.965, or 2π
   (b) 0.785, 2.356, 3.927, or 5.498
   (c) $\frac{\pi}{4}$ or $\frac{5\pi}{4}$
   (d) 0, 1.823, π, 4.460, or 2π

9. (a) Because $2\sin x \cos x = \sin(2x)$, we can write $r = 900 \sin\theta \cos\theta$ as $r = 450 \sin 2\theta$.
   (b) 13.2° and 76.8°
   (c) There are no angles that will give a range of 500 ft.
   (d) 45°

11. (a) $h(t) = -50\cos\left(\frac{2\pi}{15}t\right) + 62.5$
    (b) 5.77, 9.23, 20.77, 24.23, 35.77, 39.23, 50.77, and 54.23 min

13. (a) True
    (b) Sometimes true
    (c) True
    (d) Sometimes true
    (e) Sometimes true
    (f) Sometimes true

15. (a) The amplitude is $\pi/2$ and the period is 2π.
    (b) $f(x) = x$
    (c) $f(x) = -x + 2\pi$

Section 7.4

1. (a) $y = 0.880x + 20.42$
   (b) 0.907

3. (a)

   (b) $h = 6f$, where h is the height and f is the length of your foot
   (c) The statement does not seem accurate. The foot size is too small and the slope is too steep.

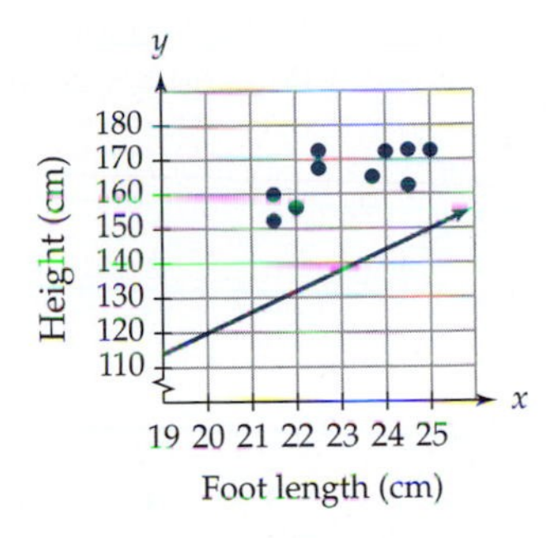

   (d) The regression equation is $h = 3.84x + 76.3$. The slope means that as foot length increases by 1 cm, height increases by about 3.84 cm.
   (e) The accuracy is fair; the correlation coefficient is only 0.68.

5. (a) The equation is $s = 0.021y + 11.88$, where y is the number of years since 1980 and t is the time.
      i. The slope represents the increase in speed per year.
      ii. The y-intercept represents the speed in the year 1800.
      iii. Yes, the regression line is very good. The correlation is 0.9875.
   (b) i. 4 min 36 sec
      ii. 3 min 44 sec
      iii. We can find the reciprocal of our output and multiply by 60 to get $t = 60/(0.021y + 11.88)$.
   (c) 1954

7. (a) These data do seem to fall in a linear pattern.

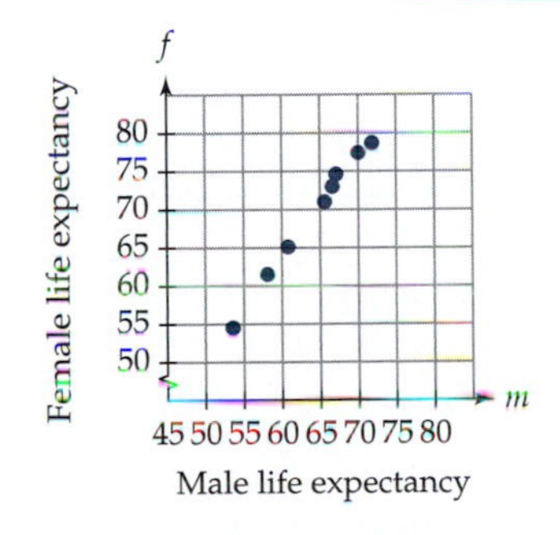

(b) The regression equation is $f = 1.35m - 17.36$.
(c) The gap is increasing. You can tell because the slope is greater than 1, which means that for every year that is added to the male life expectancy, more than one is added to the female life expectancy.

9. A correlation of zero means that there is no relationship. Coffee consumption and height are not related at all.

Section 7.5

1. (a)

x	1	2	3	4	5
$f(x) = 2^x$	2	4	8	16	32
$g(x) = \log(2^x)$	0.30	0.60	0.90	1.20	1.51
$h(x) = \ln(2^x)$	0.69	1.39	2.08	2.77	3.47

(b) The difference between successive entries in both the log and ln rows are constant.
(c) $g(x) = \log 2^x = x \log 2$. This function is a linear function with a slope of log 2 and a y-intercept of 0. A similar argument shows it for the natural log.

3. (a) The graph of $\log x$ versus $\log f(x)$ looks most linear. $f(x) = 30.20 \cdot x^{0.15}$.
(b) The graph of x versus $g(x)$ looks most linear. $g(x) = 5.51x + 26.53$.
(c) The graph of x versus $\log h(x)$ looks most linear. $h(x) = 19.5 \cdot 1.2^x$.

5. (a)

Year	1985	1987	1988	1989	1990
log(record sales) (in millions)	2.08	1.91	1.82	1.56	1.44
Year	1991	1992	1993	1994	
log(record sales) (in millions)	1.34	1.30	1.18	1.07	

(b) $\log s = -0.117y + 2.682$
(c) $s = 481 \cdot 0.76^y$

7. (a) $p = -0.412y + 58.01$
(b) $p = 60.72 \cdot 0.9894^y$
(c) The exponential function fits better.
(d) $p_u = 0.412y + 41.99, p_u = 100 - 60.72 \cdot 0.9894^y$

9. (a)

Make and Model	log(horsepower)	log(miles per gallon)
Suzuki Swift	1.90	1.48
Hyundai Accent	1.96	1.43
Saturn SCI	2	1.43
Ford Escort	2.04	1.40
Toyota Celica	2.11	1.34
Chevrolet Lumina	2.20	1.30
Ford Mustang	2.28	1.30
Lexus ES300	2.30	1.28
Cadillac Seville	2.44	1.23
Jaguar XJ8	2.46	1.23

(b) $\log m = -0.437 \log h + 2.290$
(c) This equation becomes $m = 195 \cdot h^{-0.437}$.

11. (b) $w = 8.92 \cdot a^{0.363}$
(c) $l = 19.60 \cdot a^{0.175}$
(d) $w = 8.92 \cdot (l/19.6)^{2.07}$. This function is another power function.

13. (a) $p = 480{,}173 \cdot 1.086^y$. The growth factor is 1.086.
(b) The equation is $r = 202.3 \cdot 1.075^y$. The growth factor is 1.075.

15. (a)

Term	1	2	3	4	5
Fibonacci Number	1	1	2	3	5
Term	6	7	8	9	10
Fibonacci Number	8	13	21	34	55

(b) $f = 0.488 \cdot 1.598^t$
(c) The growth factor, 1.598, is similar to the golden ratio, which is approximately 1.618.

17. (a) $r = 23.6 \cdot 0.787^n$
(b) The probability of landing tail side up is about 21%.
(c) Answers will vary.

Review Exercises

1. (a) $\pm\frac{\sqrt{33}}{2}$
(b) $\pm 5\sqrt{5}$

(c) $\dfrac{-3 \pm \sqrt{89}}{4}$

(d) $0, -\frac{1}{2}$, or -2

(e) $\dfrac{17 \pm \sqrt{913}}{6}$

(f) $-\frac{2}{3}$ or 8

3. (a) $L = \dfrac{1}{2f}\sqrt{\dfrac{T}{\rho}}$
   (b) $T = (2fL)^2\rho$
   (c) $h = \pm\sqrt{s^2 - r^2}$
   (d) $y = \pm\sqrt{b^2 - \dfrac{b^2x^2}{a^2}}$
5. (a) 11.896
   (b) 1.236
   (c) 0.606
   (d) −2.317
   (e) −0.645
7. (a) −0.000433
   (b) 80.5%, 70.7%
   (c) 2780 years
9. (a) Because the rate for Switzerland is greater than the rate for Austria, Switzerland must eventually overtake Austria.
   (b) 81 years
11. (a) 1
    (b) No solution
    (c) No solution
13. (a) $0.100 < x < 3.041$
    (b) $1.318 < x < 4.965$
    (c) $0 \le x < 1.249$, $\dfrac{\pi}{2} < x < 4.391$, or $\dfrac{3\pi}{2} < x \le 2\pi$
15. (a) $0, \dfrac{\pi}{2}, \pi, \dfrac{3\pi}{2}, 2\pi$
    (b) $\dfrac{\pi}{12}, \dfrac{5\pi}{12}, \dfrac{13\pi}{12}, \dfrac{17\pi}{12}$
17. (a) $y = 12{,}393 - 1061x$
    (b) The slope means that the car loses $1061 in value each year.
    (c) The y-intercept means that the price of a new car is about $12,393.
    (d) −0.9924
19. (a) x versus $\log f(x)$ is the most linear. $f(x) = 10.927 \cdot (1.301)^x$.
    (b) x versus $g(x)$ is the most linear. $g(x) = 10.4 + 5.457x$.
    (c) $\log x$ versus $\log h(x)$ is the most linear. $h(x) = 16.13 \cdot x^{0.4969}$.

Section 8.1

1. (a) 34 ft/sec
   (b) 26 ft/sec
   (c) 19.6 ft/sec
   (d) 18 ft/sec
   (e) 50 ft/sec
3. (a) 6.23 ft/sec
   (b) 4.36 ft/sec
   (c) 3.61 ft/sec
   (d) 2.58 ft/sec
   (e) Decreasing
5. (a) $\frac{7}{8}$
   (b) $-\frac{2}{3}$
   (c) −1
7. 160 ft/sec
9. Answers will vary.
11. (a) 1
    (b) 2
    (c) 0
    (d) 0
13. (a) 6
    (b) 4
    (c) Does not exist
    (d) 16
    (e) 16
    (f) 16
15. (a) e^2
    (b) e^3
    (c) e^n
17. (a) 1
    (b) Does not exist
19. (a) Method A describes a continuous function, and method B describes a function that is not continuous.
    (b) Method A would be less controversial because in method B, those earning slightly more than $10,000 will pay a lot more in taxes than those earning slightly less than $10,000.
21. $x = 2$ is a point of discontinuity in g.

Section 8.2

1. 4 ft/sec
3. (a) $B(2) = 92{,}160$; $B'(2) \approx 255{,}000$.
   (b) $B(2)$ indicates the number of bacteria on a square centimeter of fresh chicken after 2 days, whereas, $B'(2)$ indicates the rate of change of the number of bacteria at 2 days.

5. (a) 6
   (b) 3
   (c) 8
   (d) −8
   (e) 32
7. (a) $f'(x) = 3$
   (b) $g'(x) = 2x + 6$
   (c) $f'(x) = 3x^2$
   (d) $f'(x) = 4x - 6$
9. C, D, B, A
11. (a) At point A, the elevator is moving up.
    (b) Between points B and C, the elevator is stationary.
    (c) At point D, the elevator is moving down.
    (d) At point E, the elevator is just starting to move down.
    (e) The elevator made three stops above point A.
13. (a) 1
    (b) $\frac{1}{2}$
    (c) The baby is gaining the most weight the first few months.

15. (a)

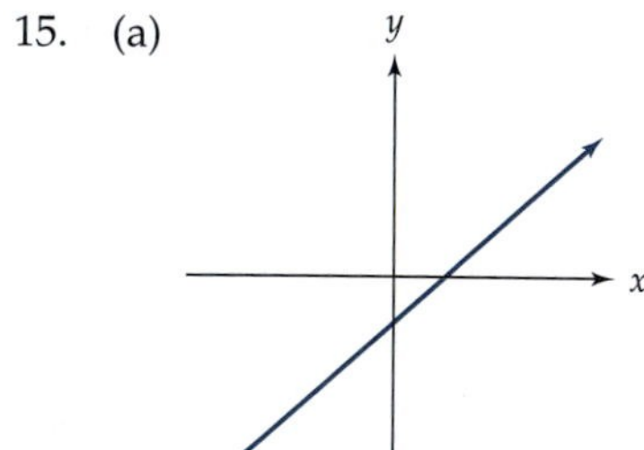

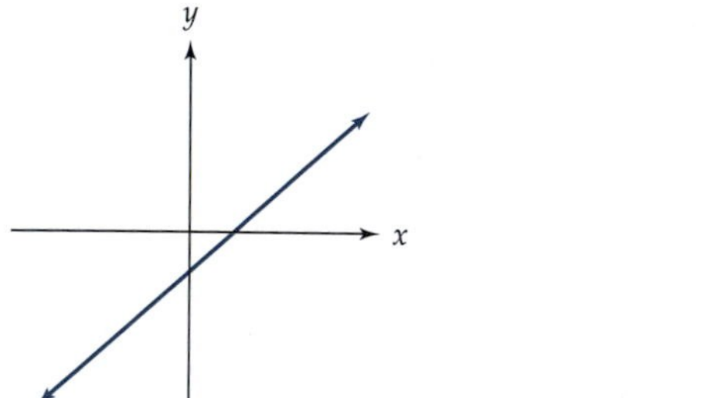

(b)

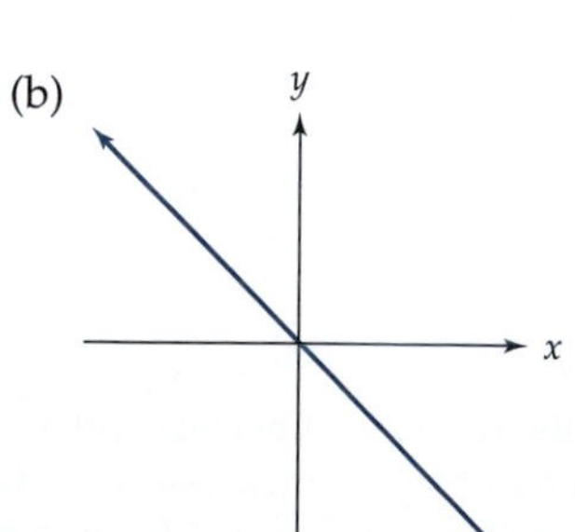

(c)

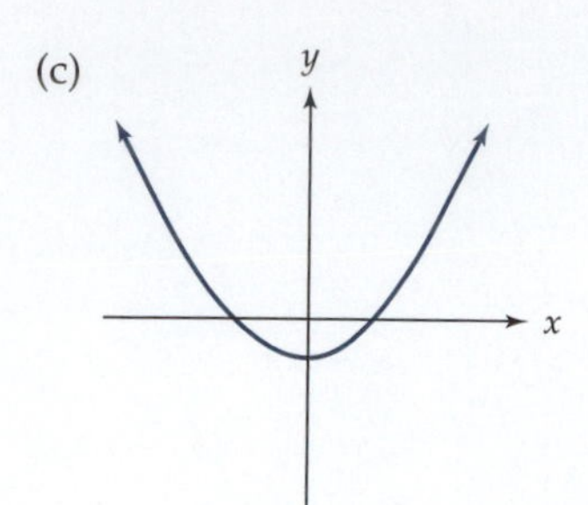

(d)

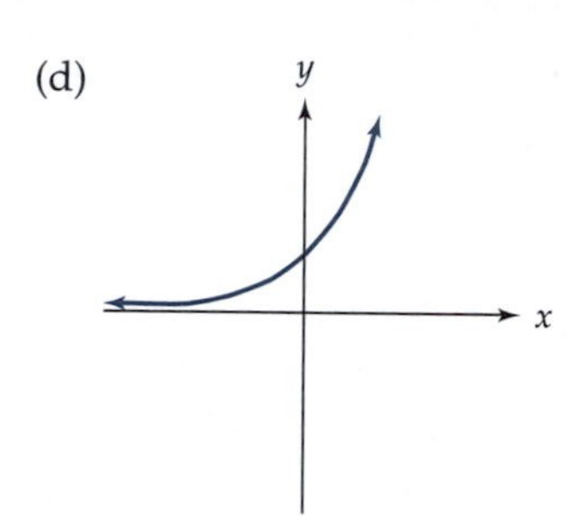

(e)

(f)

17. When the function is multiplied by a constant, its derivative is also multiplied by the same constant.

Section 8.3

1. (a) $1, 5, 9, 13, 17$
   (b) $1, 3, 9, 27, 81$
   (c) $2, 4, 7, 11, 16$
   (d) $5, 1, -5, -13, -23$
   (e) $2, 3, 5, 8, 13$
3. (a) $a_n = 4n$
   (b) $a_n = -13 + 10n$
   (c) $a_n = 12 - 14n$
5. (a) 1334
   (b) 140
   (c) -12
7. 774
9. (a) 3
   (b) ∞
   (c) $\frac{15}{4}$
11. ≈ 0.00242
13. 6.4
15. (a) 32
    (b) $\frac{14}{3}$
    (c) $\frac{4}{5}$
17. (a) $\frac{5}{9}$
    (b) 1
    (c) $\frac{25}{99}$
    (d) $\frac{271}{99}$
19. (a) $10 + 100 + 1000 + 10{,}000 + 100{,}000$
    (b) \$100,000
    (c) \$1,000,000,000
    (d) The population is less than the number of people that are supposed to be sending you money. There is no way it will work.
21. (a) 35
    (b) $n^2 + n\left(\frac{n-1}{2}\right)$
    (c) $a_1 = 1$ and $a_n = a_{n-1} + 3n - 2$
23. $\approx 1.84467 \times 10^{19}$

Section 8.4

1. Area ≈ 35 square units
3. (a) Area ≈ 22.37 square units
   (b) Area ≈ 35.52 square units
   (c) Area ≈ 8.38 square units
5. (a) Area ≈ 5.36 square units. This approximation is an underestimate.
   (b) Area ≈ 5.77 square units. This approximation is an overestimate.
   (c) Area ≈ 5.57 square units. This approximation is between the other two.
7. (a) Area $= 0$
   (b) Area $= 6$
9. (a) 72 square units
   (b) 4 square units
   (c) 2π square units
11. (a) -2 square units
    (b) 0 square units
    (c) -3 square units
13. (a) 2 square units
    (b) 0 square units
    (c) 4 square units
    (d) 2 square units
15. (a) $v(t) = 30 + 5t$
    (b) 212.5 mi
    (c) The value of the integral from part (b) represents the total distance traveled from $t = 0$ to $t = 5$.

Review Exercises

1. (a) 14 ft/sec
   (b) 7.6 ft/sec
   (c) 6.16 ft/sec
   (d) 6 ft/sec
3. (a) 0
   (b) 1
   (c) 2
5. No
7. (a) 2
   (b) -1
   (c) 32
9. (a) $f'(x) = 5$
   (b) $g'(x) = 4x$
   (c) $h'(x) = 3x^2 - 5$
11. (a) $y = 3x + 6$
    (b) $y = -8x - 4$
13. (a) 42
    (b) 2,621,440
    (c) 1540
15. (a) 1305
    (b) 2887.5
    (c) 16
    (d) 6
17. (a) 22.5
    (b) 54
    (c) $\frac{9}{4}\pi$

19. (a) -4
    (b) 0

21. (a) $r(t) = 5$
    (b) 50
    (c) After 5 min, 50 gal have flowed through the pipe.

Section 9.1

1. (a)

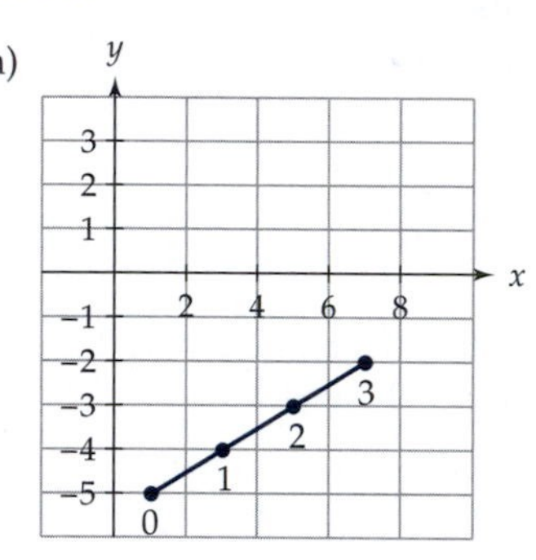

(b)

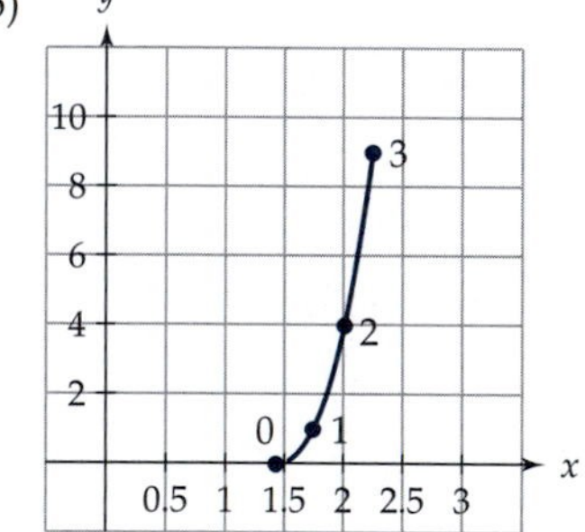

(c)

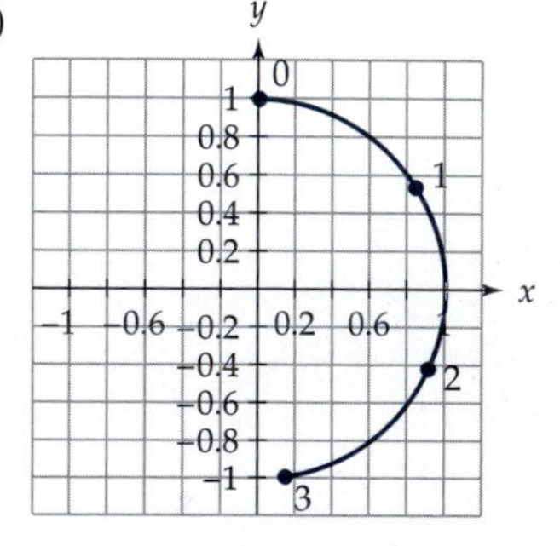

3. (a) The first one traces more quickly than the second.
   (b) They start at different places.
   (c) One traces in the opposite direction from the other.

5. (a) These parametric equations describe an ellipse.

7. $x = 10\cos(-(\pi/30)t)$ and $y = 10\sin(-(\pi/30)t)$ for $0 \le t \le 60$

9. $x = 100t\cos 40°$ and $y = 100t\sin 40° - 16t^2 + 3$; 311 ft

11. (a) $y = -12\cos\theta + 12$ and $x = -12\sin\theta + 12\theta$
    (b) $x = -12\sin(6\pi t) + 72\pi t$ and $y = -12\cos(6\pi t) + 12$
    (c) $0 \le t \le \frac{4}{3}$

13. (a)

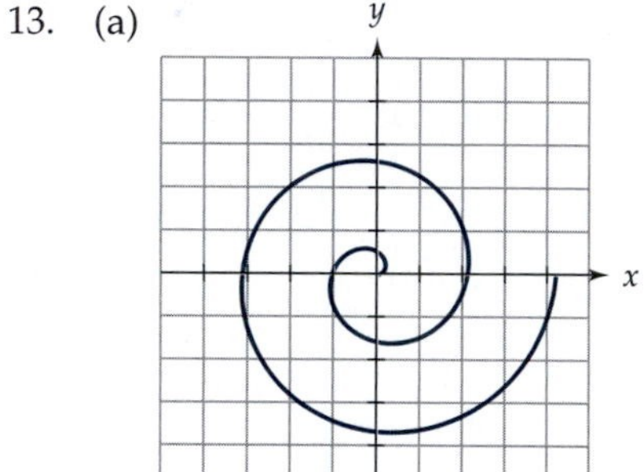

    (b) The equations give the formula for a circle with a changing radius. As θ increases, the circle is drawn and the radius spirals outward.

15. (a) No
    (b) $\theta \approx 56.44°$
    (c) The balls hit each other when $x_2 = 6$. The balls will hit each other as long as x_2 is less than the point where the first ball hits the ground, approximately 26.

17. (a) The calculator evaluates the function at different values of t and then "connects the dots." If T_{step} is set to 60, then it will only evaluate 6 points as it goes around the circle, giving us six straight lines instead of a circle.
    (b) i. $T_{step} = 120$
        ii. $T_{step} = 90$
        iii. $T_{step} = 72$
        iv. $T_{step} = \dfrac{360}{n}$

Section 9.2

1. (a) $7\mathbf{i} + 10\mathbf{j}$
   (b) $-8\mathbf{i} - 5\mathbf{j}$
   (c) $4\mathbf{i} - 7\mathbf{j}$
   (d) $7\mathbf{i} + 8\mathbf{j}$
3. (a) Quadrant II
   (b) Quadrant I
   (c) Quadrant III
   (d) Quadrant IV
5. (a) $2.57\mathbf{i} + 3.06\mathbf{j}$
   (b) $-3\mathbf{i} + 5.2\mathbf{j}$
   (c) $-2.82\mathbf{i} - 1.03\mathbf{j}$
   (d) $1.88\mathbf{i} - 0.68\mathbf{j}$
7. (a) $\mathbf{u} + \mathbf{v} = 4\mathbf{i} + 7\mathbf{j}$
   (b) $\mathbf{u} + \mathbf{v} = 4\mathbf{i} - 16\mathbf{j}$
   (c) A vector with magnitude of 9.85 at an angle of 52.0°
   (d) A vector with magnitude of 12.12 at an angle of 90°
9. 131.1 mph at 62.4° south of due west
11. 242.9 miles at 54° west of north
13. (a) 90°
   (b) The smallest force would be of 4 lb, when the direction of the second vector is 270°.
   (c) The magnitude is approximately 6.32 lb and the direction is 71.6° or 251.6°.
   (d) −9.6° or 189.6°
15. No
17. (d) $n = 106.5$, $m = 70.9$
19. 45 lb

Section 9.3

1. (a) 19
   (b) 31
   (c) $9 + 2y$
   (d) $x^2 + 6$
3. (a) −18.22
   (b) $W(10, t) = -21.23 + 1.23t$
   (c) $W(10, t)$ is a linear function
5. (a) $I(90, h) = 86.6459477 + 0.07300256h + 0.00581763h^2$. This function is a quadratic function.
   (b) $I(90, 80) \approx 129.72°$

7. (a)

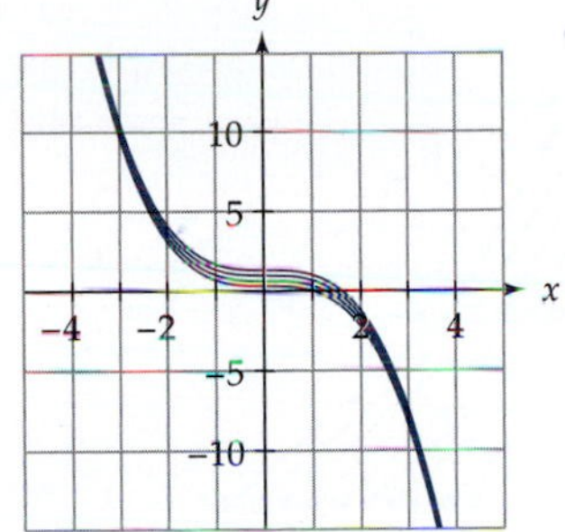

(b)

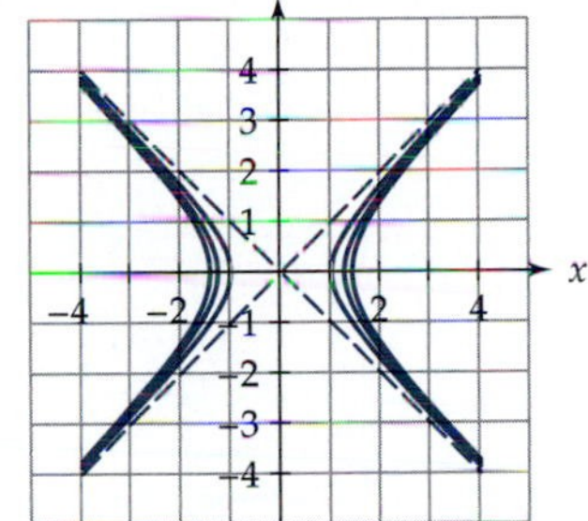

(c)

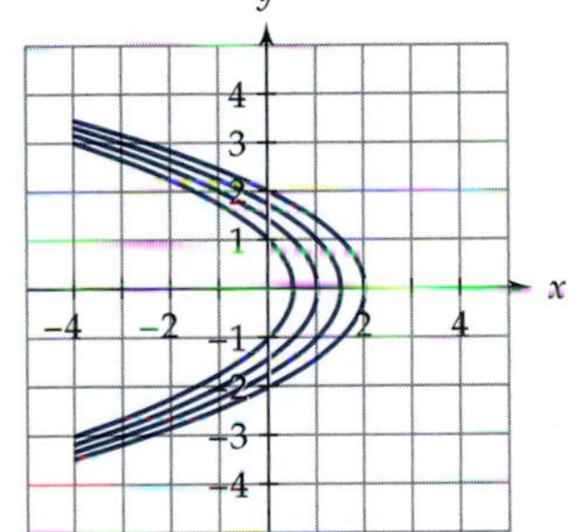

(d)

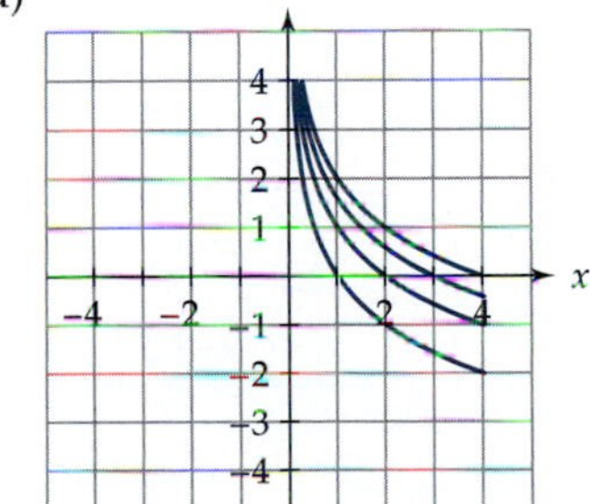

9.

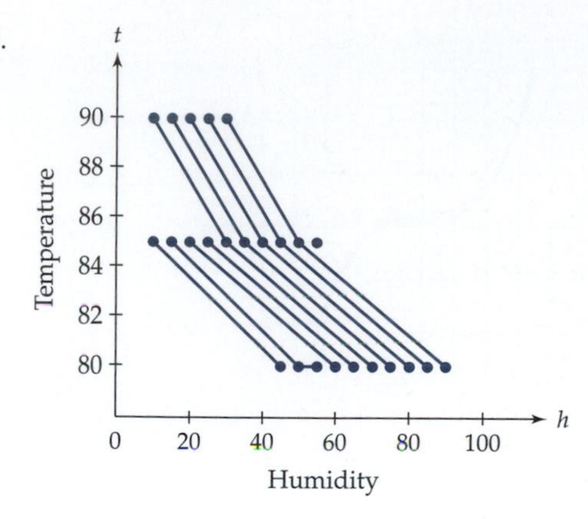

11. (a)

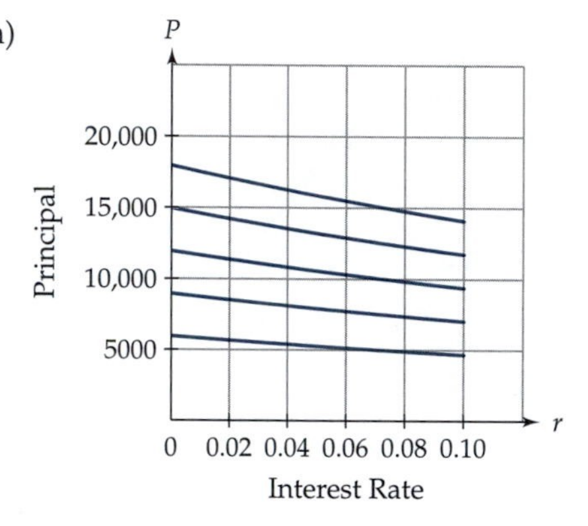

(b) About 10,300

13. (a) $w(w_0, d) = (1.58 \times 10^7) \cdot w_0 \cdot (D + 3960)^{-2}$

(b)

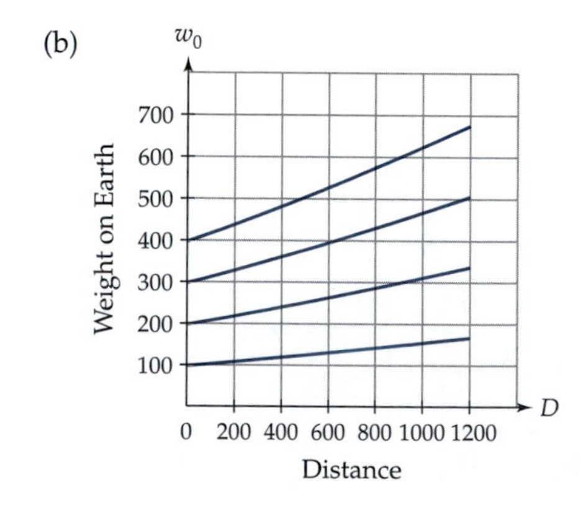

(c) The object weighs 208.65 lb on the surface of Earth.

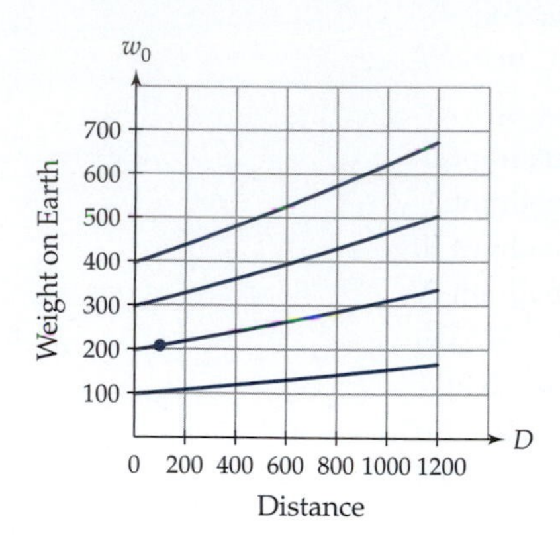

Review Exercises

1. $x = 60\sqrt{2}t$ and $y = 60\sqrt{2}t - 16t^2$, 450 ft.
3. (a) $x = -25\sin((\pi/15)t)$ and $y = -25\cos((\pi/15)t) + 30$
   (b) The equation $x = -25\sin((\pi/15)t)$ changes sign to become $x = 25\sin((\pi/15)t)$.
5. (a) $-2\mathbf{i} + 10\mathbf{j}$
   (b) $7\mathbf{i} - 2\mathbf{j}$
   (c) $-2\mathbf{i} + 9\mathbf{j}$
7. (a) $\frac{3\sqrt{2}}{2}\mathbf{i} + \frac{3\sqrt{2}}{2}\mathbf{j}$
   (b) $-4\mathbf{i} + 4\sqrt{3}\mathbf{j}$
   (c) $2\mathbf{i} - 2\sqrt{3}\mathbf{j}$
9. 108.48 mph at 19.98° west of due south
11. 187.5 lb
13. (a) $V(3, 6) = 54\pi$ is the volume of a cylinder that has a radius of 3 and a height of 6.
    (b) $V(3, h) = 9\pi h$ is the volume of a cylinder that has a radius of 3 and a height of h.
    (c) $V(r, 6) = 6\pi r^2$ is the volume of a cylinder that has a radius of r and a height of 6.

INDEX